Cyclic Nucleotide Phosphodiesterases in Health and Disease

edited by

Joseph A. Beavo
Sharron H. Francis
Miles D. Houslay

CRC Press
Taylor & Francis Group
Boca Raton   London   New York

CRC Press is an imprint of the
Taylor & Francis Group, an informa business

Cover art was kindly provided by Dr. Robert X. Xu, Structural Sciences RTP, GlaxoSmithKline, Research Triangle Park, NC.

CRC Press
Taylor & Francis Group
6000 Broken Sound Parkway NW, Suite 300
Boca Raton, FL 33487-2742

© 2007 by Taylor & Francis Group, LLC
CRC Press is an imprint of Taylor & Francis Group, an Informa business

First issued in paperback 2019

No claim to original U.S. Government works

ISBN 13: 978-0-367-45332-9 (pbk)
ISBN 13: 978-0-8493-9668-7 (hbk)

Library of Congress Cataloging-in-Publication Data

Cyclic nucleotide phosphodiesterases in health and disease / editors, Joseph Beavo, Sharron Francis, and Miles Houslay.
    p. ; cm.
    Includes bibliographical references and index.
    ISBN-13: 978-0-8493-9668-7 (alk. paper)
    ISBN-10: 0-8493-9668-9 (alk. paper)
    1. Cyclic nucleotide phosphodiesterases. I. Beavo, Joseph. II. Francis, Sharron H. III. Houslay, Miles D.
    [DNLM: 1. Phosphoric Diester Hydrolases--physiology. 2. Nucleotides, Cyclic--physiology. 3. Phosphodiesterase Inhibitors--therapeutic use.   QU 136 C9953 2006]

QP609.C92C93 2006
612'.01513--dc22
                          2006018995

Visit the Taylor & Francis Web site at
http://www.taylorandfrancis.com

and the CRC Press Web site at
http://www.crcpress.com

Preface

The establishment of a Gordon Research Conference devoted to cyclic nucleotide phospho-diesterases has had an enormous impact on the field by facilitating collaborations and friendships and promoting research in this important area. The final impetus that led to the generation of this book was nurtured from this source. Apocryphally one can probably trace its beginning to an afternoon on the veranda at the 2004 Gordon Conference in Italy where Joe and Miles had been taking in the ambiance of the site and therefore were susceptible to Sharron's "great idea" that it was time for another "PDE Book." So, with such an auspicious start, and after Sharron agreed to help, we began this project for an all-encompassing book on "PDEology."

The last major book compendium dedicated to cyclic nucleotide PDEs was published over 15 years ago.* During this time an enormous amount of progress has occurred in the PDE field so a revised update on this topic is clearly needed. For example, the number of PDE families in mammals has increased from five to eleven: three families are cAMP-specific; three families are cGMP-specific; and five families hydrolyze both cAMP and cGMP well. Characterizations of genetic knockouts for several PDEs have started to appear. The crystal structures of catalytic domains for nine PDEs (some complexed with inhibitors or product) and one regulatory domain have been published. Intracellular binding partners for several PDEs have been found and evidence for select pools of cyclic nucleotides that modulate a variety of processes has led to an appreciation of the regulation of specific cyclic nucleotide pools by distinct PDEs. From a physiological perspective, we are now beginning to unravel at least some of the unique cellular roles for different PDEs. From a medical perspective several PDE inhibitors are now approved for treatment of a number of different pathological conditions, and the remarkable medical and financial success of the PDE5-selective inhibitors marketed as Viagra®, Levitra®, and Cialis® has elevated these medications to blockbuster status.

We have been very pleased that the authors for the various chapters in this volume have been willing to set aside time to contribute in such a positive way. In developing this project, we have tried to not only cover each PDE family but also to have chapters that highlight key therapeutic and biomedical areas. Inevitably we have missed out on some due to space or deadline limitations and we apologize for this. It has been our privilege to work in such an exciting field and interact with a large community of enthusiastic and talented scientists from all over the world. We are very grateful to all who have contributed both directly and indirectly and trust that you will find this book to be a valuable reference source. We would particularly like to thank our "better halves" for their forbearance when we have each been locked away editing, collating, phoning, cajoling, and writing over weekends and evenings. Enjoy!!

Miles, Joseph, and Sharron

---

*Beavo, J.A. and Houslay, M.D., 1990. Cyclic nucleotide phosphodiesterases; structure, function, regulation, and drug action, in *Molecular Pharmacology of Cell Regulation*, 1st ed., Houslay, M.D., 344 pp. Chichester: John Wiley & Sons.

Editors

Joseph A. Beavo is a professor of pharmacology at the University of Washington. He obtained a BS from Stetson University in Deland, Florida and then a PhD from Vanderbilt University in Nashville, Tennessee. He has been involved in research concerning cyclic nucleotide phosphodiesterases since his graduate work with Drs. Joel Hardman and Earl Sutherland in the 1960s. He is an active member of the American Society for Pharmacology and Experimental Therapeutics and the National Academy of Sciences, having served on the editorial boards of journals published by both organizations. His current research focuses on the structural mechanisms of regulation of cyclic nucleotide phosphodiesterases and also on the functional roles played by the different phosphodiesterase gene products.

Sharron H. Francis is a research professor of molecular physiology and biophysics at Vanderbilt University School of Medicine in Nashville, Tennessee. She obtained her undergraduate degree with emphasis in biology and chemistry from the Western Kentucky State in Bowling Green; studies for her doctoral degree in physiology were performed at Vanderbilt University under the guidance of Jane Park, PhD. She did postdoctoral study with Dr. Herman Eisen at Washington University in St. Louis, Missouri and was a research fellow in the Laboratory of Biochemistry at the National Heart and Lung Institute in Bethesda, Maryland under the guidance of Earl Stadtman, PhD. She is an active member of the American Society for Biochemistry and Molecular Biology, served the Scientific Advisory Board for Cell Pathways, Inc., and also has served as a consultant for numerous pharmaceutical companies. Her research interests focus on mechanisms involved in cGMP signaling, molecular characteristics of cGMP-dependent protein kinase, and phosphodiesterase-5, which are targets of cGMP, and the features that contribute to potent inhibition of phosphodiesterase-5 by selective inhibitors.

Miles D. Houslay is a Gardiner professor of biochemistry at the University of Glasgow, Scotland, UK. He obtained his first degree in biochemistry at the University of Wales in Cardiff and then his PhD from King's College, Cambridge. He has held faculty positions at the Universities of Cambridge and Manchester. He is a fellow of the Royal Society of Edinburgh, Colworth Medal Holder of the British Biochemical Society, has served as member/chair of numerous grant agencies, and has consulted widely in the pharmaceutical industry. He has been involved in cell signaling research since its inception. His current research interest is focused on the role of phosphodiesterase-4 isoforms in underpinning cAMP compartmentalization and cross-talk processes and potential for identifying novel therapeutic opportunities.

Contributors

Einar M. Aandahl
University of Oslo
Oslo, Norway

Aniella Abi-Gerges
INSERM, University of Paris
Châtenay-Malabry, France

Joseph A. Beavo
University of Washington
Seattle, Washington

Andrew T. Bender
University of Washington
Seattle, Washington

Emmanuel P. Bessay
Vanderbilt University School
   of Medicine
Nashville, Tennessee

Mitsi A. Blount
Vanderbilt University School
   of Medicine
Nashville, Tennessee

Graeme B. Bolger
University of Alabama
   at Birmingham
Birmingham, Alabama

Laurence L. Brunton
School of Medicine
University of California at San Diego
La Jolla, California

Liliana R.V. Castro
INSERM, University of Paris
Châtenay-Malabry, France

Marco Conti
Stanford University School of Medicine
Stanford, California

Jackie D. Corbin
Vanderbilt University School of Medicine
Nashville, Tennessee

Rick H. Cote
University of New Hampshire
Durham, New Hampshire

Shireen-A. Davies
University of Glasgow
Glasgow, Scotland

Jonathan P. Day
University of Glasgow
Glasgow, Scotland

Eva Degerman
Lund University
Lund, Sweden

Sara A. Epperson
School of Medicine
University of California at San Diego
La Jolla, California

Laure Favot
Pôle Biologie Santé UFR SFA
Poitiers, France

Rodolphe Fischmeister
INSERM, University of Paris
Châtenay-Malabry, France

Sharron H. Francis
Vanderbilt University School of Medicine
Nashville, Tennessee

Mark A. Giembycz
University of Calgary
Institute of Infection, Immunity
   and Inflammation
Calgary, Alberta, Canada

Lena Stenson Holst
Lund University
Lund, Sweden

Miles D. Houslay
University of Glasgow
Glasgow, Scotland

Kye-Im Jeon
University of Rochester School
  of Medicine and Dentistry
Rochester, New York

S.-L. Catherine Jin
Stanford University School
  of Medicine
Stanford, California

Junichi Kambayashi
Otsuka Maryland Medicinal
  Laboratories LLC
Rockville, Maryland

Hengming Ke
The University of North Carolina
Chapel Hill, North Carolina

Thérèse Keravis
Université Louis Pasteur
  (Strasbourg I)
Illkirch, France

Jun Kotera
Tanabe Seiyaku Co. Ltd.
Saitama, Japan

Christina Kruuse
University of Copenhagen
Glostrup, Denmark

Adam Lerner
Boston Medical Center
Boston, Massachusetts

Yongge Liu
Otsuka Maryland Medicinal
  Laboratories, LLC
Rockville, Maryland

Claire Lugnier
CNRS, Université Louis Pasteur
  (Strasbourg I)
Illkirch, France

M.R. MacLean
University of Glasgow
Glasgow, Scotland

Vincent Manganiello
National Institutes of Health
Bethesda, Maryland

Sergio E. Martinez
University of Washington
Seattle, Washington

Donald H. Maurice
Queen's University
Kingston, Ontario, Canada

Jeffrey M. McKenna
Celgene Corporation
San Diego, California

Frank S. Menniti
Pfizer Global Research and
  Development
Groton, Connecticut

Tamar Michaeli
Albert Einstein College of Medicine
  of Yeshiva University
Bronx, New York

Eun-Yi Moon
Korea Institute of Bioscience
  and Biotechnology
Taejon, Korea

Matthew A. Movsesian
VA Salt Lake City Health Care System
University of Utah
Salt Lake City, Utah

George W. Muller
Celgene Corporation
Summit, New Jersey

F. Murray
School of Medicine
University of California at San Diego
La Jolla, California

David J. Nagel
University of Rochester School
of Medicine and Dentistry
Rochester, New York

James M. O'Donnell
West Virginia University Health
Sciences Center
Morgantown, West Virginia

Kenji Omori
Tanabe Seiyaku Co. Ltd.
Saitama, Japan

N.J. Pyne
University of Strathclyde
Glasgow, Scotland

Wito Richter
Stanford University School of Medicine
Stanford, California

Francesca Rochais
INSERM, University of Paris
Châtenay-Malabry, France

Christopher J. Schmidt
Pfizer Global Research and Development
Groton, Connecticut

Thomas Seebeck
University of Berne
Berne, Switzerland

Yasmin Shakur
Otsuka Maryland Medicinal
Laboratories, LLC
Rockville, Maryland

Antonio P. Silva
Novartis Pharma AG
Basel, Switzerland

Carolyn J. Smith
Medsn Inc.
Jersey City, New Jersey

Christine A. Strick
Pfizer Global Research
and Development
Groton, Connecticut

Kjetil Taskén
University of Oslo
Oslo, Norway

Rothewelle Tate
University of Strathclyde
Glasgow, Scotland

Douglas G. Tilley
Clinical Research Laboratory
Duke University
Durham, North Carolina

Sanjay Tiwari
University Hospital Schleswig-Holstein
Luebeck, Germany

Grégoire Vandecasteele
INSERM, University of Paris
Châtenay-Malabry, France

Valeria Vasta
University of Washington
Seattle, Washington

Huanchen Wang
The University of North Carolina
Chapel Hill, North Carolina

Laurent Wentzinger
University of Berne
Berne, Switzerland

Oliwia Witczak
University of Oslo
Oslo, Norway

Chen Yan
University of Rochester School
of Medicine and Dentistry
Rochester, New York

Han-Ting Zhang
West Virginia University Health
Sciences Center
Morgantown, West Virginia

Kam Y.J. Zhang
Plexxikon, Inc.
Berkeley, California

Allan Z. Zhao
University of Pittsburgh
Pittsburgh, Pennsylvania

Roya Zoraghi
Vanderbilt University
  School of Medicine
Nashville, Tennessee

Table of Contents

Introduction

Chapter 1 Cyclic Nucleotide Phosphodiesterase Superfamily 3

Joseph A. Beavo, Miles D. Houslay, and Sharron H. Francis

Chapter 2 Phosphodiesterase Isoforms—An Annotated List 19

Graeme B. Bolger

Section A
Specific Phosphodiesterase Families—Regulation, Molecular
and Biochemical Characteristics

Chapter 3 Calmodulin-Stimulated Cyclic Nucleotide Phosphodiesterases 35

Andrew T. Bender

Chapter 4 PDE2 Structure and Functions 55

Sergio E. Martinez

Chapter 5 Phosphodiesterase 3B: An Important Regulator
of Energy Homeostasis 79

Eva Degerman and Vincent Manganiello

Chapter 6 Cellular Functions of PDE4 Enzymes 99

Graeme B. Bolger, Marco Conti, and Miles D. Houslay

Chapter 7 Phosphodiesterase 5: Molecular Characteristics Relating to Structure,
Function, and Regulation 131

*Sharron H. Francis, Roya Zoraghi, Jun Kotera, Hengming Ke,
Emmanuel P. Bessay, Mitsi A. Blount, and Jackie D. Corbin*

Chapter 8 Photoreceptor Phosphodiesterase (PDE6):
A G-Protein-Activated PDE Regulating Visual
Excitation in Rod and Cone Photoreceptor Cells 165

Rick H. Cote

Chapter 9 PDE7 195

Tamar Michaeli

Chapter 10   cAMP-Phosphodiesterase 8 Family...205

Valeria Vasta

Chapter 11   PDE9...221

Jun Kotera and Kenji Omori

Chapter 12   PDE10A: A Striatum-Enriched, Dual-Substrate Phosphodiesterase............237

Christine A. Strick, Christopher J. Schmidt, and Frank S. Menniti

Chapter 13   PDE11...255

Kenji Omori and Jun Kotera

Section B
Nonmammalian Phosphodiesterases

Chapter 14   Protozoal Phosphodiesterases 277

Laurent Wentzinger and Thomas Seebeck

Chapter 15   Studies of Phosphodiesterase Function Using Fruit
Fly Genomics and Transgenics 301

Shireen-A. Davies and Jonathan P. Day

Section C
Phosphodiesterases Functional Significance: Gene-Targeted
Knockout Strategies

Chapter 16   Insights into the Physiological Functions of PDE4
from Knockout Mice...323

S.-L. Catherine Jin, Wito Richter, and Marco Conti

Chapter 17   Regulation of cAMP Level by PDE3B—Physiological
Implications in Energy Balance and Insulin Secretion 347

Allan Z. Zhao and Lena Stenson Holst

Section D
Compartmentation in Cyclic Nucleotide Signaling

Chapter 18   Heart Failure, Fibrosis, and Cyclic Nucleotide
Metabolism in Cardiac Fibroblasts...365

Sara A. Epperson and Laurence L. Brunton

Chapter 19   Role of A-Kinase Anchoring Proteins in the
Compartmentation in Cyclic Nucleotide Signaling 377

Oliwia Witczak, Einar M. Aandahl, and Kjetil Taskén

Chapter 20   Role of Phosphodiesterases in Cyclic Nucleotide
Compartmentation in Cardiac Myocytes 395

*Aniella Abi-Gerges, Liliana R.V. Castro, Francesca Rochais,
Grégoire Vandecasteele, and Rodolphe Fischmeister*

Section E
Phosphodiesterases as Pharmacological Targets in Disease Processes

Chapter 21   Role of PDEs in Vascular Health and Disease: Endothelial
PDEs and Angiogenesis..417

Thérèse Keravis, Antonio P. Silva, Laure Favot, and Claire Lugnier

Chapter 22   Regulation of PDE Expression in Arteries: Role
in Controlling Vascular Cyclic Nucleotide Signaling 441

Donald H. Maurice and Douglas G. Tilley

Chapter 23   Regulation and Function of Cyclic Nucleotide
Phosphodiesterases in Vascular Smooth Muscle
and Vascular Diseases 465

Chen Yan, David J. Nagel, and Kye-Im Jeon

Chapter 24   Role of Cyclic Nucleotide Phosphodiesterases
in Heart Failure and Hypertension 485

Matthew A. Movsesian and Carolyn J. Smith

Chapter 25   Molecular Determinants in Pulmonary Hypertension:
The Role of PDE5..501

N.J. Pyne, F. Murray, Rothewelle Tate, and M.R. MacLean

Chapter 26   Role of PDE5 in Migraine 519

Christina Kruuse

Chapter 27   Phosphodiesterase-4 as a Pharmacological Target
Mediating Antidepressant and Cognitive Effects on Behavior 539

Han-Ting Zhang and James M. O'Donnell

Chapter 28   Role of Phosphodiesterases in Apoptosis.......................................559

Adam Lerner, Eun-Yi Moon, and Sanjay Tiwari

Section F
Development of Specific Phosphodiesterase Inhibitors as Therapeutic Agents

Chapter 29   Crystal Structure of Phosphodiesterase Families
and the Potential for Rational Drug Design 583

Kam Y.J. Zhang

Chapter 30 Structure, Catalytic Mechanism, and Inhibitor
Selectivity of Cyclic Nucleotide Phosphodiesterases 607

Hengming Ke and Huanchen Wang

Chapter 31 Bench to Bedside: Multiple Actions of the PDE3
Inhibitor Cilostazol 627

Junichi Kambayashi, Yasmin Shakur, and Yongge Liu

Chapter 32 Reinventing the Wheel: Nonselective Phosphodiesterase
Inhibitors for Chronic Inflammatory Diseases... 649

Mark A. Giembycz

Chapter 33 Medicinal Chemistry of PDE4 Inhibitors....................................... 667

Jeffrey M. McKenna and George W. Muller

Index 701

Introduction

1 Cyclic Nucleotide Phosphodiesterase Superfamily

Joseph A. Beavo, Miles D. Houslay, and Sharron H. Francis

CONTENTS

1.1 Background 3
1.2 Multiple Forms 7
1.3 Phosphodiesterase Inhibitors 11
1.4 *In Vitro* Selectivity 11
1.5 *In Vivo* Selectivity 12
1.6 Concluding Remarks 15
References 15

Signaling pathways employing adenosine $3',5'$-cyclic monophosphate (cAMP) or guanosine $3',5'$-monophosphate (cGMP) (Figure 1.1) play a critical role in the control of myriad processes across a broad spectrum of organisms. The actions of many physiological signals including hormones, neurotransmitters, odorants, and light are mediated by very specific and highly controlled changes in the cellular levels of these second messengers. These changes in either global cyclic nucleotide content or that constrained to specific cellular compartments within the cell, in turn, control a wide range of physiological processes as diverse as vision, olfaction, synaptic function, muscle contraction, water and electrolyte homeostasis, immune responses, hemostasis, carbohydrate and lipid metabolism, and gene expression. The amplitude and duration of the cyclic nucleotide responses are determined by the balance between their rates of synthesis by adenylyl or guanylyl cyclases, and their rates of degradation by one or more cyclic nucleotide phosphodiesterases (PDEs).[1] These PDEs catalyze the insertion of a water-derived hydroxyl group at the phosphorous end in the cyclic phosphate ring (Figure 1.1), thereby hydrolyzing the $3'$ bond of either cAMP or cGMP and converting it to its $5'$ counterpart. While both cyclic nucleotides can be transported out of the cell at a low rate, evidence indicates that, quantitatively, the catalytic action of PDEs provides the only pathway to rapidly lower cellular cyclic nucleotide content at least in the majority of cell types. As a result, the members of the PDE superfamily are well placed to be the targets for pharmacological interventions in a wide variety of pathological conditions impinging on these processes and in pathogenic organisms that inflict health and economic hardships.

1.1 BACKGROUND

The second messengers, cAMP and cGMP, were first identified by Sutherland and colleagues almost 50 years ago. Very soon after the discovery of these cyclic nucleotides, an enzymatic activity was found that could degrade them by catalyzing the hydrolysis of the $3'$ cyclic phosphate bond. As the cyclic phosphate structure is strained, this reaction is highly

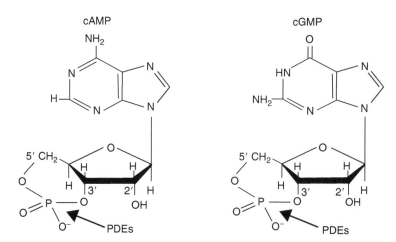

FIGURE 1.1 Structures of cAMP and cGMP showing the location of the 3′ phosphate bond hydrolyzed by class I phosphodiesterases.

exothermic (>10 kcal/mol), making it essentially irreversible in the cell.[2,3] This rapid and essentially irreversible reaction provides a nearly perfect mechanism for the control of a high-affinity second messenger, particularly, in instances where the levels of the messenger must oscillate rapidly. Indeed, very recently it has been demonstrated that cAMP undergoes rapid oscillations in beta cell lines akin to those long noted for Ca^{2+}.[4]

Two classes of PDEs (class I and class II) that can hydrolyze cAMP and cGMP are known to exist in nature. In mammals and flies, only class I enzymes have been identified whereas yeast and protozoans contain both class I and class II PDEs. Little is known about the group of proteins designated as class III PDEs, since they have not been biochemically characterized or shown to be specific for cyclic nucleotides. It is interesting to note that multiple genes encoding cyclic nucleotide PDEs are present in nearly all eukaryotic species. However, in general, the complexity and number of different gene products increase as the complexity of the species increases. As shown in Figure 1.2, simple species like yeast appear to contain only one gene of class I PDE. More complex organisms like *Caenorhabditis elegans* contain at least five genes of this class. Interestingly, most of the mammalian *PDE* genes have relatively close homologs in simpler organisms and several genes found in lower species are most homologous to some of the more recently discovered mammalian PDEs. Examples include the Zebra fish PDE7 homologs and the *Drosophila* homologs of PDE8 and PDE11. DMPDE "32948" is a Drosophila PDE that has a class I PDE consensus catalytic sequence that is most homologus to mammalian PDE9. However, it contains little other sequence that is homologus to any mammalian PDE. It will be of importance to determine if these homologs are present due to mere chance or perhaps indicative of some important highly conserved functions that these PDEs have in common. Similarly, homologs of PDE1C are expressed in nearly all species.

Class I PDEs are the most extensively studied groups of PDEs. In mammalian tissues, three of the eleven PDE families selectively hydrolyze cAMP, three families selectively hydrolyze cGMP, and five families hydrolyze both cyclic nucleotides with varying efficacies (Figure 1.3). It is clear from Figure 1.2 that there appears to be a close relationship between the PDE's and PDE3 (top shaded region) and also as expected among most of the PDEs that are regulated by GAF domains (middle shaded region). Given the similarity in substrate specificity, it is not surprising that the PDE4s and PDE7s are homologus. It is perhaps a bit unexpected that the PDE8s appear to be less related, however, they may just be a bit more ancient. Class I PDEs share a highly conserved catalytic domain (Figure 1.4); in most of the class I PDEs, the

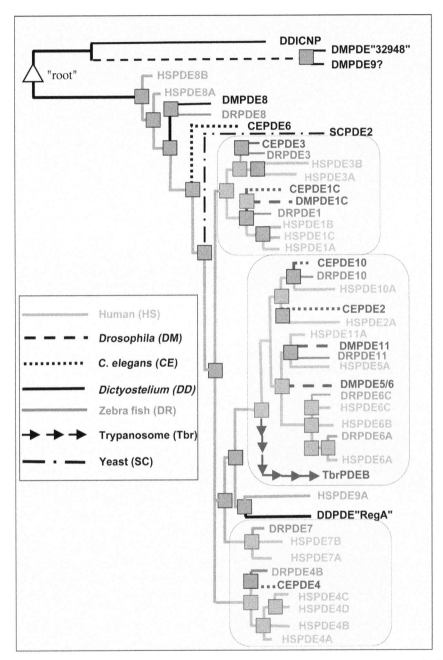

FIGURE 1.2 (See color insert following page 274.) This diagram shows a phenagram depiction of primary amino acid sequence relationships among PDE holoenzymes expressed by seven organisms widely studied in biology. They are color coded by species as indicated. The protein sequences of the A1 or B1, etc., variant of all the major PDE gene products were aligned using the program Multalin as accessed from the Expasy Tools web page (http://us.expasy.org/tools/). Default values for all alignment parameters were used and the results were plotted as a phenagram. This allows the relationship of each PDE from a lower species to be compared to the closest mammalian counterpart. A slightly different set of sequence relationships results if one uses the catalytic domain sequences for the alignment.

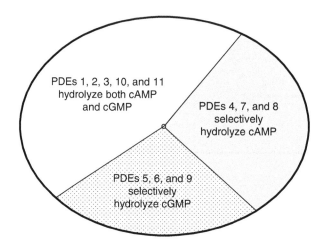

FIGURE 1.3 Cartoon of the three major types of mammalian PDEs classified according to their substrate selectivity. PDEs 1, 2, 3, 10, and 11 efficiently catalyze the hydrolysis of both cAMP and cGMP. At low substrate levels, PDEs 5, 6, and 9 show great selectivity towards cGMP whereas PDEs 4, 7, and 8 preferentially hydrolyze cAMP.

catalytic domain is located in the carboxyl-terminal portion of the protein, but in one PDE from *Trypanosoma cruzi* the catalytic domain is located in the midportion of the polypeptide (see Chapter 14). Diverse regulatory features including cyclic-nucleotide-binding sites that are associated with GAF domains, phosphorylation sites, domains that provide for interaction with specific organelles, membranes, or other proteins are commonly found in more amino-terminal portions of class I PDEs (Figure 1.4). However, in several instances, features in more carboxyl-terminal portions of PDEs are also involved in the regulation by phosphorylation or association with membranes or proteins (Figure 1.4). Remarkably, almost half of the class I PDE families contain one or more GAF domains that provide for dimerization, cyclic nucleotide binding, regulatory functions, or other unidentified roles; among mammalian PDEs, the subgroup of GAF-containing PDEs is comprised of PDEs 2, 5, 6, 10, and 11, and the close evolutionary relationship of this subgroup of PDEs is shown in Figure 1.2. The role of GAF domains differs among PDEs in which they have been studied (see Chapter 4, Chapter 7, and Chapter 8), and for some GAF-domain-containing PDEs, no function has yet been found (see Chapter 12 and Chapter 13). Since GAF domains are rarely found in other mammalian proteins, the abundance of these domains in the PDE superfamily is particularly remarkable, and it is notable that some PDEs from lower forms also contain GAF domains (see Chapter 14).

For mammalian PDE families, the amino acid sequence identity among the catalytic domains ranges from 25% to 52% (Figure 1.5). Fourteen amino acids are invariant among these catalytic domains and are likely to provide the key features required for the catalytic function of these enzymes. The catalytic domains of all class I PDEs contain a consensus sequence ($HNX_2HNX_NE/D/QX_{10}HDX_2HX_{25}E$) that is a double version of a common metal-binding motif (italicized) present in a wide variety of proteases and nucleotide-binding enzymes of all phyla and kingdoms. This motif forms a novel metal-binding site, a portion of which is associated with Zn^{2+} binding in unrelated proteins; with the exception of PDE3, at least one molecule of Zn^{2+} is bound in the catalytic site of all class I PDEs that have been examined. There is also a similar, but modified metal-binding motif found in class II PDEs (see Chapter 14), and evidence suggests that Zn^{2+} is a similar, but critical part of some of these PDEs as well.[5]

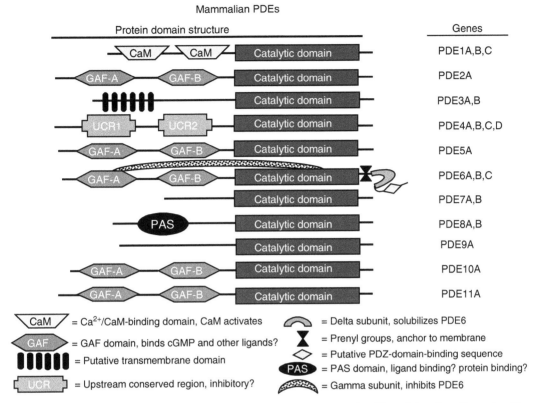

FIGURE 1.4 (See color insert following page 274.) Cartoon of the 11 families of class I cyclic nucleotide phosphodiesterases found in mammals. A conserved catalytic domain showing approximately 25%–52% amino acid sequence identity between any two different families is shown in blue. Regulatory domains that bind Ca^{2+}/calmodulin (CaM) are in yellow; regulatory GAF domains are in red; upstream conserved regions (UCRs) are in green; and a PAS domain of unknown function is shown in black. Some family members are represented by only one gene; others contain as many as four genes. Many of the genes have alternative splicing or alternative start sites to generate a large number of different mRNA products. The longest known form for each PDE family is depicted here. However, many families contain variants that differ substantially in the length of the regulatory domain and the functional domains contained therein. These shorter variants are fully described in the chapters devoted to each family.

1.2 MULTIPLE FORMS

Class I PDEs are expressed in all eukaryotic genomes that have been sequenced. Now that the sequences of several mammalian genomes have been determined, it appears that there are 11 different PDE families. At present, 21 genes that encode proteins containing active PDE catalytic domains have been identified (Figure 1.4). Many mammalian PDE gene families contain more than one gene, and most of the individual genes can produce multiple mRNAs and protein sequences by utilizing alternative transcriptional start sites or alternative splicing that can occur at either the amino-terminus or the carboxyl-terminus. This mRNA expression and processing are regulated in a cell-type-specific manner. However, the actual number of active PDE mRNAs and proteins remains unclear, as our ability to predict all possible splice variants and alternative start sites is imperfect. It is also not known if all variants that have been detected occur in all species. The total number of PDE mRNAs in any one species is likely to exceed 100. Chapter 2 has a comprehensive list of all PDEs currently described in

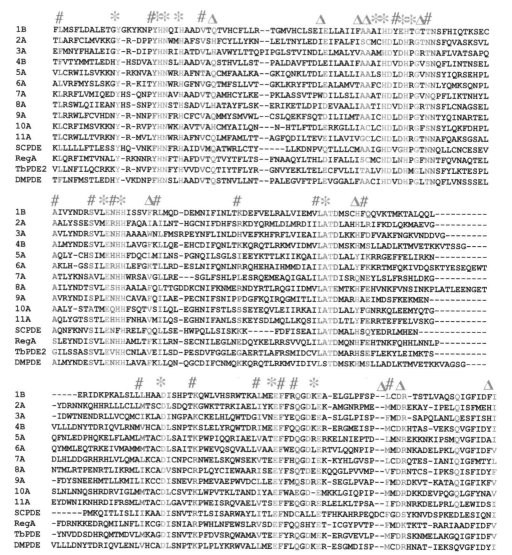

FIGURE 1.5 (See color insert following page 274.) Clustal W sequence alignment of amino acid sequences of Class I PDEs. The 14 amino acids that are invariant among all Class I PDEs are indicated in red and highlighted by an asterisk. Amino acids colored blue (#) are highly conserved (e.g., a tyrosine/phenylalanine or a leucine/isoleucine type of substitution), and positions in which the amino acids are colored green (Δ) have similar residues that are more distantly related. Amino acids that are invariant in mammalian PDE families (in addition to the residues that are invariant across all class I PDEs) are in purple. SCPDE = yeast PDE2, RegA = Dictyostelium PDE, TbPDE = Trypanosome TbrPDEB. IB (beginning at Phe-204) is human PDE1B (NP_000915), 2A (beginning at Thr-638) is human PDE2A (AAC51320), 3A (beginning at Glu-634) is human PDE3A (NP_000912), 4B (beginning at Thr-215) is human PDE4B (NP_002591), 5A (beginning at Val-594) is human PDE5 (NP_001074), 6A (beginning at Ala-541) is human PDE6A (NP_000431), 7A (beginning at Lys-209) is human PDE7A (NP_002594), 8A (beginning at Thr-421) is human PDE8A (NP_002596), 9A (beginning at Thr-294) is human PDE9A (AAC39778), 10A (beginning at Lys-497) is human PDE10A (NP_006652), 11A (beginning at Thr-202) is human PDE11A (BAB16371), Yeast PDE2 is PDE2 of *S. cerevisiae* (CAA999689), RegA from *D. discoideum* (AAB03508), TbPDE2A is PDE2A from *T. brucei* (AAG23160), and Drosophila is (AAC51320). (Figure provided courtesy of Roya Zoraghi.)

Nomenclature

Currently there are two generally accepted nomenclature systems being used in the mammalian PDE field. The systems are very similar and most journals accept the first one, but a few ask that the second be used. In the first system, a distinct PDE name would be constructed as follows: for example HsPDE4A1[1]. The Hs signifies the species of origin, *Homo sapiens*; PDE denotes that it is a 3′,5′-cyclic nucleotide PDE; the Arabic numeral 4 signifies that it is a member of the PDE4 gene family; the capital A signifies it is the A gene. Initially a gene was given the letter A if it was the first reported in the literature for any species and the B gene the second, etc. However, once most PDE genes had been identified for any species, then subsequent homologs in a second or third species are usually (ideally) named for their previously named homolog, not their absolute order of discovery. That is, in such cases the nomenclature defers to the sequence alignment for naming, in order not to add confusion when comparing PDEs among different species.

A second system that is a slightly modified version of this nomenclature is required by some journals (e.g., *Biology of Reproduction*) because these journals adhere to the general mouse gene nomenclature system listed on the Jackson Laboratories web site (http://www.informatics.jax.org/mgihome/nomen/gene.shtml). In this nomenclature system, the same PDE would be written mostly in lower case letters in italics as *Pde4a_vl* when referring to the gene and in nonitalicized capital letters as PDE4A_V1 when referring to the protein. The letter *A* or *a* denotes the gene within the PDE4 family and the letter *V* or *v* stands for the splice or alternative start variant of the gene. Given the similarity between the systems, it would seem that either should be acceptable and either can be used without undue confusion among reports in different journals. In this book we use almost exclusively the first system. Occasionally, authors will affix a numerical suffix to the name to indicate the molecular weight of the expressed protein product, e.g., HSPDE3A-136.

This system of nomenclature can break down when applied to PDEs from a number of "lower species" mostly because the homologies are not universally well conserved. Therefore, several more specialized nomenclature systems have been adopted or commonly used for reference to cyclic nucleotide PDEs in these "lower" species. For example see the one now used for kinetoplast PDEs (trypanosomes) described by Seebeck in this volume.

GenBank. It is to be expected that this list will continue to expand as more information on splicing and alternative start sites become available. There are also several web sites that have periodically provided this type of information and may continue to be updated in the coming years (www.xpharm.com/, www.signaling-gateway.org/, http://depts.washington.edu/pde/Nomenclature.html).

Increasingly, efforts are being made to determine the functional importance of the presence of so many different PDE genes in mammals and their protein products that exhibit such diverse functional characteristics. Twenty years ago many scientists in the field were baffled by the functional relevance of some PDEs with K_m values that were much higher than levels expected for either cAMP or cGMP within the cell. In addition, many scientists felt that multiple isoforms of a particular PDE most likely represented either degradation products or, at best, redundant functions. Actually, this perception holds true for most key regulatory enzyme systems in biology. Today, the perception in the PDE field has changed to be almost diametrically opposite. With increased understanding of alternative splicing, alternative promoters, differential expression, differential binding to scaffolding proteins, selective compartmentation, differential synthesis and degradation rates, and differential control by multiple

signaling pathways, most scientists have begun to realize that understanding the roles played by multiple forms of an enzyme is absolutely essential for our understanding of regulatory biology. Indeed, many chapters in this book provide details about our emerging understanding of specific roles played by the protein products of the different PDE genes, and in some instances, even for particular splice variants of the individual genes. In addition, selective anchoring of PDEs within the subcellular milieu is now well documented for many PDEs. In some instances, such as the retinal PDEs, the anchoring is fairly static since the PDEs are restricted to one area of the cell and, within that area, they are almost exclusively associated with the membrane. In other instances, anchoring of PDEs is a dynamic process that changes with different regulatory signals. These topics are well covered in several chapters (Chapter 6 for example, Chapter 19, and Chapter 20), which are devoted to particular PDE families.

One consequence of an organism having the flexibility to use alternative splicing and alternative start sites to generate different forms of a particular PDE family is to provide molecular mechanisms that allow selective subcellular localization of different forms of that enzyme. A consensus is emerging that, for cyclic nucleotide functions, it is not only the regulation of long-term steady levels that are important, but also the regulation of the amplitude and duration of these second messenger signals that are critical to the proper control of many physiological processes. We now know that in most tissues, cyclic nucleotides can oscillate on a minute-to-minute timescale and in several sensory tissues even on a milliseconds timescale. For example in the retina, the total pool of cGMP can turnover within 15 ms after a moderate light flash.[6] In olfactory neurons, cAMP level peaks within 50 ms and falls to near basal levels within another 150 ms.[7] The rapidity of these responses allows the animal to react quickly to sensory input, and this is made possible, in part, by very high concentrations of PDE6 in the photoreceptor outer segments and of PDE1C in the cilia of olfactory neurons, respectively. Somewhat slower oscillations also occur in other cell types. A recent paper on changes in cAMP levels in a pancreatic beta cell line shows that the cAMP level in these cells can oscillate multiple times within a 10 min timescale, and that these levels closely parallel intracellular Ca^{2+} oscillations.[4] The spikes and valleys of cellular cAMP content are made possible in large part by the oscillation of the activity state of particular PDEs between less active and more active forms; the mechanisms mediating the transition between different activity states of PDEs are not fully understood, but such transitions are likely to be common among PDEs. Interestingly, the cells apparently can respond not only to the frequency and amplitude of the oscillations, but also to the average combined signal. In fact, it is this combined signal that we normally measure experimentally.

It is increasingly clear that in order to understand fully how PDEs can control the amplitude, duration, and frequency of cyclic nucleotide oscillations, we will have to be able to measure changes in PDE activity in real time with spatial as well as temporal resolution. The relatively recent emergence of techniques, utilizing fluorescent reporters for Ca^{2+} and particularly for cAMP and cGMP, has just begun to have sufficient sensitivity and speed to allow these measurements. Currently, it would appear that only those sensors that make use of the rapid responses of cyclic-nucleotide-gated channels have sufficiently rapid off rates to approach this problem. However, it seems likely that better probes will become available in the near term.[8]

In addition to the rapid oscillations that can occur, we also are beginning to appreciate that there are very likely to be either pools or clouds of cAMP and cGMP in subcellular regions of the cell. This idea has been placed on much firmer ground in recent years by the discoveries of an increasing number of anchoring or scaffolding proteins that sequester specific splice variants of PDEs, as well as cyclases, protein kinases, phosphoprotein phosphatases, channels, A-kinase anchor proteins (AKAPs), substrates, etc., in particular locations within the cell. Many chapters in this volume discuss specific examples of this type of regulation. To date most of the data regarding these types of regulation have come from the study of the members of the PDE4 family (Chapter 6). However, there is no reason to suppose that it is limited to this particular family and, indeed, recent evidence discussed in other chapters

highlight examples of PDEs 1, 2, 3, 5, 6, and 7 targeted to specific subcellular localizations and interacting with scaffolding proteins.

Why are there so many different PDE isozymes? On the one hand, it is amazing that PDEs comprise such a complex superfamily of enzymes that are designed solely to hydrolyze either cAMP or cGMP and that so many representatives of these enzymes have been conserved throughout evolution, withstanding considerable selective pressure. Not only that, but even in families with complex isoform complements, such as PDE4, the conservation between humans and mouse is remarkable. This conservation strongly suggests that there is strong selective advantage for a species to tightly regulate cyclic nucleotide levels through the process of differential control of a given combination of multiple PDEs. On the other hand, it is important to point out that in the PDE field, and also possibly for many other regulatory enzymes, the role for a particular PDE family or variants within a family in one species may not be the same as it is for another. For example, PDE1C is upregulated in human smooth muscle in response to proliferative signals.[9] However, this variant of PDE1 may not be expressed in bovine, rat, or mouse smooth muscle cells. Similar examples are seen in the population of PDEs expressed in testis from different animals, and in this case the PDEs even vary between rats and mice.[10] On the one hand, this variation gives one pause in extrapolating experimental results derived from various gene disruption and transgenic animal models to human physiology. On the other hand, all species are different, and it is just as easy to imagine a simple mutation in a promoter region of a gene as important to speciation as to hundreds or thousands of mutations in the coding regions. Similarly, it is then equally easy to imagine that different species might find it efficacious to utilize different promoters to change the expression level or subcellular localization of a PDE to best suit its biological needs.

1.3 PHOSPHODIESTERASE INHIBITORS

One of the greatest advances in the PDE field in the last 15 years has been the increased availability and more recently the clinical use of family-selective inhibitors. Clinically, the huge therapeutic and financial success of the PDE5-selective inhibitors, Viagra (sildenafil), Cialis (tadalafil), and Levitra (vardenafil) have focused renewed energy into this area of research. A number of investigators have adopted new approaches in identifying small molecules that could serve as potential platforms for the synthesis of a new generation of selective PDE inhibitors (see Chapter 29 and Chapter 30). Currently, several highly selective PDE inhibitors either are approved or in the late-stage clinical trials for the treatment of a variety of pathologies. In addition to the PDE5-selective inhibitors including the recently approved udenafil (Zydena), these inhibitors include the PDE3 inhibitor (cilostazol) for the treatment of intermittent claudication (see Chapter 31) and the PDE4 inhibitor Ariflo (cilomilast) for the treatment of chronic obstructive pulmonary disease (see Chapter 28), and Aggrenox (extended release of dipyridamole with aspirin) for the treatment of recurrent stroke. Furthermore, the range of uses for the existing PDE inhibitors is expanding rapidly. For example, the PDE5 inhibitors are used clinically or are in trial for the treatment of pulmonary arterial hypertension and high-altitude sickness and likely to be tried shortly for cardiac hypertrophy (see Chapter 24 and Chapter 25). PDE4 inhibitors are also being considered for treatment of central nervous system (CNS) diseases including memory and schizophrenia as well as possibly acting as antidepressants (see Chapter 27).

1.4 *IN VITRO* SELECTIVITY

There are several facets of selectivity that should be considered when using PDE inhibitors either in laboratory experiments or evaluating them when reading manuscripts describing their use. Perhaps the most important is *in vitro* specificity. The term specific is to be strongly

discouraged because all currently characterized PDE inhibitors have some degree of cross-reactivity with other families of PDEs. Most researchers in the field agree that a compound should show at least a 30-fold and preferably greater than 100-fold, selectivity over all other PDEs in order to be called truly selective in an *in vitro* experiment. This means that the K_i should be 100-fold lower (or the IC_{50} if determined at low substrate concentration) for the PDE in question compared to all other PDEs. With 100-fold selectivity, one should easily be able to define assay conditions that achieve 90% inhibition of the PDE in question with 10% or less effect on other PDEs. Unfortunately, many of the inhibitors described in the literature as being selective or specific show less discrimination than the criteria mentioned above. Another problem in comparative studies is that several PDEs have quite different K_m values for cAMP or cGMP. This means that while true K_i values are comparative, IC_{50} values determined at a single concentration of either cAMP or cGMP can be potentially misleading if one enzyme is near saturation with the substrate and the other is not. It is possible in some tissues to still obtain meaningful date (e.g., to determine the portion of the measured PDE activity that is due to a particular isozyme) even with semi- or partially selective inhibitors. However, to be rigorous, this condition requires additional prior information regarding the various PDE isoforms present in the tissue. As it is particularly difficult to find comparable *in vitro* data on all of the PDE families, difficulties will likely persist.

1.5   *IN VIVO* SELECTIVITY

The difficulties concerning interpretation of selectivity are compounded when PDE inhibitors are used in intact cells or whole animals. Ideally, it would also be important to have some knowledge of the local cyclic nucleotide concentration in the vicinity of the PDE in question, but currently technologies that can experimentally assess the condition are not available. Moreover, in order to be interpreted confidently, it needs to be established that the compound enters the cell with minimal difficulty and that it exhibits the same IC_{50} in the intact cell as it does in broken cell preparations. Unfortunately, this is seldom tested in the literature. Interpretation of results is also complicated by the fact that the degree of inhibition of a PDE activity that is required to cause physiological and pharmacological effects is often not clear. As it has been known since the early experiments of Earl Sutherland and colleagues that low doses of PDE inhibitors can greatly potentiate low doses of adenylyl cyclase agonists, this is a particularly vexing problem. For example, it is often not clear how to interpret a small effect of a drug *in vivo* on a process when on the one hand it may penetrate the cell poorly, but on the other hand full inhibition of one or more PDEs may not be required to show an effect. It is expected that these issues will continue to complicate the interpretation of experiments based solely on the use of PDE inhibitors in the foreseeable future, but readers are urged to judiciously consider the problem when interpreting the data derived from studies employing PDE inhibitors.

Since all the subsequent chapters of this book are either specific to a particular PDE or a particular tissue or process, in this introductory chapter we have included a table of commonly used PDE inhibitors, their relative selectivities, and IC_{50} values (Table 1.1). Special emphasis is placed on those compounds that are commercially available. Where the data are available, each compound listed is thought to be at least 30-fold selective for the PDE family under which it is listed. We have also included a second table describing a number of commonly used, but only partially selective, PDE inhibitors along with brief descriptions of their use and selectivity (Table 1.2). It is anticipated that many additional new proprietary company compounds will become available in the near future. Finally, it should be noted that clinically, it may be well known that drugs that inhibit a particular combination of PDEs may be more efficacious and possibly less toxic than drugs that are selective against individual PDEs. This possibility is discussed in more detail in Chapter 32. It is also useful to know that

TABLE 1.1
Selective Inhibitors of Various PDE Families

PDE	Inhibitors	Source	IC_{50}	Usage	Ref.
PDE1	IC224	ICOS	80 nM	Vinpocetine and Sch51866 are used mostly *in vitro*; IC224 may be the most permeable in intact cells	11
	SCH51866	Schering	13–100 nM		12
	Vinpocetine	W-A	14 μM		13
PDE2	EHNA	W-A	1 μM	Investigated for improving memory and decreasing endothelial permeability under inflammatory conditions	14
	PDP	ICOS	0.6 nM		15
	IC933	Bayer	4 nM		11
	Bay 60-7550		4.7 nM		16
PDE3	Trequinsin	W-A	300 pM	Milrinone is currently approved for short-term treatment of congestive heart failure	17
	OPC-33540	Otsuka	~1 nM		18
	Cilostamide	W-A	20 nM	Cilastazol is used for intermittent claudication. It also inhibits adenosine uptake with an IC_{50} in the 5–10 μM range	19
	Milrinone	W-A	150 nM		20
	Cilastazol	Otsuka	200 nM		21
PDE4	Roflumilast	Altana	0.8 nM	Many compounds have been developed to treat COPD, but most have also caused serious side effects. PDE4 inhibitors are also under investigation for other inflammatory conditions and for CNS disorders including depression, schizophrenia, and memory loss	22
	AWD12-281	Elbion	10 nM		23
	SCH351591	Schering	60 nM		24
	Cilomilast	GSK	120 nM		25
	Rolipram	W-A	1 μM		26
	Ro 20-1724	W-A	2 μM		27
PDE5	Vardenafil	Bayer	0.1 nM	Sildenafil, vardenafil, and tadalafil are in usage as erectile dysfunction drugs. These compounds are in trials for other indications such as pulmonary hypertension and benign prostatic hyperplasia. See note on selectivity of the PDE5 inhibitors toward PDEs 6 and 11 in Table 1.2	28
	Udenafil	Dong-A	6 nM		29
	Sildenafil	Pfizer	4 nM		30
	Tadalafil	Lilly-ICOS	2 nM		31
PDE6	No selective inhibitors available	—	—	No truly selective small molecule PDE6 inhibitors are available. PDE6 inhibition may be a source of visual side effects related to sildenafil use. Zaprinast and dipyridamole show some selectivity toward PDE6	—

continued

TABLE 1.1 (continued)
Selective Inhibitors of Various PDE Families

PDE	Inhibitors	Source	IC_{50}	Usage	Ref.
PDE7	BRL 50481	GSK	260 nM	PDE7-selective inhibitors investigated as anti-inflammatory agents as yet have shown limited utility *in vivo*	33
	IC242	ICOS	370 nM		34
PDE8	No selective inhibitors available	—	—	—	—
PDE9	BAY 73-6691	Bayer	55 nM	BAY 73-6691 is in preclinical development for the treatment of Alzheimer's disease	35
PDE10	No selective inhibitors available	—	—	No truly selective inhibitors are yet available. Dipyridamole shows some selectivity but is not specific for PDE10	—
PDE11	No selective inhibitors available	—	—	Most interest in PDE11 relates to observation that it is also inhibited by tadalafil, which is thus a potential cause for side effects	—

W-A, widely available; EHNA, *erythro*-9-(2-hydroxy-3-nonyl)adenine; and IBMX, 3-isobutyl-1-methylxanthine.

TABLE 1.2
Isozyme Cross-Reactivity of Commonly Used PDE Inhibitors

Inhibitor	PDE	IC_{50}
Sildenafil	PDE1	350 nM
	PDE6	12–50 nM
Tadalafil	PDE6	2 μM
	PDE11	73 nM
Vardenafil	PDE6	11 nM
Zaprinast	PDE1	6 μM
	PDE5	0.13 μM
	PDE10	22 μM
	PDE11	12 μM
Zardaverine	PDE3	0.5–2 μM
	PDE4	0.8–4 μM
Dipyridamole	PDE5	0.9 μM
	PDE6	0.13 μM
	PDE7	0.60 μM
	PDE8	9 μM
	PDE10	1 μM
	PDE11	0.4 μM
Trequinsin	PDE4	0.2–0.8 μM
	PDE2	~1 μM
IBMX	Multiple	2–50 μM

Although many inhibitors are called selective, this is a relative term. Compounds are termed selective based on the fact that other isoforms are only inhibited at higher concentrations. Some compounds such as dipyridamole and zardaverine are selective for a limited number of isoforms. However, inhibitors that target multiple isoforms may be useful because of this property. For example, dipyridamole is a component of the stroke-treatment drug, Aggrenox.

many of the nonselective PDE inhibitors have greatly different efficacy toward some PDEs compared to others. For example, the widely used 3-isobutyl-1-methylxanthine (IBMX) has an IC_{50} for PDE4s in the range of 5 µM but has little or no efficacy toward PDE 8 or 9 (see Chapter 10 and Chapter 11).

1.6 CONCLUDING REMARKS

Since the publication of the last book in this series, an enormous amount of information on the regulation and roles of PDEs has been accumulated. Investigators in the PDE field have now largely completed the cataloging of the various PDE gene family members in humans and have made a good start in determining their functions in the cell. We still have much work ahead in order to understand how many different splice and alternative start variants exist in any one tissue or species and, of course, the physiological uses for which the cell utilizes these variants is only just beginning to be understood. Progress has been made in determining the three-dimensional structures of the catalytic and some of the regulatory domains of PDEs, although we still await the structure for a PDE holoenzyme. Major advances continue to be made in the design and synthesis of inhibitors that are potent and highly selective for particular PDEs, and major efforts have been initiated to develop inhibitors that target pathogens that adversely impact health and economic welfare of humans and their domestic animals. Finally, we are now beginning to see highly effective drugs come to market that make use of the early understanding of the regulation and roles of specific isozymes of PDEs. It is clear that much needs to be done but the future seems bright and challenging. We hope that you will enjoy reading the chapters of this book and possibly take away some good ideas for your own research projects.

REFERENCES

1. Beavo, J.A., M. Conti, and R.J. Heaslip. 1994. Multiple cyclic nucleotide phosphodiesterases. *Mol Pharmacol* 46 (3):399–405.
2. Greengard, P., S.A. Rudolph, and J.M. Sturtevant. 1969. Enthalpy of hydrolysis of the 3' bond of adenosine 3',5'-monophosphate and guanosine 3',5'-monophosphate. *J Biol Chem* 244 (17): 4798–800.
3. Hayaishi, O., P. Greengard, and S.P. Colowick. 1971. On the equilibrium of the adenylate cyclase reaction. *J Biol Chem* 246 (18):5840–3.
4. Dyachok, O., Y. Isakov, J. Sagetorp, and A. Tengholm. 2006. Oscillations of cyclic AMP in hormone-stimulated insulin-secreting beta-cells. *Nature* 439 (7074):349–52.
5. Londesborough, J. 1985. Evidence that the peripheral cyclic AMP phosphodiesterase of rat liver plasma membranes is a metalloenzyme. *Biochem J* 225 (1):143–7.
6. Goldberg, N.D., A.A.D. Ames, J.E. Gander, and T.F. Walseth. 1983. Magnitude of increase in retinal cGMP metabolic flux determined by [18]O incorporation into nucleotide alpha-phosphoryls corresponds with intensity of photic stimulation. *J Biol Chem* 258 (15):9213–9.
7. Breer, H., I. Boekhoff, and E. Tareilus. 1990. Rapid kinetics of second messenger formation in olfactory transduction. *Nature* 345 (6270):65–8.
8. Shaner, N.C., P.A. Steinbach, and R.Y. Tsien. 2005. A guide to choosing fluorescent proteins. *Nat Methods* 2 (12):905–9.
9. Rybalkin, S.D., K.E. Bornfeldt, W.K. Sonnenburg, I.G. Rybalkina, K.S. Kwak, K. Hanson, E.G. Krebs, and J.A. Beavo. 1997. Calmodulin-stimulated cyclic nucleotide phosphodiesterase (PDE1C) is induced in human arterial smooth muscle cells of the synthetic, proliferative phenotype. *J Clin Invest* 100 (10):2611–21.
10. Rossi, P., R. Pezzotti, M. Conti, and R. Geremia. 1985. Cyclic nucleotide phosphodiesterases in somatic and germ cells of mouse seminiferous tubules. *J Reprod Fertil* 74 (2):317–27.
11. Snyder, P.B., J.M. Esselstyn, K. Loughney, S.L. Wolda, and V.A. Florio. 2005. The role of cyclic nucleotide phosphodiesterases in the regulation of adipocyte lipolysis. *J Lipid Res* 46 (3):494–503.

12. Watkins, R.W., H.R. Daris Jr., A. Fawzi, H.S. Ahn, J. Cook, R. Cleven, L. Hoos, D. McGregor, R. McLeod, K. Pula, R. Tedesco, and D. Tulshian. 1995. Antihypertensive, hemodynamic, and vascular protective effects of SCH-51866, an inhibitor of cGMP hydrolysis. *FASEB J* 9:A342.

13. Hagiwara, M., T. Endo, and H. Hidaka. 1984. Effects of vinpocetine on cyclic nucleotide metabolism in vascular smooth muscle. *Biochem Pharmacol* 33 (3):453–7.

14. Podzuweit, T., P. Nennstiel, and A. Muller. 1995. Isozyme selective inhibition of cGMP-stimulated cyclic nucleotide phosphodiesterases by *erythro*-9-(2-hydroxy-3-nonyl) adenine. *Cell Signal* 7 (7):733–8.

15. Seybold, J., D. Thomas, M. Witzenrath, S. Boral, A.C. Hocke, A. Burger, A. Hatzelmann, H. Tenor, C. Schudt, M. Krull, H. Schutte, S. Hippenstiel, and N. Suttorp. 2005. Tumor necrosis factor-alpha-dependent expression of phosphodiesterase 2: Role in endothelial hyperpermeability. *Blood* 105 (9):3569–76.

16. Boess, F.G., M. Hendrix, F.J. van der Staay, C. Erb, R. Schreiber, W. van Staveren, J. de Vente, J. Prickaerts, A. Blokland, and G. Koenig. 2004. Inhibition of phosphodiesterase 2 increases neuronal cGMP, synaptic plasticity and memory performance. *Neuropharmacology* 47 (7):1081–92.

17. Tanaka, T., T. Ishikawa, M. Hagiwara, K. Onoda, H. Itoh, and H. Hidaka. 1988. Effects of cilostazol, a selective cAMP phosphodiesterase inhibitor on the contraction of vascular smooth muscle. *Pharmacology* 36 (5):313–20.

18. Sudo, T., K. Tachibana, K. Toga, S. Tochizawa, Y. Inoue, Y. Kimura, and H. Hidaka. 2000. Potent effects of novel anti-platelet aggregatory cilostamide analogues on recombinant cyclic nucleotide phosphodiesterase isozyme activity. *Biochem Pharmacol* 59 (4):347–56.

19. Hidaka, H., and T. Endo. 1984. Selective inhibitors of three forms of cyclic nucleotide phosphodiesterase—Basic and potential clinical applications. *Adv Cyclic Nucleotide Protein Phosphorylation Res* 16 (245):245–59.

20. Harrison, S.A., D.H. Reifsnyder, B. Gallis, G.G. Cadd, and J.A. Beavo. 1986. Isolation and characterization of bovine cardiac muscle cGMP-inhibited phosphodiesterase: A receptor for new cardiotonic drugs. *Mol Pharmacol* 29:506–14.

21. Ruppert, D., and K.U. Weithman. 1982. HL 725, an extremely potent inhibitor of platelet phosphodiesterase and induced platelet aggregation in vitro. *Life Sci* 31:2037–43.

22. Hatzelmann, A., and C. Schudt. 2001. Anti-inflammatory and immunomodulatory potential of the novel PDE4 inhibitor roflumilast in vitro. *J Pharmacol Exp Ther* 297 (1):267–79.

23. Schmidt, D.T., N. Watson, G. Dent, E. Ruhlmann, D. Branscheid, H. Magnussen, and K.F. Rabe. 2000. The effect of selective and non-selective phosphodiesterase inhibitors on allergen- and leukotriene C-4-induced contractions in passively sensitized human airways. *Br J Pharmacol* 131 (8):1607–18.

24. Billah, M., G.M. Buckley, N. Cooper, H.J. Dyke, R. Egan, A. Ganguly, L. Gowers, A.F. Haughan, H.J. Kendall, C. Lowe, M. Minnicozzi, J.G. Montana, J. Oxford, J.C. Peake, C.L. Picken, J.J. Piwinski, R. Naylor, V. Sabin, N.Y. Shih, and J.B.H. Warneck. 2002. 8-Methoxyquinolines as PDE4 inhibitors. *Bioorg Med Chem Lett* 12 (12):1617–9.

25. Griswold, D.E., E.F. Webb, A.M. Badger, P.D. Gorycki, P.A. Levandoski, M.A. Barnette, M. Grous, S. Christensen, and T.J. Torphy. 1998. SE 207499 (Ariflo), a second generation phosphodiesterase 4 inhibitor, reduces tumor necrosis factor alpha and interleukin-4 production in vivo. *J Pharmacol Exp Ther* 287 (2):705–11.

26. Schwabe, U., M. Miyake, Y. Ohga, and J.W. Daly. 1976. 4-(3-Cyclopentyloxy-4-methoxyphenyl)-2-pyrrolidone (ZK 62711): A potent inhibitor of adenosine cyclic 3′,5′-monophosphate phosphodiesterases in homogenates and tissue slices from rat brain. *Mol Pharmacol* 12 (6):900–10.

27. Sheppard, H., and W.H. Tsien. 1975. Alterations in the hydrolytic activity, inhibitor sensitivity and molecular size of the rat erythrocyte cyclic AMP phosphodiesterase by calcium and hypertonic sodium chloride. *J Cyclic Nucleotide Res* 1 (4):237–42.

28. Bischoff, E. 2004. Potency, selectivity, and consequences of nonselectivity of PDE inhibition. *Int J Impot Res* 16 (Suppl. 1):S11–4.

29. Oh, T.Y., K.K. Kang, B.O. Ahn, M. Yoo, and W.B. Kim. 2000. Erectogenic effect of the selective phosphodiesterase type 5 inhibitor, DA-8159. *Arch Pharmacol Res* 23 (5):471–6.

30. Boolell, M., M.J. Allen, S.A. Ballard, S. Gepi-Attee, G. Muirhead, A.M. Naylor, I.H. Osterloh, and C. Gingell. 1996. Sildenafil: An orally active type 5 cyclic GMP-specific phosphodiesterase inhibitor for the treatment of penile erectile dysfunction. *Int J Impot Res* 8 (2):47–52.

31. Padma-Nathan, H., J.G. McMurray, W.E. Pullman, J.S. Whitaker, J.B. Saoud, K.M. Ferguson, and R.C. Rosen. 2001. On-demand IC351 (Cialis) enhances erectile function in patients with erectile dysfunction. *Int J Impot Res* 13 (1):2–9.
32. Lugnier, C., P. Schoeffter, B.A. Le, E. Strouthou, and J.C. Stoclet. 1986. Selective inhibition of cyclic nucleotide phosphodiesterases of human, bovine and rat aorta. *Biochem Pharmacol* 35 (10):1743–51.
33. Smith, S.J., L.B. Cieslinski, R. Newton, L.E. Donnelly, P.S. Fenwick, A.G. Nicholson, P.J. Barnes, M.S. Barnette, and M.A. Giembycz. 2004. Discovery of BRL 50481 [3-(N,N'-dimethylsulfonamido)-4-methyl-nitrobenzene], a selective inhibitor of phosphodiesterase 7: In vitro studies in human monocytes, lung macrophages, and CD8[+] T-lymphocytes. *Mol Pharmacol* 66 (6):1679–89.
34. Lee, R., S. Wolda, E. Moon, J. Esselstyn, C. Hertel, and A. Lerner. 2002. PDE7A is expressed in human B-lymphocytes and is up-regulated by elevation of intracellular cAMP. *Cell Signal* 14 (3):277–84.
35. Wunder, F., A. Tersteegen, A. Rebmann, C. Erb, T. Fahrig, and M. Hendrix. 2005. Characterization of the first potent and selective PDE9 inhibitor using a cGMP reporter cell line. *Mol Pharmacol* 68 (6):1775–81.

2 Phosphodiesterase Isoforms— An Annotated List

Graeme B. Bolger

CONTENTS

2.1 Introduction 19
References 25

2.1 INTRODUCTION

One of the characteristic features of the phosphodiesterase (PDE) superfamily is its incredible diversity of isoforms. This diversity has been generated in evolution by two mechanisms. The first is gene duplication followed by gene divergence. Indeed, it is now clear that all 21 mammalian PDE genes are sufficiently similar and that they all evolved from a single ancestral gene in this way. The second mechanism is the generation of multiple isoforms from any individual gene, through the use of alternative mRNA splicing or the use of multiple transcriptional start sites for each isoform. Some PDE gene families, such as the PDE4 family, make extensive use of both mechanisms, whereas others, such as the PDE6 family, have a relatively limited number of different isoforms.

Table 2.1 presents a list of all known mammalian PDE isoforms, as defined by their mRNAs and the proteins that they encode. Only sequences that have been deposited in GenBank or allied databases are included. Every attempt has been made to include every isoform that encodes a unique protein. On occasion, some cDNAs have been submitted to GenBank that differ from other entries only in their nonprotein-encoding regions (i.e., 3$'$ or 5$'$ untranslated regions [UTRs]) and they have also been included.

For conciseness, GenBank entries that are generated from expressed sequence tags (ESTs) have been excluded, unless they encode full-length proteins. Genomic sequences have also been excluded, with the exception of a few genomic entries that, in the opinion of GenBank staff, or the submitting scientist, clearly encode genuine mRNAs. In most cases, these mRNAs encode proteins that are strongly homologous to those of another species (e.g., rat versus mouse) and were felt to be sufficiently reliable to be worthy of inclusion.

The number of PDE isoforms in the databases that have been defined solely on the basis of their genomic DNA sequence have exploded in the past few years, as more mammalian genomes have been sequenced (e.g., the chimpanzee, the dog, and other recent examples), and are likely to increase substantially in the near future. However, to date, very few, if any, of the isoforms from these newly sequenced organisms have been identified as experimentally isolated cDNAs. Instead, they have been identified by bioinformatic approaches from genomic sequence, almost always as homologs of experimentally isolated cDNAs from other organisms. For this reason, they have been excluded from this table. If all these sequences were included in this table, it would be even longer than it is already.

TABLE 2.1
Mammalian Cyclic Nucleotide Phosphodiesterase Isoforms

Gene	Human Isoforms				Equivalent Mammalian Isoforms			
	Name	GenBank	Size (aa)	Ref.	Name	GenBank	Size (aa)	Ref.
PDE1A	PDE1A1	AB038224; AF110235; AF110238	529	1, 2	MMPDE1A1	AF023529	529	3
					BTPDE1A1	L34069	514	4
	PDE1A2	AF110238; AF110236; NM_005019	545	2, 5	MMPDE1A2	AK043647	545	6
					BTPDE1A2	M90358	530	
	PDE1A3	U40370; AF110236	535	2, 7	MMPDE1A?	U56649	565*	8, 9
	PDE1A4	AB038224; AJ401610; AF110235	519	1, 2, 5				
	PDE1A5	(Same as 1A2)						
	PDE1A6	(Same as 1A3)						
	PDE1A7	(Now PDE1A10)			MMPDE1A7	AF159298	456	9
	PDE1A8	AB038224	536	1	MMPDE1A8	AY845863	?	9
	PDE1A9	AB038224	552	1	MMPDE1A9	AY845864	465	9
	PDE1A10	AB038224	501	1				
	PDE1A11	AB038224	511	1				
	PDE1A12	AB038224	518	1				
PDE1B	PDE1B1	U56976	536	10, 11	MMPDE1B1	L01695	535	106
					RNPDE1B1	M94537	535	12
					BTPDE1B1	M94867	534	13
	PDE1B2	AJ401609	516	5	MMPDE1B2	AK004772	516	
PDE1C	PDE1C1	U40371	634	7	MMPDE1C1	L76944	631	14
	PDE1C2	AK091734	769		RNPDE1C2	L41045	768	15
	PDE1C3	U40372	709	7	MMPDE1C3	AK030423	706	
	PDE1C4	(Not yet identified)			MMPDE1C4	L76947	654	14
	PDE1C5	(Not yet identified)			MMPDE1C5	L76946	654	14
	PDE1C6	(Not yet identified)			MMPDE1C6	AK029531	603	
	PDE1C7	(Not yet identified)			MMPDE1C7	AK014887	617	
	PDE1C8	(Not yet identified)			MMPDE1C8	AK504499	501	
PDE2A	PDE2A1	AK092278	786*		BTPDE2A1	M73512	921	16
					MMPDE2A2	NM_284323	928	
	PDE2A2	AK095024	950		RNPDE2A2	U21101	928	17
					BTPDE2A2	L49503	942	18

Family	Isoform	GenBank accession	aa	Ref.	Ortholog	GenBank accession	aa	Ref.
	PDE2A3	U67733; NM_002599	941	19	MMPDE2A3	NM_018779	938	
	PDE2A4	AY495087	932					
PDE3A	PDE3A1	M91667; U36798; NM_000921	1141	20, 21	MMPDE3A1		1141	
	PDE3A2		985?	21	SSPDE3A2	AF161582	985	22
	PDE3A3		?	21				
PDE3B	PDE3B1	AY459346	1112	23	MMPDE3B1	NM_011055	1110	24
					RNPDE3B1	Z22867	1108	26
PDE4A	PDE4A1	U97584	647	25	MMPDE4A1	AF142645	610	
					RNPDE4A1	M26715; J04554; L27062	610	27, 28
					RNPDE4A?	M25348; M28411	358*	29
	PDE4A2	(Does not exist)			RNPDE4A2	M6717; J04554	493*	27
	PDE4A3	(Does not exist)			RNPDE4A3	M26716; J04544	585*	27
	PDE4A4	L20965	886	30	MMPDE4A5	AF142643	844	26
					RNPDE4A5	L20757	844	28
	PDE4A6	M37744	775*	31	(Does not exist)			
	PDE4A7	U18087	686*	32	(Does not exist)			
	PDE4A8	U18088	323*	32, 33	RNPDE4A6	L36467	763	34, 35
	PDE4A9	AY593872	864	(Unpublished)	(Does not exist)			
	PDE4A10	AF073745	825	36				
	PDE4A11	AY618547; L20967	860	30, 37				
PDE4B	PDE4B1	L20966	736	30	RNPDE4B1	AF202372; J04563	736	38, 39
	PDE4B2	L20971; M97515	564	30, 40	MMPDE4B2	BC023751	564	
		L12686	606*	41	RNPDE4B2	L27058	564	28
					MMPDE4B2	M25350; M28413	360*	29, 42
	PDE4B3	U85048	721	43	MMPDE4B3	AF326555; AF326556; AF208023	721	44, 45
					RNPDE4B3	U95748	721	43
	PDE4B4	(Genomic only)		39	RNPDE4B4	AF202733	659	39
PDE4C	PDE4C1	L20968	251*	30	RNPDE4C1	M25347; M28410	359*	29
		Z46632	712*	46		L27061	536*	28
	PDE4C2	U88712	606	47				
	PDE4C3	U88713	700*	47				
	PDE4C4	U66346	791*	48				
	PDE4C5	U66347	426*	48				
	PDE4C6	U66348	518*	48				

continued

TABLE 2.1 (continued)

Gene	Name	Human Isoforms GenBank	Size (aa)	Ref.	Name	Equivalent Mammalian Isoforms GenBank	Size (aa)	Ref.
PDE4D	PDE4C7	U66349	427*	48				
	PDE4D1	U50157	585	49	MMPDE4D1	AK081466	584	51, 52
		U79571	205*	50	RNPDE4D1	U09455; M25349; M28412	584	28
						L27960	551*	
	PDE4D2	U50158	507	49	RNPDE4D2	U09456; M25349	506	51, 53
		AF012074	507	50				
	PDE4D3	L20970; U50159	673	30, 49	RNPDE4D3	U09457	672	51, 53
		U02822	604*	54		L27059	578*	28
	PDE4D4	L20969	809	30, 50	RNPDE4D4	AF031373	803	55
	PDE4D5	AF012073	745	50				
	PDE4D6	AF536976	518	56	RNPDE4D6	AF536974	517	56
	PDE4D7	AF536976	748	56	MMPDE4D7	AF536978	747	56
					RNPDE4D7	AF536979	747	56
	PDE4D8	AF536977	687	56				
	PDE4D9	AY245867; AY388960	679	57	MMPDE4D9	AY388962	678	
					RNPDE4D9	AY388961	678	
PDE5	PDE5A1	AF043731; D89094; AJ004865; AB001635	875	58–60	MMPDE5A1	AF541937	865	61
					BTPDE5A1	L16545	865	61
	PDE5A2	AF043732	833	58	RNPDE5A2	D89093	833	62
	PDE5A3	AY264918	823	63				
PDE6A	PDE6A1	M26061	859	64	MMPDE6A1	X60664	859	65
					BTPDE6A1	M27541	859	66
					BTPDE6A2	M26043	859	64
					CFPDE6A1	AJ233677	643*	67
					RPPDE6A1	AY044174	866	68
PDE6B	PDE6B1	X66142	854	69	MMPDE6B1	X55968; X60133	856	65, 70
					BTPDE6B1	J05553	853	71
PDE6C	PDE6C1	U31973	858	72	MMPDE6C1	AF411063	861	73
					BTPDE6C1	M37838	855	73
					BTPDE6C1	M33140	195*	74

PDE7A	PDE7A1	L12052	482	75				
	PDE7A2	U67932	456	76, 77	MMPDE7A2	U68171	456	76
	PDE7A3	AF332652	424	78				
PDE7B	PDE7B1	AB038040; AJ251860	450	79, 80	MMPDE7B1	AF190639; AK035385	446	80, 81
	PDE7B2	(Not yet identified)			RNPDE7B1/2	AB057409	446	82
	PDE7B3	(Not yet identified)			RNPDE7B3	AB057410	359	82
	PDE7B4	(Not yet identified)			RNPDE7B4	AB057411	459	82
PDE8A	PDE8A1	AF056490; AF388183; AF332653	829	78, 83, 84	MMPDE8A1	AF067806	823	85
	PDE8A2	AF388184	783	84	RNPDE8A1	AB092696	823	86
	PDE8A3	AF388185	449	84	RNPDE8A2	AB092696	761	86
	PDE8A4	AF388186	582	84				
	PDE8A5	AF388187	582	84				
PDE8B	PDE8B1	AF079529; AB085824	885	87, 88	MMPDE8B1	BC032286	730*	
	PDE8B2 or 3	AB085825	788	88				
	PDE8B2	AY129948	838	87				
	PDE8B4	AB085827	865	88				
	PDE8B5	AB085826	830	88				
	PDE8B?	AX959103	?					
PDE9A	PDE9A1	AF067223	593	89	MMPDE9A1	AF031147	534	90
	PDE9A2	AF048837; AF067224	533	89, 91	MMPDE9A2	AF068247	534	89
	PDE9A3	AF067225	466	89				
	PDE9A4	AF067226	465	89				
	PDE9A5	AY242121	?	92				
	PDE9A5	AY196299	?	93				
	PDE9A6	AY196300	?	93				
	PDE9A7	AY196301	?	93				
	PDE9A8	AY196302	?	93				
	PDE9A9	AY196303	?	93				
	PDE9A10	AY196304	?	93				
	PDE9A11	AY196305	?	93				
	PDE9A12	AY196306	?	93				
	PDE9A13	AY196307	?	93				
	PDE9A14	AY196308	?	93				

continued

TABLE 2.1 (continued)

Gene	Human Isoforms				Equivalent Mammalian Isoforms			
	Name	GenBank	Size (aa)	Ref.	Name	GenBank	Size (aa)	Ref.
	PDE9A15	AY196309	?	93				
	PDE9A16	AY196310	?	93				
	PDE9A17	AY196311	?	93				
	PDE9A18	AY196312	?	93				
	PDE9A19	AY196313	?	93				
	PDE9A20	AY196314	?	93				
PDE10A	HSPDE10A1	AB020593; AF127479; AF041798	779	94-96	MMPDE10A1	AF110507; AC104323	779	97
	HSPDE10A2	AF127480; AB026816	789	95, 98	MMPDE10A2	AY360383	796	99
					RNPDE10A2	AB027155	794	100
					MMPDE10A3	AY039249	790	99
					RNPDE10A3	AB027156; AY462095	788	100, 101
					RNPDE10A4	(Not in GenBank)		
					RNPDE10A5	(Not in GenBank)		
					RNPDE10A6	(Not in GenBank)		
	PDE10A7	AB041798	731	96	RNPDE10A7	(Not in GenBank)		
	PDE10A8	AB041798	747	96	RNPDE10A8	(Not in GenBank)		
	PDE10A9	AB041798	737	96	RNPDE10A9	(Not in GenBank)		
	PDE10A10	AB041798	699	96				
					RNPDE10A11	AY426091	852	101
					RNPDE10A12	AY426092	883	101
					RNPDE10A13	AY462093	714	101
					RNPDE10A14	AY462094	653	101
PDE11A	PDE11A1	AJ251509	489	102				
	PDE11A2	AF281865	576	103	RNPDE11A2	AB059360	581	104
	PDE11A3	AJ278682; AB038041	684	103, 105	RNPDE11A3	AB059361	685	104
	PDE11A4	AB036704	934	105	RNPDE11A4	AB059362	935	104

Also omitted are GenBank entries of the form NM_*nnnnnn* (where *n* is a single digit). These entries have all been curated by GenBank staff from primary sequence entries. The accession number for these entries frequently changes with time, as material in GenBank undergoes additional annotation and editing. In addition, the links from these entries to the primary sequence data can be difficult to follow, especially for the uninitiated. Every attempt, therefore, has been to include primary, investigator-submitted sequence data, and the corresponding references in the literature.

Historically, the generation of full-length cDNAs for many PDE isoforms was a potentially error-prone process. As a result, a number of sequences were submitted to GenBank that, at least in retrospect, did not encode full-length proteins, or contain artifacts of cloning. These entries are potential sources of confusion to the uninitiated. In this table, such entries are indicated by an asterisk in the column (size [aa]).

The entries in the table follow the present standard PDE nomenclature. This nomenclature first indicates the PDE family name (i.e., PDE1 versus PDE4). It then indicates the locus or gene name, which follows the nomenclature used in the genome project (i.e., PDE4A versus PDE4B). This is followed by a numeral that indicates the number of transcript (i.e., PDE4B1 versus PDE4B3). Different isolates of a single transcript are sometimes indicated by letters (i.e., PDE4A1A versus PDE4A1B), but as these different isolates encode identical proteins, this differentiation is omitted from this table. Finally, entries are typically prefixed by the species name (e.g., HSPDE4A1 for the human [*Homo sapiens*] PDE4A1 isoform). To improve the readability of the table, this prefix has been omitted from the entries of human isoforms, but is included for those of other species (MM for *Mus musculus*, i.e., mice; RN for *Rattus norvegicus*, i.e., rats; BT for *Bos taurus*, i.e., cattle; SS for *Sus scrofa*, i.e., pigs).

Using the table: One of the most important aspects of the table is that the orthologous isoforms are all grouped on the same row. The first column of the table indicates the gene name (e.g., PDE1). The next four columns describe the various human isoforms encoded by that gene, one isoform per row. The next further four columns describe the corresponding mammalian isoforms. For example, under PDE1A, the human PDE1A1 isoform is entered in the human isoform columns. The corresponding mouse, rat, and bovine isoforms (MMPDE1A1, RNPDE1A1, and BTPDE1A1, respectively) are entered on the same row, in the four columns marked Equivalent mammalian isoforms. This feature of the table allows the easy identification of those isoforms that are orthologous to any given human isoform.

REFERENCES

1. Michibata, H., N. Yanaka, Y. Kanoh, K. Okumura, and K. Omori. 2001. Human Ca^{2+}/calmodulin-dependent phosphodiesterase PDE1A: Novel splice variants, their specific expression, genomic organization, and chromosomal localization. *Biochim Biophys Acta* 1517:278–287.
2. Snyder, P.B., V.A. Florio, K. Ferguson, and K. Loughney. 1999. Isolation, expression and analysis of splice variants of a human Ca^{2+}/calmodulin-stimulated phosphodiesterase (PDE1A). *Cell Signal* 11:535–544.
3. Sonnenburg, W.K., S.D. Rybalkin, K.E. Bornfeldt, K.S. Kwak, I.G. Rybalkina, and J.A. Beavo. 1998. Identification, quantitation, and cellular localization of PDE1 calmodulin-stimulated cyclic nucleotide phosphodiesterases. *Methods* 14:3–19.
4. Sonnenburg, W.K., D. Seger, K.S. Kwak, J. Huang, H. Charbonneau, and J.A. Beavo. 1995. Identification of inhibitory and calmodulin-binding domains of the PDE1A1 and PDE1A2 calmodulin-stimulated cyclic nucleotide phosphodiesterases. *J Biol Chem* 270:30989–31000.

5. Fidock, M., M. Miller, and J. Lanfear. 2002. Isolation and differential tissue distribution of two human cDNAs encoding PDE1 splice variants. *Cell Signal* 14:53–60.

6. Sonnenburg, W.K., D. Seger, and J.A. Beavo. 1993. Molecular cloning of a cDNA encoding the "61-kDa" calmodulin-stimulated cyclic nucleotide phosphodiesterase. Tissue-specific expression of structurally related isoforms. *J Biol Chem* 268:645–652.

7. Loughney, K., T.J. Martins, E.A. Harris, K. Sadhu, J.B. Hicks, W.K. Sonnenburg, J.A. Beavo, and K. Ferguson. 1996. Isolation and characterization of cDNAs corresponding to two human calcium, calmodulin-regulated, $3',5'$-cyclic nucleotide phosphodiesterases. *J Biol Chem* 271:796–806.

8. Yan, C., A.Z. Zhao, W.K. Sonnenburg, and J.A. Beavo. 2001. Stage and cell-specific expression of calmodulin-dependent phosphodiesterases in mouse testis. *Biol Reprod* 64:1746–1754.

9. Vasta, V., W.K. Sonnenburg, C. Yan, S.H. Soderling, M. Shimizu-Albergine, and J.A. Beavo. 2005. Identification of a new variant of PDE1A calmodulin-stimulated cyclic nucleotide phosphodiesterase expressed in mouse sperm. *Biol Reprod* 73:598–609.

10. Jiang, X., J. Li, M. Paskind, and P.M. Epstein. 1996. Inhibition of calmodulin-dependent phosphodiesterase induces apoptosis in human leukemic cells. *Proc Natl Acad Sci USA* 93:11236–11241.

11. Yu, J., S.L. Wolda, A.L. Frazier, V.A. Florio, T.J. Martins, P.B. Snyder, E.A. Harris, K.N. McCaw, C.A. Farrell, B. Steiner, J.K. Bentley, J.A. Beavo, K. Ferguson, and R. Gelinas. 1997. Identification and characterisation of a human calmodulin-stimulated phosphodiesterase PDE1B1. *Cell Signal* 9:519–529.

12. Repaske, D.R., J.V. Swinnen, S.L. Jin, J.J. Van Wyk, and M. Conti. 1992. A polymerase chain reaction strategy to identify and clone cyclic nucleotide phosphodiesterase cDNAs. Molecular cloning of the cDNA encoding the 63-kDa calmodulin-dependent phosphodiesterase. *J Biol Chem* 267:18683–18688.

13. Bentley, J.K., A. Kadlecek, C.H. Sherbert, D. Seger, W.K. Sonnenburg, H. Charbonneau, J.P. Novack, and J.A. Beavo. 1992. Molecular cloning of cDNA encoding a "63"-kDa calmodulin-stimulated phosphodiesterase from bovine brain. *J Biol Chem* 267:18676–18682.

14. Yan, C., A.Z. Zhao, J.K. Bentley, and J.A. Beavo. 1996. The calmodulin-dependent phosphodiesterase gene PDE1C encodes several functionally different splice variants in a tissue-specific manner. *J Biol Chem* 271:25699–25706.

15. Yan, C., A.Z. Zhao, J.K. Bentley, K. Loughney, K. Ferguson, and J.A. Beavo. 1995. Molecular cloning and characterization of a calmodulin-dependent phosphodiesterase enriched in olfactory sensory neurons. *Proc Natl Acad Sci USA* 92:9677–9681.

16. Sonnenburg, W.K., P.J. Mullaney, and J.A. Beavo. 1991. Molecular cloning of a cyclic GMP-stimulated cyclic nucleotide phosphodiesterase cDNA. Identification and distribution of isozyme variants. *J Biol Chem* 266:17655–17661.

17. Yang, Q., M. Paskind, G. Bolger, W.J. Thompson, D.R. Repaske, L.S. Cutler, and P.M. Epstein. 1994. A novel cyclic GMP stimulated phosphodiesterase from rat brain. *Biochem Biophys Res Commun* 205:1850–1858.

18. Juilfs, D.M., H.J. Fulle, A.Z. Zhao, M.D. Houslay, D.L. Garbers, and J.A. Beavo. 1997. A subset of olfactory neurons that selectively express cGMP-stimulated phosphodiesterase (PDE2) and guanylyl cyclase-D define a unique olfactory signal transduction pathway. *Proc Natl Acad Sci USA* 94:3388–3395.

19. Rosman, G.J., T.J. Martins, W.K. Sonnenburg, J.A. Beavo, K. Ferguson, and K. Loughney. 1997. Isolation and characterization of human cDNAs encoding a cGMP-stimulated $3',5'$-cyclic nucleotide phosphodiesterase. *Gene* 191:89–95.

20. Meacci, E., M. Taira, M. Moos Jr., C.J. Smith, M.A. Movsesian, E. Degerman, P. Belfrage, and V. Manganiello. 1992. Molecular cloning and expression of human myocardial cGMP-inhibited cAMP phosphodiesterase. *Proc Natl Acad Sci USA* 89:3721–3725.

21. Wechsler, J., Y.H. Choi, J. Krall, F. Ahmad, V.C. Manganiello, and M.A. Movsesian. 2002. Isoforms of cyclic nucleotide phosphodiesterase PDE3A in cardiac myocytes. *J Biol Chem* 277:38072–38078.

22. Choi, Y.H., D. Ekholm, J. Krall, F. Ahmad, E. Degerman, V.C. Manganiello, and M.A. Movsesian. 2001. Identification of a novel isoform of the cyclic-nucleotide phosphodiesterase PDE3A expressed in vascular smooth-muscle myocytes. *Biochem J* 353:41–50.
23. Miki T., M. Taira, S. Hockman, F. Shimada, J. Lieman, M. Napolitano, D. Ward, H. Makino, and V.C. Manganiello. 1996. Characterization of the cDNA and gene encoding human PDE3B, the cGIP1 isoform of the human cyclic GMP-inhibited cyclic nucleotide phosphodiesterase family. *Genomics* 36:476–485.
24. Taira, M., S.C. Hockman, J.C. Calvo, P. Belfrage, and V.C. Manganiello. 1993. Molecular cloning of the rat adipocyte hormone-sensitive cyclic GMP-inhibited cyclic nucleotide phosphodiesterase. *J Biol Chem* 268:18573–18579.
25. Sullivan, M., G. Rena, F. Begg, L. Gordon, A.S. Olsen, and M.D. Houslay. 1998. Identification and characterization of the human homologue of the short PDE4A cAMP-specific phosphodiesterase RD1 (PDE4A1) by analysis of the human HSPDE4A gene locus located at chromosome 19p13.2. *Biochem J* 333:693–703.
26. Olsen, A.E., and G.B. Bolger. 2000. Physical mapping and promoter structure of the murine cAMP-specific phosphodiesterase PDE4A gene. *Mamm Genome* 11:41–45.
27. Davis, R.L., H. Takayasu, M. Eberwine, and J. Myres. 1989. Cloning and characterization of mammalian homologs of the *Drosophila* dunce+ gene. *Proc Natl Acad Sci USA* 86:3604–3608.
28. Bolger, G.B., L. Rodgers, and M. Riggs. 1994. Differential CNS expression of alternative mRNA isoforms of the mammalian genes encoding cAMP-specific phosphodiesterases. *Gene* 149:237–244.
29. Swinnen, J.V., D.R. Joseph, and M. Conti. 1989. Molecular cloning of rat homologues of the *Drosophila melanogaster* dunce cAMP phosphodiesterase: Evidence for a family of genes. *Proc Natl Acad Sci USA* 86:5325–5329.
30. Bolger, G., T. Michaeli, T. Martins, St.T. John, B. Steiner, L. Rodgers, M. Riggs, M. Wigler, and K. Ferguson. 1993. A family of human phosphodiesterases homologous to the dunce learning and memory gene product of *Drosophila melanogaster* are potential targets for antidepressant drugs. *Mol Cell Biol* 13:6558–6571.
31. Livi, G.P., P. Kmetz, M.M. McHale, L.B. Cieslinski, G.M. Sathe, D.P. Taylor, R.L. Davis, T.J. Torphy, and J.M. Balcarek. 1990. Cloning and expression of cDNA for a human low-K_m, rolipram-sensitive cyclic AMP phosphodiesterase. *Mol Cell Biol* 10:2678–2686.
32. Sullivan, M., M. Egerton, Y. Shakur, A. Marquardsen, and M.D. Houslay. 1994. Molecular cloning and expression, in both COS-1 cells and *S. cerevisiae*, of a human cytosolic type-IVA, cyclic AMP specific phosphodiesterase (hPDE-IVA-h6.1). *Cell Signal* 6:793–812.
33. Horton, Y.M., M. Sullivan, and M.D. Houslay. 1995. Molecular cloning of a novel splice variant of human type IVA (PDE-IVA) cyclic AMP phosphodiesterase and localization of the gene to the p13.2–q12 region of human chromosome 19 [published erratum appears in *Biochem J* 1995; 312 (Pt 3):991]. *Biochem J* 308:683–691.
34. Bolger, G.B., I. McPhee, and M.D. Houslay. 1996. Alternative splicing of cAMP-specific phosphodiesterase mRNA transcripts. Characterization of a novel tissue-specific isoform, RNPDE4A8. *J Biol Chem* 271:1065–1071.
35. Naro, F., R. Zhang, and M. Conti. 1996. Developmental regulation of unique adenosine $3',5'$-monophosphate-specific phosphodiesterase variants during rat spermatogenesis [published erratum appears in *Endocrinology* 1996; 137 (10):4114]. *Endocrinology* 137:2464–2472.
36. Rena, G., F. Begg, A. Ross, C. MacKenzie, I. McPhee, L. Campbell, E. Huston, M. Sullivan, and M.D. Houslay. 2001. Molecular cloning, genomic positioning, promoter identification, and characterization of the novel cyclic AMP-specific phosphodiesterase PDE4A10. *Mol Pharmacol* 59:996–1011.
37. Wallace, D.A., L.A. Johnston, E. Huston, D. MacMaster, T.M. Houslay, Y.F. Cheung, L. Campbell, J.E. Millen, R.A. Smith, I. Gall, R.G. Knowles, M. Sullivan, and M.D. Houslay. 2005. Identification and characterization of PDE4A11, a novel, widely expressed long isoform encoded by the human PDE4A cAMP phosphodiesterase gene. *Mol Pharmacol* 67:1920–1934.
38. Colicelli, J., C. Birchmeier, T. Michaeli, K. O'Neill, M. Riggs, and M. Wigler. 1989. Isolation and characterization of a mammalian gene encoding a high-affinity cAMP phosphodiesterase. *Proc Natl Acad Sci USA* 86:3599–3603.

39. Shepherd, M., T. McSorley, A.E. Olsen, L.A. Johnston, N.C. Thomson, G.S. Baillie, M.D. Houslay, and G.B. Bolger. 2003. Molecular cloning and subcellular distribution of the novel PDE4B4 cAMP-specific phosphodiesterase isoform. *Biochem J* 370:429–438.

40. McLaughlin, M.M., L.B. Cieslinski, M. Burman, T.J. Torphy, and G.P. Livi. 1993. A low-K_m, rolipram-sensitive, cAMP-specific phosphodiesterase from human brain. Cloning and expression of cDNA, biochemical characterization of recombinant protein, and tissue distribution of mRNA. *J Biol Chem* 268:6470–6476.

41. Obernolte, R., S. Bhakta, R. Alvarez, C. Bach, P. Zuppan, M. Mulkins, K. Jarnagin, and E.R. Shelton. 1993. The cDNA of a human lymphocyte cyclic-AMP phosphodiesterase (PDE IV) reveals a multigene family. *Gene* 129:239–247.

42. Swinnen, J.V., K.E. Tsikalas, and M. Conti. 1991. Properties and hormonal regulation of two structurally related cAMP phosphodiesterases from the rat Sertoli cell. *J Biol Chem* 266:18370–18377.

43. Huston, E., S. Lumb, A. Russell, C. Catterall, A.H. Ross, M.R. Steele, G.B. Bolger, M.J. Perry, R.J. Owens, and M.D. Houslay. 1997. Molecular cloning and transient expression in COS7 cells of a novel human PDE4B cAMP-specific phosphodiesterase, HSPDE4B3. *Biochem J* 328:549–558.

44. Fehr, C., R.L. Shirley, J.K. Belknap, J.C. Crabbe, and K.J. Buck. 2002. Congenic mapping of alcohol and pentobarbital withdrawal liability loci to a <1 centimorgan interval of murine chromosome 4: Identification of Mpdz as a candidate gene. *J Neurosci* 22:3730–3738.

45. Cherry, J.A., B.E. Thompson, and V. Pho. 2001. Diazepam and rolipram differentially inhibit cyclic AMP-specific phosphodiesterases PDE4A1 and PDE4B3 in the mouse. *Biochim Biophys Acta* 1518:27–35.

46. Engels, P., M. Sullivan, T. Muller, and H. Lubbert. 1995. Molecular cloning and functional expression in yeast of a human cAMP-specific phosphodiesterase subtype (PDE IV-C). *FEBS Lett* 358:305–310.

47. Owens, R.J., S. Lumb, M.K. Rees, A. Russell, D. Baldock, V. Lang, T. Crabbe, M. Ballesteros, and M.J. Perry. 1997. Molecular cloning and expression of a human phosphodiesterase 4C. *Cell Signal* 9:575–585.

48. Obernolte, R., J. Ratzliff, P.A. Baecker, D.V. Daniels, P. Zuppan, K. Jarnagin, and E.R. Shelton. 1997. Multiple splice variants of phosphodiesterase PDE4C cloned from human lung and testis. *Biochim Biophys Acta* 1353:287–297.

49. Nemoz, G., R. Zhang, C. Sette, and M. Conti. 1996. Identification of cyclic AMP-phosphodiesterase variants from the PDE4D gene expressed in human peripheral mononuclear cells. *FEBS Lett* 384:97–102.

50. Bolger, G.B., S. Erdogan, R.E. Jones, K. Loughney, G. Scotland, R. Hoffmann, I. Wilkinson, C. Farrell, and M.D. Houslay. 1997. Characterization of five different proteins produced by alternatively spliced mRNAs from the human cAMP-specific phosphodiesterase PDE4D gene. *Biochem J* 328:539–548.

51. Sette, C., E. Vicini, and M. Conti. 1994. The rat PDE3/IVd phosphodiesterase gene codes for multiple proteins differentially activated by cAMP-dependent protein kinase [published erratum appears in *J Biol Chem* 1994; 269 (32):20806]. *J Biol Chem* 269:18271–18274.

52. Swinnen, J.V., D.R. Joseph, and M. Conti. 1989. The mRNA encoding a high-affinity cAMP phosphodiesterase is regulated by hormones and cAMP. *Proc Natl Acad Sci USA* 86:8197–8201.

53. Jin, S.L., J.V. Swinnen, and M. Conti. 1992. Characterization of the structure of a low K_m, rolipram-sensitive cAMP phosphodiesterase. Mapping of the catalytic domain. *J Biol Chem* 267:18929–18939.

54. Baecker, P.A., R. Obernolte, C. Bach, C. Yee, and E.R. Shelton. 1994. Isolation of a cDNA encoding a human rolipram-sensitive cyclic AMP phosphodiesterase (PDE IVD). *Gene* 138:253–256.

55. Iona, S., M. Cuomo, T. Bushnik, F. Naro, C. Sette, M. Hess, E.R. Shelton, and M. Conti. 1998. Characterization of the rolipram-sensitive, cyclic AMP-specific phosphodiesterases: Identification and differential expression of immunologically distinct forms in the rat brain. *Mol Pharmacol* 53:23–32.

56. Wang, D., C. Deng, B. Bugaj-Gaweda, M. Kwan, C. Gunwaldsen, C. Leonard, X. Xin, Y. Hu, A. Unterbeck, and M. De Vivo. 2003. Cloning and characterization of novel PDE4D isoforms PDE4D6 and PDE4D7. *Cell Signal* 15:883–891.

57. Gretarsdottir, S., G. Thorleifsson, S.T. Reynisdottir, A. Manolescu, S. Jonsdottir, T. Jonsdottir, T. Gudmundsdottir, S.M. Bjarnadottir, O.B. Einarsson, H.M. Gudjonsdottir, M. Hawkins, G. Gudmundsson, H. Gudmundsdottir, H. Andrason, A.S. Gudmundsdottir, M. Sigurdardottir, T.T. Chou, J. Nahmias, S. Goss, S. Sveinbjornsdottir, E.M. Valdimarsson, F. Jakobsson, U. Agnarsson, V. Gudnason, G. Thorgeirsson, J. Fingerle, M. Gurney, D. Gudbjartsson, M.L. Frigge, A. Kong, K. Stefansson, and J.R. Gulcher. 2003. The gene encoding phosphodiesterase 4D confers risk of ischemic stroke. *Nat Genet* 35:131–138.

58. Loughney, K., T.R. Hill, V.A. Florio, L. Uher, G.J. Rosman, S.L. Wolda, B.A. Jones, M.L. Howard, L.M. McAllister-Lucas, W.K. Sonnenburg, S.H. Francis, J.D. Corbin, J.A. Beavo, and K. Ferguson. 1998. Isolation and characterization of cDNAs encoding PDE5A, a human cGMP-binding, cGMP-specific $3',5'$-cyclic nucleotide phosphodiesterase. *Gene* 216:139–147.

59. Yanaka, N., J. Kotera, A. Ohtsuka, H. Akatsuka, Y. Imai, H. Michibata, K. Fujishige, E. Kawai, S. Takebayashi, K. Okumura, and K. Omori. 1998. Expression, structure and chromosomal localization of the human cGMP-binding cGMP-specific phosphodiesterase PDE5A gene. *Eur J Biochem* 255:391–399.

60. Stacey, P., S. Rulten, A. Dapling, and S.C. Phillips. 1998. Molecular cloning and expression of human cGMP-binding cGMP-specific phosphodiesterase (PDE5). *Biochem Biophys Res Commun* 247:249–254.

61. McAllister Lucas, L.M., W.K. Sonnenburg, A. Kadlecek, D. Seger, H.L. Trong, J.L. Colbran, M.K. Thomas, K.A. Walsh, S.H. Francis, J.D. Corbin, J.A. Beavo. 1993. The structure of a bovine lung cGMP-binding, cGMP-specific phosphodiesterase deduced from a cDNA clone. *J Biol Chem* 268:22863–22873.

62. Kotera, J., N. Yanaka, K. Fujishige, Y. Imai, H. Akatsuka, T. Ishizuka, K. Kawashima, and K. Omori. 1997. Expression of rat cGMP-binding cGMP-specific phosphodiesterase mRNA in Purkinje cell layers during postnatal neuronal development. *Eur J Biochem* 249:434–442.

63. Sopory, S., T. Kaur, and S.S. Visweswariah. 2004. The cGMP-binding, cGMP-specific phospho-diesterase (PDE5): Intestinal cell expression, regulation and role in fluid secretion. *Cell Signal* 16:681–692.

64. Pittler, S.J., W. Baehr, J.J. Wasmuth, D.G. McConnell, M.S. Champagne, P. vanTuinen, D. Ledbetter, and R.L. Davis. 1990. Molecular characterization of human and bovine rod photo-receptor cGMP phosphodiesterase alpha-subunit and chromosomal localization of the human gene. *Genomics* 6:272–283.

65. Baehr, W., M.S. Champagne, A.K. Lee, and S.J. Pittler. 1991. Complete cDNA sequences of mouse rod photoreceptor cGMP phosphodiesterase alpha- and beta-subunits, and identification of β'-, a putative beta-subunit isozyme produced by alternative splicing of the beta-subunit gene. *FEBS Lett* 278:107–114.

66. Ovchinnikov, Y.A., V.V. Gubanov, K.A. Khramtsov, V.M. Ischenko, V.E. Zagranichny, K.G. Muradov, T.M. Shuvaeva, and V.M. Lipkin. 1987. Cyclic GMP phosphodiesterase from bovine retina: Amino acid sequence of the alpha-subunit and the nucleotide sequence of the corresponding cDNA. *FEBS Lett* 223:169–173.

67. Petersen-Jones, S.M., D.D. Entz, and D.R. Sargan. 1999. cGMP phosphodiesterase-alpha muta-tion causes progressive retinal atrophy in the Cardigan Welsh corgi dog. *Invest Ophthalmol Vis Sci* 40:1637–1644.

68. Yamazaki, M., N. Li, V.A. Bondarenko, R.K. Yamazaki, W. Baehr, and A. Yamazaki. 2002. Binding of cGMP to GAF domains in amphibian rod photoreceptor cGMP phosphodiesterase (PDE). Identification of GAF domains in PDE alphabeta subunits and distinct domains in the PDE gamma subunit involved in stimulation of cGMP binding to GAF domains. *J Biol Chem* 277:40675–40686.

69. Khramtsov, N.V., E.A. Feshchenko, V.A. Suslova, B.E. Shmukler, B.E. Terpugov, T.V. Rakitina, N.V. Atabekova, and V.M. Lipkin. 1993. The human rod photoreceptor cGMP phosphodiesterase beta-subunit. Structural studies of its cDNA and gene. *FEBS Lett* 327:275–278.

70. Bowes, C., T. Li, M. Danciger, L.C. Baxter, M.L. Applebury, and D.B. Farber. 1990. Retinal degeneration in the rd mouse is caused by a defect in the beta subunit of rod cGMP-phosphodies-terase. *Nature* 347:677–680 [see comments].

71. Lipkin, V.M., N.V. Khramtsov, I.A. Vasilevskaya, N.V. Atabekova, K.G. Muradov, V.V. Gubanov, T. Li, J.P. Johnston, K.J. Volpp, and M.L. Applebury. 1990. Beta-subunit of bovine rod photoreceptor cGMP phosphodiesterase. Comparison with the phosphodiesterase family. *J Biol Chem* 265:12955–12959.

72. Viczian, A.S., N.I. Piriev, and D.B. Farber. 1995. Isolation and characterization of a cDNA encoding the alpha′ subunit of human cone cGMP-phosphodiesterase. *Gene* 166:205–211.

73. Li, T.S., K. Volpp, and M.L. Applebury. 1990. Bovine cone photoreceptor cGMP phosphodiesterase structure deduced from a cDNA clone. *Proc Natl Acad Sci USA* 87:293–297.

74. Charbonneau, H., R.K. Prusti, H. LeTrong, W.K. Sonnenburg, P.J. Mullaney, K.A. Walsh, and J.A. Beavo. 1990. Identification of a noncatalytic cGMP-binding domain conserved in both the cGMP-stimulated and photoreceptor cyclic nucleotide phosphodiesterases. *Proc Natl Acad Sci USA* 87:288–292.

75. Michaeli, T., T.J. Bloom, T. Martins, K. Loughney, K. Ferguson, M. Riggs, L. Rodgers, J.A. Beavo, and M. Wigler. 1993. Isolation and characterization of a previously undetected human cAMP phosphodiesterase by complementation of cAMP phosphodiesterase-deficient *Saccharomyces cerevisiae*. *J Biol Chem* 268:12925–12932.

76. Bloom, T.J., and J.A. Beavo. 1996. Identification and tissue-specific expression of PDE7 phosphodiesterase splice variants. *Proc Natl Acad Sci USA* 93:14188–14192.

77. Han, P., X. Zhu, and T. Michaeli. 1997. Alternative splicing of the high affinity cAMP-specific phosphodiesterase (PDE7A) mRNA in human skeletal muscle and heart. *J Biol Chem* 272: 16152–16157.

78. Glavas, N.A., C. Ostenson, J.B. Schaefer, V. Vasta, and J.A. Beavo. 2001. T cell activation up-regulates cyclic nucleotide phosphodiesterases 8A1 and 7A3. *Proc Natl Acad Sci USA* 98: 6319–6324.

79. Sasaki, T., J. Kotera, K. Yuasa, and K. Omori. 2000. Identification of human PDE7B, a cAMP-specific phosphodiesterase. *Biochem Biophys Res Commun* 271:575–583.

80. Gardner, C., N. Robas, D. Cawkill, and M. Fidock. 2000. Cloning and characterization of the human and mouse PDE7B, a novel cAMP-specific cyclic nucleotide phosphodiesterase. *Biochem Biophys Res Commun* 272:186–192.

81. Hetman, J.M., S.H. Soderling, N.A. Glavas, and J.A. Beavo. 2000. Cloning and characterization of PDE7B, a cAMP-specific phosphodiesterase. *Proc Natl Acad Sci USA* 97:472–476.

82. Sasaki, T., J. Kotera, and K. Omori. 2002. Novel alternative splice variants of rat phosphodiesterase 7B showing unique tissue-specific expression and phosphorylation. *Biochem J* 361:211–220.

83. Fisher, D.A., J.F. Smith, J.S. Pillar, St.S.H. Denis, and J.B. Cheng. 1998. Isolation and characterization of PDE8A, a novel human cAMP-specific phosphodiesterase. *Biochem Biophys Res Commun* 246:570–577.

84. Wang, P., P. Wu, R.W. Egan, and M.M. Billah. 2001. Human phosphodiesterase 8A splice variants: Cloning, gene organization, and tissue distribution. *Gene* 280:183–194.

85. Soderling, S.H., S.J. Bayuga, and J.A. Beavo. 1998. Cloning and characterization of a cAMP-specific cyclic nucleotide phosphodiesterase. *Proc Natl Acad Sci USA* 95:8991–8996.

86. Kobayashi, T., M. Gamanuma, T. Sasaki, Y. Yamashita, K. Yuasa, J. Kotera, and K. Omori. 2003. Molecular comparison of rat cyclic nucleotide phosphodiesterase 8 family: Unique expression of PDE8B in rat brain. *Gene* 319:21–31.

87. Hayashi, M., Y. Shimada, Y. Nishimura, T. Hama, and T. Tanaka. 2002. Genomic organization, chromosomal localization, and alternative splicing of the human phosphodiesterase 8B gene. *Biochem Biophys Res Commun* 297:1253–1258.

88. Gamanuma, M., K. Yuasa, T. Sasaki, N. Sakurai, J. Kotera, and K. Omori. 2003. Comparison of enzymatic characterization and gene organization of cyclic nucleotide phosphodiesterase 8 family in humans. *Cell Signal* 15:565–574.

89. Guipponi, M., H.S. Scott, J. Kudoh, K. Kawasaki, K. Shibuya, A. Shintani, S. Asakawa, H. Chen, M.D. Lalioti, C. Rossier, S. Minoshima, N. Shimizu, and S.E. Antonarakis. 1998. Identification and characterization of a novel cyclic nucleotide phosphodiesterase gene (PDE9A) that maps to 21q22.3: Alternative splicing of mRNA transcripts, genomic structure and sequence. *Hum Genet* 103:386–392.

90. Soderling, S.H., S.J. Bayuga, and J.A. Beavo. 1998. Identification and characterization of a novel family of cyclic nucleotide phosphodiesterases. *J Biol Chem* 273:15553–15558.

91. Fisher, D.A., J.F. Smith, J.S. Pillar, St.S.H. Denis, and J.B. Cheng. 1998. Isolation and characterization of PDE9A, a novel human cGMP-specific phosphodiesterase. *J Biol Chem* 273:15559–15564.

92. Wang, P., P. Wu, R.W. Egan, and M.M. Billah. 2003. Identification and characterization of a new human type 9 cGMP-specific phosphodiesterase splice variant (PDE9A5). Differential tissue distribution and subcellular localization of PDE9A variants. *Gene* 314:15–27.

93. Rentero, C., A. Monfort, and P. Puigdomenech. 2003. Identification and distribution of different mRNA variants produced by differential splicing in the human phosphodiesterase 9A gene. *Biochem Biophys Res Commun* 301:686–692.

94. Fujishige, K., J. Kotera, H. Michibata, K. Yuasa, S. Takebayashi, K. Okumura, and K. Omori. 1999. Cloning and characterization of a novel human phosphodiesterase that hydrolyzes both cAMP and cGMP (PDE10A). *J Biol Chem* 274:18438–18445.

95. Loughney, K., P.B. Snyder, L. Uher, G.J. Rosman, K. Ferguson, and V.A. Florio. 1999. Isolation and characterization of PDE10A, a novel human 3′,5′-cyclic nucleotide phosphodiesterase. *Gene* 234:109–117.

96. Fujishige, K., J. Kotera, K. Yuasa, and K. Omori. 2000. The human phosphodiesterase PDE10A gene genomic organization and evolutionary relatedness with other PDEs containing GAF domains. *Eur J Biochem* 267:5943–5951.

97. Soderling, S.H., S.J. Bayuga, and J.A. Beavo. 1999. Isolation and characterization of a dual-substrate phosphodiesterase gene family: PDE10A. *Proc Natl Acad Sci USA* 96:7071–7076.

98. Kotera, J., K. Fujishige, K. Yuasa, and K. Omori. 1999. Characterization and phosphorylation of PDE10A2, a novel alternative splice variant of human phosphodiesterase that hydrolyzes cAMP and cGMP. *Biochem Biophys Res Commun* 261:551–557.

99. Hu, H., E.A. McCaw, A.L. Hebb, G.T. Gomez, and E.M. Denovan-Wright. 2004. Mutant huntingtin affects the rate of transcription of striatum-specific isoforms of phosphodiesterase 10A. *Eur J Neurosci* 20:3351–3363.

100. Fujishige, K., J. Kotera, and K. Omori. 1999. Striatum- and testis-specific phosphodiesterase PDE10A isolation and characterization of a rat PDE10A. *Eur J Biochem* 266:1118–1127.

101. O'Connor, V., A. Genin, S. Davis, K.K. Karishma, V. Doyere, C.I. De Zeeuw, G. Sanger, S.P. Hunt, G. Richter-Levin, J. Mallet, S. Laroche, T.V. Bliss, and P.J. French. 2004. Differential amplification of intron-containing transcripts reveals long term potentiation-associated up-regulation of specific Pde10A phosphodiesterase splice variants. *J Biol Chem* 279:15841–15849.

102. Fawcett, L., R. Baxendale, P. Stacey, C. McGrouther, I. Harrow, S. Soderling, J. Hetman, J.A. Beavo, and S.C. Phillips. 2000. Molecular cloning and characterization of a distinct human phosphodiesterase gene family: PDE11A. *Proc Natl Acad Sci USA* 97:3702–3707.

103. Hetman, J.M., N. Robas, R. Baxendale, M. Fidock, S.C. Phillips, S.H. Soderling, and J.A. Beavo. 2000. Cloning and characterization of two splice variants of human phosphodiesterase 11A. *Proc Natl Acad Sci USA* 97:12891–12895.

104. Yuasa, K., T. Ohgaru, M. Asahina, and K. Omori. 2001. Identification of rat cyclic nucleotide phosphodiesterase 11A (PDE11A): Comparison of rat and human PDE11A splicing variants. *Eur J Biochem* 268:4440–4448.

105. Yuasa, K., J. Kotera, K. Fujishige, H. Michibata, T. Sasaki, and K. Omori. 2000. Isolation and characterization of two novel phosphodiesterase PDE11A variants showing unique structure and tissue-specific expression. *J Biol Chem* 275:31469–31479.

106. Polli, J.W., R.L., Kineaid. 1992. Molecular clearing of DNA encoding a calmodulin-dependent phospho diesterase enriched in striatum. *Proc. Natl. Acad. Sci. USA*. 89:11079–11083.

Section A

Specific Phosphodiesterase Families—Regulation, Molecular and Biochemical Characteristics

3 Calmodulin-Stimulated Cyclic Nucleotide Phosphodiesterases

Andrew T. Bender

CONTENTS

3.1 Introduction 35
3.2 Biochemical Characteristics 36
    3.2.1 Primary Structure and Activation by Ca^{2+}/CaM 36
    3.2.2 Kinetic Properties 38
    3.2.3 Regulation by Phosphorylation 39
    3.2.4 Inhibitors 40
3.3 PDE1 Genetics 41
    3.3.1 PDE1A 42
    3.3.2 PDE1B 42
    3.3.3 PDE1C 43
3.4 Tissue-Specific Expression and Functional Roles of PDE1 44
    3.4.1 Cardiovascular System 44
    3.4.2 Nervous System 45
    3.4.3 Immune System 47
    3.4.4 Testis and Sperm 48
3.5 Conclusions 49
Acknowledgments 50
References 50

3.1 INTRODUCTION

The PDE1 family of enzymes has a relatively long history and was one of the first PDE families to be discovered.[1,2] It is one of the most studied and well characterized of the 11 PDE families. PDE1 enzymes were initially isolated from animal tissues, particularly bovine, and characterized by their unique biochemical properties. Calmodulin-stimulated PDE activity was originally identified in bovine[1,3] and rat brain.[2] Subsequent work found that PDE1 enzymes hydrolyze both cAMP and cGMP and distinguished them from other PDEs as having the unique property of being activated by the binding of a complex of calcium and calmodulin (CaM). To date, three PDE1 isoforms PDE1A, PDE1B, and PDE1C have been discovered and are the products of separate genes. The advent of the molecular biology revolution facilitated the cloning and study of the different PDE1 isoforms. Molecular biology advances have also allowed the discovery of the many variants of the three PDE1 genes and aided the study of their expression and localization.

An irksome difficulty of the PDE1 research field has been cryptic nomenclature of the enzymes. Initially, PDEs were classified based on their regulatory properties and PDE1 was simply referred to as CaM-stimulated PDE as researchers were unaware of the multitude of

families that were yet to be discovered. As new isoforms were discovered, they were named based on their molecular mass and tissue of discovery, i.e., PDE1B was the "63 kDa bovine brain CaM-stimulated PDE". The adoption of the current PDE nomenclature system in 1995[4] has alleviated much of the confusion and instituted a nomenclature system based on PDE primary structure. The current nomenclature scheme uses the first two letters to represent the species, the next three letters plus an Arabic numeral to designate the cyclic nucleotide phosphodiesterase gene family, the next letter to represent the individual gene product within the family, and the final Arabic numeral to represent the variant. Thus, the "63 kDa bovine brain CaM-stimulated PDE" is now identified as BtPDE1B1. In this chapter, separate gene products will be referred to as isoforms and individual mRNA or protein products of the different genes arising from alternative splicing or alternative transcriptional starts will be called variants.

As with many of the PDE families, little is known about the physiological function of the PDE1 enzymes. Study of the function of PDE1 has been hindered by the lack of a readily available PDE1-selective inhibitor that will rapidly enter cells. However, a great deal of work has described the expression pattern of PDE1s in different tissues and cell types and how the expression can change in the course of physiological and pathological processes. These data together with some recent genetic approaches for studying PDE1 have yielded clues about PDE1 function. This chapter will provide only a brief review of the biochemical character-istics of the PDE1s, as their characteristics have been well reviewed previously.[5–7] Instead, the main emphasis will be genetic regulation and tissue-specific expression of PDE1 as these areas have been the focus of recent research. Additionally, what has been discovered about the functional roles of PDE1 will be discussed.

3.2   BIOCHEMICAL CHARACTERISTICS

3.2.1   PRIMARY STRUCTURE AND ACTIVATION BY Ca^{2+}/CAM

As is common with most PDE families, the PDE1 enzymes have a primary structure consisting of a somewhat conserved C-terminal catalytic domain and an N-terminal regulatory domain. The unique distinguishing feature of the PDE1 family is the existence of two regulatory CaM-binding sites at the N-termini of the enzymes. The enzymes are known to exist as dimers,[8–10] although the nature or functional significance of dimerization has not been elucidated. The molecular weight of the enzymes ranges from 58 to 86 kDa per monomer.

The binding of an activated Ca^{2+}/CaM complex serves to stimulate PDE1 enzymatic activity. The activation by CaM is an effect on V_{max} and not on K_m.[11] The domain organi-zation of the enzymes includes a C-terminal catalytic domain, two N-terminal CaM-binding domains, and an inhibitory domain situated between the two CaM-binding sites[12] (see Figure 3.1A). The inhibitory domain is conserved in all the PDE1 enzymes[13] and may interact with a site near the catalytic domain that maintains the enzyme in an inactive conformation. It was demonstrated with PDE1A that upon binding of Ca^{2+}/CaM a conformational change occurs where the inhibition is released and full enzyme activity is allowed (Figure 3.1B).[12] CaM-mediated displacement of an autoinhibitory domain has been shown to be an activation mechanism for other proteins such as CaM kinase II, myosin light chain kinase, CaM kinase kinase, and calcineurin.[14] However, exactly how the inhibitory domain interacts with the rest of the protein to prevent PDE1 enzyme activity has not been elucidated.

The different PDE1 isoforms vary in sensitivity to activation by Ca^{2+}/CaM (see Figure 3.2). In the presence of excessive CaM, the EC_{50} for activation by Ca^{2+} varies from 0.27 μM (PDE1A1) to 3.02 μM (PDE1C1).[15] However, the affinity is not uniquely high or low for all the variants of each isoform. Variation in amino acid sequence at the N-terminus of some isoforms confers sequence changes to the first CaM-binding site that can alter PDE1 affinity

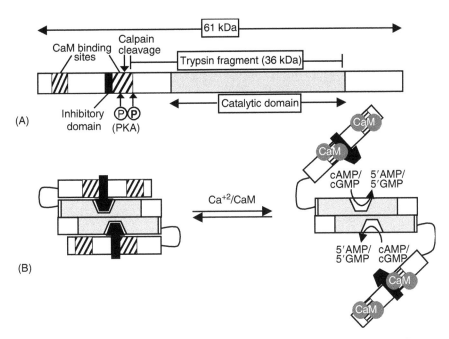

FIGURE 3.1 PDE1 domain organization and regulatory sites. (A) The PDE1A2 domain structure is diagrammed with the calmodulin-binding sites (striped box), inhibitory domain (solid box), and catalytic domain (gray) as shown. The m-calpain and trypsin cleavage sites are labeled as they are the PKA phosphorylation sites with the site that affects calmodulin affinity shown in bold. (B) PDE1 has been shown to exist as a dimer and in the absence of calmodulin, its activity is thought to be inhibited by an interaction between the inhibitory domain and the catalytic domain. Upon binding of calmodulin, the inhibition is relieved and the catalysis is allowed to occur.

for CaM. For instance, the PDE1A1 variant has an N-terminus that is distinct from that of PDE1A2 and consequently has a nearly 10-fold lower EC_{50} for the activation by CaM. These properties produce an array of PDE1 enzymes that would respond to different degrees to changes in Ca^{2+} levels in the cell. Thus, different cell types may express individual PDE1 enzymes to regulate specific calcium-mediated processes. Additionally, phosphorylation is a posttranslational mechanism that can reversibly decrease the affinity for Ca^{2+}/CaM (Figure 3.2). The variations in Ca^{2+}/CaM affinity produced by phosphorylation and expression of unique N-terminal sequences are likely to be important mechanisms for imparting a highly specific regulation of PDE1 activity by Ca^{2+} *in vivo* and the two concepts will be discussed in more detail in later sections. Exactly how the biochemical properties of PDE1 regulation are related to the regulation of the enzyme by Ca^{2+} signals in intact cells is beginning to be investigated.[16]

Wide variations in the fold stimulation of PDE1 activity by CaM have been reported. One reason for the variable activation is that proteolysis can cleave off the inhibitory domain thereby irreversibly activating the enzyme. Limited proteolysis of the 59 kDa bovine brain PDE1 *in vitro* by α-chymotrypsin produced a 45 kDa fragment that could not be activated by the addition of CaM, but had activity equal to native enzyme in the presence of CaM.[11] Similarly, cleavage of PDE1A2 with trypsin was used to generate a 36 kDa fragment that was fully active in the absence of CaM.[17] By amino acid sequence analysis, the identity of the fragment was shown to contain the catalytic domain (see Figure 3.1A). The proteolytic activation may be physiologically relevant in the brain as m-calpain was found to cleave bovine brain PDE1A2 between residues [126]Gln and [127]Ala. This cleavage site is C-terminal to the first CaM-binding site (residues 24–45), but resides in what is thought to be the second CaM-binding

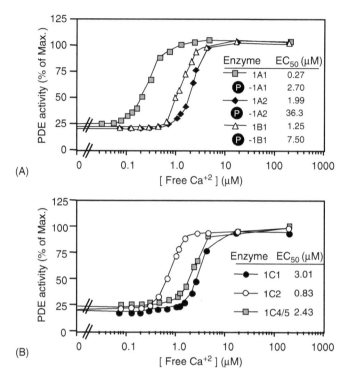

(A)

(B)

FIGURE 3.2 PDE1 enzymes vary in their sensitivity to activation by Ca^{2+}/CaM. The sensitivity for activation by Ca^{2+}/CaM was determined for PDE1 enzymes transiently expressed in COS-7 cells as reported in Ref. [15]. PDE activity was measured in the presence of $4\,\mu g/mL$ CaM with varying concentrations of Ca^{2+} using $1\,\mu M$ cGMP as substrate. (A) PDE1A1 and PDE1A2 have unique N-termini and disparate Ca^{2+} EC_{50} values. Phosphorylation of PDE1A1[27] and PDE1A2 [25] by PKA or PDE1B1 by CaM kinase II[20] decreases the affinity for Ca^{2+}/CaM and increases the EC_{50} for the activation. The values shown for the phosphorylated enzymes were calculated based on the fold changes determined in the original reports. (B) The PDE1C2 variant has a large N-terminal sequence, which is not present in PDE1C1 or PDE1C4/5 that likely alters its affinity for Ca^{2+}/CaM.

site (residues 108–138)[12] (see Figure 3.1A). Cleavage by calpain generates a 45 kDa fragment that is constitutively active even in the absence of CaM.[18] This could serve as an irreversible long-term mechanism for activating PDE1A2 in contrast to the activation by Ca^{2+}/CaM-binding, which would be short term and reversible. Cleavage of PDE1 by calpain has been suggested to occur *in vivo* in rat hearts subjected to ischemia and reperfusion.[19]

3.2.2 KINETIC PROPERTIES

The three PDE1 isoforms have distinct enzymatic properties as they vary in specificity for cAMP or cGMP as substrate and in affinity for Ca^{2+}/CaM. Catalytic properties of many of the PDE1 isoforms and variants have been determined and are listed in Table 3.1 with references. The PDE1A family is characterized by a strong preference for cGMP as substrate as the enzymes have a 27- to 42-fold higher affinity (lower K_m) for cGMP than cAMP. The PDE1B family also prefers cGMP as a substrate, although the difference (four- to eightfold lower K_m for cGMP than cAMP) is not as pronounced as with the PDE1A family. In contrast, the PDE1C family does not display a preference for cGMP or cAMP as substrate and has nearly equal K_m values for the two nucleotides. Regardless of the *in vitro* substrate specificity of the PDE1 isoforms, they may conceivably function in the cell to regulate either cAMP or cGMP.

TABLE 3.1
PDE1 Enzyme Kinetic Properties[a]

Enzyme	kDa	K_m (μM)		V_{max}	Reference
		cGMP	cAMP	cGMP/cAMP	
HsPDE1A1	60.6	2.62	72.7	0.61	41
HsPDE1A2/5	62.3	2.20	93.1	0.54	41
HsPDE1A3/6	61.2	3.15	10.1	0.53	41
HsPDE1A4	59.6	3.51	12.4	0.56	41
HsPDE1B1	61.4	5.90	23.8	1.61	50
HsPDE1B2	59.0	1.21	9.96	2.37	50
MmPDE1C1	71.7	2.2	3.5	0.77	15
RnPDE1C2	86.7	1.1	1.2	0.83	15
HsPDE1C3	80.8	0.57	0.33	1.22	45
MmPDE1C4/5	74.0	1.0	1.1	1.0	15

[a]Kinetic constants for PDE activity of the PDE1 isoforms and their variants were determined from recombinantly expressed proteins. PDE1A variants were expressed in Sf9 cells, PDE1B variants were expressed in HEK cells, and PDE1C variants were expressed in COS-7 cells except for HSPDE1C3, which was expressed in yeast.

Despite the major differences in K_m values for various PDE1 isoforms, they all have a high specific activity for cyclic nucleotide hydrolysis, although there is some variability in the values reported for the different isoforms. CaM-activated purified bovine brain PDE1B was found to have V_{max} values ranging from around 20[20] to 30 μmol cGMP/min/mg.[21] A larger range of values for PDE1A have been reported as values have ranged from around 40–50 μmol cAMP/min/mg for purified bovine lung PDE1A[22] and purified brain PDE1A2[18] to higher values of 200–300 μmol cAMP/min/mg for purified brain PDE1A[23] and recombinant bovine PDE1A1 and PDE1A2.[12] Specific activity values for purified PDE1C have not been reported. All the PDE1 enzymes have nearly equal V_{max} values for cGMP and cAMP and these ratios are listed in Table 3.1.

3.2.3 REGULATION BY PHOSPHORYLATION

Phosphorylation has been biochemically characterized as a mechanism for the regulation of both PDE1A and PDE1B. However, phosphorylation has also been recently suggested to regulate PDE1C.[24] cAMP-dependent protein kinase (PKA) was first reported to phosphorylate bovine brain PDE1A2,[25] but has also been reported to phosphorylate bovine heart PDE1A1.[26,27] Phosphorylation of PDE1A results in the incorporation of 2 mol of phosphate per mol of PDE1 monomer.[25,26] Bovine brain PDE1A2 phosphorylation results in a reduction of calmodulin affinity and shifts the EC$_{50}$ for the activation by calmodulin from 0.51 to 9.3 nM.[25] Phosphorylation is blocked by the binding of CaM and can be reversed by the phosphoprotein phosphatase calcineurin. In a later report using proteolysis and mass spectrometry, Ser120 and Ser138 were identified as the principal PDE1A2 residues phosphorylated.[26] However, only phosphorylation of Ser120 was found to reduce the affinity for CaM; the importance of this site is indicated in Figure 3.1A by bold letters. The modification of CaM-binding sensitivity by PKA-mediated phosphorylation could serve as a feedback mechanism for regulating the cross talk between the cAMP- and Ca^{2+}-signaling systems. Recently, evidence has been reported that PDE1C may be regulated by PKA phosphorylation as well. In AtT20 neuroendocrine cells, treatment with CPT-cAMP was found to reduce PDE1 activity in a PKA-dependent manner.[24]

Furthermore, this effect was potentiated by the phosphoprotein phosphatase inhibitor calyculin A and was demonstrated to be due to a reduced ability of CaM to stimulate activity. The effect could be reproduced *in vitro* with partially purified PDE1C from the cells and added PKA.

Unlike PDE1A, PDE1B is not phosphorylated by PKA but is instead phosphorylated by CaM kinase II. Similar to PDE1A, phosphorylation of PDE1B decreases the enzyme's affinity for CaM.[21] Phosphorylation of purified bovine brain PDE1B by CaM kinase II increased by sixfold the EC_{50} for the activation by CaM.[20] However, the binding of CaM to the PDE blocks the phosphorylation and change in enzyme kinetics. Peptide mapping of tryptic digests of phosphorylated PDE1B identified two major phosphorylated peptides that were suggested to be responsible for the shift in CaM affinity. Like PDE1A, PDE1B phosphorylation is also regulated by calcineurin as it was found to dephosphorylate CaM kinase II phosphorylated enzyme and reverse the reduction in affinity for CaM.[21] It has been hypothesized that the significance of the regulation of PDE1B by phosphorylation in this manner serves as a timing loop for regulating cAMP.[5] Initial Ca^{2+} entry into the cell could activate the Ca^{2+}/CaM-sensitive adenylyl cyclase and CaM kinase II resulting in the phosphorylation and desensitization of PDE1B and allowing the accumulation of cAMP to occur. As calcium levels rise further, calcineurin is activated and dephosphorylates PDE1B, allowing the PDE to become sensitive to activation by Ca^{2+}/CaM. The activated PDE1B then hydrolyzes the cAMP thereby terminating the signal. This would function as a negative feedback process to modulate cAMP levels. Similar interactions may also exist in cGMP signal transduction as Ca^{2+} levels can be regulated by cGMP and in turn may feedback to affect the PDE1B phosphorylation state and control the duration and amplitude of cGMP levels.

3.2.4 INHIBITORS

Although a number of compounds have been found to inhibit PDE1 activity *in vitro*, no truly selective PDE1 inhibitors are available. PDE1 inhibitors may block activity by either preventing the interaction of Ca^{2+}/CaM with the enzyme or by interfering with cyclic nucleotide binding at the catalytic site. The list of compounds that inhibit PDE1 by interfering with Ca^{2+}/CaM interaction with the PDE is quite extensive and includes compounds that interfere with Ca^{2+}, such as EGTA, and compounds that interfere with CaM, such as trifluoperazine and W7. In general, these types of compounds have limited utility as research probes for PDE1 in a cellular context, as they will additionally inhibit a large number of other Ca^{2+}/CaM-regulated proteins.

Compounds that interfere with the catalytic site of PDE1s have been characterized, but few provide a high degree of selectivity over other PDE isoforms. Nonselective PDE inhibitors such as theophylline and IBMX (isobutylmethylxanthine) inhibit members of the PDE1 family. Derivatization of IBMX at the C-8 position to 8-methoxymethyl-IBMX does not affect the IC_{50} of the compound for PDE1 (low micromolar) but increases the selectivity for PDE1 over other PDEs by 30- to 50-fold.[28] Some selectivity for different PDE1 isoforms can be seen with other compounds. For example, vinpocetine is more selective for PDE1A and PDE1B in comparison to PDE1C.[15] Newer compounds are being developed that more specifically target PDE1. SCH51866 is a compound that inhibits the PDE1s in the low to mid-nanomolar range,[15,29] although it also inhibits other cGMP-hydrolyzing PDEs. Recently, ICOS reported a compound, IC224, that inhibits PDE1 with an IC_{50} of 80 nM and has a 127-fold selectivity for the next most sensitive PDE.[30]

Inhibition of PDE1 may be relevant in some clinical contexts. Clinically used drugs such as the erectile dysfunction (ED) drugs sildenafil and vardenafil have been reported to be PDE1 inhibitors. After PDE5, the PDE1 and PDE6 enzymes have the lowest IC_{50} values for

sildenafil and vardenafil, although the PDE1 IC_{50} values are higher than normal clinically used doses.[31,32] Other clinically used drugs that can act as PDE1 inhibitors include the anti-Parkinsonian agents amantadine[33] and deprenyl,[34] and calcium-channel blockers.[35] However, it has not been demonstrated that inhibition of PDE1 is involved in the mechanism of action of any of these drugs or is responsible for their side effects. Nevertheless, the possibility exists that these drugs may accumulate at high levels in selective compartments or cell types and produce physiological effects consequential to inhibition of PDE1. This idea may be especially germane to the ED drugs as many of them are currently being investigated for the treatment of a wide variety of pathologies beyond ED. The existence of naturally occurring PDE1 inhibitors has also been reported; for instance, ginsenoids have been found to be potent inhibitors of some PDE1 enzymes.[36] Also, the existence of a protein inhibitor of PDE1 was reported in glioblastoma multiforme, but not in normal human cerebral tissue.[37] Additionally, differentiation-inducing factor-1 (DIF-1) from *Dictyostelium discoideum*, which is a potent antiproliferative agent in mammalian cells, was suggested to inhibit PDE1 as its mechanism of action. DIF-1 was demonstrated to elevate cAMP levels in mammalian cells and inhibit PDE1 at concentrations where PDE3A, PDE3B, PDE8A, and the CaM-dependent enzymes, calcineurin and myosin light chain kinase, were not inhibited.[38]

3.3   PDE1 GENETICS

The PDE1 family is comprised of three separate genes: PDE1A, PDE1B, and PDE1C. However, each of these three genes has alternative promoters and is alternatively spliced to give rise to a multitude of disparate mRNAs and protein products (see Figure 3.3). The number of known unique variants of each gene has significantly increased recently as the advances in molecular biology and sequencing technologies have greatly facilitated the identification of new PDE1 variant mRNAs. However, the protein products of many of these mRNAs in tissues have yet to be documented. The regulation of PDE-splicing patterns is

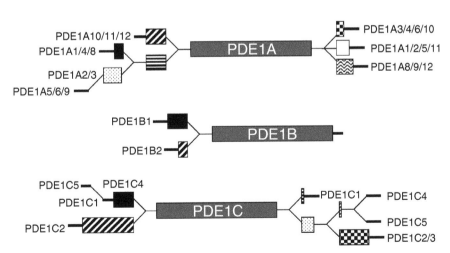

FIGURE 3.3 Organization of mRNA and protein products of PDE1A/B/C variants. Many PDE1 isoform variants are produced by the use of alternative start sites or alternative splicing. Common regions for each isoform are shown as dark boxes and labeled with the isoform identity, but are not shown to scale. The boxes with assorted shading patterns represent unique amino acid sequences and are shown to scale with each other. Thick darkened lines at the termini represent untranslated nucleotide sequence. Only the human variants are shown for PDE1A and PDE1B whereas the variants from multiple species are shown for PDE1C (see Table 3.1 for species designations).

complex and not well understood. Nonetheless, it appears that the production of splice variants is a regulated process and results in variant production in a cell- and tissue-specific manner. The purpose of generating variants with unique N- or C-termini is not always apparent as many of the variants have highly similar enzymatic properties. The unique mRNA sequences may have important regulatory features or the unique amino acid sequences may be important for PDE localization or protein–protein interactions.

3.3.1 PDE1A

Early work identified PDE1A variants in bovine tissues such as brain and heart.[39] Later studies identified more variants in bovine as well as other species. To date, the PDE1A isoform has proven to be the most complex of the three PDE1 isoforms. The human PDE1A gene spans over 120 kb of genomic DNA on chromosome 2q32.[40] For the human PDE1A gene alone, three different N-termini and three different C-termini have been identified and are matched in various combinations to give rise to multiple different protein products[40–42] (see Figure 3.3) with proteins ranging in molecular weight from roughly 59 to 62 kDa. In addition, some mRNA variants exist that have different untranslated regions but still encode identical proteins. At the N-terminus, there are several separate transcriptional start sites, resulting in three separate N-terminal amino acid sequences. The mRNA for PDE1A1/4/8 encodes 18 amino acids that are different from a unique 34 amino acid stretch of PDE1A2/3/5/6/9 whereas PDE1A10/11/12 have a 38 amino acid sequence that is distinct from either of the other two N-termini. The mRNAs encoding three different C-terminal amino acid sequences have also been identified. At the C-terminus, alternative splicing can occur producing three alternative 3' sequences.[40] The PDE1A variants differ greatly in their tissue expression. PDE1A1 and PDE1A4 mRNA were found to have a broad tissue distribution. However, mRNAs of other variants were found to have highly specific expression as PDE1A5 and PDE1A6 were detected only in brain[40] or thyroid[42] and PDE1A10 was detected only in testis.[40] Other PDE1A-variant mRNAs were not detected by Northern blot in appreciable amounts in any of the tissues tested.

The variation in primary sequences of PDE1A variants can translate into significant changes in enzymatic properties, specifically in their affinity for Ca^{2+}/CaM. This is not surprising given that the N-terminal variations are in regions of the first CaM-binding site. Bovine PDE1A1 and PDE1A2 have different N-terminal sequences and correspondingly have different affinities for CaM. PDE1A1 has a 10-fold higher affinity for CaM than PDE1A2.[43,44] The Ca^{2+}/CaM-binding affinities of most of the other N-terminal variants have not been characterized, but as the regions of variation are in or around the CaM-binding domains differences in affinity may be large. Many of the other kinetic properties of the variants such as inhibitor potencies, cAMP/cGMP K_m, and cGMP/cAMP V_{max} ratios have been found to be highly similar.[5,41,45]

3.3.2 PDE1B

In contrast to PDE1A, only two variants of PDE1B have been identified. PDE1B1 was first identified more than 10 years ago and has been extensively studied. The variant has been characterized in bovine,[46,47] human,[48] and mouse.[49] PDE1B2 was first reported in 2002 for human[42] and only one other report[50] concerning the enzyme has been published. The difference between the two variants lies at the N-terminus where PDE1B1 has a unique 38 amino acid sequence and PDE1B2 has an 18 amino acid sequence that is divergent from that of PDE1B1 (see Figure 3.3 and Figure 3.5). The longer N-terminal sequence of PDE1B1 gives the protein a larger molecular mass (61.4 kDa) compared to the PDE1B2 variant (59.0 kDa).

Alignment of the PDE1B variants' mRNA sequences with the genomic sequence of PDE1B has revealed that these unique N-terminal sequences are products of separate first exons located large distances apart $(11.5\,kb)$[50] (see diagram in Figure 3.5A). The fact that the PDE1B1 first exon is located a large distance upstream from the second exon is also true for mouse PDE1B1 and is underscored by the fact that researchers originally cloning and characterizing the mouse PDE1B1 gene failed to detect the first exon.[49] The PDE1B1 and PDE1B2 unique first exons were each found to have separate functional promoters, allowing separate transcriptional regulation of the two variants.[50] This may be significant as it was discovered that the PDE1B1 and PDE1B2 mRNAs are regulated differently. During mono-cyte to macrophage differentiation, the PDE1B2 promoter is activated and produces an mRNA that is translated into functional PDE1B2 protein. In contrast, in the same cells, PDE1B1 mRNA is likely subjected to a form of posttranslational repression as the PDE1B1 mRNA is present and even induced, but no protein can be detected.[50]

The unique N-terminal protein sequences do not appreciably alter the enzymatic charac-teristics of the PDE1B variants as similar K_m values and cGMP/cAMP V_{max} ratios were noted for recombinant PDE1B1 and PDE1B2. However, PDE1B2 has a roughly threefold lower EC_{50} for activation by CaM than PDE1B1.[50] It is not known if this difference in affinity has a functional consequence in the cell. As PDE1B1 and PDE1B2 were found to have separate promoters; the reason for the existence of the two variants may be for highly regulated tissue-specific expression, which will be discussed later in Section 3.4.3. Alternatively, the unique amino acid sequences may be required for cellular targeting or protein–protein interactions.

3.3.3 PDE1C

The PDE1C family is distinguished from the other two PDE1 families in that it does not prefer cGMP/cAMP as substrate and hydrolyzes each nucleotide equally well. PDE1C enzymes are the largest of the PDE1 family and their predicted molecular masses range in size from 71 to 86 kDa. Five variants of PDE1C encoding four different proteins have been identified from several different species. PDE1C1/4/5 were isolated from mouse whereas PDE1C2 and PDE1C1/3 have been isolated from rat and human cDNA libraries, respect-ively.[15,45,51] The unique protein and untranslated regions for the PDE1C variants are dia-grammed in Figure 3.3. PDE1C2 has a unique 98 amino acid sequence at its N-terminus that is distinct from the 38 amino acids at the N-terminus of the other four variants.[51] Although the PDE1C/3/4/5 variants encode proteins with identical N-termini, there is a great deal of heterogeneity in the C-terminal portions of these variants.[15] PDE1C1 has the shortest unique C-terminus (four amino acids) whereas PDE1C2/3 have the longest (79 amino acids). PDE1C4 and PDE1C5 encode identical C-terminal protein products, but differ in their $3'$ untranslated sequences. The purpose of having unique $3'$ UTR regions for the PDE1C variants has not been explored, but it may be a regulatory feature of the individual mRNA products.

The different PDE1C variants were found to have highly similar enzymatic properties, inhibitor profiles, and specificity for cAMP versus cGMP.[15] However, the N-terminal differ-ence of PDE1C2 from the other PDE1C variants may be reflected in the fact that it has a roughly three- to fourfold lower EC_{50} for the activation by Ca^{2+}/CaM.[15] One notable difference in the PDE1C variants is their tissue expression pattern. RNase protection and in situ hybridization demonstrated that the mRNA expression for different PDE1C variants is specific to certain tissues and can even be highly localized to structures or cell types within a tissue. For example, PDE1C2 was found to be uniquely expressed at high levels in sensory neurons of the olfactory epithelium.[51] Other PDE1C variants have been localized to different regions of the brain, cardiovascular system, and testis as is described later.

3.4   TISSUE-SPECIFIC EXPRESSION AND FUNCTIONAL ROLES OF PDE1

One of the challenging aspects of PDE research is elucidating a role for individual PDEs *in vivo*. This task is especially difficult for the PDE1 family enzymes as a highly selective inhibitor is currently unavailable. Alternatively, genetic approaches have recently been used to investigate PDE1 function. An initial step in studying PDE function is identifying which PDEs are expressed in individual tissues and cell types. Numerous studies have been undertaken to this end and have demonstrated highly specific expression of PDE1 isoforms and variants. Furthermore, PDE1 expression patterns have been found to change in several physiological and pathological settings, thus yielding clues to what processes the PDE1s may regulate.

3.4.1   CARDIOVASCULAR SYSTEM

All three of the PDE1 isoforms are expressed in tissues of the cardiovascular system. However, their prevalence is highly variable among tissues and cell types. Furthermore, in specific tissues there can be significant differences among species in which PDE1 isoform is expressed. In cardiac bovine tissue, PDE1A1 is highly expressed.[52] Likewise, in rat heart PDE1 constitutes about 60% of the total cGMP PDE activity with PDE1A1 and PDE1C representing 85% and 15% of that activity, respectively. However, some studies suggest that PDE1 is not expressed in myocytes themselves, but is instead expressed in smooth muscle of arteries supplying the muscle.[53] In contrast, another report found that PDE1C mRNA is expressed not only in rat ventricular tissue, but also in isolated myocytes as well.[54] Myocyte-specific overexpression of a Ca^{2+}/CaM-regulated adenylate cyclase (AC8) in transgenic mice was found to change the ratio of the PDE1 isoforms in homogenized hearts. While cAMP PDE1 activity was increased 71%, cGMP PDE1 activity was reduced 28%.[55] This may suggest a switch in the PDE1 isoforms that are expressed to compensate for the higher cAMP levels, which are likely to result from increased AC8 expression. It was suggested that PDE1 isoforms, along with other PDE isoforms in the heart, exist to modify PKA regulation of L-type Ca^{2+} channels and other cardiac targets.[55] Similarly, when rat aortas were banded to model pressure overload, an upregulation of CaM-stimulated cGMP PDE1 activity was found in left ventricular tissue.[56] However, only a slight increase in PDE1A mRNA was detected by Northern blotting and the increase in activity may be due to a posttranscriptional or posttranslational mechanism. This PDE1 upregulation may function to offset the increased synthesis of natriuretic peptides.

All three PDE1 isoforms have been identified in vascular tissue. Interestingly, PDE1 expression is highly variable and exhibits both tissue- and species-specific expression patterns. Smooth muscle cells (SMCs) isolated from human and monkey aorta express PDE1A and PDE1B whereas SMCs from rat aorta express only PDE1A.[57] PDE1C was absent from quiescent SMCs, but found to be induced in human SMCs proliferating in culture. However, PDE1C was not found in proliferating SMCs from a number of other species and may be unique to human SMC.[58] PDE1C was also identified in SMC isolated from a human atherosclerotic lesion. PDE1C expression was well correlated with human SMC of the synthetic proliferative phenotype and both pharmacological PDE1 inhibitors and antisense oligonucleotide treatment against PDE1C inhibited human SMC proliferation.[59] These results indicate that PDE1C is required for SMC proliferation, but is dispensable for quiescent SMC function. Thus, inhibition of PDE1C could be a specific and useful target in treating disease states, such as atherosclerosis, where SMC proliferation is problematic.

Another medically relevant finding concerning PDE1 was the induction of PDE1A1 in nitrate-tolerant rats. Nitroglycerin (NTG) is a therapeutic agent used for the treatment of angina pectoris due to its activity as a vasodilator via its ability to stimulate cGMP

production by the activation of soluble guanylyl cyclase (GC). In vascular SMCs, cGMP activates PKG, which can mediate relaxation by several mechanisms including decreasing Ca^{2+} release, increasing Ca^{2+} sequestration, and activating myosin light chain phosphatase.[60] However, NTG efficacy is reduced during chronic treatment due to the development of tolerance. In aortas of rats where nitrate tolerance was induced by NTG treatment, induction of PDE1A1 mRNA, protein, and activity was found.[61] The partially selective PDE1 inhibitor vinpocetine was able to partly reverse the tolerance to NTG when vasorelaxation and cGMP accumulation were measured. Furthermore, the induction of PDE1A1 was suggested to mediate the nitrate-tolerance-induced supersensitivity to vasoconstrictors such as angiotensin II[61] (see model in Figure 3.4). Angiotensin II treatment was found to reduce the accumulation of cGMP in response to atrial natriuretic factor (ANF) treatment. This reduction is likely mediated by the effect of angiotensin II to elevate Ca^{2+} that can activate PDE1A1. Functionally, the role of PDE1 in vascular smooth muscle may be to modulate contraction by reducing cGMP in response to a Ca^{2+} signal generated by vasoconstrictors (see model in Figure 3.4). This was illustrated in several earlier studies of PDE1 activation. Using a method designed to estimate the association of Ca^{2+}/CaM with PDE in intact tissue,[62] it was found that the exposure of intact coronary artery strips to histamine increased the association of PDE1 with Ca^{2+}/CaM.[63] Thus, intracellular Ca^{2+}-altering stimuli likely trigger PDE1 modulation of SMC contraction.

3.4.2 Nervous System

The brain was one of the first tissues in which PDE1 activity was discovered and a great deal of the initial characterization of PDE1 biochemistry was performed in bovine brain.[8,9,43,64]

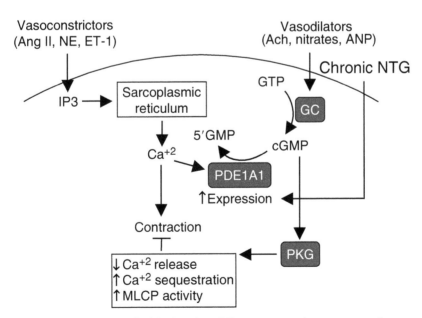

FIGURE 3.4 PDE1A1 is upregulated with chronic NTG treatment and opposes vascular smooth muscle relaxation. Chronic NTG treatment induces the increased expression of PDE1A1 in vascular smooth muscle that contributes to tolerance of the vasodilatory effects of NTG. PDE1A1 hydrolyzes cGMP generated from GC activated by vasodilators. cGMP activates PKG and induces relaxation primarily by decreasing calcium levels. PDE1A1 can be activated by a rise in Ca^{2+} that is generated by vasoconstrictors and thus potentiates contraction by reducing cGMP. (Adapted from Refs. [60,61].)

All three PDE1 isoforms are present in the central nervous system (CNS) to differing degrees. Importantly, the expression of the different PDE isoforms is highly specific and localized not only in different regions and cell types, but also in individual neurons in the same tissue. Although a great deal of work has identified the expression pattern of PDE1 isoforms in neuronal tissues, few studies have explored PDE1 function in neurons.

Olfactory neurons are one neuronal cell type that exhibits highly specific PDE1 isoform expression. PDE1C2 was found to be highly enriched in the sensory neurons of the rat olfactory epithelium[51] where both PDE1A and PDE1B were absent.[65] PDE1C2 expression is highly localized as it was not found in all neurons. Furthermore, PDE1C2 has a distinct subcellular distribution as it is localized to olfactory cilia and is not present in the axons or cell bodies.[66] Odorant sensing is a rapid cAMP-dependent process that likely requires highly tuned regulation by a high-turnover-responsive PDE such as PDE1C2. Hypothetically, PDE1C2 could function as a cAMP regulator when an odorant activates adenylyl cyclase (G_{olf}-activated type III adenylyl cyclase) and triggers the production of cAMP that activates a cAMP-regulated cation channel. The channel initially allows Ca^{2+} entry that generates the electrical neuronal signal, but later Ca^{2+} could activate PDE1C2 that hydrolyzes cAMP and in part terminates the odorant signal.[51] Thus, the Ca^{2+} signal not only mediates the odorant response, but also elicits a feedback pathway to trigger its termination. It is likely that in neurons PDE1 would function to reduce cyclic nucleotides in response to a Ca^{2+} signal. In another example, PDE1 has also been suggested to reduce cGMP accumulation in cerebellar granule neurons in response to increased intracellular Ca^{2+}.[67] It was suggested that physiological agonists, such as *N*-methyl-D-aspartate (NMDA), which increase Ca^{2+} entry into the cell can activate PDE1 in neurons and thus reduce cyclic nucleotide levels.

Although both PDE1A and PDE1B are highly expressed in the brain, they have different expression patterns. In situ hybridization studies of mouse brain found that PDE1A and PDE1B mRNA are expressed in unique and mostly nonoverlapping regions of the mouse brain.[65] PDE1B is widely expressed in many regions, but is found at highest levels in the caudate-putamen, nucleus accumbens, olfactory tubercle, and dentate gyrus of the hippocampus. In contrast, PDE1A has a more limited distribution and is expressed at highest levels in the cerebral cortex and in the pyramidal cells of the hippocampus. Both PDE1A and PDE1B appeared to be localized to the cell bodies of neurons. PDE1 expression in the brain not only varies by region, but also can even differ among neurons in the same region. Purkinje neurons of the mouse cerebellum express high levels of PDE1B, but only in a subset of the cells.[68] In contrast, PDE5 was expressed in all Purkinje neurons. The highly specific and differential PDE1 isoform expression patterns suggest that their different regulatory and enzymatic properties make the isoforms highly specialized for specific functions in neurons. For example, the expression pattern of PDE1B was generally homologous to that of dopamine receptors, suggesting that it may modulate dopamine function in regions such as the caudate-putamen, nucleus accumbens, or olfactory tubercle.[65]

As suggested by the localization studies,[65] using PDE1B knockout mice, Reed et al.[69] found that PDE1B plays a role in dopaminergic signaling. PDE1B expression was ablated by disruption of the gene in the catalytic region and the absence of PDE1B mRNA in brain confirmed the success of the knockout.[69] PDE1B knockout mice were identical in appearance and in most behaviors to wild-type mice. However, they exhibited increased locomotor activity. The increased locomotor activity was exaggerated in mice treated with the dopamine agonist methamphetamine. On a molecular level, the role of PDE1B in dopaminergic function was manifested as the knockout mice exhibiting increased phosphorylation of DARP-32 and GluR1 in response to D1 agonists. The data suggest that in the absence of PDE1B, cAMP or cGMP levels are elevated and increase the phosphorylation state of signaling proteins in response to dopamine agonists. PDE1B knockout mice also showed decreased learning and

memory as demonstrated by their performance in Morris water maze tests. PDE1B knockout mice may serve as a tool in future experiments for the identification of other processes that are regulated by PDE1B.

3.4.3 IMMUNE SYSTEM

Cyclic nucleotides, particularly cAMP, are established regulators of immune cell function. Both cAMP and cGMP are generally considered to be anti-inflammatory and PDE inhibitors can inhibit inflammatory functions such as cytokine release.[70] However, most of these effects are ascribed to the inhibition of PDE4 activity.[70] Nevertheless, studies have demonstrated the increased expression of PDE1 in several immune cell types upon activation or differentiation, suggesting that PDE1 may play a requisite role in the immune response. Recent results have suggested what role PDE1 may play in the function of immune cells.

The first evidence of PDE1 expression in leukocytes was the report of induction of CaM-stimulated PDE activity in phytohemagglutinin (PHA)-stimulated mononuclear cells. PDE1 activity and protein were found in PHA-activated bovine mononuclear cells, but not in untreated cells.[71] PHA induced a PDE1 protein with a molecular mass of 63 kDa that was detected by Western blotting with an anti-PDE1 antibody, and the protein had an HPLC elution profile similar to that of 63 kDa CaM-stimulated bovine brain PDE1, suggesting that the PDE1 isoform was PDE1B. Later studies also reported low PDE1 activity in CD4$^+$ and CD8$^+$ cells,[72] but an induction of PDE1 activity and PDE1B mRNA was observed upon the activation of human T cells with either PHA or CD3/CD28 treatment.[73] PHA-activated human T cells were found to upregulate both the PDE1B1 and PDE1B2 variants[50,74] (see Figure 3.5D). The treatment of PHA- or CD3/CD28-activated T cells with 8-methoxymethyl-IBMX inhibited the production of the B-cell activating cytokine IL-13.[73] PDE1B may be important not only for the activation of lymphoid cells, but also for their survival. Indeed, the treatment of the human lymphoblastoid cell line RPMI-8392 with either vinpocetine or antisense oligonucleotides against PDE1B reduced PDE1 activity and induced apoptosis.[74] Thus, PDE1B induction likely plays a role in T-cell activation and survival.

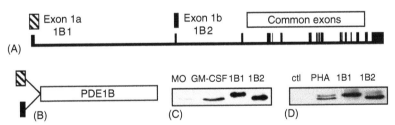

FIGURE 3.5 PDE1B1 and PDE1B2 have distinct N-termini and are upregulated in activated human immune cells. The unique N-terminal amino acid sequences of PDE1B1 and PDE1B2 are the products of separate first exons that have different transcriptional start sites whereas the remaining C-terminal portions of the proteins are identical. (A) The exon and intron organization of the *PDE1B* gene is shown and (B) the protein products are diagrammed with the unique N-terminal sequences of PDE1B1 and PDE1B2 shown as the cross-hatched (*PDE1B1*) or darkened (*PDE1B2*) areas. (C) Recombinant PDE1B1 and PDE1B2 as well as PDE1 from monocytes (MO) and GM-CSF-treated monocytes (*GM-CSF*) were immunoprecipitated. The immunoprecipitates were separated by SDS-PAGE and Western blot, using an antibody recognizing a portion of the common C-terminal region of PDE1B. Only a band with the same migration as recombinant PDE1B2 is found in GM-CSF-treated monocytes. (D) The same anti-PDE1B C-terminal antibody detected both PDE1B variants in immunoprecipitates from primary T cells treated with PHA but not in control untreated T cells (*ctl*). (Modified from Ref. [50].)

PDE1 activity is very low in circulating monocytes, but is greatly induced when monocytes are differentiated *in vitro* into macrophages.[75,76] Likewise, PDE1 activity has been found in primary macrophages isolated from humans as well as a number of other species.[77,78] One of the interesting features of macrophages is their remarkable heterogeneity of phenotypes depending on their location in the body. PDE1 expression has been found to vary greatly among macrophages of different phenotypes and different species. Monocytes can be differentiated into macrophages of diverse phenotypes by the exposure to different cytokines. It was found that differentiating monocytes with the cytokine granulocyte–macrophage colony-stimulating factor (GM-CSF) highly induced PDE1B expression whereas the differentiation with macrophage colony-stimulating factor (M-CSF) induced only a small increase in PDE1B.[76] Monocytes differentiated to macrophages with GM-CSF phenotypically resemble alveolar macrophages whereas monoctyes differentiated with M-CSF resemble peritoneal macrophages. The large induction of PDE1 is highly specific as GM-CSF differentiation in the presence of IL-4 directs the differentiation to a dendritic cell instead of a macrophage and suppresses the induction of PDE1.[50] Immunoprecipitation of PDE1 from monocytes differentiated with GM-CSF shows a band that migrates with the same molecular mass as recombinant PDE1B2 by Western blotting, but not recombinant PDE1B1 (see Figure 3.5). Thus, only the PDE1B2 and not the PDE1B1 variant is induced with differentiation. Correspondingly, high PDE1 activity was also found in isolated human alveolar macrophages[78] whereas peritoneal macrophages isolated from rat[77] as well as mouse[79] did not express PDE1. What role PDE1 plays in macrophage biology is at this time unclear. The fact that PDE1B is upregulated with macrophage differentiation suggests that it plays a role in differentiation, determination of the final phenotype of the macrophage, or regulation of a cellular function that is specific to the alveolar macrophage type.

3.4.4 TESTIS AND SPERM

Recent work has demonstrated the importance of cyclic nucleotides in the reproductive system, particularly in sperm and testis. cAMP has been demonstrated to regulate several important processes in sperm migration and fusion with the oocyte such as capacitation, motility, hyperactivation, and the acrosome reaction.[80] Thus, tight regulation of cyclic nucleotide levels is required and reflected in the wealth of PDE isoforms that are expressed. Treatment of sperm with PDE inhibitors can increase cAMP-stimulated motility.[81,82] Clinically, nonspecific PDE inhibitors have been used in *in vitro* fertilization to increase the procedure's success rate.[81] The semiselective PDE1 inhibitor 8-methoxymethyl-IBMX was found to increase the acrosome reaction with no apparent effect on sperm motility whereas the opposite was true for PDE4 inhibitors.[83] This suggests that the PDE1 isoforms in sperm may have highly specialized functions and are involved in the regulation of sperm activation.

The PDE1 family is well represented in testis and sperm as multiple variants of the PDE1A and PDE1C isoforms are expressed. PDE1 activity was detected and characterized in mouse and rat male germ cells and was found to constitute a large portion of the PDE activity in these cells.[84,85] The PDE1A and PDE1C genes were found to be the most highly expressed in mouse testis[86] as PDE1B expression is low as determined by in situ hybridization (see Figure 3.6A). PDE1A and PDE1C mRNA were strongly detected only in the seminiferous tubule but not in the interstitium. Interestingly, the detected signal varied between tubules, indicating that the expression of the two enzymes varies during spermatogenesis. As implied, PDE1A and PDE1C expression patterns were cell-type- and stage-specific and differed in developing versus mature sperm with PDE1C expression occurring earlier than PDE1A (Figure 3.6B).[86] PDE1A mRNA was found in round to elongated spermatids whereas the mRNA for PDE1C was found prominently during early meiotic prophase and throughout meiotic and

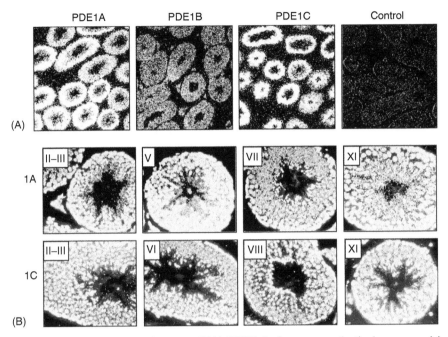

FIGURE 3.6 (See color insert following page 274.) PDE1 isoforms are selectively expressed in mouse testis. (A) In situ hybridization was used to detect mRNA for PDE1A, PDE1B, and PDE1C in cross sections of mouse testis. A control panel is shown where a sense probe for PDE1A was used. While PDE1B expression is uniformly low, PDE1A and PDE1C show strong expression in the seminiferous tubules with variable expression between different tubules. (B) High magnification images of PDE1A (1A) and PDE1C (1C) mRNA in adult mouse testis are shown at various stages of the seminiferous cycle as indicated by the roman numerals labeling each panel. Nuclei are shown in blue and silver grains indicate mRNA localization. The PDE1A and PDE1C mRNA show distinct localization and temporal expression patterns. (Data adapted from Ref. [8].)

postmeiotic stages. Immunocytochemistry revealed PDE1A protein expression in the tails of elongating and elongated spermatids. PDE1C protein distribution was similar to the mRNA expression as PDE1C protein was found in pachytene spermatocytes and round, elongating, and elongated spermatids and was localized to the cytoplasm. Furthermore, it was recently reported that a newly identified variant of PDE1A, PDE1A7, is expressed in mouse sperm.[87] This variant was localized to the sperm tail only using immunocytochemical staining. Taken together, the above data suggest that PDE1s may play a role not only in mature sperm function, but also in sperm development as well.

3.5 CONCLUSIONS

A great deal of work has characterized the biochemical properties and regulation of PDE1 on a molecular level. Recent studies have begun to put this work into a cellular context.[16] An important aspect of these studies has been the elucidation of the biochemical and regulatory properties unique to the different PDE1 isoforms and variants. This point is emphasized by the highly specific expression of the different enzymes in unique cell types and tissues. Future experiments will hopefully better characterize the genetics of the different variants to determine how their expression is regulated. As PDE1 has been implicated to be involved in multiple clinically relevant processes, it is an attractive therapeutic target. The biochemical differences between the PDE1 isoforms and variants may be exploited to develop highly

specific inhibitors. Consequently, the discovery of selective inhibitors for PDE1 enzymes may provide tools for treating pathological conditions and could also serve as important research tools. Overall, study of the PDE1 family has continued to advance in recent years and future advances hold great potential for medical benefit.

ACKNOWLEDGMENTS

The author would like to recognize Joseph Beavo, Valeria Vasta, and Sergei Rybalkin for their advice and proofreading of this chapter. Support for this work was provided by NIH grant DK21723 to Joseph Beavo. Also, Andrew Bender is a trainee under University of Washington Cardiovascular Pathology Training Grant NIH HL-07312.

REFERENCES

1. Cheung, W.Y. 1970. Cyclic $3',5'$-nucleotide phosphodiesterase. Demonstration of an activator. *Biochem Biophys Res Commun* 38 (3):533–8.
2. Kakiuchi, S., and R. Yamazaki. 1970. Calcium dependent phosphodiesterase activity and its activating factor (PAF) from brain studies on cyclic $3',5'$-nucleotide phosphodiesterase (3). *Biochem Biophys Res Commun* 41 (5):1104–10.
3. Cheung, W.Y. 1971. Cyclic $3',5'$-nucleotide phosphodiesterase. Evidence for and properties of a protein activator. *J Biol Chem* 246 (9):2859–69.
4. Beavo, J.A., M. Conti, and R.J. Heaslip. 1994. Multiple cyclic nucleotide phosphodiesterases. *Mol Pharmacol* 46 (3):399–405.
5. Kakkar, R., R.V. Raju, and R.K. Sharma. 1999. Calmodulin-dependent cyclic nucleotide phosphodiesterase (PDE1). *Cell Mol Life Sci* 55 (8–9):1164–86.
6. Sonnenburg, W.K., G.A. Wayman, D.R. Storm, and J.A. Beavo. 1998. Cyclic nucleotide regulation by calmodulin. In *Calmodulin and signal transduction*, eds. L.J. Van Eldik and D.M. Watterson, 237–87. San Diego: Academic Press.
7. Wang, J.H., R.K. Sharma, and M.J. Mooibroek. 1990. Calmodulin-stimulated cyclic nucleotide phosphodiesterases. In *Cyclic nucleotide phosphodiesterases: Structure, regulation, and drug action*, eds. J.A. Beavo and M.D. Houslay, 19–60. New York: John Wiley & Sons.
8. Sharma, R.K., A.M. Adachi, K. Adachi, and J.H. Wang. 1984. Demonstration of bovine brain calmodulin-dependent cyclic nucleotide phosphodiesterase isozymes by monoclonal antibodies. *J Biol Chem* 259 (14):9248–54.
9. Shenolikar, S., W.J. Thompson, and S.J. Strada. 1985. Characterization of a Ca^{2+}–calmodulin-stimulated cyclic GMP phosphodiesterase from bovine brain. *Biochemistry* 24 (3):672–8.
10. Rossi, P., M. Giorgi, R. Geremia, and R.L. Kincaid. 1988. Testis-specific calmodulin-dependent phosphodiesterase. A distinct high affinity cAMP isoenzyme immunologically related to brain calmodulin-dependent cGMP phosphodiesterase. *J Biol Chem* 263 (30):15521–7.
11. Kincaid, R.L., I.E. Stith-Coleman, and M. Vaughan. 1985. Proteolytic activation of calmodulin-dependent cyclic nucleotide phosphodiesterase. *J Biol Chem* 260 (15):9009–15.
12. Sonnenburg, W.K., D. Seger, K.S. Kwak, J. Huang, H. Charbonneau, and J.A. Beavo. 1995. Identification of inhibitory and calmodulin-binding domains of the PDE1A1 and PDE1A2 calmodulin-stimulated cyclic nucleotide phosphodiesterases. *J Biol Chem* 270 (52):30989–1000.
13. Zhao, A.Z., C. Yan, W.K. Sonnenburg, and J.A. Beavo. 1997. Recent advances in the study of Ca^{2+}/CaM-activated phosphodiesterases: Expression and physiological functions. *Adv Second Messenger Phosphoprotein Res* 31:237–51.
14. Hoeflich, K.P., and M. Ikura. 2002. Calmodulin in action: Diversity in target recognition and activation mechanisms. *Cell* 108 (6):739–42.
15. Yan, C., A.Z. Zhao, J.K. Bentley, and J.A. Beavo. 1996. The calmodulin-dependent phosphodiesterase gene PDE1C encodes several functionally different splice variants in a tissue-specific manner. *J Biol Chem* 271 (41):25699–706.

16. Goraya, T.A., and D.M. Cooper. 2005. Ca^{2+}–calmodulin-dependent phosphodiesterase (PDE1): Current perspectives. *Cell Signal* 17 (7):789–97.
17. Charbonneau, H., S. Kumar, J.P. Novack, D.K. Blumenthal, P.R. Griffin, J. Shabanowitz, D.F. Hunt, J.A. Beavo, and K.A. Walsh. 1991. Evidence for domain organization within the 61-kDa calmodulin-dependent cyclic nucleotide phosphodiesterase from bovine brain. *Biochemistry* 30 (32):7931–40.
18. Kakkar, R., R.V. Raju, and R.K. Sharma. 1998. In vitro generation of an active calmodulin-independent phosphodiesterase from brain calmodulin-dependent phosphodiesterase (PDE1A2) by m-calpain. *Arch Biochem Biophys* 358 (2):320–8.
19. Kakkar, R., D.P. Seitz, R. Kanthan, R.V. Rajala, J.M. Radhi, X. Wang, M.K. Pasha, R. Wang, and R.K. Sharma. 2002. Calmodulin-dependent cyclic nucleotide phosphodiesterase in an experimental rat model of cardiac ischemia-reperfusion. *Can J Physiol Pharmacol* 80 (1):59–66.
20. Hashimoto, Y., R.K. Sharma, and T.R. Soderling. 1989. Regulation of Ca^{2+}/calmodulin-dependent cyclic nucleotide phosphodiesterase by the autophosphorylated form of Ca^{2+}/calmodulin-dependent protein kinase II. *J Biol Chem* 264 (18):10884–7.
21. Sharma, R.K., and J.H. Wang. 1986. Calmodulin and Ca^{2+}-dependent phosphorylation and dephosphorylation of 63-kDa subunit-containing bovine brain calmodulin-stimulated cyclic nucleotide phosphodiesterase isozyme. *J Biol Chem* 261 (3):1322–8.
22. Sharma, R.K., and J.H. Wang. 1986. Purification and characterization of bovine lung calmodulin-dependent cyclic nucleotide phosphodiesterase. An enzyme containing calmodulin as a subunit. *J Biol Chem* 261 (30):14160–6.
23. Hansen, R.S., H. Charbonneau, and J.A. Beavo. 1988. Purification of calmodulin-stimulated cyclic nucleotide phosphodiesterase by monoclonal antibody affinity chromatography. *Methods Enzymol* 159:543–57.
24. Ang, K.L., and F.A. Antoni. 2002. Reciprocal regulation of calcium dependent and calcium independent cyclic AMP hydrolysis by protein phosphorylation. *J Neurochem* 81 (3):422–33.
25. Sharma, R.K., and J.H. Wang. 1985. Differential regulation of bovine brain calmodulin-dependent cyclic nucleotide phosphodiesterase isoenzymes by cyclic AMP-dependent protein kinase and calmodulin-dependent phosphatase. *Proc Natl Acad Sci USA* 82 (9):2603–7.
26. Florio, V.A., W.K. Sonnenburg, R. Johnson, K.S. Kwak, G.S. Jensen, K.A. Walsh, and J.A. Beavo. 1994. Phosphorylation of the 61-kDa calmodulin-stimulated cyclic nucleotide phosphodiesterase at serine 120 reduces its affinity for calmodulin. *Biochemistry* 33 (30):8948–54.
27. Sharma, R.K. 1991. Phosphorylation and characterization of bovine heart calmodulin-dependent phosphodiesterase. *Biochemistry* 30 (24):5963–8.
28. Wells, J.N., and J.R. Miller. 1988. Methylxanthine inhibitors of phosphodiesterases. *Methods Enzymol* 159:489–96.
29. Vemulapalli, S., R.W. Watkins, M. Chintala, H. Davis, H.S. Ahn, A. Fawzi, D. Tulshian, P. Chiu, M. Chatterjee, C.C. Lin, and E.J. Sybertz. 1996. Antiplatelet and antiproliferative effects of SCH 51866, a novel type 1 and type 5 phosphodiesterase inhibitor. *J Cardiovasc Pharmacol* 28 (6):862–9.
30. Snyder, P.B., J.M. Esselstyn, K. Loughney, S.L. Wolda, and V.A. Florio. 2005. The role of cyclic nucleotide phosphodiesterases in the regulation of adipocyte lipolysis. *J Lipid Res* 46 (3):494–503.
31. Card, G.L., B.P. England, Y. Suzuki, D. Fong, B. Powell, B. Lee, C. Luu, M. Tabrizizad, S. Gillette, P.N. Ibrahim, D.N. Artis, G. Bollag, M.V. Milburn, S.H. Kim, J. Schlessinger, and K.Y. Zhang. 2004. Structural basis for the activity of drugs that inhibit phosphodiesterases. *Structure* (*Camb*) 12 (12):2233–47.
32. Bischoff, E. 2004. Potency, selectivity, and consequences of nonselectivity of PDE inhibition. *Int J Impot Res* 16 (Suppl. 1):S11–4.
33. Kakkar, R., R.V. Raju, A.H. Rajput, and R.K. Sharma. 1997. Amantadine: An antiparkinsonian agent inhibits bovine brain 60 kDa calmodulin-dependent cyclic nucleotide phosphodiesterase isozyme. *Brain Res* 749 (2):290–4.
34. Kakkar, R., R.V. Raju, A.H. Rajput, and R.K. Sharma. 1996. Inhibition of bovine brain calmodulin-dependent cyclic nucleotide phosphodiesterase isozymes by deprenyl. *Life Sci* 59 (21): PL337–41.

35. Sharma, R.K., J.H. Wang, and Z. Wu. 1997. Mechanisms of inhibition of calmodulin-stimulated cyclic nucleotide phosphodiesterase by dihydropyridine calcium antagonists. *J Neurochem* 69 (2):845–50.
36. Sharma, R.K., and J. Kalra. 1993. Ginsenosides are potent and selective inhibitors of some calmodulin-dependent phosphodiesterase isozymes. *Biochemistry* 32 (19):4975–8.
37. Lal, S., R.V. Raju, and R.K. Sharma. 1998. Novel protein inhibitor of calmodulin-dependent cyclic nucleotide phosphodiesterase from glioblastoma multiforme. *Neurochem Res* 23 (4):533–8.
38. Shimizu, K., T. Murata, T. Tagawa, K. Takahashi, R. Ishikawa, Y. Abe, K. Hosaka, and Y. Kubohara. 2004. Calmodulin-dependent cyclic nucleotide phosphodiesterase (PDE1) is a pharmacological target of differentiation-inducing factor-1, an antitumor agent isolated from *Dictyostelium*. *Cancer Res* 64 (7):2568–71.
39. Novack, J.P., H. Charbonneau, J.K. Bentley, K.A. Walsh, and J.A. Beavo. 1991. Sequence comparison of the 63-, 61-, and 59-kDa calmodulin-dependent cyclic nucleotide phosphodiesterases. *Biochemistry* 30 (32):7940–7.
40. Michibata, H., N. Yanaka, Y. Kanoh, K. Okumura, and K. Omori. 2001. Human Ca^{2+}/calmodulin-dependent phosphodiesterase PDE1A: Novel splice variants, their specific expression, genomic organization, and chromosomal localization. *Biochim Biophys Acta* 1517 (2):278–87.
41. Snyder, P.B., V.A. Florio, K. Ferguson, and K. Loughney. 1999. Isolation, expression and analysis of splice variants of a human Ca^{2+}/calmodulin-stimulated phosphodiesterase (PDE1A). *Cell Signal* 11 (7):535–44.
42. Fidock, M., M. Miller, and J. Lanfear. 2002. Isolation and differential tissue distribution of two human cDNAs encoding PDE1 splice variants. *Cell Signal* 14 (1):53–60.
43. Hansen, R.S., and J.A. Beavo. 1986. Differential recognition of calmodulin–enzyme complexes by a conformation-specific anti-calmodulin monoclonal antibody. *J Biol Chem* 261 (31):14636–45.
44. Sharma, R.K., and J. Kalra. 1994. Characterization of calmodulin-dependent cyclic nucleotide phosphodiesterase isoenzymes. *Biochem J* 299 (Pt 1):97–100.
45. Loughney, K., T.J. Martins, E.A. Harris, K. Sadhu, J.B. Hicks, W.K. Sonnenburg, J.A. Beavo, and K. Ferguson. 1996. Isolation and characterization of cDNAs corresponding to two human calcium, calmodulin-regulated, 3',5'-cyclic nucleotide phosphodiesterases. *J Biol Chem* 271 (2):796–806.
46. Bentley, J.K., A. Kadlecek, C.H. Sherbert, D. Seger, W.K. Sonnenburg, H. Charbonneau, J.P. Novack, and J.A. Beavo. 1992. Molecular cloning of cDNA encoding a "63"-kDa calmodulin-stimulated phosphodiesterase from bovine brain. *J Biol Chem* 267 (26):18676–82.
47. Repaske, D.R., J.V. Swinnen, S.L. Jin, J.J. Van Wyk, and M. Conti. 1992. A polymerase chain reaction strategy to identify and clone cyclic nucleotide phosphodiesterase cDNAs. Molecular cloning of the cDNA encoding the 63-kDa calmodulin-dependent phosphodiesterase. *J Biol Chem* 267 (26):18683–8.
48. Yu, J., S.L. Wolda, A.L. Frazier, V.A. Florio, T.J. Martins, P.B. Snyder, E.A. Harris, K.N. McCaw, C.A. Farrell, B. Steiner, J.K. Bentley, J.A. Beavo, K. Ferguson, and R. Gelinas. 1997. Identification and characterisation of a human calmodulin-stimulated phosphodiesterase PDE1B1. *Cell Signal* 9 (7):519–29.
49. Reed, T.M., J.E. Browning, R.L. Blough, C.V. Vorhees, and D.R. Repaske. 1998. Genomic structure and chromosome location of the murine PDE1B phosphodiesterase gene. *Mamm Genome* 9 (7):571–6.
50. Bender, A.T., C.L. Ostenson, E.H. Wang, and J.A. Beavo. 2005. Selective up-regulation of PDE1B2 upon monocyte-to-macrophage differentiation. *Proc Natl Acad Sci USA* 102 (2):497–502.
51. Yan, C., A.Z. Zhao, J.K. Bentley, K. Loughney, K. Ferguson, and J.A. Beavo. 1995. Molecular cloning and characterization of a calmodulin-dependent phosphodiesterase enriched in olfactory sensory neurons. *Proc Natl Acad Sci USA* 92 (21):9677–81.
52. Hansen, R.S., and J.A. Beavo. 1982. Purification of two calcium/calmodulin-dependent forms of cyclic nucleotide phosphodiesterase by using conformation-specific monoclonal antibody chromatography. *Proc Natl Acad Sci USA* 79 (9):2788–92.
53. Bode, D.C., J.R. Kanter, and L.L. Brunton. 1991. Cellular distribution of phosphodiesterase isoforms in rat cardiac tissue. *Circ Res* 68 (4):1070–9.

54. Verde, I., G. Vandecasteele, F. Lezoualc'h, and R. Fischmeister. 1999. Characterization of the cyclic nucleotide phosphodiesterase subtypes involved in the regulation of the L-type Ca^{2+} current in rat ventricular myocytes. *Br J Pharmacol* 127 (1):65–74.
55. Georget, M., P. Mateo, G. Vandecasteele, L. Lipskaia, N. Defer, J. Hanoune, J. Hoerter, C. Lugnier, and R. Fischmeister. 2003. Cyclic AMP compartmentation due to increased cAMP-phosphodiesterase activity in transgenic mice with a cardiac-directed expression of the human adenylyl cyclase type 8 (AC8). *Faseb J* 17 (11):1380–91.
56. Yanaka, N., Y. Kurosawa, K. Minami, E. Kawai, and K. Omori. 2003. cGMP-phosphodiesterase activity is up-regulated in response to pressure overload of rat ventricles. *Biosci Biotechnol Biochem* 67 (5):973–9.
57. Rybalkin, S.D., K.E. Bornfeldt, W.K. Sonnenburg, I.G. Rybalkina, K.S. Kwak, K. Hanson, E.G. Krebs, and J.A. Beavo. 1997. Calmodulin-stimulated cyclic nucleotide phosphodiesterase (PDE1C) is induced in human arterial smooth muscle cells of the synthetic, proliferative phenotype. *J Clin Invest* 100 (10):2611–21.
58. Rybalkin, S.D., C. Yan, K.E. Bornfeldt, and J.A. Beavo. 2003. Cyclic GMP phosphodiesterases and regulation of smooth muscle function. *Circ Res* 93 (4):280–91.
59. Rybalkin, S.D., I. Rybalkina, J.A. Beavo, and K.E. Bornfeldt. 2002. Cyclic nucleotide phosphodiesterase 1C promotes human arterial smooth muscle cell proliferation. *Circ Res* 90 (2):151–7.
60. Yan, C., D. Kim, T. Aizawa, and B.C. Berk. 2003. Functional interplay between angiotensin II and nitric oxide: Cyclic GMP as a key mediator. *Arterioscler Thromb Vasc Biol* 23 (1):26–36.
61. Kim, D., S.D. Rybalkin, X. Pi, Y. Wang, C. Zhang, T. Munzel, J.A. Beavo, B.C. Berk, and C. Yan. 2001. Upregulation of phosphodiesterase 1A1 expression is associated with the development of nitrate tolerance. *Circulation* 104 (19):2338–43.
62. Miller, J.R., and J.N. Wells. 1988. Estimating the association of phosphodiesterase with calmodulin in intact cells. *Methods Enzymol* 159:594–604.
63. Saitoh, Y., J.G. Hardman, and J.N. Wells. 1985. Differences in the association of calmodulin with cyclic nucleotide phosphodiesterase in relaxed and contracted arterial strips. *Biochemistry* 24 (7):1613–8.
64. Strada, S.J., M.W. Martin, and W.J. Thompson. 1984. General properties of multiple molecular forms of cyclic nucleotide phosphodiesterase in the nervous system. *Adv Cyclic Nucleotide Protein Phosphorylation Res* 16:13–29.
65. Yan, C., J.K. Bentley, W.K. Sonnenburg, and J.A. Beavo. 1994. Differential expression of the 61 kDa and 63 kDa calmodulin-dependent phosphodiesterases in the mouse brain. *J Neurosci* 14 (3 Pt 1):973–84.
66. Juilfs, D.M., H.J. Fulle, A.Z. Zhao, M.D. Houslay, D.L. Garbers, and J.A. Beavo. 1997. A subset of olfactory neurons that selectively express cGMP-stimulated phosphodiesterase (PDE2) and guanylyl cyclase-D define a unique olfactory signal transduction pathway. *Proc Natl Acad Sci USA* 94 (7):3388–95.
67. Baltrons, M.A., S. Saadoun, L. Agullo, and A. Garcia. 1997. Regulation by calcium of the nitric oxide/cyclic GMP system in cerebellar granule cells and astroglia in culture. *J Neurosci Res* 49 (3):333–41.
68. Shimizu-Albergine, M., S.D. Rybalkin, I.G. Rybalkina, R. Feil, W. Wolfsgruber, F. Hofmann, and J.A. Beavo. 2003. Individual cerebellar Purkinje cells express different cGMP phosphodiesterases (PDEs): *in vivo* phosphorylation of cGMP-specific PDE (PDE5) as an indicator of cGMP-dependent protein kinase (PKG) activation. *J Neurosci* 23 (16):6452–9.
69. Reed, T.M., D.R. Repaske, G.L. Snyder, P. Greengard, and C.V. Vorhees. 2002. Phosphodiesterase 1B knock-out mice exhibit exaggerated locomotor hyperactivity and DARPP-32 phosphorylation in response to dopamine agonists and display impaired spatial learning. *J Neurosci* 22 (12):5188–97.
70. Essayan, D.M. 2001. Cyclic nucleotide phosphodiesterases. *J Allergy Clin Immunol* 108 (5):671–80.
71. Hurwitz, R.L., K.M. Hirsch, D.J. Clark, V.N. Holcombe, and M.Y. Hurwitz. 1990. Induction of a calcium/calmodulin-dependent phosphodiesterase during phytohemagglutinin-stimulated lymphocyte mitogenesis. *J Biol Chem* 265 (15):8901–7.
72. Tenor, H., L. Staniciu, C. Schudt, A. Hatzelmann, A. Wendel, R. Djukanovic, M.K. Church, and J.K. Shute. 1995. Cyclic nucleotide phosphodiesterases from purified human $CD4^+$ and $CD8^+$ T lymphocytes. *Clin Exp Allergy* 25 (7):616–24.

73. Kanda, N., and S. Watanabe. 2001. Regulatory roles of adenylate cyclase and cyclic nucleotide phosphodiesterases 1 and 4 in interleukin-13 production by activated human T cells. *Biochem Pharmacol* 62 (4):495–507.

74. Jiang, X., J. Li, M. Paskind, and P.M. Epstein. 1996. Inhibition of calmodulin-dependent phosphodiesterase induces apoptosis in human leukemic cells. *Proc Natl Acad Sci USA* 93 (20):11236–41.

75. Gantner, F., R. Kupferschmidt, C. Schudt, A. Wendel, and A. Hatzelmann. 1997. In vitro differentiation of human monocytes to macrophages: Change of PDE profile and its relationship to suppression of tumour necrosis factor-alpha release by PDE inhibitors. *Br J Pharmacol* 121 (2):221–31.

76. Bender, A.T., C.L. Ostenson, D. Giordano, and J.A. Beavo. 2003. Differentiation of human monocytes in vitro with granulocyte–macrophage colony-stimulating factor and macrophage colony-stimulating factor produces distinct changes in cGMP phosphodiesterase expression. *Cell Signal* 16 (3):365–74.

77. Kobialka, M., H. Witwicka, J. Siednienko, and W.A. Gorczyca. 2003. Metabolism of cyclic GMP in peritoneal macrophages of rat and guinea pig. *Acta Biochim Pol* 50 (3):837–48.

78. Tenor, H., A. Hatzelmann, R. Kupferschmidt, L. Stanciu, R. Djukanovic, C. Schudt, A. Wendel, M.K. Church, and J.K. Shute. 1995. Cyclic nucleotide phosphodiesterase isoenzyme activities in human alveolar macrophages. *Clin Exp Allergy* 25 (7):625–33.

79. Tenor, H., and C. Schudt. 1996. Analysis of PDE isoenzyme profiles in cells and tissues by pharmacological methods. In *Phosphodiesterase inhibitors*, eds. C. Schudt, G. Dent, and K. Rabe, 21–40. London: Academic Press.

80. Harrison, R.A. 2003. Cyclic AMP signalling during mammalian sperm capacitation—still largely terra incognita. *Reprod Domest Anim* 38 (2):102–10.

81. Henkel, R.R., and W.B. Schill. 2003. Sperm preparation for ART. *Reprod Biol Endocrinol* 1 (1):108.

82. Wennemuth, G., A.E. Carlson, A.J. Harper, and D.F. Babcock. 2003. Bicarbonate actions on flagellar and Ca^{2+}-channel responses: Initial events in sperm activation. *Development* 130 (7):1317–26.

83. Fisch, J.D., B. Behr, and M. Conti. 1998. Enhancement of motility and acrosome reaction in human spermatozoa: Differential activation by type-specific phosphodiesterase inhibitors. *Hum Reprod* 13 (5):1248–54.

84. Geremia, R., P. Rossi, D. Mocini, R. Pezzotti, and M. Conti. 1984. Characterization of a calmodulin-dependent high-affinity cyclic AMP and cyclic GMP phosphodiesterase from male mouse germ cells. *Biochem J* 217 (3):693–700.

85. Geremia, R., P. Rossi, R. Pezzotti, and M. Conti. 1982. Cyclic nucleotide phosphodiesterase in developing rat testis. Identification of somatic and germ-cell forms. *Mol Cell Endocrinol* 28 (1):37–53.

86. Yan, C., A.Z. Zhao, W.K. Sonnenburg, and J.A. Beavo. 2001. Stage and cell-specific expression of calmodulin-dependent phosphodiesterases in mouse testis. *Biol Reprod* 64 (6):1746–54.

87. Vasta, V., W.K. Sonnenburg, C. Yan, S.H. Soderling, M. Shimizu-Albergine, and J.A. Beavo. 2005. Identification of a new variant of PDE1A calmodulin-stimulated cyclic nucleotide phosphodiesterase expressed in mouse sperm. *Biol Reprod* 73:598–609.

4 PDE2 Structure and Functions

Sergio E. Martinez

CONTENTS

4.1 Introduction 55
4.2 Enzymology 56
    4.2.1 Kinetics 56
    4.2.2 Catalytic Site Inhibitors 59
4.3 Gene Organization and Splice Variants 60
    4.3.1 Gene Organization 60
    4.3.2 Splice Variants 60
4.4 Structure 61
    4.4.1 Regulatory Segment 61
    4.4.2 Allosteric cGMP-Binding Pocket: Structure, Cyclic
           Nucleotide Affinity, and Specificity.. 62
    4.4.3 Catalytic Domain 65
4.5 Physiological Roles 65
    4.5.1 Atrial Natriuretic Peptide and Adrenal Steroidogenesis 66
    4.5.2 Endothelial Cells.. 67
    4.5.3 Platelet Aggregation 67
    4.5.4 PDE2 in Neurons: NO-cGMP Pathway.. 68
    4.5.5 PDE2 in Fibroblasts 69
    4.5.6 Role in Chloride Transport 69
    4.5.7 Calcium Currents in the Heart 69
    4.5.8 Localization 70
    4.5.9 Other Likely Roles for PDE2 71
           4.5.9.1 Olfaction 71
           4.5.9.2 Immunology 71
           4.5.9.3 Prolactin Release 72
4.6 Concluding Remarks 72
Acknowledgment.. 72
References 72

4.1 INTRODUCTION

Phosphodiesterase-2 (PDE2) is one of the 11 identified families of mammalian $3',5'$-cyclic nucleotide phosphodiesterases (PDEs). It is also known as the cGMP-stimulated PDE, whose hydrolysis of cAMP can be stimulated up to 30-fold by binding of cGMP to the GAF-B

regulatory domain.* PDE2 has been characterized at the gene and protein levels and x-ray crystal structures are available for fragments consisting of the catalytic domain or the cGMP-binding regulatory segment.[1,2] Early studies on PDE2 focused on the partial or complete purification of the enzyme from various tissues and determination of its kinetic properties.[3–19] The discovery that the adenosine deaminase inhibitor, *erythro*-9-(2-hydroxy-3-nonyl)-adenine (EHNA), was also a selective inhibitor for PDE2, began a widespread change in research focus into the signaling pathways regulated by PDE2. A recurrent theme for the physiological roles of PDE2 revolves around the ability of PDE2 to, in many cases, integrate the cGMP and cAMP signaling pathways, particularly in those cases where they oppose each other's actions.

4.2 ENZYMOLOGY

4.2.1 KINETICS

PDE2 is a homodimer of two monomers (~103 kDa each) composed of a regulatory segment that contains two GAF domains followed by a catalytic domain (Figure 4.1). Both cAMP and cGMP are hydrolyzed by PDE2A with different kinetics. For the bovine cardiac enzyme, the apparent K_m is 36 μM for cAMP and 11 μM for cGMP (Table 4.1). Either cAMP or cGMP can bind to the GAF-B domain in the regulatory segment,[20] but with somewhat different effects. With either cAMP or cGMP as a substrate, the enzyme is positively cooperative with Hill coefficients of 1.9 or 1.3, respectively. At high-substrate concentrations the V_{max} is approximately the same for both nucleotides (~120 μmol/min/mg). As can be seen in Figure 4.2, cGMP at concentrations of 1–5 μM stimulates the hydrolysis of cAMP by a factor of 5–30 by shifting the substrate or velocity curve to the left, and converting it into a noncooperative Michaelis–Menten function.[10,11,21,22] On this basis, PDE2A has been called the cGMP-stimulated PDE. *In vitro* binding studies show that cGMP binds to the PDE2 holoenzyme with

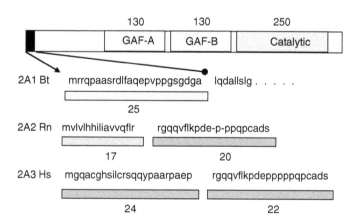

FIGURE 4.1 (See color insert following page 274.) PDE2 is a homodimer of 103 kD subunits, each containing two GAF domains followed by a catalytic domain. PDE2 has three splice variants: PDE2A1, PDE2A2, and PDE2A3. The splice sequences are shown from bovine, rat, and human, respectively. Rectangles of different colors underlie portions that are different among splice variants. Rectangles of the same color underlie portions that are nearly identical in sequence.

---

*GAF domains are small molecule-binding regulatory and dimerization domains found in a wide variety of prokaryotic and a few eukaryotic enzymes. In mammals, most known GAF domains are on PDEs and half of those bind cyclic nucleotides. The acronym GAF is taken from the first letters of the initial three proteins in which they were first identified, G, cGMP-stimulated PDE; A, Anabaena adenylyl cyclase; F, Fhla transcription factor. There now exist over 900 proteins in which GAF domains have been identified, mostly in plants and prokaryotes.

TABLE 4.1
Kinetic Properties of PDE2 Purified from Bovine Heart and Calf Liver

Property	Bovine Cardiac	Calf Liver
Reference number	10	19
K_m, μM		
cAMP	36	33
cGMP	11	15
V_{max}, μmol/min/mg		
cAMP	120	170
cGMP	120	200
Hill coefficient		
cAMP	1.9	1.6–1.8
cGMP	1.3	1.2–1.6
K_a for cGMP, μM	—	0.5

an affinity constant of approximately 10–30 nM.[20] Cyclic AMP binds to PDE2 with an affinity about 30 times weaker than cGMP binding. Therefore, various investigators consider that, at least *in vivo*, some cGMP and cAMP are always likely to be bound to the GAF-B domain because the basal concentration of both cyclic nucleotides in the cell is in the 10^{-7}–10^{-6} M concentration range. For example, for the enzyme purified from bovine adrenal tissue, 100–120 nM unlabeled cGMP or 4.7–5 μM unlabeled cAMP could displace 50% of 50 nM [^{3}H]-cGMP.[11] Similarly, when unlabeled forms of either cAMP or cGMP were used to compete with ^{3}H-cGMP binding to baculovirus-expressed bovine PDE2A1, the IC_{50} for cGMP was 22 nM, but nearly 600 nM for cAMP.[20] Despite these high binding affinities, the K_a for activation of cGMP-stimulated cAMP hydrolysis has been observed to be 0.5 μM,[19] which is a much higher value than the ~20 nM K_d value observed for the binding of cGMP to GAF-B.[20] Clearly, we have more to learn about the molecular mechanisms that relate binding to activation in PDE2.

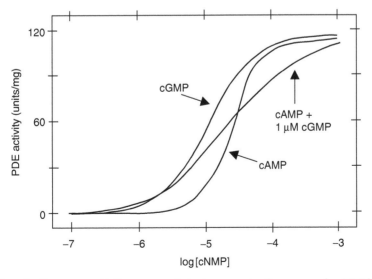

FIGURE 4.2 Phosphodiesterase activity versus substrate concentration curves for PDE2 purified from bovine heart. (Redrawn from PhD thesis of T.J. Martins [Ref. 96].)

One hint may lie in an interesting autoinhibitory effect of the GAF-A domain on cyclic nucleotide binding by the GAF-B fragment. In this case (control), IC$_{50}$ values of 7 for cGMP and 146 for cAMP were seen.[20] Recombinant constructs that additionally featured either the GAF-A domain or the GAF-A domain, together with the catalytic domain, showed little effect on the apparent affinity for cGMP, but weakened the apparent affinity for cAMP. This may be reflected in an increase in the cooperativity of cAMP binding in the complete holoenzyme.

The structural basis for activation upon occupancy of the GAF domains is unclear. Most models of allosteric cooperativity require contacts between subunits of an oligomer. One puzzle is that the GAF-A domain appears to provide the dimerization site for the holoenzyme and very likely does not bind either cGMP or cAMP.[2,20] GAF-B alone does not confer dimerization and the crystal structure of the catalytic domain shows no dimerization interface.[1] The GAF-A dimer may then serve to create oligomer contacts among the two GAF-Bs and two catalytic domains, creating a complex with two regulatory and two catalytic cGMP-binding sites. Some guesses at an allosteric mechanism might be made from the probable linear arrangement of domains, as in 27 Å electron micrograph maps of PDE5 and PDE6.[23,24] It may be that the two active sites in the dimer are mutually blocked by the catalytic domains when GAF-B is unoccupied, and more exposed when GAF-B binds cGMP. Alternatively, contacts with the GAF-B domains may hinder or open access to the catalytic sites.

It has been presumed that these kinetic changes are due to conformational changes in the protein. In fact, an apparent conformational change has been detected indirectly by chymotryptic cleavage of bovine heart PDE2A.[25] In this study, it was shown that the 103 kDa native enzyme is relatively resistant to chymotryptic cleavage in the absence of cGMP, with fragments larger than ~90 kDa being formed. In the presence of 1 μM cGMP, which should saturate the GAF-B site, the enzyme was more susceptible to proteolysis and several additional chymotryptic fragments of 60, 36, 21, and 1 kDa were produced. One major cleavage site was C-terminal to Y553, which we now know is located in the short, 15 residue sequence between the last residue of the PDE2A GAF-AB x-ray crystal model (Y555 in MmPDE2A, or Y562 in HsPDE2A) and the first residue of the PDE2A catalytic domain model (S578 in HsPDE2A). This result strongly suggests that the short bridge between GAF-B and the catalytic domain may change its conformation upon cGMP binding to GAF-B.

An interesting phenomenon is the apparent stimulation of PDE2 activity by low concentrations of certain compounds normally thought to be competitive inhibitors.[15,19,22] In general, a competitive inhibitor can increase the V_{max} of a positively cooperative enzyme if the substrate and inhibitor are both at low concentration and the inhibitor can overcome active site inhibition by stimulating the enzyme at the allosteric site.[26] For PDE2 partially purified from rat liver and activated with 3 μM cGMP, the hydrolysis of cAMP followed normal Michaelis–Menten kinetics.[22] Up to 0.2 mM 3-isobutyl-1-methylxanthine (IBMX) showed simple linear competitive inhibition with respect to cAMP ($K_i = 18$ μM), or 5–100 μM cGMP as substrate. In the absence of cGMP, 3–50 μM IBMX stimulated basal cAMP hydrolysis only at 3–5 μM cAMP, with maximal stimulation at 50 μM IBMX. Other inhibitors used in early PDE work have also shown similar activating properties. PDE2 purified to homogeneity from calf liver was inhibited by IBMX, dipyridamole, and papaverine when activated with 1 μM cGMP, with IC$_{50}$s of 4–13 μM.[19] In the absence of cGMP, with 0.5 μM cAMP as substrate, IBMX, dipyridamole, and papaverine increased activity, but cilostamide was inhibitory. The increase was again dependent on low-drug and low-substrate concentration with maximal stimulation (150%–180%) at 0.5–2.5 μM cAMP and 100 μM IBMX or dipyridamole, or 30 μM papaverine. At higher cAMP, the three drugs were inhibitory. Evidence that these inhibitors are mimicking the activation of PDE2 by cGMP binding to GAF-B is that papaverine, IBMX, or dipyridamole reduced the Hill coefficient for cAMP hydrolysis from 1.8 to a value between 1.1 and 1.2.

To the author's knowledge, no one has measured the affinity of any of these classical PDE inhibitors for isolated PDE GAF-B or AB domains. This would be a valuable step to gain further insight into the molecular mechanism of this phenomenon. High-affinity inhibitors designed to be specific for the active site might not bind to the structurally different GAF-B binding site, thus avoiding the need to account for allosteric stimulation by inhibitors.

4.2.2 CATALYTIC SITE INHIBITORS

Many of the more recent studies of signaling pathways impacted by PDE2 have relied on PDE2-selective inhibitors, as genetic knockout animals are not yet available. The first compound reported to show substantial selectivity for PDE2 inhibition was EHNA (Figure 4.3). EHNA has an IC_{50} for human myocardium PDE2 of approximately 0.8 μM.[27] However, EHNA is also a high-affinity inhibitor of adenosine deaminase, therefore requiring the use of controls such as other specific inhibitors of adenosine deaminase that do not inhibit either PDE2 or adenosine uptake to be useful for work with intact cells. Recently, various pharmaceutical companies have developed several newer inhibitors with increased potency and improved selectivity compared to EHNA.[28–31] One of these compounds, 9-(6-phenyl-2-oxohex-3-yl)-2-(3,4-dimethoxybenzyl)-purin-6-one (PDP), has an IC_{50} of 0.6 nM and a greater than 1000-fold selectivity for PDE2 when compared with other PDEs.[31] A PDE2-selective inhibitor from Bayer, BAY 60-7750, has an IC_{50} of 4.7 nM and a selectivity of at least 50-fold against other PDEs.[29] Recently, Pfizer also has reported oxindole as a PDE2-selective inhibitor, with an IC_{50} of 40 nM and a selectivity of approximately 290-fold.[30]

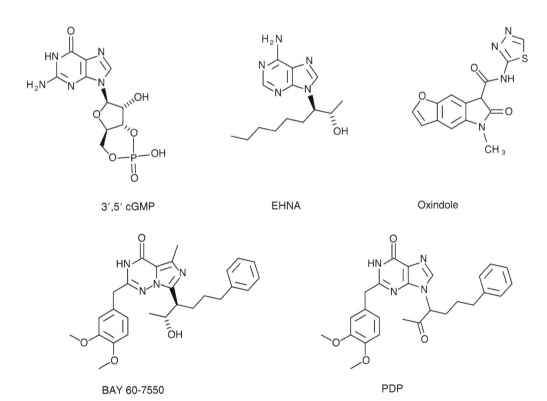

FIGURE 4.3 PDE2-selective inhibitors that have been described in the literature. cGMP, a substrate of PDE2, is shown for comparison.

Finally, a new scaffold for PDE2 inhibitors, based on the 2-morpholinochromones, has been reported.[28] However, at present, only EHNA (Sigma) and BAY 60-7750 (Alexis Biochemicals) are available commercially.

4.3   GENE ORGANIZATION AND SPLICE VARIANTS

4.3.1   Gene Organization

There is a single human PDE2 gene composed of 31 exons, significantly more than for the human GAF-PDEs PDE5A, 6B, 10A, or 11A.[32] The most closely related PDE gene is that for PDE10A. The gene is located on mouse chromosome 7 and human chromosome 11 (11q13.4).

4.3.2   Splice Variants

Three splice variants for the PDE2A gene are known from cDNAs of three mammalian species (Figure 4.1). They differ in the N-terminal 25–46 residues. PDE2A1 was first completely sequenced at the amino acid level from bovine heart.[33] A cDNA sequence was soon obtained from a bovine adrenal cDNA library.[34] PDE2A1 is considered to be a soluble, cytoplasmic enzyme, as the majority of cGMP-stimulated PDE activity in bovine heart and adrenal cortex is in the supernatant fraction.[10] A second PDE2 splice variant, PDE2A2, was cloned from rat[35] and mouse (D. Juilfs, unpublished) brain cDNA libraries. PDE2A3 was likewise cloned from bovine and mouse brain cDNA libraries (D. Juilfs, unpublished) and a human brain cDNA library.[36]

A 62-nucleotide intron, specific to bovine PDE2A1, is inserted downstream of the methionine start site for rat PDE2A2, resulting in a frame shift and initiation from a methionine downstream of that for PDE2A2. Human PDE2A3 also lacks this intron, and has a unique start methionine. The first 24 residues of PDE2A3 are different from PDE2A2, but the next 22 are nearly identical to PDE2A2, but not shared with PDE2A1 (Figure 4.1). In the rat genome, there is no apparent intron at the splice junction between PDE2A2 and PDE2A3, so editing must occur through some other mechanism at the mRNA or protein level.

To date, there are no known differences in kinetic behavior between the PDE2A splice variants. There is no information on selective tissue distribution of splice variants. However, some evidence has been compiled to indicate that the different N-terminal sequences may mediate different subcellular localizations, and it is also possible that they contribute to interactions with different binding partners in the cell. In the brain, PDE2 activity partitions into both particulate and soluble fractions.[14,35,37] In both the rabbit and bovine brain, greater than 75% of the cGMP-stimulated PDE activity was associated with particulate fractions.[14,18] Both PDE2A2 and PDE2A3 are therefore thought to be largely membrane associated, with the N-termini splice regions functioning as membrane-targeting domains.

In the SMART database of protein domains,[38,39] the rat PDE2A2 N-terminal residues 50–189 (just before the GAF-AB region) show weak homology (E-value = 0.44) to the consensus GAF sequence. Given the high sequence identity among BtPDE2A1, RnPDE2A2, and HsPDE2A3 after the splice junctions, if such a GAF domain exists in RnPDE2A2, it should be present in all mammalian PDE2 enzymes. When BtPDE2A1 was expressed in insect sf9 cells and treated with chymotrypsin, an N-terminal 25 kDa fragment was generated that crystallized and diffracted to 9 Å resolution (unpublished data). These results strongly suggest the N-terminal region before GAF-A folds into a compact domain.

After NGF-stimulated differentiation of PC12 cells to a neuronal phenotype, but not in undifferentiated cells, several phosphoproteins were found to co-immunoprecipitate with an epitope-tagged PDE2A2 construct.[40] The identity of these proteins or the physiological significance of these intriguing interactions remains to be determined.

4.4 STRUCTURE

4.4.1 REGULATORY SEGMENT

The original determination of the boundaries of the PDE2 catalytic domain, and regulatory segment containing GAFs, was deduced through amino acid sequence comparisons[33,41] and photolabeling experiments.[42,43] Unfortunately, an experimental structure for the PDE2A holoenzyme is not yet available (nor for any other PDE holoenzyme). However, separate x-ray crystal structures have been determined for the catalytic domain at 1.7 Å,[1] and the GAF-containing portion of the mouse PDE2A regulatory segment at 2.9 Å.[2] The latter structure appeared first and revealed the location of the cGMP-binding site, clarified the stoichiometry of binding, and identified the dimerization contacts for the holoenzyme. Two PDE2A regulatory segments form a dimer related by twofold symmetry (Figure 4.4). A single subunit extends over 100 Å in its longest dimension, with the centers of the two GAF domains in the same monomer separated by about 65 Å. The first, GAF-A domain, is followed by a connecting helix, a short linker, and the second, GAF-B domain. The subunits cross at the connecting helix. The dimer interface is comprised of amino acids from GAF-A as well as contacts derived from the first seven turns of the connecting helix, burying a total of 2814 Å[2] of solvent-accessible surface. The interactions between the two GAF-A domains alone bury 1338 Å[2] of solvent-accessible surface. In addition, the interaction between the connecting helices buries 1476 Å[2] and involves residues from one side of the first seven turns. The monomers diverge after Cys386 because of a bend that causes the last three turns of the connecting helices to move further and further away from each other, thus making the centers of the two GAF-B domains 65 Å apart. Cys386, located near the end of the helix–helix interface, forms a disulfide bridge with its symmetry partner (Figure 4.5) despite the continued presence of high level of reducing agent in the crystallization milieu. Cys386 is not conserved in other PDE

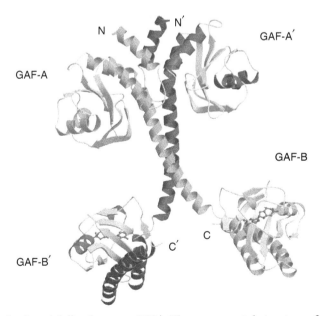

FIGURE 4.4 (See color insert following page 274.) The x-ray crystal structure of the GAF-AB dimer from mouse PDE2A. One monomer has orange helices and yellow β sheets. The other has purple helices and cyan β sheets. Loops are in gray. The bound cGMP in each GAF-B domain is in red. (Figure made with MOLSCRIPT [Ref. 97].)

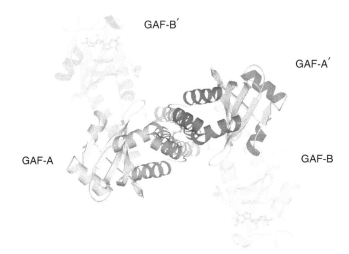

FIGURE 4.5 (See color insert following page 274.) A view down the dimer twofold axis of mouse PDE2A GAF-AB, from GAF-A to GAF-B. The disulfide bond between Cys386 and Cys386′ is shown in red. (Figure made with MOLSCRIPT [Ref. 97].)

families. GAF-A and GAF-B both have an antiparallel β sheet with strand order 3-2-1-6-5-4 (Figure 4.6). On one side, an irregular collection of helices and loops define the walls of the cGMP-binding pocket. On the other side, a bundle of two or three helices pack against the β sheet. The overall structural features of GAF-A and GAF-B are very similar, but they also have distinct conformational differences as well.

4.4.2 Allosteric cGMP-Binding Pocket: Structure, Cyclic Nucleotide Affinity, and Specificity

Cyclic GMP is bound in the GAF-B pocket in an anti-conformation (Figure 4.6 and Figure 4.7). The cGMP is entirely buried with more than 99% of its surface excluded from contact with solvent by the protein. This is reflected in the very high affinity of binding of cGMP to GAF-B. In filter-binding competition assays for [^{3}H]-cGMP with a fragment of GAF-B alone, the IC_{50} values for nonlabeled cGMP or cAMP are 7 and 146 nM, respectively.[20] Therefore, GAF-B must exist in a more open conformation before binding cGMP. In GAF-A, there is neither electron density nor room for a bound cGMP in this conformation. In particular, the key cGMP-binding side chains in GAF-B are poorly conserved in GAF-A, suggesting that PDE2A GAF-A cannot bind cyclic nucleotides and serves only as a dimerization domain.[2] However, an unknown ligand for GAF-A may be a possibility and we do not as yet have any information on the structure of the GAF domains in the absence of added cGMP.

How does GAF-B bind cGMP, and with what specificity? The binding site divides contacts among the three parts of a 3′,5′-cyclic nucleotide: contacts to the invariant ribose ring and cyclic phosphate group, and to the guanine ring that determine specificity (Figure 4.7). In contrast to high-affinity cyclic nucleotide binding sites in other proteins, the most distinctive feature of cyclic nucleotides, the cyclic phosphate, is not bound by interactions with side chains. Instead, the negative charge on the phosphate interacts with the positive end of a helix dipole from the two-turn helix α3 (Figure 4.6 and Figure 4.7). The two backbone amides of Ile458 (on α3) and Tyr481 make 2.5 and 3.0 Å H-bonds to the nonbridging oxygens, O2A and O1A, respectively. For the ribose ring, only side chain contacts are made. The 2′ hydroxyl group forms H-bonds with the hydroxyl group of Thr492 and the water, HOH2. The

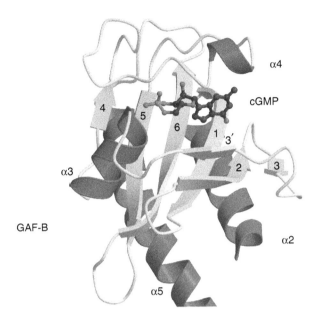

FIGURE 4.6 (See color insert following page 274.) The GAF-B domain from mouse PDE2A. The cGMP sits in a binding pocket and is shown as a ball and stick model in CPK colors. Secondary structure elements are labeled. (Figure made with MOLSCRIPT [Ref. 97].)

3′ oxygen in the ribose ring interacts with a water HOH3, which itself H-bonds to Glu512, the only charged residue in the binding pocket. The side chain of Val484 is in close proximity to several ribose atoms.

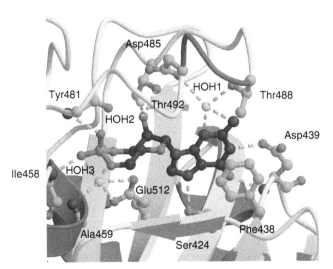

FIGURE 4.7 (See color insert following page 274.) The cGMP-binding site in GAF-B. Shown are six side chains that form hydrogen bonds (pink dashed lines) with the cGMP either directly (Asp439, Thr492, Ser424) or through a bound water (Thr488, Asp485, Glu512), one that stacks with the guanine ring (Phe438), and four residues that make backbone amide contacts (Asp439, Ile458, and Tyr481 directly to cGMP, and Ala459 through a bound water). Helix α4 is depicted as a coil for visibility of two contacts. (Figure made with MOLSCRIPT [Ref. 97].)

In the guanine-binding pocket, one residue appears to provide PDE2A with at least some of its ability to discriminate between cGMP and cAMP. The carboxylate side chain of Asp439 forms a 2.6 Å H-bond with the protonated N1 of the guanine (Figure 4.7). This contact would not form with cAMP as the N1 of adenine is not protonated. The main chain amide nitrogen forms a 2.8 Å H-bond with O6 of the guanine base. However, this amide hydrogen would clash with the NH_2 group of the cAMP in the same position (assuming cAMP binds in a very similar manner as cGMP). A third, possibly weaker specificity determinant is the water HOH1, which mediates 2.8 Å H-bonds between the hydroxyl of Thr488 and the N2 amino group of cGMP, which is absent in cAMP. Other determinants of affinity, but presumably not specificity, include Ser424, whose hydroxyl is 2.7 Å from the unprotonated N7, and the side chain of Phe438, which is base stacked with the guanine ring. On the other side of the guanine ring, Ile422 is the nearest side chain but it is likely too far removed to interact strongly.

When each amino acid in proximity to the cGMP was mutated to alanine or other residues, it was revealed that those residues contacting the phosphate or ribose moieties are most critical for binding.[20] Point mutations in residues that interact with the guanine ring lowered apparent cGMP affinity without abolishing it and in some cases actually increased the apparent cAMP affinity by up to eightfold. Thus, it appears that specificity for cGMP versus cAMP is determined largely by the purine-binding pocket, whereas the high overall affinity binding is provided largely by the phosphate- and ribose-binding pockets.

Given that there are three contacts which are seemingly incompatible with cAMP binding to GAF-B, the fact remains that cAMP binds with fairly high affinity to a GAF-B fragment (146 versus 7 nM for cGMP).[20] Asp439, the source of two of these contacts to cGMP, is on a loop between the β2 and β3 strands. This loop varies in length and sequence among both cGMP- and cAMP-binding GAF domains, with a necessarily different conformation in each case.[44] It could be that this loop has inherent flexibility and might move Asp439 away from bound cAMP. Alternately, perhaps cAMP does not bind in the same overall position as cGMP. A crystal structure with cAMP bound is needed to test these hypotheses.

Sequence alignments of GAF domains of PDEs 2, 5, 6, 10, and 11 and a cyanobacterial adenylyl cyclase show a motif N[KR]X(5–24)FX(3)DE, the NKFDE motif, which is perfectly conserved in all cases. Mutagenesis of these conserved residues caused essentially complete loss of cGMP binding in PDE2A.[20] However, these residues do not line the cGMP-binding pocket as originally expected from the results of mutagenesis studies. Rather, the structure of the PDE2A GAF domains show that the NKFDE motif is contained in a loop on the opposite side of the β sheet from the cGMP-binding site. From these observations and the x-ray model, it has been suggested that the NKFDE residues are part of a switching mechanism that closes the α4-helix over the cGMP.[44]

A later 1.9 Å crystal structure of the GAF domains from a cyanobacterial, cAMP-stimulated adenylyl cyclase from the gene cyaB2[44] confirmed similar features in its cAMP-binding sites to those found in the PDE2A cGMP-binding site. Three important differences are that in the adenylyl cyclase both GAF-A and GAF-B bound cAMP, the adenine-binding pockets are quite different from that of the guanine-binding pocket of PDE2 GAF-B, and the monomers within the GAF-AB dimer of cyaB2 have an antiparallel arrangement, as opposed to the parallel dimer of PDE2A. Whereas in the latter, only GAF-A dimerizes and GAF-B domains are 65 Å apart, in the adenylyl cyclase the GAF-A in one subunit dimerizes with GAF-B from the other subunit. Nevertheless, substitution of the cGMP-stimulated PDE2A GAF-AB into the cAMP-activated cyclase creates a cGMP-stimulated chimera.[45] Determining the regulatory mechanism of both natural holoenzymes and various chimeras will likely require determination of their structures.

4.4.3 CATALYTIC DOMAIN

The ligand-free crystal structure of the human PDE2A catalytic domain[1] showed it to be similar to that of other published PDE domains. The crystal packing arrangement gives no indication of a physiologically relevant dimer. Thus, the GAF-A contacts may provide the sole dimerization interface in the holoenzyme. The catalytic domain is composed of 15 α-helices and 6 3_{10}-helices arranged in a compact fold. Three subdomains are evident in the protein fold of the catalytic domain. The active site lies mainly within the third subdomain. The overall structure of the PDE2A catalytic domain is very similar to that of PDE3B, PDE4D, and PDE5A. The domain sequence identities among PDE2A and PDE3B, PDE4D, and PDE5A are 21.3%, 25.1%, and 32.8%, respectively. In spite of the modest sequence similarities, the overall fold is highly conserved. More than 230 equivalent Cα atoms from PDE2A could be superimposed on structures of PDE4 (RMSD 1.4 Å), PDE3 (RMSD 1.2 Å), and PDE5 (RMSD 1.0 Å). For more information about the structure of this and other PDE catalytic domains, refer to Chapter 29.

4.5 PHYSIOLOGICAL ROLES

The physiological roles of and the signaling pathways regulated by PDE2 are slowly beginning to be deciphered. Until recently progress has been slow as no PDE2 gene disruptions, or dominant negative mutations, have yet been reported and there has been a paucity of selective inhibitors to achieve chemical knockout. There are also no known diseases that are caused due to PDE2 dysfunction. Most progress in this area therefore has been due to inferences deduced initially from localization studies or the recent use of selective PDE2 inhibitors in tissues, suspected of being regulated by PDE2 because of high expression in specific cells, as discussed later in this chapter.

Three common themes for PDE2 function are beginning to emerge. First, in a few tissues, it is clear that cells can take advantage of the fact that cGMP binding to the regulatory GAF-B domain will allow a rapid increase in the rate of cAMP hydrolysis in the cell. The first and perhaps the best example of this role is discussed below for ANP or cGMP inhibition of aldosterone secretion from the adrenal cortex.[46,47] It is possible that PDE2 will be found to be a major PDE expressed in most tissues in which cGMP opposes the physiological effects of cAMP. Several other examples of this role are discussed in later chapters in this book for effects in heart[48–56] (Chapter 20) and endothelial cells[31,57–62] (Chapter 21).

A second and related role for PDE2 is its role as part of a more complicated regulatory mechanism also involving the cGMP-inhibited PDE3s. In cells expressing both of these PDEs, evidence[53,55,59,63–66] is beginning to suggest mechanisms requiring both families of PDEs that allow the timing and duration of the cAMP signal in the cell to be regulated by cGMP. This regulation takes advantage of the fact that PDE2 can be activated by cGMP and PDE3 can be inhibited by it. In these cases it is speculated that initial cGMP binding to the catalytic site of PDE3 first inhibits cAMP degradation, thereby increasing the amplitude and duration of the cAMP signal. At later times, at higher concentrations and perhaps slightly different subcellular locations, cGMP can also activate PDE2, thereby rapidly terminating the cAMP signal. It is also likely that this second event, i.e., activation of PDE2 by cGMP, will shorten the duration of the cGMP signal as cGMP is also efficiently hydrolyzed by PDE2. Examples involving this classic negative feedback regulation are thought to be present in cardiac cells[53,55] (Chapter 20), endothelial cells[59] (Chapter 21), and platelets[63–66] (Chapter 22).

Finally, it is likely that in some cases, PDE2 functions to modulate the regional or spatial distribution of both cAMP and cGMP in the cell. One idea is that it controls the spread of a cloud of cAMP that is locally formed in response to regional stimuli in the cell. Again, the

best examples of this role to date are for cardiac myocytes.[48,67] It should be noted that these three roles are not mutually exclusive and in fact are all likely to be present to some degree in most tissues that express PDE2. Abbreviated discussions of each of these roles are discussed in the following sections.

4.5.1 Atrial Natriuretic Peptide and Adrenal Steroidogenesis

One of the first signaling pathways shown to be regulated by cGMP binding to PDE2 was aldosterone secretion from the zona glomerulosa cells of bovine adrenal cortex.[47] In response to adrenocorticotropic hormone (ACTH) these cells increase cAMP and aldosterone secretion (Figure 4.8). They also respond to atrial natriuretic peptide (ANP) by activating cGMP synthesis that antagonizes the ACTH-induced increase in aldosterone. As more than 90% of PDE activity in the adrenal cortex was found to be due to the cGMP-stimulated PDE2, it was a likely choice for tying these two hormone-signaling pathways together. This idea was reinforced when it was found that most of the PDE2 in the adrenal cortex was localized, apparently at very high concentrations, in the zona glomerulosa cells. Using analogs of cAMP, only some of which could be hydrolyzed by PDE2, to stimulate aldosterone synthesis, it was shown that ANP was able to inhibit aldosterone secretion elicited by cAMP analogs that were PDE2 substrates and not by analogs that were not PDE2 substrates.

A later study with rat and human glomerulosa cells[46] also showed that PDE2 activity controls the cAMP signal induced by ACTH. These authors further suggest that when cAMP reaches its plateau phase, cAMP itself stimulates PDE2 to decrease cAMP to a basal state of production. PDE2 was found to be the main cAMP-PDE activity in a tumor cell line from rat medulla (pheochromocytoma PC12), and as in glomerulosa cells, appeared to cause a decrease in cAMP in response to ANP.[37]

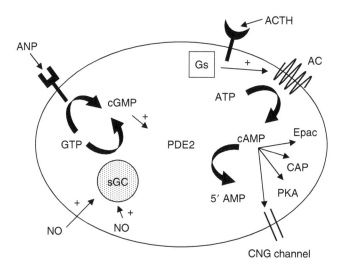

FIGURE 4.8 General scheme for PDE2 signaling pathways. PDE2 ties together cAMP (on the right) and cGMP (on the left) signaling pathways. An effector, such as the hormone atrial natriuretic peptide (ANP), binds to a receptor that generates cGMP. Alternatively, an effector can generate nitric oxide downstream, which activates nitric oxide–sensitive guanylyl cyclase and generates cGMP. This cGMP activates PDE2, which hydrolyzes cAMP generated by effectors (such as adrenocorticotropic hormone, ACTH) that activate G-coupled receptors. cAMP regulates several targets inside the cell, including protein kinase A, cAMP-regulated ion channels, the catabolite activator protein, and guanine nucleotide exchange factors that are activated by cAMP, such as Epac.

Most recently, adrenal glomerulosa cells have been used for real-time monitoring of PDE2 activity using an adenovirus-encoded cAMP sensor.[68] This sensor is a chimeric protein consisting of domains from yellow and cyan fluorescent proteins flanking the cAMP-binding domain from the guanine nucleotide exchange factor, Epac2. Binding of cAMP causes the Y and C domains to move closer together, enhancing fluorescence resonance energy transfer (FRET) and increasing the longer wavelength yellow emission at the expense of the shorter wavelength cyan emission. This sensor can detect cAMP over a range of 100 nM to 20 μM.[69] Using this system it was shown that increased PDE2 activity due to ANP could block either forskolin or ACTH-induced increases in cAMP with a half-life of approximately 15 s. Thus, this system can come close to monitoring PDE2 activity in real time. These authors repeated the basic observations of the earlier studies and further suggested that the high rate of cyclic nucleotide hydrolysis by PDE2 may be at least partly responsible for creating cAMP compartments within the cell.

4.5.2 ENDOTHELIAL CELLS

The role of PDE2 in endothelial cell function is at present an area of intense interest. Roles of PDE2 in endothelial permeability, and also in VEGF stimulation of endothelial cell migration and proliferation have also been proposed[31,57,58,61,62] and are discussed in Chapter 21 and Chapter 22, and so will not be addressed in this chapter. It is important to note that the exact roles and quantitative importance of PDE2 in these cells may vary among species, origin of the cells (e.g., small vessel, large vessel, pulmonary vessel, etc.), and conditions of culture.

In an interesting recent paper, Seybold and colleagues[31] have suggested that induction of PDE2 by TNF-α plays a major role in regulation of endothelial cell hyperpermeability and consequent lung edema. In this example, TNF-α, a cytokine produced in a variety of inflammatory situations, increases endothelial permeability by inducing endothelial PDE2 expression.[31] These investigators found that in cultured human umbilical vein endothelial cells (HUVECs), TNF-α can increase PDE2 mRNA and activity, but not PDE3 or PDE4 activity. Moreover, transfection of PDE2A3 into HUVECs was as effective in destabilizing endothelial barrier function as TNF-α stimulation of PDE2 activity. They also showed that inhibition of PDE2 with 1 μM PDP, a high-affinity PDE2-selective inhibitor, was protective for both transfected and untransfected HUVECs.

Finally, at least one study has implicated the activity of PDE2 in pulmonary vasoconstriction caused by low oxygen in the perfused rat lung.[70] EHNA-inhibited PDE2 activity was shown to be present in the smooth muscle of rat lung and at least partly responsible for this response.

4.5.3 PLATELET AGGREGATION

Platelet aggregation can be inhibited by activators of either adenylyl cyclase, such as prostacyclin (PGI2), or activators of nitric oxide–sensitive (previously known as soluble) guanylyl cyclase, such as NO donors (Figure 4.9). Aggregation also can be retarded by inhibitors of cyclic nucleotide PDEs. It has been known for over 20 years that the major cAMP-PDEs in platelets are PDE2 and PDE3; PDE5, a cGMP-selective enzyme is also abundant in platelets. Therefore, investigators have studied the role of these PDEs in platelet function.[7,63–66,71–73] In aggregate, these studies suggest that inhibition of platelet aggregation is most directly mediated by increased cAMP, not cGMP, and that the activity of PDE3 is most responsible for regulation of this cAMP (see Chapter 31). Only under conditions where cGMP is elevated[66] does PDE2 appear to play a major regulatory role. Activation of PDE2 by cGMP binding to the GAF-B domain appears to limit the rise of cAMP both directly by hydrolyzing the cAMP and indirectly by more rapidly removing the cGMP that is also

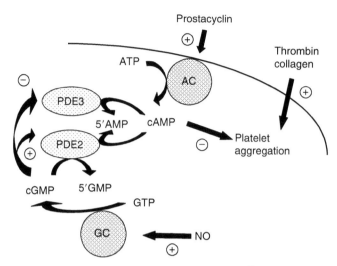

FIGURE 4.9 A tentative role for PDE2 in platelet aggregation.[66] In wounds, exposed collagen or thrombin causes platelets to clump and form a clot. Clumping can be prevented by prostacyclin, which raises cAMP. At low cAMP, PDE3 is the main cAMP-PDE. An NO donor can raise the level of cGMP, inhibiting PDE3 but stimulating PDE2 activity. Alternatively, high cAMP might also stimulate PDE2 directly.

inhibiting the PDE3. For example, platelets incubated with EHNA achieved a 4.1-fold increase in cAMP, whereas those incubated with milrinone (PDE3 inhibitor) alone only achieved a 1.7-fold increase.[66] However, only milrinone could inhibit, in a concentration-dependent manner, platelet aggregation from a variety of agonists including thrombin and collagen. The PDE3 inhibitors, motapizone and cilostamide, both raised cAMP and inhibited release of serotonin from thrombin-activated platelets, which was not affected by treatment with PDP, a PDE2 inhibitor.[65] These results also suggest that some pools of cAMP differ for PDE2 and PDE3, at least in the absence of NO.

4.5.4   PDE2 in Neurons: NO-cGMP Pathway

PDE2 has been implicated in several neuronal NO-cGMP pathways. One early study determined which PDEs are responsible for hydrolysis of N-methyl-D-aspartate (NMDA)-induced changes in cAMP and cGMP in primary neuronal cultures from the cerebral cortex of newborn rat pups.[74] In this case NMDA did not increase cAMP, but did increase cGMP to 400% of control. However, in the presence of 0.5 mM IBMX, NMDA increased cAMP to 600% and cGMP to 3500%. To determine which PDEs hydrolyzed the cAMP or cGMP, NMDA dose–response curves were carried out in the presence of 10 μM PDE inhibitors. Only rolipram produced a significant increase in cAMP and only EHNA increased cGMP, up to 2000%. The NMDA-induced increases in cAMP by rolipram or cGMP with EHNA were antagonized by 10 μM MK-801, a noncompetitive NMDA receptor antagonist. This study suggests that both cAMP and cGMP are involved in NMDA receptor-mediated signaling in cerebral cortical neuronal cultures. It also strongly suggests that in these neurons, PDE2 is predominantly involved in the regulation of cGMP at least in response to NMDA. Presumably this is due to differential compartment of the two nucleotides as PDE2 can efficiently hydrolyze both cyclic nucleotides *in vitro*.

More recently, in the hippocampus the PDE2-selective inhibitor BAY 60-7550 has been shown to enhance long-term potentiation in the CA1 region and improve social recognition memory and the memory of new objects in both rats and mice.[29] A suggestion that here too PDE2 might be working via the NMDA receptor pathway comes from the observation that BAY 60-7550 reversed MK-801 induced deficits in performance in a maze.

In rat striatal cells, PDE2 was found to be the main cGMP-hydrolyzing activity after exposure to NO.[75] The details of how PDE2 is involved in these processes remain to be determined.

4.5.5   PDE2 IN FIBROBLASTS

In rat ventricular fibroblasts, NO-stimulated cGMP production mediates most of the effects of interleukin-1β (IL-1β) on cAMP accumulation, in large part by activating PDE2.[76] In this study, IL-1β induced iNOS and a 10-fold increase in NO production. Moreover, 1 mM N^G-monomethyl arginine (L-NMMA), an inhibitor of the nitric oxide-sensitive guanylyl cyclase (sGC), completely inhibited the NO increase and restored cAMP accumulation if cells were treated with 1 μM isoproterenol or 30 μM forskolin after the IL-1β treatment. NO donors could also reversibly attenuate isoproterenol-stimulated cAMP accumulation. In the presence of 1 mM sodium nitroprusside, either 1 mM IBMX or 0.2 mM EHNA completely restored isoproterenol-stimulated cAMP accumulation to control levels, suggesting that the effect of NO is attributable at least in part to PDE2 activation by cGMP. In contrast, when cells were treated with IL-1β, IBMX or EHNA only partially restored isoproterenol-induced cAMP accumulation to control levels. The PDE inhibitor effect was from 50% to 15% inhibition of maximal isoproterenol-stimulated cAMP accumulation. These results suggest that some of the effect of iNOS induction is not dependent on acute cGMP production and PDE2 activation.

4.5.6   ROLE IN CHLORIDE TRANSPORT

In the kidney, one component of the nephron that is regulated by cAMP is the thick ascending limb (THAL). Hormones that increase cAMP production in the THAL stimulate Cl⁻ absorption and agents that increase cGMP inhibit Cl⁻ secretion. Ortiz and Garvin hypothesized[77] that endogenous NO stimulates sGC and activates PDE2, thereby decreasing cAMP and inhibiting THAL Cl⁻ absorption. Using isolated rat THALs, these investigators showed that NO donors increase cGMP and decrease THAL Cl⁻ absorption. Importantly 50 μM EHNA blocked most of this decrease, as did LY-83583, an inhibitor of sGC, and 10 μM db-cAMP, a nonhydrolyzable cAMP analog.

4.5.7   CALCIUM CURRENTS IN THE HEART

Hormonal regulation of cardiac function is mediated at least in part through changes in cAMP and cGMP. An important example is the β-adrenergic stimulation of the cardiac L-type Ca^{2+} channel and its calcium current, I_{Ca}. In the heart, cAMP activates PKA, which phosphorylates the L-type Ca^{2+} channel, increasing the incidence of the open state. The role of cGMP is less well characterized. Several studies on cardiac myocytes have reported an inhibition of this calcium current by cGMP or NO donors acting via guanylyl cyclase.[48–56] This has been attributed to PDE2 (also see Chapter 20).

An initial study was on isolated, voltage-clamped frog ventricular myocytes.[50] IBMX partially reduced the inhibitory effect of cGMP on I_{Ca}. This pointed to a cAMP-PDE activity regulating I_{Ca}. Suggestive evidence that this was the cGMP-stimulated PDE2 in these cells came from another study.[54] In a particulate fraction from frog ventricular myocytes, the

half-maximal concentration of cGMP for cGMP-stimulated PDE activity was 1.1 μM, very similar to the 0.7 μM for half-maximal reduction of cAMP-elevated I_{Ca} in patch clamp experiments. IBMX inhibited cGMP-stimulated PDE activity with an IC_{50} of 20 μM, whereas 100 μM IBMX almost entirely abolished the decrease in I_{Ca} caused when 10 μM cGMP is perfused after 30 μM cAMP. Fischmeister and Hartzell also used patch clamp and perfusion pipette techniques for similar experiments.[49] After isoprenaline treatment or intracellular perfusion with cAMP, 20 μM cGMP reversibly reduced stimulated I_{Ca} by 67% with a similar IC_{50} of 0.6 μM. Intracellular perfusion with 0.1–20 μM cGMP did not change basal I_{Ca}.

A later patch clamp study with frog ventricular myocytes made use of EHNA, a PDE2-specific inhibitor, to more positively identify PDE2.[52] The NO donors SIN-1 (3-morpholino-sydnonymine) and SNP abolished ~50% of I_{Ca} induced over its basal level with isoprenaline (1 and 0.1 μM, respectively). These NO donors could be totally blocked by 30 μM EHNA (IC_{50} ~ 3 μM). EHNA had no effect on basal I_{Ca} over 0.1–30 μM or in cells perfused with 5 or 10 μM cAMP. These experiments suggest that NO-induced increase in cGMP activates PDE2, which then hydrolyzes cAMP and deactivates the PKA that phosphorylates and activates the L-type Ca^{2+} channel. By testing different inhibitors, PDE2 appears to be the major cAMP-PDE activity regulating the I_{Ca}. Effects of 30 μM EHNA and 500 μM IBMX on I_{Ca} stimulated by 10 μM cAMP were not additive.

In an attempt to gauge how tightly the L-type Ca^{2+} channels are associated with PDE2, one study used a double voltage clamp technique.[48] One pipette was placed at each end of a frog ventricular myocyte and a double-barreled microperfusion system was placed against the cell. Using NO donors and scavengers, and EHNA, they demonstrated a strong local depletion of cAMP on the side of the cell where PDE2 was activated, but a modest reduction in the other half. This might be explained by a tight coupling between L-type Ca^{2+} channels and local PDE2 molecules also anchored in or near the membrane.

Species differences in the relative importance of PDEs 2, 3, and 4 become apparent from rat ventricular myocytes.[56] By RT-PCR, rat ventricular cells had mRNA for PDEs 1, 2, 3, and 4. PDEs 2 and 4 were in greatest abundance. Selective inhibitors for PDEs 1, 2, 3, and 4 inhibitors (MIMX, EHNA, cilostamide, RO 20-1724, respectively) were tried alone or in double combinations for effect on basal I_{Ca}. As in frog ventricular myocytes, EHNA did not affect basal I_{Ca}. The only stimulation was 47% from 0.1 μM cilostamide plus 0.1 μM RO-1724. Adding 10 μM EHNA increased this to 107%, almost as much as did 100 μM IBMX (120%). For these three PDE inhibitors, the percent increase in I_{Ca} versus concentration showed the rank order of potency PDE4 > PDE3 > PDE2 > PDE1.

In human atrial myocytes, basal calcium current is controlled by PDE2.[53] Unlike rat and frog ventricular myocytes, EHNA alone (10 μM) stimulated basal I_{Ca}. Cyclic GMP regulates I_{Ca} by controlling the intracellular cAMP concentration through opposing actions on PDE3 and PDE2.[55] Intracellular perfusion of 0.5 or 5 μM cGMP increased I_{Ca} by 64% and 36%, respectively. The authors hypothesized that PDE3 is inhibited at the lower cGMP concentration, thereby raising cAMP, activating PKA, and phosphorylating a subunit of the Ca^{2+} channel, whereas at higher concentrations cGMP-stimulated PDE2 is activated, which then lowers cAMP. PDE2 also appears to regulate the pacemaker current I_f in response to β-adrenergic stimulation from norepinephrine in guinea pig in sinoatrial cells.[78]

4.5.8 LOCALIZATION

Several studies, mostly on brain, reported using riboprobes or immunocytochemistry to localize PDE2A in tissues. Western and Northern blots showed PDE2A expressed in a variety of human tissue types, most heavily in brain[36] and especially the neocortex.[79] An

in situ mRNA hybridization study also has been carried out in rat brain at different developmental stages. Riboprobes against PDE2, PDE5, and PDE9 showed that PDE2 was strongly transcribed in neuronal cell bodies of the limbic system, which controls learning, memory, and emotion.[80] An earlier study of PDE2 in adult rat brain gave similar results.[81] In rat and mouse hippocampal slices, PDE2 was indirectly detected by a rise in cGMP in the presence of EHNA. The increase in cGMP was detected in the mouse tissue slices by cGMP immunocytochemistry,[82] as well as by radioimmunoassay (RIA) in rat.[83] PDE2 mRNA also was detected in rat and mouse hippocampus using antisense probes.[82] An immunohistochemical study in mouse brain found PDE2 in structures of the limbic system, as well as the habenula, basal ganglia, cerebral cortex, thalamic and hypothalamic nuclei, cerebellum, and midbrain.[84]

Two studies have attempted to further localize PDE2 activity in the mammalian brain. A cGMP-stimulated cAMP-PDE activity is a component of bovine brain coated vesicles immunoprecipitated with anti-clathrin polyclonal antibodies.[85] Noyama and Maekawa studied soluble, Triton X-100 solubilized, and membrane raft fractions prepared from rat cerebral cortex.[86] PDE activity was ascribed to PDE2 and PDE10 based on EHNA and dipyridamole inhibition, and the lack of inhibition by other PDE inhibitors, including zaprinast. PDE2 may be the dominant species judging from the K_m values for cGMP and cAMP. Western blotting confirmed PDE2 in all fractions.

4.5.9 OTHER LIKELY ROLES FOR PDE2

4.5.9.1 Olfaction

In humans, the main olfactory epithelium contains about 40 million olfactory sensory neurons, each of which expresses one of about 1000 olfactory receptor genes. These neurons project through the cribriform plate in the skull to about 2000 synapse points called glomeruli in the olfactory bulb. Hints of a small, distinct group of olfactory neurons having a different signaling regulation came with the discovery of a subset of olfactory neurons containing an olfactory-specific membrane guanylyl cyclase, GC-D,[87] and later PDE2. In mice, PDE2 could be detected in GC-D containing neurons by immunocytochemical staining with PDE2 antisera, or in situ hybridization of mRNA.[88,89] These neurons did not contain PDE1C as did most others in the olfactory epithelium. These neurons most likely function to detect a distinct set of odorants or pheromones. There is a separate, much smaller patch of olfactory epithelium in mammals, the septal organ, which also contains a small number of olfactory neurons that stain positive for PDE2.[90] The exact role of PDE2 in either the olfactory neurons in the septal organ or main olfactory epithelium is not known.

4.5.9.2 Immunology

PDE2 has been detected in thymocytes and macrophages, but the signaling pathways it regulates have not been determined. In mouse thymocytes,[91] PDE4 activity contributed about 80% of the total basal cAMP-PDE activity in a cytosolic extract, but on addition of 10 μM cGMP, PDE2 accounted for 80% of the activity. PDE2 activity is rapidly but transiently decreased on treatment of these thymocytes with the lectin, phytohemagglutinin.

To avoid problems in interpretation caused by the many different macrophage phenotypes in blood, one study used U937 cells, a histiocytic leukemia cell line that can be induced to differentiate to a macrophage-like phenotype with phorbol-12-myristate-13-acetate (PMA) treatment.[92] PMA treatment caused a large increase in PDE2 activity and mRNA levels.

PDE2 is also expressed in mouse glycolate-elicited peritoneal macrophages,[93] and a possible PDE2 activity was detected in rat peritoneal macrophages.[94]

4.5.9.3 Prolactin Release

In the anterior pituitary gland, NO inhibits prolactin release and decreases the concentration of cAMP.[95] This effect appears to be mediated by activation of both PDE2 and protein kinase G (PKG).

4.6   CONCLUDING REMARKS

Many pathways involving PDE2 action remain to be worked out in greater detail. Highly selective and potent PDE2 inhibitors have now been generated and when freely available will provide a unique resource through which to study these pathways; this will be a great improvement over studies that utilized inhibitors that were nonspecific (e.g., IBMX) or also inhibited another enzyme (e.g., EHNA). Further customization of active site inhibitors using the PDE2 catalytic domain structure, or possibly ligands that activate or inhibit the GAF domain, should lead to even better tools for studying PDE2 function and might lead to medical application. The recent demonstration of a role for PDE2 in cardiac[48–56] and endothelial function[31,57–62] strongly suggest this possibility. Finally, the crystal structure of PDE2 GAF-AB identified for the first time the molecular nature of the cyclic nucleotide-binding site of a PDE GAF domain. However, structures of holoenzymes will likely be needed to determine the structural basis through which cGMP binding to the GAF domain regulates PDE2 catalytic activity.

ACKNOWLEDGMENT

This work was supported by NIH Grants DK21723 and HL44948 to Joseph A. Beavo.

REFERENCES

1. Iffland, A., D. Kohls, S. Low, J. Luan, Y. Zhang, M. Kothe, Q. Cao, A.V. Kamath, Y.H. Ding, and T. Ellenberger. 2005. Structural determinants for inhibitor specificity and selectivity in PDE2A using the wheat germ *in vitro* translation system. *Biochemistry* 44 (23): 8312–8325.
2. Martinez, S.E., A.Y. Wu, N.A. Glavas, X.B. Tang, S. Turley, W.G. Hol, and J.A. Beavo. 2002. The two GAF domains in phosphodiesterase 2A have distinct roles in dimerization and in cGMP binding. *Proc Natl Acad Sci USA* 99 (20): 13260–13265.
3. Beavo, J.A., J.G. Hardman, and E.W. Sutherland. 1970. Hydrolysis of cyclic guanosine and adenosine 3′,5′-monophosphates by rat and bovine tissues. *J Biol Chem* 245 (21): 5649–5655.
4. Beavo, J.A., J.G. Hardman, and E.W. Sutherland. 1971. Stimulation of adenosine 3′,5′-monophosphate hydrolysis by guanosine 3′,5′-monophosphate. *J Biol Chem* 246 (12): 3841–3846.
5. Egrie, J.C., and F.L. Siegel. 1977. Adrenal medullary cyclic nucleotide phosphodiesterase. Subcellular distribution, partial purification, and regulation of enzyme activity. *Biochim Biophys Acta* 483 (2): 348–366.
6. Franks, D.J., and J.P. Macmanus. 1971. Cyclic GMP stimulation and inhibition of cyclic AMP phosphodiesterase from thymic lymphocytes. *Biochem Biophys Res Commun* 42 (5): 844–849.
7. Hidaka, H., and T. Asano. 1976. Human blood platelet 3′:5′-cyclic nucleotide phosphodiesterase. Isolation of low-K_m and high-K_m phosphodiesterase. *Biochem Biophys Acta* 429 (2):485–497.
8. Hurwitz, R.L., R.S. Hansen, S.A. Harrison, T.J. Martins, M.C. Mumby, and J.A. Beavo. 1984. Immunologic approaches to the study of cyclic nucleotide phosphodiesterases. *Adv Cyclic Nucleotide Protein Phosphorylation Res* 16: 89–106.

9. Klotz, U., and K. Stock. 1972. Influence of cyclic guanosine-3′,5′-monophosphate on the enzymatic hydrolysis of adenosine-3′,5′-monophosphate. *Naunyn Schmiedebergs Arch Pharmacol* 274 (1): 54–62.

10. Martins, T.J., M.C. Mumby, and J.A. Beavo. 1982. Purification and characterization of a cyclic GMP-stimulated cyclic nucleotide phosphodiesterase from bovine tissues. *J Biol Chem* 257 (4): 1973–1979.

11. Miot, F., P.J. Van Haastert, and C. Erneux. 1985. Specificity of cGMP binding to a purified cGMP-stimulated phosphodiesterase from bovine adrenal tissue. *Eur J Biochem* 149 (1): 59–65.

12. Moss, J., V.C. Manganiello, and M. Vaughan. 1977. Substrate and effector specificity of a guanosine 3′:5′-monophosphate phosphodiesterase from rat liver. *J Biol Chem* 252 (15): 5211–5215.

13. Mumby, M.C., T.J. Martins, M.L. Chang, and J.A. Beavo. 1982. Identification of cGMP-stimulated cyclic nucleotide phosphodiesterase in lung tissue with monoclonal antibodies. *J Biol Chem* 257 (22): 13283–13290.

14. Murashima, S., T. Tanaka, S. Hockman, and V. Manganiello. 1990. Characterization of particulate cyclic nucleotide phosphodiesterases from bovine brain: Purification of a distinct cGMP-stimulated isoenzyme. *Biochemistry* 29 (22): 5285–5292.

15. Pyne, N.J., M.E. Cooper, and M.D. Houslay. 1986. Identification and characterization of both the cytosolic and particulate forms of cyclic GMP-stimulated cyclic AMP phosphodiesterase from rat liver. *Biochem J* 234 (2): 325–334.

16. Russell, T.R., W.L. Terasaki, and M.M. Appleman. 1973. Separate phosphodiesterases for the hydrolysis of cyclic adenosine 3′,5′-monophosphate and cyclic guanosine 3′,5′-monophosphate in rat liver. *J Biol Chem* 248 (4): 1334–1340.

17. Terasaki, W.L., and M.M. Appleman. 1975. The role of cyclic GMP in the regulation of cyclic AMP hydrolysis. *Metabolism* 24 (3): 311–319.

18. Whalin, M.E., S.J. Strada, and W.J. Thompson. 1988. Purification and partial characterization of membrane-associated type II (cGMP-activatable) cyclic nucleotide phosphodiesterase from rabbit brain. *Biochim Biophys Acta* 972 (1): 79–94.

19. Yamamoto, T., V.C. Manganiello, and M. Vaughan. 1983. Purification and characterization of cyclic GMP-stimulated cyclic nucleotide phosphodiesterase from calf liver. Effects of divalent cations on activity. *J Biol Chem* 258 (20): 12526–12533.

20. Wu, A.Y., X.B. Tang, S.E. Martinez, K. Ikeda, and J.A. Beavo. 2004. Molecular determinants for cyclic nucleotide binding to the regulatory domains of phosphodiesterase 2A. *J Biol Chem* 279 (36): 37928–37938.

21. Erneux, C., D. Couchie, J.E. Dumont, J. Baraniak, W.J. Stec, E.G. Abbad, G. Petridis, and B. Jastorff. 1981. Specificity of cyclic GMP activation of a multi-substrate cyclic nucleotide phosphodiesterase from rat liver. *Eur J Biochem* 115 (3): 503–510.

22. Erneux, C., F. Miot, J.M. Boeynaems, and J.E. Dumont. 1982. Paradoxical stimulation by 1-methyl-3-isobutylxanthine of rat liver cyclic AMP phosphodiesterase activity. *FEBS Lett* 142 (2): 251–254.

23. Kajimura, N., M. Yamazaki, K. Morikawa, A. Yamazaki, and K. Mayanagi. 2002. Three-dimensional structure of non-activated cGMP phosphodiesterase 6 and comparison of its image with those of activated forms. *J Struct Biol* 139 (1): 27–38.

24. Kameni Tcheudji, J.F., L. Lebeau, N. Virmaux, C.G. Maftei, R.H. Cote, C. Lugnier, and P. Schultz. 2001. Molecular organization of bovine rod cGMP-phosphodiesterase 6. *J Mol Biol* 310 (4): 781–791.

25. Stroop, S.D., and J.A. Beavo. 1991. Structure and function studies of the cGMP-stimulated phosphodiesterase. *J Biol Chem* 266 (35): 23802–23809.

26. Segel, I.H. 1975. *Enzyme kinetics: Behavior and analysis of rapid equilibrium and steady state enzyme systems*. New York: John Wiley & Sons.

27. Podzuweit, T., P. Nennstiel, and A. Muller. 1995. Isozyme selective inhibition of cGMP-stimulated cyclic nucleotide phosphodiesterases by *erythro*-9-(2-hydroxy-3-nonyl)adenine. *Cell Signal* 7 (7): 733–738.

28. Abbott, B.M., and P.E. Thompson. 2004. PDE2 inhibition by the PI3 kinase inhibitor LY294002 and analogues. *Bioorg Med Chem Lett* 14 (11): 2847–2851.

29. Boess, F.G., M. Hendrix, F.J. van der Staay, C. Erb, R. Schreiber, W. van Staveren, J. de Vente, J. Prickaerts, A. Blokland, and G. Koenig. 2004. Inhibition of phosphodiesterase 2 increases neuronal cGMP, synaptic plasticity and memory performance. *Neuropharmacology* 47 (7): 1081–1092.

30. Chambers, R.J., K. Abrams, N.Y. Garceau, A.V. Kamath, C.M. Manley, S.C. Lilley, D.A. Otte, D.O. Scott, A.L. Sheils, D.A. Tess, A.S. Vellekoop, Y. Zhang, and K.T. Lam. 2006. A new chemical tool for exploring the physiological function of the PDE2 isozyme. *Bioorg Med Chem Lett* 16 (2): 307–310.

31. Seybold, J., D. Thomas, M. Witzenrath, S. Boral, A.C. Hocke, A. Burger, A. Hatzelmann, H. Tenor, C. Schudt, M. Krull, H. Schutte, S. Hippenstiel, and N. Suttorp. 2005. Tumor necrosis factor-alpha-dependent expression of phosphodiesterase 2: Role in endothelial hyperpermeability. *Blood* 105 (9): 3569–3576.

32. Yuasa, K., Y. Kanoh, K. Okumura, and K. Omori. 2001. Genomic organization of the human phosphodiesterase PDE11A gene. Evolutionary relatedness with other PDEs containing GAF domains. *Eur J Biochem* 268 (1): 168–178.

33. Trong, H.L., N. Beier, W.K. Sonnenburg, S.D. Stroop, K.A. Walsh, J.A. Beavo, and H. Charbonneau. 1990. Amino acid sequence of the cyclic GMP stimulated cyclic nucleotide phosphodiesterase from bovine heart. *Biochemistry* 29 (44): 10280–10288.

34. Sonnenburg, W.K., P.J. Mullaney, and J.A. Beavo. 1991. Molecular cloning of a cyclic GMP-stimulated cyclic nucleotide phosphodiesterase cDNA. Identification and distribution of isozyme variants. *J Biol Chem* 266 (26): 17655–17661.

35. Yang, Q., M. Paskind, G. Bolger, W.J. Thompson, D.R. Repaske, L.S. Cutler, and P.M. Epstein. 1994. A novel cyclic GMP stimulated phosphodiesterase from rat brain. *Biochem Biophys Res Commun* 205 (3): 1850–1858.

36. Rosman, G.J., T.J. Martins, W.K. Sonnenburg, J.A. Beavo, K. Ferguson, and K. Loughney. 1997. Isolation and characterization of human cDNAs encoding a cGMP-stimulated $3',5'$-cyclic nucleotide phosphodiesterase. *Gene* 191 (1): 89–95.

37. Whalin, M.E., J.G. Scammell, S.J. Strada, and W.J. Thompson. 1991. Phosphodiesterase II, the cGMP-activatable cyclic nucleotide phosphodiesterase, regulates cyclic AMP metabolism in PC12 cells. *Mol Pharmacol* 39 (6): 711–717.

38. Letunic, I., R.R. Copley, B. Pils, S. Pinkert, J. Schultz, and P. Bork. 2006. SMART 5: Domains in the context of genomes and networks. *Nucleic Acids Res* 34 (Database issue): D257–D260.

39. Schultz, J., F. Milpetz, P. Bork, and C.P. Ponting. 1998. SMART, a simple modular architecture research tool: Identification of signaling domains. *Proc Natl Acad Sci USA* 95 (11): 5857–5864.

40. Bentley, J.K., D.M. Juilfs, and M.D. Uhler. 2001. Nerve growth factor inhibits PC12 cell PDE2 phosphodiesterase activity and increases PDE2 binding to phosphoproteins. *J Neurochem* 76 (4): 1252–1263.

41. Charbonneau, H., R.K. Prusti, H. LeTrong, W.K. Sonnenburg, P.J. Mullaney, K.A. Walsh, and J.A. Beavo. 1990. Identification of a noncatalytic cGMP-binding domain conserved in both the cGMP-stimulated and photoreceptor cyclic nucleotide phosphodiesterases. *Proc Natl Acad Sci USA* 87 (1): 288–292.

42. Stroop, S.D., H. Charbonneau, and J.A. Beavo. 1989. Direct photolabeling of the cGMP-stimulated cyclic nucleotide phosphodiesterase. *J Biol Chem* 264 (23): 13718–13725.

43. Tanaka, T., S. Hockman, M. Moos Jr., M. Taira, E. Meacci, S. Murashima, and V.C. Manganiello. 1991. Comparison of putative cGMP-binding regions in bovine brain and cardiac cGMP-stimulated phosphodiesterases. *Second Messengers Phosphoproteins* 13 (2–3): 87–98.

44. Martinez, S.E., S. Bruder, A. Schultz, N. Zheng, J.E. Schultz, J.A. Beavo, and J.U. Linder. 2005. Crystal structure of the tandem GAF domains from a cyanobacterial adenylyl cyclase: Modes of ligand binding and dimerization. *Proc Natl Acad Sci USA* 102 (8): 3082–3087.

45. Kanacher, T., A. Schultz, J.U. Linder, and J.E. Schultz. 2002. A GAF-domain-regulated adenylyl cyclase from Anabaena is a self-activating cAMP switch. *Embo J* 21 (14): 3672–3680.

46. Cote, M., M.D. Payet, E. Rousseau, G. Guillon, and N. Gallo-Payet. 1999. Comparative involvement of cyclic nucleotide phosphodiesterases and adenylyl cyclase on adrenocorticotropin-induced increase of cyclic adenosine monophosphate in rat and human glomerulosa cells. *Endocrinology* 140 (8): 3594–3601.

47. MacFarland, R.T., B.D. Zelus, and J.A. Beavo. 1991. High concentrations of a cGMP-stimulated phosphodiesterase mediate ANP-induced decreases in cAMP and steroidogenesis in adrenal glomerulosa cells. *J Biol Chem* 266 (1): 136–142.

48. Dittrich, M., J. Jurevicius, M. Georget, F. Rochais, B. Fleischmann, J. Hescheler, and R. Fischmeister. 2001. Local response of L-type Ca^{2+} current to nitric oxide in frog ventricular myocytes. *J Physiol* 534 (Pt 1): 109–121.

49. Fischmeister, R., and H.C. Hartzell. 1987. Cyclic guanosine $3',5'$-monophosphate regulates the calcium current in single cells from frog ventricle. *J Physiol* 387: 453–472.

50. Hartzell, H.C., and R. Fischmeister. 1986. Opposite effects of cyclic GMP and cyclic AMP on Ca^{2+} current in single heart cells. *Nature* 323 (6085): 273–275.

51. Jurevicius, J., V.A. Skeberdis, and R. Fischmeister. 2003. Role of cyclic nucleotide phosphodiesterase isoforms in cAMP compartmentation following beta2-adrenergic stimulation of ICa, L in frog ventricular myocytes. *J Physiol* 551 (Pt 1): 239–252.

52. Mery, P.F., C. Pavoine, F. Pecker, and R. Fischmeister. 1995. *Erythro*-9-(2-hydroxy-3-nonyl)adenine inhibits cyclic GMP-stimulated phosphodiesterase in isolated cardiac myocytes. *Mol Pharmacol* 48 (1): 121–130.

53. Rivet-Bastide, M., G. Vandecasteele, S. Hatem, I. Verde, A. Benardeau, J.J. Mercadier, and R. Fischmeister. 1997. cGMP-stimulated cyclic nucleotide phosphodiesterase regulates the basal calcium current in human atrial myocytes. *J Clin Invest* 99 (11): 2710–2718.

54. Simmons, M.A., and H.C. Hartzell. 1988. Role of phosphodiesterase in regulation of calcium current in isolated cardiac myocytes. *Mol Pharmacol* 33 (6): 664–671.

55. Vandecasteele, G., I. Verde, C. Rucker-Martin, P. Donzeau-Gouge, and R. Fischmeister. 2001. Cyclic GMP regulation of the L-type Ca(2+) channel current in human atrial myocytes. *J Physiol* 533 (Pt 2): 329–340.

56. Verde, I., G. Vandecasteele, F. Lezoualc'h, and R. Fischmeister. 1999. Characterization of the cyclic nucleotide phosphodiesterase subtypes involved in the regulation of the L-type Ca^{2+} current in rat ventricular myocytes. *Br J Pharmacol* 127 (1): 65–74.

57. Favot, L., T. Keravis, V. Holl, A. Le Bec, and C. Lugnier. 2003. VEGF-induced HUVEC migration and proliferation are decreased by PDE2 and PDE4 inhibitors. *Thromb Haemost* 90 (2): 334–343.

58. Favot, L., T. Keravis, and C. Lugnier. 2004. Modulation of VEGF-induced endothelial cell cycle protein expression through cyclic AMP hydrolysis by PDE2 and PDE4. *Thromb Haemost* 92 (3): 634–645.

59. Keravis, T., N. Komas, and C. Lugnier. 2000. Cyclic nucleotide hydrolysis in bovine aortic endothelial cells in culture: Differential regulation in cobblestone and spindle phenotypes. *J Vasc Res* 37 (4): 235–249.

60. Lugnier, C., and V.B. Schini. 1990. Characterization of cyclic nucleotide phosphodiesterases from cultured bovine aortic endothelial cells. *Biochem Pharmacol* 39 (1): 75–84.

61. Netherton, S.J., and D.H. Maurice. 2005. Vascular endothelial cell cyclic nucleotide phosphodiesterases and regulated cell migration: Implications in angiogenesis. *Mol Pharmacol* 67 (1): 263–272.

62. Suttorp, N., S. Hippenstiel, M. Fuhrmann, M. Krull, and T. Podzuweit. 1996. Role of nitric oxide and phosphodiesterase isoenzyme II for reduction of endothelial hyperpermeability. *Am J Physiol* 270 (3 Pt 1): C778–C785.

63. Colman, R.W. 2004. Platelet cyclic adenosine monophosphate phosphodiesterases: Targets for regulating platelet-related thrombosis. *Semin Thromb Hemost* 30 (4): 451–460.

64. Dickinson, N.T., E.K. Jang, and R.J. Haslam. 1997. Activation of cGMP-stimulated phosphodiesterase by nitroprusside limits cAMP accumulation in human platelets: Effects on platelet aggregation. *Biochem J* 323 (Pt 2): 371–377.

65. Dunkern, T.R., and A. Hatzelmann. 2005. The effect of Sildenafil on human platelet secretory function is controlled by a complex interplay between phosphodiesterases 2, 3, and 5. *Cell Signal* 17 (3): 331–339.

66. Manns, J.M., K.J. Brenna, R.W. Colman, and S.B. Sheth. 2002. Differential regulation of human platelet responses by cGMP inhibited and stimulated cAMP phosphodiesterases. *Thromb Haemost* 87 (5): 873–879.

67. Mongillo, M., C.G. Tocchetti, A. Terrin, V. Lissandron, Y.F. Cheung, W.R. Dostmann, T. Pozzan, D.A. Kass, N. Paolocci, M.D. Houslay, and M. Zaccolo. 2005. Compartmentalized phosphodiesterase-2 activity blunts β-adrenergic cardiac inotropy via an NO/cGMP-dependent pathway. *Circ Res* 98 (2): 226–234.

68. Nikolaev, V.O., S. Gambaryan, S. Engelhardt, U. Walter, and M.J. Lohse. 2005. Real-time monitoring of the PDE2 activity of live cells: Hormone-stimulated cAMP hydrolysis is faster than hormone-stimulated cAMP synthesis. *J Biol Chem* 280 (3): 1716–1719.

69. Nikolaev, V.O., M. Bunemann, L. Hein, A. Hannawacker, and M.J. Lohse. 2004. Novel single chain cAMP sensors for receptor-induced signal propagation. *J Biol Chem* 279 (36): 37215–37218.

70. Haynes, J. Jr., D.W. Killilea, P.D. Peterson, and W.J. Thompson. 1996. *Erythro*-9-(2-hydroxy-3-nonyl)adenine inhibits cyclic-3′,5′-guanosine monophosphate-stimulated phosphodiesterase to reverse hypoxic pulmonary vasoconstriction in the perfused rat lung. *J Pharmacol Exp Ther* 276 (2): 752–757.

71. Haslam, R.J., N.T. Dickinson, and E.K. Jang 1999. Cyclic nucleotides and phosphodiesterases in platelets. *Thromb Haemost* 82 (2): 412–423.

72. Grant, P.G., A.F. Mannarino, and R.W. Colman. 1990. Purification and characterization of a cyclic GMP-stimulated cyclic nucleotide phosphodiesterase from the cytosol of human platelets. *Thromb Res* 59 (1): 105–119.

73. Jensen, B.O., F. Selheim, S.O. Doskeland, A.R. Gear, and H. Holmsen. 2004. Protein kinase A mediates inhibition of the thrombin-induced platelet shape change by nitric oxide. *Blood* 104 (9): 2775–2782.

74. Suvarna, N.U., and J.M. O'Donnell. 2002. Hydrolysis of *N*-methyl-D-aspartate receptor-stimulated cAMP and cGMP by PDE4 and PDE2 phosphodiesterases in primary neuronal cultures of rat cerebral cortex and hippocampus. *J Pharmacol Exp Ther* 302 (1): 249–256.

75. Wykes, V., T.C. Bellamy, and J. Garthwaite. 2002. Kinetics of nitric oxide-cyclic GMP signalling in CNS cells and its possible regulation by cyclic GMP. *J Neurochem* 83 (1): 37–47.

76. Gustafsson, A.B., and L.L. Brunton. 2002. Attenuation of cAMP accumulation in adult rat cardiac fibroblasts by IL-1beta and NO: Role of cGMP-stimulated PDE2. *Am J Physiol Cell Physiol* 283 (2): C463–C471.

77. Ortiz, P.A., and J.L. Garvin. 2001. NO inhibits NaCl absorption by rat thick ascending limb through activation of cGMP-stimulated phosphodiesterase. *Hypertension* 37 (2 Pt 2): 467–471.

78. Herring, N., L. Rigg, D.A. Terrar, and D.J. Paterson. 2001. NO-cGMP pathway increases the hyperpolarisation-activated current, *I*(f), and heart rate during adrenergic stimulation. *Cardiovasc Res* 52 (3): 446–453.

79. Sadhu, K., K. Hensley, V.A. Florio, and S.L. Wolda. 1999. Differential expression of the cyclic GMP-stimulated phosphodiesterase PDE2A in human venous and capillary endothelial cells. *J Histochem Cytochem* 47 (7): 895–906.

80. van Staveren, W.C., H.W. Steinbusch, M. Markerink-Van Ittersum, D.R. Repaske, M.F. Goy, J. Kotera, K. Omori, J.A. Beavo, and J. De Vente. 2003. mRNA expression patterns of the cGMP-hydrolyzing phosphodiesterases types 2, 5, and 9 during development of the rat brain. *J Comp Neurol* 467 (4): 566–580.

81. Repaske, D.R., J.G. Corbin, M. Conti, and M.F. Goy. 1993. A cyclic GMP-stimulated cyclic nucleotide phosphodiesterase gene is highly expressed in the limbic system of the rat brain. *Neuroscience* 56 (3): 673–686.

82. van Staveren, W.C., H.W. Steinbusch, M. Markerink-van Ittersum, S. Behrends, and J. de Vente. 2004. Species differences in the localization of cGMP-producing and NO-responsive elements in the mouse and rat hippocampus using cGMP immunocytochemistry. *Eur J Neurosci* 19 (8): 2155–2168.

83. van Staveren, W.C., M. Markerink-van Ittersum, H.W. Steinbusch, and J. de Vente. 2001. The effects of phosphodiesterase inhibition on cyclic GMP and cyclic AMP accumulation in the hippocampus of the rat. *Brain Res* 888 (2): 275–286.

84. Juilfs, D.M., S. Soderling, F. Burns, and J.A. Beavo. 1999. Cyclic GMP as substrate and regulator of cyclic nucleotide phosphodiesterases (PDEs). *Rev Physiol Biochem Pharmacol* 135: 67–104.

85. Silva, W.I., W. Schook, T.W. Mittag, and S. Puszkin. 1986. Cyclic nucleotide phosphodiesterase activity in bovine brain coated vesicles. *J Neurochem* 46 (4): 1263–1271.
86. Noyama, K., and S. Maekawa. 2003. Localization of cyclic nucleotide phosphodiesterase 2 in the brain-derived Triton-insoluble low-density fraction (raft). *Neurosci Res* 45 (2): 141–148.
87. Fulle, H.J., R. Vassar, D.C. Foster, R.B. Yang, R. Axel, and D.L. Garbers. 1995. A receptor guanylyl cyclase expressed specifically in olfactory sensory neurons. *Proc Natl Acad Sci USA* 92 (8): 3571–3575.
88. Juilfs, D.M., H.J. Fulle, A.Z. Zhao, M.D. Houslay, D.L. Garbers, and J.A. Beavo. 1997. A subset of olfactory neurons that selectively express cGMP-stimulated phosphodiesterase (PDE2) and guanylyl cyclase-D define a unique olfactory signal transduction pathway. *Proc Natl Acad Sci USA* 94 (7): 3388–3395.
89. Meyer, M.R., A. Angele, E. Kremmer, U.B. Kaupp, and F. Muller. 2000. A cGMP-signaling pathway in a subset of olfactory sensory neurons. *Proc Natl Acad Sci USA* 97 (19): 10595–10600.
90. Ma, M., X. Grosmaitre, C.L. Iwema, H. Baker, C.A. Greer, and G.M. Shepherd. 2003. Olfactory signal transduction in the mouse septal organ. *J Neurosci* 23 (1): 317–324.
91. Michie, A.M., M. Lobban, T. Muller, M.M. Harnett, and M.D. Houslay. 1996. Rapid regulation of PDE-2 and PDE-4 cyclic AMP phosphodiesterase activity following ligation of the T cell antigen receptor on thymocytes: Analysis using the selective inhibitors *erythro*-9-(2-hydroxy-3-nonyl)-adenine (EHNA) and rolipram. *Cell Signal* 8 (2): 97–110.
92. Bender, A.T., C.L. Ostenson, D. Giordano, and J.A. Beavo. 2004. Differentiation of human monocytes *in vitro* with granulocyte-macrophage colony-stimulating factor and macrophage colony-stimulating factor produces distinct changes in cGMP phosphodiesterase expression. *Cell Signal* 16 (3): 365–374.
93. Tenor, H., and C. Schudt. 1996. Analysis of PDE isoenzyme profiles in cells and tissues by pharmacological methods. In *Phosphodiesterase Inhibitors*, eds. C. Schudt, G. Dent, and K.F. Rabe. San Diego: Academic Press, pp. 21–40.
94. Witwicka, H., M. Kobialka, and W.A. Gorczyca. 2002. Hydrolysis of cyclic GMP in rat peritoneal macrophages. *Acta Biochim Pol* 49 (4): 891–897.
95. Velardez, M.O., A. De Laurentiis, M. del Carmen Diaz, M. Lasaga, D. Pisera, A. Seilicovich, and B.H. Duvilanski. 2000. Role of phosphodiesterase and protein kinase G on nitric oxide-induced inhibition of prolactin release from the rat anterior pituitary. *Eur J Endocrinol* 143 (2): 279–284.
96. Martins, T.J. 1984. Cyclic GMP-stimulated cyclic nucleotide phosphodiesterase: Purification from bovine tissues and characterization. PhD diss., University of Washington.
97. Kraulis, P. 1991. MOLSCRIPT: A program to produce both detailed and schematic plots of protein structures. *J Appl Cryst* 24: 946–950.

5 Phosphodiesterase 3B: An Important Regulator of Energy Homeostasis

Eva Degerman and Vincent Manganiello

CONTENTS

5.1 Background.. 79
5.2 The PDE3 Gene Family... 80
    5.2.1 Kinetic Properties and Inhibitor Sensitivities 80
    5.2.2 Structural Organization... 81
    5.2.3 Crystal Structure of PDE3B.. 83
    5.2.4 Tissue- and Cell-Specific Expression and Function......................... 83
    5.2.5 Subcellular Localization 85
    5.2.6 Insulin-Induced Formation of Macromolecular Complexes
          Containing PDE3B... 85
5.3 PDE3B in Physiological Contexts.. 86
    5.3.1 Role and Regulation of PDE3B 86
5.4 Role of PDE3B in Diabetes 88
Acknowledgments 91
References 91

5.1 BACKGROUND

Hydrolysis of intracellular 3′,5′-cyclic adenosine monophosphate (cAMP) and 3′,5′-cyclic guanosine monophosphate (cGMP) is catalyzed by a large, diverse, and complex group of structurally related cyclic nucleotide phosphodiesterases (PDEs). Eleven major *PDE* gene families (PDE1–11) have been identified; they differ in primary structures, affinities for cAMP and cGMP, responses to specific effectors, sensitivities to specific inhibitors, and mechanisms by which they are regulated.[1–3] All families are composed of more than one member and, in many cases, the members are the products of more than one gene. The PDE3 family[4] contains two subfamilies, PDE3A and PDE3B, which are encoded by different but related genes. PDE3A and PDE3B isoforms exhibit distinct, but overlapping patterns of expression. For example, PDE3A is more highly expressed than PDE3B in cells of the cardiovascular system and PDE3B is highly expressed in cells of importance for glucose and lipid metabolism. With the development of PDE family-selective inhibitors, it became apparent that PDE3s regulated cAMP pools which are important in platelet and myocardial physiology, i.e., specific PDE3 inhibitors inhibited platelet aggregation and augmented myocardial contractility.[1–4] Consistent with the expression of PDE3A in platelets, studies

with platelets from PDE3A and PDE3B knockout (KO) mice indicated that the PDE3 inhibitor, cilostamide, inhibited the aggregation of platelet preparations from PDE3B, but not PDE3A, KO mice (unpublished results). Functional roles of PDE3A and PDE3B isoforms in the cardiovascular system have not been dissected, although recent reports have suggested a critical role for murine PDE3A in the pathophysiology of cardiomyocyte apoptosis and cardiac failure.[5,6] The functional roles of PDE3 in the cardiovascular system will be covered in Chapter 21 through Chapter 25.

In this chapter, we present a general overview of the PDE3 gene family, and then focus on the physiological roles and regulation of PDE3B, and the role of PDE3A in oocyte maturation. Among PDE3Bs, the adipocyte PDE3B has been extensively studied since the early 1970s. It is now established that this enzyme plays a key role in mediating insulin-induced inhibition of lipolysis.[4] More recently, a role for PDE3B in the regulation of insulin-induced glucose uptake, GLUT-4 translocation, and lipogenesis has been demonstrated.[7,8] PDE3B also seems to be important in mediating the antiglycogenolytic effects of insulin/IGF-1 in hepatocytes,[9] and in regulating inhibitory effects of IGF-1 and leptin, as well as the stimulatory effects of glucon-like peptide 1 (GLP-1) and glucose, on insulin secretion in beta cells.[10–12] We will also discuss the role of PDE3B in the context of energy homeostasis, diabetes, and treatment of diabetes, drawing in part, on our studies with PDE3B KO mice. Chapter 17 discusses PDE3B from the perspective of obesity and diabetes. In that chapter, the metabolic characterization of mice that specifically overexpresses PDE3B in beta cells is described. These KO and transgenic mice, together with studies described in this chapter, indeed support an important role for PDE3B in the regulation of energy homeostasis.

5.2   THE PDE3 GENE FAMILY

5.2.1   KINETIC PROPERTIES AND INHIBITOR SENSITIVITIES

PDE3s can be distinguished from other PDEs by their high affinities for both cAMP and cGMP, with K_m values in the range of 0.1 to 0.8 μM, and maximal velocity (V_{max}) for cAMP is higher (approximately 4- to 10-fold) than for cGMP.[4] PDE3 isoforms hydrolyze cAMP and cGMP in a mutually competitive manner. Some biologic effects of endogenous cGMP may be mediated by the inhibition of PDE3, which results in increased cAMP levels and activation of protein kinase A (PKA), or the activation of other cAMP targets such as exchange proteins activated by cAMP or cAMP-GTP exchange factors (GEF) (EPACs).[13] For example, in rabbit platelets,[14,15] mouse thymocytes,[16] human atrial and frog ventricle myocytes,[17] and vascular smooth muscle,[18,19] an increase in cGMP brought about by nitrovasodilators (which release nitric oxide and activate guanylyl cyclase) and other cGMP-elevating agents allows for more effective competition of cGMP with cAMP at the PDE3 catalytic site; the inhibition of PDE3 is believed to contribute to increases in cAMP. PDE3 isoforms are also characterized by their sensitivity to a number of specific inhibitors and drugs, including cilostamide, cilostazol, enoximone, and lixazinone,[19–24] compounds that are relatively selective for PDE3, with K_i and IC_{50} (concentration that inhibits 50%) values at least 30- to 100-fold lower for PDE3 than for other PDE families. The availability of family-specific PDE inhibitors, especially for PDEs 3, 4, and 5, has increased the understanding of some functions of individual PDEs in the regulation of cyclic nucleotide-mediated processes in specific cellular environments, e.g., specific inhibition of PDE3s results in, presumably by inhibiting PDE3A, the stimulation of myocardial contractility, inhibition of platelet aggregation, relaxation of vascular and airway smooth muscle, inhibition of oocyte maturation, and inhibition of proliferation of cultured vascular smooth muscle cells (VSMC).[4,24,25] Furthermore, the inhibition of PDE3B is assumed to result in the inhibition of the antilipolytic action of insulin and potentiation of insulin secretion.[9,26,27]

The pharmaceutical industry has exhibited considerable interest in developing specific inhibitors of individual PDE families and subfamilies as therapeutic agents to replace non-selective PDE inhibitors, such as theophylline, which have narrow therapeutic windows due to the interaction with multiple PDEs and other biological targets. Because of their potent inotropic effects, several clinical trials were initiated to study the potential therapeutic utility of PDE3 inhibitors in the treatment of cardiac failure. Although these drugs provided immediate hemodynamic benefits, prolonged use led to increased mortality. At present, milrinone is used only on an acute basis for stabilization and acute treatment of cardiac failure in a hospital setting.[28] Cilostazol, another PDE3 inhibitor, is, however, approved as a treatment for intermittent claudication, a peripheral vascular disease.[29] As for the potential use of PDE3 inhibitors in the therapy for obesity and type 2 diabetes (T2D), one has to consider both the expected beneficial effects regarding insulin secretion, as well as the nonbeneficial effects in target tissues for insulin action and in heart, since current PDE3 inhibitors are not selective for the PDE3A or PDE3B isoforms. Some of these issues are discussed in this chapter.

5.2.2 Structural Organization

Using an affinity matrix composed of the *N*-(2-isothiocyanato)ethyl derivative of cilosta-mide, a specific inhibitor of PDE3, as well as other methods, we and others have purified PDE3s to homogeneity from several tissues, including rat[30] and bovine[31] adipose tissues, bovine aortic smooth muscle,[32] human platelets,[33] human placenta,[34] rat liver,[35,36] bovine heart,[37] and human platelets.[38] Molecular cloning revealed in several of these tissues the presence of two PDE3 subfamilies, PDE3A and PDE3B, which are the products of different but related genes.[39,40] Analysis of PDE3-deduced sequences indicates that PDE3As (or PDE3B) from different species are more closely related than are PDE3A and PDE3B from the same species. The *PDE3A* gene is located on human chromosome 12, and *PDE3B*, on chromosome 11.[41–44] As seen in Figure 5.1, the murine and human *PDE3* genes are large, ranging from mouse *PDE3B* at ~121 kb to human *PDE3A* at ~312 kb. The genomic organization

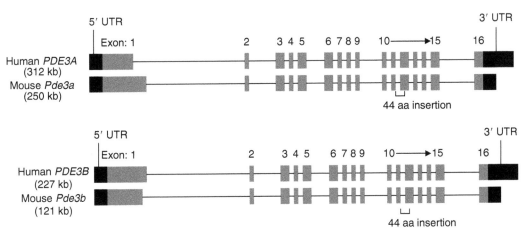

FIGURE 5.1 (See color insert following page 274.) Structure of PDE3 genes. The human *PDE3A* and *PDE3B*, and mouse *Pde3a* and *Pde3b*, gene coordinates were downloaded from the NCBI Gene database, and used to determine the gene size, exon size, and the exon/intron boundaries (also cf. Refs. [38–42] for human *PDE3A*, and rat and human *PDE3B*). The 44 amino acid insert was identified by aligning the protein sequences of PDE3s with those of PDE4s and locating the corresponding PDE3 exons containing the insert.

of *PDE3A* and *PDE3B* genes is identical, with 16 exons predicted in mouse and human. The *PDE3B* gene is thought to encode one PDE3B, which is primarily associated with membranes. Three PDE3A variants, including predominantly cytosolic forms, are generated from the single *PDE3A* gene by unknown mechanisms, i.e., possibly by the utilization of alternative transcription initiation sites,[43] mRNA splicing, or posttranscriptional processing.[44,45] It is also possible that smaller PDE3A variants could be generated by the proteolytic removal of membrane-association regions.

The overall predicted structural organization of PDE3A and PDE3B proteins is also identical, with a regulatory domain located in the N-terminal region, and the catalytic domain conserved among all PDEs in the C-terminal portion of the enzymes, followed by a hydrophilic C-terminal region.[4,46,47] Within the conserved domain of PDE3s, however, is an insert of 44 amino acids (encoded by exons 11 and 12, cf. Figure 5.1) that does not align with sequences in conserved domains of other PDE families, and the sequence of this segment differs in PDE3A and PDE3B isoforms. This insertion, which distinguishes PDE3 catalytic domains from those of other PDEs and may identify subfamilies within the PDE3 family, disrupts the first (of two) putative metal-binding domains present in PDEs.

The regulatory N-terminal portions of PDE3A and PDE3B are quite divergent and contain large hydrophobic regions with predicted transmembrane helical segments. As shown in Figure 5.2, the N-terminal portion of PDE3 (in this case PDE3B) can be arbitrarily divided into two regions, region 1 (aa 1–300) which contains a large hydrophobic domain (NHR1) with 5–6 predicted transmembrane helices, followed by region 2 (aa 300–500), with a smaller hydrophobic region (NHR2, ~50 aa). Subcellular fractionation and immunofluorescence localization of FLAG-tagged N-terminal truncated human PDE3A and mouse PDE3B

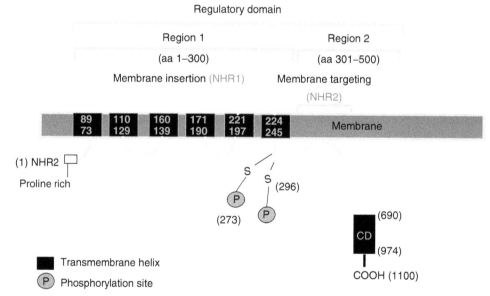

FIGURE 5.2 (See color insert following page 274.) Structural organization of PDE3B. Region 1 of the N-terminal portion of PDE3B includes NHR1, a large hydrophobic domain which contains 5–6 transmembrane helices; region 2 contains NHR2, a smaller hydrophobic region. Portions of both NHR1 and NHR2 seem to be involved in efficient membrane targeting/association. The C-terminal portion contains the catalytic domain (amino acids 690–974), which is conserved among different PDE family members. Between the membrane-association region and the catalytic domain are localized a number of serines that are phosphorylated in response to insulin and cAMP-increasing hormones; two of these serines (corresponding to murine PDE3B) are depicted. The role for the proline-rich domain in the N-terminal part of PDE3B is not known.

recombinant proteins in Sf9 insect cells and NIH 3006 or COS-7 cells, respectively, indicated that although NHR1 contains the structural determinants that are responsible for strong association with, or insertion of adipocyte PDE3B into membranes, a second domain in NHR2 seemed to be important for efficient membrane association and targeting.[46,48,49] The N-terminal region of PDE3B also appears to be important for PDE3B interactions with other proteins as judged from the analysis using gel filtration (see Section 5.2.6). As seen in Figure 5.2, downstream of NHR1 lie regulatory serine residues that are phosphorylated in intact cells in response to insulin and IGF-1, as well as to agents that increase cAMP.[4] As will be discussed in Section 5.3.1, PDE3B, as well as PDE3A, seems to be phosphorylated at multiple sites; this is believed to be important for the regulation of activity as well as protein–protein interactions.

The different myocardial PDE3A variants are thought to be identical, except for the presence of different lengths of the N-terminal regulatory domain, with PDE3A-136 containing NHR1, NHR2, and predicted phosphorylation sites, PDE3A-118 lacking NHR1 and the upstream phosphorylation site (analogous to Ser273 in PDE3B, Figure 5.2), and PDE3A-94, which lacks NHR1, NHR2, and the phosphorylation sites.[44,45,50,51] It is presumed the PDE3A variants are targeted to different compartments, where they participate in domain-specific regulation of cAMP and cGMP signaling.

5.2.3 Crystal Structure of PDE3B

More recently, the catalytic domain of human PDE3B was expressed at high levels (as much as 20 mg/L) in *Escherichia coli* and, after appropriate refolding, purified to homogeneity using a novel dihydropyridazinone affinity ligand coupled to Affi-10 gel.[52] Tota and colleagues[52–54] determined the catalytic core to span amino acids 654–1073, similar to findings in the earlier studies of rat PDE3B and human PDE3A.[47,55] The catalytic cores of all PDEs contain a histidine-rich, PDE-specific, signature sequence motif, $HD(X)_2H(X)_4N$, and two consensus metal-binding domains, $H(X)_3H(X)_{24-26}E$; amino acids from each of these motifs have been shown to contribute critical contacts to a novel binuclear divalent metal-binding site in most PDEs (Chapter 29).[56–58] As mentioned in Section 5.2.2, the 44 amino acid insert in the catalytic unit that is specific to PDE3 disrupts the first putative metal-binding sequence. Thus, the number of residues separating the second histidine from glutamic acid in the first consensus sequence is greatly increased in PDE3. The 3D structure of the catalytic core of human PDE3B, with bound IBMX (a nonselective PDE inhibitor) or MERCK1 (a novel dihydropyridazinone derivative, which inhibits the catalytic core of PDE3B with an IC_{50} value of ~0.27 nM), was solved at 2.4 Å resolution by Tota and colleagues.[54] The overall structure of the catalytic domain of PDE3B is similar to that of PDE4B, PDE4D, and PDE5A. It contains 16 α-helices, which form three subdomains; the 44 amino acid insert (which lacked any clear secondary structural organization) lies between helices 6 and 7 in the first subdomain. The two putative metal-binding sites, which are thought to bind Mg, not Zn, in the crystal structure,[54] are in a pocket at the interface of the three subdomains. The binding contacts for either inhibitor do not overlap with the metal-binding sites.[54] The 3D structures of PDE3B, in conjunction with site-directed mutagenesis studies (primarily with human PDE3A), have identified molecular structural determinants, which are important in both substrate-binding specificity for cAMP and cGMP, as well as for the structure–activity relationships of different inhibitors.[54–58] Refinement and extension of this latter information could lead to the development of more potent, specific, and clinically useful compounds.

5.2.4 Tissue- and Cell-Specific Expression and Function

In situ hybridization studies, together with Northern and Western blot analyses, demonstrated distinct, but overlapping, distribution of PDE3A and PDE3B mRNAs or proteins.[59,60]

PDE3B mRNA or protein is prominent in white and brown adipose tissues or adipocytes, hepatocytes, beta cells, renal collecting duct epithelium, and developing spermatocytes, whereas PDE3A is more abundant in platelets, heart, vascular smooth muscle, and oocytes.[4,59,60] These studies and other findings suggest that PDE3A and PDE3B are likely to exhibit cell-specific differences in properties and regulation and may serve cell-specific functions. This concept is supported by the studies in PDE3A and PDE3B KO mice. For example, as noted above, PDE3 inhibitors block aggregation of platelets from PDE3B, but not PDE3A, KO mice (unpublished results).

Another example of cell-specific function of PDE3 isoforms comes from the studies of PDE3A KO mice, a model of female sterility (Figure 5.3). Mammalian oocytes are physiologically arrested in prophase I of the first meiotic division; just prior to ovulation, the gonadotropin surge reinitiates meiosis, which progresses to metaphase II, when oocytes are competent for fertilization. Elevated oocyte cAMP maintains meiotic arrest, presumably via PKA-induced phosphorylation of specific substrates, including, perhaps, Wee1B and Cdc25b,[61] and thus preventing the activation of, and signaling by, mitogen-activated protein (MAP) kinase and maturation-promoting factor (MPF). PDE3A is the predominant cAMP PDE in murine oocytes, and PDE3 inhibitors have been reported to inhibit oocyte maturation *in vivo* and *in vitro*, without affecting ovulation.[25] Oocytes from PDE3A KO mice[62] lacked cAMP-specific PDE activity, contained increased cAMP levels, and failed to undergo spontaneous maturation (up to 48 h). Although female PDE3A KO mice were viable and ovulated normal number of oocytes, they were sterile because ovulated KO oocytes were arrested at the germinal vesicle stage, and, therefore, could not be fertilized. Meiotic maturation in PDE3A KO oocytes was restored by inhibiting PKA, which confirmed that increased cAMP-PKA signaling was most likely responsible for the meiotic blockade. Oocytes from PDE3A KO mice that underwent germinal vesicle breakdown and meiotic maturation exhibited the activation of MAP kinase and MPF, and became competent for fertilization *in vitro* by spermatozoa. These findings provided genetic evidence indicating that the resumption of meiosis *in vivo* and *in vitro* requires PDE3A activity, and described an *in vivo* model where

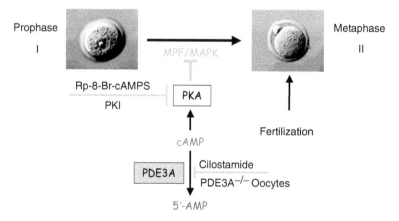

FIGURE 5.3 (See color insert following page 274.) Role of PDE3A in oocyte maturation and female infertility. PDE3A[-/-] mice represent the first genetic model indicating that disruption of cAMP signaling *in vivo* can lead to female infertility, and that PDE3A is necessary for meiotic resumption and fertilization of mouse oocytes.[62] Studies with PKA inhibitors, Rp-8-Br-cAMPs and PKI (protein kinase inhibitor peptide), are consistent with PKA playing a major role in maintaining meiotic arrest at prophase I.

meiotic maturation and ovulation are dissociated.[62] Female PDE3B KO mice, on the other hand, are fertile.[63] Other phenotypic characteristics of PDE3B KO mice will be further discussed in this chapter and other chapters. It is noteworthy that PDE3A and PDE3B KO mice show no compensatory upregulation of PDE4. Lack of compensatory upregulation of other PDEs has also been noted in PDE4A, PDE4B, and PDE4D KO mice (cf. Chapters 6 and 16), indicating that each PDE, at least at the subfamily level, has a unique role.

5.2.5 SUBCELLULAR LOCALIZATION

PDE3 family members exhibit differences in their subcellular localization. PDE3B is predominantly associated with membranes in adipocytes, hepatocytes, and pancreatic beta cells. Subcellular fractionation and immunofluorescence studies indicate that the N-terminal hydrophobic domains contain the structural determinants responsible for membrane association of adipocyte PDE3B.[46,48,49] PDE3A is cytosolic in platelets,[32,37,64,65] but found in the cytosol and in association with sarcoplasmic reticulum in myocardium.[44,50] Studies in VSMC indicate that PDE3A and PDE3B are present in cytoplasmic and membrane fractions, respectively,[45,51] and there seems to be differential regulation of the expression of PDE3A and PDE3B isoforms by cAMP in contractile or quiescent and synthetic or activated VSMC.[66] Whether cytosolic PDE3 isoforms are generated by proteolytic removal of the membrane-associated regions, the utilization of alternative transcription initiation sites,[43,45] or mRNA splicing is unknown.

Precise identification of the membranes with which PDE3B associates with adipocytes is not complete. Subcellular fractionation and immunohistochemistry show PDE3B localization to endoplasmic reticulum as well as plasma membranes in rat adipocytes and mouse 3T3-L1 adipocytes.[48,49,67,68] Furthermore, adipocyte PDE3B is localized to detergent-resistant parts of the plasma membrane, including lipid rafts and caveolae.[68] Caveolae are special forms of lipid rafts observed as small flask-shaped 50–100 nm invaginations of the plasma membrane.[69] Although caveolae are observed in many cell types, they are particularly abundant in adipocytes. They have a high content of cholesterol and sphingolipids and are stabilized by one or more isoforms of caveolin. The possibility of the functional interactions between PDE3B and caveolin is of particular interest, since in adipocytes, PDE3B is phosphorylated and activated in response to insulin as well as to cAMP-increasing hormones. Caveolae-signaling platforms are thought to be important in both cAMP- and insulin-regulated pathways, e.g., especially lipolysis and glucose uptake.[70] The finding that PDE3B is localized to both plasma membranes as well as the endoplasmic reticulum is also of special interest, since caveolae seem to have a role in membrane trafficking and lipid transport between plasma membrane and endoplasmic reticulum. In different studies, it has been shown that caveolae transport cholesterol from plasma membrane to endoplasmic reticulum and vice versa, probably to supply the plasma membrane with newly synthesized cholesterol.[71] One could speculate that PDE3B can also be transported in caveolae between the endoplasmic reticulum and plasma membrane. Interestingly, HMG-CoA reductase, an enzyme critical for cholesterol biosynthesis, is associated with the endoplasmic reticulum and may be regulated by cAMP.[72] The specific roles for different intracellular pools of PDE3B are not yet known.

5.2.6 INSULIN-INDUCED FORMATION OF MACROMOLECULAR COMPLEXES CONTAINING PDE3B

During gel filtration chromatography, PDE3B elutes as a much larger molecule than would be expected from its amino acid content. For example, gel filtration of detergent-treated

membranes from rat adipocytes and mouse 3T3-L1 adipocytes using Superose 6 chromatography shows the elution of PDE3B in high molecular weight (MW) complexes, ~670 kDa (in unstimulated rat and 3T3-L1 adipocytes) and >4000 kDa (in unstimulated rat adipocytes and insulin-treated 3T3-L1 adipocytes)[68] (unpublished data). The N-terminal region of PDE3B appears to be necessary for the presence of PDE3B in these large complexes.[46,48,49] Studies with 3T3-L1 adipocytes show that these complexes apparently contain phosphorylated or activated PDE3B (the complexes are enriched in phosphorylated species) and various signaling molecules possibly involved in its activation by insulin, i.e., insulin receptor substrate (IRS)-1, phosphatidylinositol-3-kinase (PI3K) p85, protein kinase B (PKB), heat shock protein (HSP)-90, phosphoprotein phosphatase (PP)-2A, and 14-3-3 proteins[68] (Ahmad et al., unpublished data). Phosphorylation or activation of PDE3B and recruitment of PDE3B and other signaling molecules into the complexes can be blocked by the treatment with wortmannin, a PI3K inhibitor, prior to insulin.

Several interacting partners for PDE3 have been identified, e.g., PDE3B and the insulin receptor in human adipocytes,[73] PDE3B and PKB in 3T3-L1 adipocytes,[74] PDE3B and PDE3A with 14-3-3,[75,76] PDE3B with a 50 kDa protein,[77] PDE3A and the leptin receptor in human platelets,[78] PDE3B with PI3K gamma,[79] and, as described above, PDE3B and caveolin-1 (direct or indirect via caveolae) in primary rat adipocytes,[69] suggesting that PDE3 regulates cAMP signaling in multiple intracellular compartments. In FDCP2 myeloid cells, IL-4 activated PDE3 via janus kinase (JAK)–induced tyrosine phosphorylation of IRS-2 and subsequent activation of PI3K and PKB.[80] A similar pathway has been recently suggested for leptin-induced activation of platelet PDE3A, which may be central to leptin's prothrombotic action and its ability to activate platelets.[78] In platelets, a constituitive leptin receptor or PDE3A molecular complex may be involved in the activation of PDE3A and its regulation of a cAMP pool that modulates platelet aggregation.[78]

Serine 296/302 (mouse/rat) in PDE3B (phosphorylated by insulin as well as isoproterenol) is localized within a classical 14-3-3-binding phosphopeptide motif and has been reported as a potential site for interaction of PDE3B with 14-3-3 proteins.[75] With respect to PKB, the structural determinants that are responsible for the interaction between PDE3B and PKB seem to reside in the N-terminal portion of PDE3B, since N-terminal deletion mutants in which the first 302 and 604 amino acids were deleted did not coimmunoprecipitate with commercially available recombinant PKB (Ahmad et al., unpublished results).

5.3   PDE3B IN PHYSIOLOGICAL CONTEXTS

5.3.1   ROLE AND REGULATION OF PDE3B

The role of adipocyte PDE3B has been extensively studied, starting from the early 1970s. It is now established that PDE3B has a key role in mediating insulin-induced inhibition of lipolysis[4] (Figure 5.4). More recently, a role for PDE3B in the regulation of insulin-induced glucose uptake, GLUT-4 translocation, and lipogenesis has been demonstrated; it is of interest that the pools of cAMP regulated by PDE3B seem to mediate these effects via PKA-independent mechanisms (Figure 5.4).[7] PDE3B has also been shown to have important roles in hepatocytes, pancreatic beta cells, and hypothalamic cells. Activation of PDE3B by IGF-1/insulin[10] and leptin[11] in pancreatic beta cells and hepatocytes,[9] and by leptin in the hypothalamus,[81] seems to be important for the inhibitory effects of IGF-1 and leptin on cAMP-mediated insulin secretion,[10,11] and inhibitory effects of leptin on glycogenolysis[9] and feeding behavior.[81] These data are in agreement with data from PDE3B KO (discussed in Section 5.4) and transgenic PDE3B mice specifically overexpressing PDE3B in beta cells (Chapter 17).

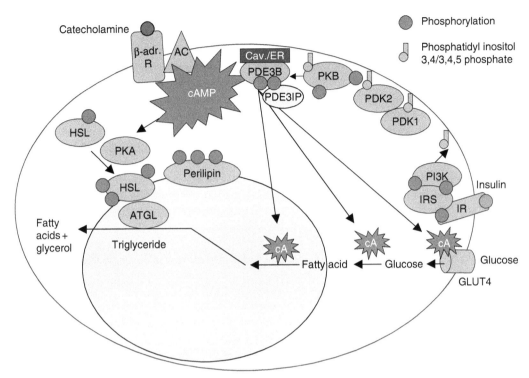

FIGURE 5.4 (See color insert following page 274.) Roles of PDE3B in adipocyte metabolic functions. Activation of adenylate cyclase (AC) results in increases in 3′,5′-cyclic adenosine monophosphate (cAMP) and activation of cAMP-dependent protein kinase A (PKA), which in turn leads to phosphorylation of perilipin (which allows access of hormone-sensitive lipase (HSL) to the surface of triglyceride droplets), and to phosphorylation and activation of HSL, and its translocation to fat droplets. HSL is phosphorylated under basal conditions; PKA induces phosphorylation at additional sites. Another lipase, adipose triglyceride lipase (ATGL) is believed to contribute to basal lipolysis. Insulin-stimulation of adipocytes results in phosphorylation and activation of phosphodiesterase 3B (PDE3B), which is believed to be the major step in insulin-induced inhibition of lipolysis. Activation of PDE3B leads to the reduction of cAMP and, thereby, the activity of PKA. PKA-mediated phosphorylation and activation of HSL is reduced, resulting in the inhibition of lipolysis. The signaling pathway from the insulin receptor to PDE3B is only partially understood. Phosphatidylinositol-3-kinase (PI3K) is likely to be involved and protein kinase B (PKB), presumably phosphorylated/activated by phosphoinositide-dependent kinases (PDK1, 2), may be the major insulin-sensitive PDE3B kinase; protein phosphatase2A (not shown) may be important in dephosphorylation of PDE3B. PDKs are activated by phosphoinositides (PI-3,4,5-P3) generated by activated PI3K. The role for endoplasmic reticulum (ER) and caveolae (cav.) localization of PDE3B and for PDE3B-interacting proteins (PDE3IP, see text) (under basal as well as hormone-induced conditions) is under investigation. In addition to having a key role in the antilipolytic action of insulin, PDE3B also contributes to the regulation of insulin-induced glucose uptake and lipogenesis.

Phosphorylation and activation of PDE3 induced by cAMP-increasing hormones and insulin and IGF-1 has been demonstrated in intact adipocytes, hepatocytes, and platelets.[4] Mechanisms involved in hormone-induced PDE3 phosphorylation have been elucidated in most detail for the adipocyte PDE3B. Insulin-induced phosphorylation and activation of PDE3B is a key event in the antilipolytic effect of insulin.[82–84] Activation of PDE3B leads to increased hydrolysis of cAMP and, consequently, inhibition of catecholamine-induced lipolysis, a process that involves PKA-dependent phosphorylation of hormone-sensitive lipase and perilipin.[4,85,86] Insulin-mediated phosphorylation and activation of PDE3B involves tyrosine

phosphorylation of IRS proteins catalyzed by the activated insulin receptor tyrosine kinase (IRTK), activation of PI3K, and increased production of phosphatidyl inositol 3,4/3,4,5 phosphates.[4,87] This leads to activation of PKB, which is believed to be one of the important kinases phosphorylating and activating PDE3B.[74,88,89] However, based on a complex phosphorylation pattern of PDE3B (Nilsson et al., unpublished results), additional insulin-activated kinases most likely are involved in the phosphorylation and activation of PDE3B.

The activation and phosphorylation of PDE3B by cAMP-increasing hormones is thought to be important in feedback regulation of cAMP- and cAMP-mediated responses. Furthermore, there seems to be a cross talk between isoproterenol- and insulin-mediated signaling in rat adipocytes at the level of PDE3B; in the presence of submaximal concentrations of isoproterenol, insulin more efficiently phosphorylates and activates PDE3B.[83] Very little information has been presented concerning PKA-mediated regulation of PDE3B, although the current hypothesis suggests that PKA mediates the activation of PDE3B induced by cAMP-increasing hormones. Corbin and coworkers demonstrated that PKA could activate PDE3 in adipocyte microsomes,[90] and Rascón et al.[91] and Rahn et al.[92] demonstrated that PKA could phosphorylate PDE3B. However, phosphorylation was not coupled to activation of the enzyme in these experiments.

Regarding the identity of sites phosphorylated in PDE3B in response to hormone stimulation, the initial data have suggested that multiple sites are phosphorylated, and that the regulation of phosphorylation is complex. In primary rat adipocytes, Rahn et al.[92] found that both insulin and isoproterenol induced phosphorylation of PDE3B on serine 302. On the other hand, in murine 3T3-L1 adipocytes, Kitamura et al.[74] identified serine 296 (corresponding to serine 302 in rat) as the site phosphorylated by isoproterenol and serine 273 (serine 279 in rat) as the site phosphorylated in response to insulin. In addition, two-dimensional tryptic phosphopeptide mapping of PDE3B indicates a number of additional sites phosphorylated in adipocytes and this becomes especially obvious in the presence of phosphatase inhibitors.[93] The limiting factor for the identification has so far been the small amounts of tryptic phosphopeptides that can be obtained from endogenous adipocyte PDE3B. In order to be able to identify additional insulin- and cAMP-mediated phosphorylation sites in PDE3B, we have overexpressed PDE3B in primary rat adipocytes and cultured H42E hepatoma cells. Stimulation of these cells that overexpress PDE3B with insulin or isoproterenol indeed demonstrates multisite phosphorylation of PDE3B and identification of the sites is ongoing. Exactly which kinases are involved in the regulation of PDE3B is not known. Although PKB[74,88,89] and PKA are good candidates, the complex phosphorylation pattern indicates the involvement of additional kinases. Multisite phosphorylation (five sites) of PDE3A in response to phorbol myristate acetate (PMA) in HeLa cells has been recently reported.[76] Phosphorylation of Ser428, not Ser312, correlated with binding of 14-3-3 to PDE3A, suggesting that phosphorylation of specific sites regulated interactions with specific partners. Finally, as discussed above, insulin-induced phosphorylation of PDE3B in 3T3-L1 adipocytes seems to induce the formation of a complex containing, among other proteins, phosphorylated or activated PDE3B and PKB. Complex formation most likely contributes to the overall regulation of activity and localization of PDE3A and PDE3B, and PDE3 downstream effects.

5.4 ROLE OF PDE3B IN DIABETES

A loss of normal regulation of adipose tissue metabolism could play a key role in the development of the hyperglycemia and dyslipidemia that characterize patients with T2D. For example, patients with obesity and T2D show increased levels of nonesterified free fatty acids (FFA), which are known to induce systemic insulin resistance.[94–98] PDE3B is a key component in the antilipolytic action of insulin. Thus, the genetic alteration of PDE3B

leading to reduced enzyme activity or an inability of insulin to induce phosphorylation and activation of the enzyme could result in increased lipolysis and increased circulating FFA, leading to the disruption of body fuel homeostasis. Early studies on adipocytes from diabetic patients demonstrated reduced PDE activity in adipose tissue.[99] Also, downregulation of PDE3B activity and decreased mRNA expression levels have been reported in adipose tissue of JCR: LA-cp rats, a strain that develops obesity, insulin resistance, and vasculopathy.[100] Furthermore, in those rats PDE3A was upregulated in VSMC, where, presumably by reducing cAMP, it might be important in stimulating proliferation of VSMC in diabetes-related vasculopathy. In obese, diabetic KKAy mice, PDE3B was shown to be downregulated in adipose tissue but upregulated in liver; the treatment of the KKAy mice with pioglitazone, an insulin-sensitizing, antidiabetic drug, increased adipocyte PDE3B expression, and its responsiveness to insulin.[101–103] We recently reported that, in 3T3-L1[104,105] and human[106] adipocytes, downregulation of PDE3B by TNF-α, perhaps mediated via MAP kinase signaling,[106,107] contributes to TNF-α- and ceramide-induced lipolysis, an effect that could be reversed by treating 3T3-L1 adipocytes with troglitazone.[105] Excess production of TNF-α has been shown to enhance the rate of adipose tissue lipolysis, hence increasing the concentration of circulating FFA and contributing to TNF-α-induced systemic insulin resistance.

In an effort to examine the physiological roles of PDE3B in the intact mouse, PDE3B-deficient 129/SvJ mice (*Pde3b* KO mice) were generated through partial reduction of the first exon in the *Pde3b* gene.[63] Characterization of the phenotype of the PDE3B KO mice demonstrated multiple alterations in insulin signaling and signs of insulin resistance, consistent with metabolic dysregulation associated with downregulation of PDE3B in the animal and tissue culture models discussed earlier.

In PDE3B KO mice, the expression of PDE3B protein and activity was virtually absent in adipose tissue, pancreatic beta cells, and liver. PDE3A mRNA content was not altered in heart and liver of KO mice, suggesting no compensatory increase in this isoform. PDE4 activity was also retained. The KO mice did not exhibit altered behavior, fertility, or life span. Whole body weight development was similar in KO and wild-type (WT) control mice, although KO mice tended to be slightly heavier. However, when animals were fed a standard diet or a high-fat diet, the weight of the gonadal adipose tissue was lower and a significant decrease in adipocyte size was observed, indicating substantial alterations in lipid metabolism.

Intraperitoneal injection of cAMP-elevating catecholamines such as isoproterenol (a general beta adrenergic receptor agonist) or CL 316,243 (a beta-3-specific agonist) significantly increased adipose tissue lipolysis, measured as release of FFA and glycerol, to a greater extent in intact PDE3B KO mice relative to WT controls. Basal (nonstimulated) serum levels of FFA and glycerol, measured in the fed state, were not different in KO mice compared to WT, but after a 20 h fast both components were increased in KO mice, indicating an elevated lipolytic drive in these mice. Catecholamine-stimulated lipolysis in adipocytes isolated from the KO mice was significantly increased compared to WT adipocytes. Furthermore, insulin did not inhibit catecholamine-stimulated lipolysis in PDE3B KO adipocytes, thus substantiating the hypothesis that the activation of PDE3B is a key event in the antilipolytic action of insulin. Insulin-stimulated lipogenesis was significantly enhanced in the adipocytes of PDE3B KO mice and paralleled an increase in the expression of fatty acid synthase.[63]

Consistent with previously reported involvement of PDE3B in the secretion of insulin from pancreatic beta cells, intraperitoneal or intravenous administration of CL 316,243 increased serum insulin to a greater extent in KO than in WT mice. In KO mice, the effect of CL 316,243 on serum insulin was greater than that of isoproterenol. Since insulin-secreting beta cells of the pancreas are believed not to express beta-3 receptors, the stimulatory effect of CL 316,243 is not likely to be exerted directly on beta cells, but rather reflects an indirect action, most likely via components from adipose tissue, the major site of beta-3 receptor expression.

This finding suggests cross talk between adipose tissue and pancreatic beta cells, the nature of which has not been elucidated. FFA secreted from adipocytes are known to affect insulin secretion and could serve as such a link, but FFA were similar in serum of the KO mice treated with isoproterenol or CL 316,243, despite the larger increase in serum insulin in the latter. Either an adipocyte-derived cytokine (adipokine) is released to a greater extent in response to beta-3 stimulation in the KO mice, or the beta cells in KO mice exhibit a higher responsiveness to the agent, or both. Pancreatic islets isolated from PDE3B KO mice released significantly more insulin than did wild-type islets upon stimulation with glucose, either alone or in combination with GLP-1.[63] This finding further strengthens the concept that a PDE3B-regulated pool of cAMP plays an important role in regulating nutrient-stimulated insulin secretion (see Chapter 17).

Several signs of insulin resistance were identified in the PDE3B KO mice. A significantly reduced capacity to dispose of blood glucose as well as serum FFA was apparent during insulin tolerance tests. Further, in spite of greatly increased serum insulin levels observed after injection of CL 316,243, glucose was not more efficiently eliminated from the circulation of KO mice. Moreover, hyperinsulinemic–euglycemic clamp experiments showed a reduced responsiveness to insulin and a decreased capacity of insulin to inhibit endogenous glucose production in the liver. It thus appeared that the PDE3B KO mice with time (after at least 3–4 months) develop insulin resistance, possibly with a prime location in the liver. *In vitro* analyses of the livers from KO mice did reveal several alterations; for example, increased content of cAMP, increased phosphorylation of PKA substrates, increased expression of some gluconeogenetic enzymes, increased triglyceride accumulation, as well as changes in inflammatory cytokine expression and insulin-dependent signaling pathways.[63]

However, since there were no indications of hyperglycemia, the PDE3B KO animals did not, during the test conditions, develop diabetes. This may reflect a combination of metabolic dysregulation in pancreas, liver, and adipose tissue, i.e., altered regulation of insulin secretion due to absence of PDE3B in pancreatic beta cells, reduced ability of insulin to inhibit endogenous glucose production due to absence of PDE3B in the liver, and various alterations in insulin signaling in adipocytes lacking PDE3B. In addition, it is conceivable that the endocrine function of adipocytes of the KO mice is affected. Adiponectin, an important adipokine secreted from adipose tissue and believed to increase insulin sensitivity, was significantly increased in the serum of KO mice. The higher production and secretion of adiponectin might help counteract factors promoting insulin resistance, as might the diminished size of KO adipocytes.[63]

In summary, the phenotypic characteristics of PDE3B KO mice indicate that PDE3B plays a unique role not only in adipocytes but also in pancreatic beta cells and hepatocytes and is not apparently functionally compensated for, by other PDE gene families or by the other PDE3 family member, PDE3A. Although not frankly diabetic, PDE3B KO mice demonstrate alterations in the regulation of energy homeostasis, including signs of insulin resistance and a number of other alterations in the regulation of energy homeostasis. Consistent with our findings in PDE3B KO mice, the administration of PDE3 inhibitors to intact rats and mice (Table 5.1) increased lipolysis and insulin secretion, and blocked insulin-induced suppression of endogenous glucose production.[108–116] Thus, absence or inhibition (e.g., via milrinone) of PDE3B could be associated with the development of insulin resistance, perhaps related to altered insulin signaling and dysregulation of hepatic glucose output. On the other hand, absence or inhibition of beta cell PDE3B potentiates insulin secretion, which could be beneficial for patients with T2D. This is also in agreement with *in vitro* studies on the role of PDE3B in beta cells, and with data from transgenic PDE3B mice, which demonstrate that overexpressing PDE3B in beta cells results in defective insulin secretion, leading to glucose intolerance (see Chapter 17).

TABLE 5.1
Comparison of Effects of PDE3 Inhibitor and Phenotypic Characteristics of PDE3-Null Mice

	PDE3 Inhibitor[108–116]	PDE3B KO[63]
Fatty acid levels	Increased	Increased
Insulin response	Increased	Increased
Glucose levels	Increased, NA[a]	Increased, NA
Effects of insulin		
Total body glucose uptake	NA	NA
Suppression of lipolysis	Blocked	Blocked
Suppression of glucose production	Blocked	Blocked

[a]NA, not affected.

Thus, regarding the use of PDE3B as a target for drugs in the treatment of T2D, one has to keep in mind the dynamic interplay among multiple tissues expressing PDE3B (Table 5.1). Tissue-specific delivery systems might be useful, although delivery of a PDE3B inhibitor to hepatocytes and adipocytes could result in worsening of glucose disposal (hepatocyte) and glucotoxicity, or in the development of insulin resistance (increased lipolysis and FFA), delivery to pancreatic beta cells might be beneficial (by increasing insulin secretion).

Although a recent study found no association of several polymorphisms in the promoter region or exon 4 of the human *PDE3B* gene with T2D in Japanese patients,[117] our results do raise the question as to whether the *PDE3B* gene could function as a susceptibility or modifier gene for the development of specific types of diabetes. Crossbreeding of PDE3B KO mice with mice that exhibit other disruptions in insulin-signaling pathways might offer clues in this area. For example, given the apparent effects of PDE3B deficiency on insulin-regulated hepatic glucose output, and the recent report describing downregulation of PDE4 in skeletal muscle of MIRKO mice,[118] it would be of interest to analyze insulin resistance and perform hyperinsulinemic–euglycemic clamps in combined PDE3B/PDE4 KO mice or in PDE3B KO mice receiving a specific PDE4 inhibitor such as rolipram.

ACKNOWLEDGMENTS

This work was supported by Swedish Medical Research Council Grant 3362 (E.D.); the Medical Faculty, Lund University; and the following foundations: Swedish Diabetes Association, Stockholm; Albert Påhlsson, Malmö, Sweden; and Novo Nordisk, Copenhagen, Denmark. We acknowledge the close collaborative efforts of Dr. Marco Conti and his laboratory in elucidating the reproductive phenotype of the PDE3A KO mice. We also thank Dr. Faiyaz Ahmad for communicating his unpublished results.

REFERENCES

1. Francis, S.H., I.V. Turko, and J.D. Corbin. 2001. Cyclic nucleotide phosphodiesterases: Relating structure and function. *Prog Nucleic Acid Res Mol Biol* 65:1.
2. Manganiello, V., and E. Degerman. 2004. Cyclic nucleotide phosphodiesterases. In *Encyclopedia of Biological Chemistry*, eds. W. Lennarz, and M.D. Lane. New York: Elsevier Science, pp. 501–505.
3. Conti, M., and S.L. Jin. 1999. The molecular biology of cyclic nucleotide phosphodiesterases. *Prog Nucleic Acid Res Mol Biol* 63:1.

4. Shakur, Y., L. Stenson Holst, T. Rahn Landström, et al. 2001. Regulation and function of the cyclic nucleotide phosphodiesterase (PDE3) gene family. *Prog Nucleic Acid Res Mol Biol* 66:241.

5. Ding, B., J. Abe, H. Wei, et al. 2005. A positive feedback loop of phosphodiesterase 3 (PDE3) and inducible cAMP early repressor (ICER) leads to cardiomyocyte apoptosis. *Proc Natl Acad Sci USA* 102:14771.

6. Ding, B., J. Abe, H. Wei, et al. 2005. Functional role of phosphodiesterase 3 in cardiomyocyte apoptosis. *Circulation* 111:2469.

7. Zmuda-Trzebiatowska, E., A. Oknianska, V. Manganiello, et al. 2006. Role of PDE3B in insulin-induced glucose uptake, GLUT-4 translocation and lipogenesis in primary rat adipocytes. *Cell Signal* 3:382.

8. Eriksson, J.W., C. Wesslau, and U. Smith. 1994. The cGMP-inhibitable phosphodiesterase modulates glucose transport activation by insulin. *Biochim Biophys Acta* 1189:163.

9. Zhao, A.Z., M.M. Shinohara, D. Huang, et al. 2000. Leptin induces insulin-like signaling that antagonizes cAMP elevation by glucagon in hepatocytes. *J Biol Chem* 275:11348.

10. Zhao, A.Z., H. Zhao, J. Teague, et al. 1997. Attenuation of insulin secretion by insulin-like growth factor 1 is mediated through activation of phosphodiesterase 3B. *Proc Natl Acad Sci USA* 94:3223.

11. Zhao, A.Z., K.E. Bornfeldt, and J.A. Beavo. 1998. Leptin inhibits insulin secretion by activation of phosphodiesterase 3B. *J Clin Invest* 102:869.

12. Furman, B., N. Pyne, P. Flatt, et al. 2004. Targeting beta-cell cyclic $3',5'$ adenosine monophosphate for the development of novel drugs for treating type 2 diabetes mellitus. A review. *J Pharm Pharmacol* 56 (12):1477.

13. Springett, G.M., H. Kawasaki, and D.R. Spriggs. 2004. Non-kinase second-messenger signaling: New pathways with new promise. *Bioessays* 26 (7):730.

14. Maurice, D.H., and R.J. Haslam. 1990. Molecular basis of the synergistic inhibition of platelet function by nitrovasodilators and activators of adenylate cyclase: Inhibition of cyclic AMP breakdown by cyclic GMP. *Mol Pharmacol* 37:671.

15. Fisch, A., J. Michael-Hepp, J. Meyer, et al. 1995. Synergistic interaction of adenylate cyclase activators and nitric oxide donor SIN-1 on platelet cyclic AMP. *Eur J Pharmacol* 289:455.

16. Marcoz, P., A.F. Prigent, M. Lagarde, et al. 1993. Modulation of rat thymocyte proliferative response through the inhibition of different cyclic nucleotide phosphodiesterase isoforms by means of selective inhibitors and cGMP-elevating agents. *Mol Pharmacol* 44:1027.

17. Kirstein, M., M. Rivet-Bastide, S. Hatem, et al. 1995. Nitric oxide regulates the calcium current in isolated human atrial myocytes. *J Clin Invest* 95:794.

18. Maurice, D.H., and R.J. Haslam. 1990. Nitroprusside enhances isoprenaline-induced increases in cAMP in rat aortic smooth muscle. *Eur J Pharmacol* 191:471.

19. Komas, N., C. Lugnier, and J.C. Stoclet. 1991. Endothelium-dependent and independent relaxation of the rat aorta by cyclic nucleotide phosphodiesterase inhibitors. *Br J Pharmacol* 104:495.

20. Weishaar, R.E., M.H. Cain, and J.A. Bristol. 1985. A new generation of phosphodiesterase inhibitors: Multiple molecular forms of phosphodiesterase and the potential for drug selectivity. *J Med Chem* 28:537.

21. Alvarez, R., G.L. Banerjee, J.J. Bruno, et al. 1986. A potent and selective inhibitor of cyclic AMP phosphodiesterase with potential cardiotonic and antithrombotic properties. *Mol Pharmacol* 29:554.

22. Manganiello, V., E. Degerman, and M. Elks. 1988. Selective inhibitors of specific phosphodiesterases in intact adipocytes. *Methods Enzymol* 159:504.

23. Beavo, J.A., and D.H. Reifsnyder. 1990. Primary sequence of cyclic nucleotide phosphodiesterase isozymes and the design of selective inhibitors. *Trends Pharmacol Sci* 11:150.

24. Komas, N., S. Movsesian, E. Kedev, et al. 1996. cGMP-inhibited phosphodiesterase (PDE3). In *Phosphodiesterase inhibitors*, eds. C. Schudt, G. Dent, and K.F. Rabe, 89. London: Academic Press.

25. Conti, M., C.B. Andersen, F.J. Richard, et al. 1998. Role of cyclic nucleotide phosphodiesterases in resumption of meiosis. *Mol Cell Endocrinol* 145:9.

26. Elks, M.L., V.C. Manganiello, and M. Vaughan. 1983. Hormone-sensitive particulate cAMP phosphodiesterase activity in 3T3-L1 adipocytes. Regulation of responsiveness by dexamethasone. *J Biol Chem* 258:8582.

27. Eriksson, H., M. Ridderstråle, E. Degerman, et al. 1995. Evidence for the key role of the adipocyte cGMP-inhibited cAMP phosphodiesterase in the antilipolytic action of insulin. *Biochim Biophys Acta* 1266:101.

28. Movsesian, M.A. 2000. Therapeutic potential of cyclic nucleotide phosphodiesterase inhibitors in heart failure. *Expert Opin Investig Drug* 9:963.

29. Wang, S., J. Cone, M. Fong, et al. 2001. Interplay between inhibition of adenosine uptake and phosphodiesterase type 3 on cardiac function by cilostazol, an agent to treat intermittent claudication. *J Cardiovasc Pharmacol* 38 (5):775.

30. Degerman, E., P. Belfrage, A.H. Newman, et al. 1987. Purification of the putative hormone sensitive cyclic AMP phosphodiesterase from rat adipose tissue using a derivative of cilostamide as a novel affinity ligand. *J Biol Chem* 262:5797.

31. Degerman, E., V.C. Manganiello, A.H. Newman, et al. 1988. Purification, properties and polyclonal antibodies for the particulate, low K_m cAMP phosphodiesterase from bovine adipose tissue. *Second Messengers Phosphoproteins* 12:171.

32. Rascón, A., S. Lindgren, L. Stavenow, et al. 1992. Purification and properties of the cGMP inhibited cAMP phosphodiesterase from bovine aortic smooth muscle. *Biochim Biophys Acta* 1134:149.

33. Degerman, E., M. Moos Jr., A. Rascón, et al. 1994. Single-step affinity purification, partial structure and properties of human platelet cGMP inhibited cAMP phosphodiesterase. *Biochim Biophys Acta* 1205:189.

34. LeBon, T.R., J. Kasuya, R.J. Paxton, et al. 1992. Purification and characterization of guanosine 3′,5′-monophosphate-inhibited low K_m adenosine 3′,5′-monophosphate phosphodiesterase from human placental cytosolic fractions. *Endocrinology* 130:3265.

35. Pyne, N.J., M.E. Cooper, and M.D. Houslay. 1987. The insulin- and glucagon-stimulated 'dense-vesicle' high-affinity cyclic AMP phosphodiesterase from rat liver. Purification, characterization and inhibitor sensitivity. *Biochem J* 242:33.

36. Boyes, S., and E.G. Loten. 1988. Purification of an insulin-sensitive cyclic AMP phosphodiesterase from rat liver. *Eur J Biochem* 174:303.

37. Harrison, S.A., D.H. Reifsnyder, B. Gallis, et al. 1986. Isolation and characterization of bovine cardiac muscle cGMP-inhibited phosphodiesterase: A receptor for new cardiotonic drugs. *Mol Pharmacol* 29:506.

38. Grant, P.G., and R.W. Colman. 1984. Purification and characterization of a human platelet cyclic nucleotide phosphodiesterase. *Biochemistry* 23:1801.

39. Meacci, E., M. Taira, M. Moos Jr., et al. 1992. Molecular cloning and expression of human myocardial cGMP-inhibited cAMP phosphodiesterase. *Proc Natl Acad Sci USA* 89:3721.

40. Taira, M., S.C. Hockman, J.C. Calvo, et al. 1993. Molecular cloning of the rat adipocyte hormone sensitive cyclic GMP-inhibited cyclic nucleotide phosphodiesterase. *J Biol Chem* 268:18573.

41. Miki, T., M. Taira, S. Hockman, et al. 1996. Characterization of the cDNA and gene encoding human PDE3B, the cGIP1 isoform of the human cyclic GMP-inhibited cyclic nucleotide phosphodiesterase family. *Genomics* 36:476.

42. Löbbert, R.W., A. Winterpacht, B. Seipel, et al. 1996. Molecular cloning and chromosomal assignment of the human homologue of the rat cGMP-inhibited phosphodiesterase 1 (*PDE3A*)—a gene involved in fat metabolism located at 11p 15.1. *Genomics* 37:211.

43. Kasuya, J., H. Goko, and Y. Fujita-Yamaguchi. 1995. Multiple transcripts for the human cardiac form of the cGMP-inhibited cAMP phosphodiesterase. *J Biol Chem* 270:14305.

44. Hambleton, R., J. Krall, E. Tikishvili, et al. 2005. Isoforms of cyclic nucleotide phosphodiesterase PDE3 and their contribution to cAMP hydrolytic activity in subcellular fractions of human myocardium. *J Biol Chem* 280:39168.

45. Choi, Y.H., D. Ekholm, J. Krall, et al. 2001. Identification of a novel isoform of the cyclic nucleotide phosphodiesterase PDE3A expressed in vascular smooth-muscle myocytes. *Biochem J* 353:41.

46. Leroy, M.-J., E. Degerman, M. Taira, et al. 1996. Characterization of two recombinant PDE3 (cGMP-inhibited cyclic nucleotide phosphodiesterase) isoforms, RcGIP1 and HcGIP2, expressed in NIH 3006 murine fibroblasts and Sf9 insect cells. *Biochemistry* 35:10194.

47. He, R., N. Komas, D. Ekholm, et al. 1998. Expression and characterization of deletion recombinants of two cGMP-inhibited cyclic nucleotide phosphodiesterases (PDE-3). *Cell Biochem Biophys* 29:89.
48. Kenan, Y., T. Murata, Y. Shakur, et al. 2000. Functions of the N-terminal region of cyclic nucleotide phosphodiesterase 3 (PDE3) isoforms. *J Biol Chem* 275:12331.
49. Shakur, Y., K. Takeda, Y. Kenan, et al. 2000. Membrane localization of cyclic nucleotide phosphodiesterase 3 (PDE3). Two N-terminal domains are required for the efficient targeting to, and association of, PDE3 with endoplasmic reticulum. *J Biol Chem* 275:38749.
50. Smith, C.J., J. Krall, V.C. Manganiello, et al. 1993. Cytosolic and sarcoplasmic reticulum-associated low K_m, cGMP-inhibited cAMP phosphodiesterase in mammalian myocardium. *Biochem Biophys Res Commun* 190:516.
51. Liu, H., and D.H. Maurice. 1998. Expression of cyclic GMP-inhibited phosphodiesterases 3A and 3B (PDE3A and PDE3B) in rat tissues: Differential subcellular localization and regulated expression by cyclic AMP. *Br J Pharmacol* 125:1501.
52. Varnerin, J.P., C.C. Chung, S.B. Patel, et al. 2004. Expression, refolding, and purification of recombinant human phosphodiesterase 3B: Definition of the N-terminus of the catalytic core. *Protein Expr Purif* 35 (2):225.
53. Scapin, G., S.B. Patel, C. Chung, et al. 2004. Crystal structure of human phosphodiesterase 3B: Atomic basis for substrate and inhibitor specificity. *Biochemistry* 43 (20):6091.
54. Cheung, P.P., H. Xu, M.M. McLaughlin, et al. 1996. Human platelet cGI-PDE: Expression in yeast and localization of the catalytic domain by deletion mutagenesis. *Blood* 88 (4):1321.
55. Colman, R.W. 2004. Platelet cyclic adenosine monophosphate phosphodiesterases: Targets for regulating platelet-related thrombosis. *Semin Thromb Hemost* 30 (4):451.
56. Zhang, W., H. Ke, and R.W. Colman. 2002. Identification of interaction sites of cyclic nucleotide phosphodiesterase type 3A with milrinone and cilostazol using molecular modeling and site-directed mutagenesis. *Mol Pharmacol* 62 (3):514.
57. Zhang, W., H. Ke, A.P. Tretiakova, et al. 2001. Identification of overlapping but distinct cAMP and cGMP interaction sites with cyclic nucleotide phosphodiesterase 3A by site-directed mutagenesis and molecular modeling based on crystalline PDE4B. *Protein Sci* 10 (8):1481.
58. Ke, H. 2004. Implications of PDE4 structure on inhibitor selectivity across PDE families. *Int J Impot Res* 16 (Suppl. 1):24.
59. Reinhardt, R.R., E. Chin, J. Zhou, et al. 1995. Distinctive anatomical patterns of gene expression for cGMP-inhibited cyclic nucleotide phosphodiesterases. *J Clin Invest* 95:1528.
60. Reinhardt, R.R., and C.A. Bondy. 1996. Differential cellular pattern of gene expression for two distinct cGMP-inhibited cyclic nucleotide phosphodiesterases in developing and mature rat brain. *Neuroscience* 72:567.
61. Han, S.J., and M. Conti. 2006. New pathways from PKA to the Cdc2/cyclin B complex in oocytes: Wee1B as a potential PKA substrate. *Cell Cycle* 5 (3):227.
62. Masciarelli, S., K. Horner, C. Liu, et al. 2004. Cyclic nucleotide phosphodiesterase 3A-deficient mice as a model of female infertility. *J Clin Invest* 114 (2):196.
63. Hun Choi, Y., S. Park, S. Hockman, et al. Alterations in regulation of energy homeostasis in cyclic nucleotide phosphodiesterase 3B (mPDE3B)-null mice. *J. Clin Invest* (in press).
64. Grant, P.G., A.F. Mannarino, and R.W. Colman. 1988. cAMP-mediated phosphorylation of the low-K_m cAMP phosphodiesterase markedly stimulates its catalytic activity. *Proc Natl Acad Sci USA* 85:9071.
65. Macphee, C.H., D.H. Reifsnyder, T.A. Moore, et al. 1988. Phosphorylation results in activation of a cAMP phosphodiesterase in human platelets. *J Biol Chem* 263:10353.
66. Maurice, D.H., D. Palmer, D.G. Tilley, et al. 2003. Cyclic nucleotide phosphodiesterase activity, expression, and targeting in cells of the cardiovascular system. *Mol Pharmacol* 64:533.
67. Kono, T., F.W. Robinson, and J.A. Sarver. 1975. Insulin-sensitive phosphodiesterase. Its localization, hormonal stimulation, and oxidative stabilization. *J Biol Chem* 250 (19):7826.
68. Nilsson, R., F. Ahmad, K. Swärd, et al. 2006 Plasma membrane cyclic nucleotide phosphodiesterase 3B (PDE3B) is associated with caveolae in primary rat adipocytes. *Cell Signal* 18 (10):1713.

69. Cohen, A.W., T.P. Combs, P.E. Scherer, et al. 2003. Role of caveolin and caveolae in insulin signaling and diabetes. *Am J Physiol Endocrinol Metab* 285 (6):E1151.

70. Razani, B., T.P. Combs, X.B. Wang, et al. 2002. Caveolin-1-deficient mice are lean, resistant to diet-induced obesity, and show hypertriglyceridemia with adipocyte abnormalities. *J Biol Chem* 277:8635.

71. Liu, P., M. Rudick, and R.G. Anderson. 2002. Multiple functions of caveolin-1. *J Biol Chem* 277:41295.

72. Botham, K.M. 1992. Cyclic AMP and the regulation of cholesterol metabolism. *Biochem Soc Trans* 2:454.

73. Rondinone, C.M., E. Carvalho, T. Rahn, et al. 2000. Phosphorylation of PDE3B by phosphatidylinositol 3-kinase associated with the insulin receptor. *J Biol Chem* 275:10093.

74. Kitamura, T., Y. Kitamura, S. Kuroda, et al. 1999. Insulin-induced phosphorylation and activation of cyclic nucleotide phosphodiesterase 3B by the serine–threonine kinase. *Akt Mol Cell Biol* 9:6286.

75. Onuma, H., H. Osawa, K. Yamada, et al. 2002. Identification of the insulin-regulated interaction of phosphodiesterase 3B with 14-3-3 beta protein. *Diabetes* 51:3362.

76. Pozuelo Rubio, M., D.G. Campbell, N.A. Morrice, et al. 2005. Phosphodiesterase 3A binds to 14-3-3 proteins in response to PMA-induced phosphorylation of Ser428. *Biochem J* 392 (Pt 1):163.

77. Onuma, H., H. Osawa, T. Ogura, et al. 2005. A newly identified 50 kDa protein, which is associated with phosphodiesterase 3B, is phosphorylated by insulin in rat adipocytes. *Biochem Biophys Res Commun* 337:976.

78. Elbatarny, H.S., and D.H. Maurice. 2005. Leptin-mediated activation of human platelets: Involvement of a leptin receptor and phosphodiesterase 3A-containing cellular signaling complex. *Am J Physiol Endocrinol Metab* 289 (4):695.

79. Patrucco, E., A. Notte, L. Barberis, et al. 2004. PI3K-gamma modulates the cardiac response to chronic pressure overload by distinct kinase-dependent and -independent effects. *Cell* 118 (3):375.

80. Ahmad, F., G. Gao, L.M. Wang, et al. 1999. IL3 and IL4 activate cyclic nucleotide phosphodiesterases 3 (PDE3) and 4 (PDE4) by different mechanisms in FDCP2 myeloid cells. *J Immunol* 162:4864.

81. Zhao, A.Z., J.N. Huan, S. Gupta, et al. 2002. A phosphatidylinositol 3-kinase phosphodiesterase 3B-cyclic AMP pathway in hypothalamic action of leptin on feeding. *Nat Neurosci* 8:727.

82. Degerman, E., C.J. Smith, H. Tornqvist, et al. 1990. Evidence that insulin and isoprenaline activate the cGMP-inhibited low-K_m cAMP phosphodiesterase in rat fat cells by phosphorylation. *Proc Natl Acad Sci USA* 87:533.

83. Smith, C.J., V. Vasta, E. Degerman, et al. 1991. Hormone-sensitive cyclic GMP-inhibited cyclic AMP phosphodiesterase in rat adipocytes. Regulation of insulin- and cAMP-dependent activation by phosphorylation. *J Biol Chem* 266:13385.

84. Smith, C.J., and V.C. Manganiello. 1989. Role of hormone-sensitive low K_m cAMP phosphodiesterase in regulation of cAMP-dependent protein kinase and lipolysis in rat adipocytes. *Mol Pharmacol* 35:381.

85. Holm, C., D. Langin, V. Manganiello, et al. 1997. Regulation of hormone-sensitive lipase activity in adipose tissue. *Methods Enzymol* 286:45.

86. Londos, C., R.C. Honnor, and G.S. Dhillon. 1985. cAMP-dependent protein kinase and lipolysis in rat adipocytes. III. Multiple modes of insulin regulation of lipolysis and regulation of insulin responses by adenylate cyclase regulators. *J Biol Chem* 260:15139.

87. Rahn, T., M. Ridderstråle, H. Tornqvist, et al. 1994. Essential role of phosphatidylinositol 3-kinase in insulin-induced activation and phosphorylation of the cGMP-inhibited cAMP phosphodiesterase in rat adipocytes. Studies using the selective inhibitor wortmannin. *FEBS Lett* 350:314.

88. Wijkander, J., T. Rahn Landström, V. Manganiello, et al. 1998. Insulin-induced phosphorylation and activation of phosphodiesterase 3B in rat adipocytes: Possible role for protein kinase B but not mitogen-activated protein kinase or p70 S6 kinase. *Endocrinology* 139:219.

89. Ahmad, F., L.N. Cong, L. Stenson Holst, et al. 2000. Cyclic nucleotide phosphodiesterase 3B is a downstream target of protein kinase B and may be involved in regulation of effects of protein kinase B on thymidine incorporation in FDCP2 cells. *J Immunol* 164 (9):4678.

90. Gettys, T.W., A.J. Vine, M.F. Simonds, et al. 1988. Activation of the particulate low K_m phosphodiesterase of adipocytes by addition of cAMP-dependent protein kinase. *J Biol Chem* 263:10359.

91. Rascón, A., E. Degerman, M. Taira, et al. 1994. Identification of the phosphorylation site in vitro for cAMP-dependent protein kinase on the rat adipocyte cGMP-inhibited cAMP phosphodiesterase. *J Biol Chem* 269:11962.

92. Rahn, T., L. Rönnstrand, M.-J. Leroy, et al. 1996. Identification of the site in the cGMP-inhibited phosphodiesterase phosphorylated in adipocytes in response to insulin and isoproterenol. *J Biol Chem* 271:11575.

93. Resjö, S., A. Oknianska, S. Zolnierowicz, et al. 1999. Phosphorylation and activation of phosphodiesterase type 3B (PDE3B) in adipocytes in response to serine/threonine phosphatase inhibitors: Deactivation of PDE3B in vitro by protein phosphatase type 2A. *Biochem J* 341:839.

94. Boden, G. 1997. Role of fatty acids in the pathogenesis of insulin resistance and NIDDM. *Diabetes* 46:3.

95. McGarry, J.D., and R.L. Dobbins. 1999. Fatty acids, lipotoxicity and insulin secretion. *Diabetologia* 42:128.

96. Reaven, G.M. 1995. The fourth musketeer—from Alexandre Dumas to Claude Bernard. *Diabetologia* 38:3.

97. Frayn, K.N. 1993. Insulin resistance and lipid metabolism. *Curr Opin Lipidol* 4:197.

98. Frayn, K.N., C.M. Williams, and P. Arner. 1996. Are increased plasma non-esterified fatty acid concentrations a risk marker for coronary heart disease and other chronic diseases? *Clin Sci (Lond)* 90:243.

99. Engfeldt, P., P. Arner, J. Bolinder, et al. 1982. Phosphodiesterase activity in human subcutaneous adipose tissue in insulin- and noninsulin-dependent diabetes mellitus. *J Clin Endocrinol Metab* 55:983.

100. Nagaoka, T., T. Shirakawa, T.W. Balon, et al. 1998. Cyclic nucleotide phosphodiesterase 3 expression in vivo: Evidence for tissue-specific expression of phosphodiesterase 3A or 3B mRNA and activity in the aorta and adipose tissue of atherosclerosis-prone insulin-resistant rats. *Diabetes* 47:1135.

101. Tang, Y., H. Osawa, H. Onuma, et al. 1999. Improvement in insulin resistance and the restoration of reduced phosphodiesterase 3B gene expression by pioglitazone in adipose tissue of obese diabetic KKAy mice. *Diabetes* 48:1830.

102. Tang, Y., H. Osawa, H. Onuma, et al. 2001. Adipocyte-specific reduction of phosphodiesterase 3B gene expression and its restoration by JTT-501 in the obese, diabetic KKAy mouse. *Eur J Endocrinol* 145:93.

103. Tang, Y., H. Osawa, H. Onuma, et al. 2001. Phosphodiesterase 3B gene expression is enhanced in the liver but reduced in the adipose tissue of obese insulin resistant db/db mouse. *Diabetes Res Clin Pract* 54:145.

104. Rahn Landström, T., J. Mei, M. Karlsson, et al. 2000. Down-regulation of cyclic-nucleotide phosphodiesterase 3B in 3T3-L1 adipocytes induced by tumour necrosis factor alpha and cAMP. *Biochem J* 346 (Pt 2):337.

105. Mei, J., L. Stenson Holst, T. Rahn Landström, et al. 2002. C(2)-ceramide influences the expression and insulin-mediated regulation of cyclic nucleotide phosphodiesterase 3B and lipolysis in 3T3-L1 adipocytes. *Diabetes* 51:631.

106. Zhang, H.H., M. Halbleib, F. Ahmad, et al. 2002. Tumor necrosis factor-alpha stimulates lipolysis in differentiated human adipocytes through activation of extracellular signal-related kinase and elevation of intracellular cAMP. *Diabetes* 51:2929.

107. Ryden, M., A. Dicker, V. van Harmelen, et al. 2002. Mapping of early signaling events in tumor necrosis factor-alpha-mediated lipolysis in human fat cells. *J Biol Chem* 277:1085.

108. Degerman, E., V. Manganiello, J.J. Holst, et al. 2004. Milrinone efficiently potentiates insulin secretion induced by orally but not intravenously administered glucose in C57BL6J mice. *Eur J Pharmacol* 498:319.

109. Cheung, P., G. Yang, and G. Boden. 2003. Milrinone, a selective phosphodiesterase 3 inhibitor, stimulates lipolysis, endogenous glucose production, and insulin secretion. *Metabolism* 52:1496.

110. Cases, J.A., I. Gabriely, X.H. Ma, et al. 2001. Physiological increase in plasma leptin markedly inhibits insulin secretion in vivo. *Diabetes* 50:348.
111. Cheung, P., G. Yang, and G. Boden. 2001. Milrinone, a selective phosphodiesterase 3 inhibitor, stimulates lipolysis, endogenous glucose production, and insulin secretion. *Metabolism* 52:1496.
112. El-Metwally, M., R. Shafiee-Nick, N.J. Pyne, et al. 1997. The effect of selective phosphodiesterase inhibitors on plasma insulin concentrations and insulin secretion in vitro in the rat. *Eur J Pharmacol* 324:227.
113. Nakaya, Y., A. Minami, S. Sakamoto, et al. 1999. Cilostazol, a phosphodiesterase inhibitor, improves insulin sensitivity in the Otsuka Long-Evans Tokushima Fatty Rat, a model of spontaneous NIDDM. *Diabetes Obes Metab* 1:37.
114. Parker, J.C., M.A. VanVolkenburg, N.A. Nardone, et al. 1997. Modulation of insulin secretion and glycemia by selective inhibition of cyclic AMP phosphodiesterase III. *Biochem Biophys Res Commun* 236:665.
115. Shafiee-Nick, R., N.J. Pyne, and B.L. Furman. 1995. Effects of type-selective phosphodiesterase inhibitors on glucose-induced insulin secretion and islet phosphodiesterase activity. *Br J Pharmacol* 115:1486.
116. Yang, G., and L. Li. 2003. In vivo effects of phosphodiesterase III inhibitors on glucose metabolism and insulin sensitivity. *J Chin Med Assoc* 66:210.
117. Osawa, H., T. Niiya, H. Onuma, et al. 2003. Systematic search for single nucleotide polymorphisms in the 5′ flanking region of the human phosphodiesterase 3B gene: Absence of evidence for major effects of identified polymorphisms on susceptibility to Japanese type 2 diabetes. *Mol Genet Metab* 79:43.
118. Yechoor, V.K., M.-E. Patti, K. Ueki, et al. 2004. Distinct pathways of insulin-regulated versus diabetes-regulated gene expression: An in vivo analysis in MIRKO mice. *Proc Natl Acad Sci USA* 101:16525.

6 Cellular Functions of PDE4 Enzymes

Graeme B. Bolger, Marco Conti, and Miles D. Houslay

CONTENTS

6.1   Introduction 100
6.2   Transcriptional Regulation of the *PDE4* Genes as a Feedback
      Mechanism in Signaling 101
6.3   Phosphorylation of PDE4 Enzymes by PKA and ERK 102
      6.3.1   Phosphorylation of PDE4s by PKA 102
      6.3.2   Phosphorylation of PDE4 Isoforms by ERK Kinases.....................103
6.4   Dimerization, Multimerization, and Intramolecular Domain Interactions.............104
      6.4.1   Interactions between UCR1 and UCR2 104
      6.4.2   Dimerization of PDE4 Enzymes and Its Regulatory and
              Pharmacological Effects 105
      6.4.3   Dimerization, Multimerization, Phosphorylation, and the
              High-Affinity Rolipram-Binding Site..105
6.5   Interaction of PDE4A5 with the Immunophilin XAP2 106
      6.5.1   PDE4A5 Interacts Specifically with XAP2 106
      6.5.2   Functional and Pharmacological Implications of the XAP2–PDE4A5
              Interaction...107
6.6   Interaction of PDE4 Isoforms with Members of the SRC Family of Protein
      Tyrosine Kinases...108
      6.6.1   PDE4 Proteins Interact with SRC, FYN, and LYN 109
      6.6.2   Functional and Pharmacological Implications of the PDE4-SH3
              Domain Interaction..109
6.7   Role of the PDE4A1 Amino-Terminal Region in Membrane Association.............110
      6.7.1   Functional Implications of PDE4A1 Membrane Localization 110
6.8   Interaction of PDE4 Isoforms with Myomegalin...110
      6.8.1   PDE4D Isoforms Interact with Myomegalin.....................................110
      6.8.2   Implications of the PDE4D–Myomegalin Interaction 111
6.9   Interaction of PDE4 Isoforms with the Disrupted in Schizophrenia (DISC1)
      Protein...112
      6.9.1   The DISC1 Protein Is a Multifunctional Scaffold Protein.....................112
      6.9.2   PDE4B Encodes Isoforms That Interact with DISC and May Itself Be
              a Schizophrenia Susceptibility Gene 112
      6.9.3   Functional Implications of the Association between PDE4
              Isoforms and DISC1 113
6.10 The Role of PDE4 Isoforms in Regulating Signaling through the β-Adrenergic
      Receptor...113
      6.10.1 Interaction of PDE4 Isoforms with β-Arrestins.........................113

      6.10.2 Role of the PDE4D5–β-Arrestin Interaction on Phosphorylation
              of the β₂AR...114
      6.10.3 Role of the PDE4D5 Amino-Terminal Region in Its Interaction
              with β-Arrestin..116
      6.10.4 Role of RACK1 in the PDE4D5–β-Arrestin Interaction 116
      6.10.5 Functional Implications of the PDE4D5–β-Arrestin-β₂AR Interaction.....117
6.11   Associations between PDE4 Isoforms and AKAPs...118
      6.11.1 Interactions between PDE4D3 and mAKAP..118
      6.11.2 ERK5 Associates with PDE4D3 119
      6.11.3 Interactions between PDE4D3, mAKAP, and Epac1 119
      6.11.4 Functional Consequences of the ERK5–PDE4D3-Epac1-mAKAP
              Interaction..119
      6.11.5 Interaction of PDE Isoforms with AKAPs in the Centrosome..................119
      6.11.6 Summary—AKAPs and PDE4 Function..120
6.12   Conclusion and Pharmacological Implications 120
References 121

6.1 INTRODUCTION

The phosphodiesterase 4 (PDE4) enzymes, also termed cAMP-specific phosphodiesterases or high-affinity cAMP PDEs, specifically hydrolyze cAMP with high affinity and are the targets for a class of drugs with antidepressant, memory-enhancing, anti-inflammatory, and immunomodulatory functions. Recent developments in the biochemistry and biology of the PDE4 enzymes have provided new insights into their roles in cells and have clarified much older work on the actions of PDE4-specific inhibitors. This chapter will discuss recent developments in the regulation of PDE4s in cells, with special emphasis on their phosphorylation and their interactions with other proteins. Other chapters in this volume will discuss the atomic structure of the PDE4s, their roles in intact organisms, and recent developments in PDE4 inhibitor discovery. Additional aspects of PDE4 biology have been discussed in other comprehensive reviews.[1–4]

The PDE4s can be distinguished from other classes of PDEs, including the PDE7 and PDE8 classes, which are also specific for cAMP, in the structure of the catalytic region of the proteins and also in their ability to be inhibited by specific compounds, of which the prototype is rolipram. Rolipram and other PDE4-specific inhibitors act at the catalytic site of the PDE4 enzymes and thereby act as competitive inhibitors of enzyme function. The PDE4 enzymes are also distinguished from the other PDE classes by the presence of two regions of amino acid sequence, first identified by Bolger and co-workers, called upstream conserved sequences (UCRs) 1 and 2 (UCR1 and UCR2; in Ref. [5]), which are located in the amino-terminal half of the protein, distinct from the catalytic region (Figure 6.1). UCR1 and UCR2 are conserved in all vertebrate PDE4 homologues and also in the PDE4 homologues of several model organisms, including the *dunce* gene of *Drosophila melanogaster*[6] and the PDE4 homologues of *Aplysia kurodai*[7] and *Caenorhabditis elegans*. UCR1 and UCR2, like other amino-terminal regions of the PDE4 proteins, regulate various functions of the PDE4 enzymes, as described further in the chapter.

Extensive molecular cloning studies have identified at least 20 different PDE4 isoforms (see separate chapter for a detailed list of all mammalian PDE isoforms). Mammalian PDE4s are encoded by four genes (*PDE4A*, *PDE4B*, *PDE4C*, and *PDE4D* in humans), with additional diversity generated by the use of alternative mRNA splicing and the use of different promoters. The various isoforms encoded by each gene can be divided into the long forms, which contain

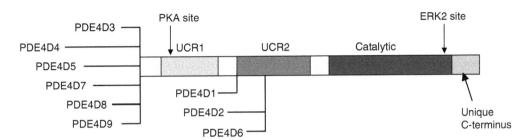

FIGURE 6.1 (See color insert following page 274.) Schematic representation of the human PDE4D isoforms. (Modified from Refs. [18, 116, 139]) The *PDE4D* gene is one of four human genes (*PDE4A*, *PDE4B*, *PDE4C* and *PDE4D*) encoding PDE4 isoforms. PDE4D encodes nine different proteins (PDE4D1 through PDE4D9), through the use of alternative mRNA splicing and by initiation of transcription from promoters specific to each isoform. The heavy bar in the figure shows regions of sequence common to all, or most, PDE4D proteins and which are also homologous to sequences in isoforms encoded by all four human *PDE4* genes and by the dunce gene of *Drosophila melanogaster*. These homologous regions include the catalytic regions of the proteins and upstream conserved regions 1 and 2 (UCR1 and UCR2; from Ref. [5]). The unique C-terminus is seen in all nine PDE4D isoforms but not in those encoded by other *PDE4* genes. The thinner lines projecting from the heavy bar indicate regions of sequence unique to an individual isoform. Although this figure shows only the PDE4D isoforms, those encoded by the other three *PDE4* genes have a similar overall organization.[1]

both UCR1 and UCR2, the short forms, which contain only UCR2, and the super-short forms, which contain only half of UCR2 (Figure 6.1). The various isoforms encoded by a single gene can differ in their enzymatic properties and their distribution in different tissues. For example, five different isoforms, all encoded by the *PDE4A* gene, have widely different patterns of tissue expression: PDE4A1, which is expressed exclusively in brain, particularly the cerebellum,[8] PDE4A4 or PDE4A5, PDE4A10, and PDE4A11, which are expressed widely in tissues,[9–12] and PDE4A7 or PDE4A8, which is expressed almost exclusively in testis.[13,14] Similarly, each of the nine PDE4D isoforms has a specific pattern of tissue expression.[9,15–19] The various PDE4 isoforms encoded by individual genes may also differ in their intracellular targeting, interactions with other proteins, and regulation by phosphorylation, as described in detail later in this chapter.

6.2 TRANSCRIPTIONAL REGULATION OF THE *PDE4* GENES AS A FEEDBACK MECHANISM IN SIGNALING

Given the complex structure of the *PDE4* genes, which is a feature common to all mammalian *PDE4* genes as well as orthologs in model organisms such as *Drosophila*, it is likely that multiple mechanism of regulation of *PDE4* gene transcription are operating in most cells. In addition to cell- and tissue-specific expression, the regulation of PDE4D and PDE4B mRNAs by cAMP was explored soon after the cloning of the rodent PDE4D and PDE4B cDNAs. PDE4D1 or PDE4D2 (formerly rPDE3) and PDE4B2 (formerly rPDE4) are induced upon stimulation of endocrine cells with ligands that activate GPCRs.[20] The large induction of these mRNAs (up to 100-fold) is caused by a cAMP-dependent increase in transcription of the *PDE4D* and *PDE4B* genes.[21] Although other promoters have been implicated in the control of PDE4D transcripts (see last paragraph of this section), this large increase in cAMP-dependent PDE4 mRNA transcription depends

primarily on the activation of internal promoters located downstream of the exons coding for UCRs and upstream of the first exons included in the short PDE4D1 and PDE4B2 forms.[22,23] This region of these genes has been further characterized by deletion mutagenesis, and putative cAMP regulatory elements (CREs) have been identified.[22,23] Further evidence for the involvement of a CRE and CREB in this regulation is the finding that dominant negative CREB constructs abolish cAMP transcriptional regulation in cortical neurons.[22] Thus, an increase in cAMP in many cells causes a large activation of transcription of PDE4B and PDE4D short forms. Similar regulation has been observed in endometrium,[24] vascular smooth muscle,[25] Jurkat cells,[26,27] granulosa cells,[28] and thyroid cells,[29] suggesting a ubiquitous role for this regulation.

Factors other than cAMP levels have been implicated in the transcriptional regulation of the PDE4 short forms. Initial studies were performed in human monocytes and the human monocytic cell line Mono-MAC6. Exposure of these inflammatory cells to lipopolysaccharide (LPS) produced an increase in steady-state mRNA levels and increased PDE4 activity, particularly of PDE4B2.[30] More detailed studies clearly demonstrated that PDE4B2 is the predominant phosphodiesterase isoform expressed in monocytes and that LPS induces transcription of this splice variant,[31] whereas this isoform is constitutively expressed in neutrophils.[32] Similar LPS stimulation of PDE4B2 has been demonstrated in mouse PBMCs and macrophages.[33] Although the mechanisms of LPS-induced PDE4B2 transcription have not been elucidated, several canonical Nf–kB sites that are present in these intronic promoters in proximity to the CREs are likely to play an important role in this regulation. The toll-like receptors (TLRs), which are the cell–surface receptors for LPS, are also likely to play an important role.

Cyclic AMP regulation of PDE4 transcription is not restricted to the internal promoter regulating the short PDE4D1 and PDE4B2 isoforms. The PDE4D5 promoter is also sensitive to cAMP, and a 2–3-fold increase in PDE4D5 mRNA transcription rates has been observed in human airway smooth muscle cells upon stimulation with either forskolin or isoproterenol.[34] This regulation of the PDE4D5 promoter is again dependent on one or more CRE elements, as their ablation obliterates the effect of cAMP.[34] Exposure of Jurkat cells to cAMP analogs also up-regulates PDE4A4 and PDE4D3 mRNAs, suggesting that other PDE4 promoters may be sensitive to cAMP regulation, although the transcription rate was not investigated in these studies. It is therefore most likely that myriad transcriptional regulators impinge on the more than 20 promoters that regulate PDE4 transcripts. Although largely unexplored, these regulatory mechanisms may play an important role in cell adaptation under physiological and pathological conditions.

6.3 PHOSPHORYLATION OF PDE4 ENZYMES BY PKA AND ERK

6.3.1 PHOSPHORYLATION OF PDE4S BY PKA

Early biochemical studies demonstrated that elevation of cAMP levels produces phosphorylation and activation of cAMP-hydrolyzing PDEs in hepatocytes and other cells (in Ref. [35]; see Ref. [1] for details of early research). Subsequently, Conti and colleagues demonstrated that PKA can phosphorylate and activate the long PDE4D isoform, PDE4D3, and that phosphorylation occurs at two typical PKA consensus phosphorylation sites: S13, located within its unique amino-terminal region, and at S54, located at the very beginning of UCR1 (Figure 6.1, Refs. [36–40]). This second PKA site is conserved in the long PDE4 isoforms encoded by all four mammalian *PDE4* genes [5,41] (see Ref. [1] for an alignment). Conversely,

the short PDE4 isoforms, which do not contain these phosphorylation sites, are not activated by PKA.[36,38]

PKA phosphorylation of PDE4D3 has a number of enzymatic and pharmacological consequences. Phosphorylation at S54 clearly activates PDE4D3, increasing its relative V_{max} by approximately 6–8-fold and also increasing its ability to be inhibited by rolipram.[36] Phosphorylation and changes in PDE4D3 enzymatic activity could be produced by activation of PKA by TSH in the FRTL-5 cell line and also by prostaglandin E2 treatment of U937 cells.[38,39,42] Mutation of the S54 in PDE4D3 to alanine (S54A) blocked the ability of the enzyme to be phosphorylated by PKA and abolished the effects of PKA on enzyme activity and rolipram sensitivity.[40]

Houslay and colleagues have studied the effect of various mutations within the PKA consensus site of PDE4D3.[43] They show that the S54A mutation has no direct effect on enzyme activity, but causes sufficient change in the PDE4D3 catalytic region to change the sensitivity of the enzyme to inhibition by rolipram. The S54D mutation mimicked both the effect of PKA on increasing PDE4D3 activity and in enhancing its sensitivity to inhibition by rolipram.[43] In contrast, no change in either property was detected with the S54T mutation. In contrast, the E53A mutation mimics the activation of PDE4D3 by phosphorylation but does not change rolipram sensitivity.[43] These changes demonstrate that relatively small changes in charge or configuration of the amino-terminal regions of the long PDE4 forms can produce detectable changes in the conformation of the catalytic region of the protein.

Other long PDE4 isoforms also undergo PKA phosphorylation, both *in vitro* and upon elevation of cAMP levels in cells. In all these isoforms, phosphorylation occurs at the serine that corresponds to S54 in UCR1 of PDE4D3. PKA phosphorylation increases the activity of the long isoforms PDE4A8, PDE4B1, PDE4C2, PDE4D5, PDE4D7, PDE4D8, and PDE4D9 by approximately 50% but, in contrast to the case for PDE4D3, does not appear to change their ability to be inhibited by rolipram.[18,19,44] The exact physiological consequences of these changes are now becoming clearer, as discussed in subsequent sections.

6.3.2   PHOSPHORYLATION OF PDE4 ISOFORMS BY ERK KINASES

PDE4B, PDE4C, and PDE4D isoforms have a typical MAPK kinase consensus phosphorylation site (the first S in the sequence PQSPSP) within the catalytic region of the enzyme (S579 in PDE4D3; see Ref. [1] for complete alignments). Houslay and colleagues have determined the regulation of ERK2 phosphorylation of PDE4D3 and some of its biochemical consequences.[45] Phosphorylation of PDE4D3 profoundly reduces the activity of the enzyme (greater than 75% reduction in activity). Phosphorylation and activity change were abolished upon mutation of S579 to alanine (S579A), and mutation of S579 to aspartic acid (S579D) mimicked the change in activity produced by ERK2 phosphorylation.[45,46] Treatment of COS1 cells transfected to express PDE4D3 with epidermal growth factor (EGF) produced phosphorylation and inhibition of PDE4D3 and elevation of cAMP levels, all of which could be blocked with the MEK inhibitor PD98059. Similarly, treatment of HEK293 cells and F442A cells with EGF also produced changes in the activity of PDE4D3 and PDE4D5, which were also blocked with PD98059. Although the S579A mutant blocked EGF-induced inhibition of PDE4D3, it did not affect the ability of PDE4D3 to be phosphorylated by PKA. As EGF-mediated inhibition of PDE4D3 was transient, it is possible that PKA-induced activation of PDE4D3 would counterbalance the effects of ERK2 phosphorylation.[45]

MAPK kinases bind to their substrates via two docking sites, of the sequence KIM and FQF. Both sites are present in the PDE4 catalytic region, are conserved among all PDE4 isoforms (having the sequences **VETKKVTSSGVLLL** and **FQF**), and straddle the ERK2 phosphorylation site.[1] Structural analysis of the PDE4 catalytic region indicates that these

two sites are located on the surface of the protein, where they are available for interaction with other proteins (in Refs. [47,48]; see also the relevant chapters in this volume). Mutation of either or both of these two sites to alanine raises the K_m of ERK2 for PDE4D3, thereby attenuating PDE4D3 phosphorylation *in vitro*, and blocks phosphorylation of PDE4D3 by ERK2 in intact cells.[49] It also blocks the co-immunoprecipitation of PDE4D3 and ERK2 from cells.[49] These findings expand and confirm the concept that ERK2 phosphorylation of PDE isoforms occurs in cells in a physiological fashion. The exact physiological consequences of these events, particularly in intact organisms, require further investigation. However, it has been shown that ERK regulation of PDE4 *in vivo* affects learning and memory in rats[50] and may also play a role in NGF-mediated neurite growth.[51]

6.4 DIMERIZATION, MULTIMERIZATION, AND INTRAMOLECULAR DOMAIN INTERACTIONS

Key to understanding many of the pharmacological and regulatory aspects of PDE4 function are recent insights into how the various domains of the PDE4 proteins can interact in intramolecular and intermolecular interactions. In this section, we will first describe domain–domain interactions and then discuss mechanisms for interactions between different PDE4 polypeptides.

6.4.1 INTERACTIONS BETWEEN UCR1 AND UCR2

The identification of UCR1 and UCR2 as intrinsic components of PDE4 proteins[5] posed a number of interesting questions about their role in PDE4 function. UCR1 and UCR2 are located in the amino-terminal half of PDE4 proteins and are clearly separate from the catalytic regions (Figure 6.1). Studies of PDE4 deletion constructs have shown that UCR1 and UCR1 play no direct role in catalysis (reviewed in Ref. [1]). However, their strong evolutionary conservation suggested that they play an important role in PDE4 biology, as regions of amino acid sequence that are highly conserved in evolution are usually of functional significance.

The groups of Bolger, Conti and Houslay have independently studied the interaction of UCR1 and UCR2, using a variety of methods, including two-hybrid assays and pull-down experiments with these regions expressed as fusions.[52,53] They have demonstrated that both UCR1 and UCR2 act as authentic protein domains, i.e., each can fold into independent modules that appear to retain their structure in a variety of contexts. This behavior of UCR1 and UCR2 is reminiscent of many other protein domains, such as SH2 and SH3 domains, that undergo self-folding and are able to participate in protein–protein interactions.[54,55]

These studies also showed that UCR1 and UCR2 were capable of interacting with each other and that this interaction requires electrostatic interactions between specifically charged amino acids in both UCR1 and UCR2.[53] The interaction also appears to be dependent on the phosphorylation status of the PKA site that is located within UCR1 (Figure 6.1). Phosphorylation of UCR1 with PKA attenuated its ability to interact with UCR2.[53] Mutation of the phosphorylation site to aspartic acid (S54D) or alanine (S54A) also affects the interaction, although the exact direction of this effect remains uncertain.[52,53,56] Removal of UCR1, treatment with antibodies directed against UCR1, or treatment with the acidic phospholipids, phosphatidic acid or phosphatidylcholine produced an increase in the activity of the enzyme that was similar to, and not additive with, PKA activation.[52,57] These data show that a PKA-induced interaction between UCR1 and UCR2 leads to an alteration in the catalytic region that changes the activity of enzyme and also changes its sensitivity to rolipram.

UCR1 and UCR2 also modulate the effects of ERK phosphorylation on PDE4 enzyme activity. UCR1 and UCR2 seem to amplify the effects of ERK2 inhibition, as truncation of UCR1 and UCR2 partially attenuate the ability of ERK2 to inhibit enzymatic activity, as compared to the full-length long-form enzyme.[49,58] In contrast, the lone UCR2 that is present in the short PDE4B2 and PDE4D2 isoforms (Figure 6.1) increases the ERK-induced activation of this isoform.[49,58] Collectively, these data demonstrate that the *PDE4B*, *PDE4C*, and *PDE4D* genes encode a series of isoforms that are differentially inhibited or activated by PKA, ERK2, or a combination of both enzymes,[58] thereby providing a mechanism for the selective regulation of particular isoforms present in various tissues or intracellular compartments.

6.4.2   DIMERIZATION OF PDE4 ENZYMES AND ITS REGULATORY AND PHARMACOLOGICAL EFFECTS

It has also been shown that UCR1 and UCR2 mediate dimerization of PDE4 enzymes and that this has specific regulatory consequences. Using a variety of approaches, Conti and colleagues have shown that the long PDE4D3 isoform, which has both UCR1 and UCR2 (Figure 6.1), behaves as a dimer in cells, whereas the short PDE4D2 isoform, which lacks UCR1, is a monomer.[56] Internal deletions within either UCR1 or UCR2 abolish the dimerization of PDE4D3, demonstrating a role for these regions in dimerization. Intriguingly, the precise amino acids within UCR1 and UCR2 that mediate dimerization are different from those that mediate the intramolecular interaction between UCR1 and UCR2.[53,56]

Dimerization of PDE4D3 appears to alter its enzymatic properties and its ability to be inhibited by rolipram. Mutants of PDE4D3 that block dimerization also abolish the ability of the enzyme to be activated by PKA or by phosphatidic acid.[59] They also reduce the sensitivity of the enzyme to inhibition by rolipram. Therefore, dimerization of the long PDE4 isoforms may be an additional function of UCR1 or UCR2 that provides an additional explanation for the differential regulation of the long and short PDE4 isoforms in cells.[56] The domain interactions described in these studies are reminiscent of those that occur in other PDE isoforms, such as PDE5 and PDE6, where dimerization is mediated by GAF domains located at the amino terminus of these isoforms and where oligomerization is required for the regulation of catalytic activity (see relevant chapters by Francis and Beavo in this volume).

6.4.3   DIMERIZATION, MULTIMERIZATION, PHOSPHORYLATION, AND THE HIGH-AFFINITY ROLIPRAM-BINDING SITE

One important aspect of recent studies on the dimerization of the PDE4s is the new insight that it provides into a long-standing controversy in PDE4 pharmacology: the exact nature of what has been widely termed the high-affinity rolipram-binding site (HARBS). HARBS was originally defined as the ability of a subset of PDE4 inhibitors to bind at high affinity to sites in a variety of tissues, especially the brain (see Ref. [1] for a more detailed discussion). In contrast, in a number of tissues, binding occurred at lower affinities and this was attributed to low-affinity rolipram-binding sites (LARBS). However, compounds binding to HARBS would invariably inhibit enzyme activity over a broad range of concentrations, depending on the tissue source or the system used to express recombinant enzyme. Initially, it was speculated that HARBS could involve inhibitor binding at a site different from the catalytic region of the PDE4 proteins, possibly even on a separate polypeptide. However, Torphy and colleagues[60] then showed that recombinant PDE4 isoforms could provide HARBS when expressed in cells, demonstrating that HARBS was an intrinsic property of PDE4 proteins. Subsequent studies have demonstrated that virtually any long PDE4 isoform, when expressed as a recombinant protein in an appropriate system, can provide HARBS. Studies of deletion constructs strongly suggest that HARBS requires the presence of not only the catalytic region

of the protein, but also UCR2.[60] However, as an isolated UCR2 has no ability to bind to PDE4 inhibitors, it is unlikely that UCR2 forms a distinct binding site, but instead is required for the catalytic region of the enzyme to assume a conformation that binds PDE4 inhibitors with high affinity. This suggests that any long PDE4 isoform can create HARBS in the appropriate cellular context (see also Ref. [61]). Indeed, Souness and Rao have proposed the term high- and low-affinity PDE4 conformers as a preferred alternative to the terms LARBS and HARBS.

Clearly, dimerization is one of the ways to generate HARBS. Mutants of PDE4D3 that block dimerization reduce the sensitivity of the enzyme to inhibition by rolipram.[59] Binding assays suggest that dimerization is accompanied by an increase in HARBS, although at least some HARBS are seen even when dimerization does not occur. These data suggest that, although dimerization is not an absolute requirement for the generation of HARBS, it stabilizes the enzyme in a HARBS conformation.[59]

Dimerization or multimerization may also contribute to ability of PDE4 isoforms to aggregate when expressed at high levels in cells.[63] Treatment of cells overexpressing recombinant PDE4A4 or PDE4A5 with various PDE4 inhibitors causes PDE4A4 or PDE4A5 to relocalize to form accretion foci. This process seemed to be independent of PKA phosphorylation (as it was not affected by the PKA inhibitor H89), and was strongly dependent on the specific PDE4 inhibitor tested. Deletion of a portion of the unique amino-terminal region of PDE4A4 or PDE4A5 altered foci formation, indicating that this region was essential for this process. Although the exact mechanisms for the generation of these foci, and any potential physiological relevance, remains to be determined, these studies demonstrate an intriguing effect of pharmacological PDE4 inhibitors on PDE4s in cells and may provide a novel assay for subcategorizing PDE4 inhibitors.

6.5   INTERACTION OF PDE4A5 WITH THE IMMUNOPHILIN XAP2

The interaction of PDE4 proteins with molecular chaperones and other proteins that modulate protein folding in cells, such as immunophilins, has the potential to influence diverse aspects of PDE4 function. Modification of the conformation of PDE4 enzymes in particular has the potential to influence domain interactions or dimerization or multimerization, and thereby influence any of the phenomena described in Section 6.4. It also has the potential to influence the regulation of PDE4s by phosphorylation or by other protein–protein interactions. The ubiquity of these proteins in cells suggests that they have the potential to act on most PDE4 isoforms. To date, however, most of the observations have focused on the long isoform, PDE4A4 or PDE4A5 (i.e., human PDE4A4 and rat PDE4A5, which are strongly homologous, even in their unique amino-terminal regions).

6.5.1   PDE4A5 INTERACTS SPECIFICALLY WITH XAP2

Bolger and colleagues identified immunophilin XAP2 as an interacting partner of PDE4A5 in a two-hybrid screen.[64] XAP2 (previously also known as ARA9 or AIP) is a member of the immunophilin protein family, whose members are all characterized by the presence of a region of amino acid sequence called the immunophilin domain. In many immunophilins, this domain has *cis–trans* peptidyl-prolyl isomerase activity (i.e., is a PPIase) and is a target for immunosuppressive drugs such as cyclosporine, tacrolimus (FK506), or rapamycin.[65] However, a number of members of the immunophilin family, including XAP2, do not appear to have either of these functions. However, it is likely that all immunophilins play an important role in protein folding. In addition to its immunophilin domain, XAP2 has at least one tetratricopeptide repeat (TPR) domain, which is located in the carboxyl-terminal

half of the protein.[66-68] TPR domains are found in a large number of proteins and mediate protein–protein interactions.[69,70]

XAP2 has a number of cellular functions and was first identified as a protein that interacts with the X protein of hepatitis B virus.[71] It has been studied extensively in its role as a member of a multiprotein complex containing the aryl hydrocarbon (dioxin) receptor, a member of the steroid-hormone nuclear receptor superfamily.[66-68,72,73] XAP2 is also part of a similar complex associated with the peroxisome proliferator-activated receptor α (PPARα; Ref. [74]). XAP2 appears to be essential for the nuclear trafficking of these receptors and their transcriptional activation activity.[73] Other members of both of these complexes include Hsp90 and several other proteins with chaperone-like function. The TPRs of XAP2 mediate its interaction with Hsp90, as they bind to a specific motif, of amino acid sequence EEVD, located at the extreme-carboxyl-terminal end of Hsp90.[73] The XAP2 TPRs are similar in amino acid sequence to those of other proteins, such as PP5A, Hop1, Cyp40, and FKBP52, that also interact with Hsp90 and have been subjected to x-ray crystallographic study.[75-79] XAP2 therefore appears to be a protein that interacts with a variety of proteins, and in turn can modify their structure or localization in cells.

Two-hybrid and pull-down studies showed that PDE4A5 interacted specifically with XAP2 and not related immunophilins, and that XAP2 in turn interacted only with PDE4A5 and not other PDE4 isoforms. The unique amino-terminal region of PDE4A5 was necessary for it to interact with XAP2, thereby explaining its specificity of interaction with XAP2. However, a second region of PDE4A5, located within UCR2 and having the amino acid sequence EELD, was also essential for the interaction. This sequence was similar to the motif in the carboxyl-terminal half of Hsp90 that interacts with the XAP2 TPRs, suggesting that the EELD motif of PDE4A5 also interacts with the XAP2 TPRs. Consistent with this prediction, mutations of single amino acids within the XAP2 TPR blocked its interaction with PDE4A5, and mutations of amino acids within the EELD motif of PDE4A5 blocked its interaction with XAP2.

The interaction of PDE4A5 and XAP2 has three biochemical effects.[64] The first is that XAP2 produces a marked inhibition of the enzymatic activity of PDE4A5. This action of XAP2 is dose-dependent but is maximally saturable at about 60% of activity. This suggests that the effect of XAP2 is noncompetitive and is regulated by regions of PDE4A5 outside of the catalytic region, which is consistent with the mutagenesis and binding studies described above. It is also consistent with work described elsewhere in this review, that UCR1, UCR2, and the unique amino-terminal regions of PDE4 family members can regulate their enzymatic activity. A second effect of interaction is that XAP2 increases the sensitivity of the enzyme to inhibition by rolipram. This effect is again consistent with the idea that changes in the conformation of UCR1, UCR2, or the unique amino-terminal region of PDE4A5 produces changes in the conformation of the catalytic region, which in this case is detectable by changes in rolipram sensitivity. The third effect is that XAP2 attenuates the ability of PDE4A5 to be phosphorylated by PKA. The exact mechanism of this third effect is unclear. One possibility is that XAP2 blocks access to the PKA site, located within UCR1 (Figure 6.1), by virtue of its binding to UCR2 and the unique amino-terminal region. Alternatively, XAP2 could block the interaction of UCR2 with UCR1 (see the above section) and thereby induce a conformational change within UCR1 that affects its ability to be phosphorylated.

6.5.2 FUNCTIONAL AND PHARMACOLOGICAL IMPLICATIONS OF THE XAP2–PDE4A5 INTERACTION

One of the most noteworthy aspects of the PDE4A5–XAP2 interaction is that it directly changes several enzymatic and pharmacological characteristics of PDE4A5. This interaction demonstrates that protein–protein interactions may significantly change the conformation of

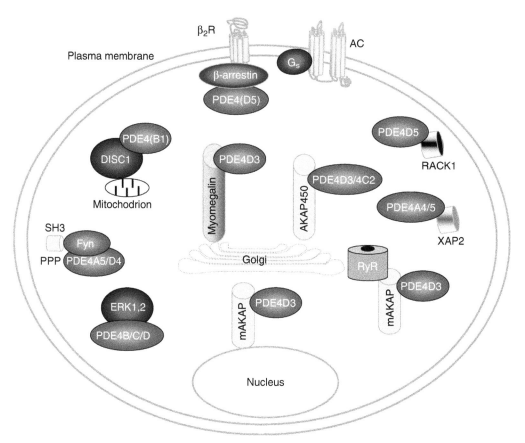

FIGURE 6.2 (See color insert following page 274.) PDE4 isoforms and their anchoring or scaffold protein partners. Various PDE4 isoforms are shown, with each of its known partner proteins and intracellular location.

PDE4 isoforms and thereby affect their properties as pharmacological targets within cells. For example, interactions with proteins that are likely to modulate the folding of PDE4 proteins, such as the immunophilins, may contribute to their dimerization, multimerization, or ability to act as a HARBS (see Section 6.4). Each protein partner is likely to have different effects on the pharmacological properties of a given PDE4 isoform, i.e., the same PDE4 isoform may be inhibited at significantly lower concentrations of a PDE4 inhibitor when it is associated with a given protein partner than with another. As the protein partners of any given PDE4 isoform will differ in different subcellular compartments, or in different tissues, such an isoform will be inhibited differently in each of these different cellular contexts. As different PDE4 inhibitors may preferentially interact with different PDE4 conformers (see Section 6.4), this effect is likely to be both isoform-specific and inhibitor-selective (Figure 6.2).

6.6   INTERACTION OF PDE4 ISOFORMS WITH MEMBERS OF THE SRC FAMILY OF PROTEIN TYROSINE KINASES

Houslay and colleagues have demonstrated that several PDE4 isoforms can interact with proteins containing SH3 domains, such as the SRC, FYN, and LYK protein tyrosine

kinases.[80–82] SH3 domains are discrete, self-folding globular modules first identified in the SRC protein tyrosine kinase and which are present in a wide variety of proteins.[54] They mediate interactions between proteins and therefore are required for the association of numerous proteins in a large variety of signaling complexes. SH3 domains interact with several different motifs, all of which share core proline-rich motifs, with specificity of these interactions determined by differences in the regions surrounding the SH3 domain and the proline-rich regions.

6.6.1 PDE4 Proteins Interact with SRC, FYN, and LYN

The SH3-domain-containing proteins SRC, FYN, and LYN interact with human PDE4A4 and its rat homologue, PDE4A5. These proteins have a highly conserved unique amino-terminal region that contains a series of Class 1 SH3 domains of sequence PxxPxxR or PxxRxR (at prolines 4, 37, and 61 in both rat and human; Refs. [5,9]). Removal of this unique amino-terminal region from rat PDE4A5 blocked its association with SH3-domain-containing proteins.[80] The interaction of human PDE4A4 with SH3 domains is partially attenuated by the removal of this region, but this interaction also requires a proline-rich region located in LR2 (i.e., between UCR2 and the catalytic region of the enzyme, Figure 6.1; Ref. [81]). The sequences of human PDE4A4 and rat PDE4A5 LR2 are quite divergent, with that of human PDE4A4 being more proline-rich,[5,9] which presumably explains the higher avidity of human PDE4A4 for SH3 domains. Indeed, the human PDE4A10 and PDE4A11 isoforms, which lack the PDE4A4 or PDE4A5 unique amino-terminal region, but have an identical LR2, are capable of interacting with LYN, although the avidity of this interaction is considerably less than that of PDE4A4.[12,83]

Pull-down studies and phage display analysis have shown that there is considerable specificity for the interaction of PDE4 isoforms with specific SH3 domains. These analyses have shown that rat PDE4A5 and human PDE4A4 both have a marked preference for SRC protein tyrosine kinases, in the order LYN > FYN > SRC.[80,84] In contrast, PDE4D4, which has a unique amino-terminal domain that also interacts with SH3 domains, has a broader range of interactions, with no selectivity between LYN and FYN and similar avidity for the SH3 domains of fodrin and ABL.[82]

6.6.2 Functional and Pharmacological Implications of the PDE4-SH3 Domain Interaction

A number of studies have shown that the interaction of PDE4A isoforms with SH3-domain-containing proteins does not affect the catalytic activity of the enzyme.[80–82] However, the interaction of PDE4A5 with LYN may affect its subcellular targeting. Deletion of the first 9 amino acids of PDE4A5 blocked its ability to interact with LYN and also changed its ability to target membranes, as determined by cell fractionation experiments, and also to produce foci at the cell margin on immunofluorescence studies.[84]

The interaction of human PDE4A4, but not rat PDE4D5, with LYN profoundly changes the ability of the enzyme to be inhibited by rolipram.[81] This effect is presumably mediated by the proline-rich sequences within the PDE4A4 LR2, which are located next to Helix 0 of the catalytic region of the enzyme (see other chapters in this volume on the structure of the PDE4 catalytic region). This effect differs between PDE4-selective inhibitors—i.e., it is seen with R-rolipram but not with its enantiomer, S-rolipram.[81] This effect is consistent with SH3-mediated switching of PDE4A4 from a LARBS to a HARBS and may contribute to the varying proportions of LARBS and HARBS in different cell types or tissues.

6.7   ROLE OF THE PDE4A1 AMINO-TERMINAL REGION IN MEMBRANE ASSOCIATION

PDE4A1 (formerly known as RD1) is a super-short PDE4 isoform, lacking UCR1 and having a truncated UCR2, which possesses a unique 25 amino acid amino-terminal region that is completely conserved in mammals.[85,86] PDE4A1 is unique in that it is entirely membrane associated and this is due to its isoform specific N-terminal region.[8] Removal of this region generates a fully soluble and active enzyme and, conversely, fusion of this region to a variety of normally soluble proteins renders them entirely membrane associated, with an intracellular distribution which parallels that of PDE4A1.[8,87,88]

Proton NMR study of the unique PDE4A1 amino-terminal region shows that it has two helical regions separated by a mobile hinge.[89] Study of constructs in which various regions of the amino-terminal region were fused to green fluorescent protein (GFP) or chloramphenicol acetyl transferase (CAT) showed that membrane association was determined through the action of helix 2 (sequence [14]PWLVGWWDQFKR[25]), also called TPAS1.[90] Helix 2 becomes membrane associated via penetration of the membrane bilayer.[90] The core insertion module is formed by the W19:W20 pairing, facilitated by L16:V17 pairing. For this module to function, Ca^{2+} is required to bind to D21, which provides a molecular switch facilitating insertion in less than 10 ms. This module also exhibits a preference for interacting with phosphatidic acid, as opposed to other phospholipids. It has been proposed that the basis for this selectivity is a propensity for one surface of helix 2 to achieve charge neutrality.[90] When Ca^{2+} binds to D21, phosphatidic acid is the only lipid to exhibit a charge of -2 at physiological pH value, which is required to effect overall neutrality in the presence of Ca^{2+}. Consistent with this model, the D21A mutation allows bilayer insertion to occur in a Ca^{2+}-independent fashion but destroys selectivity for phosphatidic acid.

6.7.1   FUNCTIONAL IMPLICATIONS OF PDE4A1 MEMBRANE LOCALIZATION

PDE4A1 is typically found associated with Golgi and vesicles that traffic out to the cell surface plasma membrane.[90] However, this localization is perturbed when selectivity for phosphatidic acid is destroyed. Preliminary studies also indicate that helix 1 (amino acids 1–12) plays a role in Golgi retention of this isoform. PDE4A1 is found exclusively in the brain[8,9,91] and undergoes upregulation in frontal and parietal cortex on chronic administration of antidepressants,[91] implying that it may have a role in mood disorders. Study of PDE4A mouse knockouts, which are in progress, should provide additional data on its functional significance.

6.8   INTERACTION OF PDE4 ISOFORMS WITH MYOMEGALIN

6.8.1   PDE4D ISOFORMS INTERACT WITH MYOMEGALIN

Conti and Colleagues identified myomegalin as a novel protein that interacts with PDE4D isoforms. As its name implies, myomegalin is a large (2324 amino acids) protein that is enriched in skeletal and cardiac muscles. Several smaller isoforms, produced by alternative mRNA splicing, expressed in heart and testis, as described in more detail later in the chapter. A related protein, called MMGL (myomegalin-like), has also been identified and, like myomegalin itself, is a large gene that encodes multiple isoforms via alternative mRNA splicing and is preferentially expressed in heart and skeletal muscles.[92]

Myomegalin is a complex protein with a number of interesting structural features, as predicted from its amino acid sequence. It contains several extensive regions of coiled-coil

α helix. There is a leucine zipper located very close to the amino terminus, which is homologous to that present in the centrosomin protein of *D. melanogaster*. There is also a region that is homologous to the centractin or ARP1-binding domain of dynactin, a microtubule-associated protein that is also involved in vesicular transport. It also has a separate region that has a loop–helix–loop structure that is strongly conserved in several other proteins. The shorter, testis-expressed form of myomegalin corresponds to the carboxyl-terminal one-third of the muscle-specific form and contains almost exclusively α-helical and coiled-coil domains.

Subcellular fractionation and immunoblotting studies demonstrate that the heart, skeletal muscle, and testis forms of myomegalin are soluble only under conditions that yielded myofibril components and also are strongly associated with particulate components. Immunofluorescence studies on COS7 cells demonstrated that myomegalin localizes to an inner Golgi or centrosome region. Immunofluorescence studies on COS7 cells transfected with the FLAG-tagged testis isoform of myomegalin also showed that it localized to this region. Immunofluorescence studies of testis showed myomegalin immunoreactivity in the Golgi complex of seminiferous tubules. In contrast, studies of skeletal and cardiac muscle cells showed that myomegalin is present in a periodic pattern along muscle fibrils, corresponding to the z-band of skeletal muscle. Immunofluorescence for PDE4D isoforms demonstrated a pattern that colocalized with myomegalin in these cells.

Studies of various PDE4D isoforms, and various truncation constructs, in two-hybrid and pull-down studies, demonstrated that the PDE4D UCR2 was necessary for their interaction with myomegalin. Consistent with this analysis, PDE4D3 and PDE4D2 could be co-immunoprecipitated with myomegalin in extracts from COS7 cells transfected with constructs encoding these PDE4 isoforms and the testis isoform of myomegalin. In contrast, the PDE4D1 isoform, which lacks UCR2, was incapable of interacting with myomegalin in these assays.

6.8.2 IMPLICATIONS OF THE PDE4D–MYOMEGALIN INTERACTION

Myomegalin is a very large protein with multiple potential functional domains. It also exists in multiple isoforms, each with a characteristic pattern of tissue expression and subcellular localization. Each of these isoforms could have distinct functions in cells. In addition, a close homologue of myomegalin, MMGL, has been identified, which, like myomegalin, is preferentially expressed in muscle cells but, unlike myomegalin, is both cytoplasmic and nuclear in location.[92] Also, myomegalin has structural features that are homologous to those seen in centrosomins, which are components of the centrosome and mitotic spindle. It is also weakly related to members of the golgin family, which are proteins that are components of the Golgi apparatus. Like the golgins, myomegalin can localize to the Golgi or centrosome.

Another aspect of myomegalin biology has emerged from the study of chromosomal translocations associated with various myeloproliferative disorders. Myomegalin has been identified as a portion of a fusion protein generated by a specific chromosomal translocation present in these disorders, t(1;5)(q23;q33). Note that myomegalin maps to chromosome 1, region 1q23, whereas MMGL maps to chromosome 1, region 1q1.[92,93] This translocation generated a fusion between the first 905 amino acids of the testis form of myomegalin and the transmembrane and tyrosine kinase domains of the platelet-derived growth factor receptor β (PDGFRB; Ref. [93]). Myomegalin may function in this fusion as an activator of PDGFRB, and the coiled-coil domains of myomegalin may mediate this activation. This study therefore provides additional insight into the potential regulatory functions of myomegalin.

It is reasonable to predict that myomegalin could associate with specific PDE4D isoforms and target them to specific subcellular structures. Given the diverse subcellular localization of the various myomegalin isoforms, and the tissue-specific expression of these isoforms, these

subcellular structures could vary between tissues. In turn, these subcellular structures may contain additional proteins that regulate PDE4D function or act as downstream targets of cAMP action. For example, it may bring PDE4D isoforms closer to PKA subunits that are anchored by AKAPs specific to a particular tissue. It is likely that these regulatory elements will be both tissue specific and cell-compartment specific.

6.9   INTERACTION OF PDE4 ISOFORMS WITH THE DISRUPTED IN SCHIZOPHRENIA (DISC1) PROTEIN

6.9.1   THE DISC1 PROTEIN IS A MULTIFUNCTIONAL SCAFFOLD PROTEIN

Numerous lines of evidence have shown that genetic factors play an important role in the predisposition of humans to schizophrenia, a devastating and frequently occurring disorder of mood and cognitive functioning. For example, twin studies have demonstrated that genetic factors and early childhood environment are strong predisposing factors.[94] The *DISC1* gene was first identified at the breakpoint of a chromosomal translocation [t(1;11)(q42.1;q14.3)] that segregated with schizophrenia and related disorders in a family.[95] Extensive genetic and neuropsychiatric studies have demonstrated that mutations in *DISC1* clearly predispose humans to schizophrenia, rather than merely serving as markers for the disease (see Ref. [96] for an extensive review). Parallel genetic studies have also identified a number of other genes that, when mutated, may predispose individuals to schizophrenia and which have become fertile subjects for functional analysis.[96] *DISC1* encodes an 854-amino acid protein that is selectively expressed in brain and that has been shown to interact with a number of other brain proteins, such as FEZ1 and NUDEL.[97–99]

6.9.2   PDE4B ENCODES ISOFORMS THAT INTERACT WITH DISC AND MAY ITSELF BE A SCHIZOPHRENIA SUSCEPTIBILITY GENE

A PDE4B protein was identified as an interacting partner of DISC1 in a two-hybrid screen.[100] The interaction between DISC1 and PDE4 isoforms was confirmed by pull-down, co-immunoprecipitation, and colocalization studies.[100] Analysis of PDE4B and DISC1 truncations indicated that a region in the amino-terminal region of DISC1 interacts with UCR2 of PDE4B. UCR2 is strongly conserved among all PDE4 isoforms[1] and it was shown that DISC1 was indeed capable of binding to representative isoforms encoded by all four *PDE4* genes.[100] Phosphorylation of PDE4B by PKA in neuronal cells dramatically reduced the amount of PDE4B that associated with DISC1. This is consistent with a model in which PKA phosphorylation disrupts the UCR1–UCR2 interaction (see the above sections), thereby exposing a region on UCR2 that is then able to interact with DISC1.[100] Therefore, the interaction between DISC1 and PDE4B may be dynamic and also influenced by cAMP signaling through PKA.

Further evidence for a functional interaction between PDE4B isoforms and DISC1 was provided by the identification of a chromosomal translocation (t(1;16)(p31.2;q21)) that involves the *PDE4B* gene and which predisposes to schizophrenia and related disorders.[100] This translocation, which to date has only been identified in a single pedigree, disrupts an intron located between two different exons that encode portions of the unique amino-terminus of the PDE4B1 isoform.[100] The translocation therefore disrupts only expression of the PDE4B1 isoform, producing, as expected, a 50% reduction in PDE4B1 expression in carriers of the translocation.[100] If these data are confirmed in other pedigrees, they would provide important genetic confirmation of the functional data described above and also

stimulate search for associations between schizophrenia and other similar disorders and other *PDE4* genes.

6.9.3 FUNCTIONAL IMPLICATIONS OF THE ASSOCIATION BETWEEN PDE4 ISOFORMS AND DISC1

The converging genetic and functional data described above imply that PDE4 isoforms and DISC1 comprise an important regulatory system. It is possible that DISC1 tethers PDE4 to important cellular structures much in the same way that A-kinase activating proteins (AKAPs) tether PKA and other proteins to specific cell regions and thereby direct PKA action to specific substrates (see further sections in this chapter for additional discussion on AKAPs). Tethering of PDE4 isoforms to specific cellular regions can modulate cAMP levels in these local environments and thus regulate downstream cAMP-mediated signaling events. The exact downstream effectors of the DISC1–PDE4 module remain to be determined—indeed, it is not even certain whether these effects would be modulated by PKA or by EPACs.

6.10 THE ROLE OF PDE4 ISOFORMS IN REGULATING SIGNALING THROUGH THE β-ADRENERGIC RECEPTOR

6.10.1 INTERACTION OF PDE4 ISOFORMS WITH β-ARRESTINS

Many heptahelical G-protein coupled receptors (GPCRs), such as the β_2-adrenergic receptor (β_2AR), act by coupling though the G-protein G_s to stimulate adenylyl cyclase and thus generate cAMP at the cell membrane.[101] One important regulator of this process is the rapid uncoupling of β_2AR from adenylyl cyclase, which occurs when the β_2AR is phosphorylated by the G-protein receptor associated kinase2 (GRK2). Phosphorylation of the β_2AR in turn recruits β-arrestins from the cytosol, which bind to phosphorylated, ligand-occupied β_2AR and in turn have multiple effects on β_2AR function. One important function of β-arrestins is to block coupling of the β_2AR to G_s and thus prevent stimulation of adenylyl cyclase by the β_2AR. The β-arrestins also target the β_2AR to endosomes, where they can be targeted for degradation or recycled rapidly to the cell membrane.[102] These β-arrestins are also versatile proteins that have many other functions in cells, including interacting with numerous other GPCRs and also serving as scaffold or adaptor proteins for numerous intracellular signaling proteins.[103,104]

Recent studies have demonstrated that PDE4 isoforms interact directly with β-arrestin and are recruited to the β_2AR on agonist stimulation.[105–108] Stimulation of Hek293B2 cells and cardiac myocytes with the β_2AR agonist isoproterenol recruits β-arrestin to the membrane, and this is accompanied by the recruitment of PDE4D5, and, to a lesser extent, PDE4D3, to the membrane, and specifically to the β_2AR itself.[105–108] Mouse embryonic fibroblasts derived from β-arrestin2 knockout mice[109] lacked the ability to recruit PDE4D5 to the membrane on agonist stimulation.[105] The recruitment of PDE4D5 to the β_2AR has the potential to lower cAMP levels in the local vicinity of the β_2AR and thus initiate a short-feedback loop: initially, activation of the β_2AR activates adenylyl cyclase and increases cAMP levels in the vicinity of the β_2AR. However, recruitment of PDE4D5 to the β_2AR by β-arrestin increases PDE4 activity in the vicinity of the β_2AR and thereby lowers cAMP levels. This short-feedback loop is synergistic with the ability of β-arrestin to block coupling of the β_2AR to G_s, thereby preventing stimulation of adenylyl cyclase.[110]

6.10.2 ROLE OF THE PDE4D5–β-ARRESTIN INTERACTION ON PHOSPHORYLATION OF THE β₂AR

Increases in cAMP levels in the local environment the β₂AR activate protein kinase A (PKA) and have the potential to increase the phosphorylation of numerous substrates. Among the substrates of PKA at the plasmid membrane is the β₂AR itself, and PKA phosphorylation of the β₂AR switches coupling of the β₂AR from G_s to G_i. This in turn has two functional consequences: it inhibits adenylyl cyclase activity and couples the β₂AR to the activation of ERK1 or ERK2 kinases.[111,112] The activation of the ERK pathway by the β₂AR also requires SRC and other regulators of ERK signaling, such as RAS and RAF.[113–115]

It has been shown recently that the recruitment of PDE4D5 to the β₂AR is a key regulator of β₂AR phosphorylation by PKA.[106–108] β-Arrestin-mediated recruitment of PDE4D5 to the agonist-occupied, GRK2-phosphorylated β₂AR lowers cAMP levels in the local environment of the β₂AR, reduces PKA activity, and thereby attenuates PKA phosphorylation of the β₂AR. Through this action, the recruited PDE4 serves to regulate the ability of the β₂AR to switch between the G_s-mediated activation of adenylyl cyclase and the G_i-mediated activation of ERK2 signaling.[106–108]

Key to the dissection of the functional implications of the PDE4D5–β-arrestin-β₂AR interaction is the use of a mutant, PDE4D5-D556A (Refs. [105–108]). This mutation replaces an aspartic acid in the catalytic region of the enzyme that is involved in metal-ion binding and is essential for catalysis (Ref. [47]; see also appropriate chapters in this volume on PDE4 structure). When overexpressed in cells, this mutant displaces endogenous PDE4 isoforms from their protein partners and thus acts as a dominant negative. The mutation does not affect the ability of PDE4D5 to bind to β-arrestin. Overexpression of this mutant blocked the interaction of endogenous PDE4 isoforms with β-arrestin, reduced the recruitment of PDE4 isoforms to the β₂AR on agonist stimulation, and enhanced membrane-associated PKA activity.

Biochemical analysis of the effects of β-arrestin-mediated recruitment of PDE4D5 to the membrane has focused on its effects on PKA phosphorylation of the β₂AR.[106–108] Agonist-mediated recruitment of PDE4D5 to the β₂AR increased its phosphorylation by PKA, and this effect increases with treatment of the cells with rolipram and also overexpression of PDE4D5-D556A.[106] Recruitment of PDE4D5 is also accompanied by phosphorylation of ERK1 and ERK2, an effect that is agonist dependent and also augmented by treatment of the cells with rolipram or overexpression of PDE4D5-D556A.[106] Treatment of the cells with pertussis toxin, which is a specific inhibitor of G_i, also blocked ERK phosphorylation, demonstrating that this effect requires a switch in β₂AR coupling from G_s to G_i.[106] These effects were seen in both Hek293B2 cells and neonatal cardiac myocytes. These studies indicate that β-arrestin-mediated recruitment of PDE4D5 to the β₂AR has specific functional implications related to ERK signaling (Figure 6.3).

SiRNA-mediated knockdown of PDE4 isoforms has extended and confirmed these studies.[108] In these studies, siRNA was directed against a region located near the 3′ end of PDE4D mRNAs; as this region is common to all PDE4D isoforms (Figure 6.1), it would be expected to knockdown mRNAs encoding all PDE4D isoforms, which was indeed confirmed by immunoblotting. Similarly, siRNA was directed against the corresponding region of PDE4B mRNAs, which have an overall structure that is similar to those encoded by *PDE4D*.[1] Knockdown of PDE4D, but not PDE4B, isoforms enhanced the agonist-mediated phosphorylation of the β₂AR and enhanced phosphorylation and activation of ERK. The magnitude of this effect was similar to the treatment of cells by rolipram. This effect was specific to PDE4D5, as knockdown of PDE4D5, but not PDE4D3 (directed against the unique 5′ ends of their specific mRNAs), augmented agonist-mediated β₂AR phosphorylation and ERK activation.

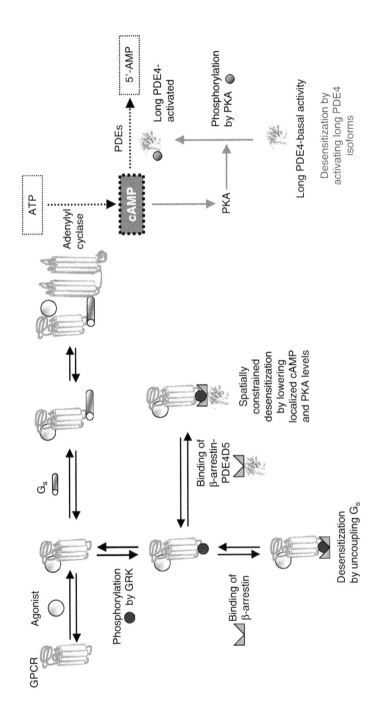

FIGURE 6.3 (See color insert following page 274.) Regulation of β₂-adrenergic signaling by PDE4D5 and β-arrestins. The β-arrestin recruits PDE4D5 to the agonist-occupied, GRK-phosphorylated β₂-adrenergic receptor. The recruitment of PDE4D5 to the β₂-adrenergic receptor lowers cAMP levels in the vicinity of the receptor and thus serves as a short-feedback regulatory loop. See Section 6.10 for details. Activation of PKA also produces phosphorylation and activation of long PDE4 isoforms, which also serves as a feedback mechanism. See Section 6.3 and Section 6.10 for details.

6.10.3 ROLE OF THE PDE4D5 AMINO-TERMINAL REGION IN ITS INTERACTION WITH β-ARRESTIN

All PDE4 isoforms have the potential to interact with β-arrestin, which is mediated by a conserved region at the carboxyl-terminal end of the PDE4 catalytic region.[105,107] However, PDE4D5, compared to other PDE4 isoforms, is preferentially recruited to the β$_2$AR by β-arrestins.[107,108] These observations were first made by immunoprecipitation of endogenous PDE4D isoforms from a variety of cell types, including neonatal cardiac myocytes, in which the proportion of PDE4D5, as compared to other PDE4D isoforms, that was recruited to the membrane upon β$_2$AR stimulation was always greater than its relative abundance in lysates from the corresponding cell type.[105–107] Similarly, PDE4D5 was more abundant in β-arrestin immunoprecipitates from these cells.[107]

The greater avidity of β-arrestins for PDE4D5 is mediated by its interaction with the PDE4D5 unique amino-terminal region (Figure 6.1). This region of PDE4D5 consists of 88 amino acids that have no detectable homology to those of any other PDE4 isoform (Figure 6.1, Ref. [116]). Two-hybrid assays and pull-down experiments showed that PDE4D5 interacted more avidly with β-arrestin than did other PDE4D isoforms.[107] Deletion of the PDE4D5 amino-terminal region reduced the avidity of its interaction with β-arrestin 2 to the level seen with PDE4D3, another long PDE4 isoform that has a different amino-terminal region (Figure 6.1). Mutation of a number of individual amino acids in the PDE4D5 amino-terminal region had a similar effect. These data show that the PDE4D5 amino-terminal region is essential for mediating the selectivity of its interaction with β-arrestin 2.[107]

Parallel studies showed that deletion of the extreme carboxyl-terminal region of PDE4D5 also abolished its interaction with β-arrestin 2.[107] As PDE4D5 shares this region with all PDE4D isoforms (Figure 6.1), it is likely that this region is responsible for mediating interactions between all PDE4D isoforms, especially PDE4D3 and β-arrestin. Because the interaction between these isoforms and β-arrestin is less avid than the interactions between PDE4D5 and β-arrestin, the strength of the interaction mediated by the PDE4D carboxyl-terminal region is likely to be less than that mediated by the unique PDE4D5 amino-terminal region.

There are two regions on β-arrestin 2 that are necessary for it to interact with PDE4D5.[107] These are located at the extreme amino-terminus of β-arrestin 2 and within its carboxyl-terminal half. These regions are clearly different from those implicated in the interaction of β-arrestin 2 with some of its other partners, such as clathrin, adaptor protein 2 (AP2), or JNK3 kinase.[107] The structure of β-arrestin, as deduced by x-ray crystallography,[117] shows that these two regions do not juxtapose each other. It is possible that each of these regions may interact separately with one of the two regions on PDE4D5, which we have shown to be necessary for the interaction. However, further study of the interaction, preferably with additional structural and interaction data, will be required to validate this hypothesis.

6.10.4 ROLE OF RACK1 IN THE PDE4D5–β-ARRESTIN INTERACTION

RACK1 is a seven-bladed, WD-repeat, β-propeller protein that has been implicated as an interacting partner of a number of signal transduction proteins, including various PKC isoforms, β-integrins, IGF1, the G$_{s\gamma}$ subunit, and SRC (see Refs. [118–120] for recent developments and Ref. [121] for a review). The exact physiological implications of each of these interactions are currently an active subject of investigation in a number of laboratories.

Bolger, Houslay and colleagues have demonstrated previously that RACK1 interacts specifically with PDE4D5. These groups initially identified RACK1 as a partner of PDE4D5 in a two-hybrid screen and confirmed the interaction by co-immunoprecipitation of the two proteins from untransfected cell lines.[122] Co-immunoprecipitation and two-hybrid

studies showed that RACK1 interacted specifically with PDE4D5 and not with other PDE4 isoforms.[122] Conversely, PDE4D5 does not interact with $G_{s\beta}$ or any other WD-repeat proteins that have been tested.[122] Analysis of truncation constructs demonstrated that the unique amino-terminal region of PDE4D5 was required for it to interact with RACK1,[122] consistent with its specific interaction with PDE4D5 and not other PDE4 isoforms. Further study of the PDE4D5 amino-terminus identified a 38-amino acid region, which we have called RAID1, for RACK1 interaction domain 1, which was necessary for the interaction.[123] A synthetic peptide corresponding to this region bound to RACK1 with an affinity (K_d, 6 nm) that was very similar to that of PDE4D5 itself.[123] Study of deletion and point mutations in RACK1 implicated four of the seven β-propeller blades of RACK1 in the interaction.[124] RAID1 appears to have an amphipathic helical structure that may interact with the β-propeller blades of RACK1 in a manner akin to the interaction of $G_{s\gamma}$ with $G_{s\beta}$.[123]

As RACK1 does not significantly affect the enzymatic properties of PDE4D5,[122] its physiological effects on PDE4D5 were initially uncertain. However, the subsequent discovery that β-arrestin also interacts with the PDE4D5 amino-terminus (see previous section) suggested that RACK1 and β-arrestin would compete for interaction to PDE4D5 by virtue of their binding to overlapping regions on PDE4D5. Using three-hybrid, pull-down, and peptide display approaches, Bolger et al. have now shown that RACK1 and β-arrestin do indeed compete for interaction with overlapping regions of PDE4D5.[125] Not surprisingly, the ability of RACK1 to modulate the β-arrestin–PDE4D5 interaction has predictable physiological consequences. siRNA-mediated knockdown of RACK1 increases the proportion of PDE4D5 that interacts with β-arrestin.[125] Furthermore, siRNA-mediated RACK1 knockdown increases β-arrestin-mediated targeting of PDE4D5 to the β_2AR on isoproterenol stimulation, significantly attenuates PKA phosphorylation of the β_2AR, and markedly reduces ERK activation.[125] Therefore, it is clear that RACK1 has important physiological effects on PDE4D5 that in turn modulate the functioning of the β_2AR.

6.10.5 FUNCTIONAL IMPLICATIONS OF THE PDE4D5–β-ARRESTIN-β_2AR INTERACTION

Agonist-mediated recruitment of PDE4D5 to the β_2AR by β-arrestin modulates PKA activity in the vicinity of the β_2AR. One obvious question here is precisely what substrates of PKA are present in the vicinity of the β_2AR that in turn could be regulated by these activities. Clearly, the β_2AR is one important PKA substrate that is regulated by these interactions, as described in detail above. However, it is certainly possible that there are additional PKA substrates in the vicinity of the β_2AR that are also affected by the recruitment of PDE4D5 to the β_2AR. There are a number of potential PKA substrates at the plasma membrane, including ion channels, cytoskeletal proteins, and other kinases. However, it is not known which of these potential substrates is tethered sufficiently closely to the β_2AR so that it could be regulated by the recruitment of PDE4D5 to the β_2AR. In this context, the tethering of PKA itself to various cellular structures by A-kinase activating proteins (AKAPs) would determine whether it is located sufficiently close to the β_2AR so that it could be regulated in this fashion (see the separate section, further in the chapter, for detailed discussion on the potential role of AKAPs in PDE4-mediated signaling). AKAP79 is associated with the β_2AR in a number of cell types, including Hek293 cells, and appears to be essential for the regulation of β_2AR phosphorylation by PKA,[108] but whether this is true in other systems remains to be determined.

One potential membrane-associated PKA substrate is Csk kinase, which phosphorylates and inhibits Srk. Csk has important functions in numerous cells, including T-lymphocytes, where it serves as an effector of CD28-mediated signaling. β-arrestin recruitment of PDE4D5

to CD28 attenuates PKA phosphorylation of Csk.[126] The exact physiological consequences of these events are under active investigation.

A related issue is what cellular and organismal phenotypes are affected by activation of the ERK pathway in cells and intact organisms. In all of the studies described above, activation of the ERK pathway was assessed almost exclusively by autophosphorylation of ERK. Activation of ERK signaling has been associated with numerous different functions, including cell proliferation and DNA replication, accumulation of lipids and other metabolic functions of cells, as well as a variety of phenotypes in the central nervous system. In the future, we predict that the functional role of PDE4D5–β-arrestin-$β_2$AR interaction will be investigated in a variety of such systems. Many of the experimental tools described above, such as the dominant-negative PDE4D5 mutant and siRNA, will probably be essential tools in these efforts.

6.11   ASSOCIATIONS BETWEEN PDE4 ISOFORMS AND AKAPs

AKAPs are a very diverse family of proteins that anchor PKA subunits to various subcellular structures.[54,127,128] One of the most important aspects of AKAP function is that they can bind to multiple components of a signaling pathway[129] and thereby facilitate their functional interaction in both space and time. The diversity of AKAPs provides the potential for their interacting with different PDE isoforms in different tissues and in different subcellular compartments.

6.11.1   INTERACTIONS BETWEEN PDE4D3 AND mAKAP

Scott and colleagues have demonstrated that PDE4D3 interacts directly with mAKAP, a high–molecular weight AKAP that is enriched in skeletal and cardiac muscles.[130–132] The interaction between PDE4D3 and mAKAP is highly specific, in that PDE4D3 interacts only weakly with other AKAPs and there is little interaction between mAKAP and other PDE4 isoforms. mAKAP binds directly to the unique amino-terminal region of PDE4D3 (Figure 6.1). This portion of PDE4D3 contains a PKA phosphorylation site (S13), which is distinct from the PKA phosphorylation site located within UCR1 common to all long PDE4 isoforms (S54 in PDE4D3). Binding of mAKAP to the unique amino-terminal region of PDE4D3 is enhanced at least fivefold on phosphorylation of S13, but not by phosphorylation of S54.[131] In contrast, phosphorylation of S54, but not S13, activates PDE4D3 activity (see also above, PKA phosphorylation). PDE4D3 interacts with a specific region of the mAKAP protein (amino acids 1286–1401 of the 2301 amino acid protein), which is clearly distinct from the region that binds the RII subunit of PKA.[130]

The interaction between PDE4D3 and mAKAP has a number of immediate physiological consequences. The ability of mAKAP to tether PKA and PDE4D3 lowers cAMP levels in the vicinity of mAKAP and thereby lowers PKA activity. Inhibition of PDE4D3 increases PKA activity. For example, PKA activity co-immunoprecipitates with mAKAP from cardiac myocytes, and the activity of PKA can be augmented in these immunoprecipitates by inhibition of PDE4D3 activity by rolipram.[130] One substrate for PKA in this complex is, of course, PDE4D3 itself, and stimulation of PKA activity produces enhanced phosphorylation of PDE4D3 in cells, which can be augmented by transfection of the cells by mAKAP and is abolished by mutation of the PKA phosphorylation sites to alanine (S13A and S54A). As S13 phosphorylation is necessary for PDE4D3 to interact with mAKAP, mAKAP requires anchored PKA to interact with PDE4D3 efficiently in cells.[131] A mutation in the PKA-binding region of mAKAP (I2062P) that blocks its ability to interact with the RII subunit of PKA also blocked the phosphorylation of PDE4D3 phosphorylation and its interaction mAKAP.[131] These observations provide evidence for a short-feedback loop

that regulates PDE4D3 activity in cells that ultimately regulates PKA activity on substrates in the immediate vicinity.

6.11.2 ERK5 ASSOCIATES WITH PDE4D3

PDE4B, PDE4C, and PDE4D isoforms, including PDE4D3, bind to, and are phosphorylated by, ERK kinases (see section above). Scott and colleagues have demonstrated recently that mAKAP-associated PDE4D3 is preferentially associated with ERK5, and that ERK5 could be co-immunoprecipitated with mAKAP.[132] However, ERK5 could not be co-immunoprecipitated with mAKAP in cells overexpressing PDE4D3 with a mutant (the KIM/FQF double mutant, as described above) that blocks its association with ERKs. This demonstrates that the interaction of mAKAP and ERK5 is mediated by their mutual binding to PDE4D3.[132] As expected, mAKAP-associated PDE4D3 activity could be inhibited by serum (which activates ERKs) and this effect could be blocked by the MEK inhibitor PD98059.

6.11.3 INTERACTIONS BETWEEN PDE4D3, mAKAP, AND EPAC1

Recently, Scott and colleagues have demonstrated that PDE4D3 interacts directly with Epac1.[132] Epac1 (exchange protein directly activated by cAMP) is a cAMP-dependent G-protein exchange factor specific for the small–molecular weight G-protein Rap1 (see Ref. [133] for a review). It is a downstream effector of cAMP signaling that is independent of PKA. Scott and colleagues were led to investigate the role of Epac1 in mAKAP-tethered signaling because they detected a cAMP-dependent block of ERK5 activity in mAKAP immunoprecipitates that was not blocked by treatment with H89, KT5720, or any other tested specific PKA inhibitor.[132] However, it was blocked by a compound (8-CPT-2'-O-Me-cAMP, also called 007) that is a specific inhibitor of Epac1.[134] Overexpression of RapGap (an upstream regulator of Rap1, which in turn regulates Epac1) in cells also augmented ERK activity in mAKAP immunoprecipitates, and this was not affected by cAMP stimulation or PKA inhibitors. These data demonstrate that ERK5 forms a complex with mAKAP that is mediated by its interaction with PDE4D3 and that allows it to be regulated by Epac1.

6.11.4 FUNCTIONAL CONSEQUENCES OF THE ERK5–PDE4D3-EPAC1-mAKAP INTERACTION

The experiments summarized in this section demonstrate that mAKAP mediates the interaction of a large number of proteins involved in cAMP-mediated signaling. Furthermore, other studies have identified additional proteins that associate with mAKAP, including protein phosphatase (PP2A) subunits A and C and the cardiac ryanodine receptors.[71,135] Given the complexity of this complex, as well as the possibility that its components will vary between different tissue types, the precise functional consequences of these interactions are likely to be important but not easily predicted. Nonetheless, it is clear that this pathway has the potential ability to facilitate crosstalk between cAMP and ERK kinase signaling in numerous contexts.

6.11.5 INTERACTION OF PDE ISOFORMS WITH AKAPS IN THE CENTROSOME

A number of groups have demonstrated that PDE isoforms localize to the centrosome along with PKA RII and that specific AKAPs mediate the localization of these proteins in this area. In the Sertoli cells of the testis, PDE4D3 is largely insoluble and can be shown by immunofluorescence studies to target centrosomes in a distribution that is similar to that of AKAP450.[136] PDE4D3, the PKA RII subunit, and AKAP450 can be co-immunoprecipitated

from these cells. These interactions do not appear to be mediated by myomegalin.[136] In COS1 cells, PDE4D3 and PDE4C2, but not other PDE4 isoforms, can co-immunoprecipitate with AKAP250 and AKAP450.[137] Expression in COS1 cells of dominant-negative mutants of PDE4D3 and PDE4C2 (i.e., mutations in these isoforms of the amino acid that is homologous to the D556A mutation of PDE4D5 [see the above section]) renders them constitutively phosphorylated and activated by PKA. In contrast, while the corresponding mutants of PDE4A4 and PDE4B1 were phosphorylated in a manner similar to their unmutated counterparts; SiRNA-mediated knockdown of PKA RII or treatment of the cells with the peptide Ht31, which disrupts the binding of PKA RII to AKAPs, prevented the constitutive phosphorylation of the dominant-negative mutants of PDE4D3 and PDE4C2. These studies suggest that AKAP450 may tether certain PDE4 isoforms, such as PDE4D3 and PDE4C2, in the centrosome so that they are preferentially phosphorylated by PKA.[136,137]

6.11.6 SUMMARY—AKAPS AND PDE4 FUNCTION

Rapid progress is made in determining the specific associations of various PDE4 isoforms and AKAPs. Given the diversity of both AKAPs and members of the PDE4 family, it is likely that additional associations will be uncovered in the near future. Study of the ERK5-PDE4D3-Epac1-mAKAP complex, in particular, has suggested a number of potential regulatory interactions. Intriguingly, it is clear that the key regulatory implication of the ERK5–PDE4D3–Epac1–mAKAP complex—i.e., its ability to facilitate crosstalk between cAMP and ERK kinase signaling—is also the key attribute of the interactions between PDE4D5–β-arrestin-β_2-adrenergic-receptor complex that was discussed in Section 6.10.

Since all the components of both the ERK5-PDE4D3-Epac1-mAKAP complex and the PDE4D5–β-arrestin-β_2-adrenergic-receptor complex are present in cardiac myocytes, it is highly likely that either or both of these pathways play a critical role in cardiac function. They may play a role in cardiac remodeling after injury, in the development of cardiac hypertrophy in response to various stresses, or in the generation of normal or abnormal cardiac rhythms. Although some of these questions can be examined in cell culture, it is likely that such questions will be more clearly answered in animal models. Recent work on the role of PDE4D in the heart, using PDE4D knockout mice developed by Conti and colleagues, demonstrates the value of these systems (in Ref. [138]; see separate chapter in this volume for additional discussion). In particular, generation of animal models that overexpress a dominant-negative PDE4 mutant (i.e., the D556A mutation in PDE4D5, or the equivalent mutations in other PDE4 isoforms) has particular promise, as such mutations have the ability to dissect out the relative contributions of specific PDE4–protein interactions in a living animal.

6.12 CONCLUSION AND PHARMACOLOGICAL IMPLICATIONS

Extraordinary progress has been made in the past few years in determining the mechanisms by which PDE4s are regulated in cells. It is however likely that many additional regulatory mechanisms have yet to be discovered. There remains a large number of PDE4 isoforms—and additional isoforms are likely to be identified in the near future—for which protein partners have yet to be identified. Additionally, many PDE4 isoforms have distinctive patterns of expression in tissues, and it is entirely possible that many isoforms are regulated by proteins or entire pathways that are highly tissue-specific, and which have yet to be discovered.

An important practical question is how studies on the cellular regulation of PDE4 isoforms should influence drug development. Traditionally, the PDE4 inhibitor development has focused on compounds that interact at, or close to, the active site of the enzyme. It has now become abundantly clear that the conformation of the PDE4 catalytic site can be

influenced by phosphorylation and by protein–protein interactions, as described in detail in many sections above. It is likely that many of the cell-specific or tissue-specific actions of PDE4 inhibitors reflect differences in the binding partners of the PDE4 isoforms expressed in those cells or tissues. Indeed, the studies described in this chapter have the potential to clarify a number of long-standing issues in PDE4 inhibitor development, for example, the nature of phenomena such as HARBS and on targeting inhibitor action at specific tissues, especially the CNS. Therefore, we would predict that, over the next few years, we should see the development of novel PDE4 inhibitors that act at the catalytic site but can be targeted to specific isoforms by virtue of specific conformational changes in the catalytic regions of these isoforms that are in turn dictated by specific protein–protein interactions. These newer PDE4 inhibitors should have a higher degree of isoform and tissue selectivity than those present in the currently available inhibitors.

However, the studies discussed in this chapter also suggest that a significant fraction of PDE4 inhibitor development should be targeted at the development of compounds that act outside the catalytic regions of PDE4 enzymes. These novel inhibitors could block any of the novel protein–protein interactions that have been identified in these studies, or could interfere with the interaction of several kinases, such as PKA or ERK, with their specific phosphorylation sites on the PDE4 proteins (i.e., not with the catalytic regions of these kinases). Admittedly, the generation of these compounds might be more difficult than the development of those targeted to the catalytic site and they might be less desirable as drugs. For example, drugs specifically targeted to protein–protein interactions might be more hydrophobic (i.e., less soluble and more protein-bound) than would be desirable for therapeutic use. However, they would also have the potential to be vastly more tissue specific and cell-compartment specific than the agents currently available.

REFERENCES

1. Houslay, M.D., M. Sullivan, and G.B. Bolger. 1998. The multienzyme PDE4 cyclic adenosine monophosphate-specific phosphodiesterase family: Intracellular targeting, regulation, and selective inhibition by compounds exerting anti-inflammatory and antidepressant actions. *Adv Pharmacol* 44:225–342.
2. Baillie, G.S., and M.D. Houslay. 2005. Arrestin times for compartmentalised cAMP signalling and phosphodiesterase-4 enzymes. *Curr Opin Cell Biol* 17:129–134.
3. Conti, M, W. Richter, C. Mehats, G. Livera, J.Y. Park, and C. Jin. 2003. Cyclic AMP-specific PDE4 phosphodiesterases as critical components of cyclic AMP signaling. *J Biol Chem* 278:5493–5496.
4. Houslay, M.D., P. Schafer, and K.Y. Zhang. 2005. Keynote review: Phosphodiesterase-4 as a therapeutic target. *Drug Discov Today* 10:1503–1519.
5. Bolger, G, T. Michaeli, T. Martins, T. St. John, B. Steiner, L. Rodgers, M. Riggs, M. Wigler, and K. Ferguson. 1993. A family of human phosphodiesterases homologous to the dunce learning and memory gene product of *Drosophila melanogaster* are potential targets for antidepressant drugs. *Mol Cell Biol* 13:6558–6571.
6. Qiu, Y.H., C.N. Chen, T. Malone, L. Richter, S.K. Beckendorf, and R.L. Davis. 1991. Characterization of the memory gene dunce of *Drosophila melanogaster*. *J Mol Biol* 222:553–565.
7. Park, H., J.A. Lee, C. Lee, M.J. Kim, D.J. Chang, H. Kim, S.H. Lee, Y.S. Lee, and B.K. Kaang. 2005. An Aplysia type 4 phosphodiesterase homolog localizes at the presynaptic terminals of Aplysia neuron and regulates synaptic facilitation. *J Neurosci* 25:9037–9045.
8. Shakur, Y., M. Wilson, L. Pooley, M. Lobban, S.L. Griffiths, A.M. Campbell, J. Beattie, C. Daly, and M.D. Houslay. 1995. Identification and characterization of the type-IVA cyclic AMP-specific phosphodiesterase RD1 as a membrane-bound protein expressed in cerebellum. *Biochem J* 306:801–809.
9. Bolger, G.B., L. Rodgers, and M. Riggs. 1994. Differential CNS expression of alternative mRNA isoforms of the mammalian genes encoding cAMP-specific phosphodiesterases. *Gene* 149:237–244.

10. McPhee, I., L. Pooley, M. Lobban, G. Bolger, and M.D. Houslay. 1995. Identification, characterization, and regional distribution in brain of RPDE-6 (RNPDE4A5), a novel splice variant of the PDE4A cyclic AMP phosphodiesterase family. *Biochem J* 310:965–974.

11. McPhee, I., S. Cochran, and M.D. Houslay. 2001. The novel long PDE4A10 cyclic AMP phosphodiesterase shows a pattern of expression within brain that is distinct from the long PDE4A5 and short PDE4A1 isoforms. *Cell Signal* 13:911–918.

12. Wallace, D.A., L.A. Johnston, E. Huston, D. MacMaster, T.M. Houslay, Y.F. Cheung, L. Campbell, J.E. Millen, R.A. Smith, I. Gall, R.G. Knowles, M. Sullivan, and M.D. Houslay. 2005. Identification and characterization of PDE4A11, a novel, widely expressed long isoform encoded by the human PDE4A cAMP phosphodiesterase gene. *Mol Pharmacol* 67:1920–1934.

13. Bolger, G.B., I. McPhee, and M.D. Houslay, 1996. Alternative splicing of cAMP-specific phosphodiesterase mRNA transcripts. Characterization of a novel tissue-specific isoform, RNPDE4A8. *J Biol Chem* 271:1065–1071.

14. Naro, F., R. Zhang, and M. Conti. 1996. Developmental regulation of unique adenosine 3',5'-monophosphate-specific phosphodiesterase variants during rat spermatogenesis. *Endocrinology* 137:2464–2472. [Published erratum appears in *Endocrinology* 1996 Oct;137(10): 4114.]

15. Swinnen, J.V., D.R. Joseph, and M. Conti. 1989. Molecular cloning of rat homologues of the Drosophila melanogaster dunce cAMP phosphodiesterase: Evidence for a family of genes. *Proc Natl Acad Sci USA* 86:5325–5329.

16. Miro, X., S. Perez-Torres, P. Puigdomenech, J.M. Palacios, and G. Mengod. 2002. Differential distribution of PDE4D splice variant mRNAs in rat brain suggests association with specific pathways and presynaptical localization. *Synapse* 45:259–269.

17. Salanova, M., S.Y. Chun, S. Iona, C. Puri, M. Stefanini, and M. Conti. 1999. Type 4 cyclic adenosine monophosphate-specific phosphodiesterases are expressed in discrete subcellular compartments during rat spermiogenesis. *Endocrinology* 140:2297–2306.

18. Wang, D., C. Deng, B. Bugaj-Gaweda, M. Kwan, C. Gunwaldsen, C. Leonard, X. Xin, Y. Hu, A. Unterbeck, and M. De Vivo. 2003. Cloning and characterization of novel PDE4D isoforms PDE4D6 and PDE4D7. *Cell Signal* 15:883–891.

19. Richter, W., S.L. Jin, and M. Conti. 2005. Splice variants of the cyclic nucleotide phosphodiesterase PDE4D are differentially expressed and regulated in rat tissue. *Biochem J* 388:803–811.

20. Swinnen, J.V., D.R. Joseph, and M. Conti. 1989. The mRNA encoding a high-affinity cAMP phosphodiesterase is regulated by hormones and cAMP. *Proc Natl Acad Sci USA* 86:8197–8201.

21. Swinnen, J.V., K.E. Tsikalas, and M. Conti. 1991. Properties and hormonal regulation of two structurally related cAMP phosphodiesterases from the rat Sertoli cell. *J Biol Chem* 266:18370–18377.

22. D'Sa, C., L.M. Tolbert, M. Conti, and R.S. Duman. 2002. Regulation of cAMP-specific phosphodiesterases type 4B and 4D (PDE4) splice variants by cAMP signaling in primary cortical neurons. *J Neurochem* 81:745–757.

23. Vicini, E., and M. Conti. 1997. Characterization of an intronic promoter of a cyclic adenosine 3',5'-monophosphate (cAMP)-specific phosphodiesterase gene that confers hormone and cAMP inducibility. *Mol Endocrinol* 11:839–850.

24. Mehats, C., G. Tanguy, E. Dallot, B. Robert, R. Rebourcet, F. Ferre, and M.J. Leroy. 1999. Selective up-regulation of phosphodiesterase-4 cyclic adenosine 3',5'- monophosphate (cAMP)-specific phosphodiesterase variants by elevated cAMP content in human myometrial cells in culture. *Endocrinology* 140:3228–3237.

25. Rose, R.J., H. Liu, D. Palmer, and D.H. Maurice. 1997. Cyclic AMP-mediated regulation of vascular smooth muscle cell cyclic AMP phosphodiesterase activity. *Br J Pharmacol* 122:233–240.

26. Seybold, J., R. Newton, L. Wright, P.A. Finney, N. Suttorp, P.J. Barnes, I.M. Adcock, and M.A. Giembycz. 1998. Induction of phosphodiesterases 3B, 4A4, 4D1, 4D2, and 4D3 in Jurkat T-cells and in human peripheral blood T-lymphocytes by 8-bromo-cAMP and Gs-coupled receptor agonists. Potential role in β₂-adrenoreceptor desensitization. *J Biol Chem* 273:20575–20588.

27. Yamashita, N., M. Yamauchi, J. Baba, and A. Sawa. 1997. Phosphodiesterase type 4 that regulates cAMP level in cortical neurons shows high sensitivity to rolipram. *Eur J Pharmacol* 337:95–102.

28. Park, J.Y., F. Richard, S.Y. Chun, J.H. Park, E. Law, K. Horner, S.L. Jin, and M. Conti. 2003. Phosphodiesterase regulation is critical for the differentiation and pattern of gene expression in granulosa cells of the ovarian follicle. *Mol Endocrinol* 17:1117–1130.
29. Takahashi, S.I., T. Nedachi, T. Fukushima, K. Umesaki, Y. Ito, F. Hakuno, J.J. Van Wyk, and M. Conti. 2001. Long-term hormonal regulation of the cAMP-specific phosphodiesterases in cultured FRTL-5 thyroid cells. *Biochim Biophys Acta* 1540:68–81.
30. Verghese, M.W., R.T. McConnell, J.M. Lenhard, L. Hamacher, and S.L.C. Jin. 1995. Regulation of distinct cyclic AMP-specific phosphodiesterase (phosphodiesterase type 4) isozymes in human monocytic cells. *Mol Pharmacol* 47:1164–1171.
31. Ma, D., P. Wu, R.W. Egan, M.M. Billah, and P. Wang. 1999. Phosphodiesterase 4B gene transcription is activated by lipopolysaccharide and inhibited by interleukin-10 in human monocytes. *Mol Pharmacol* 55:50–57.
32. Beasley, S.C., N. Cooper, L. Gowers, J.P. Gregory, A.F. Haughan, P.G. Hellewell, D. Macari, J. Miotla, J.G. Montana, T. Morgan, R. Naylor, K.A. Runcie, B. Tuladhar, and J.B. Warneck. 1998. Synthesis and evaluation of a novel series of phosphodiesterase IV inhibitors. A potential treatment for asthma. *Bioorg Med Chem Lett* 8:2629–2634.
33. Jin, S.L., and M. Conti. 2002. Induction of the cyclic nucleotide phosphodiesterase PDE4B is essential for LPS-activated TNF-alpha responses. *Proc Natl Acad Sci USA* 99:7628–7633.
34. Le Jeune, I.R., M. Shepherd, G. Van Heeke, M.D. Houslay, and I.P. Hall. 2002. Cyclic AMP-dependent transcriptional upregulation of phosphodiesterase 4D5 in human airway smooth muscle cells. Identification and characterisation of a novel PDE4D5 promoter. *J Biol Chem* 277:35980–35989.
35. Marchmont, R.J., and M.D. Houslay. 1980. Insulin trigger, cyclic AMP-dependent activation and phosphorylation of a plasma membrane cyclic AMP phosphodiesterase. *Nature* 286:904–906.
36. Sette, C., E. Vicini, and M. Conti. 1994. The ratPDE3/IVd phosphodiesterase gene codes for multiple proteins differentially activated by cAMP-dependent protein kinase. *J Biol Chem* 269:18271–18274. [Published erratum appears in *J Biol Chem* 1994 Aug 12; 269 (32):20806].
37. Sette, C., E. Vicini, and M. Conti. 1994. Modulation of cellular responses by hormones: Role of cAMP specific, rolipram-sensitive phosphodiesterases. *Mol Cell Endocrinol* 100:75–79.
38. Sette, C., S. Iona, and M. Conti. 1994. The short-term activation of a rolipram-sensitive, cAMP-specific phosphodiesterase by thyroid-stimulating hormone in thyroid FRTL-5 cells is mediated by a cAMP-dependent phosphorylation. *J Biol Chem* 269:9245–9252.
39. Alvarez, R., C. Sette, D. Yang, R.M. Eglen, R. Wilhelm, E.R. Shelton, and M. Conti. 1995. Activation and selective inhibition of a cyclic AMP-specific phosphodiesterase, PDE-4D3. *Mol Pharmacol* 48:616–622.
40. Sette, C., and M. Conti. 1996. Phosphorylation and activation of a cAMP-specific phosphodiesterase by the cAMP-dependent protein kinase. Involvement of serine 54 in the enzyme activation. *J Biol Chem* 271:16526–16534.
41. Engels, P., M. Sullivan, T. Muller, and H. Lubbert. 1995. Molecular cloning and functional expression in yeast of a human cAMP-specific phosphodiesterase subtype (PDE IV-C). *FEBS Lett* 358:305–310.
42. Oki, N., S.I. Takahashi, H. Hidaka, and M. Conti. 2000. Short-term feedback regulation of cAMP in FRTL-5 thyroid cells. Role of PDE4D3 phosphodiesterase activation. *J Biol Chem* 275:10831–10837.
43. Hoffmann, R., I.R. Wilkinson, J.F. McCallum, P. Engels, and M.D. Houslay. 1998. cAMP-specific phosphodiesterase HSPDE4D3 mutants which mimic activation and changes in rolipram inhibition triggered by protein kinase A phosphorylation of Ser-54: Generation of a molecular model. *Biochem J* 333:139–149.
44. MacKenzie, S.J., G.S. Baillie, I. McPhee, C. MacKenzie, R. Seamons, T. McSorley, J. Millen, M.B. Beard, G. Van Heeke, and M.D. Houslay. 2002. Long PDE4 cAMP specific phosphodiesterases are activated by protein kinase A-mediated phosphorylation of a single serine residue in Upstream Conserved Region 1 (UCR1). *Br J Pharmacol* 136:421–433.

45. Hoffmann, R., G.S. Baillie, S.J. MacKenzie, S.J. Yarwood, and M.D. Houslay. 1999. The MAP kinase ERK2 inhibits the cyclic AMP-specific phosphodiesterase HSPDE4D3 by phosphorylating it at Ser579. *EMBO J* 18:893–903.

46. Lenhard, J.M., D.B. Kassel, W.J. Rocque, L. Hamacher, W.D. Holmes, I. Patel, C. Hoffman, and M. Luther. 1996. Phosphorylation of a cAMP-specific phosphodiesterase (HSPDE4B2B) by mitogen-activated protein kinase. *Biochem J* 316 (Pt 3):751–758.

47. Xu, R.X., A.M. Hassell, D. Vanderwall, M.H. Lambert, W.D. Holmes, M.A. Luther, W.J. Rocque, M.V. Milburn, Y. Zhao, H. Ke, and R.T. Nolte. 2000. Atomic structure of PDE4: Insights into phosphodiesterase mechanism and specificity. *Science* 288:1822–1825.

48. Lee, M.E., J. Markowitz, J.O. Lee, and H. Lee. 2002. Crystal structure of phosphodiesterase 4D and inhibitor complex(1). *FEBS Lett* 530:53–58.

49. MacKenzie, S.J., G.S. Baillie, I. McPhee, G.B. Bolger, and M.D. Houslay. 2000. ERK2 mitogen-activated protein kinase binding, phosphorylation, and regulation of the PDE4D cAMP-specific phosphodiesterases. The involvement of COOH-terminal docking sites and NH2-terminal UCR regions. *J Biol Chem* 275:16609–16617.

50. Zhang, H.T., Y. Zhao, Y. Huang, N.R. Dorairaj, L.J. Chandler, and J.M. O'Donnell. 2004. Inhibition of the phosphodiesterase 4 (PDE4) enzyme reverses memory deficits produced by infusion of the MEK inhibitor U0126 into the CA1 subregion of the rat hippocampus. *Neuropsychopharmacology* 29:1432–1439.

51. Gao, Y., E. Nikulina, W. Mellado, and M.T. Filbin. 2003. Neurotrophins elevate cAMP to reach a threshold required to overcome inhibition by MAG through extracellular signal-regulated kinase-dependent inhibition of phosphodiesterase. *J Neurosci* 23:11770–11777.

52. Lim, J., G. Pahlke, and M. Conti. 1999. Activation of the cAMP-specific phosphodiesterase PDE4D3 by phosphorylation. Identification and function of an inhibitory domain. *J Biol Chem* 274:19677–19685.

53. Beard, M.B., A.F. Olsen, R.E. Jones, S. Erdogan, M.D. Houslay, and G.B. Bolger. 2000. UCR1 and UCR2 domains unique to the cAMP-specific phosphodiesterase family form a discrete module via electrostatic interactions. *J Biol Chem* 275:10349–10358.

54. Pawson, T., and J.D. Scott. 1997. Signaling through scaffold, anchoring, and adaptor proteins. *Science* 278:2075–2080.

55. Cohen, G.B., R. Ren, and D. Baltimore. 1995. Modular binding domains in signal transduction proteins. *Cell* 80:237–248.

56. Richter, W., and M. Conti. 2002. Dimerization of the type 4 cAMP-specific phosphodiesterases is mediated by the upstream conserved regions (UCRs). *J Biol Chem* 277:40212–40221.

57. Grange, M., C. Sette, M. Cuomo, M. Conti, M. Lagarde, A.F. Prigent, and G. Nemoz. 2000. The cAMP-specific phosphodiesterase PDE4D3 is regulated by phosphatidic acid binding. Consequences for cAMP signaling pathway and characterization of a phosphatidic acid binding site. *J Biol Chem* 275:33379–33387.

58. Baillie, G.S., S.J. MacKenzie, I. McPhee, and M.D. Houslay. 2000. Sub-family selective actions in the ability of erk2 MAP kinase to phosphorylate and regulate the activity of PDE4 cyclic AMP-specific phosphodiesterases. *Br J Pharmacol* 131:811–819.

59. Richter, W., and M. Conti. 2004. The oligomerization state determines regulatory properties and inhibitor sensitivity of type 4 cAMP-specific phosphodiesterases. *J Biol Chem* 279:30338–30348.

60. Jacobitz, S., M.M. McLaughlin, G.P. Livi, M. Burman, and T.J. Torphy. 1996. Mapping the functional domains of human recombinant phosphodiesterase 4A: Structural requirements for catalytic activity and rolipram binding. *Mol Pharmacol* 50:891–899.

61. Laliberte, F., Y. Han, A. Govindarajan, A. Giroux, S. Liu, B. Bobechko, P. Lario, A. Bartlett, E. Gorseth, M. Gresser, and Z. Huang. 2000. Conformational difference between PDE4 apoenzyme and holoenzyme. *Biochemistry* 39:6449–6458.

62. Souness, J.E., and S. Rao. 1997. Proposal for pharmacologically distinct conformers of PDE4 cyclic AMP phosphodiesterases. *Cell Signal* 9:227–236.

63. Terry, R., Y.F. Cheung, M. Praestegaard, G.S. Baillie, E. Huston, I. Gall, D.R. Adams, and M.D. Houslay. 2003. Occupancy of the catalytic site of the PDE4A4 cyclic AMP phosphodiesterase by

rolipram triggers the dynamic redistribution of this specific isoform in living cells through a cyclic AMP independent process. *Cell Signal* 15:955–971.

64. Bolger, G.B., A.H. Peden, M.R. Steele, C. MacKenzie, D.G. McEwan, D.A. Wallace, E. Huston, G.S. Baillie, and M.D. Houslay. 2003. Attenuation of the activity of the cAMP-specific phospho-diesterase PDE4A5 by interaction with the immunophilin XAP2. *J Biol Chem* 278:33351–33363.

65. Marks, A.R. 1996. Cellular functions of immunophilins. *Physiol Rev* 76:631–649.

66. Ma, Q., and J.P. Whitlock, Jr. 1997. A novel cytoplasmic protein that interacts with the Ah receptor, contains tetratricopeptide repeat motifs, and augments the transcriptional response to 2,3,7,8-tetrachlorodibenzo-p-dioxin. *J Biol Chem* 272:8878–8884.

67. Carver, L.A., and C.A. Bradfield. 1997. Ligand-dependent interaction of the aryl hydrocarbon receptor with a novel immunophilin homolog in vivo. *J Biol Chem* 272:11452–11456.

68. Meyer, B.K., M.G. Pray-Grant, J.P. Vanden Heuvel, and G.H. Perdew. 1998. Hepatitis B virus X-associated protein 2 is a subunit of the unliganded aryl hydrocarbon receptor core complex and exhibits transcriptional enhancer activity. *Mol Cell Biol* 18:978–988.

69. Sikorski, R.S., M.S. Boguski, M. Goebl, and P. Hieter. 1990. A repeating amino acid motif in CDC23 defines a family of proteins and a new relationship among genes required for mitosis and RNA synthesis. *Cell* 60:307–317.

70. Uzawa, S., I. Samejima, T. Hirano, K. Tanaka, and M. Yanagida. 1990. The fission yeast cut1 + gene regulates spindle pole body duplication and has homology to the budding yeast ESP1 gene. *Cell* 62:913–925.

71. Kuzhandaivelu, N., Y.S. Cong, C. Inouye, W.M. Yang, and E. Seto. 1996. XAP2, a novel hepatitis B virus X-associated protein that inhibits X transactivation. *Nucleic Acids Res* 24:4741–4750.

72. Kazlauskas, A., L. Poellinger, and I. Pongratz. 2002. Two distinct regions of the immunophilin-like protein XAP2 regulate dioxin receptor function and interaction with hsp90. *J Biol Chem* 277:11795–11801.

73. Hollingshead, B.D., J.R. Petrulis, and G.H. Perdew. 2004. The aryl hydrocarbon (Ah) receptor transcriptional regulator hepatitis B virus X-associated protein 2 antagonizes p23 binding to Ah receptor-Hsp90 complexes and is dispensable for receptor function. *J Biol Chem* 279:45652–45661.

74. Sumanasekera, W.K., E.S. Tien, R. Turpey, J.P. Vanden Heuvel, and G.H. Perdew. 2003. Evidence that peroxisome proliferator-activated receptor alpha is complexed with the 90-kDa heat shock protein and the hepatitis virus B X-associated protein 2. *J Biol Chem* 278:4467–4473.

75. Das, A.K., P.W. Cohen, and D. Barford. 1998. The structure of the tetratricopeptide repeats of protein phosphatase 5: Implications for TPR-mediated protein–protein interactions. *EMBO J* 17:1192–1199.

76. Scheufler, C., A. Brinker, G. Bourenkov, S. Pegoraro, L. Moroder, H. Bartunik, F.U. Hartl, and I. Moarefi. 2000. Structure of TPR domain-peptide complexes: Critical elements in the assembly of the Hsp70–Hsp90 multichaperone machine. *Cell* 101:199–210.

77. Taylor, P., J. Dornan, A. Carrello, R.F. Minchin, T. Ratajczak, and M.D. Walkinshaw. 2001. Two structures of cyclophilin 40: Folding and fidelity in the TPR domains. *Structure (Camb)* 9:431–438.

78. Wu, B., P. Li, Y. Liu, Z. Lou, Y. Ding, C. Shu, S. Ye, M. Bartlam, B. Shen, and Z. Rao. 2004. 3D structure of human FK506-binding protein 52: Implications for the assembly of the glucocorticoid receptor/Hsp90/immunophilin heterocomplex. *Proc Natl Acad Sci USA* 101:8348–8353.

79. Yang, J., S.M. Roe, M.J. Cliff, M.A. Williams, J.E. Ladbury, P.T. Cohen, and D. Barford. 2005. Molecular basis for TPR domain-mediated regulation of protein phosphatase 5. *EMBO J* 24:1–10.

80. O'Connell, J.C., J.F. McCallum, I. McPhee, J. Wakefield, E.S. Houslay, W. Wishart, G. Bolger, M. Frame, and M.D. Houslay. 1996. The SH3 domain of Src tyrosyl protein kinase interacts with the N-terminal splice region of the PDE4A cAMP-specific phosphodiesterase RPDE-6 (RNPDE4A5). *Biochem J* 318:255–262.

81. McPhee, I., S.J. Yarwood, G. Scotland, E. Huston, M.B. Beard, A.H. Ross, E.S. Houslay, and M.D. Houslay. 1999. Association with the SRC family tyrosyl kinase LYN triggers a conformational change in the catalytic region of human cAMP-specific phosphodiesterase HSPDE4A4B. Consequences for rolipram inhibition. *J Biol Chem* 274:11796–11810.

82. Beard, M.B., J.C. O'Connell, G.B. Bolger, and M.D. Houslay. 1999. The unique N-terminal domain of the cAMP-specific phosphodiesterase PDE4D4 allows for interaction with specific SH3 domains. *FEBS Lett* 460:173–177.

83. Rena, G., F. Begg, A. Ross, C. MacKenzie, I. McPhee, L. Campbell, E. Huston, M. Sullivan, and M.D. Houslay. 2001. Molecular cloning, genomic positioning, promoter identification, and characterization of the novel cyclic amp-specific phosphodiesterase PDE4A10. *Mol Pharmacol* 59:996–1011.

84. Beard, M.B., E. Huston, L. Campbell, I. Gall, I. McPhee, S. Yarwood, G. Scotland, and M.D. Houslay. 2002. In addition to the SH3 binding region, multiple regions within the N-terminal noncatalytic portion of the cAMP-specific phosphodiesterase, PDE4A5, contribute to its intracellular targeting. *Cell Signal* 14:453–465.

85. Davis, R.L., H. Takayasu, M. Eberwine, and J. Myres, 1989. Cloning and characterization of mammalian homologs of the Drosophila dunce + gene. *Proc Natl Acad Sci USA* 86:3604–3608.

86. Sullivan, M., G. Rena, F. Begg, L. Gordon, A.S. Olsen, and M.D. Houslay. 1998. Identification and characterization of the human homologue of the short PDE4A cAMP-specific phosphodiesterase RD1 (PDE4A1) by analysis of the human HSPDE4A gene locus located at chromosome 19p13.2. *Biochem J* 333:693–703.

87. Scotland, G., and M.D. Houslay. 1995. Chimeric constructs show that the unique N-terminal domain of the cyclic AMP phosphodiesterase RD1 (RNPDE4A1A; rPDE-IVA1) can confer membrane association upon the normally cytosolic protein chloramphenicol acetyltransferase. *Biochem J* 308:673–681.

88. Pooley, L., Y. Shakur, G. Rena, and M.D. Houslay. 1997. Intracellular localization of the PDE4A cAMP-specific phosphodiesterase splice variant RD1 (RNPDE4A1A) in stably transfected human thyroid carcinoma FTC cell lines. *Biochem J* 321:177–185.

89. Smith, K.J., G. Scotland, J. Beattie, I.P. Trayer, and M.D. Houslay. 1996. Determination of the structure of the N-terminal splice region of the cyclic AMP-specific phosphodiesterase RD1 (RNPDE4A1) by 1-H NMR and identification of the membrane association domain using chimeric constructs. *J Biol Chem* 271:16703–16711.

90. Baillie, G.S., E. Huston, G. Scotland, M. Hodgkin, I. Gall, A.H. Peden, C. MacKenzie, E.S. Houslay, R. Currie, T.R. Pettitt, A.R. Walmsley, M.J. Wakelam, J. Warwicker, and M.D. Houslay. 2002. TAPAS-1, a novel microdomain within the unique N-terminal region of the PDE4A1 cAMP-specific phosphodiesterase that allows rapid, Ca^{2+}-triggered membrane association with selectivity for interaction with phosphatidic acid. *J Biol Chem* 277:28298–28309.

91. D'Sa, C., A.J. Eisch, G.B. Bolger, and R.S. Duman. 2005. Differential expression and regulation of the cAMP-selective phosphodiesterase type 4A splice variants in rat brain by chronic antidepressant administration. *Eur J Neurosci* 22:1463–1475.

92. Soejima, H., S. Kawamoto, J. Akai, O. Miyoshi, Y. Arai, T. Morohka, S. Matsuo, N. Niikawa, A. Kimura, K. Okubo, and T. Mukai. 2001. Isolation of novel heart-specific genes using the BodyMap database. *Genomics* 74:115–120.

93. Wilkinson, K., E.R. Velloso, L.F. Lopes, C. Lee, J.C. Aster, M.A. Shipp, and R.C. Aguiar. 2003. Cloning of the t(1;5)(q23;q33) in a myeloproliferative disorder associated with eosinophilia: Involvement of PDGFRB and response to imatinib. *Blood* 102:4187–4190.

94. Sullivan, P.F., K.S. Kendler, and M.C. Neale. 2003. Schizophrenia as a complex trait: Evidence from a meta-analysis of twin studies. *Arch Gen Psychiatry* 60:1187–1192.

95. Millar, J.K., J.C. Wilson-Annan, S. Anderson, S. Christie, M.S. Taylor, C.A. Semple, R.S. Devon, D.M. Clair, W.J. Muir, D.H. Blackwood, and D.J. Porteous. 2000. Disruption of two novel genes by a translocation cosegregating with schizophrenia. *Hum Mol Genet* 9:1415–1423.

96. Harrison, P.J., and D.R. Weinberger. 2005. Schizophrenia genes, gene expression, and neuropathology: On the matter of their convergence. *Mol Psychiatry* 10:40–68.

97. Miyoshi, K., A. Honda, K. Baba, M. Taniguchi, K. Oono, T. Fujita, S. Kuroda, T. Katayama, and M. Tohyama. 2003. Disrupted-In-Schizophrenia 1, a candidate gene for schizophrenia, participates in neurite outgrowth. *Mol Psychiatry* 8:685–694.

98. Hayashi, M.A., F.C. Portaro, M.F. Bastos, J.R. Guerreiro, V. Oliveira, S.S. Gorrao, D.V. Tambourgi, O.A. Sant'Anna, P.J. Whiting, L.M. Camargo, K. Konno, N.J. Brandon, and A.C. Camargo. 2005. Inhibition of NUDEL (nuclear distribution element-like)-oligopeptidase activity by disrupted-in-schizophrenia 1. *Proc Natl Acad Sci USA* 102:3828–3833.

99. Brandon, N.J., E.J. Handford, I. Schurov, J.C. Rain, M. Pelling, B. Duran-Jimeniz, L.M. Camargo, K.R. Oliver, D. Beher, M.S. Shearman, and P.J. Whiting. 2004. Disrupted in Schizophrenia 1 and Nudel form a neurodevelopmentally regulated protein complex: Implications for schizophrenia and other major neurological disorders. *Mol Cell Neurosci* 25:42–55.

100. Millar, J.K., B.S. Pickard, S. Mackie, R. James, S. Christie, S.R. Buchanan, M.P. Malloy, J.E. Chubb, E. Huston, G.S. Baillie, P.A. Thomson, E.V. Hill, N.J. Brandon, J.C. Rain, L.M. Camargo, P.J. Whiting, M.D. Houslay, D.H. Blackwood, W.J. Muir, and D.J. Porteous. 2005. DISC1 and PDE4B are interacting genetic factors in schizophrenia that regulate cAMP signaling. *Science* 310:1187–1191.

101. Sunahara, R.K., C.W. Dessauer, and A.G. Gilman. 1996. Complexity and diversity of mammalian adenylyl cyclases. *Annu Rev Pharmacol Toxicol* 36:461–480.

102. Shenoy, S.K., and R.J. Lefkowitz. 2003. Trafficking patterns of β-arrestin and G protein-coupled receptors determined by the kinetics of β-arrestin deubiquitination. *J Biol Chem* 278:14498–14506.

103. Lefkowitz, R.J., and S.K. Shenoy. 2005. Transduction of receptor signals by β-arrestins. *Science* 308:512–517.

104. Gainetdinov, R.R., R.T. Premont, L.M. Bohn, R.J. Lefkowitz, and M.G. Caron. 2004. Desensitization of G protein-coupled receptors and neuronal functions. *Annu Rev Neurosci* 27:107–144.

105. Perry, S.J., G.S. Baillie, T.A. Kohout, I. McPhee, M.M. Magiera, K.L. Ang, W.E. Miller, A.J. McLean, M. Conti, M.D. Houslay, and R.J. Lefkowitz. 2002. Targeting of cyclic AMP degradation to β₂-adrenergic receptors by β-arrestins. *Science* 298:834–836.

106. Baillie, G.S., A. Sood, I. McPhee, I. Gall, S.J. Perry, R.J. Lefkowitz, and M.D. Houslay. 2003. β-Arrestin-mediated PDE4 cAMP phosphodiesterase recruitment regulates β-adrenoceptor switching from Gs to Gi. *Proc Natl Acad Sci USA* 100:940–945.

107. Bolger, G.B., A. McCahill, E. Huston, Y.F. Cheung, T. McSorley, G.S. Baillie, and M.D. Houslay. 2003. The unique amino-terminal region of the PDE4D5 cAMP phosphodiesterase isoform confers preferential interaction with β-arrestins. *J Biol Chem* 278:49230–49238.

108. Lynch, M.J., G.S. Baillie, A. Mohamed, X. Li, C. Maisonneuve, E. Klussmann, G. Van Heeke, and M.D. Houslay. 2005. RNA silencing identifies PDE4D5 as the functionally relevant cAMP phosphodiesterase interacting with β-arrestin to control the protein kinase A/AKAP79-mediated switching of the β₂-adrenergic receptor to activation of ERK in HEK293B2 cells. *J Biol Chem* 280:33178–33189.

109. Bohn, L.M., R.J. Lefkowitz, R.R. Gainetdinov, K. Peppel, M.G. Caron, and F.T. Lin. 1999. Enhanced morphine analgesia in mice lacking β-arrestin 2. *Science* 286:2495–2498.

110. Brunton, L.L. 2003. PDE4: Arrested at the border. *Sci STKE* 2003:E44.

111. Daaka, Y., L.M. Luttrell, and R.J. Lefkowitz. 1997. Switching of the coupling of the β₂-adrenergic receptor to different G proteins by protein kinase A. *Nature* 390:88–91.

112. Luttrell, L.M., F.L. Roudabush, E.W. Choy, W.E. Miller, M.E. Field, K.L. Pierce, and R.J. Lefkowitz. 2001. Activation and targeting of extracellular signal-regulated kinases by β-arrestin scaffolds. *Proc Natl Acad Sci USA* 98:2449–2454.

113. McDonald, P.H., C.W. Chow, W.E. Miller, S.A. Laporte, M.E. Field, F.T. Lin, R.J. Davis, and R.J. Lefkowitz. 2000. β-Arrestin 2: A receptor-regulated MAPK scaffold for the activation of JNK3. *Science* 290:1574–1577.

114. Luttrell, L.M., S.S. Ferguson, Y. Daaka, W.E. Miller, S. Maudsley, R.G. Della, F. Lin, H. Kawakatsu, K. Owada, D.K. Luttrell, M.G. Caron, and R.J. Lefkowitz. 1999. β-Arrestin-dependent formation of β₂ adrenergic receptor-Src protein kinase complexes. *Science* 283:655–661.

115. Cao, W., L.M. Luttrell, A.V. Medvedev, K.L. Pierce, K.W. Daniel, T.M. Dixon, R.J. Lefkowitz, and S. Collins. 2000. Direct binding of activated c-Src to the β₃-adrenergic receptor is required for MAP kinase activation. *J Biol Chem* 275:38131–38134.

116. Bolger, G.B., S. Erdogan, R.E. Jones, K. Loughney, G. Scotland, R. Hoffmann, I. Wilkinson, C. Farrell, and M.D. Houslay. 1997. Characterization of five different proteins produced by alternatively spliced mRNAs from the human cAMP-specific phosphodiesterase PDE4D gene. *Biochem J* 328:539–548.

117. Han, M., V.V. Gurevich, S.A. Vishnivetskiy, P.B. Sigler, and C. Schubert. 2001. Crystal structure of β-arrestin at 1.9 A: Possible mechanism of receptor binding and membrane Translocation. *Structure (Camb)* 9:869–880.

118. Chen, S., F. Lin, and H.E. Hamm. 2005. RACK1 binds to a signal transfer region of G βγ and inhibits phospholipase C β$_2$ activation. *J Biol Chem* 280:33445–33452.

119. Lopez-Bergami, P., H. Habelhah, A. Bhoumik, W. Zhang, L.H. Wang, and Z. Ronai. 2005. RACK1 mediates activation of JNK by protein kinase C [corrected]. *Mol Cell* 19:309–320.

120. Kiely, P.A., M. Leahy, D. O'Gorman, and R. O'Connor. 2005. RACK1-mediated integration of adhesion and insulin-like growth factor I (IGF-I) signaling and cell migration are defective in cells expressing an IGF-I receptor mutated at tyrosines 1250 and 1251. *J Biol Chem* 280:7624–7633.

121. McCahill, A., J. Warwicker, G.B. Bolger, M.D. Houslay, and S.J. Yarwood. 2002. The RACK1 scaffold protein: A dynamic cog in cell response mechanisms. *Mol Pharmacol* 62:1261–1273.

122. Yarwood, S.J., M.R. Steele, G. Scotland, M.D. Houslay, and G.B. Bolger. 1999. The RACK1 signaling scaffold protein selectively interacts with the cAMP-specific phosphodiesterase PDE4D5 isoform. *J Biol Chem* 274:14909–14917.

123. Bolger, G.B., A. McCahill, S.J. Yarwood, M.S. Steele, J. Warwicker, and M.D. Houslay. 2002. Delineation of RAID1, the RACK1 interaction domain located within the unique N-terminal region of the cAMP-specific phosphodiesterase, PDE4D5. *BMC Biochem* 3:24.

124. Steele, M.R., A. McCahill, D.S. Thompson, C. MacKenzie, N.W. Isaacs, M.D. Houslay, and G.B. Bolger. 2001. Identification of a surface on the β-propeller protein RACK1 that interacts with the cAMP-specific phosphodiesterase PDE4D5. *Cell Signal* 13:507–513.

125. Bolger, G.B., G.S. Baillie, X. Li, M.J. Lynch, P. Herzyka, A. Mohamed, L.H. Mitchell, A. McCahill, C. Hundsrucker, E. Klussmann, D.R. Adams, and M.D. Houslay. 2006. Scanning peptide array analyses identify overlapping binding sites for the signalling scaffold proteins, beta-arrestin and RACK1 in cAMP-specific phosphodiesterase PDE4D5, *Biochem J* 398:23–36.

126. Abrahamsen, H., G. Baillie, J. Ngai, T. Vang, K. Nika, A. Ruppelt, T. Mustelin, M. Zaccolo, M. Houslay, and K. Tasken. 2004. TCR- and CD28-mediated recruitment of phosphodiesterase 4 to lipid rafts potentiates TCR signaling. *J Immunol* 173:4847–4858.

127. Tasken, K., and E.M. Aandahl. 2004. Localized effects of cAMP mediated by distinct routes of protein kinase A. *Physiol Rev* 84:137–167.

128. Wong, W., and J.D. Scott. 2004. AKAP signalling complexes: Focal points in space and time. *Nat Rev Mol Cell Biol* 5:959–970.

129. Coghlan, V.M., B.A. Perrino, M. Howard, L.K. Langeberg, J.B. Hicks, W.M. Gallatin, and J.D. Scott. 1995. Association of protein kinase A and protein phosphatase 2B with a common anchoring protein. *Science* 267:108–111.

130. Dodge, K.L., S. Khouangsathiene, M.S. Kapiloff, R. Mouton, E.V. Hill, M.D. Houslay, L.K. Langeberg, and J.D. Scott. 2001. mAKAP assembles a protein kinase A/PDE4 phosphodiesterase cAMP signaling module. *EMBO J* 20:1921–1930.

131. Carlisle Michel, J.J., K.L. Dodge, W. Wong, N.C. Mayer, L.K. Langeberg, and J.D. Scott. 2004. PKA-phosphorylation of PDE4D3 facilitates recruitment of the mAKAP signalling complex. *Biochem J* 381:587–592.

132. Dodge-Kafka, K.L., J. Soughayer, G.C. Pare, J.J. Carlisle Michel, L.K. Langeberg, M.S. Kapiloff, and J.D. Scott. 2005. The protein kinase A anchoring protein mAKAP coordinates two integrated cAMP effector pathways. *Nature* 437:574–578.

133. Bos, J.L. 2003. Epac: A new cAMP target and new avenues in cAMP research. *Nat Rev Mol Cell Biol* 4:733–738.

134. Enserink, J.M., A.E. Christensen, J. de Rooij, M. van Triest, F. Schwede, H.G. Genieser, S.O. Doskeland, J.L. Blank, and J.L. Bos. 2002. A novel Epac-specific cAMP analogue demonstrates independent regulation of Rap1 and ERK. *Nat Cell Biol* 4:901–906.

135. Kapiloff, M.S., N. Jackson, and N. Airhart. 2001. mAKAP and the ryanodine receptor are part of a multi-component signaling complex on the cardiomyocyte nuclear envelope. *J Cell Sci* 114:3167–3176.
136. Tasken, K.A., P. Collas, W.A. Kemmner, O. Witczak, M. Conti, and K. Tasken. 2001. Phosphodiesterase 4D and protein kinase A type II constitute a signaling unit in the centrosomal area. *J Biol Chem* 276:21999–22002.
137. McCahill, A., T. McSorley, E. Huston, E.V. Hill, M.J. Lynch, I. Gall, G. Keryer, B. Lygren, K. Tasken, G. Van Heeke, and M.D. Houslay. 2005. In resting COS1 cells a dominant negative approach shows that specific, anchored PDE4 cAMP phosphodiesterase isoforms gate the activation, by basal cyclic AMP production, of AKAP-tethered protein kinase A type II located in the centrosomal region. *Cell Signal* 17:1158–1173.
138. Lehnart, S.E., X.H. Wehrens, S. Reiken, S. Warrier, A.E. Belevych, R.D. Harvey, W. Richter, S.L. Jin, M. Conti, and A.R. Marks. 2005. Phosphodiesterase 4D deficiency in the ryanodine-receptor complex promotes heart failure and arrhythmias. *Cell* 123:25–35.
139. Gretarsdottir, S., G. Thorleifsson, S.T. Reynisdottir, A. Manolescu, S. Jonsdottir, T. Jonsdottir, T. Gudmundsdottir, S.M. Bjarnadottir, O.B. Einarsson, H.M. Gudjonsdottir, M. Hawkins, G. Gudmundsson, H. Gudmundsdottir, H. Andrason, A.S. Gudmundsdottir, M. Sigurdardottir, T.T. Chou, J. Nahmias, S. Goss, S. Sveinbjornsdottir, E.M. Valdimarsson, F. Jakobsson, U. Agnarsson, V. Gudnason, G. Thorgeirsson, J. Fingerle, M. Gurney, D. Gudbjartsson, M.L. Frigge, A. Kong, K. Stefansson, and J.R. Gulcher. 2003. The gene encoding phosphodiesterase 4D confers risk of ischemic stroke. *Nat Genet* 35:131–138.

7 Phosphodiesterase 5: Molecular Characteristics Relating to Structure, Function, and Regulation

Sharron H. Francis, Roya Zoraghi, Jun Kotera, Hengming Ke, Emmanuel P. Bessay, Mitsi A. Blount, and Jackie D. Corbin

CONTENTS

7.1   Introduction 132
7.2   Tissue Distribution 133
7.3   Characteristics of the Gene Encoding PDE5, Properties of the Various
      Isoforms, and Gene Regulation 134
7.4   Physical Characteristics 135
      7.4.1   Domain Organization 135
      7.4.2   Regulatory Domain 136
             7.4.2.1   Overall Structural and Functional Features 136
             7.4.2.2   Cyclic GMP Binding 137
             7.4.2.3   GAF Functions 138
             7.4.2.4   Dimerization 141
      7.4.3   Catalytic Domain 141
             7.4.3.1   Overall Structural Features 141
             7.4.3.2   Metal-Binding Site 142
             7.4.3.3   Substrate or Inhibitor Binding 143
             7.4.3.4   Crystal Structures of PDE5 147
7.5   Inhibitor Specificity 148
7.6   Different Conformers of PDE5 148
7.7   Impact of PDE5 Regulatory Mechanisms on cGMP Signaling and Action
      of Inhibitors 149
      7.7.1   Short-Term Regulation of cGMP Signaling 150
      7.7.2   Impact of Concentration and Affinity of cGMP Receptors
            on cGMP Signaling 152
      7.7.3   Negative Feedback Pathway for cGMP Signaling 153

7.7.4 Involvement of PDE5 in Erectile Dysfunction
        and Pulmonary Hypertension.. 154
7.8 PDE5 as a Drug Target in the Treatment of Disease Processes 155
7.9 Concluding Remarks 155
References 155

7.1 INTRODUCTION

Phosphodiesterase 5 (PDE5) was discovered in 1976 as a cGMP-binding protein that comigrated with a cGMP-specific phosphodiesterase (PDE) activity on diethylaminoethyl (DEAE) cellulose chromatography.[1,2] It was eventually shown that the cGMP binding and cGMP-hydrolytic activities were associated with the same protein;[3] the protein was first called the cGMP-binding protein PDE (cG-BPP)[3] and has subsequently been labeled as the cGMP-binding cGMP-specific PDE (cGMP B-PDE, cGMP-BPDE), PDE-V, and PDE5. Studies of the initial purification and characterization of PDE5 and its functions have been previously summarized.[3–11] PDE5 was the first PDE shown to directly bind cGMP at allosteric sites.[3] It was also the first mammalian receptor shown for cAMP or cGMP other than the cAMP- and cGMP-dependent protein kinases (PKA and PKG, respectively). For years, interest in PDE5 languished in large part due to the uncertainty of the role of cGMP in cellular signaling processes. Interest in PDE5 was re-energized in the mid- to late 1980s by accumulating evidence that cGMP plays a prominent role in several physiological processes (Figure 7.1).[12–19] Since PDE5 is particularly abundant in the vascular smooth muscle, it increasingly became a protein of interest as a potential pharmacological target for the treatment of hypertension and other vascular diseases.[4,10,20–27]

Development of inhibitors with improved potency and selectivity for PDE5, such as zaprinast (Figure 7.2), advanced insight into some PDE5 functions. Introduction of sildenafil (Viagra), vardenafil (Levitra), tadalafil (Cialis), and udenafil (zydena) highly selective and

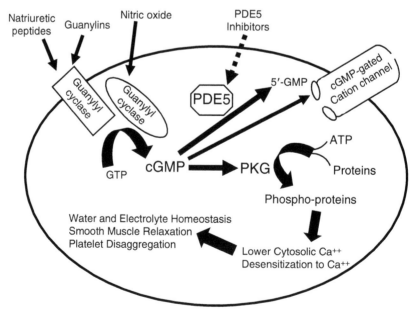

FIGURE 7.1 General scheme for the role of PDE5 and PKG in cGMP-signaling pathways. PKG, cGMP-dependent protein kinase; PDE5, phosphodiesterase-5.

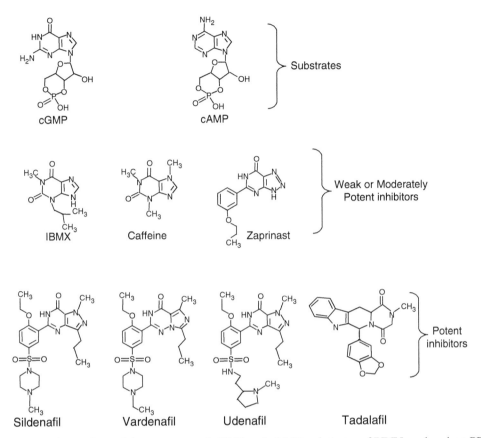

FIGURE 7.2 Comparison of the structures of cGMP and cAMP, substrates of PDE5, and various PDE5 inhibitors. 3-Isobutyl-1-methylxanthine (IBMX) and caffeine are weak inhibitors; zaprinast is a moderately potent inhibitor; sildenafil, vardenfil, tadalafil, and udenafil are potent PDE5 inhibitors.

potent inhibitors of PDE5 (Figure 7.2), to the commercial market and the remarkable success of these medications in the treatment of erectile dysfunction (ED) continues to energize interest in PDE5.[9,20,24,28,29] PDE5 is now recognized to play a role in modulating smooth muscle tone in general and penile corpus cavernosal smooth muscle tone in particular (Figure 7.1).[9,30,31] It is also implicated to have a role in platelet function,[2,32–34] pulmonary vascular function,[35–37] recovery from stroke,[38,39] control of water and electrolyte homeostasis by the intestine,[40] neuronal cell function,[41] Purkinje fiber physiology,[42–44] cardiac function,[45,46] and serotonin transport.[47] Advances in understanding the biochemical properties of PDE5 have uncovered seminal features that are common to class I PDEs, including the role of Zn^{2+} in catalysis[7,48–51] and the functions of invariant amino acids.[50–53] Studies using PDE5 have also enhanced understanding of the complex function and regulation of cGMP-binding PDEs,[54–60] and expanded the appreciation of regulatory features that impact PDE functions under physiological conditions and pharmacological regimens.[9,28,61–64]

7.2 TISSUE DISTRIBUTION

PDE5 protein occurs in high levels in most smooth muscle, lung, platelets, kidney, gastro-intestinal epithelial cells, pulmonary artery endothelial cells, and Purkinje neurons and has been detected in proximal renal tubules, collecting renal ducts, Skene periurethral glands,

vaginal epithelium, and epithelial cells of pancreatic ducts.[2,36,40,65–67] PDE5 is the major 'cGMP-hydrolyzing PDE in lung,[68] pulmonary arterial endothelial cells,[36] platelets,[31,69,70] and penile corpus cavernosum of human, dog, and rabbit.[20,21,24,71] PDE5 mRNA has been detected in many more tissues, but the abundance of the protein in these tissues is unknown[2,6,26,42,66,72]. PDE5 was first purified and characterized from rat lung[4,23] and subsequently cloned from several species.[67,73–76]

7.3 CHARACTERISTICS OF THE GENE ENCODING PDE5, PROPERTIES OF THE VARIOUS ISOFORMS, AND GENE REGULATION

PDE5 is the product of a single gene that is ~8.5 kb in length, contains 23 exons (Figure 7.3), and is located on human chromosome 4q25–27.[67] PDE5 cDNAs have been isolated from bovine, dog, human, mouse, and rat.[67,73,74,76,77] PDE5 proteins from different species have a high degree of identity; bovine and human PDE5A1 enzymes are 96% identical after

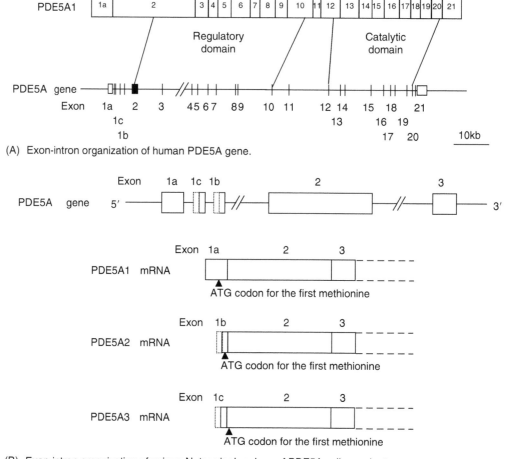

(A) Exon-intron organization of human PDE5A gene.

(B) Exon-intron organization of unique N–terminal regions of PDE5A splice variants.

FIGURE 7.3 Exon–intron organization of human PDE5 gene and of unique N-terminal regions of PDE5A splice variants. (A) The upper panel shows the structure of the PDE5A1 protein with corresponding exon numbers. The lower panel shows the exon–intron organization of the human PDE5A gene. Vertical lines are exons; white boxes in the lower panel indicate untranslated regions. (B) The upper panel shows the exon–intron organization of the human PDE5A gene. The white boxes indicate exons or mRNAs.

excluding an insertion of a 10 amino acid[72] glutamine-rich segment near the N-terminus in the human enzyme.[67] Three alternatively spliced variants, PDE5A1, PDE5A2, and PDE5A3, have been identified.[67,73,74] These differ only in the 5′ end of the respective mRNAs and the corresponding amino acid sequence at the extreme N-terminus in the protein products. Alternative first exons arranged in the order A1–A3–A2 account for these variants; however, there is no start methionine codon in the first exon of PDE5A3 cDNA (Figure 7.3). PDE5A3 is translated from the first codon for methionine of exon 2, which is the first common exon of the PDE5A variants. PDE5A2 is the most widely distributed isoform.[78] Human PDE5A1, PDE5A2, and PDE5A3 are comprised of 875 amino acids (~100 kDa), 833 amino acids (~95 kDa), and 823 amino acids (~95 kDa), respectively. A fourth form (PDE5A4) that was reported several years ago was apparently the result of a cloning error. A single base pair change (T152C) was detected in clones from T84 cells and causes replacement of Val51 in PDE5A2 by alanine (GenBank accession number AY264918).[40] No functional effects have thus far been associated with this change.

Factors that regulate expression of PDE5 are poorly understood; angiotensin II or androgens are reported to increase PDE5 expression.[79–81] Both cGMP and cAMP have been implicated in modulating transcription of PDE5, but there are contradictory reports.[73,82–84] In mouse neuroblastoma cells, elevation of cAMP is associated with increased PDE5 mRNA expression[82]; in rat vascular smooth muscle cells, cAMP increases expression of PDE5A2 mRNA, but not PDE5A1.[73] A short gene promoter (139 bp) is located upstream from the PDE5A1-specific first exon, but extensions on either side (308 bp upstream containing AP2- and Sp1-binding sites and 156 bp downstream containing Sp1-binding sequences) are reported to contribute to the regulation of PDE5 gene transcription.[84] Another promoter that contains both Sp1- and AP2-binding sites is located in an intron upstream from the PDE5A2-specific first exon. One of the Sp1-binding sites is reported to be important for both basal and cAMP- or cGMP-stimulated promotion of PDE5A2 expression.[84] In contrast, the treatment of myometrial cells isolated from pregnant or nonpregnant women with either human chorionic gonadotropin or dibutyryl cAMP causes a reduction in PDE5 expression and activity.[85] The basis for these discrepancies is unknown and will require more extensive studies to resolve.

To date, no significant functional differences among PDE5A1, PDE5A2, and PDE5A3 have been detected. PDE5s from various tissues exhibit similar K_m values for cGMP and IC_{50} values for inhibitors, although some modest differences have been reported.[86] This might reflect different proportions of PDE5 isoforms in a given tissue or different biochemical status of the enzyme, e.g., varied proportions of the phospho- and unphospho-PDE5s or the highly active and less active forms described later. Tissue distribution of PDE5A1, PDE5A2, and PDE5A3 varies considerably. PDE5A1 and PDE5A2 are coexpressed in a variety of tissues, but PDE5A2 is apparently expressed most widely; PDE5A2 is the more abundant isoform in human corpus cavernosum and a number of other tissues. PDE5A3 may be selectively expressed in smooth muscle cells.[78] In most tissues examined so far, PDE5 is largely cytosolic; however, some PDE5 is associated with the particulate fraction in certain tissues.[36,68] The low level of PDE5 in cardiomyocytes from healthy animals is primarily localized to the Z band, but under pathological conditions, it is dispersed throughout the cell,[87] and in pulmonary arterial smooth muscle cells, a significant portion of PDE5 is in the particulate fraction.[35]

7.4   PHYSICAL CHARACTERISTICS

7.4.1   DOMAIN ORGANIZATION

PDE5 isoforms are comprised of two identical subunits (subunit M_r ~95,000–100,000). PDE5 is a chimeric protein comprised of two major domains, i.e., a more N-terminal regulatory (R) domain and a more C-terminal catalytic (C) domain that are approximately equal in size (Figure 7.4). Ongoing communication between PDE5 R and C domains in PDE5 holoenzyme

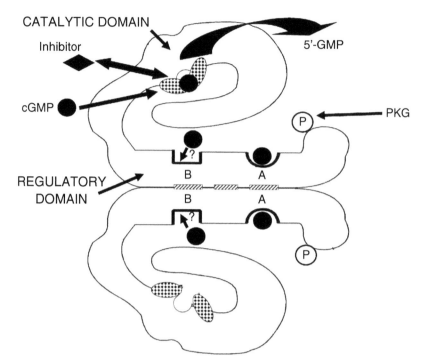

FIGURE 7.4 Working model of PDE5. The regulatory domain is located in the N-terminal portion of PDE5 and contains the phosphorylation site at Ser102, two GAFs (A and B), and dimerization contacts shown as crosshatched bars. Cyclic GMP binds to GAF-A in the regulatory domain, but whether GAF-B also binds cGMP is still a question as indicated by (?). The catalytic domain is located in the C-terminal portion of the molecule and contains the catalytic pocket, which includes metal-binding sites, shown as a doublet of black and white ovals and the substrate, cGMP- (black circle) or inhibitor (black diamond)-binding site.

contributes importantly to the regulation of enzyme function; the literature is replete with evidence of this communication, but the molecular mechanisms involved are still unknown.[3,237,33,34,54,61,103,104] The majority of the sequence in the R domain, all of the C domain sequence, and the phosphorylation consensus site for PKG or PKA are highly conserved in all PDE5 isoforms. PDE5 monomers dimerize tightly through complex interactions between R domains.[10,59,88,89] Interdomain communication between the R and C domains provides critical reciprocal regulation of enzyme functions. When the R and C domains are expressed separately, each domain retains the salient features of function as determined in the holoenzyme.[59,88,90,91] The C domains of PDE5 are most similar in sequence to the C domains of PDE6, a cGMP-specific PDE family (~44%),[92] and PDE11, a PDE with dual-substrate specificity (~50%).[93-95] The PDE5 catalytic site exhibits ~100-fold preference for cGMP over cAMP, and all known inhibitors directly compete with substrate for access to the catalytic pocket (Figure 7.4). The R domain is homologous to other members of the GAF-containing subgroup of class I PDEs, i.e., PDE2, PDE6, PDE10, and PDE11, and the bulk of the similarity resides in the GAFs.

7.4.2 REGULATORY DOMAIN

7.4.2.1 Overall Structural and Functional Features

The R domain of PDE5 is comprised of a number of elements: a single phosphorylation site (Ser102 or Ser92 in human or bovine PDE5A1, respectively) that can be phosphorylated by

either PKG or PKA,[96] a complex set of dimerization contacts, autoinhibitory elements, and two GAFs (A and B) that are arranged in tandem (Figure 7.4). The GAF acronym was derived from the names of the first three classes of proteins recognized to contain these sequences, i.e., cGMP-binding PDEs, *Anabaena* adenylyl cyclase, and *Escherichia coli* Fh1A.[97,98] GAFs occur in only three known groups of chordate proteins, including the latent-transforming growth-factor-binding protein 4 from humans[99] and a hypothetical EF-hand containing protein in mouse;[100] among these, the GAF motif is by far the most abundant in the PDE superfamily.[60,101] GAFs in the PDEs are comprised of a sequence of ~110 amino acids that are highly homologous and predicted to have a common evolutionary origin. The GAF-A sequence in PDE5 is 52% identical to that in PDE11 and 48% identical to GAF-B in PDE2. However, the amino acid sequences among GAFs in human PDEs do not have a pattern of stronger homology among GAF-A or GAF-B sequences. The GAFs in PDE5 are evolutionarily, structurally, and biochemically distinct from the catalytic site of PDE5 as well as from other cyclic nucleotide (cN)-binding sites, including those in cN-dependent protein kinases, cN-gated cation channels, and cN-regulated guanine-nucleotide exchange factors.[102] Both GAFs in PDE5 contain the characteristic NKXD motif that is common among a subgroup of GAFs.[58,76]

GAFs are commonly described as small ligand-binding motifs, but direct evidence for ligand binding is available for only a few proteins; an equally important function may be their involvement in protein–protein interactions. Both functions are represented in PDE5, but there are distinct differences among PDEs as evidenced when comparing GAF functions in PDE2 and PDE5.[58,60,101]

7.4.2.2 Cyclic GMP Binding

One prominent feature of the PDE5 R domain is the binding of cGMP at allosteric sites that shows high selectivity (~100- to 300-fold) for cGMP over cAMP (Figure 7.4).[3,4,10,59] In PDE5 holoenzyme, occupation of the catalytic site by cGMP or inhibitors increases cGMP binding at allosteric sites in the R domain;[3,10] as would be predicted based on the principle of reciprocity,[105] allosteric cGMP binding increases affinity of the catalytic site for cGMP (not shown) and inhibitors (Figure 7.5).[61] Cyclic GMP occupation of the allosteric sites on PDE5 holoenzyme (Figure 7.6A), the isolated R domain (Figure 7.6B), or mutants containing only GAF-A (not shown) causes an apparent elongation that can be detected as slowed migration of the proteins on nondenaturing polyacrylamide gel electrophoresis.[5,59,62,106] The relationship of this apparent elongation to activation of PDE5 by allosteric cGMP binding remains to be determined.

The stoichiometry of allosteric cGMP binding for either PDE5 holoenzyme or isolated R domain is ~0.6–0.8 mol/monomer. The binding pattern contains a high-affinity component and a low-affinity component (Figure 7.7A), whose rates differ by ~10-fold.[10,59,62,106,107] Biphasic cGMP-binding kinetics are found in native PDE5, recombinant PDE5, His-tagged PDE5, glutathione-*S*-transferase (GST)-fusion or His-tagged constructs of the isolated R domain, and His-tagged constructs containing only GAF-A.[59,62,106] The high- and low-affinity components appear to be in equilibrium since the high-affinity component accounts for 100%, 70%, or 40% of total cGMP binding among different preparations; in addition, this distribution can be experimentally changed by processes such as phosphorylation (Figure 7.7B).[10,59,76,90,106] Point mutations of a number of amino acids in the GAFs of bovine PDE5 selectively alter either the high- or low-affinity cGMP binding[107]; thus, it was proposed that GAF-A and GAF-B provide for the high- and low-affinity cGMP-binding components, respectively. More recent studies[108] have established that GAF-A binds cGMP with high affinity, but a role for GAF-B in contributing to the heterogeneity of cGMP binding is now in question.

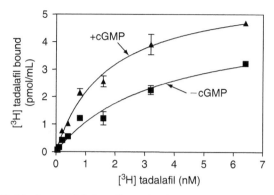

FIGURE 7.5 Cyclic GMP binding to PDE5 allosteric sites in the R domain increases catalytic site function. Effect of cGMP (10 μM) on [³H]tadalafil binding at 4°C using the Millipore vacuum filtration method. (Adapted from Ref. [61]. With permission.)

7.4.2.3 GAF Functions

Both GAFs in PDE5 play important roles in the regulation of enzyme functions. Both appear to contribute to the dimerization of PDE5 although confirmation of this role will require the elucidation of the x-ray crystal structure of the R domain.[59,89,108]

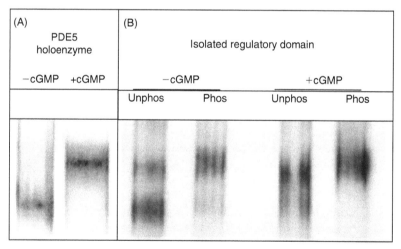

FIGURE 7.6 PDE5 holoenzyme or isolated R domain exist in multiple conformations. (A) Cyclic GMP binding to PDE5 holoenzyme slows migration of the enzyme on nondenaturing gel electrophoresis which is consistent with an apparent elongation of the molecule. (B) Either cGMP binding, phosphorylation, or the combination of cGMP binding and phosphorylation of PDE5 R domain slows migration of the protein on nondenaturing gel electrophoresis. Phospho-PDE5 R domain was stoichiometrically phosphorylated using purified PKA C subunit. Unphospho- and phospho-PDE5 R domains were incubated in the absence or presence of 20 mM cGMP at 4°C, and aliquots were subjected to native gel electrophoresis; buffer for gel shown in the left panel lacked cGMP, buffer in gel shown in right panel contained 1 mM cGMP. (From Ref. [90]. With permission.)

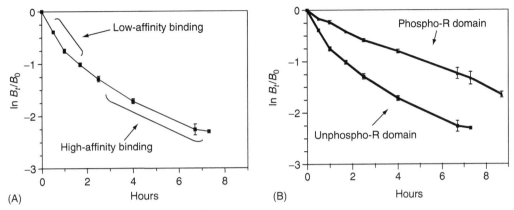

FIGURE 7.7 Exchange dissociation of [³H]cGMP from isolated unphospho-PDE5 R domain is kinetically heterogeneous and phosphorylation converts this to a linear high affinity pattern. Unphospho- and phospho-R domain proteins were incubated for 60 min in a [³H]cGMP-binding mixture containing 1 μM [³H]cGMP to allow for saturation of the allosteric cGMP-binding sites. Aliquots were removed and subjected to vacuum filtration to determine binding at zero time (B_0). A 100-fold excess of unlabeled cGMP was added and aliquots were removed at various times (B_t) for determination of [³H]cGMP remaining bound. (A) Exchange dissociation pattern for unphospho-R domain of PDE5. (B) Comparison of exchange dissociation for unphospho-and phospho-R domain of PDE5. (Adapted from Ref. [90]. With permission.)

7.4.2.3.1   GAF-A

GAF-A binds cGMP with high affinity and may account entirely for PDE5 allosteric cGMP binding (K_D ~ 100–200 nM).[59,106,109] Allosteric cGMP binding increases catalytic site affinity for cGMP, enhances catalytic activity (V_{max}), and converts PDE5 into a stably activated form.[33,104,110] Evidence suggests that this activation is mediated by cGMP binding to GAF-A, since an anti-GAF-A antibody blocks the effect.[104] Mutant proteins containing only GAF-A exhibit ~300-fold higher selectivity for cGMP over cAMP, perhaps slightly more so than the selectivity (~100-fold) exhibited by holoenzyme[59] and have higher cGMP-binding affinity (K_D < 40 nM) than PDE5 holoenzyme (K_D = ~100–200 nM) or the isolated R domain containing both GAFs (K_D ~110 nM). This suggests that the sequence containing GAF-B and its surrounding amino acids has an autoinhibitory effect on cGMP-binding affinity of GAF-A. A truncation mutant containing the entire N-terminus through GAF-A (Met1–Glu321) has high-affinity, biphasic cGMP-binding kinetics consistent with the interpretation that the kinetic heterogeneity of cGMP binding derives from structural heterogeneity of GAF-A. The heterogeneity of cGMP binding by GAF-A may reflect asymmetry in the dimeric GAF-A,[59] i.e., the monomers within a dimer differ, or it could be that different conformers of the entire GAF-A dimer exist, i.e., GAF-As within a dimer are identical but differ from the conformations of GAF-A found in other dimers. A recent report used NMR to define the backbone ¹H, ¹³C, ¹⁵N resonance assignments to the backbone atoms of the GAF-A dimer and provided the first structural information on PDE5 GAF-A.[89]

7.4.2.3.2   GAF-B

GAF-B plays a role in the regulation of a number of PDE5 functions: it modulates allosteric cGMP binding by GAF-A;[59] it is important in the sequestration of Ser102 from phosphorylation by cN-dependent protein kinases in the absence of cGMP (see Section 7.4.2.3.4);[59] it appears to contribute significantly to PDE5 dimerization (see Section 7.4.2.4);[59] and it impacts selectivity of

PDE5 for vardenafil versus sildenafil or tadalafil (see Section 7.4.3.3) (Blount et al., unpublished results). Significant cGMP binding by PDE5 GAF-B has not been demonstrated; however, this possibility cannot be excluded because low-affinity cGMP binding may not be detected by the methods that have been used (Figure 7.4).[59] PDE5 GAF-B also lacks portions of the signature sequence $(SX_{(13-18)}FDX_{(18-22)}IAX_{(21)}[Y/N]X_{(2)}VDX_{(2)}TX_{(3)}TX_{(19)}[E/Q])$ that provides for cGMP binding in PDE2 GAF-A.[101,111]

7.4.2.3.3 Kinetic Heterogeneity

The molecular basis for the kinetic heterogeneity of cGMP binding in PDE5 is not under-stood, but the heterogeneity is likely to be related to conformational differences either among PDE5 dimers or between monomers within a given dimer. In a GST-fusion protein containing only PDE5 GAF-A, a single high-affinity cGMP-binding site was found (K_D ~ 30 nM);[106,109] these results are complicated by the fact that the GST tag may influence the kinetics. Three His-tagged constructs of varying length and containing GAF-A, but not GAF-B, exhibit high-affinity cGMP binding (K_D ~ 28–45 nM) that is kinetically heterogeneous.[59] Phosphor-ylation of Ser102 increases the affinity of allosteric cGMP binding ~10-fold in either PDE5 holoenzyme or the isolated R domain in the isolated R domain: phosphorylation converts the pattern of kinetic heterogeneity of cGMP binding to a single high-affinity species (Figure 7.7B).[54,90] Other unidentified processes may produce a similar effect, thereby accounting for the differences seen in various preparations of PDE5, the isolated R domain, and constructs containing only GAF-A. Interconversion of the high- and low-affinity cGMP-binding com-ponents in PDE5 R domain may have an important role in modulating physiological functioning of PDE5.[33,104]

7.4.2.3.4 Regulation by Phosphorylation or Dephosphorylation

Occupation of the allosteric sites by cGMP in either the PDE5 holoenzyme or the isolated R domain induces an apparent elongation and exposes Ser102 for phosphorylation by PKG or PKA (Figure 7.6).[11,90] In R domain C-terminal truncation mutants containing only GAF-A and N-terminal sequences extending through the phosphorylation site at Ser102, phosphoryl-ation is not modulated by cGMP.[59] This indicates that the sequence containing GAF-B and flanking amino acids is critical for cGMP regulation of Ser102 phosphorylation by cN-dependent protein kinases. The apparent elongation induced by phosphorylation can be detected as slowed migration on nondenaturing gels; in the presence of cGMP, the phospho-R domain exhibits a migration pattern that is distinct from either the cGMP-bound unphosphoprotein or the phos-phoprotein alone (Figure 7.6),[62] but the migration of phospho-PDE5 and cGMP-bound PDE5 holoenzyme does not differ perceptibly (Bessay et al., unpublished results).

The consensus sequence containing the phosphorylation site, Ser102, is a ~10-fold better substrate for PKG over PKA[96] and in studies of purified PDE5, this is the only residue modified.[11,56] Phosphorylation is absent in a point mutant in which Ser102 is replaced with a nonphosphorylatable amino acid.[106] The preference for PKG over PKA is mediated by a phenylalanine that is located four amino acids C-terminal to the phosphorylation site and provides a negative determinant for PKA, but not for PKG.[96] The isolated R domain is phosphorylated by PKG with a K_D of ~3 μM compared to a K_D of ~68 μM for the synthetic peptide (RKISASEFDRPLR) containing this sequence; this indicates that the affinity of PKG for PDE5 is enhanced by contacts outside the phosphorylation consensus sequence.[106] Studies in intact cells suggest that phosphorylation of PDE5 is largely, if not entirely, mediated by PKG.[27,33,103]

Phosphorylation of Ser102 increases affinity of PDE5 allosteric cGMP binding ~10-fold,[54,62] affinity of catalytic site for cGMP (i.e., lower K_m) with little effect on V_{max},[27,54,103]

and affinity for inhibitors even in the absence of cGMP (Bessay et al., unpublished results). Interaction with small proteins related to the Pγ-inhibitory subunit of PDE6 reportedly blocks activation of PDE5 by PKA phosphorylation in crude extracts, but the mechanism of this effect and physiological relevance are unknown.[112] Inhibition of PDE5 catalytic activity by Pγ has also been reported, but this effect is controversial since other investigators have not found an effect in the studies using purified PDE5 and Pγ.[113] After cGMP elevation, phosphorylation of PDE5 has been reported to occur in vascular smooth muscle cells, uterine smooth muscle, gastrointestinal epithelial cells, human colonic T84 cells, and platelets.[27,34,40,103] *In vitro* studies using purified PDE5 and studies in intact smooth muscle cells indicate that phosphoprotein phosphatase 1 is a major determinant of PDE5 dephosphorylation and that dephosphorylation is not regulated by either cGMP or PDE5 inhibitors.[103,177]

7.4.2.4 Dimerization

PDE5 monomers dimerize through strong contacts in the R domain,[10,50,59,88] but the role of dimerization in enzyme function is not known. Multiple point mutations of a leucine zipper motif in the PDE5 R domain did not measurably destabilize the dimer.[88] PDE5 truncation mutants containing only GAF-A or GAF-B also form high-affinity homodimers.[59] Two GAF-A modules dimerize with $K_D < 30$ nM, and GAF-B modules form a dimer with a $K_D = 1–20$ pM. In addition, the components in the sequence (Thr322–Asp403) between the GAFs add to dimer stability (Figure 7.4).[59] Whether all these contacts occur in the holoenzyme is yet to be determined.

7.4.3 CATALYTIC DOMAIN

The C domain hydrolyzes cGMP with a $K_m \sim 1–5$ μM, $V_{max} \sim 2–5$ μmol/min/mg, and k_{cat} $\sim 4–10$ s^{-1}; cAMP and cGMP are hydrolyzed at similar rates, but the affinity for cGMP is ~ 100 times greater than that for cAMP ($K_m \sim 300$ μM). The low affinity for cAMP makes it unlikely that cAMP would be hydrolyzed by PDE5 under physiological conditions. Catalytic site function has been studied in considerable detail using both the holoenzyme and isolated C domain (amino acids 535–875, human). Isolated C domain of PDE5A1 is monomeric and retains the salient catalytic properties (K_m, k_{cat}) of the holoenzyme.[88] Site-directed mutagenesis of amino acids within the C domain and elucidation of several x-ray crystal structures of the isolated C domain have identified residues that are critical in binding metals, cGMP, or inhibitors, or that impact catalytic turnover rate. These details will be summarized in the following section.

7.4.3.1 Overall Structural Features

The PDE5 C domain contains three helical subdomains as occurs in other PDE C domains (Figure 7.8) (see Chapter 29 and Chapter 30).[51,52] The C domains of most PDEs contain 16 α-helices. However, the residues that are located in helices H8 and H9 in other PDEs form different structures in PDE5. In the unliganded PDE5, these form a coil, but when sildenafil is bound, they form a 3_{10} helix.[175] The catalytic site is formed by the juxtaposition of different regions of the protein;[50,52,114] the substrate- or inhibitor-binding site has a volume of ~ 330 Å^3, is about 10 Å deep, and has a narrow aperture at the protein–solvent surface. Sung et al. designated four functional regions: (1) the metal-binding site (M site), (2) the core (Q) pocket lined by Gln817, Phe820, Val782, and Tyr612, (3) the hydrophobic (H) pocket (formed by Phe786, Ala783, Leu804, and Val782), and (4) the lid (L) region composed of Tyr664,

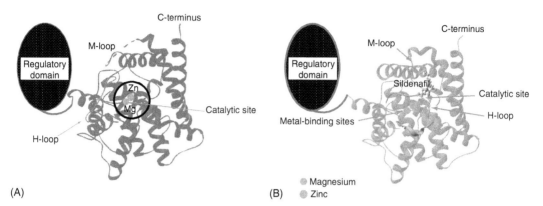

FIGURE 7.8 (See color insert following page 274.) Cartoon depicting the ribbon diagram of the overall structure of PDE5 catalytic domain in the absence and presence of sildenafil. (A) The divalent metal-binding site that forms part of the catalytic site is concomitantly occupied by zinc (pink ball) and magnesium (purple ball). Critical loops (H and M) in the protein structure are indicated. H-loop is shown in purple. (B) Sildenafil stick model, shown in gold, is bound in the catalytic pocket near to the metal-binding sites. H-loop shown in gold has moved from a position on the left in (A) to cover the catalytic site.

Met816, Ala823, and Gly819, which narrows the opening to the catalytic site. Two loops designated as the H-loop (Val660–His684, human) and M-loop (Phe787–Lys812) have subsequently been described (Figure 7.8).[50,114,175]

7.4.3.2 Metal-Binding Site

PDE5, like other PDEs, requires divalent cations for catalytic function and perhaps for maintaining catalytic site structure.[51,115,116] Recognition of conserved Zn^{2+}-binding motifs ($HX_3HX_n(E/D)$) in all class I PDEs, their involvement in catalysis, and direct evidence for Zn^{2+} binding and support of catalytic activity at submicromolar Zn^{2+} in mammalian PDEs were first demonstrated using PDE5.[7]

As predicted from the results of enzymological studies using site-directed mutagenesis of the PDE5 C domain, the metal-binding site is novel and involves multiple amino acids.[7,48] The Zn^{2+} that is visualized in the crystal structure (Figure 7.8) is coordinated to the side chains of His617, Asp654, His653, Asp764, and two water molecules; these six coordinates form an octahedron.[50,52] Site-directed mutagenesis of each of these amino acids profoundly decreases catalytic activity.[48,50,52,53] A point mutation of one of these histidines (His643 in bovine PDE5, His653 in human PDE5) or Asp764 was particularly deleterious to catalytic efficiency.[53] The second metal ion, presumably Mg^{2+}, but not yet identified, also forms an octahedron involving contacts with Asp654 and five water molecules. Polarization of water that bridges the two cations by a negatively charged amino acid such as Asp764 may enhance the nucleophilicity of the oxygen to generate the hydroxyl catalyst that is implicated in breaking the cyclic phosphate ring.[6] A point mutation replacing this aspartic acid in bovine PDE5 causes a 2000-fold loss in catalytic efficiency (k_{cat}/K_m), a result that is consistent with its central role in catalytic function.[53] The results from x-ray crystal structures agree well with biochemical data, indicating that the phosphohydrolase action of class I PDEs involves an in-line nucleophilic substitution of a solvent hydroxyl at the phosphorous that disrupts the P–O bond at the $3'$ oxygen of the ribose and causes inversion of the configuration at

the phosphorous.[117–120] It is likely that Zn^{2+} binding also provides for structural integrity since ethylene diaminetetraacetate (EDTA) ablates binding of high-affinity inhibitors.[115]

7.4.3.3 Substrate or Inhibitor Binding

Contacts between the PDE5 catalytic site and the substrate, cGMP, have been inferred from the cocrystal structure of PDE5 in the complex with three potent inhibitors (sildenafil, tadalafil, and vardenafil), two weak inhibitors (3-isobutyl-1-methylxanthine (IBMX) and icarasid II), and the product (5′ GMP); many of the amino acids that interact with sildenafil are depicted in Figure 7.9. Early studies comparing the effects of site-directed mutagenesis on the affinity of PDE5 for cGMP, sildenafil, zaprinast, and the sildenafil-related compound UK122764 revealed that the more potent the inhibitor in the series, i.e., sildenafil, the more closely the effect of the mutations mimics changes in affinity for cGMP.[64] More recent studies have identified amino acids that are important for cGMP affinity, but not for sildenafil affinity (see later) (Zoraghi et al., unpublished results). Many of the contacts between PDE5 and cGMP or inhibitors are largely provided by amino acids in the Q pocket (lined by Gln817, Phe820, Val782, and Tyr612) and the H pocket (formed by Phe786, Ala783, Leu804, and Val782) (Figure 7.9).[50,52,114] Tadalafil binding utilizes many of the same contacts as those identified for sildenafil and vardenafil, but it also exploits other contacts. The cyclic phosphate ring of cGMP extends into the metal-binding pocket based on the position of the 5′ phosphate in 5′ GMP. However, the precise positioning of this critical component of cGMP in the catalytic pocket is not known because 5′ GMP is a low-affinity product poised to dissociate.

Recent studies implicate an important role for the H-loop in providing high-affinity interaction with cGMP, but not sildenafil.[175] Unique structural features in the inhibitors undoubtedly

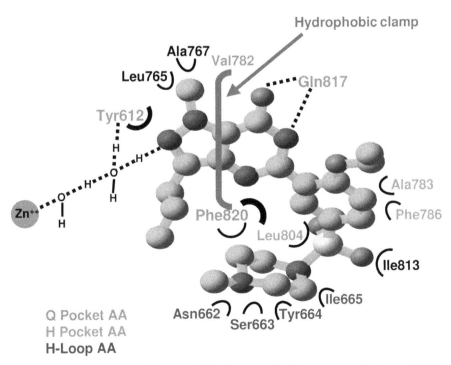

FIGURE 7.9 (See color insert following page 274.) Cartoon depicting contacts between PDE5 catalytic site amino acids and sildenafil. Heavy bars indicate particularly strong interactions.

make contacts that are absent in the cGMP–PDE5 complex and influence the positioning of these ligands in the catalytic site. With these caveats in mind, a number of insights into important contacts have been made.

7.4.3.3.1 Ligand Interactions

Contacts with Phe820 and Val782. The heterocyclic rings of 5′-GMP, sildenafil, vardenafil, tadalafil, and IBMX (Figure 7.2) are sandwiched between the hydrophobic side chains of Val782 and Phe820 (Figure 7.9); this has been described as a hydrophobic clamp. The five-member rings in each ligand form a face-to-face π–π stacking interaction with the phenyl ring of Phe820; the five-member rings of sildenafil and vardenafil also interact with Tyr612 (see later).[52,121] Site-directed mutagenesis of Phe820 (F820A) causes pronounced loss in the affinity for cGMP (60-fold), sildenafil (70-fold), vardenafil (450-fold), tadalafil (140-fold), or IBMX (90-fold), whereas the deleterious effects of mutation of Val782 (V782A) are less remarkable (6-, 10-, 23-, 3-, and 12-fold, respectively) (Zoraghi et al., unpublished results). A stacking interaction involving the amino acid that is homologous to Phe820 may be a common contact between substrates or inhibitors in all PDEs.

Contacts with Tyr612. Site-directed mutagenesis studies indicate that Tyr612 is critical to high affinity of PDE5 for cGMP.[53,59] In the crystal structure of PDE5, Tyr612 interacts with the five-membered ring of sildenafil (Figure 7.9), vardenafil, or 5′-GMP. One of the nitrogens in the pyrazole of sildenafil or the imidazole of vardenafil forms a hydrogen bond (H-bond) with a water molecule that in turn forms a H-bond bridge to both the hydroxyl groups of Tyr612 and another water molecule that coordinates to the catalytic site Zn^{2+}.[52] Substitution of alanine for this tyrosine produces a ~120-fold loss in affinity for vardenafil versus a 15-fold loss in affinity for cGMP, and a 30- or 50-fold loss in affinity for sildenafil or tadalafil, respectively. This implies that the contacts between Tyr612 and the inhibitors are much more important than for substrate affinity. Substitution of phenylalanine for Tyr612 only slightly reduces the affinity for cGMP (twofold) or tadalafil (threefold), but affinity for sildenafil, vardenafil, or IBMX is improved approximately twofold. V_{max} values for these proteins are unaffected. This indicates that the Tyr612 hydroxyl group contributes substantially to the difference in potency between vardenafil and sildenafil or tadalafil, and the hydrophobicity of this position is critical for the potency of both vardenafil and sildenafil.[121]

Contacts with Gln817. In the cocrystal structures of PDE5 and 5′ GMP, vardenafil, sildenafil, tadalafil, or IBMX, each ligand interacts directly with the invariant Gln817 (Figure 7.9 and Figure 7.10).[52] In PDE4, the side chain of this glutamine is immobilized by a H bond with another amino acid and forms a bidentate H bond with cAMP involving donation of a hydrogen from the Gln N-ε atom to the N-1 atom of adenine and from the C-6 exocyclic amino group of the adenine ring to the Gln O-ε atom.[51] Xu et al. suggested that the side chain of the invariant glutamine could be rotated ~180°, thereby reversing the positions of the O-ε and N-ε atoms. The side chain could then form a similar bidentate H bond with the guanine of cGMP, which has different H-bonding potential at its N-1 proton and the C-6 carbonyl substituent compared to cAMP. It was predicted that in cGMP-specific PDEs the side chain of the invariant glutamine would be fixed in the alternate orientation, thereby providing for cGMP-specific catalysis.[51]

In a cocrystal structure of PDE5 and 5′-GMP, the side chain of Gln817 forms a bidentate H bond with the N-1 proton and C-6 carbonyl oxygen of guanine (Figure 7.10); it is held in place by a H bond with Gln775, which orients the side chain of Gln817 optimally for interaction with the guanine ring of cGMP and should discriminate against interaction with cAMP (Figure 7.10).[50,52,114] The interpretation that the orientation of the side chain of this invariant glutamine determines cN selectivity in PDEs quickly gained acceptance, and it was referred to as the glutamine switch or the cN-selectivity switch.[114] If this mechanism is

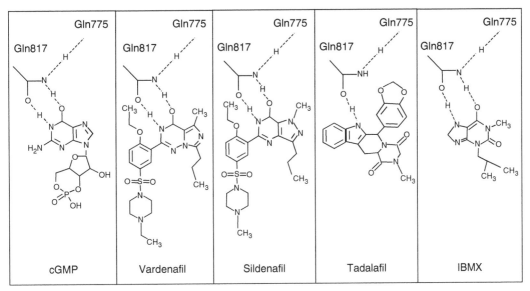

FIGURE 7.10 Schematic depiction of the interactions between the O-ε and N-ε atoms of invariant glutamine with substrate and inhibitors in PDEs. Dotted lines depict the bidentate hydrogen bond that has been detected in x-ray cocrystal structures of PDE5 and between the invariant catalytic site Gln817 in PDE5 and cGMP or inhibitors. Orientation of the O-ε and N-ε atoms of the Gln817 side chain is immobilized by H bond to Gln775.

correct, the site-directed mutagenesis of Gln775 (Q775A) should release rotational constraints on Gln817, impair interaction with cGMP, and improve interaction with cAMP. However, results do not support this prediction. In the Q775A mutant, the affinity for cGMP is lowered (~20-fold) as would be expected from the glutamine switch model, but the affinity for cAMP is not affected. If Gln775 determines the orientation of the Gln817 side chain as predicted from the x-ray crystal structures, then our data indicate that freeing the rotational constraint on that side chain does not improve affinity for cAMP. Therefore, contrary to current dogma, structural features other than Gln817 in the PDE5 catalytic site provide for cN selectivity. The 20-fold effect on cGMP affinity in the Q775A mutant may suggest that the role of Gln775 is not restricted to limitation of rotation of the Gln817 side chain.[122]

The results from the studies described above along with results from other groups question the role of the glutamine switch. The side chain of the invariant glutamine in PDE9, a cGMP-specific PDE, is not tethered.[123] Chimeras of PDE4 and PDE5, created by switching the amino acids that tether the side chain of the invariant glutamines between these enzymes, lose cN selectivity, but this is due primarily to loss of affinity for the natural substrate rather than a gain in affinity for the other cN.[114] The proper alignment of the Gln817 side chain is the key to the high-affinity interaction of cGMP with the PDE5 catalytic site, but does not impact cAMP affinity.[122] Therefore, the experimental evidence in PDE5 does not support the putative glutamine switch as a determinant of cN selectivity. Whether this theory applies in other PDEs that are highly selective for either cAMP or cGMP remains to be experimentally determined.

Vardenafil and sildenafil also form bidentate H bonds with Gln817 involving positions in their heterocyclic ring that are analogous to those in 5'-GMP (Figure 7.10).[52] IBMX also forms a bidentate H bond, but the contacts differ (Figure 7.10).[50,124] In contrast, tadalafil forms a single H bond with Gln817. Substitution of an alanine for Gln817 weakens affinity

for cGMP (60-fold), sildenafil (43-fold), vardenafil (500-fold), tadalafil (95-fold), and IBMX (60-fold).[122] Maximum catalytic activity for either cGMP or cAMP is not compromised.[122]

The effect of the Q817A mutation on the potencies of vardenafil and sildenafil emphasizes that the contribution of Gln817 to vardenafil potency is far greater than for sildenafil and that this residue contributes importantly to the selectivity of PDE5 for vardenafil over sildenafil, which declined from 29- to 2-fold in the Q817A mutant. The similar magnitude of the effect of the Q817A substitution on potency for sildenafil, tadalafil, or IBMX suggests that despite the absolute difference in potencies of these inhibitors and H-bonding pattern, the relative importance of Gln817 for potency of each inhibitor is similar. The affinity of the Q775A mutant for sildenafil or vardenafil is not impaired despite the purported importance of the proper orientation of the Gln817 side chain in the cocrystal structures. These results further emphasize differences in cGMP, sildenafil, and vardenafil binding to PDE5 catalytic site, which are likely to be critical in strategies to design inhibitors with improved potency and specificity.

The loss of free energy of binding (ΔG) in the Q817A mutant for cGMP, sildenafil, vardenafil, IBMX, and tadalafil is equivalent to ~1 H bond.[122] This result is hard to reconcile with the two H bonds reported in the x-ray cocrystal structures of PDE5 C domain with 5' GMP, sildenafil, or vardenafil; notably, there is a similar loss in free energy of binding with tadalafil, which only forms one H bond with Gln817 in the x-ray crystal structure (Zoraghi et al., unpublished results).[52,114]

Influences from other residues. Mutation of a number of amino acids that are in close proximity with 5'-GMP or inhibitors in the x-ray cocrystal structures with PDE5 C domain has modest effects on the affinity for these ligands (Figure 7.9). Substitution of alanine for Leu765, Phe786, or His613 weakens affinity for cGMP by ~10-, 10-, and 2-fold, respectively. The affinity of L765A for sildenafil, vardenafil, tadalafil, or IBMX is weakened 3-, 4-, 3-, and 0-fold; affinity of F786A for sildenafil, vardenafil, tadalafil, or IBMX is weakened 6-, 7-, 17-, and 12-fold; affinity of H613A for these same inhibitors is weakened 4-, 7-, 37-, and 12-fold (Zoraghi et al., unpublished results).

7.4.3.3.2 R Domain Influence on Inhibitor Potency

Recent results indicate that affinity of PDE5 holoenzyme for vardenafil-based compounds differs from that of the isolated C domain by ~10-fold, but affinity of the two proteins for cGMP, or for other inhibitors, i.e., sildenafil-based compounds, tadalafil, or IBMX, does not differ.[176] Using truncation mutagenesis, 46 amino acids in the N-terminal portion of GAF-B have been shown to provide for potent inhibition of PDE5 by vardenafil-based compounds and for the potency difference between vardenafil and sildenafil or tadalafil.[176] This documents an unexpected influence of the R domain on catalytic site interaction with certain classes of inhibitors and further emphasizes the importance of studying holoenzymes when assessing inhibitor potencies and defining critical contacts. Similar effects with different classes of inhibitors may occur in other PDEs, particularly in families such as the PDE4 family in which the length of the R domain among members varies markedly.

7.4.3.3.3 H-Loop Influences

A role for amino acids in the H-loop (Val660–His683) in catalytic site function has recently been revealed. In the original crystal structures of PDE5, the H-loop was not visible,[52] but in more recent structures, the H-loop can be traced (Figure 7.8A and Figure 7.8B).[50,175] The cocrystal of PDE5 with 5'-GMP has been considered to be a good model for interactions between the enzyme and the substrate cGMP, but in this cocrystal, the H-loop is chimerically replaced with the equivalent sequence in PDE4 and no contacts between the H-loop and 5' GMP are evident.[114] Despite this, mutations involving the H-loop cause 25- to 30-fold loss in

affinity for cGMP, with less impact on the turnover rate of the enzyme.[175] These mutations have little, if any, effect on the affinity for sildenafil; effect on the potency of other inhibitors has not yet been tested (Francis et al., unpublished results), but binding of inhibitors to the PDE5 catalytic site causes pronounced movement of the H-loop that is inhibitor-specific. When sildenafil is bound to PDE5, the H-loop migrates as much as 25 Å to cover the catalytic pocket (Figure 7.8B). Although mutation of H-loop amino acids to alanine or deletion of the H-loop does not substantially affect affinity for sildenafil, several residues (Asn662, Ser663, Tyr664, Ile665) contact the methylpiperazine ring of sildenafil in the crystal structure (Figure 7.9).[175]

Gly659, which immediately precedes the H-loop and is invariant in all class I PDEs, is critical to enzyme function. The mutant (G659A) has ~24-fold weaker affinity for cGMP and a 17-fold decrease in V_{max} compared to wild type PDE5.[175] It seems likely that the flexibility provided by this invariant glycine provides important functional characteristics to all class I PDEs. Notably, the affinity of the G659A mutant for sildenafil is not significantly affected (Francis et al., unpublished results). This emphasizes further differences in the binding characteristics of cGMP and inhibitors.

7.4.3.4 Crystal Structures of PDE5

Each x-ray crystal structure of the PDE5 C domain has significantly advanced understanding of the topography of the PDE5 catalytic site, identified ligand interactions, and provided important new information. However, investigators interested in the fine details of the PDE5 catalytic site, and in particular the interactions with inhibitors, should consider the various structures very carefully. All enzymes used for the determination of the x-ray crystal structures have been expressed in *E. coli*. The enzyme in the Sung study[52] is ~1000-fold less active ($V_{max} = 2$ nmol/min/mg) than that previously reported for isolated recombinant C domain ($k_{cat} \sim 3$ s^{-1}), and native or recombinant holoenzyme ($V_{max} = 2$–5 μmol/min/mg or $k_{cat} \sim 2$–4 s^{-1})[10,57,88] and affinities for the inhibitors were not reported.[52] No activity values were reported in the study of the PDE5/PDE4 chimera, but this construct is likely to have lost some subtle features of the natural PDE5 catalytic site.[114] The protein studied by Huai et al.[50] and Wang et al.[175] has the native sequence and exhibits kinetic properties [$K_m = 5$ μM, $V_{max} = 2$–4 μmol/min/mg ($k_{cat} \cong 1.3$–2.6 s^{-1})] and IC$_{50}$ values like that of native and recombinant PDE5. Not surprisingly, the crystal structure of the PDE5 in complex with IBMX determined by Huai et al. showed some conformational differences from those of PDE5A in the Sung study.[50] In studies by Wang et al.,[175] the cocomplex of PDE5 with sildenafil is similar to the structure reported by Sung et al., but the orientation of the piperazine ring of the inhibitor and the contacts of this moiety with the catalytic site amino acids differ (Figure 7.9). In the Sung et al. study,[52] amino acids 665–675 are disordered and not visualized, and in the studies by Zhang et al.,[114] this segment which had been replaced by the homologous sequence from PDE4 is traceable, but does not contact 5′ GMP. Conversely, in the Huai et al. studies, this sequence is traceable and has been designated as a part of the H-loop (Val660–His683).[50] Importantly, the H-loop sequence is a critical component for binding cGMP, and while it comes into close contact with the piperazine ring of sildenafil, it appears to have no effect on the affinity for sildenafil. Another region that displays a difference between the structures involves the M-loop (Asp788–Asn811 in human PDE5), which is ordered in the Sung et al. study but disordered in the Huai structure (code for the PDB is 1RKP) of PDE5A complexed with IBMX;[50] the M-loop assumes different traceable conformations when complexed with sildenafil or icarasid II, another inhibitor with a different structure. These differences are important considerations for those engaged in designing potent PDE5 inhibitors.

7.5   INHIBITOR SPECIFICITY

Spatial features of PDE5 catalytic site were originally assessed using a battery of cGMP analogs[125,126,127] and it was demonstrated that substitutions at several points on the molecule varying in size and chemical properties can be accommodated. Information about the biochemical properties of the catalytic site have also been advanced with the availability of inhibitors that are progressively more potent and selective for PDE5. Early studies depended on the use of nonspecific inhibitors such as IBMX or caffeine ($IC_{50} > 10,000$ nM) and papaverine, but the introduction of more potent methylxanthine derivatives, zaprinast (IC_{50} ~ 100–300 nM), sildenafil (IC_{50} ~ 4 nM), vardenafil (IC_{50} ~ 0.4 nM), and tadalafil (IC_{50} ~ 2 nM) revolutionized PDE5 research (Figure 7.2).[20,126,128–130]

Because of the role of cGMP in the regulation of smooth muscle tone and the abundance of PDE5 in the vascular smooth muscle, PDE5 was considered to be an excellent pharmacological target for the treatment of hypertension, and the development of sildenafil targeted this disease. All known PDE5 inhibitors show competitive kinetics for the inhibition of cGMP hydrolysis, and sildenafil, vardenafil, and tadalafil interact with the PDE5 catalytic site in a mutually exclusive manner (Figure 7.4).[61] The examination of the structures of the inhibitors (Figure 7.2) emphasizes that the catalytic pocket of PDE5[115] can accommodate molecules of considerable size and varied design. Unlike cGMP, which interacts with both the allosteric cGMP-binding sites and the catalytic site of PDE5, the inhibitors do not interact with the cGMP-binding sites.[28,52,64] Additional novel contacts provide for the 1000- to 10,000-fold higher affinity of sildenafil, tadalafil, or vardenafil compared to cGMP or IBMX; sildenafil is estimated to make approximately 17 contacts with PDE5 compared to five contacts made by IBMX. The high potency of these compounds and their selectivity for PDE5 catalytic site has been exploited to study PDE5 as will be described later; for many of these studies use of [^{3}H]inhibitors has proved to be invaluable.[61,115]

7.6   DIFFERENT CONFORMERS OF PDE5

Multiple lines of evidence suggest that PDE5 exists in different conformers that relate to the activation state of the enzyme,[5,27,33,34,61,62,104,115] and the balance between the activity states can be altered by events initiated in either the R or C domains. Activation of PDE5 catalytic activity and allosteric cGMP-binding activity by phosphorylation of Ser102 or activation of catalysis by allosteric cGMP binding suggests that the enzyme is in equilibrium between a less active autoinhibited state, also known as the taut (T) state, and a more active or relaxed (R) states according to the model for allosteric transitions[131] (Figure 7.11), but to date, no region providing for autoinhibition of catalytic site function has been identified. Purified PDE5 appears to exist in two distinct conformers that can be resolved on nondenaturing gels,[90] and exchange and dissociation studies of radiolabeled inhibitors demonstrate that the catalytic site of PDE5 exhibits high- and low-affinity states (Figure 7.12);[61,115] the low-affinity state is converted to the high-affinity state with sustained occupation of the catalytic site (Blount et al., unpublished results). The isolated C domain of PDE5 also exhibits two affinities for inhibitors, but the low-affinity state is not stably converted to the high-affinity state. This indicates that the R domain of PDE5 plays a critical role in modulating the flux between the two states; while the effect can be mediated by catalytic site-specific ligands, the influence of the R domain is required. Results also indicate that this requires an intact GAF-B.[176,177] Modulation of the equilibrium between these two states most likely occurs through a similar mechanism in unphospho-, phospho-, and cGMP-bound PDE5s.[177]

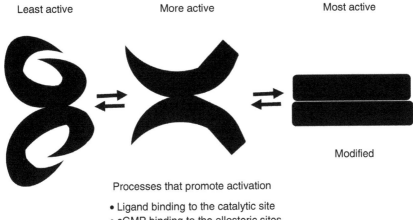

FIGURE 7.11 Working model of conformational changes in PDE5 that relate to different states of activation of affinity for inhibitors.

7.7   IMPACT OF PDE5 REGULATORY MECHANISMS ON cGMP SIGNALING AND ACTION OF INHIBITORS

PDE5 is subjected to long-term, developmental, and rapid onset regulation by a variety of biological processes (Table 7.1).[26,43,73,83,85,132,133] Mechanisms associated with long-term modulation of PDE5 gene expression and protein level are poorly understood, but a number of signals including angiotensin[81] and androgens[79,80] have been reported to induce increased production of PDE5. Potential mechanisms for the rapid regulation of the catalytic and

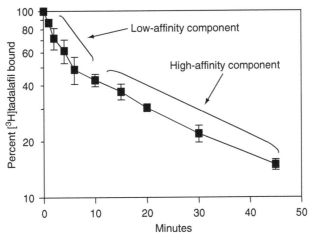

FIGURE 7.12 Exchange dissociation of [³H]tadalafil from PDE5. Purified bovine PDE5 (0.35 nM) was added to a reaction mixture containing [³H]inhibitor (30 nM), 0.2 mg/mL histone IIA-S, 10 mM potassium phosphate, pH 6.8, 25 mM 2-mercaptoethanol, and incubated at 4°C for 1 h. An aliquot was removed at zero time (B_0) to determine binding of inhibitor to the enzyme. Unlabeled inhibitor was then added (final concentration of 7 µM), and aliquots were removed at indicated times (B_t) and subjected to Millipore vacuum filtration to determine [³H]inhibitor remaining bound. (Adapted from Ref. [61]. With permission.)

TABLE 7.1
Processes Associated with PDE5 Role in Negative Feedback Control of cGMP Signaling

1. Substrate (cGMP) availability at the catalytic site
2. Direct induction of a more active catalytic conformation by cGMP binding to the catalytic site
3. Stimulation of catalytic site activity by allosteric cGMP binding
4. Ligand binding at the catalytic site increases allosteric cGMP binding which stimulates catalytic site function
5. Phosphorylation stimulated by ligand binding at the catalytic site stimulates catalytic site function
6. Cyclic GMP binding at the catalytic site increases allosteric cGMP binding, phosphorylation, and catalytic site function
7. Increased phosphorylation promotes allosteric cGMP binding, increases catalytic site activity
8. Increased phosphorylation promotes allosteric cGMP binding which stimulates further phosphorylation
9. Increased sequestration of cGMP by allosteric sites
10. Increased PDE5 gene transcription and protein level

allosteric cGMP-binding activities of the enzyme involve multiple processes including changes in cGMP level,[33,104,133] phosphorylation and dephosphorylation events,[27,40,54,90,103] oxidation and reduction processes (Francis et al., unpublished results), degradation by caspases,[134,135] and equilibrium between different conformations (Figure 7.11).[61,115]

7.7.1 SHORT-TERM REGULATION OF cGMP SIGNALING

Cyclic GMP binding to the allosteric sites in the PDE5 R domain increases catalytic activity.[110] Elevation of cGMP by atrial natriuretic peptide (ANP) treatment of vascular smooth muscles causes an approximately fourfold change in PDE5 activity measured under V_{max} conditions (10 micromolar cGMP) and the effect is reversible.[27] The Beavo and Koesling groups have also shown that elevation of cGMP causes a direct activation of PDE5 catalytic activity that is time-dependent and reversible.[33,104] In platelets where PDE5 is abundant, Koesling's group has provided compelling evidence that nitric oxide–induced elevation of cGMP causes a rapid (5–10 s) and sustained activation of PDE5 that persists for more than an hour. As a result, cGMP elevation in platelets following the exposure to nitric oxide peaks quickly and returns to near-basal levels after 30 to 40 s (Figure 7.13). Evidence suggests that this activation of PDE5 catalytic activity is due to cGMP binding to allosteric sites in the R domain and is likely to account for the well-known desensitization to nitric oxide that occurs in platelets. A subsequent challenge with nitric oxide produces a blunted elevation of cGMP. These observations can be recapitulated when the nitric-oxide-sensitive guanylyl cyclase and PDE5 are stably transfected into HEK-293 cells, suggesting that these two proteins are sufficient for this characteristic response to a nitric oxide challenge. Phosphorylation of PDE5 is not required for nor does it enhance the extent of activation of PDE5 that is observed when cGMP is elevated; however, phosphorylation substantially prolongs PDE5 activation. The half-life for the activated state of unphospho-PDE5 is ~3 min compared to ~20 min for phospho-PDE5. Since cGMP binding causes persistent activation of PDE5 and phosphorylation substantially increases cGMP-binding affinity and slows cGMP dissociation, it is reasonable to infer that the effect of phosphorylation is mediated via the increased affinity for cGMP at the allosteric sites.

Rybalkin et al.[104] reported similar effects of cGMP on recombinant PDE5 expressed in HEK cells. Activation of PDE5 by cGMP binding occurs only in fresh enzyme preparations; with storage, catalytic activity progressively increases, affinity for sildenafil increases, and cGMP stimulation of activity wanes. A monoclonal anti-GAF-A antibody blocks activation of freshly isolated PDE5 by cGMP but has no effect on the aged PDE5 that has assumed the

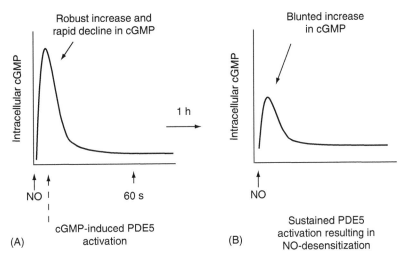

FIGURE 7.13 Desensitization of platelets to nitric oxide is associated with prolonged activation of PDE5. (A) Treatment of naïve platelets with nitric oxide produces a robust increase in intracellular cGMP that quickly declines. The elevation of cGMP is accompanied by activation of PDE5, which causes the rapid return of cGMP to near-basal levels. (B) Rechallenge of platelets that had been treated with nitric oxide ~1 h earlier with a fresh supply of nitric oxide caused a much lower increase in cGMP. This desensitization to a rechallenge with nitric oxide was primarily due to activation of PDE5 that was preserved from the initial treatment. (Adapted from Ref. [174]. With permission.)

more active state. This could be due to structural changes occurring in either the R domain or the C domain to achieve a more active conformation of the enzyme. In the study by Rybalkin et al., activation of PDE5 by cGMP lowers the K_m by approximately threefold and increases the V_{max} approximately threefold.[104] The comparison of the V_{max} of this cGMP-activated form of PDE5 to that for purified PDE5 is not available.[27] The question arises as to how the cGMP-mediated conversion of PDE5 to a highly activated form compares to the inhibitor-mediated conversion of PDE5 to a high-affinity state described earlier (Blount et al., unpublished results). The cGMP-activated PDE5 studied by Rybalkin et al. has approximately threefold increased affinity for sildenafil, which is similar in magnitude to the effect elicited by prolonged incubation with catalytic site-specific inhibitors.[104] An improved understanding of the mechanisms that mediate this interconversion is necessary for complete understanding of PDE5 functions.

These observations have several important implications: (1) the powerful activation of PDE5 by cGMP binding is consistent with the interconversion of different conformations of the enzyme; (2) powerful autoinhibitory influences that are present in the less active form which compromise both the affinity of the enzyme for catalytic site ligands and the maximal turnover rate are relieved upon cGMP binding to the allosteric site; (3) phosphorylation of Ser102 (human PDE5) stabilizes the active form of PDE5 by directly enhancing allosteric cGMP-binding affinity and catalytic site function (Bessay et al., unpublished results); and (4) with aging, the enzyme is spontaneously converted from a less active to a more active state. The molecular mechanisms that provide for these effects are unknown; it is possible that the activated form of the enzyme is energetically favored and with aging either GAF-A or the C domain may undergo a slow conformational change to produce the more active form. If so, does this occur in the cell and are these different forms of PDE5 physiologically or pharmacologically relevant?

In combination, in tissues where PDE5 is present, these myriad regulatory processes have the potential to provide a powerful response to changes in cellular cGMP, PDE inhibitors, or

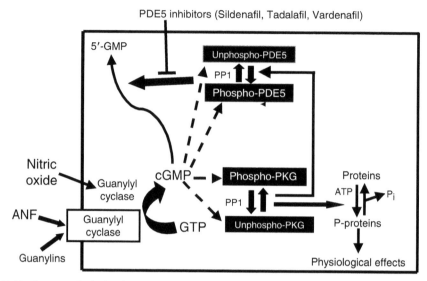

FIGURE 7.14 Cartoon depicting interplay among unphospho-PKG, phospho-PKG, unphospho-PDE5, and phospho-PDE in cGMP signaling pathway. PKG, cGMP-dependent protein kinase and PDE5, phosphodiesterase 5. Larger boxes surrounding phospho-PDE5 and phospho-PKG indicate that allosteric cGMP-binding sites in these forms have higher affinity for cGMP than do the unphospho-forms shown in the smaller boxes. Dotted arrows indicate cGMP binding to allosteric sites on both proteins. Larger box for phospho-PDE5 also indicates higher catalytic activity for the hydrolysis of cGMP.

to other relevant stimuli, such as apoptotic signals.[36,136–138] Under normal physiological conditions, elevation of cellular cGMP elicits a rapid PDE5-mediated negative feedback regulation of cGMP (Figure 7.13 and Figure 7.14);[27,33,104] a chronic decline in cGMP may cause downregulation of the enzyme and hypersensitivity to subsequent changes in cGMP.[132] The same mechanisms that provide for the rapid response to elevation of cGMP foster a feed-forward action to increase potency of inhibitors of PDE5 such as sildenafil, vardenafil, or tadalafil (Figure 7.14) the high concentration of PDE5 in smooth muscle cells and the high affinity of the enzyme for potent inhibitors in likely to promote rapid entry of these inhibitors into the cell and prolonged cellular retention of these inhibitors. The possibility that prolonged elevation of cGMP could elicit increased PDE5 expression should be considered when PDE5 inhibitors are used chronically; this includes medication regimens for the treatment of pulmonary hypertension and excessive use of these compounds for the enhancement of erectile function.

7.7.2 IMPACT OF CONCENTRATION AND AFFINITY OF cGMP RECEPTORS ON cGMP SIGNALING

Under basal metabolic conditions excluding the possible effects of compartmentation, cellular cGMP concentration has been estimated to be relatively low ($<0.1\ \mu$M)[14,22,71,139,140] compared to the K_m of PDE5 for cGMP, K_D of the allosteric cGMP-binding sites for cGMP, and total cGMP-binding sites represented in PDE5 and PKG (Figure 7.15). Under basal conditions, cGMP occupancy of either type of site is predicted to be low. In addition, the allosteric cGMP-binding sites on PDE5 are similar in affinity to the allosteric cGMP-binding sites of PKG, so that the competition of PKGs and PDE5 for cGMP would be significant (Figure 7.15).[63,71,141] As cGMP synthesis increases in response to agonists, cellular cGMP would also increase; since PDE5 is predicted to typically function at sub-K_m substrate concentrations, increased substrate availability would accelerate cGMP hydrolysis by mass

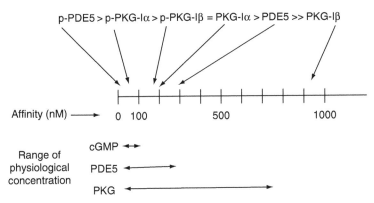

FIGURE 7.15 Schematic depicting the affinities of the allosteric cGMP-binding sites of unphosphos-PDE5 and phospho-PDE5 (p-PDE5), unphospho-PDE5 and phospho-PKG-Iα (p-PKG-Iα), and unphospho-PKG-Iβ and phospho-PKG-Iβ (p-PKG-Iβ) in relationship to intracellular concentration range for cGMP, PDE5, and PKGs.

action (Table 7.1). Increased occupation of the catalytic site by substrate or substrate analogs promotes interaction of cGMP with the allosteric sites in the PDE5 R domain.[3,10,57] Cyclic GMP binding to the PDE5 allosteric sites in turn increases affinity of the catalytic site for cGMP and induces a conformational change that exposes Ser102 (human PDE5) for efficient phosphorylation by PKG.[11,33,62,104] Phosphorylation of Ser102 increases affinity of both the allosteric and catalytic sites for cGMP, thereby increasing catalytic function at subsaturating cGMP levels[54,63,90] (Bessay, unpublished results). Recent results indicate that catalytic site-specific ligands such as sildenafil, vardenafil, or tadalafil can also promote phosphorylation of Ser102 *in vitro*.[177] PDE5 is phosphorylated in intact cells in response to elevation of cGMP; this occurs rapidly and is largely, if not exclusively, mediated by PKG.[27,32,34,103] Whether PDE5 inhibitors also promote phosphorylation in intact cells in the absence of significant cGMP elevation is not known.

Koesling's group has shown that in platelets, nitric oxide–induced increases in cGMP reach or exceed ~50 μM; in this instance, both the PDE5 allosteric cGMP-binding sites and the catalytic site would be saturated and catalysis would be at V_{max}. At any point in time, cellular cGMP will be distributed among (1) allosteric sites in cGMP-binding PDEs, such as PDE5 or PDE2, (2) allosteric sites in the PKG R domain, (3) cGMP-hydrolyzing catalytic sites such as these in the C domains of PDEs 1, 2, 3, 5, 9, and 11, and (4) other cGMP-binding proteins such as the cN-gated cation channels.[9,63,71,102,141] The balance in this dynamic competition for cGMP will be influenced by the amount of each protein, the affinity of the respective sites for cGMP, the modulation of these affinities by processes such as phosphorylation, and the compartments in which they are located (Figure 7.14 and Figure 7.15). Cyclic GMP bound in allosteric sites of either PDEs or PKG is temporally sequestered from potential hydrolysis by the catalytic sites.[63,141,142] When cGMP is bound to allosteric sites in PDE5 or PDE2, it fosters increased breakdown of cGMP by the respective catalytic sites, thereby blunting cGMP signaling, but when cGMP levels begin to decline, cGMP that dissociates could potentially serve as a reservoir for prolonging cGMP signaling.

7.7.3 NEGATIVE FEEDBACK PATHWAY FOR cGMP SIGNALING

The mechanisms described earlier provide for classical negative feedback to dampen or terminate the magnitude of the cellular response to cGMP signaling. The balance of cGMP distribution among the sites on PDE5 and PKG in vascular smooth muscle cells has

important implications; no other cGMP-binding proteins are known to exist in significant concentrations in these cells where cGMP signaling plays such a prominent role. Estimates of cGMP and allosteric cGMP-binding sites on PKG at basal level suggest that the cGMP-binding sites on PKG would be ~15% to 30% saturated. Thus, a three-to sixfold increase in cGMP would almost fully activate catalysis and maximize activity in the cGMP-signaling pathway.[14,71,140] In fact, in pig coronary artery strips, a threefold elevation in cGMP by the atrial natriuretic peptide (ANP) treatment produces maximum relaxation of the tissue.[140] Excessive accumulation of cGMP could slow recovery from the agonist signal and obliterate the fine-tuned response that has been demonstrated in other signaling pathways.[143] In rabbit corpus cavernosum under basal conditions, the molar ratio of PDE5 allosteric cGMP-binding sites to cGMP is ~10-fold, so that these sites could clearly bind a significant portion of cGMP, even after cGMP elevation.[71] Phosphorylation events alter the function of all these sites; when cGMP is elevated, PKG undergoes autophosphorylation, which increases its affinity for cGMP, and PDE5 is phosphorylated by PKG, resulting in increased affinity of PDE5 allosteric sites for cGMP and increased cGMP hydrolysis by the catalytic site (Figure 7.14 and Figure 7.15). Such an array of mechanisms for negative feedback control of cGMP signaling suggests that cells cannot tolerate excessive and prolonged activation of PKG. These simplistic considerations have not addressed the potentially powerful effects of selective compartmentation of the various components.

7.7.4 INVOLVEMENT OF PDE5 IN ERECTILE DYSFUNCTION AND PULMONARY HYPERTENSION

In individuals with normal erectile function, the increase in cGMP synthesis in response to sexual arousal is accompanied by countervailing activation of PDE5 to dampen or terminate cGMP signaling. Despite this delicate balance, in the normal erectile response the increase in cGMP is sufficient to cause the vasodilatation required to achieve penile tumescence. Inadequate levels of PDE5 may disturb the balance in this pathway, contributing to excessive cGMP accumulation and maladies such as priapism.[132] Conversely, excess PDE5 activity compared to guanylyl cyclase activity will produce ED. Pharmacological inactivation of PDE5 activity by sildenafil, vardenafil, or udenafil shifts the balance to favor cGMP accumulation and thereby improves the quality of the erectile response (Figure 7.14). A similar scenario occurs in pulmonary arterial hypertension where there is an imbalance between cGMP synthesis and hydrolysis by PDE5.[144] As in the treatment of ED, the use of sildenafil to block PDE5 action shifts the balance in favor of cGMP accumulation, promotes increased vasodilatation, and lowers arterial pressure. In contrast to the vasculature of the penis, the pulmonary vascular bed appears to steadily produce nitric oxide, which provides the agonist drive for activation of the nitric oxide–sensitive guanylyl cyclase and cGMP synthesis (see Chapter 25). In this setting, blocking PDE5 catalytic activity has an immediate effect on vascular smooth muscle tone. PDE5 inhibitors have no effect in the vasculature of the penis in the absence of external stimuli that promote increased release of nitric oxide from either neuronal or endothelial sources.

Activation of PDE5 by phosphorylation, allosteric cGMP binding, or ligand-induced conformational changes *in vitro* improves interaction of the PDE5 catalytic site with the three medications currently used for the treatment of ED. Sildenafil (Viagra), tadalafil (Cialis), vardenafil (Levitra) or udenafil (Zydena) are most effective for the relief of ED when taken 0.5 to 2 h before sexual activity.[28,129,145–148] This allows sufficient time for the inhibitors to be absorbed, enter vascular smooth muscle cells, and bind to the PDE5 catalytic site. Based on *in vitro* studies with purified PDE5, it is plausible that the following events occur within the cell following ingestion of one of these inhibitors. First, inhibitor binding produces a conformational change in PDE5 that transforms it into a high-affinity

binding partner for the inhibitors and may also foster phosphorylation of Ser102 (human).[177] Second, when the cGMP level in the cell begins to rise in response to a nitric oxide signal, cGMP binds to the allosteric sites on PDE5, which further enhances affinity of the catalytic site for the inhibitors. Third, phosphorylation of Ser102 that is stimulated by allosteric cGMP binding or occupation of the catalytic site by inhibitors enhances affinity for the inhibitors. This combination of events should cause the inhibitors to block the negative feedback pathway and concomitantly increase their own potencies and efficiencies this is a classical feed-forward regulation of function. In other words, these regulatory mechanisms produce physiological negative feedback regulation of the cGMP pathway, but the same mechanisms produce pharmacological enhancement of PDE5 inhibitor action.

7.8 PDE5 AS A DRUG TARGET IN THE TREATMENT OF DISEASE PROCESSES

PDE5-selective inhibitors have proved to be highly efficacious in treating ED and pulmonary hypertension, both of which are maladies originating primarily from compromised function of the vasculature. Sildenafil (Viagra) was originally approved for the treatment of ED, but under the trademark name of Revatio, it has recently been approved[149] in the United States for the treatment of pulmonary hypertension.[144,150–152] Recent studies have also implicated a potentially beneficial role for PDE5-selective inhibitors in several forms of endothelial dysfunction,[153–155] Raynaud's syndrome,[156] chronic obstructive pulmonary disease (COPD),[157] pressure overload–induced cardiac hypertrophy,[46] cognition,[158–160] cerebral vascular reactivity,[161] alveolar development,[162] chronic renal disease,[163] protection against ischemia and reperfusion injuries,[164–166] and certain forms of female sexual dysfunction.[167] It is hoped that treatment of some or all of these maladies will be improved in the future by the insights developed through the use of PDE5-selective inhibitors.

Myriad studies have established that sildenafil has minimal effects on cardiac function in either healthy individuals or those suffering from many different forms of cardiovascular disease, and its effect on either systemic arterial pressure or cardiac output is very slight.[168–173] Despite these findings and the fact that PDE5 level is very low in heart,[68] recent evidence indicates that in rats, sildenafil prevents cardiac hypertrophy in response to pressure overload and also reverses cardiac hypertrophy induced by pressure overload.[46] Furthermore, a recent report by this same group demonstrated that the acute administration of sildenafil blunts adrenergic-stimulated cardiac contractility in healthy men.[45] This effect is poorly understood in light of the lack of significant cardiovascular effects of sildenafil in most clinical studies, but it could be an important consideration in some patients.

7.9 CONCLUDING REMARKS

Numerous advances have recently been made in understanding the basic biochemical features of catalysis and mechanisms that regulate PDE5 function. Pharmacological targeting of PDE5 for the treatment of ED and other vascular diseases such as pulmonary hypertension has proved highly successful and encourages more vigorous efforts not only to develop improved PDE5 inhibitors, but also to discover appropriate inhibitors that will selectively target other PDEs for the treatment of a variety of dysfunctions.

REFERENCES

1. Lincoln, T.M., C.L. Hall, C.R. Park, and J.D. Corbin, 1976. Guanosine 3′,5′-cyclic monophosphate binding proteins in rat tissues. *Proc Natl Acad Sci USA* 73:2559–2563.

2. Coquil, J.F., D.J. Franks, J.N. Wells, M. Dupuis, and P. Hamet. 1980. Characteristics of a new binding protein distinct from the kinase for guanosine 3',5'-monophosphate in rat platelets. *Biochim Biophys Acta* 631:148–165.

3. Francis, S.H., T.M. Lincoln, and J.D. Corbin. 1980. Characterization of a novel cGMP binding protein from rat lung. *J Biol Chem* 255:620–626.

4. Francis, S.H. 1985. Effectors of rat lung cGMP binding protein-phosphodiesterase. *Curr Top Cell Regul* 26:247–262.

5. Francis, S.H., D.M. Chu, M.K. Thomas, A. Beasley, K. Grimes, J.L. Busch, I.V. Turko, T.L. Haik, and J.D. Corbin. 1998. Ligand-induced conformational changes in cyclic nucleotide phosphodiesterases and cyclic nucleotide-dependent protein kinases. *Methods* 14 (1):81–92.

6. Francis, S.H., I.V. Turko, and J.D. Corbin. 2001. Cyclic nucleotide phosphodiesterases: Relating structure and function. *Prog Nucleic Acid Res Mol Biol* 65:1–52.

7. Francis, S.H., J.L. Colbran, L.M. McAllister-Lucas, and J.D. Corbin. 1994. Zinc interactions and conserved motifs of the cGMP-binding cGMP-specific phosphodiesterase suggest that it is a zinc hydrolase. *J Biol Chem* 269:22477–22480.

8. Francis, S.H., M.K. Thomas, and J.D. Corbin. 1990. Cyclic GMP-binding cyclic GMP-specific phosphodiesterase from lung. In *Cyclic nucleotide phosphodiesterases: Structure, regulation, and drug action*, eds. J. Beavo and M.D. Housley, 117–140. New York: John Wiley & Sons.

9. Corbin, J.D., and S.H. Francis. 1999. Cyclic GMP phosphodiesterase 5: Target for sildenafil. *J Biol Chem* 274:13729–13732.

10. Thomas, M.K., S.H. Francis, and J.D. Corbin. 1990. Characterization of a purified bovine lung cGMP-binding cGMP phosphodiesterase. *J Biol Chem* 265:14964–14970.

11. Thomas, M.K., S.H. Francis, and J.D. Corbin. 1990. Substrate- and kinase-directed regulation of phosphorylation of a cGMP-binding phosphodiesterase by cGMP. *J Biol Chem* 265 (25):14971–14978.

12. Forte, L.R., W.J. Krause, and R.H. Freeman. 1988. Receptors and cGMP signalling mechanism for *E. coli* enterotoxin in opossum kidney. *Am J Physiol* 255:F1040–F1046.

13. Forte, L.R., W.J. Krause, and R.H. Freeman. 1989. *Escherichia coli* enterotoxin receptors: Localization in opossum kidney, intestine, and testis. *Am J Physiol* 257:F874–F881.

14. Francis, S.H., B.D. Noblett, B.W. Todd, J.N. Wells, and J.D. Corbin. 1988. Relaxation of vascular and tracheal smooth muscle by cyclic nucleotide analogs that preferentially activate purified cGMP-dependent protein kinase. *Mol Pharmacol* 34 (4):506–517.

15. Hofmann, F. 1985. The molecular basis of second messenger systems for regulation of smooth muscle contractility. *J Hypertens* (Suppl. 3):S3–S8.

16. Ignarro, L.J., and P.J. Kadowitz. 1985. The pharmacological and physiological role of cyclic GMP in vascular smooth muscle relaxation. *Annu Rev Pharmacol Toxicol* 25:171–191 [Review].

17. Raeymaekers, L., F. Hofmann, and R. Casteels. 1988. Cyclic GMP-dependent protein kinase phosphorylates phospholamban in isolated sarcoplasmic reticulum from cardiac and smooth muscle. *Biochem J* 252 (1):269–273.

18. Lincoln, T.M., T.L. Cornwell, S.S. Rashatwar, and R.M. Johnson. 1988. Mechanism of cyclic-GMP-dependent relaxation in vascular smooth muscle. *Biochem Soc Trans* 16:497–499.

19. Murad, F. 1988. Role of cyclic GMP in the mechanism of action of nitrovasodilators, endothelium-dependent agents and atrial natriuretic peptide. *Biochem Soc Trans* 16:490–492.

20. Ballard, S.A., C.J. Gingell, K. Tang, L.A. Turner, M.E. Price, and A.M. Naylor. 1998. Effects of sildenafil on the relaxation of human corpus cavernosum tissue in vitro and on the activities of cyclic nucleotide phosphodiesterase isozymes. *J Urol* 159 (6):2164–2171.

21. Boolell, M., M.J. Allen, S.A. Ballard, S. Gepi-Attee, G.J. Muirhead, A.M. Naylor, I.H. Osterloh, and C. Gingell. 1996. Sildenafil: An orally active type 5 cyclic GMP-specific phosphodiesterase inhibitor for the treatment of penile erectile dysfunction. *Int J Impot Res* 8:47–52.

22. Bush, P.A., W.J. Aronson, G.M. Buga, J. Rajfer, and L.J. Ignarro. 1992. Nitric oxide is a potent relaxant of human and rabbit corpus cavernosum. *J Urol* 147 (6):1650–1655.

23. Francis, S.H., and J.D. Corbin. 1988. Purification of cGMP-binding protein phosphodiesterase from rat lung. *Methods Enzymol* 159:722–729.

24. Jeremy, J.Y., S.A. Ballard, A.M. Naylor, M.A. Miller, and G.D. Angelini. 1997. Effects of sildenafil, a type-5 cGMP phosphodiesterase inhibitor, and papaverine on cyclic GMP and cyclic AMP levels in the rabbit corpus cavernosum in vitro. *Br J Urol* 79 (6):958–963.
25. Lugnier, C., P. Schoeffter, A. Le Bec, E. Strouthou, and J.C. Stoclet. 1986. Selective inhibition of cyclic nucleotide phosphodiesterases of human, bovine and rat aorta. *Biochem Pharmacol* 35 (10):1743–1751.
26. Sanchez, L.S., S.M. de la Monte, G. Filippov, R.C. Jones, W.M. Zapol, and K.D. Bloch. 1998. Cyclic-GMP-binding, cyclic-GMP-specific phosphodiesterase (PDE5) gene expression is regulated during rat pulmonary development. *Pediatr Res* 43 (2):163–168.
27. Wyatt, T.A., A.J. Naftilan, S.H. Francis, and J.D. Corbin. 1998. ANF elicits phosphorylation of the cGMP phosphodiesterase in vascular smooth muscle cells. *Am J Physiol Heart Circ Physiol* 274 (2 Pt 2):H448–H455.
28. Francis, S.H., and J.D. Corbin. 2003. Molecular mechanisms and pharmacokinetics of phosphodiesterase-5 antagonists. *Curr Urol Rep* 4 (6):457–465.
29. Padma-Nathan, H., W.D. Steers, and P.A. Wicker. 1998. Efficacy and safety of oral sildenafil in the treatment of erectile dysfunction: A double-blind, placebo-controlled study of 329 patients. Sildenafil Study Group. *Int J Clin Pract* 52 (6):375–379.
30. Sebkhi, A., J.W. Strange, S.C. Phillips, J. Wharton, and M.R. Wilkins. 2003. Phosphodiesterase type 5 as a target for the treatment of hypoxia-induced pulmonary hypertension. *Circulation* 107 (25):3230–3235.
31. Wallis, R.M., J.D. Corbin, S.H. Francis, and P. Ellis. 1999. Tissue distribution of phosphodiesterase families and the effects of sildenafil on tissue cyclic nucleotides, platelet function, and the contractile responses of trabeculae carneae and aortic rings in vitro. *Am J Cardiol* 83:3–12.
32. Friebe, A., and D. Koesling. 2003. Regulation of nitric oxide-sensitive guanylyl cyclase. *Circ Res* 93 (2):96–105.
33. Mullershausen, F., A. Friebe, R. Feil, W.J. Thompson, F. Hofmann, and D. Koesling. 2003. Direct activation of PDE5 by cGMP: Long-term effects within NO/cGMP signaling. *J Cell Biol* 160 (5):719–727.
34. Mullershausen, F., M. Russwurm, W.J. Thompson, L. Liu, D. Koesling, and A. Friebe. 2001. Rapid nitric oxide-induced desensitization of the cGMP response is caused by increased activity of phosphodiesterase type 5 paralleled by phosphorylation of the enzyme. *J Cell Biol* 155 (2):271–278.
35. Wharton, J., J.W. Strange, G.M. Moller, E.J. Growcott, X. Ren, A.P. Franklyn, S.C. Phillips, and M.R. Wilkins. 2005. Antiproliferative effects of phosphodiesterase type 5 inhibition in human pulmonary artery cells. *Am J Respir Crit Care Med* 172 (1):105–113.
36. Zhu, B., S. Strada, and T. Stevens. 2005. Cyclic GMP-specific phosphodiesterase 5 regulates growth and apoptosis in pulmonary endothelial cells. *Am J Physiol Lung Cell Mol Physiol* 289 (2):L196–L206.
37. Zhu, B., L. Vemavarapu, W.J. Thompson, and S.J. Strada. 2005. Suppression of cyclic GMP-specific phosphodiesterase 5 promotes apoptosis and inhibits growth in HT29 cells. *J Cell Biochem* 94 (2):336–350.
38. Zhang, R., Y. Wang, L. Zhang, Z. Zhang, W. Tsang, M. Lu, L. Zhang, and M. Chopp. 2002. Sildenafil (Viagra) induces neurogenesis and promotes functional recovery after stroke in rats. *Stroke* 33 (11):2675–2680.
39. Zhang, L., R.L. Zhang, Y. Wang, C. Zhang, Z.G. Zhang, H. Meng, and M. Chopp. 2005. Functional recovery in aged and young rats after embolic stroke: Treatment with a phosphodiesterase type 5 inhibitor. *Stroke* 36 (4):847–852.
40. Sopory, S., T. Kaur, and S.S. Visweswariah. 2004. The cGMP-binding, cGMP-specific phosphodiesterase (PDE5): Intestinal cell expression, regulation and role in fluid secretion. *Cell Signal* 16 (6):681–692.
41. Van Staveren, W.C., H.W. Steinbusch, M. Markerink-Van Ittersum, D.R. Repaske, M.F. Goy, J. Kotera, K. Omori, J.A. Beavo, and J. De Vente. 2003. mRNA expression patterns of the cGMP-hydrolyzing phosphodiesterases types 2, 5, and 9 during development of the rat brain. *J Comp Neurol* 467 (4):566–580.

42. Bender, A.T., and J.A. Beavo. 2004. Specific localized expression of cGMP PDEs in Purkinje neurons and macrophages. *Neurochem Int* 45 (6):853–857.
43. Kotera, J., N. Yanaka, K. Fujishige, Y. Imai, H. Akatsuka, T. Ishizuka, K. Kawashima, and K. Omori. 1997. Expression of rat cGMP-binding cGMP-specific phosphodiesterase mRNA in Purkinje cell layers during postnatal neuronal development. *Eur J Biochem* 249 (2):434–442.
44. Shimizu-Albergine, M., S.D. Rybalkin, I.G. Rybalkina, R. Feil, W. Wolfsgruber, F. Hofmann, and J.A. Beavo. 2003. Individual cerebellar Purkinje cells express different cGMP phosphodiesterases (PDEs): In vivo phosphorylation of cGMP-specific PDE (PDE5) as an indicator of cGMP-dependent protein kinase (PKG) activation. *J Neurosci* 23 (16):6452–6459.
45. Borlaug, B.A., V. Melenovsky, T. Marhin, P. Fitzgerald, and D.A. Kass. 2005. Sildenafil inhibits beta-adrenergic-stimulated cardiac contractility in humans. *Circulation* 112 (17):2642–2649.
46. Takimoto, E., H.C. Champion, M. Li, D. Belardi, S. Ren, E.R. Rodriguez, D. Bedja, K.L. Gabrielson, Y. Wang, and D.A. Kass. 2005. Chronic inhibition of cyclic GMP phosphodiesterase 5A prevents and reverses cardiac hypertrophy. *Nat Med* 11 (2):214–222.
47. Zhu, C.B., W.A. Hewlett, S.H. Francis, J.D. Corbin, and R.D. Blakely. 2004. Stimulation of serotonin transport by the cyclic GMP phosphodiesterase-5 inhibitor sildenafil. *Eur J Pharmacol* 504 (1–2):1–6.
48. Francis, S.H., I.V. Turko, K.A. Grimes, and J.D. Corbin. 2000. Histidine-607 and histidine-643 provide important interactions for metal support of catalysis in phosphodiesterase-5. *Biochemistry* 39:9591–9596.
49. He, F., A. Seryshev, C.W. Cowan, and T.G. Wensel. 2000. Multiple zinc binding in retinal rod cGMP phosphodiesterase, PDE6Balphabeta. *J Biol Chem* 275 (27):20572–20577.
50. Huai, Q., Y. Liu, S.H. Francis, J.D. Corbin, and H. Ke. 2004. Crystal structures of phosphodiesterases 4 and 5 in complex with inhibitor 3-isobutyl-1-methylxanthine suggest a conformation determinant of inhibitor selectivity. *J Biol Chem* 279 (13):13095–13101.
51. Xu, R.X., A.M. Hassell, D. Vanderwall, M.H. Lambert, W.D. Holmes, M.A. Luther, W.J. Rocque, et al. 2000. Atomic structure of PDE4: Insights into phosphodiesterase mechanism and specificity. *Science* 288:1822–1825.
52. Sung, B.J., K.Y. Hwang, Y.H. Jeon, J.I. Lee, Y.S. Heo, J.H. Kim, J. Moon, et al. 2003. Structure of the catalytic domain of human phosphodiesterase 5 with bound drug molecules. *Nature* 425 (6953):98–102.
53. Turko, I.V., S.H. Francis, and J.D. Corbin. 1998. Potential roles of conserved amino acids in the catalytic domain of the cGMP-binding cGMP-specific phosphodiesterase. *J Biol Chem* 273 (11):6460–6466.
54. Corbin, J.D., I.V. Turko, A. Beasley, and S.H. Francis. 2000. Phosphorylation of phosphodiesterase-5 by cyclic nucleotide-dependent protein kinase alters its catalytic and allosteric cGMP-binding activities. *Eur J Biochem* 267:2760–2767.
55. Granovsky, A.E., M. Natochin, R.L. McEntaffer, T.L. Haik, S.H. Francis, J.D. Corbin, and N.O. Artemyev. 1998. Probing domain functions of chimeric PDE6alpha'/PDE5 cGMP-phosphodiesterase. *J Biol Chem* 273 (38):24485–24490.
56. Turko, I.V., S.H. Francis, and J.D. Corbin. 1998. Binding of cGMP to both allosteric sites of cGMP-binding cGMP-specific phosphodiesterase (PDE5) is required for its phosphorylation. *Biochem J* 329 (Pt 3):505–510.
57. Turko, I.V., T.L. Haik, L.M. McAllister-Lucas, F. Burns, S.H. Francis, and J.D. Corbin. 1996. Identification of key amino acids in a conserved cGMP-binding site of cGMP-binding phosphodiesterases. A putative NKXnD motif for cGMP binding. *J Biol Chem* 271 (36):22240–22244.
58. Wu, A.Y., X.B. Tang, S.E. Martinez, K. Ikeda, and J.A. Beavo. 2004. Molecular determinants for cyclic nucleotide binding to the regulatory domains of phosphodiesterase 2A. *J Biol Chem* 279 (36):37928–37938.
59. Zoraghi, R., E.P. Bessay, J.D. Corbin, and S.H. Francis. 2005. Structural and functional features in human PDE5A1 regulatory domain that provide for allosteric cGMP binding, dimerization, and regulation. *J Biol Chem* 280 (12):12051–12063.

60. Zoraghi, R., J. Corbin, and S. Francis. 2004. Properties and functions of GAF domains in cyclic nucleotide phosphodiesterases and other proteins. *Mol Pharmacol* 65 (2):267–278.

61. Blount, M.A., A. Beasley, R. Zoraghi, K.R. Sekhar, E.P. Bessay, S.H. Francis, and J.D. Corbin. 2004. Binding of tritiated sildenafil, tadalafil, or vardenafil to the phosphodiesterase-5 catalytic site displays potency, specificity, heterogeneity, and cGMP stimulation. *Mol Pharmacol* 66 (1):144–152.

62. Francis, S.H., K.A. Grimes, L. Liu, W.J. Thompson, and J.D. Corbin. 2001. Phosphorylation of isolated human PDE5 regulatory domain increases cGMP-binding affinity and induces an apparent conformational change. *J Biol Chem* 277 (49):47581–47587.

63. Kotera, J., S.H. Francis, K.A. Grimes, A. Rouse, M.A. Blount, and J.D. Corbin. 2004. Allosteric sites of phosphodiesterase-5 sequester cyclic GMP. *Front Biosci* 9:378–386.

64. Turko, I.V., S.A. Ballard, S.H. Francis, and J.D. Corbin. 1999. Inhibition of cyclic GMP-binding cyclic GMP-specific phosphodiesterase (type 5) by sildenafil and related compounds. *Mol Pharmacol* 56:124–130.

65. D'Amati, G., C.R. di Gioia, M. Bologna, D. Giordano, M. Giorgi, S. Dolci, and E.A. Jannini. 2002. Type 5 phosphodiesterase expression in the human vagina. *Urology* 60 (1):191–195.

66. Kotera, J., K. Fujishige, and K. Omori. 2000. Immunohistochemical localization of cGMP-binding cGMP-specific phosphodiesterase (PDE5) in rat tissues. *J Histochem Cytochem* 48:685–693.

67. Loughney, K., T.R. Hill, V.A. Florio, L. Uher, G.J. Rosman, S.L. Wolda, B.A. Jones, et al. 1998. Isolation and characterization of cDNAs encoding PDE5A, a human cGMP-binding cGMP-specific 3′,5′-cyclic nucleotide phosphodiesterase. *Gene* 216:137–147.

68. Corbin, J.D., A. Beasley, M.A. Blount, and S.H. Francis. 2005. High lung PDE5: A strong basis for treating pulmonary hypertension with PDE5 inhibitors. *Biochem Biophys Res Commun* 334 (3):930–938.

69. Berkels, R., T. Klotz, G. Sticht, U. Englemann, and W. Klaus. 2001. Modulation of human platelet aggregation by the phosphodiesterase type 5 inhibitor sildenafil. *J Cardiovasc Pharmacol* 37 (4):413–421.

70. Chiu, P.J., S. Vemulapalli, M. Chintala, S. Kurowski, G.G. Tetzloff, A.D. Brown, and E.J. Sybertz. 1997. Inhibition of platelet adhesion and aggregation by E4021, a type V phosphodiesterase inhibitor, in guinea pigs. *Naunyn Schmiedebergs Arch Pharmacol* 355 (4):463–469.

71. Gopal, V.K., S.H. Francis, and J.D. Corbin. 2001. Allosteric sites of phosphodiesterase-5 (PDE5). A potential role in negative feedback regulation of cGMP signaling in corpus cavernosum. *Eur J Biochem* 268:3304–3312.

72. Mancina, R., S. Filippi, M. Marini, A. Morelli, L. Vignozzi, A. Salonia, F. Montorsi, et al. 2005. Expression and functional activity of phosphodiesterase type 5 in human and rabbit vas deferens. *Mol Hum Reprod* 11 (2):107–115.

73. Kotera, J., K. Fujishige, Y. Imai, E. Kawai, H. Michibata, H. Akatsuka, N. Yanaka, and K. Omori. 1999. Genomic origin and transcriptional regulation of two variants of cGMP-binding cGMP-specific phosphodiesterases. *Eur J Biochem* 262:866–872.

74. Lin, C.S., A. Lau, R. Tu, and T.F. Lue. 2000. Expression of three isoforms of cGMP-binding cGMP-specific phosphodiesterase (PDE5) in human penile cavernosum. *Biochem Biophys Res Commun* 268:628–635.

75. Lin, C.S., Z.C. Xin, G. Lin, and T.F. Lue. 2003. Phosphodiesterases as therapeutic targets. *Urology* 61 (4):685–691.

76. McAllister-Lucas, L.M., W.K. Sonnenburg, A. Kadlecek, D. Seger, H. LeTrong, J.L. Colbran, M.K. Thomas, et al. 1993. The structure of a bovine lung cGMP-binding, cGMP-specific phosphodiesterase deduced from a cDNA clone. *J Biol Chem* 268:22863–22873.

77. Burns, F., A.Z. Zhao, and J.A. Beavo. 1996. Cyclic nucleotide phosphodiesterases: Gene complexity, regulation by phosphorylation, and physiological implications. *Adv Pharmacol (New York)* 36:29–48 [Review].

78. Lin, C.S., S. Chow, A. Lau, R. Tu, and T.F. Lue. 2002. Human PDE5A gene encodes three PDE5 isoforms from two alternate promoters. *Int J Impot Res* 14 (1):15–24.

79. Morelli, A., S. Filippi, R. Mancina, M. Luconi, L. Vignozzi, M. Marini, C. Orlando, et al. 2004. Androgens regulate phosphodiesterase type 5 expression and functional activity in corpora cavernosa. *Endocrinology* 145 (5):2253–2263.

80. Zhang, X.H., A. Morelli, M. Luconi, L. Vignozzi, S. Filippi, M. Marini, G.B. Vannelli, R. Mancina, G. Forti, and M. Maggi. 2005. Testosterone regulates PDE5 expression and in vivo responsiveness to tadalafil in rat corpus cavernosum. *Eur Urol* 47 (3):409–416; Discussion 416.

81. Kim, D., T. Aizawa, H. Wei, X. Pi, S.D. Rybalkin, B.C. Berk, and C. Yan. 2005. Angiotensin II increases phosphodiesterase 5A expression in vascular smooth muscle cells: A mechanism by which angiotensin II antagonizes cGMP signaling. *J Mol Cell Cardiol* 38 (1):175–184.

82. Giordana, D., M. Giorgi, C. Sette, S. Biagioni, and G. Augusti-Tocco. 1999. Expression of cGMP-binding cGMP-specific phosphodiesterase (PDE5) in mouse tissues and cell lines using an antibody against the enzyme amino-terminal domain. *FEBS Lett* 446:218–222.

83. Lin, C.S., S. Chow, A. Lau, R. Tu, and T.F. Lue. 2001. Identification and regulation of human PDE5A gene promoter. *Biochem Biophys Res Commun* 280:684–692.

84. Lin, C.S., S. Chow, A. Lau, R. Tu, and T.F. Lue. 2001. Regulation of human PDE5A2 intronic promoter by cAMP and cGMP: Identification of a critical Sp1-binding site. *Biochem Biophys Res Commun* 280:693–699.

85. Belmonte, A., C. Ticconi, S. Dolci, M. Giorgi, A. Zicari, A. Lenzi, E.A. Jannini, and E. Piccione. 2005. Regulation of phosphodiesterase 5 expression and activity in human pregnant and non-pregnant myometrial cells by human chorionic gonadotropin. *J Soc Gynecol Investig* 12 (8):570–577.

86. Wang, P., P. Wu, J.G. Myers, A. Stamford, R.W. Egan, and M.M. Billah. 2001. Characterization of human, dog and rabbit corpus cavernosum type 5 phosphodiesterases. *Life Sci* 68 (17):1977–1987.

87. Senzaki, H., C.J. Smith, G.J. Juang, T. Isoda, S.P. Mayer, A. Ohler, N. Paolocci, G.F. Tomaselli, J.M. Hare, and D.A. Kass. 2001. Cardiac phosphodiesterase 5 (cGMP-specific) modulates beta-adrenergic signaling in vivo and is down-regulated in heart failure. *FASEB J* 15 (10):1718–1726.

88. Fink, T.L., S.H. Francis, A. Beasley, K.A. Grimes, and J.D. Corbin. 1999. Expression of an active, monomeric catalytic domain of the cGMP-binding cGMP-specific phosphodiesterase (PDE5). *J Biol Chem* 274 (49):34613–34620.

89. Sekharan, M.R., P. Rajagopal, and R.E. Klevit. 2005. Backbone ^{1}H, ^{13}C, and ^{15}N resonance assignment of the 46 kDa dimeric GAF A domain of phosphodiesterase 5. *J Biomol NMR* 33 (1):75.

90. Francis, S.H., E.P. Bessay, J. Kotera, K.A. Grimes, L. Liu, W.J. Thompson, and J.D. Corbin. 2002. Phosphorylation of isolated human phosphodiesterase-5 regulatory domain induces an apparent conformational change and increases cGMP binding affinity. *J Biol Chem* 277 (49):47581–47587.

91. Liu, L., T. Underwood, H. Li, R. Pamukcu, and W.J. Thompson. 2001. Specific cGMP binding by the cGMP binding domains of cGMP-binding cGMP-specific phosphodiesterase. *Cell Signal* 13:1–7.

92. Charbonneau, H., N. Beier, K.A. Walsh, and J.A. Beavo. 1986. Identification of a conserved domain among cyclic nucleotide phosphodiesterases from diverse species. *Proc Natl Acad Sci USA* 83:9308–9312.

93. Fawcett, L., R. Baxendale, P. Stacey, C. McGrouther, I. Harrow, S. Soderling, J. Hetman, J.A. Beavo, and S.C. Phillips. 2000. Molecular cloning and characterization of a distinct human phosphodiesterase gene family: PDE11A. *Proc Natl Acad Sci USA* 97:3702–3707.

94. Yuasa, K., T. Ohgaru, M. Asahina, and K. Omori. 2001. Identification of rat cyclic nucleotide phosphodiesterase 11A (PDE11A): Comparison of rat and human PDE11A splicing variants. *Eur J Biochem* 268 (16):4440–4448.

95. Yuasa, K., Y. Kanoh, K. Okumura, and K. Omori. 2001. Genomic organization of the human phosphodiesterase PDE11A gene. Evolutionary relatedness with other PDEs containing GAF domains. *Eur J Biochem* 268 (1):168–178.

96. Colbran, J.L., S.H. Francis, A.B. Leach, M.K. Thomas, H. Jiang, L.M. McAllister, and J.D. Corbin. 1992. A phenylalanine in peptide substrates provides for selectivity between cGMP- and cAMP-dependent protein kinases. *J Biol Chem* 267 (14):9589–9594.

97. Aravind, L., and C.P. Ponting. 1997. The GAF domain: An evolutionary link between diverse phototransducing proteins. *Trends Biochem Sci* 22:458–459.

98. Charbonneau, H. 1990. Structure–function relationships among cyclic nucleotide phosphodiesterases. In *Cyclic nucleotide phosphodiesterases: Structure, regulation and drug action*, eds. J. Beavo and M.D. Houslay, 267–296. New York: Wiley.

99. Giltay, R., G. Kostka, and R. Timpl. 1997. Sequence and expression of a novel member (LTBP-4) of the family of latent transforming growth factor-beta binding proteins. *FEBS Lett* 411 (2–3):164–168.

100. Okazaki, Y., M. Furuno, T. Kasukawa, J. Adachi, H. Bono, S. Kondo, I. Nikaido, et al. 2002. Analysis of the mouse transcriptome based on functional annotation of 60,770 full-length cDNAs. *Nature* 420 (6915):563–573.

101. Martinez, S.E., W.G. Hol, and J. Beavo. 2002. GAF domains: Two-billion-year-old molecular switches that bind cyclic nucleotides. *Mol Interv* 2 (5):317–323.

102. Francis, S.H., M.A. Blount, R. Zoraghi, and J.D. Corbin. 2005. Molecular properties of mammalian proteins that interact with cGMP: Protein kinases, cation channels, phosphodiesterases, and multi-drug anion transporters. *Front Biosci* 10:2097–2117.

103. Rybalkin, S.D., I.G. Rybalkina, R. Feil, F. Hofmann, and J.A. Beavo. 2002. Regulation of cGMP-specific phosphodiesterase (PDE5) phosphorylation in smooth muscle cells. *J Biol Chem* 277 (5):3310–3317.

104. Rybalkin, S.D., I.G. Rybalkina, M. Shimizu-Albergine, X.B. Tang, and J. Beavo. 2003. PDE5 is converted to an activated state upon cGMP binding to the GAF A domain. *EMBO J* 22 (3):469–478.

105. Weber, G. 1975. Energetics of ligand binding to protein. *Adv Protein Chem* 29:1–83.

106. Liu, L., T. Underwood, H. Li, R. Pamukcu, and W.J. Thompson. 2002. Specific cGMP binding by the cGMP binding domains of cGMP-binding cGMP specific phosphodiesterase. *Cell Signal* 14 (1):45–51.

107. McAllister-Lucas, L.M., T.L. Haik, J.L. Colbran, W.K. Sonnenburg, D. Seger, I.V. Turbo, J.A. Beavo, S.H. Francis, and J.D. Corbin. 1995. An essential aspartic acid at each of two allosteric cGMP-binding sites of a cGMP phosphodiesterase. *J Biol Chem* 270 (51):30671–30679.

108. Zoraghi, R., E.P. Bessay, J.D. Corbin, and S.H. Francis. 2005. Structural and functional features in human PDE5A1 regulatory domain that provide for allosteric cGMP binding, dimerization, and regulation. *J Biol Chem* 280 (12):12051–12063.

109. Sopory, S., S. Balaji, N. Srinivasan, and S.S. Visweswariah. 2003. Modeling and mutational analysis of the GAF domain of the cGMP-binding, cGMP-specific phosphodiesterase, PDE5. *FEBS Lett* 539 (1–3):161–166.

110. Okada, D., and S. Asakawa. 2002. Allosteric activation of cGMP-specific, cGMP-binding phosphodiesterase (PDE5) by cGMP. *Biochemistry* 41 (30):9672–9679.

111. Martinez, S.E., A.Y. Wu, N.A. Glavas, X.B. Tang, S. Turley, W.G. Hol, and J.A. Beavo. 2002. The two GAF domains in phosphodiesterase 2A have distinct roles in dimerization and in cGMP binding. *Proc Natl Acad Sci USA* 99 (20):13260–13265.

112. Lochhead, A., E. Nekrasova, V.Y. Arshavsky, and N.J. Pyne. 1997. The regulation of the cGMP-binding cGMP phosphodiesterase by proteins that are immunologically related to gamma subunit of the photoreceptor cGMP phosphodiesterase. *J Biol Chem* 272 (29):18397–18403.

113. Granovsky, A.E., R. McEntaffer, and N.O. Artemyev. 1998. Probing functional interfaces of rod PDE gamma-subunit using scanning fluorescent labeling. *Cell Biochem Biophys* 28 (2–3):115–133.

114. Zhang, K.Y., G.L. Card, Y. Suzuki, D.R. Artis, D. Fong, S. Gillette, D. Hsieh, et al. A glutamine switch mechanism for nucleotide selectivity by phosphodiesterases. *Mol Cell* 15 (2):279–286.

115. Corbin, J.D, M.A. Blount, J.L. Weeks, A. Beasley, K.P. Kuhn, Y.S. Ho, L.F. Saidi, J.H. Hurley, J. Kotera, and S.H. Francis. 2003. [3H]Sildenafil binding to phosphodiesterase-5 is specific, kinetically heterogenous, and stimulated by cGMP. *Mol Pharmacol* 63 (6):1364–1372.

116. Robison, G.A., R.W. Butcher, and E.W. Sutherland. 1971. *Cyclic AMP*. New York: Academic Press.

117. Braumann, T., C. Erneux, G. Petridis, W.D. Stohrer, and B. Jastorff. 1986. Hydrolysis of cyclic nucleotides by a purified cGMP-stimulated phosphodiesterase: Structural requirements for hydrolysis. *Biochim Biophys Acta* 871:199–206.

118. Burgers, P.M., F. Eckstein, D.H. Hunneman, J. Baraniak, R.W. Kinas, K. Lesiak, and W.J. Stec. 1979. Stereochemistry of hydrolysis of adenosine 3′:5′-cyclic phosphorothioate by the cyclic phosphodiesterase from beef heart. *J Biol Chem* 254 (20):9959–9961.

119. Goldberg, N.D., T.F. Walseth, J.H. Stephenson, T.P. Krick, and G. Graff. 1980. ^{18}O-Labeling of guanosine monophosphate upon hydrolysis of cyclic guanosine 3′:5′-monophosphate by phosphodiesterase. *J Biol Chem* 255:10344–10347.

120. Jarvest, R.L., G. Lowe, J. Baraniak, and W.J. Stec. 1982. A stereochemical investigation of the hydrolysis of cyclic AMP and the (Sp)- and (Rp)-diastereoisomers of adenosine cyclic 3′:5′-phosphorothioate by bovine heart and baker's-yeast cyclic AMP phosphodiesterases. *Biochem J* 203 (2):461–470.

121. Corbin, J., S. Francis, and R. Zoraghi. 2006. Tyrosine-612 in PDE5 contributes to higher affinity for vardenafil over sildenafil. *Int J Impot Res* 18 (3):251–257.

122. Zoraghi, R., J.D. Corbin, and S.H. Francis. 2006. Phosphodiesterase-5 GLN-817 is critical for cGMP, vardenafil, or sildenafil affinity: Its orientation impacts cGMP but not cAMP affinity. *J Biol Chem* 281 (9):5553–5558.

123. Huai, Q., H. Wang, W. Zhang, R.W. Colman, H. Robinson, and H. Ke. 2004. Crystal structure of phosphodiesterase 9 shows orientation variation of inhibitor 3-isobutyl-1-methylxanthine binding. *Proc Natl Acad Sci USA* 101 (26):9624–9629.

124. Huai, Q., J. Colicelli, and H. Ke. 2003. The crystal structure of AMP-bound PDE4 suggests a mechanism for phosphodiesterase catalysis. *Biochemistry* 42 (45):13220–13226.

125. Thomas, M.K., S.H. Francis, S.J. Beebe, T.W. Gettys, and J.D. Corbin. 1992. Partial mapping of cyclic nucleotide sites and studies of regulatory mechanisms of phosphodiesterases using cyclic nucleotide analogues. *Adv Second Messenger Phosphoprotein Res* 25:45–53.

126. Sekhar, K.R., P. Grondin, S.H. Francis, and J.D. Corbin. 1996. Design and synthesis of xanthines and cyclic GMP analogues as potent inhibitors of PDE5. In *Phosphodiesterase inhibitors*, eds. C. Schudt, G. Dent, and K.F. Rabe, 135–146. New York: Academic Press.

127. Beltman, J., D.E. Becker, E. Butt, G.S. Jensen, S.D. Rybalkin, B. Jastorff, and J.A. Beavo. 1995. Characterization of cyclic nucleotide phosphodiesterases with cyclic GMP analogs: Topology of the catalytic domains. *Mol Pharmacol* 47:330–339.

128. Griffith, T.M., D.H. Edwards, M.J. Lewis, and A.H. Henderson. 1985. Evidence that cyclic guanosine monophosphate (cGMP) mediates endothelium-dependent relaxation. *Eur J Pharmacol* 112 (2):195–202.

129. Padma-Nathan, H., J.G. McMurray, W.E. Pullman, J.S. Whitaker, J.B. Saoud, K.M. Ferguson, and R.C. Rosen. 2001. On-demand IC351 (Cialis) enhances erectile function in patients with erectile dysfunction. *Int J Impot Res* 13 (1):2–9.

130. Porst, H., R. Rosen, H. Padma-Nathan, I. Goldstein, F. Giuliano, E. Ulbrich, and T. Bandel. 2001. The efficacy and tolerability of vardenafil, a new, oral, selective phosphodiesterase type 5 inhibitor, in patients with erectile dysfunction: The first at-home clinical trial. *Int J Impot Res* 13 (4):192–199.

131. Monod, J., J. Wyman, and J.P. Changeaux. 1965. On the nature of allosteric transitions: A plausible model. *J Mol Biol* 12 (1):88–118.

132. Champion, H.C., T.J. Bivalacqua, E. Takimoto, D.A. Kass, and A.L. Burnett. 2005. Phosphodiesterase-5A dysregulation in penile erectile tissue is a mechanism of priapism. *Proc Natl Acad Sci USA* 102 (5):1661–1666.

133. Macpherson, J.D., T.D. Gillespie, H.A. Dunkerley, D.H. Maurice, and B.M. Bennett. 2005. Inhibition of phosphodiesterase 5 selectively reverses nitrate tolerance in the venous circulation. *J Pharmacol Exp Ther* 317 (1):188–195.

134. Frame, M., K.F. Wan, R. Tate, P. Vandenabeele, and N.J. Pyne. 2001. The gamma subunit of the rod photoreceptor cGMP phosphodiesterase can modulate the proteolysis of two cGMP binding cGMP-specific phosphodiesterases (PDE6 and PDE5) by caspase-3. *Cell Signal* 13 (10):735–741.

135. Frame, M.J., R. Tate, D.R. Adams, K.M. Morgan, M.D. Houslay, P. Vandenabeele, and N.J. Pyne. 2003. Interaction of caspase-3 with the cyclic GMP binding cyclic GMP specific phosphodiesterase (PDE5a1). *Eur J Biochem* 270 (5):962–970.

136. Whitehead, C.M., K.A. Earle, J. Fetter, S. Xu, T. Hartman, D.C. Chan, T.L. Zhao, et al. 2003. Exisulind-induced apoptosis in a non-small cell lung cancer orthotopic lung tumor model augments docetaxel treatment and contributes to increased survival. *Mol Cancer Ther* 2 (5):479–488.

137. Chan, D.C., K.A. Earle, T.L. Zhao, B. Helfrich, C. Zeng, A. Baron, C.M. Whitehead, et al. 2002. Exisulind in combination with docetaxel inhibits growth and metastasis of human lung cancer and prolongs survival in athymic nude rats with orthotopic lung tumors. *Clin Cancer Res* 8 (3):904–912.

138. Li, H., L. Liu, M.L. David, C.M. Whitehead, M. Chen, J.R. Fetter, G.J. Sperl, R. Pamukcu, and W.J. Thompson. 2002. Pro-apoptotic actions of exisulind and CP461 in SW480 colon tumor cells involve beta-catenin and cyclin D1 down-regulation. *Biochem Pharmacol* 64 (9):1325–1336.

139. Eigenthaler, M., C. Nolte, M. Halbrugge, and U. Walter. 1992. Concentration and regulation of cyclic nucleotides, cyclic-nucleotide-dependent protein kinases and one of their major substrates in human platelets. Estimating the rate of cAMP-regulated and cGMP-regulated protein phosphorylation in intact cells. *Eur J Biochem* 205:471–481.

140. Jiang, H., J.L. Colbran, S.H. Francis, and J.D. Corbin. 1992. Direct evidence for cross-activation of cGMP-dependent protein kinase by cAMP in pig coronary arteries. *J Biol Chem* 267 (2):1015–1019.

141. Kotera, J., K.A. Grimes, J.D. Corbin, and S.H. Francis. 2003. cGMP-dependent protein kinase protects cGMP from hydrolysis by phosphodiesterase-5. *Biochem J* 372 (Pt 2):419–426.

142. Haddox, M.K., J.H. Stephenson, M.E. Moser, D.B. Glass, J.G. White, B. Holmes-Gray, and N.D. Goldberg. 1979. Ascorbic acid modulation of splenic cell cyclic GMP metabolism. *Life Sci* 24:1555–1566.

143. Keely, S.L., and J.D. Corbin. 1977. Involvement of cAMP-dependent protein kinase in the regulation of heart contractile force. *Am J Physiol* 233:H269–H275.

144. Galie, N., H.A. Ghofrani, A. Torbicki, R.J. Barst, L.J. Rubin, D. Badesch, T. Fleming, et al. 2005. Sildenafil citrate therapy for pulmonary arterial hypertension. *N Engl J Med* 353 (20):2148–2157.

145. Ormrod, D., S.E. Easthope, and D.P. Figgitt. 2002. Vardenafil. *Drugs Aging* 19 (3):217–227; Discussion 228–229.

146. Carson, C.C., and T.F. Lue. 2005. Phosphodiesterase type 5 inhibitors for erectile dysfunction. *BJU Int* 96 (3):257–280.

147. Montorsi, F., A. Briganti, A. Salonia, P. Montorsi, and P. Rigatti. 2004. The use of phosphodiesterase type 5 inhibitors for erectile dysfunction. *Curr Opin Urol* 14 (6):357–359.

148. Montorsi, F., A. Salonia, F. Deho, A. Cestari, G. Guazzoni, P. Rigatti, and C. Stief. 2003. Pharmacological management of erectile dysfunction. *BJU Int* 91 (5):446–454.

149. Bhatia, S., R.P. Frantz, C.J. Severson, L.A. Durst, and M.D. McGoon. 2003. Immediate and long-term hemodynamic and clinical effects of sildenafil in patients with pulmonary arterial hypertension receiving vasodilator therapy. *Mayo Clin Proc* 78 (10):1207–1213.

150. Michelakis, E., W. Tymchak, D. Lien, L. Webster, K. Hashimoto, and S. Archer. 2002. Oral sildenafil is an effective and specific pulmonary vasodilator in patients with pulmonary arterial hypertension: Comparison with inhaled nitric oxide. *Circulation* 105 (20):2398–2403.

151. Michelakis, E.D., W. Tymchak, M. Noga, L. Webster, X.C. Wu, D. Lien, S.H. Wang, D. Modry, and S.L. Archer. 2003. Long-term treatment with oral sildenafil is safe and improves functional capacity and hemodynamics in patients with pulmonary arterial hypertension. *Circulation* 108 (17):2066–2069.

152. Oliveira, E.C., and C.F. Amaral. 2005. Sildenafil in the management of idiopathic pulmonary arterial hypertension in children and adolescents. *J Pediatr* (*Rio J*) 81 (5):390–394.

153. Katz, S.D., K. Balidemaj, S. Homma, H. Wu, J. Wang, and S. Maybaum. 2000. Acute type 5 phosphodiesterase inhibition with sildenafil enhances flow-mediated vasodilation in patients with chronic heart failure. *J Am Coll Cardiol* 36 (3):845–851.

154. Hryniewicz, K., C. Dimayuga, A. Hudaihed, A.S. Androne, H. Zheng, K. Jankowski, and S.D. Katz. 2005. Inhibition of angiotensin-converting enzyme and phosphodiesterase type 5 improves endothelial function in heart failure. *Clin Sci* (*Lond*) 108 (4):331–338.

155. Sommer, F., and W. Schulze. 2005. Treating erectile dysfunction by endothelial rehabilitation with phosphodiesterase 5 inhibitors. *World J Urol* 23(6):385–392.

156. Boin, F., and F.M. Wigley. 2005. Understanding, assessing and treating Raynaud's phenomenon. *Curr Opin Rheumatol* 17 (6):752–760.

157. Alp, S., M. Skrygan, W.E. Schmidt, and A. Bastian. 2005. Sildenafil improves hemodynamic parameters in COPD—an investigation of six patients. *Pulm Pharmacol Ther*.

158. Prickaerts, J., A. Sik, F.J. van der Staay, J. de Vente, and A. Blokland. 2005. Dissociable effects of acetylcholinesterase inhibitors and phosphodiesterase type 5 inhibitors on object recognition memory: Acquisition versus consolidation. *Psychopharmacology* (*Berlin*) 177 (4):381–390.

159. Prickaerts, J., A. Sik, W.C. van Staveren, G. Koopmans, H.W. Steinbusch, F.J. van der Staay, J. de Vente, and A. Blokland. 2004. Phosphodiesterase type 5 inhibition improves early memory consolidation of object information. *Neurochem Int* 45 (6):915–928.

160. Rutten, K., J.D. Vente, A. Sik, M.M. Ittersum, J. Prickaerts, and A. Blokland. 2005. The selective PDE5 inhibitor, sildenafil, improves object memory in Swiss mice and increases cGMP levels in hippocampal slices. *Behav Brain Res* 164 (1):11–16.

161. Rosengarten, B., R.T. Schermuly, R. Voswinckel, M.G. Kohstall, H. Olschewski, N. Weissmann, W. Seeger, M. Kaps, F. Grimminger, and H.A. Ghofrani. 2005. Sildenafil improves dynamic vascular function in the brain: Studies in patients with pulmonary hypertension. *Cerebrovasc Dis* 21 (3):194–200.

162. Ladha, F., S. Bonnet, F. Eaton, K. Hashimoto, G. Korbutt, and B. Thebaud. 2005. Sildenafil improves alveolar growth and pulmonary hypertension in hyperoxia-induced lung injury. *Am J Respir Crit Care Med* 172 (6):750–756.

163. Rodriguez-Iturbe, B., A. Ferrebuz, V. Vanegas, Y. Quiroz, F. Espinoza, H. Pons, and N.D. Vaziri. 2005. Early treatment with cGMP phosphodiesterase inhibitor ameliorates progression of renal damage. *Kidney Int* 68 (5):2131–2142.

164. Kukreja, R.C., R. Ockaili, F. Salloum, C. Yin, J. Hawkins, A. Das, and L. Xi. 2004. Cardioprotection with phosphodiesterase-5 inhibition—a novel preconditioning strategy. *J Mol Cell Cardiol* 36 (2):165–173.

165. Kukreja, R.C., F. Salloum, A. Das, R. Ockaili, C. Yin, Y.A. Bremer, P.W. Fisher, et al. 2005. Pharmacological preconditioning with sildenafil: Basic mechanisms and clinical implications. *Vascul Pharmacol* 42 (5–6):219–232.

166. Bremer, Y.A., F. Salloum, R. Ockaili, E. Chou, W.B. Moskowitz, and R.C. Kukreja. 2005. Sildenafil citrate (Viagra) induces cardioprotective effects after ischemia/reperfusion injury in infant rabbits. *Pediatr Res* 57 (1):22–27.

167. Mayer, M., C.G. Stief, M.C. Truss, and S. Uckert. 2005. Phosphodiesterase inhibitors in female sexual dysfunction. *World J Urol* 23(6):393–397.

168. Padma-nathan, H., I. Eardley, R.A. Kloner, A.M. Laties, and F. Montorsi. 2002. A 4-year update on the safety of sildenafil citrate (Viagra). *Urology* 60 (2 Suppl. 2):67–90.

169. Vardi, Y., L. Klein, S. Nassar, E. Sprecher, and I. Gruenwald, I. 2002. Effects of sildenafil citrate (Viagra) on blood pressure in normotensive and hypertensive men. *Urology* 59 (5):747–752.

170. Conti, C.R., C.J. Pepine, and M. Sweeney. 1999. Efficacy and safety of sildenafil citrate in the treatment of erectile dysfunction in patients with ischemic heart disease. *Am J Cardiol* 83 (5A):29C–34C.

171. Jackson, G., N. Benjamin, N. Jackson, and M.J. Allen. 1999. Effects of sildenafil citrate on human hemodynamics. *Am J Cardiol* 83 (5A):13C–20C.

172. Herrmann, H.C., G. Chang, B.D. Klugherz, and P.D. Mahoney. 2000. Hemodynamic effects of sildenafil in men with severe coronary artery disease. *N Engl J Med* 342 (22):1622–1626.

173. Kloner, R.A., M. Brown, L.M. Prisant, and M. Collins. 2001. Effect of sildenafil in patients with erectile dysfunction taking antihypertensive therapy. Sildenafil Study Group. *Am J Hypertens* 14 (1):70–73.

174. Koesling, D., F. Mullershausen, A. Lange, A. Friebe, E. Mergia, C. Wagner, and M. Russwurm. 2005. Negative feedback in NO/cGMP signalling. *Biochem Soc Trans* 33 (Pt 5):1119–1122.

175. Wang, H., Y.Liu, O. Huai, J.Cai, R. Zoranghi, S.H. Francis, J.D. Corbin, H. Robinson, A.Xin, G. Lin, and H.Ke. 2006. Multiple conformation of phosphodieserase-5 implications for enzyme function and drug development. *J.Biol Chem* 281(30):21469–214790.

176. Blount, M.A., R. Zoraghi, J.D.Corbin, and S.H. Francis. 2006. A 46-amino acid segment in phosphodiesterase-5 GAF-B domain provides for high Vardenafil potency over sildenafil and tadalafil and is involved in PDE5 dimerization. *Mol Pharamacol* (In press).

177. Bessay, E.P., Zoraghi, R., Blount, M.A., Grimes, K.A., Beasley, A., Francis, S.H., and Corbin, J.D. 2007. Phosphorylation of phosphodiesterase-5 is promoted by a conformational change induced by sildanafil, vardanafil, or tadalafil. *Front Biosci* 12:1899–1910.

8 Photoreceptor Phosphodiesterase (PDE6): A G-Protein-Activated PDE Regulating Visual Excitation in Rod and Cone Photoreceptor Cells

Rick H. Cote

CONTENTS

8.1   Overview 166
8.2   Vertebrate Phototransduction Signaling Pathway 166
     8.2.1   The Photoreceptor Cell is Compartmentalized for Light Capture
            and Signal Propagation.. 167
     8.2.2   Major Steps in Visual Excitation in Rod Photoreceptors............................... 168
     8.2.3   Recovery of the Dark-Adapted State and Light Adaptation........................... 169
     8.2.4   cGMP Metabolic Flux during Visual Transduction 170
8.3   Tissue and Subcellular Distribution of PDE6.. 171
8.4   Structure of PDE6 Catalytic and Inhibitory Subunits 171
     8.4.1   Catalytic Subunit 173
         8.4.1.1   cGMP Binding Regulatory GAF Domains 173
         8.4.1.2   Catalytic Domain.. 174
         8.4.1.3   C-Terminal Prenylation 175
     8.4.2   Inhibitory γ-Subunit 175
         8.4.2.1   Structure and Domain Organization of the γ-Subunit 176
8.5   Catalytic Properties of PDE6... 177
8.6   Pharmacology 177
     8.6.1   Effects of PDE Inhibitors on Photoreceptor and Retinal Physiology 177
     8.6.2   Potency and Selectivity of PDE Inhibitors for PDE6 *in Vitro*...................... 178
8.7   Allosteric Role of the cGMP-Binding GAF Domains of PDE6............................. 179
8.8   Regulation of PDE6 Activation by the G-Protein, Transducin 180
8.9   PDE6 Interacting Proteins 181
     8.9.1   Glutamic Acid–Rich Protein-2 (GARP2) 181
     8.9.2   Prenyl Binding Protein/PDEδ (PrBP/δ).. 181
8.10   Retinal Diseases due to Defects in PDE6 182

8.11   Conclusion 183
Acknowledgments 183
References 184

8.1   OVERVIEW

Rod and cone photoreceptors of the vertebrate retina utilize cyclic nucleotides as the primary intracellular messenger for converting photic stimuli into an electrical response. Cyclic nucleotide levels in a dark-adapted photoreceptor are maintained by the balance of cyclic nucleotide synthesis (catalyzed by cyclases) and degradation (catalyzed by phosphodiesterases, PDEs). Upon light stimulation, the rapid activation of the photoreceptor PDE, classified as PDE6, triggers a dramatic drop in cyclic nucleotide levels that lead to cell hyperpolarization. The lifetime of activated PDE6 is also a highly regulated process. PDE6 inactivation is temporally coordinated with activation of the photoreceptor cyclase so that cyclic nucleotide levels recover quickly to restore the dark-adapted state of the photoreceptor.

It is not surprising, therefore, that this member of the superfamily of PDEs has properties that make it uniquely suited for the demands of serving as the central enzyme of the visual transduction signaling cascade. In this chapter, the physiological context in which PDE6 operates to control the initial events of vision is presented first, followed by a review of the structural properties of the PDE6 holoenzyme and its catalytic and inhibitory subunits. After presenting the enzymological and pharmacological properties of PDE6, a summary of the various proteins (most notably the G-protein, transducin) that contribute to the precise regulation of PDE6 activation and inactivation during visual transduction will be given. The chapter concludes with a discussion of the involvement of PDE6 in diseases of the retina.

8.2   VERTEBRATE PHOTOTRANSDUCTION SIGNALING PATHWAY

The initial events of visual perception occur in the retina where two distinct classes of photoreceptor cells are responsible for converting photon absorption into an electrical response that is propagated to second-order retinal neurons and eventually to the optic nerve. Rod photoreceptors function at dim levels of illumination, and can reliably detect single photons of light. However, during daylight illumination conditions, the rod photoresponse is saturated and the visual system depends on cone photoreceptors. Cones are much less-sensitive light detectors, but possess the important property that their photoresponses do not saturate even at the brightest illumination conditions. In addition, most vertebrates have two or more types of cone photoreceptors that differ in the spectral sensitivity of their visual pigments. This confers the potential to discriminate color by comparing the output of each spectrally distinct group of cone photoreceptors.[1,2]

Although the light responsiveness of rods and cones differ in many respects, both classes of photoreceptors possess the same overall biochemical machinery for transforming photic stimuli into altered neurotransmitter release at the photoreceptor synapse. Some of the differences in rod and cone cell physiology are likely because of the functional differences in the distinct rod and cone isoforms found for the central phototransduction proteins (including rod and cone PDE6 catalytic and inhibitory subunits). Other factors contributing to the differences in sensitivity and temporal responsiveness of rods and cones may include the following: differences in cellular anatomy (Figure 8.1), presence of proteins uniquely expressed in rods or cones, and differences in the relative abundance of certain proteins found in both cell types.[3] Because of the difficulty in obtaining sufficient quantities of isolated

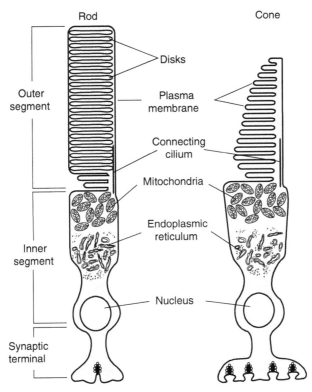

FIGURE 8.1 Schematic diagram of vertebrate rod and cone photoreceptors. The phototransducing outer segment is joined to the mitochondria-rich ellipsoid portion of the inner segment by the connecting cilium. Light-induced changes in membrane potential in the outer segment are transmitted through the inner segment to alter neurotransmitter release at the synaptic terminal that communicates with second-order retinal neurons.

cone cells for biochemical studies, the better-understood rod-phototransduction pathway will be the primary focus of this review.

The highly complex biochemical pathways that constitute the visual transduction process can be categorized, somewhat artificially, into the following stages: (1) activation or excitation, (2) inactivation or termination, and (3) light adaptation or desensitization. A major accomplishment in the field of signal transduction has been the quantitative modeling of the activation pathway and its correlation with the amplitude and kinetics of the electrical photoresponse of isolated photoreceptors.[4] In addition, the major steps in the termination pathway are now known, including the importance of calcium as a feedback regulator for reactions involved in response termination. However, at the biochemical level, we still lack a complete understanding of how the gain and kinetics of the photoresponse are controlled during the process of light adaptation.[4–9]

8.2.1 THE PHOTORECEPTOR CELL IS COMPARTMENTALIZED FOR LIGHT CAPTURE AND SIGNAL PROPAGATION

Rod and cone photoreceptors are highly specialized sensory neurons consisting of several structurally and functionally distinct compartments (Figure 8.1). Phototransduction is initiated in the outer segment, an organelle consisting of either a stack of disk membranes (rods)

or continuous invaginations of the plasma membrane (cones). In both cases, the outer segment membranes contain high levels of the visual pigment to optimize photon capture, as well as the other central phototransduction proteins. The close packing of outer segment membranes (30 nm repeat distance) restricts free diffusion in the cytoplasm, resulting in localization of the photoexcitation process.

Adjacent to the outer segment is the metabolic compartment of the cell (the inner segment), which is connected to the outer segment by a nonmotile cilium. The connecting cilium is responsible for regulating the transport of proteins and metabolites from their site of synthesis in the inner segment to the phototransducing outer segment. Mitochondria are localized within the inner segment near the connecting cilium, whereas organelles dedicated to protein biosynthesis are centrally located near the nucleus. The photoreceptor synaptic terminus adjoins the inner segment, and it is here that the light-induced changes in membrane potential generated in the outer segment lead to a slowing of neurotransmitter release, and transmission of the light stimulus to other classes of retinal neurons.

8.2.2 Major Steps in Visual Excitation in Rod Photoreceptors

The first step in vertebrate vision is the photoisomerization of 11-*cis* retinal to the all-*trans* isomer that is covalently bound to the G-protein-coupled receptor, rhodopsin (Figure 8.2). Activated rhodopsin (specifically, metarhodopsin II) undergoes a conformational change in its cytoplasmic loops that allow this transmembrane receptor to bind the heterotrimeric G-protein, transducin. Following GDP–GTP exchange on the α-subunit of transducin and

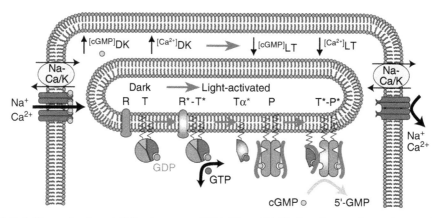

FIGURE 8.2 (See color insert following page 274.) The cGMP signaling pathway in vertebrate rod photoreceptors. In dark-adapted rod photoreceptors (left side of diagram), cytoplasmic cGMP (green circles) and calcium concentrations are high, and some of the cGMP-gated cation channels in the plasma membrane (purple tetramer) are fully liganded and in their open state. Upon absorption of a photon by rhodopsin (R; pink integral disk membrane protein), isomerization of the 11-*cis* retinal chromophore occurs to activate the receptor (R*). This leads to binding of transducin (T; pie-shaped heterotrimer) to R*, guanine nucleotide exchange of GDP (gray circle) for GTP (red circle), and formation of the activated transducin α-subunit with bound GTP (Tα*). The Tα* species then binds PDE6 holoenzyme (P; blue αβ catalytic dimer with red γ-subunits), causing de-inhibition by the γ-subunit (Tα*-P*), and a large acceleration of catalysis of cGMP to 5′-GMP at the active site (green arrow). The light-induced drop in cGMP concentration (right side of diagram) causes the ligand-gated ion channel to close, causing membrane hyperpolarization. Ongoing extrusion of calcium by the Na^+–Ca^{2+}/K^+ exchanger (gray) in the absence of calcium influx through the channel also causes intracellular calcium concentration to decline, which is vital for the recovery process. (Adapted from Ref. [9].)

dissociation from the transducin βγ dimer, the activated α-GTP then binds to the inactive PDE6 holoenzyme. Displacement of the inhibitory γ-subunit of PDE6 by activated transducin (to be described in detail further in the chapter) accelerates cGMP hydrolysis at the active site of PDE6, resulting in a rapid drop in cytoplasmic cGMP levels. cGMP-gated ion channels in the plasma membrane that were open in the dark-adapted state then dissociate their liganded cGMP, causing channel closure and membrane hyperpolarization. High concentrations of rhodopsin, transducin, and PDE6 confined to the surface of the rod disk membrane serve to increase the sensitivity, speed, and amplification properties of these initial events in phototransduction. It is the rapid decline in cellular cGMP levels that transmits by diffusion the photoactivation signal to the plasma membrane where the ion channels are localized.

As the central effector enzyme in the visual excitation pathway, PDE6 activation and inactivation must be precisely regulated. The visual excitation process is dominated by the >100-fold activation of PDE6 by transducin, along with a catalytic efficiency for PDE6 that operates at rates approaching the diffusion-controlled limit. The speed and extent of PDE6 activation ensures that cGMP levels in the photoreceptor can be regulated on the millisecond timescale required for this sensory neuron. (For further reviews, see Refs. [4,8,10].)

8.2.3 RECOVERY OF THE DARK-ADAPTED STATE AND LIGHT ADAPTATION

An opposing set of reactions is responsible for controlling the duration of each step in the excitation pathway. Active turnoff mechanisms are necessary to precisely control the kinetics of the electrical response, as well as to accelerate the restoration of the dark-adapted state (thereby enhancing temporal resolution of the visual system). Although it has not been determined whether cGMP contributes directly to response termination by means of feedback regulation of cGMP-binding proteins (e.g., PDE6 itself or cyclic nucleotide-dependent protein kinases), it is well documented that calcium signaling controls several aspects of photoresponse termination.[7,11,12] Inactivation of metarhodopsin II results from phosphorylation by rhodopsin kinase (GRK1), whose activity is regulated by interaction with the calcium binding protein recoverin or S-modulin. Phosphorylation of rhodopsin and its subsequent binding to visual arrestin effectively quenches the ability of rhodopsin to activate transducin. The inactivation of transducin depends on the intrinsic GTPase activity on the α-subunit, which is not subject to regulation by calcium. Rather, the GTPase rate is modulated by a multiprotein complex consisting of the regulator of G-protein signaling-9 (RGS9), the type 5 G-protein β-subunit (Gβ5), the RGS9 anchoring protein (R9AP), and the γ-subunit of PDE6. Inactivation of transducin restores the ability of the PDE6 γ-subunit to rebind and inhibit the active site of PDE6. The relative affinity of γ-subunit binding to transducin α-subunit versus the PDE6 catalytic subunits may be determined by the state of cGMP occupancy on the regulatory GAF domains of PDE6.

Reinhibition of PDE6 catalysis is accompanied by stimulation of photoreceptor guanylate cyclase (GC) activity to restore cGMP levels to their dark-adapted state. Calcium regulation of GC activity is mediated by guanylate cyclase activating proteins (GCAPs) that bind to and activate GC when calcium levels drop in the outer segment. This calcium-regulated stimulation of cGMP synthesis is the dominant reaction controlling the response amplitude and the kinetics of the recovery phase following flash illumination of a rod photoreceptor. (For reviews, see Refs. [5,8,13].)

In addition to the above-mentioned reactions involved in controlling the amplitude and kinetics of the photoresponse recovery, photoreceptors have an extraordinary ability to adapt to an enormous range of illumination conditions. The reactions involved in excitation and termination must therefore be subject to desensitization, so that incremental changes in light intensity can be detected in the presence of constant illuminance. Two defining physiological

characteristics of light adaptation are the compression of the response amplitude and the acceleration of the kinetics of the electrical response to an incremental change in light stimulation. Multiple biochemical mechanisms are responsible for the desensitization of the photoresponse, many of which serve to precisely regulate cGMP metabolism. Calcium is a central regulator of light adaptation in rods and cones, and targets for calcium regulation include GC and rhodopsin kinase (mentioned above) as well as calcium–calmodulin regulation of the cGMP-gated ion channel. The extent of PDE6 activation in light-adapted photoreceptors will affect the kinetics and amplitude of the response to an incremental change in light intensity; in this regard, cGMP metabolic flux provides a calcium-independent mechanism for adaptation. Of particular relevance to cone photoreceptors (which must function at high levels of illuminance) is the process of bleaching adaptation, in which depletion of visual pigment capable of photoactivation reduces the gain of the visual transduction pathway. Finally, some aspects of light adaptation currently lack biochemical correlates, and it is plausible that novel aspects of PDE6 regulation may account for some of these physiological phenomena. (For reviews, see Refs. [7,12,14,15].)

8.2.4 cGMP Metabolic Flux during Visual Transduction

The photoreceptor cell does not make artificial distinctions between excitation, recovery, and adaptation, but rather responds to a light stimulus with a coordinated activation of the intersecting cGMP and calcium signaling pathways described above. As the final electrical output of the photoreceptor depends on the open or closed state of the cGMP-gated ion channels in the outer segment, it is the dynamics of cGMP metabolic flux (i.e., PDE6 and GC activities) that determine the photoresponse.

In the outer segment of a dark-adapted rod photoreceptor, the metabolically active cGMP concentration is several micromolar[16] and the free calcium concentration is several hundred nanomolar.[7] The total cellular content of each second messenger is actually much higher because binding sites for cGMP and calcium exist in the outer segment that buffer the cytoplasmic free concentration. In the case of cGMP, the total cellular content of cGMP in the amphibian rod outer segment is ~60 μM.[17] However, most of the cGMP in the outer segment is metabolically inactive on the timescale of the flash photoresponse, as judged by the small, maximal light-induced decline (<20%) in total cGMP following flash illumination.[17,18] Electrophysiological measurements of the extent of cGMP-gated channel activation in dark-adapted rods also provide estimates of 2–4 μM free cytoplasmic cGMP.[19,20] Quantitative measurements of cGMP-binding sites in frog rod outer segments identified two classes of binding sites, one with high affinity ($K_D = 60$ nM; site concentration ~40 μM), which is likely to be the cGMP-binding sites in the PDE6 GAF domain, and a second class of lower affinity sites ($K_D = 7$ μM) whose identity is unknown.[21] Thus the majority of the cellular cGMP content in the rod outer segment is not metabolically active, at least during the initial stages of the light response.[22]

The cytoplasmic cGMP concentration in dark-adapted rod outer segments is maintained by a balance of cGMP hydrolysis and cGMP synthesis by nonactivated PDE6 and GC (whose activity is neither inhibited nor activated at this calcium concentration), respectively. cGMP metabolic flux in darkness causes the entire pool of cytoplasmic cGMP to turn over very rapidly (every 1–2 s[23,24]), even though the PDE6 hydrolytic rate is only 0.1% of its catalytic potential.[25,26]

Immediately following illumination, PDE6 is activated by transducin, and cGMP levels in the outer segment drop rapidly to cause channel closure. However, there is a lag between PDE6 activation and calcium-dependent GC activation that occurs because cytoplasmic calcium levels do not fall until after channel closure has occurred. Even after calcium levels drop and GC is thereby activated, the catalytic power of activated PDE6 exceeds the synthetic rate of activated GC. Although the cytoplasmic cGMP concentration remains low during this

period where both the enzymes are activated, cGMP metabolic flux is accelerated 10-fold.[23,24,27,28] This acceleration of cGMP metabolic flux in the light is an important mechanism of increasing the operating range of the photoreceptor to allow light-adapted photoresponses to occur in the presence of constant background illumination.[14]

After the light stimulus is extinguished, PDE6 inactivation is a prerequisite for photoresponse recovery. Once PDE6 activity has returned to its basal level, cGMP levels can be rapidly restored by activated GC, cGMP-gated channels reopen, cytoplasmic calcium levels rise, and cGMP metabolic flux returns to its dark-adapted rate.

8.3 TISSUE AND SUBCELLULAR DISTRIBUTION OF PDE6

The expression of PDE6 is much more restricted than most of the other vertebrate PDE families. PDE6 catalytic subunits are found primarily in the outer segments of rods and cones in the retina. Rod photoreceptors express the α and β catalytic subunits of PDE6, whereas cones express the α'-subunit. Likewise, distinct rod γ- and cone γ'-subunit isoforms are exclusively expressed in rods and cones, respectively. The PDE6 holoenzyme content in photoreceptor cells is quite high compared to the cellular abundance of most other PDE families. There is approximately 1 PDE6 per 10 transducin and 300 rhodopsin molecules on the disk membrane surface of both amphibian[21,29,30] and mammalian[31] rod outer segments, equivalent to a cytoplasmic concentration of PDE6 holoenzyme of 20 μM. The relative ratio of the proteins responsible for initial steps in the visual cascade has probably evolved to optimize the signal-to-noise properties of phototransduction.[5]

Outside the retina, photoreceptive cells in the pineal gland contain all the components of the retinal phototransduction pathway.[32,33] In mammalian neonates, cone (but not rod) PDE6 catalytic subunits are present,[32,34] whereas in the chicken pineal gland the cone α'-subunit, rod β-subunit, and both γ- and γ'-subunits are detected.[35] Interestingly, primary retinoblastoma cultures and some retinoblastoma cell lines contain transcripts for both rod and cone PDE6 catalytic subunits and can produce functional cone PDE6,[36] but other retinoblastoma-derived cell lines (e.g., Y79) contain PDE6 transcripts but virtually no functional enzyme is produced.[37,38]

Low levels of PDE6 catalytic subunit transcript have been reported in nonphotoreceptive tissues.[39,40] Upstream promoter regions for rod and cone PDE6 catalytic subunit genes have been identified that promote transcription primarily in photoreceptor cells and, to a much lesser extent, in brain.[41–43] With one exception,[44,45] there are no reports of protein expression of PDE6 catalytic subunits in mammalian tissues other than retina, pineal gland, and retina-derived tumors.

In contrast, several studies have shown that the PDE6 inhibitory γ-subunit may be widely distributed in nonphotoreceptive tissues.[46–51] The role of the γ-subunit in these tissues presumably lacking PDE6 catalytic subunits is unclear. Although regulation of PDE5 activity by the γ-subunit has been reported in partially purified preparations of the enzyme,[46,47] purified, recombinant PDE5 does not bind γ-subunit nor are its cGMP binding or catalytic properties affected by the γ-subunit.[52] A number of potential binding partners for the γ-subunit have been identified, many of which contain SH3 domains.[51,53] Furthermore, γ-subunit interactions with this class of proteins may be regulated by reversible phosphorylation.[48,50] The significance of these observations awaits further study.

8.4 STRUCTURE OF PDE6 CATALYTIC AND INHIBITORY SUBUNITS

The PDE6 holoenzyme differs in fundamental ways from the other members of the class I PDEs. For rod PDE6 (with the possible exception of chicken rod PDE6),[54] the catalytic subunits are coded by two distinct genes, α and β, which form a catalytic heterodimer.[55–58]

TABLE 8.1
Catalytic and Inhibitory Subunits of Human PDE6

Common Name	Gene Name	NCBI Gene ID	Human Chromosome	NCBI Protein RefSeq	Number of a.a.	Molecular Weight (Da)
Rod α	*PDE6A*	5145	5q31.2-q34	NP_000431	860	99504
Rod β	*PDE6B*	5158	4p16.3	NP_000274	854	98276
Cone α'	*PDE6C*	5146	10q24	NP_006195	858	98957
Rod γ	*PDE6G*	5148	17q25	NP_002593	87	9512
Cone γ'	*PDE6H*	5149	12p13	NP_006196	83	8943

In contrast, cone PDE6 is similar to all other class I PDEs in that it is a homodimeric enzyme composed of α' subunits.[59,60] A second major distinctive trait of PDE6 (this one shared by rod and cone isoforms) is the association of high-affinity inhibitory rod or cone γ-subunits to the corresponding catalytic dimer[61,62] Table 8.1 lists the rod and cone PDE6 catalytic and inhibitory subunits identified in humans. The rod PDE6 holoenzyme has been conclusively shown to exist as a tetramer,[57,63] and the cone enzyme is assumed to be the same but experimental confirmation is currently lacking.

Although the x-ray crystal structure is yet to be solved for the PDE6 holoenzyme or for individual subunits or folding domains, a low-resolution representation of the PDE6 holoenzyme has been obtained using electron microscopic and image analysis techniques.[64,65] As seen in Figure 8.3, the two catalytic subunits form a dimer, with each subunit organized into three structural domains. Comparison of the volume of each domain with the predicted mass of the GAFa, GAFb, and catalytic domains, as well as with the volumes of the PDE2 GAFa-GAFb structure [66] or the PDE5 catalytic domain crystal structure[67] support this assignment of the molecular organization of PDE6 as seen in Figure 8.3.[64] The ability of individual GAF domains to bind cGMP,[54,66,68–70] along with the ability of the monomeric catalytic domain of PDE5 to hydrolyze cyclic nucleotides[71] argues that the observed structures fold

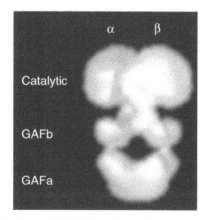

FIGURE 8.3 Molecular organization of PDE6 based on electron microscopy and image analysis at 2.8 nm resolution.[64] This three-dimensional model shows the organization of the rod PDE6 catalytic αβ dimer into three distinct domains, representing the catalytic domain and the two tandem GAF domains. Low molecular weight subunits are not discriminated at this level of resolution.

into independent, functional domains. Note that this low-resolution structural model of PDE6 does not allow visualization of the γ-subunit.

8.4.1 CATALYTIC SUBUNIT

The primary sequence of the α, β, and α' catalytic subunits of PDE6[72–74] consist of an N-terminal region of unknown function, two regulatory GAF domains (GAFa and GAFb), the catalytic domain, and a C-terminal motif that is subject to isoprenylation (Figure 8.4).

8.4.1.1 cGMP Binding Regulatory GAF Domains

The tandem GAF domains (SMART accession number, SM00065)[75,76] of PDE6 catalytic subunits are highly conserved in amino acid sequence within the PDE6 family as well as show strong sequence similarity to the other four GAF-containing PDE families (PDE2, PDE5, PDE10, and PDE11). For example, the human GAFa domains of the α- and β-subunits are 68% identical, whereas the rod and cone PDE6 GAFa domains share ~54% amino acid identity; PDE5 GAFa has only about 30% amino acid identity with the three PDE6 catalytic subunits. The sequence identity within the GAFb domains of PDE6 is significantly greater than for GAFa domains: 83%, ~77%, and ~22% amino acid identity for α-versus β-subunits, rod versus cone subunits, and PDE5 versus PDE6, respectively.

The GAF domains found in class I PDEs serve several functions: (1) one of the two tandem GAF domains consists of a functional cyclic nucleotide binding pocket; (2) GAF domains undergo an allosteric transition upon ligand binding that is communicated to the catalytic domain; (3) for those PDEs which have been studied to date (i.e., PDE2, PDE5, and PDE6),

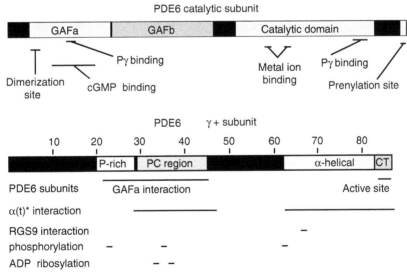

FIGURE 8.4 Domain organization of the catalytic and inhibitory subunits of PDE6. The catalytic subunit of PDE6 (top) consists of two tandem GAF domains, a catalytic domain, and a prenylated C-terminus. The 10 kDa γ-subunit has several functional domains: a proline-rich (P-rich) region and a polycationic region (PC region) that both interact with the cGMP binding site on PDE6, an α-helical region, and the C-terminal (CT) residues, which bind to the active site of PDE6. The regions known to interact with transducin and RGS9 are indicated, along with putative sites for phosphorylation (at T22, T35, T52) and ADP-ribosylation (at R33 and R36).

the GAF domains are a primary structural element involved in catalytic subunit dimerization.[66,70,77] The PDE6 GAF domains serve an additional, unique function, namely to bind the γ-subunit in a reciprocal, allosteric manner (discussed in Section 8.7).

In its nonactivated state, the rod PDE6 holoenzyme (αβγ₂) binds two cGMP molecules with high but unequal affinity[78–80] (see Figure 8.4); another class of low-affinity cGMP-binding sites found in rod outer segments[21] might also represent cGMP binding to the PDE6 holoenzyme, as four potential cGMP-binding sites exist in the GAFa and GAFb domains of the catalytic dimer. The cGMP-binding affinity for PDE6 is somewhat weaker for amphibian rod PDE6 and for mammalian cone PDE6 isoforms compared to mammalian rod PDE6.[60,79] The cGMP binding sites on PDE6 catalytic subunits discriminate cGMP over cAMP by ~10^6-fold, with the N1/C6 region of the purine ring of cGMP being a major determinant of nucleotide specificity.[54,81] The cGMP binding site in PDE6 has been localized to the GAFa domain in rod and cone isoforms, and several amino acid residues important for high-affinity interactions with cGMP have been defined.[54,82] In particular, several residues within or close to the so-called H4 helix (found in cGMP binding PDEs but not other classes of GAF-containing proteins) are critical for stabilizing high-affinity cGMP binding to PDE6.[54,82]

The GAFa domains of rod and cone PDE6 catalytic subunits serve as the primary dimerization domain between two individual subunits. This is evident in the structural model of the PDE6 holoenzyme[64] (Figure 8.3). An essential region for dimerization of rod PDE6 was defined by mutagenesis in the N-terminal region of the GAFa domain; this work also demonstrated strong selectivity for αβ heterodimer formation over αα or ββ homodimer species,[77] consistent with biochemical evidence for rod heterodimers.[58]

The GAFa domain is also a major site of interaction with the inhibitory γ-subunit of PDE6. Direct cross-linking of the proline-rich region of the γ-subunit (see Section 8.4.2.1) to the M138/G139 residues in GAFa (bovine PDE6 α-subunit) has been reported.[83] In addition, mutagenesis of PDE6-specific amino acids in the vicinity of this site can abolish binding of the γ-subunit to the catalytic dimer.[82] The proximity of γ-subunit interacting sites with the cGMP binding pocket in GAFa provides structural support for the positive cooperativity between cGMP and γ-subunit binding (discussed in Section 8.7). Differences in the binding of the γ-subunit to the α- and β-subunit reveal potential structural differences between α-GAFa and β-GAFa that may account for heterogeneity in cGMP or γ-subunit affinity for the holoenzyme.[84]

Relatively little is known about the function of the GAFb domain of PDE6. Although it contains the signature residues (NFKDE) for a GAF domain, the GAFb domain of PDE6 has a 29-amino acid insert in this region, which is lacking in GAFa and which may impair cGMP binding. Mutation of a critical asparagine residue in the PDE6 GAFb of a chimeric PDE5 or PDE6 construct eliminated cGMP binding, which might be due to conformational change that indirectly destabilized cGMP binding in GAFa.[52] GAFb may also contribute to stabilizing the formation of catalytic dimers, but is not the major determinant of dimer formation.[77]

8.4.1.2 Catalytic Domain

The PDE6 catalytic domain (SMART accession number, SM00471) contains the same invariant catalytic site residues that typify all class I PDEs.[85,86] Multiple sequence alignment of the catalytic domains of human PDE6 and PDE5 reveals 84% amino acid identity between the two rod subunit isoforms, ~75% amino acid identity between rod and cone PDE6 subunits, and 43% identity between PDE5 and PDE6. Because of the difficulty in expressing functional PDE6 catalytic subunits in heterologous systems,[37,52,87–89] site-directed mutagenesis has not been feasible to date. Instead, much of the recent progress in understanding the

structure and function of the PDE6 catalytic domain has been obtained by constructing and expressing chimeric proteins containing both PDE5 and PDE6 sequences and by resorting to structural homology modeling based on the crystal structure of the PDE5 catalytic domain.

The structural basis for the very high catalytic efficiency of cGMP hydrolysis in PDE6 compared with that in PDE5 is not well understood, but is likely to result from differences in the metal-binding pocket of the active site. Substituting two PDE5 residues for their PDE6 counterparts in the divalent cation binding region (i.e., A608G and A612G; bovine PDE5 numbering, corresponding to bovine cone PDE6 a.a. 562 and 566) increased the k_{cat} by 10-fold.[90] Replacement of the H-loop (adjacent to the metal binding site) of PDE5 with the corresponding PDE6 sequence also introduced a greater PDE6-like character to the recombinantly expressed catalytic domain protein.[89] Other structural determinants of the 500-fold greater k_{cat} for PDE6 versus PDE5 await further work. The histidine residues responsible for binding divalent cations in the active site of PDE5[91,92] are present in PDE6 and bind zinc with high affinity and magnesium with lower affinity.[93] These divalent cations are not only critical for the catalytic mechanism but also confer structural stability to the enzyme.

As mentioned earlier, PDE6 is unique among the class I PDEs in having its catalytic activity regulated by an extrinsic protein, the γ-subunit. The γ-subunit exerts its inhibitory action by direct binding of its C-terminal residues (see Section 8.4.2.1) to the catalytic pocket, thereby blocking access of cGMP to its binding site.[94,95] Using recombinantly expressed PDE5 and PDE6 chimeras in which the γ-interacting region of PDE6 was inserted into the PDE5 catalytic domain sequence, specific γ-interacting amino acid residues (PDE6 α′: M758, Q752, F777, and F781) have been identified within the M-loop of the catalytic domain structure.[89,90,96]

8.4.1.3 C-Terminal Prenylation

PDE6 is the only vertebrate phosphodiesterase family containing a C-terminal CAAX (A, aliphatic; X, any amino acid) motif that is posttranslationally modified by sequential isoprenylation, proteolysis, and carboxymethylation. The mammalian rod PDE6 α-subunit is farnesylated at the cysteine residue (indicated in italics) of the terminal sequence *C*C(I/V)Q, whereas the rod β-subunit has a geranylgeranyl group bound to the cysteine of the terminal sequence *C*(C/R)IL.[97,98] On the basis of the CLML sequence for the mammalian enzymes, cone PDE6 is likely to be geranylgeranylated at its C-terminus. The prenylated, carboxymethylated C-termini are hydrophobic (especially the geranylgeranylated moieties) and are responsible for anchoring PDE6 to the outer segment disk membrane,[99] thereby facilitating two-dimensional collisions with transducin during visual excitation. PDE6 is membrane bound, except when the 17 kDa PrBP/δ protein (originally termed PDEδ) is present to bind to the prenyl groups and solubilize PDE6 from the membrane (see Section 8.9.2 for discussion). Disk membrane association of PDE6 can be disrupted by altering the ionic, nucleotide, divalent cation, or illumination conditions, which are useful for purification of the enzyme.[100]

8.4.2 Inhibitory γ-Subunit

Considering its small size (~10 kDa), the γ-subunit of rod and cone PDE6[101,102] serves a remarkable number of functions (Figure 8.4): (1) A primary function of the γ-subunit is to block catalysis by competing with cGMP at the active site of PDE6;[95] (2) the γ-subunit enhances the binding affinity of cGMP to the regulatory GAFa domain;[103] (3) activated transducin interacts at two distinct sites with the γ-subunit,[104] leading to de-inhibition of PDE6 at its active site; (4) during deactivation, the γ-subunit participates in a protein complex with the RGS9 and other proteins to accelerate the GTPase activity of activated transducin;[105–107] (5) the γ-subunit contains a recognition sequence for binding to SH3

domain-containing proteins;[51] (6) the γ-subunit is also a substrate for several posttranslational modifications.[108,109]

8.4.2.1 Structure and Domain Organization of the γ-Subunit

It is likely that the γ-subunit exists in solution as a natively unfolded (or intrinsically disordered) protein,[110] because of its small size, low hydrophobicity, and minimal secondary structure.[111,112] This structural feature of the γ-subunit may account for this small protein's ability to span the distance from the GAFa domain to the active site on the catalytic domain of the catalytic dimer,[65,84] as well as explain how the γ-subunit can serve several distinct functions, i.e., moonlighting.[113]

The N-terminal region of the γ-subunit has the greatest variability in amino acid sequence, particularly when comparing the rod and cone γ-subunit sequences from different animals. However, beginning at amino acid G19 (all amino acid positions refer to the mammalian rod γ-subunit sequence in this section), there are very few amino acid substitutions in the remainder of the protein (100% a.a. identity for all mammalian rod γ-subunits reported to date), suggesting the structural and functional importance of the remaining 75% of the protein.

A proline-rich region (a.a. 20–28; Figure 8.4) represents an interaction site for proteins containing SH3 domains.[51] This region also contains a consensus sequence for proline-directed kinases, and both cdk5 kinase and MAP kinase have been shown to phosphorylate T22 *in vitro*.[114–116] Proline-23, when chemically modified, cross-links with the GAFa domain of rod PDE6, demonstrating specific interaction of this region of the γ-subunit.[83] This region is also known, from peptide competition studies, to stabilize cGMP binding to GAFa.[26]

The central portion of the γ-subunit (a.a. 20–50) has high hydrophilicity,[117] especially in its polycationic region (a.a. 29–45) in which 50% of the residues are either lysine or arginine and only two hydrophobic amino acids occur. This polycationic region is a major site of interaction with activated transducin α-subunit.[104,118–122] In addition, a consensus sequence for protein kinase-A (and related serine or threonine kinases) is present at T35.[116,123,124] The primary effect of phosphorylation at either T22 or T35 is to reduce the ability of the γ-subunit to interact with activated transducin, but the physiological occurrence of these *in vitro* phosphorylation reactions and their relevance for the visual signaling pathway is uncertain.[116] Two asparagines (N33 and N36) are potential sites for ADP ribosylation, and could modulate γ-subunit interactions with transducin.[109,125]

The combined proline-rich and polycationic regions of the γ-subunit (comprising P20 through K45) are a primary interaction site of the γ-subunit with the PDE6 catalytic dimer.[118,126–129] Peptide mapping studies showed that the proline-rich and polycationic regions bind to catalytic dimers with 50-fold greater affinity than peptides that bind to the α-helical and C-terminal regions.[26] Cross-linking studies identified P23, S30, and S40 as individual sites of interaction with catalytic dimers.[84]

The amino acid sequence between the polycationic region and the α-helical region of the γ-subunit contains a high content of glycine residues, which may impart flexibility and a low extent of secondary structure. The ability of the γ-subunit to interact with RGS9 to accelerate transducin inactivation is localized to V66 of the γ-subunit sequence.[107] In addition, a consensus site for phosphorylation by G-protein-coupled receptor kinase-2 (distinct from rhodopsin kinase, GRK1) has been identified at T52,[48] although the relevance to phototransduction is not known.

The α-helical region near the C-terminus of the γ-subunit (a.a. 62–83) is a major site of interaction with activated transducin α-subunit, and is primarily responsible for the GTPase accelerating activity of the γ-subunit.[107,130,131] Leucine 76, C68, and W70 of the γ-subunit appear to be key contacts for transducin binding.[132–136] Mutagenesis and cross-linking studies have demonstrated the importance of amino acids F73, N74, H75, and L78 of

the γ-subunit to interact with the PDE6 catalytic subunits,[84,136] even though the α-helical region itself does not directly block the active site. The emerging picture is that transducin α-subunit binds to one face of the γ-subunit α-helical region, and disrupts contacts between the opposite face of the α-helical domain and the catalytic domain of PDE6.

The last five C-terminal residues (a.a. 83–87) directly bind to the active site to cover the catalytic pocket and block cyclic nucleotide entry.[94–96,118,120,131] They also interact with the α-subunit of transducin during the relief of inhibition caused by transducin binding to the γ-subunit.[107]

Although most of the above discussion does not distinguish differences in γ-subunit interactions with the α- and β-subunits of rod PDE6, both biochemical and structural evidence clearly indicate that the two γ-subunits bind differentially to the catalytic dimer of rod PDE6.[26,80,84,137,138] Because the GAFa domains of the α- and β-subunit of PDE6 are more dissimilar (68% amino acid identity) than the GAFb (83%) or catalytic domains (84%), a reasonable hypothesis is that most of the differences in γ-subunit binding reside in the GAFa domains.

8.5   CATALYTIC PROPERTIES OF PDE6

In rod outer segments, the concentration of the γ-subunit is equal to that of the PDE6 catalytic subunits.[138] The very high affinity of γ-subunit for the catalytic dimer ensures that virtually all PDE6 exists as the nonactivated holoenzyme ($\alpha\beta\gamma_2$) in its dark-adapted state.[63,139] Biochemical estimates of the dark-adapted (basal) activity of rod PDE6 indicates that only 1 out of 2200 catalytic dimers lack bound γ-subunit and are fully active,[25] consistent with electrophysiological measurements of PDE6 activity in dark-adapted rods.[140]

At the other extreme, removal of the γ-subunit from the catalytic dimer by limited proteolysis,[62] by competition with polycationic proteins,[55] or by extraction of the γ-subunit by binding to activated transducin[141] relieves γ-subunit inhibition and stimulates catalysis. (The mechanism of activation by transducin is discussed in Section 8.8.) Activated rod PDE6 catalytic dimer has a ~1000-fold higher turnover number for cGMP (bovine, $k_{cat} = 5440$ s^{-1}; frog, $k_{cat} = 7500$ s^{-1}; see Refs. [26,138]) than for the closely related PDE5 enzyme. cAMP is a very poor substrate for mammalian rod PDE6 catalysis; the specificity constants (k_{cat}/K_M) for the two substrates differ by ~100-fold, primarily due to differences in the K_M values for cGMP ($K_M = 14$ μM) and cAMP ($K_M = 900$ μM).[26] Wide variations in the literature for the K_M for PDE6[4] are now known to result from restricted substrate diffusion when the enzyme was studied in rod outer segment membrane preparations.[30,142] The k_{cat}/K_M value of 4×10^8 M^{-1} s^{-1} for bovine rod PDE6 catalysis of cGMP[26] approaches the diffusion-controlled limit for a bimolecular collision,[143] making PDE6 a nearly perfectly designed catalytic machine for rapidly lowering cGMP levels in the photoreceptor cell during light stimulation.

8.6   PHARMACOLOGY

8.6.1   Effects of PDE Inhibitors on Photoreceptor and Retinal Physiology

Before the discovery of the rod cGMP-gated ion channels,[144] electrophysiological studies in which isolated retinas were perfused with solutions containing isobutylmethylxanthine (IBMX) provided strong evidence for PDE6 regulating the light sensitivity and photoresponse kinetics.[145–147] Suction electrode recordings of isolated photoreceptors[148] have exploited PDE inhibitors for the purpose of measuring changes in PDE6 catalytic activity in response to illumination.[23] Exposure of intact photoreceptor cells to IBMX[149] or zaprinast or

dipyridamole[150] has also provided mechanistic insights into how PDE6 activation sets the light sensitivity and controls the electrical response of the photoreceptor. cGMP levels in intact rod outer segment suspensions are elevated upon exposure to PDE inhibitors,[151] but inhibitor potency is greatly reduced compared to purified PDE6 and not all of the PDE6 can be inhibited. These observations may be explained by several factors reducing the effectiveness of PDE inhibitors: high cellular PDE6 concentration, competition of drug with the γ-subunit for the PDE6 active site, and compensating feedback mechanisms that reduce cGMP synthesis.

Application of PDE inhibitors to isolated retinas to suppress PDE6 activity provides insights into the various mechanisms leading to photoreceptor cell death and retinal degeneration. Treatment with the nonspecific PDE inhibitor IBMX was shown to elevate cGMP levels in the retina and lead to photoreceptor cell degeneration in both rod- and cone-dominant retinas,[152,153] as well as in human retinas.[154–156]

More recent work with PDE5-targeted inhibitors (most of which also potently inhibit PDE6) has revealed alterations in retinal physiology,[157–160] some of which may reflect involvement of other retinal neurons besides photoreceptors. In cases where the retina is susceptible to retinal degenerative diseases, use of PDE5-targeted inhibitors may have greater effects on retinal function.[161] Overall, currently available evidence suggests no major adverse or irreversible effects on retinal function or health with short-term, intermittent administration of PDE5-targeted drugs in healthy individuals; however, PDE5 inhibitor therapy has been contraindicated for humans with retinitis pigmentosa (for review, see Ref. [162]). Long-term use of PDE5-targeted inhibitors may downregulate PDE6 expression in the retina, but further work is needed in this area to confirm the currently available evidence.[163,164]

8.6.2 POTENCY AND SELECTIVITY OF PDE INHIBITORS FOR PDE6 *IN VITRO*

Pharmacological studies of purified PDE6 give differing results for inhibitor potency depending on whether the PDE6 catalytic dimer or holoenzyme is studied. In the few instances where the mechanism of inhibition has been carefully evaluated, PDE6 catalytic dimers display classical competitive inhibition by zaprinast[165] and E4021.[166] A classical competitive inhibition mechanism is also supported by the crystal structures of PDE inhibitors bound to the active site of the structurally homologous PDE5 catalytic domain.[67] For the PDE6 holoenzyme, however, competition between the γ-subunit and PDE5-targeted inhibitors (but not IBMX) greatly reduces the apparent affinity of drug for the PDE6 enzyme.[151,165,166] Although the γ-interacting residues are distinct from the amino acids which contact bound drug molecules, it is reasonable to assume that binding of these radically different types of inhibitors is mutually exclusive.

Purified PDE6 has a pharmacological profile most similar to that of PDE5, in that most compounds inhibit both PDE families with similar potencies.[88,151,165] For example, vardenafil is the most potent inhibitor of PDE6 reported to date ($K_I = 0.7$ and 0.3 nM for rod and cone PDE6, respectively[151]), but shows little selectivity versus PDE5 ($K_I = 0.25$ nM). Therefore, most drugs targeting PDE5 are more aptly categorized as PDE5/6-selective inhibitors. One notable exception is tadalafil, a bona fide PDE5-selective inhibitor that is 300–600-fold more potent toward PDE5 ($K_I = 3$ nM) than rod PDE6 ($K_I = 2100$ nM).[151,167] Although the structure of tadalafil bound to the PDE5 catalytic pocket has been determined,[67] structural homology modeling of the PDE6 catalytic domain (in the absence of solved crystal structures) fails to reveal an obvious structural basis for the drug selectivity (also see Ref. [168]). Although the rod and cone isoforms of PDE6 are generally very similar in their inhibitor sensitivities,[88,151,165] xanthine-based inhibitors (e.g., IBMX and 8-methoxymethyl-IBMX) have a three- to fourfold greater potency for cone PDE6 compared to the rod enzyme.[151] With the advent of new therapeutic uses for PDE5 inhibitors requiring long-term

administration of the drug (e.g., pulmonary hypertension), a greater understanding of the difference in the pharmacologic profiles of PDE5 and the rod and cone PDE6 isoforms is needed.

8.7   ALLOSTERIC ROLE OF THE cGMP-BINDING GAF DOMAINS OF PDE6

For the case of two GAF-PDEs (PDE2 and PDE5), cGMP binding to the GAF domains induces conformational changes that relieve inhibition of catalysis in the active site of the enzyme.[169–173] Furthermore, cGMP binding to PDE5 induces a change in the shape of the enzyme, stimulates phosphorylation of PDE5, and alters inhibitor binding affinity at the active site.[174–176] Although intramolecular allosteric communication between the GAF and catalytic domains is predicted for the PDE6 catalytic dimer based on its structural similarities to PDE5, no evidence supports this hypothesis at present. The cGMP occupancy of the GAF domains fails to alter cAMP hydrolytic activity at the active site,[25,26,177] although there are inherent experimental difficulties in quantifying ligand binding when the regulatory ligand can also be hydrolyzed at the active site. Effects of cGMP binding on PDE6 catalytic subunit phosphorylation[123] or on changes in the inhibition constant of PDE inhibitors have not been examined.

One well-documented allosteric effect of cGMP binding to the PDE6 GAF domains is the enhancement of the affinity of the γ-subunit to the PDE6 catalytic dimer (Figure 8.5). This is seen as a decrease in the basal activity of PDE6 holoenzyme when cGMP occupies the GAF domain[25]. Interestingly, cGMP binding to PDE6 GAFa domain increases the intrinsic γ-subunit affinity for one, but not both, of the γ-subunit interaction sites on the catalytic dimer; the second γ-subunit binding site retains the same affinity for PDE6 regardless of the

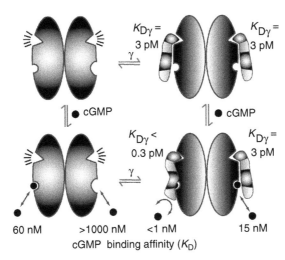

FIGURE 8.5 Summary of the reciprocal positive cooperativity between cGMP and γ-subunit binding to the rod PDE6 catalytic dimer. The top left represents the PDE6 αβ catalytic dimer, with the notch representing the active site and the half-circle representing the regulatory cGMP binding site on each subunit. Addition of γ-subunit (shown with its catalytic-interacting and GAF-interacting subdomains) causes binding of one γ-subunit per catalytic subunit (left to right reactions). Addition of cGMP results in cGMP occupancy of binding sites on the GAFa domains of the catalytic dimer (top to bottom reactions). Dissociation constants for the γ-subunit ($K_{D\gamma}$) and cGMP (K_D; listed at bottom) are for bovine rod PDE6 (see text for details).

state of occupancy of the GAF domains by cGMP.[26] On transducin activation of PDE6, the γ-subunit remains associated with the PDE6 catalytic dimer when cGMP is present, but is released in a complex with transducin α-subunit when the GAF domains are unoccupied.[177,178]

The allosteric change in the GAFa domain is reciprocal, in that addition of γ-subunit to PDE6 catalytic dimers increases the binding affinity for cGMP to the GAF domains.[79,103] In this case as well, cGMP binds with high affinity to one site on the PDE6 catalytic dimer lacking bound γ-subunit; addition of stoichiometric amounts of γ-subunit converts a second low-affinity site into a high-affinity cGMP binding site[80,138] (Figure 8.5).

8.8   REGULATION OF PDE6 ACTIVATION BY THE G-PROTEIN, TRANSDUCIN

Transducin activation of PDE6 results from the binding of the activated transducin α-subunit ($α_t$-GTP) to multiple sites on the nonactivated PDE6 holoenzyme ($αβγ_2$), leading to displacement of the γ-subunit from its site of inhibition at the entrance to the PDE6 catalytic site. The prevailing model of visual excitation asserts that $α_t$-GTP binds to each catalytic subunit, displacing both γ-subunits and activating cGMP hydrolysis at each active site in the catalytic dimer.[5,179] However, evidence for the stoichiometry of $α_t$-GTP binding to PDE6 and the extent to which PDE6 can be activated remains uncertain.[8] Under some conditions, a single α-subunit of transducin was able to maximally activate one PDE6 catalytic dimer, suggesting that either PDE6 catalytic dimer has only one functional active site, or that a single $α_t$-GTP can relieve inhibition of both γ-subunits at both active sites.[180–182] Furthermore, it is typically observed that transducin activates PDE6 to about one-half of the trypsin-activated level in frog[138,183] and bovine[181,184] rod outer segments.

In addition to relieving inhibition by the C-terminal domains of the γ-subunit at the PDE6 active site, transducin activation also reduces γ-subunit affinity to one of the two GAFa domains. This causes 10-fold accelerated cGMP dissociation from one binding site and a concomitant dissociation of one γ-subunit from the holoenzyme. The second GAFa domain retains high affinity for cGMP, and the second γ-subunit remains associated with the PDE6 catalytic dimer.[79,138]

Taking transducin activation together with the reciprocal positive cooperativity and the binding heterogeneity with which cGMP and the γ-subunit interact with the PDE6 catalytic dimer (see Section 8.7), a potentially unique function for the cGMP binding GAF domains of PDE6 emerges. In dark-adapted photoreceptors, cytoplasmic-free cGMP levels are several micromolar[21] (see Section 8.2.4), both GAFa domains are occupied with cGMP, and two γ-subunits block catalysis at the active site. Upon light activation of PDE6, displacement of a γ-subunit by transducin will relieve inhibition at the active site and lower cGMP affinity for the GAFa domain. For transient light activation, cGMP dissociation from the GAFa domains is unlikely because recovery of cGMP levels is fast, thereby promoting tight reassociation of the γ-subunit and the return of PDE6 holoenzyme to its dark-adapted state. For prolonged illumination (i.e., during light adaptation) during which cGMP levels remain low, cGMP dissociation from the GAFa domain might occur, lowering the γ-subunit affinity for its PDE6 catalytic subunit and permitting the γ-subunit (complexed with transducin α-subunit and RGS9) to serve as a GTPase accelerating factor.[105] This would increase the rate of transducin inactivation and help restore PDE6 to its nonactivated state. In this way, the GAFa domains of PDE6 might be sensors of cytoplasmic cGMP, and respond to sustained decreases in cGMP levels with a negative feedback mechanism to help restore the ability to detect light stimuli. An alternative hypothesis that the GAF domains buffer cellular cGMP and release it during photoresponse recovery[21,185] has not been supported by the kinetics of cGMP binding and dissociation with the GAFa domains.[28,138]

8.9 PDE6 INTERACTING PROTEINS

8.9.1 GLUTAMIC ACID–RICH PROTEIN-2 (GARP2)

Alternative splicing of the β-subunit of the rod cGMP-gated ion channel (CNGB1) in its N-terminal domain generates two truncated proteins, one of which is called glutamic acid–rich protein-2 (GARP2). GARP2 is an ~300 a.a. protein with a unique 8-amino acid C-terminal extension, a high content of proline and glutamate residues, and exists in a natively unfolded state.[186–189] GARP2 is specifically expressed in rod photoreceptors,[188] is abundant at the rims of the outer segment disk membranes,[187,188] and is present in roughly stoichiometric amounts relative to PDE6.[31,189]

Binding partners for GARP2 were initially identified by peptide affinity chromatography (PDE6, guanylate cyclase, and the ATP-binding cassette transporter [ABCR];[188]), by gel filtration chromatography (PDE6[188]), or by immunoprecipitation (peripherin[190]). Korschen et al. (1999) reported that addition of a recombinant GARP2 fusion protein reversed the activation of PDE6 by transducin, but GARP2 had no effect on the PDE6 holoenzyme or on the catalytic dimer.[188] More recent investigations of GARP2–PDE6 interactions have confirmed the high affinity with which GARP2 binds to PDE6, and novel chromatographic procedures for isolating and purifying native GARP2 from bovine rod outer segments (ROS) have been developed.[189,191] Using purified, native GARP2, Pentia et al. failed to confirm an inhibitory effect of GARP2 on transducin-activated PDE6;[31] this inhibitory action of GARP2 on activated PDE6 is now attributed to the fusion tag present on the recombinant GARP2 used in the earlier study.[192]

However, Pentia et al. showed that GARP2 can inhibit the basal activity of the PDE6 holoenzyme by up to 80%.[31] This opens the possibility that GARP2 serves to reduce spontaneous activation of PDE6, and thereby enhances the signal-to-noise ratio of rod photoreceptors under very dim (e.g., single photon) illumination conditions. Other potential roles for GARP2 at the rims of the outer segment disk membrane have been proposed,[188–190] and additional work is needed to define whether the PDE6–GARP2 interactions described above are physiologically relevant in regulating PDE6 during visual transduction.

8.9.2 PRENYL BINDING PROTEIN/PDEδ (PRBP/δ)

The 17 kDa prenyl binding protein/PDEδ (PrBP/δ) was originally identified as the PDE6 δ-subunit because it copurified with soluble rod and cone PDE6 from bovine retinal extracts.[193,194] The structure of PrBP/δ consists of an immunoglobulin-like domain that folds into a hydrophobic pocket[195] capable of binding both farnesylated and geranylgeranylated proteins.[196,197] Unlike most photoreceptor proteins whose tissue distribution is restricted to the retina, PrBP/δ is present in numerous tissues.[194,198] PrBP/δ can interact (at least *in vitro*) with numerous other proteins besides PDE6, including opsin kinases (GRK1 and GRK7), transducin βγ dimers, the retinitis pigmentosa GTPase regulator (RPGR), members of the ras GTPase superfamily, and Arl2 and Arl3 (for review, see Ref. [199]).

Attempts to understand the function of PrBP/δ by defining its subcellular localization in the retina have been inconclusive to date. PrBP/δ immunofluorescence has been observed in the rod outer segment,[194] in both inner and outer segments of rods and cones,[197,200] in the inner segment of a single cone class in bullfrogs,[201] and in the connecting cilium region of rods and cones.[201] Localization of PrBP/δ in the inner segment and in the connecting cilium region of the photoreceptor is intriguing because several of the above-mentioned PrBP/δ interacting proteins (i.e., RPGR,[202] Arl3,[203] and Rab8[204]) are also found here. Because the connecting cilium is the conduit for macromolecules that are produced in the inner segment and then transported to the outer segment, the colocalization of PrBP/δ and several potential binding

partners is consistent with PrBP/δ functioning as a chaperone for the transport of prenylated proteins. Further support for a transport role for PrBP/δ comes from recent work with a transgenic mouse in which the *PrBP/δ* gene has been deleted; in this situation, opsin kinase is partially or completely absent in the outer segments of rods or cones, respectively, with corresponding defects in the photoresponse.[200]

Biochemical studies of PDE6 regulation by PrBP/δ have clearly demonstrated that PrBP/δ binds to the farnesylated and geranylgeranylated C-termini of rod PDE6 catalytic subunits. Association of PrBP/δ with PDE6 causes the complex to be released from the disk membranes.[194,196,201] As a consequence, the maximal extent of PDE6 activation by transducin is lowered.[201,205] Furthermore, PrBP/δ binding to mammalian PDE6 causes a reduction in the binding affinity of cGMP for one of the PDE6 GAFa domains.[80] In the absence of PrBP/δ, one of the two cGMP binding sites on bovine PDE6 is essentially nonexchangeable ($K_D < 1$ nM), whereas the other site has a binding affinity of ~20 nM. For PDE6 complexed with PrBP/δ, the nonexchangeable, high-affinity site is converted to a readily exchangeable form with a similar affinity as the second cGMP site.[80]

These effects of PrBP/δ on PDE6 *in vitro* led to the hypothesis that PrBP/δ acts as a negative feedback mechanism during light adaptation to impair transducin activation of PDE6.[205] However, the PrBP/δ content in ROS is only 10% of the PDE6 concentration, arguing against PrBP/δ as a PDE6 regulatory protein or subunit.[201] The substoichiometric amounts of PrBP/δ relative to PDE6, in conjunction with its presence in the inner segment (where PDE6 is virtually absent) and connecting cilium make it more likely that PrBP/δ transiently associates with PDE6 during transport to the outer segment, but does not participate in regulation of the phototransduction pathway itself.

8.10   RETINAL DISEASES DUE TO DEFECTS IN PDE6

Degenerative diseases affecting the retina constitute a family of retinal dystrophies that result in loss of vision. These diseases can result from genetic or nongenetic factors. Retinal diseases are characterized by loss of photoreceptor function, often accompanied by the death of rods and cones. Often, the genetic basis of retinal degeneration can be ascribed to a defect (mutation) in a component of the visual transduction pathways in rod and cone photoreceptors. Updated information on the genetic basis of retinal diseases can be found at RetNet (www.sph.uth.tmc.edu/RetNet/).

It is well established that biochemical defects in cGMP metabolism account for some forms of inherited retinal degenerative diseases. In a mutant mouse (*rd*) that develops retinitis pigmentosa, first the rods and then the cones degenerate as the animal matures or grows. The disease causes elevated levels of cGMP in the retina before photoreceptor cell death.[206,207] The elevation of cGMP levels in the *rd* retina results from a deficiency of PDE6 in rod cells, specifically a mutation in the PDE6 β-subunit.[208–213] A similar genetic defect in Irish setter dogs (*rcd*) causing abnormal cGMP metabolism is also because of a mutation in the PDE6 β-subunit.[214–216]

In humans, mutations in the rod PDE6 catalytic subunit genes account for ~5% of recessively inherited cases of retinitis pigmentosa.[217,218] Defects in the β-subunit of rod PDE6 (cataloged in NCBI Online Mendelian Inheritance in Man, OMIM 180072) have been shown to cause autosomal recessive retinitis pigmentosa[219–221] or congenital stationary night blindness.[222–224] Mutations in the α-subunit of PDE6 (OMIM 180071) also lead to retinitis pigmentosa in humans[225] and progressive retinal atrophy in dogs.[218,226] Targeted disruption of the mouse rod γ-subunit (OMIM 180073) causes rapid retinal degeneration because of disruptions in cGMP metabolism,[227] but no naturally occurring mutations leading

to retinal disease have been reported.[228–230] To date, the cone α'-subunit has not been reported to undergo mutations that cause retinal disease.[231,232] Mutations in the cone γ'-subunit are apparently rare, but in one case a mutation in the 5'-untranslated region of the gene was correlated with a cone dystrophy.[233] Aside from the PDE6 subunits, mutations in the aryl hydrocarbon receptor-interacting protein-like 1 (AIPL1, OMIM 604392) have been shown to cause defective transport of rod PDE6 from inner to outer segment, resulting in Leber's congenital amaurosis.[234,235] In all of these disease processes, reduced functioning of PDE6 presumably causes an elevation of the circulating dark current, but the underlying cellular mechanisms that cause photoreceptor cell death are currently not understood.

8.11 CONCLUSION

Considering the fact that the visual signaling pathway is arguably the best-understood G-protein coupled signaling pathway, it is perhaps a bit surprising that there remain so many gaps in our knowledge of the PDE6 family. Regarding the mechanism of transducin activation of rod PDE6, we still lack a step-by-step mechanism by which activated transducin interacts with the PDE6 holoenzyme and deinhibits the enzyme by the binding to the PDE6 inhibitory γ-subunit. The physiological significance of the regulatory cGMP binding sites of PDE6 for conformational changes in the catalytic dimer and for the allosteric control of γ-subunit binding remains to be elucidated. Some aspects of the electrical responses of rod photoreceptors, particularly during light adaptation, cannot be explained by known photo-transduction pathways; in this regard, novel mechanisms of PDE6 regulation involving PDE6-interacting proteins may be involved in desensitizing the photoresponse of rods. Furthermore, although cone photoreceptors display numerous differences in their light responsiveness compared to rod photoreceptors, relatively little is known about the extent to which cone PDE6 is regulated in a different manner than its rod counterpart. Finally, although there is a clear link between disruptions of PDE6 function and retinal disease, therapies have yet to be developed that can intervene to regulate cGMP metabolism in photoreceptors and thereby prevent the degeneration of photoreceptor cells that leads to vision loss and blindness.

Apart from its central role in vision, PDE6 is also a subject of increased scrutiny in other areas as well. With the increasing use of PDE5-selective inhibitors for a growing number of therapeutic applications, some of which involve long-term drug administration, potential adverse effects on PDE6 function must be considered; in addition, a better understanding of differences in the molecular architecture of the actives sites of PDE5 and PDE6 will form the basis for enhancing inhibitor selectivity. There is limited evidence to date that PDE6 subunits may be found outside the retina, and may be differentially expressed during development. The significance of having the G-protein-coupled PDE6 expressed in nonphotoreceptive cells is extremely intriguing and merits further investigation of other signal transduction pathways in which this highly regulated enzyme may participate.

ACKNOWLEDGMENTS

Work from the author's laboratory has been supported in part by grants from the National Institutes of Health (National Eye Institute) and the New Hampshire Agricultural Experiment Station. This is scientific contribution number 2308 from the New Hampshire Agricultural Experimental Station. I am grateful to the past and present members of my laboratory for their scientific and intellectual contributions and to Sherry Palmer for expert assistance with the illustrations.

REFERENCES

1. Rodieck, R.W. 1998. *The first steps in seeing.* Sunderland, MA: Sinauer Press.
2. Fain, G.L. 2003. *Sensory transduction.* Sunderland, MA: Sinauer Press.
3. Ebrey, T., and Y. Koutalos. 2001. Vertebrate photoreceptors. *Prog Retin Eye Res* 20:49.
4. Pugh, E.N. Jr., and T.D. Lamb. 1993. Amplification and kinetics of the activation steps in phototransduction. *Biochim Biophys Acta* 1141:111.
5. Pugh, E.N., and T.D. Lamb. 2000. Phototransduction in vertebrate rods and cones: Molecular mechanisms of amplification, recovery, and light adaptation. In *Molecular mechanisms in visual transduction,* eds. D.G. Stavenga, W.J. DeGrip, and E.N. Pugh, 183. New York: Elsevier Science B.V.
6. Burns, M.E., and D.A. Baylor. 2001. Activation, deactivation, and adaptation in vertebrate photoreceptor cells. *Annu Rev Neurosci* 24:779.
7. Fain, G.L., et al. 2001. Adaptation in vertebrate photoreceptors. *Physiol Rev* 81:117.
8. Arshavsky, V.Y., T.D. Lamb, and E.N. Pugh Jr. 2002. G proteins and phototransduction. *Annu Rev Physiol* 64:153.
9. Zhang, X., and R.H. Cote. 2005. cGMP signaling in vertebrate retinal photoreceptor cells. *Front Biosci* 10:1191.
10. Stavenga, D.G., W.J. DeGrip, and E.N. Pugh Jr. 2000. *Molecular mechanisms in visual transduction.* Amsterdam: Elsevier Science.
11. Nakatani, K., et al. 2002. Calcium and phototransduction. *Adv Exp Med Biol* 514:1.
12. Burns, M.E., and V.Y. Arshavsky. 2005. Beyond counting photons: Trials and trends in vertebrate visual transduction. *Neuron* 48:387.
13. Pugh, E.N. Jr., et al. 1997. Photoreceptor guanylate cyclases: A review. *Biosci Rep* 17:429.
14. Pugh, E.N. Jr., S. Nikonov, and T.D. Lamb. 1999. Molecular mechanisms of vertebrate photoreceptor light adaptation. *Curr Opin Neurobiol* 9:410.
15. McBee, J.K., et al. 2001. Confronting complexity: the interlink of phototransduction and retinoid metabolism in the vertebrate retina. *Prog Retin Eye Res* 20:469.
16. Pugh, E.N. Jr., and T.D. Lamb. 1990. Cyclic GMP and calcium: The internal messengers of excitation and adaptation in vertebrate photoreceptors. *Vision Res* 30:1923.
17. Cote, R.H., et al. 1984. Light-induced decreases in cGMP concentration precede changes in membrane permeability in frog rod photoreceptors. *J Biol Chem* 259:9635.
18. Blazynski, C., and A.I. Cohen. 1986. Rapid declines in cyclic GMP of rod outer segments of intact frog photoreceptors after illumination. *J Biol Chem* 261:14142.
19. Cobbs, W.H., and E.N. Pugh Jr. 1985. Cyclic GMP can increase rod outer-segment light-sensitive current 10-fold without delay of excitation. *Nature* 313:585.
20. Yau, K.-W., and K. Nakatani. 1985. Light-suppressible, cyclic GMP-sensitive conductance in the plasma membrane of a truncated rod outer segment. *Nature* 317:252.
21. Cote, R.H., and M.A. Brunnock. 1993. Intracellular cGMP concentration in rod photoreceptors is regulated by binding to high and moderate affinity cGMP binding sites. *J Biol Chem.* 268:17190.
22. Corbin, J.D., et al. 2003. Regulation of cyclic nucleotide levels by sequestration. In *Handbook of Cell Signaling,* eds. R. Bradshaw and E. Dennis, 465. San Diego: Academic Press.
23. Hodgkin, A.L., and B.J. Nunn. 1988. Control of light-sensitive current in salamander rods. *J Physiol (Lond)* 403:439.
24. Dawis, S.M., et al. 1988. Regulation of cyclic GMP metabolism in toad photoreceptors. *J Biol Chem* 263:8771.
25. D'Amours, M.R., and R.H. Cote. 1999. Regulation of photoreceptor phosphodiesterase catalysis by its noncatalytic cGMP binding sites. *Biochem J* 340:863.
26. Mou, H., and R.H. Cote. 2001. The catalytic and GAF domains of the rod cGMP phosphodiesterase (PDE6) heterodimer are regulated by distinct regions of its inhibitory γ subunit. *J Biol Chem* 276:27527.
27. Koutalos, Y., et al. 1995. Characterization of guanylate cyclase activity in single retinal rod outer segments. *J Gen Physiol* 106:863.
28. Calvert, P.D., et al. 1998. Onset of feedback reactions underlying vertebrate rod photoreceptor light adaptation. *J Gen Physiol* 111:39.

29. Hamm, H.E., and M.D. Bownds. 1986. Protein complement of rod outer segments of frog retina. *Biochem* 25:4512.
30. Dumke, C.L., et al. 1994. Rod outer segment structure influences the apparent kinetic parameters of cyclic GMP phosphodiesterase. *J Gen Physiol* 103:1071.
31. Pentia, D.C., S. Hosier, and R.H. Cote. 2006. The glutamic acid–rich protein-2 (GARP2) is a high-affinity rod photoreceptor phosphodiesterase (PDE6) binding protein that modulates its catalytic properties. *J Biol Chem* 281:5500.
32. Blackshaw, S., and S.H. Snyder. 1997. Developmental expression pattern of phototransduction components in mammalian pineal implies a light-sensing function. *J Neurosci* 17:8074.
33. Holthues, H., and L. Vollrath. 2004. The phototransduction cascade in the isolated chick pineal gland revisited. *Brain Res* 999:175.
34. Carcamo, B., et al. 1995. The mammalian pineal expresses the cone but not the rod cyclic GMP phosphodiesterase. *J Neurochem* 65:1085.
35. Morin, F., et al. 2001. Expression and role of phosphodiesterase 6 in the chicken pineal gland. *J Neurochem* 78:88.
36. Hurwitz, R.L., et al. 1990. Expression of the functional cone phototransduction cascade in retinoblastoma. *J Clin Invest* 85:1872.
37. Piriev, N.I., et al. 2003. Expression of cone photoreceptor cGMP-phosphodiesterase α'-subunit in Chinese hamster ovary, 293 human embryonic kidney, and Y79 retinoblastoma cells. *Mol Vis* 9:80.
38. White, J.B., W.J. Thompson, and S.J. Pittler., 2004. Characterization of 3′,5′-cyclic nucleotide phosphodiesterase activity in Y79 retinoblastoma cells: Absence of functional PDE6. *Mol Vis* 10:738.
39. Weber, B., et al. 1991. Genomic organization and complete sequence of the human gene encoding the β-subunit of the cGMP phosphodiesterase and its localisation to 4p16.3. *Nucleic Acids Res* 19:6263.
40. Collins, C., et al. 1992. The human β-subunit of rod photoreceptor cGMP phosphodiesterase: Complete retinal cDNA sequence and evidence for expression in brain. *Genomics* 13:698.
41. Ogueta, S.B., et al. 2000. The human cGMP-PDE β-subunit promoter region directs expression of the gene to mouse photoreceptors. *Invest Ophthalmol Vis Sci* 41:4059.
42. Taylor, R.E., et al. 2001. A PDE6A promoter fragment directs transcription predominantly in the photoreceptor. *Biochem Biophys Res Commun* 282:543.
43. Viczian, A.S., et al. 2004. Conserved transcriptional regulation of a cone phototransduction gene in vertebrates. *FEBS Lett* 577:259.
44. Ahumada, A., et al. 2002. Signaling of rat Frizzled-2 through phosphodiesterase and cyclic GMP. *Science* 298:2006.
45. Wang, H.,Y. Lee, and C.C. Malbon. 2004. PDE6 is an effector for the Wnt/Ca(2$^+$)/cGMP-signalling pathway in development. *Biochem Soc Trans* 32:792.
46. Lochhead, A., et al. 1997. The regulation of the cGMP-binding cGMP phosphodiesterase by proteins that are immunologically related to the γ-subunit of the photoreceptor cGMP phospho-diesterase. *J Biol Chem* 272:18397.
47. Tate, R.J., et al. 1998. The γ-subunit of the rod photoreceptor cGMP-binding cGMP-specific PDE is expressed in mouse lung. *Cell Biochem Biophys* 29:133.
48. Wan, K.F., et al. 2001. The inhibitory γ-subunit of the type 6 retinal cyclic guanosine monopho-sphate phosphodiesterase is a novel intermediate regulating p42/p44 mitogen-activated protein kinase signaling in human embryonic kidney 293 cells. *J Biol Chem* 276:37802.
49. Tate, R.J., V.Y. Arshavsky, and N.J. Pyne. 2002. The identification of the inhibitory γ-subunits of the type 6 retinal cyclic guanosine monophosphate phosphodiesterase in nonretinal tissues: Dif-ferential processing of mRNA transcripts. *Genomics* 79:582.
50. Wan, K.F., et al. 2003. The inhibitory γ-subunit of the type 6 retinal cGMP phosphodiesterase functions to link c-src and G-protein-coupled receptor kinase 2 in a signaling unit that regulates p42/p44 mitogen-activated protein kinase by epidermal growth factor. *J Biol Chem* 278:18658.
51. Morin, F., et al. 2003. A proline-rich domain in the γ-subunit of phosphodiesterase 6 mediates interaction with SH3-containing proteins. *Mol Vis* 9:449.

52. Granovsky, A.E., et al. 1998. Probing domain functions of chimeric PDE6α'/PDE5 cGMP-phosphodiesterase. *J Biol Chem* 273:24485.

53. Houdart, F., et al. 2005. The regulatory subunit of PDE6 interacts with pacsin in photoreceptors. *Mol Vis* 11:1061.

54. Huang, D., et al. 2004. Molecular determinants of cGMP-binding to chicken cone photoreceptor phosphodiesterase. *J Biol Chem* 279:48143.

55. Miki, N., et al. 1975. Purification and properties of the light-activated cyclic nucleotide phosphodiesterase of rod outer segments. *J Biol Chem* 250:6320.

56. Baehr, W., M.J. Devlin, and M.L. Applebury. 1979. Isolation and characterization of cGMP phosphodiesterase from bovine rod outer segments. *J Biol Chem* 254:11669.

57. Fung, B.K.K., et al. 1990. Subunit stoichiometry of retinal rod cGMP phosphodiesterase. *Biochem* 29:2657.

58. Artemyev, N.O., et al. 1996. Subunit structure of rod cGMP-phosphodiesterase. *J Biol Chem* 271:25382.

59. Hurwitz, R.L., et al. 1985. cGMP phosphodiesterase in rod and cone outer segments of the retina. *J Biol Chem* 260:568.

60. Gillespie, P.G., and J.A. Beavo. 1988. Characterization of a bovine cone photoreceptor phosphodiesterase purified by cyclic GMP-sepharose chromatography. *J Biol Chem* 263:8133.

61. Dumler, I.L., and R.N. Etingof. 1976. Protein inhibitor of cyclic adenosine 3',5'-monophosphate phosphodiesterase in retina. *Biochim Biophys Acta* 429:474.

62. Hurley, J.B., and L. Stryer. 1982. Purification and characterization of the γ regulatory subunit of the cyclic GMP phosphodiesterase from retinal rod outer segments. *J Biol Chem* 257:11094.

63. Deterre, P., et al. 1988. cGMP phosphodiesterase of retinal rods is regulated by two inhibitory subunits. *Proc Natl Acad Sci USA* 85:2424.

64. Kameni Tcheudji, J.F., et al. 2001. Molecular organization of bovine rod cGMP-phosphodiesterase 6. *J Mol Biol* 310:781.

65. Kajimura, N., et al. 2002. Three-dimensional structure of nonactivated cGMP phosphodiesterase 6 and comparison of its image with those of activated forms. *J Struct Biol* 139:27.

66. Martinez, S.E., et al. 2002. The two GAF domains in phosphodiesterase 2A have distinct roles in dimerization and in cGMP binding. *Proc Natl Acad Sci USA* 99:13260.

67. Sung, B.J., et al. 2003. Structure of the catalytic domain of human phosphodiesterase 5 with bound drug molecules. *Nature* 425:98.

68. Ho, Y.-S.J., L.M. Burden, and J.H. Hurley. 2000. Structure of the GAF domain, a ubiquitous signaling motif and a new class of cyclic GMP receptor. *EMBO J* 19:5288.

69. Wu, A.Y., et al. 2004. Molecular determinants for cyclic nucleotide binding to the regulatory domains of phosphodiesterase 2A. *J Biol Chem* 279:37928.

70. Zoraghi, R., et al. 2005. Structural and functional features in human PDE5A1 regulatory domain that provide for allosteric cGMP binding, dimerization, and regulation. *J Biol Chem* 280:12051.

71. Fink, T.L., et al. 1999. Expression of an active, monomeric catalytic domain of the cGMP-binding cGMP-specific phosphodiesterase (PDE5). *J Biol Chem* 274:34613.

72. Ovchinnikov, Y.A., et al. 1987. Cyclic GMP phosphodiesterase from bovine retina: Amino acid sequence of the α-subunit and nucleotide sequence of the corresponding cDNA. *FEBS Lett* 223:169.

73. Lipkin, V.M., et al. 1990. β-subunit of bovine rod photoreceptor cGMP phosphodiesterase: Comparison with the phosphodiesterase family. *J Biol Chem* 265:12955.

74. Li, T., K. Volpp, and M.L. Applebury. 1990. Bovine cone photoreceptor cGMP phosphodiesterase structure deduced from a cDNA clone. *Proc Natl Acad Sci USA* 87:293.

75. Aravind, L., and C.P. Ponting. 1997. The GAF domain: An evolutionary link between diverse phototransducing proteins. *Trends Biochem Sci* 22:458.

76. Zoraghi, R., J.D. Corbin, and S.H. Francis. 2004. Properties and functions of GAF domains in cyclic nucleotide phosphodiesterases and other proteins. *Mol Pharmacol* 65:267.

77. Muradov, K.G., et al. 2003. The GAFa domains of rod cGMP-phosphodiesterase 6 determine the selectivity of the enzyme dimerization. *J Biol Chem* 278:10594.

78. Gillespie, P.G., and J.A. Beavo. 1989. cGMP is tightly bound to bovine retinal rod phosphodiesterase. *Proc Natl Acad Sci USA* 86:4311.

79. Cote, R.H., M.D. Bownds, and V.Y. Arshavsky. 1994. cGMP binding sites on photoreceptor phosphodiesterase: Role in feedback regulation of visual transduction. *Proc Natl Acad Sci USA* 91:4845.

80. Mou, H., et al. 1999. cGMP binding to noncatalytic sites on mammalian rod photoreceptor phosphodiesterase is regulated by binding of its γ- and δ-subunits. *J Biol Chem* 274:18813.

81. Hebert, M.C., et al. 1998. Structural features of the noncatalytic cGMP binding sites of frog photoreceptor phosphodiesterase using cGMP analogs. *J Biol Chem* 273:5557.

82. Muradov, H., K.K. Boyd, and N.O. Artemyev. 2004. Structural determinants of the PDE6 GAFa domain for binding the inhibitory δ-subunit and noncatalytic cGMP. *Vision Res* 44:2437.

83. Muradov, K.G., et al. 2002. Direct interaction of the inhibitory γ-subunit of rod cGMP phosphodiesterase (PDE6) with the PDE6 GAFa domains. *Biochem* 41:3884.

84. Guo, L.W., et al. 2005. Asymmetric interaction between rod cyclic GMP phosphodiesterase γ-subunits and αβ-subunits. *J Biol Chem* 280:12585.

85. Conti, M. 2004. A view into the catalytic pocket of cyclic nucleotide phosphodiesterases. *Nat Struct Biol* 11:809.

86. Zhang, K.Y., et al. 2004. A glutamine switch mechanism for nucleotide selectivity by phosphodiesterases. *Mol Cell* 15:279.

87. Qin, N., and W. Baehr. 1994. Expression and mutagenesis of mouse rod photoreceptor cGMP phosphodiesterase. *J Biol Chem* 269:3265.

88. Zhang, J., et al. 2004. Differential inhibitor sensitivity between human recombinant and native photoreceptor cGMP-phosphodiesterases (PDE6s). *Biochem Pharmacol* 68:867.

89. Muradov, H., K.K. Boyd, and N.O. Artemyev. Analysis of PDE6 function using chimeric PDE5/6 catalytic domains. *Vision Res* 46:860.

90. Granovsky, A.E., and N.O. Artemyev. 2001. Partial reconstitution of photoreceptor cGMP phosphodiesterase characteristics in cGMP phosphodiesterase-5. *J Biol Chem* 276:21698.

91. Francis, S.H., et al. 2000. Histidine-607 and histidine-643 provide important interactions for metal support of catalysis in phosphodiesterase-5. *Biochem* 39:9591.

92. Francis, S.H., et al. 1994. Zinc interactions and conserved motifs of the cGMP-binding cGMP-specific phosphodiesterase suggest that it is a zinc hydrolase. *J Biol Chem* 269:22477.

93. He, F., et al. 2000. Multiple zinc binding sites in retinal rod cGMP phosphodiesterase, PDE6αβ. *J Biol Chem* 275:20572.

94. Artemyev, N.O., et al. 1996. Mechanism of photoreceptor cGMP phosphodiesterase inhibition by its γ-subunits. *Proc Natl Acad Sci USA* 93:5407.

95. Granovsky, A.E., M. Natochin, and N.O. Artemyev. 1997. The γ-subunit of rod cGMP-phosphodiesterase blocks the enzyme catalytic site. *J Biol Chem* 272:11686.

96. Granovsky, A.E., and N.O. Artemyev. 2000. Identification of the γ-subunit interacting residues on photoreceptor cGMP phosphodiesterase, PDE6α'. *J Biol Chem* 275:41258.

97. Qin, N., S.J. Pittler, and W. Baehr. 1992. In vitro isoprenylation and membrane association of mouse rod photoreceptor cGMP phosphodiesterase α- and β-subunits expressed in bacteria. *J Biol Chem* 267:8458.

98. Anant, J.S., et al. 1992. In vivo differential prenylation of retinal cyclic GMP phosphodiesterase catalytic subunits. *J Biol Chem* 267:687.

99. Catty, P., and P. Deterre. 1991. Activation and solubilization of the retinal cGMP-specific phosphodiesterase by limited proteolysis: Role of the C-terminal domain of the β-subunit. *Eur J Biochem* 199:263.

100. Kuhn, H. 1980. Light- and GTP-regulated interaction of GTPase and other proteins with bovine photoreceptor membranes. *Nature* 283:587.

101. Ovchinnikov, Y.A., et al. 1986. Cyclic GMP phosphodiesterase from cattle retina: Amino acid sequence of the γ-subunit and nucleotide sequence of the corresponding cDNA. *FEBS Lett* 204:288.

102. Hamilton, S.E., and J.B. Hurley. 1990. A phosphodiesterase inhibitor specific to a subset of bovine retinal cones. *J Biol Chem* 265:11259.

103. Yamazaki, A., et al. 1982. Reciprocal effects of an inhibitory factor on catalytic activity and noncatalytic cGMP binding sites of rod phosphodiesterase. *Proc Natl Acad Sci USA* 79:3702.

104. Artemyev, N.O., et al. 1992. Sites of interaction between rod G-protein α-subunit and cGMP-phosphodiesterase γ-subunit. Implications for phosphodiesterase activation mechanism. *J Biol Chem* 267:25067.

105. Arshavsky, V.Y., and M.D. Bownds. 1992. Regulation of deactivation of photoreceptor G protein by its target enzyme and cGMP. *Nature* 357:416.

106. He, W., C.W. Cowan, and T. Wensel. 1998. RGS9, a GTPase accelerator for phototransduction. *Neuron* 20:95.

107. Slep, K.C., et al. 2001. Structural determinants for regulation of phosphodiesterase by a G protein at 2.0 A. *Nature* 409:1071.

108. Hayashi, F. 1994. Light-dependent in vivo phosphorylation of an inhibitory subunit of cGMP-phosphodiesterase in frog rod photoreceptor outer segments. *FEBS Lett* 338:203.

109. Bondarenko, V.A., et al. 1997. Residues within the polycationic region of cGMP phosphodiesterase γ-subunit crucial for the interaction with transducin α-subunit. Identification by endogenous ADP-ribosylation and site-directed mutagenesis. *J Biol Chem* 272:15856.

110. Uversky, V.N., 2002. Natively unfolded proteins: A point where biology waits for physics. *Protein Sci* 11:739.

111. Berger, A.L., R.A. Cerione, and J.W. Erickson. 1997. Real time conformation changes in the retinal phosphodiesterase γ-subunit monitored by resonance energy transfer. *J Biol Chem* 272:2714.

112. Uversky, V.N., et al. 2002. Effect of zinc and temperature on the conformation of the gamma subunit of retinal phosphodiesterase: A natively unfolded protein. *J Proteome Res* 1:149.

113. Tompa, P., C. Szasz, and L. Buday. 2005. Structural disorder throws new light on moonlighting. *Trends Biochem Sci* 30:484.

114. Tsuboi, S., et al. 1994. Phosphorylation of an inhibitory subunit of cGMP phosphodiesterase in *Rana catesbiana* rod photoreceptors. I. Characterization of the phosphorylation. *J Biol Chem* 269:15016.

115. Matsuura, I., et al. 2000. Phosphorylation by cyclin-dependent protein kinase 5 of the regulatory subunit of retinal cGMP phosphodiesterase. I. Identification of the kinase and its role in the turnoff of phosphodiesterase in vitro. *J Biol Chem* 275:32950.

116. Paglia, M.J., H. Mou, and R.H. Cote. 2002. Regulation of photoreceptor phosphodiesterase (PDE6) by phosphorylation of its inhibitory γ-subunit reevaluated. *J Biol Chem* 277:5017.

117. Kyte, J., and R.F. Doolittle. 1982. A simple method for displaying the hydropathic character of a protein. *J Mol Biol* 157:105.

118. Lipkin, V.M., et al. 1988. Active sites of the cyclic GMP phosphodiesterase γ-subunit of retinal rod outer segments. *FEBS Lett* 234:287.

119. Morrison, D.F., et al. 1989. Interaction of the γ-subunit of retinal rod outer segment phosphodiesterase with transducin. *J Biol Chem* 264:11671.

120. Brown, R.L. 1992. Functional regions of the inhibitory subunit of retinal rod cGMP phosphodiesterase identified by site-specific mutagenesis and fluorescence spectroscopy. *Biochem* 31:5918.

121. Artemyev, N.O., et al. 1993. A site on transducin α-subunit of interaction with the polycationic region of cGMP phosphodiesterase inhibitory subunit. *J Biol Chem* 268:23611.

122. Skiba, N.P., H. Bae, and H.E. Hamm. 1996. Mapping of effector binding sites of transducin α-subunit using $G\alpha_t/G\alpha_{i1}$ chimeras. *J Biol Chem* 271:413.

123. Udovichenko, I.P., et al. 1993. Phosphorylation of bovine rod photoreceptor cyclic GMP phosphodiesterase. *Biochem J* 295:49.

124. Xu, L.X., et al. 1998. Phosphorylation of the γ-subunit of the retinal photoreceptor cGMP phosphodiesterase by the cAMP-dependent protein kinase and its effect on the γ-subunit interaction with other proteins. *Biochem* 37:6205.

125. Bondarenko, V.A., et al. 1999. Suppression of GTP/Tα-dependent activation of cGMP phosphodiesterase by ADP-ribosylation by its γ-subunit in amphibian rod photoreceptor membranes. *Biochem* 38:7755.

126. Artemyev, N.O., and H.E. Hamm. 1992. Two-site high-affinity interaction between inhibitory and catalytic subunits of rod cyclic GMP phosphodiesterase. *Biochem J* 283:273.

127. Takemoto, D.J., et al. 1992. Domain mapping of the retinal cyclic GMP phosphodiesterase γ-subunit. Function of the domains encoded by the three exons of the γ-subunit gene. *Biochem J* 281:637.

128. Lipkin, V.M., et al. 1993. Site-directed mutagenesis of the cGMP phosphodiesterase γ-subunit from bovine rod outer segments: Role of separate amino acid residues in the interaction with catalytic subunits and transducin α-subunit. *Biochim Biophys Acta* 1176:250.

129. Natochin, M., and N.O. Artemyev. 1996. An interface of interaction between photoreceptor cGMP phosphodiesterase catalytic-subunits and inhibitory γ-subunits. *J Biol Chem* 271:19964.

130. Arshavsky, V.Y., et al. 1994. Regulation of transducin GTPase activity in bovine rod outer segments. *J Biol Chem* 269:19882.

131. Skiba, N.P., N.O. Artemyev, and H.E. Hamm. 1995. The carboxyl terminus of the γ-subunit of rod cGMP phosphodiesterase contains distinct sites of interaction with the enzyme catalytic subunits and the α-subunit of transducin. *J Biol Chem* 270:13210.

132. Otto-Bruc, A., et al. 1993. Interaction between the retinal cyclic GMP phosphodiesterase inhibitor and transducin. Kinetics and affinity studies. *Biochem* 32:8636.

133. Slepak, V.Z., et al. 1995. An effector site that stimulates G-protein GTPase in photoreceptors. *J Biol Chem* 270:14319.

134. Liu, Y., V.Y. Arshavsky, and A.E. Ruoho. 1996. Interaction sites of the COOH-terminal region of the γ-subunit of cGMP phosphodiesterase with the GTP-bound α-subunit of transducin. *J Biol Chem* 271:26900.

135. Tsang, S.H., et al. 1998. Role for the target enzyme in deactivation of photoreceptor G protein in vivo. *Science* 282:117.

136. Granovsky, A.E., and N.O. Artemyev. 2001. A conformational switch in the inhibitory γ-subunit of PDE6 upon enzyme activation by transducin. *Biochem* 40:13209.

137. Whalen, M.M., and M.W. Bitensky. 1989. Comparison of the phosphodiesterase inhibitory subunit interactions of frog and bovine rod outer segments. *Biochem J* 259:13.

138. Norton, A.W., et al. 2000. Mechanism of transducin activation of frog rod photoreceptor phosphodiesterase: Allosteric interactions between the inhibitory γ-subunit and the noncatalytic cGMP binding sites. *J Biol Chem* 275:38611.

139. Wensel, T.G., and L. Stryer. 1986. Reciprocal control of retinal rod cyclic GMP phosphodiesterase by its γ-subunit and transducin. *Proteins* 1:90.

140. Rieke, F., and D.A. Baylor. 1996. Molecular origin of continuous dark noise in rod photoreceptors. *Biophys J* 71:2553.

141. Yamazaki, A., et al. 1983. Activation mechanism of rod outer segment cyclic GMP phosphodiesterase: Release of inhibitor by the GTP/GTP binding protein. *J Biol Chem* 258:8188.

142. Leskov, I.B., et al. 2000. The gain of rod phototransduction: Reconciliation of biochemical and electrophysiological measurements. *Neuron* 27:525.

143. Fersht, A. 1999. *Structure and mechanism in protein science.* New York: W.H. Freeman & Co.

144. Fesenko, E.E., S.S. Kolesnikov, and A.L. Lyubarsky. 1985. Induction by cyclic GMP of cationic conductance in plasma membrane of retinal rod outer segment. *Nature* 313:310.

145. Lipton, S.A., H. Rasmussen, and J.E. Dowling. 1977. Electrical and adaptive properties of rod photoreceptors in Bufo marinus. *J Gen Physiol* 70:771.

146. Capovilla, M., L. Cervetto, and V. Torre. 1982. Antagonism between steady light and phosphodiesterase inhibitors on the kinetics of the rod photoresponses. *Proc Natl Acad Sci USA* 79:6698.

147. Capovilla, M., L. Cervetto, and V. Torre. 1983. The effect of phosphodiesterase inhibitors on the electrical activity of toad rods. *J Physiol (Lond)* 343:277.

148. Baylor, D.A., T.D. Lamb, and K.W. Yau. 1979. The membrane current of single rod outer segments. *J Physiol (Lond)* 288:589.

149. Cervetto, L., and P.A. McNaughton. 1986. The effects of phosphodiesterase inhibitors and lanthanum ions on the light-sensitive current of toad retinal rods. *J Physiol (Lond)* 370:91.

150. Rispoli, G., P.G. Gillespie, and P.B. Detwiler. 1990. Comparative effects of phosphodiesterase inhibitors on detached rod outer segment function. In *Sensory Transduction*, eds. A. Borsellino, L. Cervetto, and V. Torre, 157. New York: Plenum Press,.

151. Zhang, X., Q. Feng, and R.H. Cote. 2005. Efficacy and selectivity of phosphodiesterase-targeted drugs in inhibiting photoreceptor phosphodiesterase (PDE6) in retinal photoreceptors. *Invest Ophthalmol Vis Sci* 46:3060.

152. Lolley, R.N., et al. 1977. Cyclic GMP accumulation causes degeneration of photoreceptor cells: Simulation of an inherited disease. *Science* 196:664.

153. Williams, D.S., N.J. Colley, and D.B. Farber. 1987. Photoreceptor degeneration in a pure-cone retina. Effects of cyclic nucleotides, and inhibitors of phosphodiesterase and protein synthesis. *Invest Ophthalmol Vis Sci* 28:1059.

154. Ulshafer, R.J., C.A. Garcia, and J.G. Hollyfield. 1980. Sensitivity of photoreceptors to elevated levels of cGMP in the human retina. *Invest Ophthalmol Vis Sci* 19:1236.

155. Ulshafer, R.J., S.J. Fliesler, and J.G. Hollyfield. 1984. Differential sensitivity of protein synthesis in human retina to a phosphodiesterase inhibitor and cyclic nucleotides. *Curr Eye Res* 3:383.

156. Sandberg, M.A., S. Miller, and E.L. Berson. 1990. Rod electroretinograms in an elevated cyclic guanosine monophosphate-type human retinal degeneration. Comparison with retinitis pigmentosa. *Invest Ophthalmol Vis Sci* 31:2283.

157. Estrade, M., et al. 1998. Effect of a cGMP-specific phosphodiesterase inhibitor on retinal function. *Eur J Pharmacol* 352:157.

158. Barabás, P., Z. Riedl, and J. Kardos. 2003. Sildenafil, N-desmethyl-sildenafil and zaprinast enhance photoreceptor response in the isolated rat retina. *Neurochem Int* 43:591.

159. Mochida, H., et al. 2004. T-0156, a novel phosphodiesterase type 5 inhibitor, and sildenafil have different pharmacological effects on penile tumescence and electroretinogram in dogs. *Eur J Pharmacol* 485:283.

160. Luke, M., et al. 2005. Effects of phosphodiesterase type 5 inhibitor sildenafil on retinal function in isolated superfused retina. *J Ocul Pharmacol Ther* 21:305.

161. Behn, D., and M.J. Potter. 2001. Sildenafil-mediated reduction in retinal function in heterozygous mice lacking the γ-subunit of phosphodiesterase. *Invest Ophthalmol Vis Sci* 42:523.

162. Laties, A.M., and E. Zrenner. 2002. Viagra (sildenafil citrate) and ophthalmology. *Prog Retin Eye Res* 21:485.

163. Gonzalez, C.M., et al. 1999. Sildenafil causes a dose- and time-dependent downregulation of phosphodiesterase type 6 expression in the rat retina. *Int J Impot Res* 11:S9-S14.

164. Heywood, R., I.H. Osterloh, and S.C. Phillips. 2000. Sildenafil causes a dose- and time-dependent downregulation of phosphodiesterase type 6 expression in the rat retina: Response. *Int J Impot Res* 12:241.

165. Gillespie, P.G., and J.A. Beavo. 1989. Inhibition and stimulation of photoreceptor phosphodiesterases by dipyridamole and M&B 22,948. *Mol Pharmacol* 36:773.

166. D'Amours, M.R., et al. 1999. The potency and mechanism of action of E4021, a PDE5-selective inhibitor, on the photoreceptor phosphodiesterase depends on its state of activation. *Mol Pharmacol* 55:508.

167. Maw, G.N., et al. 2003. Design, synthesis, and biological activity of β-carboline-based type-5 phosphodiesterase inhibitors. *Bioorg Med Chem Lett* 13:1425.

168. Manallack, D.T., R.A. Hughes, and P.E. Thompson. 2005. The next generation of phosphodiesterase inhibitors: Structural clues to ligand and substrate selectivity of phosphodiesterases. *J Med Chem* 48:3449.

169. Martins, T.J., M.C. Mumby, and J.A. Beavo. 1982. Purification and characterization of a cyclic GMP-stimulated cyclic nucleotide phosphodiesterase from bovine tissues. *J Biol Chem* 257:1973.

170. Yamamoto, T., V.C. Manganiello, and M. Vaughan. 1983. Purification and characterization of cyclic GMP-stimulated cyclic nucleotide phosphodiesterase from calf liver. *J Biol Chem* 258:12526.

171. Corbin, J.D., et al. 2000. Phosphorylation of phosphodiesterase-5 by cyclic nucleotide-dependent protein kinase alters its catalytic and allosteric cGMP-binding activities. *Eur J Biochem* 267:2760.

172. Okada, D., and S. Asakawa. 2002. Allosteric activation of cGMP-specific, cGMP-binding phosphodiesterase (PDE5) by cGMP. *Biochem* 41:9672.

173. Rybalkin, S.D., et al. 2003. PDE5 is converted to an activated state upon cGMP binding to the GAFa domain. *EMBO J* 22:469.

174. Francis, S.H., et al. 1998. Ligand-induced conformational changes in cyclic nucleotide phosphodiesterases and cyclic nucleotide-dependent protein kinases. *Methods* 14:81.

175. Francis, S.H., et al. 2002. Phosphorylation of isolated human phosphodiesterase-5 regulatory domain induces an apparent conformational change and increases cGMP binding affinity. *J Biol Chem* 277:47581.

176. Corbin, J.D., et al. 2003. [³H]sildenafil binding to phosphodiesterase-5 is specific, kinetically heterogeneous, and stimulated by cGMP. *Mol Pharmacol* 63:1364.

177. Arshavsky, V.Y., C.L. Dumke, and M.D. Bownds. 1992. Noncatalytic cGMP binding sites of amphibian rod cGMP phosphodiesterase control interaction with its inhibitory γ-subunits. A putative regulatory mechanism of the rod photoresponse. *J Biol Chem* 267:24501.

178. Yamazaki, A., et al. 1990. Interactions between the subunits of transducin and cyclic GMP phosphodiesterase in *Rana catesbeiana* rod photoreceptors. *J Biol Chem* 265:11539.

179. Wensel, T.G., and L. Stryer. 1990. Activation mechanism of retinal rod cyclic GMP phosphodiesterase probed by fluorescein-labeled inhibitory subunit. *Biochem* 29:2155.

180. Bruckert, F., et al. 1994. Activation of phosphodiesterase by transducin in bovine rod outer segments: Characteristics of the successive binding of two transducins. *Biochem* 33:12625.

181. Melia, T.J., et al. 2000. Enhancement of phototransduction protein interactions by lipid surfaces. *J Biol Chem* 275:3535.

182. Yamazaki, M., et al. 2002. Binding of cGMP to GAF domains in amphibian rod photoreceptor cGMP phosphodiesterase (PDE). *J Biol Chem* 277:40675.

183. Whalen, M.M., M.W. Bitensky, and D.J. Takemoto. 1990. The effect of the γ-subunit of the cyclic GMP phosphodiesterase of bovine and frog (*Rana catesbiana*) retinal rod outer segments on the kinetic parameters of the enzyme. *Biochem J* 265:655.

184. Gillespie, P.G. 1990. Phosphodiesterases in visual transduction by rods and cones. In *Cyclic Nucleotide Phosphodiesterases: Structure, Regulation, and Drug Action*, eds. J. Beavo and M.D. Houslay, 163. New York: John Wiley & Sons.

185. Yamazaki, A., et al. 1996. Possible stimulation of retinal rod recovery to dark state by cGMP release from a cGMP phosphodiesterase noncatalytic site. *J Biol Chem* 271:32495.

186. Sugimoto, Y., et al. 1991. The amino acid sequence of a glutamic acid–rich protein from bovine retina as deduced from the cDNA sequence. *Proc Natl Acad Sci USA* 88:3116.

187. Colville, C.A., and R.S. Molday. 1996. Primary structure and expression of the human β-subunit and related proteins of the rod photoreceptor cGMP-gated channel. *J Biol Chem* 271:32968.

188. Korschen, H.G., et al. 1999. Interaction of glutamicacid–rich proteins with the cGMP signalling pathway in rod photoreceptors. *Nature* 400:761.

189. Batra-Safferling, R., et al. 2006. Glutamic acid–rich proteins of rod photoreceptors are natively unfolded. *J Biol Chem* 281:1449.

190. Poetsch, A., L.L. Molday, and R.S. Molday. 2001. The cGMP-gated channel and related glutamic acid–rich proteins interact with peripherin-2 at the rim region of rod photoreceptor disc membranes. *J Biol Chem* 276:48009.

191. Pentia, D.C., et al. 2005. Purification of PDE6 isozymes from mammalian retina. *Methods Mol Biol* 307:125.

192. Kaupp, U.B., and R. Seifert. 2002. Cyclic nucleotide-gated ion channels. *Physiol Rev* 82:769.

193. Gillespie, P.G., et al. 1989. A soluble form of bovine rod photoreceptor phosphodiesterase has a novel 15 kDa subunit. *J Biol Chem* 264:12187.

194. Florio, S.K., R.K. Prusti, and J.A. Beavo. 1996. Solubilization of membrane-bound rod phosphodiesterase by the rod phosphodiesterase recombinant δ-subunit. *J Biol Chem* 271:1.

195. Hanzal-Bayer, M., et al. 2002. The complex of Arl2-GTP and PDEδ: From structure to function. *EMBO J* 21:2095.

196. Cook, T.A., et al. 2000. Binding of the δ-subunit to rod phosphodiesterase catalytic subunits requires methylated, prenylated C-termini of the catalytic subunits. *Biochem* 39:13516.

197. Zhang, H.B., et al. 2004. Photoreceptor cGMP phosphodiesterase δ-subunit (PDEδ) functions as a prenyl-binding protein. *J Biol Chem* 279:407.

198. Marzesco, A.M., et al. 1998. The rod cGMP phosphodiesterase δ-subunit dissociates the small GTPase Rab13 from membranes. *J Biol Chem* 273:22340.
199. Zhang, H., et al. 2005. Assay and functional properties of PrBP(PDEδ), a prenyl binding protein interacting with multiple partners. *Methods Enzymol* 403:42.
200. Baehr, W., et al. 2005. The physiological role of PDEδ in mouse photoreceptors. *Invest Ophthalmol Vis Sci* 46: (E-abstract) 1699.
201. Norton, A.W., et al. 2005. Evaluation of the 17 kDa prenyl binding protein as a regulatory protein for phototransduction in retinal photoreceptors. *J Biol Chem* 280:1248.
202. Hong, D.H., et al. 2003. RPGR isoforms in photoreceptor connecting cilia and the transitional zone of motile cilia. *Invest Ophthalmol Vis Sci* 44:2413.
203. Grayson, C., et al. 2002. Localization in the human retina of the X-linked retinitis pigmentosa protein RP2, its homologue cofactor C, and the RP2 interacting protein Arl3. *Hum Mol Genet* 11:3065.
204. Moritz, O.L., et al. 2001. Mutant rab8 impairs docking and fusion of rhodopsin-bearing post-Golgi membranes and causes cell death of transgenic *Xenopus* rods. *Mol Biol Cell* 12:2341.
205. Cook, T.A., et al. 2001. The δ-subunit of type 6 phosphodiesterase reduces light-induced cGMP hydrolysis in rod outer segments. *J Biol Chem* 276:5248.
206. Farber, D.B., and R.N. Lolley. 1974. Cyclic guanosine monophosphate elevation in degenerating photoreceptor cells of the C3H mouse retina. *Science* 186:449.
207. Lolley, R.N., and D.B. Farber. 1976. A proposed link between debris accumulation, guanosine $3',5'$ cyclic monophosphate changes and photoreceptor cell degeneration in retina of RCS rats. *Exp Eye Res* 22:477.
208. Schmidt, S.Y., and R.N. Lolley. 1973. Cyclic-nucleotide phosphodiesterase: An early defect in inherited retinal degeneration of C3H mice. *J Cell Biol* 57:117.
209. Farber, D.B., and R.N. Lolley. 1976. Enzymic basis for cyclic GMP accumulation in degenerative photoreceptor cells of mouse retina. *J Cyclic Nucleotide Res* 2:139.
210. Farber, D.B., and R.N. Lolley. 1977. Light-induced reduction in cyclic GMP of retinal photo-receptor cells in vivo: Abnormalities in the degenerative disease of RCS rats and *rd* mice. *J Neurochem* 28:1089.
211. Bowes, C., et al. 1990. Retinal degeneration in the *rd* mouse is caused by a defect in the β-subunit of rod cGMP-phosphodiesterase. *Nature* 347:677.
212. Pittler, S.J., and W. Baehr. 1991. Identification of a nonsense mutation in the rod photoreceptor cGMP phosphodiesterase β-subunit gene of the *rd* mouse. *Proc Natl Acad Sci USA* 88:8322.
213. Lem, J., et al. 1992. Retinal degeneration is rescued in transgenic *rd* mice by expression of the cGMP phosphodiesterase β-subunit. *Proc Natl Acad Sci USA* 89:4422.
214. Aguirre, G., et al. 1978. Rod-cone dysplasia in Irish setters: A defect in cyclic GMP metabolism in visual cells. *Science* 201:1133.
215. Suber, M.L., et al. 1993. Irish setter dogs affected with rod/cone dysplasia contain a nonsense mutation in the rod cGMP phosphodiesterase β-subunit gene. *Proc Natl Acad Sci USA* 90:3968.
216. Clements, P.J.M., et al. 1993. Confirmation of the rod cGMP phosphodiesterase β-subunit (PDEβ) nonsense mutation in affected rcd-1 Irish setters in the UK and development of a diagnostic test. *Curr Eye Res* 12:861.
217. McLaughlin, M.E., et al. 1995. Mutation spectrum of the gene encoding the β subunit of rod phosphodiesterase among patients with autosomal recessive retinitis pigmentosa. *Proc Natl Acad Sci USA* 92:3249.
218. Dryja, T.P., et al. 1999. Frequency of mutations in the gene encoding the α subunit of rod cGMP-phosphodiesterase in autosomal recessive retinitis pigmentosa. *Invest Ophthalmol Vis Sci* 40:1859.
219. McLaughlin, M.E., et al. 1993. Recessive mutations in the gene encoding the β-subunit of rod phosphodiesterase in patients with retinitis pigmentosa. *Nature Genet* 4:130.
220. Bayes, M., et al. 1995. Homozygous tandem duplication within the gene encoding the β-subunit of rod phosphodiesterase as a cause for autosomal recessive retinitis pigmentosa. *Hum Mutat* 5:228.
221. Danciger, M., et al. 1995. Mutations in the *PDE6B* gene in autosomal recessive retinitis pigmentosa. *Genomics* 30:1.

222. Gal, A., et al. 1994. Heterozygous missense mutation in the rod cGMP phosphodiesterase β-subunit gene in autosomal dominant stationary night blindness. *Nature Genet* 7:64.
223. Gal, A., et al. 1994. Gene for autosomal dominant congenital stationary night blindness maps to the same region as the gene for the β-subunit of the rod photoreceptor cGMP phosphodiesterase (PDEB) in chromosome 4p16.3. *Hum Mol Genet* 3:323.
224. Muradov, K.G., A.E. Granovsky, and N.O. Artemyev. 2003. Mutation in rod PDE6 linked to congenital stationary night blindness impairs the enzyme inhibition by its γ-subunit. *Biochem* 42:3305.
225. Huang, S.H., et al. 1995. Autosomal recessive retinitis pigmentosa caused by mutations in the α-subunit of rod cGMP phosphodiesterase. *Nature Genet* 11:468.
226. Petersen-Jones, S.M., D.D. Entz, and D.R. Sargan. 1999. cGMP phosphodiesterase-α mutation causes progressive retinal atrophy in the Cardigan Welsh corgi dog. *Invest Ophthalmol Vis Sci* 40:1637.
227. Tsang, S.H., et al. 1996. Retinal degeneration in mice lacking the γ-subunit of the rod cGMP phosphodiesterase. *Science* 272:1026.
228. Cotran, P.R., et al. 1991. Genetic analysis of patients with retinitis pigmentosa using a cloned cDNA probe for the human γ-subunit of cyclic GMP phosphodiesterase. *Exp Eye Res* 53:557.
229. Hahn, L.B., E.L. Berson, and T.P. Dryja. 1994. Evaluation of the gene encoding the γ-subunit of rod phosphodiesterase in retinitis pigmentosa. *Invest Ophthalmol Vis Sci* 35:1077.
230. Dekomien, G., and J.T. Epplen. 2003. Analysis of *PDE6D* and *PDE6G* genes for generalised progressive retinal atrophy (gPRA) mutations in dogs. *Genet Sel Evol* 35:445.
231. Semple-Rowland, S.L., and D.A. Green. 1994. Molecular characterization of the α′-subunit of cone photoreceptor cGMP phosphodiesterase in normal and *rd* chicken. *Exp Eye Res* 59:365.
232. Gao, Y.Q., et al. 1999. Screening of the gene encoding the α′-subunit of cone cGMP-PDE in patients with retinal degenerations. *Invest Ophthalmol Vis Sci* 40:1818.
233. Piri, N., et al. 2005. A substitution of G to C in the cone cGMP-phosphodiesterase γ-subunit gene found in a distinctive form of cone dystrophy. *Ophthalmology* 112:159.
234. Liu, X., et al. 2004. AIPL1, the protein that is defective in Leber congenital amaurosis, is essential for the biosynthesis of retinal rod cGMP phosphodiesterase. *Proc Natl Acad Sci USA* 101:13903.
235. Ramamurthy, V., et al. 2004. Leber congenital amaurosis linked to AIPL1: A mouse model reveals destabilization of cGMP phosphodiesterase. *Proc Natl Acad Sci USA* 101:13897.

9 PDE7

Tamar Michaeli

CONTENTS

9.1 Introduction 195
9.2 PDE7A...195
    9.2.1 PDE7A1 196
    9.2.2 PDE7A2 199
    9.2.3 PDE7A3 199
9.3 PDE7B...199
    9.3.1 PDE7B1...200
9.4 Perspectives 201
Acknowledgment..202
References 202

9.1 INTRODUCTION

PDE7 is a family of high-affinity, cAMP-specific PDEs comprised of two genes—*PDE7A* and *PDE7B*. PDE7A was isolated by functional complementation of phosphodiesterase-deficient yeast and identified as a novel PDE based on its amino acid sequence, kinetics, substrate selectivity, and pharmacological properties.[1] Identification of PDE7A raised the possibility that other PDEs have evaded earlier biochemical or pharmacological detection. Consequently, database searches for EST sequences encoding previously unidentified PDEs lead to the identification of PDE7B and the novel PDE families—PDE8, PDE9, PDE10, and PDE11.

Catalytic domains of PDE7 family members exhibit 70% residue identity, but only 25%–40% amino acid identity to those of catalytic domains of other PDE families. PDE7A and PDE7B catalytic activity can be distinguished from that of other cAMP-specific PDEs by sensitivity to the nonselective PDE inhibitor IBMX and resistance to the PDE4-selective drug, rolipram.

Significant progress has been accomplished since the initial discovery in identification of PDE7A and PDE7B transcripts and proteins, their unique and provocative expression patterns, their domain structure, complex effects on cAMP signaling, and in development of family-selective inhibitors. Despite progress in deciphering PDE7 function, physiological roles of PDE7 proteins remain elusive, unresolved or lacking, and are the major focus of current research.

9.2 PDE7A

Three PDE7A transcripts and protein products have been identified as alternate splice variants utilizing different 5′ or 3′ coding exons.[2,3] Three different proteins, PDE7A1, PDE7A2, and PDE7A3, each with a unique combination of N′ and C′ termini are the products of this splicing

FIGURE 9.1 (See color insert following page 274.) Diagram and partial amino acid sequence of PDE7A encoded proteins. A central domain common to all PDE7A splice variants is depicted in red. Unique N' and C' terminal residues are presented in black. Splice variant encoding each of these termini are indicated above and below the sequences in green (PDE7A1, 7A2 or 7A3). Two repeats found in PDE7A1 and PDE7A3 N-termini are underlined. PKA pseodusubstrate sites within each repeat are in bold. Amino acid residue numbers at sites of N-terminal splicing are noted above or below their sequences.

pattern (Figure 9.1). Alternative N' termini of PDE7A proteins affect intracellular localization and interactions with cellular proteins, whereas alterations of C' termini affect catalytic activity.

Kinetics of recombinant PDE7A1 and PDE7A2 proteins from multiple sources demonstrated a high-affinity, cAMP-specific PDE activity with K_m values of 0.2 and 0.1 μM, respectively.[1,2,4–7] Purification of the recombinant, soluble PDE7A catalytic domain provided an estimate of a V_{max} value of 2.29 ± 0.14 μmoles/min/mg (H. Ke, personal communication). PDE7A activity requires Mg^{2+} for optimal activity at pH 7.5. PDE7A activity is reduced strongly at pH values below 6.5 and above 8.5.[5] The crystal structure of the PDE7A catalytic domain is given in detail in other chapters of this volume, as is identification of residues that confer its resistance to the PDE4-selective inhibitor, rolipram.[7] Both the PDE7A catalytic domain crystal structure and the chromatographic mobility of this domain after renaturation indicate a monomeric structure. The potency and effects of several PDE7A-selective inhibitors are described in other chapters of this volume.[8–14]

Insoluble recombinant PDE7A catalytic domain can be renatured into an active state exhibiting an identical K_m value, and a V_{max} value of 0.46 μmoles/min/mg, both within the range of values obtained for the soluble recombinant PDE7A catalytic domain.[15] Additionally, renatured PDE7A proteins exhibit pH sensitivity and a monomeric structure similar to those of soluble-recombinant PDE7A proteins and thus appear to be a valuable current source of purified PDE7A proteins until methods for large-scale expression and purification of soluble, full-length PDE7A proteins from native or recombinant sources are fully developed. Full-length renatured PDE7A1 proteins have a V_{max} value of 0.95 μmoles/min/mg, suggesting that PDE7A catalytic activity is not altered drastically by the PDE7A1 regulatory domain.[16]

9.2.1 PDE7A1

Human and murine PDE7A1 is a 57 kDa protein of 482 amino acids encoded by a 4.2 kb mRNA whose complete sequence has not been fully obtained.[1,5] Available cDNAs contain the entire open reading frame, but lack about 0.8 kb of its 5′ untranslated sequence. Regulatory functions of the long 5′ or 3′ untranslated regions of this transcript are not known although destabilizing ARE sequences have been identified downstream of the termination codon and a CpG island with methylated cytosines have been identified in the PDE7A locus at the site of the currently known G-C rich 5′ end of the PDE7A1 transcript. In the human genome this putative 5′ end maps to an endogenous Not I restriction endonuclease cleavage site.[17]

Expression of PDE7A1 RNA has been detected on Northern blots and by RT PCR in many human and murine tissues including the brain, testis, spleen, thymus, pancreas, lung, kidney, placenta, $CD4^+$ and $CD8^+$ T lymphocytes, B lymphocytes, and proinflammatory cells.[1,3–5,18,19] In situ hybridization details enrichment of PDE7A RNA in deep layers of the cerebral cortex, its presence in both neuronal and nonneuronal cells, and in the

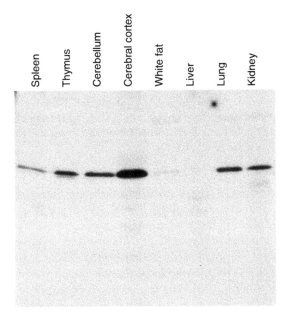

FIGURE 9.2 Relative abundance of PDE7A1 proteins in mouse tissues. Immunodetection of PDE7A1 in soluble extracts of mouse spleen, thymus, cerebellum, cerebral cortex, adipose, liver, lung and kidney. A western blot probed with affinity-purified antibodies directed against the PDE7A1 C-terminus is shown.

above-mentioned tissues.[20,21] Expression of the 57 kDa PDE7A1 protein in lymphocytes and proinflammatory cells has been described[4,19,22] and its relative abundance in mouse tissues depicts the cerebral cortex as the site of highest PDE7A1 expression among the analyzed tissues (Figure 9.2). Drastic induction of PDE7A1 expression has been noted in human peripheral, naïve T cells upon their costimulation with low concentrations of αCD3 and αCD28 monoclonal antibodies.[18] Increased expression of PDE7A1 has also been observed upon transfer of monocytes to tissue culture conditions,[23] and in some leukemias where expression is increased along with elevations in intracellular cAMP levels.[22] Although a CpG-rich sequence at the site of the currently cloned 5′ end of the transcript has been implicated in promoting transcription, full promoter sequences used for constitutive or inducible PDE7A1 transcription have not been identified yet.

Recently, it has been demonstrated that PDE7A1 is a bifunctional protein whose regulatory domain serves as a powerful inhibitor of PKA signaling.[16] The PDE7A1 N-terminus (a.a. 1–146) contains two repeated sequences, each 18 residues long bearing PKA pseudosubstrate sites, RRGAI (Figure 9.1). PKA pseudosubstrate sites are found exclusively in sequences of PKA inhibitors—PKI and the PKA regulatory subunit RI. Like PKI and RI, the PDE7A1 N-terminus interacts directly with the catalytic subunit of PKA and inhibits kinase activity with high affinity. *In vivo*, PDE7A1 associates with the dissociated C subunit of PKA, inhibits its kinase activity efficiently, and colocalizes with tetrameric PKA II complexes. In yeast, the PDE7A1 regulatory domain actually suppresses C-dependent, cAMP-independent, physiological responses. Thus, PDE7A1 is a bifunctional inhibitor of cAMP signaling, exerting its effects both via catalytic degradation of cAMP and via direct inhibition of PKA kinase activity (Figure 9.3). Inhibition exerted by the PDE7A1 regulatory domain is actually more potent than that exerted by its catalytic domain when these PDE7A1-derived peptides are overexpressed in CHO-K1 cells. The relative contribution of each of the PDE7A1 domains to termination of endogenous cAMP signals in cells where PDE7A1 is naturally expressed remains to be determined.

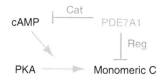

FIGURE 9.3 (See color insert following page 274.) Bi-functional inhibition of cAMP signaling by PDE7A1. A diagram depicting inhibitory targets along the cAMP signaling pathway for PDE7A1 domains. The catalytic C-terminus (cat) degrades cAMP, while the regulatory N-terminus (reg) inhibits the catalytic subunit of PKA (monomeric C).

PDE7A1 proteins have been localized mainly to the outer face of the Golgi apparatus and are also distributed throughout the cytoplasm.[24] PDE7A1 proteins are solubilized easily upon cellular fractionation in isotonic buffered solutions leaving 2%–4% of the cellular PDE7A1 content as particulate, salt, and nonionic detergents insoluble material. PDE7A1 association with the Golgi apparatus is disrupted easily upon isotonic cell extraction suggesting dynamic, labile interactions of PDE7A1 with Golgi localized proteins. Interestingly, PDE7A1 colocalizes with the regulatory subunit of PKA II to the Golgi, potentially via interactions with the PKA anchor protein MTG. MTG binds to both PDE7A1 and PDE4A via a similar mechanism that utilizes three independent fragments of MTG encompassing 85% of its sequence. It therefore appears that additional studies are required to determine whether multiple PDEs bind MTG via shared structural elements or via nonspecific associations, and whether these PDEs actually colocalize with MTG *in vivo*. The functional significance of PDE7A1 localization to the outer face of the Golgi apparatus, and to other intracellular sites, remains to be determined.

The acute induction of PDE7A1 expression upon costimulation of the T cell receptor suggested its possible involvement in the regulation of T cell proliferation.[18] The functional relevance of this observation, however, is in dispute as evidence for and against PDE7A1 involvement in T cell proliferation has been obtained. The observation that antisense oligonucleotides electroporated into human T cells reduced T cell proliferation implies PDE7A1 involvement in this process.[18] By contrast, a targeted disruption of the PDE7A catalytic domain revealed no changes in T cell proliferation.[25] Further, PDE7-selective inhibitors did not affect proliferation of human $CD8^+$ T cells even though they reduce TNFα release from aged human monocytes.[23] Multiple factors may account for the discrepancies between these observations including species differences, long-term compensatory mechanisms in mice, ineffectiveness of PDE7-selective inhibitors *in vivo*, T cell purification methods, nonspecific targeting of the antisense oligonucleotides, and the independent capacity of each of the PDE7A1 domains to block cAMP signaling. Of these, only compensation via increases in activities of other PDEs present in T cells was ruled out by quantitative pharmacological analysis of PDE activities present in T cells of wild-type and PDE7A catalytic knockout mice. Of particular concern are observations of effectiveness of PDE7-selective inhibitors in increasing intracellular cAMP content only at high concentrations and in combination with other family-selective inhibitors, whereas direct evidence for their *in vivo* selective inhibition of PDE7A1 activity in T cells is lacking.[23,26] It thus appears that additional studies are required to discern between the bona fide T cell response and the effects of all the above mentioned factors. Intriguing, however, is the strategy used to generate the PDE7A catalytic knockout mice that ablated all RNA encoding the catalytic domain while preserving RNA encoding the PDE7A1 regulatory domain. It is therefore possible that the PDE7A1 N-terminus, a potent inhibitor of PKA, is expressed in the PDE7A catalytic knockout mice, thus blocking effects of the disruption of the catalytic domain. Such that despite

increases in cAMP content and the activation of PKA in these mice, the expressed regulatory domain of PDE7A1 interferes with phenotypic manifestation of effects of reductions in phosphodiesterase activity and with determination of PDE7A1 function.

9.2.2 PDE7A2

Human PDE7A2 is a 50 kDa protein of 456 amino acids encoded by a 3.8 kb mRNA whose sequence has not been fully obtained.[2] A cDNA containing a termination codon upstream of the initiator methionine contains the entire open reading frame but lacks parts of its unique 5′ untranslated sequences. PDE7A2 3′ untranslated sequences resemble those of PDE7A1.

The unique N-terminus of PDE7A2 is 20 residues long, is hydrophobic, and confers an association of PDE7A2 with particulate membrane fractions including the plasma membrane (T. Michaeli, unpublished observations). This hydrophobic N-terminus is spliced onto the middle of the second PKA pseudosubstrate site of PDE7A1. PDE7A2 is therefore not anticipated to interact with the PKA catalytic subunit, as binding to C requires at least one intact repeat. Consequently, PDE7A2 is thought to block cAMP signaling primarily via the localization of its catalytic activity to the vicinity of sites of cAMP synthesis.

In agreement with the distribution of PDE7A mRNA, PDE7A2 proteins are detected easily in detergent solubilized extracts derived from skeletal and cardiac muscles.[23] In lymphoid and proinflammatory cells where a fifth of the PDE7A mRNA is that of PDE7A2, PDE7A2 proteins are not observed, while PDE7A1 proteins are detectable.[23] It is possible that differences in mRNA availability for translation, or competition between mRNAs bearing different 5′ untranslated sequences affect their translational efficiency and are noticeable in cells expressing multiple PDE7A mRNAs.

9.2.3 PDE7A3

Human PDE7A3 is a 50 kDa protein of 425 amino acids encoded by a transcript with an uncharacterized 5′ end.[3] The PDE7A3 protein N-terminus appears to be identical to that of PDE7A1 but its C-terminus is altered. As a result of alternative splicing that removes 62 residues from the PDE7A1 C-terminus and replaces them with a novel sequence of nine amino acids, PDE7A3 lacks phosphodiesterase activity. As PDE7A3 lacks part of helix 15 and helix 16 of the catalytic domain structure, it appears that the contributions of these helices to the integrity of the catalytic domain are essential. PDE7A3 thus appears to retain only the capacity of PDE7A1 to interact and inhibit the catalytic subunit of PKA. Consistent with its identity with the N-terminus of PDE7A1 and its expression from the same promoter, expression of PDE7A3 RNA and protein product is induced in CD4[+] T lymphocytes following costimulation of the T cell receptor.

Although PDE7A1 and PDE7A3 are expressed in the same cells, chromatographic fractionation of phosphodiesterases of HUT78 lymphoma cells identified PDE7A1 and PDE7A3 as two independent protein peaks.[3] This observation indicates that PDE7A1 and PDE7A3 do not form heterodimers and that PDE7A3 is not anticipated to have antagonistic, dominant-negative effects on PDE7A1 phosphodiesterase activity. It thus appears that the induction of PDE7A3 expression in CD4[+] T lymphocytes following costimulation of the T cell receptor can augment PDE7A1-mediated reductions in PKA activity in a time-dependent manner, but cannot directly affect PDE7A1 hydrolytic activity.

9.3 PDE7B

Four alternatively spliced PDE7B transcripts have been identified, of which PDE7B1, PDE7B2, and PDE7B3 are depicted in the literature and an additional variant is found in GenBank.[6,27,28] No endogenous protein products have been detected yet. Kinetics of

recombinant PDE7B1 proteins from insect and mammalian cells estimate K_m values of 0.03–0.2 μM that are consistent with it being a high-affinity, cAMP-specific phosphodiesterase.[6,27,28] PDE7B1 exhibits sensitivity in the low micromolar range to the nonselective phosphodiesterase inhibitor IBMX and to dipyridamole (PDEs 5, 6, 8, and 10 inhibitor), but not to inhibitors selective for PDEs 3, 4, or 5. IC_{50} values for inhibition of PDE7B by PDE7A-selective inhibitors have not been specified yet.

A PDE7B mRNA band of 5.5–5.6 kb is common to human, mouse, and rat tissues, with high abundance in heart, liver, skeletal muscle, and brain.[6,27,28] This mRNA is most abundant in the striatum, enriched in some neurons.[29,30] An additional, abundant mRNA band of 2 kb is present in rat testis, but is not observed in human or mouse testis.[29] It is therefore possible that the 5.5–5.6 kb PDE7B mRNA band contains multiple alternatively spliced mRNAs or that some PDE7B transcripts evade detection because of their low abundance or because of the analysis of a small number of samples. Dot blot analysis of human RNA reveals, in addition, the presence of PDE7B RNA in the pancreas, eye, thyroid, ovary, liver, caudate nucleus, and putamen.[6,28]

The open reading frames of three full-length alternatively spliced rat PDE7B cDNAs are depicted in Figure 9.4. PDE7B1 is widely distributed in heart, brain, spleen, lung, skeletal muscle, and kidney.[29] Abundant in rat testis is PDE7B2, which encodes N-terminally truncated PDE7B1 with a novel, 5′ untranslated region. Of lower abundance in skeletal muscle and heart is PDE7B3, which is identical to PDEB1 with the exception of addition of an exon encoding 13 amino acids between residues 27–28 of PDE7B1. The additional splice variant found in Genbank encodes a novel, not fully identified, N-terminus that is contiguous to the PDE7B1 sequence from residues 28 and further. Functional consequences of this alternative splicing pattern are not known yet, although changes in content of PKA phosphorylation sites have been noted as given in detail further in the chapter.

A putative PKA phosphorylation site is present on the N-terminus of PDE7B1 and PDE7B3.[29] A second putative PKA phosphorylation site is present on the C-terminus shared by all three murine PDE7B splice variants, but is not present on known human PDE7B protein sequences. All these sites can be phosphorylated by PKA *in vitro* and in forskolin-stimulated COS7 cells expressing recombinant PDE7B proteins.[29] The functional consequences of PKA phosphorylation on phosphodiesterase activity of all three splice variants, however, are not known.

9.3.1 PDE7B1

PDE7B1 has been implicated in negative feedback regulation of dopaminergic signaling in rat striatal neurons through up-regulation of PDE7B1 expression.[31] In these cells, PDE7B1 mRNA expression is increased about twofold within 6 h of stimulation with dopamine.

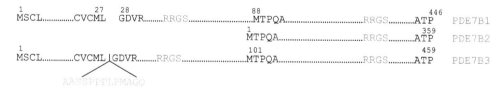

FIGURE 9.4 (See color insert following page 274.) Diagram and partial amino acid sequence of PDE7B encoded proteins. Partial amino acid sequence of rat PDE7B1, 7B2 and 7B3 is presented as indicated on their right hand side in red. Residue numbers are indicated above each sequence. PKA phosphorylation site sequences RRGS are depicted (blue). The sequence of an exon coding for 13 amino acids found within the PDE7B3 sequence is indicated in green.

The transcriptional induction of PDE7B1 appears to be mediated by the dopamine D1 receptor, cAMP, PKA, and the transcription factor CREB. The PDE7B1 promoter sequences bear several transcription factor–binding sites including a CREB binding site that mediates PDE7B1 promoter responsiveness to cAMP, but is not required for basal expression. Like other phosphodiesterase promoters, the PDE7B1 promoter lacks a TATA and a CCAAT box. Although no functional link between PDE7B1 and dopaminerginc signaling has been established yet, these observations are of great potential therapeutic relevance since loss of dopaminergic neurotransmission in striatum causes neurodegenerative disease. PDE7B1 may thus be a target for intervention with the progression of diseases like Parkinson and Huntington.

9.4 PERSPECTIVES

Knowledge of the PDE7 family of cAMP-specific PDEs has clearly expanded since the discovery of this novel, previously undetected, family of PDEs, and with it the number of new questions that have arisen. PDE7 splice variants possess an intriguing distribution among human and murine tissues. Inducible expression of some PDE7 splice variants suggests their involvement in important cellular functions and potentially also in human disease. Subcellular localization and domain structure and function reveal the complex interactions of PDE7 splice variants with cellular architecture, with the cAMP-signaling pathway, and their unique contribution to specificity of termination of cAMP or PKA signals. Strides in identification of family members, particularly PDE7A proteins and PDE7B coding sequences, have advanced considerably, and additional products are anticipated as exemplified by a new PDE7B splice variant depicted in Genbank. As Northern blots are not useful in predicting the number of splice variants present within a given tissue, either due to detection limits or due to the presence of multiple splice variants of similar size in observed mRNA bands, the absence of identified 5′ untranslated sequences and transcription initiation sites critically hamper the determination of actual transcript size, of mRNA stability and translational efficiency, and of promoter sequences.

Both PDE7A1 and PDE7B1 have an inducible expression pattern that suggests their involvement in T cell proliferation and in dopaminergic signaling, respectively. A functional link between these PDEs and their potential functions is lacking, highlighting the need for extensive *in vivo* physiological studies of normal tissues, of multiple genetically modified animal models and through diseased states. PDE7A1 function is particularly challenging because of its bifunctional inhibition of cAMP signaling, which requires a complete ablation of the entire protein product for functional studies. A nearly complete ablation of PDE7A is analyzed for PDE7A1 and PDE7A2 functions and may resolve differences in the interpretation of currently published observations (T. Michaeli, unpublished observations). Because of long-term compensatory processes that may occur in knockout mice, it is also necessary to examine short-term inhibition of PDE7 splice variants on their functions, pharmacologic or sRNA-mediated, to establish their physiological effects.

PDE7 proteins analyzed thus far are restricted to specific cellular compartments that may prove critical for their physiological effects. PDE7A1 colocalization with PKA II may allow fine-tuning of cAMP signals and their integration with other signaling pathways as has been recently demonstrated for protein complexes containing PKA II and PDE4D3, which generate localized pulses of cAMP.[32] In effect, colocalization of PDE4D3 with PKA II, and with additional signaling molecules, constitutes dynamic feedback loops that control basal activation of PKA as well as limit its stimulation.[32,33] It remains to be determined whether PDE7A1 and PKA II complexes play similar roles in fine-tuning the physiological responses of T cells or pancreatic β-cells.

Among PDEs, the bifunctional inhibition of cAMP signaling exerted by PDE7A1 is unique to PDE7 proteins. While increasing its impact on cAMP and PKA signaling, bifunctional inhibition renders PDE7A1 a therapeutic target that is phenotypically resistant to effects of inhibitors targeting the PDE7 catalytic domain. Development of novel drugs, or methods, to inhibit activities or expression of all PDE7A1 domains, is required to block its contribution to cAMP and PKA signaling. Currently developed PDE7-selective inhibitors, however, may in effect be used as either PDE7A2-selective inhibitors or to target PDE7B, should they prove effective against its catalytic activity.

ACKNOWLEDGMENT

Support from NIH DK58871 for this research is greatly appreciated.

REFERENCES

1. Michaeli, T., T.L. Bloom, T. Martins, K. Loughney, K. Ferguson, M. Riggs, L. Rodgers, J.A. Beavo, and M. Wigler. 1993. *J Biol Chem* 268:12925–12932.
2. Han, P., X. Zhu, and T. Michaeli. 1997. *J Biol Chem* 272:16152–16157.
3. Glavas, N.A., C. Ostenson, J.B. Schaefer, V. Vasta, and J.A. Beavo. 2001. *Proc Natl Acad Sci USA* 98:6319–6324.
4. Bloom, T.J., and J.A. Beavo. 1996. *Proc Natl Acad Sci USA* 93:14188–14192.
5. Wang, P., P. Wu, R.W. Egan, and M.M. Billah. 2000. *Biochem Biophys Res Commun* 276:1271–1277.
6. Sasaki, T., J. Kotera, K. Yuasa, and K. Omori. 2000. *Biochem Biophys Res Commun* 271:575–583.
7. Wang, H., Y. Liu, Y. Chen, H. Robinson, and H. Ke. 2005. *J Biol Chem* 280:30949–30955.
8. Martinez, A., A. Castro, C. Gil, M. Miralpeix, V. Segarra, T. Domenech, J. Beleta, J.M. Palacios, H. Ryder, X. Miro, C. Bonet, J.M. Casacuberta, F. Azorin, B. Pina, and P. Puigdomenech. 2000. *J Med Chem* 43:683–689.
9. Lorthiois, E., P. Bernardelli, F. Vergne, C. Oliveira, A.K. Mafroud, E. Proust, L. Heuze, F. Moreau, M. Idrissi, A. Tertre, B. Bertin, M. Coupe, R. Wrigglesworth, A. Descours, P. Soulard, and P. Berna. 2004. *Bioorg Med Chem Lett* 14:4623–4626.
10. Vergne, F., P. Bernardelli, E. Lorthiois, N. Pham, E. Proust, C. Oliveira, A.K. Mafroud, F. Royer, R. Wrigglesworth, J. Schellhaas, M. Barvian, F. Moreau, M. Idrissi, A. Tertre, B. Bertin, M. Coupe, P. Berna, and P. Soulard. 2004. *Bioorg Med Chem Lett* 14:4607–4613.
11. Vergne, F., P. Bernardelli, E. Lorthiois, N. Pham, E. Proust, C. Oliveira, A.K. Mafroud, P. Ducrot, R. Wrigglesworth, F. Berlioz-Seux, F. Coleon, E. Chevalier, F. Moreau, M. Idrissi, A. Tertre, A. Descours, P. Berna, and M. Li. 2004. *Bioorg Med Chem Lett* 14:4615–4621.
12. Bernardelli, P., E. Lorthiois, F. Vergne, C. Oliveira, A.K. Mafroud, E. Proust, N. Pham, P. Ducrot, F. Moreau, M. Idrissi, A. Tertre, B. Bertin, M. Coupe, E. Chevalier, A. Descours, F. Berlioz-Seux, P. Berna, and M. Li. 2004. *Bioorg Med Chem Lett* 14:4627–4631.
13. Pitts, W.J., W. Vaccaro, T. Huynh, K. Leftheris, J.Y. Roberge, J. Barbosa, J. Guo, B. Brown, A. Watson, K. Donaldson, G.C. Starling, P.A. Kiener, M.A. Poss, J.H. Dodd, and J.C. Barrish. 2004. *Bioorg Med Chem Lett* 14:2955–2958.
14. Kempson, J., W.J. Pitts, J. Barbosa, J. Guo, O. Omotoso, A. Watson, K. Stebbins, G.C. Starling, J.H. Dodd, J.C. Barrish, R. Felix, and K. Fischer. 2005. *Bioorg Med Chem Lett* 15:1829–1833.
15. Richter, W., T. Hermsdorf, T. Kronbach, and D. Dettmer. 2002. *Protein Expr Purif* 25:138–148.
16. Han, P., P. Sonati, C. Rubin, and T. Michaeli. 2006. *J Biol Chem* 281:15050–15057.
17. Torras-Llort, M., and F. Azorin. 2003. *Biochem J* 373:835–843.
18. Li, L., C. Yee, and J.A. Beavo. 1999. *Science* 283:848–851.
19. Smith, S.J., S. Brookes-Fazakerley, L.E. Donnelly, P.J. Barnes, M.S. Barnette, and M.A. Giembycz. 2003. *Am J Physiol Lung Cell Mol Physiol* 284:L279–L289.
20. Miro, X., S. Perez-Torres, J.M. Palacios, P. Puigdomenech, and G. Mengod. 2001. *Synapse* 40:201–214.
21. Perez-Torres, S., R. Cortes, M. Tolnay, A. Probst, J.M. Palacios, and G. Mengod. 2003. *Exp Neurol* 182:322–334.

22. Lee, R., S. Wolda, E. Moon, J. Esselstyn, C. Hertel, and A. Lerner. 2002. *Cell Signal* 14:277–284.
23. Smith, S.J., L.B. Cieslinski, R. Newton, L.E. Donnelly, P.S. Fenwick, A.G. Nicholson, P.J. Barnes, M.S. Barnette, and M.A. Giembycz. 2004. *Mol Pharmacol* 66:1679–1689.
24. Asirvatham, A.L., S.G. Galligan, R.V. Schillace, M.P. Davey, V. Vasta, J.A. Beavo, and D.W. Carr. 2004. *J Immunol* 173:4806–4814.
25. Yang, G., K.W. McIntyre, R.M. Townsend, H.H. Shen, W.J. Pitts, J.H. Dodd, S.G. Nadler, M. McKinnon, and A.J. Watson. 2003. *J Immunol* 171:6414–6420.
26. Nakata, A., K. Ogawa, T. Sasaki, N. Koyama, K. Wada, J. Kotera, H. Kikkawa, K. Omori, and O. Kaminuma. 2002. *Clin Exp Immunol* 128:460–466.
27. Hetman, J.M., S.H. Soderling, N.A. Glavas, and J.A. Beavo. 2000. *Proc Natl Acad Sci USA* 97:472–476.
28. Gardner, C., N. Robas, D. Cawkill, and M. Fidock. 2000. *Biochem Biophys Res Commun* 272:186–192.
29. Sasaki, T., J. Kotera, and K. Omori. 2002. *Biochem J* 361:211–220.
30. Reyes-Irisarri, E., S. Perez-Torres, and G. Mengod. 2005. *Neuroscience* 132:1173–1185.
31. Sasaki, T., J. Kotera, and K. Omori. 2004. *J Neurochem* 89:474–483.
32. Dodge-Kafka, K.L., J. Soughayer, G.C. Pare, J.J. Carlisle Michel, L.K. Langeberg, M.S. Kapiloff, and J.D. Scott. 2005. *Nature* 437:574–578.
33. McCahill, A., T. McSorley, E. Huston, E.V. Hill, M.J. Lynch, I. Gall, G. Keryer, B. Lygren, K. Tasken, G. van Heeke, and M.D. Houslay. 2005. *Cell Signal* 17:1158–1173.

10 cAMP-Phosphodiesterase 8 Family

Valeria Vasta

CONTENTS

10.1 Introduction 205
10.2 Biochemical Characteristics..206
10.3 Pharmacological Characteristics 207
10.4 Gene Organization and Alternative Transcripts 208
10.5 PDE8 Localization...210
10.6 PDE8 Structure..211
    10.6.1 REC Domain..211
    10.6.2 PAS Domain..211
        10.6.2.1 PAS Domain in Bacterial Cyclic Nucleotide
                Phosphodiesterases..212
        10.6.2.2 Is PAS Domain Involved in Regulation of PDE8?......................213
    10.6.3 Motifs 213
10.7 PDE8 Biological Functions..214
    10.7.1 PDE8A in Testis 214
    10.7.2 PDE8B in Brain...214
    10.7.3 PDE8s in Various Cell Types 214
10.8 Nonmammalian PDE8s 215
10.9 Conclusions..215
Acknowledgments 216
References 216

10.1 INTRODUCTION

The cloning of cyclic nucleotide phosphodiesterase 8A (PDE8A)[1,2] inaugurated the era of bioinformatics-guided PDE discovery that in the span of 2 years led to the fast paced identification of the second member of the family, PDE8B, as well as PDE9A, PDE10A, and PDE11A. All of these PDEs had previously eluded the classical biochemical approaches to protein identification. However, the large amount of information provided by expressed sequence tag (EST) databases rapidly led to the recognition of amino acid sequences that contained the classic consensus sequence and other features of PDEs. However, there were sufficient differences to suggest that they might represent new PDE families. In addition to having a conserved catalytic domain, both PDE8s have a unique N-terminal domain organization including a putative REC (signal receiver domain) and a PAS (acronym for PER, ARNT, and SIM proteins) domain (Figure 10.1).[3–5] Since the first identification of the PDE8 family, the biochemical, pharmacological, and genetic characteristics of the two PDE8s have been elucidated, but their biological functions still remain largely unknown.

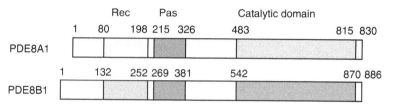

FIGURE 10.1 (See color insert following page 274.) Schematic representation of human PDE8A1 and PDE8B1 proteins. The boundaries of the REC and PAS domain have been defined using Pfam[71] and the ones of the catalytic area are defined according to Ref. [72].

PDE8A was initially identified by searching human and mouse EST databases for sequences with homology to the catalytic domain of other PDEs.[1,2] A cDNA was isolated including a partial coding region for a novel human PDE.[1] Sequence comparisons revealed that the predicted catalytic domain of this PDE shared only 20%–39% sequence identity with those of the seven previously identified PDE families. Therefore, the authors designated this protein, PDE8A, as the first member of a new family. The complete sequence of the human PDE8A transcript variant 1 (PDE8A1) was later defined and shown to encode a protein of 829 amino acids with a predicted molecular mass of 93,315 Da.[4,6] The mouse ortholog was contemporaneously recognized, sharing 85% sequence identity with the human PDE8A within the catalytic domain.[2] The full-length cDNA for mouse PDE8A contained 3678 bp and encoded a protein containing 823 amino acids with a predicted molecular mass of 93,171 Da. A phylogenic tree of PDE8s versus the other PDE families, inferred from their catalytic domain sequences, is reported in Ref. [7].

Shortly after the PDE8A sequence was reported, another partial cDNA encoding a human protein was identified and designated PDE8B because of its similarity to PDE8A.[8] The full-length human PDE8B1 was then shown to encode a protein of 885 residues and predicted molecular mass of 98,978 Da.[5,9] PDE8s orthologs are present in the genome of all mammalian species and, with the completion of various eukaryotes genomes, they are recognized in species as distant as *Drosophila melanogaster,*[9,10] *Danio rerio,* and *Caenorhabditis elegans.*

10.2 BIOCHEMICAL CHARACTERISTICS

PDE8A and PDE8B exhibit a high affinity for cAMP. The high affinity of PDE8s for cAMP may indicate that these enzymes metabolize cAMP in basal conditions or in regions of the cell where cAMP has to be maintained at low concentrations. The enzymes will neither hydrolyze cGMP appreciably nor appear to be modulated by cGMP. The kinetic parameters of human PDE8A were initially assessed with a baculovirus-expressed protein truncated in the amino-terminal 284 amino acids.[1] This enzyme displayed standard Michaelis–Menten kinetics with a K_m of 55 nM for cAMP. A similar K_m of 40 nM for cAMP was later reported for full-length human PDE8A expressed in Cos-7 cells.[11] Similarly, the mouse PDE8A expressed in Sf9 cells exhibited a K_m of 150 nM for cAMP and did not appreciably hydrolyze cGMP. Furthermore cGMP up to 100 μM, did not affect cAMP hydrolysis.[2] As all Class I PDEs, PDE8A contains two conserved metal-binding sites and the activity is dependent on the presence of divalent cations, Mg^{2+} and Mn^{2+}, in particular.[1] The V_{max} of PDE8A is not known.

Human recombinant PDE8B is also a high-affinity cAMP PDE with a K_m value of 101 nM.[11] Similar to PDE8A, the cAMP hydrolytic activity of PDE8B is not affected by cGMP and the enzyme does not hydrolyze cGMP.[8] The relative V_{max} value of PDE8B1 was twofold higher than PDE8A1 when normalized to the relative amount of the two proteins

expressed in Cos-7 cells.[11] In this cell system, the two proteins, expressed with a N-terminal tag, were equally distributed between membranes and cytosol.[11]

10.3 PHARMACOLOGICAL CHARACTERISTICS

The pharmacology of PDE8 is different from that of most other PDEs. The initial assays of PDE8A and PDE8B revealed an unusual insensitivity to 3-isobutyl-1-methyl-xanthine (IBMX), a compound that inhibits most PDEs with an IC_{50} of 2–50 μM. Therefore PDE8s are similar in this respect to PDE9A, and to Trypanosomial and Plasmodium PDEs that are also IBMX-insensitive.[7,12–14] The discovery of PDE8s and their pharmacological features resolved a controversial anomaly in the previous literature that had suggested the presence of IBMX-insensitive cAMP PDEs in liver.[15]

Moreover, most inhibitors with high affinity for other PDEs have been shown to be ineffective for PDE8s. However, dipyridamole, usually considered a weakly selective PDE5 and PDE6 inhibitor, also inhibits PDE8A and PDE8B at concentrations only 5–10 fold higher than that necessary for PDE5 and PDE6 (IC_{50} 4.5–29 μM)[1,2,11] (Table 10.1). The PDE4 inhibitors, cilomilast and L-826,141, have been reported to also inhibit PDE8A and PDE8B, respectively, with an IC_{50} of 7 μM (whereas the IC_{50} for PDE4s is in the nanomolar range).[16,17] Trequinsin, a PDE2 and PDE3 inhibitor, has been reported to inhibit PDE8B with an IC_{50} of 2 μM.[17] The PDE5 inhibitor vardenafil, unlike sildenafil or tadalafil, is also a modest inhibitor of PDE8A with an IC_{50} of 57 μM.[16] Papaverine, a nonselective PDE inhibitor, has a weak inhibitory potency, with an IC_{50} of 174 μM.[2] All the other tested common PDEs inhibitors did not inhibit PDE8s.[1,2,11,16] No specific PDE8 inhibitors have been reported to date in the literature or in patents. The lack of inhibitors has thus far hampered the study of PDE8 biological function.

The high affinity for cAMP and insensitivity to IBMX suggest that features in the substrate or inhibitor-binding pocket must be unique to PDE8 and could possibly be exploited to design specific inhibitors. Co-crystal structures of other PDEs with inhibitors have revealed a common scheme in drug binding including a hydrophobic clamp formed by conserved hydrophobic residues that sandwich the inhibitor in the active site and hydrogen bonding to an invariant glutamine.[16] These key residues for inhibitor binding are conserved in PDE8s. Similarly, PDE9A is poorly inhibited by IBMX (IC_{50} 230 μM[12]) to the PDE8 proteins. As hypothesized for PDE9A,[18] the tight binding of cAMP to PDE8A ($K_m = 40$–150 nM) might require higher

---

TABLE 10.1
Inhibitor Sensitivity of PDE8A and PDE8B

Compounds	Inhibitory Effect (IC_{50} μM) [reference number]	
	PDE8A	PDE8B
Dipyridamole	4–9[1,2,11]	23 [11]
Cilomilast	7 [16]	Not determined
Vardenafil	57 [16]	Not determined
Papaverine	174 [2]	Not determined
L-826,141	Not determined	7.2 [17]
Trequinsin	Not determined	2.1 [17]
IBMX, Vinpocetine, EHNA, Milrinone, Rolipram, Sildenafil, Tadalafil	>100 [2,11,16]	>100 [11]

concentrations of IBMX for an effective inhibition. Analysis of the crystal structure of PDE9A in complex with IBMX has suggested that a change in the side-chain conformation of the invariant glutamine making a hydrogen bond with N7 of the xanthine, as well as the absence of a hydrogen bond to the carbonyl oxygen O6 of the inhibitor, may account for the weak binding affinity of IBMX for PDE9A.[18] The elucidation of a crystal structure of PDE8s should clarify the reasons for IBMX insensitivity and facilitate the design of PDE8s-specific inhibitors. The unique domains in the N-terminal regions of the PDE8 proteins may have an allosteric regulatory function and possibly will represent a target for development of drugs, which could be more selective than those targeting the conserved catalytic region.

10.4   GENE ORGANIZATION AND ALTERNATIVE TRANSCRIPTS

The PDE8 family includes two isogenes: PDE8A and PDE8B. The gene for the human PDE8A is localized on chromosome 15q25.3 whereas the gene for PDE8B is on chromosome 5q13.3. The gene organization of PDE8A and PDE8B is similar to other PDEs and comprises 22 exons showing very similar boundaries and spanning 157 and 217 kb in their respective genes. PDE8A and PDE8B appear to be expressed in various alternative transcript forms that could potentially generate different protein isoforms (Figure 10.2). These variants arise from exon skipping or insertion events all limited to the 5′ portion upstream from the region coding

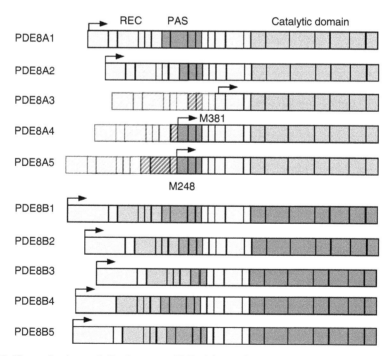

FIGURE 10.2 (See color insert following page 274.) Alternative transcript variants of human PDE8A and PDE8B. The exon organization of the various transcripts is schematized with indication of the regions coding for the REC (exons 3–6 in PDE8A1 and PDE8B1), PAS domains (exons 7–10 in PDE8A1 and PDE8B1), and catalytic area (exons 16–22). Arrows indicate translation start codons. Hatched areas indicate the proposed untranslated region for some of the PDE8A variants with their initiator methionine number relative to human PDE8A1 sequence.[4]

for the catalytic domain. The 5′ donor and 3′ acceptor splice sites are consistent with the canonical GT/AG rule.

Two transcript variants of PDE8A were initially identified in human testis: PDE8A1 and PDE8A2, which was missing a 138 nucleotide sequence corresponding to exon 8.* Three further variants arise from the skipping of exons 7 and 8 in PDE8A3 and exon 7 in PDE8A4, whereas PDE8A5 is generated by the insertion of a new exon after exon 7.[4] The expression level of these variants as determined by real-time PCR appears to be lower than the level of PDE8A1 in a large panel of tissues and it has not yet been shown if the respective protein isoforms are produced.[4] PDE8A2 has a deletion of 46 amino acids in a portion of the protein that corresponds to the PAS domain. In the other three transcript variants (PDE8A3– PDE8A5), the alternative splicing events introduce a premature stop codon due to a frame-shift in the original reading frame, thereby affecting the position of the initiation AUG codon. The protein variants predicted by conceptual translation would be truncated at the N-terminus: PDE8A3 is presumed to start at Met381 of PDE8A1 and is missing the putative REC and PAS domain; PDE8A4 and PDE8A5 transcripts would encode the same protein isoform starting at Met248 of PDE8A1 and missing the REC domain and the N-terminal part of the PAS domain[4] (Figure 10.2). Alternatively, it can also be hypothesized that proteins truncated at the C-termini could be produced or even that these mRNAs could be utilized as polycistronic mRNAs, producing both the protein upstream from the splicing-originated– stop codon and the one downstream. Indeed, instances of polycistronic transcripts generated by inclusion of a stop codon through splicing have been reported in mammals.[19] A further consideration arises from what is actually the most recognized outcome of splicing events that introduce premature stop codons. It is accepted that ORFs with stop codons more than 50–55 nucleotides upstream from an exon–exon junction are the result of routine abnormalities in the splicing process and are targeted for a process called nonsense-mediated decay.[20] This is indeed a widespread survey mechanism that blocks the production of potentially toxic or simply inactive truncated proteins.[21] Therefore, a more realistic prediction may be that PDE8A3, PDE8A4, and PDE8A5 mRNAs are unproductive and targeted for degradation. This would also explain the low expression of these variants compared to PDE8A1.[4]

PDE8B is also expressed in multiple transcript variants.[5,11] Unlike PDE8A, the splicing process producing the five variants of human PDE8B does not alter the reading frame by introducing a premature stop codon. Therefore, the putative PDE8B proteins would differ from each other due to particular amino acid deletions. The transcript variants of human PDE8B are PDE8B1 containing 22 exons; PDE8B2, which lacks exon 8 (encoding part of the PAS domain); PDE8B3, which lacks exons 8, 9, and 10 (encoding the whole PAS domain); PDE8B4 lacking exon 2; and PDE8B5* lacking exon 12 (encoding a region between the PAS domain and the catalytic area) (Figure 10.2). This nomenclature has been incorporated in the NCBI Reference Sequences at the human PDE8B gene page. All these splice variants are expressed at the mRNA level in human thyroid[5,11] but PDE8B1 is apparently the dominant PDE8B form. PDE8B1 and PDE8B3 are expressed equally in human placenta, and PDE8B3 is the most abundant form in the brain.[5] Evidence of expression of these variants at the protein level is still lacking.

The biological significance of alternative transcripts for PDE8A and PDE8B is still unclear. The various transcripts may just represent noise of the splicing process or actually serve a function. Splicing events changing the stability of mRNA could be an important mechanism for regulating the level of protein expression. Splicing events that alter the protein sequence could

---

*As renumbered herein and previously labeled as PDE8B3 in Ref. [11].

potentially alter the binding properties, intracellular localization, enzymatic activity, or stability.[22,23] All these questions remain unanswered until the predicted protein isoforms are identified.

A putative promoter for the transcription of human PDE8A1–PDE8A5 has been described as containing a TATA box and other transcription factor binding sites.[4] It appears though that the location of this TATA box is actually downstream from the transcription initiation site; therefore, this would not be a relevant promoter element. Moreover, the CG-rich nucleotide composition of the potential promoter area of PDE8A and PDE8B indicates that these genes might be transcribed from a CpG island-type promoter. CG-rich and TATA-less promoters have also been reported for other PDEs.[24–27] As mentioned above, alternative transcripts for human PDE8A and PDE8B indicate that the 5′ of the mRNAs might be variable and transcribed from different promoters, possibly in a tissue-specific manner.

10.5 PDE8 LOCALIZATION

Northern blot analysis showed that human PDE8A is expressed as an approximately 4.5 kb mRNA in a wide variety of tissues with the highest level in testis, ovary, small intestine, and colon.[1] The real-time-PCR quantification indicates that the relative expression of human PDE8A is as follows: testis > spleen > colon > small intestine > ovary > placenta > kidney.[4] A single band of 4.4 kb has been detected by Northern blot analysis in mouse tissues with the highest expression in testis, followed by eye, liver, skeletal muscle, heart, kidney, ovary, and brain in the decreasing order.[2] An elevated level of expression of PDE8A in rat testis, liver, kidney, and heart has also been reported.[9]

Northern blot analysis of PDE8B in human tissues initially indicated a more restricted expression compared with PDE8A. The 4.2 kb PDE8B mRNA was expressed at the highest levels in thyroid, and at lower levels in brain, spinal cord, and placenta.[8] Apparently, the expression of PDE8B in rat and mouse is different from that in human tissues. For instance, PDE8B is particularly abundant in rat brain whereas it is expressed at lower levels in other tissues including thyroid.[9] PDE8B seems to be expressed in a wide range of tissues in humans and in mouse.[9]

A large wealth of data on the expression of PDEs is provided by the EST and microarray databases. The expression profile of the EST listed under the human PDE8A Unigene Hs.9333 shows that this PDE has been detected in most human tissues. A similar broad expression is indicated by the profile of EST counts listed at the human PDE8B Unigene Hs.78106. The more restricted distribution observed for human PDE8A, and for PDE8B in particular, by Northern blot analysis might simply denote the limitation in sensitivity of this type of assay. On the other hand Northern blot analysis should represent a more accurate estimate of the relative abundance of an mRNA between tissues. On the whole, these data underscore the requirement of a more quantitative data on PDE8s expression as well as the need to assess which cellular components within the examined tissues contain PDE8s at the mRNA level and particularly at the protein level.

Microarray analysis has also provided a large wealth of information on PDEs in different tissues, cell types, and subtypes, as well as on the expression under specific experimental or pathological settings. These data can be accessed through repositories such as the NCBI Gene Expression Omnibus at http://www.ncbi.nlm.nih.gov/geo, the Genomics Institute of the Novartis Research Foundation at http://symatlas.gnf.org/SymAtlas, and the European Bioinformatics Institute at http://www.ebi.ac.uk/arrayexpress. This vast amount of information can provide useful indications, keeping in mind the current limits of interpretations of microarray data.[28,29] As an example, the microarray analysis of gene expression in mouse testis during the progression of spermatogenesis showed that PDE8A is induced when the

meiotic stage of maturation is reached,[30] and this data confirmed the previous results obtained by in situ hybridization histochemistry on mouse testis[2] (microarray data searchable at http://mrg.genetics.washington.edu/).

10.6 PDE8 STRUCTURE

The N-terminal portion of PDE8 contains a putative REC (signal receiver or response regulator domain) and a PAS domain. These features are unique to this PDE family and are different from the GAF domains present in various other PDEs.

10.6.1 REC DOMAIN

In various proteins, the REC domain is involved in receiving a signal from a sensor protein in a phosphotransfer signal transduction pathway commonly referred to as two-component signaling, as it involves two types of proteins: histidine kinases and response regulator proteins.[31] Such signaling pathways have been identified in prokaryotes, fungi, slime molds, and plants, but it is still unclear if they are present in mammalian cells.[31] In this pathway, the signal is conveyed by phosphorylation of an aspartyl residue in the REC domain by a sensor histidine kinase. This in turn induces a conformational change that mediates the downstream response, at times using the PAS domain as an effector. Thus, the protein containing the REC domain is generally called the response regulator. The REC domain is often found N-terminal to a DNA-binding effector domain in bacterial transcription factors.[32] On the basis of the sequence similarity between these bona fide response regulator domains and the sequences in PDE8 proteins, it has been postulated that the N-terminal region contains a REC domain.[4,9,11] Of the amino acid residues conserved between the REC domain of PDE8 and the response regulator proteins, an aspartic acid residue, preceded by conserved hydrophobic amino acids, could be the phosphorylation site.[9] This residue is conserved in PDE8s of all species. Moreover, some, but not all, of the other residues implied in the conserved mechanism of phosphorylation and propagation of the associated conformational changes in response regulator domains also appear to be conserved in PDE8s.[31] Such a level of conservation suggests that this domain might function as a response regulator and modulate the phosphodiesterase activity. Interestingly, this type of pathway regulates the cAMP-PDE RegA that controls *Dictyostelium* development, sporulation, and osmotic stress response.[33,34] Phosphorylation of the conserved aspartate residue in RegA REC domain activates the hydrolysis of cAMP at least 20-fold.[33] Although little is known about two-component pathways in mammalian cells, histidine kinase homologues have been recently identified and therefore it is possible that PDE8 might be a target for REC phosphorylation.[35,36]

10.6.2 PAS DOMAIN

The second domain recognizable in the PDE8s, downstream from the REC domain, is the PAS domain, so called because it was first identified in the *Drosophila* clock protein PER, in the transcription factor ARNT (aryl-hydrocarbon receptor nuclear translocator), and in SIM (single-minded protein) in insects. PAS domains are structural modules that can be found in proteins in all kingdoms of life.[37] These proteins are often signaling proteins and the PAS domain functions as a sensory module for oxygen tension, redox potential, or light intensity through associated cofactors.[38] Alternatively, PAS domains mediate protein–protein interactions.[38] Although the amino acid sequences of the different PAS domains show little similarity, their three-dimensional structures appear to be conserved and also show common conformational flexibility.[39] A structural model of the PDE8A PAS domain can be constructed based on the homology to FixL PAS domain crystal structure (Figure 10.3).[3] The predicted model is very similar to the classic PAS domain and folds into a hand-like structure

Protein-interaction region

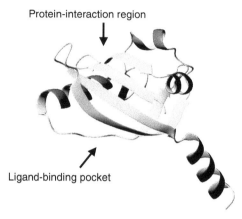

Ligand-binding pocket

FIGURE 10.3 (See color insert following page 274.) Structural model of PDE8A PAS domain based on the homology to the known structure of FiXL PAS domain. The putative protein-interaction site corresponds to the hydrophobic patch described by the surface of the β sheet. The possible ligand-binding pocket in analogy to other PAS proteins is also indicated. (Adapted from Ref. [3]. With permission from Elsevier.)

with an internal pocket framed by a five-stranded antiparallel β sheet flanked by α helices. This structure, based on an analogy to other PAS proteins, offers potential interaction sites in the inner pocket, formed by conserved hydrophobic residues, or in the hydrophobic surface patch described by the β sheet (Figure 10.3).

Because PAS domains function as sensors in signaling pathways and share a common fold, it is possible that they also have common dynamic properties, which would allow them to communicate with transducer proteins through a conserved mechanism. The structural studies suggest that glycines conserved throughout the PAS family serve as hinge points, allowing substructures in the protein to move relative to each other.[39] These glycines are also present in the PDE8s PAS domains.

10.6.2.1 PAS Domain in Bacterial Cyclic Nucleotide Phosphodiesterases

The PAS domain is also found in some bacterial PDEs containing the EAL catalytic domain (named from the conserved sequence motif Glu-Ala-Leu,[40]) that, in contrast to class I PDEs, contains only one metal-binding site.[41] Such enzymes, including PDEA1 of *Acetobacter xylinum* (AxPDEA1)[42] and *Escherichia coli* direct oxygen sensor (EcDOS),[43] catalyze the hydrolysis of the cyclic nucleotide, cyclic bis(3′–5′)diguanylic acid, an allosteric activator of cellulose synthase, and possibly of other pathways.[40] The PAS domain in AxPDEA1 and EcDOS is associated with a heme group that serves as a ligand-regulated switch for the activity. Indeed, oxygen triggers conformational changes, involving a loop in the PAS domain that leads to a decrease in the cyclic nucleotide phosphodiesterase activity.[44] A model has been proposed where the conformational changes induce a scissor-type subunit motion, altering the relative position of the catalytic area and the enzyme activity.[44] It is possible that the eukaryotic PDE8s are directly related evolutionarily to the bacterial PDEs. Alternatively, as seen in many instances, the PAS module was laterally exchanged and incorporated in a new signaling protein.[45] Although such a parallelism to bacterial PAS-regulated PDEs is appealing, it appears that the PDE8 PAS domain does not contain the conserved histidine, coordinating the heme in all oxygen-sensing PAS proteins.[46]

10.6.2.2   Is PAS Domain Involved in Regulation of PDE8?

Considering the roles of the PAS domain in other proteins, it is possible that PDE8 PAS domains associate with a cofactor or are involved in protein–protein interactions that are functionally relevant for allosteric regulation of enzyme activity. A first indication that this might indeed be the case comes from the observation that recombinant human PDE8A1 containing a partially deleted PAS domain (Δ AA215–281) exhibited a six times lower activity than the full-length form[47] although it is not clear if the K_m or V_{max} of the enzyme was affected. Therefore the PAS domain may be required for the proper assembly of the protein, stabilization of the protein, or act as an intramolecular activator. PAS might directly modulate the enzyme activity or do so secondarily to its interaction with cofactors or other proteins. Such a possibility is supported in the same study on PDE8A overexpressed in HEK293 cells by co-immunoprecipitation experiments, indicating that the PAS domain is required for an association of PDE8A with IkB proteins and that this interaction causes an activation of the enzyme.[47] IkB proteins are normally complexed with NF-kB in resting cells, maintaining the transcription factor in an inactive state. Overexpression of PDE8A apparently competes with the NF-kB p65–p50 complex for IkB binding and would be expected to release the transcription factor from inhibition. However no activation of NF-kB could be observed in the same experimental condition.[47] Therefore, the physiological role of the interaction between PDE8A and IkB proteins remains to be elucidated and more importantly, needs to be verified in cell types that normally express PDE8s. Considering that the PAS domain in some PDE8A and PDE8B isoforms is deleted, the process of alternative splicing might assume regulatory implications.

The hydrophobic core of PDE8 PAS domains could potentially associate with a small molecule and act as a ligand-regulated switch in signaling as in other PAS domain proteins. However, no evidence of such a ligand has emerged as yet. PDE8 PAS domains are unlikely to be oxygen sensors because the heme-binding histidine found in other heme–PAS proteins is not conserved. An eukaryotic PAS-kinase involved in metabolic status signaling is thought to be regulated, through binding of an as yet unidentified endogenous molecule, to its PAS domain.[48] It has been proposed that such a ligand would lead to a conformational change that propagates to a loop region of the PAS domain that normally inhibits the kinase, thus relieving the inhibition.[49] The screening of a library of organic compounds led to the identification of molecules that bind to this kinase PAS domain.[49] Therefore, a similar screening might lead to the discovery of molecules that would bind the PDE8 PAS domain, whereas biochemical approaches might eventually lead to the identification of physiological ligands. Considering the allosteric regulation usually exerted by PAS ligation in other proteins, it is tempting to speculate that interacting molecules might be utilized for drug development, leading to compounds that could be more selective than the substrate competitive PDE inhibitors developed so far. Depending on the structural relationship of the PAS domain and the catalytic region of PDE8s, such ligands could turn out to be inhibitors, as in the bacterial phosphodiesterases, or activators as in PAS kinase.

10.6.3   Motifs

Further elements of potential importance for the regulation of localization and function of PDE proteins can be identified by searches of conserved motifs present in the sequence. Putative phosphorylation sites localized between the PAS domain and the catalytic region have been described for cAMP- and cGMP-dependent protein kinases in both PDE8A and PDE8B.[4,11] Additional consensus motifs for phoshorylation by protein kinase C, casein kinase II, and tyrosine kinases have also been identified.[4] The N-terminal region

is highly conserved among species and between PDE8A and PDE8B and contains a six amino acid consensus motif for myristoylation, suggesting that PDE8s might associate with membranes.

10.7 PDE8 BIOLOGICAL FUNCTIONS

The biological functions of PDE8s are still unknown and can only be speculated based on the relevance of cAMP-regulated processes in the particular cell types in which PDE8s are found. The most important reason for such a lack in knowledge is that no specific inhibitors or transgenic models have become available to address PDE8s physiological roles.

10.7.1 PDE8A in Testis

Northern analysis for PDE8A expression in human, mouse, and rat tissues showed a high expression in testis. In situ hybridization[2] and microarray analysis[30] showed that PDE8A expression is regulated temporally and spatially in the seminiferous tubules. PDE8A, in particular, appeared to be most highly expressed in middle-to-late pachytene spermatocytes. Recently, immunofluorescence analysis of mature mouse sperm has indicated that PDE8A is localized at the level of the midpiece, a region of the flagellum where the mitochondria are localized.[50] cAMP is an important second messenger for mature sperm function[51,52] and PDE8A might regulate cAMP-responsive pathways in these cells. Therefore PDE8A modulating drugs might be evaluated for male contraception or fertility intervention.

10.7.2 PDE8B in Brain

PDE8B localization in human brain has been investigated by in situ hybridization using 45 nucleotide probes.[53] PDE8B mRNA was expressed in different regions of the brain and predominantly in the basal ganglia and in the granule cell layer of the dentate gyrus in the hippocampal formation.[53] Conversely no specific hybridization signal could be detected for PDE8A. In this study, brains from Alzheimer patients were also examined and an increase in the expression of the PDE8B in cortical areas and parts of the hippocampal formation was observed. However, considering that PDE8B mRNA expression showed a positive correlation with age, the study did not conclusively establish if the increase, in expression of PDE8B, observed in Alzheimer brains is a function of age rather than the disease. PDE8B is also expressed in rat brain.[9] Northern blot analysis showed expression in various areas of the brain with the exception of cerebellum whereas in situ hybridization analysis demonstrated positive staining in the neurons of the frontal cortex region.[9]

10.7.3 PDE8s in Various Cell Types

Upregulation of PDE8A expression has been shown to occur during human CD4+ T cell activation *in vitro*.[6] The increase in PDE8A and the parallel induction in PDE7A expression are potentially relevant in releasing the cAMP inhibitory action on the T cell activation.[54]

   PDE8A mRNA expression has been detected in the human osteosarcoma cell line MG-63[55] and in the mouse osteoblastic cell line MC3T3-E1.[56] As an increase in cAMP levels can promote osteoblastic differentiation,[57] it is possible that PDE8 inhibitors could be utilized for the treatment of metabolic bone disease. Conversely, as bone-loss induced by long-term use of glucocorticoid might involve an increase in adenylate cyclase activity,[58,59] an activator of PDE8A could counteract this effect.

   cAMP is a crucial second messenger in the regulation of thyroid function as thyrocytes largely depend on cAMP signaling for replication and differentiation.[60] Moreover, a common

pathophysiological mechanism of hyperthyroidism, thyroid, and pituitary adenomas is the constitutive activation of the cAMP-dependent mitogenic pathway.[60,61] Given the abundant and fairly selective expression of PDE8B in thyroid, it has been proposed that an inhibitor might be beneficial to treat hypothyroidism.[11] PDE upregulation seems to be a common physiological feedback mechanism opposing the chronic increase of cAMP in endocrine disorders.[62–64] In pituitary adenomas, a significant component of PDE activity has been found to be IBMX-insensitive and is most likely because of an increase in PDE8B that is not expressed in normal pituitary.[63] Drugs that would activate PDE8B might potentiate this effect and have therapeutic value in endocrine disorders.

In endometrial stromal cells, an increase in cAMP favors the uterine decidualization process and therefore supports the egg implantation. PDE8B is expressed in these cells and might be targeted as an alternative, nonsteroidal therapy for endometrial development and implantation in subfertile women.[65]

10.8 NONMAMMALIAN PDE8s

The completion of sequencing of the genomes of many different organisms has led to identification of orthologs of many mammalian PDEs. PDE8s appear to be encoded in the genome of *Drosophila melanogaster*,[9,10] *Caenorhabditis elegans* (NP_490787) and *Caenorhabditis briggsae* (CAE71587), *Ciona intestinalis*, *Danio rerio*, *Tetraodon nigrovirdis*, and *Takifugu rubripes*.

Five different transcripts encoding four different polypeptides for PDE8 have been identified in *Drosophila melanogaster*.[10] Three of the proteins include the REC and PAS domain and have differing N-terminal portions. The fourth protein does not have these domains and starts near the N-terminal of the catalytic domain. Only one of the isoforms maintains the N-terminal myristoylation motif. Mutagenesis of *Drosophila* or RNA interference living-cell microarrays[66] might allow the analysis of PDE8 function *in vivo*.

Caenorhabditis elegans is a particularly advantageous model for deciphering gene function on a genome-wide scale taking advantage of the feasibility of the RNA interference approach to knock down the expression of any gene in this system.[67] The power of the technique is illustrated by the similarity of the phenotype resulting from RNA interference on *C. elegans* PDE4D ortholog to the phenotype of the mouse knockout.[68] Indeed, a fertility problem has been detected in both the cases, although not in all tested samples for *C. elegans*.

RNA interference experiments directed toward the *C. elegans* PDE8 (NM_058386.1) did not show an obvious phenotype in one study[69] (results searchable at http://www.wormbase.org/db/seq/rnai?name=Y95B8A.10), whereas in another study evidence of embryonic lethality was observed in one experimental sample, although with poor reproducibility[70] (http://www.worm.mpi-cbg. de/phenobank2/cgi-bin/MenuPage.py). These studies were limited to the analysis of gene function in early embryo development. Further studies on adult organisms might lead to identification of other functions.

10.9 CONCLUSIONS

PDE8s are high-affinity cAMP-specific PDEs potentially involved in the control of basal cAMP levels or maintaining low cAMP in specific cell regions. The particular structure of these PDEs suggests that these enzymes might be regulated either by phosphorylation, cofactors, or by protein–protein interactions. Although PDE8s appear to be broadly expressed, the quantitative differences among tissues and the expression at the protein level and subcellular localization still remain to be investigated. As for the biological functions of

PDE8s, future studies will depend on the development of specific inhibitors suitable to study these enzymes. The unique domain structure suggests that drugs specific for PDE8s might be developed. Meanwhile, the use of transgenic models and RNA interference might compensate for the lack of specific inhibitors and lead to a better understanding of the biological and pharmacological potential of PDE8s. The localization of PDE8s, although apparently not restricted to single cell types as for ideal drug targets, indicates that these enzymes might be important targets for disease treatments.

ACKNOWLEDGMENTS

I thank Drs. Joseph Beavo, Enrico Patrucco, and James Surapisitchat for reviewing the manuscript and for helpful discussion. This study was supported by NIH grants DK 21723 and U54 HD 42454.

REFERENCES

1. Fisher, D.A., et al. 1998. Isolation and characterization of PDE8A, a novel human cAMP-specific phosphodiesterase. *Biochem Biophys Res Commun* 246 (3): 570–7.
2. Soderling, S.H., S.J. Bayuga, and J.A. Beavo. 1998. Cloning and characterization of a cAMP-specific cyclic nucleotide phosphodiesterase. *Proc Natl Acad Sci USA* 95 (15): 8991–6.
3. Soderling, S.H., and J.A. Beavo. 2000. Regulation of cAMP and cGMP signaling: New phosphodiesterases and new functions. *Curr Opin Cell Biol* 12 (2): 174–9.
4. Wang, P., et al. 2001. Human phosphodiesterase 8A splice variants: Cloning, gene organization, and tissue distribution. *Gene* 280 (1–2): 183–94.
5. Hayashi, M., et al. 2002. Genomic organization, chromosomal localization, and alternative splicing of the human phosphodiesterase 8B gene. *Biochem Biophys Res Commun* 297 (5): 1253–8.
6. Glavas, N.A., et al. 2001. T cell activation up-regulates cyclic nucleotide phosphodiesterases 8A1 and 7A3. *Proc Natl Acad Sci USA* 98 (11): 6319–24.
7. Yuasa, K., et al. 2005. PfPDE1, a novel cGMP-specific phosphodiesterase, from the human malaria parasite *Plasmodium falciparum. Biochem J* 392 (Pt 1): 221–9.
8. Hayashi, M., et al. 1998. Molecular cloning and characterization of human PDE8B, a novel thyroid-specific isozyme of 3′,5′-cyclic nucleotide phosphodiesterase. *Biochem Biophys Res Commun* 250 (3): 751–6.
9. Kobayashi, T., et al. 2003. Molecular comparison of rat cyclic nucleotide phosphodiesterase 8 family: Unique expression of PDE8B in rat brain. *Gene* 319: 21–31.
10. Day, J.P., et al. 2005. Cyclic nucleotide phosphodiesterases in *Drosophila melanogaster. Biochem J* 388 (Pt 1): 333–42.
11. Gamanuma, M., et al. 2003. Comparison of enzymatic characterization and gene organization of cyclic nucleotide phosphodiesterase 8 family in humans. *Cell Signal* 15 (6): 565–74.
12. Fisher, D.A., et al. 1998. Isolation and characterization of PDE9A, a novel human cGMP-specific phosphodiesterase. *J Biol Chem* 273 (25): 15559–64.
13. Rascon, A., et al. 2002. Cloning and characterization of a cAMP-specific phosphodiesterase (TbPDE2B) from *Trypanosoma brucei. Proc Natl Acad Sci USA* 99 (7): 4714–9.
14. Zoraghi, R., and T. Seebeck. 2002. The cAMP-specific phosphodiesterase TbPDE2C is an essential enzyme in bloodstream form *Trypanosoma brucei. Proc Natl Acad Sci USA* 99 (7): 4343–8.
15. Lavan, B.E., T. Lakey, and M.D. Houslay. 1989. Resolution of soluble cyclic nucleotide phosphodiesterase isoenzymes, from liver and hepatocytes, identifies a novel IBMX-insensitive form. *Biochem Pharmacol* 38 (22): 4123–36.
16. Card, G.L., et al. 2004. Structural basis for the activity of drugs that inhibit phosphodiesterases. *Structure (Camb)* 12 (12): 2233–47.
17. Liu, S., et al. 2005. Dynamic activation of cystic fibrosis transmembrane conductance regulator by type 3 and type 4D phosphodiesterase inhibitors. *J Pharmacol Exp Ther* 314 (2): 846–54.

18. Huai, Q., et al. 2004. Crystal structure of phosphodiesterase 9 shows orientation variation of inhibitor 3-isobutyl-1-methylxanthine binding. *Proc Natl Acad Sci USA* 101 (26): 9624–9.

19. Blumenthal, T. 1998. Gene clusters and polycistronic transcription in eukaryotes. *Bioessays* 20 (6): 480–7.

20. Lejeune, F., and L.E. Maquat. 2005. Mechanistic links between nonsense-mediated mRNA decay and pre-mRNA splicing in mammalian cells. *Curr Opin Cell Biol* 17 (3): 309–15.

21. Lewis, B.P., R.E. Green, and S.E. Brenner. 2003. Evidence for the widespread coupling of alternative splicing and nonsense-mediated mRNA decay in humans. *Proc Natl Acad Sci USA* 100 (1): 189–92.

22. Lareau, L.F., et al. 2004. The evolving roles of alternative splicing. *Curr Opin Struct Biol* 14 (3): 273–82.

23. Stamm, S., et al. 2005. Function of alternative splicing. *Gene* 344: 1–20.

24. Vicini, E., and M. Conti. 1997. Characterization of an intronic promoter of a cyclic adenosine 3′,5′-monophosphate (cAMP)-specific phosphodiesterase gene that confers hormone and cAMP inducibility. *Mol Endocrinol* 11 (7): 839–50.

25. Fujishige, K., et al. 2000. The human phosphodiesterase PDE10A gene genomic organization and evolutionary relatedness with other PDEs containing GAF domains. *Eur J Biochem* 267 (19): 5943–51.

26. Torras-Llort, M. and F. Azorin. 2003. Functional characterization of the human phosphodiesterase 7A1 promoter. *Biochem J* 373 (Pt 3): 835–43.

27. Liu, H., et al. 2005. Identification of promoter elements in 5′-flanking region of murine cyclic nucleotide phosphodiesterase 3B gene. *Methods Mol Biol* 307: 109–24.

28. Marshall, E. 2004. Getting the noise out of gene arrays. *Science* 306 (5696): 630–1.

29. Sherlock, G. 2005. Of fish and chips. *Nat Methods* 2 (5): 329–30.

30. Shima, J.E., et al. 2004. The murine testicular transcriptome: Characterizing gene expression in the testis during the progression of spermatogenesis. *Biol Reprod* 71 (1): 319–30.

31. Robinson, V.L., D.R. Buckler, and A.M. Stock. 2000. A tale of two components: a novel kinase and a regulatory switch. *Nat Struct Biol* 7 (8): 626–33.

32. Elsen, S., et al. 2004. RegB/RegA, a highly conserved redox-responding global two-component regulatory system. *Microbiol Mol Biol Rev* 68 (2): 263–79.

33. Thomason, P.A., et al. 1999. The RdeA-RegA system, a eukaryotic phospho-relay controling cAMP breakdown. *J Biol Chem* 274 (39): 27379–84.

34. Ott, A., et al. 2000. Osmotic stress response in *Dictyostelium* is mediated by cAMP. *Embo J* 19 (21): 5782–92.

35. Steeg, P.S., et al. 2003. Histidine kinases and histidine phosphorylated proteins in mammalian cell biology, signal transduction and cancer *Cancer Lett* 190 (1): 1–12.

36. Besant, P.G., E. Tan, and P.V. Attwood. 2003. Mammalian protein histidine kinases. *Int J Biochem Cell Biol* 35 (3): 297–309.

37. Ponting, C.P., and L. Aravind. 1997. PAS: A multifunctional domain family comes to light. *Curr Biol* 7 (11): R674–7.

38. Taylor, B.L., and I.B. Zhulin. 1999. PAS domains: Internal sensors of oxygen, redox potential, and light. *Microbiol Mol Biol Rev* 63 (2): 479–506.

39. Vreede, J., et al. 2003. PAS domains. Common structure and common flexibility. *J Biol Chem* 278 (20): 18434–9.

40. Romling, U., M. Gomelsky, and M.Y. Galperin. 2005. C-di-GMP: The dawning of a novel bacterial signalling system. *Mol Microbiol* 57 (3): 629–39.

41. Sasakura, Y., et al. 2002. Characterization of a direct oxygen sensor heme protein from *Escherichia coli*. Effects of the heme redox states and mutations at the heme-binding site on catalysis and structure. *J Biol Chem* 277 (26): 23821–7.

42. Tal, R., et al. 1998. Three cdg operons control cellular turnover of cyclic di-GMP in *Acetobacter xylinum*: Genetic organization and occurrence of conserved domains in isoenzymes. *J Bacteriol* 180 (17): 4416–25.

43. Delgado-Nixon, V.M., G. Gonzalez, and M.A. Gilles-Gonzalez. 2000. Dos, a heme-binding PAS protein from *Escherichia coli*, is a direct oxygen sensor. *Biochemistry* 39 (10): 2685–91.

44. Kurokawa, H., et al. 2004. A redox-controlled molecular switch revealed by the crystal structure of a bacterial heme PAS sensor. *J Biol Chem* 279 (19): 20186–93.

45. Aravind, L., V. Anantharaman, and L.M. Iyer. 2003. Evolutionary connections between bacterial and eukaryotic signaling systems: A genomic perspective. *Curr Opin Microbiol* 6 (5): 490–7.
46. Tomita, T., et al. 2002. A comparative resonance Raman analysis of heme-binding PAS domains: Heme iron coordination structures of the BjFixL, AxPDEA1, EcDos, and MtDos proteins. *Biochemistry* 41 (15): 4819–26.
47. Wu, P., and P. Wang. 2004. Per-Arnt-Sim domain-dependent association of cAMP-phosphodiesterase 8A1 with IkappaB proteins. *Proc Natl Acad Sci USA* 101 (51): 17634–9.
48. Rutter, J. 2002. Essay: Amersham biosciences and science prize. PAS domains and metabolic status signaling. *Science* 298 (5598): 1567–8.
49. Amezcua, C.A., et al. 2002. Structure and interactions of PAS kinase N-terminal PAS domain: Model for intramolecular kinase regulation. *Structure (Camb)* 10 (10): 1349–61.
50. Baxendale, R.W., and L.R. Fraser. 2005. Mammalian sperm phosphodiesterases and their involvement in receptor-mediated cell signaling important for capacitation. *Mol Reprod Dev* 71 (4): 495–508.
51. Esposito, G., et al. 2004. Mice deficient for soluble adenylyl cyclase are infertile because of a severe sperm-motility defect. *Proc Natl Acad Sci USA* 101 (9): 2993–8.
52. Nolan, M.A., et al. 2004. Sperm-specific protein kinase A catalytic subunit Calpha2 orchestrates cAMP signaling for male fertility. *Proc Natl Acad Sci USA* 101 (37): 13483–8.
53. Perez-Torres, S., et al. 2003. Alterations on phosphodiesterase type 7 and 8 isozyme mRNA expression in Alzheimer's disease brains examined by in situ hybridization. *Exp Neurol* 182 (2): 322–34.
54. Vang, T., et al. 2003. Combined spatial and enzymatic regulation of Csk by cAMP and protein kinase a inhibits T cell receptor signaling. *J Biol Chem* 278 (20): 17597–600.
55. Ahlstrom, M., et al. 2005. Cyclic nucleotide phosphodiesterases (PDEs) in human osteoblastic cells; the effect of PDE inhibition on cAMP accumulation. *Cell Mol Biol Lett* 10 (2): 305–19.
56. Wakabayashi, S., et al. 2002. Involvement of phosphodiesterase isozymes in osteoblastic differentiation. *J Bone Miner Res* 17 (2): 249–56.
57. Qin, L., L.J. Raggatt, and N.C. Partridge. 2004. Parathyroid hormone: A double-edged sword for bone metabolism. *Trends Endocrinol Metab* 15 (2): 60–5.
58. Rizzoli, R., V. von Tscharner, and H. Fleisch. 1986. Increase of adenylate cyclase catalytic-unit activity by dexamethasone in rat osteoblast-like cells. *Biochem J* 237 (2): 447–54.
59. Wong, M.M., et al. 1990. Long-term effects of physiologic concentrations of dexamethasone on human bone-derived cells. *J Bone Miner Res* 5 (8): 803–13.
60. Dremier, S., et al. 2002. The role of cyclic AMP and its effect on protein kinase A in the mitogenic action of thyrotropin on the thyroid cell. *Ann NY Acad Sci* 968: 106–21.
61. Spiegel, A.M. 1996. Mutations in G proteins and G protein-coupled receptors in endocrine disease. *J Clin Endocrinol Metab* 81 (7): 2434–42.
62. Persani, L., et al. 2000. Induction of specific phosphodiesterase isoforms by constitutive activation of the cAMP pathway in autonomous thyroid adenomas. *J Clin Endocrinol Metab* 85 (8): 2872–8.
63. Persani, L., et al. 2001. Relevant cAMP-specific phosphodiesterase isoforms in human pituitary: Effect of Gs(alpha) mutations. *J Clin Endocrinol Metab* 86 (8): 3795–800.
64. Wattel, S., et al. 2005. Gene expression in thyroid autonomous adenomas provides insight into their physiopathology. *Oncogene* 24 (46): 6902–16.
65. Bartsch, O., B. Bartlick, and R. Ivell. 2004. Phosphodiesterase 4 inhibition synergizes with relaxin signaling to promote decidualization of human endometrial stromal cells. *J Clin Endocrinol Metab* 89 (1): 324–34.
66. Wheeler, D.B., et al. 2004. RNAi living-cell microarrays for loss-of-function screens in *Drosophila melanogaster* cells. *Nat Methods* 1 (2): 127–32.
67. O'Rourke, S.M., and B. Bowerman. 2005. Genomics: Frontiers of gene function. *Nature* 434 (7032): 444–5.
68. Park, J.Y., et al. 2003. Phosphodiesterase regulation is critical for the differentiation and pattern of gene expression in granulosa cells of the ovarian follicle. *Mol Endocrinol* 17 (6): 1117–30.
69. Fraser, A.G., et al. 2000. Functional genomic analysis of *C. elegans* chromosome I by systematic RNA interference. *Nature* 408 (6810): 325–30.

70. Sonnichsen, B., et al. 2005. Full-genome RNAi profiling of early embryogenesis in *Caenorhabditis elegans*. *Nature* 434 (7032): 462–9.
71. Bateman, A., et al. 2004. The Pfam protein families database. *Nucleic Acids Res* 32 (Database issue): D138–41.
72. Xu, R.X., et al. 2000. Atomic structure of PDE4: Insights into phosphodiesterase mechanism and specificity. *Science* 288 (5472): 1822–5.

11 PDE9

Jun Kotera and Kenji Omori

CONTENTS

11.1   Introduction 221
11.2   Cloning...222
11.3   Genomic Organization and Splice Variants 223
11.4   Enzymatic Profiles and Sensitivity to Inhibitors 227
11.5   Crystal Structure 228
11.6   Tissue-Expression Patterns in Humans and Mice 229
11.7   *In Situ* Hybridization Analysis of PDE9A in Rat Brain.........................230
11.8   Subcellular Localization of PDE9A Variants 231
11.9   PDE9A Inhibitors 232
11.10   Conclusions 233
References 233

11.1 INTRODUCTION

Cyclic guanosine monophosphate (cGMP) is produced via activation of guanylyl cyclases in response to extracellular signals activated by signal-transducing agents, such as nitric oxide (NO), natriuretic peptides, and guanylins.[1,2] cGMP activates cGMP-dependent protein kinases (PKG) and cGMP-gated ion channels. Activated PKG phosphorylates many intracellular proteins that are involved in modulating intracellular calcium levels; elevation of cGMP as a whole causes lowering of intracellular calcium and desensitization of a number cellular signaling processes to calcium. The cGMP-gated ion channel regulates calcium influx into the cell. The combined effects of cGMP on PKG and the cGMP-gated ion channel lead to a variety of physiological events, such as relaxation of vascular smooth muscles, airway distention, inhibition of cell proliferation, inhibition of platelet aggregation, neuronal transmission, visual signaling, and apoptosis.[1–3]

cGMP is hydrolyzed by 3′,5′-cyclic nucleotide phosphodiesterases (PDEs), and as a result, intracellular cyclic nucleotides return to basal levels. Currently, the mammalian PDE superfamily consists of 11 families classified on the basis of amino acid sequence homology, substrate specificity, inhibitor sensitivity, and structural domains.[3–5] Among these 11 families, 5 (PDE2A, PDE5A, PDE6, PDE10A, and PDE11A) are categorized as GAF (cGMP-binding and stimulated phosphodiesterases, *Anabaena* adenylyl cyclases, and *Escherichia coli* FhlA) PDEs, and the others are known as non-GAF PDEs.[6,7] At least one GAF domain in PDE2A, PDE5A, and the PDE6 isoforms acts as an allosteric cGMP-binding domain.

Phosphodiesterase 9A (PDE9A) cDNA was discovered in 1998 and classified as the ninth PDE family using bioinformatics.[8,9] PDE9A as a member of the non-GAF PDEs is cGMP-specific and, unlike any other PDE including GAF PDEs, has an N-terminal region that contains no protein domain. A single gene encodes PDE9 and there are currently 21 known splice variants. The tissue-expression pattern of PDE9A mRNA is relatively broad in humans

with some variants exhibiting unique tissue-expression patterns. Although PDE9A has been considered as a therapeutic target for the treatment of a variety of conditions, there are only few reports that describe the physiological roles of PDE9A. More recently, a potent and selective PDE9A inhibitor was reported in the literature.[10] In this chapter, the molecular and enzymatic characteristics, crystal structure, tissue expression, and subcellular localization of PDE9A are described.

11.2   CLONING

cDNAs for the ninth PDE family were retrieved from the Incyte human EST database and the mouse EST database using human PDE4B cDNA and mouse PDE1A2 cDNA sequences as queries, respectively.[8,9] Human PDE9A cDNA (designated herein as PDE9A1, GenBank accession number AF048837) amplified from human brain cDNA encodes a full-length protein of 593 amino acids including 261 amino acids in the C-terminal region that composes the catalytic domain of PDE. Cloned human PDE9A possesses the HD domain (HD[L/I/V/M/F/Y]XHX[A/G]-(X)2-NX[L/I/V/M/F/Y]) as 3',5'-cyclic nucleotide phosphodiesterase signature in the C-terminal catalytic domain.[11] The amino acid sequence of the PDE9 catalytic domain shows similar homology (~30% identity) to all other mammalian PDE catalytic domains. Percentages of identical amino acids between the catalytic domain in human PDE9A and the catalytic domain in the other 10 PDE families is as follows: PDE1A (32%), PDE2A (31%), PDE3A (28%), PDE4A (33%), PDE5A (30%), PDE6A (31%), PDE7A (34%), PDE8A (34%), PDE10A (28%), and PDE11A (29%). This degree of similarity indicates that cloned human PDE9A is not an isozyme of the other known PDE families because similarity in the catalytic domains between PDE isozymes in the same PDE family is more than 65% (e.g., 93% identity between PDE4A and PDE4B, 68% identity between PDE7A and PDE7B, and 79% identity between PDE8A and PDE8B). Phylogenetic tree analysis of mammalian PDEs demonstrated that PDE9A has evolved independently of other PDEs (Figure 11.1). Unlike other PDEs, the N-terminal regions of PDE9A and PDE7A contain no known specific domain motifs. Although PROSITE searches revealed that Thr-96, Thr-235, and Thr-260 of human PDE9A1 are putative phosphorylation sites by PKG and cAMP-dependent protein kinase (PKA), no study has so far investigated the *in vitro* or *in vivo* phosphorylation at these sites.

Mouse PDE9A cDNA (designated herein as mouse PDE9A2, GenBank accession number AF031147) isolated from mouse kidney cDNA by PCR[9] contains an open reading frame encoding a full-length protein of 534 amino acids. The amino acid sequence of rat PDE9A cDNA (rat PDE9A2) has also been reported[12,13] and is listed in GenBank (accession nos. AF372654 and AY145898). Amino acid sequences of rat PDE9A2 and mouse PDE9A2 show 97.6% identity and are, respectively, 91.7% and 92.3% identical to human PDE9A2 (Figure 11.2). Amino acid sequences particularly in the C-terminal regions of the rat and mouse PDE9A2 are more diverse than human PDE9A2, whereas amino acid sequences in the other regions of human, rat, and mouse PDE9A2 are homologous.

Besides mammalian PDEs, the PDE9A catalytic domain is most homologous to RegA, a *Dictyostelium discoideum* PDE. Soderling et al.[9] have reported that the C-terminal part of the catalytic domain of mouse PDE9A is 41% identical to that of RegA. This identity percent is higher than that between the corresponding C-terminal region of PDE9A catalytic domain and other mammalian PDEs (29%–36%).[9] However, PDE9A amino acid sequence outside the catalytic domain has little or no similarity with that of RegA. With regard to enzymatic properties, RegA is cAMP specific with a K_m value of 5 μM,[14] whereas PDE9A is cGMP specific.

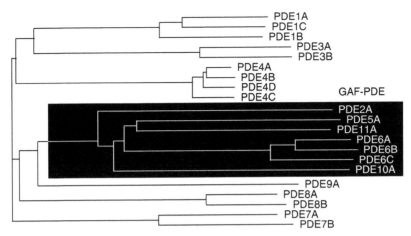

FIGURE 11.1 Phylogenetic tree of the PDE superfamily. The tree is generated using NJ algorithm of PHILIP based on multiple alignment of amino acid sequences containing PDEase_I (Pfam accession number PF00233) motif in the catalytic domain of human PDEs analyzed with CLUSTALW. The tree of PDE families containing GAF sequence is shown in a closed box. PDE1A, P54750; PDE1B, Q01064; PDE1C, Q14123; PDE2A, O00408; PDE3A, Q14432; PDE3B, Q13370; PDE4A, P27815; PDE4B, Q07343; PDE4C, Q08493; PDE4D, Q08499; PDE5A, D89094; PDE6A, P16499; PDE6B, P35913; PDE6C, X94354; PDE7A, Q13946; PDE7B, AB038040; PDE8A, AF056490; PDE8B, AF079529; PDE9A, AF048837; PDE10A, AB020593; and PDE11A, AB036704.

11.3   GENOMIC ORGANIZATION AND SPLICE VARIANTS

The human *PDE9A* gene consists of 22 exons (exons 1–3, 3b, 4, 4b, and 5–20) scattered in 125 kb of human genomic sequence (Figure 11.3)[15,16] (http://www.ncbi.nlm.nih.gov/entrez/query.fcgi?db = gene&cmd = Retrieve&dopt = full_report&list_uids = 5152) (GenBank accession nos. NC_000021 and AB017602). Currently, 21 splice variants of human PDE9A mRNAs are registered in the NCBI database (Table 11.1, Figure 11.3). Unfortunately, the nomenclature designations of the human PDE9A splice variants are rather complicated and have been confusing (Table 11.1). Therefore, a common nomenclature for PDE9A variants is necessary for clarity of communication in future studies. Human PDE9A cDNA clone (GenBank accession number AF048837) was first reported in the literature as PDE9A2[8] and conversely mouse PDE9A clone (GenBank accession number AF031147) was termed PDE9A1.[9] Currently, however, the human PDE9A cDNA clone is known as PDE9A1 (NCBI database), whereas the mouse PDE9A clone has been renamed PDE9A2 (Table 11.1). This new nomenclature is used hereafter in this book.

Human PDE9A1, which is the longest form among all variants, consists of 20 exons (exons 1–20 except for exons 3b and 4b) coding for a protein of 593 amino acid residues,[8,15,16] where the catalytic domain is encoded by exons 12–18 (Figure 11.3). Cloning studies have reported as many as 20 splice variants for human PDE9A all of which possess exons 1, 7, and 9–20.[15,16] Exon 3b is unique to PDE9A5, and exon 4b is specific to PDE9A19 and PDE9A21. Therefore, some PDE9A exons can be translated in different reading frames. For instance, the absence of exon 6 in PDE9A4 causes alternative translation initiation in exon 1 in a reading frame different from that in PDE9A3, which possesses exon 6 (Figure 11.3). PDE9A1, PDE9A2, PDE9A4, PDE9A5, PDE9A9, PDE9A13, and PDE9A18 have the same initiator methionine in exon 1, whereas PDE9A3, PDE9A6, PDE9A10, PDE9A12, PDE9A16, and

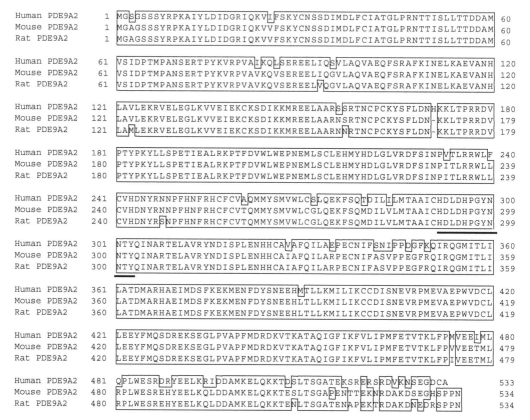

FIGURE 11.2 Alignment of human, rat, and mouse PDE9As (PDE9A2). Amino acid sequences of PDE9As are shown in one-letter code. Identical amino acid residues are boxed. The thick line indicates 3′,5′-cyclic nucleotide phosphodiesterase signature.

PDE9A17 use another methionine in exon 1 as translation initiation site. PDE9A7, PDE9A8, PDE9A14, PDE9A19, and PDE9A20 have the first methionine in exon 8, whereas the first methionine of PDE9A11 and PDE9A15 is situated in exon 7. Recently, the 21st splice variant (PDE9A21) has been listed in NCBI database. As PDE9A21 possesses exon 3b but lacks exon 5, its first methionine is located in exon 3b. In spite of the considerable number of splice variants for PDE9A, there are only few reports on the functional characteristics, tissue distribution, and subcellular localization of each variant. Particularly, analysis of PDE9A promoter has not been performed, although it might be assumed that all known PDE9A variants, which have exon 1 in common, are regulated by a common promoter. The 21 splice variants of PDE9A have been reported at mRNA level, however, most reports have focused only on 5′-diversities of PDE9A mRNAs. According to the literature[8,15,16] only human PDE9A1 and PDE9A2 mRNAs have been clearly confirmed as contiguous fragments coding for the putative coding region. Therefore, there is little direct evidence that full-length PDE9A splice variants are functionally expressed in tissues. Clearly, further studies on the complete expression of PDE9A splice variants at mRNA levels and proteins are needed.

    The human *PDE9A* gene is located on human chromosome 21q22.3,[15] an area in the genome that is related to two genetic diseases, i.e., nonsyndromic hereditary deafness and bipolar affective disorder.[17–19] Several studies have indicated that bipolar affective disorder is

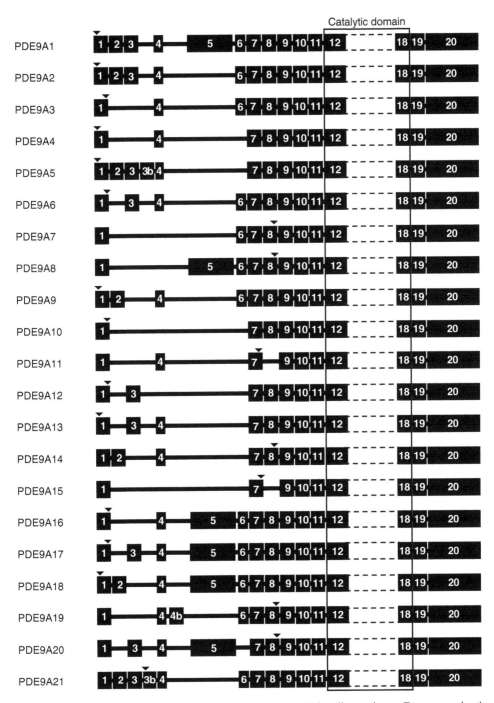

FIGURE 11.3 Structural organization of predicted human PDE9A splice variants. Exon organization of PDE9A splice variants is illustrated. Boxes with numbers represent exons and their numbers. Arrow heads indicate the position of initiation methionine.

TABLE 11.1
GenBank Accession Numbers of PDE9A Splice Variants in Humans and Mice

Variant	Amino Acid Residue	Human[8]	Human[15]	Human[16]	Human[24]	Human (NCBI Database)	Human (NCBI Database)	Mouse[9]
PDE9A1	593	AF048837[a]	AF067223			NM_002606		N/A
PDE9A2	533		AF067224			NM_001001567		AF031147[a]
PDE9A3	466		AF067225			NM_001001568		N/A
PDE9A4	465		AF067226			NM_001001569		N/A
PDE9A5	540			AY196299		NM_001001570		N/A
PDE9A6	492			AY196300	AY242121[a]	NM_001001571		N/A
PDE9A7	386			AY196301		NM_001001572		N/A
PDE9A8	386			AY196302		NM_001001573		N/A
PDE9A9	507			AY196303		NM_001001574		N/A
PDE9A10	433			AY196304		NM_001001575		N/A
PDE9A11	376			AY196305		NM_001001576		N/A
PDE9A12	459			AY196306		NM_001001577		N/A
PDE9A13	491			AY196307		NM_001001578		N/A
PDE9A14	386			AY196308		NM_001001579		N/A
PDE9A15	376			AY196309		NM_001001580		N/A
PDE9A16	526			AY196310		NM_001001581		N/A
PDE9A17	552			AY196311		NM_001001582		N/A
PDE9A18	567			AY196312		NM_001001583		N/A
PDE9A19	386			AY196313		NM_001001584		N/A
PDE9A20	386			AY196314		NM_001001585		N/A
PDE9A21	485						AY701187	N/A

[a]AF048837, AY242121, and AF031147 were named PDE9A2, PDE9A5, and PDE9A1 in Refs. [8], [24], and [9], respectively. In this book, these clones are renamed PDE9A1, PDE9A6, and PDE9A2, respectively.

related to proteins that regulate cyclic nucleotide metabolism.[20–22] In particular, cGMP-specific phosphodiesterase activity has been reported to be lower in amniotic fluid cells with trisomy 21 than in normal cells.[23] Triplication of genes, including *PDE9A*, located in chromosome 21 is associated with phenotypes of Down syndrome.[24] On the other hand, single nucleotide polymorphisms in coding region (cSNPs), haplotype-tagged SNPs (htSNPs), and other direct evidence of genetic relatedness to diseases including nonsyndromic hereditary deafness and bipolar affective disorder have not been reported yet.

Mouse PDE9A gene, *Pde9a* (approximately 95 kb) is located on chromosome 17 and consists of at least 19 exons corresponding to exons 1–4 and exons 6–20 of human *PDE9A*[16] (http://www.ncbi.nlm.nih.gov/entrez/query.fcgi?db = gene&cmd = Retrieve&dopt = full_report&list_uids = 18585) (GenBank accession number NC_000083). Although a few reports on mouse PDE9A cDNA have been published, exons corresponding to exons 3b, 4b, and 5 of the human *PDE9A* gene have not yet been identified in the mouse *Pde9a* gene. All mouse PDE9A clones registered in the GenBank (GenBank accession number AF031147, BC061163, AF068247, AK030509, NM_008804, AK168143, BC046915 (partial), and BC038675 (partial)) are derived from PDE9A2, which lacks the exon corresponding to exon 5 of the human *PDE9A* gene.

11.4 ENZYMATIC PROFILES AND SENSITIVITY TO INHIBITORS

The characteristics of recombinant PDE9A have been reported.[8,9,25,26] Human PDE9A with a $5'$ FLAG tag produced in transfected Sf9 cells shows ~1400-fold selectivity for cGMP versus cAMP, and the affinity of PDE9A for cGMP is high relative to cAMP, with a K_m value of 0.17 μM for cGMP compared to a K_m value of 230 μM for cAMP (Table 11.2).[8] However, there is only one report that describes the enzymatic properties of PDE9A splice variants.[25] The report demonstrates that K_m values of $3'$ V5-6xHis-tagged human PDE9A1 and PDE9A6 (termed PDE9A5 in the report) proteins produced in transfected HEK293 cells for cGMP are 0.25 and 0.39 μM, respectively, and that V_{max} values of these variants are 0.96 μmol/min/mg protein and 2.55 μmol/min/mg protein, respectively. Thus, the V_{max} values of these variants are slightly different whereas the K_m values are almost equal. PDE9A catalytic domain containing a His-tag and produced in *E. coli* for a crystallographic study has been shown to exhibit a K_m value of 0.14 μM and a V_{max} value of 1.53 μmol/min/mg for cGMP with an IC_{50} value, in the case of IBMX, of approximately 500 μM.[26] These findings agree with the values obtained using full-length PDE9A.[8,9,25]

PDE9A1 activity increases in the presence of Mn^{2+}, Mg^{2+}, and Ca^{2+} at concentrations ranging from 100 μM to 100 mM.[8,25] It maximally hydrolyzes cGMP in the presence of 1–10 mM Mn^{2+}, 10 mM Mg^{2+}, or 10 mM Ca^{2+} with an activity approximately 37-fold, 20-fold, or 17-fold, respectively, that observed in the absence of these metal ions. Another study has shown that PDE9A1 and PDE9A6 are similarly activated by metal ions.[25] As PDE9A6 lacks exons 2 and 5, the study suggests that these two exons are not involved in metal binding. Crystal structure analysis of PDE9A has also reported that four amino acid residues (His-316 in exon 12, His-352 in exon 13, Asp-353 in exon 13, and Asp-462 in exon 16) in human PDE9A1 catalytic domain are associated with metal, such as zinc and magnesium, binding.[26]

PDE9A's sensitivity to typical PDE inhibitors is shown in Table 11.2.[8,9,25,26] SCH51866 (a PDE1/5 dual inhibitor) and sildenafil (a PDE5-selective inhibitor) inhibit recombinant PDE9A with IC_{50} values of 1.6–3.3 and 7–11 μM, respectively. Vardenafil (a potent PDE5 inhibitor) also inhibits PDE9A with an IC_{50} value of 0.58 μM.[27] Zaprinast (which inhibits PDE1, PDE5, and PDE6 enzymes) weakly inhibits PDE9A with an IC_{50} value of 29–46 μM. PDE5 inhibitors that weakly inhibit PDE9A have chemical structures derived from cGMP.

TABLE 11.2
Enzymatic Profiles of PDE9A and IC$_{50}$ Values of PDE Inhibitors of PDE9A Activity

	r_hPDE9A1	r_hPDE9A6[24]	r_mPDE9A2[9]	r_hPDE9A2 cat.[25]
K_m value	(μM)	(μM)	(μM)	(μM)
cGMP	0.17[8], 0.25[24]	0.39	0.07	0.14
cAMP	230[8]	N/A	N/A	181
IC$_{50}$ value	(μM)	(μM)	(μM)	(μM)
IBMX	>100[8,24]	>100	>200	500
Dipyridamole	>100[8]	N/A	>38	N/A
Zaprinast	35[8], 44[24]	46	29	N/A
Rolipram	>100[8]	N/A	>200	N/A
SKF94120	>100[8]	N/A	N/A	N/A
Vinpocetine	>100[8]	N/A	N/A	N/A
EHNA	N/A	N/A	>100	N/A
Enoximone	N/A	N/A	>100	N/A
SCH51866	3.3[24]	2.9	1.55	N/A
Sildenafil	8[24]	11	7	N/A
Vardenafil	0.58[26] (r_hPDE9A)			
Tadalafil	>100[26] (r_hPDE9A)			
cAMP	N/A	N/A	>100	N/A

Note: Recombinant human PDE9A1; r_hPDE9A6, recombinant human PDE9A6; r_mPDE9A2, recombinant mouse PDE9A2; r_hPDE9A2 cat., recombinant human PDE9A2 catalytic domain, r_hPDE9A, recombinant human PDE9A (a variant type was not specified).

By contrast, tadalafil, a potent PDE5 inhibitor that is structurally different from sildenafil, vardenafil, and zaprinast does not inhibit PDE9A (IC$_{50}$ > 100 μM).[27] The IC$_{50}$ values of sildenafil, vardenafil, tadalafil, and zaprinast for PDE5A are 3.5, 0.14, 6.7,[27] and 680 nM,[28] respectively, and, therefore, the selectivity of these compounds for PDE5A is 2000-fold, 4100-fold, >14,800-fold, and 50-fold, respectively, that for PDE9A. Inhibitors of other PDEs, such as 3-isobutyl-1-methylxanthine (IBMX, a nonspecific inhibitor), dipyridamole (a PDE2, 4, 5, 10, and 11 inhibitor), rolipram (a PDE4 inhibitor), SKF94120 (a PDE3 inhibitor), vinpocetine (a partially specific PDE1 inhibitor), *erythro*-9-(2-hydroxy-3-nonyl)-adenine (EHNA, a PDE2 inhibitor), and enoximone (a PDE3 inhibitor) do not inhibit PDE9A significantly. As the K_m value of PDE9A for cAMP is very high, cAMP at a concentration of up to 100 μM does not inhibit the hydrolyzing activity of PDE9A for cGMP. Some data on PDE9A variant sensitivities to typical PDE inhibitors have also been reported.[25] PDE9A1 and PDE9A6 produced from transfected HEK293 cells show similar sensitivity to IBMX, zaprinast, sildenafil, and SCH51866 (Table 11.2).

11.5   CRYSTAL STRUCTURE

The crystal structure of PDE9A catalytic domain at 2.23 Å resolution has been reported in 2004.[26] Details of this structure are described in Chapter 29, where the catalytic domain of PDE9A2 (amino acids 181–506) with His-tag produced in *E. coli* was examined. The catalytic domain of PDE9A2 had a K_m value of 139 nM and a V_{max} of 1.53 μmol/min/mg for cGMP. The IC$_{50}$ value of IBMX, a nonspecific PDE inhibitor, was approximately 500 μM. These values are consistent with those reported in the literature for full-length PDE9A,[8,9,25]

indicating that truncated PDE9A is functional and has an enzymatic activity almost identical to that of full-length PDE9A.

As mentioned above,[8] PDE9A activity can be enhanced in the presence of metal ions, such as Mn^{2+}, Mg^{2+}, and Ca^{2+}. Accordingly, the binding mode of PDE9A2 catalytic domain to metal ions was examined in crystallographic studies. The first metal ion (Zn^{2+}) interacts with His-256, His-292, Asp-293, Asp-402, and two water molecules in an octahedral configuration, whereas the second metal ion (Mn^{2+}) forms an octahedron with Asp-293 and five water molecules.

The structure of PDE9A2–IBMX complex was then compared with that of PDE5A1–IBMX complex and PDE4D2–IBMX complex. The human PDE9A2 catalytic domain is composed of 16 α-helices with a single α-helix in residues 181–206, whereas the corresponding residues in PDE4D2 have two α-helices and a 3_{10}-helix. The α-helix in residues 272–275 of PDE9A2 corresponds to the 3_{10}-helix in PDE4D2. Superposition of PDE9A2 catalytic domain and PDE4D2 in PDE4D2–IBMX complex yielded an rms deviation of 1.5 Å for the Cα atoms of residues 207–495 in PDE9A2 or residues 115–411 in PDE4D2. This suggests that the structure of PDE9A is very similar to that of PDE4D. In contrast, the conformations of PDE9A2 and PDE5A1 catalytic domains are different because they show an rms deviation of 2.8 Å for the Cα atoms of residue 272. In addition, the portion of PDE5A1 at residues 535–567 has two helices, which is different from the corresponding region of PDE9A2 (residues 181–206). The H loop residues 301–316 in PDE9A2 and 661–676 in PDE5A1 exhibit positional differences. Although the M loop residues 425–448 in PDE9A2 are well ordered, the majority of the residues in the corresponding region of PDE5A1 are not traceable.

It had been assumed that PDE9A catalytic site resembles more that of PDE5A than that of PDE4D because both PDE9A and PDE5A hydrolyze cGMP, whereas PDE4D specifically hydrolyzes cAMP. However, analysis of the three-dimensional structures of these three PDEs reported the opposite, i.e., that the structure of PDE9A catalytic domain is more similar to that of PDE4D than that of PDE5A. These findings should help with the development of PDE9A-specific inhibitors using structure-based drug design.

11.6 TISSUE-EXPRESSION PATTERNS IN HUMANS AND MICE

PDE9A mRNA is broadly distributed in many tissues as listed in Table 11.3. Northern blot analysis with a PDE9A probe gives a single band of ~2.0 kb that is relatively abundant in the spleen, small intestine, brain, colon, prostate, kidney, and placenta.[8,15] Moderate or low expression is also observed in the heart, pancreas, lung, testis, thymus, skeletal muscles, ovary, and liver. Another report using a human multiple tissue-expression array (Clontech) has demonstrated that, in addition to the above tissues, many other tissues express PDE9A mRNA at detectable levels.[16] Moreover, expression patterns of PDE9A1–6 and PDE9A8–19 mRNAs have been confirmed in the colon, prostate, spleen, peripheral blood leukocytes, small intestine, thymus, testis, and ovary using RT-PCR analysis.[16] Most PDE9A-variant mRNAs were shown to be present in all these tissues, whereas PDE9A11, PDE9A12, PDE9A14, and PDE9A15 mRNAs were found to be specifically expressed in peripheral blood leukocytes, prostate, ovary, and thymus, respectively. Tissue distribution analysis of PDE9A1 and PDE9A6 (termed PDE9A5 in earlier literature) mRNAs by quantitative PCR analysis has revealed that these two variants are highly expressed in immune tissues (spleen, lymph node, and thymus) in addition to the prostate and spleen, and are more abundant in $CD4^+$ and $CD8^+$ T cells than in $CD19^+$ B cells, neutrophils, and monocytes.[25] Also, the expression levels of these two variants in the small intestine, brain, kidney, and colon are moderate. As for the expression ratio of PDE9A1 and PDE9A6 in human tissues, several differences have been found. Relative mRNA expression levels of PDE9A1 are reported to

TABLE 11.3
Tissue-Expression Patterns of PDE9A mRNAs in Humans and Mice

Tissue	Human[8]	Human[15]	Human[16]	Mouse[9]
Brain	+++	++	+	+
Lung	+	++	+	+
Liver	N/A	+	+	+
Kidney	N/A	+++	++	++++
Spleen	++++	+++	++	−
Small intestine	+++	++	++	N/A
Heart	+	++	+	−
Skeletal muscle	+	++	+	−
Thymus	++	+	+	N/A
Testis	+	+	+	−
Placenta	N/A	+++	+	N/A
Prostate	N/A	++++	+++	N/A
Colon	N/A	++++	+++	N/A
Jejunum	N/A	N/A	++	N/A
Pancreas	N/A	++	+	N/A
Ovary	N/A	+	+	N/A

−, Absent; +, weak; ++, moderate; +++, strong; ++++, strongest signals.

be higher than those of PDE9A6 in the spleen, small intestine, brain, kidney, prostate, colon, lymph node, thymus, CD4$^+$ T cells, and CD8$^+$ T cells. By contrast, in the lung, heart, neutrophils, and monocytes, mRNA expression levels of PDE9A1 are relatively lower than those of PDE9A6. In CD19$^+$ B cells and liver, relative mRNA expression levels of both the variants are similar. Thus, mRNA expression of PDE9A splice variants diverges among cell types.

In mice, Northern blot analysis has demonstrated that PDE9A mRNA is abundant in the kidney and that PDE9A mRNA expression levels in the liver, lung, and brain are low (Table 11.3).[9] Mouse PDE9A mRNA, which is the same size as a human PDE9A mRNA (~2.0 kb), has, according to the literature, not been detected in the heart, spleen, skeletal muscle, and testis. In addition, expression patterns of mouse PDE9A mRNA in other tested tissues have been reported to be different from that of humans.[8] However, detailed information on expression patterns of mouse PDE9A mRNA are needed to discuss species difference in PDE9A tissue distribution.

11.7 *IN SITU* HYBRIDIZATION ANALYSIS OF PDE9A IN RAT BRAIN

Two groups have reported detailed expression patterns of PDE9A mRNA in rat brain by in situ hybridization analysis.[12,13,29] Detailed methods for in situ hybridization of PDE9A mRNA in rat brain using digoxigenin-labeled riboprobes are described in the literature.[30] In the olfactory bulb, strong in situ hybridization signals can be detected in the anterior olfactory nucleus. In the cortex, cells in layers II–VI express PDE9A mRNA, and most regions of the basal ganglia and basal forebrain contain PDE9A mRNA. Expression of PDE9A mRNA has also been seen in the thalamus and hypothalamus; however, the level of this expression is relatively low compared to that in other brain regions. PDE9A mRNA is also expressed in all regions of the cerebellum, especially in Purkinje cells where PDE9A

mRNA expression is highly abundant. Thus, PDE9A mRNA is widely distributed in most regions of the rat brain whereas other cGMP-hydrolyzing PDEs such as PDE2A and PDE5A are expressed in specific regions of the brain.[12,13]

To determine which cell types express PDE9A mRNA in the brain, double staining of PDE9A antisense probe with anti–neuronal nuclei (NeuN) or anti–glial fibrillary acidic protein (GFAP) antibody was performed. PDE9A mRNAs have been shown to be predominantly located in NeuN positive cells (neurons) of the brain.[13]

Expression patterns of PDE9A mRNA during development from embryonic stage (E15) until postnatal day 21 (P21) have also been reported.[29] PDE9A mRNA expression in brain regions does not seem to be altered dramatically during postnatal development. In contrast, PDE2A and PDE5A mRNA expression levels change during postnatal development, especially in the cerebellum.[29,31] PDE2A mRNA is expressed in the molecular layer, Purkinje cell layer, and internal granule layer at P21, P10–21, and P21, respectively. PDE5A mRNA expression increases dramatically after birth in the cerebellum, hippocampus, cortex, and caudate putamen.

The brain expresses multiple cGMP-hydrolyzing PDEs, paranthesis (PDE1A, PDE1B, PDE1C, PDE2A, PDE3, PDE5A, PDE9A, PDE10A, and PDE11A).[4,5,12,13,28,31,42] Among these PDEs, PDE9A hydrolyzes cGMP with the highest affinity (K_m, 0.07–0.17 μM). The K_m values of other cGMP-hydrolyzing PDEs are as follows: K_m, 1–5 μM for PDE1, 15–30 μM for PDE2, 0.1–0.8 μm for PDE3, 1–5 μM for PDE5, 3–7 μM for PDE10, and 1 μM for PDE11. Therefore, each PDE family is likely to play a specific role in intracellular signal transduction. Considering the high affinity of PDE9A for cGMP, PDE9A may be involved in maintaining basal levels of cGMP in neuronal cells. Moreover, as PDE9A variants show unique subcellular localization (as described in Section 11.8), PDE9A might participate in regulating compartmentation of cGMP signaling in cells.

11.8   SUBCELLULAR LOCALIZATION OF PDE9A VARIANTS

Subcellular localization of the PDE9A splice variants PDE9A1 and PDE9A6 has been investigated using recombinant proteins produced in HEK293 cells.[25] Although PDE9A6 was termed PDE9A5 in a report published in 2003,[25] it was designated as PDE9A6 in another report published the same year,[16] as shown in Table 11.1. It has been reported that recombinant PDE9A6 protein is localized in the cytoplasm whereas recombinant PDE9A1 protein is observed only in the nucleus.[25] PDE9A1 is found to have a pat7 potential nuclear localization signal ([105]PLRD[109]R[110]RV), which is encoded by exon 5. The PDE9A1 mutant with EE or AA substitutions at [109]R[110]R is localized in the cytoplasm of transfected HEK293 cells, suggesting that the pat7 nuclear localization signal is involved in the nuclear localization of PDE9A1. Thus, the pat7 potential nuclear localization signal in exon 5 of PDE9A1 causes nuclear localization of recombinant proteins. This concept suggests that PDE9A variants such as PDE9A16 and PDE9A17, which possess exon 5 encoding the PLRDRRV sequence in their reading frame, might also localize into the nucleus.

To examine subcellular localization of native PDE9A protein, Western blot analysis was performed using affinity-purified anti-human PDE9A antibody raised against (C)RERSRDVKNSEGD (Cys residue in parentheses is introduced to produce a conjugate with keyhole-limpet hemocyanin), which is derived from the C-terminal sequence of human PDE9A, and recombinant PDE9A1 protein as positive control. Immunoreactive protein, which has a molecular mass corresponding to recombinant PDE9A1, is present in the nucleus of blood CD4[+] T cells, where PDE9A mRNA is highly expressed.[25] By contrast, immunoreactive protein corresponding to PDE9A1 has not been detected in the cytoplasm of T cells.[25] Thus, the presence of native PDE9A protein in human tissues has been suggested for the first

time and its subcellular localization has been shown to be the same as that of recombinant PDE9A1. PDE9A1 also contains the G2 myristoylation site, and by metabolic labeling, Wang et al.[25] has demonstrated that PDE9A1 is indeed myristoylated. However, recombinant PDE9A1 protein in the nucleus does not anchor to the nuclear membrane, but is present in the soluble fraction.

Several PDEs and their splice variants have been found in the nucleus[32,33] and perinuclear region.[34–42] Accordingly, compartmentation of cAMP signaling has been discussed in a number of reports. For instance, the PDE4 family has many splice variants that exhibit unique subcellular localization.[36–41] Some of these variants interact with A-kinase anchoring proteins (AKAPs),[40,43] indicating that these PDEs could regulate cAMP signaling in a local space. Regarding cGMP signaling, one PKG subtype (PKG-I α) binds to GKAP42 (42-kDa cGMP-dependent protein kinase anchoring protein) and anchors to the perinuclear region.[44] PKG, activated by intracellular cGMP, is known to regulate transcription of several genes.[45,46] PDE9A1, which is localized in the nucleus, might be associated with compartmentation of cGMP signaling, and, therefore, regulate nuclear signaling via cGMP. As some PDE9A splice variants exhibit unique tissue expression, unique combination of PDE9A variants occurs in each tissue. Therefore, other PDE9A variants as well as PDE9A1 and PDE9A6 are likely to show unique subcellular localization, and the biological roles of PDE9A in cell signal transduction are expected to be divergent among cells.

11.9 PDE9A INHIBITORS

Recently, a PDE9A-specific inhibitor BAY 73-6691 (1-(2-chlorophenyl)-6-[(2R)-3,3,3-tri-fluoro-2-methylpropyl]-1,5-dihydro-4H-pyrazolo[3,4-d]pyrimidine-4-one) (Figure 11.4) has been reported.[10] BAY 73-6691 inhibits recombinant human and mouse PDE9A with IC_{50} values of 55 and 100 nM, respectively. PDE1C and PDE11A, which are the next most susceptible PDEs, have IC_{50} values of 1400 and 2600 nM, respectively. BAY 73-6691 exhibits very low inhibitory effects on PDE2A, PDE3B, PDE4B, PDE7B, PDE8A, and PDE10A ($IC_{50} \gg 4000$ nM) and shows weak inhibition of PDE5A ($IC_{50} > 4000$ nM). Therefore, the selectivity of this compound for PDE9A is more than 25-fold that for all other PDEs.

The method used for identifying the PDE9A inhibitor involved a unique assay system for monitoring intracellular cGMP signaling.[47] This method involved indirect monitoring of intracellular cGMP levels by converting the cGMP signal into a calcium signal. This was accomplished by using the olfactory cyclic nucleotide-gated channel (CNGA2) and the phosphoprotein aequorin (calcium-sensing protein) stably transfected into CHO cells (cGMP-reporter cell line). This allowed measurement of intracellular cGMP levels via

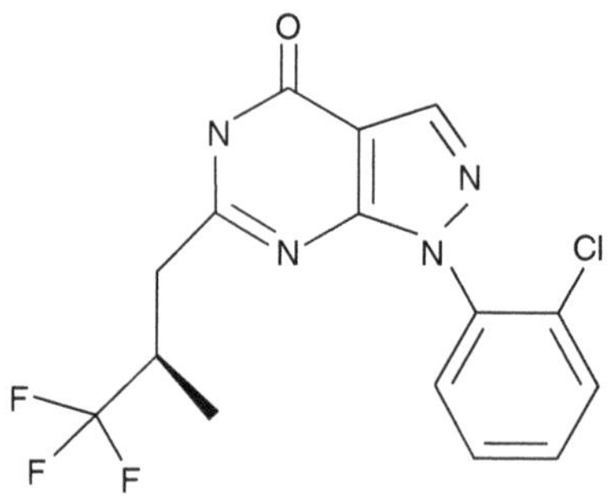

FIGURE 11.4 Chemical structure of the PDE9A inhibitor BAY 73-6691.

aequorin luminescence in response to calcium elevation induced by CNGA2 activation. The cGMP-reporter cell line was transfected with PDE9A cDNA, and the cellular activity and cell permeability of PDE9A inhibitors were measured via aequorin luminescence. BAY 73-6691 was selected as a compound that increases aequorin luminescence in the presence of compounds that activate a soluble (or nitric oxide–sensitive) guanylyl cyclase. This method is very useful as an ultrahigh throughput screen for the discovery of novel compounds that inhibit PDE9A activity.

PDE9A mRNA has been detected in many regions of the brain as described above, suggesting that PDE9A might regulate neuronal cGMP-mediated signaling. In addition, it has been reported that nitric oxide–cGMP signaling in neurons is involved in the process of learning memory and that PDE9A inhibitors improve long-term potentiation in several animal models of cognition.[48]

11.10 CONCLUSIONS

In this chapter we reported the molecular profiles of PDE9As and their tissue distribution. Currently, 21 splice variants of human PDE9A have been reported at mRNA level; however, only few splice variants have been confirmed as complete mRNAs coding for the predicted open reading frames. Cellular localization of PDE9A has not been reported except in the brain and immune tissues. To elucidate the roles of PDE9A in each tissue, PDE9A-expressing cells should be identified and detailed analyses of PDE9A splice variants should be performed. Synthesized PDE9A inhibitors have been patented and a report describing the *in vivo* functions of PDE9A using specific inhibitors has just been published in the patent. It is therefore expected that experimental and clinical evidence using PDE9A-deficient mice and PDE9A-specific inhibitors will elucidate the physiological roles of PDE9A and its association with diseases. However, such active-site-directed inhibitors will inhibit the entire PDE9 family and thus it will be important to identify the roles of specific isoforms in key cell types.

REFERENCES

1. Francis, S.H., and J.D. Corbin. 1999. Cyclic nucleotide-dependent protein kinases: Intracellular receptors for cAMP and cGMP action. *Crit Rev Clin Lab Sci* 36:275.
2. Beavo, J.A., and L.L. Brunton. 2002. Cyclic nucleotide research—still expanding after half a century. *Nat Rev Mol Cell Biol* 3:710.
3. Soderling, S.H., and J.A. Beavo. 2000. Regulation of cAMP and cGMP signaling: New phosphodiesterases and new functions. *Curr Opin Cell Biol* 12:174.
4. Beavo, J.A. 1995. Cyclic nucleotide phosphodiesterases: Functional implications of multiple isoforms. *Physiol Rev* 75:725.
5. Francis, S.H., I.V. Turko, and J.D. Corbin. 2001. Cyclic nucleotide phosphodiesterases: Relating structure and function. *Prog Nucleic Acid Res Mol Biol* 65:1.
6. Aravind, L., and C.P. Ponting. 1997. The GAF domain: An evolutionary link between diverse phototransducing proteins. *Trends Biochem Sci* 22:458.
7. Yuasa, K., et al. 2001. Genomic organization of the human phosphodiesterase PDE11A gene: Evolutionary relatedness with other PDEs containing GAF domains. *Eur J Biochem* 268:168.
8. Fisher, D.A., et al. 1998. Isolation and characterization of PDE9A, a novel human cGMP-specific phosphodiesterase. *J Biol Chem* 273:15559.
9. Soderling, S.H., S.J. Bayuga, and J.A. Beavo. 1998. Identification and characterization of a novel family of cyclic nucleotide phosphodiesterases. *J Biol Chem* 273:15553.
10. Wunder, F., et al. 2005. Characterization of the first potent and selective PDE9 inhibitor using a cGMP reporter cell line. *Mol Pharmacol* 68:1775.

11. Aravind, L., and E.V. Koonin. 1998. The HD domain defines a new superfamily of metal-dependent phosphohydrolases. *Trends Biochem Sci* 23:469.

12. Andreeva, S.G., et al. 2001. Expression of cGMP-specific phosphodiesterase 9A mRNA in the rat brain. *J Neurosci* 21:9068.

13. van Staveren, W.C., et al. 2002. Cloning and localization of the cGMP-specific phosphodiesterase type 9 in the rat brain. *J Neurocytol* 31:729.

14. Thomason, P.A., et al. 1998. An intersection of the cAMP/PKA and two-component signal transduction systems in *Dictyostelium*. *EMBO J* 17:2838.

15. Guipponi, M., et al. 1998. Identification and characterization of a novel cyclic nucleotide phosphodiesterase gene (PDE9A) that maps to 21q22.3: Alternative splicing of mRNA transcripts, genomic structure and sequence. *Hum Genet* 103:386.

16. Rentero, C., A. Monfort, and P. Puigdomenech. 2003. Identification and distribution of different mRNA variants produced by differential splicing in the human phosphodiesterase 9A gene. *Biochem Biophys Res Commun* 301:686.

17. Bonne-Tamir, B., et al. 1996. Linkage of congenital recessive deafness (gene DFNB10) to chromosome 21q22.3. *Am J Hum Genet* 58:1254.

18. Veske, A., et al. 1996. Autosomal recessive nonsyndromic deafness locus (DFNB8) maps on chromosome 21q22 in a large consanguineous kindred from Pakistan. *Hum Mol Genet* 5:165.

19. Straub, R.E., et al. 1994. A possible vulnerability locus for bipolar affective disorder on chromosome 21q22.3. *Nat Genet* 8:291.

20. Belmaker, R.H., J. Zohar, and R.P. Ebstein. 1980. Cyclic nucleotides in mental disorder. *Adv Cyclic Nucleotide Res* 12:187.

21. Ebstein, R.P., et al. 1988. Lithium modulation of second messenger signal amplification in man: Inhibition of phosphatidylinositol-specific phospholipase C and adenylate cyclase activity. *Psychiatry Res* 24:45.

22. Mathews, R., et al. 1997. Increased G alpha q/11 immunoreactivity in postmortem occipital cortex from patients with bipolar affective disorder. *Biol Psychiatry* 41:649.

23. Karlsson, J.O., et al. 1990. Cyclic guanosine monophosphate metabolism in human amnion cells trisomic for chromosome 21. *Biol Neonate* 57:343.

24. Antonarakis, S.E., et al. 2004. Chromosome 21 and Down syndrome: From genomics to pathophysiology. *Nat Rev Genet* 5:725.

25. Wang, P., et al. 2003. Identification and characterization of a new human type 9 cGMP-specific phosphodiesterase splice variant (PDE9A5). Differential tissue distribution and subcellular localization of PDE9A variants. *Gene* 314:15.

26. Huai, Q., et al. 2004. Crystal structure of phosphodiesterase 9 shows orientation variation of inhibitor 3-isobutyl-1-methylxanthine binding. *Proc Natl Acad Sci USA* 101:9624.

27. Gbekor, E., et al. 2002. Selectivity of sildenafil and other phosphodiesterase type 5 (PDE5) inhibitors against all human phosphodiesterase families. *Eur Urol* 1 (Suppl 1):63.

28. Yanaka, N., et al. 1998. Expression, structure and chromosomal localization of the human cGMP-binding cGMP-specific phosphodiesterase PDE5A gene. *Eur J Biochem* 255:391.

29. van Staveren, W.C., et al. 2003. mRNA expression patterns of the cGMP-hydrolyzing phosphodiesterases types 2, 5, and 9 during development of the rat brain. *J Comp Neurol* 467:566.

30. van Staveren, W.C. and M. Markerink-van Ittersum. 2005. Localization of cyclic guanosine 3',5'-monophosphate-hydrolyzing phosphodiesterase type 9 in rat brain by nonradioactive in situ hybridization. *Methods Mol Biol* 307:75.

31. Kotera, J., et al. 1997. Expression of rat cGMP-binding cGMP-specific phosphodiesterase mRNA in Purkinje cell layers during postnatal neuronal development. *Eur J Biochem* 249:434.

32. Lupidi, G., et al. 1990. 3'–5' cAMP phosphodiesterase in pig liver nuclei. *Ital J Biochem* 39:30.

33. Lugnier, C., et al. 1999. Characterization of cyclic nucleotide phosphodiesterase isoforms associated to isolated cardiac nuclei. *Biochim Biophys Acta* 1472:431.

34. Benelli, C., et al. 1988. Changes in low-K_m cAMP phosphodiesterase activity in liver Golgi fractions from hyper- and hypoinsulinemic rats. *Diabetes* 37:717.

35. Shakur, Y., et al. 1995. Identification and characterization of the type-IVA cyclic AMP-specific phosphodiesterase RD1 as a membrane-bound protein expressed in cerebellum. *Biochem J* 306:801.

36. Pooley, L., et al. 1997. Intracellular localization of the PDE4A cAMP-specific phosphodiesterase splice variant RD1 (RNPDE4A1A) in stably transfected human thyroid carcinoma FTC cell lines. *Biochem J* 321:177.

37. Jin, S.L., et al. 1998. Subcellular localization of rolipram-sensitive, cAMP-specific phosphodiesterases. Differential targeting and activation of the splicing variants derived from the PDE4D gene. *J Biol Chem* 273:19672.

38. Geoffroy, V., et al. 1999. Activation of a cGMP-stimulated cAMP phosphodiesterase by protein kinase C in a liver Golgi-endosomal fraction. *Eur J Biochem* 259:892.

39. Taskén, K.A., et al. 2001. Phosphodiesterase 4D and protein kinase a type II constitute a signaling unit in the centrosomal area. *J Biol Chem* 276:21999.

40. Dodge, K.L., et al. 2001. mAKAP assembles a protein kinase A/PDE4 phosphodiesterase cAMP signaling module. *EMBO J* 20:1921.

41. Verde, I., et al. 2001. Myomegalin is a novel protein of the Golgi/centrosome that interacts with a cyclic nucleotide phosphodiesterase. *J Biol Chem* 276:11189.

42. Kotera, J., et al. 2004. Subcellular localization of cyclic nucleotide phosphodiesterase type 10A variants, and alteration of the localization by cAMP-dependent protein kinase-dependent phosphorylation. *J Biol Chem* 279:4366.

43. Asirvatham, A.L., et al. 2004. A-kinase anchoring proteins interact with phosphodiesterases in T lymphocyte cell lines. *J Immunol* 173:4806.

44. Yuasa, K., K. Omori, and N. Yanaka. 2000. Binding and phosphorylation of a novel male germ cell-specific cGMP-dependent protein kinase-anchoring protein by cGMP-dependent protein kinase Iα. *J Biol Chem* 275:4897.

45. Gudi, T., et al. 1996. Regulation of gene expression by cGMP-dependent protein kinase. Transactivation of the c-fos promoter. *J Biol Chem* 271:4597.

46. Chan, S.H., et al. 2004. Nitric oxide regulates c-fos expression in nucleus tractus solitarii induced by baroreceptor activation via cGMP-dependent protein kinase and cAMP response element-binding protein phosphorylation. *Mol Pharmacol* 65:319.

47. Wunder, F., et al. 2005. A cell-based cGMP assay useful for ultra-high-throughput screening and identification of modulators of the nitric oxide/cGMP pathway. *Anal Biochem* 339:104.

48. Hendrix, M. 2005. Selective inhibitors of cGMP phosphodiesterases as procognitive agents. *BMC Pharmacol* 5 (Suppl 1):S5.

12 PDE10A: A Striatum-Enriched, Dual-Substrate Phosphodiesterase

Christine A. Strick, Christopher J. Schmidt,
and Frank S. Menniti

CONTENTS

12.1 Introduction 237
12.2 Identification and Characterization of a Novel PDE Family 238
    12.2.1 Cloning of PDE10A..238
    12.2.2 PDE10A Primary Structure and Splice Variants 238
    12.2.3 Transcriptional Start Site...239
    12.2.4 Genomic Organization of PDE10A 241
12.3 PDE10A Localization 241
12.4 Activity and Function of PDE10A 244
    12.4.1 Enzymatic and Kinetic Properties...244
    12.4.2 Regulation of PDE10A 245
    12.4.3 Physiological Regulation of Cyclic Nucleotides....................248
    12.4.4 Physiological Functions and Therapeutic Implications..........249
12.5 Summary 251
Acknowledgments 251
References 252

12.1 INTRODUCTION

Cyclic nucleotide signaling is an integral part of synaptic communication in the mammalian central nervous system (CNS). Research to date has been focused primarily on the role of cAMP signaling; however, elements of the cGMP-signaling cascade are particularly highly expressed in the brain. In addition, phosphodiesterases (PDEs) are highly expressed in the CNS and, in fact, the highest expression of a number of different PDEs occurs in brain. Despite this, with the exception of members of the PDE4 family, there has been relatively little investigation of the role of PDEs in shaping cyclic nucleotide responses in the CNS. In this regard, PDE10A is particularly striking in that it is expressed almost exclusively in the CNS. Further in the chapter (Section 12.2), we review our current understanding of the structure, localization, and function of this enzyme.

12.2 IDENTIFICATION AND CHARACTERIZATION OF A NOVEL PDE FAMILY

12.2.1 CLONING OF PDE10A

The discovery of a new PDE, PDE10A, was reported simultaneously in mid-1999 by three laboratories, each of which used a bioinformatics approach to identify this novel PDE family. Soderling et al.[1] identified a murine EST clone (ID 760844) that contained sequences similar to the putative GAF- or cGMP-binding domains of PDEs 2, 5, and 6. Both Loughney et al.[2] and Fujishige et al.[3] searched the expressed sequence tag (EST) database against known PDEs and identified a novel human EST (W04835) that contained the sequence motif common to the catalytic region of all known PDEs. The original EST sequence information was used to generate a full-length cDNA by screening human fetal brain[2] or fetal lung[3] cDNA libraries or mouse testis cDNA.[1]

12.2.2 PDE10A PRIMARY STRUCTURE AND SPLICE VARIANTS

The nucleotide sequences of the first two published human PDE10A cDNAs code for identical molecules, and define PDE10A1, a protein of 779 amino acids with a predicted molecular mass of 88,412 Da. The deduced amino acid sequence of human PDE10A1 is shown in Figure 12.1. Both mouse and rat PDE10A,[4] which was published shortly after the original mouse and human publications and was also cloned in our laboratory,[5] are highly

```
                          *
MRIEERKSQH LTGLTDEKVK AYLSLHPQVL DEFVSESVSA ETVEKWLKRK  50

NNKSEDESAP KEVSRYQDTN MQGVVYELNS YIEQRLDTGG DNQLLLYELS 100

SIIKIATKAD GFALYFLGEC NNSLCIFTPP GIKEGKPRLI PAGPITQGTT 150

VSAYVAKSRK TLLVEDILGD ERFPRGTGLE SGTRIQSVLC LPIVTAIGDL 200

IGILELYRHW GKEAFCLSHQ EVATANLAWA SVAIHQVQVC RGLAKQTELN 250
                    Regulatory (GAF)
DFLLDVSKTY FDNIVAIDSL LEHIMIYAKN LVNADRCALF QVDHKNKELY 300

SDLFDIGEEK EGKPVFKKTK EIRFSIEKGI AGQVARTGEV LNIPDAYADP 350

RFNREVDLYT GYTTRNILCM PIVSRGSVIG VVQMVNKISG SAFSKTDENN 400

FKMFAVFCAL ALHCANMYHR IRHSECIYRV TMEKLSYHSI CTSEEWQGLM 450

QFTLPVRLCK EIELFHFDIG PFENMWPGIF VYMVHRSCGT SCFELEKLCR 500

FIMSVKKNYR RVPYHNWKHA VTVAHCMYAI LQNNHTLFTD LERKGLLIAC 550

LCHDLDHRGF SNSYLQKFDH PLAALYSTST MEQHHFSQTV SILQLEGHNI 600
                  Catalytic
FSTLSSSEYE QVLEIIRKAI IATDLALYFG NRKQLEEMYQ TGSLNLNNQS 650

HRDRVIGLMM TACDLCSVTK LWPVTKLTAN DIYAEFWAEG DEMKKLGIQP 700

IPMMDRDKKD EVPQGQLGFY NAVAIPCYTT LTQILPPTEP LLKACRDNLS 750

QWEKVIRGEE TATWISSPSV AQKAAASED                        779
```

FIGURE 12.1 The deduced amino acid sequence of human PDE10A1. The (*) marks the start of residues common to most PDE10A splice variants. The regulatory or GAF domain is underlined and labeled in the N-terminal portion of the protein. The catalytic domain is underlined and labeled in the C-terminal portion of the protein. Sequence from AB020593, AF127479.

homologous to the human protein, with 95% amino acid identity overall, and 98% within the catalytic region. The catalytic domain of PDE10A is contained within the C-terminal portion of the protein and is most similar (40%–47% amino acid identity) to that of PDEs 2A, 5A, 6A–C, and 11A.[6] The N-terminal portion of PDE10A protein contains sequences homologous to the regulatory or GAF domains present in this same subset of PDEs. The initially identified mouse, rat, and human sequences diverge at the extreme N-terminus, representing alternately spliced forms of the protein.

Numerous splice variants derived from the single PDE10A gene have been identified in human, mouse, and rat. These variants define at least 15 unique N-terminal amino acid sequences. Fujishige et al.[7] also found two unique C-terminal sequences in human PDE10A cDNAs; this same group identified only a single C-terminus for rat PDE10A variants.[4] PDE10A splice variants are summarized in Table 12.1. Splice variants 1–14 are numbered as described in GenBank or literature references; other splice variants that have also been documented but not named are designated PDE10A15–18. There are two published rat PDE10A5 sequences that differ slightly but are obviously related. These are listed as PDE10A5[4] and PDE10A5′.[8] The N-terminus of PDE10A6 was reported as a truncated sequence,[4] but nucleotides that encode this sequence are also found in the 5′ untranslated region of PDE10A13.[8] Translation of PDE10A13 is proposed to start at the first downstream Met, which would be at amino acid 71 within the PDE10A1 sequence. If this is the case for PDE10A6, then its N-terminus may be identical to that of PDE10A13. This possibility is also listed in Table 12.1. None of these variants (with the exception of rat PDE10A14, whose predicted N-terminus starts within GAFA) is altered in either the regulatory domain or the catalytic domain, so the proteins encoded by these splice variants would not be expected to differ in their catalytic properties. In support of this, Kotera et al.[9] expressed PDE10A1 and PDE10A2 in COS-7 cells and found that K_m values for cAMP and cGMP were very similar for both isoforms.

PDE10A2 is the predominant form in some rat and human tissues, including brain[4,9] and has also been identified in mouse (AY360383). Interestingly, the second most highly expressed human transcript, PDE10A1, which results from insertion of a 128-base fragment into the N-terminal region of PDE10A2, was not detected in rat brain, and the second most highly expressed rat transcript, PDE10A3, contains a distinct N-terminal amino acid sequence not found in humans.[4] These three N-terminal splice variants are reported to be differentially localized within the cell and other PDE10A splice variants display differential patterns of expression within the brain (see Section 12.3).

12.2.3 TRANSCRIPTIONAL START SITE

Fujishige et al.[2] used 5′ rapid amplification of cDNA ends (RACE) and human caudate nucleus RNA to locate a major transcriptional start site for PDE10A at a guanosine residue 179 bases upstream of the initiating ATG codon of PDE10A2. The genomic sequence upstream from the transcriptional start site is highly GC rich and contains a GC box, but not a typical TATA or CAAT box motif, at least within 779 bases upstream of the initiator ATG. Although these are features common to the regulatory region of constitutively expressed housekeeping genes,[10] multiple elements that suggest a more dynamic regulation are also found in this region, including myeloid zinc-finger-1 (MZF-1), AP-2, Sp1, and GATA-1 sites. It is not known if some combination of these elements or some yet to be determined regulatory element is responsible for the distinctive tissue-specific expression of PDE10A. Other rat PDE10A splice variants, including those whose expression is increased in response to synaptic activity in hippocampal neurons,[8] originate from a unique promoter far upstream from that for PDE10A2. As the exact transcription start sites for these variants have not been identified, regulatory elements within this promoter that might respond to synaptic activity have not been identified.

TABLE 12.1
Known Splice Variants of PDE10A

Variant	Species	N-Terminus	N-Terminus Joins 10A1 Sequence at AA	C-Terminus Diverges from 10A1 at AA	Reference
10A1	Human	MRIEERKSQHLTG	(14) LTDEK.	(717) LGFYN...AASED	AB020593
					AF127479
10A2	Human	MEDGPSNNASCFRRLTECFLSPS		(717) LGFYN...AASED	AB026816
	Rat		(14) LTDEK.	(769) ATSKS...RKVDD	AF127480
	Mouse			(769) GPAPS...VKVED	AB027155
					AY360383
10A3	Rat	MSNDSPEGAVGSCNATG	(14) LTDEK.	(769) ATSKS...RKVDD	AB027156
10A4	Rat	MDAYEGWQVMDWKPPETAAC	(14) LTDEK.	(769) ATSKS...RKVDD	4
10A5	Rat	MSKKRKALEGGGGGGGEPQLPEEEPTAWFGRSSEEPG	(14) LTDEK.	(769) ATSKS...RKVDD	4
10A5′	Rat	MSKKRKALEGGGGGGGEPQLPEEEPTAWFGRSSEEPAG CLPITFKGGSKGPALLALRNRTDSRGQ	(14) LTDEK.	(769) ATSKS...RKVDD	8
10A6	Rat	*TGNEAERSFVLEDTLAGM	(14) LTDEK. (72) QGVVY.	(769) ATSKS...RKVDD	4
10A7	Human	MNPQSFENY	(57) ESAPK.	(717) LGFYN...AASED	9
10A8	Human	MRIEERKSQHLTG	(14) LTDEK.	(717) FLHGG...PRSLF	9
10A9	Human	MEDGPSNNASCFRRLTECFLSPS	(14) LTDEK.	(717) FLHGG...PRSLF	9
10A10	Human	MNPQSFENY	(14) LTDEK.	(717) FLHGG...PRSLF	9
10A11	Rat	MSKKRKALEGGGGGGGEPQLPEEEPTAWFGGSSEE PAGCLPITFKGGSKGPALLALRNRTDSRGQMSNDS PEGAVGSCNATG	(14) LTDEK.	(769) ATSKS...RKVDD	AY462091
10A12	Rat	MSKKRKALEGGGGGGGEPQLPEEEPTAWFGGSSEEPAG CLPITFKGGSKGPALLALRNRTDSRGQMSNDSPE GAVGSCNATGSTGSTGELGKEFHTPPRRKSASDSRL ALCMG	(14) LTDEK.	(769) ATSKS...RKVDD	AY462092
10A13	Rat	M	(72) QGVVY.	(769) ATSKS...RKVDD	AY462093
10A14	Rat	M	(133) KEGQP.	(769) ATSKS...RKVDD	AY462094
10A15	Mouse	MEKLYG	(14) LTDEK.	(769) GPAPS...VKVED	AF110507
10A16	Rat	MSG	(14) LTDEK.	(769) ATSKS...RKVDD	5
10A17	Mouse	MELGGFQOAQLCFGFPSPSATTQG	(14) LTDEK.	(769) GPAPS...VKVED	ACI04323
10A18	Mouse	MSNDSTEGTVGSCNATG	(14) LTDEK.	(769) GPAPS...VKVED	AK039249

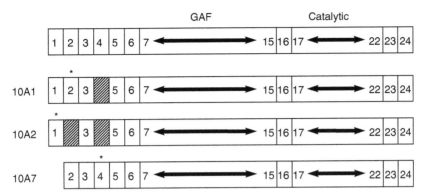

FIGURE 12.2 Exon structure of human PDE10A showing generation of major 5' splice variants by exon skipping. The (*) indicates translational start.

12.2.4 GENOMIC ORGANIZATION OF PDE10A

Human PDE10A maps to chromosome 6q26–27.[2,3] The human genomic sequence spans more than 200 kb and includes at least 24 exons.[7] Fujishige et al. have proposed that the three distinct N-terminal variants of human PDE10A are generated through an exon skipping process (Figure 12.2), and the two C-termini result from an alteration in the usage of the 5' acceptor splice site in exon 23. The regulatory region of PDE10A is encoded in exons 7–15, whereas the catalytic domain is contained within exons 17–23. The mapping of additional rat splice variants[4,8] has expanded the PDE10A genomic region on rat chromosome 1 upstream by approximately another 300 kb from exons 1–24 that are equivalent to those in the human gene and has introduced additional complexity into the generation of rat splice variants. A portion of the unique sequence at the 5' end in rat-variant PDE10A11 can be found in the human genome about 90 kb upstream of the 5' end of PDE10A2 (P. Zagouras, personal communication, 2005), suggesting that the human PDE10A locus may need to be expanded as well, although to date no human PDE10A cDNAs containing these sequences have been reported.

Although the GAF-containing PDEs (PDE2A, 5A, 6A–C, 10A, and 11A) form a subgroup based on the degree of amino acid identity within their catalytic domains, their exon organization suggests that all these genes do not share a common ancestor gene. Although the intron–exon organization of the genes for PDEs 5, 6, and 11 is closely related, the exon organizations for PDE2A and PDE10A are completely different from this group and from each other, suggesting that those genes were established separately.[11]

12.3 PDE10A LOCALIZATION

Initial surveys using Northern blot and dot blot analyses indicated that PDE10A mRNA is highly expressed only in brain and testis.[2–4] In human brain, Fujishige et al.[3] noted high expression in two particular regions, the caudate nucleus and putamen. This group subsequently showed using in situ hybridization that PDE10A mRNA is localized within the rodent brain to the caudate nucleus and olfactory tubercle. These regions are part of a contiguous group of subcortical brain nuclei collectively referred to as the striatum, a name that derives from the striate appearance imparted by bundles of myelinated primary motor axons that traverse this tissue. Quantitative PCR analysis of different microdissected brain regions by Seeger et al.[5] indicated that PDE10A mRNA is expressed in rodent striatum at

TABLE 12.2
PDE10A mRNA Expression (Rat)

Tissue	PDE10A mRNA Expression (Relative to Prefrontal Cortex)
Total brain	1.1
Prefrontal cortex	1
Striatum	13.1
Hippocampus	0.62
Cerebellum	0.54
Cortex	0.89
Thalamus	0.34
Spinal cord	0.18
Liver	0.0004
Lung	0.0024
Testis	0.50
Heart	0.033
Kidney	0.013

Note: PDE10A mRNA levels were determined by quantitative PCR and normalized to expression seen in prefrontal cortex.

levels more than an order of magnitude higher than for any other brain region (Table 12.2). The expression of PDE10A protein in rat brain was mapped by immunohistochemistry and Western blot analysis.[5] Consistent with mRNA expression patterns, PDE10A protein is expressed at high levels in the caudate nucleus, nucleus accumbens, and olfactory tubercle in the rat brain. The expression pattern of PDE10A protein in other mammalian species is essentially the same as in rat, as illustrated for primate in Figure 12.3.

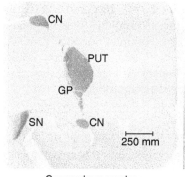
Cynomolgus monkey

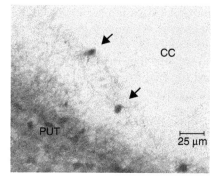
Human

FIGURE 12.3 (See color insert following page 274.) Expression of PDE10A protein in primate and human brain. Left panel: A coronal section through the cynomolgus monkey brain immunostained with PDE10A antibody (brown reaction product) and counterstained with hematoxylin. Dense staining is observed in caudate nucleus (CN), putamen (PUT), globus pallidus (GP), and substantia nigra (SN). Scale bar = 250 mm. Right panel: A high magnification image at the intersection of the putamen (PUT) and corpus callosum (CC) of human brain. The arrows point to two putative medium spiny neurons expressing PDE10A immunoreactivity throughout the cell bodies and proximal processes. Scale bar = 25 μm.

The expression pattern of PDE10A mRNA and protein indicates that the enzyme is highly expressed in the medium spiny neurons of the striatum, which comprise 90%–95% of the total neuronal population in the striatum.[12] Although the cell bodies of these neurons are relatively small (~12–20 μm), they elaborate extensive dendritic trees that form spheres of diameter as much as 300–500 μm surrounding the cell bodies. The presence of dense PDE10A-like immunoreactivity throughout the parenchyma in these brain regions is consistent with the protein being transported throughout the dendritic arborizations of these neurons. The medium spiny neurons project to the globus pallidus or substantia nigra. The observation of dense PDE10A-like immunoreactivity in both of these regions as well as the fibers of the striatonigral pathway, in the virtual absence of PDE10A mRNA, is consistent with the protein being transported to the axon terminal regions of these neurons.[5] The striatum also contains several small populations of interneurons, which, despite the low numbers, have a profound influence on striatal function.[12] However, PDE10A is not expressed in any of these interneuron populations, based on lack of colocalization of PDE10A-like immunoreactivity with immunological markers for each of the interneuron types.[13] The high level of expression of PDE10A in striatal medium spiny neurons suggests that a principal physiological function of the enzyme is in the regulation of signal transduction in these neurons.

PDE10A protein is also expressed at lower levels in a number of other neuronal populations, consistent with the presence of low levels of PDE10A mRNA. These include neurons in most cortical layers, hippocampal pyramidal neurons, dentate and cerebellar granule neurons.[5] Interestingly, the distribution of the enzyme in these latter neuronal populations appears to differ significantly from that in the medium spiny neurons. In neurons outside the striatum, PDE10A appears to be restricted to neuronal nuclei, based on immunohistochemistry (Figure 12.4). The specificity of immunoreactivity was confirmed by immunoadsorption with recombinant PDE10A as well as the presence of a band of the predicted molecular weight in extracts of brain regions that contain exclusively nuclear expressed PDE10A.

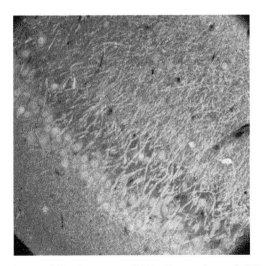

FIGURE 12.4 (See color insert following page 274.) Expression of PDE10A and microtubule-associated protein (MAP2) in rat hippocampal pyramidal neurons. Confocal image of rat hippocampal neurons stained for the neuronal marker MAP2 (green) and PDE10A (red). Overlap of the two markers is in orange. MAP staining is apparent throughout the cell bodies and dendritic trees of these neurons (extending in the northeast direction). PDE10A staining overlaps with MAP2 staining only in the perinuclear region (orange).

The significance of this apparently differential distribution of PDE10A in the medium spiny compared to other neurons is not yet known. However, the mechanism may be related to the expression of different N-terminal splice variants. Kotera et al.[14] report that the major splice variants of PDE10A expressed in brain (PDE10A2 and PDE10A3 [rat] or PDE10A1 [human]) differ in the degree of membrane association when expressed in PC12h cells. We find that in rat striatal tissue, PDE10A is primarily membrane associated,[13] consistent with the major expressed isoform being PDE10A2. A scan of the amino acid sequence of rat PDE10A2 using PSORT, a program which detects subcellular localization signals,[15] predicts a single transmembrane region at amino acids 198–214. This region is within the N-terminal regulatory domain and so would be found in all known PDE10A splice variants. Although this predicted hydrophobic region could account for the association of PDE10A2 protein with membrane fractions, there is no experimental evidence that PDE10A2 protein spans the membrane. In fact, the report that the phosphorylation state of the unique N-terminus of PDE10A2 can alter the membrane association of the protein[14] argues against membrane association through a transmembrane domain. Other PDE10A splice variants may predominate outside the striatum. A subset of variants described by O'Connor et al.[8] (PDE10A3 or 10A6, PDE10A5 or 10A11) are expressed specifically in cerebellum, olfactory bulb, dentate gyrus, areas CA1–CA3 of the hippocampal formation, and cortex. Expression of these variants can be up-regulated in the dentate gyrus in response to long-term potentiation (LTP), whereas the expression of PDE10A2 does not change. It is possible that these splice variants contain motifs that restrict the distribution of the enzyme within these neurons, accounting for the differential subcellular distribution of PDE10A in striatal neurons compared to neurons throughout the rest of the brain.

PDE10A mRNA expression is relatively low in tissues outside the CNS. The highest level of expression of mRNA in peripheral tissues is in testis.[1–4] Quantitative PCR indicates that mRNA in this tissue in rat is ~20-fold lower than in striatum, based on a comparison of total tissue levels (Table 12.2). It appears that, unlike the CNS, expression of PDE10A protein in testis may vary across species. Using monoclonal antibody 24F3.F11, which recognizes an epitope in the conserved N-terminal region of the protein,[5] PDE10A protein is evident in dog testis and to a lesser extent in monkey testis (Figure 12.5). PDE10A containing the 24F3.F11 epitope is below the level of detection in rat and human testis by IHC. These findings do not rule out the possibility that splice variants of PDE10A not detectable by 24F3.F11 are expressed in testis of different species. Expression levels of PDE10A mRNA in other peripheral tissues are an order of magnitude or more lower than in testis (Table 12.2).

12.4 ACTIVITY AND FUNCTION OF PDE10A

12.4.1 ENZYMATIC AND KINETIC PROPERTIES

Characterization of recombinant PDE10A in a number of different laboratories indicates that the enzyme can hydrolyze both cAMP and cGMP.[1–4] The structure of the PDE10A catalytic domain has been determined using x-ray crystallography.[16] The key glutamine residue involved in substrate specificity, Q716, has no apparent restriction for rotation, thus allowing binding of either substrate (Figure 12.6). Lack of restriction of rotation for this conserved glutamine residue has also been proposed to account for dual substrate specificity in other PDEs.[17] The affinity of PDE10A for cAMP is considerably higher than for cGMP (cAMP, $K_m = 0.05$ μM, cGMP, $K_m = 3$ μM), whereas the V_{max} is approximately fourfold greater for cGMP.[1] These characteristics led to the initial hypothesis that PDE10A may be a cAMP-inhibited cGMP PDE.

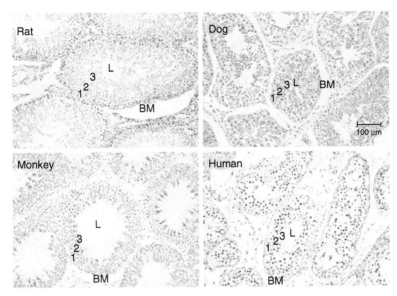

FIGURE 12.5 (See color insert following page 274.) Expression of PDE10A protein in testis of different species. PDE10A-like immunoreactivity appears as diffuse brown immunoperoxidase reaction product in monkey and dog but not in rat and human. Sections are counterstained with hematoxylin. 1, Sertoli cells and spermatogonia; 2, spermatocytes; 3, spermatids; L, lumen; BM, Basement membrane. Scale bar = 100 μm.

The most potent and specific PDE10A inhibitor identified to date is papaverine.[18,19] Papaverine is a benzylisoquinoline derived from the opium poppy that has been in therapeutic use for more than 100 years. The compound, which is a smooth muscle relaxant, was considered a relatively nonselective PDE inhibitor and appears to have a number of other pharmacological activities. However, we find that papaverine is a relatively potent inhibitor of PDE10A (K_i of approximately 30 nM). Furthermore, the potency for inhibition of PDE10A is 10-fold or greater than for inhibition of members of the other 10 PDE families. Dipyridamole ($K_i \approx 1$ μM) and zaprinast ($K_i \approx 22$ μM) also inhibit PDE10A, although with less potency and specificity. Thus, papaverine at present represents the best widely available pharmacological tool to investigate the physiology of the enzyme. The development of more potent and specific PDE10A inhibitors is underway in a number of different laboratories.

12.4.2 REGULATION OF PDE10A

The regulation of PDE10A is just beginning to be investigated. PDE10A is one of the groups of PDEs that contain GAF domains. The GAF acronym[20] derives from the types of proteins in which they were first identified (cGMP-specific and -stimulated PDEs, *Anabaena* adenylate cyclases and *Escherichia coli* FhlA). GAFs have been shown to be allosteric cyclic nucleotide-binding sites in some PDEs. An alignment of amino acids from the most highly conserved regions of GAFA and GAFB domains of representative members of the PDE2, 5, 6, 10, and 11 families is shown in Figure 12.7. Direct binding of cGMP to the GAF domain has been established for PDE6,[21] PDE2,[22] and PDE5,[23–25] and Turko et al.[24] performed mutational analysis to identify key amino acids that support cGMP binding within the highly conserved NKFDE motif in the GAF domain of bovine PDE5A. Several of these residues are missing from GAFA in PDE10A but are present in GAFB. Recently, the crystal structure of the N-terminal regulatory segment of murine PDE2A was solved to reveal that the PDE2A

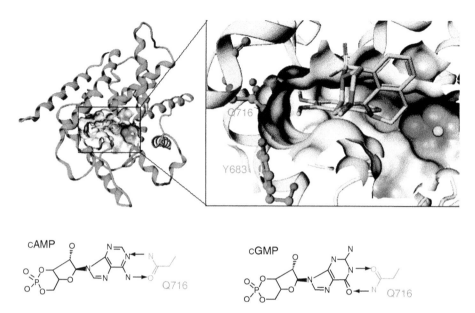

FIGURE 12.6 (See color insert following page 274.) Crystal structure of rat PDE10A catalytic domain. Overall, the catalytic domain of PDE10A folds into a characteristic all-α-helical fold, with a metal ion bound at the interface of three structural subdomains, seen in all the other PDE catalytic domain crystal structures (upper left structure). The substrate-binding site in PDE10A consists of a deep hydrophobic cleft, which is seen here occupied by the dimethoxy quinazolyl moiety of a potent PDE10A inhibitor ($IC_{50} = 40$ nM; Ref. [16]). The binding site, seen in greater detail in the upper right structure, has a conserved glutamine residue (Q716) at one end, and the metal-binding sites at the other. In this crystal structure, Q716 is oriented in such a manner that the amino nitrogen makes hydrogen-bond (H-bond) interactions with the methoxy oxygens of the inhibitor and is an H-bond donor to the hydroxyl of tyrosine (Y683). This orientation for the side chain of Q716 is consistent with a cAMP-binding mode, and is the orientation seen for the corresponding glutamine in the crystal structure of PDE4B–AMP complex.[17] Consistent with the prediction about dual-specificty PDEs,[17] the side chain of Q716 is free to rotate by 180°, so as to adopt a cGMP-binding orientation. In this orientation, the carboxyl oxygen of PDE10A would accept a hydrogen bond from Y683, and the amino nitrogen would interact with the bound water. This orientation of the conserved glutamine has been seen before in the crystal structure of the PDE5A–GMP complex.[17] Both cGMP and cAMP can be docked into the PDE10A-binding pocket with no steric clashes.

GAFA domain serves as a site for dimerization, whereas cGMP binds to a deeply buried site in GAFB.[26] Unexpectedly, the cGMP-binding pocket does not contain the highly conserved NKFDE motif that was assumed to be directly involved in cGMP binding. These residues may, instead, function to stabilize the cyclic nucleotide-binding pocket.[27,28] Martinez et al.[26] proposed a cGMP-binding motif based on residues in the cGMP-binding pocket of the PDE2A GAFB domain that contacts the cyclic nucleotide. Although 10 of 11 of these residues are conserved in the GAFB domain of PDE10A, with some minor shift in spacing, neither Soderling et al.[1] nor Fujishige et al.[3] was able to demonstrate cGMP binding *in vitro* under conditions in which cGMP binding to PDE2A or 5A was easily detected. However, Schultz and colleagues used a chimeric protein containing the GAF domains of PDE10A coupled to a bacterial adenylate cyclase to demonstrate that the PDE10A GAF domains bind to cAMP and that cAMP binding is transduced into altered cyclase activity.[29] These results suggest that cAMP may regulate PDE10A activity through a GAF-mediated allosteric effect, but this remains to be directly demonstrated.

```
      S X(13-18)                              F D X(18-22)
S R C C L L L V S E D N L Q L S C K V I G D - K V - - - L G    PDE2A GAFA
C S V F L L D Q N - - - - - E L V A K V F D G G V V D D - -          GAFB
Y S L F L V C E D S S N D K F L I S R L F D V A E G S T L E    PDE5A GAFA
D F S F S S V F H - - - - - - - - M E C - E E L E - K S S D          GAFB
S L F M Y R T R N G - - I A E L A T R L F N V H K D A V L E    PDE6A GAFA
E I N F Y K V I D - - - - - - Y I L H G - - K E - D I K V            GAFB
A L Y F L G E C N N - - - S L C I F T P P G I K E G K P R L    PDE10AGAFA
A D R C A L F Q V D H K N K E L Y S D L F D I G E E K E G K          GAFB
C S L F L V E G A A A G K K T L V S K F F D V H A G T P L L    PDE11AGAFA
K C S A D A E N S - - - - - - - - F K - - E S M E - K S S -          GAFB


                                    I A X(21)
- - - - - - - - - - - E E V S F P L - T G C L G Q V V E D K K    PDE2A GAFA
- - - - E S - - - - - Y E I R I P A D Q G I A G H V A T T G Q           GAFB
- - E V S N - - - - - N C I R L E W N K G I V G H V A A L G E    PDE5A GAFA
- - T L T R - - - - - E H D A N K I N Y M Y A Q Y V K N T M E           GAFB
D C L V M P - - - - D Q E I V F P L D M G I V G H V A H S K K    PDE6A GAFA
- - I P N P - - - - - P P D H W A L V S G L P A Y V A Q N G L           GAFB
- - I P A G - - - - - P I T Q G T T V S - - - A Y V A K S R K    PDE10AGAFA
P V - - - F K K T - K E I R F S I E K G I A G Q V A R T G E           GAFB
P C S S T E N S N E V Q V P W G K G I I G Y V G E H G E T V    PDE11AGAFA
- - - - - - - - - - - - Y S D W L I N N S I A E L V A S T G L           GAFB


                      Y/N X(2)V D X(2)T X(3)  T X(19)
S I Q L K D L T S E D - - V Q Q L Q S M L G C E L Q A M L C    PDE2A GAFA
I L N I P D A Y A H P L F Y R G V D D S T G F R T R N I L C           GAFB
P L N I K D A Y E D P R F N A E V D Q I T G Y K T Q S I L C    PDE5A GAFA
P L N I P D V S K D K R F P W T T E N T G N V N Q Q C I R S           GAFB
I A N V P N T E E D E H F C D F V D I L T E Y K T K N I L A    PDE6A GAFA
I C N I M N A P A E D F F A F Q K E P L D E S G W M I K N V           GAFB
T L L V E D I L G D E R F P R G T G L E S G T R I Q S V L C    PDE10AGAFA
V L N I P D A Y A D P R F N R E V D L Y T G Y T T R N I L C           GAFB
- - N I P D A Y Q D R R F N D E I D K L T G Y K T K S L L C    PDE11AGAFA
P V N I S D A Y Q D P R F D A E A D Q I S G F H I R S V L C           GAFB


              E/Q              NK/R X(5-14)
V P V I S R A T D Q V V A L A C A F - - - N K L E G - - D L    PDE2A GAFA
F P I K N E N Q E V I G V A E L V - - - - N K I N G - - P W           GAFB
M P I K N H R E E V V G V A Q A I - - - - N K K S G N G G T    PDE5A GAFA
L L C T P I K N G K K N K V I G V C Q L V N K M E E N T G K           GAFB
S P I M N G K D V V A I I M A V - - - - N K V D G - - S H    PDE6A GAFA
L S M P I V N K K E E I V G V A T F Y - - N R K D G - - - -           GAFB
L P I V T A I G D L I G I L E L Y R H W G K K p - - - - - -    PDE10AGAFA
M P I V S R G S - V I G V V Q M V - - - - N K I S G - - E A           GAFB
M P I R S S D G E I I G V A Q A I - - - - N K I P - - - S A    PDE11AGAFA
V P I W N S N H Q I I G V A Q V L - - - - N R L D G E G A P           GAFB


      F X(3) D E
- - - F T D E D E H V I                                        PDE2A GAFA
- - - F S K F D E D L A                                              GAFB
- - - F T E K D E K D F                                        PDE5A GAFA
V K P F N R N D E                                                    GAFB
- - - F T K R D E E I L                                        PDE6A GAFA
- - - F D E M D E T L M                                              GAFB
- - - F C L S H Q E V A                                        PDE10AGAFA
- - - F S K T D E N N F                                              GAFB
- - - F T E D D E K V M                                        PDE11AGAFA
- K P F D D A D Q R L                                                GAFB
```

FIGURE 12.7 GAF domains of PDEs. GAF domains of human PDEs were aligned using DNASTAR (Clustal W) adjusted for maximum alignment of motifs. Conserved amino acids that comprise the proposed cyclic nucleotide binding motif and the GAF motif are shown above the sequences and are given in bold in the figure. Accession numbers for sequences used are PDE2A, U67733; PDE5A, AB001635; PDE6A, M26061; PDE10A, AB020593; PDE11A, AB036704.

In addition to the GAF domain, the primary amino acid sequence of PDE10A1 contains a number of consensus phosphorylation motifs that are identified by Prosite.[30] As phosphorylation has been shown to regulate the activity of many PDEs,[31] the presence of such sites in the PDE10A sequence offer the potential that this PDE may be similarly modulated. Interestingly, the single PKA consensus site predicted in the unique N-terminal region of PDE10A2 can be phosphorylated by PKA *in vitro*.[9] Phosphorylation at this site is accompanied by a shift in the localization of PDE10A2 from the membrane to cytosol[14] in recombinant PDE10A2 expressed in PC12 cells. The authors did not report whether this phosphorylation produced a change in activity. Prosite also predicts a number of consensus PKC and CK2 phosphorylation sites, but none of these has been shown to be utilized.

Regulation of protein levels of PDE10A is also beginning to be elucidated. Hebb et al.[32] report that PDE10A mRNA and protein levels are decreased in striatum obtained postmortem from human Huntington's disease patients. There is also a decrease in striatal expression of PDE10A in two mouse models of Huntington's disease, the R6/1 and R6/2 mice. This decrease is an early event that precedes the development of motor symptoms in these mice. In a subsequent study, it was found that mutant huntingtin, the putative pathogenic agent in Huntington's disease, causes a decrease in transcription initiation rate of the *PDE10A* gene.[33] It is also interesting to note that PDE10A is a substrate for caspase 3 cleavage.[34] Thus, it is possible that in diseased medium spiny neurons, activation of caspase cleavage cascades and degradation of PDE10A protein may also contribute to the early loss of this enzyme in the brains of patients with Huntington's disease. In hippocampus, the expression of some forms of PDE10A is increased in response to the induction of LTP.[8] The mechanisms for transcriptional regulation of PDE10A expression in striatum and hippocampus remain to be determined.

12.4.3 Physiological Regulation of Cyclic Nucleotides

The high level of PDE10A expression in striatal medium spiny neurons suggests that a principal physiological function of the enzyme is in the regulation of cyclic nucleotide signaling within these neurons. Thus, we have begun to investigate the regulation of cyclic nucleotide metabolism by PDE10A in the medium spiny neurons.

In primary cultures of rat striatal neurons, pharmacological activation of guanylyl cyclase by sodium nitroprusside increases cGMP, and PDE10A inhibitors potentiate this increase. Inhibition of PDE10A also increases the magnitude and time course of the elevation of cGMP in response to a physiological stimulus such as activation of *N*-methyl D-aspartate (NMDA) receptors expressed by these neurons.[35] Consistent with this observation in cultured neurons, systemic administration of the PDE10A inhibitor papaverine to mice also causes a rapid and substantial increase in cGMP level in striatum.[19] This increase in striatal cGMP is almost completely attenuated by genetic deletion of neuronal nitric oxide synthase (nNOS). The observation that inhibition of PDE10A causes a robust increase in cellular and tissue levels of cGMP suggests that the enzyme regulates cGMP signaling in the striatal medium spiny neurons.

There is also evidence that PDE10A inhibitors alter cAMP metabolism in the medium spiny neurons. PDE10A inhibition potentiates a forskolin-stimulated increase in cAMP in striatal neurons in culture, although the potency of the inhibitors is approximately fourfold lower than that required to potentiate a sodium nitroprusside–induced increase in cGMP. However, there is little or no effect of PDE10A inhibition on the increase in cAMP in response to stimulation of NMDA receptors in the cultures. There is also no effect of PDE10A inhibition on striatal tissue level of cAMP *in vivo*. This lack of effect of PDE10A inhibition on total tissue levels of cAMP in the cultures and *in vivo* may be because of the fact that there are multiple pools of cAMP in the medium spiny neurons, the bulk of which are unaffected by PDE10A inhibition. In contrast,

using microdialysis of rat striatum to sample the efflux of cyclic nucleotides into the extracellular space, we observe that PDE10A inhibition causes an increase in both cGMP and cAMP.[19] To further examine a role for PDE10A in the regulation of cAMP signaling in striatal neurons, we investigated the effects of PDE10A inhibition on phosphorylation of proteins downstream of the cyclic nucleotide–regulated kinases.

cAMP response element binding (CREB) protein is a transcription factor activated by phosphorylation in a wide variety of cell types in response to a variety of cellular signaling events. In primary cultures of rat striatal neurons, CREB is phosphorylated in response to NMDA-receptor activation. Using automated fluorescence microscopy and image analysis, we observe that the magnitude and duration of the NMDA-induced nuclear phosphoCREB (pCREB) response is enhanced in the presence of PDE10A inhibitors.[35] Selective inhibition of other PDEs does not alter the NMDA-induced increase in pCREB, suggesting a preferential role for PDE10A in the regulation of this response. The increase in CREB phosphorylation is blocked by inhibitors of PKA. However, this response is not affected by inhibition of cGMP formation with the nNOS inhibitor L-NAME. Consistent with these findings, PDE10A inhibition also causes an increase in CREB phosphorylation in mouse striatum *in vivo*.[19] Genetic deletion of nNOS, which almost completely attenuates the PDE10A inhibition-induced increase in striatal cGMP, has no effect on the increase in CREB phosphorylation. Thus, in both striatal neurons in primary culture and *in vivo*, PDE10A inhibition causes an increase in CREB phosphorylation that is not affected by abrogation of cGMP signaling. These data, together with the results obtained with microdialysis, suggest that PDE10A regulates cAMP signaling in the striatal neurons. Thus, PDE10A appears to function as a true dual substrate PDE in the striatal medium spiny neurons. A question that will be interesting to address is whether PDE10A is regulating cGMP and cAMP signaling in distinct compartments as PDE10A is distributed throughout the medium spiny neurons, yet CREB signaling is largely confined to the nucleus.

The role of PDE10A in the regulation of cyclic nucleotide signaling in other neuronal populations is not yet known. In contrast to the observed effects in striatal neurons, there is no effect of PDE10A inhibition on cGMP in mouse brain regions outside the striatum. This is perhaps not surprising, given the much lower levels of protein expression and the more restricted subcellular distribution. Investigation of the effects of PDE10A inhibitors on discreet cellular events downstream of changes in cyclic nucleotide levels may be the most fruitful approach to reveal the role of the enzyme in nonstriatal neurons.

12.4.4 Physiological Functions and Therapeutic Implications

The high level of expression and unique subcellular distribution of PDE10A in the striatal medium spiny neurons suggest that the enzyme has a particularly prominent role in the function of these neurons. The striatal medium spiny neurons are the principal input site and the first site for information integration in the basal ganglia circuit of the mammalian brain. The basal ganglia are a series of interconnected subcortical nuclei that integrate widespread cortical input with dopaminergic signaling to plan and execute relevant motor and cognitive patterns while suppressing unwanted or irrelevant patterns.[36,37] Significant insight into basal ganglia function has been garnered from human diseases in which dysfunction in this circuit is implicated. These include neuropsychiatric disorders such as schizophrenia, obsessive–compulsive disorders, and addictions. Medium spiny neurons projecting to the external globus pallidus express high levels of dopamine D2 receptors and blockade of these receptors accounts in large part for the antipsychotic efficacy of currently clinically used agents such as haloperidol and ziprasidone.[38] Disruption of basal ganglia function is also a prominent feature of the neurodegenerative disorders such as Parkinson's disease and

Huntington's disease. The clinical manifestation of Parkinson's disease is caused by the degeneration of the dopaminergic input to the striatal medium spiny neurons.[39] These clinical symptoms are ameliorated early in the disease by the pharmacological restoration of dopamine receptor stimulation on these neurons. In Huntington's disease, the medium spiny neurons degenerate[40] and loss of PDE10A mRNA and protein expression appears to be an early event in the degenerative process.[32,33] There are a number of animal behavioral models that have been used to probe the basal ganglia dysfunction that underlies these human diseases and to predict the therapeutic utility of new approaches to treat these diseases. We have begun to use some of these models to investigate the physiological function of PDE10A and the therapeutic implications of pharmacological manipulation of this enzyme. Manipulation of PDE10A activity was accomplished using the PDE10A inhibitor papaverine and a mouse line in which the gene for PDE10A is disrupted.

A prominent phenotype of the PDE10A knockout mouse is a reduction of motor activity.[41] These mice show reduced exploratory activity when placed in a novel environment. These mice also have a blunted response to the stimulatory effects of the NMDA-receptor antagonists, phencyclidine and MK-801. In contrast, the PDE10A knockout mice respond similarly to wild-type mice in response to amphetamine and methamphetamine, which cause a hyperlocomotor response by increasing dopamine release in the striatum. The effects of pharmacological inhibition of PDE10A with papaverine are qualitatively similar to that produced by genetic deletion of the enzyme.[19] Specifically, papaverine causes a decrease in both spontaneous locomotor activity and that induced by phencyclidine. Papaverine also has some inhibitory effect on amphetamine-induced hyperlocomotor activity. These results are consistent with the hypothesis that a physiological function of PDE10A is in the regulation of the excitability of the striatal medium spiny neurons. In general terms, increased activity of the striatal medium spiny neurons tends to decrease overall behavioral responsivity. The observation that deletion or pharmacological inhibition of PDE10A causes a decrease in spontaneous behavioral activity suggests that inhibition of the enzyme causes an activation of these neurons. Both NMDA-receptor antagonists and dopamine-releasing agents cause a suppression of the activity of the medium spiny neurons, and this produces the behavioral hyperactivity observed after the administration of these agents. The observation that PDE10A deletion or pharmacological inhibition reduces the hyperactivity caused by NMDA-receptor antagonists suggests that PDE10A inhibition permits the medium spiny neurons to maintain activity in the face of reduced NMDA-receptor signaling. However, on the basis of more limited effect of PDE10A inhibition on amphetamine-induced hyperlocomotion, it would appear that PDE10A inhibition is unable to prevent the inactivation of the medium spiny neurons in response to dopamine-receptor signaling. This suggests that PDE10A may regulate medium spiny neuron activity via a signaling cascade more tightly coupled to NMDA-receptor signaling than to dopamine-receptor signaling. However, this hypothesis clearly requires further research.

The behavioral effects of PDE10A inhibition are reminiscent of those of dopamine D2 receptor antagonists.[38,42] These drugs, which include haloperidol, risperdol, olanzapine, and ziprasidone among others, are first-line therapies for the treatment of psychosis in schizophrenia and other disorders. Thus, it was of interest to investigate the possible utility of PDE10A inhibitors as antipsychotic agents. It is hypothesized that psychosis in schizophrenia may result from aberrant attribution of salience to irrelevant environmental stimuli.[43] Conditioned avoidance responding (CAR) in rodents is a measure of the ability of a compound to reduce stimulus salience and all clinically active antipsychotic agents are active in animal conditioned avoidance paradigms.[44] We find that in both rats and mice, papaverine produces a dose-dependent inhibition of CAR.[19] Furthermore, the PDE10A knockout mice acquire the CAR task at a slower rate than wild-type mice.[41] One possible explanation for this is that

the salience of the training stimuli are reduced in the knockout mice compared to wild types. Significantly, once the PDE10A knockout mice have been trained in the CAR task, dopamine D2 receptor antagonists suppress CAR. In contrast, papaverine is without effect, confirming that the effect of papaverine on CAR in normal animals results from PDE10A inhibition. Thus, the ability of PDE10A inhibitors to inhibit stimulant-induced hyperlocomotion and to suppress CAR suggests that such compounds may be a novel therapeutic approach to the treatment of psychosis.

Rodefer et al.[45] report that the prototype PDE10A inhibitor, papaverine, ameliorates a deficit in extradimensional set shifting produced in rats by subchronic administration of the NMDA-receptor antagonist, phencyclidine. There was no effect in NMDA antagonist-naive animals. Treatment of rodents with NMDA-receptor antagonists such as phencyclidine models the glutamatergic hypofunction that is hypothesized to underlie schizophrenia.[46] Thus, the fact that the PDE10A inhibitors are active in the set-shifting task only in the phencyclidine-treated animals but not in untreated controls suggests that these compounds may have a specific beneficial effect on executive function in patients with schizophrenia. This effect may result from facilitation of the interaction between the dorsolateral striatum (where PDE10A is highly expressed) and the dorsolateral prefrontal cortex, a circuit involved in mediating this aspect of executive function.[47]

The role of PDE10A in regulation of cyclic nucleotide signaling outside the striatum is not as well understood. It has been observed that expression of some splice variants of PDE10A is increased in hippocampus in response to the induction of LTP.[8] Preliminary results also indicate that PDE10A inhibitors increase CREB phosphorylation in primary cultures of rat cortical neurons (R.J. Kleiman and L. Kimmel, personal communication). Such an effect fits well with the perinuclear localization of PDE10A in these neurons. Given that CREB phosphorylation is implicated in regulating LTP, these combined results suggest that PDE10A may play a role in regulating this form of synaptic plasticity.

12.5   SUMMARY

PDE10A is one of the group of PDEs identified in the late 1990s by molecular genetic techniques and a picture is now starting to emerge as to the physiological function of this enzyme. Perhaps the most salient feature of PDE10A is its high expression in a single population of neurons in the brain, the striatal medium spiny projection neurons. Evidence is accumulating that PDE10A plays a significant role in regulating the activity of these neurons and that inhibition of this enzyme may have therapeutic utility in the treatment of psychosis. PDE10A is expressed at lower levels in a variety of other neurons throughout the brain with a subcellular distribution that is much more restricted than that in the medium spiny neurons. The function of PDE10A in these other neuronal populations is also under study and may include a role in the regulation of neuronal plasticity. The mechanisms of the regulation of PDE10A expression and activity are also beginning to be elucidated. It will be of considerable interest to further investigate the physiology of PDE10A, as this will surely shed new light on mechanisms of cyclic nucleotide signaling in the CNS.

ACKNOWLEDGMENTS

We thank Dr. Diane Stephenson for providing examples of immunohistochemical local-ization of PDE10A in brain and testis, Dr. Robin Kleiman for sharing data on the function of PDE10A in primary neuronal culture models, Dr. Jay Pandit for the information on the

crystal structure of PDE10A, and Dr. Panayiotis Zagouras for bioinformatics support (all from Pfizer Global Research and Development). We also appreciate their helpful comments on the manuscript.

REFERENCES

1. Soderling, S.H., S.J. Bayuga, and J.A. Beavo. 1999. Isolation and characterization of a dual-substrate phosphodiesterase gene family: PDE10A. *Proc Natl Acad Sci USA* 96:7071.
2. Loughney, K., et al. 1999. Isolation and characterization of PDE10A, a novel human 3',5'-cyclic nucleotide phosphodiesterase. *Gene* 234:109.
3. Fujishige, K., et al. 1999. Cloning and characterization of a novel human phosphodiestease that hydrolyzes both cAMP and cGMP (PDE10A). *J Biol Chem* 274:18438.
4. Fujishige, K., J. Kotera, and K. Omori. 1999. Striatum- and testis-specific phosphodiesterase PDE10A. Isolation and characterization of a rat PDE10A. *Eur J Biochem* 226:118.
5. Seeger, T.F., et al. 2003. Immunohistochemical localization of PDE10A in the rat brain. *Brain Res* 985:113.
6. Altschul, S.F., et al. 1997. Gapped BLAST and PSI-BLAST: A new generation of protein database search programs. *Nuc Acids Res* 25:3389.
7. Fujishige, K., et al. 2000. The human phosphodiesterase PDE10A gene. Genomic organization and evolutionary relatedness with other PDEs containing GAF domains. *Eur J Biochem* 267:5943.
8. O'Connor, V., et al. 2004. Differential amplification of intron-containing transcripts reveals long term potentiation-asociated up-regulation of specific PDE10A phosphodiesterase splice variants. *J Biol Chem* 279:15841.
9. Kotera, J., et al. 1999. Characterization and phosphorylation of PDE10A2, a novel alternative splice variant of human phosphodiesterase that hydrolyzes cAMP and cGMP. *Biochem Biophys Res Commun* 261:551.
10. Dynan, W.S. 1986. Promoters for housekeeping genes. *Trends Genet* 2:196.
11. Yausa, K., et al. 2001. Genomic organization of the human phosphodiesterase PDE11A gene. Evolutionary relatedness with other PDEs containing GAF domains. *Eur J Biochem* 268:168.
12. Kawaguchi, Y. 1997. Neostriatal cell subtypes and their functional roles. *Neurosci Res* 27:1.
13. Xie, Z., W.O. Adamowicz, W.D. Eldred, A.B. Jakowski, R.J. Kleiman, D.G. Morton, D.T. Stephenson, C.A. Strick, R.D. Williams, and F.S. Menniti. 2006. Cellular and subcellular localization of PDE10A, a striatal-specific phosphodiesterase. *Neurosci* 139:597.
14. Kotera, J., et al. 2004. Subcellular localization of cyclic nucleotide phosphodiesterase type 10A variants, and alteration of the localization by cAMP-dependent protein kinase-dependent phosphorylation. *J Biol Chem* 279:4366.
15. Nakai, K., and P. Horton. 1999. PSORT: A program for detecting sorting signals in proteins and predicting their subcellular localization. *TIBS* 24:34.
16. Pandit, J., et al. The x-ray crystal structure of the catalytic domain of rat recombinant PDE10A. Manuscript in preparation.
17. Kam Y.J., et al. 2004. A glutamine switch mechanism for nucleotide selectivity by phosphodiesterases. *Mol Cell* 15:279.
18. Schmidt, C.J., et al. 2003. The neurochemical and behavioral effects of papaverine *in vivo* suggest PDE10 inhibition is "antipsychotic." *Schizophr Res* 60:114.
19. Siuciak, J.A., D.S. Chapin, J.F. Harms, L.A. Lebel, L.C. James, S.A. McCarthy, L.K. Chambers, A. Shrikehande, S.K. Wong, F.S. Menniti, and C.J. Schmidt. 2006. Inhibition of the striatum-enriched phosphodiesterase PDE10A: A novel approach to the treatment of psychosis. *Neuropharmacology* 51:386.
20. Arvind, L., and C.P. Ponting. 1997. The GAF domain: An evolutionary link between diverse phototransducing proteins. *TIBS* 22:458.
21. Gillespie, P.G., and J.A. Beavo. 1989. cGMP is tightly bound to bovine retinal rod phosphodiesterase. *Proc Natl Acad Sci USA* 86:4311.

22. Stroop, S.D., H. Charbonneau, and J.A. Beavo. 1989. Direct photolabeling of the cGMP-stimulated cyclic nucleotide phosphodiesterase. *J Biol Chem* 264:13718.
23. McAllister-Lucas, L.M., et al. 1995. An essential aspartic acid at each of two allosteric cGMP-binding sties of a cGMP-specific phosphodiesterase. *J Biol Chem* 270:30671.
24. Turko, I.V., et al. 1996. Identification of key amino acids in a conserved cGMP-binding site of a cGMP-binding phosphodiesterase. A putative NKX$_n$D motif for cGMP binding. *J Biol Chem* 271:22240.
25. Liu, L., T. Underwood, H. Li, R. Pamukcu, and W.J. Thompson. Specific cGMP binding by the cGMP binding domains of cGMP-binding cGMP specific phosphodiesterase. *Cell Signal* 14:45.
26. Martinez, S.E., A.Y. Wu, N.A. Glavas, X.B. Tang, S. Turley, W.G.J. Hol, and J.A. Beavo. 2002. The two GAF domains in phosphodiesterase 2A have distinct roles in dimerization and in cGMP binding. *Proc Natl Acad Sci USA* 99:13260.
27. Martinez, S.E., S. Bruder, A. Schultz, N. Zheng, J.E. Schultz, J.A. Beavo, and J.U. Linder. 2005. Crystal structure of the tandem GAF domains from a cyanobacterial adenylyl cyclase: Modes of ligand binding and dimerization. *Proc Natl Acad Sci USA* 102:3082–3087.
28. Wu, A.Y., X.B. Tang, S.E. Martinez, K. Ikeda, and J.A. Beavo., 2004. Molecular determinants for cyclic nucleotide binding to the regulatory domains of phosphodiesterase 2A. *J Biol Chem* 279:37928.
29. Gross-Langenhoff, M., K. Hofbauer, J. Weber, A. Schultz, and J.E. Schultz. 2006. cAMP is a ligand for the tandem GAF domain of human phosphodiesterase 10 and cGMP for the tandem GAF domain of phosphodiesterase 11. *J Biol Chem* 281:2841.
30. Sigrist, C.J.A., L. Cerutti, N. Hulo, A. Gattiker, L. Falquet, M. Pagni, A. Bairoch, and P. Bucher. 2002. PROSITE: A documented database using patterns and profiles as motif descriptors. *Brief Bioinform* 3:265.
31. Burns, F., A.Z. Zhao, and J.A. Beavo. 1996. Cyclic nucleotide phosphodiesterases: Gene complexity, reglation by phosphorylation, and physiological implications. *Adv Pharmacol* 36:29.
32. Hebb, A.L., H.A. Robertson, and E.M. Denovan-Wright. 2004. Striatal phosphodiesterase mRNA and protein levels are reduced in Huntington's disease transgenic mice prior to the onset of motor symptoms. *Neuroscience* 123:967.
33. Hu, H., E.A. McCaw, A.L.O. Hebb, G.T. Gomez, and E.M. Denovan-Wright. 2004. Mutant huntingtin affects the rate of transcription of striatum-specific isoforms of phosphodiesterase 10A. *Eur J Neurosci* 20:3351.
34. Frame, M., K.F. Wan, R. Tate, P. Vandenabeele, and N.J. Pyne. 2001. The γ-subunit of the rod photoreceptor cGMP phosphodiesterase can modulate the proteolysis of two cGMP binding cGMP-specific phosphodiesterases (PDE6 and PDE5) by caspase-3. *Cell Signal* 13:735.
35. Kleiman, R.J., L. Kimmel, M.A. Collins, and F.S. Menniti. Inhibitors of PDE10 potentiate NMDA-induced phosphorylation of CREB in striatal and cortical neurons. Manuscript in preparation.
36. Wilson, C.J. 1998. Basal ganglia. In *The synaptic organization of the brain*, 4th edition, ed. G.M. Sheperd, pp. 329–376. New York: Oxford University Press.
37. Graybiel, A.M. 2000. The basal ganglia. *Curr Biol* 10:R509.
38. Kapur, S., and D. Mamo. 2003. Half a century of antipsychotics and still a central role for dopamine D2 receptors. *Prog Neuro psychopharm Biol Psychiatry* 27:1081.
39. Beal, M.F. 2001. Experimental models of Parkinson's disease. *Nat Revs Neuroscience* 2:325.
40. Graveland, G.A., R.S. Williams, and M. DiFiglia. 1985. Evidence for degenerative and regenerative changes in neostriatal spiny neurons in Huntington's Disease. *Science* 227:770.
41. Siuciak, J.A., S.A. McCarthy, D.S. Chapin, R.A. Fujiwara, L.C. James, R.D. Williams, J.L. Stock, J.D. McNeish, C.A. Strick, F.S. Menniti, and C.J. Schmidt. 2006. Genetic deletion of the striatum-enriched phosphodiesterase PDE10A: Evidence for altered striatal function. *Neuropharmacology* 51:374.
42. Civelli, O., J.R. Bunzow, and D.K. Grandy 1993. Molecular diversity of the dopamine receptors. *Ann Rev Pharmacol Toxicol* 33:281
43. Kapur, S. 2003. Psychosis as a state of aberrant salience: A framework linking biology, phenomenology, and pharmacology in schizophrenia. *Am J Psychiatry* 160:13.

44. Wadenberg, M.-L.G., and P.B. Hicks. 1999. The conditioned avoidance response test re-evaluated: Is it a sensitive test for the detection of potentially atypical antipsychotics? *Neurosci Biobehav Rev* 23:851.
45. Rodefer J.S., E.R. Murphy, and M.G. Baxter. 2005. The PDE10A inhibitor papaverine reverses subchronic PCP-induced deficits in attentional set-shifting. *Eur J Neurosci* 21:1070.
46. Jentsch, J.D., and R.H. Roth. 1999. The Neuropsychopharmacology of phencyclidine: From NMDA receptor hypofunction to the dopamine hypothesis of schizophrenia. *Neuropsychopharmacology* 20:201.
47. Dalley, J.W., R.N. Cardinal, and T.W. Robbins. 2004. Prefrontal executive and cognitive functions in rodents: Neural and neurochemical substrates. *Neurosci Biobehav Rev* 28:771.

13 PDE11

Kenji Omori and Jun Kotera

CONTENTS

13.1 Introduction 255
13.2 Primary Structure..256
13.3 Enzymatic Properties...257
13.4 Sensitivity to Known Phosphodiesterase Inhibitors.................................259
13.5 Other Protein Features...261
13.6 Tissue Distribution of PDE11A mRNA 261
13.7 Detection and Localization of PDE11A Protein......................................263
13.8 Genomic Organization 264
13.9 Single Nucleotide Polymorphisms...269
13.10 Physiological Role of PDE11A and Effects of Its Inhibition.................269
13.11 Discussion 271
References 273

13.1 INTRODUCTION

Cyclic adenosine $3',5'$-monophosphate (cAMP) and cyclic guanosine $3',5'$-monophosphate (cGMP) are second messengers that regulate many physiological functions in mammalians. cAMP participates in various cellular events, such as gene expression, cell growth and differentiation, hormonal control, olfaction, neurotransmission, smooth muscle regulation, and reproduction.[1,2] cGMP signaling is also linked to regulation of many cellular activities and functions, including cell proliferation, apoptosis, visual and neuronal signal transduction, platelet function, vascular resistance, water and electrolyte homeostasis, airway distension, and bone formation.[3] Intracellular cyclic nucleotide levels are dependent on a balance in cAMP and cGMP production and degradation mediated by adenylyl and guanylyl cyclases,[4,5] and $3',5'$-cyclic nucleotide phosphodiesterases (PDEs), respectively.[2,6–8] Downstream effector proteins for cyclic nucleotide signaling are cAMP-dependent protein kinase (PKA), cGMP-dependent protein kinase (PKG), cAMP-guanine nucleotide exchange factors (GEFs), and cyclic nucleotide-gated ion channels. Activation of these effector proteins is dependent on the availability of cyclic nucleotides, which must be strictly controlled by PDEs as negative regulators. In fact, intracellular compartmentation of PDEs with signaling molecules in some instances underlies the molecular mechanism governing the hydrolysis of cyclic nucleotides related to certain signaling processes,[9] and the fundamentals of this compartmentation have been established with six major PDEs (PDE1–6). On the other hand, the discovery of novel PDEs has provided hints for explaining unknown regulation of cyclic nucleotide signaling, which could not be fully illustrated by the six major PDEs.

All mammalian PDEs are class I PDEs. The main characteristic of these PDEs is the HD domain HD[L/I/V/M/F/Y]XHX[A/G]-(X)₂-NX[L/I/V/M/F/Y], which contains highly conserved

histidine and aspartic acid residues situated in the C-terminal catalytic region.[10] The N-terminal parts of class I PDEs contain distinct protein domains that are associated with either isoform-specific regulation of enzymatic activity or protein localization, such as calmodulin binding (PDE1), cGMP binding (PDEs 2, 5, and 6), N-terminal membrane targeting (PDE4), hydrophobic membrane association (PDE3), and PKA- or PKG-dependent phosphorylation (PDEs 1–7, 10, and 11). Judged from amino acid sequence similarities, enzymatic properties, and sensitivity to inhibitors, 11 PDE families composed of 21 genes have been identified in mammals. PDE11, which is literally the 11th and the latest member of the PDE superfamily, has GAF (cGMP-binding and stimulated PDEs, *Anabaena* adenylyl cyclases, and *Escherichia coli* FhlA protein) domains[11] in its N-terminal region, and therefore, belongs to the GAF-PDE subgroup of families, as do PDE2, PDE5, PDE6, and PDE10.[12] PDE11 consists of one gene (*PDE11A*), but includes several splicing variants with unique structural features generated from the gene. Cloning of human PDE11A1 cDNA, one of the PDE11A splice variants (see Section 13.8), was first achieved using bioinformatics and molecular biology approaches, and recombinant PDE11A was found to catalyze the hydrolysis of both cAMP and cGMP, thereafter defined as dual substrate PDE.[13]

Although the first report on PDE11A was published more than 6 years ago, today there are only few publications on this PDE family. Among them, reports of analyses on mRNA and protein distribution of PDE11A variants and studies on PDE11A knockout mice largely contributed to a better understanding of PDE11A family. In addition to these reports, reports on biochemistry and pharmacology of a PDE5A and PDE11A dual inhibitor were intriguing and important for knowing the physiological roles of PDE11A. However, there are significant discrepancies between the findings in some reports, and controversial interpretations have been made. This chapter deals with the molecular basis of, and current inconsistencies in the observations on, PDE11A.

13.2 PRIMARY STRUCTURE

The human PDE11 family is composed of four splice variants, namely PDE11A1 (GenBank accession no. AJ251509), PDE11A2 (GenBank accession no. AF281865), PDE11A3 (GenBank accession nos. AB038041 and AJ278682), and PDE11A4 (GenBank accession no. AB036704) (Figure 13.1 and Table 13.1). The deduced amino acid sequences of these isoforms show that all of them are N-terminal variants with various GAF domain structures together with a common C-terminal catalytic domain sequence. PDE11A1, which is the shortest form among PDE11A variants, is predicted as a protein of 490 amino acids that includes the C-terminal region half of the GAF-B domain sequence located at its N-terminus.[13] PDE11A2 is a calculated protein of 576 amino acids that contains a complete GAF-B domain and a small C-terminal part of the GAF-A domain sequence.[14] PDE11A3 is a putative protein of 684 amino acids that include a complete GAF-B domain sequence and a C-terminal portion of the GAF-A domain sequence with a unique N-terminal flanking extension.[14,15] PDE11A4, which is the longest form among PDE11A variants, is a predicted protein of 934 amino acids with two complete GAF domain sequences (GAF-A and GAF-B) and an N-terminal flanking sequence that contains two consensus sequences for PKA and PKG phosphorylation.[15] Variant forms with distinct GAF domains have not yet been identified in other GAF-PDEs, and from this point of view, PDE11A is a unique GAF-PDE.

As for experimental animals, cDNAs for three rat PDE11A variants (PDE11A2, PDE11A3, and PDE11A4) have been cloned. Rat PDE11A variants are highly similar to human PDE11A2, PDE11A3, and PDE11A4, respectively.[16] For example, rat PDE11A4 is composed of 935 amino acids, and is overall 94% identical to human PDE11A4. In the mouse, DNA sequences potentially coding for mouse PDE11As are found in the mouse genome database

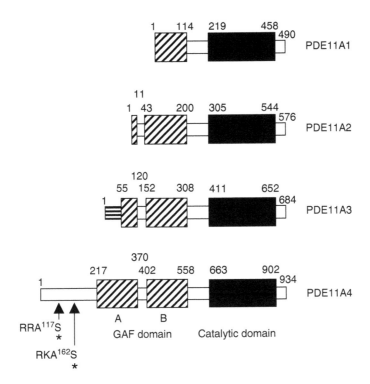

FIGURE 13.1 Schematic representation and enzymatic characterization of the four PDE11A variants in human. Catalytic domains identified by Pfam PDEase I motif Pfam accession no. PF00233) are indicated by a black box, and GAF domain sequences identified by sequence similarity to other GAF-PDEs (Pfam accession no. PF01590) are shown in crosshatched boxes. The N-terminal sequence of PDE11A3 indicated by horizontal stripe pattern is unique and not included in other PDE11A variants. The positions of two phosphorylation sites in PDE11A4 are shown by the arrows, and phosphorylation residues in the motifs are indicated by asterisks.

(http://www.ncbi.nlm.nih.gov/entrez/query.fcgi?db=gene&cmd=Retrieve&dopt=Graphics& list_uids = 241489). However, currently there is no report describing the characteristics of mouse PDE11A gene products.

13.3 ENZYMATIC PROPERTIES

PDE11A is a cAMP- and cGMP-hydrolyzing enzyme although it is the most homologous among PDEs to PDE5A, a cGMP-PDE (43% identical over 721 amino acids). Reported K_m values of recombinant human PDE11A1, PDE11A2, PDE11A3, and PDE11A4 range from 1.0 to 5.7 μM for cAMP and 0.5 to 4.2 μM for cGMP (Table 13.1).[13–17] As compared to the K_m values, the relative V_{max} values for cAMP and cGMP hydrolysis are remarkably different among the PDE11A variants.[16] PDE11A4 shows the highest relative V_{max} values for cAMP and cGMP (100-fold higher than those of PDE11A1). However, ratios of V_{max} values for cAMP and cGMP are similar among these human variants. In rats, the enzymatic properties of PDE11A4 are similar to those of human PDE11A4 (K_m values of 3.9 ± 0.13 μM for cAMP and 1.6 ± 0.08 μM for cGMP).[16] Rat variants PDE11A2, PDE11A3, and PDE11A4 also show similar characteristics to human PDE11A variants in regard to K_m and relative V_{max} values. A cDNA corresponding to rat PDE11A1 has not been identified. Artificially

TABLE 13.1
Molecular Characteristics and Enzymatic Properties of Human PDE11A Variants

		PDE11A1	PDE11A2	PDE11A3	PDE11A4	Ref.
Predicted molecular weight		55,786.81	65,766.54	78,133.86	104,810.71	
Amino acid residue number		490	576	684	934	
N-terminal tag		Histidine-tagged	Histidine-tagged	Histidine-tagged	Histidine-tagged	
K_m values	Substrate	(μM)	(μM)	(μM)	(μM)	
	cAMP	1.04	N/A	N/A	N/A	[13]
		N/A	3.3	5.7	N/A	[14]
		N/A	N/A	3 ± 0.28	3 ± 0.26	[15]
		3.2 ± 0.14	2 ± 0.01	2 ± 0.03	2.7 ± 0.03	[16]
		N/A	N/A	N/A	2.4 ± 0.96	[17]
	cGMP	0.52	N/A	N/A	N/A	[13]
		N/A	3.7	4.2	N/A	[14]
		N/A	N/A	1.5 ± 0.07	1.4 ± 0.06	[15]
		2.1 ± 0.09	0.95 ± 0.01	0.99 ± 0.02	1.3 ± 0.07	[16]
		N/A	N/A	N/A	0.97 ± 0.08	[17]
Relative V_{max}[a]	3.5 μM cAMP	1	31	13	97	[16]
	1.3 μM cGMP	1	35	15	109	[16]
V_{max} ratio	V_{max} (cAMP)/V_{max} (cGMP)	2.6	2.3	2.3	2.3	[16]

N/A, Not available.

[a] PDE activities of recombinant PDE11A proteins produced in COS cells are normalized by immunoreactive signals obtained by anti-Xpress antibody which detects recombinant proteins carrying on N-terminal leader peptide (Xpress epitope) encoded by the mammalian expression vector pcDNA4/His Max.

generated rat PDE11A1 showed a profile similar to human PDE11A1. Thus, PDE11A4 is the most active form among the four PDE11A variants in humans and rats. PDE11A4 is a typical GAF-PDE having two complete GAF domains. One of the functions of the GAF domain is cGMP binding, which is related to allosteric control of certain PDEs. This type of allosteric control has been demonstrated in cGMP-activated PDE2A, cGMP-binding PDE5, and cGMP-binding PDE6s (see Chapter 4, Chapter 7, and Chapter 8). Activation of PDE2A is a particularly good example for cGMP-mediated allosteric regulation. However, in the case of PDE11A4, no potent cyclic nucleotide binding has been identified to date, and under the conditions studied thus far, PDE11A4 is not seemingly activated by either cGMP or cAMP[15] similarly to PDE10A.[18] Recently, the tandem GAF domain of PDE11A4 has been demonstrated to cause the cGMP-mediated allosteric activation on its fusion construct with cyanobacterial adenylyl cyclase,[19] and the N[K/R]-(X)$_n$-F-(X)$_3$-DE motif included in the sequence NKIPEGAPFTEDDE of the GAF-A domain has been shown to be involved in this allosteric activation. Thus, the two complete GAF domains of PDE11A4 are functional domains for a regulator of the adjacent catalytic domain. Although no report has yet exhaustively investigated the differences in GAF functions (such as cyclic nucleotide binding, resultant activation, and dimerization) among PDE11A variants having various GAF domain forms, the lack of the respective GAF domains has been demonstrated to result in significant reduction of catalytic activity as described earlier. The tandem GAF domain found in PDE11A4 may be the only functional form necessary for the allosteric regulation, and the presence of this domain form in PDE11A4 may lead to active conformation necessary for full catalytic activity of this enzyme. This seems to be the reason for the highest hydrolytic activity of PDE11A4 among PDE11A variants.

13.4 SENSITIVITY TO KNOWN PHOSPHODIESTERASE INHIBITORS

Sensitivities of the four PDE11A variants to various PDE inhibitors are summarized in Table 13.2. Although various assay conditions with different substrate concentrations (see Table 13.2) were used,[13–17,20,21] PDE11A variant sensitivity to known PDE inhibitors is fundamentally the same. The nonspecific PDE inhibitor, 3-isobutyl-1-methylxanthine (IBMX), showed weak inhibitory activity towards all four PDE11A variants with IC$_{50}$ values in the range of 25–81 μM. Vinpocetine weakly inhibited PDE11A3 with IC$_{50}$ values ranging between 49 and 68 μM, and had IC$_{50}$ values for PDE11A4 of more than 100 μM. Zaprinast inhibited all four PDE11A variants with IC$_{50}$ values in the range of 5–33 μM, whereas SCH51866 inhibited PDE11A3 and PDE11A4 (IC$_{50}$ values in the range of 8.6–25 μM) but not PDE11A2 (IC$_{50}$ > 100 μM). Dipyridamole showed potent inhibitory activity towards all PDE11A variants (IC$_{50}$ = 0.34–1.8 μM), whereas E4021 potently inhibited PDE11A3 and PDE11A4 with IC$_{50}$ values in the range of 0.88–1.8 μM. The IC$_{50}$ values of milrinone, erythro-9-(2-hydroxy-3-nonyl)-adenine (EHNA), rolipram, papaverine, enoximone, Ro 201724, and pentoxifylline were over 100 μM for at least one PDE11A variant. Sildenafil (Viagra, Pfizer Inc., New York, NY), vardenafil (Levitra, Bayer AG, Germany), and tadalafil (Cialis, Lilly ICOS LLC, Indianapolis, IN), all of which are potent PDE5 inhibitors used clinically, inhibited PDE11A1 with IC$_{50}$ values of 2.73, 0.162, and 0.037 μM, respectively[21] and PDE11A4 with IC$_{50}$ values of 3.8, 0.84, and 0.073 μM, respectively.[17] All the PDE inhibitors examined had similar inhibitory activity towards cGMP and cAMP hydrolysis. cAMP and cGMP substrates inhibited PDE11A-mediated hydrolysis of cGMP and cAMP, respectively. The IC$_{50}$ values of cAMP and cGMP inhibition were affected by substrate concentrations (Table 13.2), and those obtained at low substrate concentrations were comparable to the K_m values of this enzyme, indicating that cAMP and cGMP compete for the same catalytic site.

TABLE 13.2
Inhibitor Sensitivities of Human PDE11A Variants

		IC$_{50}$ Values (μM)				
Inhibitor	Substrate	PDE11A1	PDE11A2	PDE11A3	PDE11A4	Ref.
IBMX	0.17 μM cGMP	49.8	N/A	N/A	N/A	[13]
	0.012 μM cAMP	N/A	80 $\pm$ 43	25 $\pm$ 10	N/A	[14]
	3.5 μM cAMP	N/A	N/A	30 $\pm$ 3.9	65 $\pm$ 13	[15]
	1.3 μM cGMP	N/A	N/A	38 $\pm$ 3.5	81 $\pm$ 16	[15]
	0.03 μM cGMP	N/A	N/A	N/A	26.54 $\pm$ 9.23	[20]
Milrinone	0.17 μM cGMP	>100	N/A	N/A	N/A	[13]
	3.5 μM cAMP	N/A	N/A	>100	>100	[15]
	1.3 μM cGMP	N/A	N/A	>100	>100	[15]
	0.03 μM cGMP	N/A	N/A	N/A	>100	[20]
Vinpocetine	3.5 μM cAMP	N/A	N/A	49 $\pm$ 9.2	>100	[15]
	1.3 μM cGMP	N/A	N/A	68 $\pm$ 4	>100	[15]
	0.03 μM cGMP	N/A	N/A	N/A	>100	[20]
Rolipram	0.17 μM cGMP	>100	N/A	N/A	N/A	[13]
	0.012 μM cAMP	N/A	>200	N/A	N/A	[14]
	3.5 μM cAMP	N/A	N/A	>100	>100	[15]
	1.3 μM cGMP	N/A	N/A	>100	>100	[15]
	0.03 μM cGMP	N/A	N/A	N/A	>100	[20]
Zaprinast	0.17 μM cGMP	12	N/A	N/A	N/A	[13]
	0.012 μM cAMP	N/A	28 $\pm$ 7.6	5 $\pm$ 1.4	N/A	[14]
	3.5 μM cAMP	N/A	N/A	18 $\pm$ 10	26 $\pm$ 6.8	[15]
	1.3 μM cGMP	N/A	N/A	11 $\pm$ 3.6	33 $\pm$ 5.3	[15]
	0.03 μM cGMP	N/A	N/A	N/A	8.16 $\pm$ 2.39	[20]
Dipyridamole	0.17 μM cGMP	0.37	N/A	N/A	N/A	[13]
	0.012 μM cAMP	N/A	1.8 $\pm$ 1.1	0.82 $\pm$ 0.1	N/A	[14]
	3.5 μM cAMP	N/A	N/A	0.36 $\pm$ 0.11	0.82 $\pm$ 0.28	[15]
	1.3 μM cGMP	N/A	N/A	0.34 $\pm$ 0.09	0.72 $\pm$ 0.08	[15]
	0.1 μM cGMP	N/A	N/A	N/A	0.84 $\pm$ 0.22	[17]
	0.03 μM cGMP	N/A	N/A	N/A	0.34	[20]
Papaverine	0.012 μM cAMP	N/A	>100	N/A	N/A	[14]
EHNA	0.012 μM cAMP	N/A	>100	N/A	N/A	[14]
	3.5 μM cAMP	N/A	N/A	>100	>100	[15]
	1.3 μM cGMP	N/A	N/A	>100	>100	[15]
	0.03 μM cGMP	N/A	N/A	N/A	>100	[20]
SCH51866	0.012 μM cAMP	N/A	>100	N/A	N/A	[14]
	3.5 μM cAMP	N/A	N/A	11 $\pm$ 4.8	22 $\pm$ 1.8	[15]
	1.3 μM cGMP	N/A	N/A	8.6 $\pm$ 2.7	25 $\pm$ 5.8	[15]
E4021	3.5 μM cAMP	N/A	N/A	0.88 $\pm$ 0.13	1.8 $\pm$ 0.33	[15]
	1.3 μM cGMP	N/A	N/A	0.66 $\pm$ 0.19	1.8 $\pm$ 0.25	[15]
Enoximone	0.012 μM cAMP	N/A	>100	N/A	N/A	[14]
Sildenafil	0.012 μM cAMP	N/A	>0.5	N/A	N/A	[14]
	0.1 μM cGMP	N/A	N/A	N/A	3.8 $\pm$ 0.75	[17]
	0.03 μM cGMP	N/A	N/A	N/A	3.15 $\pm$ 0.23	[20]
	Not specified	2.73	N/A	N/A	N/A	[21]
Vardenafil	0.1 μM cGMP	N/A	N/A	N/A	0.84 $\pm$ 0.16	[17]
	Not specified	0.162	N/A	N/A	N/A	[21]
Tadalafil	0.1 μM cGMP	N/A	N/A	N/A	0.073 $\pm$ 0.0031	[17]
	Not specified	0.037	N/A	N/A	N/A	[21]

continued

TABLE 13.2 (continued)
Inhibitor Sensitivities of Human PDE11A Variants

		IC$_{50}$ Values (μM)				
Inhibitor	Substrate	PDE11A1	PDE11A2	PDE11A3	PDE11A4	Ref.
Ro 201724	0.012 μM cAMP	N/A	>200	N/A	N/A	[14]
Pentoxifylline	0.012 μM cAMP	N/A	>100	N/A	N/A	[14]
cAMP	1.3 μM cGMP	N/A	N/A	8.2 $\pm$ 0.43	9 $\pm$ 0.38	[15]
	0.1 μM cGMP	N/A	N/A	N/A	1.2 $\pm$ 0.063	[17]
cGMP	3.5 μM cAMP	N/A	N/A	5.1 $\pm$ 0.12	3.1 $\pm$ 0.16	[15]
	0.1 μM cGMP	N/A	N/A	N/A	0.56 $\pm$ 0.22	[17]

N/A, Not available.

No specific PDE11A inhibitor has been reported. However, among the PDE inhibitors tested, tadalafil shows the most potent inhibitory activity for PDE11A.[17,21] It has been discussed whether the administration of tadalafil causes physiological changes by PDE11A inhibition or not. PDE11A inhibition by tadalafil is described in Section 13.10.

13.5 OTHER PROTEIN FEATURES

PDE11A4 possesses two canonical phosphorylation sites (RRA^{117}S and RKA^{162}S, for either PKA or PKG) in the N-terminal flanking sequence of GAF-A domain (Figure 13.1). These phosphorylation sites are unique to PDE11A4 among PDE11A variants. PDE11A4 is a good substrate for PKA, PKG-Iα, and PKG-Iβ in an *in vitro* phosphorylation test, and alanine substitutions at Ser117 and Ser162 reduce the ^{32}P incorporation.[15] PKA- or PKG-dependent phosphorylation of PDEs has been reported to cause activation (PDE3, PDE4, and PDE5),[22–24] changes in the affinity for interacting proteins (PDE1),[25] and alteration of subcellular localization (PDE10).[26] The existence of these sites in the N-terminal region of PDE11A4 implies that PDE11A4 might be subjected to cAMP- or cGMP-mediated regulation. However, no effects of PKA phosphorylation on PDE11A4 enzymatic activity have been shown nor the phosphorylation of PDE11A4 has been demonstrated *in vivo*.

13.6 TISSUE DISTRIBUTION OF PDE11A mRNA

As shown in Table 13.3, several kinds of PDE11A transcripts with different sizes are present in different human tissues.[13,15] Northern blot analysis with a probe sensitive to all four PDE11A splice variants demonstrated that PDE11A transcripts are particularly dense in the prostate.[13] The testis, salivary gland, pituitary gland, kidney, and liver contain PDE11A transcripts at moderate levels. The low-level expression of PDE11A transcripts is observed in the adrenal gland, mammary gland, thyroid gland, pancreas, spinal cord, and trachea. PDE11A transcripts seem to be absent in the bladder, colon, heart, small intestine, stomach, and uterus. A probe for PDE11A3 and PDE11A4 demonstrated strong signals in the prostate, moderate signals in the testis, salivary gland, pituitary gland, thyroid gland, and liver, and no signal in the kidney.[15] Moderate expression has been confirmed in pancreas (unpublished observation). In the prostate, where PDE11A transcripts are abundant, a major band of approximately 6 kb and minor bands of 2 kb and 10–10.5 kb are present.[13,15] PDE11A transcripts in the testis and skeletal muscle are approximately 3[15] and 8.5 kb[13], respectively.

TABLE 13.3
Tissue Distribution of PDE11A mRNAs and Proteins of Human PDE11A Variants

| | PDE11A mRNA | | | PDE11A Protein | | |
| | Northern Blotting | | PCR | Immunoblotting | | |
Ref.	Fawcett et al.[13]	Yuasa et al.[15]	Yuasa et al.[12]	Fawcett et al.[13]	D'Andrea et al.[20]	Loughney et al.[31]
Preparation of cDNA probe	cDNA encoding amino acids 121–263 of PDE11A1	Nucleotides 1237–1801 of PDE11A4	—	—	—	—
Preparation of antibody	—	—	—	The synthetic peptide SAIFDRNRKDELPRL as an antigen,[a] polyclonal antibody, and affinity-purified	The synthetic peptide VATNRSKWEELHQKR as an antigen, polyclonal antibody, and affinity-purified; purchased from FabGennix (Shreveport, LA) VATNRSKWEELHQKR[b]	The C-terminal 87 amino acid protein of PDE11A as an antigen, monoclonal antibody, and epitope sequence is located within
Tissue						
Prostate	10.5 kb (high) 6.0 kb (high)	10 kb (moderate) 6.0 kb (high) 2.0 kb (low)	PDE11A4 (high)	56 kDa	100 kDa (low) 70 kDa (high)	105 kDa (high)
Skeletal muscle	8.5 kb (moderate)	N/A	PDE11A1	78 kDa (high)	100 kDa (low)	Undetectable
Testis	N/A	3 kb (moderate)	PDE11A3	65 kDa (low) 56 kDa (low)	70 kDa (low)	Undetectable
Pituitary gland	N/A	N/A	N/A	N/A	100 kDa (moderate) 70 kDa (faint)	105 kDa (moderate)
Liver	N/A	N/A	PDE11A4 (low) PDE11A1 (low)	N/A	N/A	105 kDa (low)
Heart	Undetectable	N/A	PDE11A4 (low)	N/A	N/A	105 kDa (low)

N/A, Not available.

[a] The same results have been reported with the antibodies against other two peptides NNTIMYDQVKKSWAK and VATNRSKWEELHQKR. Wayman et al.[27] and Baxendale et al.[28] have reported immunohistochemistry probably with the above antibodies.

[b] Another monoclonal antibody (epitope is situated within SAIFDRNRKDELPRL) has been obtained. However, this monoclonal antibody reacted with proteins other than PDE11A.

As for the expression of PDE11A variants in human tissues, polymerase chain reaction (PCR) analysis has shown that the levels of PDE11A1 transcripts are high in the skeletal muscle and low in the liver,[12,15] and that only PDE11A3 primers produced PCR products in the testis. PDE11A4 transcripts are abundant in the prostate and present at low levels in the liver and heart. Tissue expression pattern of PDE11A2, which was cloned from human testis cDNA library, has not yet been reported. Thus, distinct tissue expression patterns of human PDE11A are due to multiple splice variants.

In rats, tissue expression patterns of PDE11A4 are different from those in humans.[16] Transcripts of rat PDE11A4 are poorly expressed in the prostate but densely present in the brain, heart, kidney, and liver. As in humans, expression of rat PDE11A3 transcripts is limited to the testis. On the other hand, rat PDE11A2 transcripts are present in the brain, lung, skeletal muscle, spleen, testis, and prostate. Thus, PDE11A4 is the only major variant that exhibits species difference in tissue distribution between humans and rats. In knockout mice carrying the β-galactosidase gene (lacZ) in the PDE11A locus, PDE11A promoter-driven mRNA production is observed in the prostate and testis but not in the aorta.[27]

13.7   DETECTION AND LOCALIZATION OF PDE11A PROTEIN

Immunoblot analysis or immunohistochemistry of PDE11A protein has been reported in several papers.[13,20,27–31] Using immunoblot analysis with affinity-purified rabbit polyclonal antibody raised against the synthetic peptide EPH-3 (SAIFDRNSR2KDELPRL) derived from the human PDE11A catalytic domain sequence, Fawcett et al.[13] first reported an immunoreactive protein of 56 kDa, which is in good agreement with the predicted size of PDE11A1 protein, in the cytosolic fraction of human prostate (Table 13.3). This study also showed a weak signal of 56 kDa protein (PDE11A1) in the skeletal muscle with two additional signals of 65 kDa (low level) and 78 kDa (high level) proteins considered to represent PDE11A2 and PDE11A3, respectively. The same results were demonstrated with two other antibodies against the synthetic peptides EPH-2 (NNTIMYDQVKKSWAK) and EPH-4 (VATNRSKWEELHQKR) (see Table 13.3).[13]

PDE11A immunoreactive protein has been reported in human and mouse reproductive systems by immunohistochemistry with the polyclonal antibody against EPH-3. In mice, positive staining was detected in developing sperm cells including spermatogonia, spermatocytes, and spermatids within the seminiferous tubules, interstitial Leydig cells involved in testosterone production, basement membrane, and vascular smooth muscle but not in Sertoli cells and connective tissue components.[27,28] Immunoreactive signals were also identified in human germinal epithelium of the seminiferous tubules and in all the above-mentioned reproductive tissues of humans.[28] Besides male gonad, immunoreactive signals were also detected in smooth muscle cells within the human penile vasculature and bundles of smooth muscle fiber adjacent to the vascular spaces of the corpus cavernosum.[29] In human prostate, immunoreactive protein was observed in the glandular epithelial and subepithelial tissues close to the transition zone and in the epithelial layer of glandular ducts in the fibromuscular stroma.[30] The pituitary gland, where many hormones related to sex and reproduction are secreted, contains PDE11A transcripts.[13,15] Immunoreactive proteins were detected in the acidophils of the anterior lobe that secretes prolactin and growth hormone but not in the posterior lobe.[28] These findings suggest that PDE11A plays a part in the hormonal regulation of the reproductive system and its development. However, immunoblot analyses of human tissue extracts with the polyclonal antibody against EPH-3 have detected multiple immunoreactive bands or shorter immunoreactive protein than expected as described earlier. In addition to this, the molecular size of immunoreactive proteins for mouse PDE11A has not yet been investigated because of the lack of immunoblot analysis of mouse tissue extracts. In

order to make the results of immunohistochemistry with this antibody reliable, these issues should be clarified. Since Fawcett et al.[13] have reported that other two antibodies give the same results as the antibody against EHP-3; in addition to the above detailed immunoblot analyses, further experiments with these antibodies, for example, will be necessary to make clear the above results.

Loughney et al.[31] generated monoclonal antibody against a C-terminal 87 amino acid peptide of PDE11A produced in *E. coli*. The epitope sequence of the monoclonal antibody was situated within VATNRSKWEELHQK, which is almost identical to the antigen peptide sequence (EHP-4) of the antibodies used in the reports by Fawcett et al.[13] (see above) and D'Andrea et al.[20] (see later). The monoclonal antibody demonstrated the presence of an immunoreactive 105 kDa protein corresponding to PDE11A4 in human prostate (strong), pituitary gland (moderate), heart (weak), and liver (weak) on immunoblot analysis. In their report, the other three PDE11A proteins (PDE11A1–3) were not detected in tissues where mRNA expression of these variants has been confirmed. Immunohistochemistry with monoclonal antibody revealed a faint signal for PDE11A4 in neuronal cells within the parasympathetic ganglia and in a subset of nerves, no signal in cardiac myocytes of human heart, and a strong signal in the glandular epithelium of human prostate. Another monoclonal antibody generated by Loughney et al.,[31] which recognizes the EPH-3 sequence, also detected the above described PDE11A4 proteins, however, immunoreactivity to a large unrelated protein was observed with this monoclonal antibody.

Using immunoblot analysis with affinity-purified rabbit polyclonal antibody against the above described synthetic peptide EPH-4 (VATNRSKWEELHQKR), D'Andrea et al.[20] have reported the presence of immunoreactive protein of ~100 kDa in human prostate, bladder, skeletal muscle, corpus cavernosum, and testis, and that of 70 kDa in all these tissues except bladder (Table 13.3). The detection of an immunoreactive ~100 kDa protein in human prostate is matched with that reported by Loughney et al.[31] but the observation is inconsistent with that by Fawcett et al.,[13] demonstrating immunoreactive 56 kDa protein. Regarding immunoreactive proteins in human testis and skeletal muscle, the results obtained by D'Andrea et al.,[20] Loughney et al.,[31] and Fawcett et al.[13] are totally different in spite of using the antibodies recognizing the same peptide sequence (EPH-4). Immunohistochemistry by D'Andrea et al.[20] has demonstrated that immunoreactive PDE11A proteins are distributed over many human tissues, however, signals are also detected in some tissues such as colon, ovary, small intestine, and spleen, where PDE11A mRNA is not expressed at visible levels.[13,15]

As stated above, the localization of PDE11A protein has been investigated with antibodies in several major organs, however, there are a number of differences in the apparent protein size, localization, and tissue distribution of PDE11A in the reports. Currently, the observations by Loughney et al.,[31] which are supported by clear data of immunoblot analysis, seem to be the most reliable. It should be noted that the experiments of immunohistochemistry have not been reproduced by other investigators, and that immunoblot analyses of the immunoreactive proteins, which are necessary to show the protein species of PDE11A, have not always been accompanied by immunohistochemical analyses. Possible reasons for these differences are summarized and discussed in Section 13.11.

13.8 GENOMIC ORGANIZATION

The human *PDE11A* gene is located in the 480 kb genomic DNA region of chromosome 2 (2q31.2). The main body of *PDE11A* is available on the web (http://www.ncbi.nlm.nih.gov/ntrez/query.fcgi?db = gene&cmd = Retrieve&dopt = Graphics&list_uids = 50940), and further 5′ upstream sequence encoding putative promoter regions, exons, and introns for PDE11A3 and PDE11A4 can be retrieved from the GenBank AC011998 sequence (*Homo sapiens* BAC clone

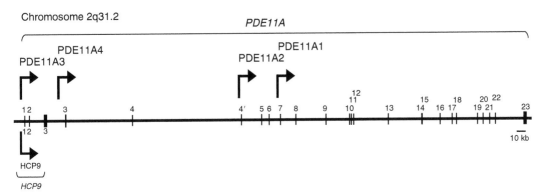

FIGURE 13.2 Structure of the human *PDE11A* and *HCP9* genes. The human *PDE11A* (upper) and *HCP9* (lower) genes are illustrated to scale. The thick horizontal line represents the genomic sequence of *PDE11A* overlapping with the testis-type cytochrome *c* pseudogene *HCP9*, and the short vertical lines and closed boxes indicate exons with exon numbers. The spaces between exons correspond to introns. Transcriptional start sites for the four PDE11A variants and a putative gene product of *HCP9* are indicated by horizontal arrows with vertical lines.

RP11-318 N5 from 2, complete sequence). Four PDE11A variants with a common cDNA sequence encoded by exons 8–23 are generated by alternative splicing (Figure 13.2 and Figure 13.3). In addition to the common sequence at the C-terminal side, cDNAs for PDE11A1, PDE11A3, and PDE11A4 contain sequences encoded by exon 7 (as a noncoding region), exons 1–2 plus exons 4–6, and exons 3–6, respectively.[12] An open reading frame of PDE11A2 is composed of exons 8–23, exons 5–6, and an unidentified exon coding for the 5' noncoding region (exon 4'). Currently, the human *PDE11A* gene is composed of 24 exons. All of the 5' donor and 3' acceptor splice sites, except for 5' splice sites of exons encoding the N-termini and exon 20, are consistent with the canonical GT/AG rule. Exons 3–5, 6, and 8–12 code for GAF domains, and exons 14–22 for the catalytic domain. With regard to exon–intron organization, the catalytic domains of PDE5A, PDE6B, and PDE11A, which are encoded in exons 14–22 of *PDE11A* and exons 12–20 of *PDE5A* and *PDE6B*, are analogous. In spite of high sequence similarity in the catalytic domain of PDE11A with PDE2A and PDE10A (42%–43%), exon organization of *PDE11A* is quite different from that of *PDE2A*[12] or *PDE10A*.[32] The protein regions including GAF domains of PDE5A, PDE6B, and PDE11A are not similar in exon–intron organization compared with the catalytic domains. Although several common exon–intron junctions are found among these PDEs, unique junctions are also present in each GAF domain region.

Transcription of mRNAs of the four PDE11A splice variants is regulated by distinct promoters. Transcription initiation sites of PDE11A1, PDE11A3, and PDE11A4 have already been determined by 5' RACE (rapid amplification of cDNA ends) reaction using cDNAs from human skeletal muscle, testis, and prostate, respectively.[12] On the other hand, transcription initiation site for PDE11A2 has not yet been reported, and therefore, is predicted from the 5' sequence of the longest form of PDE11A2 cDNA.[14] Figure 13.4 shows TRANSFAC[33] analysis of DNA elements that may contribute to the transcriptional control of each PDE11A variant. As reported by Yuasa et al.,[12] the genomic sequence upstream of the transcription initiation site of PDE11A4 is TATA-less and GC-rich, and includes an Sp1-binding sequence. Transcription of PDE11A4 is initiated at multiple transcription start sites in spite of the presence of a CCAAT box. These are the characteristics of a housekeeping gene. On the contrary, the putative promoter regions of PDE11A1, PDE11A2, and PDE11A3

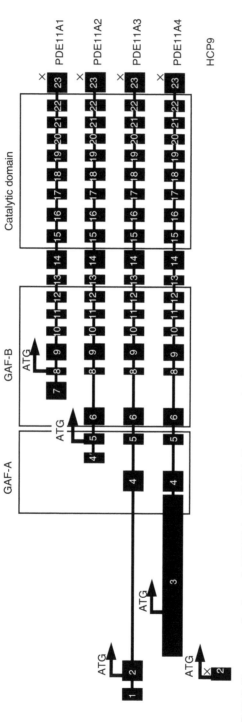

FIGURE 13.3 Exon organization of *PDE11A* and *HCP9*. Exon organization of the four PDE11A splice variants and the *HCP9* gene product are shown. Numbered boxes represent exons and their numbers. Exons encoding GAF domains and the catalytic domains are boxed. ATG with an arrow shows an initiation codons and X-mark indicates a termination codon.

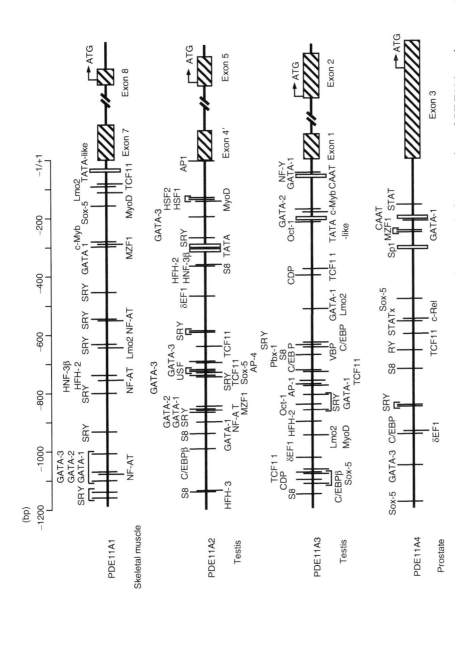

FIGURE 13.4 Putative transcriptional elements in the 5′ upstream of *PDE11A*. Major transcription start sites of PDE11A variants are designated as +1. The 5′ upstream genomic regions of the four PDE11A variants are illustrated by horizontal lines, and the positions of potential regulatory elements, which were searched (cutoff score = 90) by **TRANSFAC**,[33] are indicated by vertical lines. Open and crosshatched boxes represent the putative genomic region involved in transcription and exons, respectively.

all include potential TATA motifs. The possible binding motifs for SRY (sex-determining region Y gene product)[34] and Sox-5 (SRY-related high-mobility group box)[35] are commonly found in the potential regulatory regions of the four PDE11A variants. The binding site of myogenic regulatory factor,[36] MyoD, is located in the putative promoter region of PDE11A1, a skeletal muscle type of PDE11A. On the other hand, an androgen response element, which is present in the promoter regions of many prostate-specific genes,[37] is absent in the putative regulatory region of the prostate-specific variant PDE11A4. Further studies are necessary to reveal the mechanisms of specific expression of the PDE11A variants.

Transcription of PDE11A3 is unique in that the putative promoter of this variant is shared with the testis-specific cytochrome *c* pseudogene *HCP9* in humans[12] (Figure 13.2). The termination codon TGA is present in the open reading frame of *HCP9*, resulting in a nonfunctional protein. In addition, the nucleotide sequence encoding the C-terminal part located in intron 2 of *PDE11A* is disrupted by the *Alu* insertion (Figure 13.5). However, since *HCP9* promoter is still functional, PDE11A3 transcripts are produced and PDE11A3 protein is assumed to be translated from an alternative reading frame located in exon 2. Therefore, the expression of PDE11A3 transcripts is testis-specific.

In rodents, the rat and mouse PDE11A genes (*Pde11a*) are located on chromosome 3 (3q23) and chromosome 2 (2C3), respectively, and their sequence information is available on the Web sites (http://www.ncbi.nlm.nih.gov/entrez/query.fcgi?db = gene&cmd = retrieve& dopt = default&list_uids = 140928) and (http://www.ncbi.nlm.nih.gov/entrez/query.fcgi? db = gene&cmd = retrieve&dopt = default&list_uids = 241489), respectively. Both the rat and mouse *Pde11a* genes have almost the same structure as the human *PDE11A* gene. As in humans, a testis-specific cytochrome *c* gene (*Cyct*), which is a functional gene different from *HCP9*, is situated in the 5′ upstream of both *Pde11a*. Due to this gene organization, transcription of rat and mouse PDE11A3s occurs in a similar way to that of human PDE11A3.[16]

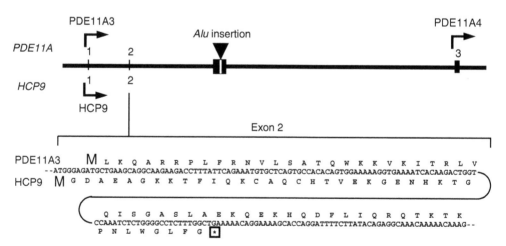

FIGURE 13.5 Comparison of the structures of *PDE11A* and *HCP9* genes. The common 5′ genomic region of *PDE11A* and *HCP9* is illustrated. The thick horizontal line represents the genomic sequence of *PDE11A* and *HCP9*, and the short vertical lines and closed boxes indicate exons with exon numbers. The spaces between exons correspond to introns. Transcriptional start sites for PDE11A3, PDE11A4, and HCP9 are shown by horizontal arrows with vertical lines. *Alu* insertion in exon 3 of *HCP9* is indicated by a white line with an arrowhead. The nucleotide and deduced amino acid sequences of the N-termini of PDE11A3 and HCP9 are also shown. The initiators (methionine) are indicated M with large font size. The position of the termination codon in HCP9 sequence is indicated by an asterisk enclosed by a box.

13.9 SINGLE NUCLEOTIDE POLYMORPHISMS

Currently no single nucleotide polymorphisms (SNPs), leading to amino acid substitution and frame shift, have been reported in-NCBI SNPs database (http://www.ncbi.nlm.nih.gov/SNP/snp_ref.cgi?locusId = 50940# GenBank(mrna)). However, one deletion of three serial nucleotides has been identified as IMS-JST080989 in a database of Japanese single nucleotide polymorphisms (JSNP database [http://snp.ims.u-tokyo.ac.jp/index.html]) in a coding region close to the PDE11A C-terminus by comparison of nucleotide sequences from expressed sequence tag (EST), cDNA, and genomic BAC clones. This deletion causes the removal of the TCC codon for Ser922 of PDE11A4, and is found in human genomic sequence deposited as AC073834 (*Homo sapiens* BAC clone RP11-250 N10 from 2, complete sequence). However, at this point, the heterozygosity of this SNP is unknown. No report has described any effects of this deletion and the location of this serine residue outside the conserved catalytic domain and close to the C-terminus suggests that it may produce no significant alteration of enzymatic characteristics.

13.10 PHYSIOLOGICAL ROLE OF PDE11A AND EFFECTS OF ITS INHIBITION

Gene disruption is one of the effective approaches to investigate the physiological roles of a gene. To study the role of PDE11A in spermatozoa physiology, PDE11A knockout (PDE11$^{-/-}$) mice have been generated by inserting the *lacZ-neo* cassette into exon 12, which is one of the exons coding for the catalytic domain of the mouse *Pde11a* gene.[27] PDE11$^{-/-}$ mice are born according to the Mendelian ratio and grow without any obvious anatomical abnormality. The fertility rate of the deficient homozygotes PDE11$^{-/-}$ is similar to that of the heterozygotes PDE11$^{+/-}$ or the wild-type PDE11$^{+/+}$ littermates. Although the tissue distribution of mouse PDE11A transcripts has not been established by Northern blot analysis, expression of PDE11A transcripts has been demonstrated in the prostate and testis by staining with LacZ derived from the *lacZ* gene inserted in the *Pde11a* locus. No LacZ staining was observed in the aorta.

Abnormalities of sperm function and spermatogenesis in PDE11$^{-/-}$ mice have been reported.[27] No significant difference in overall motility of ejaculated sperm between PDE11$^{-/-}$ and PDE11$^{-/-}$ mice has been observed, and no study has reported sperm malformation in PDE11$^{-/-}$ mice. However, PDE11$^{-/-}$ mice show a significant reduction in sperm forward progression rate (24%), sperm concentration (19%), and percentage of alive ejaculated sperm (20%) compared to PDE11$^{+/+}$ mice. In addition, the proportion of capacitated spermatozoa in the pre-ejaculated sperm of PDE11$^{-/-}$ mice is 53% higher than that in PDE11$^{+/+}$ mice, suggesting that the sperm activation is accelerated before ejaculation in PDE11$^{-/-}$ mice. Uncontrolled sperm activation in the testis and epididymides can affect the number of ejaculated sperm, their motility, and their viability. Spermatogenesis is tightly regulated by cAMP through the cAMP-responsive element modulator (CREM).[38] Disruption of cAMP signaling in spermatogenic cells impedes normal spermatogenesis and results in reduction of sperm number, production of sperm with impaired functions, or both. From the results obtained with PDE11$^{-/-}$ mice, Wayman et al.[27] have postulated a significant physiological role of PDE11A located in the testis and spermatozoa in cAMP-mediated regulation of spermatogenesis via cAMP hydrolysis, especially, in keeping sperm in uncapacitated state before ejaculation and in sperm forward progression after ejaculation. However, in spite of the presence of PDE11A transcripts in mouse prostate and a substantial role of this tissue, which is the secretion of prostatic fluid necessary during ejaculation for the increase of sperm motility, involvement of PDE11A in prostate function have not been investigated. A loss of PDE11A function in the prostate may also be related to the relevant phenotypes of PDE11$^{-/-}$ mice, and therefore, prostate function of the mice should be investigated.

No specific PDE11A inhibitors have been reported. Currently, tadalafil, a PDE5 inhibitor ($IC_{50} = 6.7$ nM), is the most effective PDE11A inhibitor with IC_{50} value of 73 nM for PDE11A4.[21] Other potent PDE5 inhibitors, i.e., sildenafil ($IC_{50} = 3.7$ nM) and vardenafil ($IC_{50} = 0.091$ nM), are less effective in inhibiting PDE11A than tadalafil (IC_{50} value of 3800 nM for sildenafil and 840 nM for vardenafil). The pharmacological effects of tadalafil are often referred in order to give explanation for the physiological roles of PDE11A, however, it should be noted that tadalafil's effects are observed under PDE5 inhibition. It cannot be simply concluded that the observations after administration of tadalafil are entirely caused by PDE11A inhibition. Regarding tadalafil's effects, material safety data sheet of tadalafil (http://www.ehs.lilly.com/msds/msds_tadalafil_tablets.html) indicates that chronic overexposure (daily for 6 to 12 months at doses of 25 mg/kg/d and above) to tadalafil causes testicular tissue changes (decrease of testis weight associated with degeneration and atrophy of the seminiferous epithelium) and decrease in sperm production in dogs. By contrast, no significant change in testicular tissue or sperm production has been reported in rats or mice. In a clinical trial, daily intake of tadalafil at doses of 10 and 20 mg for 6 months produced no change in spermatogenesis or reproductive hormones in men aged 45 years or older.[39] On the contrary, Pomara et al.[40] have reported that tadalafil causes a statistically significant decrease ($12.3\% \pm 2.2\%$ [control] vs. $9\% \pm 1.9\%$ [tadalafil], $p < 0.05$) in sperm straight-line velocity in men under 40 years.

Based on the tadalafil IC_{50} value for PDE11A, the testicular tissue changes and decrease in sperm production caused by the chronic overexposure of this drug in dogs are probably due, at least in part, to inhibition of PDE11A. However, the presence of PDE11A protein in testicular tissue is indispensable as the basis of inferring PDE11A function there, nevertheless, there is no report describing PDE11A mRNA and proteins in the canine reproductive system. Therefore, there remains a possibility that the testicular tissue changes may come from a simple side effect of chronically administered tadalafil appeared in the canine reproductive system because the compound has a unique chemical structure different from sildenafil and vardenafil. On the contrary, inhibition of PDE11A in the prostate, if canine PDE11A4 protein exists there as well as in humans, may cause reduction of sperm production through the changes in prostate function. Alterations in sperm parameters caused by tadalafil in men under 40 years, who had been diagnosed with fertility problems, may also be due to inhibitory effects of tadalafil on PDE11A. Certainly, alterations of sperm physiology by the administration of tadalafil in humans seem consistent with the changes found in PDE11$^{-/-}$ mice.[27] However, the pharmacokinetics and pharmacodynamics of tadalafil are totally different from those of sildenafil and vardenafil, i.e., tadalafil acts over a relatively long period of time as compared to sildenafil or vardenafil; and considering differences of these drug properties and tissue distribution of human PDE11A protein, the effects obtained with tadalafil in the studies mentioned above should be carefully assessed before concluding that they are simply attributable to PDE11A inhibition in the testis (that in the prostate may be possible). Especially, the existence of PDE11A protein in human testicular tissue is questionable as described in Section 13.7, and regarding PDE11$^{-/-}$ mice, a compensatory upregulation of other PDE genes might occur in tissues expressing PDE11A, which leads to underestimation of the physiological functions of PDE11A in wild-type animals. Therefore, studies with a potent and selective PDE11A inhibitor and more extensive clinical studies seem warranted and should be pursued to investigate the physiological roles of PDE11A, the contribution of PDE11A activity, and the effects of PDE11A inhibition in testicular and sperm function.

As described above, immunoreactive PDE11A proteins have been found in the corpus cavernosum and vasculature of human penis, and involvement of PDE11A in nerve-mediated penile erection has been investigated using UK-343664, a potent PDE5 specific inhibitor and tadalafil, a PDE5/11 dual inhibitor.[29] Although both compounds are potent relaxants of

corpus cavernosum, the relaxant effects of UK-343664 are more potent than those of tadalafil as reflected by IC_{50} values of these compounds for PDE5A. It is therefore concluded that PDE11A inhibition is not associated with corpus cavernosum relaxation.

13.11 DISCUSSION

Biochemical and molecular analyses of PDE11A have revealed that (1) human PDE11A family is composed of four N-terminal splice variants, (2) PDE11A is a dual substrate PDE with high sensitivity to tadalafil, and (3) PDE11A mRNA is expressed densely in the prostate and moderately in the testis, pituitary gland, skeletal muscle, liver, prostate, and heart. In addition, analysis of *PDE11A* genome has indicated that PDE11A3 transcription is promoted by a testis-specific cytochrome *c* pseudogene promoter. These biological characteristics of PDE11A are commonly accepted, however, the distribution and existence of PDE11A variant proteins are still controversial. Suggested PDE11A mRNA tissue distribution and human PDE11A variant proteins are summarized in Table 13.3.

Northern blot analysis has demonstrated that long forms of PDE11A mRNA are densely present in the prostate and that a short form of PDE11A mRNA can be found in the testis.[13,15] The difference in the size of PDE11A mRNA is likely due to the different PDE11A variants because PCR analysis has demonstrated that human PDE11A3 and PDE11A4 transcripts are testis- and prostate-specific, respectively.[12,15] The question arises if endogenous proteins equivalent to each of the four PDE11A variants are present in human tissues? Although several reports have indicated the presence of multiple PDE11A proteins in human tissues, the results of immunoblot analysis have been puzzling. The first report[13] using affinity-purified polyclonal antibody against the EPH-3 peptide has demonstrated the existence of an immunoreactive 56 kDa protein corresponding in size to PDE11A1 in the prostate, where PDE11A4 transcripts coding for a 105 kDa protein are abundant.[15] A strong signal for a 78 kDa protein and weak signals of 56 and 65 kDa, corresponding to PDE11A3, PDE11A1, and PDE11A2 in size, respectively, have been detected in skeletal muscle extracts where PDE11A1 transcripts are the most dominant forms.[12] Although it is difficult to explain these observations, one of the acceptable reasons for the absence of a full-size PDE11A4 protein in the prostate is posttranslational processing of PDE11A proteins. PDE11A4 protein may undergo a rapid proteolytic degradation, resulting in three smaller forms of PDE11A proteins during sample preparation or under physiological conditions. This idea is supported by the results of D'Andrea et al.[20] showing the existence of an immunoreactive ~100 kDa protein (corresponding to a recombinant protein of PDE11A4 in migration) and a 70 kDa protein in the prostate, skeletal muscle, corpus cavernosum, and testis of humans. Another possible reason is alternative translation of PDE11A4 from initiation codons situated downstream of the predicted first methionine codon. However, alternative translation of PDE11A4 is unlikely because a construct designed to produce a recombinant protein of PDE11A4 using its own initiation codon (but not an initiation codon in the N-terminal tag sequence) has been shown to give a protein with a predicted size of 105 kDa.[31] Regarding the detection of immunoreactive proteins with greater molecular mass than PDE11A1 in skeletal muscle,[13] it may be due to unidentified PDE11A variants, modification of PDE11A1 protein by a sugar chain or lipid, or insufficient specificity of the antibody used (this seems the most plausible explanation). Immunoblot analysis by D'Andrea et al.[20] also demonstrated the presence of some PDE11A protein species in the prostate and skeletal muscles, which do not agree with those predicted by analysis of PDE11A-variant mRNA. Particularly, the presence of an immunoreactive ~100 kDa protein in human testis,[20] where the existence of a 78 kDa protein of PDE11A3 is expected from mRNA analysis, is totally inconsistent with those made by another group, demonstrating no PDE11A-related protein in human testis[31] as described later.

Loughney et al.[31] have demonstrated that an immunoreactive protein of 105 kDa, equivalent to the molecular mass of PDE11A4, is the only variant actually detectable in human tissues. Their findings that PDE11A4 protein is present as a predicted form in the prostate (high level), pituitary gland (moderate), liver (lower), and heart (lower) strongly agree with tissue distribution pattern and levels of PDE11A4 mRNA.[12,15] The existence of other variant proteins in the testis and skeletal muscle has not been confirmed with this antibody in spite of the presence of PDE11A transcripts in these tissues. These findings indicate that the levels of PDE11A variant proteins other than PDE11A4 are very low in the testis and skeletal muscle or that these proteins are too unstable to be detected. Regarding the discrepancy of the presence of PDE11A3 mRNA[12,14,15] but the absence of the corresponding protein[31] in human testis, it has been suggested that translation of PDE11A3 might be inefficient, resulting in the production of PDE11A3 protein at undetectable level *in vivo* because translation of PDE11A3 is considered to be initiated from an unusual alternative translation initiation codon in a different reading frame situated within the coding region of the *HCP9* gene product (Figure 13.5).

Although several explanations and speculations on these discrepancies above are possible, differences in specificity and sensitivity of the antibodies used for PDE11A protein detection are the most probable causes. Confusion is also found in cellular localization of PDE11A proteins identified by immunohistochemistry. Several studies using polyclonal antibodies have revealed signals of PDE11A protein in murine testis and in human testis, prostate, penis, spleen, intestine, and adrenal gland.[20,27–30] Using monoclonal antibody, human PDE11A proteins have been shown to be present in the prostate and neuronal cell bodies of the heart.[31] Although at present the results obtained with anti-PDE11A antibodies are conflicting, tissue distribution of PDE11A4 protein shown by Loughney et al.'s study[31] using monoclonal antibody very well matches with that of the PDE11A4 mRNA, strongly indicating the high specificity of this monoclonal antibody for PDE11A and the presence of PDE11A4 protein with a calculated molecular weight in the prostate and other tissues. According to Loughney et al.'s results,[31] PDE11A3 protein is present at a very low or undetectable level in human testis. The results obtained with polyclonal antibodies against synthetic peptides[13,20,27,28] cannot be denied, however, more well-controlled studies will be required to explain discrepancies of tissue distribution between mRNA and proteins and differences in PDE11A species predicted by immunoblot and mRNA analyses.

The physiological roles of PDE11A have not been clarified. As is often the case, the physiological functions of proteins are often predicted and discussed based on the information of their tissue distribution. With regard to the relation of PDE11A with testicular and sperm functions, the presence of PDE11A3 mRNA in the testis is the basis of this idea in spite of the existence of PDE11A proteins being questionable there. For instance, changes in sperm parameters in PDE11$^{-/-}$ mice and in *in vivo* and clinical studies with tadalafil have been deemed to be due to a loss of PDE11A activity. Transcripts for PDE11A3 are present in the testis of humans and rodents definitively, but the corresponding protein is not present at a significant level in humans[31] as described above and has not yet investigated in rodents. Therefore, it is hard to conclude that the reported abnormal spermatogenesis is linked to inhibition of testicular PDE11A. Species difference suggested by Yuasa et al.[16] should also be considered. Studies focusing on PDE11A function in the prostate, where PDE11A4 protein is detected, are indispensable to explain and clarify the reported observations. At least, analyses of PDE11A protein in mouse testis and other tissues including the prostate using Loughney et al.'s monoclonal antibody are intriguing and should be carried out first to clarify the above discussion.

Francis[41] has given valuable comments in her discussion to resolve issues of discrepancy in PDE11A protein localization and physiological roles. Development of PDE11A-specific

inhibitors is indispensable to settle arguments on the physiological roles of PDE11A and the pharmacological effects of tadalafil on PDE11A inhibition. PDE11A-specific inhibitors would help understand endogenous activities of PDE11A and its involvement in cyclic nucleotide hydrolysis in tissue extracts. Such important approaches are urgently required in order to elucidate the physiological role of PDE11A in humans and to determine whether PDE11A-specific inhibitors have potential for the treatment of disease states. Aid in furthering this would be crystallographic analyses of the PDE11A protein as a key step in developing PDE11A-specific inhibitors.

REFERENCES

1. Beavo, J.A., and L.L. Brunton. 2002. Cyclic nucleotide research—Still expanding after half a century. *Nat Rev Mol Cell Biol* 3:710.
2. Houslay, M.D., and G. Milligan. 1997. Tailoring cAMP-signalling responses through isoform multiplicity. *Trends Biochem Sci* 22:217.
3. Soderling, S.H., and J.A. Beavo. 2000. Regulation of cAMP and cGMP signaling: New phosphodiesterases and new functions. *Curr Opin Cell Biol* 12:174.
4. Lucas, K.A., et al. 2000. Guanylyl cyclases and signaling by cyclic GMP. *Pharmacol Rev* 52:375.
5. Watts, V.J. 2002. Molecular mechanisms for heterologous sensitization of adenylate cyclases. *J Pharmacol Exp Ther* 302:1.
6. Beavo, J.A. 1995. Cyclic nucleotide phosphodiesterases: Functional implications of multiple isoforms. *Physiol Rev* 75:724.
7. Francis, S.H., I.V. Turko, and J.D. Corbin. 2001. Cyclic nucleotide phosphodiesterases: Relating structure and function. *Prog Nucleic Acid Res Mol Biol* 65:1.
8. Houslay, M.D. 2001. PDE4 cAMP-specific phosphodiesterases. *Prog Nucleic Acid Res Mol Biol* 69:249.
9. Michel, J.J., and J.D. Scott. 2002. AKAP mediated signal transduction. *Annu Rev Pharmacol Toxicol* 42:235.
10. Aravind, L., and E.V. Koonin. 1998. The HD domain defines a new superfamily of metal-dependent phosphohydrolases. *Trends Biochem Sci* 23:469.
11. Aravind, L., and C.P. Ponting. 1997. The GAF domain: An evolutionary link between diverse phototransducing proteins. *Trends Biochem Sci* 22:458.
12. Yuasa, K., et al. 2001. Genomic organization of the human phosphodiesterase PDE11A gene. Evolutionary relatedness with other PDEs containing GAF domains. *Eur J Biochem* 268:168.
13. Fawcett, L., et al. 2000. Molecular cloning and characterization of a distinct human phosphodiesterase gene family: PDE11A. *Proc Natl Acad Sci USA* 97:3702.
14. Hetman, J.M., et al. 2000. Cloning and characterization of two splice variants of human phosphodiesterase 11A. *Proc Natl Acad Sci USA* 97:12891.
15. Yuasa, K., et al. 2000. Isolation and characterization of two novel phosphodiesterase PDE11A variants showing unique structure and tissue-specific expression. *J Biol Chem* 275:31469.
16. Yuasa, K., et al. 2001. Identification of rat cyclic nucleotide phosphodiesterase 11A (PDE11A): Comparison of rat and human PDE11A splicing variants. *Eur J Biochem* 268:4440.
17. Weeks, J.L., et al. 2005. High biochemical selectivity of tadalafil, sildenafil and vardenafil for human phosphodiesterase 5A1 (PDE5) over PDE11A4 suggests the absence of PDE11A4 cross-reaction in patients. *Int J Impot Res* 17:5.
18. Fujishige, K., et al. 1999. Cloning and characterization of a novel human phosphodiesterase that hydrolyzes both cAMP and cGMP (PDE10A). *J Biol Chem* 274:18438.
19. Gross-Langenhoff, M., et al. 2006. cAMP is a ligand for the tandem GAF domain of human phosphodiesterase 10 and cGMP for the tandem GAF domain of phosphodiesterase 11. *J Biol Chem* 281:2841.
20. D'Andrea, M.R., et al. 2005. Expression of PDE11A in normal and malignant human tissues. *J Histochem Cytochem* 53:895.
21. Gebkor, E., et al. 2002. Selectivity of sildenafil and other phosphodiesterase type 5 (PDE5) inhibitors against all human phosphodiesterase families. *Eur Urol* 1 (Suppl. 1):63.

22. Degerman, E., P. Belfrage, and V.C. Manganiello. 1997. Structure, localization, and regulation of cGMP-inhibited phosphodiesterase (PDE3). *J Biol Chem* 272:6823.

23. Lim, J., G. Pahlke, and M. Conti. 1999. Activation of the cAMP-specific phosphodiesterase PDE4D3 by phosphorylation: Identification and function of an inhibitory domain. *J Biol Chem* 274:19677.

24. Corbin, J.D., et al. 2000. Phosphorylation of phosphodiesterase-5 by cyclic nucleotide-dependent protein kinase alters its catalytic and allosteric cGMP-binding activity. *Eur J Biochem* 267:2760.

25. Sharma, R.K., and J.H. Wang. 1985. Differential regulation of bovine brain calmodulin-dependent cyclic nucleotide phosphodiesterase isoenzymes by cyclic AMP-dependent protein kinase and calmodulin-dependent phosphatase. *Proc Natl Acad Sci USA* 82:2603.

26. Kotera, J., et al. 2004. Subcellular localization of cyclic nucleotide phosphodiesterase type 10A variants, and alteration of the localization by cAMP-dependent protein kinase-dependent phosphorylation. *J Biol Chem* 279:4366.

27. Wayman, C., et al. 2005. Phosphodiesterase 11 (PDE11) regulation of spermatozoa physiology. *Int J Impot Res* 17:216.

28. Baxendale, R.W., F. Burslem, and S.C. Phillips. 2001. Phosphodiesterase type 11 (PDE11) cellular localization: Progress towards defining a physiological role in testis and/or reproduction. *J Urol* 165 (Suppl.):340.

29. Baxendale, R.W., et al. 2001. Cellular localization of phosphodiesterase type 11 (PDE11) in human corpus cavernosum and the contribution of PDE11 inhibition on nerve-stimulated relaxation. *J Urol* 165 (Suppl.):233.

30. Ückert, S., et al. 2004. Immunohistochemical presence of phophodiesterase (PDE) 11A in the human prostate. *J Urol* 171 (Suppl.):352.

31. Loughney, K., J. Taylor, and V.A. Florio. 2005. 3′,5′-Cyclic nucleotide phosphodiesterase 11A: Localization in human tissues. *Int J Impot Res* 17:320.

32. Fujishige, K., et al. 2000. The human phosphodiesterase PDE10A gene genomic organization and evolutionary relatedness with other PDEs containing GAF domains. *Eur J Biochem* 267:5943.

33. Heinemeyer, T., et al. 1999. Expanding the TRANSFAC database towards an expert system of regulatory molecular mechanisms. *Nucleic Acids Res* 27:318.

34. Gubbay, J., et al. 1990. A gene mapping to the sex-determining region of the mouse Y chromosome is a member of a novel family of embryonically expressed genes. *Nature* 346:245.

35. Denny, P., et al. 1992. An SRY-related gene expressed during spermatogenesis in the mouse encodes a sequence-specific DNA-binding protein. *EMBO J* 11:3705.

36. Lassar, A.B., et al. 1989. MyoD is a sequence-specific DNA binding protein requiring a region of myc homology to bind to the muscle creatine kinase enhancer. *Cell* 58:823.

37. Riegman, P.H.J., et al. 1991. The promoter of the prostate-specific antigen gene contains a functional androgen responsive element. *Mol Endocrinol* 5:192.

38. Nantel, F., et al. 1996. Spermiogenesis deficiency and germ-cell apoptosis in CREM-mutant mice. *Nature* 380:159.

39. Hellstrom, W.J., et al. 2003. Tadalafil has no detrimental effect on human spermatogenesis or reproductive hormones. *J Urol* 170:887.

40. Pomara, G., et al. 2004. Effect of acute *in vivo* sildenafil and tadalafil treatments on semen parameters in patients with fertility problem, a randomized double-blind, cross over study. *J sex Med* 2 (suppl. 2):23.

41. Francis, S.H. 2005. Phosphodiesterase 11 (PDE11): Is it a player in human testicular function? *Int J Impot Res* 17:467.

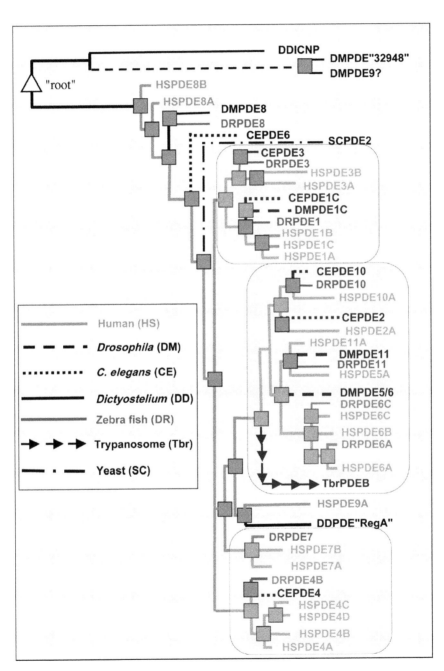

FIGURE 1.2

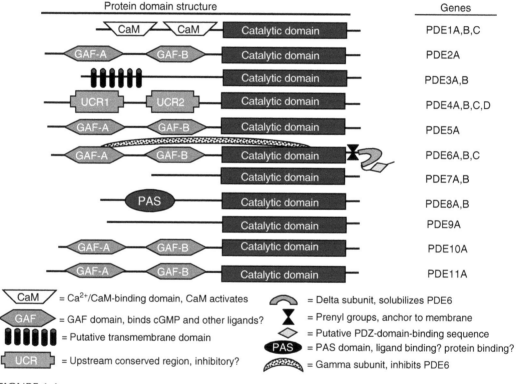

FIGURE 1.4

```
              #         *      #** *      #Δ                    Δ          ΔΔ** #**Δ#
1B      FLMSFLDALETGYGKYKNPYHNQIHAADVTQTVHCFLLR--TGMVHCLSEIELLAIIFAAAIHDYEHTGTTNSFHIQTKSEC
2A      TLARFCLMVKKGY-R-DPPYHNWMHAFSVSHFCYLLYKN--LELTNYLEDIEIFALFISCMCHDLDHRGTNNSFQVASKSVL
3A      EFMNYFHALEIGY-R-DIPYHNRIHATDVLHAVWYLTTQPIPGLSTVINDLELMALYVAAAMHDYDHPGRTNAFLVATSAPQ
4B      TFVTYMMTLEDHY-HSDVAYHNSLHAADVAQSTHVLLST--PALDAVFTDLEILAAIFAAAIHDVDHPGVSNQFLINTNSEL
5A      VLCRWILSVKKNY-RKNVAYHNWRHAFNTAQCMFAALKA--GKIQNKLTDLEILALLIAALSHDLDHRGVNNSYIQRSEHPL
6A      ALVRFMYSLSKGY-R-KITYHNWRHGFNVGQTMFSLLVT--GKLKRYFTDLEALAMVTAAFCHDIDHRGTNNLYQMKSQNPL
7A      KLRRFLVMIQEDYHS-QNPYHNVAVHAADVTQAMHCYLKE--PKLASSVTPWDILLSLIAAATHDLDHPGVNQPFLIKTNHYL
8A      TLRSWLQIIEANYHS-SNPYHNSTHSADVLHATAYFLSK--ERIKETLDPIDEVAALIAATIHDVDHPGRTNSFLCNAGSEL
9A      TLRRWLFCVHDNY-R-NNPFHNFRHCFCVAQMMYSMVWL--CSLQEKFSQTDILILMTAAICHDLDHPGYNNTYQINARTEL
10A     KLCRFIMSVKKNY-R-RVPYHNWKHAVTVAHCMYAILQN---N-HTLFTDLERKGLLIACLCHDLDHRGFSNSYLQKFDHPL
11A     TLCRWLLTVRKNY-R-MVLYHNWRHAFNVCQLMFAMLTT--AGFQDILTEVEILAVIVGCLCHDLDHRGTNNAFQAKSGSAL
SCPDE   KLLLLLLFTLESSYHQ-VNKFHNFRHAIDVMQATWRLCTY-----LLKDNPVQTLLLCMAAIGHDVGHPGTNNQLLCNCESEV
RegA    KLQRFIMTVNALY-RKNNRYHNFTHAFDVTQTVYTFLTS--FNAAQYLTHLDIFALLISCMCHDLNHPGFNNTFQVNAQTEL
TbPDE2  VLLNFILQCRKKY-R-NVPYHNFYHVVDVCQTIYTFLYR--GNVYEKLTELECFVLLITALVHDLDHMGLNNSFYLKTESPL
DMPDE   TFLNFMSTLEDHY-VKDNPFHNSLHAADVTQSTNVLLNT--PALEGVFTPLEVGGALFAACIHDVDHPGLTNQFLVNSSSEL


              #         # *#* Δ#                   #                    #*   Δ#
1B      AIVYNDRSVLENHHISSVFRLMQD-DEMNIFINLTKDEFVELRALVIEMVLATDMSCHFQQVKTMKTALQQL----------
2A      AALYSSESVMERHHFAQAIAILNT-HGCNIFDHFSRKDYQRMLDLMRDIILATDLAHHLRIFKDLQKMAEVG----------
3A      AVLYNDRSVLENHHAAAAWNLFMSRPEYNFLINLDHVEFKHFRFLVIEAILATDLKKHFDFVAKFNGKVNDDVG--------
4B      ALMYNDESVLENHHLAVGFKLLQE-EHCDIFQNLTKKQRQTLRKMVIDMVLATDMSKHMSLLADLKTMVETKKVTSSG----
5A      AQLY-CHSIMEHHFDQCLMILNS-PGNQILSGLSIEEYKTTLKIIKQAILATDLALYIKRRGEFFELIRKN----------
6A      AKLH-GSSILERHHLEFGKTLLRD-ESLNIFQNLNRRQHEHAIHMMDIAIIATDLALYFKKRTMFQKIVDQSKTYESEQEWT
7A      ATLYKNSAVLENHHWRSAVGLLRE---SGLFSHLPLESRQEMEAQIGALILATDISRQNEYLSLFRSHLDKG----------
8A      AILYNDTSVLESHHAALAFQLTTGDDKCNIFKNMERNDYRTLRQGIIDMVLATEMTKHFEHVNKFVNSINKPLATLEENGET
9A      AVRYNDISPLENHHCAVAFQILAE-PECNIFSNIPPDGFKQIRQGMITLILATDMARHAEIMDSFKEKMEN-----------
10A     AALY-STATMEQHHFSQTVSILQL-EGHNIFSTLSSSEYEQVLEIIRKAIIATDLALYFGNRKQLEEMYQTG----------
11A     AQLYGTSSTLEHHHFNHAVMILQS-EGHNIFANLSSKEYSDLMQLLKQSILATDLTLYFERRTEFFELVSKG----------
SCPDE   AQNFKNVSILENFHRELFQQLLSE-HWPQLLSISKKK-----FDFISEAILATDMALHSQYEDRLMHEN-------------
RegA    SLEYNDISVLENHHAMLTFKILRN-SECNILEGLNEDQYKELRRSVVQLILATDMQNHFEHTNKFQHHLNNLP---------
TbPDE2  GILSSASSVLEVHHCNLAVEILSD-PESDVFGGLEGAERTLAFRSMIDCVLATDMARHSEFLEKYLEIMKTS----------
DMPDE   ALMYNDESVLENHHLAVAFKLLQN-QGCDIFCNMQKKQRQTLRKMVIDIVLSTDMSKHMSLLADLKTMVETKKVAGSG----


              #    *    #          # *#  #   *          Δ#Δ                    Δ
1B      -----ERIDKPKALSLLLHAADISHPTKQWLVHSRWTKALMEEFFRQGDKEA-ELGLPFSP--LCDR-TSTLVAQSQIGFIDFI
2A      -YDRNNKQHHRLLLCLLMTSCDLSDQTKGWKTTRKIAELIYKEFFSQGDLEK-AMGNRPME--MMDREKAY-IPELQISFMEHI
3A      -IDWTNENDRLLVCQMCIKLADINGPAKCKELHLQWTDGIVNEFYEQGDEEA-SLGLPISP--FMDR-SAPQLANLQESFISHI
4B      VLLLDNYTDRIQVLRNMVHCADLSNPTKSLELYRQWTDRIMEEFFQQGDKER-ERGMEISP--MCDKHTAS-VEKSQVGFIDYI
5A      QFNLEDPHQKELFLAMLMTACDLSAITKPWPIQQRIAELVATEFFDQGDRERKELNIEPTD--LMNREKKNKIPSMQVGFIDAI
6A      QYMMLEQTRKEIVMAMMMTACDLSAITKPWEVQSQVALLVAAEFWEQGDLERTVLQQNPIP--MMDRNKADELPKLQVGFIDFV
7A      DLHLDDGRHRHLVLQMALKCADICNPCRNWELSKQWSEKVTEEFFHQGDIEK-KYHLGVSP--LCDRQTES-IANIQIGFMTYL
8A      NTMLRTPENRTLIKRMLIKCADVSNPCRPLQYCIEWAARISEEYFSQTDEEKQQGLPVVMP--VFDRNTCS-IPKSQISFIDYF
9A      -FDYSNEEHMTLLKMILIKCCDISNEVRPMEVAEPWVDCLLEEYFMQSDREK-SEGLPVAP--FMDRDKVT-KATAQIGFIKFV
10A     SLNLNNQSHRDRVIGLMMTACDLCSVTKLWPVTKLTANDIYAEFWAEGD-EMKKLGIQPIP--MMDRDKKDEVPQGQLGFYNAV
11A     EYDWNIKNHRDIFRSMLMTACDLGAVTKPWEISRQVAELVTSEFFEQGDRERLELKLTPSA--IFDRNRKDELPRLQLEWIDSI
SCPDE   ------PMKQITLISLIIKAADISNVTRTLSISARWAYLITLEFNDCALLETFHKAHRPEQDCFGDSYKNVDSPKEDLESIQNI
RegA    -FDRNKKEDRQMILNFLIKCGDISNIARPWHLNFEWSLRVSDEFFQQSHYET-ICGYPVTP--FMDKTKTT-RARIAADFIDFV
TbPDE   -YNVDDSDHRQMTMDVLMKAGDISNVTKPFDVSRQWAMAVTEEFYRQGDMEK-ERGVEVLP--MFDRSKNMELAKGQIGFIDFV
DMPDE   VLLLDNYTDRIQVLENLVHCADLSNPTKPLPLYKRWVALLMEEFFLQGDKER-ESGMDISP--MCDRHNAT-IEKSQVGFIDYI
```

FIGURE 1.5

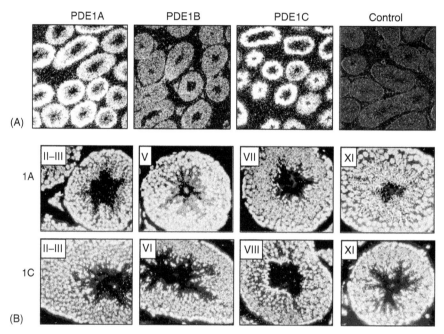

(A)

(B)

FIGURE 3.6

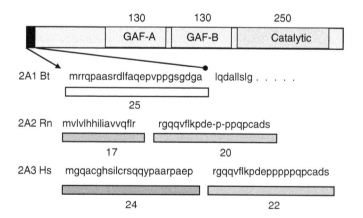

FIGURE 4.1

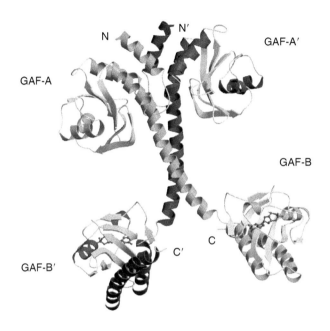

FIGURE 4.4

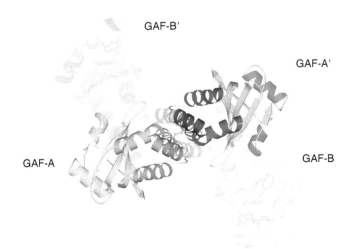

FIGURE 4.5

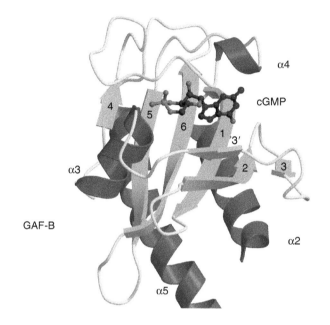

FIGURE 4.6

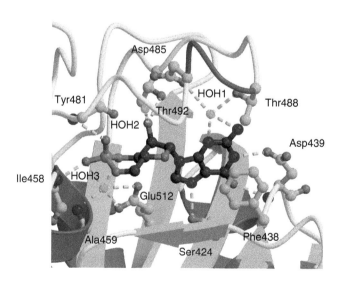

FIGURE 4.7

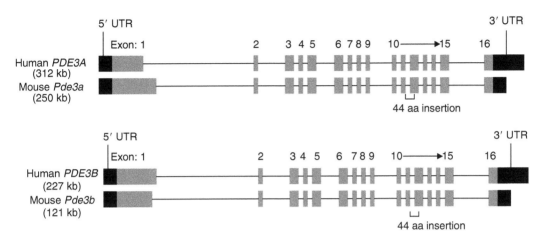

FIGURE 5.1

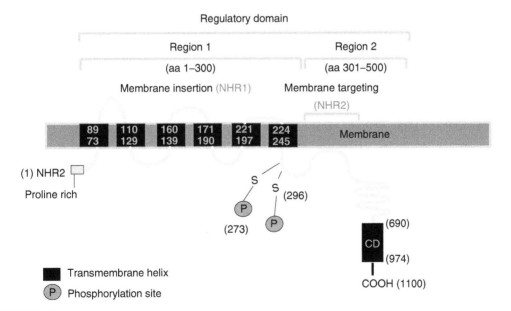

FIGURE 5.2

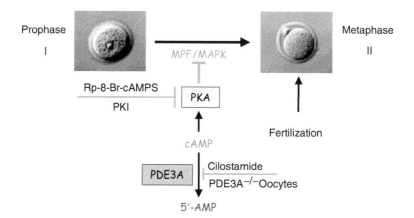

FIGURE 5.3

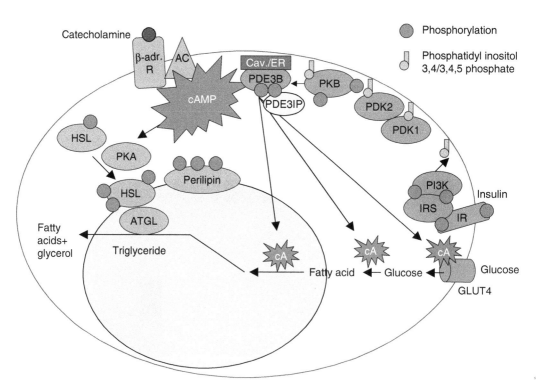

FIGURE 5.4

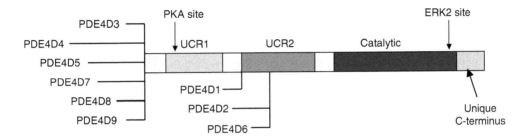

FIGURE 6.1

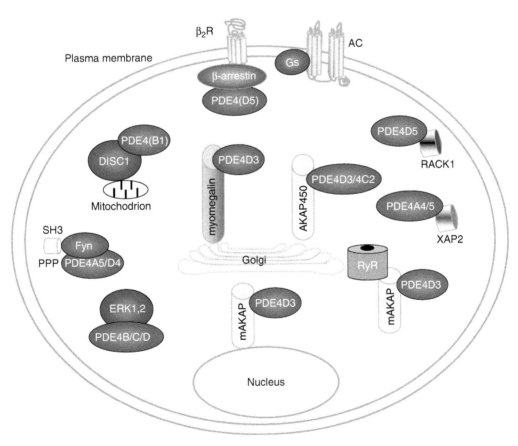

FIGURE 6.2

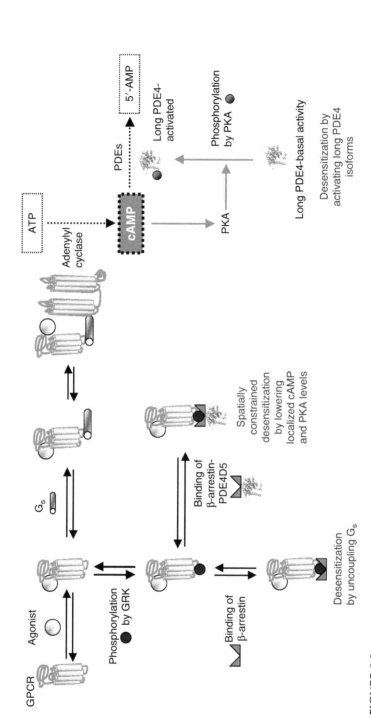

FIGURE 6.3

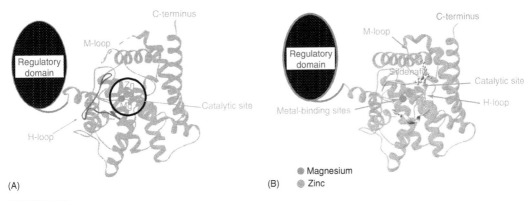

(A)

(B)

● Magnesium
● Zinc

FIGURE 7.8

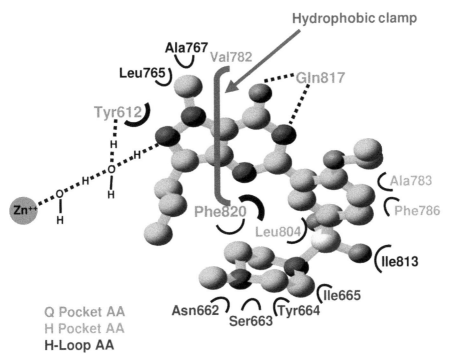

Hydrophobic clamp

Ala767 Val782
Leu765 Gln817
Tyr612
H
H
H₂O
H
Zn⁺⁺
H
Ala783
Phe786
Phe820 Leu804
Ile813
Asn662 Ser663 Tyr664 Ile665

Q Pocket AA
H Pocket AA
H-Loop AA

FIGURE 7.9

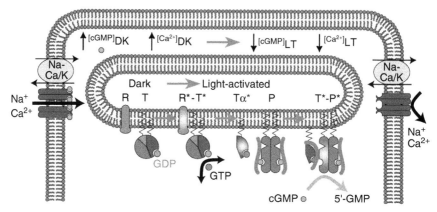

[cGMP]DK
[Ca2+]DK
[cGMP]LT
[Ca2+]LT
Na-Ca/K
Na-Ca/K
Dark
Light-activated
R T
R*-T*
Tα*
P
T*-P*
Na+ Ca2+
Na+ Ca2+
GDP
GTP
cGMP
5'-GMP

FIGURE 8.2

PDE7A1/7A3
PDE7A1/7A2
MEV...PVPQHVLSRRGAISFSSS...PNPQLSQ
46
FMTYL...............QRLS
RRGAISYDSS..............................ASIQIG
MGITLIWCLALVLIKWITSK
20
NYTYLDIAG
PDE7A2
PDE7A3

FIGURE 9.1

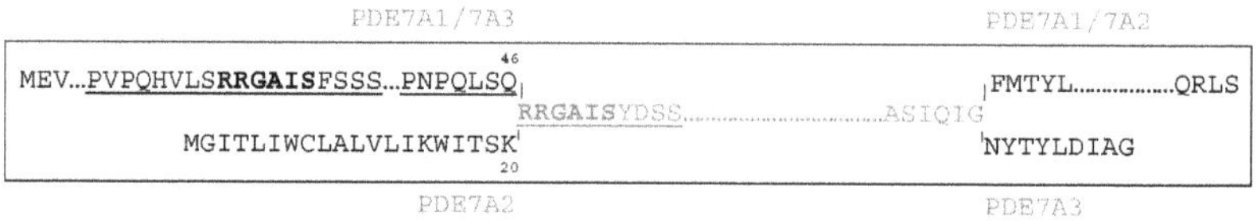

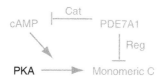

Cat
cAMP
PDE7A1
Reg
PKA
Monomeric C

FIGURE 9.3

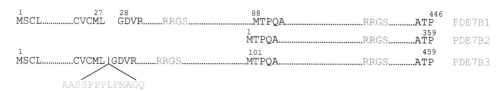

1 MSCL..............CVCML 27 28 GDVR.........RRGS.............................88 MTPQA...................................RRGS...........ATP 446 PDE7B1
1 MTPQA.......................................RRGS...........ATP 359 PDE7B2
1 MSCL..............CVCML | GDVR.........RRGS.............................101 MTPQA.......................................RRGS...........ATP 459 PDE7B3
AASSPPPLPMAGQ

FIGURE 9.4

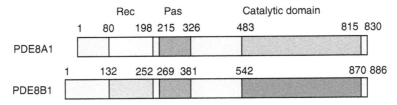

FIGURE 10.1

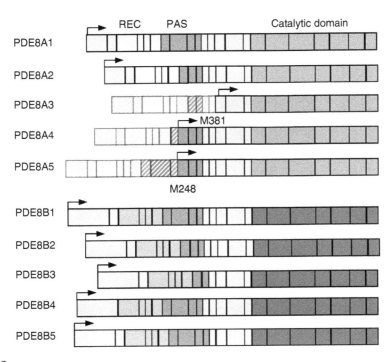

FIGURE 10.2

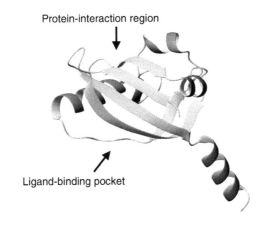

FIGURE 10.3

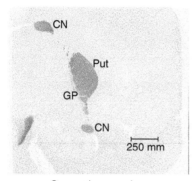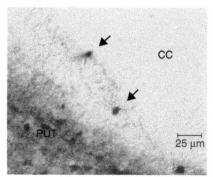

Cynomolgus monkey | Human

FIGURE 12.3

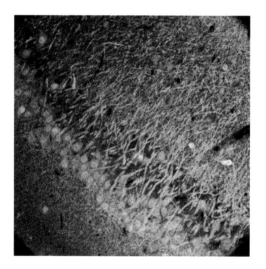

FIGURE 12.4

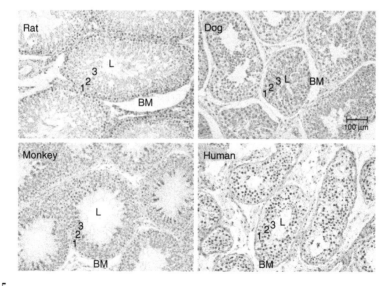

FIGURE 12.5

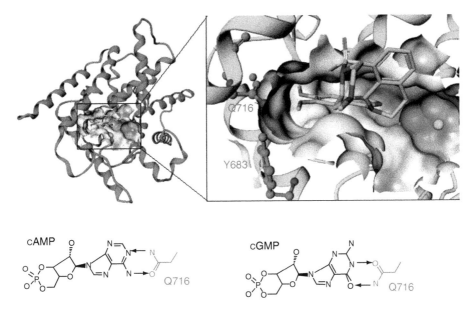

cAMP

cGMP

Q716

Q716

FIGURE 12.6

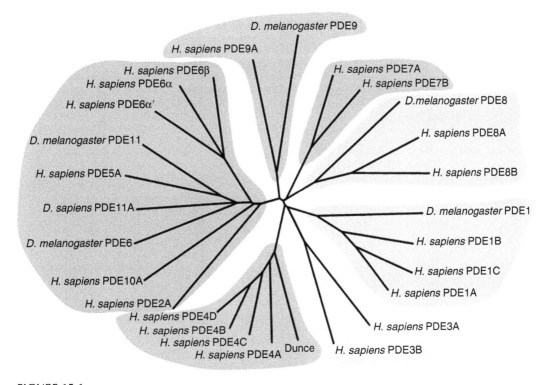

D. melanogaster PDE9

H. sapiens PDE9A

H. sapiens PDE6β
H. sapiens PDE6α

H. sapiens PDE6α'

D. melanogaster PDE11

H. sapiens PDE5A

D. sapiens PDE11A

D. melanogaster PDE6

H. sapiens PDE10A

H. sapiens PDE2A
H. sapiens PDE4D
H. sapiens PDE4B
H. sapiens PDE4C
H. sapiens PDE4A
Dunce

H. sapiens PDE7A
H. sapiens PDE7B

D.melanogaster PDE8

H. sapiens PDE8A

H. sapiens PDE8B

D. melanogaster PDE1

H. sapiens PDE1B

H. sapiens PDE1C

H. sapiens PDE1A

H. sapiens PDE3A

H. sapiens PDE3B

FIGURE 15.1

```
D.mel PDE1    222  EAFLHRVEEGYCRYRNPYHNNLHAVDVMQTIHYCLCNTGLMNWL--TDLEIFASLLAALL
H.sap PDE1B   205  MSFLDALETGYGKYKNPYTHNQIHAADVTQTVHCFLLRTGMWHCL--SEIELLAIIFAAAI
D.mel dunce   360  LNFMSTLEDHYVKD-NPFHNSLHAADVTQSTNVLLNTPALEGVF--TPLEWGGALFAACI
H.sap PDE4A   177  VTYMLTLEDHYHAD-VAYHNSLHAADVLQSTHVLLATPALDAVF--TDLEILAALFAAAI
H.sap PDE5A   596  CRWLLSVRKNYRK-NVAYHNWRHAFNTAQCMFAALKAGKIQNKL--TDLEILALLIAALS
D.mel PDE6    715  CRWLSVRKNYRP--VKYHNWRHAFNVAQTMFAMLKTGKHERFM--TDLEILGLLVACLC
H.sap PDE6b   542  VRFMYAMKKGTRK--VTYHNWRHGFNVAQTMFTLLMTGKLRCYY--SDLEAMAMVTAGFC
D.mel PDE8    769  KAWLAVIEAHYRK-SNTYHNSTHAADVMQATGAFITQLTNKDMLVMDRRMEEATALIAAAA
H.sap PDE8A   539  RSWLQIIEANYHS-SNPYHNSTHSADVLHATAYFLSKERIKETL--DPIDEVAALIAATI
D.mel PDE9    95   REWLYEVYKHYN--EVPFHNFRHCFCVAQMYAITRQANLLSRL--GDLECLILLVSCIC
H.sap PDE9A   296  RRWLFCVHDNYRN--NPFHNFRHCFCVAQMMYSMVWLCSLQEKF--SQTDILILMTAAIC
D.mel PDE11   935  CRWLLSVRKNYR--NVTYHNWRHAFNVAQTMFAILTTTQWWKIF--GEIECLALIIGCLC
H.sap PDE11A  648  CRWLLTVRKNYR--MVLYHNWRHAFNVCQLMFAMLTTAGFQDIL--TEVEILAVIVGCLC

D.mel PDE1    280  HDYEHTGTTNNFHVMSGSETALLYNDRAVLENHHASASFRLLR-EDEYNILSHLSREEFR
H.sap PDE1B   263  HDYEHTGTTNSFHIQTRSECAIVYNDRSVLENHHISSVFRLMQ-DDEMNIFINLTKDEFV
D.mel dunce   417  HDVDHPGLTNQFLVNSSSEKALMYNDRSVLENHHLAVAFKLLQ-NQGCDIFCNMQKKQRQ
H.sap PDE4A   234  HDVDHPGVSNQFLINTNSELALMYNDESVLENHHLAVGFKLLQ-EDNCDIFQNLSKRQRQ
H.sap PDE5A   653  HDLDHRGVNNSYIQRSEHPLAQLYC-HSIMEHHHFDQCLMILN-SPGNQILSGLSIEEYK
D.mel PDE6    771  HDLDHRGTNNAFQTKTESPLAILYT-TSTMEHHHFDQCVMILN-SEGNNIFQALSPEDYR
H.sap PDE6b   598  HDIDHRGTNNLYQMKSQHPLARLHG-SSYLERHHLEFGKFLLA-NESMNIFLNLNRRQHE
D.mel PDE8    828  HDVDHPGRSSAFLCNSNDALAVLYNDLTVLENHHAAITFRLTLGDDKINIFKNLDKETYK
H.sap PDE8A   596  HDVDHPGRTNSFLCNAGSELAILYNDTAVLESHHAALAFQLTTGDDKCNIFKNMERNDYR
D.mel PDE9    151  HDLDHPGYNNIYQINARTELALRYNDISPLENHHCSIAFRLLE-HPECNIFKNFSRDTFN
H.sap PDE9A   352  HDLDHPGYNNTYQINARTELAVRYNDISPLENHHCAVAFQILA-EPECNIFSNIPPDGFK
D.mel PDE11   991  HDLDHRGTNNSFQIKASSPLAQLYS-TSTMEHHHFDQCLMILN-SPGNQILANLSSDDYC
H.sap PDE11A  704  HDLDHRGTNNAFQAKSGSALAQLYGTSATLEHHHFNHAVMILQ-SEGHNIFANLSSKEYS

D.mel PDE1    339  ELRGLVIEMVLGTDMTNHFQQMKAMRQLLTLQEAT--------------------IDKQ
H.sap PDE1B   322  ELRALVIEMVLATDMSCHFQQVKTMKTALQQLER--------------------IDKP
D.mel dunce   476  TLRKMVIDIVLSTDMSKHMSLLADLKTMVETKKVAGSGVLLLDNY-----------TDRI
H.sap PDE4A   293  SLRKMVIDMVLATDMSKHMTLLADLKTMVETKKVTSSGVLLLDNY-----------SDRI
H.sap PDE5A   711  TTLKIIKQAILATDLALYIKRRGEFFELIPK--NQFNLEDPH--------------Q-KE
D.mel PDE6    829  SVMKTVESAILSTDLAMYFKKRNAFLELVLENGEFDWQGEE---------------KKD
H.sap PDE6b   656  HWIHLMDIAILATDLALYFKKRTMFQKIVDQSKIYEEQDKWVDYLS-------LETTRKE
D.mel PDE8    888  SARSTIIDMILATEMTRHFEHLARFVSVFGGEEPRDHNPQT---D----------EETSI
H.sap PDE8A   656  TLRQGIIDMVLATEMTKHFEHVNKFVNSINKPLATLEENGETDKNQEVINTMLRTPENRT
D.mel PDE9    210  MIREGIIRCILATDMARHNEILTQFMEITPIFDYSN------------------RAHIN
H.sap PDE9A   411  QIRQGMITLILATDMARHAEIMDSFKEKMENFDYS--NEE----------------HMT
D.mel PDE11   1049 RVIRVLEDAILSTDLAVYFKKRGPFLESVSQPTSYWVAEEP----------------RA
H.sap PDE11A  763  DLMQLLKQSILATDLTLYFERRTEFFELVSKGEYDWNIKN----------------HRD

D.mel PDE1    378  KVLSLVLHCCDISHPAKQWGVHHRWTMLLLEEFFRQGDLEK-ELGLPFSPLCDRNN----
H.sap PDE1B   360  KALSLLLHAADISHPTKQWLVHSRWTKALMEEFFRQGDREA-ELGLPFSPLCDRTS----
D.mel dunce   525  QVLENLVHCADLSNPTKPLPLYKRWVALLMEEFFLQGDRER-ESGMDISPMCDRHN----
H.sap PDE4A   342  QVLRNMVHCADLSNPTKPLELYRQWTDRIMAEFFQQGDRER-ERGMEISPMCDKHT----
H.sap PDE5A   754  LFLAMLMTACDLSAITKPWPIQQRIAELVATEFFDQGDRERKELNIEPTDLMNREKK---
D.mel PDE6    872  LLCGMMMTACDVSAIAKPWEVQHKVAKLVADEFFDQGDLEKLQLNTQPVAMMDRERK---
H.sap PDE6b   709  IVMAMMMTACDLSAITKPWEVQSKVALLVAAEFWEQGDLERTVLQQQPIPMMDRNKS---
D.mel PDE8    935  LMRRMLIKVADVSNPARPMQFCIEWARRIAEEYFMQTDEEKQRHLPIVMPMFDRAT----
H.sap PDE8A   716  LIKRMLIKCADVSNPCRPLQYCIEWAARISEEYFSQTDEEKQQGLPVVMPVFDRNT----
D.mel PDE9    251  LLCMILIKVADISNEARPMDVAEPWLDRLLQEFFAQSAAEKSEGLPVTPFMDPDK-----
H.sap PDE9A   452  LLKMILIKCCDISNEVRPMEVAEPWVDCLLEEYFMQSDREK-SEGLPVAPFMDRDK----
D.mel PDE11   1092 LLRAMSMTVCDLSAITKPWEIEKRWVADLVSSEFFEQGDMEKQELNITPIDIMNREKKE--
H.sap PDE11A  806  IFRSMLMTACDLGAVTKPWEISRQWAELVTSEFFEQGDRERLELKLTPSAIFDRNRK---
```

FIGURE 15.3

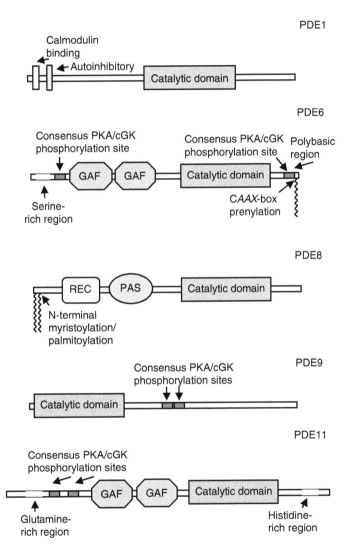

FIGURE 15.4

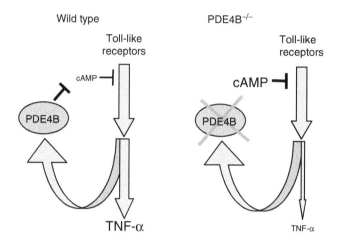

FIGURE 16.2

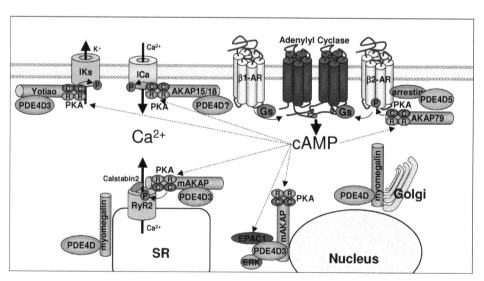

FIGURE 16.3

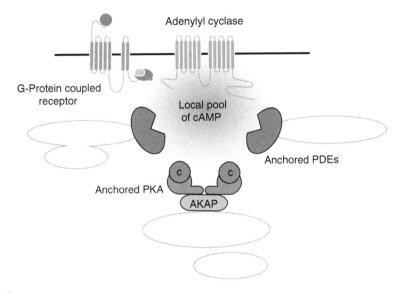

FIGURE 19.1

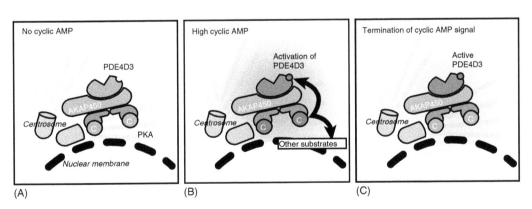

FIGURE 19.2

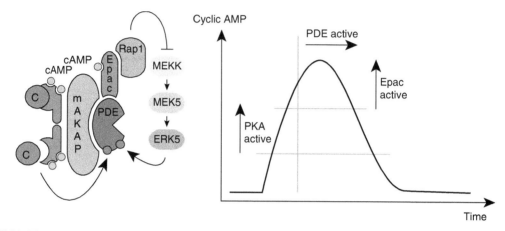

FIGURE 19.3

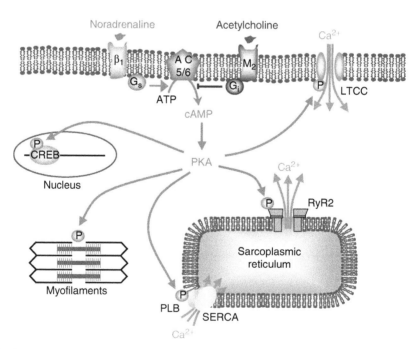

FIGURE 20.1

(A)   Double-barreled microperfusion

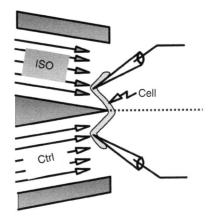

(B)   FRET-based imaging

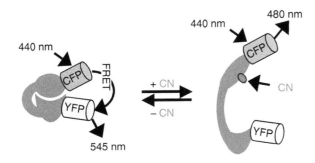

(C)   Recombinant CNG channels

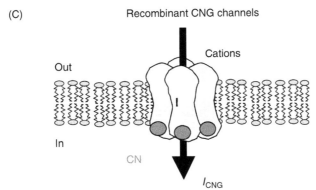

FIGURE 20.2

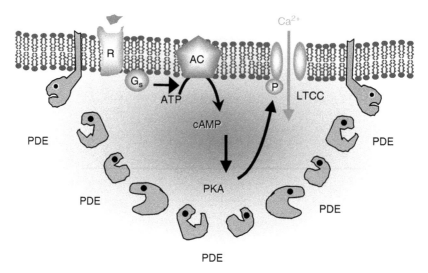

FIGURE 20.3

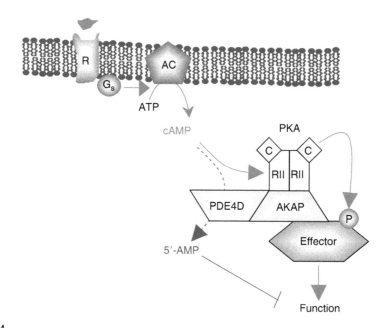

FIGURE 20.4

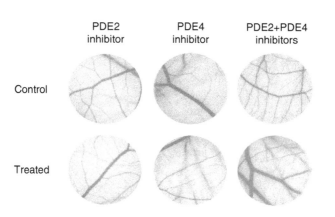

FIGURE 21.13

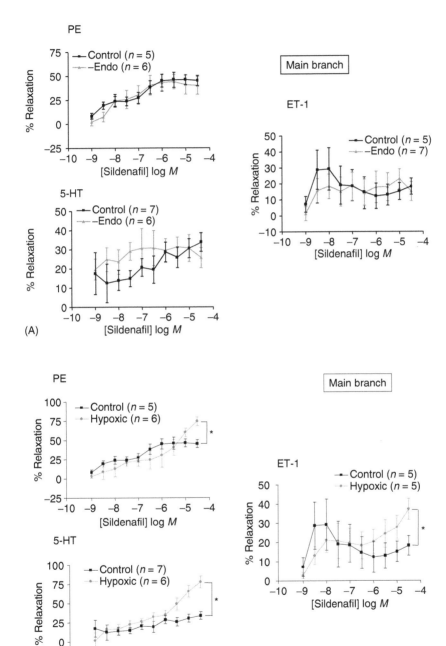

FIGURE 25.2

(*continued*)

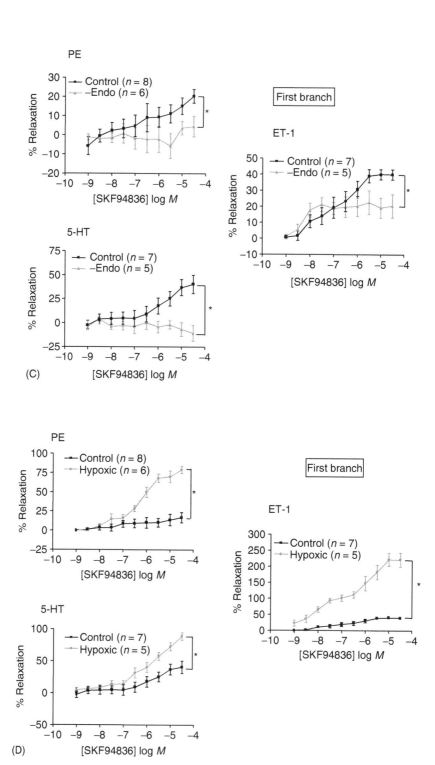

FIGURE 25.2 (continued)

Human PDE5 gene organization
Chromosome 4 – NC_000004

120777036
1

120907584
130549

NC_00004

hPDE5a1 (NM_001083)

ATG

5a

STOP

hPDE5a2 (NM_033430)

5a2

STOP

hPDE5a3 (NM_033437)

ATG

5a

STOP

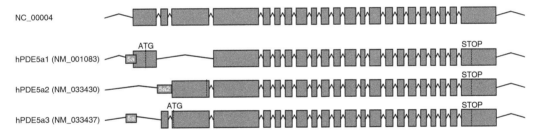

Exon
Intron
Promoter

FIGURE 25.4

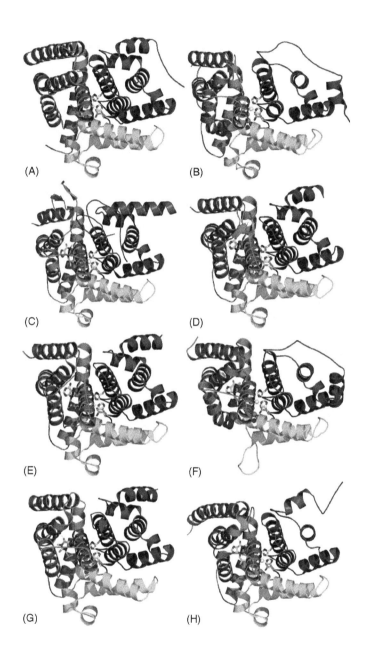

(A)

(B)

(C)

(D)

(E)

(F)

(G)

(H)

FIGURE 29.1

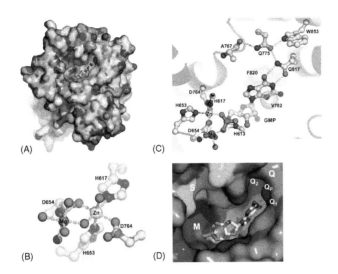

(A)   (C)

(B)   (D)

FIGURE 29.2

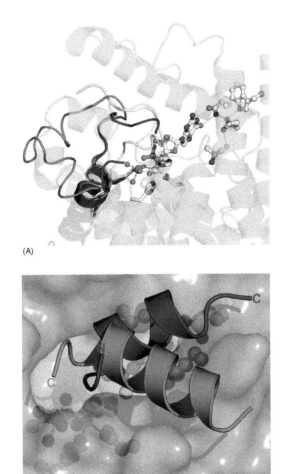

(A)

(B)

FIGURE 29.3

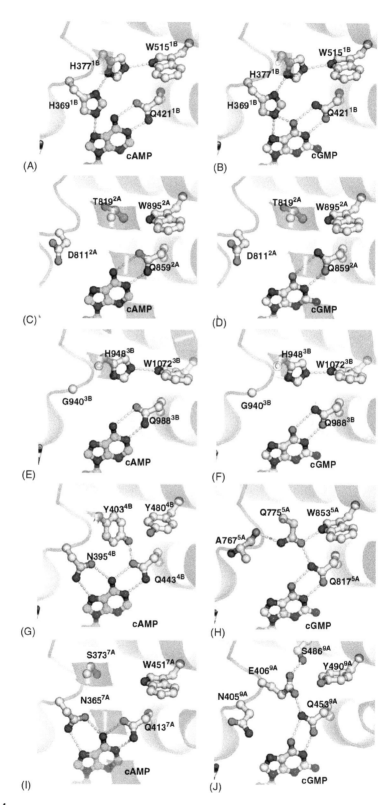

FIGURE 29.4

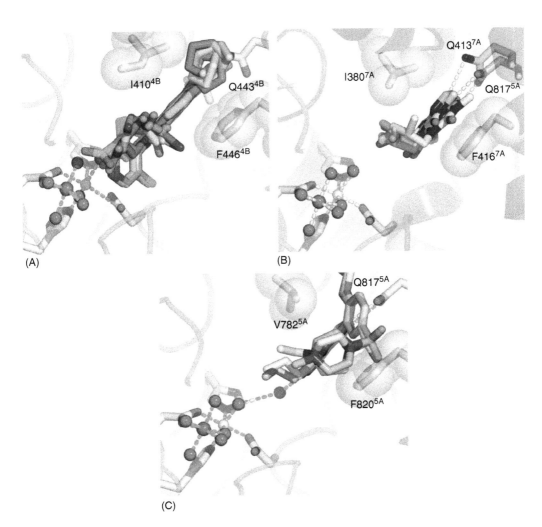

FIGURE 29.5

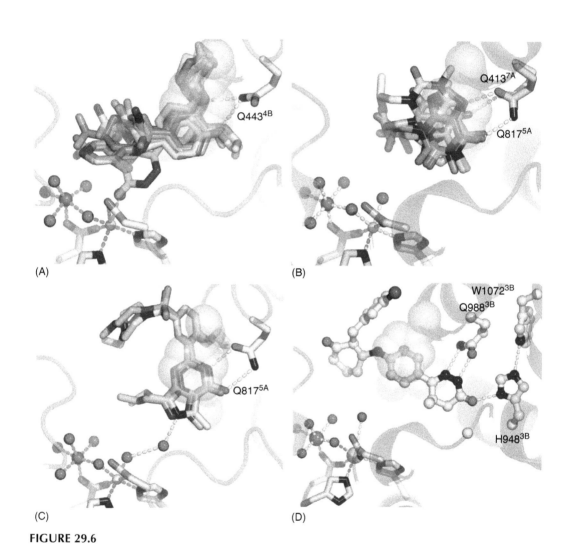

(A)   (B)   (C)   (D)

FIGURE 29.6

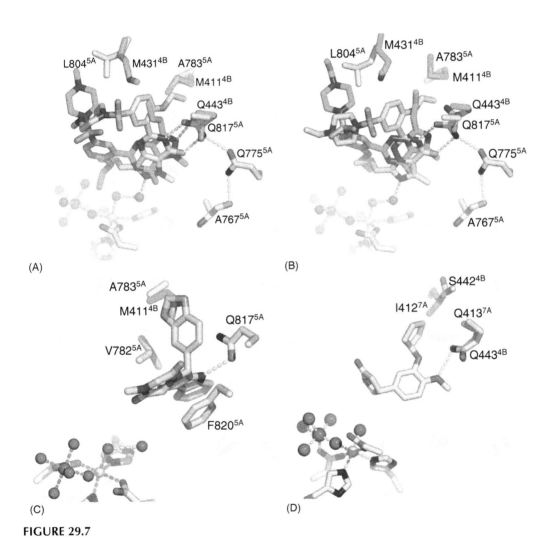

(A)

(B)

(C)

(D)

FIGURE 29.7

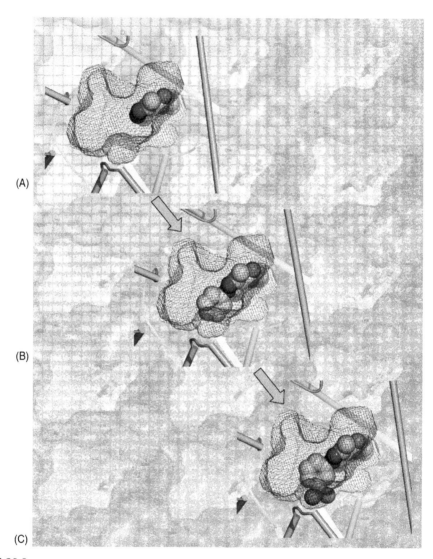

(A)

(B)

(C)

FIGURE 29.8

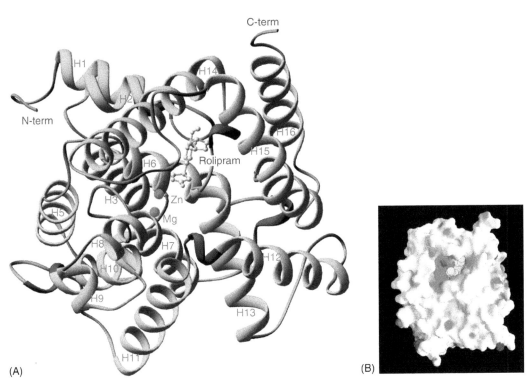

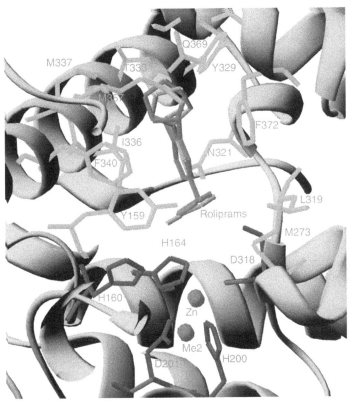

(A)

(B)

(C)

FIGURE 30.2

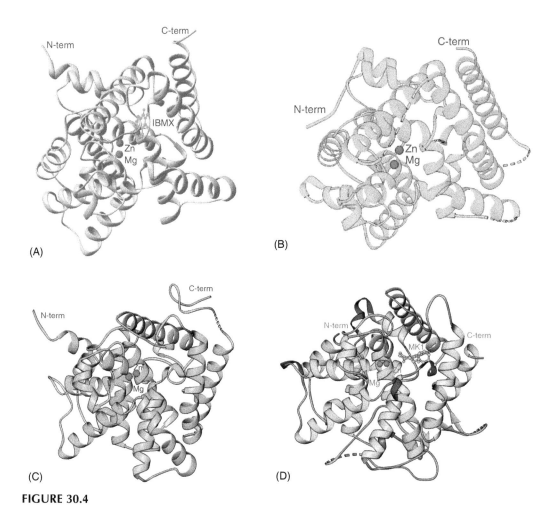

(A)

(B)

(C)

(D)

FIGURE 30.4

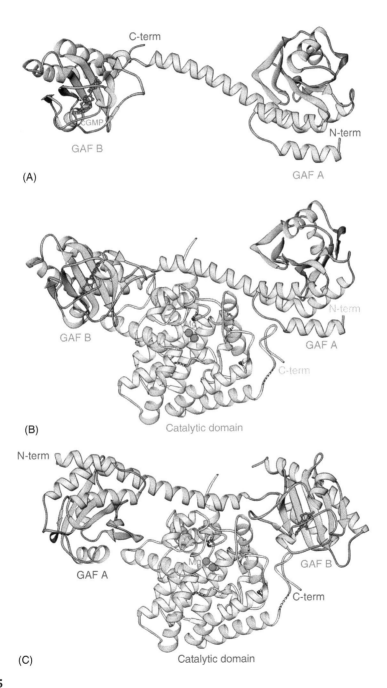

(A)

GAF B

CGMP

GAF A

C-term

N-term

(B)

GAF B

GAF A

N-term

C-term

Catalytic domain

(C)

N-term

GAF A

GAF B

Mg

C-term

Catalytic domain

FIGURE 30.5

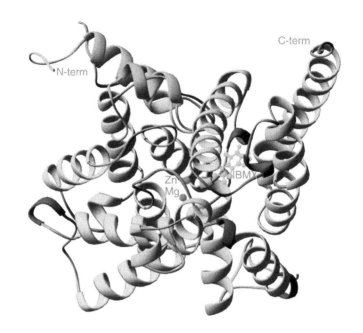

FIGURE 30.6

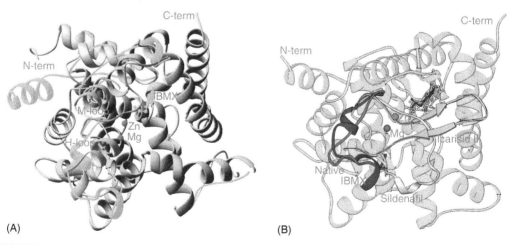

(A)

(B)

FIGURE 30.7

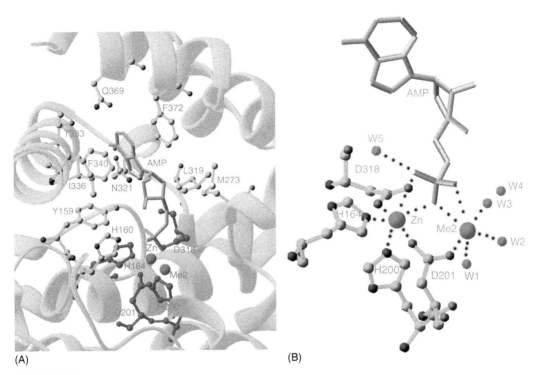

(A)

(B)

FIGURE 30.8

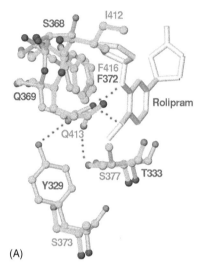

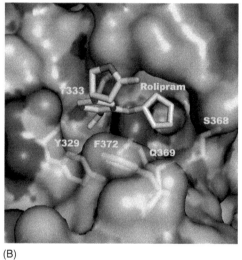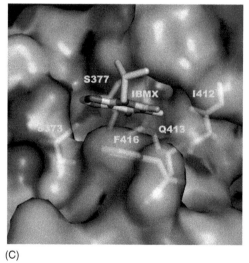

(A)

(B)   (C)

FIGURE 30.11

Section B

Nonmammalian Phosphodiesterases

14 Protozoal Phosphodiesterases

Laurent Wentzinger and Thomas Seebeck

CONTENTS

14.1 Outline 278
14.2 Fungal Phosphodiesterases 278
    14.2.1 General Remarks 278
    14.2.2 Fungal Class II Phosphodiesterases...278
        14.2.2.1 *Saccharomyces cerevisiae* ScPDE1 278
        14.2.2.2 *Schizosaccharomyces pombe* SpPDE1 280
        14.2.2.3 *Candida* spp. PDE1...280
        14.2.2.4 Other Fungi 280
        14.2.2.5 Biological Roles of Fungal PDE1 281
    14.2.3 Fungal Class I Phosphodiesterases 281
        14.2.3.1 *Saccharomyces cerevisiae* ScPDE2 281
        14.2.3.2 *Candida* spp. PDE2...282
14.3 *Dictyostelium* 283
    14.3.1 General Remarks 283
    14.3.2 DdPDE1 283
        14.3.2.1 Cell Biology of DdPDE1 284
        14.3.2.2 DdPDI—An Inhibitor of DdPDE1 285
    14.3.3 DdPDE2 285
        14.3.3.1 Cell Biology of DdPDE2 286
    14.3.4 DdPDE3 287
        14.3.4.1 Cell Biology of DdPDE3 287
    14.3.5 DdPDE4 288
    14.3.6 DdPDE5 288
        14.3.6.1 Cell Biology of DdPDE5 288
    14.3.7 DdPDE6 289
    14.3.8 DdPDE7 289
14.4 Kinetoplastida 289
    14.4.1 General Remarks 289
    14.4.2 The PDEA Family..290
        14.4.2.1 TbrPDEA of *Trypanosoma brucei* 290
        14.4.2.2 LmjPDEA of *Leishmania major* 291
    14.4.3 The PDEB Family 291
        14.4.3.1 *Trypanosoma brucei* 291
        14.4.3.2 *Leishmania major* 293
        14.4.3.3 *Trypanosoma cruzi* 293

14.4.4   The PDEC Family 294
        14.4.4.1   TcrPDEC of *T. cruzi* 294
14.5   *Plasmodium* 294
        14.5.1   General Remarks..294
        14.5.2   PfPDE1 295
                14.5.2.1   Cell Biology of PfPDE1..295
14.6   Conclusions 296
References 296

14.1   OUTLINE

The phosphodiesterases (PDEs) of protozoa, i.e., the unicellular eukaryotes, differ from their metazoan counterparts mainly in two aspects. (1) In the metazoa including humans, all PDEs belong to the class I superfamily of PDEs, which share a conserved catalytic domain. Members of the protozoal PDEs exhibit three distinct types of catalytic domains: (i) the canonical class I domain, (ii) the class II domain that is present in fungi, slime molds, and amoebae, and (iii) the related β-lactamase-type of PDE catalytic domain that was found so far only in two PDEs of the slime mold, *Dictyostelium* (Figure 14.1). (2) With respect to the regulatory domains, the protozoal PDEs contain less variety than their metazoal counterparts. While most of the protozoal PDEs do not contain any regulatory domains, those that do contain GAF domains, cNMP-binding domains that belong to the catabolite gene activator protein (CAP) family, FYVE domains, and response regulator domains.

14.2   FUNGAL PHOSPHODIESTERASES

14.2.1   General Remarks

The analysis of fungal PDEs has largely developed with the two prototypic yeasts, *Saccharomyces cerevisiae* and *Schizosaccharomyces pombe*. More recently, the PDEs of the fungal pathogens *Candida* spp. have also come into the focus of interest. The completion of a number of additional fungal genomes now allows a first overview over the fungal phosphodiesterome. Most fungi, for which complete genome sequence information is available, contain two different PDEs. One of them is a class I PDE (signature sequence: $-H(X)_3H(X)_{25-35}(D/E)-^1$), whereas the second is a member of the class II PDEs (signature sequence: (GST)-H-X-HLDH-X-X-(AGS)), so that both appear to be restricted to bacteria, fungi, and some protozoal species. Current exceptions to this two-PDE rule are the ascomycete *S. pombe* that contains only a class II PDE, and the microsporidium *Encephalitozoon cuniculi* that only contains a class I PDE. For historical reasons, the fungal class I PDEs are designated as XyPDE2 whereas the class II PDEs are designated as XyPDE1. Neither of the two PDEs is essential for cell growth under standard culture conditions, but both play distinct roles in the maintenance of cAMP homeostasis. The high-K_m, class II PDEs are the key players for responding to sensory input whereas the primary role of the low-K_m, class I PDEs be in maintaining the steady-state levels of cAMP.

14.2.2   Fungal Class II Phosphodiesterases

14.2.2.1   *Saccharomyces cerevisiae* ScPDE1

ScPDE1 was first characterized using purified enzyme and reported to be a roughly spherical homodimer with a molecular weight of 86,000.[2,3] The enzyme contains two tightly bound Zn^{2+} atoms per monomer that are essential for catalysis. Agents that interact with Zn^{2+} such

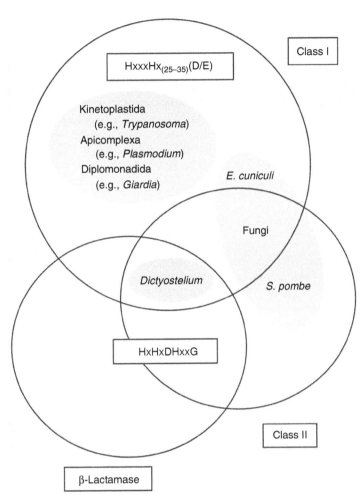

FIGURE 14.1 The three classes of PDEs found in the protozoal kingdom. The kinetoplastida, apicomplexa, diplomonadida, and presumably many other protozoa contain only class I PDEs. The fungi generally contain both, class I and class II PDEs, with some exceptions such as *Encephalitozoon cuniculi* (one class I PDE only) or *Schizosaccharomyces pombe* (one class II PDE only). The slime mold *Dictyostelium discoideum* is the only species so far that contains three classes of PDEs, class I, class II, and metallo-β-lactamase PDEs. The class II and the metallo-β-lactamase PDEs both contain the minimal signature sequence HxHxDHxxG.

as KCN (but not NaN$_3$), 1,10-phenanthroline, or thiols inactivate the enzyme. The activity can be partly reconstituted by the addition of Zn^{2+}. The enzyme remains stable in the presence of ethylenediaminetetraacetate (EDTA), demonstrating that the Zn^{2+} ions are tightly bound to the polypeptide,[3] and that ScPDE1 does not require divalent cations such as Mg^{2+} or Mn^{2+} for its activity. Interestingly, the activity of ScPDE1, i.e., its V_{max}/K_m ratio, is almost independent of temperature[2] whereas the reaction rate of the adenylyl cyclase at all ATP concentrations approximately doubles for every 10°C rise in temperature.[4] This disparate temperature dependence of the two enzymes might play a role in controlling temperature stress. ScPDE1 is a dual substrate PDE with K_m values of 250 and 160 μM for cAMP and cGMP, respectively. ScPDE1 also hydrolyzes other cyclic nucleotides, such as

cyclic inosine-, cytosine-, or uridine monophosphates, albeit with much higher K_m values (0.8–2 mM). The membrane-permeable cAMP derivative N^6,O^2-dibutyryl cAMP is not hydrolyzed to a measurable extent.[5]

The ScPDE1 gene was identified by a complementation screen for high copy-number suppressors of the heat-shock-sensitive phenotype of RAS2[val19] mutants in *S. cerevisiae*.[6] It codes for a polypeptide of 369 amino acids, which consists essentially of a class II catalytic domain (GenBank accession number P22434). Yeast cells that lack both PDE genes, ScPDE1 and ScPDE2, are perfectly viable but are heat-shock sensitive. Very low levels of any type of PDE activity are sufficient to rescue to mutant phenotype.[7] These findings have made *S. cerevisiae* ScPDE1⁻ ScPDE2⁻ deletion strains a most elegant experimental system for studying recombinant PDEs from various sources.[8]

14.2.2.2   *Schizosaccharomyces pombe* SpPDE1

SpPDE1 was identified by screening for suppressors of the phenotype conferred by the inactivation of the ran1 and protein kinase.[9] The gene recovered from this screen, *cgs2*, codes for a class II PDE of 346 amino acids, SpPDE1 (accession number P36599) that is closely similar to its homologs from *S. cerevisiae* and other fungi. In the *S. pombe* genome, *cgs2* is the only gene that codes for a recognizable PDE, and this organism does not seem to contain a class I PDE.

14.2.2.3   *Candida* spp. PDE1

The CaPDE1 gene of the human pathogen *Candida albicans* was identified by complementation cloning in a PDE-deficient *S. cerevisiae* strain.[7] Its open reading frame codes for a class II PDE of 426 amino acids (accession number XP_720545). The recombinant CaPDE1 expressed in *S. cerevisiae* is a high-K_m, dual substrate PDE, with K_m values of 490 and 250 μM for cAMP and cGMP, respectively. The V_{max} with cAMP is much higher (1172 nmol/min) than with cGMP (44 nmol/min), probably reflecting the fact that cAMP is the primary substrate. CaPDE1 activity is not stimulated by Ca^{2+}, Ca^{2+}/calmodulin, Mg^{2+} or Mn^{2+}; it is fully active in the presence of EDTA or ethylene glycol tetraacetic acid (EGTA), and its activity is inhibited by Cu^{2+} or Zn^{2+} and by thiols.

CaPDE1 homologs have also been detected in the related pathogens *Candida stellatoidea* by cross-hybridization[7] and in *Candida glabrata* by DNA sequencing (accession number CAG60064).

14.2.2.4   Other Fungi

Genome sequencing has demonstrated the presence of class II PDEs in the genomes of a number of other fungi: *Ashbya gossypii* (accession number AAS53630), *Debaryomyces hansenii* (accession number CAG89346), *Kluyveromyces lactis* (accession number CAH00671), *Aspergillus nidulans* (accession number EAA65659), *Aspergillus fumigatus* (accession number EAL90820), and *Cryptococcus neoformans* (accession number EAL18882). In all these sequences, the signature motif is well conserved, as are a number of other residues. In several other fungal species (e.g., *Gibberella zeae*, accession number EAA76096; *Magnaporthe grisea*, accession number EAA53430; *Neurospora crassa*, accession number XP_322323; and *Yarrowia lipolytica*, accession number CAG80293), the PDE1 catalytic domain is interrupted by long inserts of unknown function. In all of the fungal class II PDEs, the catalytic domain comprises essentially the entire polypeptide, and no additional functional domains have been found. The microsporidian fungus *E. cuniculi* does not contain a class II PDE at all.

14.2.2.5 Biological Roles of Fungal PDE1

In *S. cerevisiae*, neither PDE1 nor PDE2 (see Section 14.2.3.1) is essential.[6] The presence of either gene alone is sufficient to downregulate excess cAMP, even in strains with a constitutively activated adenylyl cyclase. When glucose is added to cells grown in glucose-free medium, or when the intracellular pH is temporarily reduced, yeast cells display a rapid (1–2 min) spike of intracellular cAMP. In *S. cerevisiae* mutants that lack ScPDE1, these cAMP spikes are of about a threefold higher amplitude, and of longer duration.[10]

ScPDE1 contains a cAMP-dependent protein kinase (PKA) phosphorylation motif (-RRES^{252}-) that is phosphorylated *in vivo*.[10] Phosphorylation of this residue does not alter the catalytic activity of PDE1. Nevertheless, strains that carry an S^{252}–A^{252} or an S^{252}–E^{252} mutation are no longer able to control glucose- or acidification-induced cAMP spikes. The level of ScPDE1 protein is not affected in this short-time frame, though glucose addition eventually leads to an increase in ScPDE1 mRNA level and a decrease in ScPDE1 protein level after several hours.[11] The role of S^{252} phosphorylation for the control of ScPDE1 activity remains to be resolved, and it might be specific for *S. cerevisiae*, as the S^{252} phosphorylation motif is not conserved in other fungi. Nevertheless, it is interesting to speculate, and straightforward to verify experimentally, if this phosphorylation might be involved in PDE localization, rather than in PDE activity per se. The importance of intracellular localization of PDEs for the proper functioning of cAMP signaling has become apparent in mammalian cells, and signal control by PDE localization may constitute a general paradigm in protozoa as well.

14.2.3 FUNGAL CLASS I PHOSPHODIESTERASES

14.2.3.1 *Saccharomyces cerevisiae* ScPDE2

ScPDE2 was characterized using purified enzyme from *S. cerevisiae*.[12] The enzyme was described as a monomer of 61 kDa, in good agreement with the later determination of its molecular mass from the amino acid sequence (60,999 Da). Besides the full-size 61 kDa polypeptide, the enzyme preparations also contained a nicked version of the enzyme where a single proteolytic cut produced an N-terminal 17 kDa fragment and a C-terminal 45 kDa fragment. Both remained tightly associated with each other during purification. Nicking of the 61 kDa full-size protein increased its specific activity. The increase of specific activity upon nicking suggests that the N-terminal region (the 17 kDa fragment) inhibits the C-terminal catalytic domain.

ScPDE2 contains approximately one Zn^{2+} atom per monomer. This can be removed by complexing agent such as 8-hydroxychinolin, with a concomitant inactivation of the enzyme. The activity can be restored by the addition of Zn^{2+} to the depleted enzyme. ScPDE2 activity is Mg^{2+}-dependent, but it is not stimulated by Ca^{2+} or Ca^{2+}-calmodulin, and its activity is stable against 3 M urea. In contrast to ScPDE1, ScPDE2 is highly cAMP selective, with a K_m of 0.17 μM. ScPDE2 is not an essential gene,[13] and deletion of ScPDE2 does not reduce spore viability.[14] The gene is located on chromosome 15 and codes for a class I PDE of 526 amino acids (accession number P06776). The catalytic domain is located in the C-terminal part of the polypeptide, and no other functional domains can be recognized. Despite the lack of a nuclear localization signal, ScPDE2 seems to locate exclusively to the cell nucleus.[15]

The ScPDE2 gene was independently isolated as rac1, a mutant that renders the cells sensitive to exogenous cAMP.[16] At high extracellular cAMP concentrations (4 mM), sufficient cAMP permeates the cell membrane to raise the intracellular cAMP level by 40- to 50-fold. ScPDE2$^-$ mutants are no longer able to dispose of this excess cAMP, and their intracellular cAMP signaling is severely disrupted.

14.2.3.1.1   Cell Biology of ScPDE2

The ScPDE2 mRNA is subjected to regulation by read-through translation.[15] The open reading frame of ScPDE2 terminates with the sequence CAA-TAG-CAA (stop codon underlined), a sequence that is prone to inefficient and regulatable translational termination. The efficiency of translational termination is strongly influenced by the external factors such as stress, carbon source, or temperature.[17] Read-through translation of ScPDE2 results in a protein that is extended by 20 amino acids. This extension directs the protein to the proteasomes and to rapid degradation. Since the steady-state level of cAMP is critically dependent on the level of ScPDE2 protein, diminishing the level of active ScPDE2 by increased translational readthrough leads to a rapid increase in the steady-state cAMP level.

High cAMP levels inhibit the MSN2/a stress response pathway,[18] and the deletion of ScPDE2 renders the cells hypersensitive to various types of stress.[19] ScPDE2 was also identified as a multicopy suppressor of sorbitol dependence, i.e., it suppresses the fragility of these mutants to hypo-osmotic stress,[20] and it is involved in controlling cell wall biosynthesis. These findings agree well with studies demonstrating that low PKA activity, a proxy for ScPDE2 overexpression, also confers high osmotolerance,[21,22] and that cAMP plays an important role in controlling hypha formation and filamentation of *S. cerevisiae*.[23,24]

14.2.3.2   *Candida* spp. PDE2

CaPDE2 of the human pathogen *C. albicans* was identified by database screening[25] and was found to code for a class I PDE of 571 amino acids (accession number CAA21984). The CaPDE2 gene complements the heat-shock-sensitive phenotype of an ScPDE2 deletion mutant of *S. cerevisiae*.[26]

The steady-state level of CaPDE2 mRNA increases within minutes of transferring the cells from 30°C to 37°C and exposing them to human serum, i.e., conditioning the fungus is likely to encounter upon infection of a human host. The increase in CaPDE2 mRNA is strong but transient, reaching background levels again within 60 min.[26,27] Exogenous cAMP does not affect the level of CaPDE2 mRNA.[25]

Similar to the situation in *S. cerevisiae*, CaPDE2 deletion strains exhibit a higher steady-state level of intracellular cAMP, are sensitive to nutritional starvation, are unable to enter the stationary phase, and cannot accumulate glycogen; they lose viability much more rapidly than the wild type when shifted from 30°C to 37°C.

14.2.3.2.1   Cell Biology of CaPDE2

C. albicans is a diploid fungus that frequently interconverts between yeast, hyphal, and pseudohyphal forms. These interconversions are essential for pathogenicity, and cAMP signaling exerts a major control over this differentiation. Homozygous disruption of CaPDE2 stimulates filamentation and invasive growth on nutrient agar plates.[25] Similar to what is observed with *S. cerevisiae*, CaPDE2 deletions are sensitive to the effect of exogenous cAMP, resulting in larger cell size, reduced growth rate, and lower cell viability. This observed toxicity of cAMP for CaPDE2 mutants denotes the importance of a precise temporal and spatial control of intracellular cAMP.

CaPDE2 mutants are avirulent when tested in the mouse. This lack of virulence is most likely due to the sum of the detrimental effects of excess cAMP on cell physiology, such as the inability to enter stationary phase, decreased viability at 37°C, and increased susceptibility to oxidative stress.

CaPDE2 deletion mutants exhibit increased susceptibility to the antifungal polyene antibiotic amphotericin B.[28] Amphotericin B is a polyene antibiotic from *Streptomyces nodosus* that binds preferentially to cell membrane sterols such as ergosterol. Binding disrupts the

osmotic integrity of cell membrane and leads to the uncontrolled efflux of K^+ and Mg^{2+} ions, amino acids, and other low molecular weight cell components. The CaPDE2 mutants accumulate more ergosterol in their membranes, which renders them more susceptible to the sterol-binding amphotericin B. A future combination therapy of amphotericin B with a CaPDE2 inhibitor might allow reducing the dosing of amphotericin B, and hence reduce its nephrotoxic side effects.

Similar to what is observed in *S. cerevisiae*, CaPDE2⁻ deletion mutants exhibit an altered cell wall synthesis. Their cell wall is thinner than that in wild-type cells and contains strongly reduced levels of alkali-insoluble β-glucans.[28]

14.3 DICTYOSTELIUM

14.3.1 GENERAL REMARKS

Dictyostelium discoideum has remained a supreme organism for studying the biology of cyclic nucleotides ever since the initial discovery of acrasin, the chemoattractant for *Dictyostelium* amoeba was in fact cAMP.[29] Several recent reviews have highlighted the role of cyclic nucleotide signaling in the biology of *Dictyostelium*.[30–33] Surprisingly, the investigation of the PDEs as key players in all these signaling pathways has progressed rather slowly. Much of what we know today stems from reports published in the 1970s. In that decade, the first papers appeared that demonstrated the involvement of a PDE activity in the extracellular cAMP waves that regulate cell aggregation.[34] However, it was only in 1986 that the enzyme was identified on the molecular level as pdsA/DdPDE1.[35] The major intracellular PDE was identified a full 12 years later (regA/DdPDE2).[36,37] Several additional PDEs have been discovered since, and Van Haastert and colleagues have recently proposed a unifying nomenclature[38] to which we will adhere (Figure 14.2).

DdPDE1 is coded for by the pdsA gene, is a dual-substrate class II PDE that is located both on the cell surface and in the extracellular medium; DdPDE2, encoded by the regA gene, is a cytosolic, cAMP-specific class I PDE that represents the major intracellular PDE activity; DdPDE3 is a cytosolic, low-K_m, cGMP-specific class I PDE; DdPDE4 is predicted from its DNA sequence to represent a class I PDE that contains a signal sequence, two transmembrane segments, and a catalytic domain that is exposed at the outer cell surface; DdPDE5 is a β-lactamase-type PDE that is cGMP-specific and is activated by cGMP; DdPDE6 is a dual substrate β-lactamase-type PDE that is activated by cAMP; and finally DdPDE7, which is a putative class II PDE of 425 amino acids that shares 50% sequence identity with DdPDE1. The recent completion of the *Dictyostelium* genome project[39] indicates that these seven may represent the final tally of PDEs.

14.3.2 DdPDE1

DdPDE1 is coded for by the gene pdsA. Its open reading frame codes for a class II PDE of 452 amino acids (accession number AAA68447). DdPDE1 is an extracellular PDE that is present in both a membrane-associated and a soluble form. The N-terminus of the complete polypeptide contains a highly hydrophobic signal sequence of 18 amino acids, indicating that DdPDE1 is anchored in the cell membrane through its hydrophobic N-terminus. The N-terminus of the soluble form corresponds to N^{50} of the complete polypeptide and may result from cleavage of the membrane-bound enzyme by a lysine amino peptidase between K^{49} and N^{50}.[35] Soluble DdPDE1 is a monomeric glycoprotein.[40] It is a dual-substrate PDE with similar K_m values of about 10 μM both for cAMP and cGMP[41] whose activity does not depend on divalent cations. It is inactivated by reductants such as L-cysteine, dithioerythritol,

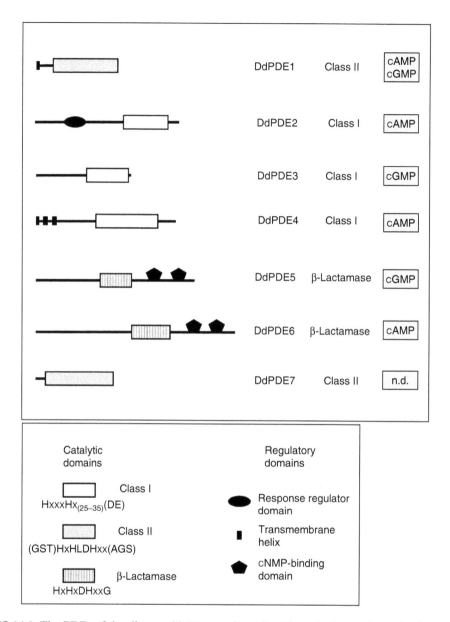

FIGURE 14.2 The PDEs of the slime mold *Dictyostelium discoideum* (n.d., not determined).

or dithithreitol, but not by β-mercaptoethanol. The broad spectrum PDE inhibitor, 3-iso-butyl-1-methylxanthine (IBMX) has no effect on the activity of DdPDE1.[42]

14.3.2.1   Cell Biology of DdPDE1

The main roles of DdPDE1 are in clearing the extracellular cAMP pulses that direct *Dictyostelium* chemotaxis, and in modulating cAMP signals during differentiation. Mutants lacking DdPDE1 do not aggregate[43,44] and cannot differentiate properly.[45] The pdsA gene is transcribed from three differentially regulated promotors:[43,45–48] (i) The vegetative promotor is located 1.9 kb upstream of the initiation codon and directs the synthesis of a 1.9 kb mRNA.

This mRNA is present in vegetative cells and disappears rapidly after the onset of starvation. The main function of DdPDE1 during vegetative growth is keeping the level of extracellular cAMP low. (ii) The aggregation promotor, located 3 kb upstream of the open reading frame, directs the synthesis of a 2.4 kb mRNA. Its expression is induced by increased levels of extracellular cAMP.[46] (iii) The late promotor is active during differentiation. It is located proximally to the open reading frame, and its primary transcript, after removing the short intron located after codon 7, results in a 2.2 kb mRNA. The activity of this promotor is regulated by the differentiation-inducing factor (DIF), a chlorinated hydrocarbon that regulates stalk cell differentiation.[49,50] During differentiation, DdPDE1 is synthesized specifically by prestalk cells, and it is involved in preventing the premature differentiation of spore cells.[45] As *Dictyostelium* has split off from the mainstream of eukaryotes very early in evolution, the concept of tissue-specific splicing of PDE transcripts may represent a very ancient mode of genetic control of PDE activity.

DdPDE1 activity is regulated at various levels: (a) differential control of the three promotors, (b) regulation of the steady-state levels of the transcripts[46] and of the active protein,[51] (c) modulation of enzyme localization (membrane-bound vs. secreted) by the shape of the incoming cAMP signal,[52] and (d) regulation of enzyme activity by the glycoprotein inhibitor DdPDI.

14.3.2.2 DdPDI—An Inhibitor of DdPDE1

The activity of the soluble, but not of the membrane-bound DdPDE1 is regulated by a specific inhibitor, DdPDI.[53] This inhibitor is a small glycoprotein (237 amino acids; accession number P22549) that contains 36 cysteine residues (17% of total amino acids), and it is coded for by a single-copy gene.[47,54] DdPDI is not expressed during vegetative growth, but it is induced immediately after the onset of starvation. Expression of DdPDI is strongly repressed by elevated levels of extracellular cAMP,[54,55] i.e., expression of the inhibitor is modulated by extracellular cAMP in the opposite sense than is the expression of its substrate PDE, DdPDE1 (see Section 3.2.1).

DdPDI is a heat- and acid-resistant glycoprotein that can be inactivated by dithiothreitol. The inhibitor forms a tight ($K_d = 10^{-10}$ M) complex with a 1:1 stoichiometry with the monomeric, soluble DdPDE1. It does not bind to, nor inactivate, the membrane-bound form of DdPDE1.[55] Binding of DdPDE1 to the inhibitor reversibly increases its K_m for cAMP or cGMP from 10 μM to several mM whereas the V_{max} of cyclic nucleotide hydrolysis remains unchanged.[41] Once the inhibitor is released, or inactivated by reduction, the K_m of DdPDE1 drops to its original low value.

14.3.3 DdPDE2

The gene coding for DdPDE2, regA, was identified through genetic screens that searched for mutants with disrupted cAMP signaling pathways during terminal differentiation.[36,37] DdPDE2 (accession number CAA06513) is a class I PDE of 793 amino acids that contains a response regulator domain[56] in its N-terminal part. It is cAMP-specific, with a K_m value of about 5 μM. This activity is not sensitive to inhibitors of human PDE4 such as rolipram or Ro20–1724, but it is moderately inhibited by IBMX, with an IC_{50} value of about 250 μM, using 1 μM cAMP as a substrate. cGMP up to 20 mM shows no effect on cAMP hydrolysis.[37]

DdPDE2 is regulated via a histidine phosphorelay cascade (Figure 14.3). The donor histidine kinase, RDEA (accession number AAB70186), is activated by autophosphorylation at histidine residue H^{65}. From this phosphoramidate, the phosphoryl group is transferred to D^{212} of the response regulator domain of DdPDE2, leading to the activation of the

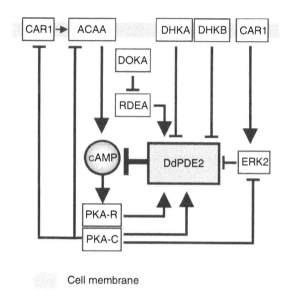

FIGURE 14.3 The regulatory circuits of DdPDE2. Arrowheads, stimulation; bars, inhibition.

downstream PDE catalytic domain.[57,58] The activity of the RDEA is negatively regulated by another histidine kinase, DOKA (accession number S71628[59]). This enzyme contains a response regulator domain with an aspartic acid residue (D^{1567}) that competes with the response regulator domain of DdPDE2 for phosphorylation by activated RDEA, and thus acts as a phosphatase for RDEA. *in vitro*, D^{212} can be phosphorylated using acetyl phosphate as a phosphate donor. The reaction is specific and Mg^{2+}-dependent, and yields a heat-labile phosphoaspartate.[37,58] The temporal stability of such mixed anhydrides on response regulators is widely different, depending on the immediate surrounding of the phosphorylated amino acid.[60] The phosphorylated state of the response regulator domain of DdPDE2 is quite stable, with a half-life of longer than 6 h.

Phosphorylation of the response regulator domain at residue D^{212} by RDEA activates DdPDE2 about 20-fold. This activation is effected via an increase in V_{max} whereas there is no change in K_m for the cAMP substrate. This contrasts with DdPDE5 (see Section 14.3.6) where the enzyme is activated primarily via decreasing the K_m for the substrate cGMP.

14.3.3.1 Cell Biology of DdPDE2

DdPDE2 is involved in every phase of *Dictyostelium* development, and it is the predominant intracellular PDE. Its mRNA remains low during vegetative growth, increases rapidly upon starvation, and is maintained at high levels throughout development.[36] During aggregation, DdPDE2 is involved in the regulatory circuit that resensitizes the cAMP receptor after every wave of the oscillatory cAMP signal; during differentiation, it is involved in the prestalk–prespore cell differentiation, and finally it controls spore formation, dormancy, and germination of the spores.[61]

In view of these diverse biological functions, it is not surprising that DdPDE2 is a virtual hub that connects several signaling cascades[62] (Figure 14.3). DdPDE2 is downregulated by extracellular signal-regulated kinase (ERK2)-mediated phosphorylation of T^{676}. The MAP kinase ERK2 is activated when the cAMP receptor CAR1 is activated by extracellular cAMP, and it is downregulated through phosphorylation by PKA. Via this loop, DdPDE2 is directly

inactivated by ERK2 and indirectly activated by PKA. In addition, PKA may also directly activate DdPDE2 by phosphorylation. ERK2-mediated phosphorylation affects DdPDE2 protein levels, rather than its enzymatic activity.[63,64] T^{676} phosphorylation promotes the association of DdPDE2 with an ubiquitinylation complex and its subsequent degradation through the ubiquitin–proteasome pathway.[65] In *Dictyostelium* mutants that express a non-phosphorylatable $T^{676}A$ mutation of DdPDE2, the steady-state level of cAMP is less than 15% of the wild-type level. Conversely, ERK2 deletion mutants exhibit constitutively high levels of DdPDE2 activity. During prespore cell differentiation, DdPDE2 is downregulated via phosphorylation of its response regulator domain by the integral membrane histidine kinase DHKA (accession number S71629). This enzyme is activated by the binding of the differentiation-specific peptide SDF-2[66] to its extracellular receptor domain. During dormancy of the spores, DdPDE2 is actively downregulated by a second histidine kinase, DHKB (accession number AAB71889). This enzyme is kept active by a differentiation-specific adenine derivative, discadenine.[67] Upon spore germination, DHKB is inactivated and thus the downregulation of DdPDE2 is terminated. The subsequent increase in DdPDE2 activity, combined with a simultaneous inactivation of the osmosensor adenylyl cyclase ACGA (accession number Q03101[68]), reduces the intracellular cAMP level. The ensuing inactivation of PKA activity leads to the germination of the spore.[61]

An additional, rather unusual mode of DdPDE2 activation has been proposed by Shaulsky et al.[69] These authors observed that DdPDE2 can physically interact with the type I regulatory subunit (accession number P05987) of PKA. This interaction leads to a 18-fold stimulation of DdPDE2 activity.

DdPDE2 is the major intracellular PDE activity, and cAMP signaling in *Dictyostelium* appears to be regulated predominantly by PDE activity. The main point of intracellular cAMP signaling appears to be the downregulation of PKA; *Dictyostelium* mutants that cannot produce cAMP cannot develop. However, a moderate level of overexpression of the catalytic subunit of PKA allows near-normal development,[70] in the complete absence of any cAMP signaling.

14.3.4 DdPDE3

The open reading frame (1822 bp) of the gene for DdPDE3 contains two introns, and it is transcribed into an mRNA of 1.9 kb. The open reading frame codes for a class I PDE of 522 amino acids (accession number AAN78319[71]). Its N-terminal part is rich in polyasparagine and polyglutamine tracts of unknown function, as they are found in many *Dictyostelium* proteins.[72] The catalytic domain of DdPDE3 is located in the C-terminal part of the molecule.

DdPDE3 is a highly cGMP-selective PDE with K_m values of 0.22 μM for cGMP and 145 μM for cAMP. DdPDE3 is absolutely dependent on divalent cations for activity, which is not modulated by cGMP or 8-Br-cGMP. Recombinant DdPDE3 is sensitive to inhibition by IBMX, with an IC_{50} of about 60 μM.

14.3.4.1 Cell Biology of DdPDE3

The gene for DdPDE3 is expressed constitutively throughout all stages of *Dictyostelium* development. A 25-fold overexpression of DdPDE3 does not result in an observable phenotype, though the overexpressed enzyme is fully active. DdPDE3 is only a minor cGMP-hydrolyzing activity, contributing about 30% of the total cGMP hydrolysis capacity, and its main role is in maintaining the low steady-state level of cGMP. Signal-induced cGMP peaks are mainly dealt with by DdPDE5 (see Section 14.3.6) whereas DdPDE3 and DdPDE6 may serve as backup activities for this function.

14.3.5 DdPDE4

DdPDE4 is only predicted from its DNA sequence.[38] It contains a hydrophobic leader sequence of 24 amino acids, followed by two transmembrane helices. It is a class I PDE with a catalytic domain that is predicted to be extracellular, and to be specific for cAMP (Peter Van Haastert, personal communication).

14.3.6 DdPDE5

The gene for DdPDE5, *pdeD/gbpA*, is located on chromosome 2 and consists of an open reading frame of 2.601 kb containing two introns of 96 and 122 bp at positions 1108 and 1565, respectively. It codes for a PDE of 867 amino acids[73] (accession number AAL06059). Its N-terminal region is rich in low-complexity sequences[72] whereas the middle part of the polypeptide represents a metallo-β-lactamase domain. The signature sequence HxHxDHxxG is characteristic for four unrelated enzyme families, the metallo-β-lactamases, glyoxalase II, a bacterial arylsulfatase, and the class II PDEs. Phylogenetic analyses indicate that the catalytic domains of DdPDE5 (and DdPDE6, see Section 14.3.7) are more closely related to the bacterial β-lactamases than to the monophyletic group of the fungal class II PDEs.[38] While the enzymatic activity of DdPDE5 cannot be predicted a priori from its sequence, deletion mapping experiments have established that the cGMP-hydrolyzing activity is indeed located in the β-lactamase domain.[73] The C-terminal part of the polypeptide contains two putative cNMP-binding domains that are homologous to the CAP-related family of cyclic nucleotide-binding sites.[74]

DdPDE5 is highly (>100-fold) selective for cGMP, and it is activated by cGMP. Since cyclic inosine-3′,5′-monophosphate (cIMP) is a good substrate, but does not bind to the activation domain, and 8-Br-cGMP is a good activator, but a poor substrate, the kinetics of enzyme activation through ligand binding to the activation domain could be explored in detail.[75] Steady-state enzyme activity is rapidly reached at the catalytic domain ($t_{1/2} < 5$ s) whereas steady-state binding of the activator to the allosteric site takes much longer time ($t_{1/2}$ of about 20 s). Activation of the enzyme by cGMP (activation constant of 0.16 μM) changes its K_m for cGMP as a substrate (5 μM for the activated enzyme vs. 20 μM for the non-activated enzyme), while leaving its V_{max} unaltered. Activation of the enzyme by 8-bromo-cGMP agrees well with the concept that the CAP-related cNMP-binding domains prefer the *syn* conformation of the ligand, which is enforced by the bulky bromo group substitution. cAMP does not activate DdPDE5, but acts as an inhibitor at high concentrations (K_i cAMP = 1.8 mM[38]). DdPDE5 appears to be insensitive to inhibition by IBMX.[71]

In many aspects, DdPDE5 resembles the human cGMP-stimulated, cGMP-specific HsPDE2, though their functionalities are obtained with different modules that share no amino acid sequence similarity. HsPDE2 has a class I catalytic domain and its cGMP regulation is mediated by a GAF domain whereas DdPDE5 is a class II and β-lactamase-type PDE and its cGMP modulation is effected via a CAP-related cNMP-binding domain.

14.3.6.1 Cell Biology of DdPDE5

DdPDE5 is transcribed into a 2.9 kb mRNA that reaches its maximum steady-state level during the aggregation phase.[73] DdPDE5 is the major cGMP hydrolytic activity in *Dictyostelium* cells, and its major function is the control of the cGMP signals generated when the cAMP receptors are stimulated by the extracellular cAMP. In wild-type cells, an exogenous cAMP pulse generates an intracellular cGMP spike that lasts for about 30 s, with a maximal amplitude of 6 pmol/10^7 cells reached 10 s after the cAMP signal. In DdPDE5⁻ deletion

mutants, the spike length is increased to 120 s, the maximal amplitude increases to 15 pmol/10^7 cells, and this maximum is reached only 20 s after the cAMP pulse.[38]

14.3.7 DdPDE6

The gene for DdPDE6, *pdeE/gbpB*, is located on chromosome 2 and represents an open reading frame of 3.288 kb that contains two introns of 202 and 80 bp. The open reading frame codes for a polypeptide of 1096 amino acids[76] (accession number AAL06060). The overall architecture of the DdPDE6 polypeptide is very similar to that of DdPDE5. The N-terminus contains stretches of low-complexity sequence, and the predicted β-lactamase catalytic domain is located in the middle of the polypeptide chain. This domain is 44% identical between DdPDE5 and DdPDE6. Other than the highly conserved signature sequence HxHxDHxxG, it has little else in common with the other class II PDEs. However, upstream of the signature sequence, a motif (FxFFxT/SxHxxPxxxxxxExxGxxxxYS/TxD) was detected that is highly conserved in a number of unidentified bacterial proteins, as well as in the 73 kDa subunits of the pre-mRNA cleavage and polyadenylation specificity factor CPSF[77] of *Homo sapiens* (CPSF73) and *S. cerevisiae* (PF1).[76] The latter two proteins are suspected 3′ nucleotidases. The C-terminal part of DdPDE6 contains two cNMP domains that are homologous in sequence to the CAP-related cyclic nucleotide-binding domains.[74] DdPDE6 is cAMP-specific, and its activity is stimulated by cAMP.[76]

14.3.8 DdPDE7

The gene for DdPDE7 is located on chromosome 5 and is predicted to code for a class II PDE. DdPDE7 consists of 425 amino acids and contains a class II catalytic domain spanning from residues 58 through 420 (accession number XP-636383). The amino acid sequence of DdPDE7 is 50% identical with DdPDE1. Its N-terminal region, including the predicted signal sequence and protease cleavage site at N^{50}, is almost completely identical with that of DdPDE1, suggesting that both represent functional homologs.

14.4 KINETOPLASTIDA

14.4.1 General Remarks

The kinetoplastids are a large order of unicellular eukaryotes, many of which are parasites, and several among them are major human pathogens. *Trypanosoma brucei* causes the fatal human sleeping sickness in Africa; *Trypanosoma cruzi* is the etiological agent of South-American Chagas disease; and several species of the genus *Leishmania* cause the complex of human leishmaniases that constitute a global problem. The PDEs of several kinetoplastids have recently gained attention as possible targets for the development of PDE-inhibitor-based antiparasitic drugs. Such developments are particularly urgent because there is essentially no satisfactory medication for any of the kinetoplastid diseases, and the few available compounds are threatened by the rapid development of drug resistance.[78–80]

The recent completion of several kinetoplastid genome projects[81–83] allows an overview over the kinetoplastid phosphodiesteromes. All kinetoplastid species appear to contain the same set of four families of predicted class I PDEs (Figure 14.4), and no class II PDEs were identified. A unified nomenclature for the kinetoplastid PDEs has recently been proposed,[84] to which we will adhere throughout the entire text. Older names are also given where appropriate.

PDEA is a PDE of about 600 amino acids that contains a class I catalytic domain in the C-terminal region. PDEB1 and PDEB2 are coded for by two tandemly arranged genes. Both

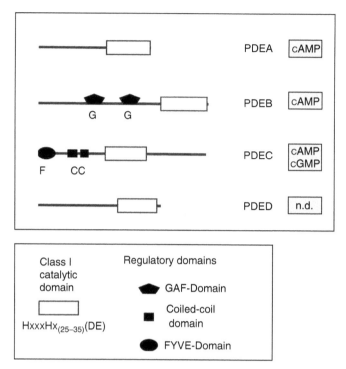

FIGURE 14.4 The PDEs of the kinetoplastida (n.d., not determined).

PDEBs consist of about 900 amino acids and contain two GAF domains[85] in their N-terminal regions. Their class I catalytic domains, which are essentially identical, are located at the C-terminal ends. PDEC is about 900 amino acids long and contains an N-terminal FYVE domain, followed by two closely spaced coiled-coil regions. The FYVE domain may serve to target PDEC to membrane-delimited organelles such as the endosomes, and thus might be involved in regulating endocytosis. Its class I catalytic domain is located in the middle of the polypeptide. And finally, PDED is a PDE of about 700 amino acids, with the class I catalytic domain at the C-terminus and no other functional domains. PDED is coded for by a single-copy gene in the genomes of *Leishmania major* and *T. brucei*, and by two dispersed copies in the genome of *T. cruzi*. An overview of the architecture of the kinetoplastid PDEs is given in Figure 14.4.

14.4.2 THE PDEA FAMILY

14.4.2.1 TbrPDEA of *Trypanosoma brucei*

TbrPDEA is a single-copy gene located on chromosome 10 that is expressed both in the bloodstream and the procyclic life-cycle stages. TbPDEA consists of 620 amino acids (accession number AAL58095) and its class I catalytic domain is located at the C-terminus.

TbrPDEA (earlier designation: TbPDE1) was initially identified by complementation screening in *S. cerevisiae*.[86] A fragment of TbrPDEA (R^{189}–T^{620}) containing the catalytic domain was expressed in *Escherichia coli*. The recombinant enzyme is cAMP-specific and exhibits an unusually high K_m for its substrate (>600 μM). cGMP neither competes as a substrate, nor does it modulate the activity of the recombinant enzyme.

When a panel of inhibitors for mammalian PDEs were tested, only four compounds (sildenafil, trequinsin, ethaverin, and dipyridamole) exhibited significant potency.[86] However, their IC_{50} values were rather high (ranging from 1 μM for sildenafil to 13 μM for dipyridamole) when compared to the potency for the specific human target enzymes.

14.4.2.1.1   Cell Biology of TbrPDEA

Homozygous deletion mutants of TbrPDEA were constructed in both procyclic and bloodstream trypanosomes. No significant phenotype was observed in culture; the differentiation from bloodstream to procyclic forms was not affected; and the infectivity for tsetse flies was not altered.[87] This study showed that TbrPDEA is not an essential gene in *T. brucei*, neither in culture nor during midgut infection of tsetse flies. The unusually high K_m of TbrPDEA, and the absence of a marked phenotype in deletion mutants, is reminiscent of the situation of ScPDE1 of *S. cerevisiae*, a high-K_m, class II PDE (see Section 2.2.1). The deletion of this enzyme also produces no appreciable phenotype, and the main function of ScPDE1 seems to be in modulating the cAMP peaks generated by external stimuli.

14.4.2.2   LmjPDEA of *Leishmania major*

LmjPDEA was identified by a *L. major* database screening. The open reading frame of LmjPDEA encodes a class I PDE of 631 amino acids (accession number AAR88144) whose catalytic domain is located in the C-terminal part. LmjPDEA shares 45.1% identity with its *T. brucei* homolog TbrPDEA, and it can complement a PDE-deficient yeast strain,[88] demonstrating that it hydrolyzes cAMP.

14.4.3   THE PDEB FAMILY

14.4.3.1   *Trypanosoma brucei*

14.4.3.1.1   TbrPDE1 of *Trypanosoma brucei*

The open reading frame of TbrPDEB1 (previously designated TbPDE2C[89]) is located on chromosome 9, and it is followed, in a distance of 2.379 kb, by the tandemly arranged open reading frame of TbrPDEB2. TbrPDEB1 (accession number AAK33016) is a class I PDE of 930 amino acids that contains two GAF domains[85] in its N-terminal region, and a class I catalytic domain at the C-terminus. TbrPDEB1 is cAMP-specific, with a K_m of 8 μM. cGMP neither competes for cAMP as a substrate, nor does it modulate the activity of TbrPDEB1. Inhibition studies showed that trequinsin, dipyridamole, ethaverin, etazolate, and sildenafil were the most potent inhibitors with IC_{50} values in the low micromolar range (see Table 14.1). Many other inhibitors of mammalian PDEs, including rolipram or IBMX, exhibited IC_{50} values of >500 μM.[89]

14.4.3.1.2   TbrPDEB2 of *Trypanosoma brucei*

The open reading frame of TbrPDEB2 (previously designated TbPDE2B[90]) is located downstream of TbrPDEB1 on chromosome 9. TbrPDEB2 (accession number XP_803815) is a class I PDE of 930 amino acids that is very similar to TbrPDEB1. It contains two tandemly arranged GAF domains, followed by the highly conserved catalytic domain. TbrPDEB2 is cAMP-specific, with a K_m of 2.4 μM.[90] Considering the high degree of sequence conservation between the catalytic domains of TbrPDEB1 and TbrPDEB2, unexpectedly significant differences in inhibitor sensitivities were observed between the two enzymes (see Table 14.1). A short, highly divergent sequence stretch in the catalytic domains (amino acids 790–820 for TbrPDEB1 and 790–821 for TbrPDEB2, respectively) may account for these differences.

TABLE 14.1
Potency of Selected Human PDE Inhibitors against *T. Brucei*,[89,90,92] *T. cruzi*,[93,97] *Leishmania*,[88] and *Plasmodium*[109] PDEs

	IC_{50} (μM) for							Selectivity for Mammalian PDE	IC_{50} (μM) for Mammalian PDE
Inhibitor	Tbr PDEA	Tbr PDEB1	Tbr PDEB2	Tcr PDEB2	Lmj PDEB2	Tcr PDEC	Pf PDE1		
Dipyridamole	13	14.6	27	17	22.6	6.9	22	PDE5/6/8/10	0.9/0.38/4.5/1.1
Trequinsin	2.5	13.3	—	—	96.6	3.9	—	PDE3	0.0003
Etazolate	25	30.6	127	—	>100	—	—	PDE4	0.55
Ethaverine	8	26.8	—	—	—	—	—	n.d.	
Sildenafil	1	42.2	>100	—	—	—	56	PDE5	0.0039
Papaverine	30	—	304	111	>100	25	>100	n.s.	2–25
IBMX	>200	1704	>1000	>1000	>100	68	>100	n.s.	2–50
Pentoxifylline	—	—	>800	—	>100	>100	>100	n.s.	45–150
Rolipram	280	—	>300	>500	>100	>100	>100	PDE4	2
Ro 20–1724	—	—	—	—	>100	>100	—	PDE4	2
Enoximone	—	—	>100	—	—	—	—	PDE3	1
Zaprinast	>100	—	>50	>500	>100	—	3.8	PDE5/6	0.76/0.15
EHNA	>100	—	>180	217	>100	>100	>100	PDE2	1
MM-IBMX	>100	—	—	—	>100	>75	—	PDE1	4
Vinpocetine	>100	—	—	134	—	—	>100	PDE1	1
Milrinone	>100	—	—	—	>100	>100	>100	PDE3	0.3
Cilostamide	>100	—	—	—	>100	>100	—	PDE3	0.005
Zardaverine	>100	—	—	—	—	>100	—	PDE3/4	0.5
Theophylline	>100	—	—	>1000	—	—	>100	n.s.	50–300

Abbreviations: EHNA, erythro-9-(2-hydroxy-3-nonyl)adenine; IBMX, 3-isobutyl-1-methylxanthine; MM-IBMX, 8-methoxymethyl-IBMX; n.s., nonselective; n.d., not determined.

A recent study[91] has demonstrated that the purified recombinant N-terminus of TbrPDEB2, containing the GAF-A domain, binds cAMP with a much higher affinity than cGMP ($K_D = 17.6$ and 289 nM, respectively). Mutating a key-binding site residue ($T^{137}A$) in the GAF-A domain resulted in a complete loss of cAMP binding. In the complete enzyme, this mutation increased the K_m value for cAMP hydrolysis fourfold (from 4.4 to 17.5 μM). Considering that the K_m for cAMP of the catalytic domain alone is ≥ 44 μM, these data indicate that binding of cAMP to the GAF-A domain increases the affinity of TbrPDEB2 for its substrate.

14.4.3.1.3   Cell Biology of TbrPDEB1 and TbrPDEB2
The observation that inhibition of TbrPDEB increases intracellular cAMP levels and blocks cell proliferation[92] indicates that the members of this family are essential for trypanosome proliferation and survival in culture. In good agreement, the homozygous deletions of TbrPDEB1 in bloodstream form trypanosomes that are not viable, and RNAi against TbrPDEB is lethal.[89] The sensitivity of bloodstream forms against any interference with PDE activity and cAMP levels were also documented by the toxicity of membrane-permeable cAMP analogs. Procyclic (insect form) trypanosomes are much less sensitive to interference with PDE activity. Here, RNAi against TbrPDEB leads to a strong increase in intracellular cAMP (about 15-fold), but the majority of the cells retain their normal shape and survive.[89]

14.4.3.2   *Leishmania major*

14.4.3.2.1   *LmjPDEB1 and LmjPDEB2 of* Leishmania major

The *L. major* genome contains two genes that code for homologs of the TbrPDEB family: LmjPDEB1 and LmjPDEB2. Both are single-copy genes, arranged in tandem on chromosome 15, with about 5 kb between their open reading frames. LmjPDEB1 (accession number AAR88146) and LmjPDEB2 (accession number AAR88145) are highly conserved class I PDEs of 941 and 931 amino acids, respectively.[88] Their sequences are very similar, with two exceptions. The first 230 amino acids share only 34% identity, and the completely conserved catalytic domains contain a stretch of 33 amino acids that is totally divergent between LmjPDEB1 and LmjPDEB2. A similar, divergent stretch of 33 amino acids is also found in the conserved catalytic domains of TbrPDEB1 and TbrPDEB2, and of TcrPDEB1 and TcrPDEB2. Its biological significance remains unclear, but it may be the reason for the different drug sensitivities exhibited by the two, or otherwise completely conserved catalytic domains of TbrPDEB1 and TbrPDEB2 (see Section 4.3.1.2). Both PDEs contain two tandemly arranged GAF domains in their N-terminal sections, separated by a short-linker region that contains a putative PKA phosphorylation site (LmjPDEB1: KKKS[404]; LmjPDEB2: KKKS[395]). This PKA site is highly conserved in the PDEB1 and PDEB2 enzymes of other kinetoplastids whereas none of the other predicted PKA sites are suggesting that it may in fact be functional.

LmjPDEB1 and LmjPDEB2 are both cAMP-specific with K_m values of 6.98 and 0.99 µM, respectively. The apparent K_m is not altered by the presence of a 100-fold excess of cGMP for both recombinant enzymes. Of a panel of mammalian PDE inhibitors, most of them were essentially inactive. Only dipyridamole ($IC_{50} = 22.6$ µM), trequinsin ($IC_{50} = 96.6$ µM), and etazolate exhibited appreciable potency (see Table 14.1).

14.4.3.2.2   *Cell Biology of LmjPDEB1 and LmjPDEB2*

Three PDE inhibitors, dipyridamole, trequinsin, and etazolate, were tested for their effect on cell proliferation in promastigote and amastigote forms in culture. The results showed that all three compounds have a strong inhibitory effect on cell survival and proliferation.[88] These observations suggest that the two PDEs are essential for cell proliferation in *L. major*, similar to the situation in *T. brucei*.

14.4.3.3   *Trypanosoma cruzi*

14.4.3.3.1   *TcrPDEB2 of* T. cruzi

A bioinformatics strategy together with a genomic DNA library screening allowed cloning of the TcrPDEB2 gene (designated TcPDE1 in the original publication[93]). TcrPDEB2 (accession number AAP49573) is a class I PDE of 929 amino acids that is very similar to its homologs from *T. brucei* and *L. major*. The polypeptide starts with a putative signal peptide (amino acids 1–22), followed by two GAF domains and a conserved class I catalytic domain. TcPDEB2 shows strong identities with the TbrPDEB family (>80% identity for the catalytic domains). The genomic organization of TcrPDEB2 is similar to what was observed in *L. major* and *T. brucei*. The gene for TcrPDEB2 is preceded by the tandemly arranged, very similar gene coding for TcrPDEB1 (accession number EAN98416), with the two open reading frames separated by 1.665 kb. TcrPDEB2 expressed in a PDE-deficient yeast strain, and exhibited a cAMP-specific activity with a K_m of 7.3 µM. The activity is two- to threefold higher in the presence of Mn^{2+} than Mg^{2+} as cofactor, and is not affected by Ca^{2+} and calmodulin. TcrPDEB2 does not hydrolyze cGMP, and cGMP neither competes as a substrate nor does it modulate the activity of the recombinant enzyme. Among a panel of

mammalian PDE inhibitors, only dipyridamole was a significantly potent inhibitor of TcrPDEB2, with an IC_{50} value of 17 μM.

14.4.3.3.2 Cell Biology of TcrPDEB2

TcrPDEB2 is preferentially located in the flagellum.[93] Within the flagellum, it appears to be associated with the paraflagellar rod structure.[94] This localization is interesting, since at least some of the numerous adenylyl cyclases of kinetoplastids localize in the flagellum[95] and can interact with paraflagellar rod proteins.[96]

14.4.4   THE PDEC FAMILY

14.4.4.1   TcrPDEC of *T. cruzi*

TcrPDEC is coded for by a single-copy gene that has two markedly different alleles.[97] TcrPDEC (accession number CAI63255 and accession number CAI63256 for the two allelic sequences) is a PDE of 924 amino acids, with the class I catalytic domain located in the middle of the polypeptide (see Figure 14.4). Recombinant TcrPDEC is a dual-substrate PDE that hydrolyzes cAMP and cGMP with similar V_{max} values and with K_m values of 32 and 78 μM, respectively. These two K_m values are similar to those of the human dual-specificity PDE, HsPDE2[98] and fungal PDE1.[3] Among a number of mammalian PDE inhibitors tested, only trequinsin, etazolate, and dipyridamole exhibited IC_{50} values below 10 μM. All the three compounds are also potent inhibitors of other trypanosome PDEs.

Unusual for a PDE, amino acids P^{10}–G^{73} of TcrPDEC constitute a FYVE domain.[99] FYVE domains are Zn^{2+}-stabilized structures that can recognize membrane-embedded phosphatidylinositol-3-phosphate (PtIns(3)P). The FYVE domain of TcrPDEC is followed by two closely spaced coiled-coil regions (D^{144}–D^{179} and K^{207}–E^{264}). The recombinant FYVE domain of TcrPDEC can form stable dimers, even in the absence of the adjacent coiled-coil regions. This dimerization of the FYVE domain may be required for the interaction of TcrPDEC with a target structure, and it might induce dimerization of PDE catalytic domains, thus influencing the catalytic properties of the enzyme.

14.5   *PLASMODIUM*

14.5.1   GENERAL REMARKS

Plasmodium falciparum and its lesser relatives, *Plasmodium ovale*, *Plasmodium vivax*, and *Plasmodium malariae*, are the causative agents of the various forms of human malaria. Despite their significance as global disease agents, little is known about their cyclic nucleotide signaling. A few reports have discussed this topic,[100–102] and four purine nucleotide cyclases have been cloned and characterized.[103] Essentially nothing is known on *P. falciparum* PDEs. In the genome database (http://www.plasmodb.org[104]), the four putative PfPDEs genes have been annotated (PFL0475w, MAL13P1.118, MAL13P1.119, and PF14_0672). All four PfPDEs contain class I catalytic domains[105] that are located in the C-terminal part of the predicted polypeptides (Figure 14.5). The amino acid sequence identities between the four catalytic domains vary from 30% to 40%, showing that they represent four distinct PDE families. Very unusually, all four PfPDEs polypeptides are predicted to contain three to six transmembrane helices in their N-terminal parts, suggesting that they are integral membrane proteins (Figure 14.5). Transmembrane helices in PDEs are rather unusual, and have only been found so far in one PDE from *Dictyostelium*, DdPDE4 (see Section 3.5), and possibly in

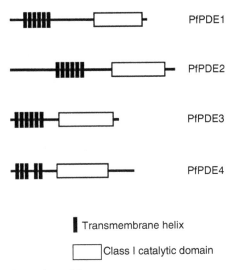

FIGURE 14.5 The PDEs of *Plasmodium falciparum*.

the human HsPDE3. The predicted topology for PfPDE2 and PfPDE4 indicates that their catalytic domains are extracellular and have the potential to be exposed to the cytoplasm of the erythrocyte. If so, the parasite would have the means to interfere with the cyclic nucleotide signaling of its host cell. Microarray-based mRNA expression profiling[106–108] has shown that various PfPDEs are differentially expressed during the life cycle of the parasite, suggesting their involvement in different physiological processes.

14.5.2 PfPDE1

Recently, Yuasa et al.[109] reported the first biochemical characterization of a PfPDE, PfPDE1, that is coded for by the previously annotated gene PFL0475w. PfPDE1 contains six putative transmembrane helices in its N-terminal part and a C-terminal class I catalytic domain. The recombinant catalytic domain is specific for cGMP, with a K_m value of 0.65 μM. cGMP hydrolysis was neither activated nor inhibited by cAMP. PDE activity was completely lost by mutagenesis of a conserved Asp^{762}, which is involved in substrate binding.[110]

Most of the inhibitors tested against the recombinant protein exhibited very low potency (IBMX, papaverine, theophylline, pentoxiphylline, vinpocetine, erythoro-9-(2-hydroxy-3-nonyl)adenine (EHNA), milrinone, and rolipram). Interestingly, compounds that inhibit human PDE5 (dipyridamole, E4021, and sildenafil) exhibited moderate activity against PfPDE1, with IC_{50} values of 22, 46, and 56 μM, respectively. Zaprinast was the most potent inhibitor tested with an IC_{50} value of 3.8 μM.[109]

14.5.2.1 Cell Biology of PfPDE1

In lysates of infected erythrocytes, most of the cGMP hydrolytic activity was detected in the membrane fraction. This localization agrees well with the presence of transmembrane helices in all the four plasmodial PDEs. This activity was potently inhibited by zaprinast.[109] Zaprinast inhibited proliferation of the parasite with an EC_{50} value of 35 μM, demonstrating the importance of zaprinast-sensitive PDE activity, including PfPDE1, for the proliferation of *Plasmodium* in infected erythrocytes.

14.6   CONCLUSIONS

The overview presented here is far from complete, as the forthcoming protozoal genome projects (e.g., *Theileria annulata*, *Theileria parva*, *Entamoeba histolytica*, *Giardia lamblia*, several additional kinetoplastids, and *Plasmodium* species) will provide ever more information about additional PDEs. However, for the time being all this information is at the level of DNA sequences. This is why we have constrained ourselves to reviewing those PDEs for which at least a shred of experimental data is available. What we have learned so far is that the paradigm of cNMP signaling may become even more complex. Compared to the relative simplicity of the metazoan world with a single class of PDE, two new types of PDE have appeared among the protozoa, the class II PDEs and the β-lactamase-type catalytic domains (Figure 14.1). Since the PDE activity of the latter cannot be predicted from sequence alone, we might see an increase of PDEs of the latter type as more of the β-lactamase enzymes are genetically or biochemically characterized. Also, additional regulatory domains such as response regulator domains of phosphorelay pathways are utilized in the protozoal PDEs, suggesting that signaling pathways might be highly individualized among species.

As many of the protozoa are medically or economically important pathogens, their PDEs are bound to gain more interest as potential drug targets. Targeting them would allow use of the enormous amount of expertise that is available in the pharmaceutical industry to explore new applications and develop novel approaches for the treatment of infectious diseases that are still major scourges of humans and their domestic animals.

REFERENCES

1. Charbonneau, H., et al. 1986. Identification of a conserved domain among cyclic nucleotide phosphodiesterases from diverse species. *Proc Natl Acad Sci USA* 83 (24):9308–12.
2. Londesborough, J., and T.M. Lukkari. 1980. The pH and temperature dependence of the activity of the high K_m cyclic nucleotide phosphodiesterase of bakers' yeast. *J Biol Chem* 255 (19):9262–7.
3. Londesborough, J., and K. Suoranta. 1983. The zinc-containing high K_m cyclic nucleotide phosphodiesterase of bakers' yeast. *J Biol Chem* 258 (5):2966–72.
4. Londesborough, J., and K. Varimo. 1979. The temperature-dependence of adenylate cyclase from baker's yeast. *Biochem J* 181 (3):539–43.
5. Fujimoto, M., A. Ichikawa, and K. Tomita. 1974. Purification and properties of adenosine 3′,5′-monophosphate phosphodiesterase from baker's yeast. *Arch Biochem Biophys* 161:54–63.
6. Nikawa, J., P. Sass, and M. Wigler. 1987. Cloning and characterization of the low-affinity cyclic AMP phosphodiesterase gene of *Saccharomyces cerevisiae*. *Mol Cell Biol* 7 (10):3629–36.
7. Hoyer, L.L. et al. 1994. A *Candida albicans* cyclic nucleotide phosphodiesterase: Cloning and expression in *Saccharomyces cerevisiae* and biochemical characterization of the recombinant enzyme. *Microbiology* 140 (Pt 7):1533–42.
8. Atienza, J.M., and J. Colicelli. 1998. Yeast model system for study of mammalian phosphodiesterases. *Methods* 14 (1):35–42.
9. DeVoti, J., et al. 1991. Interaction between ran1 + protein kinase and cAMP dependent protein kinase as negative regulators of fission yeast meiosis. *Embo J* 10 (12):3759–68.
10. Ma, P., et al. 1999. The PDE1-encoded low-affinity phosphodiesterase in the yeast *Saccharomyces cerevisiae* has a specific function in controlling agonist-induced cAMP signaling. *Mol Biol Cell* 10 (1):91–104.
11. Wera, S., P. Ma, and J.M. Thevelein. 1997. Glucose exerts opposite effects on mRNA versus protein and activity levels of Pde1, the low-affinity cAMP phosphodiesterase from budding yeast, *Saccharomyces cerevisiae*. *FEBS Lett* 420 (2–3):147–50.
12. Suoranta, K., and J. Londesborough. 1984. Purification of intact and nicked forms of a zinc-containing, Mg^{2+}-dependent, low K_m cyclic AMP phosphodiesterase from bakers' yeast. *J Biol Chem* 259 (11):6964–71.

13. Wilson, R.B., and K. Tatchell. 1988. SRA5 encodes the low-K_m cyclic AMP phosphodiesterase of *Saccharomyces cerevisiae. Mol Cell Biol* 8 (1):505–10.

14. Sass, P., et al. 1986. Cloning and characterization of the high-affinity cAMP phosphodiesterase of *Saccharomyces cerevisiae. Proc Natl Acad Sci USA* 83 (24):9303–7.

15. Namy, O., G. Duchateau-Nguyen, and J.P. Rousset. 2002. Translational readthrough of the PDE2 stop codon modulates cAMP levels in *Saccharomyces cerevisiae. Mol Microbiol* 43 (3):641–52.

16. Wilson, R.B., et al. 1993. The pde2 gene of *Saccharomyces cerevisiae* is allelic to rca1 and encodes a phosphodiesterase which protects the cell from extracellular cAMP. *FEBS Lett* 325 (3):191–5.

17. Eaglestone, S.S., B.S. Cox, and M.F. Tuite. 1999. Translation termination efficiency can be regulated in *Saccharomyces cerevisiae* by environmental stress through a prion-mediated mechanism. *EMBO J* 18 (7):1974–81.

18. Garreau, H., et al. 2000. Hyperphosphorylation of Msn2p and Msn4p in response to heat shock and the diauxic shift is inhibited by cAMP in *Saccharomyces cerevisiae. Microbiology* 146 (Pt 9):2113–20.

19. Park, J.I., C.M. Grant, and I.W. Dawes. 2005. The high-affinity cAMP phosphodiesterase of *Saccharomyces cerevisiae* is the major determinant of cAMP levels in stationary phase: Involvement of different branches of the Ras-cyclic AMP pathway in stress responses. *Biochem Biophys Res Commun* 327 (1):311–9.

20. Tomlin, G.C., et al. 2000. Suppression of sorbitol dependence in a strain bearing a mutation in the SRB1/PSA1/VIG9 gene encoding GDP-mannose pyrophosphorylase by PDE2 overexpression suggests a role for the Ras/cAMP signal-transduction pathway in the control of yeast cell-wall biogenesis. *Microbiology* 146 (Pt 9):2133–46.

21. Norbeck, J., and A. Blomberg. 2000. The level of cAMP-dependent protein kinase A activity strongly affects osmotolerance and osmo-instigated gene expression changes in *Saccharomyces cerevisiae. Yeast* 16 (2):121–37.

22. Werner-Washburne, M., et al. 1993. Stationary phase in the yeast *Saccharomyces cerevisiae. Microbiol Rev* 57 (2):383–401.

23. Mosch, H.U., et al. 1999. Crosstalk between the Ras2p-controlled mitogen-activated protein kinase and cAMP pathways during invasive growth of *Saccharomyces cerevisiae. Mol Biol Cell* 10 (5):1325–35.

24. Pan, X. and J. Heitman. 1999. Cyclic AMP-dependent protein kinase regulates pseudohyphal differentiation in *Saccharomyces cerevisiae. Mol Cell Biol* 19 (7):4874–87.

25. Bahn, Y.S., J. Staab, and P. Sundstrom. 2003. Increased high-affinity phosphodiesterase PDE2 gene expression in germ tubes counteracts CAP1-dependent synthesis of cyclic AMP, limits hypha production and promotes virulence of *Candida albicans. Mol Microbiol* 50 (2):391–409.

26. Jung, W.H., and L.I. Stateva. 2003. The cAMP phosphodiesterase encoded by CaPDE2 is required for hyphal development in *Candida albicans. Microbiology* 149 (Pt 10):2961–76.

27. Fradin, C., et al. 2003. Stage-specific gene expression of *Candida albicans* in human blood. *Mol Microbiol* 47 (6):1523–43.

28. Jung, W.H., et al. 2005. Deletion of PDE2, the gene encoding the high-affinity cAMP phosphodiesterase, results in changes of the cell wall and membrane in *Candida albicans. Yeast* 22 (4):285–94.

29. Bonner, J.T. et al. 1969. Acrasin, Acrasinase, and the sensitivity to acrasin in *Dictyostelium discoideum. Dev Biol* 20 (1):72–87.

30. Anjard, C., F. Soderbom, and W.F. Loomis. 2001. Requirements for the adenylyl cyclases in the development of *Dictyostelium. Development* 128 (18):3649–54.

31. Kimmel, A.R., and C.A. Parent. 2003. The signal to move: *D. discoideum* go orienteering. *Science* 300 (5625):1525–7.

32. Loomis, W.F., G. Shaulsky, and N. Wang. 1997. Histidine kinases in signal transduction pathways of eukaryotes. *J Cell Sci* 110 (Pt 10):1141–5.

33. Postma, M., et al. 2004. Chemotaxis: Signalling modules join hands at front and tail. *EMBO Rep* 5 (1):35–40.

34. Malchow, D., and G. Gerisch. 1974. Short-term binding and hydrolysis of cyclic 3′,5′-adenosine monophosphate by aggregating *Dictyostelium* cells. *Proc Natl Acad Sci USA* 71 (6):2423–7.

35. Lacombe, M.L., et al. 1986. Molecular cloning and developmental expression of the cyclic nucleotide phosphodiesterase gene of *Dictyostelium discoideum. J Biol Chem* 261 (36):16811–7.

36. Shaulsky, G., R. Escalante, and W.F. Loomis. 1996. Developmental signal transduction pathways uncovered by genetic suppressors. *Proc Natl Acad Sci USA* 93 (26):15260–5.
37. Thomason, P.A., et al. 1998. An intersection of the cAMP/PKA and two-component signal transduction systems in *Dictyostelium*. *MBO J* 17 (10):2838–45.
38. Bosgraaf, L., et al. 2002. Identification and characterization of two unusual cGMP-stimulated phoshodiesterases in *Dictyostelium*. *Mol Biol Cell* 13 (11):3878–89.
39. Eichinger, L., et al. 2005. The genome of the social amoeba *Dictyostelium discoideum*. *Nature* 435 (7038):43–57.
40. Carfi, A., et al. 1995. The 3D structure of a zinc metallo-beta-lactamase from *Bacillus cereus* reveals a new type of protein fold. *EMBO J* 14 (20):4914–21.
41. Kessin, R.H., et al. 1979. Binding of inhibitor alters kinetic and physical properties of extracellular cyclic AMP phosphodiesterase from *Dictyostelium discoideum*. *Proc Natl Acad Sci USA* 76 (11):5450–4.
42. Orlow, S.J., et al. 1981. The extracellular cyclic nucleotide phosphodiesterase of *Dictyostelium discoideum*. Purification and characterization. *J Biol Chem* 256 (14):7620–7.
43. Franke, J., and R.H. Kessin. 1992. The cyclic nucleotide phosphodiesterases of *Dictyostelium discoideum*: Molecular genetics and biochemistry. *Cell Signal* 4 (5):471–8.
44. Sucgang, R., et al. 1997. Null mutations of the *Dictyostelium* cyclic nucleotide phosphodiesterase gene block chemotactic cell movement in developing aggregates. *Dev Biol* 192 (1):181–92.
45. Wu, L., et al. 1995. The phosphodiesterase secreted by prestalk cells is necessary for *Dictyostelium* morphogenesis. *Dev Biol* 167 (1):1–8.
46. Faure, M., et al. 1990. The cyclic nucleotide phosphodiesterase gene of *Dictyostelium discoideum* contains three promoters specific for growth, aggregation, and late development. *Mol Cell Biol* 10 (5):1921–30.
47. Franke, J., et al. 1991. Cyclic nucleotide phosphodiesterase of *Dictyostelium discoideum* and its glycoprotein inhibitor: Structure and expression of their genes. *Dev Genet* 12 (1–2):104–12.
48. Muramoto, T., et al. 2005. Reverse genetic analyses of gamete-enriched genes revealed a novel regulator of the cAMP signaling pathway in *Dictyostelium discoideum*. *Mech Dev* 122 (5):733–43.
49. Morris, H.R., et al. 1987. Chemical structure of the morphogen differentiation inducing factor from *Dictyostelium discoideum*. *Nature* 328 (6133):811–4.
50. Thompson, C.R., and R.R. Kay. 2000. The role of DIF-1 signaling in *Dictyostelium* development. *Mol Cell* 6 (6):1509–14.
51. Mullens, I.A., et al. 1984. Developmental regulation of the cyclic-nucleotide-phosphodiesterase mRNA of *Dictyostelium discoideum*. Analysis by cell-free translation and immunoprecipitation. *Eur J Biochem* 142 (2):409–15.
52. Yeh, R.P., F.K. Chan, and M.B. Coukell. 1978. Independent regulation of the extracellular cyclic AMP phosphodiesterase-inhibitor system and membrane differentiation by exogenous cyclic AMP in *Dictyostelium discoideum*. *Dev Biol* 66 (2):361–74.
53. Riedel, V., and G. Gerisch. 1971. Regulation of extracellular cyclic-AMP-phosphodiesterase activity during development of *Dictyostelium discoideum*. *Biochem Biophys Res Commun* 42 (1): 119–24.
54. Wu, L., and J. Franke. 1990. A developmentally regulated and cAMP-repressible gene of *Dictyostelium discoideum*: Cloning and expression of the gene encoding cyclic nucleotide phosphodiesterase inhibitor. *Gene* 91 (1):51–6.
55. Franke, J., and R.H. Kessin. 1981. The cyclic nucleotide phosphodiesterase inhibitory protein of *Dictyostelium discoideum*. Purification and characterization. *J Biol Chem* 256 (14):7628–37.
56. Grebe, T.W., and J.B. Stock. 1999. The histidine protein kinase superfamily. *Adv Microb Physiol* 41:139–227.
57. Chang, W.T., et al. 1998. Evidence that the RdeA protein is a component of a multistep phosphorelay modulating rate of development in *Dictyostelium*. *EMBO J* 17 (10):2809–16.
58. Thomason, P.A., et al. 1999. The RdeA–RegA system, a eukaryotic phospho-relay controlling cAMP breakdown. *J Biol Chem* 274 (39):27379–84.
59. Ott, A., et al. 2000. Osmotic stress response in *Dictyostelium* is mediated by cAMP. *EMBO J* 19 (21):5782–92.

60. Hoch, J.A., and T.J. Silhavy. 1995. *Two-component signal transduction*. Washington: ASM Press.
61. Loomis, W.F. 1998. Role of PKA in the timing of developmental events in *Dictyostelium* cells. *Microbiol Mol Biol Rev* 62 (3):684–94.
62. Laub, M.T., and W.F. Loomis. 1998. A molecular network that produces spontaneous oscillations in excitable cells of *Dictyostelium*. *Mol Biol Cell* 9 (12):3521–32.
63. Maeda, M., et al. 1996. Seven helix chemoattractant receptors transiently stimulate mitogen-activated protein kinase in *Dictyostelium*. Role of heterotrimeric G proteins. *J Biol Chem* 271 (7):3351–4.
64. Maeda, M., et al. 2004. Periodic signaling controlled by an oscillatory circuit that includes protein kinases ERK2 and PKA. *Science* 304 (5672):875–8.
65. Mohanty, S., et al. 2001. Regulated protein degradation controls PKA function and cell-type differentiation in *Dictyostelium*. *Genes Dev* 15 (11):1435–48.
66. Anjard, C., and W.F. Loomis. 2005. Peptide signaling during terminal differentiation of *Dictyostelium*. *Proc Natl Acad Sci USA* 102 (21):7607–11.
67. Zinda, M.J., and C.K. Singleton. 1998. The hybrid histidine kinase dhkB regulates spore germination in *Dictyostelium discoideum*. *Dev Biol* 196 (2):171–83.
68. Saran, S., and P. Schaap. 2004. Adenylyl cyclase G is activated by an intramolecular osmosensor. *Mol Biol Cell* 15 (3):1479–86.
69. Shaulsky, G., D. Fuller, and W.F. Loomis. 1998. A cAMP-phosphodiesterase controls PKA-dependent differentiation. *Development* 125 (4):691–9.
70. Wang, B., and A. Kuspa. 1997. *Dictyostelium* development in the absence of cAMP. *Science* 277 (5323):251–4.
71. Kuwayama, H., et al. 2001. Identification and characterization of DdPDE3, a cGMP-selective phosphodiesterase from *Dictyostelium*. *Biochem J* 353 (Pt 3):635–44.
72. Mann, S.K., and R.A. Firtel. 1991. A developmentally regulated, putative serine/threonine protein kinase is essential for development in *Dictyostelium*. *Mech Dev* 35 (2):89–101.
73. Meima, M.E., R.M. Biondi, and P. Schaap. 2002. Identification of a novel type of cGMP phosphodiesterase that is defective in the chemotactic stmF mutants. *Mol Biol Cell* 13 (11):3870–7.
74. Harman, J.G. 2001. Allosteric regulation of the cAMP receptor protein. *Biochim Biophys Acta* 1547 (1):1–17.
75. Van Haastert, P.J., and M.M. Van Lookeren Campagne. 1984. Transient kinetics of a cGMP-dependent cGMP-specific phosphodiesterase from *Dictyostelium discoideum*. *J Cell Biol* 98 (2):709–16.
76. Meima, M.E., K.E. Weening, and P. Schaap. 2003. Characterization of a cAMP-stimulated cAMP phosphodiesterase in *Dictyostelium discoideum*. *J Biol Chem* 278 (16):14356–62.
77. Jenny, A., et al. 1996. Sequence similarity between the 73-kilodalton protein of mammalian CPSF and a subunit of yeast polyadenylation factor I. *Science* 274 (5292):1514–7.
78. Arav-Boger, R., and T.A. Shapiro. 2005. Molecular mechanisms of resistance in antimalarial chemotherapy: The unmet challenge. *Annu Rev Pharmacol Toxicol* 45:565–85.
79. Ouellette, M., J. Drummelsmith, and B. Papadopoulou. 2004. Leishmaniasis: Drugs in the clinic, resistance and new developments. *Drug Resist Updat* 7 (4–5):257–66.
80. Seebeck, T., and P. Maeser. 2006. Drug resistance in African trypanosomiasis. In *Antimicrobial drug resistance. Principle and practice for the clinic and the bench*, eds. D. Myers, et al. New Jersey: Humana Press (in press).
81. Berriman, M., et al. 2005. The genome of the African trypanosome *Trypanosoma brucei*. *Science* 309 (5733):416–22.
82. El-Sayed, N.M., et al. 2005. Comparative genomics of trypanosomatid parasitic protozoa. *Science* 309 (5733):404–9.
83. Ivens, A.C., et al. 2005. The genome of the kinetoplastid parasite, *Leishmania major*. *Science* 309 (5733):436–42.
84. Kunz, S., et al. 2006. Cyclic nucleotide specific phosphodiesterases of the kinetoplastida: A unified nomenclature. *Mol Biochem Parasitol* 145 (1):133–5.
85. Martinez, S.E., J.A. Beavo, and W.G. Hol. 2002. GAF domains: Two-billion-year-old molecular switches that bind cyclic nucleotides. *Mol Interv* 2 (5):317–23.

86. Kunz, S., et al. 2004. TbPDE1, a novel class I phosphodiesterase of *Trypanosoma brucei. Eur J Biochem* 271 (3):637–47.

87. Gong, K.W., et al. 2001. cAMP-specific phosphodiesterase TbPDE1 is not essential in *Trypanosoma brucei* in culture or during midgut infection of tsetse flies. *Mol Biochem Parasitol* 116 (2):229–32.

88. Johner, A., et al. 2006. Cyclic nucleotide specific phosphodiesterases of *Leishmania major BMC Microbiol* 6:25.

89. Zoraghi, R., and T. Seebeck. 2002. The cAMP-specific phosphodiesterase TbPDE2C is an essential enzyme in bloodstream form *Trypanosoma brucei. Proc Natl Acad Sci USA* 99 (7):4343–8.

90. Rascon, A., et al. 2002. Cloning and characterization of a cAMP-specific phosphodiesterase (TbPDE2B) from *Trypanosoma brucei. Proc Natl Acad Sci USA* 99 (7):4714–9.

91. Laxman, S., A. Rascon, and J.A. Beavo. 2005. Trypanosome cyclic nucleotide phosphodiesterase 2B binds cAMP through its GAF-A domain. *J Biol Chem* 280 (5):3771–9.

92. Zoraghi, R. et al. 2001. Characterization of TbPDE2A, a novel cyclic nucleotide-specific phosphodiesterase from the protozoan parasite *Trypanosoma brucei. J Biol Chem* 276 (15):11559–66.

93. D'Angelo, M.A., et al. 2004. Identification, characterization and subcellular localization of TcPDE1, a novel cAMP-specific phosphodiesterase from *Trypanosoma cruzi. Biochem J* 378 (Pt 1):63–72.

94. Gadelha, C., et al. 2004. Relationships between the major kinetoplastid paraflagellar rod proteins: A consolidating nomenclature. *Mol Biochem Parasitol* 136 (1):113–5.

95. Paindavoine, P., et al. 1992. A gene from the variant surface glycoprotein expression site encodes one of several transmembrane adenylate cyclases located on the flagellum of *Trypanosoma brucei. Mol Cell Biol* 12 (3):1218–25.

96. D'Angelo, M.A., et al. 2002. A novel calcium-stimulated adenylyl cyclase from *Trypanosoma cruzi*, which interacts with the structural flagellar protein paraflagellar rod. *J Biol Chem* 277 (38):35025–34.

97. Kunz, S., M. Oberholzer, and T. Seebeck. 2005. A FYVE-containing unusual cyclic nucleotide phosphodiesterase from *Trypanosoma cruzi. FEBS J* 272 (24):6412–22.

98. Francis, S.H., I.V. Turko, and J.D. Corbin. 2001. Cyclic nucleotide phosphodiesterases: Relating structure and function. *Prog Nucleic Acid Res Mol Biol* 65:1–52.

99. Birkeland, H.C., and H. Stenmark. 2004. Protein targeting to endosomes and phagosomes via FYVE and PX domains. *Curr Top Microbiol Immunol* 282:89–115.

100. Inselburg, J. 1983. Stage-specific inhibitory effect of cyclic AMP on asexual maturation and gametocyte formation of *Plasmodium falciparum. J Parasitol* 69 (3):592–7.

101. Kawamoto, F., et al. 1993. The roles of Ca^{2+}/calmodulin- and cGMP-dependent pathways in gametogenesis of a rodent malaria parasite, *Plasmodium berghei. Eur J Cell Biol* 60 (1):101–7.

102. Read, L.K., and R.B. Mikkelsen. 1991. Comparison of adenylate cyclase and cAMP-dependent protein kinase in gametocytogenic and nongametocytogenic clones of *Plasmodium falciparum. J Parasitol* 77 (3):346–52.

103. Baker, D.A., and J.M. Kelly. 2004. Purine nucleotide cyclases in the malaria parasite. *Trends Parasitol* 20 (5):227–32.

104. Gardner, M.J., et al. 2002. Genome sequence of the human malaria parasite *Plasmodium falciparum. Nature* 419 (6906):498–511.

105. Beavo, J.A. 1995. Cyclic nucleotide phosphodiesterases: Functional implications of multiple isoforms. *Physiol Rev* 75 (4):725–48.

106. Bozdech, Z., et al. 2003. The transcriptome of the intraerythrocytic developmental cycle of *Plasmodium falciparum. PLoS Biol* 1 (1):E5.

107. Le Roch, K.G., et al. 2003. Discovery of gene function by expression profiling of the malaria parasite life cycle. *Science* 301 (5639):1503–8.

108. Young, J.A., et al. 2005. The *Plasmodium falciparum* sexual development transcriptome: A microarray analysis using ontology-based pattern identification. *Mol Biochem Parasitol* 143:67–79.

109. Yuasa, K., et al. 2005. PfPDE1, a novel cGMP-specific phosphodiesterase, from the human malaria parasite *Plasmodium falciparum. Biochem J* 392 (Pt 1):221–9.

110. Zhang, W., et al. 2001. Identification of overlapping but distinct cAMP and cGMP interaction sites with cyclic nucleotide phosphodiesterase 3A by site-directed mutagenesis and molecular modeling based on crystalline PDE4B. *Protein Sci* 10 (8):1481–9.

15 Studies of Phosphodiesterase Function Using Fruit Fly Genomics and Transgenics

Shireen-A. Davies and Jonathan P. Day

CONTENTS

15.1 *Drosophila melanogaster* as a Model Organism 301
15.2 Phosphodiesterases in *Drosophila melanogaster* 302
    15.2.1 Gene Information 303
        15.2.1.1 *CG14940* 304
        15.2.1.2 *Dunce* 304
        15.2.1.3 *CG8279* 304
        15.2.1.4 *CG5411* 304
        15.2.1.5 *CG32648* 305
        15.2.1.6 *CG10231* 305
    15.2.2 Expression of Phosphodiesterases in Adult *Drosophila* 306
    15.2.3 Protein Sequence Alignments of Novel *Drosophila* Phosphodiesterases 307
    15.2.4 Structural Features of Novel *Drosophila* Phosphodiesterases 307
        15.2.4.1 PDE1 307
        15.2.4.2 PDE6 307
        15.2.4.3 PDE8 310
        15.2.4.4 PDE9 310
        15.2.4.5 PDE11 310
    15.2.5 Biochemistry and Pharmacology of *Drosophila* Phosphodiesterases 311
        15.2.5.1 Dunce 311
        15.2.5.2 Novel Phosphodiesterases—PDE1, PDE6, and PDE11 312
    15.2.6 Inhibitor Sensitivities of Novel *Drosophila* Phosphodiesterases 314
    15.2.7 Phenotypic Evidence for Functional Conservation of *Drosophila*
        and Mammalian Phosphodiesterases 315
15.3 Conclusions 317
Acknowledgments 317
References 317

15.1 *DROSOPHILA MELANOGASTER* AS A MODEL ORGANISM

The fruit fly, *Drosophila melanogaster*, has been recognized as a genetic model since the early twentieth century, when Thomas Hunt Morgan discovered the white-eyed mutation in 1910. For his work in *Drosophila* genetics, and the chromosome in heredity, Morgan was awarded the Nobel Prize in physiology or medicine in 1933.

Over the decades, it has become clear that when the fly and humans are separated by ~400 million years of evolution, many close similarities exist in gene–protein structure and function between these organisms. Thus, study of *Drosophila* development led to the identification of key developmental processes in mammals and humans, and another Nobel Prize in physiology or medicine in 1995 was awarded to Ed Lewis, Christiane Nusslein-Volhard, and Eric Weischaus.

The highly sophisticated genetics of the fruit fly has been complemented extensively by the sequenced *D. melanogaster* genome, first released in 2000. The availability of the complete genome (http://fly.ebi.ac.uk:7081/) together with other genomic resources has rapidly advanced the work in *Drosophila*. For example, the results from large-scale yeast two-hybrid experiments listing protein–protein interactions[1] are now compiled on a freely available online database (http://biodata.mshri.on.ca:80/fly_grid/servlet/SearchPage), which allows quick analysis of specific proteins of interest. Furthermore, a complete database of human disease genes and their homologs in the *Drosophila* genome has been compiled[2] (see also http://superfly.ucsd.edu/homophila/). This allows functional analysis of disease genes, many of which have unknown function, *in vivo*. Finally, the sheer range of transgenic and mutant resources available to *Drosophila* workers, including clone, mutant, and stock collections (see FlyBase home page, http://fly.ebi.ac.uk:7081/), is unparalleled in biology, allowing the studies of gene function in an organism with a complex body plan, and a range of behavioral and physiological paradigms.

15.2 PHOSPHODIESTERASES IN *DROSOPHILA MELANOGASTER*

Genetic analysis of specific loci in the late 1970s led to the discovery of regions of the *Drosophila* chromosome, which contained cyclic nucleotide phosphodiesterase (PDE)-encoding genes.[3] In particular, chromosomal region 3D4 was found to be associated with cAMP-dependent PDE activity and was necessary for fertility and oogenesis.[4,5] Biochemical analysis of adult *Drosophila* extracts showed that two distinct PDE activities were present: one, cAMP-specific (termed Form II) and the other, Form I, which catalyzed the hydrolysis of both cAMP and cGMP.[6] In 1976, Dudai et al.[7] discovered the first gene to be associated with learning defects in a newly isolated mutant, *dunce (dnc)*. In a seminal paper, the biochemical defect associated with the *dunce* mutation was shown to involve a Form II cA-PDE.[8] This work also showed that *dunce* was located on chromomere 3D4 (now further mapped to region chromosomal location 3C9-D1), strongly suggesting that *dunce* was the structural gene for the Form II enzyme. Further work revealed that the Form I but not the Form II enzyme was calcium-regulated, suggesting that neurological defects arising in *dunce* mutants were directly due to aberrant cAMP signaling in neurons, as opposed to cAMP-induced calcium influx in presynaptic transmission. Work with several alleles of *dunce* confirmed that aberrations in cA-PDE activity were directly linked to such mutations,[9] and finally, molecular analysis of the wild-type *dunce* gene showed unequivocally that this encoded a cA-PDE.[10]

The first PDE gene to be identified was *dunce*, and the discovery of *dunce*-encoded cA-PDE was instrumental in the cloning of the first mammalian PDE, RD1 thereby delineating, for the first time at the genetic level, the PDE4 family.[11] Identification of rat homologs of *dunce* uncovered four distinct gene families in 1989[12,13] encoding rolipram-sensitive cAMP-specific PDEs[14] of which RD1 is now known as PDE4A1.

The importance of *dunce* and, more specifically, cA-PDEs, in learning and behavior, long obscured the existence of other *Drosophila* PDEs. Furthermore, the general focus on cAMP signaling, as opposed to cGMP signaling, may also have partly explained the dearth of interest in *Drosophila* cG-PDEs. In recent years, however, interest in cGMP signaling *in vivo*[15,16] has led to the investigation of cG-PDEs and other novel *Drosophila* PDEs.

Indeed, the entire *Drosophila* PDE family has recently been curated using genome resources (http://fly.ebi.ac.uk:7081/.bin/fbidq.html?FBrf0159902). Polypeptide sequences of the 11 known mammalian PDEs were used as probes for putative *Drosophila* PDEs to search the Berkeley *Drosophila* Genome Project database (http://flybase.net/). Any hits were screened for the $HX_3HX_{21-23}D/E$ 3',5'-cyclic nucleotide motifs. This initial genomic survey showed that there were five PDEs encoded by the *Drosophila* genome in addition to *dunce*.

The PDE-encoding genes of *D. melanogaster* are *CG14940*, *CG8279*, *CG5411*, *CG32648*, and *CG10231*. These are PDE-encoding genes that were also confirmed in another study.[15]

In silico analysis of sequence similarity between putative *Drosophila* and human PDE sequences are shown in Figure 15.1 The dendrogram in Figure 15.1 shows distinct clustering of the PDE families, correlating closely with relevant human PDEs. *Drosophila* PDEs do not show any close similarity to human PDE2, PDE3, PDE7, or PDE10.

Sequences were aligned using ClustalW and displayed using TreeView.[17] PID accession numbers for human PDE sequences were as follows: PDE1A (NP_005010), PDE1B (NP_000915), PDE1C (Q14123), PDE2A (AAH_40974), PDE3A (QI_4432), PDE3B (NP_000913), PDE4A (NP_006193), PDE4B1 (CA123101), PDE4C (NP_000914), PDE4D (NP_006194), PDE5A (NP001074), PDE6α (P16499), PDE6α' (CAH72330), PDE6β (NP_000274), PDE7Aa (NP_002594), PDE8A1 (NP_002596), PDE9A (076083), PDE8B (NP_003710), PDE10A (CA120436), and PDE11A4 (BAB62712).

15.2.1   GENE INFORMATION

The Berkeley *Drosophila* Genome Project (BDGP) (http://www.fruitfly.org/about/) is a consortium of the *Drosophila* Genome Centre, founded by key *Drosophila* investigators in the United States. Part of the aim of the BDGP is to characterize and sequence *D. melanogaster*

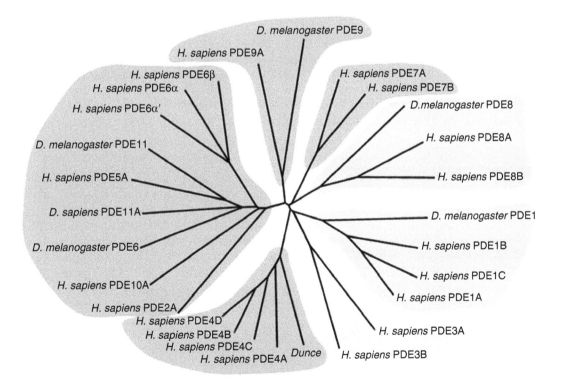

FIGURE 15.1 (**See color insert following page 274.**) Sequence similarity analysis of PDE protein sequences.

cDNAs. These are cloned from various cDNA libraries, including embryo, adult head, and adult testes libraries (http://www.fruitfly.org/EST/EST.shtml), and clones known as expressed sequence tags (ESTs) are available for the study of individual genes of interest via a comprehensive database. The availability of such cDNA clones reduces the need for library screening and cloning methods in order to obtain sequence for novel genes. Thus, use of such ESTs relevant to the novel PDEs has been instrumental in confirming, or extending, the nucleic acid sequences, and predicted amino acid sequences available from the genome database.

15.2.1.1 *CG14940*

The closest *Drosophila* homolog of mammalian PDE1 is *CG14940*. *CG14940* resides on chromosome 2L at position 33A2. Sequencing of an EST clone for *CG14940* (RE56388) shows that this gene has an open reading frame (ORF) of 1815 nucleotides, which encodes a polypeptide of 605 amino acids (aa). The transcript comprises four exons and covers 7 kb of genomic sequence. *CG14940* encodes a polypeptide, which shows 40% overall identity with human PDE1C and 63% identity in the catalytic domain.

15.2.1.2 *Dunce*

The gene *dnc* is located on the X chromosome and its exons are distributed over 148 kb of genomic DNA. Furthermore, two genes, *Pig1* and *Sgs4*, are located within the *dnc* 79 kb intron.[10] Also, the *dnc* genomic region is rich in chromosomal rearrangements. The *dnc* is a complex gene, encoding multiple RNAs arising from transcription initiation at multiple sites, alternative splicing, and generation of different 3′ ends.[18] The *dnc* encodes 14 major transcripts, which are transcribed from at least eight different start sites.[18] Deletion mutagenesis of individual transcripts, as well as phenotypic work, has shown that RNAs arising from specific transcription start sites are associated with different phenotypes, including fertility, and learning and memory.[19] Transcripts *dnc-RO* and *dnc-RB* encode full-length polypeptides of 1209 aa, dnc-PO and dnc-PB (isoform I), respectively. Polypeptides of significant length are dnc-PC (1057 aa), dnc-PD (1068 aa), dnc-PI, dnc-PJ, dnc-PK, and dnc-PJ (all others have 1070 aa; dnc-PI and dnc-PJ constitute isoform II). The polypeptides dnc-PA, dnc-PE, dnc-PF, dnc-PG, and dnc-PM are all 700–900 aa in length. Each polypeptide contains the conserved catalytic domain associated with cA-PDEs, although functional cA-PDE activity has not yet been demonstrated for all the polypeptides.

15.2.1.3 *CG8279*

BLAST searching allowed the identification of *Drosophila* gene *CG8279* as a putative homolog of mammalian PDE5 and PDE6. *CG8279* resides between positions 88C5 and 88C6 on the third chromosome. The DNA sequence predicted by the genome annotation comprises a 3393 base-pair ORF, coding for a polypeptide of 1131 aa (Table 15.1). Although an available EST (GH27433) for *CG8279* lacked 271 base pairs of the 5′ end of the ORF, a reverse transcriptase-polymerase chain reaction (RT-PCR) strategy was used to obtain the full-length ORF.[20] Northern blot analysis also demonstrated a full-length transcript of 7 kb. The predicted amino acid sequence of *CG8279* shares 28% overall sequence identity with mammalian PDE6β, but shares 51% identity within the catalytic domain. *CG8279* also shares 58% identity within the catalytic domain of PDE5.

15.2.1.4 *CG5411*

CG5411 has 19 exons and covers over 13 kb of genomic sequence as predicted on FlyBase. It resides between positions 59F3 and 59F4 on chromosome 2R. In order to confirm the

TABLE 15.1
Sequence Information for Novel PDE Genes and Deduced Proteins Identified from the
***D. Melanogaster* Genome Listed Together with Closest Human Homolog**

Gene	Human Homolog	% aa Identity (Similarity)	% aa Identity (Similarity), Catalytic Domain	Predicted Length of Polypeptide (aa)	ESTs
CG14940	PDE1	40 (56)	63 (79)	1818	RE56388
CG8279	PDE5/PDE6	28 (46)	51 (69)	1131	GH27433[a]
CG5411 transcript A	PDE8	34 (52)	60 (79)	914	SD18711
CG5411 transcript B	PDE8	35 (53)	60 (79)	904	RE31467
CG5411 transcript C	PDE8	47 (66)	60 (79)	400	GH21295
CG5411 transcript D	PDE8	37 (57)	60 (79)	805	RE35136
CG5411 transcript E	PDE8	34 (52)	60 (79)	914	RE07805
CG5411 transcript F	PDE8	23 (9)	60 (79)	400	LD46553
CG32648	PDE9	26 (34)	63 (76)	963[b]	None
CG10231	PDE11	38 (55)	77 (96)	1545	SD13096

[a]Full-length ESTs for *Drosophila* PDE genes are also listed, where available (http://www.fruitfly.org/EST/index.shtml) except for GH27433, which is not full length. The percentage identities and similarities for each gene are calculated over the length of the shorter (human) homolog; and also in relation to the catalytic domain.
[b]previously thought to be 2080 aa. Current estimate of deduced PDE9 polypeptide length from http://flybase.bio. indiana.edu/.bin/fbidq.html?FBpp0073534.
Source: Reproduced from Ref. [20]. With permission.

predicted nucleotide sequence, analysis of EST SD18711 showed that this encoded the full predicted nucleotide sequence, plus 320 bp of 5′ UTR. This results in an ORF of 3180 bp, which translates into a polypeptide of 1060 aa.

Version 3 of the *Drosophila* genome annotation (http://flybase.net/annot/) identifies five transcripts encoding four different polypeptides for PDE8, A–E, in which the ORFs of transcripts A and E are identical. Analysis of EST LD46553 suggests that a further transcript exists, identical to transcript C yet with an 800 bp extension to the 3′ UTR, designated transcript F.

15.2.1.5 *CG32648*

BLAST searching using the polypeptide sequence of mammalian PDE9 as a probe revealed significant similarity between this PDE and the conceptual translation of *Drosophila CG32648*. *CG32648* resides on the X chromosome at position 11C2–3, within a region identified by Kiger and Golanty,[3] resulting in increased cG-PDE activity when duplicated.

The ESTs for *CG32648* are not available, although in silico translation of *CG32648* results in a 963 aa polypeptide. *Drosophila* PDE9 shows 54% identity with the catalytic domain of human PDE9 but very little similarity outside this region. This may indicate that further sequence annotation for *CG32648* is required.

15.2.1.6 *CG10231*

CG10231 was identified as a homolog of mammalian PDE11. *CG10231* resides at position 37A1 on chromosome 2, where it covers nearly 9 kb of genomic sequence. The predicted nucleotide sequence of PDE11 showed that it possessed 16 exons. A full-length EST clone,

SD10396, was sequenced to verify *CG10231* and predicted to encode transcript of ~5.8 kb long. Northern blotting of adult head and body RNA confirmed the full-length transcript of *CG10231* to be 5.8 kb. *CG10231* encodes a PDE with 77% sequence identity for the catalytic domain of human PDE11. Table 15.1 provides a summary of sequence information available for the novel *Drosophila* PDEs.

15.2.2 Expression of Phosphodiesterases in Adult *Drosophila*

The emergence of *dnc* as the first gene to be associated with behavior led to intense study of sites of expression in the fly. Predictably, *dnc* is highly expressed in the larval and adult brain, including the learning and memory centers of the mushroom bodies (http://flybrain.uni-freiburg.de/Flybrain/html/terms/terms.html). The *dnc* expression has also been documented in abdominal ventral longitudinal muscle, embryonic and larval somatic muscle, and motor neurons, thus suggesting a role for *dnc* in flight and escape responses. The effect of *dnc* mutations on female sterility is also explained by high *dnc* expression in egg chamber and in nurse cells. Thus, *dnc* is widely expressed in the fly (see http://flybase.bio.indiana.edu/.bin/fbidq.html?FBgn0000479&content=phenotype).

Given the widespread importance of cyclic nucleotide signaling in tissues, global expression of other PDE genes in *Drosophila* aside from *dnc* may be expected. Indeed, expression of all novel *Drosophila* PDEs occurs in head and body tissue[20] (see also Figure 15.2) as assessed by

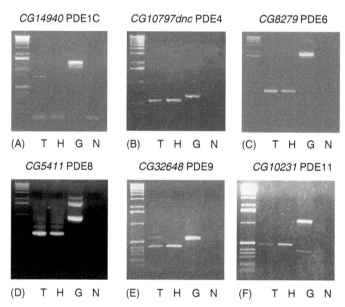

FIGURE 15.2 PDE expression in adult *D. melanogaster*. RT-PCR using cDNA templates from adult tubule (T), head (H), and control genomic DNA (G) using intron-spanning gene-specific primers for putative PDE-encoding genes. (N) indicates no template control. Assessment of band size was achieved using a 1 kb DNA ladder, shown in the first lane of each gel. PCR products obtained for all putative PDE genes are as follows: (A) *CG14940* (PDE1C), genomic-1438 bp, cDNA-743 bp; (B) *CG10797* (*dunce*) (PDE4), genomic-603 bp, cDNA-438 bp; (C) *CG8279* (PDE6), genomic-1969 bp, cDNA-406 bp; (D) *CG5411* (PDE8), genomic-998 bp, cDNA-506 bp; (E) *CG32648* (PDE9), genomic-567 bp, cDNA-391 bp; and (F) *CG10231* (PDE11), genomic-1118 bp, cDNA-472 bp. PCR products were cloned and sequenced and found to have 100% identity to PDE genes (data not shown). (Reproduced from Ref. [20]. With permission.)

RT-PCR from *Drosophila* cDNA template using gene-specific primers. Although expression of *CG14940* (PDE1) shows faint expression in head, expression of this PDE gene in head tissue is verified by the availability of PDE1 EST clones derived from head cDNA libraries.

Interestingly, the *Drosophila* Malpighian (renal) tubule expresses all identified PDEs,[20] suggesting the importance of cyclic nucleotide signaling in renal epithelia. Microarray analysis of PDE gene expression in both tubules and in the rest of adult flies via Affymetrix arrays[21] confirms the RT-PCR data. In particular, *CG8279* PDE6, *CG5411* PDE8, and *CG10231* PDE11 are all significantly enriched in tubules compared to the rest of the fly (http://www.mblab.gla.ac.uk/%7Ejulian/arraysearch.cgi).

15.2.3 PROTEIN SEQUENCE ALIGNMENTS OF NOVEL *DROSOPHILA* PHOSPHODIESTERASES

In addition to the screening of putative PDEs using the $HX_3HX_{21-23}D/E$ motif, homologs to mammalian PDE1, PDE6, PDE8, PDE9, and PDE11 were identified based on the sequence similarity within the predicted catalytic domain; as well as by the presence of sequences proposed to be regulatory and posttranslational modification sequences in the vertebrate enzymes.[22] Figure 15.3 shows the alignments of translated catalytic domains of *Drosophila* PDEs with those of their closest human homologs.

15.2.4 STRUCTURAL FEATURES OF NOVEL *DROSOPHILA* PHOSPHODIESTERASES

A schematic summary of potential regulatory and catalytic domains in the novel *Drosophila* PDEs based on in silico sequence analysis are shown in Figure 15.4 The significance of such features for each PDE will be discussed in turn. The schematic summary was deduced from ClustalW alignments for each novel *Drosophila* PDE, labeled as shown in Figure 15.4.

15.2.4.1 PDE1

The N-terminal autoinhibitory motif of mammalian PDE1A[23] is conserved between mammal and fly,[20] although only one of the two mammalian calmodulin-binding sites appears to be present in *Drosophila* PDE1 (Figure 15.4). Examination of *Drosophila* PDE1 and *Homo sapiens* PDE1C (http://www.biochemj.org/bj/388/bj3880333add.htm) revealed that the fly gene encodes an extra 13 aa at the N-terminus and is truncated by 133 aa at the C-terminus, although the significance of this is unknown.

Recently, structural studies have revealed the Q switch mechanism of PDE substrate specificity, where a conserved glutamine residue changes orientation with respect to the bound substrate.[24] Specific histidine residues surrounding this glutamine confer such rotational freedom, and this is thought to form the basis of mechanism of action of dual-specificity PDEs. The glutamine residue, which coordinates the nucleoside purine was identified in *Drosophila* PDE1 (Q439), which correlates to Q426 in PDE1C. Three critical residues surrounding the glutamine in mammalian PDE1C (H381, H373, and W496) are conserved in *Drosophila* PDE1 (H399, H391, and W539). It therefore appears that PDE1 encoded by *CG14940* possesses the structural requirements for dual-specificity PDE1 activity.

15.2.4.2 PDE6

CG8279-encoded PDE6 is the closest *Drosophila* homolog of mammalian PDE5, and a close homolog of mammalian retinal PDE6β. One of the most prominent features of the predicted protein sequence of *CG8279* is the presence of a C-terminal CAAX-box prenylation motif (http://www.biochemj.org/bj/388/bj3880333add.htm). The presence of the above mentioned motif and the presence of conserved catalytic domain residues led to its designation as a

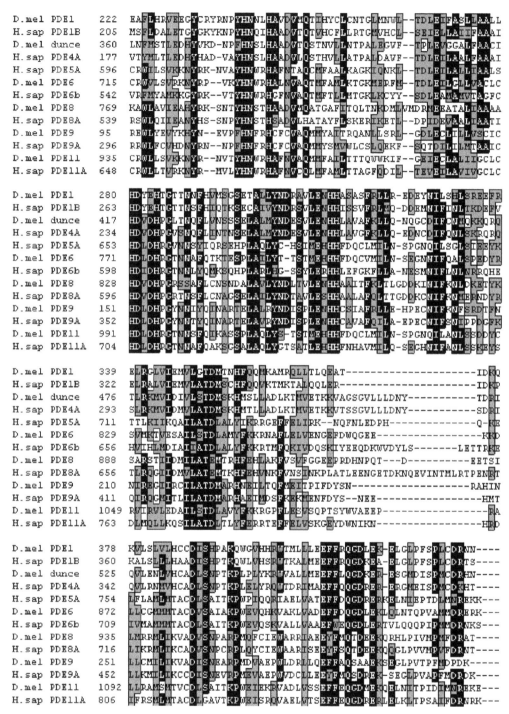

FIGURE 15.3 (See color insert following page 274.) Sequence alignments of catalytic domains of *D. melanogaster* and *H. sapiens* PDEs. These were performed using ClustalW (http://www.ebi. ac. uk/clustalw/), and drawn using BioEdit (http://www.mbio.ncsu.edu/BioEdit/bioedit.html). Comparisons of *Drosophila* and human PDEs are shown, where identical residues are shaded black; red shading indicates residues with 70% similarity. Identity or similarity was assessed using a PAM250 scoring matrix at 70% stringency. Full alignments of novel *Drosophila* PDEs with human homologs are available at http://www.biochemj.org/bj/388/bj3880333add.htm. (Reproduced from Ref. [20]. With permission.)

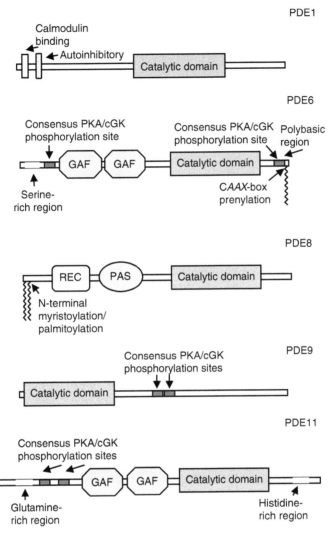

FIGURE 15.4 (See color insert follwoing page 274.) Schematic diagram of novel *Drosophila* PDEs. (Reproduced from Ref. [20]. With permission.)

PDE6 homolog in spite of the strong similarity to PDE5. *Drosophila* PDE6 contains regions similar to the tandem cGMP-binding and dimerization-mediating GAF domains[22,25] that lie within the N-terminal region of several mammalian PDEs (Figure 15.4). A polybasic region comprising four lysines, one arginine, and two serine residues is proximal to the C-terminal CAAX-box prenylation motif, which contains a consensus cGMP-dependent protein kinase (cGK) or cAMP-dependent protein kinase (PKA) phosphorylation motif (KKRS). Another cGK/PKA phosphorylation motif (KRPS) occurs at predicted aa 197–200, which may be homologous to that in mammalian PDE5. This also lies N-terminal to the GAF domains and is thought to modulate the binding of cGMP to these regulatory domains.[22] Comparison of *Drosophila* PDE6 polypeptide with *H. sapiens* PDE6β reveals an extra 99 aa at the N-terminus of *Drosophila* PDE6 and an 111 aa insertion between the end of the catalytic domain and the prenylation motif. Structural studies have uncovered the main mechanism conferring cGMP specificity in PDE5A,[24] which is the canonical vertebrate cG-PDE. This involves a glutamine

residue, Q817, which forms a clamp via hydrogen bond interactions with Q775; as well as by hydrogen bonds between the carboxyl backbone of A767 and Q775, and W853 and Q775. Each of these residues is conserved in *Drosophila* PDE6 (Q935, Q893, A885, and W970); and also in mammalian PDE6β, another cGMP-specific PDE (http://www.biochemj.org/bj/388/bj3880333add.htm), strongly suggesting that *Drosophila* PDE6 is a cGMP-specific PDE.

15.2.4.3 PDE8

The N-terminus of the *Drosophila* PDE8 homolog contains a six amino acid consensus myristoylation and palmitoylation motif, MGCAP, almost identical to that present in human PDE8A1 (http://www.biochemj.org/bj/388/bj3880333add.htm). PDE8 falls into a category of lipid-modified proteins at which modification occurs at the N-terminus; the cysteine at position 3 is a prime candidate for such posttranslational palmitoylation. Given the close conservation between mammalian and fly PDE8 at the N-terminus, it is possible that lipid modification is important for the function of PDE8, although direct experimental evidence for this has yet to be demonstrated.

PDE8 also possesses two conserved domains N-terminal to the catalytic domain: a REC (RecA) domain, identified as a phosphate-acceptor domain in the bacterial two-component system,[26] shows 38% identity and 54% similarity to human PDE8A REC domain. A PAS domain is also present with 34% identity and 53% similarity to that in human PDE8.

CG5411 encodes five transcripts (A–E), resulting in four different polypeptides (refer to Section 15.2.1.4). These different transcripts arise from alternative splicing events, with all transcripts sharing identical exons at the 3' end (FlyBase report, Gene PDE8). Transcripts A, B, C, and D all encode unique N-terminal sequences; although transcripts B, C, and D do not encode the N-terminal myristoylation/palmitoylation motif, which are not encoded by *H. sapiens* PDE8A splice forms, nor by PDE8B. Transcript C does not encode either the PAS or the REC domain.

The N-terminal regions of several PDEs have been shown to influence specific localization. PSORT (http://psort.nibb.ac.jp/) predictions for the subcellular localization of each PDE8 isoform indicated that proteins A, C, and D were most likely to be located in the cytoplasm, whereas protein B was likely to be confined to mitochondria.

Mammalian PDE8 is a high-affinity cAMP-specific PDE.[27,28] Structural comparisons of mammalian PDE8 with that of cAMP-specific PDE4B suggests that two conserved residues, Q369 and N321, make contact with the substrate, adenine. *Drosophila* PDE8 contains the equivalent of the N321 conserved residue, N729. Also, each of the four residues within the *H. sapiens* PDE8 nucleoside-binding pocket (C729, N737, Q778, and W812) are conserved in *Drosophila* N803, C811, Q851, and W885. This would suggest that *Drosophila* PDE8 is a cA-specific PDE.

15.2.4.4 PDE9

Apart from conservation of the catalytic domain with both mammalian PDE9 and with other PDEs (Figure 15.3), very little else is known about the structure of PDE9, given the possible ambiguities of the PDE9 gene sequence. However, as a putative cG-PDE, this PDE is of possible relevance to cGMP-signaling mechanisms and should be investigated further.

15.2.4.5 PDE11

Drosophila PDE11 lies in a closely related cluster with human PDE5 and PDE6β (Figure 15.1). However, *CG10231*-encoded PDE11 shows a high sequence identity (of 77%) to human PDE11 within the catalytic domain. Furthermore, the predicted *Drosophila* PDE11 polypeptide sequence contains tandem GAF domains at the N-terminal region, as does human PDE11

(Figure 15.4 and http://www.biochemj.org/bj/388/bj3880333add.htm). Human PDE11A3 and PDE11A4 isoforms each contain two GAF domains[29] and *Drosophila* PDE11 shows very close sequence identity with these isoforms of PDE11A. Therefore, it is likely that *CG10231* encodes the *Drosophila* homolog of PDE11A3/A4. PDE5 is a phosphorylation substrate for cGK[30] and contains one cGK phosphorylation motif in each subunit.[31] Interestingly, *Drosophila* PDE11 has four such cGK phosphorylation motifs (http://www.biochem-www.biochemj.org/bj/388/bj3880333add.htm) suggesting that PDE11 may be a substrate for cGK in *Drosophila*.

15.2.5   BIOCHEMISTRY AND PHARMACOLOGY OF *DROSOPHILA* PHOSPHODIESTERASES

The identification of rat homologs of *dnc* allowed initial pharmacological characterization of the mammalian PDE4 family.[14] Surprisingly, while PDE4 is sensitive to the antidepressants, rolipram and RO-20-1724, the closely related dunce product is not. Thus, although structurally very similar, different pharmacological properties are associated with mammalian and *Drosophila* enzymes, at least in the case of dunce/PDE4. However, nonspecific inhibitors of vertebrate PDEs have been used in *Drosophila* physiology in order to modulate cAMP signaling. For example, caffeine has been used to demonstrate effects on potassium currents in larval muscle fibers from *dnc* mutants[32] whereas 3-isobutyl-1-methylxanthine (IBMX) has been used to mimic the effects of dunce on calcium mobilization from L-type calcium channels in larval muscle.[33] IBMX has also been used successfully for studies of cAMP signaling, in other *Drosophila* preparations for physiological and biochemical studies, including that of the Malpighian (renal) tubule,[34] where it has been shown that IBMX inhibits cA-PDE activity. The *Drosophila* renal tubule has been established as a genetic model for transporting epithelia and is the functional equivalent of mammalian kidney and liver.[35] The tubule is critically involved in fluid transport, osmoregulation, detoxification,[36] and the immune response.[37] The available tubule phenotypes, combined with *Drosophila* tools, thus allow assessments of PDE function in an organotypic context. Fluid transport by the tubule has been shown to be sensitive to an inhibitor of vertebrate cG-PDEs, zaprinast.[38,39] The effects of either a nitric oxide donor[38] or induction of a nitric oxide synthase transgene[39] on tubule fluid transport are potentiated by treatment of tubules with zaprinast. Zaprinast has also been shown to inhibit tubule cG-PDE activity.[39] More recently, tubules have been shown to be sensitive to the PDE5 inhibitor, sildenafil.[40] Sildenafil significantly inhibits cG-PDE activity assayed at high substrate concentration (20 μM cGMP) between 10^{-9} and 10^{-5} M.[40] Thus, *Drosophila* contains high K_m, sildenafil- and zaprinast-sensitive enzymes, which will be further discussed in Section 15.2.6.

15.2.5.1   Dunce

Biochemical characterization of the *dnc*-encoded PDE showed that this was a high-affinity, cAMP-specific PDE.[14] An earlier study to characterize dunce was based on partial purification of crude adult wild-type fly homogenates by gel filtration. Subsequent assay of the purified homogenate fractions with cAMP demonstrated that dunce hydrolyzed cAMP with linear kinetics, and with a K_m of 2.2 ± 0.5 μM.[9] Interestingly, the enzyme encoded by a hypomorphic allele of *dnc*–*dnc¹* showed a similar K_m for cAMP to the wild-type enzyme (2.5 ± 0.4 μM) but a significantly reduced V_{max}. While dunce PDE is not inhibited by rolipram (see below), use of the SQ20009 compound at 100 or 140 mM is sufficient to inhibit endogenous cA-PDE activity in adult flies.[41] SQ20009 may thus provide a useful tool for pharmacological investigations for the other *Drosophila* cA-PDE, PDE8.

The reasons behind the lack of rolipram inhibition of dunce activity are still unknown. Key histidine residues, His427 and His446, in recombinant human PDE4A (rhPDE4A)[42] are not

present in dunce, which contains instead a valine and asparagine, respectively, at these positions.[10] Given that it was possible that these aa conferred rolipram resistance, generation and analysis of H427V and H446N rhPDE4A mutants was undertaken.[42] However, these mutants were found to be sensitive to rolipram, suggesting that the rolipram insensitivity of dunce is not due to the lack of equivalent His427 and His 446 residues present in PDE4. This study also investigated the influence of nonhistidine residues in rolipram sensitivity. A seven amino acid sequence conserved in PDE4 and dunce, ELALMYN (positions 491–497 in hrPDE4A), is preceded by a three-serine stretch in dunce but not in PDE4. T489S and N490S mutations in rhPDE4A did not alter the rolipram sensitivity of PDE4A.[42] Thus, presently there is no satisfactory explanation for the rolipram insensitivity of dunce PDE, given the high degree of identity and similarity between dunce and the PDE4 family.

15.2.5.2 Novel Phosphodiesterases—PDE1, PDE6, and PDE11

In order to investigate the biochemical properties of some of the uncharacterized *Drosophila* PDEs, an antibody-based approach was used to carry out the functional work on specific PDEs isolated from adult *Drosophila*.

15.2.5.2.1 PDE1

The use of an antipeptide antibody to the C-terminal epitope EQAVKDAEARALAT was validated by Western blotting of *CG14940*-transfected *Drosophila* S2 cells, which allowed identification of a 75 kDa protein. This antibody was then used to immunoprecipitate PDE1 from adult *Drosophila* head extracts, with subsequent assay of this material in PDE assays. PDE1 activity assayed at 10 μM substrate was found to hydrolyze both cAMP and cGMP and to be regulated by calcium/calmodulin[20] (Figure 15.5A) Further kinetic analysis of PDE1 showed that its K_m for cGMP was 15.3 $\pm$ 1 μM[20] (Figure 15.5B) and K_m for cAMP was

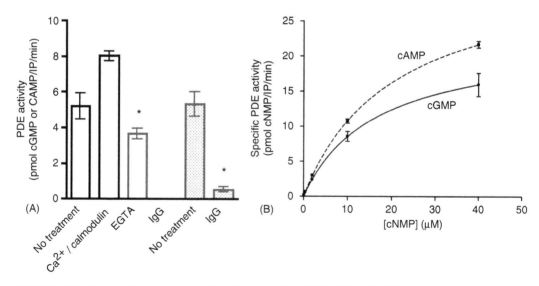

FIGURE 15.5 Biochemical characteristics of *Drosophila* PDE1. (A) cG-PDE activity (undotted bars) and cA-PDE activity (dotted bars) of immunoprecipitated PDE1. IgG indicates control immunoprecipitations (IPs), for which PDE activity was negligible. (Reproduced from Ref. [20]. With permission.) (B) Michaelis–Menten plot for cA-PDE and cG-PDE activities of PDE1. (Reproduced from Ref. [20]. With permission.)

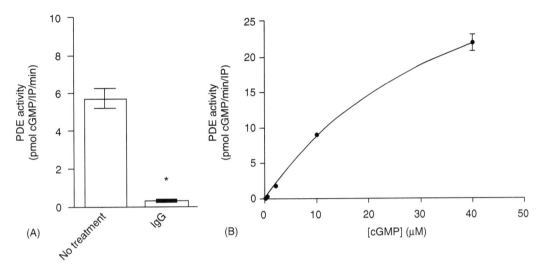

FIGURE 15.6 Biochemical characteristics of *Drosophila* PDE6. (A) cG-PDE activity of immunopreci-pitated PDE6. IgG indicates control immunoprecipitations. (B) Michaelis–Menten plot for cG-PDE activity of PDE6. (Reproduced from Ref. [20]. With permission.)

$20.5 \pm 1.5 \, \mu$M.[20] Thus, PDE1 is a dual-substrate, calcium/calmodulin-regulated enzyme as predicted from both gene and protein sequences. This is in spite of the fact that the predicted sequence for PDE1 shows that *Drosophila* PDE1 has only one calmodulin-binding site, and not two, as does human PDE1C.

In Figure 15.5A, Figure 15.6A, and Figure 15.7A, asterisk indicates data statistically significant between antibody-specific and control immunoprecipitations (IPs), assayed for cG- and cA-specific PDE activity using 10 µM substrate (pmol cGMP or cAMP/IP/min,

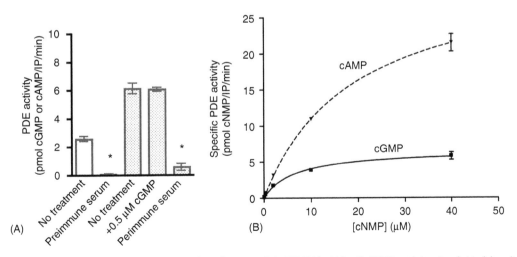

FIGURE 15.7 Biochemical characteristics of *Drosophila* PDE11. (A) cG-PDE activity (undotted bars) and cA-PDE activity (dotted bars) of immunoprecipitated PDE11. IgG indicates control immunopre-cipitations. (Reproduced from Ref. [20]. With permission.) (B) Michaelis–Menten plot for cA-PDE and cG-PDE activities of PDE11. (Reproduced from Ref. [20]. With permission.)

$\pm$ SEM, $N = 6$) where $p < 0.05$ (student's unpaired t-test). Additionally, in Figure 15.5A, asterisk indicates statistical significance ($p < 0.05$, student's unpaired t-test) between ethylene glycol tetraacetic acid (EGTA)-treated (0.1 mM EGTA) and EGTA-untreated samples assayed for cGMP-specific PDE activity.

15.2.5.2.2   PDE6

PDE6 was characterized using an antipeptide antibody to the HGSEDSHTPEHQRS epitope, which specifically recognized a 130 kDa protein in *CG8279*-transfected S2 cells. Immunoprecipitated adult head extracts were assayed for PDE activity at 10 μM substrate and this revealed that the *CG8279*-encoded protein is a cG-PDE[20] (Figure 15.6A). Furthermore, kinetic analysis shown in Figure 15.6B[20] demonstrates that PDE6 is a high K_m cG-PDE (37 $\pm$ 13 μM). This work confirmed earlier data from *CG8279*-transfected S2 cells, which showed that PDE6 is a high K_m cG-PDE.[43] The *CG8279* protein translation is closely aligned with mammalian PDE5, PDE6β, and PDE11, as well as *Drosophila* PDE11 (Figure 15.1). However, the cG-PDE specificity of this putative PDE6 homolog rules out the possibility of this as a PDE11-like enzyme. While the above condition still allows the possibility that CG8279 protein is the *Drosophila* homolog of mammalian PDE5, the presence of a CAAX prenylation motif, which is not present in PDE5, suggests that *CG8279* encodes a *Drosophila* PDE6-like enzyme.

15.2.5.2.3   PDE11

PDE11 was characterized from adult head extracts by immunoprecipitation studies using an antipeptide antibody raised to the C-terminal epitope PTSTQPSDDDNDAD, which recognized a specific protein band of 100 kDa by Western analysis of *CG10231*-transfected S2 cell extracts. Assays of PDE activity using 10 μM substrate revealed that *CG10231* encodes a PDE that hydrolyzes both cAMP and cGMP (Figure 15.7A), which hydrolyzes cGMP with a K_m of 6 $\pm$ 2 (Figure 15.7B) and cAMP with a K_m of 18.5 $\pm$ 1.5.[20] Thus, while *CG10231*-encoded PDE is closely related to mammalian PDE5, PDE6β, and *Drosophila* PDE6 (*CG8279*), this is clearly the *Drosophila* homolog of PDE11. Furthermore, *Drosophila* PDE11 is the highest affinity cG-PDE assessed so far.

In summary, the biochemical characterization of *Drosophila* PDE1, PDE6, and PDE11 showed that these are high-K_m enzymes. This was confirmed by assay of *Drosophila* tissue (Malpighian tubule) extracts, which showed cG-PDE activity with a K_m of ~15 μM.[40]

15.2.6   INHIBITOR SENSITIVITIES OF NOVEL *DROSOPHILA* PHOSPHODIESTERASES

While it is clear that *Drosophila* and mammalian PDEs share close structural conservation, preliminary evidence suggests that close functional conservation may also occur. Inhibitor studies on immunoprecipitated PDEs show that PDE1, PDE6, and PDE11 are inhibited by zaprinast and sildenafil.[20] Specifically, inhibition of PDE1 activity by zaprinast occurs at an IC_{50} value of 71 $\pm$ 39 μM (Figure 15.8A) and that by sildenafil at an IC_{50} value of 1.3 $\pm$ 0.9 μM (Figure 15.8B). Sildenafil inhibits PDE6 at nanomolar concentrations (IC_{50} value of 0.025 $\pm$ 0.005 μM, Figure 15.8B), although PDE6 is inhibited by zaprinast at an IC_{50} value of 0.65 $\pm$ 0.15 μM (Figure 15.8A). PDE11 shows sensitivity to both zaprinast and sildenafil at IC_{50} values of 1.6 $\pm$ 0.5 μM (Figure 15.8A) and 0.12 $\pm$ 0.06 μM (Figure 15.8B), respectively. Thus, the dual specificity and cG-PDEs in *Drosophila* are sensitive to inhibitors of vertebrate cG-PDEs.

Immunoprecipitated samples from *Drosophila* head lysates using antibodies specific for PDE1, PDE6, and PDE11 were assayed for cGMP-specific PDE activity (pmol cGMP hydrolyzed/IP/min $\pm$ SEM at 10 μM substrate concentration) in the absence (control) or in the presence of zaprinast (A) and sildenafil citrate (B) at varying concentrations. Data are

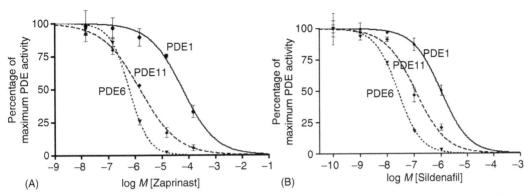

FIGURE 15.8 Inhibitor sensitivities of *Drosophila* PDEs. (Reproduced from Ref. [20]. With permission.)

expressed as percentage of maximum (control) PDE activity $\pm$ SEM ($N = 3$). Control PDE activity (pmol cGMP hydrolyzed/IP/min $\pm$ SEM) for PDE1: 2.95 $\pm$ 0.16, PDE6: 11.11 $\pm$ 0.195, and PDE11: 3.65 $\pm$ 0.15.

15.2.7 PHENOTYPIC EVIDENCE FOR FUNCTIONAL CONSERVATION OF *DROSOPHILA* AND MAMMALIAN PHOSPHODIESTERASES

Drosophila encodes several PDEs—in addition to dunce, these include two cG-PDEs (PDE6 and PDE9), one cA-PDE (PDE8A), and two dual specificity PDEs (PDE1 and PDE11). The widespread expression of all these PDEs in adult fly suggests important roles for complex cyclic nucleotide regulation in *Drosophila* physiology.

Interestingly, the *Drosophila* genome does not encode homologs of mammalian PDEs 2, 3, 7, and 10. Also, while there is no defined *Drosophila* homolog of mammalian PDE5, the *CG8279*-encoded enzyme has structural and functional features of both PDE5 and PDE6. This perhaps suggests that the *Drosophila* enzymes are the products of ancestral genes, which in vertebrates, evolved to encompass several other related PDEs. The existence of multiple additional PDEs may well confer fine-tuning of signaling processes in a necessarily more complicated body plan and increased number of specialized cell types. This is supported by the fact that simple organisms (e.g., *Drosophila*) have one gene encoding PDE4, whereas the zebra fish, *Danio rerio*, is the first organism in evolutionary terms in which multiple genes encoding PDE4 subtypes are found.[11]

An overall comparison of *Drosophila* PDEs with known biochemical characteristics and their mammalian counterparts is shown in Table 15.2. Given the close overall similarities but intriguing specific differences between the *Drosophila* and vertebrate enzymes, can functional conservation between these systems occur *in vivo*?

An *in vivo* study on dunce showed that transgenic *Drosophila* bearing either heat-shock inducible *dnc* (*hspdnc54*) or rat *RD1*, PDE4A1 (*hspRD1-12*) transgenes showed increased cA-PDE activity.[41] The *dnc* mutants, for example, the *dnc¹* allele, contain ~46% of wild-type cA-PDE activity.[41] When the *dnc* transgene is introduced into the *dnc¹* background (*hspdnc54-dnc¹* flies), cA-PDE activity is restored to wild-type levels and can eventually exceed such levels.[41] Perhaps more importantly, this effect on cA-PDE activity is also observed when the rat PDE4 construct is expressed in the *dnc* mutant background (*hspRD1-12-dnc¹* flies). Thus, functional conservation of the PDE4 enzymes occurs *in vivo*. What are the physiological consequences of this?

TABLE 15.2

Comparison of Biochemical Characteristics of *Drosophila* and Mammalian PDEs

Enzyme	K_m cGMP (μM)	K_m cAMP (μM)	IC_{50} Zaprinast (μM)	IC_{50} Sildenafil (μM)
Drosophila PDE1	15.3 ± 1	20.5 ± 1.5	71 ± 39	1.3 ± 0.9
PDE1C	2.2 ± 0.1	3.5 ± 0.3	3.5 ± 0.6	1.1 ± 0.275
	Mouse[44]	Mouse[44]	Human[45]	Bovine[46]
dunce	Undetectable at 1 mM[6]	2.2 ± 0.5[9]	—	—
RD1, PDE4A1	Undetectable at 1 mM[6]	4.1 ± 1.1[14]	—	—
	Rat brain[12]			
DPDE4, PDE4B	Undetectable at 1 mM[6]	2.9 ± 0.5[14]		
	Rat brain[47]			
DPDE4, PDE4B	>1 mM[48]	7.2 ± 2[48]	—	—
	Human[48]			
Drosophila PDE6	37 ± 13	—	0.65 ± 0.15	0.025 ± 0.005
PDE6β	17 ± 7	—	0.18 ± 0.01	0.02 ± 0.001
	Bovine[49]		Human (K_i)[50]	Human (K_i)[50]
Drosophila PDE11	6 ± 2	18.5 ± 5.5	1.6 ± 0.5	0.12 ± 0.06
PDE11A3/4	1.5 ± 0.0.07	3.0 ± 0.28	11 ± 3.6	3.8 ± 0.75
	Human	Human	Human	Human
	PDE11A3[29]	PDE11A3[29]	PDE11A3[29]	PDE11A4[51]
	1.4 ± 0.06	3.0 ± 0.28	33 ± 5.3	
	Human	Human	Human	
	PDE11A4[29]	PDE11A4[29]	PDE11A4[29]	

Values for K_m (cGMP and cAMP) and IC_{50} values for zaprinast and sildenafil are listed, where known, with appropriate references.

Source: Data for *Drosophila* PDE1, PDE6, and PDE11, as well as for mammalian PDE1C and PDE6β are reproduced from Ref. [20]. With permission.

Learning and memory processes in the fly are significantly influenced by *dnc* activity. The *dnc* mutants carrying the *dnc* or rat PDE4 transgenes were subjected to odor-avoidance tests as learning trials, where the learning index measured accounts for acquisition, retrieval, and immediate memory. Such trials showed that both the *dnc* and the rat PDE4 transgenes significantly rescued the learning deficit in heat-shocked *dnc¹* mutants, which performed better than the nonheat-shocked *dnc* and rat PDE4 transgenic lines.

Dunce activity is required for reproduction in females, and *dnc* mutant females show a sterility phenotype.[52] Rescue of *dnc* mutants with the *dnc* transgene *hspdnc54* with a heat-shock regime partially rescues this sterility phenotype to ~75% of wild-type levels, with rescued females producing eggs and progeny.[41] Although this phenotypic analysis was not performed with the rat transgene, it is entirely possible that mammalian PDE4 could rescue the fertility phenotype using appropriate gene targeting to germ cells.

Another study, using time-utilizing bovine PDE5, showed that targeted overexpression of PDE5A in transgenic *Drosophila* renal tubules results in increased cG-PDE activity and increased sensitivity to sildenafil, at substrate concentrations relevant to PDE5.[40] Thus, mammalian PDE5 displays conservation of biochemical properties and inhibitor sensitivities when expressed in *Drosophila*. PDE5 was also targeted to the apical membrane of tubules, which has been shown to be the site of endogenous PDE activity,[39] suggesting that targeting of ectopically expressed PDE to the appropriate site in situ occurs. The physiological role of

ectopically expressed PDE5 in *Drosophila* tubules is associated with modulation of basal and neuropeptide-stimulated rates of fluid transport. Given that tubules express several cG-PDEs including dual-substrate PDEs, it is likely that some of these may be specifically involved with fluid transport regulation. Thus, the PDE5 and dunce work show that remarkably functional complementation does occur between the *Drosophila* and mammalian enzymes *in vivo*.

15.3   CONCLUSIONS

Given the recent progress in targeted transgenesis in biomedically relevant model organisms like the mouse, and the advent of other cell systems, for example, stem cells, some scientists suggest that perhaps that the days of *Drosophila* research are numbered. However, this appears not to be the case. Biological research using *Drosophila* still provides us with insights into gene function, which are not easily obtainable by any other means. Furthermore, the completion of genome projects including that for mouse and human suggests that scores of novel genes exist for which there is no ascribed putative function. Even in *Drosophila*, with very well-annotated genome information, microarray analysis shows that many tissue-specific genes are still of unknown function.[21,53] Thus, if we are to uncover the biological function of key genes, *Drosophila* still has much to offer. The existence of key disease gene homologs in *Drosophila* also allows for a systematic study of human disease genes using the power of *Drosophila* genetics.

In terms of PDEs, key similarities but intriguing differences exist between the *Drosophila* and mammalian enzymes. Thus, using *Drosophila*, the scope for investigation into the function of specific PDEs in an *in vivo* context remain very promising with possible novel roles for PDEs under identification. This would apply to both the endogenous *Drosophila* PDEs as well as mammalian PDEs, given the functional conservation between these groups of enzymes. Furthermore, the structural differences between mammalian and *Drosophila* PDEs could be exploited to reveal further insights into some members of the PDE family in mammals. Finally, given the central role of cyclic nucleotide regulation in physiology, *Drosophila* and other insect-specific PDEs could well be rational potential targets for pesticide design.

ACKNOWLEDGMENTS

Work in the laboratory is supported by the UK Biotechnological and Biological Sciences Research Council. We thank J.A.T. Dow, M.D. Houslay, S. Francis, and N.J. Pyne for collaboration and helpful discussion.

REFERENCES

1. Giot, L., J.S. Bader, C. Brouwer, A. Chaudhuri, B. Kuang, Y. Li, Y.L. Hao, C.E. Ooi, B. Godwin, E. Vitols, G. Vijayadamodar, P. Pochart, H. Machineni, M. Welsh, Y. Kong, B. Zerhusen, R. Malcolm, Z. Varrone, A. Collis, M. Minto, S. Burgess, L. McDaniel, E. Stimpson, F. Spriggs, J. Williams, K. Neurath, N. Ioime, M. Agee, E. Voss, K. Furtak, R. Renzulli, N. Aanensen, S. Carrolla, E. Bickelhaupt, Y. Lazovatsky, A. DaSilva, J. Zhong, C.A. Stanyon, R.L. Finley Jr., K.P. White, M. Braverman, T. Jarvie, S. Gold, M. Leach, J. Knight, R.A. Shimkets, M.P. McKenna, J. Chant, and J.M. Rothberg. 2003. A protein interaction map of *Drosophila melanogaster*. *Science* 302 (5651):1727–1736.
2. Chien, S., L.T. Reiter, E. Bier, and M. Gribskov. 2002. Homophila: Human disease gene cognates in *Drosophila*. *Nucleic Acids Res* 30 (1):149–151.
3. Kiger, J.A. Jr., and E. Golanty. 1977. A cytogenetic analysis of cyclic nucleotide phosphodiesterase activities in *Drosophila*. *Genetics* 85 (4):609–622.

4. Kiger, J.A. Jr. 1977. The consequences of nullosomy for a chromosomal region affecting cyclic AMP phosphodiesterase activity in *Drosophila*. *Genetics* 85 (4):623–628.

5. Kiger, J.A. Jr., and E. Golanty. 1979. A genetically distinct form of cyclic AMP phosphodiesterase associated with chromomere 3D4 in *Drosophila melanogaster*. *Genetics* 91 (3):521–535.

6. Davis, R.L., and J.A. Kiger Jr. 1980. A partial characterization of the cyclic nucleotide phosphodiesterases of *Drosophila melanogaster*. *Arch Biochem Biophys* 203 (1):412–421.

7. Dudai, Y., Y.N. Jan, D. Byers, W.G. Quinn, and S. Benzer. 1976. Dunce, a mutant of *Drosophila* deficient in learning. *Proc Natl Acad Sci USA* 73 (5):1684–1688.

8. Byers, D., R.L. Davis, and J.A. Kiger Jr. 1981. Defect in cyclic AMP phosphodiesterase due to the dunce mutation of learning in *Drosophila melanogaster*. *Nature* 289 (5793):79–81.

9. Davis, R.L., and J.A. Kiger Jr. 1981. Dunce mutants of *Drosophila melanogaster*: Mutants defective in the cyclic AMP phosphodiesterase enzyme system. *J Cell Biol* 90 (1):101–107.

10. Chen, C.N., S. Denome, and R.L. Davis. 1986. Molecular analysis of cDNA clones and the corresponding genomic coding sequences of the *Drosophila* dunce[+] gene, the structural gene for cAMP phosphodiesterase. *Proc Natl Acad Sci USA* 83:9313–9317.

11. Conti, M., W. Richter, C. Mehats, G. Livera, J.Y. Park, and C. Jin. 2003. Cyclic AMP-specific PDE4 phosphodiesterases as critical components of cyclic AMP signaling. *J Biol Chem* 278 (8):5493–5496.

12. Davis, R.L., H. Takayasu, M. Eberwine, and J. Myres. 1989. Cloning and characterization of mammalian homologs of the *Drosophila* dunce+ gene. *Proc Natl Acad Sci USA* 86 (10):3604–3608.

13. Swinnen, J.V., D.R. Joseph, and M. Conti. 1989. Molecular cloning of rat homologues of the *Drosophila melanogaster* dunce cAMP phosphodiesterase: Evidence for a family of genes. *Proc Natl Acad Sci USA* 86 (14):5325–5329.

14. Henkel-Tigges, J., and R.L. Davis. 1990. Rat homologs of the *Drosophila* dunce gene code for cyclic AMP phosphodiesterases sensitive to rolipram and RO 20-1724. *Mol Pharmacol* 37 (1):7–10.

15. Morton, D.B., and M.L. Hudson. 2002. Cyclic GMP regulation and function in insects. *Adv Insect Physiol* 29:1–54.

16. Davies, S.A. 2000. Nitric oxide signalling in insects. *Insect Biochem Mol Biol* 30:1123–1138.

17. Page, R.D.M. 1996. TREEVIEW: An application to display phylogenetic trees on personal computers. *Comput Applic Biosci* 12:357–358.

18. Qiu, Y.H., C.N. Chen, T. Malone, L. Richter, S.K. Beckendorf, and R.L. Davis. 1991. Characterization of the memory gene dunce of *Drosophila melanogaster*. *J Mol Biol* 222 (3):553–565.

19. Qiu, Y., and R.L. Davis. 1993. Genetic dissection of the learning/memory gene dunce of *Drosophila melanogaster*. *Genes Dev* 7 (7B):1447–1458.

20. Day, J.P., J.A. Dow, M.D. Houslay, and S.A. Davies. 2005. Cyclic nucleotide phosphodiesterases in *Drosophila melanogaster*. *Biochem J* 388 (1):333–342.

21. Wang, J., L. Kean, J. Yang, A.K. Allan, S.A. Davies, P. Herzyk, and J.A. Dow. 2004. Function-informed transcriptome analysis of *Drosophila* renal tubule. *Genome Biol* 5 (9):R69.

22. Charbonneau, H., R.K. Prusti, H. LeTrong, W.K. Sonnenburg, P.J. Mullaney, K.A. Walsh, and J.A. Beavo. 1990. Identification of a noncatalytic cGMP-binding domain conserved in both the cGMP-stimulated and photoreceptor cyclic nucleotide phosphodiesterases. *Proc Natl Acad Sci USA* 87 (1):288–292.

23. W.K. Sonnenburg, D. Seger, K.S. Kwak, J. Huang, H. Charbonneau, and J.A. Beavo. 1995. Identification of inhibitory and calmodulin-binding domains of the PDE1A1 and PDE1A2 calmodulin-stimulated cyclic nucleotide phosphodiesterases. *J Biol Chem* 270 (52):30989–31000.

24. Zhang, K.Y., G.L. Card, Y. Suzuki, D.R. Artis, D. Fong, S. Gillette, D. Hsieh, J. Neiman, B.L. West, C. Zhang, M.V. Milburn, S.H. Kim, J. Schlessinger, and G. Bollag. 2004. A glutamine switch mechanism for nucleotide selectivity by phosphodiesterases. *Mol Cell* 15 (2):279–286.

25. Muradov, K.G., K.K. Boyd, S.E. Martinez, J.A. Beavo, and N.O. Artemyev. 2003. The GAFa domains of rod cGMP-phosphodiesterase 6 determine the selectivity of the enzyme dimerization. *J Biol Chem* 278 (12):10594–10601.

26. Wang, J. 2004. Nucleotide-dependent domain motions within rings of the RecA/AAA(+) super-family. *J Struct Biol* 148 (3):259–267.

27. Hayashi, M., K. Matsushima, H. Ohashi, H. Tsunoda, S. Murase, Y. Kawarada, and T. Tanaka. Molecular cloning and characterization of human PDE8B, a novel thyroid-specific isozyme of 3',5'-cyclic nucleotide phosphodiesterase. *Biochem Biophys Res Commun* 250 (3):751–756.

28. Soderling, S.H., S.J. Bayuga, and J.A. Beavo. 1998. Cloning and characterization of a cAMP-specific cyclic nucleotide phosphodiesterase. *Proc Natl Acad Sci USA* 95 (15):8991–8996.

29. Yuasa, K., J. Kotera, K. Fujishige, H. Michibata, T. Sasaki, and K. Omori. 2000. Isolation and characterization of two novel phosphodiesterase PDE11A variants showing unique structure and tissue-specific expression. *J Biol Chem* 275 (40):31469–31479.

30. Corbin, J.D., I.V. Turko, A. Beasley, and S.H. Francis. 2000. Phosphorylation of phosphodiesterase-5 by cyclic nucleotide-dependent protein kinase alters its catalytic and allosteric cGMP-binding activities. *Eur J Biochem* 267 (9):2760–2767.

31. Thomas, M.K., S.H. Francis, and J.D. Corbin. 1990. Substrate- and kinase-directed regulation of phosphorylation of a cGMP-binding phosphodiesterase by cGMP. *J Biol Chem* 265 (25): 14971–14978.

32. Zhong, Y., and C.F. Wu. 1993. Differential modulation of potassium currents by cAMP and its long-term and short-term effects: Dunce and rutabaga mutants of *Drosophila*. *J Neurogenet* 9 (1):15–27.

33. Bhattacharya, A., G.G. Gu, and S. Singh. 1999. Modulation of dihydropyridine-sensitive calcium channels in *Drosophila* by a cAMP-mediated pathway. *J Neurobiol* 39 (4):491–500.

34. Cabrero, P., J.C. Radford, K.E. Broderick, J. Veenstra, E. Spana, S. Davies, and J.A.T. Dow. 2002. The CRF gene of *Drosophila melanogaster* encodes a diuretic peptide that activates cAMP signalling. *J Exp Biol* 205 (24):3799–3807.

35. Dow, J.A.T., and S.A. Davies. 2003. Integrative physiology, functional genomics and epithelial function in a genetic model organism. *Physiol Rev* 83:687–729.

36. Torrie, L.S., J.C. Radford, T.D. Southall, L. Kean, A. Dinsmore, S.A. Davies, and J.A. Dow. 2004. Resolution of the insect ouabain paradox. *Proc Natl Acad Sci USA* 101 (37):13689–13693.

37. McGettigan, J., R.K. McLennan, K.E. Broderick, L. Kean, A.K. Allan, P. Cabrero, M.R. Regulski, V.P. Pollock, G.W. Gould, S.A. Davies, and J.A. Dow. 2005. Insect renal tubules constitute a cell-autonomous immune system that protects the organism against bacterial infection. *Insect Biochem Mol Biol* 35 (7):741–754.

38. Dow, J.A.T., S.H. Maddrell, S.A. Davies, N.J. Skaer, and K. Kaiser. 1994. A novel role for the nitric oxide-cGMP signaling pathway: The control of epithelial function in *Drosophila*. *Am J Physiol*. 266 (5 Pt 2):R1716–R1719.

39. Broderick, K.E., M.R. MacPherson, M. Regulski, T. Tully, J.A.T. Dow, and S.A. Davies. 2003. Interactions between epithelial nitric oxide signaling and phosphodiesterase activity in *Drosophila*. *Am J Physiol Cell Physiol* 285 (5):C1207–C1218.

40. K.E. Broderick, L. Kean, J.A.T. Dow, N.J. Pyne, and S.A. Davies. 2004. Ectopic expression of bovine type 5 phosphodiesterase confers a renal phenotype in *Drosophila*. *J Biol Chem* 279 (9):8159–8168.

41. Dauwalder, B., and R.L. Davis. 1995. Conditional rescue of the dunce learning/memory and female fertility defects with *Drosophila* or rat transgenes. *J Neurosci* 15 (5 Pt 1):3490–3499.

42. Jacobitz, S., M.D. Ryan, M.M. McLaughlin, G.P. Livi, W.E. DeWolf Jr., and T.J. Torphy. 1997. Role of conserved histidines in catalytic activity and inhibitor binding of human recombinant phosphodiesterase 4A. *Mol Pharmacol* 51 (6):999–1006.

43. Day, J.P., M.D. Houslay, and S.A. Davies. 2003. Cloning and characterisation of a novel cGMP-specific phosphodiesterase from *Drosophila melanogaster*. *BMC Meeting Abstracts: 1st International Conference on cGMP: NO/sGC Interaction and Its Therapeutic Implications*, 1 p: 0014.

44. Yan, C., A.Z. Zhao, J.K. Bentley, and J.A. Beavo. 1996. The calmodulin-dependent phosphodiesterase gene PDE1C encodes several functionally different splice variants in a tissue-specific manner. *J Biol Chem* 271 (41):25699–25706.

45. Loughney, K., T.J. Martins, E.A. Harris, K. Sadhu, J.B. Hicks, W.K. Sonnenburg, J.A. Beavo, and K. Ferguson. 1996. Isolation and characterization of cDNAs corresponding to two human calcium, calmodulin-regulated, 3',5'-cyclic nucleotide phosphodiesterases. *J Biol Chem* 271 (2):796–806.

46. Daugan, A., P. Grondin, C. Ruault, A.C. Le Monnier de Gouville, H. Coste, J.M. Linget, J. Kirilovsky, F. Hyafil, and R. Labaudiniere. 2003. The discovery of tadalafil: A novel and highly selective PDE5 inhibitor. 2:2,3,6,7,12,12a-hexahydropyrazino[1′,2′:1,6]pyrido[3,4-b]indole-1,4-dione analogues. *J Med Chem* 46 (21):4533–4542.

47. Colicelli, J., C. Birchmeier, T. Michaeli, K. O'Neill, M. Riggs, and M. Wigler. 1989. Isolation and characterization of a mammalian gene encoding a high-affinity cAMP phosphodiesterase. *Proc Natl Acad Sci USA* 86 (10):3599–3603.

48. Bolger, G., T. Michaeli, T. Martins, St. T. John, B. Steiner, L. Rodgers, M. Riggs, M. Wigler, and K. Ferguson. 1993. A family of human phosphodiesterases homologous to the dunce learning and memory gene product of *Drosophila melanogaster* are potential targets for antidepressant drugs. *Mol Cell Biol* 13 (10):6558–6571.

49. Gillespie, P.G., and J.A. Beavo. 1988. Characterization of a bovine cone photoreceptor phosphodiesterase purified by cyclic GMP-sepharose chromatography. *J Biol Chem* 263 (17):8133–8141.

50. Zhang, J., R. Kuvelkar, P. Wu, R.W. Egan, M.M. Billah, and P. Wang. 2004. Differential inhibitor sensitivity between human recombinant and native photoreceptor cGMP-phosphodiesterases (PDE6s). *Biochem Pharmacol* 68 (5):867–873.

51. Weeks, J.L., R. Zoraghi, A. Beasley, K.R. Sekhar, S.H. Francis, and J.D. Corbin. 2005. High biochemical selectivity of tadalafil, sildenafil and vardenafil for human phosphodiesterase 5A1 (PDE5) over PDE11A4 suggests the absence of PDE11A4 cross-reaction in patients. *Int J Impot Res* 17:5–9.

52. Bellen, H.J., B.K. Gregory, C.L. Olsson, and J.A. Kiger Jr. 1987. Two *Drosophila* learning mutants, dunce and rutabaga, provide evidence of a maternal role for cAMP on embryogenesis. *Dev Biol* 121 (2):432–444.

53. Andrews, J., G.G. Bouffard, C. Cheadle, J. Lu, K.G. Becker, and B. Oliver. 2000. Gene discovery using computational and microarray analysis of transcription in the *Drosophila melanogaster* testis. *Genome Res* 10 (12):2030–2043.

Section C

Phosphodiesterases Functional Significance: Gene-Targeted Knockout Strategies

16 Insights into the Physiological Functions of PDE4 from Knockout Mice

S.-L. Catherine Jin, Wito Richter, and Marco Conti

CONTENTS

16.1 Introduction 323
16.2 PDE4D Is Critical for Neonatal Survival and Growth..................................... 324
16.3 PDE4D Regulation Is Necessary for the Gonadotropin-Dependent Pattern
of Gene Expression in the Ovarian Follicle 326
16.4 PDE4D Is Involved in the Control of Airway Smooth Muscle Tone................... 327
16.5 Differential Role of PDE4s in Airway Inflammation *in Vivo* Models
of Asthma.. 328
16.6 PDE4B Induction Is Essential for Toll-Like Receptor Signaling....................... 329
16.7 PDE4B and PDE4D Are Involved in the Control of Neutrophil Recruitment
to the Lung.. 333
16.8 PDE4s and Central Nervous System Functions... 334
    16.8.1 PDE4B and PDE4D Mediate the Antidepressant Effects
of Rolipram 334
    16.8.2 PDE4D but Not PDE4B Modulates the Hypnotic Action
of Adrenoceptors: Implications for the Emetogenic Effects of PDE4
Inhibitors 335
16.9 Involvement of PDE4D in the Modulation of Excitation–Contraction
Coupling in the Heart 336
16.10 Conclusions 338
Acknowledgment.. 339
References 339

16.1 INTRODUCTION

One of the most striking features of the phosphodiesterase (PDE) superfamily is the variety of genes, transcripts, and splicing variants that have been steadily described over the years.[1–5] With the appreciation of this complexity, the idea has been taking shape that specific functions must be associated with each gene and isoenzymes. Several different approaches have been used to probe this diversity. Two decades of progress in the molecular pharmacology of PDE isozymes have paved the way to the development of inhibitors selective for individual PDE families, undoubtedly useful tools to distinguish their functions *in vitro* and *in vivo*. The development of PDE3,[6,7] PDE4,[8] and PDE5[9,10] inhibitors have provided a first clear documentation of divergent PDE function. However, this widely used pharmacological

strategy does not have the resolution to define whether individual PDE isoforms within a family serve unique and specialized functions, or whether these genes are largely redundant.

This question is above all cogent for the PDE4 family, which consists of four paralog genes (PDE4A–D), each encoding multiple variants, giving rise to at least 20 different isozymes.[5,11] These proteins are widely distributed in mammalian cells and tissues and are differentially expressed and regulated among various cell types.[12–15] Their involvement in a large array of biological processes is inferred by the use of PDE4-selective second generation inhibitors such as rolipram; yet the exact role of individual PDE4 isozymes under physiological and pathological conditions remains largely unknown. Equally unclear is the outcome of selective pharmacological inhibition of a single PDE form. Dissecting PDE4 functions either by genetic inactivation of individual genes,[16] or more recently by RNA interference,[17] has opened a new perspective to probe the function of individual PDE subtypes. Here we will summarize the data derived from the study of PDE4A, B, and D knockout (KO) mice (see Table 16.1) and, whenever possible, compare and contrast them with data obtained using PDE4 inhibitors. The PDE4 functions derived from pharmacological studies have been detailed in several excellent reviews[18–22] and elsewhere in this book (Chapter 27 and Chapter 33). Therefore, they will not be discussed in detail in this chapter.

16.2   PDE4D IS CRITICAL FOR NEONATAL SURVIVAL AND GROWTH

That a PDE serves indispensable functions in the body is underscored by the loss of viability of PDE4 KO mice. Neonatal lethality and growth retardation are the prominent phenotypes in mice with targeted disruption of the PDE4D gene. Genotyping the offspring from mating PDE4D heterozygous mice on a mixed 129/Ola and C57BL/6 genetic background revealed that approximately 40% of PDE4D KO mice die within 4 weeks after birth with most of the deaths occurring during the first 2 weeks of life.[16] This phenotype is markedly exacerbated when the PDE4D null allele is transferred to a pure C57BL/6 background, although the Mendelian inheritance of the null allele is normal in C57BL/6 fetuses at gestation days 17 and 18. No C57BL/6 PDE4D KO mice survive the first day after birth, suggesting that the penetrance of this neonatal lethality is background-dependent. This exacerbation indicates that one or more genetic modifiers functionally linked to PDE4D are absent in the C57Bl/6 genetic background. Understanding the nature of these genetic modifiers may provide novel views on the physiological role of PDE4D. The exact cause of death of PDE4D$^{-/-}$ mice is still unclear but may be due to the failure of more than one organ (unpublished observation).

PDE4D$^{-/-}$ mice are also small, and growth charts indicate that their body weight at 1–2 weeks after birth is at least 30% lower than that of their wild-type littermates.[16] Although the growth rate becomes normal after 2 weeks, on average their body size remains smaller. The decrease in weight is associated with reduced level of circulating insulin-like growth factor-I (IGF-I) measured at day 15, suggesting an impaired growth hormone-IGF-I axis in immature PDE4D null mice. Further investigation is necessary to confirm this hypothesis, but it underscores the pleiotropic function of PDE4D in the body.

Unlike PDE4D$^{-/-}$ mice, PDE4B$^{-/-}$ and PDE4A$^{-/-}$ mice appear normal in size and are viable, with body weights and growth rates similar to those of their wild-type littermates, even though a decreased viability of PDE4B$^{-/-}$ mice on a pure C57BL/6 background is beginning to emerge (see Table 16.1). These findings provide an initial indication that in mice PDE4D is functionally involved in a broader spectrum of cellular processes than PDE4A and PDE4B, most likely due to its wider expression. More importantly, such studies indicate that the functions of PDE4D cannot be fully compensated for by any other PDE gene present in the mouse genome.

TABLE 16.1

Phenotypes of PDE4A, PDE4B, and PDE4D Knockout Mice

Function	4D KO	4B KO	4A KO
General health			
Neonatal survival (on mixed 129/Ola and C57BL/6 background)	Reduced[a]	Normal [b]	Normal
Growth	Reduced by 30%	Normal	Normal
Fertility			
Female fertility	Reduced	Normal	Normal
Ovulation rate in LH/hCG-stimulated immature mice	Reduced[c]	n.d.	n.d.
Inflammation			
Airway hyperreactivity in response to methacholine	Absent[d]	Reduced	n.d.
Eosinophil recruitment to lungs after OVA sensitization and challenge	Normal	Reduced	Normal
Th2 cytokine accumulation in the airway and in lymph node cell culture	Normal	Reduced	Normal
Proliferation of tracheal lymph node T cells upon OVA stimulation	Normal	Reduced	Normal
Tracheal ring contractility in response to carbachol	Reduced	Normal	Normal
LPS-induced TNF-α production in circulating leukocytes	Normal	Reduced by 90%	Normal
LPS-induced TNF-α production in peritoneal naive macrophages	Normal	Reduced by 56%	Normal
Neutrophil recruitment to lungs after exposure to aerosolized LPS	Reduced by 48%	Reduced by 31%	Normal
Chemotaxis of splenic neutrophils in response to the chemokine KC	Reduced	Reduced	n.d.
CD18 expression in BAL neutrophils after LPS inhalation	Reduced	Reduced	n.d.
CNS			
Antidepressant-like behavior	Increased[e]	Increased	n.d.
Duration of α$_{2A}$-adrenoceptor-mediated anesthesia	Reduced[f]	Normal	n.d.
Heart			
Chronotropic effect of β$_2$-adrenoceptor signaling on cardiac myocytes	Increased	n.a.	n.a.
Exercise-induced cardiac arrhythmias	Increased	n.d.	n.d.
Heart failure after myocardial infarct	Accelerated	n.d.	n.d.
Late-onset cardiomyopathy	Yes	n.d.	n.d.

n.d., Not determined; n.a., not affected.

[a] Mice on a mixed 129/Ola and C57BL/6 background exhibit 40% neonatal lethality, whereas no mice survive when the null allele is transferred to a pure C57BL/6 background.

[b] From heterozygous mating on a pure C57BL/6 background, there is a 75% reduction in yield of PDE4B null mice.

[c] Expression of several genes involved in ovulation is decreased in PDE4D-deficient ovaries. Luteinized antral follicles with entrapped oocytes are present in the ovary of PMSG-treated PDE4D null mice.

[d] These mice develop normal airway hyperreactivity in response to serotonin.

[e] PDE4D null mice display normal anxiolytic and anxiogenic behavior, exploratory behavior, and locomotor activity.

[f] Decrease in duration of α$_{2A}$-adrenoceptor-mediated anesthesia correlates with the emetic effects of PDE4 inhibitors in humans.

16.3 PDE4D REGULATION IS NECESSARY FOR THE GONADOTROPIN-DEPENDENT PATTERN OF GENE EXPRESSION IN THE OVARIAN FOLLICLE

Gonadal function depends on the pituitary gonadotropins FSH and LH, which exert their effects through activation of cyclic nucleotide signaling.[23] In view of the established expression of PDE4s in testis and ovary,[24–26] it was expected that ablation of PDE4 would impact fertility in both female and male mice. Surprisingly, male mice that are deficient in any of the three PDE4s studied are fertile, and no disruption of testicular function could be detected. These findings in the mouse differ from those in the rat, where chronic administration of PDE4 inhibitors was demonstrated to have detrimental effects on testicular function (Conti, M., unpublished observation). An explanation of these divergent findings between mice and rats probably lies in the pattern of PDE4 expression during spermatogenesis. Whereas all four PDE4 genes are expressed at different stages of rat germ-cell development,[25] surprisingly only PDE4A is expressed during a very limited period in the final stages of the mouse spermatogenesis.[27] In mouse germ cells, PDE1A and PDE1C, but not PDE4A, are the major forms expressed.[28–30] These findings raise the possibility that germ cells from two closely related species maintain cAMP homeostasis by expressing different PDEs. The functional implications of this puzzling observation remain to be determined but indicate wide species differences in PDE expression as has been described in the heart of rodents, dogs, and humans.[31,32]

Conversely, PDE4D ablation has a major impact on female fertility in the mouse, in a manner reminiscent of that described in the *dunce* mutation of the fruit fly, *Drosophila melanogaster*.[33] In mouse, the fertility of females deficient in PDE4D is diminished, and the number of pups per litter is decreased by approximately 50%.[16] In a paradigm of superovulation, oocytes produced by immature PDE4D$^{-/-}$ females are reduced by 70%–80%.[16,34] When treated with gonadotropins, ovarian follicles of mice deficient in PDE4D are prematurely luteinized with entrapped and degenerating oocytes. Although in some cases injection of PDE4 inhibitors promotes ovulation, it often induces precocious luteinization.[35,36]

Further analysis of this fertility phenotype has uncovered an altered response of follicular granulosa cells from PDE4D$^{-/-}$ mice to gonadotropins.[34] Although cAMP accumulation is induced by luteinizing hormone (LH) in PDE4D$^{-/-}$ cells, the pattern of cyclic nucleotide accumulation is different from that of wild-type cells. Wild-type granulosa cells respond to LH treatment with a robust but transient cAMP accumulation, whereas the response of PDE4D$^{-/-}$ cells is blunted and a stable plateau of cAMP concentration is maintained for at least 3 h. Because LH stimulation in wild-type cells rapidly induces the accumulation of the PDE4D1/2 and PDE4B2 short forms,[34] it is proposed that the granulosa cells from PDE4D$^{-/-}$ mice compensate for the loss of the induction of these PDE4 isoforms by uncoupling the LH receptors from Gs protein. The response of PDE4D$^{-/-}$ granulosa cells to forskolin is comparable to that of wild-type cells, demonstrating that the adenylyl cyclase system is present but not fully activated.[34] It should be noted that PDE4B is still induced in the follicle of PDE4D null mice but cannot compensate for the absence of PDE4D. More importantly, we have demonstrated that the pattern of gene expression triggered by LH is disrupted in a subtle way by PDE4D ablation.[34] Whereas a set of genes involved in granulosa cell transition to luteal cells is induced normally, expression of genes involved in follicle rupture and ovulation is defective in cells lacking PDE4D.[34] As an example, genes coding for enzymes involved in steroidogenesis are induced normally in PDE4D$^{-/-}$ granulosa cells, together with a number of critical transcription factors including the early response genes c-*jun* and c-*fos*. These findings indicate that the LH signal can be properly translated into regulation of expression of these genes. Conversely, transcription of genes such as progesterone receptors, COX2, TSG6, HAS2, and EGF-like growth factors is greatly decreased or

absent, often with altered time courses when compared to wild-type controls.[34,37] These are the genes implicated in the phenotypic changes that lead to follicular rupture and ovulation.[38] This dissection of the differentiation of granulosa cells deficient in PDE4D allows the following conclusions on the role of PDE4s. In these endocrine cells, PDE4D represents an important regulatory mechanism controlling hormonal responses. Ablation of PDE4D, but not PDE4B, causes the loss of one of the most important desensitizing mechanisms, hence the compensatory uncoupling of LH receptors from Gs protein and adenylyl cyclase. More importantly, these data strongly suggest that PDE4D regulation serves to shape the gonado-tropin signal so that the complex programs of gene expression are activated at the appropriate time to allow ovulation and differentiation of granulosa cells. Absence of PDE4D does not interrupt cAMP signaling, but it profoundly modifies its characteristics, thus resulting in an aberrant pattern of gene expression. Together with the absence of PDE4B compensation, these findings underscore how critical specific PDE4 regulation is for proper transfer of information from the membrane to the nucleus of a cell.

16.4   PDE4D IS INVOLVED IN THE CONTROL OF AIRWAY SMOOTH MUSCLE TONE

Cyclic nucleotides play a critical role in the control of smooth muscle relaxation by intersect-ing the Ca^{2+}-signaling pathway that promotes contraction at several critical points.[39] A role for PDEs in cAMP homeostasis has been inferred by the use of selective inhibitors with PDE3 isoforms usually considered as the cAMP–PDE that is most relevant to the control of smooth muscle contractility.[39] Conversely, the use of PDE4 inhibitors in humans has not yielded consistent information on the role of PDE4 in the airway.[40,41] An initial indication for an important role of PDE4D in airway smooth muscle tone emerged during studies of a mouse model of allergic asthma.[42] When compared to their wild-type littermates, PDE4D$^{-/-}$ mice sensitized intraperitoneally and challenged intranasally with ovalbumin (OVA) do not exhibit airway hyperreactivity (AHR) to the muscarinic cholinergic agonist methacholine. More importantly, the methacholine-induced bronchoconstriction observed in the nonsensitized, naive wild-type mice is also absent in the naive PDE4D$^{-/-}$ mice.[42] These findings provide a first correlation between PDE4D activity and muscarinic cholinergic signaling in the air-way.[43,44] The lack of response to muscarinic agonists is not due to an impairment of intrinsic airway functions because the PDE4D$^{-/-}$ mice develop AHR in response to serotonin inhalation. The latter finding also indicates that PDE4D is not involved in the airway contraction evoked by serotonin signaling.

Of the five known muscarinic receptors (M_1–M_5),[45] at least two are involved in airway smooth muscle tone.[46,47] The M_3 receptor coupled to phospholipase C and Ca^{2+} signaling constitutes the primary contractile stimulus. The M_2 receptor coupled to Gαi and adenylyl cyclase inhibition is thought to contribute to bronchoconstriction by decreasing cAMP concen-tration but, in general, is thought to have only an ancillary function.[48] The absence of the airway responses in the PDE4D$^{-/-}$ mice is not associated with a downregulation of the number of muscarinic receptors, as receptor-binding studies with the muscarinic cholinergic antagonist quinuclidinyl benzilate ([^{3}H]QNB)[42] indicated similar receptor levels in wild-type and PDE4D$^{-/-}$ lungs. Instead, it is associated with a decreased inhibition of adenylyl cyclase induced by cholinergic agonists, suggesting a disruption of M_2 receptor signaling.

Further studies in an *ex vivo* model of tracheal ring contraction have pinpointed the exact steps in which PDE4D is involved in the control of airway relaxation.[49] A marked decrease in both maximal efficacy and potency of carbachol to induce tracheal contraction was observed after PDE4D ablation. Consistent with the *in vivo* studies and ruling out a generalized loss of

contractility or anatomical defects, the membrane depolarizing agent KCl evoked comparable contractile responses in tracheas from PDE4D$^{-/-}$ mice and wild-type littermates. Furthermore, the effect of PDE4D ablation appears to be specific for muscarinic cholinergic stimulation because activation of G-protein coupled receptor (GPCR) signaling by arginine vasopressin (AVP), a known potent constrictor of mouse trachea, produces a sustained contraction in both wild-type and PDE4D$^{-/-}$ tracheas. A possible explanation of the divergent effect of PDE4D ablation is that its impact is only on contraction evoked by Ca^{2+} release from intracellular stores but not that induced by Ca^{2+} entry from the extracellular space.[49] This possibility should be further pursued as it is consistent with the idea that cAMP and PDEs function in microcompartments of signaling,[50,51] in analogy with studies on the PDE4D integration in macromolecular complexes involved in the control of cardiac contractility[52] (see Section 16.9).

The blunted response in the PDE4D$^{-/-}$ trachea could be reproduced by treatment of wild-type trachea with rolipram, indicating that PDE4D ablation is not associated with developmental defects causing irreversible loss of function. It also suggests that the pharmacological effect of acute inhibition of this response by rolipram is exerted through inhibition of PDE4D, but not other PDE4s. This conclusion is supported by at least two additional observations. First, the tracheal ring contraction induced by carbachol is normal in PDE4B$^{-/-}$ – and PDE4A$^{-/-}$ mice, indicating that these PDE4s are not involved in muscarinic cholinergic signaling in this tissue. Second, PDE4D is the major PDE4 expressed in the mouse trachea.[49]

When incubated *in vitro* with the cyclooxygenase inhibitor indomethacin, the PDE4D$^{-/-}$ trachea display contractile responses to muscarinic cholinergic agonists similar to those of wild-type trachea, implying the existence of an endogenous prostanoid tone in the PDE4D$^{-/-}$ trachea, which causes relaxation.[49] This reversal of response was also observed when the PDE4D$^{-/-}$ trachea were incubated with a PKA inhibitor, Rp-cAMP, but not with the non-selective β-adrenergic antagonist, propranolol. Treatment with indomethacin had no effect on wild-type tracheal contractility, a finding consistent with the data showing that ablation of the PGE$_2$ receptor EP2 does not alter the tracheal contractile response to carbachol.[53] Subsequent studies demonstrated that the differences in carbachol responses between the wild-type and PDE4D$^{-/-}$ tracheas were not due to an increase in prostanoid release in the PDE4D$^{-/-}$ trachea, but a fivefold increase in sensitivity to the endogenous prostanoid PGE$_2$.[49] This altered sensitivity coincides with an exaggerated accumulation of cAMP in PDE4D$^{-/-}$ lungs after PGE$_2$ stimulation. Moreover, *in vivo* treatment of PDE4D$^{-/-}$ mice with indomethacin restored the airway response to methacholine, suggesting that an increased sensitivity to prostaglandin is preventing muscarinic receptor-mediated contraction also *in vivo*. Taken together, these results document that a tonic cAMP/PKA signaling present in PDE4D$^{-/-}$ airways contributes to a state of relaxation that cannot be overcome by muscarinic cholinergic stimulation, and that increased sensitivity to prostanoids is a critical factor in controlling airway relaxation *in vivo*. Thus, this phenotype implies that PDE4D is an essential component in airway smooth muscle homeostasis controlling the balance between contractile and relaxing stimuli in airway smooth muscle. These data also infer that a PDE4D-selective inhibitor may enhance the bronchodilatory response to endogenous prostanoids.

16.5   DIFFERENTIAL ROLE OF PDE4S IN AIRWAY INFLAMMATION IN *IN VIVO* MODELS OF ASTHMA

The findings summarized earlier demonstrate that the absence of methacholine-induced AHR in PDE4D$^{-/-}$ mice is caused by an impaired response to cholinergic stimulation.[42] However, responses to serotonin are increased in these mice after allergen sensitization. Moreover,

PDE4D$^{-/-}$ mice exhibit normal airway inflammation following sensitization and challenge with OVA.[42] These include dense peribronchiolar and perivascular leukocyte infiltration in the lungs, mucus accumulation in the airway, and an increased cell recruitment, mainly eosinophils, in the bronchoalveolar lavage (BAL). In addition, sensitization of PDE4D$^{-/-}$ mice with OVA also induced an elevation of circulating, OVA-specific IgE levels, as well as IL-4 production, and splenic T-cell proliferation in response to OVA challenge. These findings clearly point out that the antigen-induced inflammatory response can be dissociated from the AHR, but also indicate that PDE4D ablation *in vivo* does not have a significant impact on inflammation. This is in sharp contrast to the pharmacological data showing that PDE4 inhibitors have important anti-inflammatory effects, which have been exploited extensively.[18,19] Indeed, various PDE4 inhibitors have undergone phase III clinical trials in anticipation of receiving regulatory approval for chronic inflammatory diseases of the lung, such as asthma and chronic obstructive pulmonary disease (COPD).[8,54,55]

Analysis of the other PDE4 KO mice has resolved, at least in part, these conflicting results. In the same paradigm of allergen sensitization used for PDE4D$^{-/-}$ mice, PDE4B$^{-/-}$ mice do not develop AHR (Jin, S.-L.C. et al., unpublished results). However, these mice do respond to methacholine, excluding a defect in smooth muscle contraction.[49] Cellular infiltration is marginally affected when measuring cells in the BAL, even though some reduction in infiltration is present in histological sections of the lung (Jin, S.-L.C. et al., unpublished results). The major defect associated with PDE4B ablation is a decrease in Th-2 cytokine accumulation in the airway and in *ex vivo* culture of inflammatory cells recruited to the lung lymph nodes (Jin, S.-L.C. et al., unpublished results). Thus, it is possible that the *in vivo* effects observed with PDE4 inhibitors are mediated predominantly by PDE4B inhibition. Consistent with this hypothesis, several studies have documented that PDE4 inhibitors affect T-cell proliferation.[56–59] Moreover, PDE4B has been implicated in T-cell receptor (TCR) signaling[60,61] even though PDE4D and PDE4A may be equally recruited to the ligated TCR.[62] At present, one cannot exclude the possibility that the complex events mediating the inflammatory response need to be disrupted at multiple cellular sites, and that PDE4D and PDE4B play a distinct but complementary role in this process. Different regimens used for antigen sensitization and challenge may provide an alternative explanation for the differences between the effect of genetic PDE4 ablation and acute PDE4 inhibition. For example, Kung et al.[63] showed that rolipram causes a significant decrease in eosinophil recruitment to the airways in response to OVA. In this study, the mice were sensitized with 15 µg OVA complexed with alum, a somewhat lower dose than that used for sensitization of PDE4 null mice (50 µg OVA with alum). Further studies are required to address this issue. Nevertheless, the overall picture that is emerging is that PDE4B ablation has profound effects on inflammatory responses *in vivo*, whereas PDE4D has only limited effects.

16.6 PDE4B INDUCTION IS ESSENTIAL FOR TOLL-LIKE RECEPTOR SIGNALING

Toll-like receptors (TLR) are involved in the recognition of microbial-derived molecules, such as lipopolysaccharide (LPS) and lipoteichoic acid, and play an important role in innate immunity in organisms ranging from insects to mammals.[64,65] Stimulation of TLR triggers the expression of a large array of genes critical for immune and inflammatory responses, among which the proinflammatory cytokine, tumor necrosis factor-α (TNF-α), is of most importance. TNF-α plays a pivotal role as an initiator of the inflammatory response in the pathogenesis of inflammatory diseases as diverse as Crohn's disease, rheumatoid arthritis, and COPD.[66–70] LPS stimulation of TNF-α synthesis is suppressed by elevated cAMP level,

as shown in monocytes and macrophages, the major sources of TNF-α in the body.[71–77] This effect can be reproduced by the cAMP analog, dibutyryl cAMP,[78] as well as by inhibition of PDE4.[79–84] As demonstrated in human peripheral blood monocytes, PDE4 inhibitors markedly suppress LPS-induced TNF-α production, whereas inhibitors specific for PDE1, PDE3, or PDE5 exhibit little or no inhibitory effect,[80,81,85] suggesting a selective role of PDE4 in LPS/TLR signaling. Moreover, Verghese et al.[86] have demonstrated that LPS induces PDE4 enzymatic activity in human monocytes and Mono-Mac-6 cells, a human monocytic cell line.[86] Further studies by Wang and coworkers[87,88] have shown that PDE4B2 is the predominant PDE4 isozyme in human monocytes, in which LPS specifically increases PDE4B transcripts. This regulation of PDE4 expression and its functional consequences have been reassessed in monocytes and macrophages of PDE4 KO mice.[89,90]

Peripheral blood leukocytes isolated from wild-type mice respond to LPS stimulation with increased PDE4B mRNA levels.[89] This induction reflects the expression of PDE4B2, a short splicing variant of the PDE4B gene, as inferred by Northern blot analysis with specific probes in LPS-treated THP-1 monocytic cells and is ablated in PDE4B$^{-/-}$ cells. The induction of PDE4B mRNA in wild-type leukocytes is associated with a large increase in TNF-α protein release assessed by whole blood cultures *ex vivo* or isolated leukocytes *in vitro*.[89] A similar increase in TNF-α levels is also present in circulating leukocytes from PDE4D$^{-/-}$ mice, suggesting that PDE4D does not play a role in this response.[89] The LPS-induced TNF-α production in cells from PDE4B$^{-/-}$ mice is decreased, however, to approximately the same extent (90%) as in wild-type leukocytes treated with rolipram. The reduction of TNF-α release was associated with a more than 90% decrease in TNF-α mRNA accumulation, implicating a regulatory role of PDE4B in TNF-α production at the level of transcription and/or RNA stability. Moreover, the impaired LPS response was not due to decreased sensitivity of PDE4B$^{-/-}$ cells to LPS because higher concentrations of LPS did not restore the response.[89]

Biochemical studies using purified peritoneal naive macrophages have further defined the role of PDE4 in TLR signaling.[90] Measurement of total and rolipram-sensitive PDE activity indicated that ~40% of the total cAMP-degrading activity in naive macrophages is rolipram-sensitive (i.e., PDE4 activity). After LPS stimulation, total PDE activity is increased 2.0- to 2.5-fold. This increase is due to a 4.0- to 4.5-fold increase in PDE4 activity, whereas the rolipram-insensitive activity is unaffected. The LPS-induced PDE4 activity was further characterized by a combined immunological and genetic approach using PDE4 subtype-selective antibodies for immunoprecipitation of cell extracts of wild-type and PDE4-deficient macrophages. This procedure allows determination of the exact pattern of PDE4 activity in macrophages. Under basal conditions, all three PDE4 activities were detected in wild-type macrophages with the ratio of PDE4A:PDE4B:PDE4D activity ~1:2:1. After LPS stimulation, PDE4B activity in wild-type cells was increased approximately fivefold, whereas PDE4A or PDE4D activity was either marginally increased or not significantly affected, demonstrating that PDE4B is the major subtype induced by activation of TLR signaling. Consistent with this finding, the LPS-induced increase in PDE4 activity was ablated in PDE4B null macrophages. PDE4A and PDE4D activities in PDE4B$^{-/-}$ macrophages are comparable to those obtained in the wild-type macrophages, demonstrating a lack of compensatory increase in expression of other PDE4s in PDE4B$^{-/-}$ macrophages.

As determined for circulating monocytes, LPS-induced TNF-α production was significantly decreased in PDE4B$^{-/-}$ macrophages but not affected in PDE4A$^{-/-}$ and PDE4D$^{-/-}$ macrophages (Figure 16.1A).[90] The decrease in TNF-α production in PDE4B$^{-/-}$ cells is not caused by a generalized loss of viability or compromise in function since accumulation of other cytokines, such as IL-6, was comparable to that in wild-type cells. This suggests that the distinct signaling pathways elicited by LPS/TLR show different sensitivity to the inhibitory effects of cAMP with PDE4B specifically affecting the production of TNF-α but not that of IL-6.

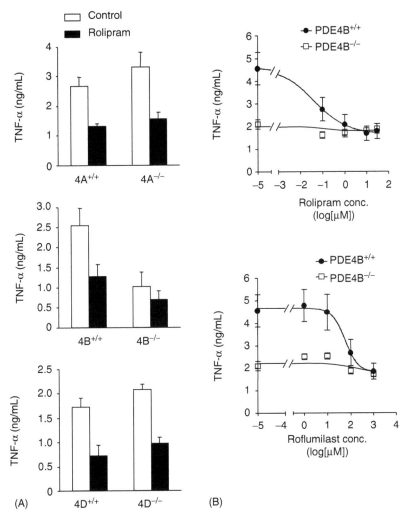

FIGURE 16.1 Effect of PDE4 inhibitors on LPS-induced TNF-α production in PDE4-deficient peritoneal macrophages. (A) Peritoneal macrophages from PDE4A$^{-/-}$, PDE4B$^{-/-}$, and PDE4D$^{-/-}$ mice and their wild-type littermates were incubated with 10 μM rolipram or vehicle (DMSO) for 30 min before LPS (100 ng/mL) stimulation for 8 h. TNF-α accumulation in the medium was determined by ELISA. No significant difference was observed between the control and rolipram-treated cells in PDE4B$^{-/-}$ mice. (B) Wild-type and PDE4B$^{-/-}$ macrophages were treated with increasing concentrations of rolipram or roflumilast for 30 min before LPS (100 ng/mL) stimulation. After 5 h, TNF-α accumulation in the medium was measured. Data are the mean $\pm$ SEM (n = 4–7 mice per group). (From Ref. [90]. With permission. Copyright 2005. The American Association of Immunologists, Inc.)

The effect of PDE4 inhibitors to effectively block TNF-α production in monocytes and macrophages is well documented in the literature.[79–84] However, no PDE4 inhibitors thus far developed can effectively distinguish between PDE4 subfamilies. Using the PDE4 KO mice, treatment of macrophages with rolipram revealed that LPS-stimulated TNF-α production is inhibited in a similar manner in wild-type cells and in PDE4A$^{-/-}$ or PDE4D$^{-/-}$ cells but is not significantly affected in PDE4B$^{-/-}$ cells (Figure 16.1A).[90] The TNF-α response in wild-type cells incubated with rolipram and LPS was comparable to that obtained in PDE4B$^{-/-}$ macrophages treated with LPS alone. Similar results were also obtained when the cells were

treated with increasing concentrations of rolipram or roflumilast, a PDE4 inhibitor currently used in phase III clinical trials (Figure 16.1B). These findings indicate that the effects of PDE4B ablation and acute rolipram inhibition on LPS stimulation of TNF-α production are not additive, and that PDE4 inhibitors exert their pharmacological effects on TNF-α production through inhibition of PDE4B alone. This highlights the pharmacological potential of compounds with PDE4B selectivity.

The blunted TNF-α response in PDE4B$^{-/-}$ macrophages was not associated with global changes in cAMP concentration because the change in cAMP level in response to LPS was not significantly different in wild-type, PDE4B$^{-/-}$, and PDE4D$^{-/-}$ cells.[90] However, the effect of PDE4B ablation on TNF-α response is cAMP-dependent, as inferred by the effects of the adenylyl cyclase activator forskolin and the PKA inhibitors H89 or Rp-cAMPs. Preincubation of wild-type macrophages with 1 μM forskolin led to a 64% decrease in LPS-induced TNF-α accumulation, reaching levels similar to those observed in PDE4B$^{-/-}$ cells incubated with LPS alone. Rolipram had minimal additional effect in forskolin-treated wild-type cells and no effect in PDE4B$^{-/-}$ cells. Moreover, blocking PKA activity with H89 restored the TNF-α response in PDE4B$^{-/-}$ cells to a level not significantly different from that in wild-type cells treated with LPS alone. These findings strongly suggest that PKA is distal to PDE4B in the cAMP-inhibitory pathway and that there is a tonic cAMP/PKA-negative constraint exerted on LPS/TLR signaling. The effects of rolipram in macrophages appeared to require an active PKA because there was little or no rolipram inhibition of TNF-α production observed in cells incubated with a combination of H89 and rolipram. Collectively, these findings, together with the absence of LPS-stimulated global cAMP changes in PDE4B$^{-/-}$ cells, are consistent with the hypothesis that compartmentalized pools of cAMP specifically regulated by PDE4B are responsible for the control of TLR signaling.

These studies, using PDE4B$^{-/-}$ cells, define the following regulatory loop that operates in the LPS/TLR signaling pathway (Figure 16.2). Cyclic AMP exerts a tonic negative constraint

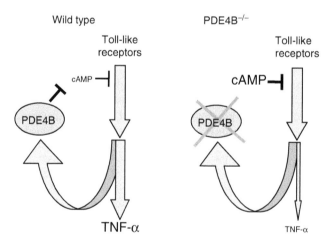

FIGURE 16.2 (See color insert following page 274.) PDE4B positive feedback loop in LPS-induced toll-like receptor signaling. Under basal conditions, cAMP exerts a tonic negative constraint on the LPS/toll-like receptor-signaling pathway. Upon LPS stimulation, expression of PDE4B mRNA and protein is induced, and the resultant increase in PDE4B activity culminates in a decrease in cAMP concentration. Consequently, the cAMP tonic constraint is removed which allows a full induction of TNF-α mRNA and protein synthesis. When PDE4B is inactivated by genetic ablation or by PDE4 inhibitors, the removal of the cAMP-negative constraint is prevented and, therefore, the LPS-induced responses are severely blunted.

on LPS-induced signaling via activation of PKA, acting as a gating mechanism.[91] Upon TLR activation by LPS, PDE4B mRNA and protein are induced, and the increase in PDE4B activity leads to a decrease in cAMP in specific microdomains. As a result, the tonic cAMP constraint is removed and a full induction of TNF-α mRNA and protein synthesis is achieved. Thus, induction of PDE4B provides a positive feedback loop that is required for the full TNF-α response. PDE4B inactivation by genetic ablation or the use of PDE4 inhibitors prevents the removal of the cAMP-negative constraint, and the LPS-induced response is severely blunted. It is likely that this positive feedback mechanism is integrated in other inflammatory signaling pathways as well, since PDE4B2 is regulated in a variety of immune cells.[60,92,93]

16.7 PDE4B AND PDE4D ARE INVOLVED IN THE CONTROL OF NEUTROPHIL RECRUITMENT TO THE LUNG

A number of reports have shown that pharmacological inhibition of PDE4 in human neutrophils suppresses several inflammatory responses, such as superoxide anion production,[94–97] degranulation,[98,99] and cell adhesion,[99,100] suggesting a role of PDE4 in neutrophil function. Using PDE4 KO mice, the involvement of individual PDE4 subtypes in neutrophil infiltration to the lung has recently been dissected using an *in vivo* mouse model of airway neutrophilia,[101] a characteristic feature of COPD. Following exposure to aerosolized LPS, the number of neutrophils recovered in BAL from PDE4B$^{-/-}$ and PDE4D$^{-/-}$ mice was decreased 31% and 48%, respectively, when compared to wild-type mice. No significant difference was observed between wild-type and PDE4A$^{-/-}$ mice, indicating that PDE4D and PDE4B, but not PDE4A, are required for neutrophil recruitment to the lung after exposure to LPS. The function of these two PDE4s has been further characterized by comparing the effects of rolipram on the neutrophil recruitment in wild-type mice and the respective PDE4 null mice. These studies show that PDE4D and PDE4B play overlapping, but not redundant, roles in the control of neutrophil recruitment, a hypothesis supported by the following observations. First, *in vivo* administration of rolipram before LPS inhalation produces a more pronounced decrease in neutrophil recruitment in wild-type mice than that which occurs with ablation of either PDE4D or PDE4B. In addition, rolipram pretreatment further decreased neutrophil recruitment in PDE4B$^{-/-}$ mice and, to a lesser extent, in PDE4D$^{-/-}$ mice. This combination of rolipram plus inactivation of one of the PDE4 genes decreased neutrophil recruitment to levels similar to those observed with rolipram in wild-type mice.[101]

The accumulation of chemokines (KC) and macrophage inflammatory protein (MIP-2) in the BAL of PDE4B$^{-/-}$ and PDE4D$^{-/-}$ mice was not significantly different from that of the wild-type mice, at least in the initial phase of exposure to LPS (2 h), suggesting that a dysfunction intrinsic to neutrophils is associated with PDE4 ablation.[101] Indeed, the reduced recruitment was associated with a significant decrease in the expression of the adhesion molecule β$_2$-integrin (CD18) in both PDE4B$^{-/-}$ and PDE4D$^{-/-}$ neutrophils compared with that in wild-type cells. No major differences were detected in the expression of two α-integrins, CD11a and CD11b. Chemotaxis in response to KC or MIP-2 was also markedly reduced in the splenic neutrophils of PDE4B$^{-/-}$ and PDE4D$^{-/-}$ mice.[101] Acute PDE4 inhibition with rolipram produced a significant decrease in KC-induced migration in wild-type neutrophils, and this decrease was comparable to that observed in each of the two PDE4-deficient neutrophils, indicating a nonadditive effect of PDE4 ablation on chemotaxis.

PDE4 is the predominant cAMP-degrading isozyme in neutrophils recovered from the BAL of wild-type mice.[101] In PDE4D$^{-/-}$ neutrophils, there is a major decrease in the PDE4 activity compared to a small, but significant, decrease in PDE4 activity in PDE4B$^{-/-}$ cells. No significant compensatory increase in rolipram-insensitive PDE activity was observed in

these cells. Immunoprecipitation of wild-type cell extracts with PDE4D- and PDE4B-specific antibodies confirmed that PDE4D is expressed at much higher levels than PDE4B in the BAL neutrophils.[101] However, since the PDE activity was measured in the neutrophils that had infiltrated the airway, it is possible that PDE4D is activated during the cell migration, whereas PDE4B is unchanged, as reported in human neutrophils in which the expression of PDE4B mRNA is not affected by LPS.[87] The PDE4B and PDE4D enzyme activities in peripheral blood neutrophils before LPS stimulation have not been compared because of the limited number of cells available for such measurements.

In summary, both PDE4B and PDE4D are involved in endotoxin-induced neutrophil recruitment to the lung. Ablation of either of the two genes affects the expression of adhesion molecules and chemotaxis. Indeed, PDE4 inhibitors have been shown to affect chemotaxis.[102] These findings provide a rationale basis for the development of more selective pharmacological strategies to manipulate neutrophil recruitment.

16.8   PDE4s AND CENTRAL NERVOUS SYSTEM FUNCTIONS

Throughout the body, the highest level of expression of PDE4s occurs in the brain. It is therefore clear that PDE4s must be involved in many regulatory pathways that utilize cAMP signaling in the brain. Here, we will review the information derived from the analysis of the PDE4 KO mice.

16.8.1   PDE4B AND PDE4D MEDIATE THE ANTIDEPRESSANT EFFECTS OF ROLIPRAM

Early clinical and basic studies have indicated that rolipram has antidepressant effects in patients with depressive disorders[103,104] as well as in animal models with depression and impaired cognition.[105–107] Indeed, cAMP signaling appears to play an important role in the pathophysiology of depression.[108,109] Long-term treatment with antidepressant drugs, such as the selective serotonin reuptake inhibitor (SSRI) fluoxetine, is associated with an activation of the cAMP-signaling pathways in certain brain regions.[110,111] In addition, increasing intracellular cAMP via stimulation of β-adrenergic receptors also produces antidepressant-like effects in animal models.[105,106,112,113] These findings strongly suggest that the antidepressant action is mediated through activation of cAMP cascades.

The PDE4A, PDE4B, and PDE4D genes are all highly expressed in human and rodent brain as shown by RT-PCR, in situ hybridization, and immunolocalization.[114–116] Each subtype exhibits a unique pattern of distribution, suggesting specialized functions for these enzymes in different sets of neurons. To define which of the PDE4 subtypes mediates the antidepressant effects of rolipram, Zhang et al.[117] examined PDE4 KO mice in several paradigms of behavior and drug sensitivity. When compared to wild-type littermates, PDE4D$^{-/-}$ mice exhibit shorter duration of immobility in tail-suspension test (TST) and forced-swim test (FST), which is indicative of antidepressant-like behavior.[118,119] To perform these tests, each mouse was either suspended by the tail from a lever (TST) or placed in a plastic cylinder, which was filled with water (FST), and the duration of immobility was recorded during the 6 min test period. For the FST, the duration of immobility is defined as floating in an upright position without additional activity other than that necessary for the animal to keep its head above water. Moreover, in the mouse PDE4D was found to be the major PDE4 subtype expressed in brain areas mediating antidepressant effects such as cerebral cortex and hippocampus.[117] Administration of the classical antidepressants fluoxetine and desipramine, a selective inhibitor of norepinephrine uptake, reduces the duration of immobility in both wild-type and PDE4D$^{-/-}$ mice in FST, even though the two mouse strains display different baselines of immobility. Thus, actions of the classical antidepressant drugs

do not require PDE4D. Conversely, repeated treatment with rolipram decreases the duration of immobility in wild-type mice, but not PDE4D$^{-/-}$ mice, indicating that PDE4D mediates the antidepressant-like effects of PDE4 inhibitors.

No differences were detected between wild-type and PDE4D$^{-/-}$ mice in the elevated plus-maze (EPM) and multicompartment chamber (MCC) tests.[117] The EPM is a paradigm widely used to evaluate anxiolytic and anxiogenic agents[120]; the MCC provides a measure of exploratory behavior and locomotor activity.[121] These data indicate that the PDE4D deficit does not produce generalized changes in behavior, but rather fairly specific behavioral effects.

Similar studies have been conducted in PDE4B$^{-/-}$ mice.[20] Preliminary results indicate that ablation of PDE4B in mice also produces antidepressant-like behavior, evidenced by decreased immobility in the FST compared to wild-type controls.[20] This antidepressant-like effect is not further affected in PDE4B$^{-/-}$ mice treated with rolipram. However, unlike PDE4D$^{-/-}$ mice, the antidepressant desipramine causes no additional decrease in immobility duration in PDE4B-deficient mice, suggesting that PDE4B is required for the antidepressant action of desipramine.

One should point out that chronic administration of antidepressants also results in an increased expression of PDE4A and PDE4B, but not PDE4D subtypes, in specific brain regions,[122–127] suggesting that PDE4A and PDE4B isozymes may modulate the effects of the antidepressant treatment. This PDE4 induction has been hypothesized to be a compensatory mechanism to counter the elevated cAMP production induced by the antidepressants studied.[127] If so, inhibitors selective to PDE4A and PDE4B may enhance or prolong the therapeutic effects of these antidepressants, a hypothesis that should be confirmed with the analysis of the PDE4A and PDE4B KO mice.

16.8.2 PDE4D but Not PDE4B Modulates the Hypnotic Action of Adrenoceptors: Implications for the Emetogenic Effects of PDE4 Inhibitors

The hypnotic action of α_2-adrenoceptor agonists is thought to be mediated at the locus coeruleus of the brain stem.[128] The α_2-adrenoceptor is negatively coupled to adenylyl cyclase via Gαi, and inhibition of the adenylyl cyclase activity is thought to play a critical role in the hypnotic effect of α_2-adrenoceptor agonists.[129] Recent observations made in ferrets indicate that PDE4 inhibitors trigger the emetic reflex through a sympathetic pathway by mimicking the pharmacological actions of presympathetic α_2-adrenoceptor antagonists.[130] This hypothesis is supported by the finding that the α_2-adrenoceptor agonist, clonidine, protects against the emetic effects of PDE4 inhibitors.[130] Interestingly, like α_2-adrenoceptor antagonists, PDE4 inhibitors also reverse the anesthetic response induced by the xylazine and ketamine combination, an α_2-adrenoceptor agonist-mediated anesthetic regimen, in rats and ferrets.[129,130] Moreover, studies in rats revealed that this anesthesia-reversing effect is functionally coupled to PDE4, specific to α_2-adrenoceptor agonist-mediated anesthesia, and relevant to emesis induced by PDE4 inhibitors.[131] Thus, it is believed that assessing the anesthesia-reversing ability is a practical and valid means of evaluating the emetic potential of PDE4 inhibitors in species, such as rodents, which do not have a vomiting reflex.

For anesthesia-reversing purpose, a combination of genetic and pharmacological approaches has been used to determine the contribution of different PDE4 subtypes in reversing α_2-adrenoceptor-mediated anesthesia in mice.[132] Following the administration of the xylazine and ketamine combination, mice deficient in PDE4D, but not PDE4B, exhibit shorter duration of anesthesia compared with their wild-type littermates.[132] Consistent with this observation, rolipram-sensitive PDE activity in the brain stem is significantly decreased in PDE4D$^{-/-}$ mice, but not in PDE4B$^{-/-}$ mice, even though both PDE4D and PDE4B were

detected in this brain area by Western blot analysis. The anesthesia-reversing effect of PDE4D deficiency appears to be specific to the α_2-adrenoceptor pathway because pentobarbital-induced anesthesia, through enhancing the GABA-mediated inhibition of synaptic transmission, is not affected in PDE4D$^{-/-}$ mice.[132] When the mice were treated with PDE4 inhibitors, the duration of anesthesia induced by xylazine and ketamine was significantly reduced in PDE4B$^{-/-}$ mice as well as in the wild-type mice but not affected in PDE4D$^{-/-}$ mice. These results indicate that PDE4B and other PDE4 subtypes, if expressed, have no impact on the duration of anesthesia in PDE4D$^{-/-}$ mice. More importantly, the reduction in the duration of anesthesia in wild-type mice after PDE4 inhibitor treatment is comparable to that produced by the single genetic inactivation of PDE4D, underscoring that PDE4D is the subtype responsible for the anesthesia-reversing effect of PDE4 inhibitors. Collectively, these results document that PDE4D, but not other PDE4s, has a modulating effect on the activity of α_{2A}-adrenoceptors, the receptor subtype responsible for the hypnotic effect of α_2-adrenoceptor agonists. Inhibition of PDE4D contributes to the anesthesia-reversing effect of PDE4 inhibitors, which implies that PDE4D is the subtype responsible for the emetic side effect induced by PDE4 inhibitors.

The nausea and emesis associated with the administration of PDE4 inhibitors constitute a major obstacle to the clinical use of these compounds.[133] Identification of the PDE4 subfamilies responsible for these adverse effects, therefore, is an example of how the genetic dissection of PDE4s may help in designing novel drugs devoid of such side effects, and, consequently, conferring an improved therapeutic window.

16.9 INVOLVEMENT OF PDE4D IN THE MODULATION OF EXCITATION–CONTRACTION COUPLING IN THE HEART

That cardiovascular functions are only marginally affected by PDE4 inhibitors has been a long-standing tenet and indeed one of the rationales for the development of this class of drugs.[18] In fact, PDE4 inhibitors have been developed as an improvement over theophylline, a nonselective PDE inhibitor, because of the absence of conspicuous cardiovascular effects. However, in rodents, and perhaps in humans, recent findings suggest that PDE4D may play a crucial role in the modulation of adrenergic responses and excitation–contraction coupling. In a first set of studies aimed at dissecting the cardiovascular functions of the different PDE4 genes, Xiang et al.[52] reported that inactivation of PDE4D disrupts subtype-specific signaling of the β_2-adrenergic receptor (β_2-AR). Using the contraction rate of cultured mouse neonatal cardiac myocytes as readout, This report shows that treatment with a PDE4-selective inhibitor such as rolipram or RS25344 produces an elevated chronotropic response to β-adrenergic agonists. Genetic ablation of PDE4A or PDE4B has no effect in this paradigm. Conversely, myocytes from PDE4D$^{-/-}$ mice show an increased chronotropic response to β-adrenergic stimulation reminiscent of the effect of rolipram on wild-type cells. In addition, pharmacological inhibition of PDE4 has no additional effect in PDE4D$^{-/-}$ myocytes, indicating that the effect of PDE4 inhibitors on the contraction rate of wild-type myocytes is mediated through inhibition of PDE4D. These data are consistent with earlier studies, demonstrating that PDE4 inhibition modulates Ca^{2+}-channel opening in cardiac myocytes from different species.[134,135]

The role of PDE4 in subtype-specific β-adrenoceptor signaling has been further dissected using knockout mice for the β_1-AR and β_2-AR, respectively.[52] The chronotropic response of cardiac myocytes to β_2-adrenergic signaling (determined in β_1-AR KO mice) was greatly elevated by PDE4 inhibition whereas the response to β_1-adrenergic stimulation (assessed

using β_2-AR KO mice) was unaffected. Moreover, whereas β_2-adrenergic signaling produces a small, transient, and PKA-independent increase in contraction rate, the response to β_2-adrenergic stimulation in the presence of PDE4 inhibitors was not only elevated, but was also longer lasting and PKA-dependent, thus resembling the response that follows β_1-AR activation. These findings strongly suggest that PDE4D, but not PDE4A or PDE4B, is an integral component of the β_2-AR-signaling complex. In this compartment, it functions to restrict the diffusion of cAMP generated by the β_2-AR, thus, allowing for subtype-specific β-adrenergic signaling. Upon pharmacological inhibition or genetic ablation of PDE4D, cAMP generated by β_2-AR signaling can more freely diffuse throughout the cell and gains access to the PKA-dependent signaling steps, by which the β_1-AR regulates myocyte contraction rate, consistent with the hypothesis that PDEs serve as a functional barrier to diffusion of cAMP.[50,51] These data are also in line with the observation that β-arrestins physically link PDE4D to the β_2-AR. Isozymes derived from the PDE4D gene, particularly PDE4D5,[17,136] have been shown to interact with β-arrestins,[137] and occupancy of β_2-AR is thought to recruit β-arrestins more efficiently than β_1-AR.[138]

In a second study probing the function of PDE4D in the heart, Lehnart et al.[139] have shown that genetic ablation of PDE4D produces a severe late-onset cardiac phenotype characterized by progressive cardiac myopathy, accelerated heart failure after myocardial infarct, and exercise-induced arrhythmias. Heart failure and cardiac arrhythmias in the PDE4D$^{-/-}$ mice result primarily from a dysregulation of the ryanodine receptor (RyR2), the major Ca^{2+}-release channel of the sarcoplasmic reticulum. This study also reports that the PDE4D splice variant, PDE4D3, is an integral component of a macromolecular complex anchored to RyR2 (see Figure 16.3), and that PDE4D deficiency in mice results in PKA hyperphosphorylation of this channel, leading to calstabin2 depletion from the RyR2 complex and a leaky RyR2 channel phenotype.

Although PDE4D3 contributes only a small fraction of the total PDE4D in the heart,[15] the deficiency of PDE4D3 in the ryanodine receptor complex is critical for the development of cardiac arrhythmias and heart failure for two reasons. First, patients with heart failure exhibit a phenotype resembling that of the PDE4D$^{-/-}$ mice. Both phenotypes show reduced levels of PDE4D3 in the RyR2 complex and PKA hyperphosphorylated and leaky RyR2 channels that have been linked to cardiac arrhythmias.[140,141] Second, and more importantly, cardiac arrhythmias and heart failure associated with pharmacological PDE4 inhibition or genetic PDE4D ablation were suppressed in mice harboring a mutated RyR2 channel that cannot be phosphorylated by PKA. Since global cAMP signaling in the PDE4D$^{-/-}$ mice was not perturbed, this study underscores how a local change in cAMP signaling (such as PKA-mediated hyperphosphorylation of RyR2 as a consequence of PDE4D3 deficiency) may be critical for the development of a major cardiac phenotype, thus pointing to the physiological relevance of cAMP microdomains. However, PDE4D ablation may not impact exclusively the RyR2/PKA homeostasis. Numerous other functional complexes involving PDE4D, such as the PDE4D-mediated regulation of β_2-adrenergic signaling described earlier, are most likely present in the heart as inferred by functional studies in cardiomyocytes or by findings in other cells (see Figure 16.3). All these complexes are likely to be disturbed by PDE4D ablation, which may contribute to the phenotype described in aging PDE4D$^{-/-}$ mice.[139]

It is very important to point out that it has long been known that there are species-specific differences in the pattern of PDEs expressed in the heart.[31,32] Thus, it is unclear whether these observations in the mouse can be directly translated to humans. Nevertheless, the systemic administration of PDE4 inhibitors currently developed for the treatment of inflammatory diseases such as asthma and COPD should be further evaluated for use in patients in light of possible cardiac effects resulting from long-term inhibition of PDE4.

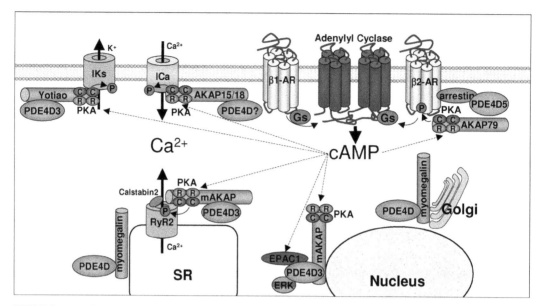

FIGURE 16.3 (See color insert following page 274.) Macromolecular complexes potentially involving PDE4D and likely affected by PDE4D ablation in the mouse heart. The subcellular space between the plasma membrane of the T tubule and the sarcoplasmic reticulum of a cardiomyocyte is presented in a schematic fashion. Black curved arrows and red circles marked with P indicate PKA-mediated phosphorylations. A macromolecular complex composed of PKA, PDE4D3, and mAKAP tethered to the nuclear membrane has been described in the rat heart.[143,144] This complex also includes EPAC1 and ERK5, both of which are bound directly to PDE4D3.[145] While PDE4D3 in the mAKAP complex binds selectively to ERK5, ERK2 has been shown to interact with and regulate the activity of various PDE4 subtypes in mammalian cell lines.[146] Both mAKAP and PDE4D3 have also been identified as components of the ryanodine receptor (RyR2) complex located in the sarcoplasmic reticulum (SR).[139,147] Based on the known interaction of mAKAP and PDE4D3, PDE4D3 is thought to be tethered through mAKAP to the RyR2 channel. The presence of PDE4D in a complex with IK channels in humans is suggested by the finding that Yotiao/AKAP 450 binds PDE4D3.[148] PDE4D5 is targeted via arrestins to β2-adrenergic receptors where it functions to control subtype-specific signaling, receptor phosphorylation, and the switch from Gs to Gi.[17,136,137] The complex of PDE4D with Ca^{2+} channels (ICa) is inferred by the identification of PDE4D/AKAP18 complex in kidney[149] and the well-established interaction of AKAP15 with $Ca_V1.2$ channels.[150] Finally, myomegalin, which binds PDE4D, has been localized in a region corresponding to the Z-band by immunocytochemistry.[151]

16.10 CONCLUSIONS

The genetic studies summarized earlier have provided novel insights into the roles of different PDE4 genes and allow a number of general conclusions on their relevance in physiological and pathological processes. At the same time, they provide a better delineation of the pharmacological potential of PDE4 inhibition.

PDE4s are involved in a plethora of physiological processes as documented by the wide array of phenotypes detected in the PDE4 KO mice. These phenotypes provide compelling evidence that these enzymes are critical for cAMP signaling. At the same time, identification of the large number of processes that are likely to be affected by PDE4 inhibition represents a formidable challenge for the development of drugs with a useful therapeutic application.

Ablation of three PDE4 genes produces largely nonoverlapping phenotypes, documenting for the first time that these genes play distinct roles in the body. PDE4D is the most widely expressed PDE4 and is involved in a large array of biological functions. This conclusion is

based on the rather severe consequences of the PDE4D ablation compared with the phenotypes, resulting from ablation of PDE4A and PDE4B. From the pharmacological standpoint, it is predicted that targeting processes involving PDE4D will be more difficult as a large number of side effects are to be expected.

In spite of the high homology of the PDE4 genes, their similar regulation, and structural homology of splicing variants, there is little compensation by other PDE4s that can be detected when a single gene is ablated. This finding again argues that PDE4s likely play functions that are in general distinct. Further studies are required to delineate the biochemical basis for these differences.

A recurrent finding from the PDE4 KO studies is that each cell type adapts to the loss of a given PDE4 with unique compensatory changes in the cAMP signaling but this adaptation is not at the level of upregulation of other PDE4 isozymes. As an example, cAMP accumulation and β-adrenergic responses are increased in lung fibroblast in spite of the compensatory downregulation of β_2-adrenergic receptors. Conversely, PDE4D ablation in granulosa cells of the ovarian follicle does not cause an overall increase in cAMP signaling but rather a compromised signal because of an adaptive GPCR uncoupling from Gs protein and adenylyl cyclase. Thus, different cell types adapt to the loss of a PDE4 using different mechanisms to maintain signaling. It will be important to further compare the adaptive changes that follow ablation of two PDE4s expressed in the same cell, as it will provide further insight into their specialized functions.

In many instances, the effect of genetic ablation of a PDE4 could be reproduced by acute inhibition using PDE4 inhibitors. Being verified in several distinct paradigms using different cells, this finding suggests that most phenotypes are the consequences of the loss of PDE hydrolytic activity. However, this is likely to be an oversimplification, as there must be divergent consequences to an acute or permanent loss of PDE4 function. Novel approaches need to be devised to explore these possibilities. Finally, the pharmacological effects of PDE4 inhibitors in many cells are lost after ablation of a single PDE4 gene. This is an observation crucial for the development of novel PDE4 inhibitors because it clearly demonstrates that inhibition of a single PDE4 subfamily is sufficient to produce a useful pharmacological effect. This concept has been documented by the selective impact of PDE4B ablation on TNF-α production and on inflammatory processes.[89,90] Thus, these genetic manipulations support the concept that subfamily-selective PDE4 inhibitors may retain their beneficial pharmacological effects and at the same time decrease the number of side effects. A major contribution of these genetic studies is that PDE4B-selective inhibitors should retain anti-inflammatory effects but should be largely devoid of the emetic or cardiac effects attributed to PDE4D. Although it is not yet clear whether the high structural homology of the different PDE4 catalytic domains will allow a pharmacological discrimination,[142] a third generation of PDE4 inhibitors with subfamily selectivity should address this possibility.

ACKNOWLEDGMENT

Work done in the authors' laboratory is supported by NIH grants HD20788, HD31544, HL67674, and U54-HD31398.

REFERENCES

1. Beavo, J.A., and L.L. Brunton. 2002. Cyclic nucleotide research—Still expanding after half a century. *Nat Rev Mol Cell Biol* 3:710–8.
2. Conti, M., and S.L. Jin. 1999. The molecular biology of cyclic nucleotide phosphodiesterases. *Prog Nucleic Acid Res Mol Biol* 63:1–38.

3. Conti, M. 2000. Phosphodiesterases and cyclic nucleotide signaling in endocrine cells. *Mol Endocrinol* 14:1317–27.
4. Francis, S.H., I.V. Turko, and J.D. Corbin. 2001. Cyclic nucleotide phosphodiesterases: Relating structure and function. *Prog Nucleic Acid Res Mol Biol* 65:1–52.
5. Houslay, M.D. 2001. PDE4 cAMP-specific phosphodiesterases. *Prog Nucleic Acid Res Mol Biol* 69:249–315.
6. Movsesian, M.A. 2003. PDE3 inhibition in dilated cardiomyopathy: Reasons to reconsider. *J Card Fail* 9:475–80.
7. Kambayashi, J., et al. 2003. Cilostazol as a unique antithrombotic agent. *Curr Pharm Des* 9:2289–302.
8. Lipworth, B.J. 2005. Phosphodiesterase-4 inhibitors for asthma and chronic obstructive pulmonary disease. *Lancet* 365:167–75.
9. Corbin, J.D. 2004. Mechanisms of action of PDE5 inhibition in erectile dysfunction. *Int J Impot Res* 16 (Suppl. 1):S4–7.
10. Corbin, J.D., et al. 2004. Vardenafil: Structural basis for higher potency over sildenafil in inhibiting cGMP-specific phosphodiesterase-5 (PDE5). *Neurochem Int* 45:859–63.
11. Conti, M., et al. 2003. Cyclic AMP-specific PDE4 phosphodiesterases as critical components of cyclic AMP signaling. *J Biol Chem* 278:5493–6.
12. Bolger, G.B., I. McPhee, and M.D. Houslay. 1996. Alternative splicing of cAMP-specific phosphodiesterase mRNA transcripts. Characterization of a novel tissue-specific isoform, RNPDE4A8. *J Biol Chem* 271:1065–71.
13. Bolger, G.B., et al. 1997. Characterization of five different proteins produced by alternatively spliced mRNAs from the human cAMP-specific phosphodiesterase PDE4D gene. *Biochem J* 328:539–48.
14. Wang, D., et al. 2003. Cloning and characterization of novel PDE4D isoforms PDE4D6 and PDE4D7. *Cell Signal* 15:883–91.
15. Richter, W., S.L. Jin, and M. Conti. 2005. Splice variants of the cyclic nucleotide phosphodiesterase PDE4D are differentially expressed and regulated in rat tissue. *Biochem J* 388:803–11.
16. Jin, S.L., et al. 1999. Impaired growth and fertility of cAMP-specific phosphodiesterase PDE4D-deficient mice. *Proc Natl Acad Sci USA* 96:11998–2003.
17. Lynch, M.J., et al. 2005. RNA silencing identifies PDE4D5 as the functionally relevant cAMP phosphodiesterase interacting with beta arrestin to control the protein kinase A/AKAP79-mediated switching of the β_2-adrenergic receptor to activation of ERK in HEK293B2 cells. *J Biol Chem* 280:33178–89.
18. Torphy, T.J. 1998. Phosphodiesterase isozymes: Molecular targets for novel antiasthma agents. *Am J Respir Crit Care Med* 157:351–70.
19. Essayan, D.M. 2001. Cyclic nucleotide phosphodiesterases. *J Allergy Clin Immunol* 108:671–80.
20. O'Donnell, J.M., and H.T. Zhang. 2004. Antidepressant effects of inhibitors of cAMP phosphodiesterase (PDE4). *Trends Pharmacol Sci* 25:158–63.
21. Banner, K.H., and M.A. Trevethick. 2004. PDE4 inhibition: A novel approach for the treatment of inflammatory bowel disease. *Trends Pharmacol Sci* 25:430–6.
22. Rose, G.M., et al. 2005. Phosphodiesterase inhibitors for cognitive enhancement. *Curr Pharm Des* 11:3329–34.
23. Richards, J.S. 2001. Perspective: The ovarian follicle—A perspective in 2001. *Endocrinology* 142:2184–93.
24. Conti, M., et al. 1981. Regulation of Sertoli cell cyclic adenosine 3′,5′ monophosphate phosphodiesterase activity by follicle stimulating hormone and dibutyrl cyclic AMP. *Biochem Biophys Res Commun* 98:1044–50.
25. Geremia, R., et al. 1982. Cyclic nucleotide phosphodiesterase in developing rat testis. Identification of somatic and germ-cell forms. *Mol Cell Endocrinol* 28:37–53.
26. Conti, M., B.G. Kasson, and A.J. Hsueh. 1984. Hormonal regulation of 3′,5′-adenosine monophosphate phosphodiesterases in cultured rat granulosa cells. *Endocrinology* 114:2361–8.
27. Welch, J.E., et al. Unique adenosine 3′,5′ cyclic monophosphate phosphodiesterase messenger ribonucleic acids in rat spermatogenic cells: Evidence for differential gene expression during spermatogenesis. *Biol Reprod* 46:1027–33.

28. Rossi, P., et al. 1985. Cyclic nucleotide phosphodiesterases in somatic and germ cells of mouse seminiferous tubules. *J Reprod Fertil* 74:317–27.
29. Yan, C., et al. 1996. The calmodulin-dependent phosphodiesterase gene PDE1C encodes several functionally different splice variants in a tissue-specific manner. *J Biol Chem* 271:25699–706.
30. Yan, C., et al. 2001. Stage and cell-specific expression of calmodulin-dependent phosphodiesterases in mouse testis. *Biol Reprod* 64:1746–54.
31. Weishaar, R.E., D.C. Kobylarz-Singer, and H.R. Kaplan. 1987. Subclasses of cyclic AMP phosphodiesterase in cardiac muscle. *J Mol Cell Cardiol* 19:1025–36.
32. Okruhlicova, L., et al. 1997. Species differences in localization of cardiac cAMP-phosphodiesterase activity: A cytochemical study. *Mol Cell Biochem* 173:183–8.
33. Davis, R.L. 1988. Mutational analysis of phosphodiesterase in *Drosophila*. *Methods Enzymol* 159:786–92.
34. Park, J.Y., et al. 2003. Phosphodiesterase regulation is critical for the differentiation and pattern of gene expression in granulosa cells of the ovarian follicle. *Mol Endocrinol* 17:1117–30.
35. Tsafriri, A., et al. 1996. Oocyte maturation involves compartmentalization and opposing changes of cAMP levels in follicular somatic and germ cells: Studies using selective phosphodiesterase inhibitors. *Dev Biol* 178:393–402.
36. McKenna, S.D., et al. 2005. Pharmacological inhibition of phosphodiesterase 4 triggers ovulation in follicle-stimulating hormone-primed rats. *Endocrinology* 146:208–14.
37. Park, J.Y., et al. 2004. EGF-like growth factors as mediators of LH action in the ovulatory follicle. *Science* 303:682–4.
38. Richards, J.S. 2005. Ovulation: New factors that prepare the oocyte for fertilization. *Mol Cell Endocrinol* 234:75–9.
39. Murthy, K.S. 2006. Signaling for contraction and relaxation in smooth muscle of the gut. *Annu Rev Physiol* 68:345–74.
40. Schmidt, D.T., et al. 2000. The effect of selective and non-selective phosphodiesterase inhibitors on allergen- and leukotriene C(4)-induced contractions in passively sensitized human airways. *Br J Pharmacol* 131:1607–18.
41. Fujii, K., et al. 1998. Novel phosphodiesterase 4 inhibitor T-440 reverses and prevents human bronchial contraction induced by allergen. *J Pharmacol Exp Ther* 284:162–9.
42. Hansen, G., et al. 2000. Absence of muscarinic cholinergic airway responses in mice deficient in the cyclic nucleotide phosphodiesterase PDE4D. *Proc Natl Acad Sci USA* 97:6751–6.
43. Nathanson, N.M. 2000. A multiplicity of muscarinic mechanisms: Enough signaling pathways to take your breath away. *Proc Natl Acad Sci USA* 97:6245–7.
44. Giembycz, M. 2000. PDE4D-deficient mice knock the breath out of asthma. *Trends Pharmacol Sci* 21:291–2.
45. Caulfield, M.P., and N.J. Birdsall. 1998. International Union of Pharmacology. XVII. Classification of muscarinic acetylcholine receptors. *Pharmacol Rev* 50:279–90.
46. Fryer, A.D., and D.B. Jacoby. 1998. Muscarinic receptors and control of airway smooth muscle. *Am J Respir Crit Care Med* 158:S154–60.
47. Ehlert, F.J. 2003. Pharmacological analysis of the contractile role of M2 and M3 muscarinic receptors in smooth muscle. *Receptors Channels* 9:261–77.
48. Stengel, P.W., et al. 2000. M(2) and M(4) receptor knockout mice: Muscarinic receptor function in cardiac and smooth muscle in vitro. *J Pharmacol Exp Ther* 292:877–85.
49. Mehats, C., et al. 2003. PDE4D plays a critical role in the control of airway smooth muscle contraction. *FASEB J* 17:1831–41.
50. Jurevicius, J., and R. Fischmeister. cAMP compartmentation is responsible for a local activation of cardiac Ca^{2+} channels by beta-adrenergic agonists. *Proc Natl Acad Sci USA* 93:295–9.
51. Zaccolo, M., and T. Pozzan. 2002. Discrete microdomains with high concentration of cAMP in stimulated rat neonatal cardiac myocytes. *Science* 295:1711–5.
52. Xiang, Y., et al. 2005. Phosphodiesterase 4D is required for β_2 adrenoceptor subtype-specific signaling in cardiac myocytes. *Proc Natl Acad Sci USA* 102:909–14.
53. Fortner, C.N., R.M. Breyer, and R.J. Paul. 2001. EP2 receptors mediate airway relaxation to substance P, ATP, and PGE2. *Am J Physiol Lung Cell Mol Physiol* 281:L469–74.

342 Cyclic Nucleotide Phosphodiesterases in Health and Disease

54. Torphy, T.J., et al. 1999. Ariflo (SB 207499), a second generation phosphodiesterase 4 inhibitor for the treatment of asthma and COPD: From concept to clinic. *Pulm Pharmacol Ther* 12:131–5.
55. Smith, V.B., and D. Spina. 2005. Selective phosphodiesterase 4 inhibitors in the treatment of allergy and inflammation. *Curr Opin Investig Drugs* 6:1136–41.
56. Essayan, D.M., et al. 1994. Modulation of antigen- and mitogen-induced proliferative responses of peripheral blood mononuclear cells by nonselective and isozyme selective cyclic nucleotide phosphodiesterase inhibitors. *J Immunol* 153:3408–16.
57. Banner, K.H., N.M. Roberts, and C.P. Page. 1995. Differential effect of phosphodiesterase 4 inhibitors on the proliferation of human peripheral blood mononuclear cells from normals and subjects with atopic dermatitis. *Br J Pharmacol* 116:3169–74.
58. Giembycz, M.A., et al. 1996. Identification of cyclic AMP phosphodiesterases 3, 4 and 7 in human CD4$^+$ and CD8$^+$ T-lymphocytes: Role in regulating proliferation and the biosynthesis of interleukin-2. *Br J Pharmacol* 118:1945–58.
59. Essayan, D.M., et al. 1997. Differential efficacy of lymphocyte- and monocyte-selective pretreatment with a type 4 phosphodiesterase inhibitor on antigen-driven proliferation and cytokine gene expression. *J Allergy Clin Immunol* 99:28–37.
60. Baroja, M.L., et al. 1999. Specific CD3 epsilon association of a phosphodiesterase 4B isoform determines its selective tyrosine phosphorylation after CD3 ligation. *J Immunol* 162:2016–23.
61. Arp, J., et al. 2003. Regulation of T-cell activation by phosphodiesterase 4B2 requires its dynamic redistribution during immunological synapse formation. *Mol Cell Biol* 23:8042–57.
62. Abrahamsen, H., et al. 2004. TCR- and CD28-mediated recruitment of phosphodiesterase 4 to lipid rafts potentiates TCR signaling. *J Immunol* 173:4847–58.
63. Kung, T.T., et al. 2000. Inhibition of pulmonary eosinophilia and airway hyperresponsiveness in allergic mice by rolipram: Involvement of endogenously released corticosterone and catecholamines. *Br J Pharmacol* 130:457–63.
64. Aderem, A., and R.J. Ulevitch. 2000. Toll-like receptors in the induction of the innate immune response. *Nature* 406:782–7.
65. Takeda, K., T. Kaisho, and S. Akira. 2003. Toll-like receptors. *Annu Rev Immunol* 21:335–76.
66. Keatings, V.M., et al. 1996. Differences in interleukin-8 and tumor necrosis factor-α in induced sputum from patients with chronic obstructive pulmonary disease or asthma. *Am J Respir Crit Care Med* 153:530–4.
67. Bell, S., and M.A. Kamm. 2000. Antibodies to tumour necrosis factor alpha as treatment for Crohn's disease. *Lancet* 355:858–60.
68. Feldmann, M., and R.N. Maini. 2001. Anti-TNF-α therapy of rheumatoid arthritis: What have we learned? *Annu Rev Immunol* 19:163–96.
69. Keating, G.M., and C.M. Perry. 2002. Infliximab: An updated review of its use in Crohn's disease and rheumatoid arthritis. *BioDrugs* 16:111–48.
70. Barnes, P.J. 2003. Cytokine-directed therapies for the treatment of chronic airway diseases. *Cytokine Growth Factor Rev* 14:511–22.
71. Kunkel, S.L., et al. Prostaglandin E2 regulates macrophage-derived tumor necrosis factor gene expression. *J Biol Chem* 263:5380–4.
72. Taffet, S.M., et al. 1989. Regulation of tumor necrosis factor expression in a macrophage-like cell line by lipopolysaccharide and cyclic AMP. *Cell Immunol* 120:291–300.
73. Spengler, R.N., et al. 1989. Dynamics of dibutyryl cyclic AMP- and prostaglandin E2-mediated suppression of lipopolysaccharide-induced tumor necrosis factor alpha gene expression. *Infect Immun* 57:2837–41.
74. Schade, F.U., and C. Schudt. 1993. The specific type III and IV phosphodiesterase inhibitor zardaverine suppresses formation of tumor necrosis factor by macrophages. *Eur J Pharmacol* 230:9–14.
75. Sinha, B., et al. 1995. Enhanced tumor necrosis factor suppression and cyclic adenosine monophosphate accumulation by combination of phosphodiesterase inhibitors and prostanoids. *Eur J Immunol* 25:147–53.
76. Gantner, F., et al. 1997. In vitro differentiation of human monocytes to macrophages: Change of PDE profile and its relationship to suppression of tumour necrosis factor-α release by PDE inhibitors. *Br J Pharmacol* 121:221–31.

77. Delgado, M., et al. 1999. Vasoactive intestinal peptide and pituitary adenylate cyclase-activating polypeptide inhibit endotoxin-induced TNF-α production by macrophages: In vitro and *in vivo* studies. *J Immunol* 162:2358–67.
78. Strieter, R.M., et al. 1988. Cellular and molecular regulation of tumor necrosis factor-α production by pentoxifylline. *Biochem Biophys Res Commun* 155:1230–6.
79. Semmler, J., H. Wachtel, and S. Endres. 1993. The specific type IV phosphodiesterase inhibitor rolipram suppresses tumor necrosis factor-α production by human mononuclear cells. *Int J Immunopharmacol* 15:409–13.
80. Prabhakar, U., et al. 1994. Characterization of cAMP-dependent inhibition of LPS-induced TNF-α production by rolipram, a specific phosphodiesterase IV (PDE IV) inhibitor. *Int J Immunopharmacol* 16:805–16.
81. Verghese, M.W., et al. 1995. Differential regulation of human monocyte-derived TNF-α and IL-1 β by type IV cAMP-phosphodiesterase (cAMP-PDE) inhibitors. *J Pharmacol Exp Ther* 272:1313–20.
82. Souness, J.E., et al. 1996. Evidence that cyclic AMP phosphodiesterase inhibitors suppress TNF-α generation from human monocytes by interacting with a 'low-affinity' phosphodiesterase 4 conformer. *Br J Pharmacol* 118:649–58.
83. Yoshimura, T., et al. 1997. Effects of cAMP-phosphodiesterase isozyme inhibitor on cytokine production by lipopolysaccharide-stimulated human peripheral blood mononuclear cells. *Gen Pharmacol* 29:633–8.
84. Barnette, M.S., et al. 1998. SB 207499 (Ariflo), a potent and selective second-generation phosphodiesterase 4 inhibitor: In vitro anti-inflammatory actions. *J Pharmacol Exp Ther* 284:420–6.
85. Seldon, P.M., et al. 1995. Suppression of lipopolysaccharide-induced tumor necrosis factor-α generation from human peripheral blood monocytes by inhibitors of phosphodiesterase 4: Interaction with stimulants of adenylyl cyclase. *Mol Pharmacol* 48:747–57.
86. Verghese, M.W., et al. 1995. Regulation of distinct cyclic AMP-specific phosphodiesterase (phosphodiesterase type 4) isozymes in human monocytic cells. *Mol Pharmacol* 47:1164–71.
87. Wang, P., et al. 1999. Phosphodiesterase 4B2 is the predominant phosphodiesterase species and undergoes differential regulation of gene expression in human monocytes and neutrophils. *Mol Pharmacol* 56:170–4.
88. Ma, D., et al. 1999. Phosphodiesterase 4B gene transcription is activated by lipopolysaccharide and inhibited by interleukin-10 in human monocytes. *Mol Pharmacol* 55:50–7.
89. Jin, S.L., and M. Conti. 2002. Induction of the cyclic nucleotide phosphodiesterase PDE4B is essential for LPS-activated TNF-α responses. *Proc Natl Acad Sci USA* 99:7628–33.
90. Jin, S.L., et al. 2005. Specific role of phosphodiesterase 4B in lipopolysaccharide-induced signaling in mouse macrophages. *J Immunol* 175:1523–31.
91. Iyengar, R. 1996. Gating by cyclic AMP: Expanded role for an old signaling pathway. *Science* 271:461–3.
92. Barber, R., et al. 2004. Differential expression of PDE4 cAMP phosphodiesterase isoforms in inflammatory cells of smokers with COPD, smokers without COPD, and nonsmokers. *Am J Physiol Lung Cell Mol Physiol* 287:L332–43.
93. Gantner, F., et al. 1997. Phosphodiesterase profiles of highly purified human peripheral blood leukocyte populations from normal and atopic individuals: A comparative study. *J Allergy Clin Immunol* 100:527–35.
94. Bessler, H., et al. 1986. Effect of pentoxifylline on the phagocytic activity, cAMP levels, and superoxide anion production by monocytes and polymorphonuclear cells. *J Leukoc Biol* 40:747–54.
95. Nielson, C.P., et al. 1990. Effects of selective phosphodiesterase inhibitors on the polymorphonuclear leukocyte respiratory burst. *J Allergy Clin Immunol* 86:801–8.
96. Rickards, K.J., et al. 2001. Differential inhibition of equine neutrophil function by phosphodiesterase inhibitors. *J Vet Pharmacol Ther* 24:275–81.
97. Jacob, C., et al. 2004. Role of PDE4 in superoxide anion generation through p44/42MAPK regulation: A cAMP and a PKA-independent mechanism. *Br J Pharmacol* 143:257–68.

98. Barnette, M.S., et al. 1996. Association of the anti-inflammatory activity of phosphodiesterase 4 (PDE4) inhibitors with either inhibition of PDE4 catalytic activity or competition for [3H]rolipram binding. *Biochem Pharmacol* 51:949–56.

99. Jones, N.A., et al. 2005. The effect of selective phosphodiesterase isoenzyme inhibition on neutrophil function in vitro. *Pulm Pharmacol Ther* 18:93–101.

100. Derian, C.K., et al. 1995. Inhibition of chemotactic peptide-induced neutrophil adhesion to vascular endothelium by cAMP modulators. *J Immunol* 154:308–17.

101. Ariga, M., et al. 2004. Nonredundant function of phosphodiesterases 4D and 4B in neutrophil recruitment to the site of inflammation. *J Immunol* 173:7531–8.

102. Kohyama, T., et al. 2002. PDE4 inhibitors attenuate fibroblast chemotaxis and contraction of native collagen gels. *Am J Respir Cell Mol Biol* 26:694–701.

103. Wachtel, H., and H.H. Schneider. 1986. Rolipram, a novel antidepressant drug, reverses the hypothermia and hypokinesia of monoamine-depleted mice by an action beyond postsynaptic monoamine receptors. *Neuropharmacology* 25:1119–26.

104. Fleischhacker, W.W., et al. 1992. A multicenter double-blind study of three different doses of the new cAMP-phosphodiesterase inhibitor rolipram in patients with major depressive disorder. *Neuropsychobiology* 26:59–64.

105. O'Donnell, J.M., 1993. Antidepressant-like effects of rolipram and other inhibitors of cyclic adenosine monophosphate phosphodiesterase on behavior maintained by differential reinforcement of low response rate. *J Pharmacol Exp Ther* 264:1168–78.

106. O'Donnell, J.M., and S. Frith. 1999. Behavioral effects of family-selective inhibitors of cyclic nucleotide phosphodiesterases. *Pharmacol Biochem Behav* 63:185–92.

107. Zhang, H.T., and J.M. O'Donnell. 2000. Effects of rolipram on scopolamine-induced impairment of working and reference memory in the radial-arm maze tests in rats. *Psychopharmacology (Berl)* 150:311–6.

108. Perez, J., et al. 2001. Protein kinase A and Rap1 levels in platelets of untreated patients with major depression. *Mol Psychiat* 6:44–9.

109. Perez, J., et al. 2002. cAMP signaling pathway in depressed patients with psychotic features. *Mol Psychiatry* 7:208–12.

110. Duman, R.S., G.R. Heninger, and E.J. Nestler. 1997. A molecular and cellular theory of depression. *Arch Gen Psychiat* 54:597–606.

111. D'Sa, C., and R.S. Duman. 2002. Antidepressants and neuroplasticity. *Bipolar Disord* 4:183–94.

112. O'Donnell, J.M., S. Frith, and J. Wilkins. 1994. Involvement of β-1 and β-2 adrenergic receptors in the antidepressant-like effects of centrally administered isoproterenol. *J Pharmacol Exp Ther* 271:246–54.

113. Zhang, H.T., et al. 2001. Comparison of the effects of isoproterenol administered into the hippocampus, frontal cortex, or amygdala on behavior of rats maintained by differential reinforcement of low response rate. *Psychopharmacology (Berl)* 159:89–97.

114. Engels, P., et al. 1995. Brain distribution of four rat homologues of the *Drosophila* dunce cAMP phosphodiesterase. *J Neurosci Res* 41:169–78.

115. Cherry, J.A., and R.L. Davis. 1999. Cyclic AMP phosphodiesterases are localized in regions of the mouse brain associated with reinforcement, movement, and affect. *J Comp Neurol* 407:287–301.

116. Perez-Torres, S., et al. 2000. Phosphodiesterase type 4 isozymes expression in human brain examined by in situ hybridization histochemistry and [3H]rolipram binding autoradiography. Comparison with monkey and rat brain. *J Chem Neuroanat* 20:349–74.

117. Zhang, H.T., et al. 2002. Antidepressant-like profile and reduced sensitivity to rolipram in mice deficient in the PDE4D phosphodiesterase enzyme. *Neuropsychopharmacology* 27:587–95.

118. Lucki, I. 1997. The forced swimming test as a model for core and component behavioral effects of antidepressant drugs. *Behav Pharmacol* 8:523–32.

119. Porsolt, R.D. 2000. Animal models of depression: Utility for transgenic research. *Rev Neurosci* 11:53–8.

120. Rodgers, R.J., and A. Dalvi. 1997. Anxiety, defence and the elevated plus-maze. *Neurosci Biobehav Rev* 21:801–10.

121. Dunn, A.J., and A.H. Swiergiel. 1999. Behavioral responses to stress are intact in CRF-deficient mice. *Brain Res* 845:14–20.
122. Ye, Y., et al. 1997. Noradrenergic activity differentially regulates the expression of rolipram-sensitive, high-affinity cyclic AMP phosphodiesterase (PDE4) in rat brain. *J Neurochem* 69:2397–404.
123. Suda, S., et al. 1998. Transcriptional and translational regulation of phosphodiesterase type IV isozymes in rat brain by electroconvulsive seizure and antidepressant drug treatment. *J Neurochem* 71:1554–63.
124. Takahashi, M., et al. 1999. Chronic antidepressant administration increases the expression of cAMP-specific phosphodiesterase 4A and 4B isoforms. *J Neurosci* 19:610–8.
125. Ye, Y., K. Jackson, and J.M. O'Donnell. 2000. Effects of repeated antidepressant treatment of type 4A phosphodiesterase (PDE4A) in rat brain. *J Neurochem* 74:1257–62.
126. Miro, X., et al. 2002. Regulation of cAMP phosphodiesterase mRNAs expression in rat brain by acute and chronic fluoxetine treatment. An in situ hybridization study. *Neuropharmacology* 43:1148–57.
127. D'Sa, C., et al. 2005. Differential expression and regulation of the cAMP-selective phosphodiesterase type 4A splice variants in rat brain by chronic antidepressant administration. *Eur J Neurosci* 22:1463–75.
128. Correa-Sales, C., B.C. Rabin, and M. Maze. 1992. A hypnotic response to dexmedetomidine, an alpha 2 agonist, is mediated in the locus coeruleus in rats. *Anesthesiology* 76:948–52.
129. Correa-Sales, C., et al. 1992. Inhibition of adenylate cyclase in the locus coeruleus mediates the hypnotic response to an alpha 2 agonist in the rat. *J Pharmacol Exp Ther* 263:1046–9.
130. Robichaud, A., et al. 2001. PDE4 inhibitors induce emesis in ferrets via a noradrenergic pathway. *Neuropharmacology* 40:262–9.
131. Robichaud, A., et al. 2002. Assessing the emetic potential of PDE4 inhibitors in rats. *Br J Pharmacol* 135:113–8.
132. Robichaud, A., et al. 2002. Deletion of phosphodiesterase 4D in mice shortens α_2-adrenoceptor-mediated anesthesia: A behavioral correlate of emesis. *J Clin Invest* 110:1045–52.
133. Giembycz, M.A. 2005. Life after PDE4: Overcoming adverse events with dual-specificity phosphodiesterase inhibitors. *Curr Opin Pharmacol* 5:238–44.
134. Vandecasteele, G., et al. 1999. Muscarinic and β-adrenergic regulation of heart rate, force of contraction and calcium current is preserved in mice lacking endothelial nitric oxide synthase. *Nat Med* 5:331–4.
135. Verde, I., et al. 1999. Characterization of the cyclic nucleotide phosphodiesterase subtypes involved in the regulation of the L-type Ca^{2+} current in rat ventricular myocytes. *Br J Pharmacol* 127:65–74.
136. Bolger, G.B., et al. 2003. The unique amino-terminal region of the PDE4D5 cAMP phosphodiesterase isoform confers preferential interaction with β-arrestins. *J Biol Chem* 278:49230–8.
137. Perry, S.J., et al. 2002. Targeting of cyclic AMP degradation to beta 2-adrenergic receptors by beta-arrestins. *Science* 298:834–6.
138. Shiina, T., et al. 2000. Interaction with beta-arrestin determines the difference in internalization behavior between beta1- and beta2-adrenergic receptors. *J Biol Chem* 275:29082–90.
139. Lehnart, S.E., X.H. Wehrens, S. Reiken, S. Warrier, A.E. Belevych, R.D. Harvey, W. Richter, S.L. Jin, M. Conti, and A.R. Marks. 2005. Phosphodiesterase 4D deficiency in the ryanodine–receptor complex promotes heart failure and arrhythmias. *Cell* 123:25–35.
140. Marx, S.O., et al. 2000. PKA phosphorylation dissociates FKBP12.6 from the calcium release channel (ryanodine receptor): Defective regulation in failing hearts. *Cell* 101:365–76.
141. Wehrens, X.H., et al. 2003. FKBP12.6 deficiency and defective calcium release channel (ryanodine receptor) function linked to exercise-induced sudden cardiac death. *Cell* 113:829–40.
142. Houslay, M.D., P. Schafer, and K.Y. Zhang. 2005. Keynote review: Phosphodiesterase-4 as a therapeutic target. *Drug Discov Today* 10:1503–19.
143. Kapiloff, M.S., et al. 1999. mAKAP: An A-kinase anchoring protein targeted to the nuclear membrane of differentiated myocytes. *J Cell Sci* 112 (Pt 16):2725–36.
144. Dodge, K.L., et al. 2001. mAKAP assembles a protein kinase A/PDE4 phosphodiesterase cAMP signaling module. *EMBO J* 20:1921–30.

145. Dodge-Kafka, K.L., J. Soughayer, G.C. Pare, J.J. Carlisle Michel, L.K. Langeberg, M.S. Kapiloff, and J.D. Scott. 2005. The protein kinase A anchoring protein mAKAP co-ordinates two integrated cAMP effector pathways. *Nature* 437:574–8.
146. Baillie, G.S., et al. 2000. Sub-family selective actions in the ability of Erk2 MAP kinase to phosphorylate and regulate the activity of PDE4 cyclic AMP-specific phosphodiesterases. *Br J Pharmacol* 131:811–9.
147. Bers, D.M. 2004. Macromolecular complexes regulating cardiac ryanodine receptor function. *J Mol Cell Cardiol* 37:417–29.
148. Tasken, K.A., et al. 2001. Phosphodiesterase 4D and protein kinase a type II constitute a signaling unit in the centrosomal area. *J Biol Chem* 276:21999–2002.
149. Meskini, N., et al. 1992. Early increase in lymphocyte cyclic nucleotide phosphodiesterase activity upon mitogenic activation of human peripheral blood mononuclear cells. *J Cell Physiol* 150:140–8.
150. Hulme, J.T., T. Scheuer, and W.A. Catterall. 2004. Regulation of cardiac ion channels by signaling complexes: Role of modified leucine zipper motifs. *J Mol Cell Cardiol* 37:625–31.
151. Verde, I., et al. 2001. Myomegalin is a novel protein of the golgi/centrosome that interacts with a cyclic nucleotide phosphodiesterase. *J Biol Chem* 276:11189–98.

17 Regulation of cAMP Level by PDE3B—Physiological Implications in Energy Balance and Insulin Secretion

Allan Z. Zhao and Lena Stenson Holst

CONTENTS

17.1 Introduction 347
17.2 The Roles of PDE3B in the Regulation of Energy Balance 348
    17.2.1 Regulation of Feeding at the Hypothalamus 348
    17.2.2 Hypothalamic Effects of cAMP on Food Intake 349
    17.2.3 Regulation of Food Intake by Leptin Requires a
           PI3K-PDE3B-cAMP Signaling Pathway 350
    17.2.4 A Working Model for Hypothalamic Control of Food
           Intake Involving Regulation of cAMP—A Perspective
           from the NPY/AgrP Neurons 352
    17.2.5 Dysregulation of cAMP in the Hypothalamus—Implication
           in Obesity 352
17.3 The Roles of PDE3B in the Regulation of Insulin Secretion—Lessons Learned
    from Cellular Studies and the Rat Insulin Promoter (RIP)-PDE3B
    Transgenic Mice 352
References 357

17.1 INTRODUCTION

The 3′,5′-cyclic adenosine monophosphate (cAMP) is a classic second messenger that is intimately involved in the regulation of food intake and body weight at the hypothalamus as well as insulin secretion from the pancreatic β-cells. In the hypothalamus, cAMP can mediate the orexigenic (gain of appetite) and anorectic (loss of appetite) effects of various peripheral hormones or neuropeptides in a region-specific and neuron-specific manner. In the pancreatic β-cells, cAMP is a classic secretagogue that stimulates insulin secretion. A series of recent studies have demonstrated that PDE3B plays an important role in mediating hormonally regulated energy balance as well as insulin secretion. A detailed account of PDE3B and its characteristics has been given in Chapter 5. In this chapter, we intend to provide an overview of recent advances in characterizing the physiological functions of PDE3B in the central nervous system (CNS) and pancreatic β-cells.

17.2   THE ROLES OF PDE3B IN THE REGULATION OF ENERGY BALANCE

17.2.1   REGULATION OF FEEDING AT THE HYPOTHALAMUS

Abundant evidence has shown that the control of food intake is primarily achieved at the hypothalamus.[1–3] Within the hypothalamus, several regions including the arcuate nucleus (ARC), ventromedial hypothalamus (VMH), dorsomedial hypothalamus (DMH), paraventricular nuclei (PVN), and the lateral hypothalamus (LH) are implicated in the regulation of food intake and body weight.[1] A variety of peripheral hormones have been demonstrated to influence energy homeostasis at these sites; these include leptin, insulin, ghrelin, cholecystokinin (CCK), and glucagon-like peptide-1 (GLP-1).[1,4–8] Among these hormones, leptin represents the long-sought adipose signal responsible for the tight control of feeding behavior and adiposity.[4] The null mutation of leptin or leptin receptor in both rodents and humans can cause severe overfeeding (hyperphagia), gross obesity, stunted linear growth, diabetes, and sterility.[4,9,10] The consistency of these phenotypes as a result of leptin or leptin-receptor mutations in both rodents and humans highlights the importance of this hormone in the long-term maintenance of body weight and energy homeostasis. The aforementioned peripheral signals are received and integrated by hypothalamic neural circuitry, which in turn regulates feeding behavior through a wide array of orexigenic and anorectic neuropeptides,[2,3,11] such as neuropeptide Y (NPY), agouti-related peptide (AgrP), melanin-concentration hormone (MCH), orexins, corticortrophin-releasing hormone (CRH), and α-melanocyte-stimulating hormone (α-MSH). cAMP plays an important role in mediating either the expression or the actions of these neuropeptides. The diagram presented in Figure 17.1 provides a limited and simplified illustration of the neuroanatomical positions of aforementioned hypothalamic regions as well as the target sites of some neuropepetides, particularly in the context of the functional roles of cAMP. The hypothalamic neurons that synthesize and secrete both NPY and AgrP are traditionally called NPY/AgrP neurons, and are almost exclusively localized in the ARC region.[12,13] NPY and AgrP are orexigenic agents, whose expression can be rapidly induced in a negative energy state (low adiposity and low leptin) and subsequently can stimulate food intake.[14–16] In a positive energy state, however, the expression of NPY and AgrP in the ARC is suppressed, which eliminates the stimulatory signals of feeding.[14,17,18] When repeatedly injected into the third ventricle, NPY and AgrP can induce hyperphagia and obesity.[7,19–22] Interestingly, gene targeting of NPY caused mild seizure but did not perturb the feeding behavior. Such observations serve as a reminder that the evolution has put multiple compensatory mechanisms in place to safeguard energy intake.[23] However, NPY-deletion in the *ob/ob* mice did lead to significant reduction in food intake and body weight, suggesting the importance of NPY in leptin-controlled energy homeostasis.[24] The pro-opiomelanocortin (POMC) neurons located in the ARC synthesize and release an anorectic peptide, α-MSH, a proteolytic product from the POMC protein.[25,26] The anorectic function of α-MSH is mediated through two of its cognate receptors, melanocortin-3R and melanocortin-4R (MC3R and MC4R, respectively), both of which are seven transmembrane G-protein-coupled receptors and highly expressed in the PVN region.[25] Gene targeting of MC4R in mice produced hyperphagia and gross obesity[27] that are phenotypically similar to the syndromes observed in the patients carrying homozygous mutations of MC4R.[28,29] Both NPY/AgrP neurons and POMC neurons project to the PVN.[30] The released AgrP serves as a specific antagonist of α-MSH to compete for the binding to the melanocortin receptors, therefore blocking the anorectic action of α-MSH.[31,32] NPY/AgrP neurons also innervate the POMC neurons, which allow NPY to exert an inhibitory effect on the activity of POMC neurons.[33]

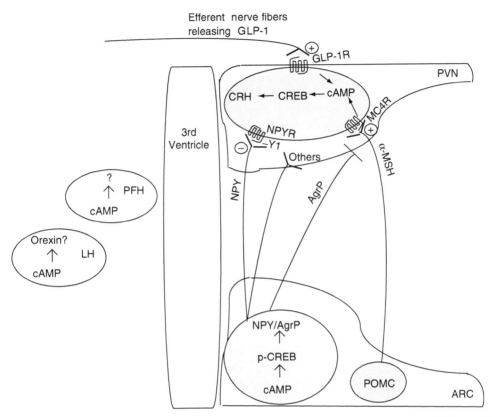

FIGURE 17.1 The differential effects of cAMP on food intake in different regions of hypothalamus. The intracellular cAMP molecules that have a stimulatory effect on food intake have been labeled in red, whereas the cAMP molecules mediating the satiety action have been labeled in black. In this case, an intracerebroventricular (ICV) injection of cAMP or forskolin will elevate the phosphorylation and activities of CREB and subsequently the expression of NPY as well as agouti-related peptide (AgrP). The anorectic peptide, α-MSH, released from the POMC-nerve fibers innervating to the PVN region (with a CRH-neuron as an example), will stimulate the production of cAMP through its Gs-coupled melanocortin receptors (MC-R, 3/4), the levels of phospho-CREB, and consequently the expression levels of anorectic neuropeptides (such as CRH). Similarly, the GLP-1 signals, released from the efferent nerve fibers projecting to the PVN region, also binds to the CRH-neurons and stimulates cAMP production, which will eventually lead to reduced food intake. Studies have shown that site-specific injection of cAMP-analogs in perifornical hypothalamus (PFH) and lateral hypothalamus (LH) also potently increases food intake, although the mechanisms of such actions in the PFH and LH are not yet clear.

17.2.2 Hypothalamic Effects of cAMP on Food Intake

As expected from the complex hypothalamic neural network involved in food-intake control, the roles of cAMP in the regulation of these processes are also multifaceted, and appear to be region-specific and neuron-specific. This topic has been extensively reviewed elsewhere.[34] The discussion here only focuses on how cAMP is regulated in the ARC, PVN, and the LH.

A direct line of evidence of demonstrating the *in vivo* effects of cAMP on the expression of orexigenic neuropeptides in the hypothalamus came from Akabayashi et al.[35] In this case, a cAMP analog, N^6-dibutyryl cAMP, was directly injected into the third ventricles of brain-cannulated rats. Through quantitative imaging analysis, the expression of NPY was found to

be sharply increased only in the ARC and medial parvocellular portion of PVN (mPVN). This region-restricted effect of cAMP on NPY expression appeared to be specific as the expression of another orexigenic peptide, galanin, was not affected at all.[35] Although we cannot rule out the possibility of cAMP stimulating NPY expression via another type of neuron, these observations, when combined with other cellular studies,[36–38] indicate a direct positive role of cAMP on the expression of NPY in the NPY/AgrP neurons in the ARC.

Some of the physiological and pharmacological evidence supporting the role of cAMP as an orexigenic second messenger came from a series of studies by Stanley and his colleagues.[39,40] Injection of 8-bromo-cAMP or cAMP-elevating agents (e.g., forskolin and IBMX) into the perifornical hypothalamus (PFH) and LH produced strong appetite-enhancing effects.[39,40] This orexigenic effect could be alleviated by co-injection of a PKA-inhibitor, H-89. Injection of the same cAMP-elevating reagents in some other hypothalamic sites (e.g., amygdala, PVN, and anterior hypothalamus) did not enhance feeding, which suggested that the effects of the cAMP analog or cAMP-elevating agents were relatively region-specific in the hypothalamus.[39,40] Although it is still not clear how cAMP enhanced feeding in these site-specific injection studies, two potential explanations can be considered (without excluding other possibilities) for these studies: (1) Injection of cAMP in the LH may have stimulated the expression and secretion of orexins that have been shown to have orexigenic effects.[41] (2) It is possible that the cAMP-analog or cAMP-elevating agents were not just confined to the specific injection sites; rather, they were able to diffuse into some adjacent regions. For example, diffusion of these agents into the ARC would be expected to stimulate the expression of NPY and AgrP and subsequently the food intake.

17.2.3   Regulation of Food Intake by Leptin Requires a PI3K-PDE3B-cAMP Signaling Pathway

Soon after the identification of the leptin receptors, it was realized that they belong to the class I cytokine receptor family.[42,43] Consequently, the janus kinase (JAK)–STAT pathway (STAT, signal transducers and activator of transcription) should be an integral part of leptin's signaling and functions, particularly for the so-called long-form variant of the leptin receptors, OB-Rb. OB-Rb is predominantly expressed in the hypothalamus.[9,43–45] Indeed, the importance of the JAK–STAT pathway is well demonstrated by the phenotypes of db/db mice in which leptin-induced JAK–STAT signaling is absent due to the lack of OB-Rb expression.[46,47] Perhaps as a further highlight of the importance of this pathway in leptin's central actions, mutant mice carrying a knock-in mutation in OB-Rb that renders it incapable of activating STAT3 developed severe hyperphagia and gross obesity.[48]

However, the JAK–STAT pathway is not the only signaling mechanism for the physiological functions of leptin. The concept of cAMP regulation by leptin was initially established in the peripheral systems including the liver and pancreatic β-cells.[49–51] In these cases, leptin triggers a PI3-kinase-dependent activation of PDE3B, which causes reduction of intracellular cAMP.[49,51] The physiological significance of this signaling mechanism is reflected by the fact that leptin can suppress GLP-1-stimulated insulin secretion and glucagon-stimulated glycogenolysis.[49–55] Only recently did evidence begin to emerge to show that this signaling scheme occurs not only in the peripheral nervous system (PNS) but also in the CNS with a significant impact on feeding behavior.

With a transgenic mouse model carrying a cAMP-responsive element (CRE)-driven *lacZ* gene, Shimizu-Albergine and colleagues found that the expression of the *lacZ* gene was drastically elevated in NPY/AgrP neurons in the ARC during fasting, which was also concomitant with the increase of NPY expression.[56] Because elevated expression of the reporter *lacZ* is often caused by increased cAMP levels, the intensity of its transcripts was

used as an indirect indicator of cAMP levels *in vivo*.[56] Interestingly, this fasting-induced signal was restricted to the NPY/AgrP neurons and was not found in the POMC neurons.[56] ICV injection of leptin reduced *lacZ* expression during fasting.[56] Importantly, injection of an inhibitor of PDE3 elevated the levels of *lacZ* transcript and this increase was restricted to the NPY/AgrP neurons in the ARC.[56] These data were reminiscent of similar findings in the periphery, and suggested that leptin may also activate the PDE3B-cAMP signaling pathway in the hypothalamus.

More definitive functional proof of this signaling mechanism in leptin-regulated energy homeostasis came from several recent studies. Our recent study has found that ICV injection of leptin significantly stimulated the activities of PI3-kinase as well as PDE3B and reduced cAMP levels in the hypothalamus.[57] Importantly, pharmacological blockade of the PDE3 activity with a selective PDE3 inhibitor, cilostamide, completely blunted the satiety and weight-reducing effects of leptin.[57] In an independent study, Schwartz and his colleagues examined the upstream component of this pathway by the administration of ICV injection to the inhibitors of PI3-kinase, wortmannin and LY294002. Both inhibitors of PI3K blocked the satiety actions of leptin.[58] Thus, the PI3K-PDE3B-cAMP signaling pathway is very likely to be an essential mechanism for leptin not only in the periphery but also in the hypothalamus. Interestingly, counterregulation of cAMP-induced cellular functions by leptin has also been found outside the context of food-intake control. Quintela et al. have recently demonstrated that leptin treatment of fetal rat neurons could completely block cAMP-induced somatostatin release.[59]

It is important to point out that the PI3K-PDE3B-cAMP pathway is not an isolated signaling mechanism for leptin. Rather, it is found to cross regulate the activity of JAK–STAT3 pathway as pharmacological blockade of PDE3B activity in the hypothalamus could eliminate leptin-stimulated phosphorylation and DNA-binding activity of STAT3.[57] These findings, although primarily based on pharmacological evidence, suggest that a reduction of cAMP through PI3K-dependent activation of PDE3B is a prerequisite step for the satiety and weight-reducing actions of leptin in the hypothalamus. Potentially, regulation of cAMP through the PI3K–PDE3B pathway can be a broad signaling theme for food-intake control. Recent studies by Schwartz and his colleagues have found that the satiety effect of insulin is also dependent on the activity of PI3-kinase.[60] In the periphery, activation of PDE3B and the subsequent reduction of cAMP by insulin have been well demonstrated to mediate the antilipolytic or antiglycogenolytic effects of insulin in the fat and liver, respectively.[61] Indeed, our recent studies have found that ICV co-injection of a PDE3 inhibitor with insulin could completely block the satiety effect of insulin (unpublished observation). The evidence gathered so far suggests that the PI3K-PDE3B-cAMP pathway is a critical signaling mechanism in the broad scheme of food-intake control. However, a complete validation of this concept will require more molecular genetic evidence, such as a CNS-specific conditional knockout of the *PDE3B* gene. In addition, PDE3A, which is structurally related to PDE3B, is also highly expressed in the CNS.[62] It remains to be determined if PDE3A might also be involved in mediating the satiety actions of leptin.

A critical issue following these studies remains to be resolved. If fasting elevates cAMP levels in the NPY/AgrP neurons, what are the hormones that are responsible for the increase of cAMP in the NP/AgrP neurons? At this stage, there is no definitive evidence to pinpoint the specific responsible harmones. Recent evidence suggests that ghrelin, an appetite-stimulating peptide hormone primarily secreted from stomach, is one strong candidate responsible for initiating the cAMP increase in the NPY/AgrP neurons. Ghrelin receptors have been mapped to the NPY/AgrP neurons, but not to the POMC neurons, which are similar to the distribution of PDE3B activity.[63–65] However, it remains to be determined if other peripheral hormones may exert similar functions in the ARC/AgrP neurons.

17.2.4 A Working Model for Hypothalamic Control of Food Intake Involving Regulation of cAMP—A Perspective from the NPY/AgrP Neurons

The evidence summarized so far strongly suggests that regulation of cAMP level at the hypothalamus is critical to the control of energy homeostasis. From the available physiological and molecular evidence, the following working model is proposed to illustrate how modulation of cAMP levels in the ARC of the hypothalamus dictates hormonal regulation of energy homeostasis (Figure 17.2A and Figure 17.2B).

In a negative energy balance state (such as fasting), the plasma concentrations of peripheral hormones like ghrelin will be elevated, which in turn will be sensed by the hypothalamus, particularly through the interaction of ghrelin with its receptor (GHS-R) on the NPY/AgrP neurons in the ARC. Upon ghrelin binding, the levels of cAMP as well as the activity of PKA will increase and consequently stimulate the expression of these two orexigenic peptides. The increased release of NPY will enhance food intake both by directly acting in the PVN and by suppressing the activity of POMC neurons. On a separate front, the secreted AgrP will act as an antagonist of α-MSH on the melanocortin receptors-3 and 4 (MC3R and MC4R) to neutralize the anorectic effect of α-MSH. As a result of these combined actions, food intake will be initiated.

In a positive energy balance state, circulating leptin and insulin signals will both rise, which in turn will be integrated by the neurons in several regions of the hypothalamus including the ARC. Upon leptin or insulin binding, the PDE3B activity in the NPY/AgrP neurons will be stimulated following the activation of PI3-kinase in the same cell population. The subsequent reduction of cAMP will cause a decrease in PKA activity as well as the phosphorylation of CREB. This decline of cAMP will suppress the expression of NPY and AgrP and thus decrease the orexigenic signals in the PVN. The reduction of cAMP levels will also relieve the inhibitory pressure on the activation of transcription factor STAT3 in the NPY or POMC neurons, which will then lead to increased production and release of the anorectic α-MSH. The binding of α-MSH to its PVN target neurons will increase cAMP synthesis through G_s-coupled MC-R, further stimulating the expression of other anorectic neuropeptides, such as CRH.

17.2.5 Dysregulation of cAMP in the Hypothalamus—Implication in Obesity

Recent studies have made us realize the importance of modulating cAMP levels within the complex hypothalamic neural network to the overall scheme of food-intake control and body weight regulation. Future studies should be directed toward understanding if cAMP regulation in the hypothalamus may have gone awry in obesity and type-2 diabetes. We already know that mutations that reduce or nullify cAMP production in the MC4R can cause an early onset of obesity and an eating disorder in humans.[66] Another recent study has found that leptin's ability to activate PDE3B is significantly diminished in the hypothalamic tissues of hyperleptinemic rats.[67] However, more genetic and physiological studies are needed, particularly those with neuron-specific knockout models, to completely elucidate the impact of cAMP-regulating pathways to the long-term control of food intake and body weight.

17.3 THE ROLES OF PDE3B IN THE REGULATION OF INSULIN SECRETION—LESSONS LEARNED FROM CELLULAR STUDIES AND THE RAT INSULIN PROMOTER (RIP)-PDE3B TRANSGENIC MICE

The presence of a PDE activity in insulin-secreting cells is long known, although neither the complexity of this enzyme nor the regulation of cyclic nucleotides in these cells has been extensively explored. Initially, the discovery of the insulinotropic effects of cAMP-elevating

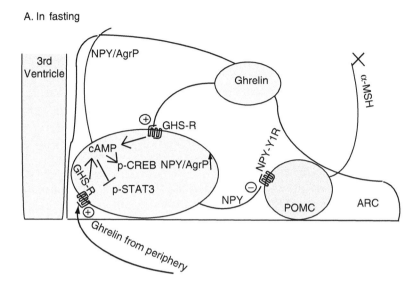

FIGURE 17.2 (A and B). A working model of hypothalamic regulation of food intake involving modulation of intracellular cAMP levels—a perspective from the arcuate nucleus (ARC). (A) In fasting, the increased plasma concentrations of ghrelin will be sensed by the GHS-receptors in the NPY/AgrP neurons. The ghrelin neurons within the hypothalamus also send out ghrelin to the NPY/AgrP neurons. The subsequent increase of cAMP in the NPY-neurons will stimulate the expression of orexigenic NPY and AgrP. The increased release of NPY will inhibit the neuronal activity of POMC neurons through the Y1-receptors of NPY (NPY-Y1R). Elevation of intracellular cAMP level will also have an inhibitory effect on the tyrosine-phosphorylation and activities of STAT3. (B) In refeeding, the rising plasma leptin signals will be integrated into the NPY as well as POMC neurons by the leptin receptors (OBR). Upon leptin binding, PI3-kinase (PI3K) and PDE3B will be activated in the NPY/AgrP neurons. The subsequent reduction of cAMP will not only shut down the expression of orexigenic NPY and AgrP but also relieve the inhibitory pressure on STAT3-phosphorylation. The decreased expression of NPY should also diminish the negative effect of NPY on POMC neurons as well as the antagonistic effect of AgrP on melanocortin receptors in the target neurons for α-melanocyte-stimulating hormone (α-MSH). Leptin also induces tyrosine-phosphorylation on STAT3 in the POMC neurons, which is expected to increase the expression of the POMC gene.

agents (e.g., forskolin) and the recognition of isobutylmethylxanthine (IBMX, a general PDE inhibitor) as a powerful amplifier of insulin secretion strongly suggested a role for cAMP in pancreatic β-cell function.[68–71] Even though it was later shown that a rise in cAMP or activation of protein kinase A is not mandatory for glucose-stimulated insulin secretion,[72,73] a vital role in the potentiation of insulin secretion by the cAMP-stimulating hormones, such as GLP-1 and glucose-dependent insulinotropic polypeptide (GIP), is undisputed. In addition to a beneficial role in insulin secretion, cAMP also positively contributes to insulin biosynthesis, β-cell growth, and survival.[74–76]

Up to now, three PDE families are known to be expressed in pancreatic β-cells, PDE1C,[77] PDE3B,[49,78] and PDE4.[79,80] It was recently reported that a combination of inhibitors selective for these three PDE families inhibits cAMP-PDE activity in clonal β-cells by around 90%,[81] suggesting that they are the dominating cAMP-degrading enzymes in β-cells. The relative contribution of these PDEs to insulin secretory events is not yet fully understood, although most data evidence favors PDE3B as the principal enzyme in this regard.[77,79,80] For example, several different PDE3-selective inhibitors can increase cAMP and enhance insulin secretion in various models, in the absence or presence of stimulatory glucose concentrations. Moreover, reducing cAMP by adenoviral overexpression of PDE3B in rat islets and INS-1 cells has been shown to significantly impair secretion and exocytosis of insulin.[82] Recent cellular studies have also demonstrated that PDE3B plays a critical role in mediating the inhibitory effects of leptin and IGF-1 on GLP-1-stimulated insulin secretion.[49,78] Figure 17.3 schematically shows the suggested mode of action of PDE3B in insulin secretion.

In an attempt to further explore the effects of cAMP reduction by overexpression of PDE3B on insulin secretion in a genetic setting, a rat insulin promoter (RIP)-driven PDE3B cDNA was used to raise transgenic mice on a C57Bl/6 × CBA background.[83] Two lines of transgenic mice exhibited β-cell-specific expression of RIP-PDE3, the RIP-PDE3B/2 and the RIP-PDE3B/7 mice overexpressing PDE3B 2–3 fold and 7–10 fold over that in the wild-type littermates, respectively.[83] In intravenous glucose tolerance testing, at the age of 4 months, a gene dose-dependent difference was seen in that RIP-PDE3B/2 mice exhibited an approximately 50% reduction of acute (within 1 min after iv injection of glucose) insulin response, whereas the secretory response of RIP-PDE3B/7 mice to the same treatment (within the same time frame) was virtually absent. Other similar dose differences have been noted both in *in vitro* and *in vivo* analyses of these two lines, strongly indicating that the effects are indeed caused by the PDE3B transgene. For reasons unclear, the phenotypes in female mice were significantly weaker than in male mice; thus, all studies so far have been conducted with the male mice.

In general agreement with the results from overexpressing PDE3B in *in vitro* models (e.g., insulinoma cells), islets isolated from both lines of RIP-PDE3B mice showed an impaired secretory response to stimulatory concentrations of glucose.[83] In accordance with previous *in vitro* data,[49,78] this impairment was most prominent when insulin secretion was enhanced by stimulation with the cAMP-elevating hormone GLP-1 in the presence of glucose. Interestingly, in intravenous glucose tolerance tests (IVGTT), the RIP-PDE3B mice also exhibited significant glucose intolerance, in addition to the dysfunction at the β-cell level. The effect was very prominent in the RIP-PDE3B/7 mice, i.e., the mice exhibiting the high level of PDE3B overexpression in β-cells. However, the RIP-PDE3B/2 mice, with relatively lower transgene expression, required a high-fat diet regimen (58% fat, calculated as energy %) to show impaired glucose disposal (see below). In IVGTT analyses of the wild-type littermates and RIP-PDE3B/2 mice at the age of 10 weeks and after high-fat feeding during 6 weeks, mice of both genotypes showed significant glucose intolerance with no apparent differences between the two lines. However, to acquire this level of glucose elimination, the high-fat-fed RIP-PDE3B/2 mice had to release more insulin relative to high-fat-fed wild-type littermates.

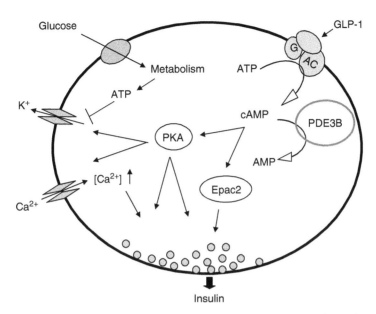

FIGURE 17.3 Suggested function of PDE3B in the regulation of insulin secretion. The main stimulator of insulin secretion is glucose, which is metabolized inside the β-cell. The subsequent increase in ATP/ADP ratio causes closure of K_{ATP}-dependent ion channels, resulting in depolarization of the plasma membrane. In consequence, L-type Ca^{2+} channels are opened, leading to influx of Ca^{2+}. The increased intracellular concentration of Ca^{2+} stimulates exocytosis of insulin. cAMP initiates processes that enhance insulin secretion, including activation of protein kinase A (PKA) and binding to exchange protein activated by cAMP 2 (Epac2). Exocytosis of insulin is then stimulated through multiple pathways, only a few of which are yet established. Glucagon-like peptide (GLP)-1 is an insulinotropic gut hormone, which acts through a G-protein-coupled receptor to activate adenylyl cyclase (AC) and thereby an increase in cAMP. PDE3B, in turn, has been shown to negatively regulate insulin secretion through its cAMP-hydrolyzing activity.

Specifically, although the acute insulin response in high-fat-fed RIP-PDE3B/2 mice was fourfold less than that in standard-diet-fed RIP-PDE3B/2 mice, it was compared to the high-fat-diet fed wild-type littermates. Together with a significant fasting hyperinsulinemia, the high-fat-fed RIP-PDE3B/2 mice clearly showed an early sign of insulin resistance. After 10 weeks of high-fat feeding, the islets of RIP-PDE3B/2 mice failed to increase insulin release adequately in IVGTT; additionally, the ability of insulin to dispose blood glucose was also severely impaired as revealed in an insulin tolerance test. These phenotypes were in sharp contrast to the wild-type mice, which displayed only moderate glucose intolerance and hyperinsulinemia on the same diet regimen. Hence, the RIP-PDE3B/2 mice appeared to be more vulnerable than the wild-type mice to the diabetogenic treatment in terms of progressive development of insulin resistance. The sites of the insulin resistance in one or several peripheral tissues remain to be defined. Preliminary data show that insulin-stimulated glucose uptake in soleus muscle and insulin-dependent lipogenesis in the adipose tissues of these mice occur to the same extent as in the wild-type mice. However, the livers of the RIP-PDE3B/2 mice fed the high-fat diet were found to be significantly more steatotic than livers of their wild-type littermates on the same diet. Because steatosis of the liver is known to be associated with hepatic insulin resistance, it is likely that the liver is a site for the insulin resistance of RIP-PDE3B/2 mice.

The reason for the early appearance of insulin resistance in high-fat-treated RIP-PDE3B/2 mice is not clear, but a connection with the increased serum levels of free fatty acids following the high-fat feeding is plausible. Dysregulation of fatty acid metabolism is known to contribute to the development of insulin resistance.[84] Although hyperleptinemia was evident in RIP-PDE3B/2 mice and their body weight was significantly higher than that of wild-type mice after a couple of weeks on the high-fat diet, the increased weight did not comprise a relative increase in fat mass but was found to represent a general rise in body weight. Furthermore, although circulating free fatty acids were elevated due to high-fat feeding, this occurred to a similar extent in transgenic and wild-type mice. Therefore, elevated circulating free fatty acids per se appear not to be the discriminating factor explaining the early and more severe insulin resistance in the RIP-PDE3B/2 mice compared to their wild-type littermates. However, in the transgenic mice, where fatty acids are elevated in a setting of impaired β-cell functions due to dysregulation of cAMP, a synergistic effect is a likely outcome; fatty acid–induced alterations in gene expression and enzyme activities in the β-cells may further aggravate the transgene-induced impairment of insulin secretion.

RIP-PDE3B/2 mice also showed early hyperglucagonemia when fed a high-fat diet. This was possibly related to a defect at the islet level; glucagon-producing α-cells were frequently seen located to the centre of transgenic islets, in contrast to a typical normal islet showing α-cells at the peripheral rim. Altered cell topography and elevated plasma glucagon is commonly associated with insulin resistance and the latter accounts in part for the increased hepatic glucose production in type 2 diabetes. Thus, the early glucose intolerance and hyperglycemia observed in high-fat-fed RIP-PDE3B/2 mice may be a combined effect of a transgene-dependent insulin secretory defect, which was further aggravated by elevated fatty acids from the high-fat diet. Furthermore, the hyperglycemic condition might synergize with the toxic effects of free fatty acids to create a severe glucolipotoxic condition, which in turn is expected to precipitate insulin resistance.

Studies of the RIP-PDE3B/7 mice, which exhibited a higher overexpression of PDE3B (than the RIP-PDE3B/2 mice), provide additional understanding of the RIP-PDE3B phenotype. For example, in IVGTT, although these mice were barely capable of responding to the glucose load, insulin was released in response to injections of GLP-1 together with glucose. However, the immediate insulin release was only ~50% of the wild-type response, an amount that appeared not sufficient to substantially improve plasma glucose disposal.

Glucose-induced insulin secretion is known to comprise two phases. This is most easily seen in experiments where isolated islets are perfused with glucose-containing buffer. The first burst of secretion is immediate, within 1–2 min, whereas the second burst is at around 10–15 min. Generally, the rapid first phase is interpreted as a result of emptying of the so-called readily releasable pool of insulin granules. The second phase, characterized by more sustained release, is understood as a result of refilling or mobilization of more remote granules. Perfusion of islets from RIP-PDE3B/7 mice demonstrated that it was primarily the first phase of glucose-stimulated insulin secretion that was impaired, indicating that the site of action of PDE3B is connected to the granules prepared for release.[83] Future investigations are still required to increase our understanding of the site of PDE3B actions in the pancreatic β-cells. We also want to caution that these conclusions were derived from a transgenic system designed for targeted overexpression of PDE3B. The specificity of these purported PDE3B functions in the β-cells derived from this transgenic model should be confirmed by evaluating the transgenic mice with similarly targeted expression of other types of PDEs (e.g., RIP-PDE7).

In summary, the transgenic RIP-PDE3B mouse model has been useful in shedding new light on the physiological significance of regulation of cAMP by PDE3B in pancreatic β-cells, particularly with respect to its roles in glucose- and GLP-1-stimulated insulin secretion. Referring to the discussion on the necessity of cAMP for nutrient-stimulated

insulin secretion, it is clear from this model that a seemingly moderate dysregulation of cAMP in pancreatic β-cells does negatively influence insulin secretion, to the extent that it affects glucose homeostasis. Furthermore, targeted PDE3B overexpression also sensitized mice to high-fat feeding so as to precipitate the diabetes-like symptoms. In adapting to a condition where insulin resistance is developing, functional compensatory increase in insulin secretion is essential. The genetic studies from the RIP-PDE3B mouse model also suggest that cAMP is important in preventing or delaying the development of fatty diet–induced insulin resistance.

REFERENCES

1. Schwartz, M.W., S.C. Woods, D. Porte Jr., R.J. Seeley, and D.G. Baskin. 2000. Central nervous system control of food intake. *Nature* 404 (6778):661–71.
2. Woods, S.C., M.W. Schwartz, D.G. Baskin, and R.J. Seeley. 2000. Food intake and the regulation of body weight. *Annu Rev Psychol* 51:255–77.
3. Williams, G., C. Bing, X.J. Cai, J.A. Harrold, P.J. King, and X.H. Liu. 2001. The hypothalamus and the control of energy homeostasis: Different circuits, different purposes. *Physiol Behav* 74 (4–5): 683–701.
4. Zhang, Y., R. Proenca, M. Maffei, M. Barone, L. Leopold, and J.M. Friedman. 1994. Positional cloning of the mouse obese gene and its human homologue [published erratum appears in *Nature* 1995 Mar 30; 374 (6521):479] (see comments). *Nature* 372 (6505):425–32.
5. Baskin, D.G., D. Figlewicz Lattemann, R.J. Seeley, S.C. Woods, D. Porte, Jr., and M.W. Schwartz. 1999. Insulin and leptin: Dual adiposity signals to the brain for the regulation of food intake and body weight. *Brain Res* 848 (1–2):114–23.
6. Tschop, M., M.A. Statnick, T.M. Suter, and M.L. Heiman. 2002. GH-releasing peptide-2 increases fat mass in mice lacking NPY: Indication for a crucial mediating role of hypothalamic agouti-related protein. *Endocrinology* 143 (2):558–68.
7. Tang-Christensen, M., N. Vrang, S. Ortmann, M. Bidlingmaier, T.L. Horvath, and M. Tschop. 2004. Central administration of ghrelin and agouti-related protein (83–132) increases food intake and decreases spontaneous locomotor activity in rats. *Endocrinology* 145 (10):4645–52.
8. Moran, T.H. 2000. Cholecystokinin and satiety: Current perspectives. *Nutrition* 16 (10):858–65.
9. Lee, G.H., R. Proenca, J.M. Montez, K.M. Carroll, J.G. Darvishzadeh, J.I. Lee, and J.M. Friedman. 1996. Abnormal splicing of the leptin receptor in diabetic mice. *Nature* 379 (6566):632–5.
10. Montague, C.T., I.S. Farooqi, J.P. Whitehead, M.A. Soos, H. Rau, N.J. Wareham, C.P. Sewter, J.E. Digby, S.N. Mohammed, J.A. Hurst, C.H. Cheetham, A.R. Earley, A.H. Barnett, J.B. Prins, and S. O'Rahilly. 1997. Congenital leptin deficiency is associated with severe early-onset obesity in humans. *Nature* 387 (6636):903–8.
11. Woods, S.C., R.J. Seeley, D. Porte, Jr., and M.W. Schwartz. 1998. Signals that regulate food intake and energy homeostasis. *Science* 280 (5368):1378–83.
12. Bi, S., B.M. Robinson, and T.H. Moran. 2003. Acute food deprivation and chronic food restriction differentially affect hypothalamic NPY mRNA expression. *Am J Physiol Regul Integr Comp Physiol* 285 (5):R1030–6.
13. Wolden-Hanson, T., B.T. Marck, and A.M. Matsumoto. 2004. Blunted hypothalamic neuropeptide gene expression in response to fasting, but preservation of feeding responses to AgrP in aging male Brown Norway rats. *Am J Physiol Regul Integr Comp Physiol* 287 (1):R138–46.
14. Schwartz, M.W., D.G. Baskin, T.R. Bukowski, J.L. Kuijper, D. Foster, G. Lasser, D.E. Prunkard, D. Porte, Jr., S.C. Woods, R.J. Seeley, and D.S. Weigle. 1996. Specificity of leptin action on elevated blood glucose levels and hypothalamic neuropeptide Y gene expression in ob/ob mice. *Diabetes* 45 (4):531–5.
15. Bertile, F., H. Oudart, F. Criscuolo, Y.L. Maho, and T. Raclot. 2003. Hypothalamic gene expression in long-term fasted rats: Relationship with body fat. *Biochem Biophys Res Commun* 303 (4): 1106–13.

16. Abbott, C.R., A.R. Kennedy, A.M. Wren, M. Rossi, K.G. Murphy, L.J. Seal, J.F. Todd, M.A. Ghatei, C.J. Small, and S.R. Bloom. Identification of hypothalamic nuclei involved in the orexigenic effect of melanin-concentrating hormone. *Endocrinology* 144 (9):3943–9.

17. Elias, C.F., C. Aschkenasi, C. Lee, J. Kelly, R.S. Ahima, C. Bjorbaek, J.S. Flier, C.B. Saper, and J.K. Elmquist. 1999. Leptin differentially regulates NPY and POMC neurons projecting to the lateral hypothalamic area. *Neuron* 23 (4):775–86.

18. Swart, I., J.W. Jahng, J.M. Overton, and T.A. Houpt. 2002. Hypothalamic NPY, AgrP, and pomc mrna responses to leptin and refeeding in mice. *am j physiol regul integr comp physiol* 283 (5):r1020–6.

19. McMinn, J.E., R.J. Seeley, C.W. Wilkinson, P.J. Havel, S.C. Woods, and M.W. Schwartz. 1998. NPY-induced overfeeding suppresses hypothalamic NPY mRNA expression: Potential roles of plasma insulin and leptin. *Regul Pept* 75–76:425–31.

20. Vettor, R., N. Zarjevski, I. Cusin, F. Rohner-Jeanrenaud, and B. Jeanrenaud. 1994. Induction and reversibility of an obesity syndrome by intracerebroventricular neuropeptide Y administration to normal rats. *Diabetologia* 37 (12):1202–8.

21. Paez, X., and R.D. Myers. 1991. Insatiable feeding evoked in rats by recurrent perfusion of neuropeptide Y in the hypothalamus. *Peptides* 12 (3):609–16.

22. Paez, X., J.W. Nyce, and R.D. Myers. 1991. Differential feeding responses evoked in the rat by NPY and NPY1–27 injected intracerebroventricularly. *Pharmacol Biochem Behav* 38 (2):379–84.

23. Erickson, J.C., K.E. Clegg, and R.D. Palmiter. 1996. Sensitivity to leptin and susceptibility to seizures of mice lacking neuropeptide Y. *Nature* 381 (6581):415–21.

24. Erickson, J.C., G. Hollopeter, and R.D. Palmiter. 1996. Attenuation of the obesity syndrome of ob/ob mice by the loss of neuropeptide Y. *Science* 274 (5293):1704–7.

25. Cone, R.D., M.A. Cowley, A.A. Butler, W. Fan, D.L. Marks, and M.J. Low. 2001. The arcuate nucleus as a conduit for diverse signals relevant to energy homeostasis. *Int J Obes Relat Metab Disord* 25 (Suppl 5): S63–7.

26. Ellacott, K.L., and R.D. Cone. 2004. The central melanocortin system and the integration of short- and long-term regulators of energy homeostasis. *Recent Prog Horm Res* 59:395–408.

27. Huszar, D., C.A. Lynch, V.Fairchild-Huntress, J.H. Dunmore, Q. Fang, L.R. Berkemeier, W. Gu, R.A. Kesterson, B.A. Boston, R.D. Cone, F.J. Smith, L.A. Campfield, P. Burn, and F. Lee. 1997. Targeted disruption of the melanocortin-4 receptor results in obesity in mice. *Cell* 88 (1):131–41.

28. Vaisse, C., K. Clement, E. Durand, S. Hercberg, B. Guy-Grand, and P. Froguel. 2000. Melano-cortin-4 receptor mutations are a frequent and heterogeneous cause of morbid obesity. *J Clin Invest* 106 (2):253–62.

29. Tao, Y.X., and D.L. Segaloff. 2003. Functional characterization of melanocortin-4 receptor muta-tions associated with childhood obesity. *Endocrinology* 144 (10):4544–51.

30. Cowley, M.A., N. Pronchuk, W. Fan, D.M. Dinulescu, W.F. Colmers, and R.D. Cone. 1999. Integration of NPY, AgrP, and melanocortin signals in the hypothalamic paraventricular nucleus: Evidence of a cellular basis for the adipostat. *Neuron* 24 (1):155–63.

31. Mizuno, T.M., H. Makimura, and C.V. Mobbs. 2003. The physiological function of the agouti-related peptide gene: The control of weight and metabolic rate. *Ann Med* 35 (6):425–33.

32. Doghman, M., P. Delagrange, A. Blondet, M.C. Berthelon, P. Durand, D. Naville, and M. Begeot. 2004. Agouti-related protein antagonizes glucocorticoid production induced through melanocortin 4 receptor activation in bovine adrenal cells: A possible autocrine control. *Endocrinology* 145 (2): 541–7.

33. Cowley, M.A., J.L. Smart, M. Rubinstein, M.G. Cerdan, S. Diano, T.L. Horvath, R.D. Cone, and M.J. Low. 2001. Leptin activates anorexigenic POMC neurons through a neural network in the arcuate nucleus. *Nature* 411 (6836):480–4.

34. Zhao, A.Z. 2005. Control of food intake through regulation of cAMP *curr top dev biol* 67: 207–24.

35. Akabayashi, A., C.T. Zaia, S.M. Gabriel, I. Silva, W.K. Cheung. and S.F. Leibowitz. 1994. Intracerebroventricular injection of dibutyryl cyclic adenosine 3′,5′-monophosphate increases hypo-thalamic levels of neuropeptide Y. *Brain Res* 660 (2):323–8.

36. Liu, J., A.I. Kahri, P. Heikkila, and R. Voutilainen. 1999. Regulation of neuropeptide Y mRNA expression in cultured human pheochromocytoma cells. *Eur J Endocrinol* 141 (4):431–5.
37. Pance, A., D. Balbi, N. Holliday, and J.M. Allen. 1995. Effect of cAMP elevation on the NPY gene transcription. *Biochem Soc Trans* 23 (1):47S.
38. May, V., C.A. Brandenburg, and K.M. Braas. 1995. Differential regulation of sympathetic neuron neuropeptide Y and catecholamine content and secretion. *J Neurosci* 15 (6):4580–91.
39. Gillard, E.R., A.M. Khan, U.H. Ahsan, R.S. Grewal, B. Mouradi, and B.G. Stanley. 1997. Stimulation of eating by the second messenger cAMP in the perifornical and lateral hypothalamus. *Am J Physiol* 273: R107–12.
40. Gillard, E.R., A.M. Khan, R.S. Grewal, B. Mouradi, S.D. Wolfsohn, and B.G. Stanley. 1998. The second messenger cAMP elicits eating by an anatomically specific action in the perifornical hypothalamus. *J Neurosci* 18 (7):2646–52.
41. Wolf, G. 1998. Orexins: A newly discovered family of hypothalamic regulators of food intake. *Nutr Rev* 56 (6):172–3.
42. White, D.W., and L.A. Tartaglia. 1996. Leptin and OB-R: Body weight regulation by a cytokine receptor. *Cytokine Growth Factor Rev* 7 (4):303–9.
43. Baumann, H., K.K. Morella, D.W. White, M. Dembski, P.S. Bailon, H. Kim, C.F. Lai, and L.A. Tartaglia. 1996. The full-length leptin receptor has signaling capabilities of interleukin 6-type cytokine receptors. *Proc Natl Acad Sci USA* 93 (16):8374–8.
44. Baskin, D.G., M.W. Schwartz, R.J. Seeley, S.C. Woods, D. Porte Jr., J.F. Breininger, Z. Jonak, J. Schaefer, M. Krouse, C. Burghardt, L.A. Campfield, P. Burn, and J.P. Kochan. 1999. Leptin receptor long-form splice-variant protein expression in neuron cell bodies of the brain and co-localization with neuropeptide Y mRNA in the arcuate nucleus. *J Histochem Cytochem* 47 (3): 353–62.
45. Tartaglia, L.A., M. Dembski, X. Weng, N. Deng, J. Culpepper, R. Devos, G.J. Richards, L.A. Campfield, F.T. Clark, J. Deeds, C. Muir, S. Sanker, A. Moriarty, K.J. Moore, J.S. Smutko, G.G. Mays, E.A. Wool, C.A. Monroe, and R.I. Tepper. 1995. Identification and expression cloning of a leptin receptor. OB-R, *Cell* 83 (7):1263–71.
46. Vaisse, C., J.L. Halaas, C.M. Horvath, J.E.J. Darnell, M. Stoffel, and J.M. Friedman. 1996. Leptin activation of Stat3 in the hypothalamus of wild-type and *ob/ob* mice but not *db/db* mice. *Nat Genet* 14 (1):95–7.
47. Ghilardi, N., S. Ziegler, A. Wiestner, R. Stoffel, M.H. Heim, and R.C. Skoda. 1996. Defective STAT signaling by the leptin receptor in diabetic mice. *Proc Natl Acad Sci USA* 93 (13):6231–5.
48. Bates, S.H., W.H. Stearns, T.A. Dundon, M. Schubert, A.W. Tso, Y. Wang, A.S. Banks, H.J. Lavery, A.K. Haq, E. Maratos-Flier, B.G. Neel, M.W. Schwartz, and M.G. Myers Jr. 2003. STAT3 signalling is required for leptin regulation of energy balance but not reproduction. *Nature* 421 (6925):856–9.
49. Zhao, A.Z., K.E. Bornfeldt, and J.A. Beavo. 1998. Leptin inhibits insulin secretion by activation of phosphodiesterase 3B. *J Clin Invest* 102 (5):869–73.
50. Fehmann, H.C., H.P. Bode, T. Ebert, A. Karl, and B. Goke. 1997. Interaction of GLP-I and leptin at rat pancreatic β-cells: Effects on insulin secretion and signal transduction. *Horm Metab Res* 29 (11):572–6.
51. Zhao, A.Z., M.M. Shinohara, D. Huang, M. Shimizu, H. Eldar-Finkelman, E.G. Krebs, J.A. Beavo, and K.E. Bornfeldt. 2000. Leptin induces insulin-like signaling that antagonizes cAMP elevation by glucagon in hepatocytes. *J Biol Chem* 275 (15):11348–54.
52. Nemecz, M., K. Preininger, R. Englisch, C. Furnsinn, B. Schneider, W. Waldhausl, and M. Roden. 1999. Acute effect of leptin on hepatic glycogenolysis and gluconeogenesis in perfused rat liver. *Hepatology* 29 (1):166–72.
53. Ceddia, R.B., G. Lopes, H.M. Souza, G.R. Borba-Murad, W.N. William Jr. R.B. Bazotte, and R. Curi. 1999. Acute effects of leptin on glucose metabolism of in situ rat perfused livers and isolated hepatocytes. *Int J Obes Relat Metab Disord* 23 (11):1207–12.
54. Aiston, S., and L. Agius. 1999. Leptin enhances glycogen storage in hepatocytes by inhibition of phosphorylase and exerts an additive effect with insulin. *Diabetes* 48 (1):15–20.

55. Kulkarni, R.N., Z.L. Wang, R.M. Wang, J.D. Hurley, D.M. Smith, M.A. Ghatei, D.J. Withers, J.V. Gardiner, C.J. Bailey, and S.R. Bloom. 1997. Leptin rapidly suppresses insulin release from insulinoma cells, rat and human islets and, *in vivo*, in mice. *J Clin Invest* 100 (11):2729–36.

56. Shimizu-Albergine, M., D.L. Ippolito, and J.A. Beavo. 2001. Downregulation of fasting-induced cAMP response element-mediated gene induction by leptin in neuropeptide Y neurons of the arcuate nucleus. *J Neurosci* 21 (4):1238–46.

57. Zhao, A.Z., J.N. Huan, S. Gupta, R. Pal, and A. Sahu. 2002. A phosphatidylinositol 3-kinase phosphodiesterase 3B-cyclic AMP pathway in hypothalamic action of leptin on feeding. *Nat Neurosci* 5 (8):727–8.

58. Niswender, K.D., G.J. Morton, W.H. Stearns, C.J. Rhodes, M.G. Myers Jr., and M.W. Schwartz. 2001. Intracellular signalling. Key enzyme in leptin-induced anorexia. *Nature* 413 (6858):794–5.

59. Quintela, M., R. Senaris, M.L. Heiman, F.F. Casanueva, and C. Dieguez. 1997. Leptin inhibits in vitro hypothalamic somatostatin secretion and somatostatin mRNA levels. *Endocrinology* 138 (12): 5641–4.

60. Niswender, K.D., C.D. Morrison, D.J. Clegg, R. Olson, D.G. Baskin, M.G. Myers Jr., R.J. Seeley, and M.W. Schwartz. 2003. Insulin activation of phosphatidylinositol 3-kinase in the hypothalamic arcuate nucleus: a key mediator of insulin-induced anorexia. *Diabetes* 52 (2):227–31.

61. Degerman, E., P. Belfrage, and V.C. Manganiello. 1997. Structure, localization, and regulation of cGMP-inhibited phosphodiesterase (PDE3). *J Biol Chem* 272 (11):6823–6.

62. Reinhardt, R.R., E. Chin, J. Zhou, M. Taira, T. Murata, V.C. Manganiello, and C.A. Bondy. 1995. Distinctive anatomical patterns of gene expression for cGMP-inhibited cyclic nucleotide phosphodiesterases [published erratum appears in *J Clin Invest* 1997 Feb 1; 99 (3):551]. *J Clin Invest* 95 (4): 1528–38.

63. Cowley, M.A., R.D. Cone, P. Enriori, I. Louiselle, S.M. Williams, and A.E. Evans., 2003. Electrophysiological actions of peripheral hormones on melanocortin neurons. *Ann N Y Acad Sci* 994: 175–86.

64. Willesen, M.G., P. Kristensen, and J. Romer. 1999. Co-localization of growth hormone secretagogue receptor and NPY mRNA in the arcuate nucleus of the rat. *Neuroendocrinology* 70 (5): 306–16.

65. Kamegai, J., H. Tamura, T. Shimizu, S. Ishii, H. Sugihara, and I. Wakabayashi. 2001. Chronic central infusion of ghrelin increases hypothalamic neuropeptide Y and agouti-related protein mRNA levels and body weight in rats. *Diabetes* 50 (11):2438–43.

66. Donohoue, P.A., Y.X. Tao, M. Collins, G.S. Yeo, S. O'Rahilly, and D.L. Segaloff. 2003. Deletion of codons 88–92 of the melanocortin-4 receptor gene: A novel deleterious mutation in an obese female. *J Clin Endocrinol Metab* 88 (12):5841–5.

67. Sahu, A., and A.S. Metlakunta. 2005. Hypothalamic phosphatidylinositol 3-kinase-phosphodiesterase 3B-cyclic AMP pathway of leptin signalling is impaired following chronic central leptin infusion. *J Neuroendocrinol* 17 (11):720–6.

68. Grill, V., K. Asplund, C. Hellerstrom, and E. Cerasi. 1975. Decreased cyclic AMP and insulin response to glucose in isolated islets of neonatal rats. *Diabetes* 24 (8):746–52.

69. Sharp, G.W. 1979. The adenylate cyclase-cyclic AMP system in islets of Langerhans and its role in the control of insulin release. *Diabetologia* 16 (5):287–96.

70. Christie, M.R., and S.J. Ashcroft. 1984. Cyclic AMP-dependent protein phosphorylation and insulin secretion in intact islets of Langerhans. *Biochem J* 218 (1):87–99.

71. Malaisse, W.J., P. Garcia-Morales, S.P. Dufrane, A. Sener, and I. Valverde. 1984. Forskolin-induced activation of adenylate cyclase, cyclic adenosine monophosphate production and insulin release in rat pancreatic islets. *Endocrinology* 115 (5):2015–20.

72. Howell, S.L., P.M. Jones, and S.J. Persaud. 1994. Regulation of insulin secretion: The role of second messengers. *Diabetologia* 37 (Suppl 2): S30–5.

73. Holst, J.J., and C. Orskov. 2001. Incretin hormones—an update. *Scand J Clin Lab Invest Suppl* 234:75–85.

74. Skoglund, G., M.A. Hussain, and G.G. Holz. 2000. Glucagon-like peptide 1 stimulates insulin gene promoter activity by protein kinase A-independent activation of the rat insulin I gene cAMP response element. *Diabetes* 49 (7):1156–64.

75. Trumper, A., K. Trumper, H. Trusheim, R. Arnold, B. Goke, and D. Horsch. 2001. Glucose-dependent insulinotropic polypeptide is a growth factor for beta (INS-1) cells by pleiotropic signaling. *Mol Endocrinol* 15 (9):1559–70.

76. Hui, H., A. Nourparvar, X. Zhao, and R. Perfetti. 2003. Glucagon-like peptide-1 inhibits apoptosis of insulin-secreting cells via a cyclic 5′-adenosine monophosphate-dependent protein kinase-A and a phosphatidylinositol 3-kinase-dependent pathway. *Endocrinology* 144 (4):1444–55.

77. Han, P., J. Werber, M. Surana, N. Fleischer, and T. Michaeli. 1999. The calcium/calmodulin-dependent phosphodiesterase PDE1C down-regulates glucose-induced insulin secretion. *J Biol Chem* 274 (32):22337–44.

78. Zhao, A.Z., H. Zhao, J. Teague, W. Fujimoto, and J.A. Beavo. 1997. Attenuation of insulin secretion by insulin-like growth factor 1 is mediated through activation of phosphodiesterase 3B. *Proc Natl Acad Sci USA* 94 (7):3223–8.

79. Parker, J.C., M.A. VanVolkenburg, R.J. Ketchum, K.L. Brayman, and K.M. Andrews. 1995. Cyclic AMP phosphodiesterases of human and rat islets of Langerhans: Contributions of types III and IV to the modulation of insulin secretion. *Biochem Biophys Res Commun* 217 (3):916–23.

80. Shafiee, N.R., N.J. Pyne, and B.L. Furman. 1995. Effects of type-selective phosphodiesterase inhibitors on glucose-induced insulin secretion and islet phosphodiesterase activity. *Br J Pharmacol* 115 (8):1486–92.

81. Pyne, N.J., and B.L. Furman. 2003. Cyclic nucleotide phosphodiesterases in pancreatic islets. *Diabetologia* 46 (9):1179–89.

82. Harndahl, L., X.J. Jing, R. Ivarsson, E. Degerman, B. Ahren, V.C. Manganiello, E. Renstrom, and L.S. Holst. 2002. Important role of phosphodiesterase 3B for the stimulatory action of cAMP on pancreatic β-cell exocytosis and release of insulin. *J Biol Chem* 277 (40):37446–55.

83. Harndahl, L., N. Wierup, S. Enerback, H. Mulder, V.C. Manganiello, F. Sundler, E. Degerman, B. Ahren, and L.S. Holst. 2004. β-cell-targeted overexpression of phosphodiesterase 3B in mice causes impaired insulin secretion, glucose intolerance, and deranged islet morphology. *J Biol Chem* 279 (15):15214–22.

84. Bergman, R.N., and M. Ader. 2000. Free fatty acids and pathogenesis of type 2 diabetes mellitus. *Trends Endocrinol Metab* 11 (9):351–6.

Section D

Compartmentation in Cyclic Nucleotide Signaling

18 Heart Failure, Fibrosis, and Cyclic Nucleotide Metabolism in Cardiac Fibroblasts

Sara A. Epperson and Laurence L. Brunton

CONTENTS

18.1 Introduction 365
18.2 Take Two Drams of Caution and Call Us in the Morning 366
18.3 The Roles of Cyclic AMP in Cardiac Hypertrophy Are
     Paradoxical and Time Dependent 367
18.4 An Additional Complexity: Cyclic AMP May Also Be a
     Proapoptotic Signal in Heart Cells 367
18.5 What PDE Activities Occur in Cardiac Fibroblasts? 367
18.6 Sildenafil: Not Just for Bedtime Anymore 368
18.7 Are Effects of PDE5 Inhibition Similar to Effects of Natriuretic
     Peptides and Cyclic GMP? 371
18.8 Summary 371
References 372

18.1 INTRODUCTION

To examine the potential roles of PDE activities in cardiac fibrosis, it is convenient to rephrase the question: Do cyclic nucleotides play any regulatory roles in the deposition of extracellular matrix (ECM) proteins by cardiac fibroblasts? The data in the literature are not unanimous, but many researchers report that a rise in cyclic AMP within cardiac fibroblasts produces a reduction in ECM deposition.[1–3] Such an action of cyclic AMP is doubly appealing because an elevation of cyclic AMP within neighboring cardiac myocytes produces a positive inotropic effect, mediated principally by the phosphorylation of phospholamban and the consequent enhancement of Ca^{2+} storage (and release) by the sarcoplasmic reticulum. Thus, one would predict that PDE inhibitors would be therapeutically useful in heart failure, that inhibition of cardiac PDE activities would both stimulate the cardiac contraction and reduce the development of fibrosis in the failing or damaged heart. As with many diseases, the situation is more complex. Within the heart, an interaction of cardiac fibroblasts and cardiac myocytes seem essential for the progression of cardiac remodeling and fibrosis.[4] Heart failure is multifactorial and involves the interaction of complex physiologic systems, including neural, paracrine, autocrine, and endocrine reflex responses to low cardiac output and the alteration of signaling pathways affecting growth, cell survival,

and apoptosis. Current concepts of heart failure emphasize that altered cardiac function and cardiac remodeling interact, one causing the other, with both sustained by neurohumoral responses.[5]

To review the effects of PDE inhibitors on cardiac fibrosis, it seems wise to approach the subject obliquely, first mentioning some caveats in interpreting data in this area of research, and then examining some roles of cyclic nucleotides in heart failure.

18.2   TAKE TWO DRAMS OF CAUTION AND CALL US IN THE MORNING

On a regular basis, a paper's title will announce that the activation of a particular G-protein or protein kinase is the cause of cardiac hypertrophy. Upon reading further, we are likely to find that the overexpression of a G-protein α-subunit or the activation of the protein kinase by a hormone or drug of unknown specificity causes neonatal rat cardiac myocytes of unknown purity, growing on a plastic culture dish in medium supplemented with 10% fetal calf serum, to change shape and get larger. The findings are not uninteresting, but the authors have probably not discovered the master key to cardiac hypertrophy.

The issues surrounding cardiac cells and hormonal responses are not as straightforward as many workers in the field suggest by their papers' titles and abstracts or by their interpretations of the data. The heart is a mixture of cell types, largely myocytes but roughly 16%–20% fibroblasts by mass, perhaps 65% fibroblasts by cell number, with endothelial cells from the endocardial lining and vasculature, and smooth muscle cells within the vasculature. There are, as well, some nerve endings and a variable amount of adipose tissue. Further, the myocytes are heterogeneous and specialized by region, as the fibroblasts may also be. There are, as a result, many possible interactions and complexities, whether we purify a cell type from isolated hearts or study various intact preparations.

There is the matter of what cell or tissue preparation is being studied, the purity and viability of that preparation, and the limitations of the preparation. For instance, freeze-clamped samples of ventricle or homogenates of great chunks of heart tissue represent an amalgam of contributions by the multiple cell types in the sample. To avoid such issues, we turn to preparations of purified cells (in which purity is too rarely assessed), which are not without their problems. As an example, adult rat ventricular myocytes undergo considerable stress during preparation (e.g., heat shock proteins are induced) and basically begin a slow process of death or dedifferentiation when isolated by enzymatic digestion and cultured under standard conditions. Cardiac fibroblasts, on the other hand, grow well in culture but their character changes in time; they need to be used in early passages. The hearts and isolated cardiac cells of species differ, even between rats and mice; as Paul Simpson has noted, "A mouse is not a small rat" (quoted in Ref. [6]). Similarly, it is inappropriate to extrapolate to human disease based on findings in cultured rodent heart cells.

Then, there is the issue of the complexity of the disease of heart failure. Isolated cell preparations of myocytes can be valuable for their purity but can scarcely be said to represent the heart. A disease as complex as heart failure with fibrosis—an interplay of multiple cell types and endocrine, paracrine, and neural factors—will not be reproduced by a single cell type in culture. As noted further in the chapter, heart failure with fibrosis is a disease of the intact organism. There is much excellent basic research being done on cardiac hypertrophy, but basic scientists should understand the nature and limitations of their model systems in relation to the conclusions being drawn. It is generally not clear to what extent animal models of heart failure and hypertrophy actually mimic the human disease.

18.3 THE ROLES OF CYCLIC AMP IN CARDIAC HYPERTROPHY ARE PARADOXICAL AND TIME DEPENDENT

A main physiological symptom of heart failure is low cardiac output, sometimes accompanied by deposition of collagen in the ventricles and by hypertrophy of the heart. At the cellular level, increased cellular cyclic AMP enhances contractility in the myocyte and generally reduces collagen production by isolated cardiac fibroblasts. Thus, one might expect that enhancing the production of cyclic AMP or reducing its degradation in cardiac myocytes and cardiac fibroblasts would stop the development of failure and improve both contractility and fibrosis. The actual result depends on the duration of treatment: β agonists produce improved cardiac function acutely but the disease will progress unimpeded. Paradoxically, blockade of cardiac β receptors, which causes acute deterioration of hemodynamic function and a worsening of symptoms, will slow the progression of the disease.[7] (This disparity between acute and chronic administration of β receptor ligands has been decisively and dramatically demonstrated in a study of their effects on airway function in a murine model of asthma; see Ref. [8]). A therapeutic role for enhancing β receptor function in heart failure has not been ruled out, however; Tevaearai and Koch have used βARKct expression to produce an upregulation of cyclic AMP signaling and improved function and reduced ventricular remodeling in various models of heart failure.[9] From their data, one may infer that the mechanism of βARKct, which acts synergistically with β blockers, involves the enhancement of additional G-protein-coupled signaling systems. Thus, although enhanced cyclic AMP alone may not be beneficial, a constellation of signals including cyclic AMP may be beneficial in treating heart failure in some models.

18.4 AN ADDITIONAL COMPLEXITY: CYCLIC AMP MAY ALSO BE A PROAPOPTOTIC SIGNAL IN HEART CELLS

Myocyte apoptosis seems to be a factor in some end-stage heart failure. From studies in neonatal rat myocytes, Ding and colleagues[10] have found that downregulation of PDE3A is associated with induction of the inducible cyclic AMP early repressor (ICER). These workers have suggested a mechanism by which PDE3A, cyclic AMP, and ICER contribute to apoptosis. The overall idea is that activation of cellular PKC by growth factors such as angiotensin II and activation of PKA by increased sympathoadrenal tone produce conditions that upregulate ICER expression. ICER transcriptionally represses PDE3A expression, elevating cyclic AMP, sustaining the activation of PKA, and forming a positive feedback loop that results in apoptosis through inhibition of CREB-mediated transcription and downregulation of the antiapoptotic factor, Bcl-2. The model presents some difficulties but also is provocative.[5] Whether the proposed PDE3A-ICER apoptosis mechanism operates in cardiac fibroblasts is not known; however, anecdotal reports suggest that sustained elevation of cellular cyclic AMP in cultured cardiac fibroblasts results in cell death (Epperson, unpublished observation).

18.5 WHAT PDE ACTIVITIES OCCUR IN CARDIAC FIBROBLASTS?

There is not a large literature on the occurrence of specific isoforms of PDEs in cardiac fibroblasts. Gustafsson and Brunton[11] have reported on the capacity of near-maximally effective concentrations of isoform-specific PDE inhibitors to enhance accumulation of cyclic AMP in response to isoproterenol (Iso) in primary cultures of rat cardiac fibroblasts.

Comparing the inhibitors to maximally effective IBMX (1 mM), accumulation to Iso + milrinone = 1%, Iso + EHNA = 8%, and to Iso + rolipram = 62%; these data indicate that PDE3 is a minor contributor to degradation following β adrenergic stimulation, PDE2 is a greater force, and PDE4 is the primary route of cyclic AMP degradation. An unidentified PDE activity, inhibited by IBMX but not by isoform-specific inhibitors, accounts for about 29% of cyclic AMP degradation. These same workers used rtPCR to examine transcripts of PDE isoforms in primary cultures of rat cardiac fibroblasts, and found message for forms 2A, 3A, 3B, 4A, 4B, and 4D, but not for 1C or 5. These workers have also noted that the capacity of NO to reduce cyclic AMP accumulation in cardiac fibroblasts is well explained by the presence of PDE2; NO stimulates the soluble guanylyl cyclase, and the resultant cyclic GMP stimulates PDE2, reducing cellular cyclic AMP accumulation by half.[12] The predominating influence of cyclic GMP–stimulated PDE2 suggests compartmentation of that activity near the site of cyclic AMP synthesis, a conclusion that Mongillo et al.[13] reached in their studies of the effect of the NO–cGMP pathway to blunt cyclic AMP-induced inotropic effects in cultured rat neonatal ventricular myocytes. It is not known whether the cardiac fibroblasts of other species have distributions of PDE activities as similar to those of rat cardiac fibroblasts.

Studies of the effects of cyclic nucleotides and PDE inhibitors on the expression of ECM proteins and matrix metalloproteases in cardiovascular tissues are in progress in a number of laboratories. From a sampling of what is known about the hormonal and transcriptional regulation of ECM proteins (Table 18.1, Refs. [1–3,53,57–60]), a complex picture emerges, indicating differences amongst laboratories and cellular preparations. In general, elevations of cyclic AMP reportedly reduce collagen synthesis; elevations of cyclic GMP produce broader antifibrotic, antihypertrophic effects.

18.6   SILDENAFIL: NOT JUST FOR BEDTIME ANYMORE

With the rising popularity of sildenafil and congeners by males for treating erectile dysfunction, we have basically conducted a large uncontrolled clinical trial on the effects of PDE5 inhibitors in humans, and this has prompted a close look at other effects of PDE5 inhibitors, using laboratory models of cardiovascular and pulmonary disease. For instance, a PDE5 inhibitor, DA-8159, was shown to attenuate development of right heart hypertrophy in a rat model of pulmonary hypertension.[46] More recently, Kass and colleagues[47] showed that chronic treatment of mice subjected to transaortic constriction (TAC) prevented the development of a hypertrophic response. The data are quite dramatic: the usual response to TAC is a doubling of cardiac mass and substantial dilation of the chambers within 3 weeks; sildenafil (100 mg/kg daily, yielding ~10 nM drug in plasma) had no effect on sham-operated controls but reduced hypertrophy and dilation in TAC subjects by two-thirds. Moreover, sildenafil treatment reversed established hypertrophy. The authors correlate these effects with increased activity of PKG1 and, in cultured neonatal rat myocytes, with suppression of three hypertrophic signaling pathways (NFAT, ERK1/2, and Akt). These provocative data not withstanding, it is premature to conclude that the modern Priapus has a healthy heart.

The relevant cellular sites of antifibrotic and antihypertrophic actions of sildenafil in the intact animal are not known. Given the complexity of the response to TAC, multiple sites of action (myocytes, nonmyocytes) in the heart and elsewhere may be involved. It seems possible that PDE5 inhibitors have additional actions as well: most PDE inhibitors that resemble adenine or adenosine (as a portion of sildenafil does) will also antagonize adenosine receptors (adenosine, acting at $A_{2a/b}$ receptors, is thought to have antifibrotic effects on cardiac fibroblasts); congeners of cyclic AMP also elevate cellular cyclic nucleotide concentrations

TABLE 18.1
Regulation of ECM Proteins, Mainly in Cardiovascular Cells

ECM Protein/Cell System	Regulation	Comments	Ref.
Collagen α1 (I)			
Tight skin (TSK) mouse versus normal mouse	AP-1	AP-1 binds negative regulatory sequence to mediate inhibition	[14]
Rabbit cardiac fibroblasts	AP-1	AP-1 necessary for T_3-induced inhibition of promoter activity	[15]
Fetal rat cardiac fibroblasts	ERK1/2	ERK activation required for gene induction	[16]
Fetal rat cardiac fibroblasts	p38MAPK	p38MAPK inhibitor (SB203580) enhances both basal and stretch-induced collagen mRNA levels	[16]
Fetal rat cardiac fibroblasts	CCAAT-binding factor (CBF/NF-Y)	Binding at the proximal promoter plus TGFβ are necessary for strain-induced promoter activation	[17]
Immortalized embryonic cardiomyocytes	TEF1 and C/EBPβ	Involved in p38αMAPK-dependent inhibition of gene transcription	[18]
Adult rat cardiac fibroblasts	Bradykinin	Bradykinin decreases collagen I mRNA	[19]
Adult rat cardiac fibroblasts	AP-1 and NFκB	Increased DNA binding in infarcted heart correlates with increased collagen mRNA levels	[20]
Collagen α2 (I)			
Mouse aorta	ERK	ERK required with TGFβ for Ang II-induced promoter activity	[21]
Mouse aorta	AP-1	AP-1 involved in Ang II-induced gene transcription	[21]
Mouse 3T3-L1	PMA	PMA treatment induces collagen mRNA levels	[22]
Collagen α1 (III)			
Rat heart myofibroblasts	Angiotensin II (Ang II)	Ang II stimulates the expression of collagen III mRNA	[23]
Adult rat cardiac fibroblasts	Bradykinin	Bradykinin decreases collagen III mRNA	[19]
Integrin α7			
Mouse clone	AP-1, AP-2, GATA	Gene contains putative binding sites	[24]
Integrin α1			
Smooth muscle cells	Nks-3.2, serum response factor, GATA-6	Form a complex to transactivate the gene	[25]
Integrin α3			
Tumor cells	Sp3	Binds GC rich motif	[26]
Laminin-5 α3			
Human mammary epithelial cells (HMECs)	CBP	Increases promoter activity	[27]

continued

TABLE 18.1 (continued)
Regulation of ECM Proteins, Mainly in Cardiovascular Cells

ECM Protein/Cell System	Regulation	Comments	Ref.
Fibronectin (FN)			
NIH/3T3 cells, JEG-3 cell line	CRE	Three CRE sites found within the 5′ flanking mediating serum and cAMP transactivation	[28–30]
NIH/3T3 cells, JEG-3 cell line	CREB	Binds CREs to promote transcription	[28–30]
Adult male rat liver, NIH/3T3 cells	ATF-2	Facilitates occupation of the CCAAT box in the promoter and activates transcription	[29,31]
Human glioblastoma cells	EGR-1	Directly activates promoter	[32]
Adult rat cardiac fibroblasts	Bradykinin	Bradykinin decreases FN mRNA	[33]
SV-40 transformed rat kidney cells	PKA, PKC, CREB	FN synthesis is associated with PKA, PKC, and CREB phosphorylation	[19]
Rat heart	Ang II	Ang II induces FN expression	[34]
Human colonic epithelial cells	Adenosine (ADO)	ADO increases FN	[35]
Porcine vascular smooth muscle cells, normal rat kidney interstitial fibroblasts (NRK 49F cells)	ERK and p38MAPK	Both kinases involved in effects of Ang II, and p38MAPK involved in effects of oxidized lipids to increase FN synthesis	[36,37]
MMPs and TIMPs			
Rat aortic smooth muscle cells	NO/cGMP	Decrease MMP2 and MMP9 activities, increase TIMP2, decrease migration	[38]
Canine cardiac fibroblasts	BNP/cGMP/PKG	Decrease collagen synthesis, increase MMP2 synthesis	[39]
Rat aortic vascular smooth muscle cells	NO/cGMP	Increase MMP9	[40]
Mice	Knockout (KO) ANP receptor/GC-A	KO increases remodeling, TNFα expression, TGFβ expression, MMP2 expression, MMP9 expression	[41]
Human umbilical vein endothelial cells	cAMP	Decrease MMP2 activity, decrease MT-MMP1	[42]
Rat cardiomyocytes	AP-1 sites and NF-B-like domain	MMP13 gene contains putative binding sites	[43]
Rat cardiac fibroblasts	AP-1	AP-1 site mediates MMP2 transcription	[44]
Rat neonatal ventricular cardiomyocytes	NF-κB	Ang II activates PKC, which activates NF-κB to increase MMP9 transcription	[45]

by competitively inhibiting cyclic nucleotide export from cells.[48–50] Indeed, there are recent data on the interaction of cyclic GMP and milrinone, an inhibitor of PDE3, with the export of cyclic AMP from isolated rabbit atria.[51] Jackson and coworkers have argued that cardiac fibroblasts express the capacity to generate adenosine from extracellular cyclic AMP, and that

adenosine thus generated can inhibit fibroblast growth (incorporation of [^{3}H] thymidine into DNA).[52] This model suggests a complex picture of several putative actions of PDE inhibitors: enhancing intracellular AMP accumulation (by inhibiting PDEs), reducing cyclic AMP export (by acting as competitive inhibitors or substrates of transport), preserving extracellular cyclic AMP (inhibiting degradation), and interfering with the interaction of adenosine at A$_2$ receptors.

18.7   ARE EFFECTS OF PDE5 INHIBITION SIMILAR TO EFFECTS OF NATRIURETIC PEPTIDES AND CYCLIC GMP?

There is a growing body of evidence that natriuretic peptides (NPs) activate pathways that protect against hypertrophy and fibrosis in the cardiac tissue. In 1998, Redondo and colleagues reported that atrial natriuretic peptide (ANP), acting via cyclic GMP and enhanced by zaprinast, reduced mitogenesis and inhibited collagen synthesis in both rat and human cardiac fibroblasts.[53] Simultaneously, Chrisman and Garbers[54] made similar observations in human dermal fibroblasts, and demonstrated the molecular basis for a reciprocal and antagonistic relationship between the mitogen-activated signaling and the activity of membrane-bound guanylyl cyclase B (GC-B, the receptor for C-type natriuretic peptide [CNP]) in these cells: mitogen-activated protein kinases stably phosphorylate and inactivate GC-B, and elevated levels of cyclic GMP block the mitogen-activated protein kinase cascade.

The bulk of NP receptors and responsiveness are in the nonmyocyte fraction of heart, especially in the fibroblasts;[55,56] thus, despite reports of effects of NPs on neonatal rat myocytes, it seems likely that the fibroblasts are a main site of direct action of NPs, especially CNP, in adult rat heart. Cultured adult rat cardiac fibroblasts secrete CNP, and CNP is more efficacious than ANP and brain NP (BNP) in stimulating cyclic GMP accumulation in these cells.[57] CNP causes inhibition of MAPK activity, collagen synthesis, and DNA synthesis, presumably through a cGMP or PKG mechanism. Nakanishi et al.[58] have recently shown that postcoronary occlusion mortality is enhanced in mice lacking the NP receptor or *GC-A* gene, and that the lack of GC-A enhances postocclusion remodeling in the ventricle. Deschepper[59,60] has provided two useful reviews that delineate the likely cell–cell interactions that need to realize the antifibrotic and antihypertrophic effects of NPs, the cellular localizations of the GC-A and GC-B receptors for the various NPs, and a number of cautionary considerations in interpreting data from various models of cardiac injury and response.

It appears that both PDE5 inhibitors and NPs are acting through cyclic GMP and PKG1 to cause alterations in similar, if not identical, pathways that result in reduction of cardiac fibroblast growth and ECM synthesis or deposition. BNP (Nesiritide) is already approved by the FDA for the treatment of dyspnea due to congestive heart failure.[7] For effects beyond those on hemodynamics and vascular tone, other NPs will likely be tested for their longer-term influence on cardiac remodeling, very possibly in combination with PDE5 inhibitors.

18.8   SUMMARY

Much data on the roles of cyclic nucleotides and signal transduction pathways in cardiac hypertrophy and fibrosis have been accumulated in model systems that bear scant resemblance to human heart failure; thus, the relevance of such data to human health is hard to determine. In clinical trials and *in vivo* experiments, PDE inhibitors produce the expected short-term improvements in cardiac function but do not alter the overall progress and downward course of the disease. In simple systems, cyclic AMP is a downregulator of collagen production as

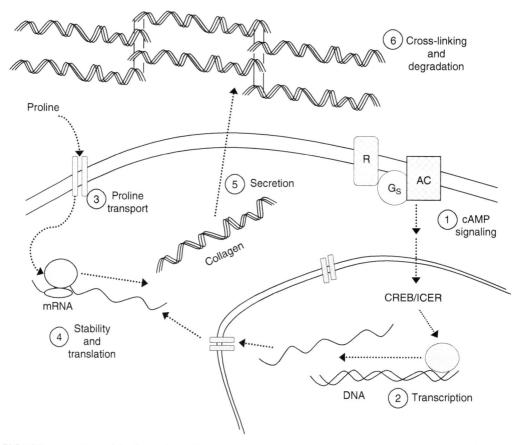

FIGURE 18.1 Potential sites of cyclic AMP-dependent regulation of proline incorporation into collagen.

assessed by incorporation of [^{3}H] proline into precipitable protein.[1–3] Yet, the scientific literature does not provide a complete picture of even the effects of cyclic AMP on the incorporation of proline into collagen in isolated cardiac fibroblasts. For instance, whether cyclic nucleotides modulate six postulated control points in collagen deposition by cardiac fibroblasts (Figure 18.1) is not known. The multifactorial nature of cardiac hypertrophy, fibrosis, and failure; the interactions of multiple cell types; and interaction of the endocrine, autocrine, and paracrine factors make experiments on physiologic systems difficult to interpret. Thus, with respect to cyclic AMP metabolism, the role of PDE activities is difficult to establish.

The picture is rosier with respect to cyclic GMP. Data in model systems showing antihypertrophic and antifibrotic effects of sildenafil and PKG activity suggest that cyclic GMP may be a key message that inhibits cardiac fibrosis and hypertrophy.[58,59] Data on the beneficial effects of BNP support such a role for cyclic GMP. Nelson Goldberg[61] would be pleased.

REFERENCES

1. Ostrom, R.S., et al. 2003. Angiotensin II enhances adenylyl cyclase signaling via Ca^{2+}/calmodulin. *J Biol Chem* 278:24461.
2. Chen, Y., et al. 2004. Functional effects of enhancing or silencing adenosine A_{2b} receptors in cardiac fibroblasts. *Am J Physiol Heart Circ Physiol* 287:H2478.

3. Swaney, J.S., et al. 2005. Inhibition of cardiac myofibroblast formation and collagen synthesis by activation and overexpression of adenylyl cyclase. *Proc Natl Acad Sci USA* 102:437.

4. Manabe, I., S. Takayuki, and R. Nagai. 2002. Gene expression in fibroblasts and fibrosis involvement in cardiac hypertrophy. *Circ Res* 91:1103.

5. Brunton, L.L. 2005. A positive feedback loop contributes to the deleterious effects of angiotensin. *Proc Natl Acad Sci USA* 102:14483.

6. Hilal-Dandan, R., J.R. Kanter, and L.L. Brunton. 2000. Characterization of G-protein signaling in ventricular myocytes from the adult mouse heart: Differences from the rat. *J Mol Cell Cardiol* 32:1211.

7. Rocco, T.P., and J.C. Fang. 2006. Pharmacotherapy of congestive heart failure. In *The Pharmacological Basis of Therapeutics, 11th ed., chap. 33, eds., L.L. Brunton, J.S. Lazo, and K.L. Parker. New York: McGraw-Hill.*

8. Callaerts-Vegh, Z., et al. 2004. Effects of acute and chronic administration of β-adrenoceptor ligands on airway function in a murine model of asthma. *Proc Natl Acad Sci USA* 101:4948.

9. Tevaearai, H.T., and W.J. Koch. 2004. Molecular restoration of β-adrenergic receptor signaling improves contractile function of failing hearts. *Trends Cardiovasc Med* 14:252.

10. Ding B, et al. 2005. A positive feedback loop of phosphodiesterase 3 (PDE3) and inducible cAMP early repressor (ICER) leads to cardiomyocyte apoptosis. *Proc Natl Acad Sci USA* 102:14771.

11. Gustafsson, Å.B., and L.L. Brunton. 2004. Interactions of the cyclic AMP and nitric oxide pathways in cardiac fibroblasts. In *Pathophysiology of cardiovascular disease*, chap. 9, eds., N.S Dhalla, H. Rupp, A. Angel, and G.N. Pierce. Boston: Kluwer Academic Publishers.

12. Gustafsson, Å.B., and L.L. Brunton. 2002. Attenuation of cAMP accumulation in adult rat cardiac fibroblasts by IL-1β and NO: Role of cGMP-stimulated PDE2. *Am J Physiol Cell Physiol* 283:C463.

13. Mongillo, M., et al. 2006. Compartmentalized phosphodiesterase-2 activity blunts β-adrenergic cardiac inotropy via an NO/cGMP-dependent pathway. *Circ Res* 98:1.

14. Philips, N., R.I. Bashey, and S.A. Jiménez. 1995. Increased α1(I) procollagen gene expression in tight skin (TSK) mice myocardial fibroblasts is due to a reduced interaction of a negative regulatory sequence with AP-1 transcription factor. *J Biol Chem* 270:9313.

15. Lee, H.-W., et al. 1998. An activator protein-1 (AP-1) response element on Pro α₁(I) collagen gene is necessary for thyroid hormone-induced inhibition of promoter activity in cardiac fibroblasts. *J Mol Cell Cardiol* 30:2495.

16. Papakrivopoulou, J., et al. 2004. Differential roles of extracellular signal-regulated kinase 1/2 and p38^{MAPK} in mechanical load-induced procollagen α₁(I) gene expression in cardiac fibroblasts. *Cardiovasc Res* 61:736.

17. Lindahl, G.E., et al. 2002. Activation of fibroblast procollagen α1(I) transcription by mechanical strain is transforming growth factor-β-dependent and involves increased binding of CCAAT-binding factor (CBF/NF-Y) at the proximal promoter. *J Biol Chem* 277:6153.

18. Ambrosino, C., et al. 2006. TEF-1 and C/EBPbeta are major p38alpha MAPK-regulated transcription factors in proliferating cardiomyocytes. *Biochem J* 396:163.

19. Singh, L.P., et al. 2001. Hexosamine-induced fibronectin protein synthesis in mesangial cells is associated with increases in cAMP responsive element binding (CREB) phosphorylation and nuclear CREB. *Diabetes* 50:2355.

20. Matsumoto, R., et al. 2004. Effects of aldosterone receptor antagonist and angiotensin II type I receptor blocker on cardiac transcriptional factors and mRNA expression in rats with myocardial infarction. *Circ J* 68:376.

21. Tharaux, P.-L., et al. 2000. Angiotensin II activates collagen I gene through a mechanism involving the MAP/ER kinase pathway. *Hypertension* 36:330.

22. Ghosh, A.K. 2002. Factors involved in the regulation of type I collagen gene expression: implication in fibrosis. *Exp Biol Med* 227:301.

23. Ghiggeri, G.M., et al. 2000. A DNA element in the α1 type III collagen promoter mediates a stimulatory response by angiotensin II. *Kidney Int* 58:537.

24. Ziober, B.L., and R.H. Kramer. 1996. Identification and characterization of the cell type-specific and developmentally regulated α7 integrin gene promoter. *J Biol Chem* 271:22915.

25. Nishida, W., et al. 2002. A triad of serum response factor and the GATA and NK families governs the transcription of smooth and cardiac muscle genes. *J Biol Chem* 277:7308.

26. Katabami, K., et al. 2006. Characterization of the promoter for the $\alpha 3$ integrin gene in various tumor cell lines: Roles of the Ets- and Sp-family of transcription factors. *J Cell Biochem* 97:530.
27. Dietze, E.C., et al. 2005. CREB-binding protein regulates apoptosis and growth of HMECs grown in reconstituted ECM via laminin-5. *J Cell Sci* 118:5005.
28. Michaelson, J.E., J.D. Ritzenthaler, and J. Roman. 2002. Regulation of serum-induced fibronectin expression by protein kinases, cytoskeletal integrity, and CREB. *Am J Physiol Lung Cell Mol Physiol* 282:L291.
29. Pesce, C.G., et al. 1999. Interaction between the (–170) CRE and the (–150) CCAAT box is necessary for efficient activation of the fibronectin gene promoter by cAMP and ATF-2. *FEBS Lett* 457:445.
30. Dean, D.C., J.J. McQuillan, and S. Weintraub. 1990. Serum stimulation of fibronectin gene expression appears to result from rapid serum-induced binding of nuclear proteins to a cAMP response element. *J Biol Chem* 265:3522.
31. Srebow, A., et al. 1993. The CRE-binding factor ATF-2 facilitates the occupation of the CCAAT box in the fibronectin gene promoter. *FEBS Lett* 327:25.
32. Liu, C., et al. 2000. The transcription factor EGR-1 directly transactivates the fibronectin gene and enhances attachment of human glioblastoma cell line U251. *J Biol Chem* 275:20315.
33. Kim, N.N., et al. 1999. Regulation of cardiac fibroblast extracellular matrix production by bradykinin and nitric oxide. *J Mol Cell Cardiol* 31:457.
34. Crawford, D.C., A.V. Chobanian, and P. Brecher, 1994. Angiotensin II induces fibronectin expression associated with cardiac fibrosis in the rat. *Circ Res* 74:727.
35. Walia, B., et al. 2004. Polarized fibronectin secretion induced by adenosine regulates bacterial–epithelial interaction in human intestinal epithelial cells. *Biochem J* 382:589.
36. Reddy, M.A., et al. 2002. The oxidized lipid and lipoxygenase product 12(*S*)-hydroxyeicosatetraenoic acid induces hypertrophy and fibronectin transcription in vascular smooth muscle cells via p38 MAPK and cAMP response element-binding protein activation. *J Biol Chem* 277:9920.
37. Sekine, S., et al. 2003. Possible involvement of mitogen-activated protein kinase in the angiotensin II-induced fibronectin synthesis in renal interstitial fibroblasts. *Arch Biochem Biophys* 415:63.
38. Gurjar, M.V., R.V. Sharma, and R.C. Bhalla. 1999. eNOS gene transfer inhibits smooth muscle cell migration and MMP-2 and MMP-9 activity. *Arterioscler Thromb Vasc Biol* 19:2871.
39. Tsuruda, T., et al. 2002. Brain natriuretic peptide is produced in cardiac fibroblasts and induces matrix metalloproteinases. *Circ Res* 91:1082.
40. Marcet-Palacios, M., et al. 2003. Nitric oxide and cyclic GMP increase the expression of matrix metalloproteinase-9 in vascular smooth muscle. *J Pharmacol Exp Ther* 307:429.
41. Vellaichamy, E., et al. 2005. Involvement of the NF-κB/matrix metalloproteinase pathway in cardiac fibrosis of mice lacking guanylyl cyclase/natriuretic peptide receptor A. *J Biol Chem* 280:19230.
42. Peracchia, F., et al. 1997. cAMP involvement in the expression of MMP-2 and MT-MMP1 metalloproteinases in human endothelial cells. *Arterioscler Thromb Vasc Biol* 17:3185.
43. Campbell, S.E., et al. 2002. The cloning and functional analysis of canine matrix metalloproteinase-13 gene promoter. *Gene* 286:233.
44. Bergman, M.R., et al. 2003. A functional activating protein 1 (AP-1) site regulates matrix metalloproteinase 2 (MMP-2) transcription by cardiac cells through interactions with JunB-Fra1 and JunB-FosB heterodimers. *Biochem J* 369:485.
45. Rouet-Benzineb, P. 2000. Angiotensin II induces nuclear factor-κB activation in cultured neonatal rat cardiomyocytes through protein kinase C signaling pathway. *J Mol Cell Cardiol* 32:1767.
46. Kang, K.K., et al. DA-8159, a new PDE5 inhibitor, attenuates the development of compensatory right ventricular hypertrophy in a rat model of pulmonary hypertension. *J Int Med Res* 31:517.
47. Takimoto, E., et al. 2005. Chronic inhibition of cyclic GMP phosphodiesterase 5A prevents and reverses cardiac hypertrophy. *Nat Med* 11:214.
48. Brunton, L.L., and S.E. Mayer. 1979. Extrusion of cyclic AMP from pigeon erythrocytes. *J Biol Chem* 254:9714.
49. Heasley, L.E., and L.L. Brunton. 1985. Prostaglandin A_1 metabolism and the inhibition of cyclic AMP extrusion by avian erythrocytes. *J Biol Chem* 260:11514.

50. Brunton, L.L., and L.E. Heasley. 1987. Cyclic AMP export and its regulation by prostaglandin A₁. In *Methods in Enzymology, vol. 159: Initiation and termination of cyclic nucleotide action*, chap. 8, eds., J.D. Corbin, and R.A. Johnson. San Diego: Academic Press, Inc.

51. Wen, J.F., et al. 2004. High and low gain switches for regulation of cAMP efflux concentration: Distinct roles for particulate GC- and soluble GC-cGMP-PDE3 signaling in rabbit atria. *Circ Res* 94:936.

52. Dubey, R.K., et al. 2001. Endogenous cyclic AMP-adenosine pathway regulates cardiac fibroblast growth. *Hypertension* 37:1095.

53. Redondo, J., J.E. Bishop, and M.R. Wilkins. 1998. Effect of atrial natriuretic peptide and cyclic GMP phosphodiesterase inhibition on collagen synthesis by adult cardiac fibroblasts. *Br J Pharmacol* 124:1455.

54. Chrisman, T.D., and D.L. Garbers. 1999. Reciprocal antagonism coordinates C-type natriuretic peptide and mitogen-signaling pathways in fibroblasts. *J Biol Chem* 274:4293.

55. Doyle, D.D., et al. 2002. Natriuretic peptide receptor-B in adult rat ventricle is predominantly confined to the nonmyocyte population. *Am J Heart Circ Physiol* 282:H2117.

56. Villegas, S., and L.L. Brunton. 1996. Cellular localization and characterization of cyclic GMP synthesis in rat heart. *Cardiovasc Pathobiol* 1:5.

57. Horio, T., et al. 2003. Gene expression, secretion, and autocrine action of C-type natriuretic peptide in cultured adult rat cardiac fibroblasts. *Endocrinology* 144:2279.

58. Nakanishi, M., et al. 2005. Role of natriuretic peptide receptor guanylyl cyclase-A in myocardial infarction evaluated using genetically engineered mice. *Hypertension* 46:441.

59. Deschepper, C.F. 2005. The many possible benefits of natriuretic peptides after myocardial infarction. *Hypertension* 46:271.

60. Deschepper, C.F. 2005. The cardiac antihypertrophic effects of cyclic GMP-generating agents: An experimental framework for novel treatments of left ventricular remodeling. *Vasc Dis Prev* 2:151.

61. Goldberg, N.D., S.B. Dietz, and A.G. O'Toole. 1969. Cyclic guanosine 3′,5′-monophosphate in mammalian tissue and urine. *J Biol Chem* 244:4458.

19 Role of A-Kinase Anchoring Proteins in the Compartmentation in Cyclic Nucleotide Signaling

Oliwia Witczak, Einar M. Aandahl, and Kjetil Taskén

CONTENTS

19.1 Introduction 377
19.2 Anchored Phosphodiesterases and Localized Pools of cAMP 378
19.3 Protein Kinase A Anchoring..379
19.4 A-Kinase Anchoring Proteins 383
19.5 A-Kinase Anchoring Proteins as Phosphodiesterase Anchoring Proteins 384
    19.5.1 AKAP450..384
    19.5.2 mAKAP 385
19.6 Signal Complexes Organized by A-Kinase Anchoring Proteins.............385
19.7 Concluding Remarks..386
Acknowledgments 387
References 387

19.1 INTRODUCTION

Cyclic AMP generation and degradation are regulated by the adenylyl cyclase and PDE families of enzymes, respectively.[1,2] The only known mechanism for the inactivation of cAMP is through degradation by a large family of cAMP-specific PDEs.[3,4] Cyclic AMP activates Epac (exchange protein directly activated by cAMP),[5,6] cAMP-regulated ion channels,[7,8] and finally, the broad substrate specificity enzyme protein kinase A (PKA) (reviewed in Ref. [9]). Compartmentalization of receptors, cyclases, and PKA by A-kinase anchoring proteins (AKAPs)[10,97] as well as generation of local pools of cAMP within the cell by the action of PDEs[11] generate a high degree of specificity in PKA-mediated signaling despite the broad substrate specificity of PKA. AKAPs contribute to specificity by targeting PKA towards specific substrates as well as versatility by assembling multiprotein signal complexes allowing signal termination by phosphoprotein phosphatases and cross talk between different signaling pathways.[9,10,12] Integrating PDEs into these anchoring complexes adds a further temporal aspect to the spatial regulation of cAMP signals (reviewed in Refs. [13,14]).

The importance of the PDEs as regulators of signaling is evident from studies of PDE-deficient mice,[15–21] siRNA-mediated PDE knockdown,[22,23] and dominant-negative PDEs.[23–25] The use of selective inhibitors has identified PDEs as important drug targets in several

377

diseases such as asthma and chronic obstructive pulmonary disease, cardiovascular diseases such as heart failure and atherosclerotic peripheral arterial disease, neurological and psychiatric disorders such as schizophrenia, and erectile dysfunction[26–31] as recently reviewed in Ref. [32].

19.2 ANCHORED PHOSPHODIESTERASES AND LOCALIZED POOLS OF cAMP

Cyclic AMP microdomains are shaped by PDEs and differ in amplitude as well as spatio-temporal dynamics. It is feasible that a cAMP gradient elicited by a distinct ligand is specifically organized to follow a distinct route of PKA signaling by reaching and activating a subset of or even a single PKA–AKAP complex to mediate a biological effect. Localized Ca^{2+} gradients and spikes are well established and are generated by controlled release and reuptake.[33–35] Similarly local domains of cAMP with discrete spatiotemporal organization are generated in response to external ligands.[36] PDEs contribute to establishing local gradients of cyclic nucleotides by localizing to subcellular compartments and by recruiting into multiprotein signaling complexes (Figure 19.1). This contributes to the temporal and spatial specificities of cyclic nucleotide signaling by regulating the availability of cAMP/cGMP to their effectors as seen, for example, in cardiomyocytes by fluorescent resonance energy transfer (FRET) imaging techniques.[37,38]

The distribution of PDEs to different subcellular localizations was proposed early on by the observation that PDE activity was found in both the soluble and particulate fractions of the cell.[39] Recent evidence further supports the above mentioned notion, and contributes to an emerging concept of a highly organized signal pathway where specific routes of cAMP signals

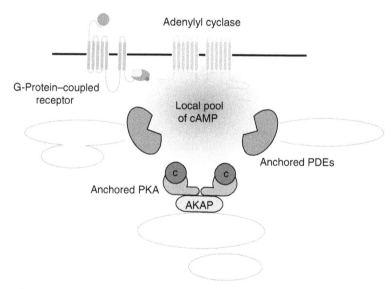

FIGURE 19.1 (See color insert following page 274.) Ligand binding to various G-protein–coupled receptors activate adenylyl cyclases in their proximity and generate pools of cAMP. The local concentration and distribution of the cAMP gradient is limited by PDEs. Particular G-protein–coupled receptors are confined to specific domains of the cell membrane in association with intracellular organelles or cytoskeletal constituents. The subcellular structures may harbor specific isozymes of PKA that through anchoring via AKAPs are localized in the vicinity of the receptor and the cyclase. PDEs are also anchored and serve to limit the extension and duration of cAMP gradients. These mechanisms serve to localize and limit the assembly and triggering of specific pathways to a defined area of the cell close to the substrate.

are formed through the localized synthesis by cell- and tissue-specific adenylyl cyclases, and where the signal is delivered to targeted effectors and terminated in a spatially and temporally defined manner by specific PDEs establishing local pools of cAMP close to the effector molecules.

Putative or established targeting domains have now been identified for most of the PDE families.[40] PDE3s are targeted to the endoplasmic reticulum by a membrane-targeting domain,[41] and PDE4D5 interacts with RACK-1, a scaffold protein which binds certain protein kinase C (PKC) isoforms.[42] PDE4D3 is targeted to the Golgi or centrosomal region through anchoring by myomegalin.[43,44] Some PDE4D and PDE4A variants bind SH3 domains, for example, Src kinases[45,46] and, via their catalytic domains, PDE4 isoforms bind to and are phosphorylated by Erk.[47] PDE4A1 contains a novel lipid-binding domain, TAPAS, with specificity for phosphatidic acid that serves to target this PDE to specific cellular membranes.[48] Most recently, the PDE4 family is reported to be recruited to activated β-adrenoreceptors through interaction with β-arrestin.[24,49] While members of the whole PDE4 family may interact with β-arrestin through their catalytic domain, PDE4D5 appears to interact with significantly higher affinity due to an additional binding site in the N-terminal region.[24,50]

In T cells, cAMP inhibits T-cell receptor-induced T-cell activation and thereby exerts important immunoregulatory functions (reviewed in Ref. [51]) through a receptor–G-protein–adenylyl cyclase–cAMP–PKA type I–C-terminal Src kinase (Csk) inhibitory pathway assembled in T-cell lipid rafts and acting on the Src family kinase Lck.[52–54] Five of the 11 families of PDEs (PDE1B, PDE3, PDE4, PDE7, and PDE8) have so far been reported to be present in primary T cells.[55–60] However, isoforms from the PDE4 subfamily account for the majority of the cAMP-hydrolyzing activity in T cells.[55,61] Concomitant stimulation of the T-cell receptor and CD28 leads to recruitment of a PDE/β-arrestin complex to T-cell lipid rafts and a profound increase in raft-associated PDE4 activity; this localized increase in cAMP PDE activity promotes degradation of cAMP and relieves the cAMP inhibition of immune function.[62] This sheds light on the process whereby costimulation mediates complete T-cell activation by a mechanism that involves release from inhibition by cAMP through recruitment and compartmentalization of a PDE. The signal elicited by CD28 that leads to recruitment of the PDE/β-arrestin complex remains, however, elusive.

Finally, AKAPs appear to provide a very interesting means of targeting PDEs as well as incorporating PDEs into integrated localized signaling complexes as discussed in the following sections.

19.3   PROTEIN KINASE A ANCHORING

PKA is a heterotetramer composed of two regulatory and two catalytic subunits. Both the regulatory (RIα, RIβ, RIIα, RIIβ) and the catalytic (Cα, Cβ, Cγ) subunits possess distinct physical and biological properties, are differentially expressed, and are able to form different isoforms of PKA holoenzymes (reviewed in Ref. [9]). Subcellular localization of PKA is mainly due to anchoring of the R subunits by AKAPs, which originally were seen as contaminants of purified PKA,[63–65] and later understood to enhance the efficiency and specificity of the signaling events. PKA type I is classically known to be largely present in the soluble fraction and was thus assumed to be mainly cytoplasmic; PKA type II is typically particulate and confined to subcellular structures and compartments anchored by cell- and tissue-specific AKAPs, a field largely pioneered by the Scott and Rubin laboratories (reviewed in Refs. [10,12,66–68,97]). However, a few dual-specific AKAPs (D-AKAPs) anchoring both PKA type I and type II as well as some AKAPs that selectively bind PKA type I have been identified more recently (see Table 19.1).

TABLE 19.1
Overview of A-Kinase Anchoring Proteins

AKAP (Gene Nomenclature Committee Name)	Tissue	Subcellular Localization	Properties/Function	Ref.
S-AKAP84/D-AKAP1/AKAP121/AKAP149 (*AKAP1*)	Testis, thyroid, heart, lung, liver, skeletal muscle, and kidney	Outer mitochondrial membrane/endoplasmic reticulum/nuclear envelope/sperm midpiece	Dual-specific AKAP. Binds lamin B and PP1. Multiple splice variants. Binds PDE7A	[81,105–110]
AKAP-KL (*AKAP2*)	Kidney, lung, thymus, and cerebellum	Actin cytoskeleton/apical membrane of epithelial cells	Multiple splice variants	[111]
AKAP110 (*AKAP3*)	Testis	Axoneme	Binds $G\alpha_{13}$	[112,113]
AKAP82/FSC1 (*AKAP4*)	Testis	Fibrous sheath of sperm tail	Potential role in sperm motility and capacitation. Multiple splice variants. Binds both RI and RII	[114–116]
AKAP75/79/150 (*AKAP5*)	Bovine/human/rat orthologs. Brain	Plasma membrane/postsynaptic density	Polybasic domains target to plasma membrane and dendrites. Binds PKC, calcineurin (PP2B), β-AR, SAP97, and PSD-95	[64,99,101,117–120]
mAKAP (*AKAP6*)	Heart (mAKAPβ, short), skeletal muscle, and brain (mAKAPα, long)	Nuclear membrane	Binds PDE4D3. Spectrin repeat domains involved in subcellular targeting. Organizes signal complex with PKA, PDE4D3, Epac, and a MEKK/MEK5/ERK5 module. Longer mAKAPα also anchors PDK1	[78,91–94,102,121]
AKAP15/18 α, β, γ, δ (*AKAP7*)	Brain, skeletal muscle, pancreas, and heart	Basolateral (α) and apical (β) plasma membrane, cytoplasm (γ), secretory vesicles (δ)	Targeted to plasma membrane via fatty acid modifications. Modulation of Na^+ and L-type Ca^{2+} channels (α). ADH-mediated translocation of AQP2 from vesicles to apical membrane in distal kidney tubules	[122–127]

AKAP95 (*AKAP8*)	Heart, liver, skeletal muscle, kidney, and pancreas	Nuclear matrix	Involved in initiation of chromosome condensation. Binds Eg7/condensin. Zinc-finger motif. Binds PDE7A	[81,128–132]
AKAP450/AKAP350/Yotiao/CG-NAP/Hyperion (*AKAP9*)	Brain, pancreas, kidney, heart, skeletal muscle, thymus, spleen, placenta, lung, and liver	Postsynaptic density/neuromuscular junction/centrosomes/Golgi	Binds PDE4D3, PP1, PP2A, PKN, and PKCε. Targets PKA and PP1 to the NMDA receptor. Multiple splice variants	[25,77,82,83, 90,133–142]
D-AKAP2 (*AKAP10*)	Liver, lung, spleen, and brain	Vesicles/peroxisomes/centrosome	Dual-specific AKAP	[143,144]
AKAP220/hAKAP220 (*AKAP11*)	Testis and brain		Binds PP1. Dual-specific AKAP	[145–148]
Gravin (*AKAP12*)	Endothelium	Actin cytoskeleton/cytoplasm	Binds PKC and β-AR. Xgravin-like (Xgl) is also a putative AKAP	[149–152]
AKAP-Lbc/Ht31/Rt31 (*AKAP13*)	Ubiquitous	Cytoplasm	Ht31 RII-binding site used in peptides to disrupt PKA anchoring. Rho-GEF that couples $G\alpha_{12}$ to Rho	[153–155]
MAP2B	Ubiquitous	Microtubules	Binds tubulin. Modulation of L-type Ca^{2+} channels	[63–65,156]
Ezrin/AKAP78	Secretory epithelia	Actin cytoskeleton	Linked to CFTR via EBP50/NHERF	[157–159]
T-AKAP80	Testis	Fibrous sheath of sperm tail		[160]
AKAP80/MAP2D	Ovarian granulosa cells		FSH-regulated protein, identified as MAP2D	[161–163]
SSeCKS (Src-suppressed C kinase substrate)	Testis, elongating spermatids	Actin remodeling	Gravin-like	[164]
Pericentrin	Ubiquitous	Centrosome	Binds dynein and γ-tubulin. Unique RII-binding domain	[165]
WAVE-1/Scar	Brain	Actin cytoskeleton	Binds Abl and Wrp. Involved in sensorimotor and cognitive function	[166,167]
Myosin VIIA	Ubiquitous	Cytoskeleton		[168]
PAP7	Steroid-producing cells (adrenal gland and gonads)	Mitochondria	Hormonal regulation of cholesterol transport into mitochondria. Binds RI *in vivo*	[169]
Neurobeachin	Brain	Golgi		[170]
AKAP28	Primary airway cells	Ciliary axonemes	Modulation of ciliary beat frequency	[171]

continued

TABLE 19.1 (continued)
Overview of A-Kinase Anchoring Proteins

AKAP (Gene Nomenclature Committee Name)	Tissue	Subcellular Localization	Properties/Function	Ref.
Myeloid translocation gene (MTG) 8 and 16b	Lymphocytes	Golgi	Binds PDE7A	[81,172,173]
AKAP140	Granulosa cells and meiotic oocytes		Upregulated by FSH in granulosa cells. Phosphorylated by CDK1 in oocytes. Not cloned	[161,162,174]
AKAP85	Lymphocytes	Golgi	Not cloned	[175]
BIG2 (Brefeldin A-inhibited guanine nucleotide-exchange protein 2)		Cytosol and Golgi	GEF for ADP-ribosylation factor GTPases. Binds RIα/RIβ and RIIα/RIIβ through three separate PKA-binding domains. Cyclic AMP-regulated translocation of BIG from cytosol to Golgi	[176]
Rab32		Mitochondria	Regulation of mitochondrial dynamics and fission	[177]
AKAP$_{CE}$	Caenorhabditis elegans		Binds to RI-like subunit. RING-finger protein with FYVE- and TGF-β receptor-binding domain	[178–180]
DAKAP550	Drosophila	Plasma membrane/cytoplasm	Contains two RII-binding sites	[181]
DAKAP200	Drosophila	Plasma membrane	Binds F-actin and Ca-calmodulin	[182,183]
Nervy	Drosophila	Axons	MTG family member. Regulates repulsive axon guidance by clustering PKA and Semaphorin 1a (Sema-1a) receptor Plexin A (PlexA)	[184]
AKAP97/radial spoke protein 3 (RSP3)	Chlamydomonas	Flagellar axonemes	Located near inner arm dyneins and possibly regulates flagellar motility	[185]

As evident from NMR structural work, the RII subunits dimerize at the N-terminus in an antiparallel fashion forming an X-type, four-helix bundle, which is necessary for both AKAP binding (N-terminal helix of both protomers) and dimerization (C-terminal helices of the bundle) through separate, but overlapping regions involved in the two events.[69–71] Dimerization is a prerequisite for AKAP binding, but deletion of residues 1–5 abolishes AKAP binding without disrupting dimer formation and branched side chains at positions 3 and 5 are critical for the interaction with the AKAPs in a hydrophobic groove, which is formed on the top of the N-terminal helices.[72,73] The RI dimerization domain contains a similar helix–turn–helix motif recently solved by NMR, which is shifted a little further from the N-terminus and encompasses amino acids 12 to 61.[74–76] The extreme N-terminus in RI is helical and believed to fold back onto the four-helix bundle and may thus contribute to differences in AKAP-binding specificity between RII and RI.

19.4   A-KINASE ANCHORING PROTEINS

The intracellular targeting and compartmentation of PKA is controlled through association with AKAPs. AKAPs are a structurally diverse family of functionally related proteins that now include more than 50 members when splice variants are included (Table 19.1). These proteins are defined on the basis of their ability to bind to PKA and to coprecipitate catalytic activity. However, the functional importance further involves targeting the enzyme to specific subcellular compartments, thereby providing spatial and temporal regulation of the PKA-signaling events. All the anchoring proteins contain a PKA-binding tethering domain and a unique targeting domain, directing the PKA–AKAP complex to defined subcellular structures, membranes, or organelles. In addition to these two domains, several AKAPs are also able to form multivalent signal transduction complexes by interaction with phosphoprotein phosphatases as well as other kinases and proteins involved in signal transduction. Through this central role in the spatial and temporal integration of effectors and substrates, AKAPs provide a high level of specificity and temporal regulation to the cAMP–PKA signaling pathway.

Direct interaction between PDE and several different AKAPs has been reported. AKAP450 targets PDE4D3 to the centrosomal region together with PKA type II in a ternary complex.[77] This was later verified using dominant negative PDEs and showed that both PDE4D2 and PDE4C2 could bind to AKAP450.[25] In cardiomyocytes, muscle AKAP (mAKAP) binds and targets both PDE4D3 and PKA type II to the perinuclear region.[78] These are the first examples of colocalized PKA/PDE complexes providing spatial control of PKA signaling by AKAP anchoring, and temporal control and termination of the cAMP signaling event by complexing PDE in the immediate vicinity. Furthermore, long PDE4 isoforms such as the PDE4D3 bound to AKAPs are activated by PKA phosphorylation.[43,79] Thus, the anchoring of both PKA and PDE in the same complex, organized by the AKAP, serves to effectively establish a negative feedback loop that terminates the cAMP signal locally (Figure 19.2). In addition to the spatial control of PDEs by subcellular compartmentation, PDE activity is also allosterically regulated, and regulated by protein–protein interactions and by posttranslational modifications, further contributing to the specificity in this signaling pathway.[40,80] More recently, studies in T-cell lines have shown that the cAMP-specific PDE7A isoform[57] binds to the Golgi-associated AKAP proteins MTG8 and MTG16b, to the nuclear envelope-associated pool of AKAP149, and to nuclear AKAP95, thus expanding the number of PDE–AKAP complexes and the spectrum of PDE isoforms that can associate with AKAPs.[81]

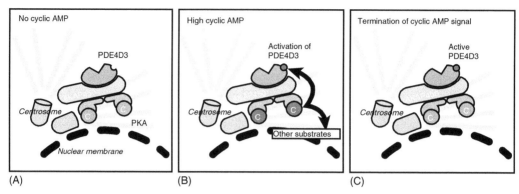

FIGURE 19.2 (See color insert following page 274.) AKAP450 targets PKA type II and PDE4D3 to the centrosomal region in Sertoli cells. A similar mechanism operates in cardiomyocytes, where mAKAP binds and targets both PDE4D3 and PKA type II to the perinuclear region. Colocalized PKA and PDE provides spatial control of PKA signaling via anchoring to the same AKAP, and temporal control and termination of cAMP signaling by a sequence of events that involve the following: (A) the effect of cAMP is mediated by PKA phosphorylation of substrate proteins; (B) PKA phosphorylates and activates the PDE4D3 (PDE4D3 and other long PDE4 isoforms are PKA substrates and phosphorylation leads to enhanced phosphodiesterase activity); and (C) the colocalized and now activated PDE4D3 degrades cAMP and terminates the signal. This serves to establish a negative feedback mechanism.

19.5   A-KINASE ANCHORING PROTEINS AS PHOSPHODIESTERASE ANCHORING PROTEINS

19.5.1   AKAP450

AKAP450 interacts with several signal transduction enzymes, in addition to PKA, including protein kinase N (PKN), protein phosphatase 1, protein phosphatase 2A, and the immature nonphosphorylated form of PKCε.[82,83] PKN is a serine/threonine kinase that associates with and phosphorylates intermediate filament proteins,[84,85] and AKAP450 targeting of PKN may thereby be important for cytoskeletal reorganization events. PKN is activated by Rho[86,87] and unsaturated fatty acids such as arachidonic acid,[88] or by truncation of its N-terminal regulatory region.[89] Interaction between AKAP450 and the nonphosphorylated form of PKCε is required for the phosphorylation-dependent maturation of PKCε. Recently, AKAP450 was reported to anchor protein kinase CK1δ, which is involved in the control of cell-cycle progression.[90] In addition, AKAP450 also anchors PDE4D3,[77] which allows tight control of the phosphorylation state of proteins regulated by cAMP signaling. Spatial control is achieved by targeting of PKA by AKAP450, whereas temporal control and inactivation of the effect of cAMP on PKA is accomplished by complexing of PDE at the same site (Figure 19.2). Controlling cAMP levels locally at centrosomes is clearly important for cell-cycle control. Overexpression of catalytically inactive, dominant negative PDE4 isoforms that displaced endogenous PDE4D3 from AKAP450 at centrosomes leads to chronic phosphorylation of the PKA phosphorylation site in PDE. In contrast to this, with catalytically active, endogenous PDE4D3, the PKA phosphorylation of this isoform only occurred when cAMP levels were elevated in response to agonist stimulation of adenylate cyclase.[25] This indicates that local cAMP levels are tightly controlled by the endogenously anchored PDE4D3 that serve as a gatekeeper for PKA activation, preventing its inappropriate occurrence in resting and unstimulated cells.

19.5.2 mAKAP

mAKAP (originally cloned and characterized as AKAP100) is a 255 kDa scaffolding protein expressed in myocytes, skeletal muscle, and brain. The mAKAP assembles a signal complex consisting of PKA and PDE4D3 at the nuclear envelope (NE), the sarcoplasmic reticulum of cardiomyocytes, and intercalated disks in adult rat heart tissue.[78,91–94] The assembly of the mAKAP signaling complex in the perinuclear region is induced by hypertrophic stimuli in rat neonatal ventriculocytes and is thought to be associated with cellular differentiation and development of a ventricular hypertrophic phenotype.[92] The induction of mAKAP expression also leads to the redistribution of RII to the NE,[92] which is interesting as PKA-mediated phosphorylation induces cAMP-responsive genes involved in propagation of cardiac hyper-trophy,[95] and the concurrent anchoring of PDE4D3 serves to establish a negative feedback loop (Figure 19.2).[78] The PKA-phosphorylation of PDE4D3 on Ser13 increases the affinity of PDE4D3 for mAKAP, suggesting that activation of mAKAP-anchored PKA enhances the recruitment of PDE4D3, facilitating quicker signal termination.[96]

19.6 SIGNAL COMPLEXES ORGANIZED BY A-KINASE ANCHORING PROTEINS

The highest level of specificity, and complexity, in cAMP–PKA signaling is accomplished by the assembly of multiprotein complexes by AKAPs. Several AKAPs with this property have been identified that provide precise spatiotemporal regulation of the cAMP–PKA pathway and integration with other signaling pathways in one-signal complex. AKAP79, AKAP450, AKAP220, Gravin, WAVE, AKAP-Lbc, and mAKAP have all been shown to scaffold signaling complexes, and it is likely that we are still in the very beginning stage of under-standing the roles AKAPs play in the orchestration of intracellular signaling events in health and disease (for studies and recent reviews, see Refs. [9,10,12,14,68,97,98]).

In addition to its role in anchoring RII, studies of AKAP79 have contributed to the evolution of the model of AKAPs as multiprotein-scaffolding proteins able to bind and anchor multiple signal transduction proteins and also regulate their enzymatic function. Although originally discovered as proteins able to bind and anchor PKA, the capacity of AKAP79 to associate with other signaling enzymes has led to an extension of the original AKAP model. By coordinating the location of PKC and the Ca^{2+}–calmodulin-dependent phosphoprotein phosphatase PP2B (calcineurin) in addition to PKA, AKAP79 positions two second-messenger-regulated kinases and a phosphatase near neuronal substrates at the post-synaptic densities.[99,100] Using an elegant experimental system that combined siRNA to knock down AKAP79 with overexpression of siRNA-resistant, mutant versions of the AKAP, selection of transfected cells and cAMP imaging using FRET reporters, Hoshi et al.[101] could show that AKAP79 coordinates different combinations of anchored enzymes to modu-late distinct signal pathways from two distinct receptors (bradykinin and muscarinic recep-tors) to two different effector systems, in this case two neuronal ion channels.

In a recent report, another signal complex involving mAKAP, PKA, PDE4D3, Epac, and a MEKK/MEK5/ERK5 module was described (Figure 19.3).[102] Whereas mAKAP targets the complex and serves as an anchoring protein for PKA and PDE4D3, PDE4D3 serves as an adaptor for Epac and ERK5 and acts as a scaffold to coordinate to a MEKK/MEK5/ERK5 module.[102] In this model, Erk phosphorylation inhibits PDE4D3[47] and thus serves to increase local cAMP. This increase in cAMP, in turn, activates PKA, which phosphorylates PDE4D3 at a different site to increase both its activity[43,79] and affinity for mAKAP[96]; this leads to decreased cAMP concentrations with a small time delay (Figure 19.3). At peak level of cAMP, Epac1,[5] a cAMP-stimulated guanine exchange protein, is also triggered and activates

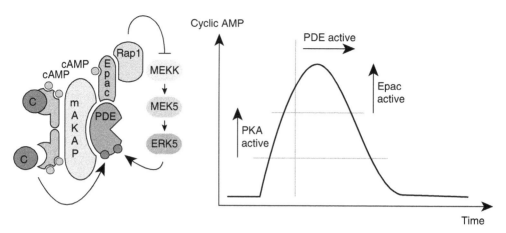

FIGURE 19.3 (See color insert following page 274.) Signal complex involving mAKAP, PKA, PDE4D3, Epac, and a MEKK/MEK5/ERK5 module. mAKAP anchors PKA and PDE4D3 whereas PDE4D3 scaffolds an Epac–Rap1 pathway that coordinates a MEKK/MEK5/ERK5 module (*left panel*). ERK5 phosphorylation inhibits PDE4D3 and increases local cAMP. PKA activated by low cAMP concentration (*right panel*) activates PDE4D3 by phosphorylation (*right panel*), leading to a small time delay to PDE activation (*right panel*). At peak cAMP concentration (*right panel*), Epac1 is triggered and activates the small Ras-like GTPase Rap-1, which inhibits MEKK and thus releases inhibition of PDE4D3 by ERK5 (*left panel*).

the small Ras-like GTPase Rap-1.[103,104] Rap-1, then in turn, inhibits MEKK and thus releases inhibition of PDE4D3 by ERK5.[102] However, as the cAMP level decreases, ERK5 activity is prominent once again, leading to inhibition of the PDE to allow for a new peak of cAMP.

Using FRET reporters, the AKAP/PKA/PDE signaling unit has been observed to generate local pulses of cAMP[102]; this is probably due to the different thresholds for activation of PKA ($K_{act} = 100–400$ nM) and PDE4D3 ($K_m = 1–4$ μM) and the time delay in PDE activation (Figure 19.3).[37] This is particularly relevant for the mAKAP signal complex that coordinates three cAMP-responsive signaling molecules and where the Epac pathway ($K_d = 4$ μM) would activate only at peak cAMP concentrations. The configuration organized by scaffolding of mAKAP and PDE4D3 itself creates a spatially organized signal complex with signal transduction molecules from several pathways that exploit the temporally regulated cAMP gradients by activating at different thresholds (Figure 19.3).[102]

19.7 CONCLUDING REMARKS

Although a number of early studies indicated possibilities of compartmentation of cAMP, the predominating view, only little more than a decade ago, was still that cAMP would be elevated throughout the cell in response to many ligands. Detailed studies of compartmentation of specific receptors and adenylyl cyclases to distinct membrane subdomains as well as live cell imaging of cAMP and unraveling of the subcellular targeting of PDEs has now made clear that physiological increases in cAMP occur in discrete microdomains. Similarly, although PKA type II was well known to be biochemically particulate and several AKAPs were known 10 years ago, the prevailing view was still that many effects of cAMP would be mediated by en bloc activation of PKA over large areas of the cell and that the C subunit would be released from a PKA holoenzyme complex and travel some distance to find its substrate. However, since then it has become clear that a large spectrum of AKAP proteins is available (more than 50 AKAPs per date when differentially targeted splice variants are included, Table 19.1). Furthermore, new AKAPs for PKA type I long thought to be primarily

cytoplasmic and freely diffusible are now increasingly reported. In addition, the requirement for anchoring of PKA in order to regulate specific substrates as well as to mediate a number of physiological effects has been extensively studied over the past decade and with few exceptions it has been shown that most cAMP/PKA-regulated physiological processes require an anchored kinase. Similarly, it has been made clear from studies discussed here that the localized cAMP microdomains require carefully compartmentalized and anchored pools of PDE enzymes. The concept described here, that has emerged over the past 10–15 years and which is now well established, is that a ligand normally will elicit a characteristic and local pool of cAMP that will follow a distinct route to reach and activate a single PKA–AKAP complex close to the substrate in order to mediate a distinct biological effect (Figure 19.1). Accordingly, each substrate (or group of selected substrates) appears to have a restricted and anchored pool of PKA and its own local spatiotemporally organized gradient of cAMP limited by PDEs. Future studies are expected to identify a number of new anchoring proteins for the plethora of PDE enzymes and to increase our understanding of PDE regulation of the temporal aspect of localized cAMP pulses.

ACKNOWLEDGMENTS

Our work is supported by grants from the Functional Genomics Programme, The Research Council of Norway, The Norwegian Cancer Society, Novo Nolchisk Foundation and the European Community.

REFERENCES

1. Sunahara, R.K., C.W. Dessauer, and A.G. Gilman. 1996. Complexity and diversity of mammalian adenylyl cyclases. *Annu Rev Pharmacol Toxicol* 36:461.
2. Soderling, S.H., and J.A. Beavo. 2000. Regulation of cAMP and cGMP signaling: New phosphodiesterases and new functions. *Curr Opin Cell Biol* 12:174.
3. Conti, M., and S.L. Jin. 1999. The molecular biology of cyclic nucleotide phosphodiesterases. *Prog Nucleic Acid Res Mol Biol* 63:1.
4. Houslay, M.D., and D.R. Adams. 2003. PDE4 cAMP phosphodiesterases: Modular enzymes that orchestrate signalling cross talk, desensitization and compartmentalization. *Biochem J* 370:1.
5. de Rooij, J., et al. 1998. Epac is a Rap1 guanine-nucleotide-exchange factor directly activated by cyclic AMP. *Nature* 396:474.
6. Kawasaki, H., et al. 1998. A family of cAMP-binding proteins that directly activate Rap1. *Science* 282:2275.
7. Kaupp, U.B., and R. Seifert. 2002. Cyclic nucleotide-gated ion channels. *Physiol Rev* 82:769.
8. Matulef, K., and W.N. Zagotta. 2003. Cyclic nucleotide-gated ion channels. *Annu Rev Cell Dev Biol* 19:23.
9. Tasken, K., and E.M. Aandahl. 2004. Localised effects of cAMP mediated by distinct routes of protein kinase A. *Physiol Rev* 84:137.
10. Michel, J.J., and J.D. Scott. 2002. AKAP mediated signal transduction. *Annu Rev Pharmacol Toxicol* 42:235.
11. Zaccolo, M., and T. Pozzan. 2002. Discrete microdomains with high concentration of cAMP in stimulated rat neonatal cardiac myocytes. *Science* 295:1711.
12. Diviani, D., and J.D. Scott. 2001. AKAP signaling complexes at the cytoskeleton. *J Cell Sci* 114:1431.
13. Smith, F.D., and J.D. Scott. 2002. Signaling complexes: Junctions on the intracellular information super highway. *Curr Biol* 12:R32.
14. Baillie, G.S., J.D. Scott, and M.D. Houslay. 2005. Compartmentalisation of phosphodiesterases and protein kinase A: Opposites attract. *FEBS Lett* 579:3264.
15. Jin, S.L., et al. 1999. Impaired growth and fertility of cAMP-specific phosphodiesterase PDE4D-deficient mice. *Proc Natl Acad Sci USA* 96:11998.

16. Hansen, G., et al. 2000. Absence of muscarinic cholinergic airway responses in mice deficient in the cyclic nucleotide phosphodiesterase PDE4D. *Proc Natl Acad Sci USA* 97:6751.

17. Jin, S.L., and M. Conti. 2002. Induction of the cyclic nucleotide phosphodiesterase PDE4B is essential for LPS-activated TNF-alpha responses. *Proc Natl Acad Sci USA* 99:7628.

18. Jin, S.L., et al. 2005. Specific role of phosphodiesterase 4B in lipopolysaccharide-induced signaling in mouse macrophages. *J Immunol* 175:1523.

19. Masciarelli, S., et al. 2004. Cyclic nucleotide phosphodiesterase 3A-deficient mice as a model of female infertility. *J Clin Invest* 114:196.

20. Robichaud, A., et al. 2002. Deletion of phosphodiesterase 4D in mice shortens alpha(2)-adreno-ceptor-mediated anesthesia, a behavioral correlate of emesis. *J Clin Invest* 110:1045.

21. Yang, G. et al. 2003. Phosphodiesterase 74-deficient mice have functional T cells. *J. Immunol* 171:6414.

22. Houslay, M.D., and G.S. Baillie. 2005. Beta-arrestin-recruited phosphodiesterase-4 desensitizes the AKAP79/PKA-mediated switching of beta2-adrenoceptor signalling to activation of ERK. *Biochem Soc Trans* 33:1333.

23. Lynch, M.J. et al. 2005. RNA silencing identifies PDE4D5 as the functionally relevant cAMP phosphochiesterase interacting with beta arrestin to control the protein kinase A/AKAP79-mediated switching of the beta2-adrenergic receptor to activation of ERK in HEK293B2 cells. *J Biol chem* 280:33178

24. Baillie, G.S., et al. 2003. Beta-arrestin-mediated PDE4 cAMP phosphodiesterase recruitment regulates beta-adrenoceptor switching from Gs to Gi. *Proc Natl Acad Sci USA* 100:940.

25. McCahill, A., et al. 2005. In resting COS1 cells a dominant negative approach shows that specific, anchored PDE4 cAMP phosphodiesterase isoforms gate the activation, by basal cyclic AMP production, of AKAP-tethered protein kinase A type II located in the centrosomal region. *Cell Signal* 17:1158.

26. Spina, D. 2003. Theophylline and PDE4 inhibitors in asthma. *Curr Opin Pulm Med* 9:57.

27. Feldman, A.M., and D.M. McNamara. 2002. Reevaluating the role of phosphodiesterase inhibitors in the treatment of cardiovascular disease. *Clin Cardiol* 25:256.

28. Grouse, J.R. 3rd, M.C. Allan, and M.B. Elam. 2002. Clinical manifestation of atherosclerotic peripheral arterial disease and the role of cilostazol in treatment of intermittent claudication. *J Clin Pharmacol* 42:1291.

29. Manji, H.K., et al. 2003. Enhancing neuronal plasticity and cellular resilience to develop novel, improved therapeutics for difficult-to-treat depression. *Biol Psyc* 53:707.

30. Millar, J.K., et al. 2005. DISC1 and PDE4B are interacting genetic factors in schizophrenia that regulate cAMP signaling. *Science* 310:1187.

31. Corbin, J.D., S.H. Francis, and D.J. Webb. 2002. Phosphodiesterase type 5 as a pharmacologic target in erectile dysfunction. *Urology* 60:4.

32. Houslay, M.D., P. Schafer, and K.Y. Zhang. 2005. Keynote review: Phosphodiesterase-4 as a therapeutic target. *Drug Discov Today* 10:1503.

33. Pozzan, T., et al. 1994. Molecular and cellular physiology of intracellular calcium stores. *Physiol Rev* 74:595.

34. Thorn, P. 1996. Spatial domains of Ca^{2+} signaling in secretory epithelial cells. *Cell Calcium* 20:203.

35. Rios, E., and M.D. Stern. 1997. Calcium in close quarters: Microdomain feedback in excitation–contraction coupling and other cell biological phenomena. *Annu Rev Biophys Biomolec Struct* 26:47.

36. Zaccolo, M., P. Magalhaes, and T. Pozzan. 2002. Compartmentalisation of cAMP and Ca^{2+} signals. *Curr Opin Cell Biol* 14:160.

37. Mongillo, M., et al. 2004. Fluorescence resonance energy transfer-based analysis of cAMP dynamics in live neonatal rat cardiac myocytes reveals distinct functions of compartmentalized phosphodiesterases. *Circ Res* 95:67.

38. Mongillo, M., et al. 2006. Compartmentalized phosphodiesterase-2 activity blunts beta-adrenergic cardiac inotropy via an NO/cGMP-dependent pathway. *Circ Res* 98:226.

39. Strada, S.J., M.W. Martin, and W.J. Thompson. 1984. General properties of multiple molecular forms of cyclic nucleotide phosphodiesterase in the nervous system. *Adv Cyclic Nucleotide Protein Phosphorylation Res* 16:13.

40. Mehats, C., et al. 2002. Cyclic nucleotide phosphodiesterases and their role in endocrine cell signaling. *Trends Endocrinol Metab* 13:29.

41. Degerman, E., P. Belfrage, and V.C. Manganiello. 1997. Structure, localization, and regulation of cGMP-inhibited phosphodiesterase (PDE3). *J Biol Chem* 272:6823.
42. Yarwood, S.J., et al. 1999. The RACK1 signaling scaffold protein selectively interacts with the cAMP-specific phosphodiesterase PDE4D5 isoform. *J Biol Chem* 274:14909.
43. Jin, S.L., et al. 1998. Subcellular localization of rolipram-sensitive, cAMP-specific phosphodiesterases. Differential targeting and activation of the splicing variants derived from the PDE4D gene. *J Biol Chem* 273:19672.
44. Verde, I., et al. 2001. Myomegalin is a novel protein of the golgi/centrosome that interacts with a cyclic nucleotide phosphodiesterase. *J Biol Chem* 276:11189.
45. Beard, M.B., et al. 1999. The unique N-terminal domain of the cAMP phosphodiesterase PDE4D4 allows for interaction with specific SH3 domains. *FEBS Lett* 460:173.
46. Beard, M.B., et al. 2002. In addition to the SH3 binding region, multiple regions within the N-terminal noncatalytic portion of the cAMP-specific phosphodiesterase, PDE4A5, contribute to its intracellular targeting. *Cell Signal* 14:453.
47. MacKenzie, S.J., et al. 2000. ERK2 mitogen-activated protein kinase binding, phosphorylation, and regulation of the PDE4D cAMP-specific phosphodiesterases—The involvement of COOH-terminal docking sites and NH2-terminal UCR regions. *J Biol Chem* 275:16609.
48. Baillie, G.S., et al. 2002. TAPAS-1, a novel microdomain within the unique N-terminal region of the PDE4A1 cAMP specific phosphodiesterase that allows rapid, Ca^{2+}-triggered membrane association with selectivity for interaction with phosphatidic acid. *J Biol Chem* 277 (31):28298.
49. Perry, S.J., et al. 2002. Targeting of cyclic AMP degradation to beta 2-adrenergic receptors by beta-arrestins. *Science* 298:834.
50. Bolger, G.B., et al. 2003. The unique amino-terminal region of the PDE4D5 cAMP phosphodiesterase isoform confers preferential interaction with beta-arrestins. *J Biol Chem* 278:49230.
51. Torgersen, K.M., et al. 2002. Molecular mechanisms for protein kinase A-mediated modulation of immune function. *Cell Signal* 14:1.
52. Skalhegg, B.S., et al. 1994. Location of cAMP-dependent protein kinase type I with the TCR–CD3 complex. *Science* 263:84.
53. Vang, T., et al. 2001. Activation of the COOH-terminal Src kinase (Csk) by cAMP-dependent protein kinase inhibits signaling through the T cell receptor. *J Exp Med* 193:497.
54. Vang, T., H. Abrahamsen, and K. Tasken. 2003. Protein kinase A intersects Src signaling in membrane microdomains. *J Biol Chem* 278 (19):17170.
55. Giembycz, M.A., et al. 1996. Identification of cyclic AMP phosphodiesterases 3, 4 and 7 in human CD4+ and CD8+ T-lymphocytes: Role in regulating proliferation and the biosynthesis of interleukin-2. *Br J Pharmacol* 118:1945.
56. Glavas, N.A., et al. 2001. T cell activation up-regulates cyclic nucleotide phosphodiesterases 8A1 and 7A3. *Proc Natl Acad Sci USA* 98:6319.
57. Li, L., C. Yee, and J.A. Beavo. 1999. CD3- and CD28-dependent induction of PDE7 required for T cell activation. *Science* 283:848.
58. Robicsek, S.A., et al. 1991. Multiple high-affinity cAMP-phosphodiesterases in human T-lymphocytes. *Biochem Pharmacol* 42:869.
59. Lee, R., et al. 2002. PDE7A is expressed in human B-lymphocytes and is up-regulated by elevation of intracellular cAMP. *Cell Signal* 14:277.
60. Bender, A.T., et al. 2005. Selective up-regulation of PDE1B2 upon monocyte-to-macrophage differentiation. *Proc Natl Acad Sci USA* 102:497.
61. Erdogan, S., and M.D. Houslay. 1997. Challenge of human Jurkat T-cells with the adenylate cyclase activator forskolin elicits major changes in cAMP phosphodiesterase (PDE) expression by up-regulating PDE3 and inducing PDE4D1 and PDE4D2 splice variants as well as down-regulating a novel PDE4A splice variant. *Biochem J* 321 (Pt 1):165.
62. Abrahamsen, H., et al. 2004. TCR- and CD28-mediated recruitment of phosphodiesterase 4 to lipid rafts potentiates TCR signaling. *J Immunol* 173:4847.
63. Theurkauf, W.E., and R.B. Vallee. 1982. Molecular characterization of the cAMP-dependent protein kinase bound to microtubule-associated protein 2. *J Biol Chem* 257:3284.

64. Sarkar, D., J. Erlichman, and C.S. Rubin. 1984. Identification of a calmodulin-binding protein that co-purifies with the regulatory subunit of brain protein kinase II. *J Biol Chem* 259:9840.

65. Lohmann, S.M., et al. 1984. High-affinity binding of the regulatory subunit (RII) of cAMP-dependent protein kinase to microtubule-associated and other cellular proteins. *Proc Natl Acad Sci USA* 81:6723.

66. Rubin, C.S. 1994. A kinase anchor proteins and the intracellular targeting of signals carried by cyclic AMP. *Biochim Biophys Acta* 1224:467.

67. Colledge, M., and J.D. Scott. 1999. AKAPs: From structure to function. *Trends Cell Biol* 9:216.

68. Dodge, K., and J.D. Scott. 2000. AKAP79 and the evolution of the AKAP model. *FEBS Lett* 476:58.

69. Newlon, M.G., et al. 1997. The A-kinase anchoring domain of type IIalpha cAMP-dependent protein kinase is highly helical. *J Biol Chem* 272:23637.

70. Newlon, M.G., et al. 1999. The molecular basis for protein kinase A anchoring revealed by solution NMR. *Nat Struct Biol* 6:222.

71. Newlon, M.G., et al. 2001. A novel mechanism of PKA anchoring revealed by solution structures of anchoring complexes. *EMBO J* 20:1651.

72. Hausken, Z.E., et al. 1994. Type II regulatory subunit (RII) of the cAMP-dependent protein kinase interaction with A-kinase anchor proteins requires isoleucines 3 and 5. *J Biol Chem* 269:24245.

73. Hausken, Z.E., et al. 1996. Mutational analysis of the A-kinase anchoring protein (AKAP)-binding site on RII. Classification of side chain determinants for anchoring and isoform selective association with AKAPs. *J Biol Chem* 271:29016.

74. Leon, D.A., et al. 1997. A stable alpha-helical domain at the N terminus of the RIalpha subunits of cAMP-dependent protein kinase is a novel dimerization/docking motif. *J Biol Chem* 272:28431.

75. Banky, P., L.J. Huang, and S.S. Taylor. 1998. Dimerization/docking domain of the type Ialpha regulatory subunit of cAMP-dependent protein kinase. Requirements for dimerization and docking are distinct but overlapping. *J Biol Chem* 273:35048.

76. Banky, P., et al. 2000. Isoform-specific differences between the type Ialpha and IIalpha cyclic AMP-dependent protein kinase anchoring domains revealed by solution NMR. *J Biol Chem* 275:35146.

77. Tasken, K.A., et al. 2001. Phosphodiesterase 4D and protein kinase a type II constitute a signaling unit in the centrosomal area. *J Biol Chem* 276:21999.

78. Dodge, K.L., et al. 2001. mAKAP assembles a protein kinase A/PDE4 phosphodiesterase cAMP signaling module. *EMBO J* 20:1921.

79. MacKenzie, S.J., et al. 2002. Long PDE4 cAMP specific phosphodiesterases are activated by protein kinase A-mediated phosphorylation of a single serine residue in upstream conserved region 1 (UCR1). *Br J Pharmacol* 136:421.

80. Conti, M. 2000. Phosphodiesterases and cyclic nucleotide signaling in endocrine cells. *Mol Endocrinol* 14:1317.

81. Asirvatham, A.L., et al. 2004. A-kinase anchoring proteins interact with phosphodiesterases in T lymphocyte cell lines. *J Immunol* 173:4806.

82. Takahashi, M., et al. 1999. Characterization of a novel giant scaffolding protein, CG-NAP, that anchors multiple signaling enzymes to centrosome and the golgi apparatus. *J Biol Chem* 274:17267.

83. Takahashi, M., et al. 2000. Association of immature hypophosphorylated protein kinase c epsilon with an anchoring protein CG-NAP. *J Biol Chem* 275:34592.

84. Matsuzawa, K., et al. 1997. Domain-specific phosphorylation of vimentin and glial fibrillary acidic protein by PKN. *Biochem Biophys Res Commun* 234:621.

85. Mukai, H., et al. 1996. PKN associates and phosphorylates the head-rod domain of neurofilament protein. *J Biol Chem* 271:9816.

86. Amano, M., et al. 1996. Identification of a putative target for Rho as the serine–threonine kinase protein kinase N. *Science* 271:648.

87. Watanabe, G., et al. 1996. Protein kinase N (PKN) and PKN-related protein rhophilin as targets of small GTPase Rho. *Science* 271:645.

88. Mukai, H., et al. 1994. Activation of PKN, a novel 120-kDa protein kinase with leucine zipper-like sequences, by unsaturated fatty acids and by limited proteolysis. *Biochem Biophys Res Commun* 204:348.

89. Takahashi, M., et al. 1998. Proteolytic activation of PKN by caspase-3 or related protease during apoptosis. *Proc Natl Acad Sci USA* 95:11566.

90. Sillibourne, J.E., et al. 2002. Centrosomal anchoring of the protein kinase CK1delta mediated by attachment to the large, coiled-coil scaffolding protein CG-NAP/AKAP450. *J Mol Biol* 322:785.

91. McCartney, S., et al. 1995. Cloning and characterization of A-kinase anchor protein 100 (AKAP100). A protein that targets A-kinase to the sarcoplasmic reticulum. *J Biol Chem* 270:9327.

92. Kapiloff, M.S., et al. 1999. mAKAP: An A-kinase anchoring protein targeted to the nuclear membrane of differentiated myocytes. *J Cell Sci* 112 (Pt 16):2725.

93. Marx, S.O., et al. 2000. PKA phosphorylation dissociates FKBP12.6 from the calcium release channel (ryanodine receptor): Defective regulation in failing hearts. *Cell* 101:365.

94. Yang, J., et al. 1998. A-kinase anchoring protein 100 (AKAP100) is localized in multiple subcellular compartments in the adult rat heart. *J Cell Biol* 142:511.

95. Zimmer, H.G., 1997. Catecholamine-induced cardiac hypertrophy: Significance of proto-oncogene expression. *J Mol Med* 75:849.

96. Carlisle Michel, J.J., et al. 2004. PKA-phosphorylation of PDE4D3 facilitates recruitment of the mAKAP signalling complex. *Biochem J* 381:587.

97. Wong, W., and J.D. Scott. 2004. AKAP signalling complexes: Focal points in space and time. *Nat Rev Mol Cell Biol* 5:959.

98. Carnegie, G.K., et al. 2004. AKAP-Lbc nucleates a protein kinase D activation scaffold. *Mol Cell* 15:889.

99. Coghlan, V.M., et al. 1995. Association of protein kinase A and protein phosphatase 2B with a common anchoring protein. *Science* 267:108.

100. Klauck, T.M., et al. 1996. Coordination of three signaling enzymes by AKAP79, a mammalian scaffold protein. *Science* 271:1589.

101. Hoshi, N., L.K. Langeberg, and J.D. Scott. 2005. Distinct enzyme combinations in AKAP signalling complexes permit functional diversity. *Nat Cell Biol* 7:1066.

102. Dodge-Kafka, K.L., et al. 2005. The protein kinase A anchoring protein mAKAP coordinates two integrated cAMP effector pathways. *Nature* 437:574.

103. Pizon, V., et al. 1988. Human Cdnas Rap1 and Rap2 homologous to the *Drosophila* gene Dras3 encode proteins closely related to Ras in the effector region. *Oncogene* 3:201.

104. Bos, J.L. 1998. All in the family? New insights and questions regarding interconnectivity of Ras, Rap1 and Ral. *EMBO J* 17:6776.

105. Lin, R.Y., S.B. Moss, and C.S. Rubin. 1995. Characterization of S-AKAP84, a novel developmentally regulated A kinase anchor protein of male germ cells. *J Biol Chem* 270:27804.

106. Chen, Q., R.Y. Lin, and C.S. Rubin. 1997. Organelle-specific targeting of protein kinase AII (PKAII). Molecular and in situ characterization of murine A kinase anchor proteins that recruit regulatory subunits of PKAII to the cytoplasmic surface of mitochondria. *J Biol Chem* 272:15247.

107. Huang, L.J., et al. 1997. Identification of a novel protein kinase A anchoring protein that binds both type I and type II regulatory subunits. *J Biol Chem* 272:8057.

108. Trendelenburg, G., et al. 1996. Molecular characterization of AKAP149, a novel A kinase anchor protein with a KH domain. *Biochem Biophys Res Commun* 225:313.

109. Huang, L.J., et al. 1999. NH2-Terminal targeting motifs direct dual specificity A-kinase-anchoring protein 1 (D-AKAP1) to either mitochondria or endoplasmic reticulum. *J Cell Biol* 145:951.

110. Steen, R.L., et al. 2000. Recruitment of protein phosphatase 1 to the nuclear envelope by A-kinase anchoring protein AKAP149 is a prerequisite for nuclear lamina assembly. *J Cell Biol* 150:1251.

111. Dong, F., et al. 1998. Molecular characterization of a cDNA that encodes six isoforms of a novel murine A kinase anchor protein. *J Biol Chem* 273:6533.

112. Mandal, A., et al. 1999. FSP95, a testis-specific 95-kilodalton fibrous sheath antigen that undergoes tyrosine phosphorylation in capacitated human spermatozoa. *Biol Reprod* 61:1184.

113. Vijayaraghavan, S., et al. 1999. Isolation and molecular characterization of AKAP110, a novel, sperm-specific protein kinase A-anchoring protein. *Mol Endocrinol* 13:705.

114. Miki, K., and E.M. Eddy. 1998. Identification of tethering domains for protein kinase A type Ialpha regulatory subunits on sperm fibrous sheath protein FSC1. *J Biol Chem* 273:34384.

115. Carrera, A., G.L. Gerton, and S.B. Moss. 1994. The major fibrous sheath polypeptide of mouse sperm: Structural and functional similarities to the A-kinase anchoring proteins. *Dev Biol* 165:272.

116. Miki, K., et al. 2002. Targeted disruption of the Akap4 gene causes defects in sperm flagellum and motility. *Dev Biol* 248:331.

117. Bregman, D.B., N. Bhattacharyya, and C.S. Rubin. 1989. High affinity binding protein for the regulatory subunit of cAMP-dependent protein kinase II-B. Cloning, characterization, and expression of cDNAs for rat brain P150. *J Biol Chem* 264:4648.

118. Carr, D.W., et al. 1992. Localization of the cAMP-dependent protein kinase to the postsynaptic densities by A-kinase anchoring proteins. Characterization of AKAP 79. *J Biol Chem* 267:16816.

119. Colledge, M., et al. 2000. Targeting of PKA to glutamate receptors through a MAGUK–AKAP complex. *Neuron* 27:107.

120. Hoshi, N., et al. 2003. AKAP150 signaling complex promotes suppression of the M-current by muscarinic agonists. *Nat Neurosci* 6:564.

121. Michel, J.J., et al. 2005. Spatial restriction of PDK1 activation cascades by anchoring to mAKA-Palpha. *Mol Cell* 20:661.

122. Gray, P.C., et al. 1997. Identification of a 15-kDa cAMP-dependent protein kinase-anchoring protein associated with skeletal muscle L-type calcium channels. *J Biol Chem* 272:6297.

123. Gray, P.C., et al. 1998. Primary structure and function of an A kinase anchoring protein associated with calcium channels. *Neuron* 20:1017.

124. Fraser, I.D., et al. 1998. A novel lipid-anchored A-kinase anchoring protein facilitates cAMP-responsive membrane events. *EMBO J* 17:2261.

125. Trotter, K.W., et al. 1999. Alternative splicing regulates the subcellular localization of A-kinase anchoring protein 18 isoforms. *J Cell Biol* 147:1481.

126. Klussmann, E., and W. Rosenthal. 2001. Role and identification of protein kinase A anchoring proteins in vasopressin-mediated aquaporin-2 translocation. *Kidney Int* 60:446.

127. Henn, V., et al. 2004. Identification of a novel A-kinase anchoring protein 18 isoform and evidence for its role in the vasopressin-induced aquaporin-2 shuttle in renal principal cells. *J Biol Chem* 279:26654.

128. Coghlan, V.M., et al. 1994. Cloning and characterization of AKAP 95, a nuclear protein that associates with the regulatory subunit of type II cAMP-dependent protein kinase. *J Biol Chem* 269:7658.

129. Eide, T., et al. 1998. Molecular cloning, chromosomal localization, and cell cycle-dependent subcellular distribution of the A-kinase anchoring protein, AKAP95. *Exp Cell Res* 238:305.

130. Collas, P., K. Le Guellec, and K. Tasken. 1999. The A-kinase-anchoring protein AKAP95 is a multivalent protein with a key role in chromatin condensation at mitosis. *J Cell Biol* 147:1167.

131. Steen, R.L., et al. 2000. A kinase-anchoring protein AKAP95 recruits human chromosome-associated protein (hCAP)-D2/Eg7 for chromosome condensation in mitotic extract. *J Cell Biol* 149:531.

132. Eide, T., et al. 2002. Distinct but overlapping domains of AKAP95 are implicated in chromosome condensation and condensin targeting. *EMBO Rep* 3:426.

133. Witczak, O., et al. 1999. Cloning and characterization of a cDNA encoding an A-kinase anchoring protein located in the centrosome, AKAP450. *EMBO J* 18:1858.

134. Schmidt, P.H., et al. AKAP350, a multiply spliced protein kinase A-anchoring protein associated with centrosomes. *J Biol Chem* 274:3055.

135. Gillingham, A.K., and S. Munro. 2000. The PACT domain, a conserved centrosomal targeting motif in the coiled-coil proteins AKAP450 and pericentrin. *EMBO Rep* 1:524.

136. Keryer, G., et al. 2003. Dissociating the centrosomal matrix protein AKAP450 from centrioles impairs centriole duplication and cell cycle progression. *Mol Biol Cell* 14 (6):2436.

137. Carlson, C.R., et al. 2001. CDK1-mediated phosphorylation of the RIIalpha regulatory subunit of PKA works as a molecular switch that promotes dissociation of RIIalpha from centrosomes at mitosis. *J Cell Sci* 114:3243.

138. Bailly, E., et al. 1992. Cytoplasmic accumulation of cyclin B1 in human cells: Association with a detergent-resistant compartment and with the centrosome. *J Cell Sci* 101 (Pt 3):529.
139. Bailly, E., et al. 1989. p34cdc2 is located in both nucleus and cytoplasm; part is centrosomally associated at G2/M and enters vesicles at anaphase. *EMBO J* 8:3985.
140. Lin, J.W., et al. 1998. Yotiao, a novel protein of neuromuscular junction and brain that interacts with specific splice variants of NMDA receptor subunit NR1. *J Neurosci* 18:2017.
141. Westphal, R.S., et al. 1999. Regulation of NMDA receptors by an associated phosphatase-kinase signaling complex. *Science* 285:93.
142. Feliciello, A., et al. 1999. Yotiao protein, a ligand for the NMDA receptor, binds and targets cAMP-dependent protein kinase II(1). *FEBS Lett* 464:174.
143. Huang, L.J., et al. 1997. D-AKAP2, a novel protein kinase A anchoring protein with a putative RGS domain. *Proc Natl Acad Sci USA* 94:11184.
144. Hamuro, Y., et al. 2002. Domain organization of D-AKAP2 revealed by enhanced deuterium exchange-mass spectrometry (DXMS). *J Mol Biol* 321:703.
145. Reinton, N., et al. 2000. Localization of a novel human A-kinase-anchoring protein, hAKAP220, during spermatogenesis. *Dev Biol* 223:194.
146. Lester, L.B., et al. 1996. Cloning and characterization of a novel A-kinase anchoring protein. AKAP 220, association with testicular peroxisomes. *J Biol Chem* 271:9460.
147. Schillace, R.V., and J.D. Scott. 1999. Association of the type 1 protein phosphatase PP1 with the A-kinase anchoring protein AKAP220. *Curr Biol* 9:321.
148. Schillace, R.V., et al. 2001. Multiple interactions within the AKAP220 signaling complex contribute to protein phosphatase 1 regulation. *J Biol Chem* 276:12128.
149. Gordon, T., et al. 1992. Molecular cloning and preliminary characterization of a novel cytoplasmic antigen recognized by myasthenia gravis sera. *J Clin Invest* 90:992.
150. Nauert, J.B., et al. 1997. Gravin, an autoantigen recognized by serum from myasthenia gravis patients, is a kinase scaffold protein. *Curr Biol* 7:52.
151. Shih, M., et al. 1999. Dynamic complexes of beta2-adrenergic receptors with protein kinases and phosphatases and the role of gravin. *J Biol Chem* 274:1588.
152. Klingbeil, P., G. Frazzetto, and T. Bouwmeester. 2001. Xgravin-like (Xgl), a novel putative A-kinase anchoring protein (AKAP) expressed during embryonic development in *Xenopus*. *Mech Dev* 100:323.
153. Carr, D.W., et al. 1992. Association of the type II cAMP-dependent protein kinase with a human thyroid RII-anchoring protein. Cloning and characterization of the RII-binding domain. *J Biol Chem* 267:13376.
154. Diviani, D., J. Soderling, and J.D. Scott. 2001. AKAP-Lbc anchors protein kinase A and nucleates Galpha 12-selective Rho-mediated stress fiber formation. *J Biol Chem* 276:44247.
155. Klussmann, E., et al. 2001. Ht31: The first protein kinase A anchoring protein to integrate protein kinase A and Rho signaling. *FEBS Lett* 507:264.
156. Davare, M.A., et al. 1999. The A-kinase anchor protein MAP2B and cAMP-dependent protein kinase are associated with class C L-type calcium channels in neurons. *J Biol Chem* 274:30280.
157. Dransfield, D.T., et al. 1997. Ezrin is a cyclic AMP-dependent protein kinase anchoring protein. *EMBO J* 16:35.
158. Sun, F., et al. 2000. E3KARP mediates the association of ezrin and protein kinase A with the cystic fibrosis transmembrane conductance regulator in airway cells. *J Biol Chem* 275:29539.
159. Sun, F., et al. 2000. Protein kinase A associates with cystic fibrosis transmembrane conductance regulator via an interaction with ezrin. *J Biol Chem* 275:14360.
160. Mei, X., et al. 1997. Cloning and characterization of a testis-specific, developmentally regulated A-kinase-anchoring protein (TAKAP-80) present on the fibrous sheath of rat sperm. *Eur J Biochem* 246:425.
161. Hunzicker-Dunn, M., J.D. Scott, and D.W. Carr. 1998. Regulation of expression of A-kinase anchoring proteins in rat granulosa cells. *Biol Reprod* 58:1496.
162. Carr, D.W., et al. 1993. Follicle-stimulating hormone regulation of A-kinase anchoring proteins in granulosa cells. *J Biol Chem* 268:20729.

163. Salvador, L.M., et al. 2004. Neuronal microtubule-associated protein 2D is a dual A-kinase anchoring protein expressed in rat ovarian granulosa cells. *J Biol Chem* 279:27621.
164. Erlichman, J., et al. 1999. Developmental expression of the protein kinase C substrate/binding protein (clone 72/SSeCKS) in rat testis identification as a scaffolding protein containing an A-kinase-anchoring domain which is expressed during late-stage spermatogenesis. *Eur J Biochem* 263:797.
165. Diviani, D., et al. 2000. Pericentrin anchors protein kinase A at the centrosome through a newly identified RII-binding domain. *Curr Biol* 10:417.
166. Westphal, R.S., et al. 2000. Scar/WAVE-1, a Wiskott–Aldrich syndrome protein, assembles an actin-associated multi-kinase scaffold. *EMBO J* 19:4589.
167. Soderling, S.H., et al. 2003. Loss of WAVE-1 causes sensorimotor retardation and reduced learning and memory in mice. *Proc Natl Acad Sci USA* 100:1723.
168. Kussel-Andermann, P., et al. 2000. Unconventional myosin VIIA is a novel A-kinase-anchoring protein. *J Biol Chem* 275:29654.
169. Li, H., et al. 2001. Identification, localization, and function in steroidogenesis of PAP7: A peripheral-type benzodiazepine receptor- and PKA (RIalpha)-associated protein. *Mol Endocrinol* 15:2211.
170. Wang, X., et al. 2000. Neurobeachin: A protein kinase A-anchoring, beige/Chediak-higashi protein homolog implicated in neuronal membrane traffic. *J Neurosci* 20:8551.
171. Kultgen, P.L., et al. 2002. Characterization of an A-kinase anchoring protein in human ciliary axonemes. *Mol Biol Cell* 13:4156.
172. Schillace, R.V., et al. 2002. Identification and characterization of myeloid translocation gene 16b as a novel A-kinase anchoring protein in T lymphocytes. *J Immunol* 168:1590.
173. Fukuyama, T., et al. 2001. MTG8 proto-oncoprotein interacts with the regulatory subunit of type II cyclic AMP-dependent protein kinase in lymphocytes. *Oncogene* 20:6225.
174. Kovo, M., et al. 2002. Expression and modification of PKA and AKAPs during meiosis in rat oocytes. *Mol Cell Endocrinol* 192:105.
175. Rios, R.M., et al. 1992. Identification of a high affinity binding protein for the regulatory subunit RII beta of cAMP-dependent protein kinase in Golgi enriched membranes of human lympho-blasts. *EMBO J* 11:1723.
176. Li, H., et al. 2003. Protein kinase A-anchoring (AKAP) domains in brefeldin A-inhibited guanine nucleotide-exchange protein 2 (BIG2). *Proc Natl Acad Sci USA* 100:1627.
177. Alto, N.M., J. Soderling, and J.D. Scott. 2002. Rab32 is an A-kinase anchoring protein and participates in mitochondrial dynamics. *J Cell Biol* 158:659.
178. Angelo, R., and C.S. Rubin. 1998. Molecular characterization of an anchor protein (AKAPCE) that binds the RI subunit (RCE) of type I protein kinase A from *Caenorhabditis elegans*. *J Biol Chem* 273:14633.
179. Angelo, R.G., and C.S. Rubin. 2000. Characterization of structural features that mediate the tethering of *Caenorhabditis elegans* protein kinase A to a novel A kinase anchor protein. Insights into the anchoring of PKAI isoforms. *J Biol Chem* 275:4351.
180. Herrgard, S., et al. 2000. Domain architecture of a *Caenorhabditis elegans* AKAP suggests a novel AKAP function. *FEBS Lett* 486:107.
181. Han, J.D., N.E. Baker, and C.S. Rubin. 1997. Molecular characterization of a novel A kinase anchor protein from *Drosophila melanogaster*. *J Biol Chem* 272:26611.
182. Rossi, E.A., et al. 1999. Characterization of the targeting, binding, and phosphorylation site domains of an A kinase anchor protein and a myristoylated alanine-rich C kinase substrate-like analog that are encoded by a single gene. *J Biol Chem* 274:27201.
183. Li, Z., et al. 1999. Generation of a novel A kinase anchor protein and a myristoylated alanine-rich C kinase substrate-like analog from a single gene. *J Biol Chem* 274:27191.
184. Terman, J.R., and A.L. Kolodkin. 2004. Nervy links protein kinase a to plexin-mediated sema-phorin repulsion. *Science* 303:1204.
185. Gaillard, A.R., et al. 2001. Flagellar radial spoke protein 3 is an A-kinase anchoring protein (AKAP). *J Cell Biol* 153:443.

20 Role of Phosphodiesterases in Cyclic Nucleotide Compartmentation in Cardiac Myocytes

Aniella Abi-Gerges, Liliana R.V. Castro, Francesca Rochais, Grégoire Vandecasteele, and Rodolphe Fischmeister

CONTENTS

20.1 Cyclic Nucleotides Regulate Cardiac Function 396
    20.1.1 Cardiac Excitation–Contraction Coupling..396
    20.1.2 Regulation of Cardiac Function by cAMP Pathways.................................397
    20.1.3 Regulation of Cardiac Function by cGMP Pathways 397
        20.1.3.1 Cardiac Nitric Oxide Signaling 397
        20.1.3.2 Cardiac Natriuretic Peptides Signaling 398
        20.1.3.3 cGMP Signaling 398
20.2 Compartmentation of Cyclic Nucleotides Signaling 398
20.3 Methods to Study Cyclic Nucleotide Compartmentation in Intact Myocytes.........399
20.4 Role of Phosphodiesterases in Cyclic Nucleotide Compartmentation 401
    20.4.1 Phosphodiesterases and Hormone Specificity 401
    20.4.2 Specific Roles of Phosphodiesterase Isoforms in Cyclic Nucleotide Compartmentation...403
        20.4.2.1 PDE2..403
        20.4.2.2 PDE3..403
        20.4.2.3 PDE4..404
        20.4.2.4 PDE5..406
        20.4.2.5 Cooperative Role of Phosphodiesterase Isoforms.....................407
20.5 Pathophysiological Role of cAMP Compartmentation 407
Acknowledgments 408
References 408

A current challenge in cellular signaling is to decipher the complex intracellular spatiotemporal organization that any given cell type has concocted to be able to discriminate among different external stimuli acting via a common signaling pathway. This obviously applies to cAMP and cGMP signaling in the heart where these cyclic nucleotides determine the regulation of cardiac function by many hormones and neuromediators.[1–5] Recent studies have identified phosphodiesterases (PDEs) as key factors in limiting the spread of cAMP and

cGMP, and in shaping and organizing intracellular-signaling microdomains. With this new role, PDEs have been promoted from the rank of a housekeeping attendant to that of an executive officer.

20.1   CYCLIC NUCLEOTIDES REGULATE CARDIAC FUNCTION

20.1.1   CARDIAC EXCITATION–CONTRACTION COUPLING

When a myocyte is depolarized by an action potential, Ca^{2+} ions enter the cell through L-type Ca^{2+} channels (LTCCs) located on the sarcolemma and generate an inward current, $I_{Ca,L}$. This $I_{Ca,L}$ triggers a subsequent release of Ca^{2+} from the sarcoplasmic reticulum (SR) through ryanodine receptors (RyR2) Ca^{2+}-release channels, a mechanism known as Ca^{2+}-induced–Ca^{2+}release (CICR).[6] Ca^{2+} ions bind to contractile proteins, such as troponin I (TnI), and induce myofilament contraction.[7] During relaxation, part of cytosolic Ca^{2+} is sequestered into the SR by an ATP-dependent Ca^{2+} pump (sarcoendoplasmic reticulum Ca^{2+}-ATPase, SERCA), which is controlled by phospholamban (PLB), thus lowering the cytosolic Ca^{2+} concentration and removing Ca^{2+} from contractile proteins (Figure 20.1).[7]

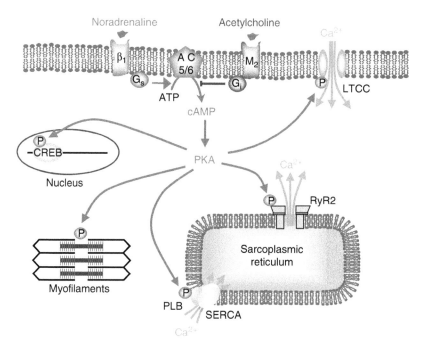

FIGURE 20.1 (See color insert following page 274.) Regulation of cardiac ECC by cAMP pathways. Upon liberation of noradrenaline by sympathetic nerve terminals, β_1-ARs activate cardiac AC isoforms (AC5 and AC6) via the stimulatory G protein (G_s). Cyclic AMP is synthesized and activates PKA, which phosphorylates different intracellular targets: LTCC phosphorylation leads to an increased Ca^{2+} influx and RyR2 phosphorylation at the SR membrane leads to an increased Ca^{2+}-induced Ca^{2+} release, both acting to enhance the force of contraction; PLB phosphorylation leads to an increased SR Ca^{2+} uptake by SERCA and TnI phosphorylation leads to a reduction in Ca^{2+} sensitivity of the myofilaments, both acting to accelerate contractile relaxation; and on the long term, CREB phosphorylation in the nucleus activates transcription. These effects are antagonized by a liberation of acetylcholine from parasympathetic nerve terminals, through activation of muscarinic M_2 receptors and inhibition of AC activity via inhibitory G proteins (G_i) (for further details, see text).

20.1.2 Regulation of Cardiac Function by cAMP Pathways

Force of contraction (inotropy) and beating frequency (chronotropy) are under the dual control of the sympathetic and parasympathetic systems. Both systems control the synthesis of cAMP in an opposite manner, and, hence, control the activity of the cAMP effectors (Figure 20.1). Catecholamines released into the synaptic cleft at sympathetic nerve terminals bind to β_1-adrenergic receptors (β_1-ARs) on the cardiac sarcolemma, activate stimulatory G proteins (G_s), which in turn activate two isoforms of adenylyl cyclase (AC), AC5 and AC6, that catalyze the conversion of ATP to cAMP. cAMP then activates the cAMP-dependent protein kinase (PKA), which initiates a series of phosphorylation processes activating excitation–contraction coupling (ECC). Phosphorylation of LTCC and RyR2 enhances their open probability, resulting in higher cytosolic Ca^{2+} concentrations and increased contractility.[7,8] Additionally, PKA phosphorylates PLB, hence stimulating SERCA activity and increasing Ca^{2+} reuptake into the SR.[9] Phosphorylation of TnI proteins decreases their sensitivity to Ca^{2+}.[10] These two latter events contribute to accelerate relaxation (lusitropic effect) during β-adrenergic stimulation. PKA also controls metabolic (glycogen synthase, phosphorylase kinase) and transcriptional activities (via cAMP-response element binding protein, CREB) in cardiac myocytes.[11] cAMP can also act in cardiomyocytes in a PKA-independent manner, through direct activation of Epac,[12] a guanine nucleotide-exchange factor for the small GTPase Rap1,[13] and HCN cyclic nucleotide–gated ion channels.[14] As a mirror image, acetylcholine released into the synaptic cleft at parasympathetic nerve terminals binds to the muscarinic M2 receptors and induces AC inhibition via inhibitory G protein (G_i), thus decreasing PKA activation (Figure 20.1).[15]

The level of intracellular cAMP is regulated by the balance between the activity of AC and the cyclic nucleotide PDEs that degrade cAMP to 5′AMP. PDE activity is found not only in cytosol, but also in a variety of membrane, nuclear, and cytoskeletal locations.[16] Cardiac PDEs fall into at least five families: PDE1, which is activated by Ca^{2+}-calmodulin; PDE2, which is stimulated by cGMP; PDE3, which is inhibited by cGMP; PDE4; and PDE5. Whereas PDE1 and PDE2 can hydrolyze both cAMP and cGMP; PDE3 preferentially hydrolyzes cAMP; PDE4 and PDE5 are specific for cAMP and cGMP, respectively.

20.1.3 Regulation of Cardiac Function by cGMP Pathways

It is generally accepted that cGMP opposes the effect of cAMP on cardiac function.[4,17] Intracellular cGMP production is achieved by two different forms of guanylyl cyclases: a soluble form (sGC), which is activated by nitric oxide (NO); a particulate form (pGC), which is activated by natriuretic peptides, such as ANP, BNP, and CNP.

20.1.3.1 Cardiac Nitric Oxide Signaling

In mammalian heart, NO can be produced by three different NO synthase (NOS) isoenzymes that are expressed in various cell types, including cardiomyocytes: the neuronal (nNOS or NOS1), the endothelial (eNOS or NOS3), and the inducible (iNOS or NOS2) isoforms.[18,19] Both eNOS and nNOS are constitutively expressed in cardiomyocytes. eNOS is located within membrane caveolae,[20] whereas nNOS is normally located in SR.[21] These isoforms undergo specific responses in the heart. For example, NO produced in SR by nNOS exerts a facilitatory effect on SERCA activity[22] and stimulates Ca^{2+} influx trough LTCC[23] whereas eNOS attenuates the inotropic response to β-adrenergic stimulation.[24] Most of the physiological effects of NO depend on cGMP synthesis that follows the binding of NO to the heme moiety of sGC. However, a direct reaction between NO and reactive thiol residues in several proteins, like RyR2, has also been shown to have significant functional consequences.[25,26]

20.1.3.2 Cardiac Natriuretic Peptides Signaling

ANP and BNP are primarily synthesized in cardiac atria and ventricles, respectively, whereas CNP is predominantly located in central nervous system, pituitary, kidney, and vascular endothelial cells.[27] CNP is secreted by endothelial cells in the heart,[28] but its role in myocardial function is less clear than that of ANP. Natriuretic peptides exert their effects by three single transmembrane natriuretic peptide receptors: NPR-A, NPR-B, and NPR-C. Both NPR-A and NPR-B have intrinsic enzymatic activity in their cytosolic domain and catalyze the synthesis of cGMP from GTP.[29] NPR-C lacks enzymatic activity, but controls local concentrations of natriuretic peptide through constitutive receptor-mediated internalization and degradation[30] and acts via G_i-dependent mechanisms.[31] All receptors can be activated by the three natriuretic peptides but NPR-A has a higher affinity for ANP and BNP whereas NPR-B is more specific for CNP.[32]

20.1.3.3 cGMP Signaling

Intracardiac cGMP acts via three main enzymes: PDE2, PDE3, and cGMP-dependent protein kinase (PKG).[4,33] Activation of PKG by cGMP reduces cardiac contractility by inhibition of LTCC activity and reduction of myofilament Ca^{2+} sensitivity.[4,34–36] At submicromolar concentrations, cGMP increases heart rate and contractility by inhibition of PDE3 and increase in cAMP levels[37–39]; at higher concentrations, cGMP exerts the opposite effect due to PDE2 activation and decrease in cAMP levels.[4,40] Due to these multiple mechanisms, the overall effect of a rise in cGMP via NO or natriuretic peptides will depend on the level of expression, the location, and the activity of PKG, PDE2, and PDE3, which may vary considerably depending on the animal species, the cardiac tissue, and the basal and stimulated levels of cAMP and cGMP.

20.2 COMPARTMENTATION OF CYCLIC NUCLEOTIDES SIGNALING

A prevailing molecular view of the receptor–enzyme–effector interaction in cells is based on the random collision theory, according to which proteins float randomly on the lipid membrane surface and contact with the target protein. A major limitation of this theory, however, is that a rapid signaling cascade involving multiple proteins may not be efficiently processed, particularly when these signaling molecules are scarce. Several observations have suggested that some components within the cyclic nucleotide-signaling pathway are colocalized to discrete regions of the plasma membrane such as caveolae[41–44] and transverse tubules,[45,46] thereby allowing rapid and preferential modulation of cAMP and cGMP production within a defined microenvironment. With the discovery of A-kinase anchoring protein (AKAP),[47,48] it has become apparent that intracellular targeting of PKA as well as the preassembly of components of signaling pathways in clusters or on scaffolds are important for the speed and organization of cAMP-signal transduction events. However, one would wonder how specificity is maintained when small diffusible molecules such as cAMP and cGMP are generated during signaling cascade. Localized cyclic nucleotide signals may be generated by interplay between discrete production sites and restricted diffusion within the cytoplasm. In addition to specialized membrane structures that may circumvent cAMP and cGMP spreading,[1,49] degradation of these cyclic nucleotides by PDEs appears critical for the formation of dynamic microdomains that confer specificity of the response.[50–53]

The first evidence for the compartmentation of cAMP signaling in heart comes from the experiments made almost 30 years ago in isolated perfused hearts.[54–56] Important differences were observed when comparing hearts perfused with different agonists activating the cAMP

cascade, particularly via β_1-AR and prostaglandin E_1 receptor (PGE$_1$-R); with isoproterenol (ISO), cAMP is elevated, the force of contraction is enhanced, soluble and particulate PKA are activated, and the activity of phosphorylase kinase and glycogen phosphorylase is increased; and with PGE$_1$, cAMP content and soluble PKA activity are also increased, but there is no change in contractile activity or in the activities of PKA substrates that regulate glycogen metabolism.[56] Similar results were reproduced in isolated myocytes.[57] The situation is even more complex if one considers that a given cardiac myocyte expresses many other G_s-coupled receptors, besides β_1-ARs and PGE$_1$-Rs, which increase cAMP but produce different effects. For instance, adult rat ventricular myocytes also express β_2-ARs, glucagon receptors (Glu-Rs), and glucagon-like peptide-1 receptors (GLP$_1$-Rs). β_2-AR stimulation increases contractile force but does not activate glycogen phosphorylase[58] and does not accelerate relaxation[59,60] (but see Ref. [58]); Glu-R stimulation activates phosphorylase and exerts positive inotropic and lusitropic effects, but the contractile effects fade with time;[61] GLP$_1$-R stimulation exerts a modest negative inotropic effect despite an increase in total cAMP comparable to that elicited by a β_1-adrenergic stimulation.[62]

The above results clearly show that the cell is able to distinguish between different stimuli acting on a common signaling cascade. One possible way to achieve that distinction is to confine the cAMP-signaling cascade to distinct intracellular compartments, which may differ depending on the stimulus used.

In the same context, several studies have shown that cGMP produced by either sGC or pGC produces different functional effects in various cell types. For instance, in human endothelial cells from umbilical vein, activation of sGC induces a more efficient relaxation than does pGC activation.[63] In airway smooth muscle cells from pig, stimulation of pGC induces relaxation exclusively by decreasing intracellular Ca^{2+} concentration, whereas sGC stimulation decreases both Ca^{2+} concentration and sensitivity of the myofilaments.[64] In human embryonic kidney, ANP, but not the NO-donor SNAP, induces a recruitment of PKG to plasma membrane and amplifies GC-A activity.[65] Differences between sGC and pGC activation have also been reported in cardiac preparations. For instance, in frog ventricular myocytes, sGC activation causes a pronounced inhibition of $I_{Ca,L}$ upon cAMP stimulation[66] whereas pGC activation has little effect.[67] In rabbit atria, pGC activation caused a larger cAMP accumulation (via PDE3 inhibition), cGMP efflux, and ANP release than activation of sGC.[68] In mouse ventricular myocytes, both pGC and sGC activation exerted similar negative inotropic effects. These effects on cell contraction were mediated by cGMP-dependent pathway involving PKG and PDEs. However, pGC activation decreased Ca^{2+} transients whereas sGC activation had marginal effects,[69] similar to what was found in pig airway smooth muscle.[64] These data suggest that pGC signaling works mainly to decrease intracellular Ca^{2+} level, whereas sGC signaling mainly decreases Ca^{2+} sensitivity.

20.3 METHODS TO STUDY CYCLIC NUCLEOTIDE COMPARTMENTATION IN INTACT MYOCYTES

During 20 years, most of the evidence supporting a compartmentation of cyclic nucleotide signaling in cardiac preparations were gathered using biochemical assays in fractionated dead tissues or cells. However, during the last decade, a number of sophisticated methods have been developed, which now allow to evaluate the role of cyclic nucleotide compartmentation in intact living cells.

The first such method combines a classical whole-cell patch-clamp recording of $I_{Ca,L}$ (as a probe for cAMP/PKA activity) with a double-barreled microperfusion system (Figure 20.2A).[50] This allows testing the effect of a local application of a receptor agonist on LTCC in the

Double-barreled microperfusion

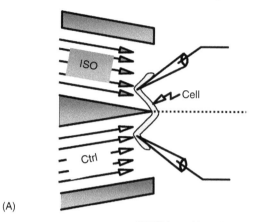

(A)

FRET-based imaging

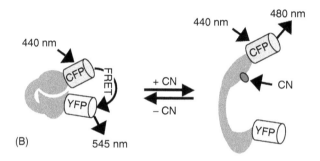

(B)

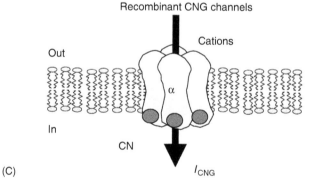

(C)

FIGURE 20.2 (See color insert following page 274.) Methods used to examine intracellular cyclic nucleotide (CN) compartmentation in intact cardiomyocytes. (A) Double-barreled microperfusion coupled with whole-cell patch-clamp technique. A cardiomyocyte is positioned transversally at the mouth of two adjacent capillaries separated by an intermediate septum of ≈5 μm thick. The cell can thus be exposed to two different solutions, for instance to ISO on one side and control (Ctrl) solution on the other side, and the activity of LTCCs can be followed separately on each side of the cell by removing Ca^{2+} ions from the other side.[50] (B) FRET-based imaging method. YFP and CFP proteins are fused to a CN-binding protein (for instance Epac[75,76] or a catalytic inactive PKG[79]). CN binding reduces FRET between CFP and YFP and the change in shape of the fluorescence emission spectrum allows CN concentrations to be visualized in real time. (C) Recombinant CNG channels. Wild-type or genetically modified α-subunits of rat olfactory CNG channel (CNGA2) form a cationic channel directly opened by CN.[49] Cardiomyocytes infected with an adenovirus encoding the native or modified channels elicit a nonselective cation current (I_{CNG}) only when CN concentration rises beneath the sarcolemmal membrane[87-89] (for further details, see text).

part of the cell exposed to the agonist and comparing it with the response of the channels located on the nonexposed part. This method provided the first evidence for a local elevation of cAMP in response to a β_2-adrenergic stimulation in frog ventricular cells as compared to a uniform elevation of cAMP in response to forskolin, a direct adenylyl cyclase activator.[50] A similar conclusion was reached using the cell-attached configuration of the patch-clamp technique in mammalian cardiomyocytes[70] and neurons[71] by applying a β_2-adrenergic agonist either inside or outside the patch pipette, while recording single LTCC activity in the patch of membrane delimited by the pipette.

More direct methods have been developed to monitor cyclic nucleotide changes using fluorescent probes and imaging microscopy. The first such probe was FlCRhR, a fluorescent indicator for cAMP, consisting of PKA in which the catalytic (C) and regulatory (R) subunits are each labeled with a different fluorescent dye, fluorescein and rhodamine, respectively.[72] Fluorescence resonance energy transfer (FRET) occurs in the holoenzyme complex R2C2, but not when cAMP binds to the R subunits and C subunits dissociate. The change in shape of the fluorescence emission spectrum allows cAMP concentrations to be visualized in real time in single living cells, as long as it is possible to microinject the cells with the labeled holoenzyme.[72] This in itself represents a major technical challenge, particularly in cardiomyocytes,[73] and has prompted the search for genetically encoded probes. A cAMP probe has been generated using the same principle as FlCRhR, but by fusing a YFP and a CFP protein to R and C subunits, respectively, instead of labeling these proteins with fluorescein and rhodamine.[74] On a similar principle, through genetic modifications of other target effectors, a number of different probes are now available for real-time measurements of cAMP (Figure 20.2B)[75–78] and cGMP[79,80] in living cells, including cardiac myocytes.[53,81–84]

A third type of approach is based on the use of recombinant cyclic nucleotide–gated (CNG) channels as cyclic nucleotide biosensors (Figure 20.2C). The methodology was developed in a series of elegant studies in model cell lines for the measurement of intracellular cAMP.[49,85,86] This method uses wild-type or genetically modified α-subunits of rat olfactory CNG channel (CNGA2), which form a cationic channel directly opened by cyclic nucleotides. Adult cardiac myocytes infected with an adenovirus encoding the native or modified channels elicit a nonselective cation current when cGMP[87] or cAMP concentration[88,89] rises beneath the sarcolemmal membrane.

20.4 ROLE OF PHOSPHODIESTERASES IN CYCLIC NUCLEOTIDE COMPARTMENTATION

20.4.1 Phosphodiesterases and Hormone Specificity

Probably the first evidence for a contribution of PDEs to intracellular cyclic nucleotide compartmentation comes from a study in guinea pig perfused hearts.[90] In that study, ISO was shown to significantly increase intracellular cAMP, cardiac contraction and relaxation, as well as phosphorylation of PLB and TnI, whereas the nonselective PDE inhibitor, 3-isobutyl-1-methylxanthine (IBMX), or the selective PDE3 inhibitor, milrinone, enhanced contraction and relaxation but had little or no effect on phosphorylation of PLB and TnI, despite a relatively large increase in tissue cAMP level.[90] These results were attributed to a functional cellular compartmentation of cAMP and PKA substrates due to a different expression of PDEs at the membrane and in the cytosol.[91] Many subsequent studies have examined the degree of accumulation of cAMP or activation of cAMP-dependent phosphorylation in particulate and soluble fractions of cardiac myocytes. In an elegant study performed in canine ventricular myocytes, Hohl and Li[92] demonstrated that cytosolic and particulate pools of cAMP are differently affected by various treatments designed to raise

intracellular cAMP. These authors have demonstrated that about 45% of the total cAMP is found in the particulate fraction in response to ISO but this fraction declined to <20% when IBMX was added to ISO, although total cAMP still increased approximately threefold. This suggests that cAMP-specific PDE activity resides predominantly in the cytosolic compartment and is responsible for the generation of particulate cAMP microdomains that cause Ca^{2+} mobilization and cardiac inotropic state through particulate PKA activation and phosphorylation of membrane and contractile proteins. Even if the cell is able to generate and accumulate cAMP that exceeds what is needed for a maximal physiologic response under PDE inhibition and forskolin stimulation, only particulate cAMP content determines the physiological response. These results show that PDEs maintain the specificity of the β-adrenergic response by limiting the amount of cAMP diffusing from membrane to cytosol.

These biochemical data are in full agreement with functional studies in frog ventricular myocytes where the effect of a local application of ISO on $I_{Ca,L}$ was tested in the presence or absence of IBMX.[50] While the $I_{Ca,L}$ response to ISO was much higher at the side of ISO applied part than in the nonexposed part of the cell, complete PDE inhibition in the presence of ISO released the cAMP signal to activate LTCCs in the remote part of the cell. Thus, these results suggest that PDE activity contributes to generate cAMP microdomains involved in the β-adrenergic stimulation of Ca^{2+} channels (Figure 20.3). A recent study using recombinant CNG channels demonstrates that this also applies to other G_s-coupled receptors (β$_1$-AR, β$_2$-AR, PGE1-R, Glu-R) with a specific pattern of PDE activity determining the specificity of the cAMP signals generated by each receptor.[89] For instance, cAMP elicited by β$_1$-AR is regulated by PDE3 and PDE4 whereas cAMP signal generated by Glu-R is exclusively regulated by PDE4. In mouse neonatal cardiomyocytes, PDE4D was shown to selectively impact cAMP signaling by β$_2$-AR while having little or no effect on β$_1$-AR signaling.[93] Indeed, while β$_2$-AR activation leads to an increase in cAMP production, the generated cAMP does not have access to the PKA-dependent signaling pathways by which the β$_1$-AR regulates the contraction rate, unless PDE4D is inhibited or its gene has been invalidated.[93]

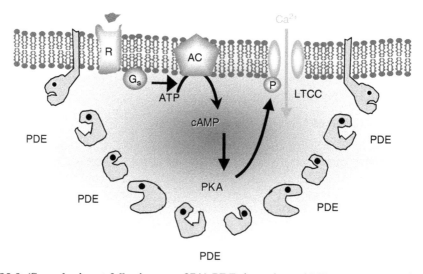

FIGURE 20.3 (See color insert following page 274.) PDE-dependent cAMP compartmentation. Activation of a G_s-coupled receptor (R) leads to AC activation and cAMP production in a compartment delimited by PDE activity. Different subsets of PDE isoforms may contribute to hormonal specificity by confining PKA phosphorylation to a limited number of substrates (LTCCs in the example shown) (for further details, see text).

20.4.2   SPECIFIC ROLES OF PHOSPHODIESTERASE ISOFORMS IN CYCLIC NUCLEOTIDE COMPARTMENTATION

The use of selective inhibitors of the dominant cardiac PDE isoforms has allowed to evaluate the contribution of four different PDE families in the compartmentation of cAMP and cGMP pathways in cardiac myocytes: PDE2, PDE3, PDE4, and PDE5. Coimmunoprecipitation experiments have further demonstrated that macromolecular complexes exist at different locations within a cardiac myocyte that include PDE3 and PDE4 isoforms, forming local signaling microdomains.

20.4.2.1   PDE2

Cyclic GMP-stimulated PDE (PDE2) hydrolyzes both cAMP and cGMP with low affinity. A single PDE2 variant, PDE2A, is expressed in cardiac tissues and isolated cardiomyocytes of several species, including rat and human.[94,95] PDE2 is found both in the cytosol and in association with functional membrane structures (plasma membrane, SR, Golgi, nuclear envelope).[16] Although PDE2 activity is relatively small compared to other cardiac PDEs, such as PDE3 and PDE4, its presence in the plasma membrane contributes to regulate the activity of cardiac LTCCs when cGMP level is increased.[4] This was first demonstrated in frog ventricular myocytes dialyzed with cAMP and cGMP, where PDE2 is able to hydrolyze cAMP and hence reduce $I_{Ca,L}$ upon application of cGMP, even when >5 μM cAMP is continuously dialyzed inside the cell via the patch pipette.[17] Increased knowledge of the contribution of PDE2 to cardiac function has accumulated after the demonstration that adenosine deaminase inhibitor, erythro-9-(2-hydroxy-3-nonyl)a (EHNA), behaves as a selective PDE2 inhibitor.[96,97] EHNA reverses the inhibitory effect of high concentration of cGMP or NO donors on $I_{Ca,L}$ in frog ventricular[96,98] and human atrial myocytes.[40,99] EHNA alone stimulates basal $I_{Ca,L}$ in isolated human atrial myocytes,[100] indicating a possible role of basal guanylyl cyclase activity in these cells.

The role of PDE2 in cyclic nucleotide compartmentation was first examined in frog cardiac myocytes, using the double-barreled microperfusion technique and local applications of NO donors or EHNA on $I_{Ca,L}$ stimulated by ISO.[98] The results of that study demonstrated that local stimulation of soluble guanylyl cyclase by NO leads to a strong local depletion of cAMP near the LTCCs due to activation of PDE2, but only to a modest reduction of cAMP in the rest of the cell. This may be explained by the existence of a tight microdomain between β-ARs, LTCC, and PDE2.[98] A similar conclusion was reached recently in rat neonatal cardiomyocytes, using the FRET-based imaging technique.[84]

PDE2 is not only involved in the control of subsarcolemmal cAMP concentration, but also controls the concentration of cGMP in that compartment. Indeed, a recent study performed in adult rat ventricular myocytes using the CNG technique compared the effects of activators of pGC (using ANP or BNP) and sGC (using NO donors) on subsarcolemmal cGMP signals, and the contribution of PDE isoforms to these signals.[87] The main result of this study is that the particulate cGMP pool is readily accessible at the plasma membrane whereas the soluble pool is not, and that the particulate pool is under the exclusive control of PDE2.[87] Therefore, differential spatiotemporal distributions of cGMP may contribute to the specific effects of natriuretic peptides and NO donors on cardiac function.

20.4.2.2   PDE3

Cyclic GMP inhibition of PDE3 can lead to cAMP increase and activation of cardiac function.[101] This mechanism accounts for the stimulatory effect of low concentrations of NO donors or cGMP on $I_{Ca,L}$ unhuman atrial myocytes.[40,99] However, a recent study

performed in perfused beating rabbit atria demonstrated that, depending on whether cGMP is produced by pGC or sGC, the effects on cAMP levels, atrial dynamics, and myocyte ANP release are different, although in both cases the effects are due to PDE3 inhibition.[68] These results suggest that cGMP–PDE3–cAMP signaling produced by pGC and sGC is compartmentalized.[68] The role of PDE3 in cyclic nucleotide compartmentation likely depends on its intracellular distribution. PDE3 is present in both cytosolic and membrane fractions of cardiac myocytes with important species and tissue differences.[16,102] For instance, in dog heart, all PDE3 activity revealed in the membrane fraction appears to be associated with the SR membrane.[103] Inhibition of PDE3 under such condition could lead to localized increases in cAMP and PKA pools, leading to increased PLB phosphorylation.

Three isoforms of PDE3 have been identified in human myocardium.[104] They appear to be generated from PDE3A gene and localize to different intracellular compartments: PDE3A-136 is present exclusively in microsomal fractions whereas PDE3A-118 and PDE3A-94 are both present in microsomal and cytosolic fractions.[105] The presence of different PDE3A isoforms in cytosolic and microsomal fractions of cardiac myocytes is especially interesting in view of the facts that cAMP metabolism in these compartments can be regulated in an independent manner and that changes in cAMP content in these compartments correlate with changes of different physiologic parameters such as intracellular Ca^{2+} homeostasis and contractility.[42,106] These observations are relevant in a physiological context since competitive inhibitors of PDE3 confer short-term hemodynamic benefits but adversely affect long-term survival in dilated cardiomyopathy.[107,108] This biphasic response is likely to result from an increase in the phosphorylation of a large number of PKA substrates, some of which may contribute to the beneficial effects (phosphorylation of PLB) whereas others contribute to the adverse effects (phosphorylation of LTCC or RyR2). If one would suppose that different isoforms regulate different proteins in response to different signals, logically agents capable of selectively activating or inhibiting individual PDE3A isoforms may have advantages over currently available nonselective PDE3 inhibitors in therapeutic applications. For instance, an agent that selectively inhibits SR-associated PDE3A-136 might preserve intracellular Ca^{2+} cycling and contractility in patients taking β-AR antagonists, without concomitant arrhythmogenic effects.[104,108]

In addition to PDE3A, cardiac myocytes also express a PDE3B isoform, at least in mouse.[109] Of particular interest is the finding that this isoform forms a complex at the cardiac sarcolemmal membrane with the G-protein-coupled, receptor-activated phosphoinositide 3-kinase γ (PI3Kγ).[109] Ablation of PI3Kγ in mice (PI3Kγ$^{-/-}$) induces an exacerbated heart failure in response to aortic constriction, which appears to be due to a PDE3B inhibition and excess cAMP. But mice carrying a targeted mutation in the PI3Kγ gene causing loss of kinase activity (PI3Kγ$^{KD/KD}$) exhibit normal cardiac contractility associated with normal cAMP levels after aortic stenosis compared to PI3Kγ$^{-/-}$. Therefore, PI3Kγ does not activate PDE3B via its kinase activity, but rather serves as an anchoring protein, which recruits PDE3B into a membrane compartment where cAMP homeostasis shapes the chronic sympathetic drive.[109]

20.4.2.3   PDE4

The PDE4 family is encoded by four genes (A, B, C, and D) that generate approximately 20 different isoforms, each of which is characterized by a unique N-terminal region.[110,111] Transcripts for PDE4A, PDE4B, and PDE4D isoforms were found in rat heart.[81,95,112,113] In the PDE4D family, mRNA for PDE4D1, PDE4D2, PDE4D3, PDE4D5, PDE4D7, PDE4D8, and PDE4D9 is present in rat heart[112,113] but only PDE4D3, PDE4D5, PDE4D8, and PDE4D9 are expressed as proteins and active enzymes.[113]

An emerging theme in PDE4 action is that individual isoforms appear to be restricted to defined intracellular microenvironments thus regulating particular sets of intracellular processes (Figure 20.4).[111,114,115] Compartmentation of PDE4 isoforms is mediated by their unique N-terminal domains, which provide the postcode for cellular localization.[116] For instance, PDE4A1 contains a lipid-binding domain, TAPAS, with specificity for phosphatidic acid that serves to target this PDE to specific cellular membranes.[117] In the heart, PDE4D3 is targeted to sarcomeric region of cardiomyocytes through binding to an anchor protein called myomegalin,[118] and to the perinuclear region through binding to muscle AKAP (mAKAP).[119] This latter complex is interesting because mAKAP not only binds PKA and PDE4D3, but also Epac1 and ERK5 kinase.[120] The three functionally distinct cAMP-dependent enzymes contained in this macromolecular complex (PKA, PDE4D3, and Epac1) respond to cAMP in different ranges of concentrations: PKA responds to nanomolar concentrations and would become activated early; PDE4D3 ($K_m = 1$–4 μM) and Epac1 ($K_d = 4$ μM) would become activated once cAMP concentrations reached micromolar levels. Conversely, inactivation of PDE4D3 and Epac1 would precede PKA holoenzyme reformation as cAMP levels decline.[120] Besides, phosphorylation of PDE4D3 by PKA on Ser54 enhances its activity[111,113] and on Ser13 increases its affinity to mAKAP[121] whereas phosphorylation by ERK5 on Ser579 suppresses its activity.[120] Therefore, when Epac1 is activated by cAMP, it mobilizes Rap1 that suppresses ERK5 activation and relieves the inhibition of PDE4D3. With such fine-tuning, this complex provides spatial control of PKA signaling by mAKAP anchoring and temporal control and termination of the cAMP-signaling event by PDE activity in the immediate vicinity.[114,119] This compartment of cAMP signaling in the perinuclear region may control the release of C subunit into the nucleus[119,122] and hence gene regulation.[114]

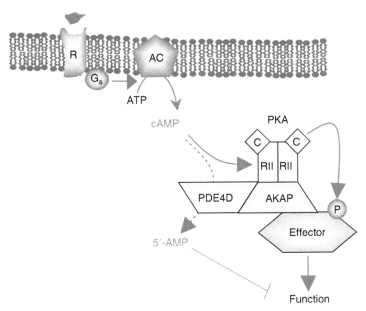

FIGURE 20.4 (See color insert following page 274.) cAMP-signaling microdomains around anchored PDE4D. PDE4D–AKAP interaction provides a molecular basis for a compartmentation of cAMP signaling. AKAP proteins bind RII subunits of PKA as well as various PKA effectors involved in cardiac function (e.g., RyR2 or LTCC). The presence of PDE4D in the same compartment limits the spread of cAMP to other compartments and controls the kinetics of the functional response (for further details, see text).

The same PDE4D3 was also found recently to be an integral component of the RyR2/Ca^{2+}-release channel complex at the SR membrane.[123] In addition to RyR2 and PDE4D3, this complex is composed of mAKAP, PKA, FKBP12.6 (calstabin2, a negative modulator or channel-stabilizing subunit of RyR2), and the protein phosphatases, PP1 and PP2A.[124,125] PKA phosphorylation of Ser2809 on RyR2 increases the open probability of the Ca^{2+}-release channel and decreases the binding affinity for the channel-stabilizing subunit calstabin2, contributing to SR Ca^{2+}-store depletion.[125] Of particular interest is the observation that heart failure in patients and animal models causes PKA hyperphosphorylation of RyR2 and leaky RyR2 channels that promote cardiac dysfunction and arrhythmias.[125] Two sets of evidence indicate that this is due to a reduction in PDE4D3 activity in the RyR2 complex:[123] first, although total PDE4 activity may be enhanced in pressure-induced congestive heart failure,[126] PDE4D3 levels in the RyR2 complex appear reduced in failing human hearts;[123] second, genetic inactivation of PDE4D in mice is associated with a cardiac phenotype comprised of a progressive, age-related cardiomyopathy, and exercise-induced arrhythmias, despite normal global cAMP signaling.[123] These results emphasize the importance of cAMP-signaling microdomains and point to the intriguing possibility that deregulation of specific compartments may lead to a disease state.

A final example of a complex around a PDE4 isoform in heart is the one formed by PDE4D5 and β-arrestins.[127] β-arrestins are scaffold proteins that initiate desensitization of β$_2$-AR (as well as several other G-protein-coupled receptors) by translocating from the cytosol to bind activated receptors at the plasma membrane. Recent studies have shown that β-arrestins can form stable complexes with all four PDE4 subfamilies in cytosol[127] but that PDE4D5 possesses a unique N-terminal region that confers preferential interaction with β-arrestins.[115,116,128,129] The specific role of this PD4D5/β-arrestin interaction in the β$_2$-AR-signaling cascade comes from a unique feature of this particular receptor, which can couple to both G$_s$ and G$_i$.[130] Upon agonist challenge, β$_2$-AR couples to G$_s$ that activates AC, thereby elevating local cAMP concentration and activating membrane PKA anchored to AKAP-79.[129] PKA in turn phosphorylates the β$_2$-AR, which triggers a shift in its coupling from G$_s$ to G$_i$, hence activating ERK through a Src-regulated pathway.[131] Therefore, recruitment by the activated β$_2$-AR of the PD4D5/β-arrestin puts a brake in the PKA phosphorylation of the receptor, and prevents its shift to G$_i$-signaling cascade; conversely, disruption of this complex enhances PKA phosphorylation of the β$_2$-AR, leading to a dramatic change in its function.[131,132]

20.4.2.4 PDE5

PDE5 is highly expressed in vascular smooth muscle, and its inhibition is a primary target for the treatment of erectile dysfunction and pulmonary hypertension.[133,134] Although the contribution of PDE5 to the regulation of cardiac function is a matter of debate,[135–137] there is evidence for PDE5 expression in cardiac myocytes, both at the mRNA[138] and protein level.[83,139] Recently, PDE5 inhibition using sildenafil (Viagra) was shown to decrease the β-adrenergic stimulation of cardiac systolic and diastolic functions in dog,[139] mouse,[83] and human,[137,140] as well as the β-adrenergic stimulation of $I_{Ca,L}$ in guinea pig ventricular myocytes.[141] In mouse ventricular myocytes, sildenafil was shown to inhibit apoptosis[142] and reduce infarct size following ischemia and reperfusion in the myocardium.[143] Moreover, chronic exposure to sildenafil was found to prevent and reverse cardiac hypertrophy in mouse hearts exposed to sustained pressure overload.[144] Most recently, PDE5 was also shown to contribute to intracellular cGMP compartmentation in cardiac myocytes.[87] Indeed, using the recombinant CNG-channel approach to measure subsarcolemmal cGMP concentration in adult rat ventricular myocytes, sildenafil produced a dose-dependent increase of the CNG

current activated by NO donors but had no effect on the current elicited by ANP. Therefore, PDE5 exerts a specific spatiotemporal control on the pool of intracellular cGMP synthesized by sGC, but not that generated by pGC which, as discussed above, is under the exclusive control of PDE2.[87] This could be either because PDE5 is more closely compartmentalized with sGC than pGC, or because PKG, which activates PDE5,[133] is compartmentalized with sGC but not pGC. Differential spatiotemporal distributions of cGMP may therefore contribute to the specific effects of natriuretic peptides and NO donors on cardiac function.[66–69,145]

20.4.2.5 Cooperative Role of Phosphodiesterase Isoforms

In many examples, more than one PDE isoform is involved in controlling the cAMP or cGMP concentration at any given intracellular location inside a cardiomyocyte. For instance, in the case of cGMP, both PDE2 and PDE5 were found to control the subsarcolemmal concentration of cGMP upon activation of sGC by NO donors in rat cardiomyocytes as demonstrated by selective inhibition of each PDE isoform. Indeed, EHNA or sildenafil used alone raised subsarcolemmal cGMP to a lower level as when the two inhibitors were applied together or when both PDEs were blocked by IBMX.[87] Similarly, the activity of cardiac LTCCs or the force of contraction is affected by the hydrolytic activity of several PDEs, since inhibition of a single PDE isoform is insufficient to raise cAMP level enough to activate these parameters.[95,146,147] Real-time measurements of cAMP in isolated cardiomyocytes using either the FRET-based or the recombinant CNG-channel method have shown that PDE4 and to a lesser extent PDE3 regulate the amplitude of cAMP response upon a β-adrenergic stimulation.[81,88] The more prominent role of PDE4 vs. PDE3 families may partly result from a larger stimulatory effect of PKA phosphorylation on the former providing a faster negative feedback regulation on cAMP concentration.[16,81,88,136,148,149] Indeed, blockade of PKA strongly increased the cAMP signal at the membrane upon β-AR stimulation of adult rat ventricular myocytes.[88]

20.5 PATHOPHYSIOLOGICAL ROLE OF cAMP COMPARTMENTATION

Cyclic AMP microdomains may be important not only for the hormonal specificity but also to prevent a global rise in cAMP, which is known to be deleterious.[125] This was illustrated in functional studies performed on the transgenic mouse line AC8TG, in which the human neuronal Ca^{2+}/calmodulin-activated type 8 adenylyl cyclase (AC8) protein is specifically expressed in cardiomyocytes.[150] In this animal model, AC and PKA activities are increased seven- and fourfold,[150] respectively, although the animals show no sign of hypertrophy or cardiomyopathy up to 3 months of age.[150,151] Isolated perfused hearts from AC8TG mice show an increased heart rate, larger amplitude of contraction, faster kinetics of contraction and relaxation, and no response to ISO as compared to nontransgenic (NTG) mice.[151] At the single-cell level, myocytes from AC8TG hearts contract faster and stronger, develop larger and faster Ca^{2+} transients, which represent the hallmarks of an improved SR function.[151] However, most surprisingly, $I_{Ca,L}$ amplitude was identical in AC8TG and NTG hearts.[151] Therefore, a compensatory mechanism must take place in AC8TG mice to prevent a continuous cAMP/PKA stimulation of cardiac LTCCs and the deleterious consequence of a Ca^{2+} overload. This mechanism appears to cause an increase in cAMP-PDE activity and a rearrangement of PDE isoforms.[152] Hence, at the single-cell level, the response of $I_{Ca,L}$ to an application of IBMX is twice larger in AC8TG vs. NTG hearts, indicating that cardiac expression of AC8 is accompanied by a strong compartmentation of the cAMP signal that shields LTCCs and protects the cardiomyocytes from Ca^{2+} overload.[152] Therefore, through

enhanced PDE activity and compartmentation, the AC8TG mouse model provides a nice example where cAMP only "makes the good, not the evil."

The concept that cardiac cAMP signaling can produce both good and bad effects, depending on where within the cell it is activated is certainly relevant to all forms of heart failure (HF), where major alterations in cAMP signaling occur. The evidence reviewed here demonstrates that physiologic cAMP signaling is confined in specific subcellular domains due to local activities of specific PDEs. We believe that the good outcomes require a strict localized control of the cAMP signaling, leading to activation of only a limited number of substrates; the bad outcomes occur when compartments are disorganized, a situation likely to exist during the morphological rearrangements that accompany hypertrophy and HF. Therefore, an in-depth analysis of cAMP signaling in pathologic hypertrophy and heart failure may provide new treatments of HF acting on localized cAMP signaling to improve heart function and clinical outcomes.

ACKNOWLEDGMENTS

This work was supported by Fondation de France (to G.V.), French Ministry of Education and Research (F.R., A.A.-G.), Association Française contre les Myopathies (to F.R.), and by European Union Contract No. LSHM-CT-2005-018833/EUGeneHeart (to R.F.).

REFERENCES

1. Steinberg, S.F., and L.L. Brunton. 2001. Compartmentation of G protein-coupled signaling pathways in cardiac myocytes. *Ann Rev Pharmacol Toxicol* 41:751.
2. Rockman, H.A., W.J. Koch., and R.J. Lefkowitz. 2002. Seven-transmembrane-spanning receptors and heart function. *Nature* 415:206.
3. Beavo, J.A., and L.L. Brunton. 2002. Cyclic nucleotide research—Still expanding after half a century. *Nat Rev Mol Cell Biol* 3:710.
4. Fischmeister, R., et al. 2005. Species- and tissue-dependent effects of NO and cyclic GMP on cardiac ion channels. *Comp Biochem Physiol A Mol Integr Physiol* 142:136.
5. Wheeler-Jones, C.P. 2005. Cell signalling in the cardiovascular system: An overview. *Heart* 91:1366.
6. Fabiato, A., and F. Fabiato. 1978. Calcium-induced release of calcium from the sarcoplasmic reticulum of skinned cells from adult human, dog, cat, rabbit, rat, and frog heart and from fetal and new-born rat ventricular. *Ann NY Acad Sci* 307:491.
7. Bers, D.M. 2002. Cardiac excitation–contraction coupling. *Nature* 415:198.
8. Brette, F., et al. 2006. Ca^{2+} currents in cardiac myocytes: Old story, new insights. *Prog Biophys Mol Biol* 91:1.
9. Mac Lennan, D.H., and E.G. Kranias. 2003. Phospholamban: A crucial regulator of cardiac contractility. *Nat Rev Mol Cell Biol* 4:566.
10. Layland, J., R.J. Solaro, and A.M. Shah. 2005. Regulation of cardiac contractile function by troponin I phosphorylation. *Cardiovasc Res* 66:12.
11. Muller, F.U., et al. 2001. Activation and inactivation of cAMP-response element-mediated gene transcription in cardiac myocytes. *Cardiovasc Res* 52:95.
12. Morel, E., et al. 2005. The cAMP-binding protein Epac induces cardiomyocyte hypertrophy. *Circ Res* 97:1296.
13. Bos, J.L. 2003. Epac: A new cAMP target and new avenues in cAMP research. *Nat Rev Mol Cell Biol* 4:733.
14. Baruscotti, M., A. Bucchi, and D. DiFrancesco. 2005. Physiology and pharmacology of the cardiac pacemaker ("funny") current. *Pharmacol Ther* 107:59.
15. Dhein, S., C.J. Van Koppen, and O.E. Brodde. 2001. Muscarinic receptors in the mammalian heart. *Pharmacol Res* 44:161.

16. Lugnier, C. 2006. Cyclic nucleotide phosphodiesterase (PDE) superfamily: A new target for the development of specific therapeutic agents. *Pharmacol Ther* 109:366.
17. Hartzell, H.C., and Fischmeister, R. 1986. Opposite effects of cyclic GMP and cyclic AMP on Ca^{2+} current in single heart cells. *Nature* 323:273.
18. Moncada, G.A., et al. 2000. Effects of acidosis and NO on nicorandil-activated K(ATP) channels in guinea-pig ventricular myocytes. *Br J Pharmacol* 131:1097.
19. Massion, P.B., and J. Balligand. 2003. Modulation of cardiac contraction, relaxation and rate by the endothelial nitric oxide synthase (ENOS): Lessons from genetically modified mice. *J Physiol* 546:63.
20. Feron, O., et al. 1996. Endothelial nitric oxide synthase targeting to caveolae—Specific interactions with caveolin isoforms in cardiac myocytes and endothelial cells. *J Biol Chem* 271:22810.
21. Xu, K.Y., et al. 1999. Nitric oxide synthase in cardiac sarcoplasmic reticulum. *Proc Natl Acad Sci USA* 96:657.
22. Khan, S.A., et al. 2003. Nitric oxide regulation of myocardial contractility and calcium cycling: Independent impact of neuronal and endothelial nitric oxide synthases. *Circ Res* 92:1322.
23. Sears, C.E., et al. 2003. Cardiac neuronal nitric oxide synthase isoform regulates myocardial contraction and calcium handling. *Circ Res* 92:e52.
24. Barouch, L.A., et al. 2002. Nitric oxide regulates the heart by spatial confinement of nitric oxide synthase isoforms. *Nature* 416:337.
25. Choi, B.R., and G. Salama. 2000. Simultaneous maps of optical action potentials and calcium transients in guinea-pig hearts: Mechanisms underlying concordant alternans. *J Physiol* 529:171.
26. Ziolo, M.T., H. Katoh, and D.M. Bers. 2001. Expression of inducible nitric oxide synthase depresses beta-adrenergic-stimulated calcium release from the sarcoplasmic reticulum in intact ventricular myocytes. *Circulation* 104:2961.
27. Fowkes, R.C., and C.A. McArdle. 2000. C-type natriuretic peptide: An important neuroendocrine regulator? *Trends Endocrinol Metab* 11:333.
28. Chen, H.H., and J.C. Burnett Jr. 1998. C-type natriuretic peptide: The endothelial component of the natriuretic peptide system. *J Cardiovasc Pharmacol* 32 (Suppl. 3):S22.
29. Wedel, B.J., and D.L. Garbers. 1998. Guanylyl cyclases: Approaching year thirty. *Trends Endocrinol Metab* 9:213.
30. Potter, L.R. 1998. Phosphorylation-dependent regulation of the guanylyl cyclase-linked natriuretic peptide receptor B: Dephosphorylation is a mechanism of desensitization. *Biochemistry* 37:2422.
31. Murthy, K.S., and G.M. Makhlouf. 2000. Heterologous desensitization mediated by G protein-specific binding to caveolin. *J Biol Chem* 275:30211.
32. Potter, L.R., S. Abbey-Hosch, and D.M. Dickey. 2006. Natriuretic peptides, their receptors and cGMP-dependent signaling functions. *Endocr Rev* 27:47.
33. Lohmann, S.M., R. Fischmeister, and U. Walter. 1991. Signal transduction by cGMP in heart. *Basic Res Cardiol* 86:503.
34. Méry, P.-F., et al. 1991. Ca^{2+} current is regulated by cyclic GMP-dependent protein kinase in mammalian cardiac myocytes. *Proc Natl Acad Sci USA* 88:1197.
35. Schroder, F., et al. 2003. Single L-type Ca^{2+} channel regulation by cGMP-dependent protein kinase type I in adult cardiomyocytes from PKG I transgenic mice. *Cardiovasc Res* 60:268.
36. Layland, J., J.M. Li, and A.M. Shah. 2002. Role of cyclic GMP-dependent protein kinase in the contractile response to exogenous nitric oxide in rat cardiac myocytes. *J Physiol* 540:457.
37. Mohan, P., et al. 1996. Myocardial contractile response to nitric oxide and cGMP. *Circulation* 93:1223.
38. Preckel, B., et al. 1997. Inotropic effects of glyceryl trinitrate and spontaneous NO donors in the dog heart. *Circulation* 96:2675.
39. Vila-Petroff, M.G., et al. 1999. Activation of distinct cAMP- and cGMP-dependent pathways by nitric oxide in cardiac myocytes. *Circ Res* 84:1020.
40. Vandecasteele, G., et al. 2001. Cyclic GMP regulation of the L-type Ca^{2+} channel current in human atrial myocytes. *J Physiol* 533:329.
41. Ostrom, R.S., and P.A. Insel. 1999. Caveolar microdomains of the sarcolemma—Compartmentation of signaling molecules comes of age. *Circ Res* 84:1110.

42. Rybin, V.O., et al. 2000. Differential targeting of β-adrenergic receptor subtypes and adenylyl cyclase to cardiomyocyte caveolae: A mechanism to functionally regulate the cAMP signaling pathway. *J Biol Chem* 275:41447.

43. Ostrom, R.S., et al. 2001. Receptor number and caveolar co-localization determine receptor coupling efficiency to adenylyl cyclase. *J Biol Chem* 276:42063.

44. Feron, O., et al. 1998. Modulation of the endothelial nitric-oxide synthase caveolin interaction in cardiac myocytes—Implications for the autonomic regulation of heart rate. *J Biol Chem* 273:30249.

45. Laflamme, M.A., and P.L. Becker. 1999. G_s and adenylyl cyclase in transverse tubules of heart: Implications for cAMP-dependent signaling. *Am J Physiol Heart Circ Physiol* 277:H1841.

46. Gao, T.Y., et al. 1997. Identification and subcellular localization of the subunits of L-type calcium channels and adenylyl cyclase in cardiac myocytes. *J Biol Chem* 272:19401.

47. Sarkar, D., J. Erlichman, and C.S. Rubin. 1984. Identification of a calmodulin-binding protein that co-purifies with the regulatory subunit of brain protein kinase II. *J Biol Chem* 259:9840.

48. Colledge, M., and J.D. Scott. 1999. AKAPs: From structure to function. *Trends Cell Biol* 9:216.

49. Rich, T.C., et al. 2000. Cyclic nucleotide–gated channels colocalize with adenylyl cyclase in regions of restricted cAMP diffusion. *J Gen Physiol* 116:147.

50. Jurevicius, J., and R. Fischmeister. 1996. cAMP compartmentation is responsible for a local activation of cardiac Ca^{2+} channels by β-adrenergic agonists. *Proc Natl Acad Sci USA* 93:295.

51. Rich, T.C., et al. 2001. A uniform extracellular stimulus triggers distinct cAMP signals in different compartments of a simple cell. *Proc Natl Acad Sci USA* 98:13049.

52. Bers, D.M., and M.T. Ziolo. 2001. When is cAMP not cAMP? Effects of compartmentalization. *Circ Res* 89:373.

53. Zaccolo, M., and T. Pozzan. 2002. Discrete microdomains with high concentration of cAMP in stimulated rat neonatal cardiac myocytes. *Science* 295:1711.

54. Corbin, J.D., et al. 1977. Compartmentalization of adenosine 3′,5′-monophosphate and adenosine 3′:5′-monophosphate-dependent protein kinase in heart tissue. *J Biol Chem* 252:3854.

55. Brunton, L.L., J.S. Hayes, and S.E. Mayer. 1979. Hormonally specific phosphorylation of cardiac troponin I and activation of glycogen phosphorylase. *Nature* 280:78.

56. Hayes, J.S., et al. 1979. Hormonally specific expression of cardiac protein kinase activity. *Proc Natl Acad Sci USA* 76:1570.

57. Hayes, J.S., et al. 1982. Evidence for selective regulation of the phosphorylation of myocyte proteins by isoproterenol and prostaglandin E1. *Biochim Biophys Acta* 714:136.

58. Bartel, S., et al. 2003. New insights into β₂-adrenoceptor signaling in the adult rat heart. *Cardiovasc Res* 57:694.

59. Xiao, R.P., and E.G. Lakatta. 1993. β₁-Adrenoceptor stimulation and β₂-adrenoceptor stimulation differ in their effects on contraction, cytosolic Ca^{2+}, and Ca^{2+} current in single rat ventricular cells. *Circ Res* 73:286.

60. Kuznetsov, V., et al. 1995. β₂-Adrenergic receptor actions in neonatal and adult rat ventricular myocytes. *Circ Res* 76:40.

61. Farah, A.E. 1983. Glucagon and the circulation. *Pharmacol Rev* 35:181.

62. Vila Petroff, M.G., et al. 2001. Glucagon-like peptide-1 increases cAMP but fails to augment contraction in adult rat cardiac myocytes. *Circ Res* 89:445.

63. Rivero-Vilches, F.J., et al. 2003. Differential relaxing responses to particulate or soluble guanylyl cyclase activation on endothelial cells: A mechanism dependent on PKG-I alpha activation by NO/cGMP. *Am J Physiol Cell Physiol* 285:C891.

64. Rho, E.H., et al. 2002. Differential effects of soluble and particulate guanylyl cyclase on Ca^{2+} sensitivity in airway smooth muscle. *J Appl Physiol* 92:257.

65. Airhart, N., et al. 2003. Atrial natriuretic peptide induces natriuretic peptide receptor-cGMP-dependent protein kinase interaction. *J Biol Chem* 278:38693.

66. Méry, P.F., et al. 1993. Nitric oxide regulates cardiac Ca^{2+} current—Involvement of cGMP-inhibited and cGMP-stimulated phosphodiesterases through guanylyl cyclase activation. *J Biol Chem* 268:26286.

67. Gisbert, M.-P., and R. Fischmeister. 1988. Atrial natriuretic factor regulates the calcium current in frog isolated cardiac cells. *Circ Res* 62:660.

68. Wen, J.F., et al. 2004. High and low gain switches for regulation of cAMP efflux concentration: Distinct roles for particulate GC- and soluble GC-cGMP-PDE3 signaling in rabbit atria. *Circ Res* 94:936.
69. Su, J., P.M. Scholz, and H.R. Weiss. 2005. Differential effects of cGMP produced by soluble and particulate guanylyl cyclase on mouse ventricular myocytes. *Exp Biol Med* (*Maywood*) 230:242.
70. Chen-Izu, Y., et al. 2000. G_i-dependent localization of β_2-adrenergic receptor signaling to L-type Ca^{2+} channels. *Biophys J* 79:2547.
71. Davare, M.A., et al. 2001. A β_2 adrenergic receptor signaling complex assembled with the Ca^{2+} channel Cav1.2. *Science* 293:98.
72. Adams, S.R., et al. 1991. Fluorescence ratio imaging of cyclic AMP in single cells. *Nature* 349:694.
73. Goaillard, J.M., P.V. Vincent, and R. Fischmeister. 2001. Simultaneous measurements of intracellular cAMP and L-type Ca^{2+} current in single frog ventricular myocytes. *J Physiol* 530:79.
74. Zaccolo, M., et al. 2000. A genetically encoded, fluorescent indicator for cyclic AMP in living cells. *Nat Cell Biol* 2:25.
75. Ponsioen, B., et al. 2004. Detecting cAMP-induced Epac activation by fluorescence resonance energy transfer: Epac as a novel cAMP indicator. *EMBO Rep* 5:1176.
76. Nikolaev, V.O., et al. 2004. Novel single chain cAMP sensors for receptor-induced signal propagation. *J Biol Chem* 279:37215.
77. Zaccolo, M. 2004. Use of chimeric fluorescent proteins and fluorescence resonance energy transfer to monitor cellular responses. *Circ Res* 94:866.
78. Nikolaev, V.O., et al. 2005. Real-time monitoring of live cell's PDE2 activity: Hormone-stimulated cAMP hydrolysis is faster than hormone-stimulated cAMP synthesis. *J Biol Chem* 280:1716.
79. Honda, A., et al. 2001. Spatiotemporal dynamics of guanosine 3′,5′-cyclic monophosphate revealed by a genetically encoded, fluorescent indicator. *Proc Natl Acad Sci USA* 98:2437.
80. Nikolaev, V.O., S. Gambaryan, and M.J. Lohse. 2006. Fluorescent sensors for rapid monitoring of intracellular cGMP. *Nat Methods* 3:23.
81. Mongillo, M., et al. 2004. Fluorescence resonance energy transfer-based analysis of cAMP dynamics in live neonatal rat cardiac myocytes reveals distinct functions of compartmentalized phosphodiesterases. *Circ Res* 95:65.
82. Warrier, S., et al. 2005. Beta-adrenergic and muscarinic receptor induced changes in cAMP activity in adult cardiac myocytes detected using a FRET based biosensor. *Am J Physiol Cell Physiol* 289:C455.
83. Takimoto, E., et al. 2005. cGMP catabolism by phosphodiesterase 5A regulates cardiac adrenergic stimulation by NOS3-dependent mechanism. *Circ Res* 96:100.
84. Mongillo, M., et al. 2006. Compartmentalized phosphodiesterase-2 activity blunts β-adrenergic cardiac inotropy via an NO/cGMP-dependent pathway. *Circ Res* 98:226.
85. Fagan, K.A., et al. 1999. Adenovirus-mediated expression of an olfactory cyclic nucleotide–gated channel regulates the endogenous Ca^{2+}-inhibitable adenylyl cyclase in C6-2B glioma cells. *J Biol Chem* 274:12445.
86. Rich, T.C., et al. 2001. In vivo assessment of local phosphodiesterase activity using tailored cyclic nucleotide–gated channels as cAMP sensors. *J Gen Physiol* 118:63.
87. Castro, L.R.V., et al. 2006. cGMP compartmentation in rat cardiac myocytes. *Circulation* 113:2221.
88. Rochais, F., et al. 2004. Negative feedback exerted by PKA and cAMP phosphodiesterase on subsarcolemmal cAMP signals in intact cardiac myocytes. An in vivo study using adenovirus-mediated expression of CNG channels. *J Biol Chem* 279:52095.
89. Rochais, F., et al. 2006. A specific pattern of phosphodiesterases controls the cAMP signals generated by different G_s-coupled receptors in adult rat ventricular myocytes. *Circ Res* 98:1081.
90. Rapundalo, S.T., R.J. Solaro, and E.G. Kranias. 1989. Inotropic responses to isoproterenol and phosphodiesterase inhibitors in intact guinea pig hearts: Comparison of cyclic AMP levels and phosphorylation of sarcoplasmic reticulum and myofibrillar proteins. *Circ Res* 64:104.
91. Weishaar, R.E., et al. 1987. Subclasses of cyclic AMP-specific phosphodiesterase in left ventricular muscle and their involvement in regulating myocardial contractility. *Circ Res* 61:539.

92. Hohl, C.M., and Q. Li. 1991. Compartmentation of cAMP in adult canine ventricular myocytes—Relation to single-cell free Ca^{2+} transients. *Circ Res* 69:1369.

93. Xiang, Y., et al. 2005. Phosphodiesterase 4D is required for β_2 adrenoceptor subtype-specific signaling in cardiac myocytes. *Proc Natl Acad Sci USA* 102:909.

94. Sadhu, K., et al. 1999. Differential expression of the cyclic GMP-stimulated phosphodiesterase PDE2A in human venous and capillary endothelial cells. *J Histochem Cytochem* 47:895.

95. Verde, I., et al. 1999. Characterization of the cyclic nucleotide phosphodiesterase subtypes involved in the regulation of the L-type Ca^{2+} current in rat ventricular myocytes. *Br J Pharmacol* 127:65.

96. Méry, P.F., et al. 1995. Erythro-9-(2-hydroxy-3-nonyl)adenine inhibits cyclic GMP-stimulated phosphodiesterase in isolated cardiac myocytes. *Mol Pharmacol* 48:121.

97. Podzuweit, T., P. Nennstiel, and A. Muller. 1995. Isozyme selective inhibition of cGMP-stimulated cyclic nucleotide phosphodiesterases by erythro-9-(2-hydroxy-3-nonyl) adenine. *Cell Signal* 7:733.

98. Dittrich, M., et al. 2001. Local response of L-type Ca^{2+} current to nitric oxide in frog ventricular myocytes. *J Physiol* 534:109.

99. Kirstein, M., et al. 1995. Nitric oxide regulates the calcium current in isolated human atrial myocytes. *J Clin Invest* 95:794.

100. Rivet-Bastide, M., et al. 1997. cGMP-stimulated cyclic nucleotide phosphodiesterase regulates the basal calcium current in human atrial myocytes. *J Clin Invest* 99:2710.

101. Kojda, G., and K. Kottenberg. 1999. Regulation of basal myocardial function by NO. *Cardiovasc Res* 41:514.

102. Muller, B., J.-C. Stoclet, and C. Lugnier. 1992. Cytosolic and membrane-bound cyclic nucleotide phosphodiesterases from guinea pig cardiac ventricles. *Eur J Pharmacol* 225:263.

103. Kauffman, R.F., et al. 1986. LY195115: A potent, selective inhibitor of cyclic nucleotide phosphodiesterase located in the sarcoplasmic reticulum. *Mol Pharmacol* 30:609.

104. Wechsler, J., et al. 2002. Isoforms of cyclic nucleotide phosphodiesterase PDE3A in cardiac myocytes. *J Biol Chem* 277:38072.

105. Hambleton, R., et al. 2005. Isoforms of cyclic nucleotide phosphodiesterase PDE3 and their contribution to cAMP-hydrolytic activity in subcellular fractions of human myocardium. *J Biol Chem* 280:39168.

106. Hayes, J.S., L.L. Brunton, and S.E. Mayer. 1980. Selective activation of particulate cAMP-dependent protein kinase by isoproterenol and prostaglandin E1. *J Biol Chem* 255:5113.

107. Movsesian, M.A. 1999. Beta-adrenergic receptor agonists and cyclic nucleotide phosphodiesterase inhibitors: Shifting the focus from inotropy to cyclic adenosine monophosphate. *J Am Coll Cardiol* 34:318.

108. Movsesian, M.A., and M.R. Bristow. 2005. Alterations in cAMP-mediated signaling and their role in the pathophysiology of dilated cardiomyopathy. *Curr Top Dev Biol* 68:25.

109. Patrucco, E., et al. 2004. PI3Kgamma modulates the cardiac response to chronic pressure overload by distinct kinase-dependent and independent effects. *Cell* 118:375.

110. Conti, M., et al. 2003. Cyclic AMP-specific PDE4 phosphodiesterases as critical components of cyclic AMP signaling. *J Biol Chem* 278:5493.

111. Houslay, M.D., and D.R. Adams. 2003. PDE4 cAMP phosphodiesterases: Modular enzymes that orchestrate signalling cross-talk, desensitization and compartmentalization. *Biochem J* 370:1.

112. Kostic, M.M., et al. 1997. Altered expression of PDE1 and PDE4 cyclic nucleotide phosphodiesterase isoforms in 7-oxo-prostacyclin-preconditioned rat heart. *J Mol Cell Cardiol* 29:3135.

113. Richter, W., S.L. Jin, and M. Conti. 2005. Splice variants of the cyclic nucleotide phosphodiesterase PDE4D are differentially expressed and regulated in rat tissue. *Biochem J* 388:803.

114. Tasken, K., and E.M. Aandahl. 2004. Localized effects of cAMP mediated by distinct routes of protein kinase A. *Physiol Rev* 84:137.

115. Baillie, G.S., J.D. Scott, and M.D. Houslay. 2005. Compartmentalisation of phosphodiesterases and protein kinase A: Opposites attract. *FEBS Lett* 579:3264.

116. Baillie, G.S., and M.D. Houslay. 2005. Arrestin times for compartmentalised cAMP signalling and phosphodiesterase-4 enzymes. *Curr Opin Biol* 17:1.

117. Baillie, G.S., et al. 2002. TAPAS-1, a novel microdomain within the unique N-terminal region of the PDE4A1 cAMP-specific phosphodiesterase that allows rapid, Ca^{2+}-triggered membrane association with selectivity for interaction with phosphatidic acid. *J Biol Chem* 277:28298.

118. Verde, I., et al. 2001. Myomegalin is a novel protein of the Golgi/centrosome that interacts with a cyclic nucleotide phosphodiesterase. *J Biol Chem* 276:11189.

119. Dodge, K.L., et al. 2001. mAKAP assembles a protein kinase A/PDE4 phosphodiesterase cAMP signaling module. *EMBO J* 20:1921.

120. Dodge-Kafka, K.L., et al. 2005. The protein kinase A anchoring protein mAKAP co-ordinates two integrated cAMP effector pathways. *Nature* 437:574.

121. Carlisle Michel, J.J., et al. 2004. PKA-phosphorylation of PDE4D3 facilitates recruitment of the mAKAP signalling complex. *Biochem J* 381:587.

122. Lugnier, C., et al. 1999. Characterization of cyclic nucleotide phosphodiesterase isoforms associated to isolated cardiac nuclei. *Biochim Biophys Acta* 1472:431.

123. Lehnart, S.E., et al. 2005. Phosphodiesterase 4D deficiency in the ryanodine receptor complex promotes heart failure and arrhythmias. *Cell* 123:23.

124. Marx, S.O., et al. 2000. PKA phosphorylation dissociates FKBP12.6 from the calcium release channel (ryanodine receptor): Defective regulation in failing hearts. *Cell* 101:365.

125. Wehrens, X.H., S.E. Lehnart, and A.R. Marks. 2005. Intracellular calcium release and cardiac disease. *Ann Rev Physiol* 67:69.

126. Takahashi, K., et al. 2002. Enhanced activities and gene expression of phosphodiesterase types 3 and 4 in pressure-induced congestive heart failure. *Heart Vessels* 16:249.

127. Perry, S.J., et al. 2002. Targeting of cyclic AMP degradation to β_2-adrenergic receptors by β-arrestins. *Science* 298:834.

128. Bolger, G.B., et al. 2003. The unique amino-terminal region of the PDE4D5 cAMP phosphodiesterase isoform confers preferential interaction with beta-arrestins. *J Biol Chem* 278:49230.

129. Lynch, M.J., et al. 2005. RNA silencing identifies PDE4D5 as the functionally relevant cAMP phosphodiesterase interacting with beta-arrestin to control the PKA/AKAP79-mediated switching of the β_2-adrenergic receptor to activation of ERK in HEK293 cells. *J Biol Chem* 280:33178.

130. Daaka, Y., L.M. Luttrell, and R.J. Lefkowitz. 1997. Switching of the coupling of the β_2-adrenergic receptor to different G proteins by protein kinase A. *Nature* 390:88.

131. Baillie, G.S., et al. 2003. β-Arrestin-mediated PDE4 cAMP phosphodiesterase recruitment regulates β-adrenoceptor switching from G_s to G_i. *Proc Natl Acad Sci USA* 100:941.

132. Houslay, M.D., and G.S. Baillie. 2003. The role of ERK2 docking and phosphorylation of PDE4 cAMP phosphodiesterase isoforms in mediating cross-talk between the cAMP and ERK signalling pathways. *Biochem Soc Trans* 31:1186.

133. Rybalkin, S.D., et al. 2003. Cyclic GMP phosphodiesterases and regulation of smooth muscle function. *Circ Res* 93:280.

134. Sastry, B.K., et al. 2004. Clinical efficacy of sildenafil in primary pulmonary hypertension: A randomized, placebo-controlled, double-blind, crossover study. *J Am Coll Cardiol* 43:1149.

135. Wallis, R.M., et al. 1999. Tissue distribution of phosphodiesterase families and the effects of sildenafil on tissue cyclic nucleotides, platelet function, and the contractile responses of trabeculae carneae and aortic rings in vitro. *Am J Cardiol* 83:3C.

136. Maurice, D.H., et al. 2003. Cyclic nucleotide phosphodiesterase activity, expression, and targeting in cells of the cardiovascular system. *Mol Pharmacol* 64:533.

137. Semigran, M.J. 2005. Type 5 phosphodiesterase inhibition: The focus shifts to the heart. *Circulation* 112:2589.

138. Kotera, J., et al. 1998. Novel alternative splice variants of cGMP-binding cGMP-specific phosphodiesterase. *J Biol Chem* 273:26982.

139. Senzaki, H., et al. 2001. Cardiac phosphodiesterase 5 (cGMP-specific) modulates β-adrenergic signaling in vivo and is down-regulated in heart failure. *FASEB J* 15:1718.

140. Borlaug, B.A., et al. 2005. Sildenafil inhibits β-adrenergic-stimulated cardiac contractility in humans. *Circulation* 112:2642.

141. Ziolo, M.T., et al. 2003. Inhibition of cyclic GMP hydrolysis with zaprinast reduces basal and cyclic AMP-elevated L-type calcium current in guinea-pig ventricular myocytes. *Br J Pharmacol* 138:986.
142. Das, A., L. Xi, and R.C. Kukreja. 2005. Phosphodiesterase-5 inhibitor sildenafil preconditions adult cardiac myocytes against necrosis and apoptosis. Essential role of nitric oxide signaling. *J Biol Chem* 280:12944.
143. Das, S., et al. 2002. Cardioprotection with sildenafil, a selective inhibitor of cyclic 3',5'-monophosphate-specific phosphodiesterase 5. *Drugs Exp Clin Res* 28:213.
144. Takimoto, E., et al. 2005. Chronic inhibition of cyclic GMP phosphodiesterase 5A prevents and reverses cardiac hypertrophy. *Nat Med* 11:214.
145. Wollert, K.C., et al. 2003. Increased effects of C-type natriuretic peptide on contractility and calcium regulation in murine hearts overexpressing cyclic GMP-dependent protein kinase I. *Br J Pharmacol* 140:1227.
146. Juan-Fita, M.J., et al. 2004. Rolipram reduces the inotropic tachyphylaxis of glucagon in rat ventricular myocardium. *Naunyn Schmiedebergs Arch Pharmacol* 370:324.
147. Juan-Fita, M.J., M.L. Vargas, and J. Hernandez. 2005. The phosphodiesterase 3 inhibitor cilostamide enhances inotropic responses to glucagon but not to dobutamine in rat ventricular myocardium. *Eur J Pharmacol* 512:207.
148. Macphee, C.H., et al. 1988. Phosphorylation results in activation of a cAMP phosphodiesterase in human platelets. *J Biol Chem* 263:10353.
149. Sette, C., and M. Conti. 1996. Phosphorylation and activation of a cAMP-specific phosphodiesterase by the cAMP-dependent protein kinase—Involvement of serine 54 in the enzyme activation. *J Biol Chem* 271:16526.
150. Lipskaia, L., et al. 2000. Enhanced cardiac function in transgenic mice expressing a Ca^{2+}-stimulated adenylyl cyclase. *Circ Res* 86:795.
151. Georget, M., et al. 2002. Augmentation of cardiac contractility with no change in L-type Ca^{2+} current in transgenic mice with a cardiac-directed expression of the human adenylyl cyclase type 8 (AC8). *FASEB J* 16 (12):1636.
152. Georget, M., et al. 2003. Cyclic AMP compartmentation due to increased cAMP-phosphodiesterase activity in transgenic mice with a cardiac-directed expression of the human adenylyl cyclase type 8 (AC8). *FASEB J* 17:1380.

Section E

*Phosphodiesterases
as Pharmacological Targets
in Disease Processes*

21 Role of PDEs in Vascular Health and Disease: Endothelial PDEs and Angiogenesis

Thérèse Keravis, Antonio P. Silva, Laure Favot, and Claire Lugnier

CONTENTS

21.1 Introduction 418
21.2 Distribution of PDEs in Various Endothelial Cells and Involvement
of Specific PDEs in Regulating cAMP and cGMP....................................419
    21.2.1 Aortic Endothelial Cells..419
    21.2.2 Pulmonary Endothelial Cells...420
    21.2.3 Human Endothelial Cells 420
21.3 Mediation of cAMP–cGMP Cross-Talk in BAEC by PDEs:
Implication of PDE4 in NO Regulation 423
21.4 Phenotypic Expression of PDEs in BAEC...424
21.5 cGMP-Induced Proliferation and Effect on PDE Expression
in the Human Permanent Endothelial Cell Line EAhy 926,
an *in Vitro* Angiogenesis Model...425
21.6 Involvement of PDEs in the Migration and Proliferation Steps
in VEGF-Stimulated HUVEC, an *in Vitro* Angiogenesis Model.............427
    21.6.1 Effect of VEGF on PDE Expression in HUVEC 427
    21.6.2 cAMP Accumulation in VEGF-Treated HUVEC 430
    21.6.3 VEGF-Induced HUVEC Proliferation 431
    21.6.4 VEGF-Induced HUVEC Cycle Progression 432
    21.6.5 Cell Cycle Protein Phosphorylation and Expression
in VEGF-Stimulated HUVEC 432
    21.6.6 VEGF-Induced HUVEC Migration 434
21.7 Effectiveness of PDE Inhibitors in the Chicken Embryo
CAM, an *in Vivo* Angiogenesis Model..435
21.8 Conclusion 435
Acknowledgments 437
References 437

21.1 INTRODUCTION

As a result of the major discovery of endothelium-dependent relaxing factor (EDRF) by Furchgott and Zawadzki[1] and its subsequent identification as nitric oxide (NO),[2] there was a great renewal in research focused on regulation of vascular contraction. The results of many studies have now clearly demonstrated that endothelial cells, which line and cover the internal face of blood vessels, play a crucial role in the physiology of blood vessels by regulating the activity of guanylyl cyclase and therefore cyclic GMP in vascular smooth muscle cells.[3,4] Endothelial cells are indeed implicated in the control of blood-formed elements and the control of vascular tone, as emphasized in Chapter 22 and Chapter 23. Also, endothelial cells play a major role in angiogenesis, which represents the formation of new blood vessels from preexisting ones. Physiological angiogenesis is active in embryonic development, menstrual cycle, and wound healing. Endothelial cells are implicated in various pathologies such as arteriosclerosis, rheumatoid arthritis, diabetic retinopathy, AIDS-associated Kaposi's sarcoma, and solid tumor growth in which angiogenesis is critical.[5] In healthy adult mammals, endothelial cells are essentially quiescent. However, in response to an increased secretion of angiogenic factors, endothelial cells can switch to a highly proliferative phenotype. Vascular endothelium growth factor (VEGF), which is for the most part synthesized and secreted by vascular smooth muscle cells, also can be secreted by solid tumors, thereby turning on an angiogenic switch. In response, the endothelial cells become highly proliferative, invade the surrounding tissues, and eventually undergo differentiation to form new blood vessels that supply the tumor with oxygen and nutrients. Angiogenesis largely depends on endothelial cell migration and proliferation. As a result, inhibition of endothelial cell growth can effectively block angiogenesis.

Studies on VEGF-induced angiogenesis, on one hand, have revealed that VEGF induces upregulation of NO synthase,[6] which consequently increases cGMP in endothelial cells.[7] On the other hand, it has been shown that elevation of cAMP has antiproliferative actions in bovine aortic endothelial cells (BAECs).[8] As PDE isozymes rapidly catabolize cAMP and cGMP to restore a basal cyclic nucleotide level, one can hypothesize that PDE isozymes might play a crucial role in angiogenesis (Figure 21.1).

This chapter is a review of endothelial cell PDEs and includes pertinent parts of our own data. Herein, we address (1) the different distribution of PDEs in various endothelial cells and

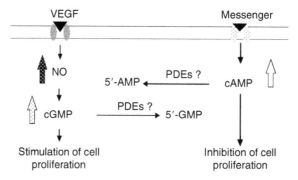

FIGURE 21.1 Potential role of PDEs in angiogenesis. cAMP and cGMP may play opposite roles in angiogenesis. On one side, VEGF-induced angiogenesis is mediated by endothelial nitric oxide synthase upregulation, and the subsequent increase in NO, by stimulating nitric oxide–sensitive (soluble) guanylyl cyclase, results in an increase of endothelial cell intracellular cGMP level and a stimulation of cell proliferation. On the other side, an increase of cAMP level in endothelial cell overcomes endothelial cell proliferation. As PDEs play a crucial role in the cAMP and cGMP balance, which may control angiogenesis, the participation and the roles of PDEs were investigated.

the identity of the PDEs that are mainly involved in regulating cyclic nucleotide levels in these cells; (2) the mediation of cAMP–cGMP crosstalk in BAEC by PDEs with the implication of PDE4 in NO regulation; (3) the phenotypic expression of PDEs in BAEC; (4) the cGMP-induced proliferation and effect on PDE expression in the human permanent endothelial cell line EAhy 926, an *in vitro* angiogenesis model; (5) the involvement of PDEs in the migration and proliferation steps in VEGF-stimulated human umbilical vein endothelial cells (HUVECs), an *in vitro* angiogenesis model; and (6) the effectiveness of PDE inhibitors in the chicken embryo chorioallantoic membrane (CAM), an *in vivo* angiogenesis model.

21.2 DISTRIBUTION OF PDEs IN VARIOUS ENDOTHELIAL CELLS AND INVOLVEMENT OF SPECIFIC PDEs IN REGULATING cAMP AND cGMP

21.2.1 AORTIC ENDOTHELIAL CELLS

Endothelial PDEs were first characterized in cultured BAEC by our group[9] at early passages (not beyond three passages), as well as in pig aortic endothelial cells (PAECs) by Souness et al.[10] by anion-exchange chromatography and by using specific inhibitors. In both species, PDE activity of aortic endothelial cells was mostly related to PDE2 (previously named cGS-PDE) and PDE4 (previously named cAMP-specific PDE) and was insensitive to Ca^{2+} and calmodulin. In BAEC, more than 80% of both cAMP- and cGMP-hydrolytic activities were cytosolic with cAMP-hydrolytic activity being about 2.5-fold greater than cGMP-hydrolytic activity at low substrate concentration (0.25 μM); however, cAMP-hydrolytic activity was 2.5-fold lower than cGMP-hydrolytic activity at higher substrate concentration (20 μM), suggesting that cAMP–PDE activity is predominant in the basal condition. Addition of 5 μM cGMP, which can increase cAMP-hydrolyzing activity by PDE2, demonstrated that PDE2 activity was present in both soluble (3.2-fold stimulation) and particulate (2-fold stimulation) fractions. HPLC resolution of cytosolic fraction (105,000 g isotonic supernatant) revealed only PDE2, which was stimulated sevenfold by cGMP, and PDE4, which was insensitive to cGMP. The cGMP-activated PDE2 and PDE4 enzymes exhibited apparent K_m values of 15 and 0.9 μM, respectively, for cAMP hydrolysis.[9] This study also showed that some PDE inhibitors can stimulate the cGMP-hydrolytic activity of PDE2 assessed using 0.25 μM cGMP as substrate, namely AAL05 (a cilostamide analogue,[11] 1.2-fold at 10 μM), dipyridamole (1.3-fold at 1 μM), IBMX (1.4-fold at 10 μM), papaverine (1.6-fold at 5 μM), trequinsin (1.2-fold at 1 μM), and zaprinast (1.5-fold at 100 μM). This is consistent with earlier results indicating that at least some of these compounds interact with the allosteric cGMP-binding site on PDE2 with different affinities and efficiencies to promote increased catalytic activity.[12] It would also suggest that the allosteric cGMP-binding site of PDE2 (presently identified as GAF B)[13] differs from the PDE5 cGMP-binding site (GAF A) in this regard.[14] The studies of Souness et al. performed in PAEC[10] using diethylamino ethanol (DEAE)-trisacryl chromatography of soluble fraction (100,000 g hypotonic supernatant) clearly confirmed that the PDE2 and PDE4 enzymes from both species display similar sensitivities to various reference PDE inhibitors. This later study was extended by addressing the effect of selective inhibitors on cyclic nucleotide accumulation in PAEC showing that rolipram, the specific inhibitor of PDE4,[15] induced cAMP accumulation only in the presence of forskolin that is known to increase cAMP level. Compounds inhibiting PDE2 as well as PDE4, such as dipyridamole and trequinsin,[9,16] induced cAMP and cGMP accumulations and potentiated forskolin-stimulated cAMP accumulation,[10] whereas no effect was observed with zaprinast, a PDE5 and PDE1 inhibitor.[15] These data indicate that PDE2 is the sole enzyme hydrolyzing cGMP in aortic endothelial cells and PDE2 activity is sufficient to prevent basal cAMP accumulation

subsequent to inhibition of PDE4 by rolipram in these cells. In contrast, when both PDE2 and PDE4 are simultaneously inhibited there is not enough remaining PDE activity to hydrolyze the cAMP, generated by adenylyl cyclase and so a net accumulation of cAMP is observed. This strengthens the above data indicating that in bovine and porcine aortic endothelial cells, the major cAMP-hydrolytic activity is related to both PDE2 and PDE4, whereas the cGMP-hydrolytic activity is only related to PDE2.

Ashikaga et al.[17] have shown that the BAEC-PDE isozyme profile may vary with cell passage number. This study showed that (1) PDE2 and a minor fraction of PDE5 accounted for cGMP hydrolysis in early passages (4–6) but these two were lost with higher passages (beyond 6); (2) cAMP was hydrolyzed by both PDE2 and PDE4 in early passages, but PDE4 was increased dramatically in higher passages; (3) a prominent PDE1 and a minor PDE3 fractions appeared in the higher passages. Furthermore, the BAEC-PDE isozyme profile may vary with culture media. In our first study,[9] BAEC were cultivated in the presence of ascorbic acid and PDE activity was resolved into PDE2 and PDE4, whereas PDE3 was expressed along PDE2 and PDE4 in the absence of ascorbic acid.[18] Altogether, these data point out that cell passage number and culture condition should be considered when studying cyclic nucleotide metabolism. It is worth to mention that PDE expression might also differ in cells in culture compared to cells freshly dissociated before getting set in culture as well as cells in intact vessels.

21.2.2 PULMONARY ENDOTHELIAL CELLS

Studies performed in pulmonary endothelial cells pointed out significant differences in PDE activities, which may result in different signal responses. Suttorp et al.[19] have shown that the use of specific inhibitors on porcine pulmonary artery endothelial cell (PPAEC) homogenate allowed the activity of PDE2, PDE3, and PDE4 to be assessed. Furthermore, a comparison of the data of Zhu et al.[20] on rat pulmonary artery endothelial cells (RPAECs) and the data of Thompson et al.[21] on rat pulmonary microvessels endothelial cell (RPMVEC) using DEAE-trisacryl chromatography revealed that RPAECs contain PDE4, PDE5, and PDE7 isozymes, whereas RPMVECs contain only PDE4, lacking cGMP-hydrolytic activity independent of cell passage number. This may explain why, contrary to RPAECs, RPMVECs are unable to overcome an intracellular cGMP increase.[20] Also, and curiously, both RPAEC and RPMVEC lack PDE2 activity in notable contrast to PPAEC, PAEC, and BAEC.

21.2.3 HUMAN ENDOTHELIAL CELLS

Two human models to study angiogenesis *in vitro* have been widely used by the scientific community: the EAhy 926 permanent cell[22] and the primary culture of HUVEC.[23] EAhy 926 cell line represents an hybridoma obtained by fusing HUVECs to epithelial cells from human pulmonary carcinoma. These cells are able to differentiate into tubular structures when cultivated on matrigel.[22]

We first used EAhy 926 cell[24] as a human model for *in vitro* angiogenesis and characterized their endothelial PDEs. HPLC chromatography of the EAhy 926 cytosol revealed only two fractions both insensitive to Ca^{2+} and calmodulin: the first fraction, which hydrolyzed both cAMP and cGMP and was potently activated by cGMP (25-fold), represents PDE2; the second fraction, which hydrolyzed only cAMP and was sensitive to rolipram and insensitive to cGMP, represents PDE4 (Figure 21.2).[25] It should be noticed that for PDE2 its cGMP–PDE activity was 12-fold greater than its basal cAMP–PDE activity. These data are similar to that obtained previously on BAEC and PAEC; therefore, this human permanent endothelial cell line was first used in the laboratory to perform studies on angiogenesis (see further).

Primary cultures of HUVEC, although is a much more difficult and time-consuming approach than using permanent cell lines, are classically used as a human model for *in vitro*

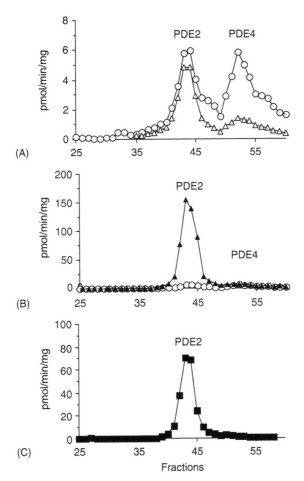

FIGURE 21.2 HPLC chromatography of EAhy 926 cell cytosolic fraction. Cell culture was performed after Edgell.[24] HPLC chromatography and PDE assay were performed as described previously.[27] (A) cAMP–PDE activity was assessed at 1 μM cAMP in presence of EGTA (○—○) or in presence of 10 μM rolipram (△–△); (B) cAMP–PDE activity was assessed at 1 μM cAMP in presence of EGTA (○—○) or in presence of 5 μM cGMP (▲–▲); (C) cGMP–PDE activity was assessed at 1 μM cGMP (■-■).

angiogenesis.[22] It has been reported that endothelial cells in culture can display alterations in their characteristics with increasing passage notably the capacity to produce NO, which is lost in BAEC and HUVEC.[26] For that reason, HUVECs were used at the second passage to study PDE profiles and cell biology. HPLC chromatography of the HUVEC cytosolic fraction resolved five peaks of PDE activity insensitive to Ca^{2+} and calmodulin.[27] The first cAMP-hydrolytic fraction, which was insensitive to rolipram and cGMP, may represent PDE7, the mRNA of which was previously identified in HUVEC by Miro et al.[28] The second cAMP-hydrolytic fraction, which also hydrolyzed cGMP and was activated by cGMP, represents PDE2. However, because a significant part of the cAMP-hydrolytic activity obtained in this fraction was inhibited by rolipram, it seems that a mixture of both PDE2 and some PDE4 is present. The major fraction of cAMP–PDE activity is obtained in the third fraction. This fraction could be subdivided into two parts based on sensitivities to rolipram and cGMP. The first part contained more PDE4 whereas the second was enriched in PDE3. Furthermore, the cGMP-hydrolytic activity was mainly restricted to one fraction eluting before the cAMP–PDE activity and represents PDE5. Altogether, HUVECs contain PDEs from five families: the

PDE activity was predominantly derived from PDE2, PDE3, and PDE4, and to a lower extent PDE5 and PDE7. In HUVEC cytosol, cAMP–PDE activity was fourfold higher than cGMP–PDE activity,[27] contrary to EAhy 926 cells, which were characterized by a 10-fold greater cGMP-hydrolytic activity than cAMP-PDE activity and contained only PDE2 and PDE4 (Figure 21.2). Although the EAhy 926 cell line originates in part from HUVECs, the differences in PDE distribution between these two cell populations pointed out possible differences in intracellular signaling between primary and permanent cell lines that must be taken into account in basic research as well as in pharmacology. The different PDE profiles may also reflect at least in part the difference between a permanent cell line in culture and primary cells; also, this could reflect that EAhy 926 cell line is an hybridoma obtained by fusing HUVEC to epithelial cells from human pulmonary carcinoma, which is a transformed cell line.

Differential expression of PDE2-protein was reported by Sadhu et al.[29] in human venous and capillary endothelial cells of tissue slices by immunostaining and confirmed by in situ hybridization. Curiously, PDE2A3 was absent from luminal endothelial cells of large vessels, such as aorta, pulmonary and renal arteries, and interestingly it was present in the endothelial cells of the vasa vasorum and a variety of microvessels in tissue slices. This study also showed that PDE2A expression was detected by Western blotting at a low level in various cultured arterial endothelial cells. The data from immunohistochemistry on vessels and those from Western blotting on cultured cells seem to be contradictory and one explanation would be that the sensitivity of the PDE2 antibody used for immunohistochemistry might be insufficient to detect a low amount of PDE2 in tissue slices; this could also reflect differences of PDE expression pattern between cells in culture and *in vivo*.

Finally, a comparison of PDE isozyme distribution in cultured human aortic endothelial cells (HAECs), human microvascular endothelial cells (HMVEC), and HUVEC has been recently reported by Netherton and Maurice.[30] The expression of PDE2, PDE3, PDE4, and PDE5 in endothelial cells from different sources were assessed by Western blot analysis and drug sensitivities. Their respective activities as well as the total cAMP–PDE and cGMP–PDE activities differed according to the cellular type. For instance, the basal cAMP–PDE versus cGMP–PDE ratio varied greatly from 3 to 21. BAEC have the greatest proportion of cGMP–PDE activity, whereas HAEC and HMEC have the lowest (2- and 14-fold less, respectively). As seen previously, BAEC have higher PDE2 activity than HUVEC.

With the recent characterization of the PDE11 family, which hydrolyzes both cAMP and cGMP, studies performed by D'Andrea et al.[31] on PDE11A expression by immunohisto-chemistry point out a high expression of PDE11A in human endothelial cells from many tissues. However, further studies are necessary to demonstrate their relative contribution to endothelial cell PDE activity.

The difference in the distribution of PDE activities (Table 21.1) in HAEC, HMVEC, and HUVEC (4 or 5 PDEs) compared to those in BAEC and PAEC (only PDE2 and PDE4) and in RPMVEC (only PDE4) suggests differences in blood vessel regulation by cyclic nucleotides (vein, artery, and microvessel), although this observation does not quite fit for EAhy 926 cells, which originate from HUVEC but contain only PDE2 and PDE4. Before studying endothelial cell signaling in a specific endothelial cell type, it would be necessary to know the complete PDE profile to define in full the nature of the cGMP-mediated crosstalk between various PDEs.[32]

All these studies show that (1) the heterogeneity of PDE distribution in various reports can be related to species, the region of the vasculature studied, as well as the sensitivity and specificity of the various methodological approaches used; (2) the cyclic nucleotide levels in endothelial cells are regulated by a major contribution from PDE2, PDE3, and PDE4 with a minor participation of PDE1, PDE5, PDE7, and possibly PDE11; (3) the knowledge of the complete PDE profile in endothelial cells under investigation is a prerequisite for studying the intracellular signaling of cyclic nucleotides.

TABLE 21.1
PDE Activities in Various Endothelial Cells

Cells	PDE1	PDE2	PDE3	PDE4	PDE5	PDE7	Ref.
BAEC	−	+	−	+	−	n.t.	[9]
	+/−	+/−	+/−	+	+/−	n.t.	[17]
	+/−	+	+/−	+	+/−	n.t.	[18]
	−	+	+	+	+	n.t.	[30]
PAEC	−	+	−	+	−	n.t.	[10]
PPAEC	−	+	+	+	−	n.t.	[19]
RPAEC	−	−	−	+	+	+	[20]
RPMVEC	−	−	−	+	−	n.t.	[21]
HAEC	−	+	+	+	+	n.t.	[30]
HMVEC	−	+	+	+	+	n.t.	[30]
HUVEC	−	+	+	+	+	+	[27]
	−	+	+	+	+	n.t.	[30]
EAhy 926	−	+	−	+	−	n.t.	[25]

Note: BAEC, bovine aorta endothelial cell; PAEC, porcine aorta endothelial cell; PPAEC, porcine pulmonary artery endothelial cell; RPAEC, rat pulmonary artery endothelial cell; RPMVEC, rat pulmonary microvascular endothelial cell; HAEC, human aorta endothelial cell; HMVEC, human microvascular endothelial cell; HUVEC, human umbilical vein endothelial cell; EAhy 926, human permanent cell line (hybridoma between HUVEC and tumoral epithelial cells); −, not detected; +, detected; +/−, presence or absence depending on the number of cell passages[17] or the cell phenotype;[18] n.t., not tested.

21.3   MEDIATION OF cAMP–cGMP CROSS-TALK IN BAEC BY PDEs: IMPLICATION OF PDE4 IN NO REGULATION

Nitric oxide (NO) is known to increase cGMP in endothelial cells,[33,34] and therefore cGMP is measured as an index of NO production. Although many studies have been performed to examine the modulation of NO production in BAEC, little is known of the intimate mechanisms regulating cAMP and cGMP levels in these cells. We have addressed the possibility that cGMP stimulation of PDE2 catalytic activity, which hydrolyzes both cAMP and cGMP, might induce a PDE2–PDE4 crosstalk. Thus, the participation of PDE2 and PDE4 in cyclic nucleotide catabolism was investigated in nonstimulated BAEC using cilostamide (normally a PDE3 inhibitor having a $K_i = 0.042$ μmol/L) as a PDE2 inhibitor ($K_i = 15$ μmol/L),[32] as BAEC do not contain PDE3,[9] and rolipram as a PDE4 inhibitor ($K_i = 0.8$ μmol/L).[15] We have measured cAMP and cGMP levels (fmol/μg DNA) in indomethacin-treated cells in the presence of NO and cGMP pathway modulators.[35] In control BAEC, the cAMP level was about fivefold greater (254 fmol/μg DNA) than the cGMP level (50 fmol/μg DNA). Pretreatment with either 100 μM L-arginine (NO synthase substrate) or 300 μM N^G-nitro-L-arginine-methyl-ester (L-NAME; NO synthase inhibitor) did not modify cyclic nucleotide levels, indicating a low basal NO and cGMP-signaling pathway activity in these cells. When treated for 5 min with a PDE2 inhibitor (50 μM cilostamide), and a PDE4 inhibitor (20 μM rolipram) alone and together independently of the pretreatment, only the combination of PDE2 and PDE4 inhibitors potently increased the cAMP level. The lack of effect of one specific PDE inhibitor on cAMP content may be related to the remaining hydrolytic activity of the other form, and the combination of both specific inhibitors may completely inhibit cAMP hydrolysis, thus inducing a marked increase in cAMP content (8.6-fold without pretreatment, $p < 0.001$; 12.9-fold with 100 μM L-arginine, $p < 0.001$; 11.3-fold with 300 μM L-NAME, $p < 0.001$).

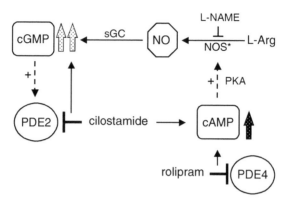

FIGURE 21.3 Implication of endothelial PDE4 in NO regulation. In presence of L-arginine (L-arg), which is the substrate for synthesis of nitric oxide (NO) by nitric oxide synthases (NOS) and subsequent increase in cGMP production by the nitric oxide–sensitive or soluble guanylyl cyclase (sGC), rolipram, a PDE4 inhibitor, significantly increased the cGMP content in BAECs. This was not observed with cilostamide alone, which was used as a PDE2 inhibitor, or in presence of L-NAME, which inhibits the production of NO by NO synthase.[35] The combination of both inhibitors caused a marked increase in cGMP level. This shows that PDE4 inhibition can increase cGMP level and suggests that this in turn increases cellular cAMP and perhaps upregulation of NO synthase[36] through the activation of PKA. The data suggest that PDE4 mediates cAMP–cGMP crosstalk in endothelial cells.

In the same samples, no change in cGMP content was observed either in control or L-NAME-treated cells. However, in the presence of L-arginine, which is the substrate for NO synthesis, the cGMP content was increased significantly by rolipram alone ($42\% \pm 2\%$ compared with L-arginine control cells, $p < 0.05$) but not by cilostamide alone.[35] This was very surprising as rolipram is involved specifically in cAMP catabolism and cilostamide inhibits PDE2, which hydrolyses both cAMP and cGMP. The combination of both inhibitors resulted in a marked increase in cGMP level ($153\% \pm 3\%$, compared with L-arginine control cells, $p < 0.001$). This shows that a PDE4 inhibitor is able to increase cGMP level and suggests a compartmentalized effect of PDE4 inhibition near the NO synthase; this interpretation is consistent with the demonstrated colocalization of NO synthase and PKA catalytic subunit in endothelial cells.[36] This result is in agreement with a cAMP-dependent phosphorylation of endothelial NO synthase[37] and regulation of the NO and cGMP pathway via a cAMP-dependent phosphorylation that enhances the NO synthase activity in endothelial cells.[38] Our data strengthen the hypothesis of cAMP activation of the NO and cGMP pathway in endothelial cells, and they show an interrelationship between cAMP and cGMP levels that is mediated by PDE4 and potentiated by PDE2 (Figure 21.3).

21.4 PHENOTYPIC EXPRESSION OF PDEs IN BAEC

BAEC are able to display two morphological and biosynthetic phenotypes in culture: (1) the cobblestone and quiescent phenotype characterized by a monolayer of closely apposed, cobblestone-shaped, polar nonproliferating cells; and (2) the spindle and proliferating phenotype characterized by elongated-shaped cells losing cell–cell contact and polarity. As intracellular cAMP elevation induced BAEC growth inhibition,[8] one would expect that PDE activity changes in the spindle and cobblestone phenotypes may be involved in phenotypic changes. Our studies performed on whole cell homogenates revealed that cAMP–PDE activity was 52% higher in the spindle phenotype compared to the cobblestone phenotype with increased PDE2, PDE3, and PDE4 activities and cGMP–PDE activity was increased by 10-fold with the induction of PDE1 and PDE5 expressions.[18] Resolution of

cytosolic cyclic nucleotide hydrolytic activities by HPLC Mono Q HR 5/5 column showed the presence of PDE2 and PDE4 along with some PDE3 in the cells with the cobblestone phenotype, whereas PDE1, PDE3, and PDE5 in addition to PDE2 and PDE4 were clearly resolved from the cytosolic fraction of the cells with the spindle phenotype. Studies at the protein and mRNA levels showed a threefold increase in a 110 kDa PDE2–protein in the cells with the spindle phenotype although a 3.9 kb PDE2-mRNA was threefold less expressed. In the cells with the spindle phenotype a 12 kb PDE3-mRNA was detected, which was absent in cells with the cobblestone phenotype, and a 7.5 kb PDE3A-mRNA was sixfold more expressed compared to cells with the cobblestone phenotype. In addition, cells with the spindle phenotype contained a twofold higher cAMP–PDE3 activity and the expression of 115 and 62 kDa PDE3-proteins that were mainly associated with the particulate fractions and a 72 kDa PDE3-protein that was essentially cytosolic. Interestingly, this protein increased significantly in the cytosol of the cells with the spindle phenotype in agreement with increased PDE3 activity revealed by HPLC. Northern blot studies showed the expression of 3.7 kb PDE4A, 3.5 kb PDE4B, and 3.6 and 7.0 kb PDE4D transcripts in cobblestone cells with the absence of PDE4C transcript. Among them, only the 7.0 kb PDE4D transcript was increased by threefold in cells with the spindle phenotype along with increased cAMP–PDE4 activity, essentially in the particulate fraction, and associated with increased expression of 100, 80, and 57 kDa PDE4-proteins; in contrast PDE4A and PDE4B-mRNAs were decreased by threefold.

If the increased cAMP–PDE activity that is associated with the spindle phenotype plays a major role in phenotype change of the endothelial cells, then specific inhibition of the elevated PDE isozymes would be predicted to prevent this change. As hypothesized, a 6-day treatment of cobblestone cells with the combination of 50 μM cilostamide (which inhibits PDE2 and PDE3) and 20 μM rolipram displayed a twofold increase in cAMP compared to untreated cells and maintained the cobblestone phenotype, whereas untreated cells displayed the spindle phenotype.[18]

Moreover, cAMP hydrolysis in the two phenotypes was differently regulated by 5 μM cGMP: 60% of the increase in total cAMP–PDE activity in the cells with the cobblestone phenotype related to cGMP stimulation of PDE2 activity; 30% of the decrease in the cells with the spindle phenotype related to cGMP inhibition of PDE3 activity. This emphasizes the roles of PDE2 and PDE3 in cAMP–cGMP crosstalk. Furthermore, it should be noticed that the great increase in cGMP–PDE activity in cells with the spindle phenotype is related to the induction of PDE1 and PDE5; this might change the cAMP and cGMP balance and thus compensate for the effect of NO synthase induction by VEGF.

These changes in PDE isozyme expression along with the crosstalk between cAMP and cGMP may well modulate NO production and consequently might participate in angiogenesis, making PDEs potential targets to modulate angiogenesis.

21.5 cGMP-INDUCED PROLIFERATION AND EFFECT ON PDE EXPRESSION IN THE HUMAN PERMANENT ENDOTHELIAL CELL LINE EAhy 926, AN *IN VITRO* ANGIOGENESIS MODEL

It has been clearly established that *in vitro* angiogenesis induced by VEGF is mediated by NO synthase upregulation, which increases intracellular cGMP level.[6,7] As cAMP is able to overcome VEGF-induced endothelial cell proliferation, we first investigated the PDE participation in *in vitro* angiogenesis (Figure 21.1). We have studied the participation of PDE2 and PDE4 in EAhy 926 cell proliferation using dipyridamole, papaverine, and trequinsin; these compounds are effective on both PDE2 and PDE4 (Table 21.2). When EAhy 926 cells were treated with these compounds for 48 h, these compounds dose-dependently and significantly inhibit EAhy 926 cell proliferation (Figure 21.4). Papaverine was the most potent compound

TABLE 21.2
Potency of PDE Inhibitors on Vascular PDE Isozymes (K_i, μM)[32]

PDEs	PDE1	PDE2	PDE3	PDE4	PDE5
Substrate	cGMP	cAMP	cAMP	cAMP	cGMP
Cilostamide	114	15	0.042	80	30
Dipyridamole	46	1.9	18	2	0.21
Trequinsin	0.4	0.8	0.005	0.3	13
Papaverine	92	5.7	2.7	2.8	2.9

PDE1, PDE3, PDE4, and PDE5 were isolated from bovine aortic vascular smooth muscle.[15] PDE2 was isolated from BAEC.[9] K_i values were determined by Lineweaver–Burk plots using cGMP in presence of Ca^{2+} and calmodulin for PDE1, cAMP in presence of 5 μM cGMP for PDE2.

with an IC_{50} value of 0.5 μM. Trequinsin was about 50-fold less potent than papaverine with an IC_{50} value of 24 μM. Furthermore, we showed that a 48 h treatment with VEGF (10 ng/ml) or 8-Br-cGMP (100 μM), a nonhydrolyzable and cell-permeant cGMP analog,

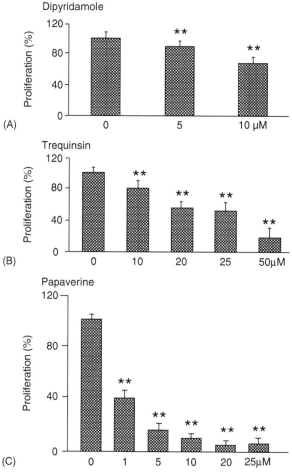

FIGURE 21.4 Effect of dipyridamole, trequinsin and papaverine, on EAhy 926 cell proliferation. Cells were treated for 48 h with increasing doses of dipyridamole (A), trequinsin (B), or papaverine (C). Cell proliferation was measured as described previously.[27]

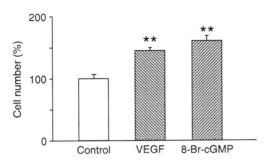

FIGURE 21.5 Effect of VEGF and 8-Br-cGMP on EAhy 926 cell proliferation. Cells were treated for 48 h with either VEGF (10 ng/ml) or 8-Br-cGMP (100 μM). Cell proliferation was measured as described previously.[27]

increased EAhy 926 cell proliferation by 45% and 59%, respectively (Figure 21.5), in agreement with the thesis that VEGF-induced angiogenesis is mediated by cGMP. At the same time, 8-Br-cGMP increased cAMP–PDE activity in the cytosolic fraction by 27% and the particulate fraction by 38%, indicating that cAMP-hydrolytic activity is increased in prolif-erative phenotype as previously seen.[18] When resolved by HPLC (Figure 21.6), no change in PDE isoform profile in terms of isoform number was seen. However, 8-Br-cGMP treat-ment induced a sevenfold increase in PDE4 activity, a threefold increase in basal PDE2 activity, and a 1.8-fold increase in cGMP-activated PDE2 activity. Northern blot analysis revealed 3.8 kb PDE2, 3.6 kb PDE4B, and 3.7 kb PDE4D-mRNA transcripts in EAhy 926 cells, and no change in mRNA expression was observed when these cells were treated with 8-Br-cGMP, indicating that the increases in PDE2 and PDE4 activities are related to post-trancriptional regulation.

These changes in cAMP–PDE and cGMP–PDE activities associated with cGMP-induced proliferation have opened a new area of research in angiogenesis focused on the changes in PDE2 and PDE4 activities.

21.6 INVOLVEMENT OF PDEs IN THE MIGRATION AND PROLIFERATION STEPS IN VEGF-STIMULATED HUVEC, AN *IN VITRO* ANGIOGENESIS MODEL

The EAhy 926 cell line,[24] which originates from a human pulmonary carcinoma, is highly proliferative per se and is not so much appropriated to study endothelial cell proliferation. Although more difficult to obtain and to cultivate, HUVECs were chosen as an *in vitro* model for angiogenesis to study PDE implications.

21.6.1 EFFECT OF VEGF ON PDE EXPRESSION IN HUVEC

VEGF secreted by smooth muscle cells or tumors represents the major signal inducing proliferation and migration of endothelial cells, thus initiating angiogenesis. We have reported that VEGF treatment of HUVEC (20 ng/ml) for 48 h induced a significant increase (+26%) in total cAMP–PDE activity in the cytosolic fraction with no change in total cGMP–PDE activity (Figure 21.7). This change in cAMP–PDE activity was related to an increase in PDE2 and PDE4 activities with no change in PDE3 activity (Figure 21.8). PDE2-, PDE3-, and PDE4-mRNAs were detected by Northern blot analysis,[39] showing the expression of 3.9 kb PDE2; 7.5 kb PDE3A, and no detectable PDE3B transcripts; 3.7 kb PDE4A, 3.7 kb PDE4B, no detectable PDE4C, and 7.8 and 3.9 kb PDE4D transcripts. PDE2, PDE3, and PDE4 were also detected by Western blot analysis, showing 93 and 67 kDa PDE2 signals;

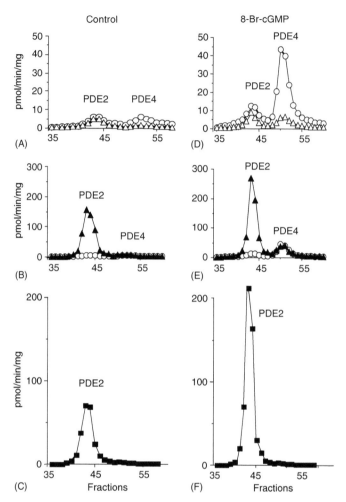

FIGURE 21.6 Effect of 8-Br-cGMP on EAhy 926 cell PDE activity. Cells were treated for 48 h with or without 100 μM 8-Br-cGMP and the effect was analyzed on cytosolic PDE activity resolved by HPLC. (A) and (D), cAMP–PDE activity was assessed at 1 μM cAMP in presence of EGTA (○—○) or in presence of 10 μM rolipram (△—△), in cells treated with (D) or without (A) 8-Br-cGMP; (B) and (E), cAMP–PDE activity was assessed at 1 μM cAMP in presence of EGTA (○—○) or in presence of 5 μM cGMP (▲—▲), in cells treated with (E) or without (B) 8-Br-cGMP; (C) and (F), cGMP–PDE activity was assessed at 1 μM cGMP (■—■) in cells treated with (F) or without (C) 8-Br-cGMP.

100 kDa PDE3A and 139 kDa PDE3B signals, although no PDE3B transcript was detected by Northern blot; 88, 62 and 59 kDa PDE4A signals; 69 and 66 kDa PDE4B signals; 117 and 70 kDa PDE4D signals (Figure 21.9). Treatment with VEGF showed (1) a significant (47%) increase in PDE2-mRNA with no statistically significant increase in PDE2 protein signals; (2) no modification of either PDE3A-mRNA or PDE3-protein; (3) significant transcript increases for PDE4A (3.5-fold), PDE4B (3.5-fold), PDE4D (+61% for the 7.8 kb and +76% for the 3.9 kb) along with significant increases in PDE4B-protein (+40% for the 69 kDa and 63% for the 66 kDa proteins) (Figure 21.10), but no statistically significant changes in PDE4A- or PDE4D- protein (data not shown).[39]

Interestingly, the increase of cAMP–PDE activity in HUVECs treated with VEGF, related to PDE2 and PDE4 activities, is similar to that obtained for EAhy 926 cells treated with

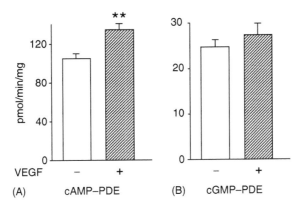

FIGURE 21.7 Effect of VEGF on HUVEC PDE activity. Cells were treated for 48 h with or without 20 ng/ml VEGF and the effect on PDE activities in the cytosolic fraction was analyzed at 1 μM cAMP or cGMP. Total cAMP–PDE (A) and cGMP–PDE (B) were assessed as described previously.[39]

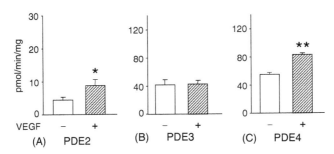

FIGURE 21.8 Effect of VEGF on HUVEC cAMP–PDE activity. Cells were treated for 48 h with or without 20 ng/ml VEGF and the effect was analyzed on total cytosolic cAMP–PDE activity. PDE2 (A), PDE3 (B), and PDE4 (C) activities were assessed as described.[39]

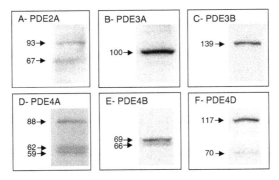

FIGURE 21.9 Characterization of PDE2, PDE3, and PDE4 in HUVEC extracts by Western blot. HUVEC extracts were prepared and analyzed by Western blot as described.[39] PDE2A was revealed with PDE2A3 antibody from Dr. Hermann Tenor, PDE3A with regulatory domain–PDE3A antibody from Dr. Vincent Manganiello, PDE3B with PDE3B antibody from Dr. Eva Degerman, PDE4A with AC55-PDE4A antibody from Dr. Marco Conti, PDE4B with K118-PDE4B antibody from Dr. Marco Conti, PDE4D with PDE4D antibody from FabGennix.

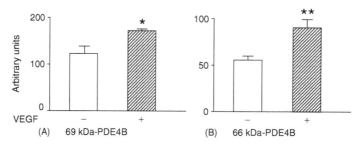

FIGURE 21.10 Effect of VEGF on PDE4B expression at protein level. Cells were treated for 48 h with or without 20 ng/ml VEGF and HUVEC extracts were prepared and analyzed by Western blot as described.[39] Densitometry was performed using the GeneGenius with GeneTool software from Syngene. Data are expressed as mean $\pm$ SEM of three experiments and analyzed by unpaired t test: $p < 0.05$.

8-Br-cGMP. This confirms the possible major role of increased cAMP–PDE activity in angiogenesis. The main difference between these human endothelial cells relates to cGMP–PDE activity, which was increased in EAhy 926 cells, whereas it was unchanged in HUVEC. This may be because of the lack of PDE5 in EAhy 926 cells, which then cannot overcome an increase in cGMP–PDE2 activity; this is in contrast to VEGF-stimulated HUVEC in which the increase in cGMP–PDE2 activity was overcome by a decrease in PDE5 activity. Furthermore, in contrast to HUVECs, no change in mRNA–PDE transcripts was seen in EAhy 926 cells. These differences point out that this human permanent cell line differs from primary cells at the level of PDE expression and is not a particularly good model for studying angiogenesis *in vitro*.

Thus, the cytosolic cAMP–PDE activity was the major hydrolytic activity and was mainly because of the activity of PDE2, PDE3, and PDE4. As PDE2 is activated by cGMP, whereas PDE3 is inhibited, and only the expression of PDE2 and PDE4 is increased in VEGF-stimulated HUVEC, we focused our investigation on PDE2, which can be activated by NO via cGMP, and on PDE4, which is insensitive to cGMP.

21.6.2 cAMP Accumulation in VEGF-Treated HUVEC

We have investigated the participation of PDE2 and PDE4 isozymes in the regulation of intracellular cAMP by measuring the effect of EHNA (a selective PDE2 inhibitor)[40] and RP73401 (a selective PDE4 inhibitor)[41] on intracellular cAMP level in control and VEGF-treated HUVEC (Figure 21.11). It is to be noticed that EHNA showed a 19-fold higher selectivity for the cGMP-activated PDE2 ($IC_{50} = 2$ μM) than for the basal PDE2 ($IC_{50} = 38$ μM), when assessed on PDE2 purified from human platelets and using 1 μM cAMP as substrate.[42] In control HUVECs, the PDE2 inhibitor did not increase cAMP, and this was recently confirmed in various vascular endothelial cells by Netherton and Maurice,[30] whereas the PDE4 inhibitor alone and in association with a PDE2 inhibitor markedly and significantly increased cAMP level by 73% and 40%, respectively. After 18 h of VEGF treatment, the intracellular cAMP content of HUVECs was not modified significantly, which might be due to either upregulation or activation of PDE2 and PDE4. When HUVEC were stimulated by VEGF, use of either a PDE2 inhibitor or a PDE4 inhibitor induced a 33% and 40% increase in cAMP, respectively. The combination of the two inhibitors had no greater effect on cAMP level, suggesting that the increase in intracellular cAMP had reached its maximal level on inhibition of either of these PDEs.

It is well established that exposure of HUVEC to VEGF is followed by an intracellular increase of cGMP related to NO production.[6] This increase in cGMP may reveal the inhibitory

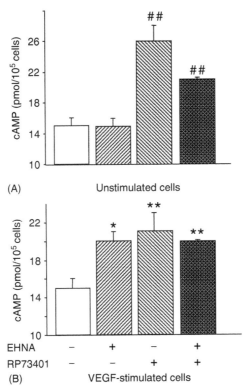

FIGURE 21.11 Effect of EHNA and RP73401 on HUVEC intracellular cAMP level. Cells were treated for 18 h with 20 μM EHNA, a PDE2-selective inhibitor, and 10 μM RP73401, a PDE4-selective inhibitor, in absence (A) or in presence (B) of 20 ng/ml VEGF. cAMP was assessed as described.[27] Data are expressed as mean ± SEM of four experiments. $^{\#\#}$, $p < 0.01$ versus control in absence of VEGF; *, $p < 0.05$ versus control in presence of VEGF; **, $p < 0.01$ versus control in presence of VEGF.

effect of EHNA toward PDE2 activity, and it might also competitively inhibit PDE3 activity, therefore increasing cAMP level markedly. In the absence of VEGF, intracellular cGMP level may not be sufficient to activate PDE2, which in its basal state is much less sensitive to EHNA.[42] The EHNA effects we observed are certainly not related to its inhibitory action on adenosine deaminase,[43] as the AMP-deamination pathway is marginal in HUVEC[44] and as adenosine itself stimulates proliferation.[45] The variations of cAMP level induced by PDE2 and PDE4 inhibitors attest to the participation of PDE2 and PDE4 inhibition in the regulation of cAMP level in nonstimulated and in VEGF-stimulated HUVEC and thus validate the use of EHNA and RP73401 to study the various processes involved in angiogenesis.

21.6.3 VEGF-INDUCED HUVEC PROLIFERATION

In view of previous reports showing that sustained elevation of cAMP inhibits the proliferation of many cell types including BAEC,[8,46,47] we first investigated the effect of PDE2 and PDE4 inhibitors on VEGF-induced HUVEC proliferation.[27] Exposure of HUVEC to VEGF for 72 h induced a 35% ($p < 0.01$) increase in cell number. Inhibition of PDE2 or PDE4 reduced VEGF-induced proliferation by 30% ($p < 0.01$) and 10% ($p < 0.05$), respectively (data not shown). In agreement with the cAMP data shown in Figure. 21.11, no greater inhibition of cell proliferation was obtained when the two inhibitors were combined (data not shown).

Reduction of proliferation by 45% was also obtained in VEGF-stimulated HUVEC treated with 8-Br-cAMP, consistent with the interpretation that PDE2 and PDE4 inhibitor effects were cAMP-mediated and overcame VEGF-induced proliferation.

21.6.4 VEGF-INDUCED HUVEC CYCLE PROGRESSION

The mechanism by which 8-Br-cAMP, PDE2, and PDE4 inhibitors reduced proliferation was explored using flow cytometry to evaluate the distribution of HUVEC in the different phases of the cell cycle (Table 21.3).[27] VEGF, by decreasing cell number by 15% in G_0/G_1 phase and increasing cell number by 26% in S phase, promoted G_1–S phase transition. As expected, 8-Br-cAMP reversed VEGF effects. Addition of a PDE2 inhibitor to VEGF-stimulated cells had no significant effect on G_1–S phase transition but inhibited the S–G_2/M transition by 27%. The combination of PDE2 and PDE4 inhibitors converted the VEGF-stimulated HUVEC to a nonproliferative status (72% in G_0/G_1 phase), which was even greater than the control condition (57%). This inhibitory effect on cell cycle progression suggests that PDE2 and PDE4 participate in the control of cell cycle progression that governs cell proliferation.

21.6.5 CELL CYCLE PROTEIN PHOSPHORYLATION AND EXPRESSION IN VEGF-STIMULATED HUVEC

The antiproliferative effect of cAMP is mediated by an intricate crosstalk between PKA, the mitogen-activated protein (MAP) kinase and ERK pathway, and the cyclin-dependent kinases (CDKs).[48,49] The cell cycle is regulated by the coordinated action of CDKs in association with their specific regulator cyclin proteins, notably cyclin D1 for G_1/S transition and cyclin A for S and G_2/M phases.[50] The kinase activity of these CDK–cyclin complexes necessary for cell cycle progression is inhibited by CDK inhibitors (CDKI) including p21$^{wafl/cip1}$ and p27^{kip1}.[51] Table 21.4 illustrates the effects of VEGF treatment on ERK phosphorylation and cell cycle protein expression in HUVECs and compares the effects of 8-Br-cAMP, PDE2, and PDE4 inhibitors on VEGF-stimulated HUVEC. As expected, VEGF, which initiates cell cycle progression, induced ERK phosphorylation, upregulation of cyclin D1 and cyclin A expression, and downregulation of p21$^{wafl/cip1}$ and p27^{kip1} expression.

In VEGF-treated cells, surprisingly, 8-Br-cAMP did not modify either p42/p44 MAP kinase phosphorylation or cyclin A expression, but, as expected, decreased cyclin D1 expression with

---

TABLE 21.3
Effects of VEGF, 8-Br-cAMP, and PDE2 and PDE4 Inhibitors on Distribution of Endothelial Cells in the Different Phases of the Cell Cycle

Treatment	G_0/G_1	S	G_2/M
Control	57.3 ± 1.09	24.3 ± 1.84	18.6 ± 1.68
VEGF (V)	45.1 ± 2.09*	32.2 ± 3.66**	17.1 ± 1.36
V + 8-Br-cAMP	56.0 ± 4.83#	24.7 ± 3.82# #	13.6 ± 2.76
V + PDE2 inhibitor	52.1 ± 2.07# #	31.6 ± 2.95	14.0 ± 0.86# #
V + PDE4 inhibitor	60.2 ± 0.92#	19.7 ± 4.21,# #	16.3 ± 2.22
V+PDE2+PDE4 inhibitors	72.1 ± 2.40**,# # #	14.5 ± 1.54**,# # #	11.4 ± 2.17# #

Following growth arrest, HUVEC were stimulated by addition of 20 ng/ml VEGF and were treated for 18 h with 600 μM 8-Br-cAMP, 20 μM EHNA (PDE2 inhibitor), and 10 μM RP73401 (PDE4 inhibitor).[27,39] Data representing the proportion of cell (%) in the different phases of cell cycle are expressed as mean ± SEM of four experiments. *, $p < 0.05$; **, $p < 0.01$ versus control cells; #, $p < 0.05$; # #, $p < 0.01$; # # #, $p < 0.001$ versus nontreated VEGF-stimulated cells.

TABLE 21.4
Effects of 8-Br-cAMP and PDE2 and PDE4 Inhibitors on Phosphorylation of p42/p44 MAPK and Expression of Key Cell Cycle Proteins in VEGF-Stimulated HUVEC

	Phosphorylation	Protein Expression			
Treatment	P-p42/p44 MAPK	Cyclin D1	Cyclin A	p21$^{waf1/cip1}$	p27^{kip1}
+VEGF/−VEGF	220%**	+67%**	+44%**	−28%*	−52%*
VEGF (V)	100%	100%	100%	100%	100%
V+8-Br-cAMP	n.s.	−23%$^{\#}$	n.s.	+64%$^{\#}$	+78%$^{\#\ \#}$
V+PDE2 inhibitor	−40%$^{\#\ \#}$	−13%$^{\#}$	+34%$^{\#}$	+103%$^{\#}$	−39%$^{\#\ \#}$
V+PDE4 inhibitor	−33%$^{\#\ \#}$	−35%$^{\#\ \#}$	−42%$^{\#\ \#}$	+141%$^{\#}$	+88%$^{\#}$
V+PDE2+PDE4 inhibitors	−43%$^{\#\ \#}$	−55%$^{\#\ \#}$	−38%$^{\#\ \#}$	+ 133%$^{\#}$	+62%$^{\#\ \#}$

HUVECs were stimulated with 20 ng/ml VEGF and treated with 600 μM 8-Br-cAMP, PDE2 inhibitor (20 μM EHNA), PDE4 inhibitor (10 μM RP73401) for 30 min for measurement of MAP kinase phosphorylation and 18 h for detection of changes in expression of cell cycle proteins.[39] For the VEGF effect, data are expressed taking control cells (−VEGF) as 100%. For treatment of VEGF-stimulated cells, data are expressed taking VEGF-stimulated cells as 100%. Results are expressed as the mean of four independent experiments. n.s., not significant; *, $p < 0.05$; **, $p < 0.01$ versus nontreated cells; $^{\#}$, $p < 0.05$; $^{\#\ \#}$, $p < 0.01$ versus nontreated VEGF-stimulated cells.

a concomitant increase in p21$^{waf1/cip1}$ and p27^{kip1}. VEGF-induced p42/p44 MAP kinase phosphorylation was decreased by a PDE2 inhibitor, and a PDE4 inhibitor, but no greater significant effect was obtained when the PDE2 inhibitor was combined with the PDE4 inhibitor, suggesting that (Figure 21.12) PDE2 and PDE4 inhibitors act on the same mechanism, which excludes any effect of 8-Br-cAMP.[39] Similar to the effect of 8-Br-cAMP, PDE2 and PDE4 inhibitors significantly overcame the VEGF-induced increase in cyclin D1 expression. The significant potentiation by PDE4 inhibitor of the action of the PDE2 inhibitor effect revealed that these compounds may act at different levels of cyclin D1 regulation, suggesting compartmentation of the action of PDE2 and PDE4 in these cells. In contrast to the lack of 8-Br-cAMP effect on cyclin A upregulation, the PDE2 inhibitor significantly potentiated the VEGF effect, whereas the PDE4 inhibitor, either alone or in combination with the PDE2 inhibitor, markedly overcame the VEGF effect, thus bringing cyclin A expression back to its basal level. This suggests that PDE2 and PDE4 inhibitors act on different mechanisms regulating cyclin A expression, and also reveals that the PDE2 inhibitor may act upstream of PDE4 inhibitor, as the PDE4 inhibitor is able to completely overcome the stimulating effect of the PDE2 inhibitor. The expression of p21$^{waf1/cip1}$ was increased by PDE2 and PDE4 inhibitors alone or in combination, restoring basal level of p21$^{waf1/cip1}$ to the level in HUVEC not stimulated by VEGF, as did by 8-Br-cAMP; this indicates that 8-Br-cAMP, PDE2 and PDE4 inhibitors act on the same mechanism regulating p21$^{waf1/cip1}$ expression.[39]

Similar to 8-Br-cAMP, PDE4 inhibitors as well as the combination of PDE2 and PDE4 inhibitors increased p27^{kip1} expression, restoring the basal level of p27^{kip1} expression in HUVEC not stimulated by VEGF. However and surprisingly, the PDE2 inhibitor markedly decreased p27^{kip1} expression, and this decrease was significantly counteracted in the presence of PDE4 inhibitor, indicating that PDE2 and PDE4 inhibitors may act in a different manner or at a different level on p27^{kip1} regulation.

PDEs have different intracellular localizations such as the sarcolemma, sarcoplasmic reticulum, and nuclear envelope.[52,53] In HUVECs, PDE2 and PDE4 that are located in the cytosolic or membrane fractions may have different relationships to various functional cell

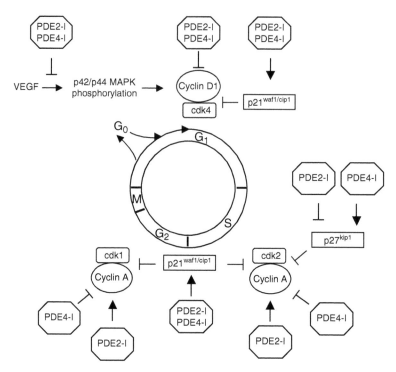

FIGURE 21.12 Effects of PDE2 inhibitor (PDE2-I) and PDE4 inhibitor (PDE4-I) on the VEGF-stimulated HUVEC cell cycle. VEGF, by phosphorylating p42/p44 MAPK, induces the expression of cyclin D1, which stimulates the G_0/G_1–S phase transition, thus initiating cell cycle progression. The complexes cyclin A/CDK2 and cyclin A/CDK1 are involved in S-phase and G_2-phase, respectively. The cyclin kinase inhibitor $p21^{waf1/cip1}$ inhibits cyclin D1/CDK4, cyclin A/CDK2, and cyclin A/CDK1 complexes. The cyclin kinase inhibitor $p^{27\ kip1}$ acts only on cyclin A/CDK2 complex. PDE2 and PDE4 inhibitors interact with some regulation points of cell cycle progression. Both PDE2 inhibitor and PDE4 inhibitor inhibit VEGF-induced p42/p44 MAPK phosphorylation and cyclin D1 expression, and they stimulate $p21^{waf1/cip1}$ expression. PDE2 and PDE4 inhibitors act differentially on cyclin A expression: stimulatory for PDE2-I and inhibitory for PDE4-I; they also act differentially on $p^{27\ kip1}$ expression: inhibitory for PDE2-I and stimulatory for PDE4-I. Arrows indicate a stimulating effect, whereas stopped lines indicate an inhibitory effect.

structures. PDE4 subtypes play a key role in compartmentalization of the cAMP and PKA signaling pathway by preventing cAMP from leaving distinct compartments and activating a local population PKA.[54] Possibly, in these cells there are cAMP pools that are distinctly regulated by either PDE2 or PDE4. These distinct pools may activate PKA subtypes that are colocalized with different cellular ultrastructures via specific A-kinase anchoring proteins (AKAPs),[55] and which may be differently associated to key cell cycle proteins, notably cyclin A and $p27^{kip1}$.

21.6.6 VEGF-Induced HUVEC Migration

Migration of endothelial cells, which is one of the steps in angiogenesis, is inhibited by elevation of PKA activity.[56,57] We have studied the effect of 8-Br-cAMP, PDE2 inhibitor, or PDE4 inhibitor on VEGF-induced migration assessed on wounded cell monolayer.[27] VEGF increased HUVEC migration by 84% ($p < 0.01$). A 24 h treatment with 8-Br-cAMP, PDE2 inhibitor, or PDE4 inhibitor reduced migration of VEGF-stimulated HUVEC by 23%

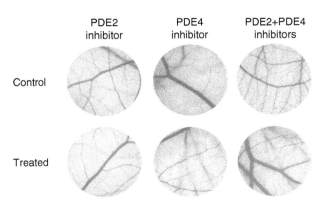

FIGURE 21.13 (See color insert following page 274.) Effects of PDE2 and PDE4 inhibitors on microvessel development of chicken embryo CAM. CAMs were treated with PDE2 inhibitor (5 μg EHNA) and PDE4 inhibitor (5 μg RP73401) and analyzed as described.[27]

($p < 0.01$), 32.5% ($p < 0.01$), and 33.7% ($p < 0.01$), respectively. The combination of both inhibitors markedly reduced cell migration (-50%, $p < 0.01$). Similar results were recently reported.[30] This study shows that PDE2 and PDE4 participate in the control of cell migration possibly at different levels of regulation governed by PKA phosphorylation.

21.7 EFFECTIVENESS OF PDE INHIBITORS IN THE CHICKEN EMBRYO CAM, AN *IN VIVO* ANGIOGENESIS MODEL

We used the chicken embryo CAM, an *in vivo* model of angiogenesis, to study the *in vivo* effects of PDE inhibitors on microvessel development.[27] The comparison of capillary total surface in the control and treated areas of CAM in the same embryo showed that PDE2 inhibitor as well as PDE4 inhibitor per se did not modify small vessel density. However, when PDE2 inhibitor was combined with the PDE4 inhibitor, the density of small vessels was reduced by 50% ($p < 0.001$) (Figure 21.13). This inhibitory effect was dose-dependent, as 0.5, 2.5, and 5 μg of compounds reduced the density of small vessels by 19% ($p < 0.01$), 25% ($p > 0.01$), and 50% ($p > 0.001$), respectively.[27]

These results document the antiangiogenic effects of PDE2 and PDE4 inhibitors in the *in vivo* CAM model. Only the association of both inhibitors was dose-dependently effective, indicating that it is necessary to inhibit various mechanisms implicated in angiogenesis *in vivo*. This result is consistent with the effect of PDE2 and PDE4 on the cell cycle, including cell cycle protein regulation, as well as on endothelial cell proliferation and migration. Nevertheless, an antiproliferative and antimigratory effect of PDE4 inhibitors on vascular smooth muscle cells could also be implicated.[58]

21.8 CONCLUSION

Most endothelial cells are characterized by the coexistence of PDE2 and PDE4 along with some PDE3, PDE5, and to a lesser extent to PDE1 and PDE7 depending on blood vessel type, phenotypic state, and culture conditions (Table 21.1). The critical role of these cells in the control of VEGF-mediated angiogenesis focused questions about PDE2 and PDE4 participation and consequently the possible development of new antiangiogenic compounds. The present data clearly show that in HUVEC, an human *in vitro* model of angiogenesis, the inhibition of both PDE2 and PDE4 increases cAMP levels, inhibits cell proliferation, cell migration, and cell cycle progression, and overcomes VEGF-induced cell cycle protein

TABLE 21.5
Effects of 8-Br-cAMP and PDE2 and PDE4 Inhibitors on EGF-Induced Processes in HUVEC

Effects	PDE2 Inhibitor	PDE4 Inhibitor	PDE2 + PDE4 Inhibitors	8-Br-cAMP
Migration	↓	↓	↓↓	↓
Proliferation	↓	↓	↓	↓
Cell cycle	G_2/M	S	S, G_2/M	S
MAP kinase	↓	↓	↓	↔
Cyclin A	↑	↓	↓	↔
Cyclin D1	↓	↓	↓	↓
p21	↑	↑	↑	↑
p27	↓	↑	↑	↑

alterations (Table 21.5). Interestingly, although 8-Br-cAMP inhibits cell migration and proliferation, 8-Br-cAMP is not able to completely mimic the effect of PDE2 and PDE4 inhibitors, because at the same concentration, it does not inhibit MAP kinase phosphorylation and cyclin A expression, indicating that PDE2 and PDE4 regulate compartmentalized cAMP pools. Further indication that PDE2 and PDE4 regulate different cAMP pools are observations derived from the use of selective inhibitors that PDE2 mediates p21[waf1/cip1] downregulation and p27[kip1] upregulation, whereas PDE4 mediates both p21[waf1/cip1] and p27[kip1] downregulation. Thus, these two PDE families participate in regulating cell cycle at different steps, which are dependent on defined patterns of intracellular targeting involving various protein interactions. This study also shows that, contrary to actions seen with 8-Br-cAMP, the inhibition of both PDE2 and PDE4 is necessary to overcome all studied alterations induced by angiogenesis, indicating that the catalytic activity of PDEs is directly implicated in regulating angiogenesis (Table 21.5). Also, it could be hypothesized that the role of PDE2 in regulating cGMP level could overcome the effect of PDE4 on cellular cAMP level, as much as the effect of PDE2 inhibitors[9,42] depends on the intracellular level of cGMP.[59] One can imagine that the increase in cGMP related to PDE2 inhibition, in contrast to PDE4, would stimulate p27[kip1] upregulation. Nevertheless, PDE2 inhibition increases cAMP level in VEGF-stimulated HUVEC more potently than PDE4 inhibition, indicating that the cAMP and cGMP control is very complex and may implicate multiple intracellular compartments playing various specific roles in intracellular signaling. Interestingly, the diversity in the PDE isozyme profile encountered in the different endothelial cells reveals a great variability in cAMP–PDE and

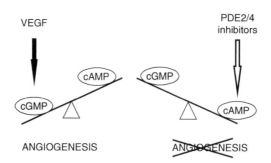

FIGURE 21.14 Role of PDE2 and PDE4 inhibitors in the cAMP and cGMP balance and the control of angiogenesis.

cGMP–PDE activity ratio, which is likely to affect the cAMP and cGMP balance and contribute to the control of angiogenesis (Figure 21.14). This emphasizes the prerequisite of a sound knowledge of the PDEs in a cell type before initiating cell signaling studies.

Our data in HUVEC clearly show that the inhibition of VEGF-upregulated PDE2 and PDE4 overcomes VEGF-stimulated migration and proliferation steps of angiogenesis and that the cAMP increases related to PDE2 and PDE4 inhibition are more effective than the cAMP increase related to 8-Br-cAMP. These studies also suggest that the cAMP and cGMP balance is of great importance in this process. This new approach in studying angiogenesis, on one hand, identifies PDE2 and PDE4 as new therapeutic targets, and it opens a new therapeutic field to prevent cancer development, atherosclerosis, and diabetes. On the other hand, PDE2 and PDE4 overexpression would be predicted to favor tissue and vascular reconstruction, but this therapeutic approach will necessitate the identification of down-regulated variants in altered tissues.

ACKNOWLEDGMENTS

We are grateful to Dr. Hermann Tenor for PDE2A3 antibody, Dr. Vincent Manganiello for PDE3A–regulatory domain antibody, Dr. Eva Degerman for PDE3B antibody, Dr. Marco Conti for PDE4A-(AC55) and PDE4B-(K118) antibodies. We also thank Dr. Cora-Jean Edgell for the EAhy 926 cell line. We acknowledge Alain Le Bec, Hélène Justiano-Basaran, and Evelyne Lacoffrette-Cronier for their skillful technical helps. This work was supported by the Ligue Régionale contre le Cancer and "ARERS."

REFERENCES

1. Furchgott, R.F., and J.V. Zawadzki. 1980. The obligatory role of endothelial cells in the relaxation of arterial smooth muscle by acetylcholine. *Nature* 288:373.
2. Palmer, R.M., A.G. Ferrige, and S. Moncada. 1987. Nitric oxide release accounts for the biological activity of endothelium-derived relaxing factor. *Nature* 327:524.
3. Arnold, W.P., et al. 1977. Nitric oxide activates guanylate cyclase and increases guanosine 3′,5′-cyclic monophosphate levels in various tissue preparations. *Proc Natl Acad Sci USA* 74:3203.
4. Furchgott, R.F., and P.M. Vanhoutte. 1989. Endothelium-derived relaxing and contracting factors. *FASEB J* 3:2006.
5. Folkman, J. 1995. Angiogenesis in cancer, vascular, rheumatoid, and other disease. *Nat Med* 1:27.
6. Papapetropoulos, A., et al. 1997. Nitric oxide production contributes to the angiogenic properties of vascular endothelial growth factor in human endothelial cells. *J Clin Invest* 100:3131.
7. Hood, J., and H.J. Granger. 1998. Protein kinase G mediates vascular endothelial growth factor-induced Raf-1 activation and proliferation in human endothelial cells. *J Biol Chem* 273:23504.
8. Leitman, D.C., R.R. Fiscus, and F. Murad. 1986. Forskolin, phosphodiesterase inhibitors, and cyclic AMP analogues inhibit proliferation of cultured bovine aortic endothelial cells. *J Cell Physiol* 127:237.
9. Lugnier, C., and V.B. Schini. 1990. Characterization of cyclic nucleotide phosphodiesterases from cultured bovine aortic endothelial cells. *Biochem Pharmacol* 39:75.
10. Souness, J.E., et al. 1990. Pig aortic endothelial-cell cyclic nucleotide phosphodiesterases. Use of phosphodiesterase inhibitors to evaluate their roles in regulating cyclic nucleotide levels in intact cells. *Biochem J* 266:127.
11. Lugnier, C., et al. 1985. Substituted carbostyrils as inhibitors of cyclic AMP phosphodiesterase. *Eur J Med Chem* 20:121.
12. Yamamoto T., et al. 1983. Complex effects of inhibitors on cyclic GMP-stimulated cyclic nucleotide phosphodiesterase. *J Biol Chem* 258:14173.
13. Wu, A.Y., et al. 2004. Molecular determinants for cyclic nucleotide binding to the regulatory domains of phosphodiesterase 2A. *J Biol Chem* 279:37928.

14. Rybalkin, S.D., et al. 2003. PDE5 is converted to an activated state upon cGMP binding to the GAF A domain. *EMBO J* 22:469.

15. Lugnier, C., et al. 1986. Selective inhibition of cyclic nucleotide phosphodiesterases of human, bovine and rat aorta. *Biochem Pharmacol* 35:1743.

16. Prigent, A.F., et al. 1988. Comparison of cyclic nucleotide phosphodiesterase isoforms from rat heart and bovine aorta. Separation and inhibition by selective reference phosphodiesterase inhibitors. *Biochem Pharmacol* 37:3671.

17. Ashikaga, T., S.J. Strada, and W.J. Thompson. 1997. Altered expression of cyclic nucleotide phosphodiesterase isozymes during culture of aortic endothelial cells. *Biochem Pharmacol* 54:1071.

18. Keravis, T., N. Komas, and C. Lugnier. 2000. Cyclic nucleotide hydrolysis in bovine aortic endothelial cells in culture: Differential regulation in cobblestone and spindle phenotypes. *J Vasc Res* 37:235.

19. Suttorp, N., et al. 1993. Role of phosphodiesterases in the regulation of endothelial permeability in vitro. *J Clin Invest* 91:1421.

20. Zhu, B., S. Strada, and T. Stevens. 2005. Cyclic GMP-specific phosphodiesterase 5 regulates growth and apoptosis in pulmonary endothelial cells. *Am J Physiol Lung Cell Mol Physiol* 289:L196.

21. Thompson, W.J., et al. 2002. Regulation of cyclic AMP in rat pulmonary microvascular endothelial cells by rolipram-sensitive cyclic AMP phosphodiesterase (PDE4). *Biochem Pharmacol* 63:797.

22. Bauer, J., et al. 1992. In vitro model of angiogenesis using a human endothelium-derived permanent cell line: Contributions of induced gene expression, G-proteins, and integrins. *J Cell Physiol* 153:437.

23. Ilan, N., S. Mahooti, and J.A. Madri. 1998. Distinct signal transduction pathways are utilized during the tube formation and survival phases of in vitro angiogenesis. *J Cell Sci* 111:3621.

24. Edgell, C.J., C.C. McDonald, and J.B. Graham. 1983. Permanent cell line expressing human factor VIII-related antigen established by hybridization. *Proc Natl Acad Sci USA* 80:3734.

25. Favot, L., T. Keravis, and C. Lugnier. 2000. Contrôle de l'angiogenèse tumorale par les inhibiteurs de phosphodiestérases (PDEs). *Arch Mal Coeur* 93:432, A 4–2.

26. Hayashi, T., et al. 1995. Estrogen increases endothelial nitric oxide by a receptor-mediated system. *Biochem Biophys Res Commun* 214:847.

27. Favot. L., et al. 2003. VEGF-induced HUVEC migration and proliferation are decreased by PDE2 and PDE4 inhibitors. *Thromb Haemost* 90:334.

28. Miro, X., et al. 2000. Phosphodiesterases 4D and 7A splice variants in the response of HUVEC cells to TNF-α. *Biochem Biophys Res Commun* 274:415.

29. Sadhu, K., et al. 1999. Differential expression of the cyclic GMP-stimulated phosphodiesterase PDE2A in human venous and capillary endothelial cells. *J Histochem Cytochem* 47:895.

30. Netherton, S., and D.H. Maurice. 2005. Vascular endothelial cell cyclic nucleotide phosphodiesterases and regulated migration: Implications in angiogenesis. *Mol Pharmacol* 67:263.

31. D'Andrea, M.R., et al. 2005. Expression of PDE11A in normal and malignant human tissues. *J Histochem Cytochem* 53:895.

32. Lugnier, C., and N. Komas. 1993. Modulation of vascular cyclic nucleotide phosphodiesterases by cyclic GMP: Role in vasodilatation. *Eur Heart J* 14:141.

33. Boulanger, C., et al. 1990. Stimulation of cyclic GMP production in cultured endothelial cells of the pig by bradykinin, adenosine diphosphate, calcium ionophore A23187, and nitric oxide. *Br J Pharmacol* 101:152.

34. Schmidt, K., W.F. Graier, and W.R. Kukovetz. 1990. EDRF-formation increases cGMP-levels in cultured endothelial cells. *Adv Second Messenger Phosphoprotein Res* 24:455.

35. Kessler, T., and C. Lugnier. 1995. Rolipram increases cyclic GMP content in L-arginine-treated cultured bovine aortic endothelial cells. *Eur J Pharmacol* 290:163.

36. Heijnen, H.F., et al. 2004. Colocalization of eNOS and the catalytic subunit of PKA in endothelial cell junctions: A clue for regulated NO production. *J Histochem Cytochem* 52:1277.

37. Butt, E., et al. 2000. Endothelial nitric-oxide synthase (type III) is activated and becomes calcium independent upon phosphorylation by cyclic nucleotide-dependent protein kinases. *J Biol Chem* 275:5179.

38. Graier, W.F., et al. 1992. Increases in endothelial cyclic AMP levels amplify agonist-induced formation of endothelium-derived relaxing factor (EDRF). *Biochem J* 288:345.
39. Favot, L., T. Keravis, and C. Lugnier. 2004. Modulation of VEGF-induced endothelial cell cycle protein expression through cyclic AMP hydrolysis by PDE2 and PDE4. *Thromb Haemost* 92:634.
40. Podzuweit, T., P. Nennstiel, and A. Muller. 1995. Isozyme selective inhibition of cGMP-stimulated cyclic nucleotide phosphodiesterases by *erythro*-9-(2-hydroxy-3-nonyl)adenine. *Cell Signal* 7:733.
41. Raeburn, D., et al. 1994. Antiinflammatory and bronchodilator properties of RP 73401, a novel and selective phosphodiesterase type IV inhibitor. *Br J Pharmacol* 113:1423.
42. Lugnier, C. 2000. Unpublished data.
43. Schaeffer, H.J., and C.F. Schwender. 1974. Enzyme inhibitors: Bridging hydrophobic and hydrophilic regions on adenosine deaminase with some 9-(2-hydroxy-3-alkyl)adenines. *J Med Chem* 17:6.
44. Smolenski, R.T., et al. 1994. Endothelial nucleotide catabolism and adenosine production. *Cardiovasc Res* 8:100.
45. Sexl, V., et al. 1995. Stimulation of human umbilical vein endothelial cell proliferation by A2-adrenosine and β_2-adrenoceptors. *Br J Pharmacol* 114:1577.
46. Hordijk, P.L., et al. 1994. cAMP abrogates the p21ras-mitogen-activated protein kinase pathway in fibroblasts. *J Biol Chem* 269:3534.
47. Koyama, H., et al. 2001. Molecular pathways of cyclic nucleotide-induced inhibition of arterial smooth muscle cell proliferation. *J Cell Physiol* 186:1.
48. Cook, S.J., and F. McCormick. 1993. Inhibition by cAMP of Ras-dependent activation of Raf. *Science* 262:1069.
49. Vadiveloo, P.K., et al. 1997. G1 phase arrest of human smooth muscle cells by heparin, IL-4 and cAMP is linked to repression of cyclin D1 and cdk2. *Atherosclerosis* 133:61.
50. Nigg, E.A. 1993. Targets of cyclin-dependent protein kinases. *Curr Opin Cell Biol* 5:187.
51. Vidal, A., and A. Koff. 2000. Cell-cycle inhibitors: Three families united by a common cause. *Gene* 247:1.
52. Lugnier, C., et al. 1999. Characterization of cyclic nucleotide phosphodiesterase isoforms associated to isolated cardiac nuclei. *Biochim Biophys Acta* 1472:431.
53. Lugnier, C. 2006. Cyclic nucleotide phosphodiesterase (PDE) superfamily: A new target for the development of specific therapeutic agents. *Pharmacol Ther* 109:366.
54. Houslay, M.D., and D.R. Adams. 2003. PDE4 cAMP phosphodiesterases: Modular enzymes that orchestrate signaling cross-talk, desensitization, and compartmentalization. *Biochem J* 370:1.
55. Dell'Acqua, M.L., and J.D. Scott. 1997. Protein kinase A anchoring. *J Biol Chem* 272:12881.
56. Kim, S., M. Harris, and J.A. Varner. 2000. Regulation of integrin α vβ 3-mediated endothelial cell migration and angiogenesis by integrin α5β 1 and protein kinase A. *J Biol Chem* 275:33920.
57. Kim, S., et al. 2002. Inhibition of endothelial cell survival and angiogenesis by protein kinase A. *J Clin Invest* 110:933.
58. Palmer, D., K. Tsoi, and D.H. Maurice. 1998. Synergistic inhibition of vascular smooth muscle cell migration by phosphodiesterase 3 and 4 inhibitors. *Circ Res* 82:582.
59. Vandecasteele G., et al. 2001. Cyclic GMP regulation of the L-type Ca^{2+} channel current in human atrial myocytes. *J Physiol* 533:329.

22 Regulation of PDE Expression in Arteries: Role in Controlling Vascular Cyclic Nucleotide Signaling

Donald H. Maurice and Douglas G. Tilley

CONTENTS

22.1 Arterial Structure–Function 441
    22.1.1 Development and Vascular Remodeling 441
    22.1.2 Cyclic Nucleotides and Their Effectors..443
    22.1.3 PDE Expression in Cells of the Cardiovascular System:
            Not All Cells Are Created Equal 444
22.2 Regulated Expression and Function of Arterial Cell PDEs....................................447
    22.2.1 PDE1 Family Variants in Arterial VSMCs..447
    22.2.2 PDE3 Family Variants in Arterial VSMCs..449
        22.2.2.1 Differential Regulation of PDE3A and PDE3B
                in Contractile and Synthetic VSMCs 451
    22.2.3 PDE4 Family Variants in Arterial VSMCs and VECs 453
        22.2.3.1 Regulated cAMP-Dependent Expression
                of *PDE4D* in VSMCs and VECs..455
22.3 Conclusions...457
References 457

22.1 ARTERIAL STRUCTURE–FUNCTION

22.1.1 DEVELOPMENT AND VASCULAR REMODELING

Healthy arteries in the cardiovascular system function to conduct and distribute oxygenated blood, nutrients, and hormones to tissues and organs. Conducting, or elastic, arteries are proximal to the heart, and, due to their great number of elastic fibers, distend to accommodate the large increases in blood flow, which accompany ejection of blood from the heart.[1] In contrast, distributing, or muscular arteries, which represent the majority of arteries, contain few elastic fibers, are innervated, and are essential for the rapid and complete distribution of blood to organs and tissues.[1] Cardiovascular disease (CVD) remains one of the leading causes of morbidity and mortality in the developed world, with over 29% of total global deaths in 2003, almost 17 million people, resulting from the major forms of CVD.[1-3] The major forms of CVD include coronary heart disease (ischaemia, heart attack), cerebrovascular disease (stroke), and hypertension, each of which can cause death or long-term disability. Vascular

remodeling, which occurs as a physiological response to alterations in blood flow, blood pressure, or arterial injury resultant from these diseases, contributes to overall long-term pathology.[1–3]

Arteries consist of three main tissue layers: the intimal, medial, and adventitial layers (Figure 22.1A). The most inner layer, the intimal layer, consists of vascular endothelial cells

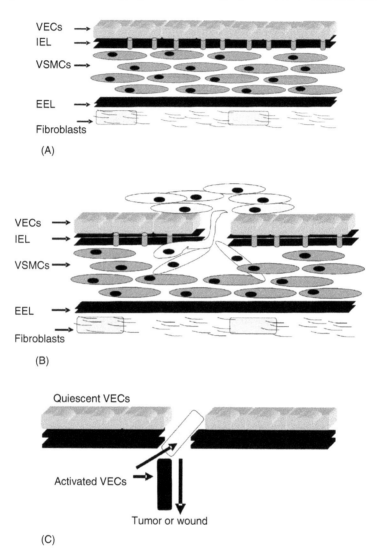

FIGURE 22.1 Structure–function of arteries. (A) Arteries are composed of three distinct layers of cells and extracellular matrix (ECM) proteins. The innermost layer, the intimal layer, is composed of vascular endothelial cells (VECs). The internal elastic lamina (IEL) separates the intimal layer from the medial layer. The medial layer is populated by vascular smooth muscle cells (VSMCs). The medial layer and the external-most layer, the adventitial layer, are separated by a lamina referred to as the external elastic lamina (EEL). The adventitial layer is sparsely populated by fibroblasts, which are encased in ECM proteins including several types of collagens. (B) In response to VEC dysfunction or vascular trauma, medial VSMCs undergo a phenotypic switch, which converts them from a contractile to a synthetic phenotype. In response to growth factors present in blood, synthetic VSMCs can migrate into the intimal space and proliferate. The neointimal VSMCs are a precursor to several vasculopathies. (C) As with VSMCs, VECs can be stimulated to migrate and proliferate. Stimuli for these behaviors in VECs include vascular endothelial growth factor released from tumors or at sites of vascular damage.

(VECs), which line the lumen and play a central role in coordinating the functions of blood vessels as well as several blood-borne cells, including leukocytes and platelets.[1] The intimal layer is separated from the middle layer, the medial layer, by a fenestrated internal elastic lamina composed of elastin. The medial layer is comprised of concentric layers of vascular smooth muscle cells (VSMCs), which are interposed by layers of VSMC-derived extracellular matrix proteins such as elastin and proteoglycans. The outermost layer, the adventitial layer, is composed primarily of fibroblast and elastic tissues.

In healthy arteries, medial layer–VSMCs control vascular patency through rhythmic contractions and relaxations and in doing so, influence the volume of blood shunted to various tissues, as well as the overall blood pressure. VSMCs of healthy blood vessels are said to be of a contractile phenotype.[1,4] In this differentiated state, VSMCs express a repertoire of proteins involved in the contractile process including smooth muscle (SM)-actin, SM–myosin heavy chain (MHC), phospholamban, and extracellular matrix (ECM) proteins including elastin and certain types of collagen.[1–5] Primarily programmed for contraction, contractile VSMCs show low rates of proliferation, migration, and synthesis of ECML proteins.[5–9] Similarly, VECs of healthy blood vessels have a low proliferative and migratory index and function both as (a) a physical barrier isolating subendothelial spaces from the blood and (b) an endocrine organ releasing numerous factors that regulate the functions of other VECs, of VSMCs, and of blood-borne cells. Several recent reviews have summarized the physiological, hemodynamic, and pharmacological factors that regulate the functions of the proteins involved in the force-producing contractile machinery of contractile VSMCs and of the impact of the VECs in controlling this function of VSMCs.[6–11]

In response to vascular damage, such as atherosclerosis or blunt injury from angioplasty, contractile VSMCs can undergo a phenotypic transition to a more synthetic phenotype (Figure 22.1B). In contrast to contractile VSMCs, the synthetic VSMCs display a more proliferative and migratory phenotype and, consistent with these functions, express a broader repertoire of proteins not involved in coordinating contractile events.[5–13] As a result of this phenotypic transition, synthetic VSMCs become sensitive to chemo-attractant and growth-promoting factors from VECs, or from plasma and blood-borne cells, and participate in vascular remodeling in response to hemodynamic and physiological demands.[5–13] The transition of VSMCs from a contractile to a synthetic phenotype also occurs when medial VSMCs are isolated and purified for cell culture.[5–13] As studies have shown that the protein expression profile of both synthetic VSMCs generated in culture and those isolated from blood vessels *in vivo* are very similar, cultured VSMCs provide a useful and convenient experimental model of synthetic VSMCs, which is used routinely to study the behavior of these cells.[11–13] Although less well defined, VECs also can be made to alter their phenotype from the quiescent state seen in stable vascular structures to one in which they adopt a more activated motile and proliferative phenotype (Figure 22.1C).[14] Migration and proliferation of these activated VECs are central events controlling the important role of these activated VECs in several processes including the initial development of the cardiovascular system and wound healing or angiogenesis in the adults (Figure 22.1C).[14]

22.1.2 CYCLIC NUCLEOTIDES AND THEIR EFFECTORS

The cyclic nucleotides, cyclic adenosine monophosphate (cAMP) and cyclic guanosine monophosphate (cGMP), regulate numerous functions in most cells, including those of the cardiovascular system.[15–28] Thus, both cAMP- and cGMP-elevating factors relax contractile VSMCs and inhibit the proliferation and migration of synthetic VSMCs.[15–36] Similarly, cAMP- and cGMP-elevating agents dynamically regulate both the barrier function of VECs, as well as the ability of the VECs to release vasoactive or chemotactic factors.[37–40] In this context, it is accepted that the main physiological regulator of VSMC functions is the intimal monolayer of VECs.[3,40] Intimal VECs are known to synthesize and release no fewer

than 20 vasoactive agents that act at neighboring VSMCs in a paracrine fashion.[37–40] Two agents synthesized and released by VECs and which act in VSMCs by increasing the levels of either cGMP or cAMP, are nitric oxide (NO) and prostacyclin (PGI$_2$).[37–40] The synthesis, release, and actions of these agents have been recently reviewed.[37–40] Briefly, diffusion of NO from VECs allows stimulation of the NO-sensitive guanylyl cyclase,[14,15] also known as soluble guanylyl cyclase (sGC). For its part, VEC-derived PGI$_2$ increases cellular cAMP by activating adenylyl cyclases largely through effects coupled to the classical Gs-coupled signal transduction system.[16,17] Each cAMP and cGMP mainly affects both contractile and synthetic VSMCs through actions mediated by protein kinases A (PKA) or G (PKG).[18,41,42] Although the cyclic nucleotides are thought to mainly exert their effects via their respective named kinases, cross-activation of PKG by cAMP is also known to occur and to allow cAMP to exert effects in cells via systems described as cGMP-dependent.[40] In addition to cAMP effects on PKG, other possible interactions between cAMP and cGMP systems have been noted. Thus, as will be described in more detail later in the chapter, the hydrolysis of cAMP can be stimulated, or inhibited, by the actions of cGMP on selected variants of the cyclic nucleotide phosphodiesterases (PDEs) family of enzymes.[43,44] These actions of cGMP on cAMP hydrolysis allow cGMP-elevating agents to either dampen or amplify cAMP signaling in cells expressing these PDEs. For instance, in VSMCs, a cell type expressing a cGMP-inhibited cAMP PDE activity, cAMP- and cGMP-elevating agents can act synergistically to alter contractility (see below). In contrast to VSMCs, as VECs can express both a cGMP-inhibited cAMP PDE and a cGMP-stimulated cAMP PDE, the impact of cGMP on cAMP dynamics is much more complex and thus will be determined based on the absolute levels of cGMP.[43–46] The cellular events involved in coordinating cyclic nucleotide–dependent (a) relaxation of VSMCs, (b) inhibition of VSMC proliferation and migration, and (c) barrier functions of VECs, have been the subject of recent reviews[43] and underpin the recent interest in assessing the therapeutic potential of using cyclic nucleotide–elevating agents to treat CVDs. In contrast to the established dominance of the roles of the cyclic nucleotide–activated kinases in VSMCs, the more recently described exchange proteins activated by cAMP (EPACs) have been shown to play an active role in several of the important functions of VECs.[47–52] Thus, EPACs have been shown to regulate both VEC intracellular and intercellular adhesions; these cAMP effectors have also been shown to regulate the release of vasoactive factors and the release of proteins involved in hemostasis and fibrinolysis from VECs. It is largely agreed that cyclic nucleotide–mediated signaling in cells is coordinated in large part due to the selective intracellular targeting of the individual components discussed above and the ability of this targeting to limit the regional intracellular impacts of cyclic nucleotide–elevating agents.

22.1.3 PDE Expression in Cells of the Cardiovascular System: Not All Cells Are Created Equal

As indicated previously, cAMP and cGMP each regulate myriad events in both activated and nonactivated VSMCs, or VECs, and agents that increase the levels of these cyclic nucleotides have marked effects on the functions of both phenotypes of each of these cell types. Past and more recent studies continue to be consistent with the hypothesis that PDEs play a central role in coordinating the effects of cyclic nucleotides in these cells.[43] Indeed, as they are the only enzymes capable of acting as intracellular buffers against unregulated cyclic nucleotide diffusion in cells, PDEs coordinate cyclic nucleotide–mediated signaling.[43,53–58] The reader is referred to the Introduction of (Chapter 1 and Chapter 2) this book for a description of the general state of the art with respect to PDEs and to Section A, Chapter 3 through Chapter 5 inclusively, for detailed descriptions of the defining family-selective characteristics of the PDEs, which at present are known to be expressed in cells of the cardiovascular system,

TABLE 22.1
Cyclic Nucleotide Phosphodiesterases Expressed in Vascular Smooth Muscle Cells

PDE Family	Contractile Rat VSMCs	Synthetic Rat VSMCs	Contractile Human VSMCs	Synthetic Human VSMCs
PDE1	PDE1A	PDE1A	PDE1A, PDE1B	PDE1A, PDE1C[a]
PDE3	PDE3A $\gg$ PDE3B	PDE3A ~ PDE3B	PDE3A $\gg$ PDE3B	PDE3A ~ PDE3B
PDE4	PDE4B2 PDE4D3/ 8/9 PDE4D5/7	PDE4B2 PDE4D3/8/ 9 PDE4D5/7	PDE4B2 PDE4D3/ 8/9 PDE4D5/7	PDE4B2 PDE4D3/ 8/9 PDE4D5/7
PDE5	PDE5A	PDE5A	PDE5A	PDE5A

Note: PDE family isoenzyme variants indicated in bold represent those species documented to change during the process of VSMC phenotypic modulation.
[a]As indicated in the text, inhibition of PDE1C in human VSMCs is associated with altered phenotypic modulation.

namely PDE1–PDE5. In addition, the reader with a special interest in the importance of PDEs in regulating activities of cells of the cardiovascular system is also referred to the other chapters in this section (Section E).

Numerous laboratories have each investigated PDE expression in contractile and synthetic VSMCs isolated from several species including human, monkey, baboon, cow, dog, rat, and mouse. Overall, these studies have shown that contractile VSMCs invariably express variants of each of the PDE1, PDE3, PDE4, and PDE5 families of enzymes (Table 22.1).[43,59–68] Similarly, synthetic VSMCs also have been shown to express variants of the PDE1, PDE3, PDE4, and PDE5 families of enzymes, although not necessarily the same PDE family variants as those expressed in contractile VSMCs (Table 22.1 and Figure 22.2).[43,59–68] Indeed, marked

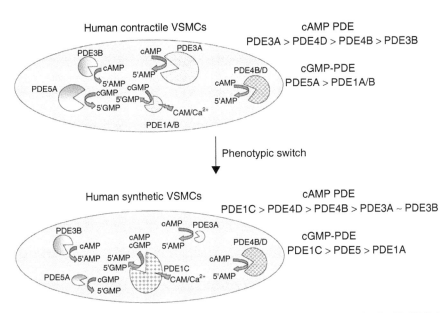

FIGURE 22.2 Phenotypic switch–associated altered expression of cAMP and cGMP PDE in human VSMCs. PDE expression is markedly affected as a result of the phenotypic switch of VSMCs from contractile to synthetic VSMC phenotypes. Relative levels of each PDE expressed in the contractile and synthetic VSMCs are indicated by their sizes within the cells and based on their rank order at the right of each cell.

differences in the specific variants expressed have been noted between contractile and synthetic VSMCs isolated from different vascular structures (Figure 22.2).[43,59] Similarly, although less extensive, a literature is emerging, which describes the PDEs expressed in various isolates of VECs from some selected species.[43,46,69–73] As with VSMCs, significant differences between isolates from distinct sources have also been reported. Thus, although human, bovine, and rodent VECs are routinely reported to express variants from each of the PDE2, PDE3, PDE4, and PDE5 families of enzymes, the levels of these activities recorded in VECs isolated from selected vascular structures between species, or from macro- and microvascular structures within species, can vary widely. For example, in one of the most comprehensive analyses thus far, bovine and human aortic VECs were shown to express distinct PDEs (Table 22.2).[46] In addition, significant differences in the PDE expression profile of human macro- and microvascular VECs were also shown in this recent study.[46] Thus, while human aortic VECs expressed very little PDE2, this activity accounted for a substantial fraction in microvascular VECs. Similarly, PDE3 activity was low in microvascular VECs but significant in macrovascular VECs. Although PDE5 activity accounted for the majority of the cGMP PDE activity in macrovascular VECs, cGMP PDE activity and PDE5 expression were very low in VECs isolated from human microvascular structures (Table 22.2).[46] Consistent with these differences in expression, PDE family-selective inhibitors differentially affected vascular endothelial growth factor–induced migration of these cells.[46] Although the factors which account for differential expression of individual PDEs within similar blood vessels between species are unknown, as macro- and microvascular structures subserve completely different functions *in vivo*, and would be expected to experience different hemodynamic forces, it is highly likely that differences in the PDEs expressed in VECs isolated from distinct vascular beds would be due, at least in part, to these different forces. On the basis of the differences in expression between cells isolated from different species, or different blood vessels, we conclude that when studying the impact of PDE inhibition on functions of cells of the cardiovascular system (a) heterogeneity of PDE expression in VSMCs, or VECs, isolated from different species will complicate the ease with which comparisons may be made and (b) VSMCs or VECs isolated from distinct vascular beds within a given species will likely respond differently.[40,43] While it is our opinion that these cautionary notes likely apply to studies of all PDEs, to our knowledge evidence for this phenomenon are presently available in the case of PDE1, PDE3, and PDE4 variants expressed in VSMCs and of PDE2 and PDE3 variants expressed in VECs. As our laboratory has focused its attention on issues related to PDE1, PDE3, and PDE4 expressions in VSMCs and VECs, in the remainder of this chapter,

TABLE 22.2
Cyclic Nucleotide Phosphodiesterases Expressed in Vascular Endothelial Cells

PDE Family	BAEC	HUVEC	HAEC	HMVEC
PDE2	PDE2A	PDE2A	PDE2A	PDE2A
PDE3	PDE3A2, PDE3B	PDE3A2, PDE3B	PDE3A2, PDE3B	PDE3A2, PDE3B
PDE4	PDE4B2, PDE4D5, PDE4D3	PDE4B2, PDE4D5, PDE4D3	PDE4B2, PDE4D5, PDE4D3	PDE4B2, PDE4D5, PDE4D3
PDE5	PDE5A1	PDE5A1	PDE5A1	PDE5A1

Note: On the basis of the results of immunoblot analyses, human VECs expressing the most of each indicated PDE family is indicated in bold. BAEC, bovine aortic endothelial cell; HUVEC, human umbilical vein endothelial cell; HAEC, human aortic endothelial cell; HMVEC, human microvascular endothelial cell.

we will limit our more detailed analysis to the roles of these enzymes. The reader is referred to Chapter 4, Chapter 7, Chapter 25, and Chapter 26 in this book for details related to PDE2 or PDE5.

22.2 REGULATED EXPRESSION AND FUNCTION OF ARTERIAL CELL PDEs

22.2.1 PDE1 FAMILY VARIANTS IN ARTERIAL VSMCs

As described in the Introduction section, Chapter 1, the PDE1 enzymes are defined by their ability to be activated upon interacting with Ca^{2+}-bound calmodulin. This regulatory mechanism, unique to PDE1s, is proposed to allow PDE1 enzymes to act as coincident detectors of Ca^{2+} and cyclic nucleotide fluxes in cells and to allow their integration. Also described in Chapter 1, the *PDE1* gene family consists of three separate genes, which each can yield numerous unique variants. Although *PDE1A* and *PDE1B* variants encode enzymes that preferentially hydrolyze cGMP, as opposed to cAMP, *PDE1C*-encoded enzymes hydrolyze both cyclic nucleotides. While several PDE1 variants are expressed in VSMCs, and these enzymes invariably play an important role in these cells under conditions of elevated intracellular Ca^{2+}, there is currently no evidence that PDE1s are involved in regulating cyclic nucleotide levels in VECs.[46,69–73]

The most comprehensive study of PDE1 expression in arterial VSMCs is that of Rybalkin et al.[64] In their report, Rybalkin and colleagues identified the PDE1 variants expressed in both contractile and synthetic VSMCs from several species including human, monkey, and rat. Overall, based on Rybalkin et al., and other studies of PDE expression in contractile VSMCs from baboon, cow, dog, or rabbit, it is clear that Ca^{2+}/calmodulin-regulated PDE activity is exclusively involved in regulating cGMP, rather than cAMP in contractile VSMCs (Figure 22.3).[43,61,64] For example, in Rybalkin,[64] the authors report that PDE1A and PDE1B,

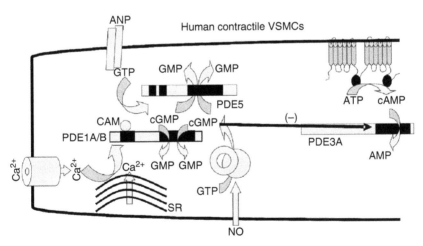

FIGURE 22.3 Scheme depicting the impact of increased intracellular Ca^{2+} on the coordinated hydrolysis of cGMP and cAMP by PDE1 and PDE3 family variants in contractile human VSMCs. Contractile human VSMCs express three distinct cGMP-hydrolyzing PDEs, namely PDE1A, PDE1B, and PDE5, and have abundant PDE3A, a cGMP-inhibited cAMP PDE. Human contractile VSMCs also express PDE4Ds (not shown). As shown, Ca^{2+} acts directly at PDE1A or PDE1B enzymes to stimulate the hydrolysis of cGMP, but not cAMP, in contractile human VSMCs. Also shown is the direct inhibition of cAMP hydrolysis by cGMP at PDE3A, an abundant cAMP-hydrolyzing PDE in contractile human VSMCs (straight arrow).

but not PDE1C, enzymes were each expressed in human, monkey, and rat aortic contractile VSMCs, albeit in markedly different species-selective proportions. Consistent with the idea that Ca^{2+}/calmodulin-activated PDEs are not involved in regulating cAMP levels in contractile VSMCs, no Ca^{2+}/calmodulin-induced hydrolysis of cAMP, nor expression of PDE1C variants, were detected in contractile VSMCs. Thus, as reviewed recently by us,[43] the data of Rybalkin and colleagues, and all other reports to similar effect, are consistent with the idea that Ca^{2+}/calmodulin-stimulated hydrolysis is involved in cGMP, but not cAMP, in contractile VSMCs (Figure 22.3). At present, the factors responsible for allowing expression of PDE1A or PDE1B in contractile VSMCs, and restricting expression of PDE1C in these cells, are not known.

As with contractile VSMCs, expression of PDE1 variants in synthetic VSMCs, isolated from distinct vascular structures within species, or the same vascular structure between species, has been shown to be somewhat variable.[43,62,64] Thus, while human and rat synthetic aortic VSMCs each express PDE1A, synthetic monkey aortic VSMCs express only PDE1B. To date, the events responsible for the species-specific-selective expression of individual *PDE1A*- or *PDE1B*-variants in synthetic VSMCs are not known. PDE1A and PDE1B variant expression in contractile VSMCs, in combination with the expression of Ca^{2+}-activated nitric-oxide synthase in these cells, may define a Ca^{2+}-dependent system, allowing coordinated control of cGMP levels (Figure 22.3). Of likely physiological importance and of potential therapeutic relevance, loss of PDE1B expression during the phenotypic transition of human aortic contractile VSMCs to more synthetic cells was associated with a marked induction of PDE1C expression in synthetic VSMCs (Figure 22.4).[62,64] Interestingly, to date no PDE1C expression has been detected in any synthetic VSMCs from any species other than human. Expression of PDE1C in synthetic human aortic VSMCs allows the Ca^{2+}/calmodulin-regulation of cAMP hydrolysis in synthetic VSMCs and may define a more global change from cGMP to cAMP sensitivity of these cells (Figure 22.4). Consistent with this expression, many cGMP-regulated enzymes are known to be down-regulated in synthetic VSMCs when compared to contractile VSMCs (Figures 22.4).[43] A molecular approach to reducing PDE1C expression in human aortic synthetic VSMCs inhibited their proliferation.[62] To date, the

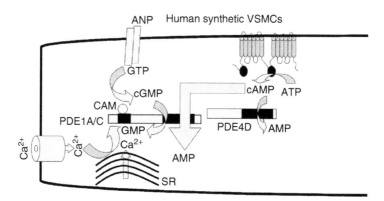

FIGURE 22.4 Scheme depicting the impact of increased intracellular Ca^{2+} on the coordinated hydrolysis of cGMP and cAMP by PDE1 family variants in synthetic human VSMCs. Synthetic human VSMCs express two distinct PDE1s, namely PDE1A and PDE1C. Although PDE1A preferentially hydrolyzes cGMP, PDE1C efficiently hydrolyzes both cGMP and cAMP. Synthetic human VSMCs express a lower level of PDE5 than do contractile human VSMCs. Synthetic human VSMCs express PDE4Ds and very low levels of PDE3A (not shown). As shown, Ca^{2+} acts directly at PDE1C to stimulate the hydrolysis of both cGMP and cAMP in synthetic human VSMCs.

absence of PDE1 inhibitors that could distinguish between PDE1A, PDE1B, and PDE1C variants has significantly hampered efforts to assess the pharmacological impact of PDE1C inhibition in cells of the cardiovascular system.

22.2.2 PDE3 Family Variants in Arterial VSMCs

As described in Section A, Chapter 3, the PDE3 family of enzymes consists of two highly homologous genes, namely *PDE3A* and *PDE3B*. The *PDE3A* gene can encode three PDE3 enzyme variants, namely a 136 kDa PDE3A1, a 118 kDa PDE3A2, and a 94 kDa PDE3A3, respectively. The *PDE3B* gene only encodes one PDE3 enzyme variant, namely a 135 kDa PDE3B.[43,57,65,67] Although both PDE3A1 and PDE3B are exclusively particulate, PDE3A2 can be found in both particulate and cytosolic fractions of cell lysates. PDE3A3 is a cytosolic enzyme in cells.[65] On the basis of an analysis of their structures, the catalytic domains of *PDE3A*- and *PDE3B*-encoded enzymes are each structurally homologous to those of the other PDE families of enzymes with one exception (Chapter 3).[67] Thus, all PDE3 enzymes contained a unique 44 amino acid insert within their catalytic core domain. Numerous mutagenesis-based studies have shown that these PDE3 inserts impact both kinetics of cAMP and cGMP hydrolysis, as well as the potency of inhibition of these enzymes by PDE3-selective inhibitors, such as cilostamide.[74,75] PDE3 enzymes have been shown to allow integration, or crosstalk, between cAMP and cGMP cellular signaling. Thus, while PDE3s bind cAMP or cGMP with high affinity ($K_m \sim 0.1 - 0.8$ μM), hydrolysis of cAMP by PDE3s is significantly faster than that of cGMP (Chapter 3).[43,67] In fact, hydrolysis of cGMP by PDE3s allows cGMP to act as a potent endogenous intracellular competitive inhibitor of cAMP hydrolysis by PDE3s, and accounts for the earlier description of PDE3s as cGMP-inhibited cAMP PDEs.[43,44,76–79] In early work in platelets and contractile VSMCs,[44,76] and more recently in cardiomyocytes,[80–82] cGMP inhibition of cAMP hydrolysis by PDE3A was shown to allow cGMP-elevating agents to synergize with activators of adenylyl cyclase to regulate cellular functions. Indeed, in an earlier work, Dr. Richard Haslam and one of us (D.H.M.) proposed that cGMP-mediated inhibition of PDE3A hydrolysis of cAMP in blood platelets, or contractile VSMCs, played a central role in allowing the important short-lived primary messengers NO and PGI$_2$ to have effects on the functions of these cells.[44,76] Although not formally tested, as PDE3s bind both cAMP and cGMP with similar affinity and cellular levels of cAMP usually far exceed those of cGMP, one might predict that cAMP could also control hydrolysis of cGMP by PDE3 and influence cGMP signaling in cells.

The full-length PDE3A1 and PDE3B share highly homologous overall structures. Thus, both contain two amino-terminal hydrophobic regions, referred to as NHR1 and NHR2, respectively, which are predicted to participate in enzyme subcellular targeting.[43,57,65,67] In addition to NHR1 and NHR2 domains, the amino-terminal sequences of the PDE3A and PDE3B variants also encode several consensus protein kinase phosphorylation sequences. Thus, PDE3A1 and PDE3B both express multiple distinct PKA consensus phosphorylation sites, and one unique PKB consensus phosphorylation site. PKA-mediated phosphorylation of PDE3A1 or PDE3B at one of these sites (S318 in human PDE3B) increase V_{max} for cAMP hydrolysis.[43,57,65,67] Similarly, PKB phosphorylation at S295 in human PDE3B increases V_{max} as well as the affinity of both PDE3A1 and PDE3B for cyclic-nucleotide binding.[43,57,65,67,83–86]

Although PKA- or PKB-catalyzed phosphorylations of PDE3s clearly activate these enzymes, evidence is emerging that these enzymes are also regulated by protein–protein interactions in cells. Thus, in as yet unpublished work from our laboratory, we have shown that PKA-catalyzed phosphorylation of human PDE3B at multiple sites, including S73, S296, and S318, promotes a dynamic association of this enzyme with members of the adaptor family

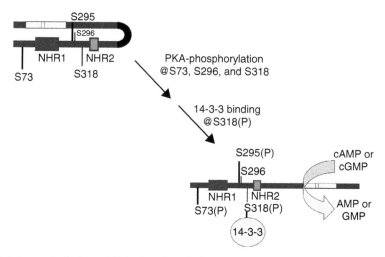

FIGURE 22.5 Scheme depicting a PKA phosphorylation-mediated configuration change in HSPDE3B, which correlates with HSPDE3B activation and 14-3-3 binding. HSPDE3B expresses three distinct PKA phosphorylation consensus sites, which are each phosphorylated in response to incubation of HSPDE3B-expressing cells with cAMP-elevating agents. When phosphorylated by PKA, one of these phospho-acceptor sites, namely S318, forms a 14-3-3 acceptor site. On the basis of the results of site-directed mutagenesis studies, 14-3-3 does not bind HSPDE3B phosphorylated by PKA on S73, S295, or by PKB at S296. *Note*: Two hydrophobic domains (NHR1 and NHR2, dark boxes) and the catalytic domain (light box) are also shown.

of proteins known as 14-3-3 proteins (D. Palmer, D.R. Raymond, L.S. Wilson, S.L. Jimmo, and D.H. Maurice, unpublished observations, Figure 22.5). Moreover, our data are consistent with the idea that interaction of human PDE3B with 14-3-3 proteins regulates intracellular trafficking of PDE3B in several cell types that express this PDE endogenously. In contrast to a previous report in murine 3T3-L1 fibroblasts,[87] in our studies, we find no evidence to support a role for PKB-mediated phosphorylation of S295 in promoting association of human PDE3B with 14-3-3 proteins. Whether this difference reflects a true species-dependent difference, or is related to methodological differences in the two studies may require further study. In data that perhaps identifies PDE3 subtype–selective regulation, a protein kinase C (PKC)-mediated phosphorylation of PDE3A2 at S428 was shown to promote association of this PDE3 variant with 14-3-3 proteins, whereas PKA-mediated phosphorylation of this PDE3 was without effect.[88] While we have not investigated the role of PKC in promoting PDE3B association with 14-3-3 proteins, our data are also consistent with the idea that PKA-mediated phosphorylation of PDE3A2 does not promote association of this PDE with 14-3-3 proteins. Although platelet PDE3A2 does not associate with 14-3-3 proteins, we recently reported that this PDE3A can participate in protein–protein interactions.[86] Thus, in this work we reported that platelet PDE3A2 existed in a complex with several proteins, including the long form of the leptin receptor, and that this association allowed leptin to promote the PKB-catalyzed phosphorylation and activation of this PDE3 (Figure 22.6). In this recent work, we proposed that these events are responsible, at least in part, for the ability of leptin to promote platelet aggregation and adhesion and may contribute to the association of obesity-associated hyperleptinemia and CVD.[86] In addition to interactions with 14-3-3 proteins, a stable association between murine PDE3B and phosphoinositide-3-kinase (PI3K) in cardiomyocytes has identified this PDE3 enzyme as potentially regulating cAMP effects on cardiomyocyte growth.[89,90] Interestingly, and again

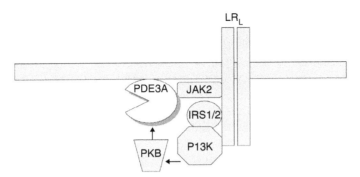

FIGURE 22.6 A human leptin receptor (LR_L)- and PDE3A-containing complex regulates leptin-mediated activation of PDE3A in human platelets. Leptin-mediated activation of its long receptor (LR_L) allows recruitment of JAK2, insulin receptor substrates 1 and 2 (IRS1/2), and phosphoinositide-3-kinase (PI3K). Localized PI3K-mediated activation of protein kinase B (PKB) allows a phosphorylation-mediated activation of human PDE3A and an increased cAMP hydrolysis. It is hypothesized that this mechanism allows leptin-mediated activation of human platelets.

suggestive of distinct regulation between PDE3A and PDE3B enzymes, the much more abundant cardiomyocyte PDE3A was not shown to associate with PI3K in murine cardiomyocytes in these studies.[89] The molecular determinants responsible for this specific interaction between PDE3B–PI3K and the molecular basis for its influence on PDE3B catalytic activity have yet to be determined.

Inhibition of PDE3 activity using selective pharmacological inhibitors, such as cilostamide, cilostazol, or milrinone, has assisted in efforts to delineate cellular functions directly regulated by these enzymes.[43,57,65,67,77] PDE3A is highly expressed in platelets, cardiomyocytes, and VSMCs.[43] In platelets, PDE3 inhibition increases intraplatelet cAMP and, via activation of PKA, prevents platelet aggregation and adhesion.[44] As such, due to its potent effects on platelets, the PDE3 inhibitor cilostazol is currently used for the control of the platelet-associated condition of intermittent claudication[91] and the basis of its utility in this condition is presented in Section F, Chapter 31, of this book. Pharmacological inhibition of PDE3 activity in synthetic VSMCs increases cellular cAMP and inhibits VSMC proliferation in cell culture, whereas PDE3 inhibition in contractile VSMCs promotes relaxation and inhibits contraction of these cells.[43,77,78,92] Pharmacological inhibition of PDE3 activity in VECs inhibits vascular endothelium growth factor–induced migration but the magnitude of the effect is dependent on the level of VEC cGMP and on the level of expression of PDE2.[46] In contrast to this effect of PDE3 inhibition on VEC migration, in as yet an unpublished study, we have shown that PDE3 inhibition in VECs in suspension promotes their adhesion to ECM proteins such as fibronectin and vitronectin. These proadhesive effects of PDE3 inhibition in suspension VECs are integrin dependent and result from PDE3-mediated, cAMP-dependent, activation of EPACs in these cells (S. Netherton, J. Sutton, and D.H. Maurice, unpublished observations).

22.2.2.1 Differential Regulation of PDE3A and PDE3B in Contractile and Synthetic VSMCs

As PDE3A is largely cytosolic and PDE3B is exclusively particulate in VSMCs, it is likely that these two PDE3s regulate distinct pools of cAMP and cGMP and, as such, distinct cellular functions. Of interest in this context, while both contractile and synthetic VSMCs express PDE3A and PDE3B variants,[77,78] the regulated expression of these PDE3s is very different during events such as the previously described phenotypic switch, or in response to prolonged

challenge of VSMCs with cAMP-elevating agents (Table 22.1).[60,93,94] Thus, during the phenotypic transition of contractile human or rat VSMCs into synthetic VSMCs *PDE3A* expression is markedly reduced (Figures 22.2).[60] Although levels of PDE3B mRNA and protein are unaffected by this phenotypic switch, PDE3A mRNA and protein levels were markedly lower in synthetic than in contractile VSMCs.[60] Indeed, levels of PDE3A were markedly lower in synthetic VSMCs derived both from cell culture and *in vivo* vascular damage –based systems. In the cell culture–based system, PDE3A expression levels were reduced after as few as five population doublings, or two rounds of subculture. In the *in vivo* context, VSMCs isolated from the neointima of damaged aortae after 14 days expressed virtually no PDE3A.[60] Although the molecular basis of this reduced PDE3A expression in synthetic VSMCs is yet unclear, it is clear that loss of PDE3A expression by synthetic VSMCs alters the ability of these cells to hydrolyze cAMP. Thus, while PDE3A plays a dominant role in hydrolyzing cAMP in contractile VSMCs, in synthetic VSMCs levels PDE3A- and PDE3B-mediated hydrolysis of cAMP were roughly equal (Figure 22.2). Also, as the expression of the other dominant cAMP PDEs expressed in contractile rat or human VSMCs, namely PDE4B and PDE4D, were unaltered during the phenotypic transition of these cells, synthetic VSMCs have an elevated PDE4 to PDE3 activity ratio than do the contractile cells (Figure 22.2). These data are of potential interest both physiologically and therapeutically. Thus, if one recalls that PDE3A is a cGMP-sensitive cAMP PDE, the loss of PDE3A during the VSMC phenotypic switch may be consistent with the observation that expression of other cGMP-regulated proteins (i.e., protein kinase G [PKG] and nitric oxide–sensitive guanylyl cyclase [sGC]) are also reduced in synthetic VSMCs.[43] As a corollary, we propose that as PDE3B is significantly less sensitive to inhibition by cGMP than PDE3A, this may account for its continued expression in synthetic VSMCs. Of course, further investigation is needed for this proposal to be validated. At a therapeutic level, these findings may provide insight into the ability of PDE3 inhibitors, such as cilostazol, to reduce neointimal formation in rodent animal models *in vivo*.[95] Thus, the finding that PDE3A levels are markedly depressed in synthetic VSMCs *in vivo* may suggest that the beneficial effects of PDE3 inhibitors in animal models of restenosis represent effects caused by inhibition of VSMC PDE3B, or alternatively, are due to non-VSMC-mediated effects. Consistent with the former findings is our earlier observation that VSMCs isolated from JCR-LA rats, a strain with elevated PDE3B activity, displayed a hyperproliferative and hypermotile phenotype that was sensitive to PDE3 inhibition.[92] Consistent with the idea that non-VSMC effects account for PDE3-inhibitor-mediated effects is the observation that PDE3 inhibitors are potent inhibitors of platelet activation and that platelet deposition at sites of vascular damage is known to be an important early event in the development of neointimal VSMC lesions.[1,2] Regrettably, the lack of suitably selective inhibitors for PDE3A or PDE3B inhibition does not allow us to formally test if selective inhibition of VSMC PDE3B would prevent neointimal formation in damaged blood vessels.

In addition to the selective reduction of PDE3A, which accompanies the VSMC phenotypic switch, PDE3A and PDE3B are also differentially affected by prolonged incubations of these cells with cAMP-elevating agents. Thus, prolonged incubation of either contractile or synthetic rat and human VSMCs with cAMP-elevating agents differentially affect PDE3A and PDE3B expression in these cells.[77,78,92–94] While forskolin, cAMP-analogues, or the β-adrenergic agonist isoproterenol each increased total PDE3 activity in VSMCs of both phenotypes, the impact of such treatments on PDE3A and PDE3B were distinct. Thus, administration of cAMP-elevating agents *in vivo* to contractile VSMCs markedly increased expression of both PDE3A and PDE3B.[43,57,65,67,96,97] In contrast, similar treatments of synthetic VSMCs in culture resulted in a selective increase in expression of PDE3B, with no change in PDE3A expression.[43,57,65,67,96,97] If the different responses of PDE3A and PDE3B to prolonged elevations of cAMP, observed in contractile and synthetic VSMCs, represent a

therapeutic avenue to regulate VSMC function will have to await the development of PDE3 family-selective inhibitors. Although the molecular basis of this difference in PDE3A and PD3B regulation in response to prolonged increases in cAMP has yet to be fully elucidated, on the basis of our recent studies of PDE4D expression in VSMCs, it may be related to differences in chromatin remodeling (see below).

22.2.3 PDE4 Family Variants in Arterial VSMCs and VECs

As described in Section A, Chapter 6, the PDE4 family is comprised of four genes, namely *PDE4A*, *PDE4B*, *PDE4C*, and *PDE4D*, each of which can encode several distinct variants.[43,54,55] Indeed, as a result of alternate splicing, or differential promoter usage, numerous PDE4 long variants (~90–120 kDa), and short variants (~60–70 kDa), can be expressed in cells. In the case of *PDE4D*s, PDE4D1, PDE4D2, and PDE4D6 are short variants whereas PDE4D3–PDE4D5, PDE4D7–PDE4D9 are each long-form gene products. As described in Chapter 6, the domain structure and regulatory characteristics of PDE4s are similar to those of other PDEs in that the carboxyl termini contain the catalytic domain and the amino termini contain the paired regulatory structures. The two amino-terminal regulatory domains in PDE4s are referred to as upstream conserved regions, UCR1 and UCR2, and these domains have been shown to create a UCR1 or UCR2 module that constitutively inhibits PDE4 long-variant activity and, in some instances regulate inhibitor sensitivity.[54,55] Also, as described in Chapter 6, PKA- or ERK-catalyzed phosphorylation of several PDE4 variants regulates their activity in numerous cell types, including VSMCs.[43,54,55,96,97] PDE4-mediated regulation of cellular cAMP has been shown to regulate a large number of physiological processes, including immune cell activation. Pharmacological inhibition of PDE4 activity has been proposed to be an effective therapeutic strategy in several pathologies including asthma and the data from these efforts have been presented in several chapters in Section E.

Contractile and synthetic VSMCs each express several distinct PDE4 enzymes (Table 22.1).[43] In contrast to the marked changes in PDE3A expression that accompanied the phenotypic switch of human or rat VSMCs from a contractile to a synthetic phenotype, this transition had no direct impact on basal levels of PDE4 expression in these cells.[43,92–94,96,97] Indeed, both contractile and synthetic VSMCs each express one PDE4B, namely PDE4B2, and at least two of the nine potential PDE4D variants. The electrophoretic mobility of the ~95 kDa VSMC anti-PDE4D antibody immunoreactive band expressed in both contractile and synthetic cells is consistent with the expression of one, or several, of PDE4D3, PDE4D8, or PDE4D9. Similarly, the mobility of the ~105 kDa VSMC anti-PDE4D immunoreactive band is consistent with the expression of either one of PDE4D5 or PDE4D7 or both of these variants. On the basis of detailed immunoblot-based analysis of VSMC lysates using a pan-PDE4D selective antiserum and several antisera raised specifically against PDE4D3, PDE4D5, PDE4D8, or PDE4D9 by Conti and colleagues, our data are consistent with the proposal that human and rat VSMCs express PDE4D3 and PDE4D5,[43,92–94,96,97] and either PDE4D8 or PDE4D9, depending on the source of VSMCs utilized (D.R. Raymond and D.H. Maurice, unpublished data). As with VSMCs, human VEC isolates have been shown to express PDE4B2 and at least two PDE4D variants (Table 22.2).[46,71] Although VECs isolated from microvascular structures had higher levels of PDE4 activity, and expressed higher levels of the anti-PDE4D ~95 kDa and ~105 kDa proteins, the identity of these species in VECs have not been unequivocally determined in these cells.

Several complex sets of regulatory systems allow cells to coordinate their initial responses to physiological or pharmacological agents and to reduce their subsequent responses (i.e., desensitize) to these effects, as required (Figure 22.7).[98–104] Although this regulation plays a

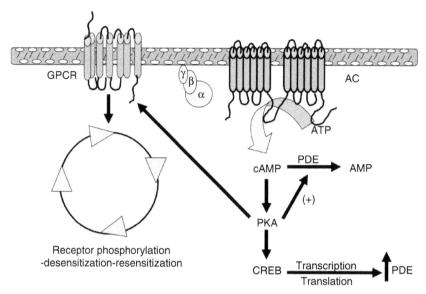

FIGURE 22.7 G-protein coupled receptor (GPCR) and nonreceptor mechanisms of desensitization to cAMP signaling in VSMCs. GPCR receptor occupancy- and PKA phosphorylation-mediated receptor internalization are involved in desensitizing cells to continued receptor activation. Co-temporally, PKA-induced phosphorylation of PKA-activatable PDEs, and of the cAMP-response element binding (CREB) protein, increases PDE-mediated hydrolysis of cAMP. In several cell types, including VSMCs, PKA-mediated effects on PDE activity play an important role in desensitizing cells to continued activation of adenylyl cyclases (AC).

central role in allowing faithful and reproducible effects of physiological agents, it often can reduce or even abolish the effectiveness of pharmacological agents used in the treatment of several maladies. While most of the work aimed at identifying the systems that allowed desensitization of cAMP-signaling focused on receptor-based mechanisms of desensitization,[98–104] both earlier work and more recent reports have shown that stimulus-dependent increases in cAMP hydrolysis by cAMP PDEs of the PDE3 and PDE4 families also contribute to desensitization in the cAMP-signaling axis (Figure 22.7). Indeed, recent reports detail some of the mechanisms that coordinate these desensitization events.[105–113] Briefly, apart from the phosphorylation-induced activation of cAMP PDEs mentioned above, and reviewed in Chapter 6, studies have revealed that intricate systems of PDE4 targeting as well as cell-type specific induction of distinct PDE4 variants can also orchestrate the desensitization process. Thus, recently it was discovered that certain PDE4 variants could interact with β-arrestin and as a result be targeted to stimulated β-adrenergic receptors to hydrolyze local pools of cAMP.[102–104] In addition to these cAMP-dependent interactions of PDE4s with β-arrestins, several recent reports have shown that PDE4s can also be found in stable intracellular complexes with cAMP effectors, including PKA and EPACs.[114–117] While these complexes have been isolated from lysates of fetal murine cardiomyocytes, and have been shown to play important roles in regulating the growth of this cell type, their existence and importance in VSMCs or VECs is currently unclear.

As stated above, it has been postulated that increased expression of several cAMP PDE isoforms in response to prolonged cAMP signaling plays a role in desensitization of cAMP effects in cells. It is generally accepted that this increased expression of PDEs results from PKA-mediated phosphorylation of CREB, the interaction of CREB with a CRE sequence

in the promoters directing expression of cAMP PDE variants and the subsequent induction of transcription and de novo synthesis.[77,78,93,94,117–120] In fact, others have reported the existence of CREs in promoters regulating expression of several PDE genes, including those encoding *PDE3B*, *PDE4A10*, *PDE4B2*, *PDE4D1/2*, *PDE4D5*, and *PDE7B1*, supporting this theory.[117–120] Recent efforts in our laboratory have elucidated some of the regulatory events allowing cell type–specific regulation of PDE4 expression in VSMCs, and shown that these events are regulated differently in contractile and synthetic VSMCs and involve differential histone acetylation.

22.2.3.1   Regulated cAMP-Dependent Expression of *PDE4D* in VSMCs and VECs

Data from our laboratory support the role of de novo protein synthesis in induction of cAMP PDE variants in VSMCs, as inclusion of actinomycin D or cycloheximide, inhibitors of transcription and translation, were able to ablate increases in PDE3A and PDE3B in contractile VSMCs and of PDE3B in synthetic VSMCs (see above). Similarly, these strategies ablated *PDE4D* expression in both contractile and synthetic VSMCs.

As described above, prolonged exposure of rat aortic VSMCs *in vivo* or of cultured human or rat aortic VSMCs *ex vivo* to cAMP-elevating agents increased the PDE4 activity and the expression of *PDE4D* variants in these cells.[77,78,93,94] Interestingly, cAMP-elevating agents increased PDE4 activity by inducing the expression of distinct *PDE4D* variants in synthetic and contractile VSMCs (Figure 22.8). Thus, when cAMP-elevating agents were administered

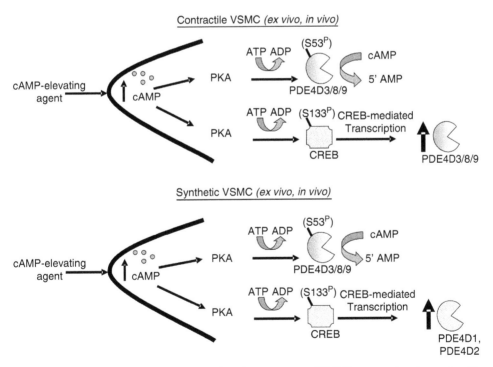

FIGURE 22.8 Differential effects of cAMP-elevating agents on PDE4D expression in contractile and synthetic VSMCs. (A) In contractile VSMCs, cAMP-elevating agents stimulate a PKA-dependent phosphorylation and activation of the major PDE4D variants in these cells (PDE4D3/8/9) and a CREB-dependent increased expression of this(ese) protein(s). (B) In synthetic VSMCs cAMP-elevating agents stimulate a PKA-dependent phosphorylation and activation of the major PDE4D variants in these cells (PDE4D3/8/9) and a CREB-dependent induction of expression of two PDE4D short variants, PDE4D1 and PDE4D2.

to contractile VSMCs *in vivo*, expression of the dominant *PDE4D* variants expressed in these cells, as detected by immunoblotting (PDE4D3/8/9 and PDE4D5/7), was significantly increased.[77,78,93,94] In contrast, cAMP-elevating agents induced expression of two previously absent *PDE4D* short forms, namely PDE4D1 and PDE4D2 (PDE4D1/2), with no change in the expression of either the *PDE4D* long-form variants, when synthetic VSMCs were similarly treated.[77,78,93,94] These data identified that *PDE4D* expression was differentially regulated by cAMP-elevating agents in contractile and synthetic VSMCs. Recently, we identified a likely molecular basis for this differential control of *PDE4D* expression in contractile and synthetic VSMCs. Thus, we proposed that a mechanism by which gene regulation was controlled through selective histone acetylation may be responsible for differences in the expression of a large number of signaling proteins in contractile and synthetic VSMCs.[77,78,93,94] Thus, while data from these studies unequivocally identified roles for each PKA, CREB, and CRE sequences in both contractile and synthetic VSMCs, they also identified that differential chromatin remodeling of a *PDE4D* intronic promoter in these cells most likely contributed to differential induction of long versus short *PDE4D* in these cells.[77,78,93,94] Indeed, using a chromatin immunoprecipitation approach, we were able to show that levels of histone acetylation of an intronic promoter regulating short *PDE4D* expression were markedly higher in synthetic VSMCs than in the contractile cells.[93,94] Consistent with this idea, increased histone acetylation correlates with promoter activation while decreases in this modification cause reduced accessibility of transcription factors for these sequences.[120–124] Two novel findings concerning cAMP signaling in VSMCs and the impact of cAMP-elevating agents on *PDE4D* expression in VSMCs emerged from these recent studies. First, we identify that contractile and synthetic VSMCs increase *PDE4D* expression in response to prolonged challenge with cAMP-elevating agents and that each VSMC phenotype utilizes PKA-dependent signaling in this process. Second, these studies show that a phenotypically dependent altered level of histone acetylation at the intronic promoter regulating *PDE4D* short forms in contractile and synthetic VSMCs likely represent the molecular basis for differential expression of selected *PDE4D* variants in these cells in response to prolonged treatment with cAMP-elevating agents. These data are consistent with numerous published reports suggesting that specific *PDE4* gene products mediate cAMP-induced desensitization to cAMP-elevating agents in individual cell types and, in fact, expand on this paradigm. Thus, in addition to confirming that *PDE4D* is a major PDE expressed in VSMCs and that prolonged cAMP elevation increases PDE4D activity, these recent findings unequivocally demonstrated that different VSMC phenotypes can respond to this challenge by inducing the expression of distinct *PDE4D* gene-derived variants. In addition to the physiological implications of the differential regulation of PDE4D variants in VSMCs, there are also two potentially important therapeutic implications that arise from this phenomenon. First, we propose that the robust increase in PDE4D short forms, which occurred upon cAMP elevation in synthetic VSMCs, would make these cells more readily able to be desensitized to the effects of prolonged cAMP-elevating agents than that might be anticipated in the contractile VSMCs. Second, we propose that development of *PDE4D* splice-variant selective inhibitors may provide greater specificity when PDE enzymes are targeted therapeutically in situations in which different effects in contractile and synthetic VSMCs are desired, such as in attempts to limit in-stent restenosis. Currently, there is substantial evidence that some PDE4 inhibitors, such as rolipram, bind to long and short PDE4 variants with different affinities, whereas other PDE4 inhibitors, such as piclamilast, do not possess this characteristic.[43] Thus, based on these differences in rolipram binding to long and short forms of PDE4D, studies are currently underway to identify PDE4 inhibitors, which might preferentially inhibit PDE4D short forms, and thus selectively augment cAMP-signaling in synthetic VSMC, but not contractile VSMC, in the same artery in animal models of in-stent restenosis. Intriguingly, a recent study showed that decreased expression of

long-PDE4D variants correlated with the increased susceptibility to carotid stroke brought on by the development of atherosclerosis, a disease marked by the very phenotypic modulation of VSMCs described in this chapter.[125] The disease-associated haplotype was reported to extend over regions of the promoters of the *PDE4D* gene, which suggests an alteration in transcriptional regulation of PDE4D variants, a phenomenon that would support a role for differential transcriptional control of PDE4D variants in phenotypically distinct VSMCs. In addition to an obvious effect on cAMP-mediated control of VSMCs, a differential induction of long and short PDE4D variants in the phenotypically distinct VSMCs may also have a broader impact on VSMC signaling. Indeed, crosstalk between ERK1 or ERK2 and PDE4D activities was recently reported to markedly influence cellular responses to both cAMP- and non-cAMP-dependent agents and expression. Indeed, taken together, these reports presented the idea that ERK1/2-mediated phosphorylation inhibited *PDE4D* long forms and translocated particulate PDE4D3 away from its targeted membrane fraction to the cytosol. In contrast, *PDE4D* short forms were activated. The importance of ERK1/2-mediated signaling in VSMC migration and proliferation and the influence of cAMP on these events make it highly likely that differential *PDE4D* induction in contractile and synthetic VSMCs may allow for cell-phenotype specific effects between the actions of growth factors and cAMP-elevating agents. Given the difficulty of isolating sufficient amounts of nonactivated VECs, further studies will be required to assess if PDE4s are also differentially sensitive to cAMP-elevating agents in VECs from stable vascular structures compared to that found in activated VECs.

22.3 CONCLUSIONS

In this chapter, we have presented data that support the idea that arterial VSMCs and VECs subserve multiple roles in the vasculature and that this phenomenon is made possible, at least in part, by the ability of these cells to utilize PDEs to dynamically regulate several cyclic nucleotide–dependent events. Moreover, we have shown that distinct expression of PDEs, in contractile and synthetic VSMCs, and in VECs isolated from distinct vascular structures, rendered difficult the facile commentary on the roles of individual enzymes based on a limited analysis. It is our wish that with continued study of the factors that regulates expression of PDEs in cells of the cardiovascular system, as well as continued study of the events that regulate the selective targeting of these enzymes in these cells, the therapeutic promise of these enzymes will soon be fulfilled.

REFERENCES

1. Pugsley, M.K., and R. Tabrizchi. 2000. The vascular system. An overview of structure and function. *J Pharmacol Toxicol Methods* 44:333.
2. World Health Organization. Cardiovascular disease: Prevention and control. http://www.who.int/dietphysicalactivity/publications/facts/cvd/en/1-5-2005.
3. Berk, B.C. 2001. Vascular smooth muscle growth: Autocrine growth mechanisms *Physiol Rev* 81:999.
4. Bauters, C., and J.M. Isner. 1997. The biology of restenosis. *Prog Cardiovasc Dis* 40:107.
5. Owens, G.K., M.S. Kumar, and B.R. Wamhoff. 2004. Molecular regulation of vascular smooth muscle cell differentiation in development and disease. *Physiol Rev* 84:767.
6. Gordon, D., M.A. Reidy, E.P. Benditt, and S.M. Schwartz. 1990. Cell proliferation in human coronary arteries. *Proc Natl Acad Sci USA* 87:4600.
7. Aikawa, M., Y. Sakomura, M. Ueda, K. Kimura, I. Manabe, S. Ishiwata, N. Komiyama, H. Yamaguchi, Y. Yazaki, and R. Nagai. 1997. Redifferentiation of smooth muscle cells after coronary angioplasty determined via myosin heavy chain expression. *Circulation* 96:82.

8. Kocher, O., and G. Gabbiani. 1986. Cytoskeletal features of normal and atheromatous human arterial smooth muscle cells. *Hum Pathol* 17:875.

9. Kocher, O., F. Gabbiani, G. Gabbiani, M.A. Reidy, M.S. Cokay, H. Peters, and I. Huttner. 1991. Phenotypic features of smooth muscle cells during the evolution of experimental carotid artery intimal thickening. Biochemical and morphologic studies. *Lab Invest* 65:459.

10. Yoshida, T., and G.K. Owens. 2005. Molecular determinants of vascular smooth muscle cell diversity. *Circ Res* 96:280.

11. Chamley-Campbell, J.H., G.R. Campbell, and R. Ross. 1979. The smooth muscle cell in culture. *Physiol Rev* 59:1.

12. Chamley-Campbell, J.H., G.R. Campbell, R. Ross. 1981. Phenotype-dependent response of cultured aortic smooth muscle to serum mitogens. *J Cell Biol* 89:379.

13. Mosse P.R., G.R. Campbell, Z.L. Wang, and J.H. Campbell. 1985. Smooth muscle phenotypic expression in human carotid arteries. I. Comparison of cells from diffuse intimal thickenings adjacent to atheromatous plaques with those of the media. *Lab Invest* 53:556.

14. Triggle, C.R., M. Hollenberg, T.J. Anderson, H. Ding, Y. Jiang, L. Ceroni, W.B. Wiehler, E.S. Ng, A. Ellis, K. Andrews, J.J. McGuire, and M. Pannirselvam. 2003. The endothelium in health and disease—a target for therapeutic intervention. *J Smooth Muscle Res* 39:249.

15. Hamad, A.M., A. Clayton, B. Islam, and A.J. Knox. 2003. Guanylyl cyclases, nitric oxide, natriuretic peptides, and airway smooth muscle function. *Am J Physiol Lung Cell Mol Physiol* 285:L973.

16. Bian, K., and F. Murad. 2003. Nitric oxide (NO)—biogeneration, regulation, and relevance to human diseases. *Front Biosci* 8:d264.

17. Hurley, J.H. 1999. Structure, mechanism, and regulation of mammalian adenylyl cyclase. *J Biol Chem* 274:7599.

18. Lincoln, T.M., N. Dey, and H. Sellak. 2001. cGMP-dependent protein kinase signaling mechanisms in smooth muscle: From the regulation of tone to gene expression. *J Appl Physiol* 91:1421.

19. Krause, M., J.E. Bear, J.J. Loureiro, and F.B. Gertler. 2002. The Ena/VASP enigma. *J Cell Sci* 115:4721.

20. Chen, L., G. Daum, K. Chitaley, S.A. Coats, D.F. Bowen-Pope, M. Eigenthaler, N.R. Thumati, U. Walter, and A.W. Clowes. 2004. Vasodilator-stimulated phosphoprotein regulates proliferation and growth inhibition by nitric oxide in vascular smooth muscle cells. *Arterioscler Thromb Vasc Biol* 24:1403.

21. Ohta, Y., T. Akiyama, E. Nishida, and H. Sakai. 1987. Protein kinase C and cAMP-dependent protein kinase induce opposite effects on actin polymerizability. *FEBS Lett* 222:305.

22. Silberbach, M., and C.T. Roberts. 2001. Natriuretic peptide signalling: Molecular and cellular pathways to growth regulation. *Cell Signal* 13:221.

23. Suhasini, M., H. Li, S.M. Lohmann, G.R. Boss, and R.B. Pilz. 1998. Cyclic-GMP-dependent protein kinase inhibits the Ras/Mitogen-activated protein kinase pathway. *Mol Cell Biol* 18:6983.

24. Schmitt, J.M., and P.J. Stork. 2001. Cyclic AMP-mediated inhibition of cell growth requires the small G protein Rap1. *Mol Cell Biol* 21:3671.

25. Howe, A.K. 2004. Regulation of actin-based cell migration by cAMP/PKA. *Biochim Biophys Acta* 1692:159.

26. Howe, A.K., and R.L. Juliano. 2000. Regulation of anchorage-dependent signal transduction by protein kinase A and p21-activated kinase. *Nat Cell Biol* 2:593.

27. Lambrechts, A., A.V. Kwiatkowski, L.M. Lanier, J.E. Bear, J. Vandekerckhove, C. Ampe, and F.B. Gertler. 2000. cAMP-dependent protein kinase phosphorylation of EVL, a Mena/VASP relative, regulates its interaction with actin and SH3 domains. *J Biol Chem* 275:36143.

28. Walter, U., M. Eigenthaler, J. Geiger, and M. Reinhard. 1993. Role of cyclic nucleotide–dependent protein kinases and their common substrate VASP in the regulation of human platelets. *Adv Exp Med Biol* 344:237.

29. Lee, M.S., E.M. David, R.R. Makkar, and J.R. Wilentz. 2004. Molecular and cellular basis of restenosis after percutaneous coronary intervention: The intertwining roles of platelets, leukocytes, and the coagulation-fibrinolysis system. *J Pathol* 203:861.

30. Bode-Boger, S.M., and G. Kojda. Organic nitrates in cardiovascular disease. *Cell Mol Biol (Noisy-le-grand)* 51:307.

31. Indolfi, C., E.V. Avvedimento, E. Di Lorenzo, G. Esposito, A. Rapacciuolo, P. Giuliano, D. Grieco, L. Cavuto, A.M. Stingone, I. Ciullo, G. Condorelli, and M. Chiariello. Activation of cAMP-PKA signaling in vivo inhibits smooth muscle cell proliferation induced by vascular injury. *Nat Med* 3:775.

32. Soyombo, A.A., G.D. Angelini, and A.C. Newby. Neointima formation is promoted by surgical preparation and inhibited by cyclic nucleotides in human saphenous vein organ cultures. *J Thorac Cardiovasc Surg* 109:2.

33. Kuchulakanti, P., and R. Waksman. 2004. Therapeutic potential of oral antiproliferative agents in the prevention of coronary restenosis. *Drugs* 64:2379.

34. Sekiguchi, M., H. Hoshizaki, H. Adachi, S. Ohshima, K. Taniguchi, M. Kurabayashi. 2004. Effects of antiplatelet agents on subacute thrombosis and restenosis after successful coronary stenting: A randomized comparison of ticlopidine and cilostazol. *Circ J* 68:610.

35. Tsuchikane, E., Y. Takeda, K. Nasu, N. Awata, and T. Kobayashi. 2004. Balloon angioplasty plus cilostazol administration versus primary stenting of small coronary artery disease: Final results of COMPASS. *Catheter Cardiovasc Interv* 63:44.

36. Indolfi, C., E. Stabile, C. Coppola, A. Gallo, C. Perrino, G. Allevato, L. Cavuto, D. Torella, E. Di Lorenzo, G. Troncone, A. Feliciello, E. Avvedimento, and M. Chiariello. 2001. Membrane-bound protein kinase A inhibits smooth muscle cell proliferation *in vitro* and in vivo by amplifying cAMP-protein kinase A signals. *Circ Res* 88:319.

37. Dormond O., and C. Ruegg. 2003. Regulation of endothelial cell integrin function and angiogenesis by COX-2, cAMP and Protein Kinase A. *Thromb Haemost* 90:577.

38. Yuan, S.Y. 2002. Protein kinase signaling in the modulation of microvascular permeability. *Vascul Pharmacol* 39:213.

39. Wojciak-Stothard, B., and A.J. Ridley. 2002. Rho GTPases and the regulation of endothelial permeability. *Vascul Pharmacol* 39:187.

40. Taylor, S.S., C. Kim, D. Vigil, N.M. Haste, J. Yang, and G.S. Anand. 2005. Dynamics of signaling by PKA. *Biochem Biophys Acta* 1754:25.

41. Seino S., and T. Shibasaki. 2005. PKA-dependent and PKA-independent pathways for cAMP-regulated exocytosis. *Physiol Rev* 85:1303.

42. Bornfeldt, K.E., and E.G. Krebs. 1999. Crosstalk between protein kinase A and growth factor receptor signaling pathways in arterial smooth muscle. *Cell Signal* 11:465.

43. Maurice, D.H., D. Palmer, D.G. Tilley, H.A. Dunkerley, S.J. Netherton, D.R. Raymond, H.S. Elbatarny, and S.L. Jimmo. 2003. Cyclic nucleotide phosphodiesterase activity, expression, and targeting in cells of the cardiovascular system. *Mol Pharmacol* 64:533.

44. Maurice, D.H., and R.J. Haslam. 1990. Molecular basis of the synergistic inhibition of platelet function by nitrovasodilators and activators of adenylate cyclase: Inhibition of cyclic AMP breakdown by cyclic GMP. *Mol Pharmacol* 37:671.

45. Maurice, D.H. 2005. Cyclic nucleotide phosphodiesterase-mediated integration of cGMP and cAMP signaling in cells of the cardiovascular system. *Front Biosci* 10:1221.

46. Netherton, S.J., and D.H. Maurice. 2005. Vascular endothelial cell cyclic nucleotide phosphodiesterases and regulated cell migration: Implications in angiogenesis. *Mol Pharmacol* 67:263.

47. Oynebraten I., N. Barois, K. Hagelsteen, F.E. Johansen, O. Bakke, G. Haraldsen. 2005. Characterization of a novel chemokine-containing storage granule in endothelial cells: Evidence for preferential exocytosis mediated by protein kinase A and diacylglycerol. *J Immunol* 175:5358.

48. Basoni, C., M. Nobles, A. Grimshaw, C. Desgranges, D. Davies, M. Perretti, I.M. Kramer, and E. Genot. 2005. Inhibitory control of TGF-beta1 on the activation of Rap1, CD11b, and transendothelial migration of leukocytes. *FASEB J* 19:822.

49. Fukuhara, S., A. Sakurai, H. Sano, A. Yamagishi, S. Somekawa, N. Takakura, Y. Saito, K. Kangawa, and N. Mochizuki. 2005. Cyclic AMP potentiates vascular endothelial cadherin-mediated cell–cell contact to enhance endothelial barrier function through an Epac-Rap1 signaling pathway. *Mol Cell Biol* 25:136.

50. Cullere, X., S.K. Shaw, L. Andersson, J. Hirahashi, F.W. Luscinskas, T.N. Mayadas. 2005. Regulation of vascular endothelial barrier function by Epac, a cAMP-activated exchange factor for Rap GTPase. *Blood* 105:1950.

51. Rondaij, M.G., E. Sellink, K.A. Gijzen, J.P. ten Klooster, P.L. Hordijk, J.A. van Mourik, and J. Voorberg. 2004. Small GTP-binding protein Ral is involved in cAMP-mediated release of von Willebrand factor from endothelial cells. *Arterioscler Thromb Vasc Biol* 24:1315.

52. Kooistra, M.R., M. Corada, E. Dejana, and J.L. Bos. 2005. Epac1 regulates integrity of endothelial cell junctions through VE-cadherin. *FEBS Lett* 579:4966.

53. Beavo, J.A., and L.L. Brunton. 2002. Cyclic nucleotide research—still expanding after half a century. *Nat Rev Mol Cell Biol* 3:710.

54. Conti, M., W. Richter, C. Mehats, G. Livera, J.Y. Park, and C. Jin. 2003. Cyclic AMP-specific PDE4 phosphodiesterases as critical components of cyclic AMP signaling. *J Biol Chem* 278:5493.

55. Houslay, M.D., and D.R. Adams. 2003. PDE4 cAMP phosphodiesterases: Modular enzymes that orchestrate signaling crosstalk, desensitization, and compartmentalization. *Biochem J* 370:1.

56. Rybalkin, S.D., C. Yan, K.E. Bornfeldt, and J.A. Beavo. 2003. Cyclic GMP phosphodiesterases and regulation of smooth muscle function. *Circ Res* 93:280.

57. Manganiello, V.C., and E. Degerman. 1999. Cyclic nucleotide phosphodiesterases (PDEs): Diverse regulators of cyclic nucleotide signals and inviting molecular targets for novel therapeutic agents. *Thromb Haemost* 82:407.

58. Haslam, R.J., N.T. Dickinson, and E.K. Jang. 1999. Cyclic nucleotides and phosphodiesterases in platelets. *Thromb Haemost* 82:412.

59. Polson, J.B., and S.J. Strada. 1996. Cyclic nucleotide phosphodiesterases and vascular smooth muscle. *Ann Rev Pharmacol Toxicol* 36:403.

60. Dunkerley, H.A., D.G. Tilley, D. Palmer, H. Liu, S.L. Jimmo, and D.H. Maurice. 2002. Reduced phosphodiesterase 3 activity and phosphodiesterase 3A level in synthetic vascular smooth muscle cells: Implications for use of phosphodiesterase 3 inhibitors in cardiovascular tissues. *Mol Pharmacol* 61:1033.

61. Rybalkin, S.D., I. Rybalkina, J.A. Beavo, and K.E. Bornfeldt. 2002. Cyclic nucleotide phosphodiesterase 1C promotes human arterial smooth muscle cell proliferation. *Circ Res* 90:151.

62. Rybalkin, S.D., I.G. Rybalkina, R. Feil, F. Hofmann, and J.A. Beavo. 2002. Regulation of cGMP-specific phosphodiesterase (PDE5) phosphorylation in smooth muscle cells. *J Biol Chem* 277:3310.

63. Kim, D., S.D. Rybalkin, X. Pi, Y. Wang, C. Zhang, T. Munzel, J.A. Beavo, B.C. Berk, and C. Yan. 2001. Upregulation of phosphodiesterase 1A1 expression is associated with the development of nitrate tolerance. *Circulation* 104:2338.

64. Rybalkin, S.D., K.E. Bornfeldt, W.K. Sonnenburg, I.G. Rybalkina, K.S. Kwak, K. Hanson, E.G. Krebs, and J.A. Beavo. 1997. Calmodulin-stimulated cyclic nucleotide phosphodiesterase (PDE1C) is induced in human arterial smooth muscle cells of the synthetic, proliferative phenotype. *J Clin Invest* 100:2611.

65. Movsesian, M.A. 2002. PDE3 cyclic nucleotide phosphodiesterases and the compartmentation of cyclic nucleotide–mediated signalling in cardiac myocytes. *Basic Res Cardiol* 97 (Suppl 1):I83.

66. Wechsler, J., Y.H. Choi, J. Krall, F. Ahmad, V.C. Manganiello, and M.A. Movsesian. 2002. Isoforms of cyclic nucleotide phosphodiesterase PDE3A in cardiac myocytes. *J Biol Chem* 277:38072.

67. Degerman, E., P. Belfrage, and V.C. Manganiello. 1997. Structure, localization, and regulation of cGMP-inhibited phosphodiesterase (PDE3). *J Biol Chem* 272:6823.

68. Hambleton, R., J. Krall, E. Tikishvili, M. Honeggar, F. Ahmad, V.C. Manganiello, and M.A. Movsesian. 2005. Isoforms of cyclic nucleotide phosphodiesterase PDE3 and their contribution to cAMP-hydrolytic activity in subcellular fractions of human myocardium. *J Biol Chem* 280:39168.

69. Orallo, F., M. Camina, E. Alvarez, H. Basaran, C. Lugnier. 2005. Implication of cyclic nucleotide phosphodiesterase inhibition in the vasorelaxant activity of the citrus-fruits flavonoid (+/-)-naringenin. *Planta Med* 71:99.

70. Favot L., T. Keravis, C. Lugnier. 2003. Modulation of VEGF-induced endothelial cell cycle protein expression through cyclic AMP hydrolysis by PDE2 and PDE4. *Thromb Haemost* 92:634.

71. Favot, L., T. Keravis, V. Holl, A. Le Bec, C. Lugnier. 2003. VEGF-induced HUVEC migration and proliferation are decreased by PDE2 and PDE4 inhibitors. *Thromb Haemost* 90:334.

72. Keravis, T., N. Komas, C. Lugnier. 2000. Cyclic nucleotide hydrolysis in bovine aortic endothelial cells in culture: Differential regulation in cobblestone and spindle phenotypes. *J Vasc Res* 37:235.

73. Lugnier, C., T. Keravis, A. Eckly-Michel. 1999. Crosstalk between NO and cyclic nucleotide phosphodiesterases in the modulation of signal transduction in blood vessel. *J Physiol Pharmacol* 50:639.

74. Zhang, W., H. Ke, and R.W. Colman. 2002. Identification of interaction sites of cyclic nucleotide phosphodiesterase type 3A with milrinone and cilostazol using molecular modeling and site-directed mutagenesis. *Mol Pharmacol* 62:514.

75. Tang, K.M., E.K. Jang, and R.J. Haslam. 1997. Expression and mutagenesis of the catalytic domain of cGMP-inhibited phosphodiesterase (PDE3) cloned from human platelets. *Biochem J* 323:217.

76. Maurice, D.H., and R.J. Haslam. 1990. Nitroprusside enhances isoprenaline-induced increases in cAMP in rat aortic smooth muscle. *Eur J Pharmacol* 191:471.

77. Liu, H., and D.H. Maurice. 1998. Expression of cyclic GMP-inhibited phosphodiesterases 3A and 3B (PDE3A and PDE3B) in rat tissues: Differential subcellular localization and regulated expression by cyclic AMP. *Br J Pharmacol* 125:1501.

78. Palmer, D., and D.H. Maurice. 2000. Dual expression and differential regulation of phosphodiesterase 3A and phosphodiesterase 3B in human vascular smooth muscle: Implications for phosphodiesterase 3 inhibition in human cardiovascular tissues. *Mol Pharmacol* 58:247.

79. Kenan, Y., T. Murata, Y. Shakur, E. Degerman, and V.C. Manganiello. 2000. Functions of the N-terminal region of cyclic nucleotide phosphodiesterase 3 (PDE 3) isoforms. *J Biol Chem* 275:12331.

80. Jurevicius, J., V.A. Skeberdis, R. Fischmeister. 2003. Role of cyclic nucleotide phosphodiesterase isoforms in cAMP compartmentation following beta2-adrenergic stimulation of ICa,L in frog ventricular myocytes. *J Physiol* 551:239.

81. Vandecasteele, G., I. Verde, C. Rucker-Martin, P. Donzeau-Gouge, R. Fischmeister. 2001. Cyclic GMP regulation of the L-type Ca^{2+} channel current in human atrial myocytes. *J Physiol* 533:329.

82. Verde, I., G. Vandecasteele, F. Lezoualc'h, R. Fischmeister. 1999. Characterization of the cyclic nucleotide phosphodiesterase subtypes involved in the regulation of the L-type Ca^{2+} current in rat ventricular myocytes. *Br J Pharmacol* 127:65.

83. Inoue, Y., K. Toga, T. Sudo, K. Tachibana, S. Tochizawa, Y. Kimura, Y. Yoshida, and H. Hidaka. 2000. Suppression of arterial intimal hyperplasia by cilostamide, a cyclic nucleotide phosphodiesterase 3 inhibitor, in a rat balloon double-injury model. *Br J Pharmacol* 130:231.

84. Zhao, A.Z., K.E. Bornfeldt, and J.A. Beavo. 1998. Leptin inhibits insulin secretion by activation of phosphodiesterase 3B. *J Clin Invest* 102:869.

85. Zhao, A.Z., M.M. Shinohara, D. Huang, M. Shimizu, H. Eldar-Finkelman, E.G. Krebs, J.A. Beavo, and K.E. Bornfeldt. 2000. Leptin induces insulin-like signaling that antagonizes cAMP elevation by glucagon in hepatocytes. *J Biol Chem* 275:11348.

86. Elbatarny, H.S., and D.H. Maurice. 2005. Leptin-mediated activation of human platelets: Involvement of a leptin receptor and phosphodiesterase 3A-containing cellular signaling complex. *Am J Physiol Endocrinol Metab* 289:E695.

87. Onuma, H., H. Osawa, K. Yamada, T. Ogura, F. Tanabe, D.K. Granner, H. Makino. 2002. Identification of the insulin-regulated interaction of phosphodiesterase 3B with 14-3-3 beta protein. *Diabetes* 51:3362.

88. Pozuelo Rubio, M., D.G. Campbell, N.A. Morrice, C. Mackintosh. 2005. Phosphodiesterase 3A binds to 14-3-3 proteins in response to PMA-induced phosphorylation of Ser428. *Biochem J* 392:163.

89. Patrucco, E., A. Notte, L. Barberis, G. Selvetella, A. Maffei, M. Brancaccio, S. Marengo, G.Russo, O. Azzolino, S.D. Rybalkin, L. Silengo, F. Altruda, R. Wetzker, M.P. Wymann, G.Lembo, E. Hirsch. 2004. PI3Kgamma modulates the cardiac response to chronic pressure overload by distinct kinase-dependent and -independent effects. *Cell* 118:375.

90. Voigt, P., M.B. Dorner, M. Schaefer. 2006. Characterization of P87Pikap, a novel regulatory subunit of phosphoinositide 3-kinase gamma that is highly expressed in heart and interacts with PDE3B. *J Biol Chem* 281:9977.

91. Barnett, A.H., A.W. Bradbury, J. Brittenden, B. Crichton, R. Donnelly, S. Homer-Vanniasinkam, D.P. Mikhailidis, and G. Stansby. 2004. The role of cilostazol in the treatment of intermittent claudication. *Curr Med Res Opin* 20:1661.

92. Netherton, S.J., S.L. Jimmo, D. Palmer, D.G. Tilley, H.A. Dunkerley, D.R. Raymond, J.C. Russell, P.M. Absher, E.H. Sage, R.B. Vernon, and D.H. Maurice. Altered phosphodiesterase 3-mediated cAMP hydrolysis contributes to a hypermotile phenotype in obese JCR:LA-cp rat aortic vascular smooth muscle cells: Implications for diabetes-associated cardiovascular disease. *Diabetes* 51:1194.

93. Tilley, D.G., and D.H. Maurice. 2002. Vascular smooth muscle cell phosphodiesterase (PDE) 3 and PDE4 activities and levels are regulated by cyclic AMP in vivo. *Mol Pharmacol* 62:497.

94. Tilley, D.G., and D.H. Maurice. 2005. Vascular smooth muscle cell phenotype-dependent phosphodiesterase 4D short form expression: Role of differential histone acetylation on cAMP-regulated function. *Mol Pharmacol* 68:596.

95. Park, S.W., C.W. Lee, H.S. Kim, N.H. Lee, D.Y. Nah, M.K. Hong, J.J. Kim, and S.J. Park. 2000. Effects of cilostazol on angiographic restenosis after coronary stent placement. *Am J Cardiol* 86:499.

96. Liu, H., D. Palmer, S.L. Jimmo, D.G. Tilley, H.A. Dunkerley, S.C. Pang, and D.H. Maurice. 2000. Expression of phosphodiesterase 4D (PDE4D) is regulated by both the cyclic AMP-dependent protein kinase and mitogen-activated protein kinase signaling pathways. A potential mechanism allowing for the coordinated regulation of PDE4D activity and expression in cells. *J Biol Chem* 275:26615.

97. Liu, H., and D.H. Maurice. 1999. Phosphorylation-mediated activation and translocation of the cyclic AMP-specific phosphodiesterase PDE4D3 by cyclic AMP-dependent protein kinase and mitogen-activated protein kinases. A potential mechanism allowing for the coordinated regulation of PDE4D activity and targeting. *J Biol Chem* 274:10557.

98. Rockman, H.A., W.J. Koch, and R.J. Lefkowitz. 2002. Seven-transmembrane-spanning receptors and heart function. *Nature* 415:206.

99. Slotkin, T.A., J.T. Auman, and F.J. Seidler. 2003. Ontogenesis of β-adrenoceptor signaling: Implications for perinatal physiology and for fetal effects of tocolytic drugs. *J Pharmacol Exp Ther* 306:1.

100. Pitcher, J.A., N.J. Freedman, and R.J. Lefkowitz. 1998. G protein-coupled receptor kinases. *Annu Rev Biochem* 67:653.

101. Perry, S.J., and R.J. Lefkowitz. 2002. Arresting developments in heptahelical receptor signaling and regulation. *Trends Cell Biol* 12:130.

102. Perry, S.J., G.S. Baillie, T.A. Kohout, I. McPhee, M.M. Magiera, K.L. Ang, W.E. Miller, A.J. McLean, M. Conti, M.D. Houslay, and R.J. Lefkowitz. 2002. Targeting of cyclic AMP degradation to β$_2$-adrenergic receptors by β-arrestins. *Science* 298:834.

103. Lynch, M.J., G.S. Baillie, A. Mohamed, X. Li, C. Maisonneuve, E. Klussmann, G. van Heeke, and M.D. Houslay. 2005. RNA silencing identifies PDE4D5 as the functionally relevant cAMP phosphodiesterase interacting with {β}arrestin to control the protein kinase A/AKAP79-mediated switching of the {β}$_2$-adrenergic receptor to activation of ERK in HEK293B2 cells. *J Biol Chem* 280:33178.

104. Baillie, G.S., A. Sood, I. McPhee, I. Gall, S.J. Perry, R.J. Lefkowitz, and M.D. Houslay. 2003. β-Arrestin-mediated PDE4 cAMP phosphodiesterase recruitment regulates β-adrenoceptor switching from Gs to Gi. *Proc Natl Acad Sci USA* 100:940.

105. Erdogan, S., and M.D. Houslay. 1997. Challenge of human Jurkat T-cells with the adenylate cyclase activator forskolin elicits major changes in cAMP phosphodiesterase (PDE) expression by up-regulating PDE3 and inducing PDE4D1 and PDE4D2 splice variants as well as down-regulating a novel PDE4A splice variant. *Biochem J* 321:165.

106. Kostic, M.M., S. Erdogan, G. Rena, G. Borchert, B. Hoch, S. Bartel, G. Scotland, E. Huston, M.D. Houslay, and E.G. Krause. 1997. Altered expression of PDE1 and PDE4 cyclic nucleotide phosphodiesterase isoforms in 7-oxo-prostacyclin-preconditioned rat heart. *J Mol Cell Cardiol* 29:3135.

107. Palmer, W.K., and S. Doukas. 1984. Dibutyryl cyclic AMP increases phosphodiesterase activity in the rat heart. *Can J Physiol Pharmacol* 62:1225.

108. Seybold, J., R. Newton, L. Wright, P.A. Finney, N. Suttorp, P.J. Barnes, I.M. Adcock, and M.A. Giembycz. 1998. Induction of phosphodiesterases 3B, 4A4, 4D1, 4D2, and 4D3 in Jurkat T-cells and in human peripheral blood T-lymphocytes by 8-bromo-cAMP and Gs-coupled receptor agonists. Potential role in beta2-adrenoreceptor desensitization. *J Biol Chem* 273:20575.

109. Verghese, M.W., R.T. McConnell, J.M. Lenhard, L. Hamacher, and S.L. Jin. 1999. Regulation of distinct cyclic AMP-specific phosphodiesterase (phosphodiesterase type 4) isozymes in human monocytic cells. *Mol Pharmacol* 47:1164.

110. Mehats, C., G. Tanguy, E. Dallot, B. Robert, R. Rebourcet, F. Ferre, and M.J. Leroy. 1999. Selective up-regulation of phosphodiesterase-4 cyclic adenosine $3',5'$-monophosphate (cAMP)-specific phosphodiesterase variants by elevated cAMP content in human myometrial cells in culture. *Endocrinology* 140:3228.

111. Mehats, C., G. Tanguy, E. Dallot, D. Cabrol, F. Ferre, and M.J. Leroy. 2001. Is up-regulation of phosphodiesterase 4 activity by PGE2 involved in the desensitization of β-mimetics in late pregnancy human myometrium? *J Clin Endocrinol Metab* 86:5358.

112. Oger, S., C. Mehats, E. Dallot, F. Ferre, and M.J. Leroy. 2002. Interleukin-1β induces phosphodiesterase 4B2 expression in human myometrial cells through a prostaglandin E2- and cyclic adenosine $3',5'$-monophosphate-dependent pathway. *J Clin Endocrinol Metab* 87:5524.

113. Rose, R.J., H. Liu, D. Palmer, and D.H. Maurice. 1997. Cyclic AMP-mediated regulation of vascular smooth muscle cell cyclic AMP phosphodiesterase activity. *Br J Pharmacol* 122:233.

114. Dodge-Kafka K.L., J. Soughayer, G.C. Pare, J.J. Carlisle Michel, L.K. Langeberg, M.S. Kapiloff, J.D. Scott. 2005. The protein kinase A anchoring protein mAKAP coordinates two integrated cAMP effector pathways. *Nature* 437:574.

115. Baillie, G.S., J.D. Scott, M.D. Houslay. 2005. Compartmentalisation of phosphodiesterases and protein kinase A: Opposites attract. *FEBS Lett* 579:3264.

116. Dodge K.L., S. Khouangsathiene, M.S. Kapiloff, R. Mouton, E.V. Hill, M.D. Houslay, L.K. Langeberg, J.D. Scott. 2001. mAKAP assembles a protein kinase A/PDE4 phosphodiesterase cAMP signaling module. *EMBO J* 20:1921.

117. Vicini, E., and M. Conti. 1997. Characterization of an intronic promoter of a cyclic adenosine $3',5'$-monophosphate (cAMP)-specific phosphodiesterase gene that confers hormone and cAMP inducibility. *Mol Endocrinol* 11:839.

118. D'Sa, C., L.M. Tolbert, M. Conti, and R.S. Duman. 2002. Regulation of cAMP-specific phosphodiesterases type 4B and 4D (PDE4) splice variants by cAMP signaling in primary cortical neurons. *J Neurochem* 81:745.

119. Rena, G., F. Begg, A. Ross, C. MacKenzie, I. McPhee, L. Campbell, E. Huston, M. Sullivan, and M.D. Houslay. 2001. Molecular cloning, genomic positioning, promoter identification, and characterization of the novel cyclic AMP-specific phosphodiesterase PDE4A10. *Mol Pharmacol* 59:996.

120. Sasaki, T., J. Kotera, and K. Omori. 2004. Transcriptional activation of phosphodiesterase 7B1 by dopamine D1 receptor stimulation through the cyclic AMP/cyclic AMP-dependent protein kinase/cyclic AMP-response element binding protein pathway in primary striatal neurons. *J Neurochem* 89:474.

121. Grunstein, M. 1997. Histone acetylation in chromatin structure and transcription. *Nature* 389:349.

122. Qiu, P., and L. Li. 2002. Histone acetylation and recruitment of serum responsive factor and CREB-binding protein onto SM22 promoter during SM22 gene expression. *Circ Res* 90:858.

123. Kumar, M.S., and G.K. Owens. 2003. Combinatorial control of smooth muscle-specific gene expression. *Arterioscler Thromb Vasc Biol* 23:737.

124. Grant, P.A. 2001. A tale of histone modifications. *Genome Biol* 2:3.

125. Gretarsdottir, S., G. Thorleifsson, S.T. Rey nisdottir, A. Manolescu, S. Jonsdottir, T. Jonsdottir, T. Gudmundsdottir, S.M. Bjarnadottir, O.B. Einarsson, H.M. Gudjonsdottir, M. Hawkins, G. Gudmundsson, H. Gudmundsdottir, H. Andrason, A.S. Gudmundsdottir, M. Sigurdardottir, T.T. Chou, J. Nahmias, S. Goss, S. Sveinbjornsdottir, E.M. Valdimarsson, F. Jakobsson, U. Agnarsson, V. Gudnason, G. Thorgeirsson, J. Fingerle, M. Gurney, D. Gudbjartsson, M.L. Frigge, A. Kong, K. Stefansson, and J.R. Gulcher. 2003. The gene encoding phosphodiesterase 4D confers risk of ischemic stroke. *Nat Genet* 35:131.

23 Regulation and Function of Cyclic Nucleotide Phosphodiesterases in Vascular Smooth Muscle and Vascular Diseases

Chen Yan, David J. Nagel, and Kye-Im Jeon

CONTENTS

23.1 Introduction 466
23.2 Role of cAMP and cGMP in Vascular Smooth Muscle 466
23.3 Multiple Structurally and Functionally Different Phosphodiesterases
in Vascular Smooth Muscle 467
    23.3.1 PDE1 as a Mediator in the Cross Talk between Ca^{2+}
    and Cyclic Nucleotides 467
    23.3.2 PDE3 in the Cross-Talk between cAMP and cGMP Signaling.................468
    23.3.3 PDE4 as a Negative Feedback Regulator of cAMP Signaling..................470
    23.3.4 PDE5 as a Negative Feedback Regulator of cGMP Signaling.................470
23.4 Roles of Phosphodiesterases in the Regulation of Vascular
Smooth Muscle Cell Functions 471
    23.4.1 Phosphodiesterases and Vascular Smooth Muscle Tone 471
        23.4.1.1 Role of PDE1 in Vascular Smooth Muscle Tone......................471
        23.4.1.2 Role of PDE5 in Vascular Smooth Muscle Tone......................471
        23.4.1.3 Role of PDE3 and PDE4 in Smooth Muscle Tone....................472
    23.4.2 PDEs and Vascular Smooth Muscle Growth 473
        23.4.2.1 Role of PDE1 in Vascular Smooth Muscle Growth 473
        23.4.2.2 Role of PDE5 in Smooth Muscle Cell Growth 475
        23.4.2.3 Role of PDE3 and PDE4 in Vascular
        Smooth Muscle Growth 475
23.5 Vascular Disease-Associated Alteration of Phosphodiesterases
and the Therapeutic Potential of Phosphodiesterase Inhibitors...........................476
    23.5.1 Nitrate Tolerance..476
    23.5.2 Pulmonary Hypertension..477
    23.5.3 Atherosclerosis and Restenosis 478
23.6 Conclusion and Perspective...479
References 480

23.1 INTRODUCTION

Cyclic nucleotides (cAMP and cGMP) regulate a wide variety of processes in vascular smooth muscle cells (SMCs) from short-term action in smooth muscle relaxation to long-term processes such as gene expression and cell growth. The versatility and specificity of cyclic nucleotide function is likely to be associated with different spatiotemporal patterns of cyclic nucleotide responses. This regulation is thought to be dependent on the organization of multiple divergent macromolecular complexes containing unique cyclases, phosphodiesterases (PDEs), kinases, and anchoring proteins. PDEs, by catalyzing the hydrolysis of cAMP and cGMP to 5′AMP and 5′GMP, play very important roles in the regulation of the amplitude, duration, and compartmentation of intracellular cyclic nucleotide signaling. PDEs constitute a large superfamily of enzymes grouped into 11 broad families based on distinct structural, kinetic, regulatory, and inhibitory properties.[1] The existence of multiple PDE isoforms with different structures, catalytic properties, and subcellular localizations in a single cell guarantees heterogeneity in the kinetics (amplitude and duration) and spatial patterns of increased cyclic nucleotides, which enable cyclic nucleotides to mediate multiple cellular processes. It is indeed reported that different PDE isozymes play distinct roles within the same cell type. For example, in mesangial cells, inhibition of PDE3 suppressed mitogenesis whereas inhibition of PDE4 had no effect.[2,3] However, inhibiting PDE4, but not PDE3, suppressed reactive oxygen species generation.[2] Similarly, in cardiomyocytes, chronic inhibition of PDE3, but not PDE4, induces cardiomyocyte apoptosis.[4] These data suggest that cells possess functionally compartmentalized pools of cAMP that are regulated by distinct PDE isozymes.

There are many pieces of experimental evidence demonstrating that cAMP exists in spatially restricted domains within numerous cell types (see detailed discussion in Section D of this book). The most direct evidence of cAMP compartmentation came from experiments using fluorescence resonance energy transfer (FRET) techniques, in which real-time cAMP dynamics were monitored in live cardiomyocytes transfected with a cAMP-dependent protein kinase (PKA)-derived fluorescent cAMP indicator.[5] In these studies, it was shown that the β-adrenergic agonist-stimulated increase in cAMP was transient and mainly localized to multiple microdomains (≈ 1 μm) corresponding to the transverse tubule and junctional sarcoplasmic reticulum (SR) membrane in rat neonatal cardiomyocytes. Free diffusion of cAMP to the cytosol is prevented by PDE activity.[5,6] These data also demonstrated that an increase in cAMP in certain cellular compartments, induced by β-AR stimulation, selectively activated a subset of PKA molecules anchored in proximity to the transverse tubule. However, PDE inhibition dissipated the cAMP gradients and led to global activation of cellular PKA.[5] These results strongly demonstrate that PDE activity plays a fundamental role in regulating not only the intensity and duration, but also the specific location of cyclic nucleotides. This is critical in guaranteeing effective, specific cyclic nucleotide signaling and function. In this chapter, we focus on the roles of different PDE isozymes in the regulation of cyclic nucleotide signaling and function in vascular SMCs. We will also review several vascular diseases that are associated with alterations in PDEs.

23.2 ROLE OF cAMP AND cGMP IN VASCULAR SMOOTH MUSCLE

A normal artery consists of quiescent arterial SMCs covered by a monolayer of endothelial cells (ECs) lining the interior surface of the blood vessel. The endothelium plays a critical role in regulating smooth muscle function. Normal endothelium synthesizes and secretes factors that both contract and relax vascular SMCs. Those that relax vascular SMCs and inhibit vascular SMC growth include prostacyclin (PGI₂), nitric oxide (NO), and C-type natriuretic peptide (CNP).[7,8] These factors regulate many SMC functions, including SMC relaxation,

proliferation, migration, inflammatory responses, and production of extracellular matrix, primarily through regulating the second messengers, cAMP and cGMP. All of these functions have been implicated in various cardiovascular diseases. PGI_2 specifically stimulates adenylyl cyclase (AC), which produces cAMP. In vascular SMCs, cAMP functions as a vasodilator as well as an inhibitor of vascular SMC activation (phenotypic modification and proliferation).[9–11] Most of these cAMP effects appear to be mediated by the activation of PKA.

cGMP is an important effector molecule of NO and natriuretic peptides (NPs), such as ANP and CNP, that act through NO-sensitive (largely soluble) or membrane-bound (particulate) guanylyl cyclases, respectively. It is well known that increases in cGMP derived from NO or CNP promote smooth muscle relaxation as well as inhibit several key events, such as SMC proliferation, migration, and the expression of inflammatory molecules in vascular SMCs during atherogenesis.[12,13] Although cGMP-dependent protein kinase (PKG) is a well-known cGMP target, cGMP also interacts with several other molecules in the cell, including cGMP-gated cation channels and the cyclic nucleotide ras guanine nucleotide exchange factor (CNrasGEF).[14,15] cGMP may also regulate cAMP levels through modulating the activities of the cGMP-stimulated PDE (PDE2) or cGMP-inhibited PDE (PDE3), resulting in altered levels of cAMP and PKA activity.[9,16,17] It has also been shown that high concentrations of cGMP may directly stimulate PKA.[18] PKG appears to be the primary receptor protein for cGMP in vascular SMCs and is believed to be a major mediator of cGMP-dependent regulation in vascular SMC relaxation as cGMP-induced relaxation of aortic smooth muscle is completely abolished in PKG-I-deficient mice.[19] cGMP and activation of PKG induce vascular relaxation by lowering intracellular Ca^{2+} and reducing sensitivity to Ca^{2+}, thereby decreasing smooth muscle tone through a variety of cGMP/PKG-dependent mechanisms.[20] However, there are other cGMP-mediated vascular SMC effects that appear to be PKG-independent. For example, recent experimental observations suggest that cGMP-dependent activation of PKA, but not PKG, might be critical in cGMP-mediated attenuation of vascular SMC growth[17] and inflammatory molecule expression[9] in vascular SMCs. This cGMP-dependent activation of PKA is largely mediated by cGMP-dependent inhibition of PDE3 activity and subsequent elevation of cAMP.[9,17]

23.3 MULTIPLE STRUCTURALLY AND FUNCTIONALLY DIFFERENT PHOSPHODIESTERASES IN VASCULAR SMOOTH MUSCLE

Early studies using diethylamino ethanol (DEAE) ion exchange chromatography together with PDE assays defined four distinct types of PDE activities present in smooth muscles from different vascular beds in various species including human, bovine, porcine, rabbit, and rat.[11,21] These PDEs were identified as the Ca^{2+}/calmodulin-stimulated PDE1 family, cGMP-inhibited PDE3 family, cAMP-specific PDE4 family, and cGMP-specific PDE5 family. The cAMP-hydrolyzing activity was mainly due to PDE3 and PDE4 activities, and the cGMP-hydrolyzing activity was mainly due to PDE1 and PDE5, although the relative contribution of each PDE may vary depending upon the species, vascular bed, and the status of cells (such as basal versus stimulated cells). For example, PDE1-dependent cGMP-hydrolyzing activity is predominant under higher calcium conditions induced by norepinephrine (NE).[22] Therefore, it is likely that different PDEs play distinct roles in the regulation of cAMP and cGMP in vascular SMCs.

23.3.1 PDE1 AS A MEDIATOR IN THE CROSS TALK BETWEEN CA^{2+} AND CYCLIC NUCLEOTIDES

Ca^{2+}/calmodulin-stimulated PDEs constitute a large family of enzymes (PDE1 family), encoded by three distinct genes, PDE1A, PDE1B, and PDE1C. Multiple N-terminal or

C-terminal splice variants have also been identified for each gene. Currently, at least 14 PDE1As, 2 PDE1Bs, and 5 PDE1Cs transcripts have been described (see a detailed review of the PDE1 family isozymes in Chapter 1) *in vitro*, the activity of all PDE1 family members can be stimulated up to 10 folds by Ca^{2+} in the presence of calmodulin.[23] However, they differ in their regulatory properties, substrate affinities, specific activities, Ca^{2+} sensitivities, and tissue and cell distributions. *in vitro* enzymatic assays have shown that PDE1A and PDE1B isozymes hydrolyze cGMP with much higher affinities than cAMP; however, PDE1C isozymes hydrolyze both cAMP and cGMP with equally high affinity. PDE1A has been detected in large vessels from many different species including mouse, rat, rabbit, cow, dog, monkey, and human.[11,24,25] PDE1B has only been reported in cultured aortic smooth muscle from monkeys and baboons.[11,26] PDE1C has been found in growing SMCs in human fetal aorta, human vascular lesions, and in cultured human SMCs.[26]

PDE1 isozymes, due to the unique nature of Ca^{2+}/calmodulin stimulation, play an important role in Ca^{2+}-mediated regulation of intracellular cyclic nucleotide levels.[27] The association of Ca^{2+}/calmodulin with PDE1 is believed to be a reversible process that is dependent on intracellular Ca^{2+} concentration. As Ca^{2+} signals are spatially and temporally complex and diverse, this provides a mechanism to dynamically regulate PDE1 activity. It has been shown that Ca^{2+}/calmodulin-stimulated PDE activity (primarily PDE1A) was rapidly stimulated in rabbit arterial strips and in cultured rat aortic SMCs, by Ca^{2+}-elevating reagents such as angiotensin II (Ang II).[27–29] It was further shown that Ang II markedly decreased atrial natriuretic peptide (ANP)-induced cGMP accumulation.[27] The partially selective PDE1 inhibitor, vinpocetine, significantly blocked the inhibitory effect of Ang II on cGMP, suggesting that PDE1A in SMCs plays a major role in mediating the inhibitory effect of Ang II on cGMP accumulation.[27] Antagonizing cGMP accumulation appears to be necessary for Ang II function because cGMP negatively regulates the Ang II-signaling pathway at multiple levels.[20] Such an attenuation of ANP- or nitroglycerine (NTG)-evoked cGMP accumulation has also been observed with many other Ca^{2+}-raising vasoconstrictors such as endothelin-1 (ET-1) and KCl.[28,29] These observations also indicate that PDE1A plays important roles in the cross talk between Ca^{2+} and cGMP signaling (Figure 23.1).

23.3.2 PDE3 in the Cross-Talk between cAMP and cGMP Signaling

PDE3 family members have similar affinities for both cAMP and cGMP with K_m values in the submicromolar range. However, they have very different V_{max} values for cAMP and cGMP, hydrolyzing cAMP with a rate ~10-fold greater than that of cGMP. Because of its high affinity and low rate of cGMP hydrolysis, it was suggested that cGMP could behave as a competitive inhibitor of cAMP hydrolysis.[30] For this reason, PDE3 family members are also referred to as cGMP-inhibited PDEs (see a detailed review of the PDE3 family isozymes in Chapter 5). Two subfamilies of PDE3 (PDE3A and PDE3B) encoded by separate, but related, genes have been identified. PDE3A2 and PDE3B1 isozymes are the two major PDE3 isozymes detected in human and rat vascular SMCs.[31–33] PDE3A2 is primarily localized in the cytosolic fraction, whereas PDE3B1 is found in the particulate fraction of cultured rat aortic SMCs.[33] PDE3A2 expression levels were dramatically downregulated in cultured vascular SMCs compared with smooth muscle from intact aortas, whereas PDE3B1 expression was not significantly altered.[34] A similar reduction in PDE3A expression was also detected in growing, neointimal SMCs (see the Chapter 22 for more details).[34] The functional difference between these two PDE3 isozymes in vascular SMCs is not yet clear, due to the lack of isoform-selective inhibitors.

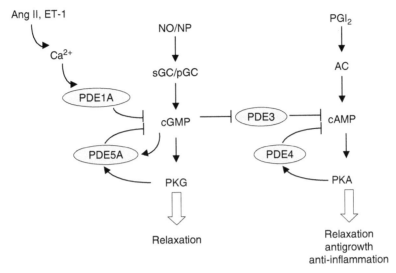

FIGURE 23.1 Functional interaction of multiple PDE isozymes in vascular SMCs. Members of at least four different PDE families (PDE1, PDE3, PDE4, and PDE5) coexist in vascular SMCs. These enzymes work in concert to regulate cAMP and cGMP signaling in vascular SMCs. Nitric oxide (NO) and natriuretic peptides (NP) stimulate NO-sensitive guanylyl cyclase (also known as soluble guanylyl cyclase, sGC) and the membrane-bound or particulate guanylyl cyclase (pGC), respectively, increase cGMP levels, which is followed by PKG activation and smooth muscle relaxation. Elevated cGMP and activated PKG interact with the N-terminal regulatory region of PDE5 to stimulate PDE5 cGMP-hydrolytic activity. Activated PDE5 acts through a negative feedback pathway to decrease intracellular cGMP level, thereby terminating cGMP signaling. Vasoconstrictors [such as angiotensin II (Ang II), norepinephrine (NE), and endothelin-1 (ET-1)] activate PDE1A via increasing intracellular Ca^{2+} concentration. Activation of PDE1A leads to increased cGMP hydrolysis, thereby attenuating cGMP accumulation. Prostacyclin (PGI2) stimulates adenylyl cyclase (AC), which leads to increased cellular cAMP, and PKA activation. Activation of PKA leads to SMC relaxation, inhibition of SMC growth, and decreased expression of inflammatory molecules. PKA activation also activates PDE4 by phosphorylating its N-terminal regulatory region, which accelerates cAMP hydrolysis and terminates cAMP signaling. cGMP, by inhibiting PDE3 cAMP-hydrolytic activity, increases cAMP accumulation. Thus, PDE3 provides one mechanism for agents that increase cGMP to regulate cAMP signaling.

The property of cGMP-mediated inhibition of PDE3 allows agents (such as NO and CNP) that elevate cGMP to increase the level of cAMP and synergize with agents that stimulate AC (i.e., PGI$_2$) (Figure 23.1). Many studies have shown that cGMP can regulate vascular SMC function through inhibition of PDE3, resulting in a subsequent increase in cellular cAMP. For example, it has been shown that the inhibitory effects of NO on SMC growth and inflammation are not mediated by PKG, but rather by PKA-dependent mechanisms. Activation of PKA by NO is mediated by cGMP-dependent inhibition of PDE3 activity, which leads to an increased cAMP level.[9,17] Moreover, nitrate-based vasodilators potentiated the relaxant effects of isoproterenol in rat aorta, via cGMP-dependent inhibition of PDE3.[35] In a manner similar to NO donors, PDE3 inhibitors enhanced cAMP accumulation stimulated by forskolin, an activator of AC,[36] or the β-adrenergic agonist, isoproterenol.[37] Thus, PDE3 provides one mechanism for agents that increase cGMP to raise cAMP level and subsequently activate PKA signaling (Figure 23.1). PDE3 inhibition can also enhance cAMP accumulation produced by activators of AC. This evidence demonstrates that PDE3 mediates a substantial amount of cross talk between cAMP and cGMP in vascular SMCs (Figure 23.1).

23.3.3 PDE4 as a Negative Feedback Regulator of cAMP Signaling

PDE4 belongs to a large family of enzymes that specifically hydrolyze cAMP with high affinity and are inhibited by rolipram and Ro-20-1724. A detailed review of the PDE4 family isozymes can be found in Chapter 6. Four PDE4 genes (*PDE4A*, *PDE4B*, *PDE4C*, and *PDE4D*) have been identified, and multiple products (variants) of each of these genes are generated through alternative promoters or splicing. PDE4 isozymes are found in almost all cell types,[38] and each cell often expresses more than one PDE4 isozyme. The expression and function of PDE4s in blood vessels have not been studied extensively because PDE4 inhibitors have modest vasorelaxant effects. It was recently reported that at least two PDE4D isoforms (PDE4D3 and PDE4D5) are expressed in rat and human vascular SMCs,[39,40] and their activity and expression are regulated in a cAMP/PKA-dependent manner (see Chapter 22 for more information).[39–41] For example, short-term exposure to cAMP-elevating agents (2–30 min) stimulated PDE4D3 and PDE4D5 activity through PKA-dependent phosphorylation in intact arteries as well as cultured SMCs.[39,41] However, long-term treatment (1–4 h) with cAMP-elevating agents markedly induced PDE4D1 and PDE4D2 expression in cultured SMCs.[40] Alternatively, PDE4D3 expression, but not PDE4D1 or PDE4D2 expression, was induced in intact vessels.[40] Nevertheless, activation and induction of PDE4 upon cAMP elevation attenuates the subsequent cAMP response both *in vitro*[32,39,40] and *in vivo*,[41] which indicates that PDE4 functions as a negative feedback regulator of cAMP signaling (Figure 23.1). PDE4 inhibitors potentiate cAMP signaling by increasing cAMP accumulation, which is synergistic in the presence of cAMP-elevating agents, cGMP-elevating agents, or PDE3 inhibition.

23.3.4 PDE5 as a Negative Feedback Regulator of cGMP Signaling

PDE5 is also known as the cGMP-binding, cGMP-specific PDE because this family of enzymes has high-affinity cGMP-binding allosteric sites in the regulatory domains and because it specifically hydrolyzes cGMP at low substrate levels. Chapter 7, as well as numerous other reviews, explains the enzymatic properties and regulatory mechanisms of PDE5 activity.[42] Only one PDE5 gene (*PDE5A*) has been cloned. To date, three PDE5A variants have been reported, PDE5A1, PDE5A2, and PDE5A3. They differ only at their N-terminal ends[43–45] and are probably generated by alternative splicing or alternative promoters.[45] PDE5A1 appears to be the predominant form expressed in most tissues, including vascular SMCs.

Bovine PDE5 is phosphorylated on Ser92 (Ser102 in human PDE5A1), which leads to increased catalytic activity.[46] *In vitro* studies showed that PDE5 could be phosphorylated by both PKA and PKG.[46,47] However, recent studies indicated that PDE5 phosphorylation *in vivo* is predominately mediated by cGMP/PKG, not cAMP/PKA.[48] The N-terminal regulatory region of PDE5 also contains two homologous domains, defined as GAF-A and GAF-B, based on their sequence homology with GAF motifs present in a large group of proteins.[49] A recent study demonstrated that cGMP binds to the regulatory GAF-A domain with high affinity, which led to substantial activation of PDE5 catalytic activity independent of phosphorylation at Ser92.[50] Because it has been shown that cGMP binding to the GAF domain is necessary for maximal phosphorylation of PDE5 by PKG *in vitro* and that phosphorylation increases the apparent affinity for cGMP binding,[46] it is hypothesized that phosphorylation of PDE5 stabilizes the cGMP-bound (activated) state of PDE5. Nevertheless, these results strongly indicate that PDE5 has low intrinsic catalytic activity at basal states when cGMP levels are low, but when cGMP levels increase and interact with the regulatory domain of PDE5, a dramatic stimulation of PDE5 cGMP-hydrolyzing activity occurs. Activation of

PDE5 by cGMP leads to a rapid decline in intracellular cGMP. Indeed, PDE5 has been shown to be an important regulator of intracellular cGMP. For example, in platelets, NO donors induce a rapid elevation followed by a rapid decline of intracellular cGMP. This rapid decline of cGMP is caused by increased PDE5 activity that correlates with increased PDE5 phosphorylation.[51] Therefore, PDE5 functions as a negative feedback regulator of cGMP signaling (Figure 23.1). PDE5 inhibitors increase and prolong intracellular cGMP accumulation, which is more profound in the presence compared with absence of NO or NP stimulation because these substances increase PDE5 activity.

23.4 ROLES OF PHOSPHODIESTERASES IN THE REGULATION OF VASCULAR SMOOTH MUSCLE CELL FUNCTIONS

Under normal conditions, vascular SMCs residing in the media of vessels are quiescent and contractile, and function principally to maintain vascular tone.[13,52] If the vessel is injured biologically (lipidemia, diabetes, smoking, viral infection) or mechanically (angioplasty), vascular SMCs respond by changing their state from a highly differentiated contractile phenotype to a dedifferentiated synthetic one.[13] By downregulating contractile proteins and acquiring the capacity to proliferate, migrate, and produce extracellular matrix proteins, these synthetic SMCs contribute to neointimal formation.[53] Cyclic nucleotides regulate the functions of both contractile and synthetic SMCs, by stimulating relaxation and inhibiting growth. Here, we will focus on the roles of PDEs in the regulation of vascular smooth muscle tone and growth.

23.4.1 PHOSPHODIESTERASES AND VASCULAR SMOOTH MUSCLE TONE

23.4.1.1 Role of PDE1 in Vascular Smooth Muscle Tone

The PDE1 partially selective inhibitor, vinpocetine, increased cGMP levels and dilated pre-constricted rabbit and rat aortas in *ex vivo* organ culture.[22,27,54–56] These data suggest that PDE1 (preferentially PDE1A) isozymes in rabbit vascular smooth muscle in intact vessels are important in regulating cGMP signaling and smooth muscle relaxation. Most vasoconstrictors, such as NE, Ang II, and ET-1, increase intracellular Ca^{2+}, which is thought to be the major mechanism of vasoconstrictor-mediated smooth muscle contraction. cGMP is a negative regulator of intracellular Ca^{2+} elevation and vasoconstriction.[20] Therefore, it is logical that vasoconstrictors increase the activity of PDE1A via increased Ca^{2+}, which then decreases cGMP levels and promotes vasoconstriction (Figure 23.1). It has indeed been shown that Ca^{2+}/calmodulin-stimulated PDE activity in the vessel wall can be rapidly enhanced by vasoconstrictors *in vivo*.[28,29] Ca^{2+}/calmodulin-stimulated PDE activity represents the major enzymatic activity responsible for the hydrolysis of cGMP in rabbit aorta stimulated with vasoconstrictors like NE.[22]

23.4.1.2 Role of PDE5 in Vascular Smooth Muscle Tone

PDE5 has been found in almost all types of vascular and visceral SMCs. PDE5 specifically hydrolyzes cGMP, a critical mediator of vasodilators such as NO and NPs in SMCs. PDE5 inhibitors, by inhibiting the hydrolytic breakdown of cGMP, potentiate and prolong the action of cGMP. This results in augmented smooth muscle relaxation. The best documented physiological function of PDE5 is the regulation of smooth muscle tone, as seen by successful clinical use of the PDE5 inhibitor sildenafil (Viagra, Pfizer), in the treatment of erectile dysfunction. This syndrome is caused by impaired ability of the corpus cavernosum smooth

muscle to vasodilate. The erectile response is dependent on relaxation of the smooth muscle of the penile vasculature in response to NO release from ECs and noncholinergic nonadrenergic (NANC) neurons surrounding the penile artery and the arteries of the corpora cavernosa. PDE5 inhibitors facilitate a higher accumulation of cGMP in response to NO, thus enhancing the erectile response.[57,58]

PDE5 also represents the predominant cGMP-hydrolyzing PDE in the lung.[59] In agreement with the predominance of PDE5 in the lung, PDE5 inhibitors show dramatic effects in pulmonary arteries and are therefore a particularly promising long-term therapy for treating pulmonary arterial hypertension (PAH).[60,61] However, sildenafil has shown modest effects on reducing systemic blood pressure, which is generally clinically insignificant when taken alone. On the other hand, sildenafil is able to greatly potentiate the effects of NO-generating compounds. Concomitant use of sildenafil and NTG or other organic nitrates may lead to severe hypotension.[62] Therefore, sildenafil is contraindicated for most patients who also use organic nitrates.

The vasorelaxant effects of PDE5 inhibitors vary in different vascular beds, which may be due to differences in local NO release. For example, the beneficial effect of Viagra on erectile function is dependent on increased NO release from cavernous NANC nerves. In addition, as mentioned above, Viagra has modest vasorelaxant effects on systemic vasculature when used alone, but organic nitrates greatly potentiate vasorelaxation by Viagra. These observations are consistent with the fact that PDE5 has low intrinsic catalytic activity when cGMP levels are low, but has high activity when cGMP levels elevate and interact with the regulatory domain of PDE5. The different vasorelaxant effects of PDE5 inhibition could also be related to variable PDE5 expression and activity, as well as the relative contribution of PDE5 to the total amount of cGMP-hydrolyzing activity in different vascular beds. For example, PDE5 expression and activity in the lung are approximately equal to that of the penile corpus cavernosum.[59]

23.4.1.3   Role of PDE3 and PDE4 in Smooth Muscle Tone

It is well known that PDE3 inhibitors promote smooth muscle relaxation in various vascular smooth muscle preparations,[21] suggesting the involvement of PDE3 in the regulation of vascular smooth muscle contractility. The effects of PDE3 inhibitors on smooth muscle relaxation are not dependent on the function of the endothelium,[63,64] which is different from PDE4 inhibitors (discussed below). *in vivo*, PDE3 inhibitors also show powerful vaso-dilatory effects.[21] For example, milrinone (a selective PDE3 inhibitor), in addition to its well-known positive inotropic effects on the heart, also reduces total peripheral resistance, enhances coronary blood flow, and reduces pulmonary vascular resistance. Both the cardiac inotropic and vasodilatory properties of milrinone contribute to its effectiveness in the acute treatment of congestive heart failure. Cilostazol (another PDE3 inhibitor), via its vasodilatory and antithrombotic effects, is an effective drug for treating chronic peripheral arterial occlusions. It is generally believed that the main mechanism by which PDE3 inhibitors relax vascular smooth muscle is through attenuating cAMP hydrolysis, thus activating the cAMP-signaling pathway. Similar to their effects on cAMP elevation (discussed above), PDE3 inhibitors have also shown synergistic effects on vascular relaxation with AC activators such as forskolin[36,65] and isoproterenol.[37,66]

In contrast to PDE3 inhibitors, PDE4 inhibitors, when used alone, are relatively poor relaxants in isolated large arteries.[21] It has been found that PDE4 inhibitors relaxed rat aortic rings much better in the presence of a functional endothelium.[64] The relaxation in the presence of endothelium was inhibited by NG-monomethyl L-arginine (L-NMMA), an inhibitor of NO synthase, and restored by the NOS substrate, L-arginine, suggesting that NO was necessary

for the effect.[64] Moreover, the combination of PDE4 inhibitors and PDE3 inhibitors produced synergistic effects. These observations suggest that PDE3 and PDE4 regulate distinct, but overlapping, cAMP pools in vascular SMCs.

It appears that the effects of PDE3 and PDE4 on smooth muscle contractility vary among different vessels. For example, in the dog, the PDE3 inhibitors, milrinone and amrinone, were more potent relaxants of coronary arteries than cerebral or renal arteries.[67] However, PDE4 has been reported to represent the predominant PDE that regulates vascular tone mediated by cAMP hydrolysis in cerebral vessels. For example, Willette et al.[68] found that in a canine model of acute cerebral vasospasm, PDE4 inhibitors such as denbufylline and rolipram, potently and completely reversed the basilar artery spasm produced by autologous blood without altering mean arterial blood pressure. In contrast, siguazodan (a PDE3 inhibitor) produced only weak relaxation of the basilar artery.[68] The favorable effect of PDE4 in cerebral vasculature may be relevant to the finding that the gene encoding PDE4D confers increased risk of ischemic stroke.[69] In view of their differential effects on different blood vessels, the *in vivo* effects of PDE3 and PDE4 inhibitors conceivably depend on the combined effects of the type of vascular bed, interaction with endogenous stimulators of AC/cAMP, as well as the interaction with endogenous GC/cGMP signaling.

23.4.2 PDEs and Vascular Smooth Muscle Growth

23.4.2.1 Role of PDE1 in Vascular Smooth Muscle Growth

In human SMCs, Rybalkin et al.[26] found that PDE1C was differentially expressed in synthetic and contractile SMCs. PDE1C expression was very low or absent in quiescent contractile smooth muscle located in the medial layer of intact adult human aorta, but expression was high in smooth muscle of human fetal aorta that contains a large number of proliferating cells. PDE1C was also markedly induced in cultured, growing human SMCs.[26] Induction of PDE1C in synthetic SMCs appears to be restricted to humans because it was not detected in several other species that were examined.[26] PDE1C expression in human cultured SMC is regulated by the cell cycle. For example, PDE1C is upregulated when SMCs enter the cell cycle (G_1–S phase), but is downregulated when SMCs exit the cell cycle (G_o phase) or become completely quiescent.[70] These observations suggest that induction of PDE1C may be required for cell-cycle progression and cell proliferation. The role of PDE1C in human SMCs proliferation was then determined by using the PDE inhibitor 8-methoxymethyl-IBMX (8MM-IBMX), at concentrations that selectively inhibit PDE1 family isozymes.[70] Moreover, antisense oligonucleotides that specifically target PDE1C were used to further verify that inhibition of PDE1C function attenuated SMC proliferation.[70] These results strongly indicate that induction of PDE1C is necessary for human arterial SMCs to grow, probably via lowering of cAMP and cGMP levels that block SMC growth. The precise molecular mechanism by which induction of PDE1C effects cell-cycle regulation deserves further investigation.

In contrast to PDE1C, PDE1A is expressed in SMCs from various species both *in vitro* and *in vivo*. PDE1A in normal, intact aortas is expressed predominantly in the cytoplasm of SMCs,[71] which is consistent with its functional role in the regulation of vascular tone.[27] We have recently studied a critical role for nuclear-localized PDE1A in vascular SMC proliferation and growth (Figure 23.2).[71] Using several models of vascular remodeling, primary cultured rat aortic SMCs, and human coronary artery SMCs, where we could manipulate growth and differentiated phenotypes, we found that PDE1A is predominantly located in the cytoplasm of contractile vascular SMCs, but is mostly nuclear in synthetic vascular SMCs. Importantly, we saw this phenomenon in three different species, indicating

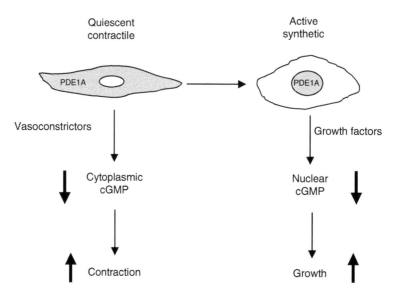

FIGURE 23.2 Potential roles of PDE1A in the regulation of vascular SMC function. PDE1A localizations are represented by gray shading, which are distinct between vascular SMCs with different phenotypes. PDE1A is expressed predominantly in the cytoplasm of vascular SMCs with quiescent, contractile phenotypes. Cytoplasmic PDE1A, in quiescent vascular SMCs, is principally responsible for regulating smooth muscle contractility. For example, vasoconstrictors stimulate cytoplasmic PDE1A activity by increasing intracellular Ca^{2+} concentration. Activation of cytoplasmic PDE1A decreases cytoplasmic cGMP, a negative regulator of vasoconstriction, and thereby promotes vasoconstriction. However, the subcellular location of PDE1A in active and synthetic vascular SMCs is predominantly nuclear. This nuclear form of PDE1A is critical for SMC growth and survival. For example, activation of nuclear PDE1A (by growth factors) decreases nuclear cGMP, thus removing an inhibitor of cell growth, and promotes SMC proliferation.

that the change in subcellular location is not species-specific. Moreover, when nuclear PDE1A expression was blocked via small interference RNA or activity was inhibited using a PDE1 inhibitor, we observed increased cell-cycle arrest and apoptosis, concurrent with increased intracellular cGMP. Furthermore, we demonstrated that in subcultured SMCs that are redifferentiated by growth on collagen gels, cytoplasmic PDE1A regulates myosin light-chain phosphorylation with little effect on apoptosis whereas nuclear PDE1A regulates cell apoptosis without a significant effect on myosin light-chain phosphorylation.[71] These observations suggest that cytoplasmic PDE1A in the contractile SMCs of normal vessels may function as a regulator of smooth muscle contractility whereas nuclear PDE1A may play a critical role in regulating synthetic SMC growth and neointima formation (Figure 23.2).

These findings suggest that PDE1A and PDE1C are both important in proliferating vascular SMCs. However, the molecular mechanisms by which PDE1A and PDE1C regulate vascular SMC proliferation are likely different. For example, using immunohistochemistry, we found that PDE1A and PDE1C are expressed in the same human atherosclerotic lesions. However, PDE1A and PDE1C exhibited different subcellular locations; PDE1A was expressed mainly in the nucleus whereas PDE1C was expressed in the perinuclear region and cytoplasm (Nagel et al., unpublished data, 2005). In addition, PDE1A and PDE1C may regulate different types of cyclic nucleotide in SMCs. For example, in rat aortic SMCs, elevated levels of cGMP but not cAMP were detected after inhibiting PDE1 (primarily PDE1A) with vinpocetine or IC86340, and downregulating PDE1A expression via small

RNA interference.[27,71] This suggests that PDE1A preferentially regulates cGMP in vascular SMCs. In human SMCs where both PDE1A and PDE1C exist, inhibition of PDE1 activity with 8-MM-IBMX caused an increase in both cAMP and cGMP, suggesting that PDE1C is capable of regulating cAMP in vascular SMCs.[70] Although cAMP and cGMP are both known to inhibit SMC proliferation,[10] it has been shown that cAMP and cGMP analogs differentially regulate cell-cycle progression.[72] These observations suggest that PDE1A and PDE1C play distinct roles in regulating vascular SMC cAMP and cGMP. The mechanisms by which PDE1A and PDE1C regulate cell growth in human cells are likely to be different; however, this remains to be determined. Attenuating the contribution of PDE1A and PDE1C to pathological vascular SMC proliferation may be helpful in reducing the neointimal formation observed in restenosis and atherosclerosis.

23.4.2.2 Role of PDE5 in Smooth Muscle Cell Growth

Endothelium-derived NO plays a critical role in attenuating SMC hyperplasia. Abnormal endothelial responses can lead to reduced bioavailability of NO. The role of PDE5 in NO/cGMP signaling might be extended to other abnormal vascular features associated with changes in cGMP signaling due to endothelial dysfunction, such as SMC hyperplasia. Most studies of PDE5 function on SMC growth have used cultured SMCs isolated from different sources. For example, it has been shown that in human pulmonary artery smooth muscle cells (PMSMCs) where 80% of cGMP hydrolysis is attributed to PDE5, inhibiting PDE5 with sildenafil significantly elevates cGMP and attenuates PMSMC growth. Sildenafil showed synergistic effects when given in the presence of NO donors and a GC activator BAY41-2272.[73] Similarly, in bovine coronary artery SMCs, sildenafil reduced platelet-derived growth factor (PDGF)-induced DNA synthesis and acted synergistically with NO donors.[17]

Recently, we found that Ang II upregulated PDE5 expression in rat aortic SMCs.[74] Ang II-mediated inhibition of cGMP signaling and SMC growth was blocked when PDE5 activity was decreased by selective PDE5 inhibitors, suggesting that upregulation of PDE5A expression is an important mechanism for Ang II antagonism of cGMP signaling and stimulation of SMC growth.[74] However, it remains to be determined if chronic use of PDE5 inhibitors can improve vascular remodeling by inhibiting SMC growth *in vivo*. Reduced bioavailability of NO also affects platelet aggregation and vascular wall inflammation, which may contribute to vascular diseases and are potentially regulated by PDE5.

23.4.2.3 Role of PDE3 and PDE4 in Vascular Smooth Muscle Growth

Recent evidence indicates that the antimitotic effects of cGMP-elevating agents are mediated by cGMP-dependent inhibition of PDE3, a subsequent increase in cAMP, and activation of PKA.[9,17] PDE3 inhibitors have been shown to inhibit DNA synthesis in cultured pig aortic SMCs *in vitro*.[75] Treatment with the PDE3 inhibitor, cilostazol, increased intracellular concentrations of cAMP, and inhibited DNA synthesis and cell growth (stimulated by various growth factors) in cultured rat aortic SMCs.[76,77] The effectiveness of PDE3 inhibition was also confirmed in animal restenosis models *in vivo*. For example, in a mouse model of photochemically induced vascular injury, oral administration of milrinone (a selective PDE3 inhibitor) for 3–21 days at 0.3–30 mg/kg suppressed intimal thickening by up to 56%, in a dose- and time-dependent manner.[78] In addition, long-term oral administration of a different PDE3 inhibitor (cilostamide) also suppressed arterial intimal hyperplasia by 83% in a rat balloon injury model.[77] These results suggest that PDE3 inhibitors may be applicable in preventing intimal hyperplasia following procedures such as percutaneous transluminal coronary angioplasty and intraluminal stent placement. In addition synergistic effects of

PDE3 inhibitors with other cAMP-elevating agents or PDE4 inhibitors have been observed to inhibit SMC growth. For example, selective inhibition of PDE3 potentiated the antiproliferative effects of forskolin, a direct activator of AC.[75]

Similar to the effects on vascular relaxation in isolated large arteries, PDE4 inhibition by itself had poor effects on attenuating cultured synthetic SMC proliferation. Combining cAMP-elevating agents or PDE3 inhibitors with PDE4 inhibitors elicited synergistic effects on inhibiting synthetic SMC migration.[79] These data are consistent with the concept that PDE4 inhibitors (when used alone) produce a minor increase in cAMP in cultured SMC, but are much more effective when combined with another cAMP- or cGMP-elevating stimulus.

23.5 VASCULAR DISEASE-ASSOCIATED ALTERATION OF PHOSPHODIESTERASES AND THE THERAPEUTIC POTENTIAL OF PHOSPHODIESTERASE INHIBITORS

Cyclic nucleotide degradation by PDEs is a highly regulated process that is controlled by different mechanisms in different physiological and pathological circumstances. Alteration of PDE activity and expression may lead to pathological consequences, which may be reversed by PDE activators and inhibitors. In addition, therapeutically targeting PDE might be useful for pathological conditions that involve alteration of other molecules in the cAMP- and cGMP-signaling pathway.

23.5.1 NITRATE TOLERANCE

When given in the short term, NTG (an NO donor) has potent vasodilator capacities. The efficacy of chronic NTG administration, however, is limited by the rapid development of tolerance (called nitrate tolerance).[80] Nitrate tolerance involves decreased vascular sensitivity and diminished cGMP elevation in vascular SMCs in response to continued nitrate treatment. Several mechanisms have been proposed to account for nitrate tolerance, including neurohumoral counterregulation[81] and mechanisms intrinsic to the vascular tissue itself, such as downregulation of NTG biotransformation and desensitized response to NO/cGMP signaling.[82] Chronic NTG treatment is also associated with an increase in sensitivity to vasoconstrictors such as catecholamines, Ang II, and ET-1.[83] All of these mechanisms may compromise the vasodilatory effects of NTG, thereby contributing to tolerance.

The possibility that one or more PDEs might play a role in nitrate tolerance development was first suggested by studies showing that nitrate tolerance in animals was partially reversed by zaprinast, an inhibitor of both PDE1 and PDE5.[84,85] It therefore seemed possible that the effect of zaprinast on the reversal of nitrate tolerance may be due to inhibition of both PDE1 and PDE5 in vascular SMC. Since inhibiting PDE5 potentiates the effects of organic nitrates, downregulating PDE5 may prevent nitrate tolerance although PDE5 may not be the primary molecule altered in the setting of nitrate tolerance.

The potential importance of vascular PDE1A (a Ca^{2+}/calmodulin-stimulated PDE) in nitrate tolerance is supported by our recent findings of PDE1A1 upregulation in a rat model of nitrate tolerance.[27] We found that continuous NTG infusion for 3 d induced nitrate tolerance, increased PDE1A expression and activity in rat aortas, but did not change PDE5A levels.[27] A PDE1 partially selective inhibitor, vinpocetine, partially restored the sensitivity of tolerant vessels to subsequent NTG exposure.[27] An increase in PDE1A would reduce cGMP accumulation and thereby decrease sensitivity of the vasculature to subsequent NTG (Figure 23.3). An induction of PDE1A would also explain the phenomenon of supersensitivity to vasoconstrictors in NTG-tolerant vessels because most vasoconstrictors, such as Ang II, NE, and ET-1, increase intracellular Ca^{2+} concentration and rapidly stimulate PDE1A activity.

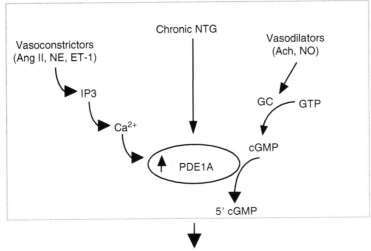

FIGURE 23.3 Potential role of PDE1A in nitrate tolerance. Endothelium-dependent and independent nitrovasodilators such as acetylcholine (Ach) and nitroglycerin (NTG) activate the NO-sensitive (largely soluble) guanylyl cyclase (sGC), which stimulates cGMP formation. Accumulation of cGMP leads to vasodilation by decreasing intracellular Ca^{2+} concentration. Vasoconstrictors such as angiotensin II (Ang II), norepinephrine (NE), and endothelin (ET)-1 cause an increase in intracellular Ca^{2+}, leading to vasoconstriction. cGMP is a key player in regulating vascular tone because it controls Ca^{2+} levels, which regulate vascular contractility. An increase in PDE1A activity by chronic NTG treatment leads to reduced cGMP accumulation in response to further stimulation with vasodilators, thereby causing decreased sensitivity of the vasculature to NTG. Increased expression of PDE1A also increases the magnitude of vasoconstrictor-mediated attenuation of cGMP accumulation, which may contribute to the phenomenon of supersensitivity to vasoconstrictors in NTG-treated vessels. (Modified from Ref. [27]. With permission.)

Thus, upregulated PDE1A expression would increase the magnitude of vasoconstrictor-mediated attenuation of cGMP accumulation, and increase sensitivity to vasoconstrictors (Figure 23.3). These findings suggest that upregulation of PDE1A may be one of the mechanisms involved in the development of nitrate tolerance and that PDE1A inhibition could be a novel therapeutic tool to limit nitrate tolerance. However, the exact roles of PDE1A in human nitrate-tolerant vessels remain to be determined.

23.5.2 PULMONARY HYPERTENSION

PAH is characterized by a progressive increase in pulmonary vascular resistance. PAH is a multifactorial process involving pulmonary vasoconstriction and vascular remodeling, and eventually leads to right ventricular failure and death. Pulmonary vascular tone and remodeling is regulated by a balance between the effects of vasodilators/antiproliferative agents such as NO and PGI2, and vasoconstrictors/mitogens such as serotonin and ET-1.[86] Because NO/cGMP plays a critical role in the regulation of pulmonary vascular relaxation, modulation of NO/cGMP signaling has been proposed as a potential therapeutic target. NO inhalation exerts beneficial effects in patients at various stages of PAH.[87] However, due to the short half-life of NO, multiple administrations are required for NO therapy, and long-term results have been disappointing.[88] Another strategy for increasing NO-dependent cGMP

signaling is through inhibition of cGMP breakdown by PDE5 inhibitors because PDE5 activity is the major cGMP-hydrolyzing PDE in the pulmonary vasculature. As expected, PDE5 inhibitors elicit extensive vasorelaxant effects on pulmonary vasculature.[61] In clinical studies of humans with severe primary pulmonary hypertension, oral administration of sildenafil showed a drastic decrease in pulmonary arterial pressure and reduced pulmonary vascular resistance.[60,61] Sildenafil also potentiated and prolonged the effects of other vasodilators such as NO inhalation when these treatments were combined.[89] Therefore, because sildenafil is taken by oral administration and has a favorable side-effect profile, it is particularly promising as a long-term therapy for the treatment of PAH.

There are several observations that suggest that PDE1, PDE3, PDE4, and PDE5 may play important roles in the development of pulmonary hypertension. For example, PDE activity is increased in different pulmonary arteries from rats with chronic hypoxia-induced pulmonary hypertension.[90] Increased cAMP–PDE activity, associated with increased PDE3 activity, was found in the main, first branch, and intrapulmonary arteries of hypoxic rats, but no changes were observed in resistance arteries.[90] Increased amounts of total cGMP–PDE were due to increased PDE1 activity in the main pulmonary artery and PDE5 activity in first branch and intrapulmonary arteries.[90] These changes in PDE activities are associated with decreased intracellular cAMP and cGMP levels. Moreover, increased PDE3 mRNA expression was found in pulmonary artery rings obtained from Spraque–Dawley rats suffering from hypoxia-induced pulmonary hypertension.[91] In addition, inhibiting PDE3 and PDE4 activities can significantly improve pulmonary artery relaxation in hypoxia-induced pulmonary hypertension. These findings suggest that multiple PDEs play important roles in the development and maintenance of hypoxia-induced pulmonary hypertension.

23.5.3 ATHEROSCLEROSIS AND RESTENOSIS

As mentioned earlier, when endothelium in the vessel wall malfunctions or is injured, vascular SMCs respond by changing from the quiescent contractile to the active synthetic phenotype.[13] Synthetic vascular SMCs contribute to cardiovascular proliferative disorders such as atherosclerosis, postangioplasty restenosis, bypass vein graft failure, and cardiac allograft vasculopathy.[92,93] Since cyclic nucleotides regulate the proliferative response of SMCs, altered PDE activity may create abnormal cAMP and cGMP levels and contribute to vascular proliferative disorders. For example, we found that PDE1A accumulated in the nuclei of synthetic SMCs *in vitro* and *in vivo*, and that inhibition of nuclear PDE1A function attenuates SMC growth and promotes SMC apoptosis.[71] These observations suggest that induction of nuclear PDE1A is essential for SMC proliferation and survival and may contribute to neointimal formation in restenosis and atherosclerosis. In addition, PDE1C has been found to be specifically expressed in human synthetic SMCs, but is not detectable in human contractile SMCs.[26] Inhibition of PDE1C function resulted in suppressed proliferation of cultured human SMCs.[70] Given that PDE1C is absent or very low in contractile SMCs, it is plausible that PDE1C inhibitors may selectively target proliferating SMCs in atherosclerotic lesions and reduce neointimal formation during restenosis in human patients.

NO is a well-known antiatherosclerotic molecule, which acts in part by inhibiting SMC proliferation and migration, and partly by reducing inflammatory responses. For example, in animals with experimentally induced atherosclerosis, NO synthase inhibitors accelerate[94,95] whereas L-arginine[96] and overexpression of endothelial NO synthase (eNOS)[97] attenuate the development of atherosclerotic lesions. Deletion of the eNOS gene on apolipoprotein $E^{-/-}$ (ApoE$^{-/-}$)-deficient background results in hypertension and increased atherosclerosis.[98] It has been shown that the inhibitory effect of NO in vascular SMC mitogenesis is mediated by cGMP-dependent inhibition of PDE3 activity, increased cAMP level, and activation of

PKA.[17] A similar mechanism has also been shown to be responsible for the inhibitory effects of NO/cGMP on inflammatory molecule expression in vascular SMCs.[9] This is corroborated by evidence that PDE3 inhibitors not only inhibited cultured SMC growth *in vitro*,[75–77] but also significantly attenuated neointima formation *in vivo* in animal models of vascular injury.[77,78] The PDE3 inhibitor, cilostazol, is an effective drug for the clinical treatment of symptoms of intermittent claudication caused by peripheral arterial disease, which itself is believed to be a manifestation of systemic atherosclerosis.[99] Furthermore, it has been reported that PDE3 activity and expression are significantly increased in aortas of atherosclerosis-prone insulin-resistant rats.[100] These findings support the important role of PDE3 in vascular diseases such as atherosclerosis and restenosis and suggest that PDE3 inhibitors may be therapeutically useful for the treatment and prevention of atherosclerosis and restenosis.

23.6   CONCLUSION AND PERSPECTIVE

Vascular SMCs express multiple PDEs that are structurally and functionally different. By concertedly and differentially regulating the amplitude, duration, and compartmentation of cAMP and cGMP, different PDE isoforms play critical roles in modifying vascular smooth muscle function. Most of our knowledge regarding PDE regulation and function in vascular SMC is limited to four PDE family members (PDE1, PDE3, PDE4, and PDE5). This is largely due to the availability of research tools for these families of enzymes, such as family-specific antibodies and family-selective inhibitors. It is very likely that vascular SMCs also express members of more recently identified PDE families (PDE7–PDE11). Molecular biology tools such as antisense and small RNA interference may be useful for exploring the functions of these PDE isoforms in vascular SMCs.

The precise balance of cyclic nucleotide levels is critical for achieving and maintaining normal physiological function of vascular smooth muscle. Alterations of PDE expression and activity disturb the balance of cAMP and cGMP, which may lead to pathological consequences and the development of vascular diseases. Most successful clinical therapies that target PDEs utilize PDE inhibitors to reduce PDE activities and increase cyclic nucleotide levels. The best known example is the use of PDE5 inhibitors to treat erectile dysfunction and pulmonary hypertension. Since NO/cGMP signaling is compromised in the settings of erectile dysfunction and pulmonary hypertension, PDE5 inhibition is frequently capable of restoring NO/cGMP signaling. Although PDE activators have not been extensively used, they may also be useful for pathological circumstances in which PDE expression or activity is downregulated.

It is worth noting that decreasing PDE expression or activity may actually be pathogenic. For example, in frog cardiac cells, cAMP compartmentation is required for local activation of cardiac Ca^{2+} channels by β-adrenergic agonists. PDE inhibition caused a reduction in cAMP-mediated localized effects due to disruption of proper cAMP subcellular localization.[101] It has also been found that increases of cAMP in specific compartments upon β-AR stimulation are critical for specific activation of a subset of PKA molecules that are anchored in proximity to the transverse tubule.[5] PDE inhibition dissipates cAMP gradients, leading to imprecise activation of PKA throughout the cell.[5] Moreover, we have recently found that PDE3A expression was significantly downregulated in failing hearts, and that chronic inhibition of PDE3 activity by PDE3 inhibitors induced cardiomyocyte apoptosis.[4] Furthermore, depletion of the PDE4D gene in mice caused abnormal PKA phosphorylation of the cardiac ryanodine receptor (RyR2), Ca^{2+} leakage, and cardiac dysfunction.[102] These observations may provide an explanation for the adverse effects seen with chronic use of PDE3 inhibitors in patients with heart failure, and may affect therapeutic strategies using PDE4 inhibitors for

the treatment of asthma and stroke. Together, these findings strongly suggest that down-regulation of PDE expression in disease states, or chronic inhibition of PDE activity, may result in dispersed cyclic nucleotide responses that attenuate the desired physiological function and enhance undesired toxic effects by broadly activating cellular signaling molecules. Therefore, while we believe that PDE inhibitors may have substantial therapeutic potential for various vascular diseases, caution should be exercised because of toxic side effects that may occur when PDE activity that is vital for normal physiologic function is inhibited.

REFERENCES

1. Soderling, S.H., and J.A. Beavo. 2000. Regulation of cAMP and cGMP signaling: New phosphodiesterases and new functions. *Curr Opin Cell Biol* 12:174.
2. Chini, C.C., et al. 1997. Compartmentalization of cAMP signaling in mesangial cells by phosphodiesterase isozymes PDE3 and PDE4. Regulation of superoxidation and mitogenesis. *J Biol Chem* 272:9854.
3. Cheng, J., et al. 2004. Differential regulation of mesangial cell mitogenesis by cAMP phosphodiesterase isozymes 3 and 4. *Am J Physiol Renal Physiol* 287:F940.
4. Ding, B., et al. 2005. Functional role of phosphodiesterase 3 in cardiomyocyte apoptosis: Implication in heart failure. *Circulation* 111:2469.
5. Zaccolo, M., and T. Pozzan. 2002. Discrete microdomains with high concentration of cAMP in stimulated rat neonatal cardiac myocytes. *Science* 295:1711.
6. Mongillo, M., et al. 2004. Fluorescence resonance energy transfer-based analysis of cAMP dynamics in live neonatal rat cardiac myocytes reveals distinct functions of compartmentalized phosphodiesterases. *Circ Res* 95:67.
7. Suga, S., et al. 1992. Endothelial production of C-type natriuretic peptide and its marked augmentation by transforming growth factor-beta. Possible existence of vascular natriuretic peptide system. *J Clin Invest* 90:1145.
8. Komatsu, Y., et al. 1992. Vascular natriuretic peptide. *Lancet* 340:622.
9. Aizawa, T., et al. 2003. Role of phosphodiesterase 3 in NO/cGMP-mediated antiinflammatory effects in vascular smooth muscle cells. *Circ Res* 93:406.
10. Koyama, H., et al. 2001. Molecular pathways of cyclic nucleotide-induced inhibition of arterial smooth muscle cell proliferation. *J Cell Physiol* 186:1.
11. Rybalkin, S.D., and K.E. Bornfeldt. 1999. Cyclic nucleotide phosphodiesterases and human arterial smooth muscle cell proliferation. *Thromb Haemost* 82:424.
12. Lloyd-Jones, D.M., and K.D. Bloch. 1996. The vascular biology of nitric oxide and its role in atherogenesis. *Annu Rev Med* 47:365.
13. Ross, R. 1993. The pathogenesis of atherosclerosis: A perspective for the 1990s. *Nature* 362:801.
14. Pham, N., et al. 2000. The guanine nucleotide exchange factor CNrasGEF activates ras in response to cAMP and cGMP. *Curr Biol* 10:555.
15. Murad, F. 1996. The 1996 Albert Lasker Medical Research Awards. Signal transduction using nitric oxide and cyclic guanosine monophosphate. *JAMA* 276:1189.
16. Vandecasteele, G., et al. 2001. Cyclic GMP regulation of the L-type Ca^{2+} channel current in human atrial myocytes. *J Physiol* 533:329.
17. Osinski, M.T., et al. 2001. Antimitogenic actions of organic nitrates are potentiated by sildenafil and mediated via activation of protein kinase A. *Mol Pharmacol* 59:1044.
18. Sausbier, M., et al. 2000. Mechanisms of NO/cGMP-dependent vasorelaxation. *Circ Res* 87:825.
19. Hofmann, F., et al. 2000. Rising behind NO: cGMP-dependent protein kinases. *J Cell Sci* 113:1671.
20. Yan, C., et al. 2003. Functional interplay between angiotensin II and nitric oxide: Cyclic GMP as a key mediator. *Arterioscler Thromb Vasc Biol* 23:26.
21. Polson, J.B., and S.J. Strada. 1996. Cyclic nucleotide phosphodiesterases and vascular smooth muscle. *Annu Rev Pharmacol Toxicol* 36:403.

22. Chiu, P.J., et al. 1988. Comparative effects of vinpocetine and 8-Br-cyclic GMP on the contraction and 45Ca-fluxes in the rabbit aorta. *Am J Hypertens* 1:262.

23. Beavo, J.A. 1995. Cyclic nucleotide phosphodiesterases: Functional implications of multiple isoforms. *Physiol Rev* 75:725.

24. Sonnenburg, W.K., et al. 1995. Identification of inhibitory and calmodulin-binding domains of the PDE1A1 and PDE1A2 calmodulin-stimulated cyclic nucleotide phosphodiesterases. *J Biol Chem* 270:30989.

25. Loughney, K., et al. 1996. Isolation and characterization of cDNAs corresponding to two human calcium, calmodulin-regulated, 3′,5′-cyclic nucleotide phosphodiesterases. *J Biol Chem* 271:796.

26. Rybalkin, S.D., et al. 1997. Calmodulin-stimulated cyclic nucleotide phosphodiesterase (PDE1C) is induced in human arterial smooth muscle cells of the synthetic, proliferative phenotype. *J Clin Invest* 100:2611.

27. Kim, D., et al. 2001. Upregulation of phosphodiesterase 1A1 expression is associated with the development of nitrate tolerance. *Circulation* 104:2338.

28. Jaiswal, R.K. 1992. Endothelin inhibits the atrial natriuretic factor stimulated cGMP production by activating the protein kinase C in rat aortic smooth muscle cells. *Biochem Biophys Res Commun* 182:395.

29. Molina, C.R., et al. 1987. Effect of *in vivo* nitroglycerin therapy on endothelium-dependent and independent vascular relaxation and cyclic GMP accumulation in rat aorta. *J Cardiovasc Pharmacol* 10:371.

30. Leroy, M.J., et al. 1996. Characterization of two recombinant PDE3 (cGMP-inhibited cyclic nucleotide phosphodiesterase) isoforms, RcGIP1 and HcGIP2, expressed in NIH 3006 murine fibroblasts and Sf9 insect cells. *Biochemistry* 35:10194.

31. Choi, Y.H., et al. 2001. Identification of a novel isoform of the cyclic-nucleotide phosphodiesterase PDE3A expressed in vascular smooth-muscle myocytes. *Biochem J* 353:41.

32. Palmer, D., and D.H. Maurice. 2000. Dual expression and differential regulation of phosphodiesterase 3A and phosphodiesterase 3B in human vascular smooth muscle: Implications for phosphodiesterase 3 inhibition in human cardiovascular tissues. *Mol Pharmacol* 58:247.

33. Liu, H., and D.H. Maurice. 1998. Expression of cyclic GMP-inhibited phosphodiesterases 3A and 3B (PDE3A and PDE3B) in rat tissues: Differential subcellular localization and regulated expression by cyclic AMP. *Br J Pharmacol* 125:1501.

34. Dunkerley, H.A., et al. 2002. Reduced phosphodiesterase 3 activity and phosphodiesterase 3A level in synthetic vascular smooth muscle cells: Implications for use of phosphodiesterase 3 inhibitors in cardiovascular tissues. *Mol Pharmacol* 61:1033.

35. Maurice, D.H., and R.J. Haslam. 1990. Nitroprusside enhances isoprenaline-induced increases in cAMP in rat aortic smooth muscle. *Eur J Pharmacol* 191:471.

36. Tanaka, T., et al. 1988. Effects of cilostazol, a selective cAMP phosphodiesterase inhibitor on the contraction of vascular smooth muscle. *Pharmacology* 36:313.

37. Lindgren, S.H., et al. 1990. Effects of isozyme-selective phosphodiesterase inhibitors on rat aorta and human platelets: Smooth muscle tone, platelet aggregation and cAMP levels. *Acta Physiol Scand* 140:209.

38. Conti, M., et al. 2003. Cyclic AMP-specific PDE4 phosphodiesterases as critical components of cyclic AMP signaling. *J Biol Chem* 278:5493.

39. Liu, H., and D.H. Maurice. 1999. Phosphorylation-mediated activation and translocation of the cyclic AMP-specific phosphodiesterase PDE4D3 by cyclic AMP-dependent protein kinase and mitogen-activated protein kinases. A potential mechanism allowing for the coordinated regulation of PDE4D activity and targeting. *J Biol Chem* 274:10557.

40. Liu, H., et al. 2000. Expression of phosphodiesterase 4D (PDE4D) is regulated by both the cyclic AMP-dependent protein kinase and mitogen-activated protein kinase signaling pathways. A potential mechanism allowing for the coordinated regulation of PDE4D activity and expression in cells. *J Biol Chem* 275:26615.

41. Tilley, D.G., and D.H. Maurice. 2002. Vascular smooth muscle cell phosphodiesterase (PDE) 3 and PDE4 activities and levels are regulated by cyclic AMP *in vivo*. *Mol Pharmacol* 62:497.

42. Rybalkin, S.D., et al. 2003. Cyclic GMP phosphodiesterases and regulation of smooth muscle function. *Circ Res* 93:280.

43. Loughney, K., et al. 1998. Isolation and characterization of cDNAs encoding PDE5A, a human cGMP-binding, cGMP-specific 3′,5′-cyclic nucleotide phosphodiesterase. *Gene* 216:139.

44. Stacey, P., et al. 1998. Molecular cloning and expression of human cGMP-binding cGMP-specific phosphodiesterase (PDE5). *Biochem Biophys Res Commun* 247:249.

45. Yanaka, N., et al. 1998. Expression, structure and chromosomal localization of the human cGMP-binding cGMP-specific phosphodiesterase PDE5A gene. *Eur J Biochem* 255:391.

46. Corbin, J.D., et al. 2000. Phosphorylation of phosphodiesterase-5 by cyclic nucleotide-dependent protein kinase alters its catalytic and allosteric cGMP-binding activities. *Eur J Biochem* 267:2760.

47. Thomas, M.K., et al. 1990. Characterization of a purified bovine lung cGMP-binding cGMP phosphodiesterase. *J Biol Chem* 265:14964.

48. Rybalkin, S.D., et al. 2002. Regulation of cGMP-specific phosphodiesterase (PDE5) phosphorylation in smooth muscle cells. *J Biol Chem* 277:3310.

49. Aravind, L., and C.P. Ponting. 1997. The GAF domain: An evolutionary link between diverse phototransducing proteins. *Trends Biochem Sci* 22:458.

50. Rybalkin, S.D., et al. 2003. PDE5 is converted to an activated state upon cGMP binding to the GAF A domain. *EMBO J* 22:469.

51. Mullershausen, F., et al. 2001. Rapid nitric oxide-induced desensitization of the cGMP response is caused by increased activity of phosphodiesterase type 5 paralleled by phosphorylation of the enzyme. *J Cell Biol* 155:271.

52. Owens, G.K. 1995. Regulation of differentiation of vascular smooth muscle cells. *Physiol Rev* 75:487.

53. Lincoln, T.M., et al. 2001. Invited review: cGMP-dependent protein kinase signaling mechanisms in smooth muscle: From the regulation of tone to gene expression. *J Appl Physiol* 91:1421.

54. Hagiwara, M., et al. 1984. Effects of vinpocetine on cyclic nucleotide metabolism in vascular smooth muscle. *Biochem Pharmacol* 33:453.

55. Ahn, H.S., et al. 1989. Effects of selective inhibitors on cyclic nucleotide phosphodiesterases of rabbit aorta. *Biochem Pharmacol* 38:3331.

56. Souness, J.E., et al. 1989. Role of selective cyclic GMP phosphodiesterase inhibition in the myorelaxant actions of M&B 22,948, MY-5445, vinpocetine and 1-methyl-3-isobutyl-8-(methylamino)xanthine. *Br J Pharmacol* 98:725.

57. Boolell, M., et al. 1996. Sildenafil, a novel effective oral therapy for male erectile dysfunction. *Br J Urol* 78:257.

58. Ballard, S.A., et al. 1998. Effects of sildenafil on the relaxation of human corpus cavernosum tissue *in vitro* and on the activities of cyclic nucleotide phosphodiesterase isozymes. *J Urol* 159:2164.

59. Corbin, J.D., et al. 2005. High lung PDE5: A strong basis for treating pulmonary hypertension with PDE5 inhibitors. *Biochem Biophys Res Commun* 334:930.

60. Prasad, S., et al. 2000. Sildenafil in primary pulmonary hypertension. *N Engl J Med* 343:1342.

61. Steiner, M.K., et al. 2005. Pulmonary hypertension: Inhaled nitric oxide, sildenafil and natriuretic peptides. *Curr Opin Pharmacol* 5:245.

62. Kloner, R.A. 2000. Cardiovascular risk and sildenafil. *Am J Cardiol* 86:57F.

63. Lugnier, C., and N. Komas. 1993. Modulation of vascular cyclic nucleotide phosphodiesterases by cyclic GMP: Role in vasodilatation. *Eur Heart J* 14 (Suppl. I):141.

64. Komas, N., et al. 1991. Endothelium-dependent and independent relaxation of the rat aorta by cyclic nucleotide phosphodiesterase inhibitors. *Br J Pharmacol* 104:495.

65. Lindgren, S., et al. 1991. Relaxant effects of the selective phosphodiesterase inhibitors milrinone and OPC 3911 on isolated human mesenteric vessels. *Pharmacol Toxicol* 64:440.

66. Maurice, D.H., et al. 1991. Synergistic actions of nitrovasodilators and isoprenaline on rat aortic smooth muscle. *Eur J Pharmacol* 192:235.

67. Harris, A.L., et al. 1989. Phosphodiesterase isozyme inhibition and the potentiation by zaprinast of endothelium-derived relaxing factor and guanylate cyclase stimulating agents in vascular smooth muscle. *J Pharmacol Exp Ther* 249:394.

68. Willette, R.N., et al. 1997. Identification, characterization, and functional role of phosphodiesterase type IV in cerebral vessels: Effects of selective phosphodiesterase inhibitors. *J Cereb Blood Flow Metab* 17:210.

69. Gretarsdottir, S., et al. 2003. The gene encoding phosphodiesterase 4D confers risk of ischemic stroke. *Nat Genet* 35:131.

70. Rybalkin, S.D., et al. 2002. Cyclic nucleotide phosphodiesterase 1C promotes human arterial smooth muscle cell proliferation. *Circ Res* 90:151.

71. Nagel, D.J., et al. 2006. Role of nuclear Ca^{2+}/calmodulin-stimulated phosphodiesterase 1A in vascular smooth muscle cell growth and survival. *Circ Res* 98 (6):777.

72. Fukumoto, S., et al. 1999. Distinct role of cAMP and cGMP in the cell cycle control of vascular smooth muscle cells: cGMP delays cell cycle transition through suppression of cyclin D1 and cyclin-dependent kinase 4 activation. *Circ Res* 85:985.

73. Wharton, J., et al. 2005. Antiproliferative effects of phosphodiesterase type 5 inhibition in human pulmonary artery cells. *Am J Respir Crit Care Med* 172:105.

74. Kim, D., et al. 2005. Angiotensin II increases phosphodiesterase 5A expression in vascular smooth muscle cells: A mechanism by which angiotensin II antagonizes cGMP signaling. *J Mol Cell Cardiol* 38:175.

75. Souness, J.E., et al. 1992. Inhibition of pig aortic smooth muscle cell DNA synthesis by selective type III and type IV cyclic AMP phosphodiesterase inhibitors. *Biochem Pharmacol* 44:857.

76. Takahashi, S., et al. 1992. Effect of cilostazol, a cyclic AMP phosphodiesterase inhibitor, on the proliferation of rat aortic smooth muscle cells in culture. *J Cardiovasc Pharmacol* 20:900.

77. Inoue, Y., et al. 2000. Suppression of arterial intimal hyperplasia by cilostamide, a cyclic nucleotide phosphodiesterase 3 inhibitor, in a rat balloon double-injury model. *Br J Pharmacol* 130:231.

78. Kondo, K., et al. 1999. Milrinone, a phosphodiesterase inhibitor, suppresses intimal thickening after photochemically induced endothelial injury in the mouse femoral artery. *Atherosclerosis* 142:133.

79. Palmer, D., et al. 1998. Synergistic inhibition of vascular smooth muscle cell migration by phosphodiesterase 3 and phosphodiesterase 4 inhibitors. *Circ Res* 82:852.

80. Munzel, T., et al. 1996. Dissociation of coronary vascular tolerance and neurohormonal adjustments during long-term nitroglycerin therapy in patients with stable coronary artery disease. *J Am Coll Cardiol* 27:297.

81. Packer, M., et al. 1987. Prevention and reversal of nitrate tolerance in patients with congestive heart failure. *N Engl J Med* 317:799.

82. Munzel, T., et al. 1996. New insights into mechanisms underlying nitrate tolerance. *Am J Cardiol* 77:24C.

83. Munzel, T., et al. 1995. Evidence for a role of endothelin 1 and protein kinase C in nitroglycerin tolerance. *Proc Natl Acad Sci USA* 92:5244.

84. De Garavilla, L., et al. 1996. Zaprinast, but not dipyridamole, reverses hemodynamic tolerance to nitroglycerin in vivo. *Eur J Pharmacol* 313:89.

85. Pagani, E.D., et al. 1993. Reversal of nitroglycerin tolerance *in vitro* by the cGMP-phosphodiesterase inhibitor zaprinast. *Eur J Pharmacol* 243:141.

86. Humbert, M., et al. 2004. Treatment of pulmonary arterial hypertension. *N Engl J Med* 351:1425.

87. Pepke-Zaba, J., et al. 1991. Inhaled nitric oxide as a cause of selective pulmonary vasodilatation in pulmonary hypertension. *Lancet* 338:1173.

88. Taylor, R.W., et al. 2004. Low-dose inhaled nitric oxide in patients with acute lung injury: A randomized controlled trial. *JAMA* 291:1603.

89. Michelakis, E., et al. 2002. Oral sildenafil is an effective and specific pulmonary vasodilator in patients with pulmonary arterial hypertension: Comparison with inhaled nitric oxide. *Circulation* 105:2398.

90. Maclean, M.R., et al. 1997. Phosphodiesterase isoforms in the pulmonary arterial circulation of the rat: Changes in pulmonary hypertension. *J Pharmacol Exp Ther* 283:619.

91. Wagner, R.S., et al. 1997. Phosphodiesterase inhibition improves agonist-induced relaxation of hypertensive pulmonary arteries. *J Pharmacol Exp Ther* 282:1650.

92. Ross, R. 1999. Atherosclerosis—an inflammatory disease. *N Engl J Med* 340:115.

93. Dzau, V.J., et al. 2002. Vascular proliferation and atherosclerosis: New perspectives and therapeutic strategies. *Nat Med* 8:1249.
94. Cayatte, A.J., et al. 1994. Chronic inhibition of nitric oxide production accelerates neointima formation and impairs endothelial function in hypercholesterolemic rabbits. *Arterioscler Thromb* 14:753.
95. Naruse, K., et al. 1994. Long-term inhibition of NO synthesis promotes atherosclerosis in the hypercholesterolemic rabbit thoracic aorta. PGH2 does not contribute to impaired endothelium-dependent relaxation. *Arterioscler Thromb* 14:746.
96. Creager, M.A., et al. 1992. L-arginine improves endothelium-dependent vasodilation in hypercholesterolemic humans. *J Clin Invest* 90:1248.
97. Kawashima, S., et al. 2001. Endothelial NO synthase overexpression inhibits lesion formation in mouse model of vascular remodeling. *Arterioscler Thromb Vasc Biol* 21:201.
98. Knowles, J.W., et al. 2000. Enhanced atherosclerosis and kidney dysfunction in eNOS(−/−) ApoE(−/−) mice are ameliorated by enalapril treatment. *J Clin Invest* 105:451.
99. Smith, J.A. 2002. Measuring treatment effects of cilostazol on clinical trial endpoints in patients with intermittent claudication. *Clin Cardiol* 25:91.
100. Nagaoka, T., et al. 1998. Cyclic nucleotide phosphodiesterase 3 expression in vivo: Evidence for tissue-specific expression of phosphodiesterase 3A or 3B mRNA and activity in the aorta and adipose tissue of atherosclerosis-prone insulin-resistant rats. *Diabetes* 47:1135.
101. Jurevicius, J., and R. Fischmeister. 1996. cAMP compartmentation is responsible for a local activation of cardiac Ca^{2+} channels by beta-adrenergic agonists. *Proc Natl Acad Sci USA* 93:295.
102. Lehnart, S.E., et al. 2005. Phosphodiesterase 4D deficiency in the ryanodine–receptor complex promotes heart failure and arrhythmias. *Cell* 123:25.

24 Role of Cyclic Nucleotide Phosphodiesterases in Heart Failure and Hypertension

Matthew A. Movsesian and Carolyn J. Smith

CONTENTS

24.1 Introduction 485
24.2 PDE3 Cyclic Nucleotide Phosphodiesterases and Heart Failure 486
    24.2.1 PDE3 Cyclic Nucleotide Phosphodiesterases in
        Cardiac and Vascular Smooth Muscle 486
    24.2.2 The Role of PDE3 Isoforms in the Regulation
        of Cyclic Nucleotide–Mediated Signaling..487
    24.2.3 The Use of PDE3 Inhibitors in the Treatment of Heart Failure...............489
    24.2.4 Inferences and New Directions..490
24.3 PDE5 Cyclic Nucleotide Phosphodiesterases in Pulmonary
    Hypertension and Heart Failure 492
    24.3.1 PDE5 Cyclic Nucleotide Phosphodiesterases in
        Cardiac and Vascular Smooth Muscles...492
    24.3.2 PDE5 Isoforms and the Regulation of Cyclic
        Nucleotide–Mediated Signaling..493
    24.3.3 The Use of PDE5 Inhibitors in the Treatment of
        Pulmonary and Systemic Hypertension...494
    24.3.4 The Use of PDE5 Inhibitors in the Treatment of Heart Failure...............494
    24.3.5 Inferences and New Directions..495
Acknowledgments 495
References 495

24.1 INTRODUCTION

A large number of cyclic nucleotide phosphodiesterases have been identified in cardiac muscle. Pharmacologic inhibitors of two families of cyclic nucleotide phosphodiesterases, PDE3 and PDE5, have been used in the treatment of heart failure and pulmonary hypertension, and this chapter is therefore focused on these two families of enzymes.

PDE3 inhibitors reduce cAMP hydrolysis and raise intracellular cAMP content in cardiac and vascular smooth muscles, thereby increasing myocardial contractility and decreasing vascular resistance. These drugs have been used as inotropic and vasodilating agents in the treatment of dilated cardiomyopathy, and are effective in the short term. In most patients, however, these drugs' long-term use is associated with increased mortality through

mechanisms that are just beginning to be elucidated. The use of PDE3 inhibitors in combination with β-adrenergic receptor antagonists and the development of inhibitors selective for individual PDE3 isoforms may provide opportunities for improving on these results.

PDE5 inhibitors reduce cGMP hydrolysis and raise intracellular cGMP content in cardiac and vascular smooth muscle. The consequent pulmonary and systemic vasodilation induced by these drugs may be useful in the treatment of heart failure. Recent studies in animals also point to a possible role for PDE5 inhibitors in limiting and reversing pathologic myocardial hypertrophy. Whether these latter actions are relevant to the use of these drugs in the treatment of heart failure remains to be determined.

24.2 PDE3 CYCLIC NUCLEOTIDE PHOSPHODIESTERASES AND HEART FAILURE

24.2.1 PDE3 CYCLIC NUCLEOTIDE PHOSPHODIESTERASES IN CARDIAC AND VASCULAR SMOOTH MUSCLE

PDE3 cyclic nucleotide phosphodiesterases are characterized by their high affinity for both cAMP and cGMP. Under comparable conditions, these enzymes have higher affinities (lower K_m values) for cGMP than for cAMP (30 μM versus 90 μM), but have much higher turnover rates (k_{cat} values) for cAMP than for cGMP (30 min^{-1} versus 6 min^{-1}).[1,2] This suggests that these enzymes, which were at one time referred to as cGMP-inhibited cAMP phosphodiesterases, function principally as cAMP-hydrolyzing enzymes, with cGMP serving as a regulator of this activity rather than as a physiologically meaningful substrate. The stimulation of renin secretion by cGMP, for example, is mediated through the inhibition of the cAMP-hydrolytic activity of PDE3.[3] Inhibition of cAMP hydrolysis by cGMP may contribute to cGMP-mediated signaling in vascular smooth muscle cells and in sinoatrial cells.[4-7] cGMP analogues can raise intracellular cAMP content in cardiac myocytes by inhibiting PDE3 activity,[8] and this may be involved in the potentiation of L-type Ca^{2+} currents in cardiac myocytes by agents that increase cGMP.[9-12] To our knowledge, a role for the cGMP-hydrolytic activity of PDE3 enzymes in the physiology of human myocardium has not been established.

Two genes encoding PDE3 cyclic nucleotide phosphodiesterases have been identified and designated *PDE3A* and *PDE3B*.[13,14] Probably through alternative transcriptional processing, the *PDE3A* gene gives rise to two mRNAs, originally designated PDE3A1 and PDE3A2.[15,16] PDE3A1 is translated into a protein that has been referred to as PDE3A-136, wheras PDE3A2 is translated into two proteins referred to as PDE3A-118 and PDE3A-94 (the numerals in the protein names correspond to the apparent molecular weights of these isoforms as inferred from SDS-PAGE).[15-17] To make the PDE3 nomenclature consistent with the nomenclature for other phosphodiesterases, we are henceforth using the designation PDE3A1, PDE3A2, and PDE3A3 for the proteins, which we formerly referred to as PDE3A-136, PDE3A-118, and PDE3A-94, respectively. All three isoforms are present in cardiac muscle. PDE3A1 is present exclusively in microsomal fractions of human myocardium (Figure 24.1). It contains two membrane-association domains, NHR1 and NHR2: NHR1 contains hydrophobic loops that insert into intracellular membranes, whereas NHR2 appears to localize the enzyme by protein–protein interactions.[18,19] PDE3A1 also contains three sites for phosphorylation and activation by PKB (P1) and PKA (P2 and P3). PDE3A2, present in microsomal and cytosolic fractions of human myocardium, lacks NHR1 and the most upstream phosphorylation site, whereas PDE3A3, also present in microsomal and cytosolic fractions of human myocardium, lacks NHR1, NHR2, and all three phosphorylation sites. All three isoforms contain the full C-terminal catalytic region (CCR) and are identical with respect to catalytic activity and sensitivity to conventional PDE3 inhibitors.[1]

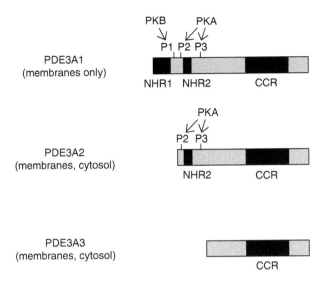

FIGURE 24.1 PDE3A isoforms identified in human cardiac and vascular smooth muscle.

In the case of PDE3B, to our knowledge, only one isoform, PDE3B1, has been identified.[14] This isoform, like PDE3A1, appears to be found exclusively in association with intracellular membranes. It has not been identified in human myocardium, but is present in mouse myocardium.[20] This and other differences between mouse and human hearts with respect to the expression of cyclic nucleotide phosphodiesterases may constitute an important limitation in our ability to draw inferences on mechanisms in human hearts based on observations in mouse models. PDE3B1 is also expressed in vascular smooth muscle myocytes, where it is recovered selectively in microsomal fractions.[21,22]

24.2.2 THE ROLE OF PDE3 ISOFORMS IN THE REGULATION OF CYCLIC NUCLEOTIDE–MEDIATED SIGNALING

PDE3 cyclic nucleotide phosphodiesterases regulate cAMP-mediated signaling in cardiac and vascular smooth muscle myocytes (Figure 24.2). In cardiac myocytes, this regulation is highly compartmentalized, i.e., cAMP content is regulated differentially in intracellular compartments represented in microsomal and cytosolic fractions of cardiac myocytes, and changes in

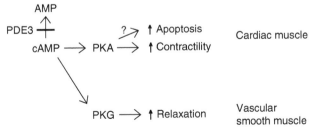

FIGURE 24.2 Overview of PDE3 inhibition in cardiac and vascular smooth muscle. Inhibition of PDE3 leads to a rise in intracellular cAMP. This activates PKA in cardiac muscle, resulting in increased contractility and possibly an increase in apoptosis. In vascular smooth muscle, the rise in cAMP activates PKG, resulting in relaxation.

the cAMP content in these compartments correlate with changes in the phosphorylation of different substrates of cAMP-dependent protein kinase (PKA) and with different physiologic responses.[23–29] Cyclic nucleotide phosphodiesterases are important in this compartmentation. In neonatal rat cardiac myocytes, for example, rises in intracellular cAMP in response to β-adrenergic receptor agonists are highly localized; when phosphodiesterase inhibitors are added, however, the distribution of the increase in intracellular cAMP becomes more diffuse.[30] Individual phosphodiesterase isoforms have specific roles in this compartmentation. In rat ventricular cardiac myocytes, PDE4 isoforms have a quantitatively greater role than PDE3 isoforms in regulating β-adrenergic receptor–induced rises in intracellular cAMP content, whereas PDE3 isoforms have a slightly greater role in regulating forskolin-induced rises in intracellular cAMP content.[31] PDE2 isoforms, which account for only a small fraction of the total cAMP-hydrolytic activity in rat cardiac myocytes, also have a major role in regulating β-adrenergic receptor–mediated rises in intracellular cAMP content; in contrast, PDE2 isoforms have little role in the regulation of forskolin-induced rises.[32]

The compartmentation of cAMP-mediated signaling is certainly important in failing human myocardium as well, but the roles of specific phosphodiesterase isoforms in this compartmentation are not clear. PDE3 isoforms are especially important in human myocardium, and as the differences among the PDE3 isoforms involve membrane-insertion domains, protein-interaction domains, and phosphorylation sites in the N-terminal regions,[15,16] these isoforms are likely to regulate cAMP content in functionally distinct intracellular compartments of cardiac myocytes in response to different upstream signals. Recent studies indicate that PDE3 isoforms comprise the majority of cAMP-hydrolytic activity in membrane-enriched fractions of failing human myocardium.[1] In cytosolic fractions of failing human hearts, PDE3 isoforms constitute the majority of cAMP-hydrolytic activity at low cAMP and Ca^{2+} concentrations. They constitute a much smaller fraction of cytosolic cAMP-hydrolytic activity at higher Ca^{2+} and cAMP concentrations, owing to increases, under these conditions, in the activity of other lower-affinity cyclic nucleotide phosphodiesterases such as PDE1, PDE2, and PDE4. cGMP inhibits cAMP hydrolysis in membrane-enriched fractions of failing human myocardium through competitive inhibition of PDE3 activity and in cytosolic fractions through competitive inhibition of both PDE3 and PDE1. At this time, little is known regarding the compartmentation of cAMP-mediated signaling in vascular smooth muscle or the possible roles of PDE3B1, PDE3A2, and PDE3A3 in any such compartmentation.

These findings on the contribution of PDE3 and PDE1 to the regulation of cAMP-mediated signaling in membrane-enriched and cytosolic fractions of human myocardium are interesting in view of the complex and seemingly contradictory actions ascribed to cGMP in cardiac myocytes. Interventions that increase myocardial cGMP content may be associated with positive inotropic effects,[33–35] no inotropic effects,[36] biphasic effects,[37] or negative inotropic effects that can antagonize the actions of cAMP-raising agents.[38–40] Some of this diversity may reflect the presence of multiple soluble (NO-sensitive) and membrane-bound (natriuretic peptide–sensitive) forms of guanylate cyclase that are regulated differentially in cardiac myocytes.[41,42] As will be discussed further in the chapter, cGMP-mediated signaling is involved in inhibiting pressure- and catecholamine-induced cardiac hypertrophy.[43–49] The high levels of cGMP-inhibited PDE1 and PDE3 cAMP-hydrolytic activity in cytosolic and microsomal fractions of failing human myocardium suggest that inhibition of cAMP hydrolysis by cGMP probably contributes to cGMP-mediated signaling in this tissue.[1] On the other hand, there is the possibility that a stimulation of PDE2 activity, as noted in rat cardiac myocytes, may contribute to a stimulation of cAMP hydrolysis by cGMP.[32] There is evidence in human atrial myocytes that the biphasic actions of cGMP may result from stimulation of PDE2 and inhibition of PDE3.[12] These considerations are likely to have therapeutic relevance because of the common concomitant use of cGMP-raising vasodilators

(organic nitrates, sodium nitroprusside) and cAMP-raising inotropic agents in the treatment of heart failure.

24.2.3   The Use of PDE3 Inhibitors in the Treatment of Heart Failure

The term heart failure has been used in reference to several patterns of cardiac pathophysiology. The syndrome of greatest importance with respect to the use of PDE3 inhibitors is dilated cardiomyopathy, which, as its name suggests, is characterized by chamber dilatation and impaired cardiac contraction. Dilated cardiomyopathy is also characterized by pathologic vasoconstriction in the systemic and pulmonary vasculature, which increases afterload and impairs emptying of the left and right ventricles. Changes in the receptor-mediated cAMP generation are part of this syndrome. These changes, which include a decrease in the density of β_1-adrenergic receptors and increase in the expressions of β-adrenergic receptor kinase, Gαi, and nucleoside diphosphate kinase, lead to an uncoupling of the occupancy of β-adrenergic receptors by their agonists from the consequent stimulation of adenylate cyclase activity.[50–55] All of these contribute to a decrease in intracellular cAMP content that is especially pronounced in cytosolic fractions of failing human myocardium.[56]

PDE3 inhibitors might be expected to be useful in the treatment of dilated cardiomyopathy for several reasons. Their ability to raise intracellular cAMP content would help to compensate for some of the molecular features of the syndrome involving reduced cAMP generation. In addition, raising intracellular cAMP content in cardiac myocytes results in inotropic responses that should have beneficial effects with respect to the clinical features of the syndrome. Several substrates of cAMP-dependent protein kinase are likely to be involved in this action. Phosphorylation of phospholamban increases Ca^{2+} sequestration by the sarcoplasmic reticulum during diastole.[57–60] Phosphorylation of ryanodine-sensitive Ca^{2+} channels increases Ca^{2+} release by the sarcoplasmic reticulum during systole.[61–63] Phosphorylation of dihydropyridine-sensitive Ca^{2+} channels increases Ca^{2+} influx during systole.[64,65] All these actions would increase the amplitude of intracellular Ca^{2+} cycling, which is attenuated in dilated cardiomyopathy.[66,67]

PDE3 inhibition also results in smooth muscle relaxation and vasodilation, which reduces resistance to ventricular emptying and thereby improves cardiac function.[68,69] The mechanisms involved in PDE inhibition may be more complex than those in cardiac myocytes. Most of the smooth muscle-relaxant effect of PDE3 inhibition seems to result from an increase in intracellular cAMP content.[70] In vascular smooth muscle myocytes, this effect seems to be attributable more to cross-activation of cGMP-dependent protein kinase (PKG) by cAMP than to the activation of cAMP-dependent protein kinase (PKA).[71] PKG phosphorylates myosin phosphatase, which decreases Ca^{2+}-activated actin–myosin ATPase activity;[72–74] it phosphorylates Ca^{2+}-dependent K^+ channels, leading to hyperpolarization of smooth muscle membranes and reduced Ca^{2+} influx;[75–79] and it phosphorylates the IP3 receptor and an IP3 receptor-associated PKG substrate, leading to reduced efflux of Ca^{2+} from the sarcoplasmic reticulum.[80–82] All of these actions reduce contraction in vascular smooth muscle. PDE3 inhibition also may raise intracellular cGMP content in vascular smooth muscle, but the extent to which this effect contributes to the vasodilatory action of PDE3 inhibition is unclear.[83] There is evidence that inhibition of cAMP hydrolysis by cGMP contributes to the vascular smooth muscle-relaxant action of nitrates.[4] Results in an animal model suggest that PDE3 inhibition may be particularly helpful in the treatment of pulmonary hypertension because of an increase in PDE3 activity in vascular smooth muscle in this condition, but this has not yet been validated in humans.[84]

In clinical trials, PDE3 inhibitors—amrinone, enoximone, milrinone, and vesnarinone— have been shown to increase myocardial contraction and reduce vascular constriction in

patients with dilated cardiomyopathy, resulting in improved hemodynamic function.[68,69,85–89] In prospective long-term trials, however, these benefits were not sustained. Instead, an increase in mortality was often observed in treated patients, and this was usually attributable to sudden cardiac death, indicating an arrhythmic etiology rather than a progressive and ultimately fatal loss of ventricular contractility.[89–95] A meta-analysis of several studies showed an overall increase in mortality in patients treated with PDE3 inhibitors of ~40%.[96]

24.2.4   INFERENCES AND NEW DIRECTIONS

What can be inferred from the results of these clinical studies? The assumption underlying the use of PDE3 inhibitors in the treatment of dilated cardiomyopathy was that the decrease in receptor-mediated cAMP generation contributes to the pathophysiology of the syndrome, and that overcoming it would be beneficial. In retrospect, it was equally possible that this decrease in receptor-mediated cAMP generation represents a compensatory response that protects the myocardium from adverse effects of catecholamine stimulation (Figure 24.3). To some degree, this latter possibility is supported by evidence in animal models that catechol-amines can induce changes in β-adrenergic receptor and Gαi expressions similar to those described in dilated cardiomyopathy in humans.[97,98] This possibility is also consistent with two other sets of clinical observations. First, increased mortality has been observed in clinical trials in which β-adrenergic receptor agonists were used to raise cAMP content in patients with dilated cardiomyopathy,[99,100] although β-adrenergic receptor agonists have been better tolerated in other trials.[101] In addition, many studies have shown that treatment with β-adrenergic receptor antagonists, which lower intracellular cAMP content in cardiac myo-cytes, has beneficial effects on survival in patients with dilated cardiomyopathy.[102–107] Sim-plistically, at least, these observations can be summarized as being consistent with the hypothesis that increases in intracellular cAMP content in cardiac myocytes, on the whole, have favorable effects on cardiac function in patients with dilated cardiomyopathy in the short term but have adverse effects on clinical outcomes in the long term.

    At the time of this writing, the specific proximate molecular mechanisms to which these adverse clinical effects can be ascribed have not been identified. Recent studies in rats suggest that inhibition of PDE3 expression and activity is associated with proapoptotic actions in cardiac myocytes.[108] In rats, these actions are associated with an increase in the expression of inducible cAMP early repressors (ICERs) that appear to participate in a positive feedback mechanism with a decrease in PDE3A expression.[109] According to this mechanism, inhibition of PDE3 leads to a sustained increase in ICER expression, which in turn leads to a decrease in

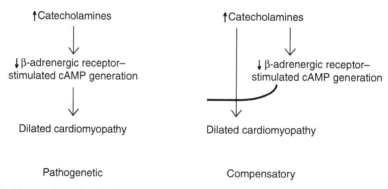

FIGURE 24.3 Alternative roles for the decrease in β-adrenergic receptor–mediated cAMP generation in the pathophysiology of dilated cardiomyopathy.

PDE3A expression; the high level of ICER would also repress the expression of CREB-regulated proteins such as Bcl-2, with proapoptotic consequences. These proapoptotic actions might be expected to contribute to the unfavorable clinical outcomes in patients with dilated cardiomyopathy treated chronically with PDE3 inhibitors. This mechanism is clearly maladaptive, and is somewhat puzzling: one would expect a cAMP-induced increase in ICER expression to serve as a compensatory mechanism to limit a cAMP-induced increase in CREB-stimulated gene expression, not to lead to a net decrease in CREB-stimulated gene expression. On the other hand, there may be CREB-independent actions of ICER that lead to the proapoptotic effects.

With regard to an etiologic role for a decrease in PDE3 activity in dilated cardiomyopathy, a reduction in the levels of some cardiac PDE3 isoforms has been observed in response to pressure load and to angiotensin I and isoproterenol in animal models.[109,110] Whether this contributes to the pathophysiology of dilated cardiomyopathy in humans is unclear. We and other investigators have found no differences between normal and failing human myocardium with respect to PDE3 activity,[111–113] but other investigators have recently reported a decrease in PDE3A1 mRNA expression and PDE3 activity in failing human myocardium.[109] The reason for this discrepancy is, however, unclear.

At another level, we noted earlier that the adverse effects on mortality in patients treated with PDE3 inhibitors were associated with an increase in sudden cardiac death. Most of these studies were conducted in the 1980s and early 1990s, i.e., before the advent of the implantation of cardiac defibrillators in patients with dilated cardiomyopathy. It is unknown whether PDE3 inhibitors increase mortality in patients with dilated cardiomyopathy in whom cardiac defibrillators have been implanted. Furthermore, the PDE3 inhibitor cilostazol has been used in the treatment of patients with peripheral vascular disease—most of whom presumably did not have dilated cardiomyopathy—without an obvious increase in sudden cardiac death.[114] These results raise the possibility that the increase in mortality seen with chronic treatment with PDE3 inhibitors occurs only in patients with dilated cardiomyopathy.

The adverse effects of chronic treatment with PDE3 inhibitors do not preclude a role for PDE3 inhibitors in the treatment of dilated cardiomyopathy. These drugs are of value in augmenting cardiac contractility and reducing vascular resistance in cases of acute cardiac failure, when the need for such treatment may be short lived, and are frequently used in such situations. PDE3 inhibitors may also be useful when contractile failure is so severe that patients have difficulty in tolerating the negative inotropic consequences of initiation of treatment with β-adrenergic receptor antagonists. Several clinical studies document the beneficial use of the combination of PDE3 inhibitors (enoximone, milrinone, or pimobendan) and β-adrenergic receptor antagonists (metoprolol, carvedilol, or esmolol) in this setting,[114–119] although to our knowledge this has yet to be demonstrated in a randomized prospective clinical trial.

The logic of treating dilated cardiomyopathy by combining agents with opposite effects on cAMP content may be questioned. In fact, however, their actions are likely not truly opposite. Some of the effects of β$_2$-adrenergic receptor occupancy are at least partially cAMP-independent, including contraction and relaxation responses to isoproterenol, activation of L-type Ca^{2+} channels and Na^+ channels, hypertrophy, and induction of a fetal gene program; these effects would not be influenced by PDE3 inhibition.[120,121] In addition, the intracellular pools of cAMP regulated by β-adrenergic receptors and PDE3 may not be identical, owing to the compartmentation of cAMP metabolism in cardiac myocytes, which is alluded to earlier. The combination of β-adrenergic receptor antagonists and PDE3 inhibitors may therefore result in an increase in the phosphorylation of some substrates of cAMP-dependent protein kinase and a decrease in the phosphorylation of others.[120] If this is the case, and if the substrates whose phosphorylation is increased contribute to the beneficial effects

of raising intracellular cAMP content in cardiac myocytes in dilated cardiomyopathy while the substrates whose phosphorylation is decreased contribute to the adverse effects, one might expect this combination to have a net beneficial effect. It should be emphasized, however, that these considerations are speculative at this time.

Another possibility has to do with the different isoforms of PDE3 in cardiac myocytes, which are likely to regulate the phosphorylation of different substrates of cAMP-dependent protein kinase. If the activity or expression of one of these isoforms could be inhibited selectively, it might be possible to increase the phosphorylation of a set of substrates that contribute to the beneficial actions of PDE3 inhibitors while avoiding an increase in the phosphorylation of substrates, which contribute to the adverse effects of PDE3 inhibitors. At this time, however, neither the specific proteins whose phosphorylation is regulated by each of the PDE3 isoforms nor the proteins whose phosphorylations cause the adverse effects of PDE3 inhibition have been identified.

24.3 PDE5 CYCLIC NUCLEOTIDE PHOSPHODIESTERASES IN PULMONARY HYPERTENSION AND HEART FAILURE

24.3.1 PDE5 CYCLIC NUCLEOTIDE PHOSPHODIESTERASES IN CARDIAC AND VASCULAR SMOOTH MUSCLES

PDE5 cyclic nucleotide phosphodiesterases bind and hydrolyze cGMP.[122] *PDE5A* is the only gene identified thus far in this PDE family. The cDNA for the cytosolic 90 kDa protein, PDE5A1, was first cloned from bovine lung.[123] Rat and human cDNAs for PDE5A1 were subsequently identified.[124–126] PDE5A1 appears to be the predominant isoform in cardiac and vascular smooth muscles, but it is not clear if it is the only isoform. Alternative splicing of PDE5A generates a second isoform, PDE5A2, in canine, rat, and human cardiovascular and neural tissues.[126–128] A third PDE5A isoform with N-terminal sequence differences, PDE5A3 (~86 kDa), was identified in cavernosal smooth muscle and other tissues that contain either a smooth muscle (bladder, uterus, prostate, urethra) or cardiac muscle component; but this isoform was not evident in brain or lung.[129] PDE5A3 may represent the 86 kDa PDE5A detected in uterine smooth muscle of pregnant rats.[130] Although all the three recombinant human isoforms have similar K_m values for cGMP (~6 μM), PDE5A2 and PDE5A3 have two- to threefold greater sensitivities to the PDE5 inhibitors zaprinast and sildenafil than does PDE5A1.[129]

PDE5 isoforms are activated by cGMP. This activation involves binding of cGMP to two allosteric regulatory (GAF) sites as well as phosphorylation by PKG. The binding of cGMP to the two N-terminal GAF sites is facilitated and stabilized by phosphorylation of serine-92 by PKG.[131] PDE5 isoforms are also activated by cGMP binding to the GAF-A domain without phosphorylation of the enzyme.[132] PDE5A1 is phosphorylated in response to atrial natriuretic peptide (ANP) in early passage cultured rat smooth muscle cells containing PKG and in response to nitric oxide (NO) in human platelets.[133,134]

The level of PDE5 in heart muscle varies from species to species, and may be influenced by the presence of heart disease. PDE5 isoforms are present in mouse hearts, and by one estimate were found to comprise 30% of the cGMP-hydrolytic activity in this tissue.[135,136] The level of expression of PDE5 is increased substantially in mouse hearts in the setting of pressure-load hypertrophy,[49] whereas it is decreased in dogs with pacing-induced dilated cardiomyopathy.[137] With respect to human myocardium, some investigators have reported the virtual absence of PDE5 from human myocardium, based on quantitation of sildenafil-sensitive cGMP-hydrolytic activity and immunoblotting with anti-PDE5 antibodies.[138] Subsequent functional studies showed low amounts of activity in human heart and meager responses to

sildenafil in this tissue.[139] Levels of PDE5 in heart were extremely low when compared to the levels in lung.[140] To our knowledge, the level of PDE5 in diseased human myocardium remains unknown.

In most tissues, PDE5 activity is predominantly cytosolic. In canine or mouse ventricular cardiac myocytes, however, PDE5A1 has been detected by cytoskeletal immunolocalization at the Z-band.[137] This localization seems to be disrupted in heart failure.[137] Expression of recombinant human PDE5A2 in COS-7 cells suggests that this isoform associates with membranes or hydrophobic proteins.[129] The association of PDE5A isoforms with other proteins or cellular structures could contribute to an intracellular compartmentation of cGMP-mediated signaling, although this has yet to be explored in cardiac or smooth muscle cells.

24.3.2 PDE5 ISOFORMS AND THE REGULATION OF CYCLIC NUCLEOTIDE–MEDIATED SIGNALING

By inhibiting cGMP hydrolysis, PDE5 potentiates the actions of cGMP and PKG in smooth muscle (Figure 24.4). PKG phosphorylates myosin phosphatase, plasma membrane K^+ channels, and proteins involved in the IP3-mediated release of Ca^{2+} from intracellular stores.[72–82,141] All of these actions reduce contraction in vascular smooth muscle, and constitute the rationale for the use of PDE5 inhibitors in the treatment of pulmonary and systemic hypertension.

The effects of cGMP and PKG in cardiac muscle are less well understood, and appear to be quite complex (Figure 24.4). As noted earlier, increasing myocardial cGMP content has been associated with diverse effects on myocardial contractility, some of which seem contradictory.[33–40] The extent to which these effects are mediated by inhibition of cAMP hydrolysis by PDE3 and PDE1 and by stimulation of cAMP hydrolysis by PDE2 is unknown.[32,142] Recently, cGMP-mediated signaling has been shown to be involved in hypertrophic responses in cardiac myocytes and heart muscle, and PDE5 inhibition prevents and reverses hypertrophic responses to pressure overload and β-adrenergic receptor agonists in animal

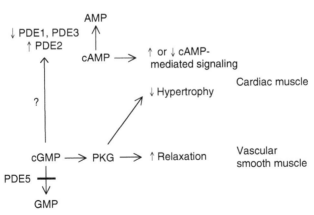

FIGURE 24.4 Overview of PDE5 inhibition in cardiac and vascular smooth muscle. Inhibition of PDE5 leads to a rise in intracellular cGMP and activation of PKG. This results in relaxation in vascular smooth muscle, and may result in antihypertrophic effects in cardiac muscle. The rise in cGMP may potentiate cAMP-mediated signaling in cardiac muscle by inhibition of PDE1 and PDE3 or inhibit cAMP-mediated signaling by stimulation of PDE2.

models.[43–49] The mechanisms involved in the antihypertrophic actions of PDE5 inhibition are as yet not fully elucidated, but may involve the calcineurin/NFAT, PI3-kinase/Akt, and ERK1/2 signaling pathways.[49]

24.3.3 THE USE OF PDE5 INHIBITORS IN THE TREATMENT OF PULMONARY AND SYSTEMIC HYPERTENSION

A number of studies in animal models have demonstrated beneficial actions of PDE5 inhibition in reducing pulmonary hypertension and pulmonary vascular remodeling, attributable to diverse causes, and in reversing the deleterious effects of increased pulmonary vascular resistance on right ventricular function.[143–153] Beneficial effects of three PDE5 inhibitors, sildenafil, tadalafil, and vardenafil, have been shown in clinical trials in humans with pulmonary hypertension of diverse etiologies.[154–160] In these clinical trials, PDE5 inhibitors had vasodilating actions that were somewhat selective for the pulmonary vasculature relative to the systemic vasculature,[160] although in some situations significant decreases in systemic vascular resistance were also observed.[158] The effects of PDE5 inhibition are synergistic with those of inhaled NO, which stimulates cGMP formation.[158,160] Long-term administration of sildenafil was associated with improvements in functional class that were sustained for at least several months as well as a decrease in right ventricular mass.[156,157,159] Because of these actions, the use of sildenafil for the treatment of pulmonary hypertension was approved in the United States in 2005.

With respect to the use of PDE5 inhibitors to treat systemic hypertension, the use of these drugs in patients treated with other antihypertensive agents, especially those that act by raising intracellular cGMP in vascular smooth muscle (e.g., organic nitrates), can cause profound systemic hypotension.[161,162] This is an important caution regarding the use of PDE5 inhibitors to treat erectile dysfunction in patients who may be taking other drugs with vasodilatory actions. Whether the vasodilating actions of PDE5 inhibition may be beneficial in the treatment of systemic hypertension remains to be determined.

24.3.4 THE USE OF PDE5 INHIBITORS IN THE TREATMENT OF HEART FAILURE

The principal rationale for the use of PDE5 inhibitors in heart failure has to do with lowering pulmonary vascular resistance and improving cardiac performance by improving right ventricular emptying rather than with direct effects on cardiac muscle.[159,163,164] Systemic vasodilation also occurs in patients with heart failure treated with PDE5 inhibitors, and the consequent reduction in left ventricular afterload may also contribute to the beneficial responses that have been observed.[158]

The possible contribution of the antihypertrophic actions of PDE5 inhibition in animal models to the beneficial effects of PDE5 inhibitors in patients with heart failure remains unknown. Some investigators have noticed a decrease in right ventricular size in patients with pulmonary hypertension treated with sildenafil,[159] but this could be attributable to the benefits of a chronic decrease in pulmonary vascular resistance and secondary effects on right ventricular hypertrophy. As noted earlier, PDE5 levels are quite low in dogs with pacing-induced dilated cardiomyopathy, whereas PDE5 levels are high in dogs with pressure overload–induced hypertrophy.[49,137] Whether PDE5 is present at significant levels in the right and left ventricles of humans with different forms of heart failure is to our knowledge unknown. Further characterization of the level and function of PDE5 in normal and failing human myocardium, together with prospective clinical trials of PDE5 inhibitors in the treatment of heart failure and pulmonary and systemic hypertensions are obvious areas for further investigation.

24.3.5 INFERENCES AND NEW DIRECTIONS

At this time, our knowledge of the role of PDE5 in hypertension and heart failure in humans is less well characterized than is the role of PDE3. As vasodilatory agents with some selectivity for the pulmonary vasculature, PDE5 inhibitors are likely to be especially useful in patients with pulmonary hypertension but low systemic blood pressures, where systemic vasodilation may be poorly tolerated. The effects these agents will have on morbidity and mortality in heart failure remain to be seen. As the many clinical trials with PDE3 inhibitors have demonstrated, hemodynamic benefits do not necessarily translate into improved long-term outcomes.

The recent reports of reduced cardiac hypertrophy in animals treated with PDE5 inhibitors raise the question of whether an increase in cGMP-mediated signaling through other mechanisms may be beneficial in patients with heart failure. As noted, this is particularly interesting given the frequent use of cGMP-raising agents as vasodilators in patients with heart failure. It may well be that these patients derive benefits from an increase in cGMP-mediated signaling in cardiac muscle. At this time, however, our knowledge of the role of myocardial cGMP-mediated signaling in heart failure in humans is, at best, rudimentary.

ACKNOWLEDGMENTS

This study was supported by Medical Research funds from the United States Department of Veterans Affairs, the American Heart Association, and the University of Utah Research Foundation (M.A.M.) and a grant (HL69061) from the National Institutes of Health (C.J.S.).

REFERENCES

1. Hambleton, R., J. Krall, E. Tikishvili, M. Honeggar, F. Ahmad, V.C. Manganiello, and M.A. Movsesian. 2005. *J Biol Chem* 280:39168.
2. Movsesian, M. 2006. Unpublished observation.
3. Kurtz, A., K.H. Gotz, M. Hamann, and C. Wagner. 1998. *Proc Natl Acad Sci USA* 95:4743.
4. Lugnier, C., T. Keravis, and A. Eckly-Michel. 1999. *J Physiol Pharmacol* 50:639.
5. Aizawa, T., H. Wei, J.M. Miano, J. Abe, B.C. Berk, and C. Yan. 2003. *Circ Res* 93:406.
6. Fung, E., and R.R. Fiscus. 2003. *J Cardiovasc Pharmacol* 41:849.
7. Shimizu, K., Y. Shintani, W.G. Ding, H. Matsuura, and T. Bamba. 2002. *Br J Pharmacol* 137:127.
8. Patel, K.N., L. Yan, A. Gandhi, P.M. Scholz, and H.R. Weiss. 2001. *Basic Res Cardiol* 96:34.
9. Lohmann, S.M., R. Fischmeister, and U. Walter. 1991. *Basic Res Cardiol* 86:503.
10. Mery, P.F., S.M. Lohmann, U. Walter, and R. Fischmeister. 1991. *Proc Natl Acad Sci USA* 88:1197.
11. Mery, P.F., C. Pavoine, L. Belhassen, F. Pecker, and R. Fischmeister. 1993. *J Biol Chem* 268:26286.
12. Vandecasteele, G., I. Verde, C. Rucker-Martin, P. Donzeau-Gouge, and R. Fischmeister. 2001. *J Physiol* 533:329.
13. Meacci, E., M. Taira, M. Moos Jr., C.J. Smith, M.A. Movsesian, E. Degerman, P. Belfrage, and V. Manganiello. 1992. *Proc Natl Acad Sci USA* 89:3721.
14. Taira, M., S.C. Hockman, J.C. Calvo, P. Belfrage, and V.C. Manganiello. 1993. *J Biol Chem* 268:18573.
15. Choi, Y.H., D. Ekholm, J. Krall, F. Ahmad, E. Degerman, V.C. Manganiello, and M.A. Movsesian. 2001. *Biochem J* 353:41.
16. Wechsler, J., Y.H. Choi, J. Krall, F. Ahmad, V.C. Manganiello, and M.A. Movsesian. 2002. *J Biol Chem* 277:38072.
17. Smith, C.J., J. Krall, V.C. Manganiello, and M.A. Movsesian. 1993. *Biochem Biophys Res Commun* 190:516.
18. Kenan, Y., T. Murata, Y. Shakur, E. Degerman, and V.C. Manganiello. 2000. *J Biol Chem* 275:12331.

19. Shakur, Y., K. Takeda, Y. Kenan, Z.X. Yu, G. Rena, D. Brandt, M.D. Houslay, E. Degerman, V.J. Ferrans, and V.C. Manganiello. 2000. *J Biol Chem* 275:38749.
20. Patrucco, E., A. Notte, L. Barberis, G. Selvetella, A. Maffei, M. Brancaccio, S. Marengo, G. Russo, O. Azzolino, S.D. Rybalkin, L. Silengo, F. Altruda, R. Wetzker, M.P. Wymann, G. Lembo, and E. Hirsch. 2004. *Cell* 118:375.
21. Tilley, D.G., and D.H. Maurice. 2002. *Mol Pharmacol* 62:497.
22. Liu, H., and D.H. Maurice. 1998. *Br J Pharmacol* 125:1501.
23. Hayes, J.S., L.L. Brunton, and S.E. Mayer. 1980. *J Biol Chem* 255:5113.
24. Hayes, J.S., N. Bowling, K.L. King, and G.B. Boder. 1982. *Biochim Biophys Acta* 714:136.
25. Rapundalo, S.T., R.J. Solaro, and E.G. Kranias. 1989. *Circ Res* 64:104.
26. Xiao, R.P., and E.G. Lakatta. 1993. *Circ Res* 73:286.
27. Xiao, R.P., C. Hohl, R. Altschuld, L. Jones, B. Livingston, B. Ziman, B. Tantini, and E.G. Lakatta. *J Biol Chem* 269:19151.
28. Kuschel, M., Y.Y. Zhou, H. Cheng, S.J. Zhang, Y. Chen, E.G. Lakatta, and R.P. Xiao. 1999. *J Biol Chem* 274:22048.
29. Rybin, V.O., X. Xu, M.P. Lisanti, and S.F. Steinberg. 2000. *J Biol Chem* 275:41447.
30. Zaccolo, M., and T. Pozzan. 2002. *Science* 295:1711.
31. Mongillo, M., T. McSorley, S. Evellin, A. Sood, V. Lissandron, A. Terrin, E. Huston, A. Hannawacker, M.J. Lohse, T. Pozzan, M.D. Houslay, and M. Zaccolo. 2004. *Circ Res* 95:67.
32. Mongillo, M., C.G. Tocchetti, A. Terrin, V. Lissandron, Y.F. Cheung, W.R. Dostmann, T. Pozzan, D.A. Kass, N. Paolocci, M.D. Houslay, and M. Zaccolo. 2005. *Circ Res* 98:226.
33. Beaulieu, P., R. Cardinal, P. Page, F. Francoeur, J. Tremblay, and C. Lambert. 1997. *Am J Physiol* 273:H1933.
34. Hirose, M., Y. Furukawa, F. Kurogouchi, K. Nakajima, Y. Miyashita, and S. Chiba. 1998. *J Pharmacol Exp Ther* 286:70.
35. Wollert, K.C., S. Yurukova, A. Kilic, F. Begrow, B. Fiedler, S. Gambaryan, U. Walter, S.M. Lohmann, and M. Kuhn. 2003. *Br J Pharmacol* 140:1227.
36. Layland, J., J.M. Li, and A.M. Shah, 2002. *J Physiol* 540:457.
37. Pierkes, M., S. Gambaryan, P. Boknik, S.M. Lohmann, W. Schmitz, R. Potthast, R. Holtwick, and M. Kuhn. 2002. *Cardiovasc Res* 53:852.
38. Balligand, J.L., R.A. Kelly, P.A. Marsden, T.W. Smith, and T. Michel. 1993. *Proc Natl Acad Sci USA* 90:347.
39. Vila-Petroff, M.G., A. Younes, J. Egan, E.G. Lakatta, and S.J. Sollott. 1999. *Circ Res* 84:1020.
40. Wegener, J.W., H. Nawrath, W. Wolfsgruber, S. Kuhbandner, C. Werner, F. Hofmann, and R. Feil. 2002. *Circ Res* 90:18.
41. Kuhn, M. 2003. *Circ Res* 93:700.
42. Padayatti, P.S., P. Pattanaik, X. Ma, and F. van den Akker. 2004. *Pharmacol Ther* 104:83.
43. Knowles, J.W., G. Esposito, L. Mao, J.R. Hagaman, J.E. Fox, O. Smithies, H.A. Rockman, and N. Maeda. 2001. *J Clin Invest* 107:975.
44. Kishimoto, I., K. Rossi, and D.L. Garbers. 2001. *Proc Natl Acad Sci USA* 98:2703.
45. Wollert, K.C., B. Fiedler, S. Gambaryan, A. Smolenski, J. Heineke, E. Butt, C. Trautwein, S.M. Lohmann, and H. Drexler. 2002. *Hypertension* 39:87.
46. Fiedler, B., S.M. Lohmann, A. Smolenski, S. Linnemuller, B. Pieske, F. Schroder, J.D. Molkentin, H. Drexler, and K.C. Wollert. 2002. *Proc Natl Acad Sci USA* 99:11363.
47. Zahabi, A., S. Picard, N. Fortin, T.L. Reudelhuber, and C.F. Deschepper. 2003. *J Biol Chem* 278:47694.
48. Hassan, M.A., and A.F. Ketat. 2005. *BMC Pharmacol* 5:10.
49. Takimoto, E., H.C. Champion, M. Li, D. Belardi, S. Ren, E.R. Rodriguez, D. Bedja, K.L. Gabrielson, Y. Wang, and D.A. Kass. 2005. *Nat Med* 11:214.
50. Bristow, M.R., R.E. Hershberger, J.D. Port, W. Minobe, and R. Rasmussen. 1989. *Mol Pharmacol* 35:295.
51. Bristow, M.R., R. Ginsburg, V. Umans, M. Fowler, W. Minobe, R. Rasmussen, P. Zera, R. Menlove, P. Shah, and S. Jamieson. 1986. *Circ Res* 59:297.
52. Ungerer, M., M. Bohm, J.S. Elce, E. Erdmann, and M.J. Lohse. 1993. *Circulation* 87:454.

53. Eschenhagen, T., U. Mende, M. Nose, W. Schmitz, H. Scholz, A. Haverich, S. Hirt, V. Doring, P. Kalmar, and W. Hoppner. 1992. *Circ Res* 70:688.
54. Feldman, A.M., A.E. Cates, W.B. Veazey, R.E. Hershberger, M.R. Bristow, K.L. Baughman, W.A. Baumgartner, and C. Van Dop. 1988. *J Clin Invest* 82:189.
55. Lutz, S., R. Mura, D. Baltus, M. Movsesian, W. Kubler, and F. Niroomand. 2001. *Cardiovasc Res* 49:48.
56. Bohm, M., B. Reiger, R.H. Schwinger, and E. Erdmann. 1994. *Cardiovasc Res* 28:1713.
57. James, P., M. Inui, M. Tada, M. Chiesi, and E. Carafoli. 1989. *Nature* 342:90.
58. Simmerman, H.K., and L.R. Jones. 1998. *Physiol Rev* 78:921.
59. Hagemann, D., and R.P. Xiao. 2002. *Trends Cardiovasc Med* 12:51.
60. Chu, G., and E.G. Kranias. 2002. *Basic Res Cardiol* 97 Suppl 1:I43.
61. Seiler, S., A.D. Wegener, D.D. Whang, D.R. Hathaway, and L.R. Jones. 1984. *J Biol Chem* 259:8550.
62. Takasago, T., T. Imagawa, and M. Shigekawa. 1989. *J Biochem (Tokyo)* 106:872.
63. Marx, S.O., and A.R. Marks. 2002. *Basic Res Cardiol* 97 Suppl 1:I49.
64. Sculptoreanu, A., T. Scheuer, and W.A. Catterall. 1993. *Nature* 364:240.
65. Kamp, T.J., and J.W. Hell. 2000. *Circ Res* 87:1095.
66. Gwathmey, J.K., L. Copelas, R. MacKinnon, F.J. Schoen, M.D. Feldman, W. Grossman, and J.P. Morgan. *Circ Res* 61:70.
67. Beuckelmann, D.J., M. Nabauer, and E. Erdmann. 1992. *Circulation* 85:1046.
68. Uretsky, B.F., T. Generalovich, P.S. Reddy, R.B. Spangenberg, and W.P. Follansbee. 1983. *Circulation* 67:823.
69. Jaski, B.E., M.A. Fifer, R.F. Wright, E. Braunwald, and W.S. Colucci. 1985. *J Clin Invest* 75:643.
70. Silver, P.J., R.E. Lepore, B. O'Connor, B.M. Lemp, L.T.R.G. Bentley, and A.L. Harris. 1988. *J Pharmacol Exp Ther* 247:34.
71. Jiang, H., J.L. Colbran, S.H. Francis, and J.D. Corbin. 1992. *J Biol Chem* 267:1015.
72. Lee, M.R., L. Li, and T. Kitazawa. 1997. *J Biol Chem* 272:5063.
73. Nakamura, M., K. Ichikawa, M. Ito, B. Yamamori, T. Okinaka, N. Isaka, Y. Yoshida, S. Fujita, and T. Nakano. 1999. *Cell Signal* 11:671.
74. Surks, H.K., N. Mochizuki, Y. Kasai, S.P. Georgescu, K.M. Tang, M. Ito, T.M. Lincoln, and M.E. Mendelsohn. 1999. *Science* 286:1583.
75. Taniguchi, J., K.I. Furukawa, and M. Shigekawa. 1993. *Pflugers Arch* 423:167.
76. Archer, S.L., J.M. Huang, V. Hampl, D.P. Nelson, P.J. Shultz, and E.K. Weir. 1994. *Proc Natl Acad Sci USA* 91:7583.
77. Lorenz, J.N., D.R. Bielefeld, and N. Sperelakis. 1994. *Am J Physiol* 266:C1656.
78. Alioua, A., J.P. Huggins, and E. Rousseau. 1995. *Am J Physiol* 268:L1057.
79. Fukao, M., H.S. Mason, F.C. Britton, J.L. Kenyon, B. Horowitz, and K.D. Keef. 1999. *J Biol Chem* 274:10927.
80. Komalavilas, P., and T.M. Lincoln. 1994. *J Biol Chem* 269:8701.
81. Komalavilas, P., and T.M. Lincoln. 1996. *J Biol Chem* 271:21933.
82. Yoshida, Y., A. Toyosato, M.O. Islam, T. Koga, S. Fujita, and S. Imai. 1999. *Mol Cell Biochem* 190:157.
83. Kauffman, R.F., K.W. Schenck, B.G. Utterback, V.G. Crowe, and M.L. Cohen. 1987. *J Pharmacol Exp Ther* 242:864.
84. Wagner, R.S., C.J. Smith, A.M. Taylor, and R.A. Rhoades. 1997. *J Pharmacol Exp Ther* 282:1650.
85. Benotti, J. R., W. Grossman, E. Braunwald, D.D. Davolos, and A.A. Alousi. 1978. *N Engl J Med* 299:1373.
86. Baim, D.S., A.V. McDowell, J. Cherniles, E.S. Monrad, J.A. Parker, J. Edelson, E. Braunwald, and W. Grossman. 1983. *N Engl J Med* 309:748.
87. Sinoway, L.S., C.S. Maskin, B. Chadwick, R. Forman, E.H. Sonnenblick, and T.H. Le Jemtel. 1983. *J Am Coll Cardiol* 2:327.
88. Seino, Y., S. Momomura, T. Takano, H. Hayakawa, and K. Katoh. 1996. *Crit Care Med* 24:1490.
89. Feldman, A.M., M.R. Bristow, W.W. Parmley, P.E. Carson, C.J. Pepine, E.M. Gilbert, J.E. Strobeck, G.H. Hendrix, E.R. Powers, R.P. Bain, et al. 1993. *N Engl J Med* 329:149.

90. DiBianco, R., R. Shabetai, B.D. Silverman, C.V. Leier, and J.R. Benotti. 1984. *J Am Coll Cardiol* 4:855.
91. Massie, B., M. Bourassa, R. DiBianco, M. Hess, M. Konstam, M. Likoff, and M. Packer. 1985. *Circulation* 71:963.
92. DiBianco, R., R. Shabetai, W. Kostuk, J. Moran, R.C. Schlant, and R. Wright. 1989. *N Engl J Med* 320:677.
93. Uretsky, B.F., M. Jessup, M.A. Konstam, G.W. Dec, C.V. Leier, J. Benotti, S. Murali, H.C. Herrmann, and J.A. Sandberg. *Circulation* 82:774.
94. Packer, M., J.R. Carver, R.J. Rodeheffer, R.J. Ivanhoe, R. DiBianco, S.M. Zeldis, G.H. Hendrix, W.J. Bommer, U. Elkayam, M.L. Kukin, et al. 1991. *N Engl J Med* 325:1468.
95. Cohn, J.N., S.O. Goldstein, B.H. Greenberg, B.H. Lorell, R.C. Bourge, B.E. Jaski, S.O. Gottlieb, F. McGrew III, D.L. DeMets, and B.G. White. 1998. *N Engl J Med* 339:1810.
96. Nony, P., J.P. Boissel, M. Lievre, A. Leizorovicz, M.C. Haugh, S. Fareh, and B. de Breyne. 1994. *Eur J Clin Pharmacol* 46:191.
97. Eschenhagen, T., U. Mende, M. Diederich, M. Nose, W. Schmitz, H. Scholz, J. Schulte am Esch, A. Warnholtz, and H. Schafer. 1992. *Mol Pharmacol* 42:773.
98. Muller, F.U., K.R. Boheler, T. Eschenhagen, W. Schmitz, and H. Scholz. 1993. *Circ Res* 72:696.
99. O'Connor, C.M., W.A. Gattis, B.F. Uretsky, K.F. Adams Jr., S.E. McNulty, S.H. Grossman, W.J. McKenna, F. Zannad, K. Swedberg, M. Gheorghiade, and R.M. Califf. 1999. *Am Heart J* 138:78.
100. The Xamoterol in Severe Heart Failure Study Group. 1990. *Lancet* 336:1.
101. Oliva, F., R. Latini, A. Politi, L. Staszewsky, A.P. Maggioni, E. Nicolis, and F. Mauri. 1999. *Am Heart J* 138:247.
102. Packer, M., W.S. Colucci, J.D. Sackner-Bernstein, C.S. Liang, D.A. Goldscher, I. Freeman, M.L. Kukin, V. Kinhal, J.E. Udelson, M. Klapholz, et al. 1996. *Circulation* 94:2793.
103. Packer, M., M.R. Bristow, J.N. Cohn, W.S. Colucci, M.B. Fowler, E.M. Gilbert, and N.H. Shusterman. 1996. *N Engl J Med* 334:1349.
104. Hjalmarson, A., S. Goldstein, B. Fagerberg, H. Wedel, F. Waagstein, J. Kjekshus, J. Wikstrand, D. El Allaf, J. Vitovec, J. Aldershvile, M. Halinen, R. Dietz, K.L. Neuhaus, A. Janosi, G. Thorgeirsson, P.H. Dunselman, L. Gullestad, J. Kuch, J. Herlitz, P. Rickenbacher, S. Ball, S. Gottlieb, and P. Deedwania. 2000. *JAMA* 283:1295.
105. Bristow, M.R., E.M. Gilbert, W.T. Abraham, K.F. Adams, M.B. Fowler, R.E. Hershberger, S.H. Kubo, K.A. Narahara, H. Ingersoll, S. Krueger, S. Young, and N. Shusterman. 1996. *Circulation* 94:2807.
106. Packer, M., A.J. Coats, M.B. Fowler, H.A. Katus, H. Krum, P. Mohacsi, J.L. Rouleau, M. Tendera, A. Castaigne, E.B. Roecker, M.K. Schultz, and D.L. DeMets. 2001. *N Engl J Med* 344:1651.
107. Goldstein, S., B. Fagerberg, A. Hjalmarson, J. Kjekshus, F. Waagstein, H. Wedel, and J. Wikstrand. 2001. *J Am Coll Cardiol* 38:932.
108. Ding, B., J.I. Abe, H. Wei, H. Xu, W. Che, T. Aizawa, W. Liu, C.A. Molina, J. Sadoshima, B.C. Blaxall, B.C. Berk, and C. Yan. 2005. *Proc Natl Acad Sci USA* 102:14771.
109. Ding, B., J. Abe, H. Wei, Q. Huang, R.A. Walsh, C.A. Molina, A. Zhao, J. Sadoshima, B.C. Blaxall, B.C. Berk, and C. Yan. 2005. *Circulation* 111:2469.
110. Smith, C.J., R. Huang, D. Sun, S. Ricketts, C. Hoegler, J.Z. Ding, R.A. Moggio, and T.H. Hintze. 1997. *Circulation* 96:3116.
111. Movsesian, M.A., C.J. Smith, J. Krall, M.R. Bristow, and V.C. Manganiello. 1991. *J Clin Invest* 88:15.
112. Schmitz, W., T. Eschenhagen, U. Mende, F.U. Muller, J. Neumann, and H. Scholz. 1992. *Basic Res Cardiol* 87 Suppl 1:65.
113. Movsesian, M.A. 1998. *Ann N Y Acad Sci* 853:231.
114. Thompson, P.D., R. Zimet, W.P. Forbes, and P. Zhang. 2002. *Am J Cardiol* 90:1314.
115. Shakar, S.F., W.T. Abraham, E.M. Gilbert, A.D. Robertson, B.D. Lowes, L.S. Zisman, D.A. Ferguson, and M.R. Bristow. 1998. *J Am Coll Cardiol* 31:1336.
116. Yoshikawa, T., A. Baba, M. Suzuki, H. Yokozuka, Y. Okada, K. Nagami, T. Takahashi, H. Mitamura, and S. Ogawa. 2000. *Am J Cardiol* 85:1495.

117. Kumar, A., G. Choudhary, C. Antonio, V. Just, A. Jain, L. Heaney, and M.A. Papp. 2001. *Am Heart J* 142:512.
118. Hauptman, P.J., D. Woods, and M.R. Prirzker. 2002. *Clin Cardiol* 25:247.
119. Metra, M., S. Nodari, A. D'Aloia, C. Muneretto, A.D. Robertson, M.R. Bristow, and L. Dei Cas. 2002. *J Am Coll Cardiol* 40:1248.
120. Movsesian, M.A. 2003. *J Card Fail* 9:475.
121. Lowes, B.D., E.M. Gilbert, W.T. Abraham, W.A. Minobe, P. Larrabee, D. Ferguson, E.E. Wolfel, J. Lindenfeld, T. Tsvetkova, A.D. Robertson, R.A. Quaife, and M.R. Bristow. 2002. *N Engl J Med* 346:1357.
122. Francis, S.H., I.V. Turko, and J.D. Corbin. 2001. *Prog Nucleic Acid Res Mol Biol* 65:1.
123. McAllister-Lucas, L.M., W.K. Sonnenburg, A. Kadlecek, D. Seger, H.L. Trong, J.L. Colbran, M.K. Thomas, K.A. Walsh, S.H. Francis, J.D. Corbin, et al. 1993. *J Biol Chem* 268:22863.
124. Yanaka, N., J. Kotera, A. Ohtsuka, H. Akatsuka, Y. Imai, H. Michibata, K. Fujishige, E. Kawai, S. Takebayashi, K. Okumura, and K. Omori. 1998. *Eur J Biochem* 255:391.
125. Kotera, J., N. Yanaka, K. Fujishige, Y. Imai, H. Akatsuka, T. Ishizuka, K. Kawashima, and K. Omori. 1997. *Eur J Biochem* 249:434.
126. Loughney, K., T.R. Hill, V.A. Florio, L. Uher, G.J. Rosman, S.L. Wolda, B.A. Jones, M.L. Howard, L.M. McAllister-Lucas, W.K. Sonnenburg, S.H. Francis, J.D. Corbin, J.A. Beavo, and K. Ferguson. 1998. *Gene* 216:139.
127. Kotera, J., K. Fujishige, Y. Imai, E. Kawai, H. Michibata, H. Akatsuka, N. Yanaka, and K. Omori. 1999. *Eur J Biochem* 262:866.
128. Kotera, J., K. Fujishige, H. Akatsuka, Y. Imai, N. Yanaka, and K. Omori. 1998. *J Biol Chem* 273:26982.
129. Lin, C.S., A. Lau, R. Tu, and T.F. Lue. 2000. *Biochem Biophys Res Commun* 268:628.
130. Buhimschi, C.S., R.E. Garfield, C.P. Weiner, and I.A. Buhimschi. 2004. *Am J Obstet Gynecol* 190:268.
131. Corbin, J.D., I.V. Turko, A. Beasley, and S.H. Francis. 2000. *Eur J Biochem* 267:2760.
132. Rybalkin, S.D., I.G. Rybalkina, M. Shimizu-Albergine, X.B. Tang, and J.A. Beavo. 2003. *Embo J* 22:469.
133. Wyatt, T.A., A.J. Naftilan, S.H. Francis, and J.D. Corbin. 1998. *Am J Physiol* 274:H448.
134. Mullershausen, F., A. Friebe, R. Feil, W.J. Thompson, F. Hofmann, and D. Koesling. 2003. *J Cell Biol* 160:719.
135. Giordano, D., M.E. De Stefano, G. Citro, A. Modica, and M. Giorgi. 2001. *Biochim Biophys Acta* 1539:16.
136. Takimoto, E., H.C. Champion, D. Belardi, J. Moslehi, M. Mongillo, E. Mergia, D.C. Montrose, T. Isoda, K. Aufiero, M. Zaccolo, W.R. Dostmann, C.J. Smith, and D.A. Kass. 2005. *Circ Res* 96:100.
137. Senzaki, H., C.J. Smith, G.J. Juang, T. Isoda, S.P. Mayer, A. Ohler, N. Paolocci, G.F. Tomaselli, J.M. Hare, and D.A. Kass. 2001. *Faseb J* 15:1718.
138. Wallis, R.M., J.D. Corbin, S.H. Francis, and P. Ellis. 1999. *Am J Cardiol* 83:3C.
139. Corbin, J., S. Rannels, D. Neal, P. Chang, K. Grimes, A. Beasley, and S. Francis. 2003. *Curr Med Res Opin* 19:747.
140. Corbin, J.D., A. Beasley, M.A. Blount, and S.H. Francis. 2005. *Biochem Biophys Res Commun* 334:930.
141. Liu, H., Z. Xiong, and N. Sperelakis. 1997. *J Mol Cell Cardiol* 29:1411.
142. Hambleton, R., J. Krall, E. Tikishvili, M. Honeggar, F. Ahmad, V.C. Manganiello, and M.A. Movsesian. 2005. *J Biol Chem* 280:39168.
143. Cohen, A.H., K. Hanson, K. Morris, B. Fouty, I.F. McMurty, W. Clarke, and D.M. Rodman. 1996. *J Clin Invest* 97:172.
144. Kodama, K., and H. Adachi. 1999. *J Pharmacol Exp Ther* 290:748.
145. Weimann, J., R. Ullrich, J. Hromi, Y. Fujino, M.W. Clark, K.D. Bloch, and W.M. Zapol. 2000. *Anesthesiology* 92:1702.
146. Oka, M. 2001. *Am J Physiol Lung Cell Mol Physiol* 280:L432.
147. Kang, K.K., G.J. Ahn, Y.S. Sohn, B.O. Ahn, and W.B. Kim. 2003. *J Int Med Res* 31:517.

148. Sebkhi, A., J.W. Strange, S.C. Phillips, J. Wharton, and M.R. Wilkins. 2003. *Circulation* 107:3230.
149. Zhao, L., N.A. Mason, J.W. Strange, H. Walker, and M.R. Wilkins. 2003. *Circulation* 107:234.
150. Pauvert, O., S. Bonnet, E. Rousseau, R. Marthan, and J.P. Savineau. 2004. *Am J Physiol Lung Cell Mol Physiol* 287:L577.
151. Yamamoto, T., A. Wada, T. Tsutamoto, M. Ohnishi, and M. Horie. 2004. *J Cardiovasc Pharmacol* 44:596.
152. Deruelle, P., T.R. Grover, and S.H. Abman. 2005. *Am J Physiol Lung Cell Mol Physiol* 289:L798.
153. Larrue, B., S. Jaillard, M. Lorthioir, X. Roubliova, G. Butrous, T. Rakza, H. Warembourg, and L. Storme. 2005. *Am J Physiol Lung Cell Mol Physiol* 288:L1193.
154. Ghofrani, H.A., R. Voswinckel, F. Reichenberger, H. Olschewski, P. Haredza, B. Karadas, R.T. Schermuly, N. Weissmann, W. Seeger, and F. Grimminger. 2004. *J Am Coll Cardiol* 44:1488.
155. Aldashev, A.A., B.K. Kojonazarov, T.A. Amatov, T.M. Sooronbaev, M.M. Mirrakhimov, N.W. Morrell, J. Wharton, and M.R. Wilkins. 2005. *Thorax* 60:683.
156. Chockalingam, A., G. Gnanavelu, S. Venkatesan, S. Elangovan, V. Jagannathan, T. Subramaniam, R. Alagesan, and S. Dorairajan. 2005. *Int J Cardiol* 99:91.
157. Wilkins, M.R., G.A. Paul, J.W. Strange, N. Tunariu, W. Gin-Sing, W.A. Banya, M.A. Westwood, A. Stefanidis, L.L. Ng, D.J. Pennell, R.H. Mohiaddin, P. Nihoyannopoulos, and J.S. Gibbs. 2005. *Am J Respir Crit Care Med* 171:1292.
158. Lepore, J. J., A. Maroo, L.M. Bigatello, G.W. Dec, W.M. Zapol, K.D. Bloch, and M.J. Semigran. 2005. *Chest* 127:1647.
159. Michelakis, E.D., W. Tymchak, M. Noga, L. Webster, X.C. Wu, D. Lien, S.H. Wang, D. Modry, and S.L. Archer. 2003. *Circulation* 108:2066.
160. Michelakis, E., W. Tymchak, D. Lien, L. Webster, K. Hashimoto, and S. Archer. 2002. *Circulation* 105:2398.
161. Webb, D.J., G.J. Muirhead, M. Wulff, J.A. Sutton, R. Levi, and W.W. Dinsmore. 2000. *J Am Coll Cardiol* 36:25.
162. Webb, D.J., S. Freestone, M.J. Allen, and G.J. Muirhead. 1999. *Am J Cardiol* 83:21C.
163. Alaeddini, J., P.A. Uber, M.H. Park, R.L. Scott, H.O. Ventura, and M.R. Mehra. 2004. *Am J Cardiol* 94:1475.
164. Preston, I.R., J.R. Klinger, J. Houtches, D. Nelson, H.W. Farber, and N.S. Hill. 2005. *Respir Med* 99:1501.

25 Molecular Determinants in Pulmonary Hypertension: The Role of PDE5

N.J. Pyne, F. Murray, Rothewelle Tate, and M.R. MacLean

CONTENTS

25.1 Pulmonary Arterial Hypertension 501
    25.1.1 Pulmonary Arterial Hypertension and Bone Morphogenetic
            Protein Receptor Mutation and Role for Modifying Factors (PDE5) 501
    25.1.2 Use of the PDE5 Inhibitor Sildenafil in Pulmonary
            Arterial Hypertension 502
    25.1.3 Hypoxia-Induced Pulmonary Vasoconstriction and
            Phosphodiesterase Inhibition 503
25.2 Phosphodiesterase Activity in Chronic Hypoxic Models
     of Pulmonary Arterial Hypertension 504
    25.2.1 Chronic Hypoxia, Cyclic Nucleotides, and Vascular Reactivity 504
    25.2.2 Role of PDE3 in Pulmonary Arterial Hypertension..................................508
    25.2.3 Genomic Organization of PDE5A Gene and
            Regulation of Expression by Chronic Hypoxia 509
    25.2.4 Role of Nuclear Factor-κB and PDE5 Expression....................................510
    25.2.5 Potential Cross-Talk Regulation between PDE5
            and PDE3 via an NFκB-Mediated Pathway 511
    25.2.6 Angiotensin II and PDE5 Expression..511
25.3 PDE5 Adaptor Proteins and Pulmonary Arterial Hypertension 512
    25.3.1 Interaction of PDEγ with PDE6 and Nonretinal Expression....................512
    25.3.2 PDEγ and Mitogenic Signaling 513
25.4 Conclusion 515
Acknowledgments 515
References 515

25.1 PULMONARY ARTERIAL HYPERTENSION

25.1.1 PULMONARY ARTERIAL HYPERTENSION AND BONE MORPHOGENETIC PROTEIN RECEPTOR MUTATION AND ROLE FOR MODIFYING FACTORS (PDE5)

The pulmonary vascular bed is a low-pressure system with a resistance approximately one-tenth that of systemic circulation. In the normal lung, pulmonary vascular tone is regulated by a balance between the effects of vasodilators/antiproliferative agents, such as prostacyclin, and vasoconstrictors/comitogens, such as serotonin and endothelin-1.[1-3] Pulmonary arterial

hypertension (PAH) is characterized by a sustained and progressive elevation in pulmonary artery pressure, right heart failure, and death. Familial PAH (fPAH) has been shown to be related to heterozygous germline mutations in the gene (BMPR2) encoding the bone mor-phogenetic protein type 2 receptor (BMPR2) and polymorphisms[4] in the gene encoding the serotonin (5-hydroxytryptamine, 5-HT) transporter (5-HTT).[5] Sporadic or idiopathic PAH (iPAH) has no demonstrable cause and PAH can also occur as a secondary consequence to many cardiorespiratory disorders. The elevated pulmonary vascular resistance is associated with remodeling of muscular pulmonary arteries and arterioles, which exhibit smooth muscle proliferation, medial hypertrophy, and fibrosis.[1] The fPAH carries an extremely poor prog-nosis. Survival has improved with epoprostenol therapy but is only 62.8% at 3 y with epoprostenol therapy and 34% without treatment. Lung transplantation is an option but is associated with a high mortality with survival of 65% at 1 y and 44% at 5 y (UNOS, US registry for transplant recipients, 1999). Hence novel therapeutic approaches are still urgently required. It is clear that PAH has a multifactorial pathobiology and it is unlikely that one factor or gene mutation will explain all forms and cases of PAH. For example, only around 60% of fPAH patients and 32% of sporadic PAH patients have the BMPR2 gene mutation. Therefore, modifying factors must interact to cause PAH. Thus, there are likely modifier genes or factors within patients carrying the BMPR2 mutations. In a small study of 33 pa-tients, Humbert et al.[6] reported that approximately 15% of patients with PAH secondary to fenfluramine ingestion also have the BMPR2 gene mutation. They concluded that the combination of this mutation and fenfluramine derivatives greatly increases the risk of developing severe PAH.

25.1.2   Use of the PDE5 Inhibitor Sildenafil in Pulmonary Arterial Hypertension

PDE5 activity has been proposed as a modifying factor in the pathobiology of PAH. PDE5 has been shown to be the main enzyme regulating cGMP hydrolysis and downstream signaling in human pulmonary artery smooth muscle cells. Indeed, sildenafil (a highly selective PDE5 inhibitor) exerts an antiproliferative effect in these cells.[7] These effects are mediated by cGMP via activation of cGMP-dependent protein kinases (PKGs). PDE5 is the major cGMP PDE subtype in the pulmonary circulation and it is also more abundant in the lung compared with other tissues.[8,9] Inhibition of PDE5 therefore offers an antiproli-ferative and vasodilator approach in the treatment of PAH. For instance, sildenafil attenu-ates hypoxia-induced PAH and pulmonary vascular remodeling in humans.[10] Experimentally, in rats exposed to hypoxia, PDE5 is localized throughout the muscularized pulmonary arteries including newly muscularized small pulmonary arteries induced by hypoxia. Sildenafil inhibits hypoxia-induced PAH and vascular remodeling.[11] Sildenafil also improved survival, increased intracellular cGMP levels, and reduced PAH and pulmonary vascular remodeling in monocrotaline-induced PAH rats[12] and partially prevented overcirculation-induced PAH in piglets.[13]

In a randomized, placebo-controlled, double-blind crossover study, sildenafil significantly improved exercise tolerance, cardiac index, and quality of life indices in patients with fPAH.[14] Hypoxia-induced PAH can also occur at altitude and secondary to chronic obstructive airway disease with alveolar hypoxia, leading to pulmonary vasoconstriction and remodeling. Silde-nafil has been shown to protect against the development of such altitude-induced PAH in man.[15,16] A more recent clinical trial has demonstrated that sildenafil is effective in the treatment of PAH (SUPER-1 study).[17] In this double-blind, placebo-controlled study, patients with symptomatic PAH (either idiopathic or associated with connective tissue disease or with repaired congenital systemic-to-pulmonary shunts) were administered sildenafil (20, 40, or 80 mg), or placebo, orally 3 times daily for 12 weeks. The 6 min walk test indicated that

all sildenafil groups had improved significantly and all sildenafil doses reduced mean pulmonary artery pressure and improved the WHO functional class. In June 2005, sildenafil (Revatio) was licensed by the FDA for the treatment of PAH.

Because of the multifactorial pathobiology of PAH, combination therapies will probably offer the optimum therapeutic strategy.[3] Sildenafil has been tested both experimentally and clinically in combination with current therapeutic strategies for treating PAH. One clinical study described the experience of a 3 y approach comparing survival of patients treated between January 2002 and December 2004 with the expected survival and with the historical survival of PAH patients treated between January 1999 and December 2001, i.e., before bosentan and sildenafil became available. Combinations of sildenafil, bosentan, and inhaled iloprost were examined in 123 patients, comparing the same dosing regime for the single and combination therapy. Survival at 1, 2, and 3 y was significantly greater than the survival of a historical control group, and greater than the expected survival. The use of combination treatment also significantly improved the combined end point of death, lung transplantation, and need for intravenous iloprost treatment.[18] Clinically, it has also been shown that combining the endothelin A/B (ETA/B) antagonist, bosentan (Tracleer) or beraprost, with sildenafil in patients with iPAH is safe and effective.[19,20] The effect of sildenafil is likewise potentiated by combination with inhaled NO in patients with PAH.[21]

In experimental models, various combinations have been examined. In the monocrotaline-induced PAH rat model, a combination of sildenafil and the prostacyclin analog beraprost attenuated the development of PAH and pulmonary vascular remodeling more than either drug alone.[22] In addition, PDE3, PDE4, and PDE5 inhibitors potentiated the pulmonary vasodilator effects of inhaled prostacyclin in a rabbit model of acute PAH.[23,24]

Sildenafil has also been shown to be extremely effective in infants born with PAH.[25] Indeed, PDE5 activity may modulate vasodilation of the pulmonary arteries at birth. Hanson et al.[26] demonstrated that at 1 h following birth, PDE5 activity and expression levels were dramatically reduced in pulmonary arteries from ovine and murine models and that this was correlated with an early transition change in pulmonary vascular resistance. However, a secondary increase in PDE5 activity was observed between 4 and 7 d, whereas pulmonary vascular resistance was maintained at a low level. These findings were interpreted in terms of a commensurate increase in nitric oxide synthase (NOS), which functions to circumvent the increase in PDE5 activity on intracellular cGMP. The secondary increase in PDE5 activity exceeded expression levels of the enzyme itself, suggesting that other regulatory factors may modulate PDE5 activity under these conditions. These results demonstrate a complex interplay between PDE5 activity or expression, NOS, and pulmonary vascular resistance, and provide support for the possibility that such complex gene expression profiling may be severely deregulated in persistent PAH of the newborn.

25.1.3 HYPOXIA-INDUCED PULMONARY VASOCONSTRICTION AND PHOSPHODIESTERASE INHIBITION

Regional vasoconstriction in response to low vascular oxygen tension, i.e., hypoxic pulmonary vasoconstriction (HPV), is present in the pulmonary vasculature. This represents the main physiological mechanism to match ventilation and perfusion in the lung. HPV diverts regional pulmonary blood flow away from poorly ventilated toward better ventilated lung regions and thus preserves ventilation–perfusion matching and systemic oxygenation.[27] Despite great efforts, the mechanism of HPV has not yet been fully understood. However, it is believed to be modulated by the endothelium and to involve oxidant and redox signaling.[28,29] As these mechanisms involve modulation of cyclic nucleotides, the role of PDE activity in HPV has been investigated. In this regard, it has been demonstrated that the PDE3 inhibitor

SCA40 can reverse the HPV observed in rat-isolated perfused lungs.[30] More recently, using an in situ rat perfused lung preparations, Phillips et al.[31] have shown that the semiselective PDE1 inhibitor vinpocetine (PDE1), cilostamide (PDE3 selective), and rolipram (PDE4 selective) all attenuate HPV. In isolated rat pulmonary arteries, the PDE3 inhibitor siguazodan, the PDE4 inhibitor rolipram, and the PDE1/PDE5 inhibitor zaprinast relax rat pulmonary arteries preconstricted by hypoxia.[32] However, as we discuss in this chapter later, these PDE inhibitors can relax pulmonary arteries preconstricted by other means such as by pretreatment with vasoconstrictors. This makes the interpretation of these *in vivo* and *in vitro* studies problematic as the inhibition of HPV could simply reflect the vasodilator activity of PDE inhibitors rather than interference of a hypoxia-induced mechanism. However, a more convincing study has implicated cGMP accumulation and PDE5 activity in the human response to hypobaric hypoxia. In this study, sildenafil was administered to healthy volunteers prior to exposure to a simulated altitude of 5000 m. Sildenafil significantly inhibited the increase in pulmonary pressures produced by hypoxia without affecting resting pulmonary pressures.[33] One interpretation of this is that HPV normally involves the transient inhibition of nitric oxide (NO) (resulting in a subsequent decline in GMP), which normally exerts a Ca^{2+}-desensitizing action.[34] It has previously been shown that inhibition of endothelial NOS potentiates HPV in rat lungs and isolated pulmonary arteries.[35] Sildenafil has, however, been shown to preserve ventilation and perfusion matching, an important consideration for its clinical use in PAH.[36]

25.2 PHOSPHODIESTERASE ACTIVITY IN CHRONIC HYPOXIC MODELS OF PULMONARY ARTERIAL HYPERTENSION

25.2.1 CHRONIC HYPOXIA, CYCLIC NUCLEOTIDES, AND VASCULAR REACTIVITY

As mentioned briefly above, acute hypoxia causes pulmonary arteriolar vasoconstriction and increased pulmonary arterial pressure. Chronic hypoxia induces a sustained increase in pulmonary arterial pressure and pulmonary vascular smooth muscle cell proliferation and the chronic hypoxic rat is widely used as a model for the study of chronic hypoxia–induced PAH.[37,38]

We have therefore used the chronic hypoxic rat models to evaluate whether cGMP PDE function is modulated in a manner that explains hyporesponsiveness of vasodilators in PAH, such as nitric oxide, which use cGMP signaling to regulate vascular tone and blood pressure. We reported that cAMP and cGMP PDE activity in the main and first-branch pulmonary arteries, but not resistance vessels is substantially increased in pulmonary arteries from chronic hypoxia–treated rats (Figure 25.1).[39] The phenotypic changes in cGMP PDE activity were correlated with decreased intracellular cGMP in these vessels. There was also a substantial increase in calcium/calmodulin-stimulated PDE1 activity in the main pulmonary artery branch and in zaprinast-sensitive cGMP PDE activity that was attributed to PDE5/PDE1, in the main, first-branch, and intrapulmonary arteries.[39]

Having established phenotypic changes in PDE activity profiles under chronic hypoxic conditions, we sought to define whether this could lead to changes in the responsiveness of arterial tissue tone to PDE inhibitors. In recent studies, we have compared vasodilator responses to sildenafil in the main and first-branch pulmonary arteries from rats exposed to normobaric and hypobaric conditions (14 d of hypoxia). The PDE5 inhibitor, sildenafil, was effective in relaxing phenylephrine-, serotonin-, and endothelin-1-induced contraction of main branch pulmonary vessels from normoxic-treated animals (Figure 25.2A). We found that the removal of the endothelium did not diminish the relaxant effect of sildenafil, suggesting that endothelial-derived nitric oxide does not contribute to the activation of an

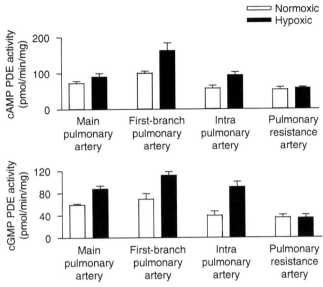

FIGURE 25.1 Effect of chronic hypoxia on cGMP and cAMP phosphodiesterase activity in different pulmonary arteries. Rats were maintained in a normbaric or hypobaric chamber after which main, first-branch, intrapulmonary, and resistance arteries were removed, homogenized, and assayed for cAMP and cGMP PDE activity using 1 μM of either [^{3}H]cAMP or [^{3}H]cGMP. Results are expressed as specific activities (pmol/min/mg) $\pm$ SD for $n = 6$ determinations. The hypobaric chamber was depressurized, over 2 d, to 550 mbar (equivalent to PO$_2$ of 110 mmHg). The temperature of the chamber was maintained at 21–22°C and the chamber was ventilated with air at approximately 45 L/min. Animals were maintained in these hypoxic and hypobaric conditions for 2 weeks.

intracellular cGMP pool that is regulated by PDE5 in pulmonary smooth muscle (Figure 25.2A). However, sildenafil was moderately more effective at relaxing phenylephrine-, serotonin-, and endothelin-1-induced contraction of main branch pulmonary vessels from hypoxic-treated animals (Figure 25.2B), suggesting that synthesis of vasorelaxants in smooth muscle such as nitric oxide that stimulate intracellular cGMP formation via soluble guanylate cyclase may be increased under hypoxic conditions. An increased expression of PDE5A2 (see later) in the main branch would counter an increase in intracellular cGMP, leading to a consequential hyporesponsiveness to agents that relax smooth muscle via a cGMP-dependent mechanism. These findings suggest that the inhibition of PDE5 with sildenafil may unmask enhanced GMP responsiveness, such that the effectiveness of endogenous relaxant agents that increase cGMP is improved.

In these studies, we also investigated the vasodilator effects of PDE3 inhibitors in pulmonary arteries from hypoxic rats. In this regard, we found that the PDE3 inhibitor, SKF94836 (siguazodan), was effective at relaxing first-branch pulmonary vessels precontracted with phenylephrine, serotonin, or endothelin-1 (Figure 25.2C). Interestingly, the relaxant effect of SKF94836 was markedly diminished in vessels in which the endothelium had been removed (Figure 25.2C). These data provide strong evidence that the endothelium releases vasoactive substances that stimulate adenylyl cyclase to induce an increase in intracellular cAMP in smooth muscle, which in turn promotes relaxation. Inhibition of PDE3 by SKF98836 appears to potentiate the relaxant effect of these putative vasoactive substances (e.g., PGI2). The relaxant effect of SKF94836 was also markedly potentiated in pulmonary vessels from chronic hypoxic rats (Figure 25.2D). This relaxant effect and that of sildenafil cannot be accounted for by increased expression of PDE3 and PDE5A, as both SKF94836 and

sildenafil are competitive inhibitors of these enzymes, respectively. On the contrary, the enhanced vasorelaxant responses are likely to be due to a hypoxic-dependent increase in the release of endogenous vasoactive substances from the endothelium and smooth muscle, and subsequent potentiation of their effects on vascular tone by PDE3/PDE5 inhibitors.

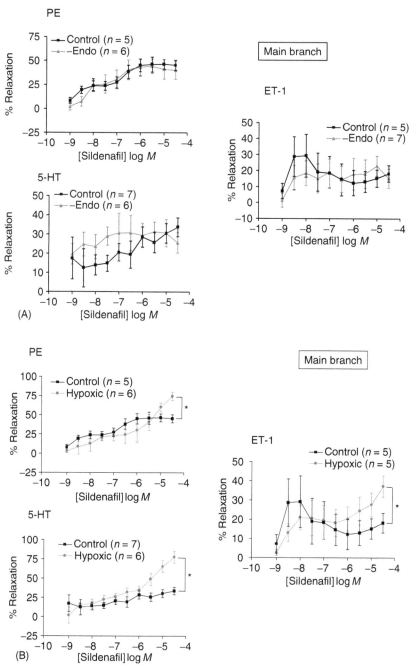

FIGURE 25.2

(*continued*)

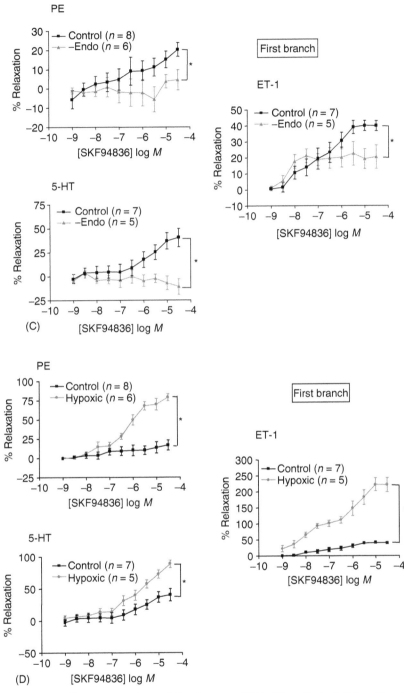

FIGURE 25.2 (continued) (See color insert following page 274.) The effect of sildenafil and SKF94836 (siguazodan) on the relaxation of pulmonary vessels. Main or first-branch pulmonary vessels were precontracted with phenylephrine (1 μM), serotonin (30 μM), or endothelin-1 (3 nM); (A) and (C) removal of endothelium; (B) and (D) chronic hypoxia (14 d in a hypobaric chamber). Results are expressed means ± SD for n = 5–8 different experiments.

Moreover, inhibition of cGMP degradation by PDE5 using sildenafil has the potential to also increase intracellular cAMP via a cGMP-mediated inhibition of PDE3, and some of its effects may occur via this mechanism and cannot be discounted.

Such models place significant importance on the potential cross-talk regulation between PDE5 and PDE3. Therefore, changes in the expression of PDE3 induced by chronic hypoxia might potentially increase the level of cross-talk regulation between intracellular cAMP and cGMP pools in terms of the regulation of vascular tone and gene expression.

25.2.2 ROLE OF PDE3 IN PULMONARY ARTERIAL HYPERTENSION

PDE3 consists of two isoforms, termed PDE3A and PDE3B, which display distinct N-terminal domains that might confer differential subcellular localization of each isoform.[40] Using RT-PCR with gene-specific primers (corresponding to positions 2989–3393 and 2912–3238 in rat PDE3A and PDE3B, respectively), we have reported that PDE3A and PDE3B transcripts are expressed in rat pulmonary vessels.[41] The alignment of the amplicons with the corresponding regions in the published rat PDE3A and PDE3B revealed 100% similarity in their nucleotide sequences, respectively, whereas there is 75% and 80.7% similarity with corresponding human PDE3A and PDE3B. Moreover, PDE3A/PDE3B transcript levels were elevated by chronic hypoxia in all the vessels and this was correlated with increased PDE3 activity in these vessels.[41] PDE3A expression and activity were also increased via a cAMP-dependent mechanism in cultured human pulmonary smooth muscle cells as evidenced by the data showing that the hypoxic-dependent increase in PDE3A expression was mimicked by treating cells with Br-cAMP and was reduced with the PKA inhibitor, H8 peptide.[41] These data provide evidence that chronic hypoxia may induce the expression of PDE3A via a mechanism that might involve protein kinase A and cAMP response element binding protein (CREB) transcription factors (see Figure 25.3).

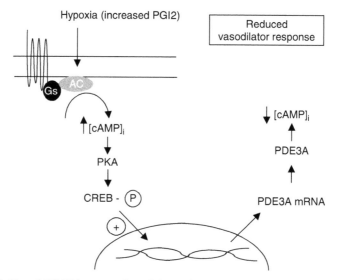

FIGURE 25.3 cAMP and PDE3A expression. Schematic showing how chronic hypoxia might induce increased levels of adenylyl cyclase activators leading to increased intracellular cAMP, which induces PDE3A expression leading to a delayed desensitization and hyporesponsiveness of vascular smooth muscle to vasorelaxants that use cAMP as a second messenger to promote a reduction in blood pressure.

25.2.3   Genomic Organization of PDE5A Gene and Regulation of Expression by Chronic Hypoxia

Bovine PDE5A was the first gene to be isolated for this isoform and the protein was shown to be composed of 865 amino acid residues.[42] Subsequently, it was reported that the N-terminal region of rat lung PDE5A was different from human and bovine PDE5A whereas the two GAF domains and the catalytic region were highly conserved among human, bovine, and rat PDE5A.[43] Indeed two forms of canine PDE5A, which represent different N-terminal variants of PDE5A cDNA have been identified.[44] The human PDE5A gene is composed of at least 21 exons with the unique N-terminal region of human PDE5A1 separated by an intron from the common regions.[45] Thus, PDE5A1 and PDE5A2 are derived from alternative splicing of the same PDE5A gene. Thus, the first two exons of PDE5A are alternative first exons encoding the isoform-specific sequences. Both PDE5A1 and PDE5A2 have similar catalytic function. However, the different N-terminal regions may confer unique regulation and subcellular localization. A third human PDE5A isoform has been identified, termed PDE5A3,[46] which appears to be expressed specifically in smooth muscle and is derived from a third alternate start site in the PDE5A gene (see Figure 25.4 for PDE5A gene organization).

The human PDE5A promoter overlaps with the A1-specific exon whereas the PDE5A2 promoter is located between the A3- and A2-specific exons. Therefore, the PDE5A promoter upstream of exon 1 may regulate expression of all three PDE5 isoforms whereas the intronic PDE5A2 promoter may drive expression of PDE5A2 alone. Sequence analysis revealed various putative regulatory elements and promoter-like sequences in the upstream and downstream regions of exons 1A and 1B in both human and rat PDE5A genes.[45] In the rat gene, several Sp1 sites were located at -2535, -1544, -1514, and -1205 upstream of exon 1A and exon 1B (the translation initiation site of PDE5A2 was designated $+1$, and the translation initiation site of rat PDE5A1 was located at -1375). The Sp1-binding sequence is critical for

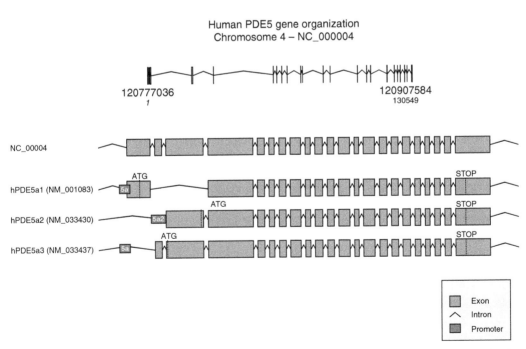

FIGURE 25.4 (See color insert following page 274.) Schematic of human PDE5A gene organization showing the location of alternate initiation sites and promoters within the gene.

basal and cGMP/cAMP-inducible promoter activities. AP-1 sites were identified at -2094 and -1709, and two AP-2 sites were situated between -4010 and -851. The cAMP response element (CRE) sites were identified at -3628, -2072, and -1634 upstream of exon 1A. Interestingly, the treatment of vascular smooth muscle cells with dibutyryl cAMP (dbcAMP) caused upregulation of the expression of PDE5A2, with no effect on the expression of PDE5A1.[45] The authors concluded that PDE5A2 is an inducible PDE5A,[45] suggesting that elements in the PDE5A2 promoter are sensitive to cAMP-dependent regulation.

The changes in PDE5 activity observed under chronic hypoxia might result from altered expression of PDE5A.[39] In this context, we have reported that PDE5A1 and PDE5A2 proteins are differentially expressed in pulmonary vascular beds of rat.[41] PDE5A2 ($M_r = 93$ kDa) was expressed in the main branch vessel whereas very low levels of PDE5A2 were found in the first branch, based upon Western blot analysis of vessel homogenates using anti-PDE5 antibody.[41] Western blot analysis also revealed that PDE5A1 ($M_r = 98$ kDa) and PDE5A2 ($M_r = 93$ kDa) were detected together in the intrapulmonary and small resistance vessels. We reported that chronic hypoxia increased PDE5A2 protein levels above basal in the main branch and first-branch pulmonary vessels.[41] Chronic hypoxia did not modulate PDE5A1/A2 protein levels in the intrapulmonary and resistance vessels.[41] The changes in PDE5A2 expression in the main and first-branch vessels can be accounted for, in part, by a chronic hypoxic–dependent increase in the transcriptional regulation of the PDE5A gene. A single PDE5A mRNA product of 300 base pairs was amplified by RT-PCR from rat main and first-branch vessel using gene-specific primers (corresponding to position 2425–2724 in rat PDE5A) and alignment of this product with the corresponding region in published rat and human PDE5A revealed 100% and 91.7% similarity in the nucleotide sequence, respectively. Significantly, PDE5 transcript levels were increased in the main and first-branch vessels by chronic hypoxia assessed by semiquantitative RT-PCR.[41] These findings raise the question as to the mechanism of chronic hypoxic–dependent upregulation of PDE5A transcriptional regulation. Since the PDE5A gene promoter contains Sp1- and CRE-binding sites, it is possible that intracellular cGMP/cAMP levels that might be temporally elevated during the early development of PAH may lead to an increase in PDE5 expression. Indeed, endothelial NO synthase (eNOS) is increased in both large and small pulmonary arteries from chronic hypoxic rats[47] and this would theoretically enhance NO-mediated signaling via a cGMP-dependent mechanism during early onset of PAH. Therefore, chronic hypoxic–dependent changes in PDE5A2 expression may be consequential, thereby leading to desensitization of NO-mediated signaling and vasorelaxation. This might contribute to increased pulmonary pressure. Previous reports have also shown that PDE5 protein expression is increased in lambs with PAH, induced by aorta-pulmonary vascular graft placement.[48] These findings may also explain the mechanism of acetylcholine-induced vasodilation, which is decreased in large proximal pulmonary arteries,[49] but actually increased in more distal resistance arteries.[50] Consistent with these findings, intracellular cGMP levels are decreased in the large pulmonary arteries but unchanged in the resistance pulmonary arteries.[51] PDE5A1 and PDEA2 expression is unchanged in the resistance arteries, thereby preserving the ability of acetylcholine to induce vasodilation, which may actually be enhanced due to increased levels of guanylate cyclase.[52] These findings are consistent with the observations of Oka[53] who showed that sildenafil selectively vasodilates the large pulmonary arteries but not the resistance arteries.

25.2.4   ROLE OF NUCLEAR FACTOR-κB AND PDE5 EXPRESSION

Nuclear factor-κB (NFκB) is a transcription factor that can be activated by hypoxia, resulting in its nuclear translocation and induction of gene products that play a significant role in

regulating vascular reactivity. For instance, NFκB regulates expression of inducible NOS (iNOS), which in turn regulates vascular tone through the action of NO and cGMP. More-over, iNOS expression is increased in the aorta of spontaneously hypertensive rats versus age-matched Wistar–Kyoto rats.[54] Additional support for a role of NFκB was evident from studies showing that the treatment of spontaneously hypertensive rats with NFκB-processing inhibitor, pyrrolidine dithiocarbamate, and the iNOS inhibitor, aminoguanidine, significantly reduced the development of hypertension and improved the diminished vascular responses to acetylcholine.[54] With respect to a role for NFκB in PAH, there is an inverse relationship between airway NO levels and NFκB activation.[55] NFκB activation has also been linked with cellular oxidative stress associated with monocrotaline-induced PAH.[56]

Given the role of NFκB in mediating hypoxic-induced stress responses on transcriptional regulation, we investigated whether PDE3 and PDE5 gene expression is regulated by this transcription factor. In this regard, we reported that the treatment of cultured human pulmonary smooth muscle cells with the inhibitor of IκB degradation, N-tosyl-L-lysine-chloromethyl ketone (TLCK), reduced the basal expression of PDE5A2.[41] Unfortunately, inhibitors used to investigate the role of NFκB are not particularly specific as they target proteosomal degradation of proteins per se. Therefore, conclusions regarding the role of NFκB using these compounds must be viewed with some circumspection. Nevertheless, the findings do highlight a potential interrelationship between NFκB and the induction of enzymes that regulate cGMP signaling (NOS and PDE5). Analysis of sequence data described by Kotera et al.[45] reveals that the promoters of PDE5A1 and PDE5A2 contain putative NFκB-binding sites at -3544 (GGGGGTTTCT), -2236 (GGGTGTTCCC), -1293 (GGGACTTTAC), -970 (GGGCCTTCCC), and -205 (GGGACTCGCC) (the translation initiation site of PDE5A2 was designated $+1$, and the translation initiation site of rat PDE5A1 was located at -1375).

25.2.5 POTENTIAL CROSS-TALK REGULATION BETWEEN PDE5 AND PDE3 VIA AN NFκB-MEDIATED PATHWAY

An NFκB-mediated upregulation of PDE5 is predicted to lead to a reduction in intracellular cGMP levels. This would effectively relieve inhibition of PDE3 by cGMP leading to conse-quential reduction in intracellular cAMP. In this regard, it is well established that cAMP via PKA inhibits NFκB activation (Figure 25.5). Therefore, it is possible that hyperactivation of NFκB might result in more pronounced upregulation of PDE5A2 expression in an amplifi-cation cascade. This will maintain PDE5 expression and would theoretically lead to dimin-ished vascular responsiveness to vasorelaxants and increased vascular smooth muscle proliferation and vascular remodeling, leading to the sustained development of PAH.

25.2.6 ANGIOTENSIN II AND PDE5 EXPRESSION

Angiotensin II can stimulate proliferation of human pulmonary artery smooth muscle cells via the AT1 receptor,[57] and endothelial angiotensin-converting enzyme (ACE) activity is increased in pulmonary arteries of patients with primary and secondary PAH.[58] In human pulmonary artery adventitial fibroblasts, hypoxia and transient overexpression of HIF-1 alpha induces an increase in ACE and ET1 expression.[59] ACE insertion (I) and deletion (D) polymorphisms can affect many disease states. The ACE DD genotype has been associ-ated with impairment in tissue oxygenation during exercise in patients with COPD[60] and also in newborns with persistent PAH.[61] Hence, an upregulation of angiotensin II responsiveness in PAH is of interest as recent studies by Kim et al.[62] have demonstrated that angiotensin II

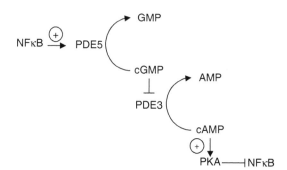

FIGURE 25.5 NFκB and PDE5–PDE3 cross-talk regulation. Schematic showing the role of NFκB in regulating PDE5 and the potential cross talk with PDE3, leading to amplification of PDE5 expression. An NFκB-mediated upregulation of PDE5 is predicted to lead to a reduction in intracellular cGMP levels. This would effectively relieve inhibition of PDE3 by cGMP, leading to consequential reduction in intracellular cAMP. The relief of cAMP-mediated inhibition of NFκB might then lead to increased NFκB-dependent PDE5A expression.

rapidly and transiently increased PDE5A mRNA transcript levels and PDE5A protein expression in rat aortic smooth muscle cells. The increase in PDE5A protein expression was mediated by the angiotensin type I receptor and involved a p42/p44 MAPK-dependent transcriptional regulation (via p62TCF, SRE, and SRF ternary complex) of the PDE5A gene. This was apparent upon blockade of angiotensin II-induced PDE5A expression by dominant negative MEK-1. The angiotensin II-stimulated upregulation of PDE5A may account for the ability of this GTP-binding protein-coupled receptor (GPCR) agonist to antagonize ANP-induced cGMP signaling and vasodilation. Such conclusions were supported by results showing that upregulation of PDE5 reduced PKG-catalyzed phosphorylation of vasodilator-stimulated phosphoprotein (VASP), an effect that was rescued using selective PDE5 inhibitors, such as sildenafil.[62] In addition, the increased expression of PDE5A may also contribute to angiotensin II-stimulated cell proliferation. This was based upon the finding that the inhibition of PDE5 with sildenafil reduced angiotensin II-stimulated cell proliferation. Relevant to this issue, there is some debate as to whether the inhibitory effects of cGMP on cell proliferation are mediated via PKGI as postnatal ablation of PKGI expression in transgenic mice reduced atherosclerotic lesions, suggesting that PKGI promotes vascular proliferation of medial vascular pulmonary smooth muscle cells.[63] Thus, the effects of sildenafil via PDE5A inhibition may be mediated via PKA and CREB or alternatively, through a consequential increased inhibition of PDE3 by cGMP, leading to ablated NFκB activation via PKA.

25.3 PDE5 ADAPTOR PROTEINS AND PULMONARY ARTERIAL HYPERTENSION

25.3.1 INTERACTION OF PDE$_\gamma$ WITH PDE6 AND NONRETINAL EXPRESSION

The phototransduction cascade (for review see Ref. [64]) involves cGMPγ phosphodiesterases (PDE6) that are expressed in rod and cone photoreceptors (termed PDE6) as tetrameric proteins composed of two catalytic subunits and two γ subunits (PDEγ). PDEγ inhibits cGMP hydrolysis at the catalytic sites of PDE6αβ. This involves interaction between a central polycationic region (amino acids 24–46) and the C-terminus of PDEγ with the PDE6 catalytic

subunits.[65,66] More recently, the polycationic region of PDEγ has been proposed to bind in an asymmetric orientation with the GAF domains of PDE6α and β subunits.[67] Light activation of rhodopsin induces GDP–GTP exchange in the G-protein, transducin and the activated GTP-bound $G_t\alpha$ then interacts with the PDE6αβγ2, removing the inhibitory influence of PDE6γ on the catalytic subunits of PDE6. PDEγ also binds to transducin via both its polycationic and C-terminal regions and the C-terminal region is involved in stimulating transducin GTPase activity.[68,69] Therefore, PDE6γ interacts with and regulates the activities of both PDE6 and transducin and can be considered a multifunctional protein.

25.3.2 PDEγ and Mitogenic Signaling

The phototransduction cascade involving rhodopsin (GPCR), G-protein, G-protein receptor-coupled kinase (GRK), and β-arrestin[64] bears many similarities with signaling by growth factors and other GPCR in nonretinal mammalian cell systems. Therefore, PDEγ might be expressed in other mammalian tissues, where it may regulate distinct PDE and receptor–G-protein-mediated pathways. Indeed, we reported that rod PDEγ is expressed in lung, kidney, testes, liver, heart, airway smooth muscle, and human embryonic kidney 293 cells and have shown its absence from these tissues in rod PDEγ knockout mice.[70–72] Moreover, we demonstrated that a peptide corresponding to the polycationic central region (amino acids, 24–46) of PDEγ reduced the ability of PDE5 to be activated by PKA in vitro.[73] We also demonstrated that PDE5 activity was coimmunoprecipitated with PDEγ from airway smooth muscle cell lysates using anti-PDEγ antibody, suggesting that these proteins form a complex. However, there is no evidence for a direct interaction between PDE5 and PDEγ and it remains to be determined whether association between PDE5 and PDEγ involves other protein–protein interactions. Nevertheless, PDEγ can modulate the susceptibility of PDE5 to proteolysis by caspase-3 in vitro,[71] suggesting that the interaction between these proteins may cause a conformational change in PDE5 that allows it to be more susceptible to caspase-3. In recent studies, we have also shown that rod and cone PDEγ (termed here PDEγ1 and PDEγ2, respectively, and which differ only in their N-terminal region) interact with the GRK2 signaling system to regulate the epidermal growth factor- and thrombin-dependent stimulation of p42/p44 MAPK in human embryonic kidney 293 cells.[74] Thrombin also stimulated the association of endogenous PDEγ1 with dynamin II, which plays a critical role in regulating endocytosis of receptor signal complexes required for the activation of p42/p44 MAPK by certain growth factors and G-protein-coupled receptor agonists. This may involve interaction of dynamin II and intermediate SH3 domain containing proteins (e.g. Grb-2) with the SH3 recognition sequence that includes amino acids 21–24 corresponding to PTVP sequence that encompasses the beginning of the central polycationic region of PDEγ. These findings demonstrated a novel role for PDEγ1/PDEγ2 in regulating endocytosis of receptor signal complexes, leading to the activation of p42/p44 MAPK and cell proliferation.[74]

As PDEγ has the potential to form complexes with PDE5, G-protein, and c-Src/GRK2, it is tempting to speculate whether PDEγ might function to localize PDE5 in GRK2 signaling complexes that are organized to stimulate the p42/p44 MAPK pathway. The association of PDE5 in this complex might then function to protect the p42/p44 MAPK from cGMP, which have been shown to inhibit mitogenic-signaling processes (see Figure 25.6). However, these interactions need to be confirmed experimentally. For instance, it is equally plausible that PDEγ and PDE5 might be placed on separate signal pathways. Nevertheless, we have been interested in establishing whether PDEγ has a functional role in PAH.[75] In this regard, we reported that PDEγ1/PDEγ2 is expressed in the main, first, intrapulmonary, and resistance pulmonary arteries.[75] Chronic hypoxia increased PDEγ1/PDEγ2 protein levels above basal in all the arteries. This was established by Western blot analysis using anti-PDEγ antibody.

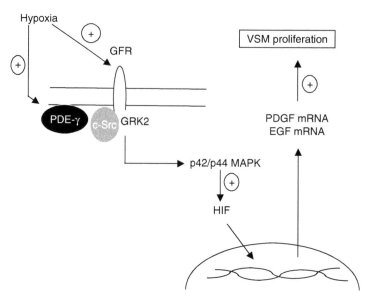

FIGURE 25.6 Schematic model showing the potential role of PDEγ in regulating growth factor-induced arterial smooth muscle remodeling in PAH. The schematic shows that PDEγ association with GRK2-c-Src may enhance signal transmission via the p42/p44 MAPK pathway, leading to increased early immediate gene expression.

RT-PCR with gene-specific primers separately amplified PDEγ1 and PDEγ2 transcripts from all the vessels,[75] although there were no differences in PDEγ transcript levels under chronic hypoxic conditions. Thus, chronic hypoxia appears to increase the expression PDEγ1/PDEγ2 via a posttranscriptional mechanism.[75] A chronic hypoxic–dependent increase in p42/p44 MAPK phosphorylation and activation status was correlated with the increase in PDEγ1/PDEγ2 expression in all the pulmonary vessels.[75] A potential link between increased PDEγ expression and enhanced p42/p44 MAPK activation was evident from results showing that the overexpression of recombinant PDEγ in HEK 293 cells enhanced EGF- and thrombin-stimulated activation of the p42/p44 MAPK pathway.[74] The activation of p42/p44 MAPK by extracellular stimuli plays a key role in regulating cell proliferation and is therefore likely to have an important role in pulmonary vessel remodeling in PAH. This conclusion is supported by several lines of previous evidence regarding the key role of p42/p44 MAPK in hypoxic-dependent PAH. For instance, others have shown that chronic and acute hypoxia–induced temporal activation of p42/p44 MAPK.[76,77] In the case of chronic treatment, the hypoxic-dependent increase in p42/p44 MAPK activation peaked at 7 d and declined thereafter. Moreover, Premkumar et al.[78] reported that hypoxia activates p42/p44 MAPK and that this resulted in c-*fos* expression via the *cis* serum response element, a critical immediate early gene involved in regulating mitogenesis. Finally, Minet et al.[79] demonstrated that hypoxic-dependent stimulation of p42/p44 MAPK was essential for hypoxia-inducible factor-1 (HIF-1) transactivation activity. The hypoxic-dependent induction of endothelial growth factor receptor, PDGF receptor, and egr-1 mRNA was correlated with p42/p44 MAPK activation. These authors also concluded that the temporal activation of the p42/p44 MAPK pathway appears to be associated with hypoxia-induced pulmonary arterial remodeling. We speculate therefore that the potential interaction between PDEγ, p42/p44 MAPK activation, and PDE5A may represent a coordinated signaling response. The possible

link between PDE5 and p42/p44 MAPK is lent weight by observations such as those by Kim et al.[62] who reported that p42/p44 MAPK plays a critical role in regulating angiotensin II-stimulated PDE5 expression.

25.4 CONCLUSION

There is an increasing body of evidence that PDE5 plays a significant role in aberrant vascular dysfunction in pulmonary hypertension. The clinical efficacy of PDE5 inhibitors suggests that there is considerable therapeutic potential for treatment of PAH, and that these drugs may in fact target one of the contributory factors to this disease.

ACKNOWLEDGMENTS

This work was supported by The British Heart Foundation and The Wellcome Trust.

REFERENCES

1. Fishman, A.P. 1998. Etiology and pathogenesis of primary pulmonary hypertension: A perspective. *Chest* 1114 (Suppl. 3):242S–247S.
2. MacLean, M.R. 1999. Endothelin and serotonin in pulmonary hypertension. *J Lab Clin Med* 134:105–114.
3. Channick, R.N., and L.J. Rubin. 2000. Combination therapy for pulmonary hypertension: A glimpse into the future. *Crit Care Med* 28:896–897.
4. Machado, R.D., et al. 2000. BMPR2 haploinsufficiency as the inherited molecular mechanism for primary pulmonary hypertension. *Am J Hum Gen* 68:92–102.
5. Eddahibi, S., et al. 2001. Serotonin transporter overexpression is responsible for PA smooth muscle hyperplasia in primary pulmonary hypertension. *J Clin Invest* 108:1141–1150.
6. Humbert, M., et al. 2002. BMPR2 germline mutations in pulmonary hypertension associated with fenfluramine derivatives. *Eur Respir J* 20:518–523.
7. Wharton, J., et al. 2005. Antiproliferative effects of phosphodiesterase type 5 inhibition in human pulmonary artery cells. *Am J Respir Crit Care Med* 172:105–134.
8. Giordano, D., et al. 2001. Expression of cGMP-binding cGMP-specific phosphodiesterase (PDE5) in mouse tissues and cell lines using an antibody against the enzyme amino-terminal domain. *Biochim Biophys Acta* 1539:16–27.
9. Corbin, J.D., et al. 2005. High lung PDE5: A strong basis for treating pulmonary hypertension with PDE5 inhibitors. *Biochem Biophys Res Commun* 334:930–938.
10. Zhao, L., et al. 2001. Sildenafil inhibits hypoxia-induced pulmonary hypertension. *Circulation* 104:424–428.
11. Sebkhi, A., et al. 2003. Phosphodiesterase type 5 as a target for the treatment of hypoxia-induced pulmonary hypertension. *Circulation* 107:3230–3235.
12. Schermuly, R.T., et al. 2004. Chronic sildenafil treatment inhibits monocrotaline-induced pulmonary hypertension in rats. *Am J Respir Crit Care Med* 169:39–45.
13. Rondelet, B., et al. 2004. Signaling molecules in overcirculation-induced pulmonary hypertension in piglets: Effects of sildenafil therapy. *Circulation* 110:2220–2225.
14. Sastry, B.K., et al. 2004. Clinical efficacy of sildenafil in primary pulmonary hypertension: A randomized, placebo-controlled, double-blind, crossover study. *J Am Coll Cardiol* 43:1149–1153.
15. Perimenis, P. 2005. Sildenafil for the treatment of altitude-induced hypoxaemia. *Expert Opin Pharmacother* 6:835–837.
16. Aldashev, A.A., et al. 2005. Phosphodiesterase type 5 and high altitude pulmonary hypertension. *Thorax* 60:683–687.
17. Galie, N., et al. 2005. Sildenafil citrate therapy for pulmonary arterial hypertension. *N Engl J Med* 353:2148–2157.

18. Hoeper, M.M., et al. 2005. Goal-oriented treatment and combination therapy for pulmonary arterial hypertension. *Eur Respir J* 26:858–863.

19. Hoeper, M.M., et al. 2004. Combination therapy with bosentan and sildenafil in idiopathic pulmonary arterial hypertension. *Eur Respir J* 24:1007–1010.

20. Ikeda, D., et al. 2005. Addition of oral sildenafil to beraprost is a safe and effective therapeutic option for patients with pulmonary hypertension. *J Cardiovasc Pharmacol* 45:286–289.

21. Preston, I.R., et al. 2005. Acute and chronic effects of sildenafil in patients with pulmonary arterial hypertension. *Respir Med* 99:1501–1510.

22. Itoh, T., et al. 2004. A combination of oral sildenafil and beraprost ameliorates pulmonary hypertension in rats. *Am J Respir Crit Care Med* 169:34–38.

23. Schermuly, R.T., et al. 2005. Lung vasodilatory response to inhaled iloprost in experimental pulmonary hypertension: Amplification by different type phosphodiesterase inhibitors. *Respir Res* 6:76.

24. Schermuly, R.T., et al. 1999. Low-dose systemic phosphodiesterase inhibitors amplify the pulmonary vasodilatory response to inhaled prostacyclin in experimental pulmonary hypertension. *Am J Resp Crit Care Med* 160:1500–1506.

25. Hon, K.L., et al. 2005. Oral sildenafil for treatment of severe pulmonary hypertension in an infant. *Biol Neonate* 88:109–112.

26. Hanson, K.A., et al. 1998. Developmental changes in lung cGMP phosphodiesterase-5 activity, protein, and message. *Am J Respir Crit Care Med* 158:279–288.

27. Von Euler, U., and G. Liljestrand. 1946. Observations on the pulmonary arterial blood pressure of the cat. *Acta Physiol Scand* 12:301–320.

28. Wolin, M.S., M. Ahmad, and S.A. Gupte. 2005. Oxidant and redox signaling in vascular oxygen sensing mechanisms: Basic concepts, current controversies, and potential importance of cytosolic NADPH. *Am J Physiol Lung Cell Mol Physiol* 289:L159–L173.

29. Ward, J.P., and T.P. Robertson. 1995. The role of the endothelium in hypoxic pulmonary vasoconstriction. *Exp Physiol* 80:793–801.

30. Crilley, T.K., J.C. Wanstall, and P.A. Bonnet. 1998. Vasorelaxant effects of SCA40 (a phosphodiesterase III inhibitor) in pulmonary vascular preparations in rats. *Clin Exp Pharmacol Physiol* 25:355–360.

31. Phillips, P.G., et al. 2005. cAMP phosphodiesterase inhibitors potentiate effects of prostacyclin analogs in hypoxic pulmonary vascular remodeling. *Am J Physiol Lung Cell Mol Physiol* 288:L103–L115.

32. Bardou, M., et al. 2001. Hypoxic vasoconstriction of rat main pulmonary artery: Role of endogenous nitric oxide, potassium channels, and phosphodiesterase inhibition. *J Cardiovasc Pharmacol* 38:325–334.

33. Ricart, A., et al. 2005. Effects of sildenafil on the human response to acute hypoxia and exercise. *High Alt Med Biol* 6:43–49.

34. Knock, G.A., et al. 2005. Modulation of PGF2{alpha}- and hypoxia-induced contraction of rat intra-pulmonary artery by p38 MAPK inhibition: A nitric oxide dependent mechanism. *Am J Physiol Lung Cell Mol Physiol* 289:L1039–L1048.

35. Emery, C.J., et al. 2003. Vasoreactions to acute hypoxia, whole lungs and isolated vessels compared: Modulation by NO. *Respir Physiol Neurobiol* 134:115–129.

36. Ghofrani, H.A., et al. 2002. Sildenafil for treatment of lung fibrosis and pulmonary hypertension: A randomised controlled trial. *Lancet* 360:895–900.

37. Hunter, C., et al. 1974. Growth of the heart and lungs in hypoxic rodents: A model of human hypoxic disease. *Clin Sci Mol Med* 46:375–391.

38. Rabinovitch, M., et al. 1979. Rat pulmonary circulation after chronic hypoxia: Haemodynamic and structural features. *Am J Physiol* 236:H818–H827.

39. MacLean, M.R., et al. 1997. Phosphodiesterase isoforms in the pulmonary arterial circulation of the rat: Changes in pulmonary hypertension. *J Pharmacol Exp Ther* 283:619–624.

40. Leroy, M.J., et al. 1996. Characterization of two recombinant PDE3 (cGMP-inhibited cyclic nucleotide phosphodiesterase) isoforms, RcGIP1 and HcGIP2, expressed in NIH 3006 murine fibroblasts and Sf9 insect cells. *Biochemistry* 35:10194–10202.

41. Murray, F., M.R. MacLean, and N.J. Pyne. 2002. Increased expression of the cGMP-inhibited cAMP-specific (PDE3) and cGMP binding cGMP-specific (PDE5) phosphodiesterases in animal and cellular models of pulmonary hypertension. *Br J Pharmacol* 137:1187–1194.

42. McAllister-Lucas, L.M., et al. 1993. The structure of the bovine lung cGMP binding, cGMP-specific phosphodiesterase deduced from a cDNA clone. *J Biol Chem* 268:22863–22873.

43. Yanaka, N., et al. 1998. Expression, structure and chromosomal localization of human cGMP binding cGMP-specific phosphodiesterase. *Eur J Biochem* 255:391–399.

44. Kotera, J., et al. 1998. Novel alternative splice variants of the cGMP binding cGMP-specific phosphodiesterase. *J Biol Chem* 273:26982–26990.

45. Kotera, J., et al. 1999. Genomic origin and transcriptional regulation of two variants of cGMP-binding cGMP-specific phosphodiesterase. *Eur J Biochem* 262:866–872.

46. Lin, C.S., et al. 2002. Human PDE5A gene encodes three PDE5A isoforms from two alternate promoters. *Int J Impotence Res* 14:15–24.

47. Le Cras, T.D., et al. 1996. Chronic hypoxia upregulates endothelial and inducible NO synthase gene and protein expression in rat lung. *Am J Physiol* 270:L164–L170.

48. Black, S.M., et al. 2001. SGC and PDE5 are elevated in lambs with increased pulmonary blood flow and pulmonary hypertension. *Am J Physiol Lung Cell Mol Physiol* 281:L1051–L1057.

49. MacLean, M.R., K.M. McCulloch, and M. Baird. 1995. Effects of pulmonary hypertension on vasoconstrictor responses to endothelin-1 and sarafotoxin S6C and on inherent tone in rat pulmonary arteries. *J Cardiovasc Pharmacol* 26:822–830.

50. MacLean, M.R., and K.M. McCulloch. 1998. Influence of applied tension and nitric oxide on responses to endothelins in rat pulmonary resistance arteries. *Br J Pharmacol* 123:991–999.

51. MacLean, M.R., et al. 1996. 5-Hydroxytryptamine receptors mediating vasoconstriction in pulmonary arteries from control and pulmonary hypertensive rats. *Br J Pharmacol* 119:917–930.

52. Li, D., N. Zhou, and R.A. Johns. 1999. Soluble guanylate cyclase gene expression and localisation in rat lung after exposure to hypoxia. *Am J Physiol* 277:L841–L847.

53. Oka, M. 2001. Phosphodiesterase 5 inhibition restores impaired Ach relaxation in hypertensive conduit pulmonary arteries. *Lung Cell Mol Pharmacol* 280:L432–L451.

54. Hong, H.J., S.H. Loh, and M.H. Yen. 2000. Suppression of the development of hypertension by the inhibitor of inducible nitric oxide synthase. *Br J Pharmacol* 131:631–637.

55. Raychaudhuri, B., et al. 1999. Nitric oxide blocks nuclear factor kappaB activation in alveolar macropharges. *Am J Respir Cell Mol Biol* 21:311–316.

56 Aziz, S.M., et al. 1997. Polyamine regulatory processes and oxidative stress in monocrotaline-treated pulmonary artery endothelial cells. *Cell Biol Int* 21:801–812.

57. Morrell, N.W., et al. 1998. Angiotensin II stimulates proliferation of human pulmonary artery smooth muscle cells via the AT1 receptor. *Chest* 114 (1 Suppl.):90S–91S.

58. Orte, C., et al. 2000. Expression of pulmonary vascular angiotensin-converting enzyme in primary and secondary plexiform pulmonary hypertension. *J Pathol* 192:379–384.

59. Krick, S., et al. 2005. Hypoxia-driven proliferation of human pulmonary artery fibroblasts: Cross-talk between HIF-1alpha and an autocrine angiotensin system. *FASEB J* 19:857–859.

60. Kanazawa, H., et al. 2002. Association between the angiotensin-converting enzyme gene polymorphisms and tissue oxygenation during exercise in patients with COPD. *Chest* 121:697–701.

61. Solari, V., and P. Pur. 2004. Genetic polymorphisms of angiotensin system genes in congenital diaphragmatic hernia associated with persistent pulmonary hypertension. *J Pediatr Surg* 39:302–306.

62. Kim, D., et al. 2005. Angiotensin II increase PDE5A expression in vascular smooth muscle: A mechanism by which angiotensin II antagonizes cGMP signaling. *J Mol Cell Cardiol* 38:175–184.

63. Wolfsgruber, W., et al. 2003. A proatherogenic role for cGMP-dependent protein kinase in vascular smooth muscle cells. *Proc Natl Acad Sci USA* 100:13519–13524.

64. Stryer, L. 1991. Visual transduction. *J Biol Chem* 266:10711–10714.

65. Natochin, A., and N.O. Artemyev. 1996. An interface of interaction between photoreceptor cGMP phosphodiesterase catalytic subunits and inhibitory gamma subunits. *J Biol Chem* 271:19964–19969.

66. Artemyev, N.O., M. Bussman, and H.E. Hamm. 1996. Mechanism of photoreceptor cGMP phosphodiesterase inhibition by its gamma-subunits. *Proc Natl Acad Sci USA* 93:5407–5412.

67. Guo, L.W., et al. 2005. Asymmetric interaction between rod cyclic GMP phosphodiesterase gamma subunits and alphabeta subunits. *J Biol Chem* 280:12585–12592.

68. Skiba, N.P., N.O. Artemyev, and H.E. Hamm. 1995. The carboxyl terminus of the gamma-subunit of rod cGMP phosphodiesterase contains distinct sites of interaction with the enzyme catalytic subunits and the alpha-subunit of transducin. *J Biol Chem* 270:13210–13215.

69. Slepak, V.Z., et al. 1995. An effector site that stimulates G-protein GTPase in photoreceptors. *J Biol Chem* 270:14319–14324.

70. Tate, R., et al. 1998. The γ subunit of the rod photoreceptor cGMP-binding cGMP-specific PDE is expressed in lung. *Cell Biochem Biophys* 29:133–144.

71. Frame, M., et al. 2001. The gamma subunit of the rod photoreceptor cGMP phosphodiesterase can modulate the proteolysis of two cGMP binding cGMP-specific phosphodiesterases (PDE6 and PDE5) by caspase-3. *Cell Signal* 13:735–741.

72. Tate, R.J., V.Y. Arshavsky, and N.J. Pyne. 2002. The identification of the inhibitory γ subunit of the type 6 retinal cyclic GMP phosphodiesterase in non-retinal tissues: Differential processing of mRNA transcripts. *Genomics* 79:582–586.

73. Lochhead, A., et al. 1997. The regulation of the cGMP binding cGMP phosphodiesterase by proteins that are immunologically related to γ subunit of the photoreceptor cGMP phosphodiesterase. *Appl. Photo Phys* 272:18397–18403.

74. Wan, K.-F., et al. 2003. The inhibitory gamma subunit of the type 6 retinal cGMP phosphodiesterase functions to link c-Src and G-protein-coupled receptor kinase 2 in a signaling unit that regulates p42/p44 mitogen-activated protein kinase by epidermal growth factor. *J Biol Chem* 278:18658–18663.

75. Murray, F., M.R. MacLean, and N.J. Pyne. 2003. An assessment of the role of the inhibitory gamma subunit of the retinal cyclic GMP phosphodiesterase and its effect on the p42/p44 mitogen-activated protein kinase pathway in animal and cellular models of pulmonary hypertension. *Br J Pharmacol* 138:1313–1319.

76. Jin, N., et al. 2000. Hypoxia activates jun-N-terminal kinase, extracellular signal-regulated protein kinases, and p38 kinase in pulmonary vessels. *Am J Respir Cell Mol Biol* 23:593–601.

77. Das, M., et al. 2001. Hypoxia-induced proliferative response of vascular adventitial fibroblast is dependent on G-protein activation of mitogen-activated protein kinases. *J Biol Chem* 276:15631–15640.

78. Premkumar, D.R., et al. 2000. Intracellular pathways linking hypoxia to activation of c-*fos* and AP-1. *Adv Exp Med Biol* 475:101–109.

79. Minet, E., et al. 2000. ERK activation upon hypoxia: Involvement of HIF-1 activation. *FEBS Lett* 468:53–58.

26 Role of PDE5 in Migraine

Christina Kruuse

CONTENTS

26.1 The Migraine Syndrome 519
26.2 Migraine Pathophysiology and the Pain Pathway 520
26.3 The Major Signaling Molecules Associated with Migraine 522
    26.3.1 Nitric Oxide...522
    26.3.2 Calcitonin Gene-Related Peptide..524
    26.3.3 Serotonin (5-HT) 524
26.4 PDE5 in the Pain Pathway of Migraine...525
26.5 Induction of Headache and Migraine by PDE5 Inhibition 526
26.6 Conclusions and Perspectives...531
Acknowledgment...532
References 532

Migraine is an episodic headache syndrome of yet unknown origin, impacting hugely not only on patients' quality of life but also on national socioeconomic considerations.[1] It has a 1 y prevalence of 10%–12% in the general population, showing a female preponderance.[2] Migraine is generally considered a neurovascular disease but the pathogenesis of the syndrome is still not fully understood. Although effective treatments of the acute migraine attack have become available, problems of treatment failures and side effects remain, and an optimal treatment for prevention of migraine attacks remains to be found. Thus, further investigations in migraine pathology are still warranted.

The signaling molecule, nitric oxide (NO), is shown to play a major part in the migraine pathophysiology along with the neuropeptide calcitonin gene-related peptide (CGRP) and serotonin (5-HT). They all appear to affect cyclic nucleotide signaling, but these downstream signal transduction pathways with possible interactions have yet to be defined, though such interactions might be interesting targets for drug development. Compounds that modulate cyclic nucleotide signaling, such as selective PDE inhibitors, provide excellent tools for investigations into these pathways both *in vitro* and *in vivo*. Recently, a selective inhibitor of the cGMP-degrading PDE5, sildenafil, has shed light on the role of cGMP and the large cerebral arteries in migraine induction.[3]

This chapter includes a short introduction to the clinical picture of migraine, the current hypothesis on migraine pathology, and proposed signaling pathways involved as well as the possible role of PDEs in migraine, focusing on the role of cGMP and PDE5.

26.1 THE MIGRAINE SYNDROME

Descriptions of migraine symptomatology can be found as far back as the 2000 B.C.,[4] but it is only within the last 30 years that an international consensus on the classification of migraine, essential for clinical research, has been reached.[5,6]

Two major subtypes of migraine exist: migraine without aura, the most common one, and migraine with aura (previously termed classic migraine) experienced by some 15%–20% of patients.[2]

The main characteristic of migraine is a throbbing, unilateral headache of moderate to severe intensity, which is aggravated by movement and light, and regularly requires bed rest for relief. The headache is self-limiting, lasting 4–72 h if untreated and accompanied by either photophobia and phonophobia or nausea.[6] The migraine syndrome may, however, also comprise other symptoms, which can be divided into various phases: (1) a premonitory phase, (2) an aura phase specific to patients suffering from migraine with aura, usually occurring before the headache phase, (3) the headache phase, and (4) a postdrome phase following the headache.[7] Premonitory symptoms typically occur hours before the onset of headache. The principal complaint is a feeling of excessive tiredness, which tends to persist during the succeeding phases, occurring also in the postdromal phase.[7] Cognitive slowing and yawning are other frequent premonitory symptoms.[7] The aura phase signals gradually developing reversible neurological symptoms reflecting cortical impulse propagation based on the spreading depression phenomena.[8] Vision is usually affected during the aura phase by induction of monochromatic flickering zigzag lights slowly moving across the visual field often followed by a scotoma. Aura may also, though less frequently, include unilateral propagating sensory symptoms and affect the speech.[6,8,9] The aura phase lasts a maximum of 60 min and may either precede the headache or start concurrent with the headache, which can be both contralateral and ipsilateral to the aura symptoms. Usually, in migraine with aura the headache is of less severe intensity.

Contributing to the enigma of migraine pathogenesis is the apparent heterogeneous appearance of some of the symptoms within the patient. Thus, some patients experience migraine attacks, both with and without aura, and in some patients, only the aura is present without accompanying headache.[6] Although usually unilateral, the migraine headache can occur bilaterally, or it may shift sides during attack or from one attack to another.

Migraine patients appear to have an increased susceptibility to a wide range of exogenous and endogenous stimuli. Although manifesting both interindividual and intraindividual variations, several factors may precipitate a migraine attack.[10,11] Some of these, e.g., sleep deprivation, irregular intake of food, and strong or flickering light, resemble the precipitating factors in epilepsy perhaps reflecting common predisposing factors in migraine and epilepsia.[12] Furthermore, physical activity or exogenous compounds such as glyceryl trinitrate (GTN), histamine, withdrawal of caffeine, and intake of alcohol may initiate attacks of migraine.[10] The frequency of attack varies and can be influenced by hormonal changes and stress. In one-third of female patients the migraine is associated with changes in the hormonal cycle.[2] Migraine frequency and headache intensity inexplicably diminish with age.[2] In patients suffering from migraine with aura, a slight, unexplained increase in ischemic stroke is reported in patients below 45 years of age.[2]

26.2 MIGRAINE PATHOPHYSIOLOGY AND THE PAIN PATHWAY

The basis for migraine pathogenesis has for some time been a matter of some controversy where divergent standpoints were taken as to whether migraine was vascular or neurogenic in origin.[13–15]

Earlier research has revealed at least three active elements in migraine pathophysiology: (1) the vessels, (2) the perivascular nerves (especially the trigeminal nerve), and (3) the central pain perception or modulation. It is now generally accepted that migraine is a neurovascular disease involving activation of the trigeminovascular pathway, where the primary sensory afferent surrounding the cerebral arteries originates from the trigeminal nerve, with cell

bodies in the trigeminal ganglion. The trigeminal afferent nerves project centrally through the trigeminal nucleus caudalis in the brain stem, which is considered to be a key relay point for the pain pathway,[16] and on to the thalamus and cerebral cortex where pain is perceived (Figure 26.1). However, there is not yet a general consensus on the initiating factor of migraine, or an account for the recurring nature of the syndrome.

Further complicating the controversy is evidence from family[17] and genetic studies[18] demonstrating that migraine most likely has a heterogeneous origin. Although migraine with aura appears largely determined by genetic factors, migraine without aura is determined to a large extent also by the environmental factors,[17] yet the question of a common gene for migraines both with and without aura remains to be resolved.

In a specific subset of migraine, the familial hemiplegic migraine (FHM), mutations have been found in a gene coding for each of (1) a pore-forming subunit of the voltage gated P/Q-type calcium channel, (2) the catalytic subunit of a sodium–potassium ATPase and, recently, (3) a voltage-gated sodium channel.[18] FHM is a rare and severe form of migraine with aura, where temporary hemiparalysis develops in addition to the other aura symptoms.[6] The genetic evidence suggests that at least the aura symptoms may be a channelopathy, albeit not limited to one type of ion channel. The cellular location and function of the involved ion channels and mutations remain to be defined.

How may migraine involve the cerebral arteries? In a remarkable study by Ray and Wolff,[19] where patients underwent neurosurgical procedures under local anesthetic, the pain sensing and headache producing structures of the brain were determined to be the large cerebral arteries, the dura in close proximity to the dural arteries and the base of the brain, the venous

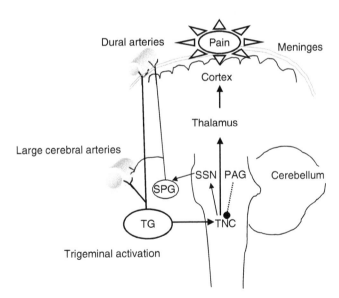

FIGURE 26.1 Migraine pain pathway. Nociceptive inputs from the large cerebral arteries, the meningeal arteries, and dura convey centrally to the CNS via the primary afferent sensory nerve, the trigeminal nerve, having its cell bodies in the trigeminal ganglion (TG) and the first-order synapses in the trigeminal nucleus caudalis (TNC). Immunoreactivity to both NOS and CGRP are found in the human trigeminal ganglion. Second-order neurons project to the thalamus and third-order neurons to the cortex where pain is perceived. The pain may be modulated by the periaquaductal grey (PAG) and hypothalamus. A parasympathetic efferent vascular response is mediated upon TNC activation through the superior salivatory nucleus (SSN) and sphenopalatine ganglion (SPG) which may cause perivascular release of NO and neuropeptides.[47]

sinuses and some of the large cranial nerves including the trigeminal and facial nerve.[19] Convergence of inputs from the skin and the arteries to the trigeminal nucleus caudalis provide the basis for the referred pain observed when stimulating the arteries.[20] In the initial vascular theory,[19] it was suggested that migraine is caused by an abnormal dilatation of the large cerebral arteries, thus stretching and activating the perivascular sensory nerves, inducing headache as a referred pain. In support of this theory, compounds relieving the headache were known to be vasoconstrictors, e.g., ergots and the serotonin (5-HT) receptor-agonists, the triptans.[21] Furthermore, the large cerebral arteries were found to dilate on the headache side during a migraine attack,[22,23] though this remains a matter of controversy.[24] The vascular theory was challenged by imaging techniques, showing that the vascular changes found were not directly linked to the headache phase;[25] further, specific brain regions were activated both during and after relieving the migraine attack with triptans.[26,27]

Is migraine then a brain disease? The presence of premonitory symptoms could suggest a central mechanism in migraine pathogenesis;[7] the observation of cutaneous allodynia, an abnormal pain sensation to nonnoxious stimuli, during migraine attacks indicated a hyper-excitability of central neurons in the pain pathway pointing to a central origin of migraine.[28] In support of this theory, a lack of central habituation to visual and auditory stimuli, which normalized during migraine attacks was found in migraine without aura.[29] The involvement of central mechanisms may be a result of any one or more of the following: (1) brain stem activation,[27] (2) a sterile neurogenic inflammation with protein extravasation from dural vessels,[30] (3) a result of the phenomena spreading depression, which activate the trigeminal afferents projecting to the brain areas associated with pain processing.[31] Any potential blood flow changes are assumed to be a result of the neurological events. Whether cortical-spreading depression itself is initiating the headache is still a matter of debate as the aura can be disassociated from the headache and there is not yet consistent evidence for the presence of cortical-spreading depression in patients with migraine without aura.[32] Clinical experience may also contradict the primary involvement of neurogenic inflammation, as compounds inhibiting neurogenic inflammation and plasma extravasation are ineffective in migraine treatment.[16]

Although the pathological mechanisms behind migraine are unclear, at least three different signaling molecules appear to be involved: NO, CGRP, and serotonin (5-HT). They may merely represent an epiphenomenon in relation to the central pain mechanisms, but from human studies it is known that 5-HT1$_{B/D}$ receptor agonists, inhibitors of NO synthase, and CGRP receptor antagonists are effective in the treatment of acute migraine attacks.

26.3 THE MAJOR SIGNALING MOLECULES ASSOCIATED WITH MIGRAINE

26.3.1 NITRIC OXIDE

Donors of NO and stimulators of endogenous NO production have long been known and used to induce headache and migraine.[33–35] A well-established human model of migraine is the GTN model,[34] where infusion of the NO donor, GTN, generates an immediate short-lasting headache in healthy subjects, concomitant with a dilatation of the large cerebral and extra cerebral arteries, with no change of cerebral blood flow.[36] With migraine patients, GTN also induces a second delayed headache resembling the patients' usual migraine attack including premonitory symptoms;[37] this occurs approximately 2–4 h after GTN infusion.[38] This delayed headache rarely occurs in healthy subjects, and then mainly occurs in those subjects with a first-degree relative suffering from migraine. The initial dilatation of the large cerebral arteries is more pronounced in migraine patients than in healthy subjects,[39] but it is

still unclear whether cerebral vasodilatation occurs in the delayed headache phase. The time delay for the second headache may suggest a modulation of the pain transduction pathway, which is specific to migraine patients, perhaps at a transcriptional level initiated either by the vasodilatation or by NO itself.[40,41] Thus, NO may have several modes of action in relation to induction of migraine symptoms, but it is generally believed to involve cGMP-related mechanisms (Figure 26.2). NO is endogenously produced by the activation of one of three NO synthases (NOS); two are Ca^{2+}-stimulated (constitutively expressed in endothelial cells, eNOS, and in neurons, nNOS) and the third is the transcriptionally regulated and inducible iNOS, highly expressed in macrophages upon stimulation.[42] The vascular effects of NO involve stimulation of the NO-sensitive (also known as the soluble) guanylate cyclase (sGC) in the vascular smooth muscle cells; levels of cGMP are increased, leading to vasodilatation with mechanical stimulation of the pain-sensitive perivascular nerve fibres,[40] causing induction of headache. In nociceptive neurons, NO and cGMP may increase excitability by

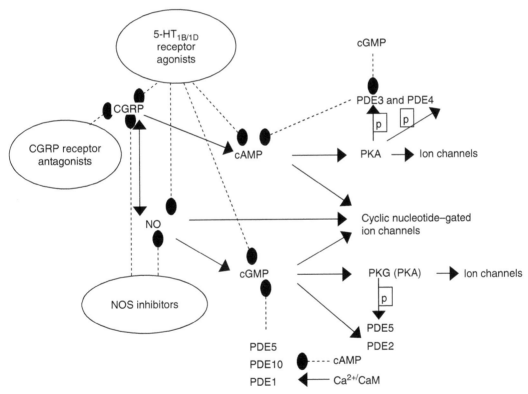

FIGURE 26.2 Possible signaling pathways in migraine. A schematic and simplified overview of the possible interactions of the major signaling molecules and the cyclic nucleotide signaling in the migraine pain transduction pathway as well as the potential effects of known antimigraine compounds. The regional and cellular locations of these events are not fully elucidated and may take place in subcellular compartments or different cell types. The cAMP- and cGMP-signaling pathways may interact at several levels. Firstly, through inhibition of PDEs with cAMP inhibiting PDE10 and cGMP inhibiting PDE3; secondly, through activation of PKA; and thirdly through activation of cyclic nucleotide–gated ion channels.[101] Several feedback mechanisms for regulating cyclic nucleotide levels are described; only the activation by phosphorylation is shown above.[71,102] For further details please see the text. NO, nitric oxide; NOS, nitric oxide synthase; PKA, cAMP-dependent protein kinase; PKG, cGMP-dependent protein kinase; punctuated line, inhibits; arrow, activates; p, phosphorylation.

affecting ion channels directly or indirectly (Figure 26.2), thus causing normal stimuli to be perceived as painful.[43] The increase in cGMP affects three molecular pathways: (1) the PDEs, (2) the cGMP-dependent protein kinase (PKG) but also at higher concentrations cAMP-dependent protein kinase (PKA),[44,45] and (3) the cyclic nucleotide–gated ion channels (CNGs) (Figure 26.2).

As a result of these diverse effects of NO and cGMP in the pain pathway, the cellular localization and mechanisms involved in migraine induction cannot be deduced by just looking at effects of NO donors. In a clinical trial, including patients with migraine without aura, subcutaneous injection of the NOS inhibitor, L-NMMA, in the thigh was shown to be effective in treating acute migraine attacks,[46] emphasizing the importance of NO as a major signaling molecule in migraine.

26.3.2 CALCITONIN GENE-RELATED PEPTIDE

The key neuropeptide involved in migraine pathophysiology is CGRP, which is present in both the human and the animal brains, as well as in the perivascular sensory nerves and trigeminal ganglion.[47] Increased levels of CGRP are found in external jugular vein blood during migraine attacks[48] and after stimulation of the trigeminal ganglion.[49] Further, infusion of CGRP elicited dilatation of large cerebral arteries and peripheral arteries causing hypotension, and symptoms of migraine in migraine patients.[50]

The cyclic nucleotide–signaling pathway affected by CGRP mainly involves cAMP. CGRP activates adenylate cyclase after binding to a G-protein-coupled CGRP-receptor,[51] though interactions with NO and cGMP signaling have been proposed (Figure 26.2).[51–53] A new concept of acute migraine treatment based on the role of CGRP has been proposed using a selective CGRP receptor antagonist. In animal models, this attenuates neurogenic-induced perivascular CGRP release, thus inhibiting the neurogenic vasodilatation and activation of the perivascular pain sensory nerves.[54] In a recent clinical trial, a proof of concept was established and the CGRP antagonist BIBN4096BS was effective in the acute treatment of migraine.[55] This is the first compound to show specific nonvascular antimigraine effect, as BIBN4096BS does not appear to affect cerebral blood flow (CBF) or large cerebral artery diameter in humans.[56]

26.3.3 SEROTONIN (5-HT)

It has been suggested that the syndrome of migraine relates to a deficiency of serotonin (5-HT) due mainly to the findings that 5-HT level in platelets is decreased during migraine attacks; serotonin infusions relieve the intensity of the attacks.[57] Treatment of acute migraine attacks was greatly improved with the development of the $5\text{-HT}_{1B/D}$-receptor agonists, also known as triptans; sumatriptan was the first developed triptan.[57] Although the main antimigraine effect of triptans appears to be constriction of dilated cranial or cerebral arteries, it also normalizes the increased plasma levels of CGRP and cGMP seen during the migraine attacks,[22,58,59] and also inhibits dural extravasation and trigeminovascular impulse transmission.[21,60]

The cyclic nucleotide effects of triptans appear to be diverse. The triptans inhibit adenylate cyclase, decrease intracellular cAMP level (Figure 26.2), and thus in smooth muscle cells induce contraction reflected in a slight constriction of the large cerebral arteries.[57] Triptans may also affect the NO/cGMP signaling pathway. In the pain signaling, the antinociceptive effects of sumatriptan were potentiated by coadministration of a NO synthase inhibitor, L-NAME.[61] In animal models, sumatriptan inhibits glutamate release[62] and attenuates the NMDA receptor–mediated, as well as the GTN-induced stimulation of nNOS in cerebral cortex and the trigeminal pathway.[63,64] Interestingly, specific serotonin-re-uptake inhibitors (SSRIs) have no effect in the treatment of migraine.[65]

26.4   PDE5 IN THE PAIN PATHWAY OF MIGRAINE

From the above, it can be hypothesized that cyclic nucleotide signaling plays an important role in the initiation, modulation, and, perhaps, cessation of a migraine pain; to define the specific actions and interactions of cGMP and cAMP is a great challenge. The intracellular levels, and thus the duration and intensity of the cyclic nucleotide signaling, are regulated largely by the balance between the rate of synthesis by cyclases and the rate of degradation by PDEs;[66] the role of efflux from the cells is not yet fully established.[67,68] Knowledge of the specific cellular and perhaps subcellular localization of the PDEs in peripheral and central structures of pain signaling might provide a basis for human studies. So far, little is known of this distribution. The main focus has been on the PDEs in the cerebral arteries because of the pain-sensitive nature of these structures and their involvement in other cerebral diseases such as stroke and cerebral vasospasms. An overview of the limited data on the distribution of PDE5 in the main structures of the pain pathway is shown in Table 26.1. Further data will be added to this table in the near future.

The presence of PDE5, PDE1A, and PDE1B mRNAs and proteins was recently found in human middle cerebral arteries[69] corresponding to the distribution previously reported in guinea pig basilar arteries.[70] A cellular localization in the vessel was not possible, although the distribution of these PDEs appeared to be related to the smooth muscle cells. It was also found that the selective PDE5 inhibitor, sildenafil, and a derivative UK-114542 inhibited the cGMP hydrolysis, although sildenafil, in contrast to UK-114542, elicited only a minor dilatory response.[69] As both the compounds are PDE5 inhibitors, inhibition of the PDE5 catalytic activity in the intact cerebral arteries appears unlikely to fully explain the difference in vasodilatory response; however, it could be a question of differential effect of the compounds on PDE5 subtypes or subcellular localization.[71] The presence of both PDE3 and PDE4 has not been shown directly, but the PDE3 inhibitor, cilostazol, and the PDE4 inhibitor, rolipram, inhibited cAMP hydrolysis in homogenates of human middle cerebral arteries.[72] Accordingly, these various data indicate activity of PDE1, PDE5, PDE3, and PDE4 in human cerebral arteries. This corresponds well with the distribution found in cerebral arteries of other animals though the function may differ.[70,72–74] High concentrations of sildenafil also inhibited cAMP hydrolysis[69] and high concentrations of cilostazol inhibited cGMP hydrolysis;[72] whether this has functional implications *in vivo* remains to be studied. In animals, where tissue is more readily available, more extensive studies have been performed on the PDEs present, their potential function, and interactions in cerebral arteries.[66,72,74,75]

TABLE 26.1
Presence of PDE5 mRNA or Protein in the Pain Transduction Pathway in Migraine

	Cerebral Arteries	Dura Mater	Trigeminal Ganglion	Trigeminal Nucleus Caudalis	Periaquaductal Grey	Thalamus	Cortex
Human	Yes	—	—	Minor	—	No	No
Guinea pig	Yes	—	—	—	—	—	—
Rat	—	—	—	—	—	No	No (protein) Yes (mRNA)
Mice	—	—	—	—	—	—	—
Dog	No	—	—	—	—	—	—
References	[69,70,76]			[80]		[81]	[81,103]

Most of such studies have characterized the presence of PDE families by using family-selective PDE inhibitors. The results appear to display differences among species and may depend on the selectivity and specificity of the inhibitors used. In general, the PDE1, PDE4, and PDE5 inhibitors show endothelium-dependent responses and PDE3 inhibitors elicit endothelium-independent vascular responses in cerebral arteries.[70,72–74]

PDE5 has been found in cerebral arteries of rodents and pigs, but, in contrast, it was found in dogs only after the induction of subarachnoid hemorrhage.[76] Zaprinast, an inhibitor of PDE1, PDE5, and PDE9, relaxed guinea pig and rat basilar arteries and mouse arterioles at high concentrations ($\sim$1 μM)[70,73,77] but was, in accordance with the PDE5 distribution, a poor dilator of canine basilar arteries.[75] The selective PDE5 inhibitor, sildenafil, evidenced a poor dilatory response in isolated guinea pig cerebral arteries corresponding to the human *in vivo* findings,[78] whereas another selective PDE5 inhibitor, UK-114542 (a derivative of sildenafil), as well as dipyridamole (a nonselective PDE5 inhibitor) was effective in eliciting relaxation of cerebral arteries, and the latter effect was also reported in humans *in vivo*.[69,70,79]

Effects of other selective PDE inhibitors in cerebral arteries have also been investigated. Vasodilator responses are found using PDE1 inhibitors,[70] but diverging intensity of dilatory effects were given for PDE3 and PDE4 inhibitors, possibly dictated by the selectivity of the inhibitor used and the species studied.[70,72,75]

In the central pain pathway, limited data is available, but high PDE5 mRNA expression in human brain samples is reported especially in the cerebellum and in the subthalamic nucleus with minor expression in the caudate nucleus, and no expression is found in cerebral cortex or thalamus.[80] There are no reports on the corresponding PDE5 activity. In situ hybridization of rat brain revealed PDE5 mRNA in the hypothalamus, but not in thalamus. In contrast to humans,[80] the PDE5 mRNA was found in neuronal cell bodies scattered in the cortex.[81]

The other specific cGMP-degrading PDE, PDE9, was, however, richly represented in the thalamic area and cortex of the rat,[81] but the function of this PDE remains to be defined. There does not appear to be any reports on the distribution of PDEs in the dura mater, trigeminal ganglion, or periaquaductal grey of either animal or human.

26.5   INDUCTION OF HEADACHE AND MIGRAINE BY PDE5 INHIBITION

Selective inhibitors of PDEs appear to provide specific tools to pinpoint the role of cyclic nucleotides in the pathophysiology of migraine. Several more or less selective PDE inhibitors are reported to induce headaches; these observations are worthy of consideration by pharmaceutical companies interested in research projects investigating headaches and migraines.

Table 26.2 presents an overview of PDE-inhibitor effects on headaches, CBF, and calculated cerebral artery diameter of both selective and nonselective PDE inhibitors.[82,83] Not all studies have investigated all parameters.

As NO donors induce migraine in migraine-prone subjects, an obvious new target to pursue after the NO investigations is the role of cGMP signaling. With the development of selective inhibitors of PDE5 for human use, tools for dissecting the cGMP-signaling pathway downstream of NO in migraine were improved.

Due to the unavailability of sildenafil in Denmark until 1998, the headache-inducing properties of dipyridamole were investigated. Apart from being a nonselective PDE5 inhibitor, dipyridamole also inhibits adenosine re-uptake; further, dipyridamole in combination with aspirin is currently recommended for secondary prevention of stroke.[82,84] Although dipyridamole is considered an inhibitor of PDE5, it may also inhibit PDE10 (present in the brain)[85] and PDE11 (present in skeletal muscle),[86] but whether this is relevant *in vivo* is unknown. The most common side effect in the clinical use of dipyridamole is the occurrence of headache in

TABLE 26.2
Effects of PDE Inhibitors on Cerebral Hemodynamics and Headache

	Pentoxifylline Nonselective Mainly cAMP↑ Adenosine Receptor Antagonist	Papavarine Nonselective Mainly cAMP↑ Adenosine Reuptake Inhibitor	Nimodipine PDE1 cGMP↑ Ca²⁺-Channel Blocker	Cilostazol PDE3 cAMP↑ Adenosine Reuptake Inhibitor	Olprinone PDE3 cAMP↑	Dipyridamole PDE5, 10, 11 Mainly cGMP↑ Adenosine Reuptake Inhibitor	Sildenafil PDE5 cGMP↑
Main PDE Inhibited	Nonselective	Nonselective	PDE1	PDE3	PDE3	PDE5, 10, 11	PDE5
cAMP/cGMP Change	Mainly cAMP↑	Mainly cAMP↑	cGMP↑	cAMP↑	cAMP↑	Mainly cGMP↑	cGMP↑
Other Effects	Adenosine Receptor Antagonist	Adenosine Reuptake Inhibitor	Ca²⁺-Channel Blocker	Adenosine Reuptake Inhibitor		Adenosine Reuptake Inhibitor	
Effects on:							
CBF	No change	↑	↑	No change	↑	No change	No change
Cerebral artery dilatation	No change	NA	NA	↓	NA	↓	No change
Headache	No	No	Yes	Yes	Not reported	Yes	Yes
Migraine in migraine patients	NA	No	No	NA	NA	Yes	Yes
References	[104]	[105,106]	[107,108]	[105,106]	[109]	[79,89,110]	[3,78]

Abbreviations: CBF, cerebral blood flow; NA, not applicable; arrow up, increase; arrow down, decrease.

Source: From Ref. [78]. With permission.

Median headache score over time

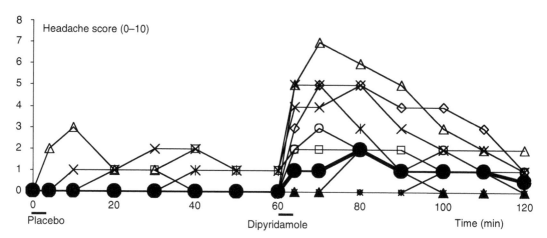

FIGURE 26.3 Headache induction by dipyridamole over time in healthy subjects. Individual headache scores are shown in thin lines. A thick line with filled circles indicates the median headache score over time ($n = 12$). Significant difference in area under the curves was seen between placebo and dipyridamole treatment ($p = 0.02$). (From Ref. [79]. With permission.)

approximately 38% of patients; this caused ~8% of all patients to discontinue treatment.[87] In a small study of healthy subjects, the headache-inducing and cerebrovascular effects of intravenous dipyridamole ($0.142 \text{ mg} \cdot \text{kg}^{-1} \cdot \text{min}^{-1}$) were studied.[79] Headache of mild to severe intensity (Figure 26.3 and Table 26.2) was induced in 8 out of 12 healthy subjects with no history of migraine. One subject manifested the symptoms of migraine without aura according to the International Headache Society[6] for a short time-interval even though the subject had no history of migraine. The dipyridamole-induced headache was accompanied by only an initial small dilatation of the large cerebral arteries, which seemed to outlast the headache response; this suggests dissociation between the initial headache and the longer-lasting vascular response. Data from an early study suggested that oral dipyridamole could induce severe headache in migraine patients though a classification of the induced headache was not performed.[88] In a recent study comparing the effects of intravenous dipyridamole in healthy subjects and in migraine patients, headache was induced in all 10 migraine patients and in 8 out of 10 healthy subjects, both groups experiencing equal pain intensity.[89] The headache evidenced the symptoms of migraine, including accompanying features in 5 of 10 migraine patients and in 1 of 10 healthy subjects with a time delay of 2–11 h postinfusion. The main difference between the two groups was the manifestation of photophobia in the migraine patients only.

With the appearance of sildenafil[90] and later vardenafil and tadalafil, more specific PDE5 inhibitors became available. Sildenafil is the only selective PDE5 inhibitor to have been used in headache and migraine investigations, although there are reports of headache as a side effect of all three compounds.[91] The first study of 10 healthy subjects, after oral administration of 100 mg sildenafil, investigated headache induction and its effects on (1) cerebral blood flow, (2) large cerebral arteries, and (3) extracerebral arteries.[78] Of particular interest, and in contrast to that previously found for other headache-inducing compounds (Table 26.2), sildenafil did not affect the cerebral arteries although it did induce headaches with a similar frequency as NO donors. The headache response was mainly monophasic and in some subjects the headache was of moderate to severe intensity (see Figure 26.4). In three subjects, the headache symptoms fulfilled the criteria

Headache score (0–10) after placebo

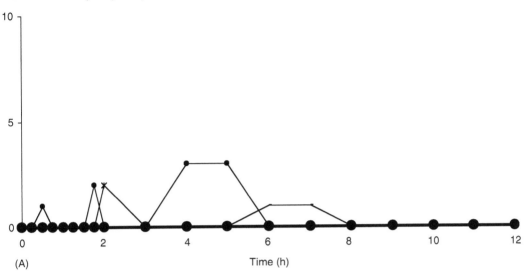

(A)   Time (h)

Headache score (0–10) after sildenafil

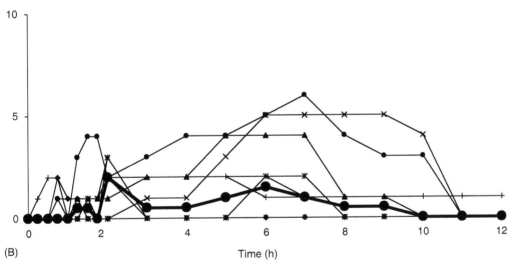

(B)   Time (h)

FIGURE 26.4 Headache induction over time after placebo (A) and sildenafil (B) in healthy subjects. Individual (thin lines) and median headache scores (thick line with filled circle) after placebo (A) and after sildenafil 100 mg (B). Nine of 10 subjects reported headache after sildenafil and 3 of 10 after placebo ($p = 0.031$). The headache was mild and transient, although after sildenafil in three subjects the symptoms fulfilled the criteria for migraine without aura after 5–7 h with moderate to severe, mostly throbbing aggravated headache, accompanied by nausea. None of the subjects had first-degree relatives suffering from migraine. (From Ref. [78]. With permission.)

for migraine with throbbing headache and accompanying symptoms (nausea), indicating a potential migraine-inducing effect even in subjects not previously prone to migraine. Although symptoms mirroring migraine were induced in healthy subjects, investigations in migraine patients were needed to fully clarify the impact of PDE5 inhibition in migraine, and a similar investigation as above was conducted in 12 migraine patients suffering from migraine without

aura.[3] Ten of 12 subjects reported headache and accompanying symptoms similar to their usual migraine attacks with a peak headache response ~4 h after sildenafil administration, but with an earlier fulfillment of migraine criteria (Figure 26.5). An element of the time delay could be ascribed to the pharmacokinetics of oral sildenafil, but an upregulation of signaling molecules cannot be excluded. The headache was slowly progressing and monophasic in the migraine

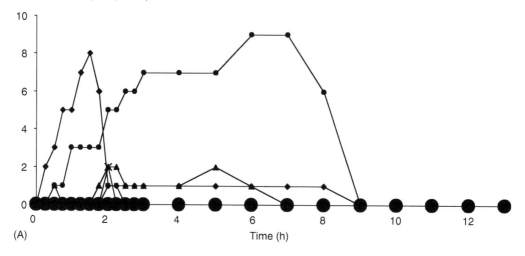

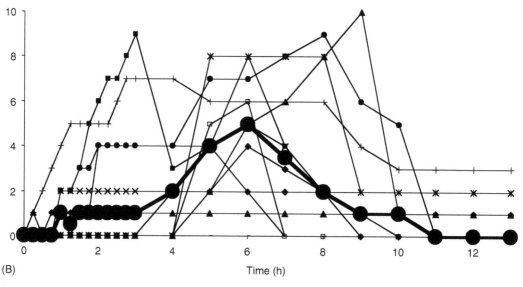

FIGURE 26.5 Headache induction after placebo (A) and sildenafil (B) in migraine patients. Individual headache scores are shown in thin lines. A thick line with filled circles indicates the median headache score over time (*n* = 12). The maximum headache score possible is 10, which equals worst possible headache. Patients were allowed to treat their migraine attack with their usual migraine medication. This accounts for the steep decreases in pain intensity seen along the individual curves. The median time to peak headache score was 4.5 h after sildenafil and the median time to fulfillment of criteria for migraine was 3.8 h. (From Ref. [3]. With permission.)

patients contrary to the biphasic GTN-induced migraine[40] perhaps reflecting a direct, non-vascular action on the initiating transduction pathway. No change in peripheral plasma levels of either cAMP, cGMP, or CGRP was found after sildenafil administration or during sildenafil-induced headache and migraine;[92] however, the efflux of cyclic nucleotides may be too small to be detected and as only cubital venous blood was investigated, a minor increase in CGRP may have escaped detection.

Sildenafil was the first compound to be used in a human headache model, which induced migraine without concomitant dilatation of the large cerebral or extra cerebral arteries, supporting evidence that initial cerebral vasodilatation is not a necessary initiation factor in migraine induction. In light of the selective effect of sildenafil on PDE5, the headache-generating effects of sildenafil are most likely due to an intracellular rise in cGMP, but the cell type involved has not been disclosed. Sildenafil is lipophilic, and there are indications that it passes the blood–brain barrier in humans.[93] A direct effect of sildenafil on CNGs has recently been proposed,[94] but whether this is possible *in vivo* remains to be investigated. Thus, the downstream mechanisms for either sildenafil or cGMP are not fully established. There are several effector molecules and pathways of cGMP, the main effectors being the PKG- and the cGMP-stimulated CNGs. Activation of CNGs may cause an increase of Ca^{2+} in cells. Depending on the cells involved, this may have opposing effects. In smooth muscle cells, the increased Ca^{2+} concentration could cause smooth muscle cell contraction, but in neuronal cells, influx of Ca^{2+} is involved in the release of excitatory amino acids such as glutamate, which could cause an increase in neuronal excitability as well as initiate cortical-spreading depression.[95] The increase in cGMP may also inhibit PDE3, thus increasing cAMP levels and, by that, activating the cAMP-stimulated protein kinase (PKA). The results with cilostazol-induced headache may favor this hypothesis.[96]

High level of cGMP may also stimulate the PKA directly although this is only likely to happen in pathophysiological conditions (see also Figure 26.2).[44] The nonvascular effects of sildenafil may be many. A recent study suggests an association of sildenafil with regulation of the antidepressant-sensitive 5-HT transporter (SERT). At high concentrations, sildenafil was found to stimulate 5-HT uptake *in vitro* through a PKG-dependent SERT activation.[97] Thus, sildenafil can decrease serotonin levels *in vivo*, which may also play a role in the induction of migraine, in light of the reduced levels of serotonin reported during migraine attacks.[57]

Considering a potential role of cortical-spreading depression in migraine induction, it is interesting that sildenafil is found not to influence cortical-spreading depression in a rat model.[98] If applicable to humans, this may suggest that the migraine-inducing potential of sildenafil is not through induction of cortical-spreading depression.

The accompanying symptoms in migraine such as photo- and phonophobia may also involve cyclic nucleotide signaling. NO has been shown to increase the spontaneous firing rate in medial vestibular neurons through a cGMP-mediated mechanism[99] and an increased excitability of auditory brain stem neurons was found to be mediated by cAMP.[100] Of relevant note, cGMP is crucial in photoreception and a side effect of PDE5 inhibitors is increased sensitivity to light.[90]

26.6 CONCLUSIONS AND PERSPECTIVES

Cyclic nucleotide signaling may be the basis for downstream pain signaling in migraine, and the migraine pathogenesis could include a cGMP-driven sensitization of the trigeminal-signaling pathway based on the apparent nonvascular migraine–inducing effect of the PDE5 inhibitor, sildenafil.

PDE5 inhibitors induce headache in patients suffering from migraine without aura; this mirrors a usual migraine attack including accompanying symptoms similar to what is seen

with NO donors. However, where NO donors have a biphasic headache response because of the immediate cerebral vasodilator effects, the PDE5 inhibitor, sildenafil, has no significant effect on the large cerebral artery diameter, a monophasic headache response, and a slightly shorter induction time. This might signify that an initial vasodilatation is not crucial for induction of migraine in a human headache model and that the pathogenesis of migraine may be located in perivascular nerves or more centrally in the pain pathway; thus the potential vasodilatation represents an epiphenomenon. PDE5 inhibitors have proved to be valuable tools in elucidating the possible migraine pathophysiology although much work remains to be done in investigating the exact signaling pathways involved downstream of cGMP and interacting with cAMP. Other selective PDE inhibitors that can be used for human studies will be valuable tools in pinpointing the cyclic nucleotides involved in the pain pathways and studies to identify the tissues involved in the pathway.

In considering the cyclic nucleotide signaling in migraine and the downstream pathways, there may be several interesting targets for investigation. cGMP appears to play an important role in modulating synaptic function and neuronal excitability either by the activation of PKG or direct actions on CNGs. Furthermore, stimulation of the cGMP cascade so far appears to have a more prominent migraine-inducing effect compared to direct stimulation of the cAMP cascade. Thus, modulation of the cGMP-signaling pathway may be a promising target for new antimigraine drugs. Discovery of compounds that activate PDE5[71] or in other ways reduce cGMP (and perhaps also cAMP level) in the brain regions of interest may prove to be beneficial in the acute, and in particular, the prophylactic treatment of migraine. However, it should be noted that one of the most important factors for the development of drugs for migraine is a specific location or function of the involved targets to provide a specific antimigraine action without unwanted side effects.

Further investigation in the distribution and function of PDEs and perhaps the downstream effectors, the protein kinases or CNGs, may provide such a specific distribution and thus be considered for potential drug development.

ACKNOWLEDGMENT

The author is supported by a grant from The Danish Medical Research Counsel.

REFERENCES

1. Rasmussen, B.K., R. Jensen, and J. Olesen. 1992. Impact of headache on sickness absence and utilization of medical services: A Danish population study. *J Epidemiol Community Health* 46 (4): 443–6.
2. Breslau, N., and B.K. Rasmussen. 2001. The impact of migraine: Epidemiology, risk factors, and comorbidities. *Neurology* 56 (6 Suppl 1): S4–12.
3. Kruuse, C., L.L. Thomsen, S. Birk, and J. Olesen. 2003. Migraine can be induced by sildenafil without changes in middle cerebral artery diameter. *Brain* 126 (Pt 1): 241–7.
4. Rapoport, A., and J. Edmeads. 2000. Migraine: The evolution of our knowledge. *Arch Neurol* 57 (8): 1221–3.
5. Anonymous. 1988. Classification and diagnostic criteria for headache disorders, cranial neuralgias, and facial pain. Headache Classification Committee of the International Headache Society. *Cephalalgia* 8 (Suppl 7): 1–96.
6. Anonymous. 2004. The international classification of headache disorders. *Cephalalgia* 24 (Suppl 1): 1–160.
7. Giffin, N.J., L. Ruggiero, R.B. Lipton, S.D. Silberstein, J.F.Tvedskov, J. Olesen, J. Altman, P.J. Goadsby, and A. Macrae. 2003. Premonitory symptoms in migraine: An electronic diary study. *Neurology* 60 (6): 935–40.

8. Lauritzen, M. 1994. Pathophysiology of the migraine aura. The spreading depression theory. *Brain* 117 (Pt 1): 199–210.
9. Russell, M.B., and J. Olesen. 1996. A nosographic analysis of the migraine aura in a general population. *Brain* 119 (Pt 2): 355–61.
10. Peatfield, R., and J. Olesen. 1993. Precipitating factors. In *The headaches*, eds. J. Olesen, P. Tfelt-Hansen, and K.M. Welch, pp. 241–45. New York: Raven Press.
11. Spierings, E.L., A.H. Ranke, and P.C. Honkoop. 2001. Precipitating and aggravating factors of migraine versus tension-type headache. *Headache* 41 (6): 554–8.
12. Bigal, M.E., R.B. Lipton, J. Cohen, and S.D. Silberstein. 2003. Epilepsy and migraine. *Epilepsy Behav* 4 (Suppl 2): S13–24.
13. Olesen, J., H.K. Iversen, and L.L. Thomsen. 1993. Nitric oxide supersensitivity: A possible molecular mechanism of migraine pain. *Neuroreport* 4 (8): 1027–1030.
14. Buzzi, M.G., and M.A. Moskowitz. 1992. The trigemino-vascular system and migraine. *Pathol Biol (Paris)* 40 (4): 313–7.
15. May, A., and P.J. Goadsby. 1999. The trigeminovascular system in humans: Pathophysiologic implications for primary headache syndromes of the neural influences on the cerebral circulation. *J Cereb Blood Flow Metab* 19 (2): 115–127.
16. Goadsby, P.J. 2005. Migraine pathophysiology. *Headache* 45 (Suppl 1): S14–24.
17. Russell, M.B., and J. Olesen. 1995. Increased familial risk and evidence of genetic factor in migraine. *Br Med J* 311 (7004): 541–4.
18. Dichgans, M., T. Freilinger, G. Eckstein, E. Babini, B. Lorenz-Depiereux, S. Biskup, M.D. Ferrari, J. Herzog, A.M. van den Maagdenberg, M. Pusch, and T.M. Strom. 2005. Mutation in the neuronal voltage-gated sodium channel SCN1A in familial hemiplegic migraine. *Lancet* 366 (9483): 371–7.
19. Ray, B., and H. Wolff. 1940. Experimental studies on headache. *Arch Surg* 41 (4): 813–56.
20. Sessle, B.J. 2005. Peripheral and central mechanisms of orofacial pain and their clinical correlates *Minerva Anestesiol* 71 (4): 117–36.
21. Villalon, C.M., D. Centurion, L.F. Valdivia, P. de Vries, and P.R. Saxena. 2003. Migraine: Pathophysiology, pharmacology, treatment, and future trends. *Curr Vasc Pharmacol* 1 (1): 71–84.
22. Friberg, L., J. Olesen, H.K. Iversen, and B. Sperling. 1991. Migraine pain associated with middle cerebral artery dilatation: Reversal by sumatriptan. *Lancet* 338 (8758): 13–7.
23. Thomsen, L.L., H.K. Iversen, and J. Olesen. 1995. Cerebral blood flow velocities are reduced during attacks of unilateral migraine without aura. *Cephalalgia* 15: 109–16.
24. Zwetsloot, C.P., J.F. Caekebeke, and M.D. Ferrari. 1993. Lack of asymmetry of middle cerebral artery blood velocity in unilateral migraine. *Stroke* 24: 1335–8.
25. Olesen, J., B. Larsen, and M. Lauritzen. 1981. Focal hyperemia followed by spreading oligemia and impaired activation of rCBF in classic migraine. *Ann Neurol* 9 (4): 344–52.
26. Weiller, C., A. May, V. Limmroth, M. Juptner, H. Kaube, R.V. Schayck, H.H. Coenen, and H.C. Diener. 1995. Brain stem activation in spontaneous human migraine attacks. *Nat Med* 1 (7): 658–60.
27. Sanchez del Rio, M., and J. Alvarez Linera. 2004. Functional neuroimaging of headaches *Lancet Neurol* 3 (11): 645–51.
28. Burstein, R., D. Yarnitsky, I. Goor-Aryeh, B.J. Ransil, and Z.H. Bajwa. 2000. An association between migraine and cutaneous allodynia. *Ann Neurol* 47 (5): 614–24.
29. Schoenen, J., A. Ambrosini, P.S. Sandor, and A. Maertens de Noordhout. 2003. Evoked potentials and transcranial magnetic stimulation in migraine: Published data and viewpoint on their pathophysiologic significance. *Clin Neurophysiol* 114 (6): 955–72.
30. Waeber, C., and M.A. Moskowitz. 2005. Migraine as an inflammatory disorder. *Neurology* 64 (10 Suppl 2): S9–15.
31. Parsons, A.A., and P.J. Strijbos. 2003. The neuronal versus vascular hypothesis of migraine and cortical spreading depression. *Curr Opin Pharmacol* 3 (1): 73–7.
32. Goadsby, P.J. 2001. Migraine, aura, and cortical spreading depression: Why are we still talking about it? *Ann Neurol* 49 (1): 4–6.
33. Sicuteri, F., E. Del Bene, M. Poggioni, and A. Bonazzi. 1987. Unmasking latent dysnociception in healthy subjects. *Headache* 27: 180–5.

34. Iversen, H.K., J. Olesen, and P. Tfelt Hansen. 1989. Intravenous nitroglycerin as an experimental model of vascular headache. Basic characteristics. *Pain* 38: 17–24.

35. Lassen, L.H., L.L. Thomsen, and J. Olesen. 1995. Histamine induces migraine via the H1-receptor. Support for the NO hypothesis of migraine. *Neuroreport* 31: 1475–9.

36. Dahl, A., D. Russell, R. Nyberg Hansen, and K. Rootwelt. 1989. Effect of nitroglycerin on cerebral circulation measured by transcranial Doppler and SPECT. *Stroke* 20: 1733–6.

37. Afridi, S.K., H. Kaube, and P.J. Goadsby. 2004. Glyceryl trinitrate triggers premonitory symptoms in migraineurs. *Pain* 110 (3): 675–80.

38. Thomsen, L.L., C. Kruuse, H.K. Iversen, and J.A. Olesen. 1994. Nitric oxide donor (nitroglycerin) triggers genuine migraine attacks. *Eur J Neurol* 1: 73–80.

39. Thomsen, L.L. 1997. Investigations into the role of nitric oxide and the large intracranial arteries in migraine headache. *Cephalalgia* 17 (8): 873–95.

40. Olesen, J., L.L. Thomsen, and H. Iversen. 1994. Nitric oxide is a key molecule in migraine and other vascular headaches. *Trends Pharmacol Sci* 15: 149– 53.

41. Reuter, U., H. Bolay, I. Jansen-Olesen, A. Chiarugi, M.S. del Rio, R. Letourneau, T.C. Theoharides, C. Waeber, and M.A. Moskowitz. 2001. Delayed inflammation in rat meninges: Implications for migraine pathophysiology. *Brain* 124 (Pt 12): 2490–2502.

42. Moncada, S., R.M. Palmer, and E.A. Higgs. 1991. Nitric oxide: Physiology, pathophysiology, and pharmacology. *Pharmacol Rev* 43: 109–42.

43. Levy, D., and A.M. Strassman. 2004. Modulation of dural nociceptor mechanosensitivity by the nitric oxide-cyclic GMP signaling cascade. *J Neurophysiol* 92 (2): 766–72.

44. Sausbier, M., R. Schubert, V. Voigt, C. Hirneiss, A. Pfeifer, M. Korth, T. Kleppisch, P. Ruth, and F. Hofmann. 2000. Mechanisms of NO/cGMP-dependent vasorelaxation. *Circ Res* 87 (9): 825–30.

45. Ahern, G.P., V.A. Klyachko, and M.B. Jackson. 2002. cGMP and S-nitrosylation: Two routes for modulation of neuronal excitability by NO. *Trends Neurosci* 25 (10): 510–7.

46. Lassen, L.H., M. Ashina, I. Christiansen, and J. Olesen. 1997. Nitric oxide synthase inhibition in migraine. *Lancet* 349: 401–2.

47. Edvinsson, L., and R. Uddman. 2005. Neurobiology in primary headaches. *Brain Res Brain Res Rev* 48 (3): 438–56.

48. Goadsby, P.J., L. Edvinsson, and R. Ekman. 1990. Vasoactive peptide release in the extracerebral circulation of humans during migraine headache. *Ann Neurol* 28: 183–7.

49. Goadsby, P.J., L. Edvinsson, and R. Ekman. 1988. Release of vasoactive peptides in the extracerebral circulation of humans and the cat during activation of the trigeminovascular system. *Ann Neurol* 23 (2): 193–6.

50. Lassen, L., P. Haderslev, V. Jacobsen, H. Iversen, B. Sperling, and J. Olesen. 2002. CGRP may play a causative role in migraine. *Cephalalgia* 22 (1): 54–61.

51. Brain, S.D., and A.D. Grant. 2004. Vascular actions of calcitonin gene-related peptide and adrenomedullin. *Physiol Rev* 84 (3): 903–34.

52. Akerman, S., D.J. Williamson, H. Kaube, and P.J. Goadsby. 2002. Nitric oxide synthase inhibitors can antagonize neurogenic and calcitonin gene-related peptide induced dilation of dural meningeal vessels. *Br J Pharmacol* 137 (1): 62–8.

53. Wei, E.P., M.A. Moskowitz, P. Boccalini, and H.A. Kontos. 1992. Calcitonin gene-related peptide mediates nitroglycerin and sodium nitroprusside-induced vasodilation in feline cerebral arterioles. *Circ Res* 70 (6): 1313–9.

54. Williamson, D.J., and R.J. Hargreaves. 2001. Neurogenic inflammation in the context of migraine. *Microsc Res Tech* 53 (3): 167–78.

55. Olesen, J., H.C. Diener, I.W. Husstedt, P.J. Goadsby, D. Hall, U. Meier, S. Pollentier, and L.M. Lesko. 2004. Calcitonin gene-related peptide receptor antagonist BIBN 4096 BS for the acute treatment of migraine. *N Engl J Med* 350 (11): 1104–10.

56. Petersen, K.A., S. Birk, L.H. Lassen, C. Kruuse, O. Jonassen, L. Lesko, and J. Olesen. 2005. The CGRP-antagonist, BIBN4096BS does not affect cerebral or systemic hemodynamics in healthy volunteers. *Cephalalgia* 25 (2): 139–47.

57. Humphrey, P.P., W. Feniuk, M.J. Perren, I.J. Beresford, M. Skingle, and E.T. Whalley. 1990. Serotonin and migraine. *Ann N Y Acad Sci* 600: 587–98; discussion 598–600.

58. Goadsby, P.J., and L. Edvinsson. 1993. The trigeminovascular system and migraine: Studies characterizing cerebrovascular and neuropeptide changes seen in humans and cats. *Ann Neurol* 33 (1): 48–56.

59. Stepien, A., and M. Chalimoniuk. 1998. Level of nitric oxide-dependent cGMP in patients with migraine. *Cephalalgia* 18 (9): 631–4.

60. Saxena, P., and P. Tfelt-Hansen. 2000. Triptans, 5-HT$_{1B/1D}$ receptor agonists in the acute treatment of migraine. In *The headaches*, 2nd ed., eds. J. Olesen, P. Tfelt-Hansen, and Welch, K. Lippincott. Philadelphia, PA: Williams & Wilkins.

61. Jain, N.K., and S.K. Kulkarni. 1999. L-NAME, a nitric oxide synthase inhibitor, modulates cholinergic antinociception. *Methods Find Exp Clin Pharmacol* 21 (3): 161–5.

62. Maura, G., and M. Raiteri. 1996. Serotonin 5-HT1D and 5-HT1A receptors respectively mediate inhibition of glutamate release and inhibition of cyclic GMP production in rat cerebellum in vitro. *J Neurochem* 66 (1): 203–9.

63. Suwattanasophon, C., P. Phansuwan-Pujito, and A. Srikiatkhachorn. 5-HT(1B/1D) serotonin receptor agonist attenuates nitroglycerin-evoked nitric oxide synthase expression in trigeminal pathway. *Cephalalgia* 23 (8): 825–32.

64. Stepien, A., M. Chalimoniuk, and J. Strosznajder. 1999. Serotonin 5HT1B/1D receptor agonists abolish NMDA receptor-evoked enhancement of nitric oxide synthase activity and cGMP concentration in brain cortex slices. *Cephalalgia* 19 (10): 859–65.

65. Moja, P., C. Cusi, R. Sterzi, and C. Canepari. 2005. Selective serotonin re-uptake inhibitors (SSRIs) for preventing migraine and tension-type headaches. *Cochrane Database Syst Rev* (3):CD002919.

66. Pelligrino, D., and Q. Wang. 1998. Cyclic nucleotide crosstalk and the regulation of cerebral vasodilation. *Prog Neurobiol* 56: 1–18.

67. Xu, H.L., V. Gavrilyuk, H.M. Wolde, V.L. Baughman, and D.A. Pelligrino. 2004. Regulation of rat pial arteriolar smooth muscle relaxation *in vivo* through multidrug resistance protein 5-mediated cGMP efflux. *Am J Physiol Heart Circ Physiol* 286 (5): H2020–7.

68. Mercapide, J., E. Santiago, E. Alberdi, and J.J. Martinez-Irujo. 1999. Contribution of phosphodiesterase isoenzymes and cyclic nucleotide efflux to the regulation of cyclic GMP levels in aortic smooth muscle cells. *Biochem Pharmacol* 58 (10): 1675–83.

69. Kruuse, C., T.S. Khurana, S.D. Rybalkin, S. Birk, U. Engel, L. Edvinsson, and J. Olesen. 2005. Phosphodiesterase 5 and effects of sildenafil on cerebral arteries of man and guinea pig. *Eur J Pharmacol* 521 (1–3): 105–14.

70. Kruuse, C., S.D. Rybalkin, T.S. Khurana, I. Jansen-Olesen, J. Olesen, and I. Edvinsson. 2001. The role of cGMP hydrolysing phosphodiesterases 1 and 5 in cerebral artery dilatation. *Eur J Pharmacol* 420 (1): 55–65.

71. Rybalkin, S.D., C. Yan, K.E. Bornfeldt, and J.A. Beavo. 2003. Cyclic GMP phosphodiesterases and regulation of smooth muscle function. *Circ Res* 93 (4): 280–91.

72. Birk, S., L. Edvinsson, J. Olesen, and C. Kruuse. 2004. Analysis of the effects of phosphodiesterase type 3 and 4 inhibitors in cerebral arteries. *Eur J Pharmacol* 489 (1–2): 93–100.

73. Rosenblum, W.I., T. Shimizu, and G.H. Nelson. 1993. Interaction of endothelium with dilation produced by inhibitors of cyclic nucleotide diesterases in mouse brain arterioles *in vivo*. *Stroke* 24 (2): 266–70.

74. Parfenova, H., M. Shibata, S. Zuckerman, R. Mirro, and C.W. Leffler. 1993. Cyclic nucleotides and cerebrovascular tone in newborn pigs *Am J Physiol* 265 (6 Pt 2): H1972–82.

75. Willette, R.N., A.O. Shiloh, C.F. Sauermelch, A. Sulpizio, M.P. Michell, L.B. Cieslinski, T.J. Torphy, and E.H. Ohlstein. 1997. Identification, characterization, and functional role of phosphodiesterase type IV in cerebral vessels: Effects of selective phosphodiesterase inhibitors. *J Cereb Blood Flow Metab* 17: 210–9.

76. Inoha, S., T. Inamura, K. Ikezaki, A. Nakamizo, T. Amano, and M. Fukui. 2002. Type V phosphodiesterase expression in cerebral arteries with vasospasm after subarachnoid hemorrhage in a canine model. *Neurol Res* 24 (6): 607–12.

77. Sobey, C.G., and L. Quan. 1999. Impaired cerebral vasodilator responses to NO and PDE V inhibition after subarachnoid hemorrhage. *Am J Physiol* 277 (5 Pt 2): H1718–24.
78. Kruuse, C., L.L. Thomsen, T.B. Jacobsen, and J. Olesen. 2002. The phosphodiesterase 5 inhibitor sildenafil has no effect on cerebral blood flow or blood velocity, but nevertheless induces headache in healthy subjects. *J Cereb Blood Flow Metab* 22 (9): 1124–31.
79. Kruuse, C., T.B. Jacobsen, L.H. Lassen, L.L. Thomsen, S.G. Hasselbalch, H. Dige-Petersen, and J. Olesen. 2000. Dipyridamole dilates large cerebral arteries concomitant to headache induction in healthy subjects. *J Cereb Blood Flow Metab* 20 (9): 1372–9.
80. Loughney, K., T.R. Hill, V.A. Florio, L. Uher, G.J. Rosman, S.L. Wolda, B.A. Jones, M.L. Howard, L.M. McAllister-Lucas, W.K. Sonnenburg, S.H. Francis, J.D. Corbin, J.A. Beavo, and K. Ferguson. 1998. Isolation and characterization of cDNAs encoding PDE5A, a human cGMP-binding, cGMP-specific 3',5'-cyclic nucleotide phosphodiesterase. *Gene* 216 (1): 139–47.
81. Van Staveren, W.C., H.W.Steinbusch, M. Markerink-Van Ittersum, D.R. Repaske, M.F. Goy, J. Kotera, K. Omori, J.A. Beavo, and J. De Vente. 2003. mRNA expression patterns of the cGMP-hydrolyzing phosphodiesterases types 2, 5, and 9 during development of the rat brain. *J Comp Neurol* 467 (4): 566–80.
82. Beavo, J.A. 1995. Cyclic nucleotide phosphodiesterases: Functional implications of multiple iso-forms. *Physiol Rev* 75: 725–48.
83. Stoclet, J., T. Karavis, N. Komas, and C. Lugnier. 1995. Cyclic nucleotide phosphodiesterases as therapeutic targets in cardiovascular diseases. *Exp Opin Invest Drugs* 4: 1081–100.
84. Fitzgerald, G.A. 1987. Dipyridamole. *N Engl J Med* 316 (20): 1247–57.
85. Fujishige, K., Kotera, J. Michibata, H. K. Yuasa, S. Takebayashi, K. Okumura, and K. Omori. 1999. Cloning and characterization of a novel human phosphodiesterase that hydrolyzes both cAMP and cGMP (PDE10A). *J Biol Chem* 274 (26): 18438–45.
86. Fawcett, L., R. Baxendale, P. Stacey, C. McGrouther, I. Harrow, S. Soderling, J. Hetman, J.A. Beavo, and S.C. Phillips. 2000. Molecular cloning and characterization of a distinct human phosphodiesterase gene family: PDE11A. *Proc Natl Acad Sci USA* 97 (7): 3702–7.
87. Anonymous. 1997. European stroke prevention study 2. Efficacy and safety data. *J Neurol Sci* 151 (Suppl): S1–77.
88. Hawkes, C.H. 1978. Dipyridamole in migraine. *Lancet* 2 (8081): 153.
89. Kruuse, C., L.H. Lassen, S. Oestergaard, H.K. Iversen, and J. Olesen. 2006. Dipyridamole may cause migraine in patients with migraine without aura. *Cephalalgia* 26(8): 925–933.
90. Morales, A., C. Gingell, M. Collins, P.A. Wicker, and I.H. Osterloh. 1998. Clinical safety of oral sildenafil citrate (Viagra) in the treatment of erectile dysfunction. *Int J Impot Res* 10:69–74.
91. Evans, R.W., and C. Kruuse. 2004. Phosphodiesterase-5 inhibitors and migraine. *Headache* 44 (9): 925–6.
92. Kruuse, C., E. Frandsen, S. Schifter, L. Thomsen, S. Birk, and J. Olesen. 2004. Plasma levels of cAMP, cGMP and CGRP in sildenafil-induced headache. *Cephalalgia* 24 (7): 547–53.
93. Pagani, S., D. Mirtella, R. Mencarelli, D. Rodriguez, and M. Cingolani. 2005. Postmortem distribution of sildenafil in histological material. *J Anal Toxicol* 29 (4): 254–7.
94. Lusche, D.F., H. Kaneko, and D. Malchow. 2005. cGMP-phosphodiesterase antagonists inhibit Ca(2+)-influx in Dictyostelium discoideum and bovine cyclic-nucleotide-gated-channel. *Eur J Pharmacol* 513 (1–2): 9–20.
95. Richter, F., O. Mikulik, A. Ebersberger, and H.G. Schaible. 2005. Noradrenergic agonists and antagonists influence migration of cortical spreading depression in rat-a possible mechanism of migraine prophylaxis and prevention of postischemic neuronal damage. *J Cereb Blood Flow Metab* 25 (9): 1225–35.
96. Birk, S., C. Kruuse, K. Petersen, O. Jonassen, P. Tfelt-Hansen, and J. Olesen. 2003. Cilostazol causes headache in healthy volunteers; evidence for invlovement of cAMP? *Cephalalgia* 7:657 (abstract).
97. Zhu, C.B., W.A. Hewlett, S.H. Francis, J.D. Corbin, and R.D. Blakely. 2004. Stimulation of serotonin transport by the cyclic GMP phosphodiesterase-5 inhibitor sildenafil. *Eur J Pharmacol* 504 (1–2): 1–6.

98. Wang, M., J. Urenjak, E. Fedele, and T.P. Obrenovitch. 2004. Effects of phosphodiesterase inhibition on cortical spreading depression and associated changes in extracellular cyclic GMP. *Biochem Pharmacol* 67 (8): 1619–27.

99. Podda, M.V., M.E. Marcocci, L. Oggiano, M. D'Ascenzo, E. Tolu, A.T. Palamara, G.B. Azzena, and C. Grassi. 2004. Nitric oxide increases the spontaneous firing rate of rat medial vestibular nucleus neurons in vitro via a cyclic GMP-mediated PKG-independent mechanism. *Eur J Neurosci* 20 (8): 2124–32.

100. Shaikh, A.G., and P.G. Finlayson. 2005. Excitability of auditory brainstem neurons, in vivo, is increased by cyclic-AMP. *Hear Res* 201 (1–2): 70–80.

101. Kaupp, U.B., and R. Seifert. 2002. Cyclic nucleotide–gated ion channels. *Physiol Rev* 82 (3): 769–824.

102. Murthy, K.S., H. Zhou, and G.M. Makhlouf. 2002. PKA-dependent activation of PDE3A and PDE4 and inhibition of adenylyl cyclase V/VI in smooth muscle. *Am J Physiol Cell Physiol* 282 (3): C508–17.

103. Kotera, J., K. Fujishige, and K. Omori. 2000. Immunohistochemical localization of cGMP-binding cGMP-specific phosphodiesterase (PDE5) in rat tissues. *J Histochem Cytochem* 48 (5): 685–93.

104. Kruuse, C., T. Jacobsen, L. Thomsen, S. Hasselbalch, E. Frandsen, H. Dige-Petersen, and J. Olesen. 2000. Effects of the nonselective phosphodiesterase inhibitor pentoxifylline on regional cerebral blood flow and large arteries in healthy subjects. *Eur J Neurol* 7:629–38.

105. McHenry, L.C. Jr., D.A. Stump, G. Howard, T.T. Novack, D.H. Bivins, and A.O. Nelson. 1983. Comparison of the effects of intravenous papaverine hydrochloride and oral pavabid HP capsulets on regional cerebral blood flow in normal individuals *J Cereb Blood Flow Metab* 3 (4): 442–7.

106. Wang, H.S., and W.D. Obrist. 1976. Effect of oral papaverine on cerebral blood flow in normals: Evaluation by the xenon-133 inhalation method. *Biol Psychiatry* 11 (2): 217–25.

107. Schmidt, J. F., and G. Waldemar. 1990. Effect of nimodipine on cerebral blood flow in human volunteers. *J Cardiovasc Pharmacol* 16 (4): 568–71.

108. Katz, A.M., and N.M. Leach. 1987. Differential effects of 1,4-dihydropyridine calcium channel blockers: Therapeutic implications. *J Clin Pharmacol* 27 (11): 825–34.

109. Yu, Y., K. Mizushige, T. Ueda, Y. Nishiyama, M. Seki, T. Aoyama, M. Ohkawa, and H. Matsuo. 2000. Effect of olprinone, phosphodiesterase III inhibitor, on cerebral blood flow assessed with technetium-99m-ECD SPECT. *J Cardiovasc Pharmacol* 35 (3): 422–6.

110. Ito, H., T. Kinoshita, Y. Tamura, J. Yokoyama, and H. Iida. 1999. Effect of intravenous dipyridamole on cerebral blood flow in humans. A pet study. *Stroke* 30 (8): 1616–20.

27 Phosphodiesterase-4 as a Pharmacological Target Mediating Antidepressant and Cognitive Effects on Behavior

Han-Ting Zhang and James M. O'Donnell

CONTENTS

27.1 Introduction 539
27.2 PDE4 Structure and Pharmacology 540
    27.2.1 Enzyme Structure 540
    27.2.2 Inhibitor Interactions...540
27.3 PDE4 in Signaling Pathways Associated with Depression and Memory.....541
    27.3.1 Noradrenergic and Serotonergic Signaling 541
    27.3.2 N-Methyl D-Aspartate Receptor–Mediated Signaling 542
27.4 Antidepressant Effects of PDE4 Inhibitors..542
    27.4.1 Preclinical Models...542
    27.4.2 Clinical Studies 544
    27.4.3 Roles of PDE4 Subfamilies 544
27.5 PDE4 and Memory 545
    27.5.1 Inhibitors in Memory Models 545
    27.5.2 Long-Term Potentiation 549
    27.5.3 Aging and Alzheimer's Disease 550
    27.5.4 Roles of PDE4 Subfamilies 551
27.6 Summary..552
Acknowledgments 553
References 553

27.1 INTRODUCTION

Phosphodiesterase-4 (PDE4) is a low K_m, cyclic AMP-selective phosphodiesterase (PDE) that is highly expressed in the central nervous system (CNS), including regions involved in the regulation of affective behavior, mood, and cognition.[1,2] Interest in its potential as a target for psychopharmacological medications began in the early 1980s with the seminal work of Wachtel that described the behavioral effects of the prototypic PDE4 inhibitor rolipram and demonstrated its activity in preclinical models sensitive to antidepressant drugs.[3,4]

539

Subsequent work demonstrated that rolipram has effects in cognitive models, antagonizing the amnesic effects of scopolamine.[5,6] Overall, this research provided an impetus for examining the potential of PDE4 inhibitors for treating depression and neuropsychiatric illnesses that involve some measure of cognitive impairment. While progress has been made in elucidating the neurochemical and molecular mechanisms that underlie the various behavioral effects of PDE4 inhibitors, it has become apparent that drug development in this area is quite challenging, in part due to an incomplete understanding of the function and regulation of PDE4 in neurons. As a consequence of this, it has proven difficult to use neurochemical and molecular indices of inhibitor interaction to predict the pattern of behavioral effects related to antidepressant and procognitive efficacy relative to side effects such as nausea, emesis, and sedation.

Several approaches have been taken to understand the function and pharmacology of PDE4 in the CNS. These involve anatomical studies, utilizing autoradiography, in situ hybridization, and immunohistochemistry, to describe the localization of PDE4 in the brain, neurochemical, and molecular studies to elucidate its involvement in signaling pathways associated with antidepressant and cognitive mechanisms, behavioral studies to define the potential of PDE4 inhibitors as psychopharmacological agents, and clinical studies to demonstrate therapeutic efficacy.[7]

27.2 PDE4 STRUCTURE AND PHARMACOLOGY

27.2.1 ENZYME STRUCTURE

PDE4 is encoded by four genes termed PDE4A, PDE4B, PDE4C, and PDE4D.[8] Of these, PDE4A, PDE4B, and PDE4D are expressed to a significant extent throughout the CNS; PDE4C, by contrast, is expressed only minimally.[1,9] Relative to most other PDEs, PDE4 exhibits a relatively low K_m (approximately 1 μM) for cyclic AMP and is responsible for a major component of cyclic AMP hydrolysis in neurons[10,11]; it is not involved in cyclic GMP hydrolysis in neurons.[12] The catalytic region of PDE4 is highly conserved among the PDE4 subfamilies. Since PDE4 inhibitors bind to the catalytic site in a competitive manner, the high level of homology among PDE4 subfamilies may account for the difficulty in developing inhibitors that exhibit marked selectivity for these individual PDE4 subfamilies (e.g., PDE4A vs. PDE4B).

The four PDE4 genes generate distinct splice variants (isoforms), five for PDE4A, four for PDE4B, three for PDE4C, and nine for PDE4D.[7,8,13] In general, these can be classified into three categories: the long-form PDE4s, which contain two highly conserved N-terminal regions (termed upstream conserved regions 1 and 2; UCR1, UCR2) that are involved in intra- and intermolecular interactions and posttranslational modification; the short-form PDE4s, which lack UCR1; and the super-short-form PDE4s, which lack UCR1 and a portion of UCR2. The catalytic sites of all PDE4s, except PDE4A, contain a phosphorylation site for extracellular signal-regulated kinase (ERK) that regulates activity in a variant-specific manner.[14]

27.2.2 INHIBITOR INTERACTIONS

The prototypic inhibitor of PDE4 is rolipram. This compound exhibits at least a 100-fold selectivity for PDE4 relative to the other 10 PDE families. It penetrates into the CNS well following systemic administration, making it useful for in vivo analyses.[15] It does not exhibit any appreciable subfamily or variant selectivity, but does show somewhat complex binding, with two binding affinity states termed the high-affinity rolipram-binding state and the

low-affinity rolipram-binding state (HARBS and LARBS). Binding of rolipram to PDE4 appears to be to the catalytic site in both instances, indicating that these are two binding conformers, rather than independent binding sites.[16] These distinct binding conformers may be due to PDE4 interaction with scaffolding and related proteins, such as A-kinase-anchoring proteins (AKAPs), receptors for activated C-kinase 1 (RACK1), and proteins that contain SH3 domains, resulting in cell-type-specific pharmacological effects.[17–20] The potency order for the effects of PDE4 inhibitors reflects these differences in that some actions such as antidepressant-like effects in rats, induction of head twitches and tremors in mice, and emesis in ferrets, corresponds to potency for interacting with the HARBS whereas others such as inhibition of mast cell degranulation and antigen-induced T-cell proliferation in guinea pigs are related to potency for the LARBS.[21–26] This differential binding appears to be of particular importance in the CNS, because PDE4 is expressed throughout both the brain and peripheral tissues, but high-affinity binding sites are detected only in the former.[27,28]

It has proven somewhat challenging to assess inhibitor interaction with the HARBS and LARBS under equivalent conditions. ^{3}H-Rolipram has been shown to be a useful ligand for labeling the HARBS, exhibiting a K_D value of approximately 1 nM; competition assays reveal the affinity of other inhibitors for this state.[27,28] The ligand ^{3}H-rolipram, however, has a relatively low affinity for the LARBS (approximately 500 nM), which makes it unsuitable for use in typical equilibrium radioligand-binding assays. While nonequilibrium assays have been used, these are somewhat cumbersome and not well suited for the assessment of inhibitor interaction with the LARBS.[16] An alternative radioligand that has proven useful is ^{3}H-piclamilast. This compound binds to both the HARBS and LARBS, with equal, high affinity (approximately 1 nM). Its use, in combination with unlabeled rolipram to isolate the LARBS, quantifies inhibitor interaction with the low-affinity conformer of PDE4.[27,28]

27.3 PDE4 IN SIGNALING PATHWAYS ASSOCIATED WITH DEPRESSION AND MEMORY

27.3.1 Noradrenergic and Serotonergic Signaling

Most proven antidepressants enhance noradrenergic or serotonergic neurotransmission, or both, in the brain. Several lines of evidence suggest that PDE4 is involved in cyclic AMP signaling mediated by adrenergic and serotonergic receptors. It has been shown that PDE4 is the enzyme responsible for the hydrolysis of cyclic AMP formed by stimulation of beta-adrenergic receptors in the cerebral cortex (Figure 27.1).[10] This receptor type has been shown to be a mediator of behavioral effects of antidepressant drugs.[29] Consistent with this, it has been found that rolipram interacts either additively or synergistically with antidepressant-like behavioral effects mediated by beta-1 or beta-2 adrenergic receptors.[30] Further, it has been shown that reducing noradrenergic function by 6-hydroxydopamine-induced noradrenergic lesions or chronic blockade of beta-adrenergic receptors with propranolol, or increasing it with chronic desipramine treatment, results in a compensatory regulation of PDE4 expression and activity.[31] This regulation has a behavioral correlate in that rats subjected to central noradrenergic lesions, which exhibit reduced PDE4 expression in the brain, are more sensitive to the antidepressant-like behavioral effects of rolipram.[32] In general, in rats, it appears that PDE4A and PDE4B are regulated to a greater extent following altered noradrenergic function than is PDE4D[33]; however, this may be species-dependent since somewhat disparate results were obtained using mice.[34] Less is known regarding PDE4 involvement in serotonin receptor–mediated signaling. However, serotonin-4, 6, and 7 receptor subtypes are positively coupled to adenylyl cyclase and indirect evidence suggests the involvement of PDE4 in their signaling pathways. Repeated treatment with serotonin reuptake inhibitors, such as

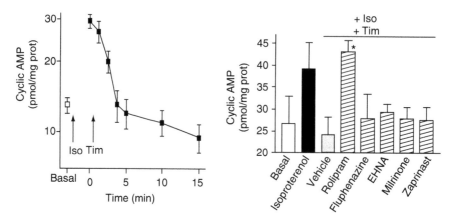

FIGURE 27.1 PDE4 hydrolyzes cyclic AMP formed by stimulation of beta-adrenergic receptors in slices of rat cerebral cortex. This is evidenced by the ability of the PDE4 inhibitor rolipram to block the hydrolysis of cyclic AMP formed following brief stimulation of beta-adrenergic receptors, produced by isoproterenol followed by timolol. Inhibitors of other PDE families have no effect on cyclic AMP hydrolysis under these conditions. (Reprinted from Ref. [10]. With permission.)

fluoxetine, which enhance serotonergic neurotransmission, increases the expression of PDE4A and PDE4B in cerebral cortex and hippocampus.[35–37]

27.3.2 *N*-Methyl d-Aspartate Receptor–Mediated Signaling

PDE4 also is involved in *N*-methyl d-aspartate (NMDA) receptor–mediated cyclic AMP signaling, which has been implicated in memory mechanisms and also may be involved in cognitive aspects of depression.[12,38] Stimulation of NMDA receptors results in increased calcium entry into neurons, which then activates calcium/calmodulin-dependent adenylyl cyclases, increasing cyclic AMP. This cyclic AMP is hydrolyzed by PDE4 (Figure 27.2).[12] By contrast, the cyclic GMP that is formed subsequent to NMDA receptor stimulation is hydrolyzed by PDE2.[12] Thus, PDE2, which hydrolyzes both cyclic AMP and cyclic GMP, does not appear to be involved in cyclic AMP hydrolysis in cerebral cortical and hippocampal neurons, suggesting distinct compartmentalization of PDE2 and PDE4.

27.4 ANTIDEPRESSANT EFFECTS OF PDE4 INHIBITORS

27.4.1 Preclinical Models

Rolipram and other PDE4 inhibitors have been shown to be active in a number of preclinical assays sensitive to antidepressant drugs. These compounds reduce the time of immobility in the forced-swim test, one of the most commonly used and extensively validated animal model for assessing antidepressant effects.[11,24] Importantly, PDE4 inhibitors produce this effect in the absence of any psychomotor-stimulant actions, which can result in false-positive findings. PDE4 inhibitors also have been shown to reduce the response rate and increase the reinforcement rate of rats under a differential-reinforcement-of-low-rate (DRL) operant schedule (Figure 27.3).[30,32,39] Such effects are consistent with those of proven antidepressants from various pharmacological classes.[40] The antidepressant-like effect of rolipram on DRL behavior persists with repeated administration; this is consistent with what has been found for tricyclic antidepressant drugs.[32,41]

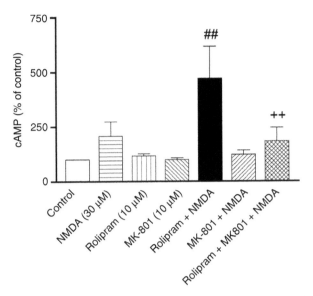

FIGURE 27.2 PDE4 hydrolyzes cyclic AMP formed by the stimulation of NMDA receptors in primary cultures of cerebral cortical neurons. This is evidenced by the ability of the PDE4 inhibitor rolipram to increase cyclic AMP concentrations following NMDA treatment; this effect is antagonized by the NMDA receptor antagonist MK-801. Inhibitors of other PDE families have no effect on NMDA receptor-stimulated cyclic AMP concentrations. Similar data are obtained using hippocampal neurons. (Reprinted from Ref. [12]. With permission.)

PDE4 inhibitors also are active in a number of other models of antidepressant sensitivity. They reverse the behavioral effects of chronic mild stress, normalize the behavior of rats of the Flinders-sensitive line as well as those that subjected to olfactory bulbectomy, both of which exhibit depression-related symptomatology.[42,43] Further, they antagonize the

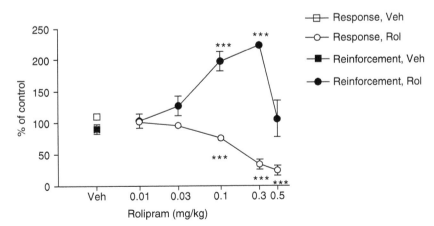

FIGURE 27.3 The PDE4 inhibitor rolipram produces antidepressant-like effects on the behavior of rats under a differential-reinforcement-of-low-rate operant schedule, reducing response rate and increasing reinforcement rate. This is the characteristic pattern produced by proven antidepressant drugs.[40] (Reprinted from Ref. [30]. With permission.)

behavioral and physiological effects of reserpine and potentiate the yohimbine-induced lethality effects shared with proven antidepressants.[4] Finally, PDE4 appears to be involved in antidepressant-induced neurogenesis in the hippocampus, which is thought to be important in the mediation of the long-term, later-developing, behavioral effects.[44]

27.4.2 CLINICAL STUDIES

In addition to the extensive body of preclinical data, clinical data support the antidepressant utility of PDE4 inhibitors; these studies focus on rolipram. Overall, the results of these studies indicate that rolipram is superior to placebo and equivalent to reference antidepressants in a number of clinical indices.[45–47] However, the dose-limiting side effects of sedation, nausea, and vomiting affect rolipram's usefulness.[48,49] Unfortunately, there has not, as yet, been a thorough psychopharmacological evaluation of a range of PDE4 inhibitors that differ structurally and pharmacologically from rolipram, in order to better assess the potential for dissociating therapeutic from side effects.

27.4.3 ROLES OF PDE4 SUBFAMILIES

It has proven somewhat difficult to assess the degree to which the different PDE4 subfamilies contribute to the mediation of the antidepressant-like behavioral effects of PDE4 inhibitors. Primarily, this results from the lack of highly subfamily-selective PDE4 inhibitors. The most well-studied inhibitors, such as rolipram and Ro 20-1724, exhibit equal potency for inhibiting the PDE4 subfamilies. While some subfamily-selective inhibitors have been developed, their selectivity is rather modest (i.e., approximately 10-fold) and not suitable for work *in vivo*. To address the roles of the PDE4 subfamilies in the mediation of antidepressant effects, knock-out mice developed by Conti and coworkers have been utilized (see Chapter 16).[50] To date, data have been reported for PDE4D- and PDE4B-deficient mice.

Mice deficient in PDE4D display an antidepressant-like behavioral profile.[11] This is evidenced by reduced immobility in the forced-swim and tail-suspension tests relative to wild-type controls (Figure 27.4). Such behavioral changes are consistent with what is observed following the administration of a proven antidepressant or PDE4 inhibitor. Examination of other behavioral measures, including general locomotor activity, anxiety-related behavior, and nociception, revealed no differences from wild-type mice. When PDE4D-deficient mice are treated with rolipram, no additional effect on behavior is observed, suggesting a role for this subfamily in the mediation of antidepressant behavioral effects of PDE4 inhibitors. By contrast, even though PDE4D-deficient mice exhibit markedly lower immobility in the forced-swim test than wild-type controls, the antidepressants desipramine and fluoxetine produce additional reductions, suggesting at least additive antidepressant actions. The data obtained using PDE4B-deficient mice suggest a different function for this subfamily.[51] While PDE4B-deficient mice show reduced sensitivity to rolipram, similar to what is observed for PDE4D-deficient mice, the antidepressant desipramine does not produce any additional effect on their behavior. This difference between the antidepressant sensitivity of PDE4D- and PDE4B-deficient mice suggests that these two PDEs may be differentially distributed in signaling pathways affected by antidepressant drugs. Loss of cyclic AMP catabolic function in a signaling pathway should increase sensitivity to presynaptically acting antidepressant drugs. This may be the case for PDE4D, but not for PDE4B, suggesting the relative importance of the former subfamily in the mediation of antidepressant effects on behavior. However, the preferential regulation of PDE4A and PDE4B, as opposed to PDE4D, in response to chronic antidepressant treatment is at odds with this interpretation.[35–37] While this discrepancy may result in

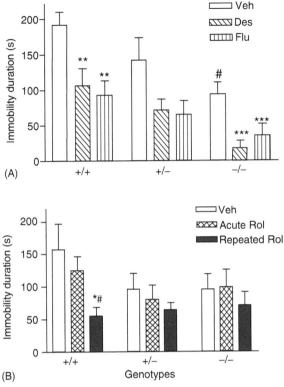

FIGURE 27.4 Mice deficient in PDE4D (−/−) exhibit an antidepressant-like behavioral profile, as evidenced by reduced immobility in the forced-swim test relative to wild-type controls (+/+); similar effects are obtained using a second model sensitive to antidepressant drugs, the tail-suspension test. Rolipram (Rol) does not produce a further reduction in the time of immobility in the forced-swim test, indicating that it works, at least in part, through this subfamily. By contrast, the antidepressants desipramine (Des) and fluoxetine (Flu) produce further antidepressant-like effects in the PDE4D-deficient mice. (Reprinted from Ref. [11]. With permission.)

part from a species difference,[34] it does highlight the need for more thorough evaluation of the behavioral phenotypes and pharmacological sensitivity of mice deficient in each of the PDE4A, PDE4B, and PDE4D subfamilies, particularly conditional knockout lines that eliminate possible developmental influences, as well as the need for inhibitors that exhibit marked subfamily selectivity.

27.5 PDE4 AND MEMORY

27.5.1 INHIBITORS IN MEMORY MODELS

The function of PDE4 in memory processes began to be recognized at the beginning of the 1980s. Studies revealed that the dunce gene in *Drosophila*, which encodes a cyclic AMP-specific PDE, is required for memory formation.[52] Mutation of this gene causes accumulation of cyclic AMP[53] and results in severe deficits in learning and memory.[54] This cyclic AMP-specific PDE was later demonstrated to be a homolog of mammalian PDE4,[55] suggesting an involvement of PDE4 in the mediation of memory. This concept was supported by the studies

of Villiger and Dunn,[56] who found that Ro 20-1724, a selective PDE4 inhibitor, facilitates the formation of long-term memory for passive avoidance conditioning in mice. Later, Randt et al.[57] reported that rolipram reverses long-term memory deficits induced by the protein synthesis inhibitor, anisomycin, in an inhibitory avoidance test in mice. In addition, the cyclic AMP level in the frontal cortex is increased after administration of rolipram, suggesting that its effect on memory results from its inhibition of PDE4 and a subsequent increase in cyclic AMP concentrations in the brain. This was the first report of the memory-enhancing effect of rolipram and its correlation with central cyclic AMP concentrations. Consistent with this, direct or indirect increases in cyclic AMP in the brain by central administration of cyclic AMP analogs or PDE4 inhibitors, respectively, improve long-term memory.[58–61]

Rolipram, when administered alone, does not always increase basal cyclic AMP concentrations in the brain, unless used at relatively high doses (up to 10 mg/kg).[57,62–64] This may account partially for some of the negative effects of rolipram on memory performance that has been reported (Table 27.1). When memory is not impaired by pharmacological or other means, rolipram is effective only in a limited number of tasks[62,65,66]; in a number of other tests, rolipram either does not alter memory[6,67–70] or even impairs it.[62,68,71,72] It is not surprising that rolipram is ineffective at doses around 0.1 mg/kg, because it does not significantly increase cyclic AMP in the brain.[62] However, it is hard to interpret the memory deficits produced by low doses of rolipram in untreated animals. One reason may be the sedative effect of rolipram,[49,73] which may result in longer escape latency in the Morris water

TABLE 27.1
Effects of PDE4 Inhibitors on Memory Performance in Untreated Animals[a]

Animal Model	PDE4 Inhibitor	Dose (mg/kg)	Efficacy	Ref.
Delayed response (monkeys)	Rolipram	0.01, IM	0, WM	[69]
Fear conditioning (mice)	Rolipram	0.1 μmol/kg	+, LTM; 0, STM	[62]
	Rolipram	0.03, SC, repeated	0, LTM; 0, STM	[68]
Fear conditioning (rabbits)	Rolipram	0.03–3 μmol/kg, SC	0	[70]
(Rats)	Rolipram	3, IP	0, LTM	[65]
		3, IP, repeated	0, LTM	[65]
		0.5, SC, continuous	+, LTM	[65]
			−, EXT	[65]
Morris water maze (rats)	Rolipram	3 and 30, IP	−, LTM	[71]
	Rolipram	3, IP, repeated	0, Acquisition; 0, LTM	[74]
	DC-AT 46	20, IP	−, LTM	[75]
(Mice)	Rolipram	0.1 μmol/kg, SC	−, LTM	[71]
	Rolipram	0.03, SC, repeated	−, LTM	[68]
Object recognition (mice)	Rolipram	0.1, IP	0, LTM	[67]
	HT0712	0.05–0.2, IP	+, LTM	
(Rats)	Rolipram	0.03–0.3, IP	+, LTM	[66]
3-Panel runway (rats)	Rolipram	0.1, IP	0, WM	[6]
Radial-arm maze (mice)	Rolipram	0.03, SC, repeated	0, Acquisition; 0, STM	[68]
Step-through (mice)	Ro 20-1274	25, 50, SC	+, LTM	[56]
	Rolipram	0.5, 1 μg, IC	−, LTM	[72]

[a]Animals were not administered drugs or manipulated physically to produce impairments in memory.

+, Enhancement; −, impairment or slowing; and 0, no effect. WM, Working memory; LTM, long-term memory; STM, short-term memory; and EXT, extinction. IP, Intraperitoneal injection; SC, subcutaneous injection; IM, intramuscular injection; and IC, intracisternal injection.

TABLE 27.2
Effects of PDE4 Inhibitors on Memory Performance in Treated Animals[a]

Animal Model	Treatment for Amnesia	PDE4 Inhibitor	Dose (mg/kg)	Efficacy	Ref.
Approach avoidance (mice)	Anisomycin	Rolipram	10, IP	+, LTM	[57]
Fear conditioning (rabbits)	L-PIA	Rolipram	3 μmol/kg	0	[70]
Morris water maze (rats)	Cerebral ischemia	Rolipram	3, IP, repeated	+, LTM; +, acquisition	[74]
Object recognition (mice)	Scopolamine	Rolipram	0.1, IP	+, WM	[66]
3-Panel runway (rats)	Scopolamine	Rolipram	0.1, IP	+, WM	[6]
	Cerebral ischemia	Rolipram	0.1, IP	+, WM	
	ECS	Rolipram	0.32, IP	+, WM	
Radial-arm maze (rats)	MK-801	Rolipram	0.1, IP	+, WM/RM	[30,38]
	Scopolamine	Rolipram	0.1, IP	+, WM/RM	[73]
	Scopolamine	Rolipram	0.02–0.2, PO	+, WM	[5]
	U0126	Rolipram	0.1, IP	+, RM	[61]
Step-through (rats)	MK801	Rolipram	0.1, IP	+, LTM	[30,38]
	U0126	Rolipram	0.1, IP	+, LTM	[61]
	U0126	Rolipram	15 μg, CA1	+, LTM	[61]
	Scopolamine	Rolipram	0.02–0.1, PO	+, LTM	[5]
	FRFS	Rolipram	5, IP	+, LTM	[105]
(Mice)	DCG-IV	Rolipram	0.5 μg, IC	+, LTM	[72]
	Scopolamine	Rolipram	10, IP	+, LTM	[6]
	Scopolamine	Rolipram	30, PO	+, LTM	[106]
	Cycloheximide	Rolipram	10, IP	+, LTM	[6]
	ECS	Rolipram	3–30, IP	+, LTM	[6]

[a]Animals were treated with drugs or physically manipulated so as to impair performance in memory tasks. Reversal by PDE4 inhibitors was assessed.

+, Enhancement; 0, no effect. IP, Intraperitoneal injection; SC, subcutaneous injection; PO, per os; CA1, bilateral hippocampal CA1 infusion; and IC, intracisternal injection. ECS, electroconvulsive shock; L-PIA, N6-(L-phenylisopropyl)adenosine; and FRFS, frequently repetitive febrile seizures. WM, Working memory; RM, reference memory; and LTM, long-term memory.

maze; this could be interpreted as memory impairment. In addition, sensitivity differences due to animal strain or line variability cannot be ruled out.

In contrast to untreated animals, pharmacologically or physically treated animals, in which memory is impaired, are sensitive to rolipram. As shown in Table 27.2, rolipram, at relatively low doses (0.02–0.3 mg/kg), reverses deficits in long-term memory and working or reference memory induced by scopolamine, a muscarinic acetylcholine receptor antagonist,[5,6,66,73] MK-801, an NMDA receptor antagonist[38,74] (Figure 27.5), DCG-IV, an agonist of group II metabotropic glutamate receptors (mGluRs),[72] U0126, a MEK inhibitor[61] (Figure 27.6), and electric convulsive shock (ECS) or cerebral ischemia.[6] These treatments produce amnesic effects via reduction of cyclic AMP/PKA signaling[38,61,72,75]; rolipram by partially restoring this signaling reverses the amnesic effects. Similarly, as shown in Table 27.3, at even lower doses, rolipram improves memory, particularly long-term memory, in aged animals, CREB-binding protein (CBP) heterozygous mice (CBP$^{+/-}$), and amyloid precursor protein

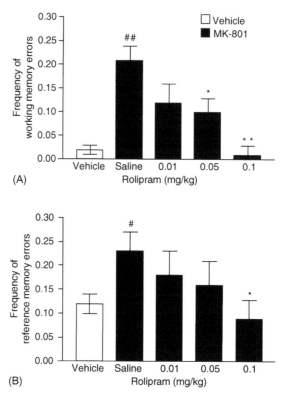

(A)

(B)

FIGURE 27.5 Rolipram antagonizes the amnesic effect of the NMDA receptor antagonist MK-801 in rats tested in the radial-arm maze. This effect is evident for both working memory and reference memory, models of short- and long-term memory, respectively. A similar effect is observed using a second model, inhibitory avoidance behavior. (Reprinted from Ref. [38]. With permission.)

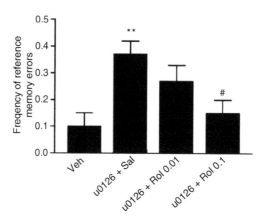

FIGURE 27.6 Rolipram antagonizes the amnesic effect of administration of the MEK inhibitor U0126 injected into the CA1 region of the hippocampi of rats tested in the radial-arm maze. A similar effect is observed using a second model, inhibitory avoidance behavior. (Reprinted from Ref. [61]. With permission.)

TABLE 27.3
Effects of PDE4 Inhibitors on Memory Performance in Aged or Transgenic Animals

Animal Model	PDE4	Dose Inhibitor	Efficacy (mg/kg)	Ref.
Barnes maze (aged mice)	Rolipram	0.05 μM, IP, repeated	+, WM	[77]
Delayed response (aged monkeys)	Rolipram	0.01, IM	−, WM	[69]
Fear conditioning (aged mice)	Rolipram	0.1 μmol/kg, SC	+, LTM	[62]
(APP/PS1 Tg mice)	Rolipram	0.03, SC, repeated	+, LTM	[68]
Morris water maze (aPP/PS1 Tg mice)	Rolipram	0.03, SC, repeated	+, LTM	[68]
Object recognition (CBP +/− mice)	Rolipram	0.1, IP	+, LTM	[67]
	HT0712	0.1, IP	+, LTM	
Radial-arm maze (APP/PS1 Tg mice)	Rolipram	0.03, SC, repeated	+, STM; +, acquisition	[68]

+, Enhancement; −, impairment. IP, Intraperitoneal injection; SC, subcutaneous injection; and IM, intramuscular injection. CBP, CREB binding protein; WM, Working memory; LTM, long-term memory; and STM, short-term memory.

(APP) and presenilin-1 (PS1) transgenic mice, an animal model of amyloid deposition that partially reproduces the cognitive deficits observed in patients with Alzheimer's disease (AD).[76] All these mice display impairment of the cyclic AMP/PKA/CREB cascade in the brain.[62,67,68,77] Therefore, the effect of rolipram appears to be dependent on changes in cyclic AMP signaling. In untreated or normal animals, cyclic AMP signaling in the brain may be maintained at a relatively high level, which is difficult to elevate by administration of rolipram alone; i.e., rolipram may only produce a small cyclic AMP difference insufficient to result in memory enhancement. Increasing rolipram doses may produce a larger cyclic AMP increase, but it may also exert a negative action on memory based on three observations. First, rolipram, at high doses, may increase cyclic AMP concentrations to extremely high levels, which mimics dunce gene mutants in *Drosophila* to produce memory deficits.[54] Second, rolipram administered at high doses may induce the expression of PDE4 and a subsequent, delayed decrease in cyclic AMP signaling and impairment of long-term memory.[71] Third, the therapeutic dose-range of rolipram is narrow. High doses of rolipram may produce sedation, impairing performance in some memory tasks.[49,73] Other PDE4 inhibitors, such as HT-0172 ((3S, 5S)-5-(3-cyclopentyloxy-4-methoxy-phenyl)-3-(3-methyl-benzyl)-piperidin-2-one) and Ro 20-1724, appear to overcome this problem and produce memory enhancement in normal animals (Table 27.1).[56,67] By contrast, in animals with memory deficits induced by pharmacological (e.g., MK-801), physical (e.g., cerebral ischemia), or genetic (e.g., APP/PS1 transgenic mice) means (i.e., treated animals), the basal cyclic AMP levels are significantly decreased. Administration of rolipram at the same doses may enhance cyclic AMP signaling to levels at or above those observed in untreated animals. This is supported by the recent work showing that rolipram enhances memory and simultaneously increases cyclic AMP signaling in the hippocampus and cerebral cortex of rats with cerebral ischemia, but not of normal rats.[74]

Taken together, the convergence of data suggests that PDE4 inhibitors, such as rolipram, improve memory via stimulation of cyclic AMP signaling.[77] They produce consistent memory-enhancing effects in animals whose memory is impaired by pharmacological, physical, or genetic means, but typically not in untreated animals exhibiting relatively normal memory performance.

27.5.2 LONG-TERM POTENTIATION

PDE4 is involved in the mediation of long-term potentiation (LTP), a cellular model of synaptic plasticity and memory.[79] It has been shown that hippocampal LTP has distinct

phases corresponding to memory formation: an early phase (E-LTP), which lasts less than 3 h and does not require new protein synthesis; and a late phase (L-LTP), which lasts longer than 3 h and requires RNA and protein synthesis.[80] E-LTP and L-LTP correspond to short-term and long-term memory, respectively. Synthesis of RNA and protein is also required for the consolidation of short-term memory to long-term memory.[81,82] Consistent with the long-term memory consolidation, E-LTP also is converted to L-LTP with facilitation of cyclic AMP/PKA signaling, which is important for LTP induction and transformation of short-term memory to long-term memory.[81] Inhibition of PDE4 with rolipram, which subsequently increases intracellular cyclic AMP concentrations, potentiates LTP induced by tetanic stimulation in slices of mouse and rat hippocampus.[62,83] Rolipram treatment also improves long-term contextual memory in both young and aged mice.[62] Interestingly, rolipram converts CA1 tetanus-stimulated E-LTP to L-LTP in slices of the rat hippocampus; this conversion is blocked by AP-5, an NMDA receptor antagonist, indicating that the rolipram action depends on the activation of NMDA receptors.[84] This supports the behavioral findings that rolipram reverses MK-801-induced deficits in memory, including working and reference memory in the radial-arm maze and long-term memory in inhibitory avoidance.

It is not known which PDE4 subfamily is involved in the mediation of memory processes. While Zhang et al.[51,61] have shown that PDE4D appears to be particularly important for long-term memory, recent studies have provided interesting clues supporting a connection between a PDE4B variant and hippocampal plasticity. LTP induction increases the expression of the PDE4B3 variant in the CA1 and dentate gyrus subregions of the rat hippocampus. Pretreatment with the NMDA receptor antagonist AP-5 antagonizes LTP-induced PDE4B3 expression. Consistently, the overall cyclic AMP concentrations in these areas are decreased.[85,86] The PDE4B3 gene is thought to be an LTP-mediated, specific plasticity gene.[85,87] However, there is no evidence indicating whether an increase in PDE4B3 alters LTP induction. Further, it is not known whether other PDE4B variants or PDE4 subfamilies are involved in hippocampal plasticity. Recently, it has been shown that DISC1 (disrupted in schizophrenia 1), a schizophrenia susceptibility gene, interacts with the UCR1 region of PDE4B, which provides additional support for a relationship between PDE4B function and cognition.[88]

27.5.3   AGING AND ALZHEIMER'S DISEASE

Age-related memory deficits have been observed in humans and a variety of animal species including rodents.[89–91] Aged rats and mice display significant impairment of hippocampus-dependent spatial memory, particularly long-term memory. Consistent with this, L-LTP in the hippocampal CA1 region also is reduced in aged animals; this is significantly and negatively correlated to the deficit in hippocampus-dependent memory.[77,91,92] Given that cyclic AMP signaling is important for the mediation of synaptic plasticity and memory, PDE4, a critical controller of intracellular cyclic AMP, may play a pivotal part in age-related memory loss. Similar to the dopamine D1/D5 receptor agonist SKF38393, which stimulates cyclic AMP signaling via activation of adenylyl cyclase, the PDE4 inhibitor rolipram attenuates the deficits in both memory and L-LTP in aged animals.[77] This suggests that PDE4 is important in the cyclic AMP signal pathway mediating memory in aged animals.[94] Inhibition of PDE4 increases cyclic AMP, which activates PKA and subsequently phosphorylates and activates CREB; this leads to improvement of memory in aged animals. This is supported by studies showing that rolipram administration reverses age-related reduction of CRE-binding activity and impairment of hippocampus-dependent memory.[77,95]

The aging process does not affect all brain regions in the same manner as observed in the hippocampus. In the prefrontal cortex of aged rats, cyclic AMP/PKA signaling is excessively

stimulated, rather than inhibited or downregulated, as evidenced by increased CRE binding and phospho-CREB expression.[69] In agreement with the biochemical changes, administration of rolipram or the PKA activator Sp-cAMPS exacerbates working memory deficits of aged rats in a delayed alternation task in the T-maze and of aged monkeys in a spatial delayed response task, in which the integrity of the prefrontal cortex is required for memory performance.[69] Consistently, direct infusion of Sp-cAMPS into the prefrontal cortex impairs working memory performance in the spatial delayed alternation task in the T-maze in rats; this is completely blocked by coinfusion of Rp-cAMPS, a cyclic AMP analog that inhibits PKA.[96] In contrast to these results, stimulation of cyclic AMP signaling in the hippocampal CA1 region enhances hippocampus-based memory and simultaneously increases hippocampal PKA activity and phospho-CREB expression.[62,97] Therefore, PDE4 may play a completely opposite role in the mediation of memory based on the status of cyclic AMP signaling in the brain of aged animals. This confirms the importance of the cyclic AMP difference for determining memory-enhancing effects of PDE4 inhibitors. However, excessive stimulation of cyclic AMP signaling may impair, rather than improve, memory, as evidenced by memory deficits observed in the dunce gene mutant *Drosophila*, which display extremely high concentrations of cyclic AMP.[53,54]

Alzheimer's disease (AD) is a progressive neurodegenerative disorder characterized by memory impairment and increased beta-amyloid peptides in the brain. Accumulation of beta-amyloid is an important etiological factor in AD.[98] Treatment of cultured cerebral cortical and hippocampal neurons with beta-amyloid$_{1-42}$ decreases KCl- and NMDA-induced CREB phosphorylation and KCl-induced expression of brain-derived neurotrophic factor (BDNF), a downstream target of CREB.[99,100] It also decreases BDNF downstream signaling by suppressing the activation of MEK/ERK1/2 or ERK, phosphatidylinositol 3-kinase (PI3-K)/Akt, and cyclic AMP/PKA pathways.[101] Consistent with this, beta-amyloid inhibits the formation of L-LTP in the mouse hippocampus and decreases CREB phosphorylation in response to glutamate.[100] In addition, mice double transgenic for APP and PS1 display impaired LTP and memory as early as 3–4 months of age; rolipram reverses these impairments but does not reverse the underlying pathophysiology.[68,100,102] Rolipram can improve memory in AD models via at least two mechanisms. First, rolipram reverses beta-amyloid-induced downregulation of cyclic AMP signaling. This is confirmed by increased phospho-CREB after rolipram treatment in hippocampal extracts and slices from APP/PS1 mice, which display a significant decrease in phospho-CREB levels.[68] Consistent with this, chronic administration of rolipram also increases expression of CREB and its downstream target BDNF in the hippocampus of normal rats.[103] Second, beta-amyloid induces suppression of ERK signaling,[101,104] although this is questioned by increased phospho-MEK1, an upstream regulator of ERK, in severe AD.[105] Since ERK phosphorylation inhibits the activity of long-form PDE4,[14,61] beta-amyloid-induced reduction of ERK may lead to disinhibition of PDE4 activity, an effect reversed by rolipram.

27.5.4 ROLES OF PDE4 SUBFAMILIES

While PDE4 has been implicated in memory processes, it is not known which subfamilies or variants are involved. It is difficult to address this question due to the lack of highly selective inhibitors of individual PDE4 subfamilies. However, studies to date have revealed some interesting indications regarding the roles of PDE4 subfamilies in memory; these include the distribution patterns and structural features of PDE4 subfamilies and the behavioral phenotype of mice lacking a specific PDE4 subfamily.

The four PDE4 subfamilies are widely but differentially distributed in mammalian tissues. PDE4A, PDE4B, and PDE4D are highly expressed throughout the brain, but PDE4C only

minimally so.[1,9] In addition, these three central PDE4 subfamilies exhibit distinct distribution patterns. For instance, PDE4A and PDE4D are expressed in the CA1 subregion of the rat hippocampus at the highest levels whereas PDE4B is expressed at a relatively low level in this structure. By contrast, PDE4B is highly expressed in the neostriatum and amygdala.[9] This indicates that, compared to PDE4A and PDE4D, PDE4B appears not to be the dominant subfamily in the mediation of hippocampal plasticity and memory, although recent studies have shown an association of LTP induction and PDE4B expression in the hippocampal CA1.[85] Distribution data suggest that PDE4B may be important for the mediation of striatum- and amygdala-based memory.

In addition to cyclic AMP, ERK signaling also is involved in the mediation of plasticity and memory.[106–108] Both signaling pathways are regulated by NMDA receptors,[109–112] which integrate PDE4 to their signaling system in mediating memory.[12,38] Given that LTP is associated with PDE4B expression[85] and ERK activation phosphorylates PDE4,[14] it appears that PDE4 is a part of the ERK signal pathway in plasticity and memory processes.

The link between PDE4 and ERK has been studied in recent years as the structure and regulation of PDE4 has become clearer. Behavioral and biochemical results suggest that PDE4 is a component of ERK signaling in the mediation of memory.[61] However, not all the PDE4 subfamilies are involved in ERK signaling-mediated memory. Only PDE4B, PDE4C, and PDE4D are phosphorylated by ERK; of these, only the PDE4B and PDE4D subfamilies are expressed appreciably in the brain.[9,14] In addition, the consequence of ERK phosphorylation of the PDE4 variants differs, depending on whether they contain a UCR1 region. ERK phosphorylation inhibits long-form PDE4 variants but, by contrast, stimulates short-form variants[14,113] Given that PDE4 activity in the hippocampus is primarily attributed to PDE4D,[9,11] ERK phosphorylation-induced changes in PDE4 activity likely involve changes in PDE4D activity predominantly. The importance of PDE4D in signal transduction also is supported by recent studies using mice deficient in PDE4D. Although PDE4 inhibition does not significantly alter intracellular basal cyclic AMP, cerebral cortical slices from PDE4D-deficient mice display a basal cyclic AMP concentration approximately 200% of those of wild-type controls,[11] indicating an important role for PDE4D in the control of intracellular cyclic AMP concentrations. Consistent with this, mice deficient in PDE4D display significant improvement of reference memory in the radial-arm maze task, long-term memory in step-down inhibitory avoidance, and spatial memory in the Morris water maze test. By contrast, mice lacking PDE4B do not show significant changes in these memory models, although they tend to exhibit a longer long-term retention in the inhibitory avoidance test.[51] In addition, recent studies have shown that both MEM1018 and MEM1091, two novel PDE4 inhibitors that reverse MK-801-induced memory impairment, exhibit modest PDE4D selectivity.[74] Furthermore, overexpression of PDE4D induced by administration of high doses of PDE4 inhibitors, which decreases cyclic AMP signaling in the hippocampus, impairs long-term spatial memory.[71] Taken together, these results suggest an important role for PDE4D in cyclic AMP hydrolysis in the hippocampus, as well as for the mediation of memory; however, roles for PDE4A and PDE4B cannot be excluded.

27.6 SUMMARY

The results of numerous studies carried out over the last two decades provide strong support that PDE4 inhibitors have antidepressant and memory-enhancing effects. Consistent with these behavioral data, neurochemical analyses have demonstrated that PDE4 is a component of neuronal-signaling pathways involved in the mediation of these behavioral actions. While

much of the behavioral work has relied on rolipram, sufficient data have been obtained using other PDE4 inhibitors to suggest that these actions are representative of this drug class. In spite of this promise of therapeutic utility, several questions and concerns remain. First, due to the lack of subfamily-selective PDE4 inhibitors, it remains unclear which of the subfamilies expressed in brain, PDE4A, PDE4B, or PDE4D, is the predominant mediator of antidepressant and cognitive effects on behavior. Some progress in this area has been made using constitutive PDE4 subfamily knockout lines (i.e., PDE4B- and PDE4D-deficient mice); results using these models suggest the potential of the PDE4D subfamily in the mediation of psychopharmacological effects of PDE4 inhibitors. However, the lack of a large body of data using these mice, the current absence of data on PDE4A-deficient mice, and inherent concerns regarding constitutive knockout lines make it difficult to draw definitive interpretations. The development of subfamily-selective inhibitors suitable for *in vivo* work will greatly accelerate progress in this area. A second issue that influences the current understanding of the neuropharmacology of PDE4 inhibitors concerns the high- and low-affinity binding conformers (HARBS and LARBS) of the enzyme. While the data obtained to date suggest the importance of the HARBS component of PDE4 in the CNS actions of PDE4 inhibitors, an incomplete understanding of molecular mechanisms underlying this complex inhibitor-binding pattern limits the extent to which rationale drug design approaches can address this problem. Finally, it has become apparent that the PDE4 variants are associated in highly unique ways with a variety of scaffolding and signaling molecules.[17–20] Understanding the functional consequences of these protein–protein interactions will help to advance both the understanding of PDE4 involvement in signal transduction mechanisms and aid in targeted drug design for neuropsychopharmacological indications.

ACKNOWLEDGMENTS

Research in the authors' laboratory is supported by grants from the National Institute of Mental Health. Debra Beery provided expert editorial assistance in the preparation of this chapter.

REFERENCES

1. Cherry, J.A. and R.L. Davis. 1999. Cyclic AMP phosphodiesterases are localized in regions of the mouse brain associated with reinforcement, movement, and affect. *J Comp Neurol* 407:287–301.
2. Iona, S., et al. 1998. Characterization of the rolipram-sensitive, cyclic AMP-specific phosphodiesterases: Identification and differential expression of immunologically distinct forms in the rat brain. *Mol Pharmacol* 53:23–32.
3. Wachtel, H. 1982. Characteristic behavioral alterations in rats induced by rolipram and other selective adenosine cyclic 3′,5′-monophosphate phosphodiesterase inhibitors. *Psychopharmacology* 77:309–316.
4. Wachtel, H. 1983. Potential antidepressant activity of rolipram and other selective cyclic adenosine 3′,5′-monophosphate phosphodiesterase inhibitors. *Neuropharmacology* 22:267–272.
5. Egawa, T., et al. 1997. Rolipram and its optical isomers, phosphodiesterase 4 inhibitors, attenuated the scopolamine-induced impairments of learning and memory in rats. *Jpn J Pharmacol* 75:275–281.
6. Imanishi, T., et al. 1997. Ameliorating effects of rolipram on experimentally induced impairments of learning and memory in rodents. *Eur J Pharmacol* 321:273–278.
7. O'Donnell, J.M. and H.T. Zhang. 2004. Antidepressant effects of inhibitors of cAMP phosphodiesterase (PDE4). *Trends Pharmacol Sci* 25:158–163.
8. Houslay, M.D. 2001. PDE4 cAMP-specific phosphodiesterases. *Prog Nucleic Acid Res* 69:249–316.
9. Perez-Torres, S., et al. 2000. Phosphodiesterase type 4 isozymes expression in human brain examined by in situ hybridization histochemistry and [³H]rolipram binding autoradiography. Comparison with monkey and rat brain. *J Chem Neuroanat* 20:349–374.

10. Ye, Y. and J.M. O'Donnell. 1996. Diminished noradrenergic stimulation reduces the activity of rolipram-sensitive, high-affinity cyclic AMP phosphodiesterase in rat cerebral cortex. *J Neurochem* 66:1894–1902.

11. Zhang, H.T., et al. 2002. Antidepressant-like profile and reduced sensitivity to rolipram in mice deficient in the PDE4D phosphodiesterase enzyme. *Neuropsychopharmacology* 27:587–595.

12. Suvarna, N.U. and J.M. O'Donnell. 2002. Hydrolysis of *N*-methyl-D-aspartate receptor-stimulated cAMP and cGMP by PDE4 and PDE2 phosphodiesterases in primary neuronal cultures of rat cerebral cortex and hippocampus. *J Pharmacol Exp Ther* 302:249–256.

13. Richter, W., et al. 2005. Splice variants of the cyclic nucleotide phosphodiesterase PDE4D are differentially expressed and regulated in rat tissue. *Biochem J* 388:803–811.

14. Baillie, G.S., et al. 2000. Sub-family selective actions in the ability of Erk2 MAP kinase to phosphorylate and regulate the activity of PDE4 cyclic AMP-specific phosphodiesterases. *Br J Pharmacol* 131:811–819.

15. Lourenco, C.M., et al. 2001. Characterization of r-(11)C-rolipram for PET imaging of phosphodiesterase-4: In vitro binding, metabolism, and dosimetry studies in rats. *Nucl Med Biol* 28:347–358.

16. Jacobitz, S., et al. 1996. Mapping the functional domains of human recombinant phosphodiesterase 4A: Structural requirements for catalytic activity and rolipram binding. *Mol Pharmacol* 50:891–899.

17. Dodge, K.L., et al. 2001. mAKAP assembles a protein kinase A/PDE4 phosphodiesterase cAMP signaling module. *EMBO J* 20:1921–1930.

18. Houslay, M.D., and D.R. Adams. 2003. PDE4 cAMP phosphodiesterases: Modular enzymes that orchestrate signaling cross-talk, desensitization and compartmentalization. *Biochem J* 370:1–18.

19. McPhee, I., et al. 1999. Association with the SRC family tyrosyl kinase LYN triggers a conformational change in the catalytic region of human cAMP-specific phosphodiesterase HSPDE4A4B. Consequences for rolipram inhibition. *J Biol Chem* 274:11796–11810.

20. Yarwood, S.J., et al. 1999. The RACK1 signaling scaffold protein selectively interacts with the cAMP-specific phosphodiesterase PDE4D5. *J Biol Chem* 274:14909–14917.

21. Barnette, M.S., et al. 1995a. The ability of phosphodiesterase IV inhibitors to suppress superoxide production in guinea pig eosinophils is correlated with inhibition of phosphodiesterase IV catalytic activity. *J Pharmacol Exp Ther* 273:674–679.

22. Barnette, M.S., et al. 1995b. Inhibitors of phosphodiesterase IV (PDE IV) increase acid secretion in rabbit isolated gastric glands: Correlation between function and interaction with high-affinity rolipram binding site. *J Pharmacol Exp Ther* 273:1396–1402.

23. Duplantier, A.J., et al. 1996. Biarylcarboxylic acids and amides: Inhibition of phosphodiesterase type IV versus [³H]rolipram binding activity and their relationship to emetic behavior in the ferret. *J Med Chem* 39:120–125.

24. Saccomano, N.A., et al. 1991. Calcium-independent phosphodiesterase inhibitors as putative antidepressants: [3-(Bicycloalkyloxy)-4-methoxyphenyl]-2-imidazolidinones. *J Med Chem* 34:291–298.

25. Schmiechen, R., et al. 1990. Close correlation between behavioural response and binding *in vivo* for inhibitors of the rolipram-sensitive phosphodiesterase. *Psychopharmacology* 102:17–20.

26. Zhang, H.T., et al. 2006. Antidepressant-like effects of PDE4 inhibitors mediated by the high-affinity rolipram binding state (HARBS) of the phosphodiesterase-4 enzyme (PDE4) in rats. *Psychopharmacology* 186:209–217.

27. Zhao, Y., et al. 2003. Inhibitor binding to type 4 phosphodiesterase (PDE4) assessed using [³H]piclamilast and [³H]rolipram. *J Pharmacol Exp Ther* 305:565–572.

28. Zhao, Y., et al. 2003. Antidepressant-induced increase in high-affinity rolipram binding sites in rat brain: Dependence on noradrenergic and serotonergic function. *J Pharmacol Exp Ther* 307:246–253.

29. Crissman, A.M., and J.M. O'Donnell. 2002. Effects of antidepressant in rats trained to discriminated centrally administered isoproterenol. *J Pharmacol Exp Ther* 302:606–611.

30. Zhang, H.T., et al. 2005. Interaction between the antidepressant-like behavioral effects of beta adrenergic agonists and the cyclic AMP PDE inhibitor rolipram. *Psychopharmacology* 182:104–115.

31. Ye, Y., et al. 1997. Noradrenergic activity differentially regulates the expression of rolipram-sensitive, high-affinity cyclic AMP phosphodiesterase (PDE4) in rat brain. *J Neurochem* 69:2397–2404.
32. O'Donnell, J.M. 1993. Antidepressant-like effects of rolipram and other inhibitors of cyclic adenosine monophosphate phosphodiesterase on behavior maintained by differential reinforcement of low response rate. *J Pharmacol Exp Ther* 264:1168–1178.
33. Farooqui, S.M., et al. 2000. Noradrenergic lesions differentially alter the expression of two subtypes of low K_m cyclic AMP-selective phosphodiesterase type 4 (PDE4A and PDE4B) subtypes in rat brain. *Brain Res* 867:52–61.
34. Dlaboga, D., et al. 2006, Regulation of phosphodiesterose-4 (PDE4) expression in mouse brain by repeated antidepressant treatement: Comparison with rolipram Brain Res 1096: 104–112.
35. Miro, X., et al. 2002. Regulation of cAMP phosphodiesterase mRNAs expression in rat brain by acute and chronic fluoxetine treatment. An in situ hybridization study. *Neuropharmacology* 43:1148–1157.
36. Takahashi, M., et al. 1999. Chronic antidepressant administration increases the expression of cAMP-specific phosphodiesterase 4A and 4B isoforms. *J Neurosci* 19:610–618.
37. Ye, Y., et al. 2000. Effects of repeated antidepressant treatment on type 4A phosphodiesterase (PDE4A) in rat brain. *J Neurochem* 74:1257–1262.
38. Zhang, H.T., et al. 2000. Inhibition of cyclic AMP phosphodiesterase (PDE4) reverses memory deficits associated with NMDA receptor antagonism. *Neuropsychopharmacology* 23:198–204.
39. O'Donnell, J.M., and S. Frith. 1999. Behavioral effects of family-selective inhibitors of cyclic nucleotide phosphodiesterases. *Pharmacol Biochem Behav* 63:185–192.
40. O'Donnell, J.M., G.J. Marek, and L.S. Seiden. 2005. Antidepressant effects assessed using behavior maintained under a differential-reinforcement-of-low-rate (DRL) operant schedule. *Neurosci Biobehav Rev* 29:785–798.
41. McGuire, P.S., and L.S. Seiden. 1980. The effects of tricyclic antidepressants on performance under a differential-reinforcement-of-low-rates schedule in rats. *J Pharmacol Exp Ther* 214:635–641.
42. Mizokawa, T., et al. 1988. The effect of a selective phosphodiesterase inhibitor, rolipram, on muricide in olfactory bulbectomized rats. *Jpn J Pharmacol* 48:357–364.
43. Overstreet, D.H., et al. 1989. Antidepressant effects of rolipram in a genetic animal model of depression: Cholinergic supersensitivity and weight gain. *Pharmacol Biochem Behav* 34:691–696.
44. Santarelli, L., et al. 2003. Requirement of hippocampal neurogenesis for the behavioral effects of antidepressants. *Science* 301:805–809.
45. Bonbon, D., et al. 1988. Is phosphodiesterase inhibition a new mechanism of antidepressant action? A double-blind double-dummy study between rolipram and desipramine in hospitalized major and/or endogenous depressives. *Eur Arch Psychiatry Neurol Sci* 238:2–6.
46. Fleischhacker, W.W., et al. 1992. A multicenter double-blind study of three different doses of the new cAMP-phosphodiesterase inhibitor rolipram in patients with major depressive disorder. *Neuropsychobiology* 26:59–64.
47. Laux, G., et al. 1988. Clinical and biochemical effects of the selective phosphodiesterase inhibitor rolipram in depressed inpatients controlled by determination of plasma level. *Pharmacopsychiatry* 21:378–379.
48. Robichaud, A., et al. 1999. Emesis induced by inhibitors of type IV cyclic nucleotide phosphodiesterase (PDE IV) in the ferret. *Neuropharmacology* 38:289–297.
49. Silvestre, J.S., et al. 1999. Preliminary evidence for an involvement of the cholinergic system in the sedative effects of rolipram in rats. *Pharmacol Biochem Behav* 64:1–5.
50. Jin, S.L., et al. 2005. Generation of PDE4 knockout mice by gene targeting. *Methods Mol Biol* 307:191–210.
51. Zhang, H.T., et al. 2003. Cyclic AMP-specific phosphodiesterase 4 (PDE4): Memory, emotion, and drug abuse. *Soc Neurosci* 9:874 (abstract).
52. Shotwell, S.L. 1983. Cyclic adenosine $3':5'$-monophosphate phosphodiesterase and its role in learning in *Drosophila*. *J Neurosci* 3:739–747.
53. Davis, R.L., and J.A. Kiger. 1981. Dunce mutants of *Drosophila melanogaster*: Mutants defective in the cyclic AMP phosphodiesterase enzyme system. *J Cell Biol* 90:101–107.

54. Tully, T., et al. 1994. Memory through metamorphosis in normal and mutant *Drosophila*. *J Neurosci* 14:68–74.
55. Henkel-Tigges, J., and R.L. Davis. 1990. Rat homologs of the *Drosophila* dunce gene code for cyclic AMP phosphodiesterases sensitive to rolipram and RO 20-1724. *Mol Pharmacol* 37:7–10.
56. Villiger, J.W., and A.J. Dunn. 1981. Phosphodiesterase inhibitors facilitate memory for passive avoidance conditioning. *Behav Neural Biol* 31:354–359.
57. Randt, C.T., et al. 1982. Brain cyclic AMP and memory in mice. *Pharmacol Biochem Behav* 17:677–680.
58. Barros, D.M., et al. 1999. Stimulators of the cAMP cascade reverse amnesia induced by intra-amygdala but not intrahippocampal KN-62 administration. *Neurobiol Learn Mem* 71:94–103.
59. Goelet. P., et al. 1986. The long and the short of long-term memory—A molecular framework. *Nature* 322:419–422.
60. Romano, A., et al. 1996. Acute administration of a permeant analog of cAMP and a phosphodiesterase inhibitor improve long-term habituation in the crab *Chasmagnathus*. *Behav Brain Res* 75:119–125.
61. Zhang, H.T., et al. 2004. Inhibition of the phosphodiesterase 4 (PDE4) enzyme reverses memory deficits produced by infusion of the MEK inhibitor U0126 into the CA1 subregion of the rat hippocampus. *Neuropsychopharmacology* 29:1432–1439.
62. Barad, M., et al. 1998. Rolipram, a type IV-specific phosphodiesterase inhibitor, facilitates the establishment of long-lasting long-term potentiation and improves memory. *Proc Natl Acad Sci USA* 95:15020–15025.
63. Schneider, H.H. 1984. Brain cAMP response to phosphodiesterase inhibitors in rats killed by microwave irradiation or decapitation. *Biochem Pharmacol* 33:1690–1693.
64. Stone, E.A., and S.M. John. 1990. In vivo measurement of extracellular cyclic AMP in the brain: Use in studies of beta-adrenoceptor function in nonanesthetized rats. *J Neurochem* 55:1942–1949.
65. Monti, B., et al. 2006. Subchronic rolipram delivery activates hippocampal CREB and Arc, enhances retention and slows down extinction of conditioned fear. *Neuropsychopharmacology* 31:278–286.
66. Rutten, K., et al. 2006. Rolipram reverses scopolamine-induced and time-dependent memory deficits in object recognition by different mechanisms of action. *Neurobiol Learn Mem* 85 (2):132–138.
67. Bourtchouladze, R., et al. 2003. A mouse model of Rubinstein–Taybi syndrome: Defective long-term memory is ameliorated by inhibitors of phosphodiesterase 4. *Proc Natl Acad Sci USA* 100:10518–10522.
68. Gong, B., et al. 2004. Persistent improvement in synaptic and cognitive functions in an Alzheimer mouse model after rolipram treatment. *J Clin Invest* 114:1624–1634.
69. Ramos, B.P., et al. 2003. Dysregulation of protein kinase a signaling in the aged prefrontal cortex: New strategy for treating age-related cognitive decline. *Neuron* 40:835–845.
70. Winsky, L., and J.A. Harvey. 1987. Effects of N6-(L-phenylisopropyl)adenosine, caffeine, theophylline and rolipram on the acquisition of conditioned responses in the rabbit. *J Pharmacol Exp Ther* 241:223–229.
71. Giorgi, M., et al. 2004. The induction of cyclic nucleotide phosphodiesterase 4 gene (PDE4D) impairs memory in a water maze task. *Behav Brain Res* 154:99–106.
72. Sato, T., et al. 2004. Inhibitory effects of group II mGluR-related drugs on memory performance in mice. *Physiol Behav* 80:747–758.
73. Zhang, H.T., and J.M. O'Donnell. 2000. Effects of rolipram on scopolamine-induced impairment of working and reference memory in the radial-arm maze tests in rats. *Psychopharmacology (Berl)* 150:311–316.
74. Zhang, H.T., et al. 2005. Effects of the novel PDE4 inhibitors MEM1018 and MEM1091 on memory in the radial-arm maze and inhibitory avoidance tests in rats. *Psychopharmacology* 179:613–619.
75. Nagakura, A., et al. 2002. Effects of a phosphodiesterase IV inhibitor rolipram on microsphere embolism-induced defects in memory function and cerebral cyclic AMP signal transduction system in rats. *Br J Pharmacol* 135:1783–1793.

76. Arendash, G.W., et al. 2001. Progressive, age-related behavioral impairments in transgenic mice carrying both mutant amyloid precursor protein and presenilin-1 transgenes. *Brain Res* 891:42–53.

77. Bach, M.E., et al. 1999. Age-related defects in spatial memory are correlated with defects in the late phase of hippocampal long-term potentiation in vitro and are attenuated by drugs that enhance the cAMP signaling pathway. *Proc Natl Acad Sci USA* 96:5280–5285.

78. Rose, G.M., et al. 2005. Phosphodiesterase inhibitors for cognitive enhancement. *Curr Pharm Des* 11:3329–3324.

79. Lynch, M.A. 2004. Long-term potentiation and memory. *Physiol Rev* 84:87–136.

80. Nguyen, P.V., et al. 1994. Requirement of a critical period of transcription for induction of a late phase of LTP. *Science* 265:1104–1107.

81. Guzowski, J.F., and J.L. McGaugh. 1997. Antisense oligodeoxynucleotide-mediated disruption of hippocampal cAMP response element binding protein levels impairs consolidation of memory for water maze training. *Proc Natl Acad Sci USA* 94:2693–2698.

82. Igaz, L.M., et al. 2002. Two time periods of hippocampal mRNA synthesis are required for memory consolidation of fear-motivated learning. *J Neurosci* 22:6781–6789.

83. Otmakhov, N., et al. 2004. Forskolin-induced LTP in the CA1 hippocampal region is NMDA receptor dependent. *J Neurophysiol* 91:1955–1962.

84. Navakkode, S., et al. 2004. The type IV-specific phosphodiesterase inhibitor rolipram and its effect on hippocampal long-term potentiation and synaptic tagging. *J Neurosci* 24:7740–7744.

85. Ahmed, T., and J.U. Frey. 2003. Expression of the specific type IV phosphodiesterase gene PDE4B3 during different phases of long-term potentiation in single hippocampal slices of rats in vitro. *Neuroscience* 117:627–638.

86. Ahmed, T., et al. 2004. Regulation of the phosphodiesterase PDE4B3-isotype during long-term potentiation in the area dentata in vivo. *Neuroscience* 124:857–867.

87. Ahmed, T., and J.U. Frey. 2005. Phosphodiesterase 4B (PDE4B) and cAMP-level regulation within different tissue fractions of rat hippocampal slices during long-term potentiation in vitro. *Brain Res* 1041:212–222.

88. Millar, J.K., et al. 2005. DISC1 and PDE4B are interacting genetic factors in schizophrenia that regulate camp signaling. *Science* 310:1187–1191.

89. Barad, M. 2003. Later developments: Molecular keys to age-related memory impairment. *Alzheimer Dis Assoc Disord* 17:168–176.

90. Grady, C.L., and F.I. Craik. 2000. Changes in memory processing with age. *Curr Opin Neurobiol* 10:224–231.

91. Rosenzweig, E.S., et al. 2003. Hippocampal map realignment and spatial learning. *Nat Neurosci* 6:609–615.

92. Etchamendy, N., et al. 2001. Alleviation of a selective age-related relational memory deficit in mice by pharmacologically induced normalization of brain retinoid signaling. *J Neurosci* 21:6423–6429.

93. Rosenzweig, E.S., and C.A. Barnes. 2003. Impact of aging on hippocampal function: Plasticity, network dynamics, and cognition. *Prog Neurobiol* 69:143–179.

94. Hsu, K.S., et al. 2002. Alterations in the balance of protein kinase and phosphatase activities and age-related impairments of synaptic transmission and long-term potentiation. *Hippocampus* 12:787–802.

95. Asanuma, M., et al. 1996. Alterations of cAMP response element-binding activity in the aged rat brain in response to administration of rolipram, a cAMP-specific phosphodiesterase inhibitor. *Brain Res Mol Brain Res* 41:210–215.

96. Taylor, J.R., et al. 1999. Activation of cAMP-dependent protein kinase A in prefrontal cortex impairs working memory performance. *J Neurosci* 19:RC23.

97. Bernabeu, R., et al. 1997. Involvement of hippocampal cAMP/cAMP-dependent protein kinase signaling pathways in a late memory consolidation phase of aversively motivated learning in rats. *Proc Natl Acad Sci USA* 94:7041–7046.

98. Suh, Y.H., et al. 2002. Amyloid precursor protein, presenilins, and alpha-synuclein: Molecular pathogenesis and pharmacological applications in Alzheimer's disease. *Pharmacol Rev* 54:469–525.

99. Tong, L., et al. 2001. Beta-amyloid-(1–42) impairs activity-dependent cAMP-response element-binding protein signaling in neurons at concentrations in which cell survival is not compromised. *J Biol Chem* 276:17301–17306.
100. Vitolo, O.V., et al. 2002. Amyloid beta-peptide inhibition of the PKA/CREB pathway and long-term potentiation: Reversibility by drugs that enhance cAMP signaling. *Proc Natl Acad Sci USA* 99:13217–13221.
101. Tong, L., et al. 2004. Beta-amyloid peptide at sublethal concentrations downregulates brain-derived neurotrophic factor functions in cultured cortical neurons. *J Neurosci* 24:6799–6809.
102. Beglopoulos, V., and J. Shen. 2006. Regulation of CRE-dependent transcription by presenilins: Prospects for therapy of Alzheimer's disease. *Trends Pharmacol Sci* 27:33–40.
103. Nibuya, M., et al. 1996. Chronic antidepressant administration increases the expression of cAMP response element binding protein (CREB) in rat hippocampus. *J Neurosci* 16:2365–2372.
104. Daniels, W.M., et al. 2001. The role of the MAP-kinase superfamily in beta-amyloid toxicity. *Metab Brain Dis* 16:175–185.
105. Zhu, X., et al. 2003. Distribution, levels, and activation of MEK1 in Alzheimer's disease. *J Neurochem* 86:136–142.
106. Di Cristo, G., et al. 2001. Requirement of ERK activation for visual cortical plasticity. *Science* 292:2337–2340.
107. Kanterewicz, B.I., et al. 2000. The extracellular signal-regulated kinase cascade is required for NMDA receptor-independent LTP in area CA1 but not area CA3 of the hippocampus. *J. Neurosci* 20:3057–3066.
108. Kelleher., R.J., et al. 2004. Translational control by MAPK signaling in long-term synaptic plasticity and memory. *Cell* 116:467–479.
109. Krapivinsky, G., et al. 2003. The NMDA receptor is coupled to the ERK pathway by a direct interaction between NR2B and RasGRF1. *Neuron* 40:775–7784.
110. Paul, S., et al. 2003. NMDA-mediated activation of the tyrosine phosphatase STEP regulates the duration of ERK signaling. *Nat. Neurosci* 6:34–42.
111. Waltereit, R., and M. Weller. 2003. Signaling from cAMP/PKA to MAPK and synaptic plasticity. *Mol. Neurobiol* 27:99–106.
112. Vanhoose, A.M., et al. 2004. Regulation of cAMP levels in area CA1 of hippocampus by Gi/o-coupled receptors is stimulus dependent in mice. *Neurosci Lett* 370:80–83.
113. MacKenzie, S.J., et al. 2000. ERK2 mitogen-activated protein kinase binding, phosphorylation, and regulation of the PDE4D cAMP-specific phosphodiesterases. The involvement of COOH-terminal docking sites and NH2-terminal UCR regions. *J Biol Chem* 275:16609–16617.
114. Chang, Y.C., et al. 2003. Febrile seizures impair memory and cAMP response-element binding protein activation. *Ann Neurol* 54:706–718.
115. Ghelardini, C., et al. 2002. DM235 (sunifiram): A novel nootropic with potential as a cognitive enhancer. *Naunyn Schmiedebergs Arch Pharmacol* 365:419–426.

28 Role of Phosphodiesterases in Apoptosis

Adam Lerner, Eun-Yi Moon, and Sanjay Tiwari

CONTENTS

28.1 Introduction 559
    28.1.1 Optimal Experimental Design 560
28.2 Proapoptotic Effects of Phosphodiesterase Inhibitors 561
    28.2.1 cAMP-Induced Lymphoid Cell Death.. 561
    28.2.2 Phosphodiesterase Inhibitor-Induced Apoptosis
            in Lymphoid Malignancies 563
    28.2.3 PDE4 Inhibitor-Induced Apoptosis in B-Cell Chronic
            Lymphocytic Leukemia 564
    28.2.4 PDE4 Inhibitor-Induced Apoptosis in Models of Acute
            Lymphoblastic Leukemia and Diffuse Large B-Cell Lymphoma 565
    28.2.5 Interrelationship between Glucocorticoid and cAMP/PDE4
            Inhibitor-Mediated Apoptosis 567
    28.2.6 Other Proposed Mechanisms for PDE4 Inhibitor-Induced
            Lymphoid Apoptosis 569
    28.2.7 PDE4 Inhibitor-Induced EPAC Activation is Antiapoptotic
            in B-Cell Chronic Lymphocytic Leukemia 571
    28.2.8 What Accounts for the Specificity of PDE4
            Inhibitor-Induced Apoptosis? 571
28.3 Antiapoptotic Effects of cAMP, cGMP, and Phosphodiesterase
    Inhibitors 572
    28.3.1 Antiapoptotic Effects of PDE4 and PDE5 Inhibitors
            in Neuronal Cells.. 573
    28.3.2 cAMP Signaling Protects Hepatocytes from Fas and Bile
            Acid-Induced Apoptosis 574
    28.3.3 cGMP Signaling and Apoptosis in Endothelial Cells, Colon
            Tumor Cell Lines, and Hepatocytes 575
    28.3.4 PDE4 Inhibitors Reduce Apoptosis and Induce
            Differentiation in Myeloid Cells 575
28.4 Do Phosphodiesterases Regulate Physiologic Apoptotic Signaling? 577
References 577

28.1 INTRODUCTION

In specific cell lineages and under specific circumstances, cyclic nucleotide phosphodiesterase (PDE) inhibitors can induce or prevent apoptosis. In the case of certain normal and transformed primary lymphocyte subsets, the prolonged elevation of cAMP by some PDE

inhibitors can induce apoptosis. In contrast, PDE inhibitor-induced cAMP or cGMP signaling augments basal cell survival or protects cells from apoptotic stimuli in a far more diverse set of cell types, including myeloid cells, neuronal cells, and hepatocytes. While the concept of exploiting the specific sensitivity of certain cell types to cAMP-induced apoptosis for medical benefit has been around since shortly after the discovery of this second messenger, it is only with the comparatively recent introduction of efficacious and clinically well-tolerated family-specific PDE inhibitors that a therapeutically feasible means by which to test these concepts has become apparent. This chapter will review what is currently known about the effects of PDE inhibitors on apoptotic signaling, with an emphasis on the potential role of PDE inhibitors in the treatment of malignancies. As discussed at the end of the chapter, in contrast to the growing literature examining the effect of PDE inhibitors on cell survival, the degree to which uninhibited PDE activities contribute to the normal physiologic regulation of apoptosis remains an interesting but thus far relatively unexplored question.

28.1.1 OPTIMAL EXPERIMENTAL DESIGN

If one goal of PDE inhibitor research is to determine whether such agents will be of benefit in the treatment of specific human diseases, the experimental models used over the years vary dramatically in the fidelity with which they mirror the complexity of treatment of humans with such agents. PDE inhibitor research has been carried out, in inverse order of frequency, in humans, animals, primary cells, and cell lines. While cell lines derived from human tumors, for instance, are useful insofar as they allow a consistent starting point for studies and maintain some of the molecular features of the original tumor, the multiple additional genetic changes that occur during the process of establishing the cell line clearly have dramatic effects on apoptotic signaling. As an example, most primary malignant human lymphoid cells do not proliferate in tissue culture whereas cell lines proliferate rapidly. Such constitutive proliferation may be of particular importance in the study of PDE inhibitor therapy for lymphoid malignancies, as early studies demonstrated clearly that PDE activity is 10- to 20-fold higher in a constitutively proliferating leukemic cell line (L1210 cells) than in normal resting peripheral blood lymphocytes.[1,2] Thus, cell lines frequently differ from the primary malignant cells that they are meant to represent both in the quantity and types (see below) of PDE and cyclic nucleotide effector proteins that they express.

In vitro studies of primary human malignant cells have their own drawbacks in that such samples can suffer from variability in genetic abnormalities, levels of contamination with normal cells, and the treatment history of the patients from which the samples are derived. *In vitro* studies of purified primary cell populations also ignore the well-documented critical involvement of stromal cells in the survival of tumor cells. If, for instance, tumor cell basal adenylyl cyclase activity is higher *in vivo* than *in vitro* because of paracrine interactions with surrounding stromal cells, primary tumor cells that appear insensitive to treatment with PDE inhibitors as single agents *in vitro* may prove to be markedly more responsive *in vivo*. Ideally, PDE inhibitor studies should be carried out both on isolated primary cells *in vitro*, for clarity of analysis, and *in vivo*, to best model the likely contribution of tumor cell and stromal cell interactions.

A wide range of pharmacologic agents have been used in studies of the effects of cyclic nucleotide signaling on cell survival, including: (a) agonist or antagonist cell permeable cyclic nucleotide analogs, some of which are specific to particular cAMP effectors such as EPAC; (b) family-specific (rolipram, cilostamide) or nonspecific (methylxanthine) PDE inhibitors; (c) the adenylyl cyclase activator forskolin; and (d) hormonal stimulants of adenylyl cyclase such as β-adrenergic agonists. While the use of hormonal stimulants mimics to some degree physiologic signaling, the interpretation of studies utilizing hormones is made more difficult in that such receptor-mediated signaling typically activates several signaling pathways rather

than cAMP-mediated signals alone. As an example, while β-adrenergic agonists are well known to induce apoptosis and cAMP-mediated signaling in specific lymphoid subsets, it remains controversial as to whether the observed apoptosis is in fact cyclic nucleotide-related.[3–5] While experiments with cyclic nucleotide analogs are at least superficially more easily interpreted, such an approach ignores the physiologic importance of PDE and cAMP effector protein subcellular organization. Family-specific PDE inhibitors, in contrast, are more likely to activate signaling pathways that are constrained by such subcellular organization. Unfortunately, while the literature examining the effects of cyclic nucleotide signaling in general on cell survival is substantial, the number of studies that have identified substantive effects of monotherapy with family-specific PDE inhibitors on cell survival is significantly more modest.

28.2 PROAPOPTOTIC EFFECTS OF PHOSPHODIESTERASE INHIBITORS

28.2.1 CAMP-INDUCED LYMPHOID CELL DEATH

Seminal studies by Gordon Tomkins and associates[6] first established that cAMP signaling could induce cell death in lymphoid cells in a cAMP-dependent protein kinase (PKA)-dependent manner. Tomkins reported that culture of S49 cells, a murine thymic lymphoma cell line, in 100 μM dibutyryl cAMP (dbcAMP) in the presence of the relatively nonspecific PDE inhibitor theophylline (200 μM), led to virtually complete cytolysis of such cells in 48 to 72 h. Such dbcAMP-induced cell death in S49 cells was preceded by arrest in G1 whereas other phases of the cell cycle were not perceptibly altered.[7] When S49 cells were grown in soft agar in the presence of gradually increasing doses of dbcAMP, sublines were isolated that were able to continue to proliferate over 48 h in the presence of dbcAMP concentrations as high as 1 mM. Such dbcAMP-resistant lines were also insensitive to growth arrest induced by β-agonists or cholera toxin. Analysis of such clones demonstrated at least three classes that varied with respect to cAMP binding, cAMP-stimulated kinase activity, and cAMP PDE induction.[8] Perhaps the best characterized class of dbcAMP-resistant S49 cells was represented by kin-A, a cell line in which a structural mutation in the regulatory subunit of PKA was inferred by demonstration that 10-fold higher concentrations of cAMP were required for comparable activation of protein kinase activity and that the mutant regulatory subunit markedly enhanced the thermostability of wild-type catalytic subunit.[9] Subsequent detailed studies by Steinberg and others[10,11] demonstrated that such cells contained charge-shifting mutations in the cAMP-binding portion of the PKA RI subunit, leading to augmented affinity of the regulatory subunit for the catalytic subunit. The demonstration that the resistance of such lines to cytolysis by dbcAMP was stable over many months and that their growth rate was equivalent to wild-type S49 cells in the absence of selection demonstrated that type 1 PKA, the predominant form of PKA in S49 cells, is largely dispensable for growth of this line *in vitro*. Nonetheless, these data demonstrate that cAMP-mediated PKA activation potently induces cell death in this T lineage cell.

Once the morphologic and biochemical hallmarks of apoptosis as defined by Wyllie and colleagues[12–14] became broadly appreciated, cAMP-induced cytolysis of lymphoid cells was recognized as a form of programmed cell death. Recognition of the sensitivity of lymphoid cells to cAMP-induced cytolysis/apoptosis was extended from cell lines to primary lymphoid cell populations including thymocytes, circulating human B lymphocytes, and chronic B-cell lymphocytic leukemia (B-CLL).[14–17] In rat thymocytes, McConkey et al.[13] reported that treatment with dbcAMP (100 μM), forskolin (50 μM), or PGE$_2$ (1 μM) induced DNA fragmentation characteristic of apoptosis within 6 h. Interestingly, such apoptosis was inhibited by prior treatment with either PMA or IL-1. Similarly, Kizaki et al.[14] reported that adenosine receptor agonists and cAMP analogs induced murine thymocyte DNA fragmentation.

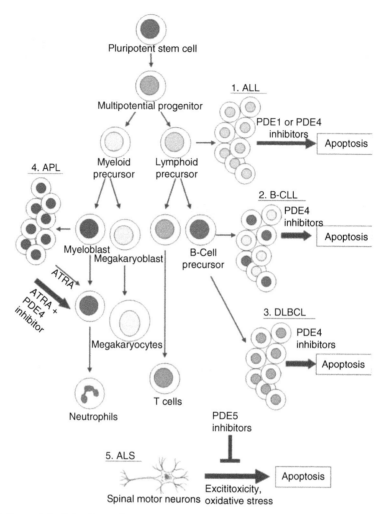

FIGURE 28.1 Potential clinical benefits of PDE inhibitor-induced regulation of cell survival. Both the pro- and antiapoptotic effects of PDE inhibitors have been proposed as therapeutically beneficial in models of hematologic malignancies and genetic disorders. PDE1 or PDE4 inhibitors have been demonstrated to induce apoptosis in models of three lymphoid malignancies: (1) ALL, (2) B-cell chronic lymphocytic leukemia (B-CLL), and (3) diffuse large B-cell lymphoma (DLBCL). The potential mechanisms by which such apoptosis occurs are summarized in subsequent figures. Interestingly, PDE4 inhibitors are not apoptotic in normal peripheral blood T cells. In contrast to their apoptotic effect in specific lymphoid malignancies, in AML, PDE4 inhibitors have been used in cell lines and in animal models to reverse the arrest in differentiation that is characteristic of this malignancy. In APL (4), a specific variant of AML, PDE4 inhibition by piclamilast potentiates the cytodifferentiating action of all-*trans*-retinoic acid (ATRA), allowing malignant promyelocytes to differentiate normally into mature neutrophils.[99] Maturing neutrophils then exit the bone marrow and have a life span of a few days. The antiapoptotic effects of PDE5 inhibitors may be of benefit in the therapy of amyotrophic lateral sclerosis (ALS) (5), a fatal neurodegenerative disease characterized by apoptosis of spinal motor neurons. Excitotoxicity or oxidative injury may play a role in neuronal apoptosis in this disease and cGMP-mediated signaling has been reported to antagonize such apoptosis. Inhibition of PDE5 with agents such as zaprinast, dipyridamole, or T-1032 has a neuroprotective effect in primary spinal motor neurons exposed to either low dose glutamate or the glutathione synthase inhibitor BSO.[69]

Importantly, in both studies, mature peripheral blood or splenic T cells were significantly less sensitive to cAMP-induced apoptosis.[13,14] Consistent with these observations, as well as prior studies demonstrated that cAMP agonists block T-cell receptor (TCR) signaling and T-cell proliferation; several groups soon after reported that treatment of T-cell hybridomas with cAMP analogs, forskolin, isobutylmethylxanthine (IBMX), or PGE$_2$ actually inhibited anti-CD3-induced cell death.[18–20] Thus in immature primary T-lineage cells, pharmacologic elevation of cAMP potently induces apoptosis whereas in more mature T-lineage cells, the same agents are significantly less active by themselves and can block the ability of T-cell receptor signaling to induce Fas-mediated apoptosis (activation-induced cell death).

28.2.2 PHOSPHODIESTERASE INHIBITOR-INDUCED APOPTOSIS IN LYMPHOID MALIGNANCIES

The recognition that cAMP signaling could induce apoptosis in primary malignant lymphoid cells such as B-CLL cells *in vitro* raised the possibility that such signaling therapy could be used in the treatment of patients with such lymphoid malignancies if a clinically tolerable pharmacologic approach to activating such signaling could be identified. With the development of both nonspecific and family-specific PDE inhibitors, several groups have now examined whether such agents induce apoptosis in either primary malignant lymphoid cell populations or cell lines derived from such cells (Figure 28.1).

The first detailed studies of PDE inhibitor-induced lymphoid apoptosis were performed by Epstein and colleagues.[21] They identified at least two forms of PDE activity in a human lymphoblastoid cell line, RPMI-8392, derived from a patient with acute lymphoblastic leukemia (ALL). One form (PDE1) hydrolyzed both cAMP and cGMP and was stimulated by calcium–calmodulin whereas the other (PDE4) was cAMP-specific.[21,22] Importantly, and in confirmation of earlier work on phytohemagglutinin (PHA)-stimulated bovine lymphocytes, PDE1B transcript was absent from resting human peripheral blood leukocytes but appeared after stimulation with PHA, a lectin known to activate T cells.[21,23] PDE1B transcript was also present in RPMI-8392 cells. Treatment of RPMI-8392 cells with vinpocetine, a known PDE1 inhibitor, at concentrations of 30 μM or greater induced apoptosis as judged by the appearance of oligonucleosome-length DNA fragmentation.[21] The PDE4 inhibitors rolipram and RO-20-1724 also induced apoptosis at concentrations of 10 μM or greater (Figure 28.1). Culture of RPMI-8392 cells with an antisense oligonucleotide against the translation initiation region of PDE1B markedly reduced PDE1B transcript levels, PDE1B enzymatic activity, and induced apoptosis whereas a scrambled control oligonucleotide did not. Of note, recent studies show that vinpocetine can inhibit other PDEs, particularly PDE4B, in the Same Concentration range that it inhibits PDE1 (10–20 μM) (Timothy Martins, personal communication). It is also of interest that mice in which the PDE1B locus has been genetically inactivated have locomoter dysfunction but no obvious hematologic abnormalities.[24] Nonetheless, these studies demonstrated for the first time that family-specific PDE inhibitors could induce apoptosis in susceptible lymphoid populations in the absence of concurrent stimulation of adenylyl cyclase.

Mentz et al.[25] reported that theophylline (100 μg/mL or 550 μM), a methylxanthine used clinically for the treatment of asthma, consistently induced apoptosis *in vitro* in B-CLL cells derived from 15 patients (mean apoptosis 87%, range 77%–96%). Methylxanthines are known to inhibit PDEs of all 11 known families with the exception of PDE8 and PDE9. Fairly marked synergy in the induction of apoptosis was observed upon combining theophylline with chlorambucil, an alkylator frequently used in the treatment of B-CLL. Bcl-2 and p53 levels fell following theophylline treatment whereas c-*myc* levels rose. (R$_p$)-8-Br-cAMPS an enantiomeric antagonist of PKA, only partially inhibited theophylline-induced apoptosis

whereas this compound completely blocked forskolin or PGE_2-induced apoptosis. Curiously, in a subsequent study the authors failed to see synergy between theophylline and forskolin in inducing B-CLL apoptosis, although the high concentrations of theophylline used in these experiments (100 μg/mL) may account for this finding, as at this concentration, theophylline alone is quite effective and may therefore mask any synergistic effect with forskolin.[26] The authors concluded that the apoptotic activity of theophylline was likely only partially dependent upon cAMP-mediated signaling.

At high serum levels, theophylline induces tremulousness, nausea, vomiting, and even seizures, limiting the therapeutic window for this drug to serum levels of 5–20 μg/mL (27.5–110 μM). Thus the concentrations of theophylline that can be safely achieved clinically are well below the concentrations required for effective inhibition of many PDEs. As an example, the IC_{50} for PDE7A, a cAMP PDE expressed in B-CLL cells, is 343.5 μM.[27] Despite these limitations, Wiernik et al.[28] carried out a phase 2 clinical trial of theophylline monotherapy (200 mg by month every 12 hours) in 25 B-CLL patients with early stage disease (Rai stage 0–1; lymphocytosis ± lymphadenopathy in the absence of hepatosplenomegaly, anemia or thrombocytopenia). One complete response was observed and 18 patients had stable disease. While this study is provocative, the nonrandomized design of the trial and the variable nature of the clinical course of patients with early stage B-CLL make it difficult to conclude whether or not the clinical response or the stable disease observed was related to theophylline therapy.

28.2.3   PDE4 Inhibitor-Induced Apoptosis in B-Cell Chronic Lymphocytic Leukemia

In an effort to determine whether family-specific PDE inhibitors might be therapeutically beneficial, our laboratory has sought to determine the relevant PDE target or targets of theophylline-induced apoptosis in B-CLL cells. While the partially specific PDE1 inhibitor vinpocetine (10 μM) induced apoptosis in B-CLL cells and PDE1B transcript was identified by RT-PCR, PDE1 protein was rarely detectable and addition of Ca^{2+}/calmodulin to cell extracts did not enhance PDE activity, suggesting that at this concentration, vinpocetine may induce apoptosis in leukemic cells by inhibiting PDE4.[29] By Western blot analysis and RT-PCR, B-CLL cells constitutively expressed PDE4A5 (130 kDa). PDE4B2 (64 kDa) and PDE4D1 (70 kDa) were induced following treatment with the PDE4 inhibitor rolipram (10 μM).[30] In addition, B-CLL cells expressed PDE3B (130 kDa) and PDE7A1 (55 kDa).[27]

Among 14 B-CLL samples tested, treatment with the PDE4 inhibitor rolipram (10 μM) induced significant apoptosis in 11 (mean 54%; range 40%–80%).[29] In keeping with earlier studies of dbcAMP-induced cytolysis in S49 cells, rolipram-induced apoptosis in B-CLL cells developed relatively slowly, peaking between 48 and 72 h. In a subsequent study, despite the presence of PDE3B enzyme in B-CLL cells, treatment with cilostamide failed to induce apoptosis.[30] By Western blot analysis, however, treatment with the PDE4 inhibitor induced an apparent compensatory upregulation of PDE3B. Consistent with a potential protective effect of such PDE3B upregulation, addition of cilostamide to rolipram augmented levels of apoptosis in five of seven rolipram-resistant leukemic samples.[30] PDE7A is present in B-CLL cells, and this enzyme is also upregulated in an apparent compensatory fashion in the human WSU-CLL cell line following treatment with the ICOS PDE7 inhibitor IC242.[27] Unfortunately, the off-target effects of the PDE7 inhibitors developed thus far have hampered efforts to determine the contribution of this enzyme to methylxanthine-induced B-CLL apoptosis. Overall, however, PDE4 is likely to be the primary target in B-CLL cells of methylxanthine-induced apoptosis (Figure 28.1). It will be of interest to determine whether treatment with PDE4 inhibitors, either alone or in combination with glucocorticoids (see below), will be of therapeutic benefit in this disease.

As might be predicted from studies cited above that utilized cyclic nucleotide analogs, rolipram treatment fails to induce apoptosis in peripheral blood T cells.[29] Peripheral blood B cells were modestly sensitive to rolipram-induced apoptosis, although such apoptosis was superimposed on a high basal apoptotic rate. Cross-linking cell surface IgM (sIgM) reduced basal apoptosis and protected peripheral blood B cells from rolipram-induced apoptosis. In contrast, cross-linking of sIgM in CLL cells failed to reduce either basal or rolipram-induced apoptosis. The failure of sIgM cross-linking to protect from rolipram-induced apoptosis is in keeping with reports that sIgM signal transduction is defective in B-CLL, perhaps due to alternative splicing of the BCR signal transduction component CD79b.[31]

Activation of B-CLL cell adenylate cyclase with forskolin (40 μM) modestly induced apoptosis as a single agent and modestly enhanced rolipram-induced apoptosis.[29] However, the relative efficacy of rolipram as a single agent both in inducing apoptosis and in raising intracellular cAMP levels is a particularly interesting aspect of these *in vitro* studies. These results suggest that under standard tissue culture conditions, basal adenylate cyclase activity is sufficiently high in B-CLL cells that inhibition of PDE4-mediated cAMP catabolism is sufficient to activate cAMP–effector pathways (flux-mediated sensitivity). In keeping with this, prolonged CREB phosphorylation and EPAC1-mediated Rap activation occur following monotherapy of B-CLL cells with rolipram (10 μM).[32] By Northern blot analysis, transcript levels for PDE4B are substantially higher in freshly isolated B-CLL cells than following tissue culture for 6 h.[29] Thus, CLL cells may prove to be more sensitive to PDE4 inhibitors *in vivo* than *in vitro*, as these data suggest that the hormonal milieu CLL cells are exposed to *in vivo* drives compensatory expression of PDE4B2, a gene whose intronic promoter is positively regulated by cAMP signaling through CREB.[33]

28.2.4 PDE4 INHIBITOR-INDUCED APOPTOSIS IN MODELS OF ACUTE LYMPHOBLASTIC LEUKEMIA AND DIFFUSE LARGE B-CELL LYMPHOMA

PDE4 inhibitors have also proven to be efficacious inducers of apoptosis in models of two other lymphoid malignancies, ALL and diffuse large B-cell lymphoma (DLBCL). As noted above, Epstein and colleagues[21] found that rolipram induced apoptosis in RPMI 8392 cells, a cell line derived from a patient with ALL. Using the T-ALL cell line CEM-CCRF as a model, Kato and colleagues[34] found that rolipram, but not PDE1 or PDE3 inhibitors, blocked proliferation of this cell line, although with a comparatively high EC$_{50}$ of 44 μM. PDE4 inhibitors also induced apoptosis of CEM cells, although the effect was most clear when the PDE4 inhibitor treatment (20 μM rolipram) was combined with adenylate cyclase activation by forskolin (20 μM). By Western blot analysis, treatment with rolipram and forskolin upregulated CEM p53 levels within 15 min and p21 WAF/CIP1, a known p53 target gene, within an hour, suggesting that the p53 and p21 proteins might play a role in this model of PDE4 inhibitor-induced apoptosis.

Shipp et al.,[35] in studies of gene chip array analyses of pathologic specimens obtained from DLBCL patients prior to treatment with standard (CHOP: Cyclophosphamide, doxorabicin, vincristine and preduisone) chemotherapy, have made the striking observation that over-expression of PDE4B transcript, among a number of other transcripts, confers a poor prognosis in such patients. To better understand the basis for this phenomenon, Smith et al.[36] examined the function and expression of PDE4B in DLBCL cell lines. Screening of normal B cells and DLBCL cell lines demonstrated that PDE4B2 was the principal PDE4B isoform expressed in such cells. RT-PCR was then performed with PDE4B2-specific oligonucleotides on an independent set of 112 newly diagnosed untreated DLBCL patients, 57 of whom were subsequently cured and 55 who developed refractory or fatal disease.

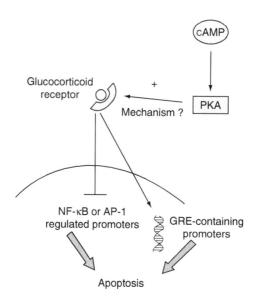

FIGURE 28.2 PDE4 inhibitors augment glucocorticoid-induced apoptosis of B-CLL cells. Although the mechanism by which PKA activity modulates glucocorticoid-induced apoptosis is not well understood, it is now clear that augmentation of PKA signaling by PDE4 inhibitor treatment enhances dexamethasone-induced apoptosis in B-CLL cells whereas inhibition of PKA signaling with (R_p)-8Br-cAMPS abrogates dexamethasone-induced apoptosis. The catalytic subunit of PKA has been reported to associate with the glucocorticoid receptor (GR) and could alter GR signaling either by phosphorylating the GR itself or associated corepressors, coactivators, or components of the transcriptional complex. Alternatively, PKA signaling has been reported to enhance transcription of the $GR\alpha$ gene and $GR\alpha$ levels may be rate limiting in determining the sensitivity of leukemic cells to glucocorticoid-mediated apoptosis. GR-mediated apoptosis could involve either transactivation of GRE-containing promoters and transcription of lysis genes or GR-mediated transrepression of NF-κB or AP-1 and repression of survival gene expression. GR-mediated transrepression occurs through a tethering mechanism that does not require a functional GR DNA-binding domain. Given a requirement for GR expression for cAMP-induced apoptosis in some lymphoid cell lines, GR may also play at least a permissive role in PDE4 inhibitor-mediated apoptosis.

This analysis confirmed the association of increased PDE4B2 expression with fatal or refractory disease. Among a series of DLBCL cell lines, levels of cAMP induced following treatment with either forskolin alone or combined with an isoform-specific PDE4B inhibitor, PLX513, varied inversely with basal expression of PDE4B. Those lines expressing low levels of PDE4B were also more sensitive to the growth inhibitory effects of PDE4B inhibitor or forskolin treatment. Reconstitution of PDE4B2 into a PDE4B low or negative DLBCL cell line, DHL6, conferred resistance to forskolin-induced apoptosis. These studies demonstrate that PDE4B2 can play a pivotal role in cAMP-induced apoptosis in this DLBCL model (Figure 28.1).

Of note, in addition to the above studies demonstrating that inhibition of PDE4 can induce apoptosis, PDE4 isoforms can also be the target of the apoptotic apparatus. Staurosporine-induced apoptosis in Rat1 fibroblasts leads to caspase 3-mediated cleavage of PDE4A5 but not other PDE4 isoforms.[37] Cleavage of the amino terminus of PDE4A5 at (69) DAVD (72) by caspase 3 leads to staurosporine-induced redistribution of PDE4A5, most likely as a result of loss of intracellular targeting by the removal of PDE4A5's amino-terminal SH3 domain. PDE5

and PDE6 have also been reported to be the target of caspase 3-mediated cleavage.[38,39] The physiologic significance of such caspase-mediated cleavage events remains to be established.

28.2.5 INTERRELATIONSHIP BETWEEN GLUCOCORTICOID AND cAMP/PDE4 INHIBITOR-MEDIATED APOPTOSIS

Treatment with cAMP analogs augments the induction of apoptosis by glucocorticoids (GCs) in susceptible lymphoid populations such as primary murine thymocytes.[40] Similarly, forskolin, either alone or in combination with PDE4 inhibitors, augments glucocorticoid receptor (GR)-mediated transcriptional activation as well as GC-mediated cell death in GC-resistant CEM-CCRF cells, a T-ALL-derived cell line.[34,41] Tiwari et al.[42] found that addition of PDE4 inhibitors augmented GC sensitivity in primary leukemic cells from B-CLL patients (Figure 28.2). In concurrent studies on CEM-CCRF cells, Tiwari et al. found no evidence of PDE4 inhibitor sensitivity whereas they confirmed prior observations that forskolin augmented GC-mediated apoptosis in these cells. Consistent with these findings, in B-CLL cells, PDE4 inhibitors augmented cAMP levels and glucocorticoid response element (GRE)-mediated transcription (as judged by GRE-containing luciferase reporter constructs) far more effectively than forskolin whereas the converse was true of CEM cells. These studies demonstrate that lymphoid cells that are sensitive to cAMP-mediated apoptosis can nonetheless differ markedly in their relative sensitivity to drugs that inhibit PDE4 or activate adenylate cyclase.

Just as augmented cAMP signaling can enhance GC-mediated apoptosis, inhibition of cAMP signaling can conversely induce GC resistance in lymphoid cells. Gruol et al.[43] selected cAMP-resistant sublines of the murine thymoma cell line WEHI-7.1 capable of growing in 100 μM dbcAMP and 75 μM IBMX. As previously described in S49 cells, such cAMP-resistant WEHI-7.1 cells had a shift in PKA activity consistent with an alteration in the affinity of the regulatory subunit for cAMP. Interestingly, while treatment of such cAMP-resistant WEHI clones with GCs still could induce apoptosis, GC-resistant clones were obtained with greater than 1000-fold higher frequency than the parental line. These doubly resistant clones did not have further alterations in PKA activity. In a subsequent study, the same group found that within the overall cell population of a cAMP-resistant WEHI-7 clone, loss of PKA activity in those cells with low basal levels of GR led concurrently to a significant reduction in sensitivity to GCs.[44] Consistent with this finding, Tiwari et al.[42] observed that the PKA antagonist (R_p)-8-Br-cAMPS almost completely inhibited GC-mediated apoptosis in primary CLL cells. Thus in certain lymphoid subtypes such as ALL and B-CLL, the level of cAMP signaling in lymphoid cells may serve as a rheostat that dictates the efficacy with which GCs induce apoptosis. While such observations have been made repeatedly since the mid-1980s, the introduction of clinically efficacious and well-tolerated PDE4 inhibitors that can induce such PKA signaling now makes testing of the therapeutic implications of this concept a realistic goal. It would be particularly appropriate to test PDE4 inhibitor and GC therapy in those lymphoid malignancies in which GCs currently play a major role, including ALL, DLBCL, and multiple myeloma.

If cAMP signaling has a major impact on GC-mediated signaling, does the GR play a role in cAMP-mediated apoptosis? Kiefer and colleagues[45] approached this question by utilizing two forms of CEM cells that differ with regard to the presence or absence of functional GRs. The parental CEM-C7 cell line is heterozygous at the GR locus, as one GR allele contains an inactivating L753F point mutation in the GR ligand-binding domain. The CEM-ICR27 line, isolated by selection of CEM-C7 cells in GCs, is GC resistant as a result of deletion of the

remaining wild-type GR allele.[46] As expected, CEM-C7 but not CEM-ICR27 cells underwent cytolysis following treatment with 10 μM methylprednisolone. Remarkably, CEM-ICR27 cells demonstrated almost complete cross-resistance to dbcAMP (100 μM) and forskolin (50 μM). Upon stable transfection of CEM-ICR27 cells with wild-type GR, sensitivity to both GCs and dbcAMP and forskolin was restored. This study demonstrates that GR function is necessary for cAMP-mediated apoptosis in this T-ALL cell-line model. Unfortunately, efforts to utilize the GR antagonist RU486 to more directly examine the role of GR signaling in cAMP-induced or PDE4 inhibitor-induced apoptosis in primary cells have been hampered by the agonist activity of RU486 (Mefipristone, a progesterone and glucocorticoid receptor antagonist) in the setting of PKA activation.[42,47]

If the GR is required for cAMP-mediated apoptosis in lymphoid cells, it is possible that cAMP signaling triggered by PDE inhibitors induces apoptosis by altering either the strength or quality of basal GR signaling such that it now surpasses a threshold level required to induce apoptosis. The ability of the GR to induce apoptosis in lymphoid cells has been variably ascribed to GRE transactivation (in studies of primary murine thymocytes) or a GR DNA-binding domain-independent inhibition of antiapoptotic NF-κB or AP-1 signaling through a tethering mechanism (in studies of leukemia or lymphoma cell lines) (Figure 28.2). In support of the former hypothesis, glucocorticoids fail to induce apoptosis in thymocytes of mice in which the wild-type GR locus has been replaced by knockin with a dimerization-defective GR that does not transactivate GREs.[48] In support of the latter hypothesis, GCs still effectively induce apoptosis in GR-negative Jurkat cells that have been stably transfected with GR mutants that are defective in GRE transactivation but maintain the ability to trans-repress NF-κB and AP-1-regulated genes.[49,50] Whether primary malignant lymphoid cells will more closely resemble cell lines or primary nonmalignant cells with regard to the mechanism by which GCs induce apoptosis awaits the development of techniques that allow investigators to address this question in such primary cells.

The molecular basis for the interrelationship of cAMP and GC-mediated apoptosis in lymphoid cells remains incompletely understood. Muller Igaz et al.[51] reported that cAMP-induced augmentation of glucocorticoid-mediated apoptosis in two T-cell hybridomas (DO.11 and 2B4) was independent of cAMP-response element (CRE) transcriptional activation as it was unaffected by cotransfection of CRE decoy oligonucleotides that effectively inhibited transcription of bona fide CRE-containing promoters. This study supports the model that PKA augments GC-induced apoptosis by directly altering GR-mediated transcription, rather than the alternate model that PKA activates a parallel CREB-mediated transcriptional pathway whose products synergize with GR-mediated transcriptional products to induce apoptosis. Zhang and Insel[52] have determined that in S49 cells, both glucocorticoid and cAMP-mediated apoptosis are associated with upregulation of the proapoptotic BH3-only Bcl-2 family member Bim. Of note, thymocytes from Bim knockout mice are insensitive to glucocorticoid-induced apoptosis.[53] These investigators noted that the mechanism that accounts for cAMP-induced upregulation of Bim expression in S49 cells is unclear as the Bim promoter does not contain CRE elements. It would therefore be of interest to determine whether cAMP upregulates Bim in GR-negative cell lines such as Jurkat, as it remains plausible that cAMP signaling upregulates Bim in a GR-dependent manner.

If PKA regulates GR signaling, how might it do so? Evans and colleagues[54] reported that PKA catalytic subunit can be coimmunoprecipitated with the glucocorticoid receptor in HEK-293 cells cotransfected to overexpress both of these proteins. *In vitro* studies have demonstrated that PKA catalytic subunit can directly phosphorylate the GR.[55] Alternatively, PKA complexed with the GR could phosphorylate GR-associated coactivators or corepressors such as GRIP-1.[56] Finally, cAMP has been shown to augment levels of glucocorticoid binding in murine lymphoma cells.[43] Depending on the model system examined, cAMP

signaling can augment GR levels as a result of either augmented transcription of the GR promoter or enhanced GR mRNA half-life.[57,58]

28.2.6 OTHER PROPOSED MECHANISMS FOR PDE4 INHIBITOR-INDUCED LYMPHOID APOPTOSIS

In addition to the GR literature cited above, other models have been proposed for the mechanism by which PDE4 inhibitors induce apoptosis in lymphoid cells. Treatment of B-CLL cells with PDE4 inhibitors induces mitochondrial depolarization, release of mitochondrial cytochrome c to the cytosol, caspase 9 and 3 activation, and poly(ADP-ribose)polymerase (PARP) cleavage (Figure 28.3).[59] Consistent with the hypothesis that such activation of an apparent mitochondrial pathway of apoptosis is a primary rather than secondary event, inhibitors of caspase 9 but not caspase 8 block rolipram-induced apoptosis in B-CLL. Two different groups have reported that 18 h after addition of rolipram (10 µM) and forskolin (40 µM) to B-CLL cells, levels of the antiapoptotic proteins Bcl-2 and Bcl-XL are downregulated in heavy membrane fractions, while there is accumulation of the proapoptotic Bcl-2 family member Bax.[59,60] Moon and colleagues further identified rolipram-induced accumulation of the proapoptotic protein BAD in B-CLL membrane fractions. In three of five B-CLL patients examined, these authors observed that rolipram and forskolin treatment induced a reduction in BAD phosphorylation at Ser112 as well as reduced association of BAD with cytosolic 14-3-3 protein.[59] As several studies have documented PP2A-induced BAD dephosphorylation, including one group that reported nonapoptotic cAMP signaling that induced PP2A-mediated BAD dephosphorylation in renal epithelial cells, Moon and colleagues examined the effects of PDE4 inhibitor treatment on PP2A activity in B-CLL cells. In PP2A immunoprecipitates from B-CLL cells, prior rolipram and forskolin treatment augmented dephosphorylation of the threonine-phosphorylated peptide KRpTIRR and such dephosphorylation was inhibited by low concentrations of okadaic acid (5 nM) that

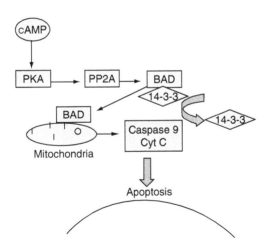

FIGURE 28.3 A proposed pathway by which PDE4 inhibitor-induced PKA signaling leads to apoptosis in B-CLL. Rolipram treatment of B-CLL leads to activation of the serine/threonine phosphatase PP2A, which in turn dephosphorylates the proapoptotic BH3-only Bcl2 family member BAD, releasing BAD from sequestration in the cytosol by 14-3-3. BAD subsequently translocates to mitochondria, where it initiates release of cytochrome c with subsequent activation of caspase 9 and apoptosis of B-CLL cells. Cotreatment of B-CLL cells with low concentrations of okadaic acid known to inhibit PP2A reduces rolipram-mediated dephosphorylation of BAD and apoptosis.

are relatively specific for PP2A activity. Total levels of B-CLL PP2Ac were also augmented 6 h after rolipram and forskolin treatment. Addition of 5 nM okadaic acid to rolipram- and forskolin-treated B-CLL cells maintained BAD Ser112 phosphorylation and association with 14-3-3 and blocked mitochondrial depolarization in four of five patient samples tested. The authors proposed that PDE4 inhibitor-induced PKA activation leads, perhaps indirectly, to PP2A activation, BAD dephosphorylation and translocation to mitochondrial membranes, and initiation of a mitochondrial pathway of apoptosis. Whether rolipram-induced upregulation of PP2Ac might occur by a transcriptional mechanism remains unexplored.

A markedly different model for PDE4 inhibitor-induced apoptosis was proposed by Smith et al.[36] in their studies of the role of PDE4B2 signaling in DLBCL cell lines (Figure 28.4). Neither the ATP binding-site PKA inhibitor H89 nor the peptide PKA inhibitor PKI blocked forskolin- and rolipram-induced growth arrest. Similarly, the EPAC-specific activator 8CPT-2Me-cAMP did not induce growth arrest and some rolipram- and forskolin-sensitive DLBCL cell lines did not even express EPAC1 transcript. Thus, the authors concluded that PDE4 inhibitors work through an as of yet unidentified PKA and EPAC-independent effector mechanism. As PDE4 inhibitors induced BAD dephosphorylation in these DLBCL cell lines, PP2A and Akt activity were examined as these enzymes were known to regulate BAD phosphorylation. While the authors found upregulated PP2A activity, they did not find any evidence that this enzyme was responsible for BAD dephosphorylation. Instead, they noted a reduction in both Akt phosphorylation and Akt activity following rolipram and forskolin

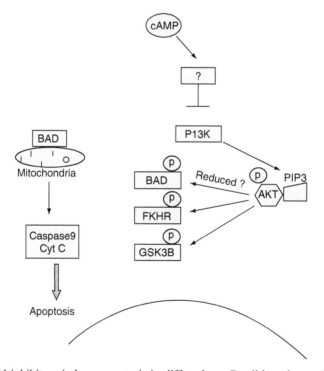

FIGURE 28.4 PDE4 inhibitors induce apoptosis in diffuse large B-cell lymphoma (DLBCL) cell lines. PDE4 inhibitors are reported to act neither through EPAC nor PKA but instead through an as yet undefined cAMP effector to inhibit PI3K. Such PDE4 inhibitor-induced PI3K inhibition results in a reduction in PKB/Akt activity and consequently reduces phosphorylation of BAD. As discussed above, dephosphorylated BAD can then induce apoptosis through a mitochondrial pathway. Alteration of the phosphorylation state of other PKB/Akt phosphorylation targets such as FKHR and GSK3B may also contribute to PDE4 inhibitor-induced apoptosis in this model.

treatment. In keeping with a role for Akt, transfection with constitutively active Akt protected the cell line from PDE4 inhibitor-induced growth arrest. Forskolin treatment of a cAMP-sensitive DLBCL cell line reduced PI3K activity towards its major substrate PIP2. DLBCL cell lines overexpressing PDE4B and resistant to cAMP-mediated growth arrest were sensitive to growth inhibition by the PI3K inhibitor LY249002. The authors concluded that PDE4B2 overexpression may render cells relatively less sensitive to cAMP-mediated inhibition of PI3K and that PI3K inhibitors may overcome such resistance to cAMP-mediated growth arrest and apoptosis.

28.2.7 PDE4 INHIBITOR-INDUCED EPAC ACTIVATION IS ANTIAPOPTOTIC IN B-CELL CHRONIC LYMPHOCYTIC LEUKEMIA

EPAC1 and EPAC2 are Rap1 GDP-exchange factors that are activated by cAMP. In an effort to determine whether selective expression in B-CLL of EPAC family members might play a role in the sensitivity of such leukemic cells to PDE4 inhibitor-induced apoptosis, Tiwari et al.[32] performed RT-PCR for EPAC1 and EPAC2 transcript on primary B-CLL cells, T-cell lymphocytic leukemia (T-CLL) cells, ALL cells, mantle cell lymphoma cells, B cells, T cells, monocytes, and neutrophils. EPAC1 transcript was detected only in B-CLL, mantle cell lymphoma, and normal human B cells whereas EPAC2 transcript was undetectable in any of these cells, although it was easily detected in the HEK-293 cell line. By pull-down analysis, treatment with rolipram and forskolin induced Rap1 activation in B-CLL cells, but not T-CLL cells, resting B cells, activated B cells, T cells, neutrophils, or monocytes. The discrepancy between the results of RT-PCR and those of the pull-down analysis with regard to normal circulating B lymphocytes and B-CLL cells was resolved by performing real-time PCR, which demonstrated that EPAC1 levels were 50-fold higher in B-CLL cells than in normal human B lymphocytes. Whether EPAC1 transcript levels are 50-fold lower in all human B cells or whether a small B-cell subpopulation expresses levels comparable to those of B-CLL cells remains to be established.

Treatment of B-CLL cells with rolipram alone (10 μM) in the absence of an exogenous stimulus for adenylate cyclase led to both robust CREB phosphorylation and Rap1 activation. In contrast, as predicted by the work of Enserink et al.[61] in other cell lineages, treatment with the EPAC-selective cAMP analog 8CPT-2Me-cAMP led to Rap1 activation but failed to induce CREB phosphorylation. While rolipram treatment induced apoptosis in B-CLL cells, treatment with 8CPT-2Me-cAMP actually reduced basal B-CLL apoptosis in a dose-dependent fashion and modestly reduced rolipram- and forskolin-induced apoptosis. The PKA antagonist (R_p)-8-Br-cAMPS antagonized rolipram- and forskolin-induced apoptosis. These data would suggest that in B-CLL cells, PDE4 inhibitors activate both a proapoptotic PKA-mediated signaling pathway and an antiapoptotic EPAC1-mediated signaling pathway, with the proapoptotic signal as the dominant effect in leukemic cells from most patients (Figure 28.5). The mechanism by which EPAC signaling confers an antiapoptotic effect in B-CLL cells is currently unknown.

28.2.8 WHAT ACCOUNTS FOR THE SPECIFICITY OF PDE4 INHIBITOR-INDUCED APOPTOSIS?

As noted above, lymphoid cells vary dramatically in their sensitivity to PDE4 inhibitor-induced apoptosis. As an example, while B-CLL cells are sensitive to PDE4 inhibitor-induced apoptosis, both normal T cells and T-CLL cells are resistant. What accounts for this variation in sensitivity? One could implicate the degree of elevation of cAMP induced following PDE4 inhibitor therapy, the specific splice isoform and type of PDE4 inhibited, or variations in

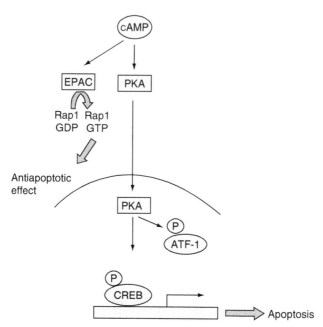

FIGURE 28.5 Signaling pathways proposed to mediate PDE4 inhibitor-induced lymphoid apoptosis. PDE4 inhibitors such as rolipram induce cell death in B-CLL cells. Rolipram treatment of B-CLL cells induces both PKA activation as judged by CREB and ATF-1 phosphorylation and EPAC activation as judged by augmented levels of GTP-bound Rap1. Despite the proapoptotic effects of PDE4 inhibitors in B-CLL, unique activation of EPAC1 by 8-CPT-2Me-cAMP is modestly antiapoptotic. Inhibition of PKA with the enantiomeric cAMP analog (R_p)-8Br-cAMPS antagonizes rolipram-mediated apoptosis in B-CLL cells, suggesting that PKA is the cAMP effector protein responsible for PDE4 inhibitor-induced apoptosis. Given that rolipram induces apoptosis in B-CLL cells, it appears that PKA-mediated proapoptotic effects predominate over EPAC-mediated antiapoptotic signaling.

the downstream effector pathways activated in these different lymphocyte populations. At least in B-CLL, neither the levels of intracellular cAMP induced nor the specific splice isoforms of PDE4 inhibited appear to account for the sensitivity of these cells to rolipram. Treatment of primary T cells with a combination of rolipram (10 μM) and forskolin (40 μM) induces high levels of cAMP but does not result in significant apoptosis.[29] Both B-CLL cells and T cells contain constitutive PDE4A long forms (PDE4A4 or PDE4A5) and inducible PDE4B2.[30,62] Instead, it appears that the specificity of PDE4 inhibitor-induced apoptosis lies in the activation of differentially expressed or regulated downstream proapoptotic targets. While the work described above has suggested that PP2A-induced BAD dephosphorylation may be one such target in B-CLL, the details of the signaling pathways that account for the relative sensitivity of different lymphoid cells to this class of drugs remain to be established.

28.3 ANTIAPOPTOTIC EFFECTS OF cAMP, cGMP, AND PHOSPHODIESTERASE INHIBITORS

cGMP-mediated signaling has antiapoptotic effects in several cell lineages. Similarly, while the studies cited above demonstrate that cAMP is proapoptotic in subsets of lymphoid populations, cAMP signaling is antiapoptotic in a wider variety of cell lineages, including cells of neuronal, hepatic, and myeloid lineage. Although the effects of PDE inhibitors have been more frequently analyzed in the proapoptotic studies described above, several groups

have recently begun to examine which PDEs regulate such antiapoptotic cyclic nucleotide signaling. The antiapoptotic effects of PDE inhibitors and cAMP- and cGMP-mediated signaling in several tissue subtypes are discussed below.

28.3.1 ANTIAPOPTOTIC EFFECTS OF PDE4 AND PDE5 INHIBITORS IN NEURONAL CELLS

cAMP and cGMP signaling induced by PDE inhibitors, hormonal stimuli, forskolin, or cyclic nucleotide analogs inhibits apoptosis in cultured spinal motor, superior cervical ganglion sympathetic, dorsal root sensory, dopaminergic, cerebellar granule, and septal cholinergic neurons.[63–68] As an example, a high basal rate of apoptosis observed following 13 days of culture of primary dopaminergic neurons is reduced by addition of forskolin (10 μM).[65] The PDE4 inhibitor rolipram (10 μM), while inactive alone, markedly enhances the antiapoptotic effects of forskolin.[65] The nonselective PDE inhibitor IBMX is also antiapoptotic whereas inhibitors of PDE2 and PDE3 show only weak activity.

Amyotrophic lateral sclerosis is a fatal neurodegenerative disease characterized by apoptosis of spinal motor neurons. Excitotoxicity or oxidative injury may play a role in this process and can be modeled by 24 h culture of rat spinal cord neurons in either low dose glutamate (10 μM) or the glutathione synthase inhibitor L-butathione-[S,R]-sulfoximine (BSO; 0.3 μM), respectively. Shimohama and colleagues[69] have reported that 8-Br-cGMP protects motor neurons against hydrogen peroxide or BSO-induced apoptosis. Consistent with this, PDE5 inhibitors (30 μM dipyridamole, 100 nM T-1032, and 1 μM zaprinast) protected spinal neurons against both acute glutamate and BSO-induced apoptosis (Figure 28.1).[70] Such neuroprotection could be abrogated by the cGMP-dependent protein kinase (PKG) inhibitor KT5823 (1 μM). Immunohistochemical staining demonstrated higher levels of PDE5 in motor than in nonmotor neurons. The investigators hypothesize that such elevated expression of PDE5 may account for the particular vulnerability of motor neurons to excitotoxic apoptosis and that this susceptibility might be ameliorated by PDE5 inhibitors. Of note, the investigators found no evidence of neuroprotection by either dbcAMP or inhibitors of PDE1-4 as vinpocetine, 8-MM-IBMX, erythro-9-(2-hydroxy-3-nonyl)-adenine (EHNA), milrinone, and rolipram were ineffective at concentrations of 10, 10, 20, 100, and 100 μM, respectively.

In related studies, nitric oxide (NO), a well-described cellular messenger in the nervous system, confers a survival signal in cerebellar granule and neuroblastoma cells by a cGMP- and PKG-dependent pathway that leads to Akt activation and subsequent CREB phosphorylation.[68,71,72] Nitric oxide–induced cGMP signaling also supports the survival of rat embryonic motor neurons cultured with brain-derived neurotrophic factor.[73]

Although dbcAMP was found to be ineffective in preventing excitotoxic apoptosis in the studies by Shimohama and colleagues described above, Hanson et al.[63] reported that treatment of almost equivalent (day 15 rather than 16) primary rat motor neuron cultures with forskolin (10 μM) and IBMX (0.1 mM) dramatically enhanced basal survival during 1 week of culture.[63] Of note, earlier work by Rydel and Greene demonstrated that different cAMP analogs have markedly different abilities to prolong survival and outgrowth of rat sympathetic and sensory neurons. While 8-(4-chlorophenylthio)-cAMP (300 μM) is efficacious in promoting survival, dbcAMP (1 mM) was largely ineffective (dbcAMP was used in the studies by Shimohama described above).[64] Regardless, Hanson's work suggests that the activity of cAMP as well as cGMP PDEs may limit the survival and growth of motor neurons under some circumstances. Consistent with this hypothesis, Nikulina et al.[74] reported that following a surgical injury to the spinal cord, the PDE4 inhibitor rolipram, delivered by mini-osmotic pumps to the spinal cord surface, enhances axonal regeneration.

The molecular mechanism by which cAMP signaling confers an antiapoptotic signal in neuronal cells remains controversial. Several laboratories have focused on the antiapoptotic

PI3K pathway and the downstream kinases Akt and GSK-3β. In HEK 293 cells, Filippa et al.[75] reported that forskolin (10 μM) or 8-(4-chlorophenylthio)-cAMP (CPT-cAMP: 1 mM) activates Akt in a PKA-dependent (i.e., inhibited by 5 μM H-89) but PI3K-independent manner (i.e., not inhibited by 100 nM wortmannin). The activation of Akt did not appear to involve direct phosphorylation by PKA and, unlike insulin-induced Akt activation, forskolin-induced Akt activation was independent of Ser473 phosphorylation. Forskolin-induced Akt activation led to GSK-3β phosphorylation and inhibition of GSK-3β activity. Despite such biochemical studies, the antiapoptotic effects of CPT-cAMP in primary sympathetic neurons are insensitive to transfection with either dominant negative PI3K or Akt.[76] Similarly, forskolin (10 μM) or CPT-cAMP (30 μM) induces GSK-3β phosphorylation in rat cerebellar granule neurons in a PKA-dependent but Akt-independent manner.[77] Inhibitors of GSK-3β reduce apoptosis in these cerebellar neurons whereas constitutively active GSK-3β transfection induces apoptosis. PKA phosphorylates GSK-3β on Ser9 *in vitro*, and transfection with a GSK-3β Ser/Ala9 mutant interferes with the ability of either forskolin or CPT-cAMP to augment neuronal survival. In conclusion, despite reports of cAMP-mediated activation of Akt in other cell lineages, cAMP-mediated antiapoptotic effects in primary neuronal cultures appear to be independent of both Akt and PI3K activity but may involve inhibition of GSK-3β.

28.3.2 cAMP SIGNALING PROTECTS HEPATOCYTES FROM FAS AND BILE ACID-INDUCED APOPTOSIS

A dose-limiting toxicity of anti-Fas directed therapy in animal models is hepatic necrosis. Fladmark et al.[78] first reported that N6-benzoyl-cAMP (100 μM) blocked induction of apoptosis in primary rat hepatocytes induced by anti-Fas treatment. Anti-Fas-induced hepatocyte apoptosis was enhanced by concurrent treatment with the PKA antagonist (R$_p$)-8-Br-cAMPS, suggesting that PKA might play a role in the antiapoptotic effect mediated by N6-benzoyl-cAMP. Webster and Anwer[79] demonstrated that dbcAMP, glucagon, or a combination of forskolin (20 μM) and the methylxanthine PDE inhibitor IBMX (20 μM) protect primary hepatocytes from apoptosis induced by treatment with the bile acid glycochenodeoxycholate (GCDC) for 2 h. The level of protection was substantial (65%) when apoptosis was measured at a 2 h end point. Such cAMP-mediated protection was abrogated by inhibitors of PKA (KT5720), PI3K (LY294002), and MAPK (PD98059). Treatment of primary rat hepatocytes with CPT-cAMP (100 μM) activates PI3K in p85 immunoprecipitates at 5–60 min after cell treatment.[80] CPT-cAMP treatment also led to Akt activation and cAMP-mediated protection was partially blocked by the Akt inhibitor SB203580.

In a subsequent study, the above investigators reported that cAMP-induced protection from apoptosis induced by GCDC (2 h), Fas ligand (3 h), or TNF-α (5 h) was largely unaffected by the PKA inhibitor (R$_p$)-8-Br-cAMPS.[81] The authors argued that their earlier studies implicating PKA may have resulted from the use of the relatively nonspecific PKA inhibitor KT5720. In contrast, the EPAC-specific cAMP analog 8CPT-2Me-cAMP (20 μM) protected rat hepatocytes from all of these apoptotic stimuli. Treatment with 8CPT-2Me-cAMP also induced Akt activation as judged by Ser473 phosphorylation whereas (R$_p$)-8-Br-cAMPS failed to block Akt activation by CPT-cAMP (100 μM). 8CPT-2Me-cAMP-induced protection of bile acid- and Fas-mediated apoptosis was inhibited by the PI3K inhibitor LY294002 whereas this was not true for TNF-α-induced apoptosis. Similarly, an earlier study by Li et al.[82] demonstrated that 8-Br-cAMP protected against TNF-α-induced apoptosis in rat hepatocytes and induced Akt activation, but that inhibition of Akt did not substantially affect the cAMP-induced protective effect. In summary, the protective effect of cAMP-induced signaling in rat

hepatocytes has been clearly established by multiple groups in studies of multiple apoptotic stimuli. Resolution of whether these antiapoptotic effects are in fact mediated through EPAC rather than PKA will require confirmatory studies and a better resolution of the mechanism underlying cAMP-mediated PI3K and Akt activation in this cell type.

28.3.3 cGMP Signaling and Apoptosis in Endothelial Cells, Colon Tumor Cell Lines, and Hepatocytes

Depending on the cell lineage examined, investigators have reported cyclic nucleotide-dependent pro- or antiapoptotic effects of atrial natriuretic peptide (ANP) signaling. ANP activates guanylyl cyclase A/B in both pulmonary arterial endothelial cells (PAECs) and pulmonary microvascular endothelial cells (PMVECs).[83] However, while treatment of PAECs with ANP leads to a transient rise in cGMP (decays within 10 min), treatment of PMVECs with ANP leads to an increase in cGMP that is sustained for more than 5 h. Of note, PDE5A1/PDE5A2 transcript, protein, and enzymatic activity are present in PAECs but not PMVECs. ANP induces growth inhibition and apoptosis in PMVECs but not in PAECs, unless the latter cells are cotreated with the PDE5 inhibitor zaprinast.[83] These studies suggest that ANP-induced cGMP signaling is proapoptotic in PMVECs but not PAECs as a result of a protective effect of PAEC PDE5.

In keeping with the work described above, inhibition of the cGMP phosphodiesterases PDE1, PDE2, and PDE5 with the drug exisulind induces activation of PKG and apoptosis in human colon tumor cell lines as well as other cancer cell-line models.[84–88] Transient transfection of the SW480 colon carcinoma cell line with constitutively active mutants of PKG1β inhibited colony formation and induced apoptosis whereas a dominant negative mutant of PKG1β grew more rapidly and was partially resistant to exisulind.[86] While antisense inhibition of PDE5A1 expression was recently found to induce growth inhibition and apoptosis in the human colon tumor cell line HT29, the effects of well characterized pharmacologic inhibitors of PDE5 on apoptosis in HT29 cells were not reported in this study.[87] In general, inhibition of cGMP PDEs with known family-specific PDE inhibitors does not induce apoptosis in such lines, suggesting that combined inhibition of multiple cGMP PDEs may be required for an apoptotic effect (W.J. Thompson, personal communication).

In contrast to the proapoptotic activity of ANP in PMVECs, an antiapoptotic effect is observed in ischemic rat livers perfused with either ANP or 8-Br-cGMP.[89] Interestingly, despite the clear antiapoptotic effect of 8-Br-cGMP, neither PKG I or II was detected in rat hepatocytes. Coperfusion of livers with a PKA inhibitor abolished the antiapoptotic effects of ANP whereas coperfusion with a PKG inhibitor (R$_p$-8-Br-pCPT-cGMPS) did not.[90] The authors conclude that PKA rather than PKG mediates the antiapoptotic effects of ANP in this model.[90] While they suggest that this may be an example of promiscuous direct cGMP-mediated activation of PKA, an alternative explanation is that cGMP-mediated inhibition of PDE3 results indirectly in augmented intracellular levels of cAMP and subsequent PKA signaling. Further studies will be required to clarify the role of PDEs in the antiapoptotic effects of ANP in the liver.

28.3.4 PDE4 Inhibitors Reduce Apoptosis and Induce Differentiation in Myeloid Cells

cAMP signaling regulates both differentiation and apoptosis in the myeloid lineage. A seminal study by Rossi et al. in 1995[91] demonstrated that forskolin, 8-Br-cAMP, and dbcAMP reduced basal apoptosis in primary neutrophils and that such an effect was abrogated by

the PKA inhibitor H-89 (100 μM). Similarly, Beebe and colleagues[92] reported that CPT-cAMP (500 μM) prolongs the time required for 50% of freshly isolated human neutrophils to undergo apoptosis from 16 to 42 h. By using low concentrations of cAMP analog pairs with known preferential affinity for sites A and B of the regulatory subunits of either type 1 or 2 PKA, these authors associated the ability of cAMP to delay neutrophil apoptosis with type 1 PKA activity. Consistent with these studies, Ottonello et al.[93] reported that either PGE$_2$ (0.1 to 10 μM) or the PDE4 inhibitor RO-20-1724 (10 μM) inhibited neutrophil apoptosis in a manner that could be blocked by H-89 (1 μM). When combined, PGE2 and RO-20-1724 had synergistic antiapoptotic effects. Although other family-specific PDEs were not examined, this report suggests that despite expression of several PDE families by neutrophils, specific inhibition of PDE4 can activate an antiapoptotic signaling pathway, even in the absence of exogenous stimulation of adenylate cyclase.

Martin et al.[94] also reported that dbcAMP (200 μM) markedly reduced basal apoptosis in human neutrophils as measured at 20 h. However, unlike the studies cited above, these authors argued that the antiapoptotic effect of dbcAMP was unlikely to be mediated by PKA as concentrations of either the PKA inhibitors H-89 (10 μM) or (R$_p$)-8-Br-cAMPS (100 μM) that reduced PKA activity had little or no effect on the antiapoptotic of dbcAMP. The antiapoptotic effects of dbcAMP, as judged either by morphologic criteria or annexin V staining, were also independent of protein translation, PI3K and MAPK signaling as concurrent treatment with cycloheximide (1 μg/mL), LY294002 (10 μM) and PD98059 (10 μM) did not restore neutrophil apoptosis to basal levels.

Studies of myeloid cell lines rather than primary myeloid cells have also demonstrated an antiapoptotic effect of cAMP signaling. A combination of forskolin (25 μM) and theophylline (1 mM) protected the promonocytic U-937 cell line from apoptosis induced by the topoisomerase II inhibitor etoposide (6 μM for 1 h), the topoisomerase 1 inhibitor camptothecin (1.0 μM for 1 h), heat treatment (30 min at 43.5°C), irradiation (20 Gy), or cadmium chloride (2 h at 125 μM).[95] Protection was not complete, but rather reduced apoptosis to levels of 50%–70% of those seen following treatment with the apoptotic stimulus alone. Theophylline and forskolin treatment led to a marked reduction in U-937 *myc* levels as judged by both Northern and Western blot analysis, suggesting that such *myc* suppression may have played a role in the antiapoptotic effects of cAMP signaling in these cells. Similarly, treatment of MC/9 cells, a murine IL-3-dependent mast cell line, with C2-ceramide (50 μM), the product of sphingomyelinase activity, induces apoptosis that can be inhibited by cotreatment with forskolin (10 μM) and IBMX (50 μM).[96] As both ceramide and forskolin/IBMX treatment led to CREB Ser133 phosphorylation, the authors concluded that such phosphorylation was insufficient to explain the pro- and antiapoptotic effects of these agents, respectively.

As in the studies on cerebellar granule and neuroblastoma cells cited above, low concentrations of nitric oxide have been reported to confer an antiapoptotic signal in a murine macrophage cell line by a cGMP-dependent pathway.[97] Inhibitors of either guanylate cyclase (LY83583 or ODQ) or protein kinase G (KT5823) reduced the cytoprotective effects of NO donors. As of yet, the cGMP phosphodiesterases regulating such a signaling pathway have not been reported.

In some contrast to the studies cited above, several recent exciting studies have suggested that agents that induce cAMP-mediated signaling, such as PDE inhibitors, can augment the ability of retinoids such as all-*trans*-retinoic acid (ATRA) to relieve the pathologic arrest of differentiation observed in acute myeloid leukemia (AML) cells, resulting in maturation of AML blasts and their subsequent physiologic death by apoptosis within several days. Addition of theophylline to arsenic trioxide, another differentiation-inducing drug, promoted blast maturation in a murine acute promyelocytic leukemia (APL) model and reversed clinical

resistance in an APL patient refractory to combined therapy with ATRA and arsenic trioxide.[98] The PDE4 inhibitor piclamilast induces PKA-mediated phosphorylation of the retinoid receptor RARα. When compared with ATRA therapy alone, addition of piclamilast to ATRA both augments maturation of the NB4 APL cell line *in vitro* and enhances survival of severe combined immunodeficient (SCID) mice in which NB4 cells have been implanted in the peritoneum.[99] Similarly, when contrasted with therapy with the rexinoid LG1069 alone, addition of the PDE4 inhibitor rolipram (100 μM) or the nonspecific PDE inhibitors theophylline (2 mM) or IBMX (500 μM) augments differentiation and apoptosis of PLB985 AML cells *in vitro*.[100] Future clinical studies will likely focus on whether addition of PDE4 inhibitors can reverse clinical resistance to retinoid therapy in a subset of non-APL patients with AML.

28.4   DO PHOSPHODIESTERASES REGULATE PHYSIOLOGIC APOPTOTIC SIGNALING?

This chapter has presented evidence from many laboratories that PDE inhibitors may prove to be therapeutically efficacious for a variety of illnesses, both malignant and nonmalignant, as a result of their ability to regulate the activity of apoptotic pathways either positively (ALL, CLL, DLBCL) or negatively (ALS, spinal cord injury). From a physiologic rather than a pharmacologic perspective, other studies have demonstrated that cyclic nucleotide-mediated signaling can regulate the survival of primary cell populations. As an example, *Drosophila* wing epidermal cells undergo programmed cell death as a result of GTP-binding protein-coupled receptor (GPCR)-coupled signaling by the insect tanning hormone, bursicon. Mutation of the bursicon receptor or loss of G protein or PKA function inhibits epidermal cell death whereas treatment with cell-permeable cyclic AMP analogs or ectopic expression of PKA induces epidermal cell death.[101] The pituitary adenylate cyclase-activating polypeptide (PAKAP) augments cerebellar granule neuron survival both *in vitro* and *in vivo* in a manner that is associated with PKA activation, is inhibited by H-89, and is mimicked by dbcAMP.[102,103]

However, it remains to be established whether PDEs play a role in regulating physiologic apoptotic pathways. Potent and specific PDE inhibitors induce a highly nonphysiologic signaling environment in cells that persist despite induction of compensatory mechanisms that normally occur following prolonged exposure to GPCR agonists such as upregulation of PDEs, regulators of G-protein signaling (RGS) proteins, and the cAMP response element modulator (CREM) gene-derived transcriptional repressor, inducible cAMP early repressor (ICER). Despite such compensatory adaptation, the persistent cyclic nucleotide effector protein activation observed following PDE inhibitor therapy is likely to be considerably more prolonged than that that would occur under even extreme physiologic conditions. Perhaps by the next edition of this book, a better understanding of the physiologic role of PDEs in apoptotic signaling will have emerged. More importantly, it is undoubtedly the hope of many investigators in this field that the next edition will also bring accounts of clinical trials that test the premise that PDE inhibitor therapy will benefit patients with the malignancies and neurologic disorders discussed above.

REFERENCES

1. Hait, W.N., and B. Weiss. 1976. *Nature* 259:321–323.
2. Epstein, P.M., J.S. Mills, C.P. Ross, S.J. Strada, E.M. Hersh, and W.J. Thompson. 1977. *Cancer Res* 37:4016–4023.
3. Yan, L., V. Herrmann, J.K. Hofer, and P.A. Insel. 2000. *Am J Physiol Cell Physiol* 279:C1665–C1674.

4. Gu, C., Y.C. Ma, J. Benjamin, D. Littman, M.V. Chao, and X.Y. Huang. 2000. *J Biol Chem* 275:20726–20733.

5. Mamani-Matsuda, M., D. Moynet, M. Molimard, H. Ferry-Dumazet, G. Marit, J. Reiffers, and M.D. Mossalayi. 2004. *Br J Haematol* 124:141–150.

6. Daniel, V., G. Litwack, and G.M. Tomkins. 1973. *Proc Natl Acad Sci USA* 70:76–79.

7. Coffino, P., J.W. Gray, and G.M. Tomkins. 1975. *Proc Natl Acad Sci USA* 72:878–882.

8. Insel, P.A., H.R. Bourne, P. Coffino, and G.M. Tomkins. 1975. *Science* 190:896–898.

9. Hochman, J., P.A. Insel, H.R. Bourne, P. Coffino, and G.M. Tomkins 1975. *Proc Natl Acad Sci USA* 72:5051–5055.

10. Steinberg, R.A., P.H. O'Farrell, U. Friedrich, and P. Coffino. 1977. *Cell* 10:381–391.

11. Hochman, J., H.R. Bourne, P. Coffino, P.A. Insel, L. Krasny, and K.L. Melmon. 1977. *Proc Natl Acad Sci USA* 74:1167–1171.

12. Wyllie, A.H., J.F.R. Kerr, and A.R. Currie. 1980. *Int Rev Cytol* 68:251–306.

13. McConkey, D.J., S. Orrenius, and M. Jondal. 1990. *J Immunol* 145:1227–1230.

14. Kizaki, H., K. Suzuki, T. Tadakuma, and Y. Ishimura. 1990. *J Biol Chem* 265:5280–5284.

15. Durant, S., and F. Homo-Delarche. 1983. *Mol Cell Endocrinol* 31:215–225.

16. Lomo, J., H.K. Blomhoff, K. Beiske, T. Stokke, and E.B. Smeland. 1995. *J Immunol* 154: 1634–1643.

17. Mentz, F., H. Merle-Beral, F. Ouaaz, and J.L. Binet. 1995. *Br J Haematol* 90:957–959.

18. Goetzl, E.J., S. An, and L. Zeng. 1995. *J Immunol* 154:1041–1047.

19. Hoshi, S., M. Furutani-Seiki, M. Seto, T. Tada, and Y. Asano. 1994. *Int Immunol* 6:1081–1089.

20. Lee, M.R., M.L. Liou, Y.F. Yang, and M.Z. Lai. 1993. *J Immunol* 151:5208–5217.

21. Jiang, X., J. Li, M. Paskind, and P.M. Epstein. 1996. *Proc Natl Acad Sci USA* 93:11236–11241.

22. Epstein, P.M., S. Moraski, and R. Hachisu. 1987. *Biochem J* 243:533–539.

23. Hurwitz, R.L., K.M. Hirsch, D.J. Clark, V.N. Holcombe, and M.Y. Hurwitz. 1990. *J Biol Chem* 265:8901–8907.

24. Reed, T.M., D.R. Repaske, G.L. Snyder, P. Greengard, and C.V. Vorhees. 2002. *J Neurosci* 22:5188–5197.

25. Mentz, F., M.D. Mossalayi, F. Ouaaz, S. Baudet, F. Issaly, S. Ktorza, M. Semichon, J.L. Binet, and H. Merle-Beral. 1996. *Blood* 88:2172–2182.

26. Mentz, F., H. Merle-Beral, and A.H. Dalloul. 1999. *Leukemia* 13:78–84.

27. Lee, R., S. Wolda, E. Moon, J. Esselstyn, C. Hertel, and A. Lerner. 2002. *Cell Signal* 14:277–284.

28. Wiernik, P.H., E. Paietta, O. Goloubeva, S.J. Lee, D. Makower, J.M. Bennett, J.L. Wade, C. Ghosh, L.S. Kaminer, J. Pizzolo, and M.S. Tallman. 2004. *Leukemia* 18:1605–1610.

29. Kim, D.H., and A. Lerner. 1998. *Blood* 92:2484–2494.

30. Moon, E., R. Lee, R. Near, L. Weintraub, S. Wolda, and A. Lerner. 2002. *Clin Cancer Res* 8:589–595.

31. Cragg, M.S., H.T. Chan, M.D. Fox, A. Tutt, A. Smith, D.G. Oscier, T.J. Hamblin, and M.J. Glennie. 2002. *Blood* 100:3068–3076.

32. Tiwari, S., K. Felekkis, E.Y. Moon, A. Flies, D.H. Sherr, and A. Lerner. 2004. *Blood* 103: 2661–2667.

33. D'Sa, C., L.M. Tolbert, M. Conti, and R.S. Duman. 2002. *J Neurochem* 81:745–757.

34. Ogawa, R., M.B. Streiff, A. Bugayenko, and G.J. Kato. 2002. *Blood* 99:3390–3397.

35. Shipp, M.A., K.N. Ross, P. Tamayo, A.P. Weng, J.L. Kutok, R.C. Aguiar, M. Gaasenbeek, M. Angelo, M. Reich, G.S. Pinkus, T.S. Ray, M.A. Koval, K.W. Last, A. Norton, T.A. Lister, J. Mesirov, D.S. Neuberg, E.S. Lander, J.C. Aster, and T.R. Golub. 2002. *Nat Med* 8:68–74.

36. Smith, P.G., F. Wang, K.N. Wilkinson, K.J. Savage, U. Klein, D.S. Neuberg, G. Bollag, M.A. Shipp, and R.C. Aguiar. 2005. *Blood* 105:308–316.

37. Huston, E., M. Beard, F. McCallum, N.J. Pyne, P. Vandenabeele, G. Scotland, and M.D. Houslay. 2000. *J Biol Chem* 275:28063–28074.

38. Frame, M., K.F. Wan, R. Tate, P. Vandenabeele, and N.J. Pyne. 2001. *Cell Signal* 13:735–741.

39. Frame, M.J., R. Tate, D.R. Adams, K.M. Morgan, M.D. Houslay, P. Vandenabeele, and N.J. Pyne. 2003. *Eur J Biochem* 270:962–970.

40. McConkey, D.J., S. Orrenius, S. Okret, and M. Jondal. 1993. *FASEB J* 7:580–585.

41. Medh, R.D., M.F. Saeed, B.H. Johnson, and E.B. Thompson. 1998. *Cancer Res* 58:3684–3693.
42. Tiwari, S., H. Dong, E.J. Kim, L. Weintraub, P.M. Epstein, and A. Lerner. 2005. *Biochem Pharmacol* 69:473–483.
43. Gruol, D.J., N.F. Campbell, and S. Bourgeois. 1986. *J Biol Chem* 261:4909–4914.
44. Gruol, D.J., F.M. Rajah, and S. Bourgeois. 1989. *Mol Endocrinol* 3:2119–2127.
45. Kiefer, J., S. Okret, M. Jondal, and D.J. McConkey. 1995. *J Immunol* 155:4525–4528.
46. Powers, J.H., A.G. Hillmann, D.C. Tang, and J.M. Harmon. 1993. *Cancer Res* 53:4059–4065.
47. Gruol, D.J., and J. Altschmied. 1993. *Mol Endocrinol* 7:104–113.
48. Reichardt, H.M., K.H. Kaestner, J. Tuckermann, O. Kretz, O. Wessely, R. Bock, P. Gass, W. Schmid, P. Herrlich, P. Angel, and G. Schutz. 1998. *Cell* 93:531–541.
49. Helmberg, A., N. Auphan, C. Caelles, and M. Karin. 1995. *EMBO J* 14:452–460.
50. Thulasi, R., D.V. Harbour, and E.B. Thompson. 1993. *J Biol Chem* 268:18306–18312.
51. Muller Igaz, L., D. Refojo, M.A. Costas, F. Holsboer, and E. Arzt. 2002. *Biochim Biophys Acta* 1542:139–148.
52. Zhang, L., and P.A. Insel. 2004. *J Biol Chem* 279:20858–20865.
53. Bouillet, P., D. Metcalf, D.C. Huang, D.M. Tarlinton, T.W. Kay, F. Kontgen, J.M. Adams, and A. Strasser. 1999. *Science* 286:1735–1738.
54. Doucas, V., Y. Shi, S. Miyamoto, A. West, I. Verma, and R.M. Evans. 2000. *Proc Natl Acad Sci USA* 97:11893–11898.
55. Haske, T., M. Nakao, and V.K. Moudgil. 1994. *Mol Cell Biochem* 132:163–171.
56. Hoang, T., I.S. Fenne, C. Cook, B. Borud, M. Bakke, E.A. Lien, and G. Mellgren. 2004. *J Biol Chem* 279:49120–49130.
57. Penuelas, I., I.J. Encio, N. Lopez-Moratalla, and E. Santiago. 1998. *J Steroid Biochem Mol Biol* 67:89–94.
58. Dong, Y., M. Aronsson, J.A. Gustafsson, and S. Okret. 1989. *J Biol Chem* 264:13679–13683.
59. Moon, E.Y., and A. Lerner. 2003. *Blood* 101:4122–4130.
60. Siegmund, B., J. Welsch, F. Loher, G. Meinhardt, B. Emmerich, S. Endres, and A. Eigler. 2001. *Leukemia* 15:1564–1571.
61. Enserink, J.M., A.E. Christensen, J. de Rooij, M. van Triest, F. Schwede, H.G. Genieser, S.O. Deskeland, J.L. Blank, and J.L. Bos. 2002. *Nat Cell Biol* 901–906.
62. Seybold, J., R. Newton, L. Wright, P.A. Finney, N. Suttorp, P.J. Barnes, I.M. Adcock, and M.A. Giembycz. 1998. *J Biol Chem* 273:20575–20588.
63. Hanson, M.G. Jr., S. Shen, A.P. Wiemelt, F.A. McMorris, and B.A. Barres. 1998. *J Neurosci* 18:7361–7371.
64. Rydel, R.E., and L.A. Greene. 1988. *Proc Natl Acad Sci USA* 85:1257–1261.
65. Yamashita, N., A. Hayashi, J. Baba, and A. Sawa. 1997. *Jpn J Pharmacol* 75:155–159.
66. D'Mello, S.R., C. Galli, T. Ciotti, and P. Calissano. 1993. *Proc Natl Acad Sci USA* 90:10989–10993.
67. Kew, J.N., D.W. Smith, and M.V. Sofroniew. 1996. *Neuroscience* 70:329–339.
68. Ciani, E., S. Guidi, R. Bartesaghi, and A. Contestabile. 2002. *J Neurochem* 82:1282–1289.
69. Urushitani, M., R. Inoue, T. Nakamizo, H. Sawada, H. Shibasaki, and S. Shimohama. 2000. *J Neurosci Res* 61:443–448.
70. Nakamizo, T., J. Kawamata, K. Yoshida, Y. Kawai, R. Kanki, H. Sawada, T. Kihara, H. Yamashita, H. Shibasaki, A. Akaike, and S. Shimohama. 2003. *J Neurosci Res* 71:485–495.
71. Ciani, E., M. Virgili, and A. Contestabile. 2002. *J Neurochem* 81:218–228.
72. Ciani, E., S. Guidi, G. Della Valle, G. Perini, R. Bartesaghi, and A. Contestabile. 2002. *J Biol Chem* 277:49896–49902.
73. Estevez, A.G., N. Spear, J.A. Thompson, T.L. Cornwell, R. Radi, L. Barbeito, and J.S. Beckman. 1998. *J Neurosci* 18:3708–3714.
74. Nikulina, E., J.L. Tidwell, H.N. Dai, B.S. Bregman, and M.T. Filbin. 2004. *Proc Natl Acad Sci USA* 101:8786–8790.
75. Filippa, N., C.L. Sable, C. Filloux, B. Hemmings, and E. Van Obberghen. 1999. *Mol Cell Biol* 19:4989–5000.
76. Crowder, R.J., and R.S. Freeman. 1999. *J Neurochem* 73:466–475.

77. Li, M., X. Wang, M.K. Meintzer, T. Laessig, M.J. Birnbaum, and K.A. Heidenreich. 2000. *Mol Cell Biol* 20:9356–9363.

78. Fladmark, K.E., B.T. Gjertsen, S.O. Doskeland, and O.K. Vintermyr. 1997. *Biochem Biophys Res Commun* 232:20–25.

79. Webster, C.R., and M.S. Anwer. 1998. *Hepatology* 27:1324–1331.

80. Webster, C.R., P. Usechak, and M.S. Anwer. 2002. *Am J Physiol Gastrointest Liver Physiol* 283:G727–G738.

81. Cullen, K.A., J. McCool, M.S. Anwer, and C.R. Webster. 2004. *Am J Physiol Gastrointest Liver Physiol* 287:G334–G343.

82. Li, J., S. Yang, and T.R. Billiar. 2000. *J Biol Chem* 275:13026–13034.

83. Zhu, B., S. Strada, and T. Stevens. 2005. *Am J Physiol Lung Cell Mol Physiol* 289:L196–L206.

84. Thompson, W.J., G.A. Piazza, H. Li, L. Liu, J. Fetter, B. Zhu, G. Sperl, D. Ahnen, and R. Pamukcu. 2000. *Cancer Res* 60:3338–3342.

85. Whitehead, C.M., K.A. Earle, J. Fetter, S. Xu, T. Hartman, D.C. Chan, T.L. Zhao, G. Piazza, A.J. Klein-Szanto, R. Pamukcu, H. Alila, P.A. Bunn Jr., and W.J. Thompson. 2003. *Mol Cancer Ther* 2:479–488.

86. Deguchi, A., W.J. Thompson, and I.B. Weinstein. 2004. *Cancer Res* 64:3966–3973.

87. Zhu, B., L. Vemavarapu, W.J. Thompson, and S.J. Strada. 2005. *J Cell Biochem* 94:336–350.

88. Deguchi, A., S.W. Xing, I. Shureiqi, P. Yang, R.A. Newman, S.M. Lippman, S.J. Feinmark, B. Oehlen, and I.B. Weinstein. 2005. *Cancer Res* 65:8442–8447.

89. Gerwig, T., H. Meissner, M. Bilzer, A.K. Kiemer, H. Arnholdt, A.M. Vollmar, and A.L. Gerbes. 2003. *J Hepatol* 39:341–348.

90. Kulhanek-Heinze, S., A.L. Gerbes, T. Gerwig, A.M. Vollmar, and A.K. Kiemer. 2004. *J Hepatol* 41:414–420.

91. Rossi, A.G., J.M. Cousin, I. Dransfield, M.F. Lawson, E.R. Chilvers, and C. Haslett. 1995. *Biochem Biophys Res Commun* 217:892–899.

92. Parvathenani, L.K., E.S. Buescher, E. Chacon-Cruz, and S.J. Beebe. 1998. *J Biol Chem* 273: 6736–6743.

93. Ottonello, L., R. Gonella, P. Dapino, C. Sacchetti, and F. Dallegri. 1998. *Exp Hematol* 26: 895–902.

94. Martin, M.C., I. Dransfield, C. Haslett, and A.G. Rossi. 2001. *J Biol Chem* 276:45041–45050.

95. Garcia-Bermejo, L., C. Perez, N.E. Vilaboa, E. de Blas, and P. Aller. 1998. *J Cell Sci* 111 (Pt 5):637–644.

96. Scheid, M.P., I.N. Foltz, P.R. Young, J.W. Schrader, and V. Duronio. 1999. *Blood* 93:217–225.

97. Yoshioka, Y., A. Yamamuro, and S. Maeda. 2003. *Br J Pharmacol* 139:28–34.

98. Guillemin, M.C., E. Raffoux, D. Vitoux, S. Kogan, H. Soilihi, V. Lallemand-Breitenbach, J. Zhu, A. Janin, M.T. Daniel, B. Gourmel, L. Degos, H. Dombret, M. Lanotte, and H. De The. 2002. *J Exp Med* 196:1373–1380.

99. Parrella, E., M. Gianni, V. Cecconi, E. Nigro, M.M. Barzago, A. Rambaldi, C. Rochette-Egly, M. Terao, and E. Garattini. 2004. *J Biol Chem* 279:42026–42040.

100. Altucci, L., A. Rossin, O. Hirsch, A. Nebbioso, D. Vitoux, E. Wilhelm, F. Guidez, M. De Simone, E.M. Schiavone, D. Grimwade, A. Zelent, H. De The, and H. Gronemeyer. 2005. *Cancer Res* 65:8754–8765.

101. Kimura, K., A. Kodama, Y. Hayasaka, and T. Ohta. 2004. *Development* 131:1597–1606.

102. Vaudry, D., B.J. Gonzalez, M. Basille, A. Fournier, and H. Vaudry. 1999. *Proc Natl Acad Sci USA* 96:9415–9420.

103. Vaudry, D., B.J. Gonzalez, M. Basille, Y. Anouar, A. Fournier, and H. Vaudry. 1998. *Neuroscience* 84:801–812.

Section F

Development of Specific
Phosphodiesterase Inhibitors
as Therapeutic Agents

29 Crystal Structure of Phosphodiesterase Families and the Potential for Rational Drug Design

Kam Y.J. Zhang

CONTENTS

29.1 Introduction 583
29.2 Phosphodiesterase Catalytic Domain Structures.......................................584
29.3 Nucleotide Selectivity..589
29.4 Hydrophobic Clamp 593
29.5 Structural Basis of Inhibitor Potency...594
    29.5.1 3-Isobutyl-1-Methylxanthine Binding to Various Phosphodiesterases 595
    29.5.2 PDE4 Inhibitors 595
    29.5.3 PDE5 Inhibitors 595
    29.5.4 PDE3 Inhibitors 597
29.6 Structural Basis of Inhibitor Selectivity 598
    29.6.1 PDE4 and PDE5 Selectivity 598
    29.6.2 PDE5 and PDE6 Selectivity 600
    29.6.3 PDE4 and PDE7 Selectivity 600
    29.6.4 PDE4B and PDE4D Selectivity..601
29.7 Potential for Structure-Based Rational Drug Design 601
29.8 Summary..602
Acknowledgments 603
References 603

29.1 INTRODUCTION

Phosphodiesterases (PDEs) are enzymes that control the level of cyclic adenosine monophosphate (cAMP) and cyclic guanosine monophosphate (cGMP), two ubiquitous second messengers that mediate biological responses to a variety of extracellular cues including hormones, neurotransmitters, chemokines, and cytokines. Increased concentration of these cyclic nucleotides results in the activation of protein kinase A (PKA) and protein kinase G (PKG). These protein kinases phosphorylate a variety of substrates, including other enzymes, receptors, transcription factors, and ion channels, which regulate myriad physiological processes, such as immune responses, cardiac and smooth muscle contraction, visual response, glycogenolysis, platelet aggregation, ion channel conductance, apoptosis, and

growth control.[1] By blocking hydrolysis of the critical cyclic phosphodiester bond, inhibition of PDE results in higher levels of cyclic nucleotides. Therefore, PDE inhibitors may have considerable therapeutic utility as anti-inflammatory agents, antiasthmatics, vasodilators, smooth muscle relaxants, cardiotonic agents, antidepressants, antithrombotics, and agents for improving memory and other cognitive functions.[2–4]

There are 21 PDE genes in the human genome, which can be categorized into 11 families. Alternative mRNA splicing of the 21 genes generates over 60 isoforms of PDEs in various human tissues. These PDEs can be divided into three groups depending on their preference for hydrolyzing either cAMP, cGMP, or both cyclic nucleotides. One group of PDEs selectively hydrolyzes cAMP (PDE4, PDE7, PDE8), a second group of enzymes is selective towards cGMP (PDE5, PDE6, PDE9), and the remaining enzymes hydrolyze both cAMP and cGMP (PDE1, PDE2, PDE3, PDE10, PDE11).[5–7] All PDEs contain a conserved catalytic domain of approximately 270 amino acids with 18%–46% sequence identity between families at the C-terminus. Regulatory domains that vary widely among the PDE families flank the catalytic core and include regions that autoinhibit the catalytic domains as well as targeting sequences that control subcellular localization.[8,9]

The crystal structures of the catalytic domains of many PDE families have been determined recently. These include PDE1B,[10] PDE2A,[11] PDE3B,[12] PDE4B,[10,13–16] PDE4D,[10,15–20] PDE5A,[10,15,20,21] PDE7B,[22] and PDE9A[23] (Figure 29.1). In addition, the structure of the heterotetrameric photoreceptor PDE6 has been determined by electron microscopy to about 29 Å resolution.[24] Although these studies suggest an arrangement for the overall association of individual monomers as well as the relationship between the regulatory and catalytic domains, the lack of details has prevented the direct comparison with the other crystal structures at atomic resolution.

There is a dearth of structural information on the regulatory domains in sharp contrast to the abundance of catalytic domain structures. The structure of the cGMP-binding and stimulated PDEs, *Anabaena* adenylyl cyclases and *Escherichia coli* FhlA protein (GAF domain) of PDE2A is the only crystal structure of regulatory domains available to date.[25] Based on this structure, a head-to-head model of PDE2 dimerization through the GAF domain has been proposed.[25] It has been demonstrated later that isolated GAF-A or GAF-B domains from PDE5 also form homodimers.[26] However, the crystal structure of the GAF domain from a cyanobacterial adenylyl cyclase has revealed a head-to-tail dimerization mode.[27] The crystal structure of the full-length PDE2A or any other GAF domain containing PDEs would shed light on this issue. Indeed, the structure of full-length PDEs is most desirable for understanding functional subtleties and inhibitor binding. However, the full-length PDE structure has eluded us so far due to various reasons including the difficulty of obtaining sufficient amounts of soluble full-length PDEs.

29.2 PHOSPHODIESTERASE CATALYTIC DOMAIN STRUCTURES

All the crystal structures of the PDE catalytic domain deposited in the Protein Data Bank to date are summarized in Table 29.1 (comments about particular structures later in the text will be annotated with the PDB identifiers). The PDE catalytic domains all share a compact α-helical structure consisting of 16 helices that can be divided into three subdomains (Figure 29.1). The active site is formed in a cradle at the junction of the three subdomains and is lined with highly conserved residues (Figure 29.2A). Located at the bottom of the active-site pocket are two metal ions coordinated by residues from all these three subdomains. These two metal ions seem to act as a linchpin that holds these three domains together. The first metal ion is a zinc ion (Zn^{2+}), which is coordinated by two histidines and two aspartates (H617[5A], H653[5A], D654[5A], D764[5A] in PDE5A) and two water molecules. These two histidines and two

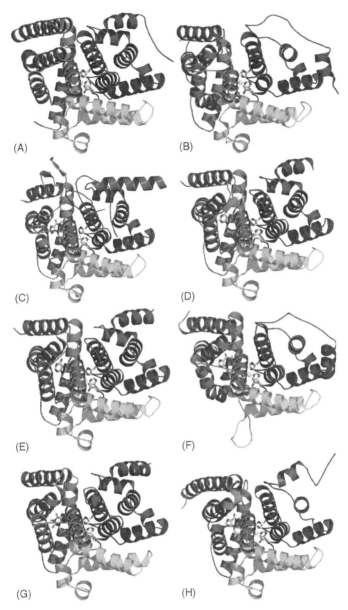

FIGURE 29.1 (See color insert following page 274.) PDE catalytic domain structures. Crystal structures of all known PDE family members are represented by ribbon diagrams with the three subdomains colored in blue (α1–α7), green (α8–α11), and red (α12–α16), respectively. (A) PDE1B, (B) PDE2A, (C) PDE3B, (D) PDE4B, (E) PDE4D, (F) PDE5A, (G) PDE7A, and (H) PDE9A.

aspartates are absolutely conserved across all the PDE family members. The second metal ion is most probably a magnesium ion (Mg^{2+}), which is coordinated by the same aspartate ($D654^{5A}$) that also coordinates to Zn^{2+}, and five water molecules, one of which bridges Mg^{2+} and Zn^{2+}. These two metal ions adopt a near-ideal octahedral coordination geometry (Figure 29.2B).

TABLE 29.1
Summary of PDE Catalytic Domain Crystal Structures

No.	PDE Isoform	Bound Inhibitor	Resolution (Å)	PDE Code
1	PDE1B	Apo	1.77	1TAZ
2	PDE2A	Apo	1.70	1Z1L
3	PDE3B	Dihydropyridazine	2.40	1SO2
4	PDE3B	IBMX	2.90	1SOJ
5	PDE4B	(R)-Mesopram	1.92	1XM6
6	PDE4B	(R)-Rolipram	2.40	1XMY
7	PDE4B	(R,S)-Rolipram	2.00	1RO6
8	PDE4B	(R,S)-Rolipram	2.31	1XN0
9	PDE4B	Cl-PhPCEE	2.40	1Y2H
10	PDE4B	NO$_2$-PhPCEE	2.55	1Y2J
11	PDE4B	8-Br-AMP	2.13	1RO9
12	PDE4B	AMP	2.00	1ROR
13	PDE4B	AMP	2.15	1TB5
14	PDE4B	Apo	1.77	1F0J
15	PDE4B	Cilomilast	2.19	1XLX
16	PDE4B	Filaminast	2.06	1XLZ
17	PDE4B	Piclamilast	2.31	1XM4
18	PDE4B	Roflumilast	2.30	1XMU
19	PDE4B	Sildenafil	2.28	1XOS
20	PDE4B	Vardenafil	2.34	1XOT
21	PDE4D	(R)-Rolipram	2.30	1Q9M
22	PDE4D	(R,S)-Rolipram	1.60	1TBB
23	PDE4D	(R,S)-Rolipram	2.00	1OYN
24	PDE4D	NH$_2$-PhPCEE	2.10	1Y2E
25	PDE4D	MeO-PhPCEE	1.70	1Y2D
26	PDE4D	NO$_2$-PhPCEE	1.36	1Y2K
27	PDE4D	PCEE	1.40	1Y2B
28	PDE4D	PhPCEE	1.67	1Y2C
29	PDE4D	AMP	1.63	1TB7
30	PDE4D	AMP	2.30	1PTW
31	PDE4D	Cilomilast	1.55	1XOM
32	PDE4D	IBMX	2.10	1ZKN
33	PDE4D	L-869298	2.00	2FM0
34	PDE4D	L-869299	2.03	2FM5
35	PDE4D	Piclamilast	1.72	1XON
36	PDE4D	Roflumilast	1.83	1XOQ
37	PDE4D	Zardaverine	1.54	1XOR
38	PDE4D	Zardaverine	2.90	1MKD
39	PDE5A	Apo	2.10	1T9R
40	PDE5A	GMP	2.00	1T9S
41	PDE5A	IBMX	2.05	1RKP
42	PDE5A	Sildenafil	1.30	1TBF
43	PDE5A	Sildenafil	2.30	1UDT
44	PDE5A	Tadalafil	1.37	1XOZ
45	PDE5A	Tadalafil	2.83	1UDU
46	PDE5A	Vardenafil	1.79	1XP0
47	PDE5A	Vardenafil	2.50	1UHO
48	PDE7A	IBMX	1.67	1ZKL
49	PDE9A	IBMX	2.23	1TBM

Abbreviations: PCEE, 3,5-dimethyl-1H-pyrazole-4-carboxylic acid ethyl ester; PhPCEE, 3,5-dimethyl-1-phenyl-1H-pyrazole-4-carboxylic acid ethyl ester; Cl-PhPCEE, 1-(2-chloro-phenyl)-3,5-dimethyl-1H-pyrazole-4-carboxylic acid ethyl ester; NO$_2$-PhPCEE, 3,5-dimethyl-1-(3-nitro-phenyl)-1H-pyrazole-4-carboxylic acid ethyl ester; NH$_2$-PhPCEE, 1-(4-amino-phenyl)-3,5-dimethyl-1H-pyrazole-4-carboxylic acid ethyl ester; MeO-PhPCEE, 1-(4-methoxy-phenyl)-3,5-dimethyl-1H-pyrazole-4-carboxylic acid ethyl ester.

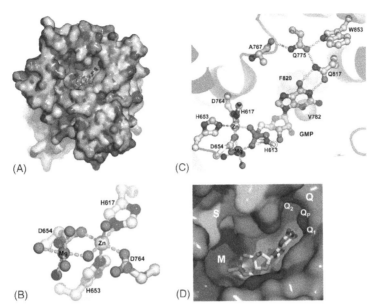

FIGURE 29.2 (See color insert following page 274.) PDE active sites. (A) The active-site pocket is shown on the PDE5A + GMP structure. (B) The metal ions Zn^{2+} and Mg^{2+} are both octahedrally coordinated. (C) Three clusters of residues that are involved in nucleotide recognition (A767, W853, Q817, Q775), formation of a hydrophobic clamp (F820 and V782), and hydrolysis are illustrated on PDE5A + GMP cocrystal structure. (D) Three regions (M, Q, and S) in the active site are indicated on the surface of PDE5A + GMP cocrystal structure.

The active site contains 12 out of 17 absolutely conserved residues in the catalytic domain across all 21 PDE gene family members.[13] There are three clusters of residues in the active site that are responsible for three distinct functions of PDEs (Figure 29.2C)[10]: (1) nucleotide recognition: a cluster of residues that controls the orientation of the amide group of an invariant glutamine for selective binding to either cAMP or cGMP in cAMP- or cGMP-selective PDEs; (2) hydrophobic clamp: two highly conserved hydrophobic residues (such as F820[5A] and V782[5A]) that sandwich the planar purine ring for substrate binding; and (3) hydrolysis: a cluster of absolutely conserved residues near the dimetal center that are responsible for cyclic nucleotide hydrolysis. The key residues involved in metal binding and hydrolysis, hydrophobic clamp, and nucleotide recognition in all 21 PDE gene family members are summarized in Table 29.2.

The active site can be subdivided into three pockets (Figure 29.2D)[15]: a metal-binding pocket (M-pocket), a solvent-filled side pocket (S-pocket), and a pocket containing the purine-selective glutamine and hydrophobic clamp (Q-pocket). The M-pocket contains the dimetal ions and highly conserved hydrophobic and polar residues that coordinate the metal ions. The S-pocket consists mainly of hydrophilic amino acids and is filled with a network of water molecules in most of the inhibitor complexes. The Q-pocket can be further divided into three distinct areas: a narrow passage (Q_P) formed by the P-clamp that only a planar structure would fit in and reach through to interact with an invariant glutamine that contacts the purine, flanked by two asymmetrical hydrophobic subpockets (Q_1 and Q_2). Unique to PDE5 is a lid region formed by a flexible H-loop, which is not shown in Figure 29.2D.[20,21]

PDE1B contains a long-sequence insertion compared to other PDEs, occurring in the loop connecting helices 15 and 16 (residues K442–V479). This loop region is mostly disordered. Consequently, residues S454–V471 in this region can be deleted to produce crystals with

TABLE 29.2

Structure-Based Sequence Correspondences for the Key Residues Involved in Metal Binding and Hydrolysis, Hydrophobic Clamp, and Nucleotide Recognition

		Key Residues											
Specificity	PDE	Metal Binding 1	Metal Binding 2	Metal Binding 3	Metal Binding 4	Clamp 1	Clamp 2	Glutamine	Position 1	Position 2	Position 3		
Dual	PDE1A	H223	H259	D260	D366	L384	F420	Q417	H369	H377	W515		
Dual	PDE1B	H227	H263	D264	D370	L388	F424	Q421	H373	H381	W496		
Dual	PDE1C	H232	H268	D269	D376	L394	F430	Q427	H379	H387	W521		
Dual	PDE2A	H660	H696	D697	D808	I826	F862	Q859	D811	T819	W895		
Dual	PDE3A	H756	H836	D837	D950	I968	F1004	Q1001	G953	H961	W1086		
Dual	PDE3B	H741	H821	D822	D937	I955	F991	Q988	G940	H948	W1072		
Dual	PDE10A	H519	H553	D554	D664	I682	F719	Q716	S667	T675	W752		
Dual	PDE11A	H224	H260	D261	D372	V390	W428	Q425	A375	S383	W461		
cAMP	PDE4A	H198	H234	D235	D352	I370	F406	Q403	N355	Y363	Y440		
cAMP	PDE4B	H238	H274	D275	D392	I410	F446	Q443	N395	Y403	Y480		
cAMP	PDE4C	H392	H428	D429	D546	I564	F600	Q597	N549	Y557	Y634		
cAMP	PDE4D	H164	H200	D201	D318	I336	F372	Q369	N321	Y329	Y406		
cAMP	PDE7A	H216	H252	D253	D362	V380	F416	Q413	N365	S373	W451		
cAMP	PDE7B	H177	H213	D214	D323	V341	F377	Q374	N326	S334	W413		
cAMP	PDE8A	H488	H524	D525	D654	I672	F709	Q706	N657	C665	W741		
cAMP	PDE8B	H84	H120	D121	D246	I264	F301	Q298	N249	C257	W333		
cGMP	PDE5A	H617	H653	D654	D764	V782	F820	Q817	A767	Q775	W853		
cGMP	PDE6A	H563	H599	D600	D716	V734	F776	Q773	A723	Q731	W809		
cGMP	PDE6B	H561	H597	D598	D714	V732	F774	Q771	A721	Q729	W807		
cGMP	PDE6C	H566	H602	D603	D719	V737	F779	Q775	A726	Q734	W812		
cGMP	PDE9A	H256	H292	D293	D402	L420	F456	Q453	N405	A413	Y490		

dramatically improved diffraction quality without affecting the overall structure and enzymatic activity (1TAZ).[10]

PDE3B has two long inserts, one is a 44-residue insert (R755–S798) between H6 and H7 that is PDE3-specific, the other is a 52-residue long insert (G1007–I1057) between H15 and H16 that is similar to one found in PDE1. Residues C767–H781 in the first loop and residues A1016–K1052 in the second loop are not visible in the electron density maps in the PDE3B crystal structures (1SO2, 1SOJ).[12] These two long and flexible loops in PDE3B could have caused the packing of multiple copies of molecules in the asymmetric unit of the crystals and the low diffraction resolution of the PDE3B crystal structures. It is conceivable that the deletion of these two flexible loops might produce crystals of higher diffraction quality, as is the case with the deletion of the loop in PDE1B.[10]

PDE5A has a flexible loop region that can adopt various conformations differing dramatically from the α-helical conformation found in all other PDEs (Figure 29.3A). For example, the corresponding residues L293–H307 in PDE4B form helix-α9 and part of helix-α10. In the apo-PDE5A structure (1T9R), this loop (L672–H685; colored yellow in Figure 29.3A) extends into the active site and blocks substrate and inhibitor binding.[10] In cocrystal structures of PDE5A complexed with sildenafil (Viagra, 1UDT), vardenafil (Levitra, 1UHO), and tadalafil (Cialis, 1UDU), this loop (colored blue in Figure 29.3A) forms part of the lid region that extends out and caps the active site upon inhibitor binding, providing further evidence for the conformational flexibility of this loop.[21] In the cocrystal structure of PDE5A in complex with 3-isobutyl-1-methylxanthine (IBMX, 1RKP), this loop (N661-H678; colored red in Figure 29.3A) adopts yet another conformation that falls into approximately the same region as helices α9 and α10 in PDE4B.[20] This loop conformation might be affected by the crystal contact that it makes with a neighboring molecule in the cocrystal structure of PDE5A + IBMX. A chimeric PDE5A protein was made in which this loop has been replaced with the corresponding region from PDE4B (1TBF, 1XP0, 1XOZ).[10] The swapped region from PDE4B adopts the same conformation in the PDE5A chimeric protein as in PDE4B, and it has no impact on the rest of the PDE5A structure (colored green as background in Figure 29.3A). Moreover, the chimeric PDE5A protein produced crystals that diffracted to a much higher resolution than the wild-type PDE5A protein in complex with the same inhibitor (Table 29.1).[10,15,21]

In the first apostructure of PDE4B (1F0J), an additional helix (H17, colored blue in Figure 29.3B) formed by residues N496–F508 was observed to pack against a neighboring molecule in a manner reminiscent of domain swapping.[13] This extra helix in the C-terminus was only observed in one of the two molecules in the asymmetric unit. In the majority of subsequent crystal structures of PDE4B in complex with AMP and various inhibitors, this H17 was not observed. The only other PDE4B structure where the H17 has been observed is that of PDE4B in complex with mesopram (1XM6).[15] In that structure, H17 (C499–L513; colored red in Figure 29.3B) is packed against the catalytic core domain in the same molecule and partially covers the active site making contact with the bound inhibitor. It seems that these residues corresponding to H17 are highly flexible, at least in the context of the isolated catalytic domain. However, it is also conceivable that this H17 region will be ordered in the full-length PDE4B, thus contributing to substrate and inhibitor binding.

29.3 NUCLEOTIDE SELECTIVITY

PDEs achieve their nucleotide selectivity through a glutamine switch mechanism.[10] The amide group of an invariant glutamine adopts one orientation to interact with cAMP and switches to another orientation by flipping 180° to interact with cGMP. This glutamine switch is controlled by an intricate network of hydrogen bonds near this absolutely conserved glutamine.

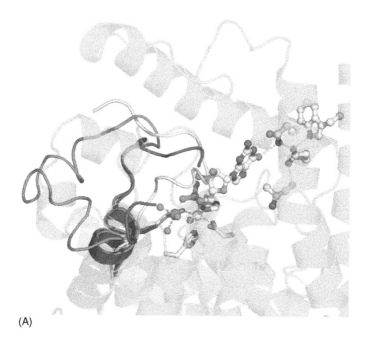

(A)

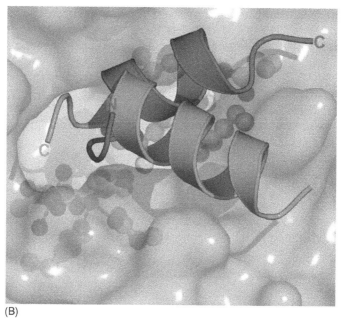

(B)

FIGURE 29.3 (See color insert following page 274.) Conformational flexibility. (A) The H-loop in PDE5A adopts various conformations depending on the context of bound inhibitor and crystal packing. Blue: PDE5A + sildenafil, red: PDE5A + IBMX, and yellow: PDE5A apostructure. The background in green is the PDE5A chimeric structure with this loop replaced by the corresponding regions in PDE4B, which is helix-8 and helix-9. The bound GMP, the dimetal ions, and several active-site residues are also shown to provide a reference point to the H- loop. (B) The helix-17 from PDE4B apostructure is shown blue and the helix-17 from PDE4B + mesopram costructure is shown in red. The N- and C-terminus of the helix-17 are labeled.

In the cAMP-selective PDEs, such as PDE4B (1TB5), the absolutely conserved Q443 forms a bidentate H-bond with the adenine moiety (Figure 29.4G). This orientation of Q443 is stabilized by H-bonding of Oε to the phenolic hydroxyl group Oη of Y329. The adenine ring also forms a bidentate H-bond with N321 in PDE4B, further stabilizing the nucleotide binding.

By contrast, in the cGMP-selective PDEs, such as PDE5A (1T9S), the orientation of the key glutamine Q817 is switched from that in PDE4B to allow H-bonding specific to the guanine ring of 5′-GMP (Figure 29.4H). This orientation of Q817 is constrained by its H-bonding with Q775. The orientation of Q775 is in turn determined by the two H-bonds that it forms with A767 and W853. This intricate network of H-bonds determines the orientation of the γ-amide group of Q817 that is favorable for cGMP but unfavorable for cAMP binding in PDE5. The role of Q775 in stabilizing the amide side chain orientation of Q817 in a guanine-preferring mode has been confirmed by a recent mutagenesis study of PDE5A, since the Q775A mutation has almost eliminated the cyclic nucleotide selectivity.[28]

The dual specificity of some PDEs such as PDE1B (1TAZ) is due to the rotational freedom of the amide group of the invariant glutamine. The key glutamine, in this case Q421, can adopt both of the orientations observed in the PDE4 and PDE5 structures, since there is no H-bond network to constrain the orientation of its γ-amide group in this enzyme (Figure 29.4A and Figure 29.4B).

This glutamine switch mechanism can account for the cyclic nucleotide selectivity for all the PDEs.[10] The key residues involved in cyclic nucleotide recognition identified through sequence alignment have revealed a pattern that fits the glutamine switch mechanism. All the cAMP-selective PDEs have an asparagine residue at position 1 (corresponding to N395[4B] in Figure 29.4G) that can form a bidentate H-bond with adenine but not with guanine (Table 29.2). All the cGMP-selective PDEs have a glutamine at position 2 (corresponding to Q775[5A] in Figure 29.4H) with the exception of PDE9A (Table 29.2). This glutamine anchors the orientation of the invariant glutamine (Q817[5A] in Figure 29.4H) to form a bidentate H-bond with guanine but not with adenine. All the dual-specific PDEs have neither asparagine at position 1 nor glutamine at position 2 to confer cyclic nucleotide selectivity (Table 29.2).

In PDE2A (1Z1L), an important purine recognition residue is Q859. The residue D811 is at position 1, T819 is at position 2, and W895 is at position 3 (Table 29.2, Figure 29.4C and Figure 29.4D). None of these neighboring residues form hydrogen bond with Q859 and therefore it can form hydrogen bonds with either cAMP or cGMP by switching the orientation of its side chain.[11] T819 might interact with W895 through a hydrogen bond and van der Waals contact. However, T819 is too far from Q859 to form a hydrogen bond. Y827 at the opposite site of T819 can form hydrogen bond with Q859 to stabilize its amide in both orientations. These have provided a structural explanation for the dual specificity of PDE2A.

In PDE3B (1SO2, 1SOJ), the purine recognition residue is Q988. The residue G940 is at position 1, H948 is at position 2, and W1072 is at position 3 (Table 29.2, Figure 29.4E and Figure 29.4F).[12] H948 forms a hydrogen bond with W1072 and this positions a carbon atom in the imidazole side chain of H948 at a closest distance of 3.4 Å from Q988, which is incapable of H-bonding. Although it is conceivable that a different side chain orientation could allow H948 to form a hydrogen bond with Q988 at the expense of the hydrogen bond between H948 and W1072, this interaction places no restriction on the orientation of the amide side chain of Q988 since the proton on the imidazole side chain of H948 can be on either nitrogen and there are no interactions to restrict the location of this proton. Consequently, the amide side chain of Q988 can freely rotate to form hydrogen bonds with either cAMP or cGMP. G940 makes no interactions with the cyclic nucleotide due to lack of a side chain.

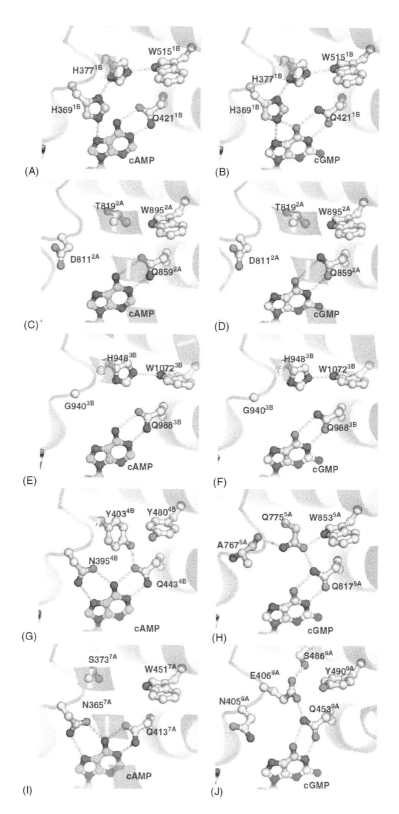

FIGURE 29.4

In PDE7A (1ZKL), the purine recognition residue is Q413. The residue N365 is at position 1, S373 is at position 2, and W451 is at position 3 (Table 29.2, Figure 29.4I).[22] Q413 forms hydrogen bond with S377 instead of S373. N365 forms bidentate H-bonds providing selectivity for adenine.

In PDE9A (1TBM), the most important purine recognition residue is Q453. The residue N405 is at position 1, A413 is at position 2, and Y490 is at position 3 based on sequence alignment (Table 29.2). This pattern is similar to the cAMP-selective PDEs. The mystery of cGMP selectivity for PDE9A was unveiled when the cocrystal structure of PDE9A bound to IBMX was reported.[23] PDE9A uses the same concept of the glutamine switch mechanism as that for PDE5A to gain selectivity for cGMP except the key residues that anchor the orientation of the invariant glutamine are different from those revealed from sequence alignment (Figure 29.4J). The mutation of Q775 in PDE5A to A413 in PDE9A has created a void in the structure that was filled by the side chain of E406, which forms H-bonds with S486 and the amide nitrogen of Q453. Therefore, E406 serves the role of anchoring the orientation of the invariant glutamine Q453 that is favorable for cGMP but not for cAMP. The structure also revealed that N405, which would be favorable for cAMP over cGMP, has swung away from the nucleotide and is not within H-bond distance from the nucleotide.

29.4 HYDROPHOBIC CLAMP

The cocrystal structures of many inhibitors in complex with various PDEs have revealed a hydrophobic clamp that anchors these inhibitors in the active site (Figure 29.5).[10,15] This hydrophobic clamp consists of a pair of highly conserved hydrophobic residues, where one jaw of the clamp is a phenylalanine in all PDE family members (e.g., F446[4B], F372[4D], F820[5A]) except PDE11, where the corresponding residue is W428[11A] (Table 29.2). A variable, but always hydrophobic, residue forms the opposite jaw of the clamp and is either valine, leucine, or isoleucine in all of the PDEs (e.g., I410[4B], I336[4D], and V782[5A]) (Table 29.2). In virtually all of the inhibitor structures, the structural elements of the hydrophobic clamp are conserved, and focused on the aromatic ring of the inhibitors presenting the H-bonding residues to the invariant glutamine. The phenylalanine engages this primary aromatic ring of the inhibitor from above with an offset face-on-face interaction[29] whereas the γ-carbon of the β-branched residue below is centered under the same ring. In the catechol diether-containing inhibitors, the face-on-face interaction centers the carbon bearing the *meta*-alkoxy substituent under the phenylalanine whereas in sildenafil and in similar analogs the polarized biaryl-linked carbon of the bicyclic aromatic group occupies a similar position. The substituents at these two types of positions thus serve a dual role, both to optimize local hydrophobic interactions with the enzymes and to modulate the position and polarity of the hydrophobic clamp. This subtle variation affords some of the inhibitors their selectivity over different PDE family members. However, the sequence variation among additional residues lining the hydrophobic pocket affords greater selectivity to the inhibitor. Interestingly, given the bicyclic nature of the cAMP

FIGURE 29.4 (See color insert following page 274.) Nucleotide recognition by PDEs. Key residues involved in cyclic nucleotide recognition are highlighted by ball-and-stick models. The cAMP and cGMP are modeled based on the crystal structures of respective PDE catalytic domain structures either apo or in complex with AMP, GMP, or inhibitors. The carbon atoms in cAMP and cGMP are colored in green and yellow, respectively. (A) cAMP recognition by PDE1B, (B) cGMP recognition by PDE1B, (C) cAMP recognition by PDE2A, (D) cGMP recognition by PDE2A, (E) cAMP recognition by PDE3B, (F) cGMP recognition by PDE3B, (G) cAMP recognition by PDE4B, (H) cGMP recognition by PDE5A, (I) cAMP recognition by PDE7A, and (J) cGMP recognition by PDE9A.

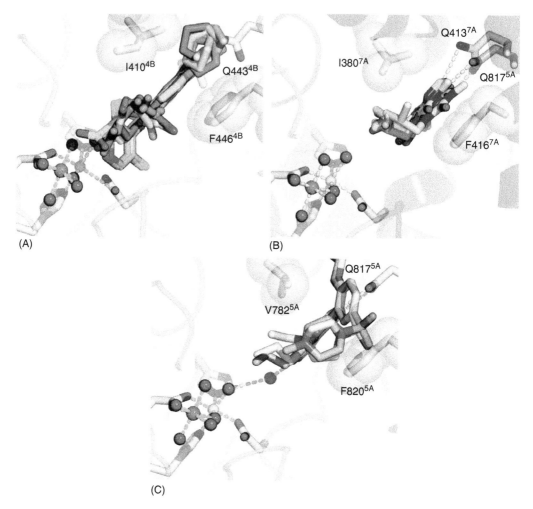

FIGURE 29.5 (See color insert following page 274.) Hydrophobic clamp. (A) Catechol diether class of PDE4 inhibitors is sandwiched by two conserved hydrophobic residues, I410^{4B} and F446^{4B}. (B) IBMX is sandwiched by two conserved hydrophobic residues, I380^{7A} and F416^{7A}. The cocrystal structures of IBMX bound to PDE3B, PDE4D, PDE5A, PDE7A, and PDE9A have been superimposed. (C) Sildenafil and vardenafil are sandwiched by two conserved hydrophobic residues, V782^{5A} and F820^{5A}.

and cGMP substrates, all of the hydrophobic interactions appear to be focused on the primary aromatic ring. In the cases of the inhibitors containing a bicyclic core, the second (fused) ring predominantly contributes productive hydrophobic interactions through the presentation of various substituents.

29.5 STRUCTURAL BASIS OF INHIBITOR POTENCY

The cocrystal structures of many inhibitors in complex with various PDEs have revealed that they share a highly conserved binding mode despite their drastically different chemotypes. There are two common features of inhibitor binding to PDEs[15]: (1) a planar ring structure of the inhibitor that is held tightly in the active site by a pair of hydrophobic residues that form a hydrophobic clamp and (2) hydrogen bond (H-bond) interactions with an

invariant glutamine residue that is essential for nucleotide selectivity.[10] These two common features can be used to define the scaffold of PDE inhibitors.

29.5.1   3-Isobutyl-1-Methylxanthine Binding to Various Phosphodiesterases

IBMX is a purine analog and is a weak and nonselective PDE inhibitor. Crystal structures of IBMX with PDE3B, PDE4D, PDE5A, PDE7A, and PDE9A have been reported (1SOJ, 1ZKN, 1RKP, 1ZKL, 1TBM).[12,20,22,23] Superposition of these cocrystal structures revealed that IBMX, a xanthine derivative and a purine analog, is sandwiched by the hydrophobic clamp and forms a hydrogen bond with the purine recognition glutamine in all the PDEs that have been studied (Figure 29.6B). IBMX binds at the common core of the active-site pocket shared by all the PDEs and interacts with highly conserved residues, which is commensurate with its non-selectivity. Moreover, IBMX flips its orientation when bound to cAMP-specific PDEs versus cGMP-specific PDEs in order to adapt to the H-bond characteristics of the nucleotide recognition glutamine. These features make IBMX a relatively weak and nonselective PDE inhibitor compared to many of the newer second and third generation selective inhibitors.

29.5.2   PDE4 Inhibitors

All the reported inhibitors bound to PDE4 represent four scaffold classes: catechol, xanthane, pyrazole, and purine analogs. The majority of these cocrystal structures are dialkoxyphenyl (catechol diether) derivatives[14,15,17,19] (Figure 29.6A) and only one is a xanthane derivative, IBMX[20] (Figure 29.6B). The superposition of the cocrystal structures of the dialkoxyphenyl family of compounds including zardaverine (1MKD, 1XOR), rolipram (1XMY, 1RO6, 1XN0, 1OYN, 1Q9M, 1TBB), filaminast (1XLZ), mesopram (1XM6), cilomilast (1XLX, 1XOM), roflumilast (1XMU, 1XOQ), piclamilast (1XM4, 1XON) with PDE4B or PDE4D revealed that the scaffold is a catechol that makes the H-bond with the purine-selective glutamine and is also sandwiched by the hydrophobic clamp (Figure 29.5A and Figure 29.6A). The catechol scaffold superposed extremely well in all of these cocrystal structures whereas the substituents showed significant variations in their binding conformation as well as in the residues with which they interact. The various substituents on the catechol scaffold explore the deep pocket close to the metal-binding site, and how well they form interactions with residues lining this pocket determines their relative binding affinity. The relatively smaller pyrrolidinone substituent in rolipram has resulted in a relatively lower binding affinity (570 nM in the case of PDE4B). The more potent dialkoxyphenyl compounds have larger substituents that could form more favorable interactions with residues lining the relatively large M-pocket. The most potent compounds, such as roflumilast and piclamilast, reach deep into the M-pocket not only interacting with residues near the metal ion but also forming a H-bond with a water molecule coordinated to the metal ion. A surprise finding from these cocrystal structures is that none of the inhibitors, with the single exception of zardaverine, directly interact with the metal ions, contrary to what was expected by many people in the field.

Recently, cocrystal structures of two enantiomeric catechol diether compounds ((+)-L-869298 and (−)-L-869299) in complex with PDE4D have been reported (2FM0, 2FM5).[30] Despite the 107-fold difference in binding affinity to PDE4D, both enantiomers interact with the same sets of residues in the active site, reminiscent of the similarity in binding of enantiomeric roliprams in the active site of PDE4B and PDE4D.[10,14,15,19]

29.5.3   PDE5 Inhibitors

The inhibitors that have cocrystal structures with PDE5A include sildenafil (1TBF, 1UDT), vardenafil (1XP0, 1UHO), tadalafil (1XOZ, 1UDU), and IBMX (1RKP).[10,15,20,21] The

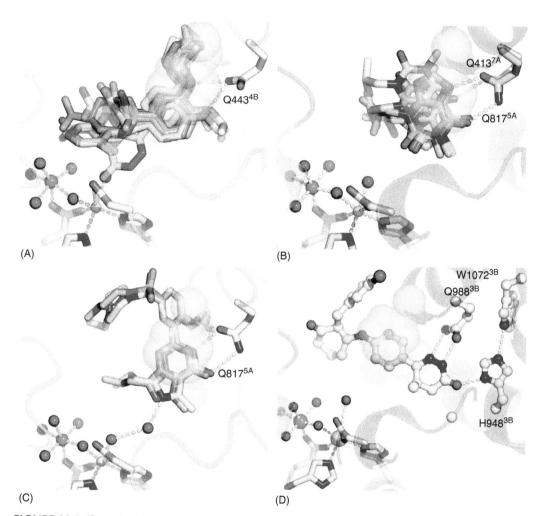

FIGURE 29.6 (See color insert following page 274.) Inhibitor potency. The H-bond between the invariant glutamine and inhibitors is indicated by dashed yellow lines. The hydrophobic clamp is indicated by semitransparent spheres. (A) Catechol diether class of PDE4 inhibitors bound to PDE4B. (B) IBMX bound to PDE3B, PDE4D, PDE5A, PDE7A, and PDE9A. IBMX binds to PDE4D and PDE7A in one orientation. IBMX binds to PDE5A and PDE9A in another orientation, which is flipped from that bound to PDE4D and PDE7A. This flipping of IBMX enables it to form bidentate H-bonds that match the orientation of the purine-selective glutamine. IBMX binds to PDE3B in an orientation similar to that in PDE5A, although it is conceivable that it might also bind in a flipped orientation. (C) Sildenafil and vardenafil bound to PDE5A. (D) The dihydropyridazinone class of compound bound to PDE3B.

chemical structures of sildenafil and vardenafil are very similar. The scaffold of sildenafil (pyrazolopyrimidinone) and vardenafil (imidazolotriazinone) mimics guanine and has the same hydrogen bond donor and acceptor characteristics. Consequently, sildenafil and vardenafil both exploit the same interactions with residues involved in nucleotide recognition.[10] The amide moiety of the pyrazolopyrimidinone group in sildenafil and the amide moiety of the imidazolotriazinone group in vardenafil both form a bidentate H-bond with the γ-amide group of Q817[5A] (Figure 29.6C). The heterocyclic nitrogen of the pyrazole moiety in sildenafil and the imidazole moiety in vardenafil forms a hydrogen bond with a water molecule, which in turn interacts with one of the waters coordinating Zn^{2+}. A large proportion of the surface

area of sildenafil and vardenafil is buried between the hydrophobic clamp residues $V782^{5A}$ and $F820^{5A}$, utilizing π–π-stacking interactions of the phenylalanine side chain to that of the pyrazolopyrimidinone and the imidazolotriazinone moiety (Figure 29.6C). Additional, more remote, hydrophobic contacts to the five-membered ring of the bicyclic core are provided by $Y612^{5A}$ and $L765^{5A}$ (residues are not shown in figures to avoid cluttering). The propyl group from the five-membered ring also forms hydrophobic contacts with $L725^{5A}$ and $F786^{5A}$ (not shown). The ethoxyphenyl group makes hydrophobic contacts with $V782^{5A}$, $F786^{5A}$, $A779^{5A}$, $I813^{5A}$, and $M816^{5A}$ (not shown). The ethylpiperazine group in vardenafil makes similar interactions with PDE5A as the methylpiperazine of sildenafil. The piperazine ring adopts a slightly different orientation that is probably due to the relative dearth of specific interactions between this group and the protein. The methylpiperazine in sildenafil and ethylpiperazine in vardenafil interact with the lid region of PDE5A, which is composed of residues Y664, M816, A823, and G819.[21] Y664 sits on an extended loop reaching towards the active site and forms a lid over the pocket. This loop in PDE5A seems to be similar to the H17 in PDE4B such that they both are potentially able to cover part of the active site. They are both flexible and multiple conformations have been observed, indicating a functional role. Indeed, these two structural elements have both been found to interact with several of the inhibitors and may even play a role in the selectivity of these inhibitors. This loop adopts a different conformation in the PDE5A complex with IBMX.[20] Moreover, this loop was found to dip into the active-site pocket, blocking substrate and inhibitor binding in the apostructure of PDE5A.[10]

Tadalafil binds to PDE5A very differently from sildenafil and vardenafil, due to its different chemical structure (Figure 29.7C). The methyne at the bridgehead between the two six-membered rings is the focal point of the hydrophobic clamp interactions with $F820^{5A}$ and $V782^{5A}$. In addition to the interactions with the hydrophobic clamp, the nearly flat four-ring moiety of pyrazinopyridoindoledione is held in place through hydrophobic interactions with side chains of $I768^{5A}$, $I778^{5A}$, $F786^{5A}$, and $L804^{5A}$ (not shown). The indole core is sandwiched between $F786^{5A}$, $L804^{5A}$, and $F820^{5A}$. The methylenedioxyphenyl group of tadalafil occupies the Q_2-pocket (Figure 29.7C) and makes hydrophobic interactions with $A783^{5A}$, $F786^{5A}$, $F787^{5A}$, $L804^{5A}$, $I813^{5A}$, and $M816^{5A}$ (not shown). Tadalafil forms only one H-bond with $Q817^{5A}$ through its NH group on the indole ring. Tadalafil makes neither direct nor water-mediated interactions with the metal ions. Despite the lack of H-bonded interactions with the protein, tadalafil makes many hydrophobic interactions that contribute to its high potency. Moreover, the rigid chemical structure of tadalafil with only one nonterminal rotatable bond also contributes to its high-binding affinity, since it loses less entropy upon protein binding compared to sildenafil and vardenafil.

29.5.4 PDE3 Inhibitors

The cocrystal structure of PDE3B in complex with a dihydropyridazine inhibitor (1SO2) showed that the inhibitor is bound in an extended conformation with the central aromatic ring overlapping with the purine ring of cyclic nucleotides (Figure 29.6D).[12] The exocyclic oxygen interacts with the side chains of T952 and H948 (not shown). The exocyclic methyl binds in a hydrophobic pocket formed by residues I938–P941 (not shown). The benzyl ring is sandwiched by the hydrophobic clamp, F991 and I955. 1,3-Hexadione sits approximately in the same area as the isobutyl of IBMX. The 2-aryl substituent extends into a nearby hydrophobic pocket formed by the side chains of F959, P975, F976, M977, L987, S990, and F991 (not shown). The dihydropyridazinone moiety is present in a very tight pocket formed by the main chain atoms of I938-P941 from H13, K950-T952 from H14, and the side chains of H948 (not shown). The dihydropyridazinone nitrogen forms hydrogen bonds to the

amide side chain of Q988 similar to those seen in the PDE3B and IBMX complex (1SOJ), but the amide side chain Q988 assumes the opposite conformation, similar to those observed in the PDE4 structures. This is the first example where both conformations of the amide group of purine recognition glutamine have been experimentally observed in dual-selective PDEs. The amide conformation of Q988 in PDE3B and IBMX complex is compatible with cGMP binding whereas the amide conformation of Q988 in PDE3B and the dihydropyridazine inhibitor complex is compatible with cAMP binding, thus demonstrating the dual specificity of PDE3B.

The analysis of the cocrystal structures of inhibitor PDE complexes has shown that the interactions with residues lining the two hydrophobic subpockets near the invariant purine-selective glutamine are important for inhibitor binding. The inhibitor potency can be further increased by exploring interactions with residues near the dimetal ion center as well as through the formation of water-mediated interactions with the metal ions.[15]

29.6 STRUCTURAL BASIS OF INHIBITOR SELECTIVITY

The catalytic domains of the 21 PDE genes share a sequence identity ranging from 18% to 87%. The lowest sequence identity is between PDE8A and PDE10A. The highest sequence identity is between the four PDE4 subtypes. Inhibitors tend to have cross-activity among closely related PDEs and especially between subtypes. This nonselectivity has contributed to the narrow therapeutic window and side effects of some PDE inhibitors. For example, the nonselectivity of PDE4 inhibitors between the four different subtypes may contribute, at least partially, to the most common side effects of nausea and emesis and thereby limit the therapeutic windows of these inhibitors as anti-inflammatory agents.[31] The popular antier-ectile dysfunction drugs, Viagra (sildenafil) and Levitra (vardenafil), are potent inhibitors of PDE5A and also cross-react with the closely related PDE6 (41%) and PDE11 (46%). Cialis (tadalafil) is also a potent PDE5 inhibitor that cross-reacts with PDE11, but not with PDE6. It is thought that this cross-reactivity with other PDEs is responsible for side effects such as blue-tinged vision and back and muscle pain that were experienced by some patients treated with these drugs.[32] Clearly, understanding of the structural basis of inhibitor selectivity at the atomic level would greatly facilitate the design of more selective inhibitors with reduced side effects and improved pharmacological profiles.

29.6.1 PDE4 and PDE5 Selectivity

There are subtle but significant differences in the active sites of PDE4B and PDE5A that could be exploited to achieve family selectivity for inhibitors. The opposite orientation of the purine-selective glutamine between PDE4B and PDE5A as well as the differences among residues near the Q_2-pocket are the major contributing factors for inhibitor selectivity between these two family members. Sildenafil and vardenafil have adopted completely different binding modes to PDE4B versus PDE5A in order to avoid some unfavorable interactions and consequently have drastically different binding affinity to these two enzymes. Tadalafil exploits the significant differences in the shape and size of the Q_2-pocket and attains both potency for PDE5A and selectivity against PDE4B.

Sildenafil and vardenafil are inhibitors against PDE4B ($IC_{50} = 20$ and $26\ \mu M$, respectively), but their potency is significantly higher for PDE5A ($IC_{50} = 0.002$ and $0.001\mu M$, respectively). The dramatic difference in binding affinity between PDE4B and PDE5A is due to the different binding mode seen in PDE4B versus PDE5A when the cocrystal structures in the presence of sildenafil or vardenafil are compared (1XOS, 1XOT, 1TBF, 1XP0) (Figure 29.7A

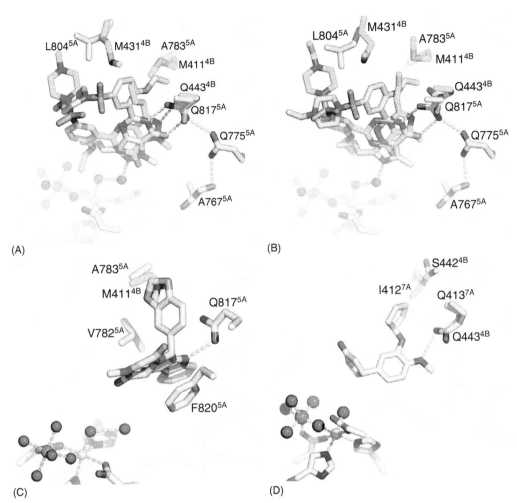

FIGURE 29.7 (See color insert following page 274.) Inhibitor selectivity. (A) Sildenafil bound to PDE5A and PDE4B. The carbon atoms of sildenafil in PDE5A and PDE4B are colored by grey and cyan, respectively. (B) Vardenafil bound to PDE5A and PDE4B. The carbon atoms of vardenafil in PDE5A and PDE4B are colored by grey and cyan, respectively. (C) Tadalafil bound to PDE5A and comparing with PDE4B. (d) Rolipram bound to PDE4B and comparing with PDE7A.

and Figure 29.7B). These different binding modes are caused by sequence changes between the active-site pocket of PDE5A and PDE4B (mainly A783[5A] to M411[4B] and L804[5A] to M431[4B]), as well as the different H-bonding patterns of Q443[4B] and Q817[5A]. Consequently, there is an almost 180° flip in the orientation along the N5-C15 axis for sildenafil (Figure 29.7A) and along the corresponding C6-N15 axis for vardenafil (Figure 29.7B). This inhibitor flipping prevents possible steric clashes between the ethoxyphenyl group and M411[4B] and M431[4B] if the compound were to remain in the position it occupies in PDE5A. Additionally, only one H-bond can now form between Q443[4B] and sildenafil, as opposed to the two formed with Q817[5A]. Concomitant with this loss of one H-bond, the flipped orientation of sildenafil has exposed many of the hydrophobic atoms in the ethoxyphenyl group to solvent, and these two factors contribute to the significant difference in the potency of sildenafil and vardenfil towards PDE5A and PDE4B.

Tadalafil is a highly selective PDE5A inhibitor (1XOZ, 1UDU). The methylenedioxyphenyl group of tadalafil occupies the Q_2-pocket (Figure 29.7C) and makes hydrophobic interactions with A783[5A], F786[5A], F787[5A], L804[5A], I813[5A], and M816[5A] (not shown). The methylenedioxyphenyl group does not fit into the relatively smaller Q_2-pocket in PDE4B and could cause steric clashes with M411[4B] (cf. to A783[5A], Figure 29.7C). This confers tadalafil not only its high potency with PDE5A but also its selectivity against PDE4B.

29.6.2 PDE5 AND PDE6 SELECTIVITY

Sildenafil and vardenafil are also potent inhibitors of PDE6 and PDE11A. This is not surprising since the sequence identity between PDE5A, PDE6, and PDE11 is very high. The sequence identity between PDE5A and PDE6 (including three subtypes, PDE6A, PDE6B, and PDE6C) is about 41% in the catalytic domain, and this is increased to 92% when only the active site residues are compared. Similarly, the sequence identity between PDE5A and PDE11A is about 46% in the catalytic domain, and this is increased to 76% when only the active site residues are compared. Of the residues within the sildenafil and vardenafil binding site, only four differences are observed between PDE5A and PDE6: S663L, L804M, M816L, and A823F. However, these residues are in the H-loop region (L-region) surrounding the phenylsulfone piperizine, which is flexible.[20,21] Nevertheless, these differences could potentially be exploited to build selectivity for other inhibitors. Tadalafil is about 200–700-fold selective for PDE5A over PDE6. However, inspection of the cocrystal structure of tadalafil with PDE5A and comparison with PDE6A sequence shed no light on the high selectivity of this compound.

29.6.3 PDE4 AND PDE7 SELECTIVITY

PDE7 has recently emerged as a target for the development of anti-inflammatory drugs. This is based on the association of PDE7A with IL-2 production and the proliferation of anti-CD3- and anti-CD28-stimulated human T-lymphocytes,[33,34] although this is still controversial.[35] Most of the reported PDE7 (PDE7A, PDE7B) inhibitors are nonselective against PDE4 due to the high degree of sequence homology (~32% sequence identity in the catalytic domain and ~72% sequence identity in the active site) between these two families. In fact many of the documented PDE7 inhibitors are either PDE4 inhibitors or close analogs thereof. The crystal structure of the catalytic domain of PDE7A bound to IBMX has recently been reported (1ZKL).[22] Comparison of PDE7 and PDE4 structures has revealed several residue differences in the active site that could be exploited for the design of selectivity between these two enzymes. The PDE4 inhibitor, rolipram, is highly selective against PDE7. The cyclopentyl group in rolipram binds in the Q_2-subpocket in PDE4B and it would clash with I412 in PDE7A, thus conferring rolipram its selectivity against PDE7A (Figure 29.7D). Several residue differences including the S442[4B]/I412[7A] mutation have made the Q_2-subpocket smaller in PDE7A than in PDE4B. Some PDE7 inhibitors have been selected to have R-groups that can be accommodated in the Q_2-subpocket of PDE7A and would still fit into that in PDE4B, thus conferring them dual inhibition of both enzymes. Based on these structural understanding of PDE4 and PDE7, the design of inhibitors selective for PDE7 over PDE4 is less challenging than, for example, the design of PDE4-subtype-selective inhibitors. In fact, PDE7-selective inhibitors, BRL-50481 and BMS-586353, have been discovered.[36] Although a selective PDE7 inhibitor, BRL-50481, was found to have no effect on the proliferation of CD8[+]-T-lymphocytes. It did however act synergistically with PDE4 inhibitors in the suppression of TNF-α release.[36] Consequently, several dual inhibitors of PDE4 and PDE7 have been

developed to explore the possibility of obtaining more efficacious drugs for the treatment of inflammatory diseases.[37,38]

29.6.4 PDE4B AND PDE4D SELECTIVITY

The design of inhibitors selective for PDE4B over PDE4D remains one of the most challenging tasks. All the inhibitors cocrystallized with PDE4 catalytic domains are nonselective between PDE4B and PDE4D. These nonselective inhibitors bound to PDE4B and PDE4D in nearly identical mode, confirming their nonselectivity. However, two PDE4D-selective inhibitors have been found. One is NVP-ABE171, which is more than 20-fold selective for PDE4D over PDE4B.[39] The other is CP-671305, which is more than 95-fold selective for PDE4D over PDE4B.[40] Perhaps the cocrystal structures of these two PDE4D-selective inhibitors will shed some light on their PDE4D selectivity and point to the way for designing PDE4B-selective inhibitors. The residues in the active site of PDE4B and PDE4D that are in contact with various inhibitors are identical. Moreover, the conformational differences in the active site between various PDE4B and PDE4D structures are very subtle. It seems that there are several possible ways of designing inhibitors with PDE4B/PDE4D selectivity. The first is to exploit the subtle differences in the conformational state of residues in the active site of PDE4B and PDE4D and the differences in their potential of induced fit in response to inhibitor binding. The second is to design inhibitors that make contact with residues beyond these from the catalytic domain, such as the C-terminal domain or the N-terminal regulatory domains, which may fold near the active site. The third is to explore potential conformational changes in the active site of the catalytic domain triggered by the presence of the N-terminal regulatory domains in the full-length enzyme.

29.7 POTENTIAL FOR STRUCTURE-BASED RATIONAL DRUG DESIGN

Structural information offers potential in formulating the rational design of more potent and selective PDE4 inhibitors. Indeed, a scaffold-based drug discovery paradigm has been applied to PDE4.[16] This paradigm begins with the screening of a low molecular weight compound library to identify low-affinity inhibitors. This is followed by high-throughput cocrystallography to select from these compounds that exhibit a dominant binding mode and have appropriate sites for substitution. Such compounds serve as scaffolds for lead optimization. Following this approach, a low-affinity 3,5-dimethyl-1H-pyrazole-4-carboxylic acid ethyl ester (PCEE) revealed the characteristic features of a potential scaffold binding to PDE4D (1Y2B, Figure 29.8A). To validate whether PCEE could serve as a scaffold for PDE4, three potential sites of substitution were identified based on the ability to make favorable chemical interactions in the available space at the active site. From this a derivative compound, 3,5-dimethyl-1-phenyl-1H-pyrazole-4-carboxylic acid ethyl ester (PhPCEE) was selected for crystallization studies with PDE4D. The cocrystal structure showed that phenyl substitution at the 1-position of the pyrazole ring does not change the binding mode of the pyrazole in PDE4, validating the PCEE moiety as a scaffold for PDE4 (1Y2C, Figure 29.8B). Based on the cocrystal structure, more than 100 compounds were computationally designed and docked into the active-site pocket. Substitutions predicted to cause undesirable interactions with residues in the active site were eliminated and 10 compounds predicted to increase binding affinity were synthesized. The cocrystal structures of several compounds in complex with PDE4B and PDE4D were obtained to facilitate and validate the chemical optimization (1Y2D, 1Y2E, 1Y2H, 1Y2J, 1Y2K). Starting from the initial scaffold a 4000-fold potency increase was obtained[16] in two rounds of chemical

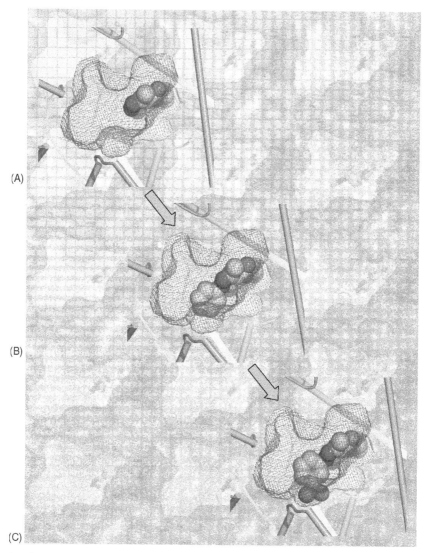

(A)

(B)

(C)

FIGURE 29.8 (See color insert following page 274.) The discovery of pyrazoles as PDE4 inhibitors by scaffold-based drug design. The blue mesh represents the active-site pocket. The spheres represent the bound inhibitor. The golden rod represents the α-helices. The background contains multiple cocrystal structures of various pyrazoles. (A) Scaffold discovery represented by PCEE. (B) Scaffold validation represented by PhPCEE. (C) Chemical optimization represented by NO_2-PhPCEE.

synthesis (Figure 29.8C), demonstrating the usefulness of this approach as a widely applicable strategy to expedite lead discovery effort for many other targets for which known small molecule modulators are limited.

29.8 SUMMARY

Recent advances in the structural biology of PDE catalytic domains have revealed the mechanism of cyclic nucleotide selectivity and structural basis of inhibitor binding. These structural insights have provided the basis for the design of potent and selective PDE inhibitors.

The cyclic nucleotide selectivity for PDEs is achieved through the glutamine switch mechanism.[10] The amide group of an invariant glutamine adopts one orientation to interact with cAMP and switches to another orientation to interact with cGMP. This glutamine switch is controlled by an intricate network of hydrogen bonds near the absolutely conserved glutamine.

Several common features of inhibitor binding to PDEs have been revealed from these cocrystal structures. The first conserved feature of inhibitor binding to PDEs is that the planar group of the inhibitor is held tightly by a hydrophobic clamp formed by two highly conserved hydrophobic residues. The second conserved feature is that the inhibitor always forms hydrogen bonds with the purine-selective glutamine. Another important feature of inhibitor binding to PDEs is that these inhibitors do not bind to the metal ions directly. Instead, they form indirect interactions with the metal ions mediated through water molecules.

As an example of how these structural insights can facilitate PDE inhibitor design, a scaffold-based inhibitor discovery strategy has been applied to the rapid discovery of a pyrazole carboxylic ester scaffold and the subsequent efficient optimization into the potent PDE4 inhibitors. The robustness and efficiency of the scaffold-based drug discovery method should make it widely applicable to expedite the lead discovery effort for many other targets for which known small molecule modulators are limited.

It remains a tremendously challenging task to design inhibitors with selectivity between closely related family members and inhibitors with selectivity among subtypes. However, the availability of these crystal structures of various families of PDEs has made these tasks more tractable. We have just begun to witness the impact of structural information on the design of PDE inhibitors and we expect more inhibitors will be designed based on these structural insights in the future.

ACKNOWLEDGMENTS

I would like to thank my colleagues at Plexxikon whose contributions have advanced our understanding of PDE structures. I would also like to thank Dr. Rick Artis, Dr. Gideon Bollag, Dr. Sam Gillette, Dr. Prabha Ibrahim, and Dr. Chao Zhang for their generous support and stimulating discussions.

REFERENCES

1. Francis, S.H., et al. 2001. Cyclic nucleotide phosphodiesterases: Relating structure and function. *Prog Nucleic Acid Res Mol Biol* 65:1–52.
2. Corbin, J.D., and S.H. Francis. 2002. Pharmacology of phosphodiesterase-5 inhibitors. *Int J Clin Pract* 56 (6):453–459.
3. Rotella, D.P. 2002. Phosphodiesterase 5 inhibitors: Current status and potential applications. *Nat Rev Drug Discov* 1 (9):674–682.
4. Souness, J.E., et al. 2000. Immunosuppressive and anti-inflammatory effects of cyclic AMP phosphodiesterase (PDE) type 4 inhibitors. *Immunopharmacology* 47 (2–3):127–162.
5. Beavo, J.A., and L.L. Brunton. 2002. Cyclic nucleotide research—Still expanding after half a century. *Nat Rev Mol Cell Biol* 3 (9):710–718.
6. Conti, M. 2000. Phosphodiesterases and cyclic nucleotide signaling in endocrine cells. *Mol Endocrinol* 14 (9):1317–1327.
7. Mehats, C., et al. 2002. Cyclic nucleotide phosphodiesterases and their role in endocrine cell signaling. *Trends Endocrinol Metab* 13 (1):29–35.
8. Sonnenburg, W.K., and J.A. Beavo. 1994. Cyclic GMP and regulation of cyclic nucleotide hydrolysis. *Adv Pharmacol* 26:87–114.

9. Houslay, M.D., and D.R. Adams. 2003. PDE4 cAMP phosphodiesterases: Modular enzymes that orchestrate signalling cross-talk, desensitization and compartmentalization. *Biochem J* 370 (Pt 1): 1–18.

10. Zhang, K.Y.J., et al. 2004. A glutamine switch mechanism for nucleotide selectivity by phosphodiesterases. *Mol Cell* 15 (2):279–286.

11. Iffland, A., et al. 2005. Structural determinants for inhibitor specificity and selectivity in PDE2A using the wheat germ in vitro translation system. *Biochemistry* 44 (23):8312–8325.

12. Scapin, G., et al. 2004. Crystal structure of human phosphodiesterase 3B: Atomic basis for substrate and inhibitor specificity. *Biochemistry* 43 (20):6091–6100.

13. Xu, R.X., et al. 2000. Atomic structure of PDE4: Insights into phosphodiesterase mechanism and specificity. *Science* 288 (5472):1822–1825.

14. Xu, R.X., et al. 2004. Crystal structures of the catalytic domain of phosphodiesterase 4B complexed with AMP, 8-Br-AMP and rolipram. *J Mol Biol* 337 (2):355–365.

15. Card, G.L., et al. 2004. Structural basis for the activity of drugs that inhibit phosphodiesterases. *Structure* 12 (12):2233–2247.

16. Card, G.L., et al. 2005. A family of phosphodiesterase inhibitors discovered by cocrystallography and scaffold-based drug design. *Nat Biotechnol* 23 (2):201–207.

17. Lee, M.E., et al. 2002. Crystal structure of phosphodiesterase 4D and inhibitor complex. *FEBS Lett* 530 (1–3):53–58.

18. Huai, Q., et al. 2003. The crystal structure of AMP-bound PDE4 suggests a mechanism for phosphodiesterase catalysis. *Biochemistry* 42 (45):13220–13226.

19. Huai, Q., et al. 2003. Three-dimensional structures of PDE4D in complex with roliprams and implication on inhibitor selectivity. *Structure (Camb)* 11 (7):865–873.

20. Huai, Q., et al. 2004. Crystal structures of phosphodiesterases 4 and 5 in complex with inhibitor 3-isobutyl-1-methylxanthine suggest a conformation determinant of inhibitor selectivity. *J Biol Chem* 279 (13):13095–13101.

21. Sung, B.J., et al. 2003. Structure of the catalytic domain of human phosphodiesterase 5 with bound drug molecules. *Nature* 425 (6953):98–102.

22. Wang, H., et al. 2005. Multiple elements jointly determine inhibitor selectivity of cyclic nucleotide phosphodiesterases 4 and 7. *J Biol Chem* 280 (35):30949–30955.

23. Huai, Q., et al. 2004. Crystal structure of phosphodiesterase 9 shows orientation variation of inhibitor 3-isobutyl-1-methylxanthine binding. *Proc Natl Acad Sci USA* 101 (26):9624–9629.

24. Kajimura, N., et al. 2002. Three-dimensional structure of non-activated cGMP phosphodiesterase 6 and comparison of its image with those of activated forms. *J Struct Biol* 139 (1):27–38.

25. Martinez, S.E., et al. 2002. The two GAF domains in phosphodiesterase 2A have distinct roles in dimerization and in cGMP binding. *Proc Natl Acad Sci USA* 99 (20):13260–13265.

26. Zoraghi, R., et al. 2005. Structural and functional features in human PDE5A1 regulatory domain that provide for allosteric cGMP binding, dimerization, and regulation. *J Biol Chem* 280 (12):12051–12063.

27. Martinez, S.E., et al. 2005. Crystal structure of the tandem GAF domains from a cyanobacterial adenylyl cyclase: Modes of ligand binding and dimerization. *Proc Natl Acad Sci USA* 102 (8): 3082–3087.

28. Zoraghi, R., et al. 2006. Phosphodiesterase-5 Gln817 is critical for cGMP, vardenafil, or sildenafil affinity: Its orientation impacts cGMP but not cAMP affinity. *J Biol Chem* 281 (9):5553–5558.

29. Burley, S.K., and G.A. Petsko. 1985. Aromatic–aromatic interaction: A mechanism of protein structure stabilization. *Science* 229 (4708):23–28.

30. Huai, Q., et al. 2006. Enantiomer discrimination illustrated by the high resolution crystal structures of type 4 phosphodiesterase. *J Med Chem* 49 (6):1867–1873.

31. Robichaud, A., et al. 2002. Deletion of phosphodiesterase 4D in mice shortens alpha(2)-adrenoceptor-mediated anesthesia, a behavioral correlate of emesis. *J Clin Invest* 110 (7):1045–1052.

32. Gresser, U., and C.H. Gleiter. 2002. Erectile dysfunction: Comparison of efficacy and side effects of the PDE-5 inhibitors sildenafil, vardenafil and tadalafil—Review of the literature. *Eur J Med Res* 7 (10):435–446.

33. Li, L., et al. 1999. CD3- and CD28-dependent induction of PDE7 required for T cell activation. *Science* 283 (5403):848–851.
34. Glavas, N.A., et al. 2001. T cell activation up-regulates cyclic nucleotide phosphodiesterases 8A1 and 7A3. *Proc Natl Acad Sci USA* 98 (11):6319–6324.
35. Yang, G., et al. 2003. Phosphodiesterase 7A-deficient mice have functional T cells. *J Immunol* 171 (12):6414–6420.
36. Smith, S.J., et al. 2004. Discovery of BRL 50481 [3-(N,N-dimethylsulfonamido)-4-methyl-nitrobenzene], a selective inhibitor of phosphodiesterase 7: In vitro studies in human monocytes, lung macrophages, and CD8$^+$ T-lymphocytes. *Mol Pharmacol* 66 (6):1679–1689.
37. Hatzelmann, A., et al. 2002. Preparation of phthalazinones as phosphodiesterase 4/7 inhibitors. In *WO2002085906A2*, 42 pp. Germany: Altana Pharma A.G.
38. Pitts, W.J., et al. 2002. Preparation of pyrimidinylaminothiazolecarboxylates and related pyrimidines as dual inhibitors of phosphodiesterases PDE7 and PDE4. In *WO2002088079A2*, 81 pp. United States: Bristol-Myers Squibb Company.
39. Trifilieff, A., et al. 2002. Pharmacological profile of a novel phosphodiesterase 4 inhibitor, 4-(8-benzo[1,2,5]oxadiazol-5-yl-[1,7]naphthyridin-6-yl)-benzoic acid (NVP-ABE171), a 1,7-naphthyridine derivative, with anti-inflammatory activities. *J Pharmacol Exp Ther* 301 (1):241–248.
40. Kalgutkar, A.S., et al. 2004. Disposition of CP-671305, a selective phosphodiesterase 4 inhibitor in preclinical species. *Xenobiotica* 34 (8):755–770.

30 Structure, Catalytic Mechanism, and Inhibitor Selectivity of Cyclic Nucleotide Phosphodiesterases

Hengming Ke and Huanchen Wang

CONTENTS

30.1 Introduction 607
30.2 Structures of PDEs...608
    30.2.1 PDE4 608
    30.2.2 PDE7A 612
    30.2.3 PDE1B..613
    30.2.4 PDE2A 614
    30.2.5 PDE3B..616
    30.2.6 PDE9A 616
    30.2.7 PDE5A 617
    30.2.8 Summary of the Catalytic Domain Structures.....................................618
30.3 Hydrolysis Mechanism...618
30.4 Dissection of Inhibitor Selectivity 620
30.5 Perspective...621
Acknowledgment..623
References 623

30.1 INTRODUCTION

Cyclic nucleotide phosphodiesterases (PDEs) hydrolyze the second messengers cyclic adenosine monophosphate (cAMP) and cyclic guanine monophosphate (cGMP) to AMP and GMP. Human genome contains 21 PDE genes that are grouped into 11 families. Alternative mRNA splicing of the 21 genes generates over 60 isoforms of PDEs in various human tissues.[1–6] The PDE molecules can be divided into three regions: an N-terminal splicing region, a regulatory domain, and a C-terminal catalytic domain (Figure 30.1). All PDEs contain a conserved catalytic domain of about 300 amino acids (AAs) and exhibit 25%–40% similarity in AA sequence among different families. However, any homology in other regions is either much weaker or nonexistent.

FIGURE 30.1 Domain structure of PDEs. The regulatory domains of PDEs contain various structural motifs: one or two calmodulin-binding domains in PDE1, two GAF domains in PDE2, PDE5, PDE6, and PDE10, one or two GAFs in PDE11, two upstream conserved regions (UCR) in PDE4, and a PAS domain in PDE8.

Each PDE family has its individual substrate preference for cAMP and cGMP. PDE families 4, 7, and 8 prefer to hydrolyze cAMP, whereas PDE5, 6, and 9 are cGMP-specific. PDE1, 2, 3, 10, and 11 show dual activities on hydrolysis of both cAMP and cGMP.[3] On the other hand, each PDE family possesses selective inhibitors that bind competitively to the conserved active sites. Selective inhibitors of PDEs have been widely studied as therapeutics for the treatment of various human diseases.[7–17] The best known example is the PDE5 inhibitor sildenafil that has been used for treatment of male erectile dysfunction. Until recently, it has been unclear how the similar active sites of PDE families distinguish subtle differences between substrates of cAMP and cGMP, and recognize structurally distinct inhibitors. This chapter will describe the current knowledge on the three-dimensional structures of the catalytic domains of the PDE families and discuss structural hints on hydrolysis mechanism, substrate specificity, and inhibitor selectivity.

30.2   STRUCTURES OF PDEs

Three-dimensional structures of the catalytic domains of seven PDE families have been published (Table 30.1): PDE1B,[18] PDE2A,[19] PDE3B,[20] PDE4B and PDE4D,[18,21–28] PDE5A,[18,26,29] PDE7A,[30] and PDE9A[31] in addition to the structures of a regulatory domain of PDE2A[32] and a fragment of PDE4D5 splicing region.[33] This chapter starts with structures of cAMP-specific families of PDE4 and PDE7, followed by dual-specific PDE1, 2, 3, and cGMP-specific PDE9 and PDE5. The PDE4D structure will be used as a reference for comparison.

30.2.1   PDE4

PDE4 is the most extensively studied PDE family with cAMP specificity. Since the first publication of the crystal structure of the catalytic domain PDE4B,[21] 33 structures of PDE4B/4D and their complexes with the product AMP or inhibitors have been deposited into the Protein Data Bank (Table 30.1). The catalytic domains of PDE4B (residues 152–528) and PDE4D2 (residues 79–438) contain 16 α-helices and no β-strands (Figure 30.2 and Figure 30.3). The 16 helices are assembled into a tight entity that may be further grouped into three subdomains.[21] The active site of PDE4 is a deep pocket formed by the juxtaposition of the three subdomains and can be divided into two major subpockets that bind divalent metals and inhibitors or substrates (Figure 30.2). Since the active site pocket has a volume of ~440 Å³ that is significantly larger than the volume of cAMP (~232 Å³),[21] it would be possible that the preferred substrate cAMP binds somewhat differently from the poor substrate cGMP and that a number of structurally distinct inhibitors fit the pocket. This characteristic is also true for other PDE families and is likely to be useful for the design of family-selective inhibitors.

Two metal ions have been identified in PDE4. The first metal has been assigned as a zinc ion and confirmed by the anomalous scattering experiment. Zinc forms six coordinations with His164, His200, Asp201, Asp318 in PDE4D2, and two water molecules. Since these four

TABLE 30.1
Structures of PDEs in Protein Data Bank

Protein	Ligand	Resolution (Å)	PDB Code	Reference
PDE1B	Unliganded	1.77	1TAZ	18
PDE2A	(GAF domain)	2.86	1MC0	32
PDE2A	Unliganded	1.70	1Z1L	19
PDE3B	Dihydropyridazine	2.4	1SO2	20
PDE3B	IBMX[a]	2.9	1SOJ	20
PDE4B2B	Unliganded	1.77	1F0J	21
PDE4B2B	(R,S)-Rolipram	2.0	1RO6	22
PDE4B2B	8-Br-AMP	2.13	1RO9	22
PDE4B2B	AMP	2.00	1ROR	22
PDE4B	AMP	2.15	1TB5	18
PDE4B	Cilomilast	2.19	1XLX	27
PDE4B	Filaminast	2.06	1XLZ	27
PDE4B	Piclamilast	2.31	1XM4	27
PDE4B	(R)-Mesopram	1.92	1XM6	27
PDE4B	Roflumilast	2.30	1XMU	27
PDE4B	(R)-Rolipram	2.40	1XMY	27
PDE4B	(R,S)-Rolipram	2.31	1XN0	27
PDE4B	Sildenafil	2.28	1XOS	27
PDE4B	Vardenafil	2.34	1XOT	27
PDE4B	Cl-PDM-PCAEE[a]	2.40	1Y2H	27
PDE4B	DMNP-PCAEE[a]	2.55	1Y2J	27
PDE4D5	(Splicing region)	NMR	1LOI	33
PDE4D	Zardaverine	2.9	1MKD	23
PDE4D2	(R,S)-Rolipram	2.0	1OYN	24
PDE4D2	(R)-Rolipram	2.3	1Q9M	24
PDE4D2	AMP	2.3	1PTW	25
PDE4D2	IBMX	2.10	1ZKN	26
PDE4D	AMP	1.63	1TB7	18
PDE4D	(R,S)-Rolipram	1.6	1TBB	18
PDE4D	Cilomilast	1.55	1XOM	27
PDE4D	Piclamilast	1.72	1XON	27
PDE4D	Roflumilast	1.83	1XOQ	27
PDE4D	Zardaverine	1.54	1XOR	27
PDE4D	DM-PCAEE[a]	1.40	1Y2B	28
PDE4D	DMP-PCAEE[a]	1.67	1Y2C	28
PDE4D	MOPDM-PCAEE[a]	1.70	1Y2D	28
PDE4D	APDM-PCAEE[a]	2.10	1Y2E	28
PDE4D	DMNP-PCAEE[a]	1.36	1Y2K	28
PDE5A	Sildenafil	2.3	1UDT	29
PDE5A	Vardenafil	2.5	1UHO	29
PDE5A	Tadalafil	2.83	1UDU	29
PDE5A	IBMX	2.05	1RKP	26
PDE5A	Unliganded	2.1	1T9R	18
PDE5A	GMP	2.0	1T9S	18
PDE5A	Sildenafil	1.30	1TBF	18
PDE5A	Tadalafil	1.37	1XOZ	27
PDE5A	Vardenafil	1.40	1XPO	27
PDE7A	IBMX	1.67	1ZKL	30
PDE9A	IBMX	2.23	1TBM	31

[a]IBMX, 3-isobutyl-1-methylxanthine; PCAEE, 1H-Pyrazole-4-carboxylic acid ethyl ester, Cl-PDM-PCAEE, 1-(2-chloro-phenyl)-3,5-dimethyl-PCAEE; DMNP-PCAEE, 3,5-Dimethyl-1-(3-nitrophenyl)-PCAEE; DM-PCAEE, 3,5-Dimethyl-PCAEE; DMP-PCAEE, 3,5-Dimethyl-1-phenyl-PCAEE; MOPDM-PCAEE, 1-(4-methoxy-phenyl)-3,5-dimethyl-PCAEE; APDM-PCAEE, 1-(4-amino-phenyl)-3,5-dimethyl-PCAEE; DMNP-PCAEE, 3,5-Dimethyl-1-(3-nitro-phenyl)-PCAEE.

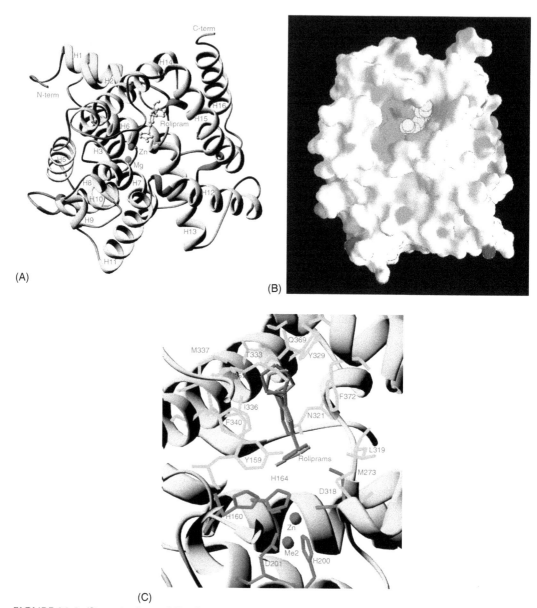

residues come from the three subdomains, one role of the zinc ion must be structural for assembly of the catalytic domain.[21] In addition, zinc is an integral element of the active site and thus plays a catalytic role. The chemical nature of the second metal could not be determined by x-ray diffraction, but is assumed to be magnesium because it is the most

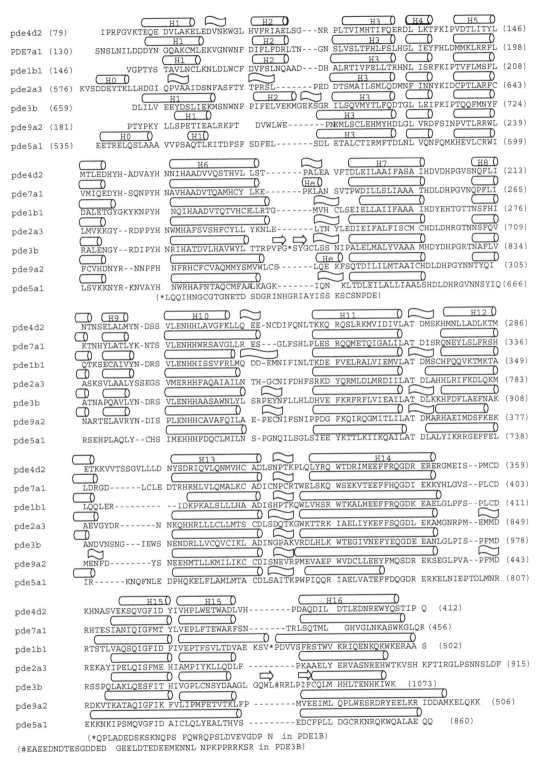

```
            ┌──H1──┐ ∿                ┌──H2──┐           ┌───H3───┐ ┌─H4─┐  ┌──H5──┐
pde4d2 (79)  IPRFGVKTEQE DVLAKELEDVNKWGL HVFRIAELSG---NR PLTVIMHTIFQERDL LKTFKIPVDTLITYL (146)
            ┌──H1──┐               ┌──H2──┐             ┌──H3──┐
PDE7a1 (130) SNSLNILDDDYN GQAKCMLEKVGNWNF DIFLFDRLTN---GN SLVSLTFHLFSLHGL IEYFHLDMMKLRRFL (198)
              ┌──H1──┐              ┌──H2──┐            ┌──H3──┐
pde1b1 (146)  VGPTYS TAVLNCLKNLDLWCF DVFSLNQAAD---DH ALRTIVFELLTRHNL ISRFKIPTVFLMSFL (208)
          ┌H0┐                       ┌───H3───┐
pde2a3 (576) KVSDDEYTKLLHDGI QPVAAIDSNFASFTY TPRSL-------PED DTSMAILSMLQDMNF INNYKIDCPTLARFC (643)
              ┌──H1──┐            ┌──H2──┐ ∿         ┌──H3──┐
pde3b (659)  DLILV EEYDSLIEKMSNWNF PIFELVEKMGEKSGR ILSQVMYTLFQDTGL LEIFKIPTQQFMNYF (724)
               ┌─H1─┐                      ┌───H3───┐
pde9a2 (181)  PTYPKY LLSPETIEALRKPT DVWLWE------PNEMLSCLEHMYHDLGL VRDFSINPVTLRRWL (239)
          ┌H0┐        ┌─H1┐              ┌──H3──┐
pde5a1 (535) EETRELQSLAAA VVPSAQTLKITDFSF SDFEL-------SDL ETALCTIRMFTDLNL VQNFQMKHEVLCRWI (599)

                     ┌───H6───┐       ∿          ┌────H7────┐         ┌─H8─┐
pde4d2    MTLEDHY-ADVAYH NNIHAADVVQSTHVL LST-------PALEA VFTDLEILAAIFASA IHDVDHPGVSNQFLI (213)
                                        ┌He┐
pde7a1    VMIQEDYH-SQNPYH NAVHAADVTQAMHCY LKE-------PKLAN SVTPWDILLSLIAAA THDLDHPGVNQPFLI (265)
pde1b1    DALETGYGKYKNPYH NQIHAADVTQTVHCELLRTG------MVH CLSEIELLAIIFAAA IHDYEHTGTTNSFHI (276)
pde2a3    LMVKKGY--RDPPYH NWMHAFSVSHFCYLL YKNLE------LTN YLEDIEIFALFISCM CHDLDHRGTNNSFQV (709)
                                        ⇒ ⇒ ∿
pde3b     RALENGY--RDIPYH NRIHATDVLHAVWYL TTRPVPG*SYGCLSS NIPALELMALYVAAA MHDYDHPGRTNAFLV (834)
                                        ┌He┐
pde9a2    FCVHDNYR--NNPFH NFRHCFCVAQMMYSMVWLCS------LQE KFSQTDILILMTAAICHDLDHPGYNNTYQI (305)
pde5a1    LSVKKNYR-KNVAYH NWRHAFNTAQCMFAALKAGK-------IQN KLTDLEILALLIAALSHDLDHRGVNNSYIQ(666)
                 (*LQQIHNGCGTGNETD SDGRINHGRIAYISS KSCSNPDE)

          ┌─H9─┐      ┌────H10────┐            ┌──────H11──────┐            ┌───H12───┐
pde4d2    NTNSELALMYN-DSS VLENHHLAVGFKLLQ EE-NCDIFQNLTKKQ RQSLRKMVIDIVLAT DMSKHMNLLADLKTM (286)
pde7a1    KTNHYLATLYK-NTS VLENHHWRSAVGLLR ES---GLFSHLPLES RQQMETQIGALILAT DISRQNEYLSLFRSH (336)
pde1b1    QTKSECAIVYN-DRS VLENHHISSVFRLMQ DD-EMNIFINLTKDE FVELRALVIEMVLAT DMSCHFQQVKTMKTA (349)
pde2a3    ASKSVLAALYSSEGS VMERHHFAQAIAILN TH-GCNIFDHFSRKD YQRMLDLMRDIILAT DLAHHLRIFKDLQKM (783)
pde3b     ATNAPQAVLYN-DRS VLENHHAASAWNLYL SRPEYNFLLHLDHVE FKRFRFLVIEAILAT DLKKHFDFLAEFNAK (908)
pde9a2    NARTELAVRYN-DIS PLENHHCAVAFQILA E-PECNIFSNIPPDG FKQIRQGMITLILAT DMARHAEIMDSFKEK (377)
pde5a1    RSEHPLAQLY--CHS IMEHHHFDQCLMILN S-PGNQILSGLSIEE YKTTLKIIKQAILAT DLALYIKRRGEFFEL (738)

          ┌────H13────┐                     ┌──────H14──────┐
pde4d2    ETKKVVTSSGVLLLD NYSDRIQVLQNMVHC ADLSNPTKPLQLYRQ WTDRIMEEFFRQGDR ERERGMEIS--PMCD (359)
pde7a1    LDRGD------LCLE DTRHRHLVLQMALKC ADICNPCRTWELSKQ WSEKVTEEFFHQGDI EKKYHLGVS--PLCD (403)
pde1b1    LQQLER--------- --IDKPKALSLLLHA ADISHPTKQWLVHSR WTKALMEEFFRQGDK EAELGLPFS--PLCD (411)
pde2a3    AEVGYDR-------N NKQHHRLLLCLLMTS CDLSDQTKGWKTTRK IAELIYKEFFSQGDL EKAMGNRPM--EMMD (849)
pde3b     ANDVNSNG---IEWS NENDRLLVCQVCIKL ADINGPAKVRDLHLK WTEGIVNEFYEQGDE EANLGLPIS--PFMD (978)
pde9a2    MENFD--------YS NEEHMTLLKMILIKC CDISNEVRPMEVAEP WVDCLLEEYFMQSDR EKSEGLPVA--PFMD (443)
pde5a1    IR------KNQFNLE DPHQKELFLAMLMTA CDLSAITKPWPIQQR IAELVATEFFDQGDR ERKELNIEPTDLMNR (807)

          ┌─H15─┐ ┌──H15──┐               ┌─────H16─────┐
pde4d2    KHNASVEKSQVGFID YIVHPLWETWADLVH--------PDAQDIL DTLEDNREWYQSTIP Q    (412)
pde7a1    RHTESIANIQIGFMT YLVEPLFTEWARFSN-------TRLSQTML GHVGLNKASWKGLQR (456)
pde1b1    RTSTLVAQSQIGFID FIVEPTFSVLTDVAE KSV*PDVVSFRSTWV KRIQENKQKWKERAA S    (502)
pde2a3    REKAYIPELQISFME HIAMPIYKLLQDLF -------PKAAELY ERVASNREHWTKVSH KFTIRGLPSNNSLDF (915)
pde3b     RSSPQLAKLQESFIT HIVGPLCNSYDAAGL GQWL#RRLPIFCQLM HHLTENHKIWK (1073)
pde9a2    RDKVTKATAQIGFIK FVLIPMFETVTKLFP --------MVEEIML QPLWESRDRYEELKR IDDAMKELQKK (506)
pde5a1    EKKNKIPSMQVGFID AICLQLYEALTHVS ---------EDCFPLL DGCRKNRQKWQALAE QQ   (860)
```

  (*QPLADEDSKSKNQPS FQWRQPSLDVEVGDP N in PDE1B)
(#EAEEDNDTESGDDED GEELDTEDEEMENNL NPKPPRRKSR in PDE3B)

FIGURE 30.3 Alignment of the sequences with secondary structures of seven PDE catalytic domains. The PDE5 active site has four different conformations. The unliganded PDE5 is used in the alignment. The secondary structures of other PDE families have no changes upon inhibitor binding.

TABLE 30.2
Alignment of AAs That Bind Rolipram and Metals

	160	164	200	273	318	321	329	332	336	340	357	368	372
PDE4D2	YH	H	HD	M	DL	N	Y	WT	IM	F	M	SQ	F
PDE4B2B	YH	H	HD	M	DL	N	Y	WT	IM	F	M	SQ	F
PDE3A1	YH	H	HD	L	DI	G	H	WT	IV	F	F	LQ	F
PDE7A	YH	H	HD	I	DI	N	S	WS	VT	F	L	IQ	F
PDE8A	YH	H	HD	M	DV	N	C	WA	IS	Y	V	SQ	F
PDE1A3A	YH	H	HD	M	DI	H	H	WT	LM	F	L	SQ	F
PDE2A3	YH	H	HD	L	DL	D	T	IA	IY	F	M	LQ	F
PDE10A2	YH	H	HD	L	DL	S	T	TA	IY	F	M	GQ	F
PDE11A	YH	H	HD	L	DL	A	S	VA	VT	F	I	LQ	W
PDE5A1	YH	H	HD	L	DL	A	Q	IA	VA	F	L	MQ	F
PDE6A	YH	H	HD	L	DL	A	Q	VA	VA	F	M	LQ	F
PDE9A1	FH	H	HD	M	DI	N	A	WV	LL	Y	F	AQ	F

common ion to produce optimal activity of PDE families. However, other divalent metal ions such as manganese also efficiently support the hydrolysis of the cyclic nucleotides and may be the preferred catalytic metal in some PDE families. For example, the activation of PDE9A by manganese was reported to be twice as effective as magnesium.[34,35] The second metal has six coordinations with Asp318 and five bound water molecules in the unliganded PDE4 structure. The six coordinations of both metal ions are arranged in a proximate octahedral configuration.

Rolipram is a PDE4-selective inhibitor and was used clinically as an antidepressant.[36] It contains a chiral carbon and thus has two enantiomeric configurations of R and S. (R)-rolipram has 10–20-fold tighter binding and 3–10-fold more effective inhibition on cAMP hydrolysis than (S)-rolipram.[37–39] However, in the x-ray structures, (R)- and (S)-roliprams bind to the active site of PDE4D2 in similar orientations and interact with the same residues of PDE4D2 (Figure 30.2 and Table 30.2). Both enantiomers of rolipram form two hydrogen bonds with the side chain of the invariant Gln369, but do not directly interact with the divalent metals. The cyclopentyloxy group of (R)- or (S)-rolipram (top in Figure 30.2C) sits in a hydrophobic pocket, interacting with residues Ile336, Met337, Phe340, Met357, Ser368, Gln369, and Phe372. The phenylmethoxy ring of (R)- or (S)-rolipram stacks against Phe372 and also contacts Tyr159, Asn321, Tyr329, Thr333, Ile336, and Gln369 via van der Waal's force. The pyrrolidone groups of (R)- and (S)-roliprams occupy similar location and interacts with the same residues of His160, Met273, Leu319, Phe340, and Phe372.[24,27]

30.2.2 PDE7A

PDE7 is also cAMP-specific and has reported K_m values of 0.03–0.2 μM for cAMP.[40,41] The isolated catalytic domain of PDE7A has K_m values of 0.2 μM for cAMP and 3.9 mM for cGMP and a specificity constant ratio $(k_{cat}/K_m)^{cAMP}/(k_{cat}/K_m)^{cGMP}$ of ~4000.[30] The entire catalytic domain of PDE7A is comparable with that of PDE4 (Figure 30.3 and Figure 30.4A), as shown by root-mean-squared (RMS) deviation of 0.93 Å for superposition of Cα atoms of PDE7A1 residues 139–455 over the equivalents of PDE4D2. However, minor differences are observed between PDE7A1 and PDE4D2. Two 3_{10}-helices in PDE4D2 correspond to α-helices in PDE7A1 (He and the C-terminal H10, Figure 30.3) and a 3_{10}-helix after H11 in

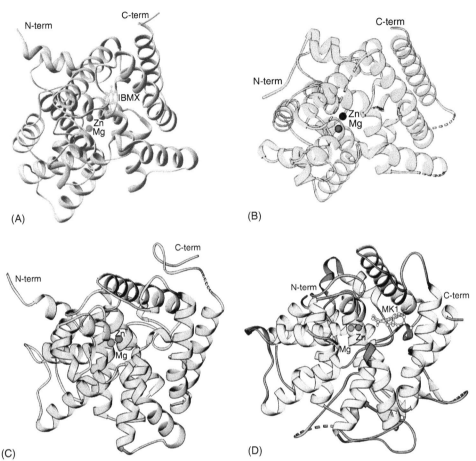

FIGURE 30.4 (See color insert following page 274.) Ribbon diagrams of PDE7A (A), PDE1B (B), PDE2A (C), and PDE3B (D). Dotted lines represent disordered portions of structures. IBMX is a nonselective PDE inhibitor. MK1 is a PDE3 selective inhibitor.

PDE4D2 (residues 271–274) becomes a coil. The most significant difference is the N-terminal portion of helix H11 (residues 304–310 of PDE7A1), which shows an average positional displacement of 2.8 Å, about 3 times the average difference (0.93 Å) for the whole catalytic domain. The difference of H11 may either reflect intrinsic conformation variation between PDE4 and PDE7, or maybe due to the fact that H11 contributes to tetramer formation in the PDE4D2 crystal, but is not involved in intermolecular interactions in PDE7A1.

30.2.3 PDE1B

PDE1 has dual specificity for both cGMP and cAMP, with K_m values of 1–30 μM for cAMP and 3 μM for cGMP.[3] The structure of almost the whole PDE1B catalytic domain,[18] which

has about 10-fold preference for cGMP, is comparable with that of PDE4, except for a few insertions or deletions (Figure 30.3 and Figure 30.4B). The superposition of 293 Cα atoms of comparable residues of PDE1B (residues 151–216, 219–242, 246–352, 257–405, 416–441, and 476–496) over the equivalents in the PDE4D–rolipram structure yields an RMS deviation of 1.6 Å. The main difference occurs in the M-loop (residues 395–416 of PDE1B or 342–363 of PDE4D2). The M-loop in PDE4D2 is ordered and is involved in binding of rolipram, but residues 406–411 of the M-loop in PDE1B are disordered. It is not clear if the M-loop in PDE1 will become ordered in the presence of substrates or inhibitors. In addition, structural comparison revealed an insertion of 38 residues (442–479) between helices H15 and H16 of PDE1B (Figure 30.3), which was not predicted by sequence alignment. A role for the inserted fragment in catalysis is not clear because it is located far away from the active site and majority of the residues (446–472) are disordered.

30.2.4 PDE2A

PDE2 hydrolyzes both cAMP and cGMP with K_m values of 30–100 μM for cAMP and 10–30 μM for cGMP.[3] The structures of the regulatory domain of mouse PDE2A (equivalent to residues 223–562 of human PDE2A3, Figure 30.5A) and the catalytic domains of human PDE2A3 (residues 576–915, Figure 30.4C) have been reported.[19,32] The regulatory domain of PDE2A contains two GAF (cGMP specific PDE, Adenylyl cyclase, and Fh1A) subdomains linked by a long α-helix (Figure 30.5A). Each GAF subdomain consists of a central five-stranded β-sheet flanked with helices. The two GAF subdomains have similar topological folding, as shown by an RMS deviation of 3 Å for the superposition of 130 Cα atoms in the domains.[32] However, cGMP binds only to the GAF B domain. No binding of cGMP to GAF A domain of PDE2A was interpreted as likely being due to the substitution of the 11 cGMP binding residues.[32] A recent report shows that cAMP preferably binds to the GAF A domain of *Trypanosoma brucei* PDE2B, which is most closely homologous to the GAF B domain of mammalian PDE2B.[42]

The folding of the catalytic domain of PDE2A3 is similar to that of PDE4D2 (Figure 30.3 and Figure 30.4C).[19] The core catalytic domain of PDE2A3 is superimposable with those of PDE4D2, as shown by an RMS deviation of 1.5 Å for backbone atoms of 293 comparable residues (597–611, 614–674, 679–720, 723–787, 790–899 in PDE2A3, Figure 30.3). The main difference is observed in the N-terminus of the catalytic domain, where PDE2A3 has an extra α-helix and two 3_{10}-helices correspond to α-helices H1 and H2 of PDE4D (Figure 30.3). These N-terminal helices have different tertiary arrangement in the two structures.

The catalytic activity of PDE2 is stimulated by cGMP binding to the regulatory domain,[43] but the mechanism is unknown. To explore the regulatory mechanism, we built a model for the assembly of the GAF and catalytic domains (Figure 30.5). Two orientations of the GAF domains are possible. In the first model, the GAF A domain interacts with residues around 705–720 (H-loop) whereas GAF B contacts the C-terminal region of the catalytic domain (Figure 30.5B). Thus, Glu278 in the GAF A domain may form a charge–charge interaction with Arg728 or Arg762. Leu389 and Tyr448 in GAF B domain may contact via van der Waal's force with Met645 and Leu836 in the catalytic domain. The second model is generated by switching the GAF domains from left to right (Figure 30.5C). In this model, Leu232 maybe involved in a hydrophobic interaction with Phe831 whereas Gln380, Asp430, and Glu446 may form hydrogen bonds with Asp652, Arg762, and Tyr719, respectively. In summary, both models predict that two GAF domains directly contact the catalytic domains, but do not directly interact with substrate or block its access to the

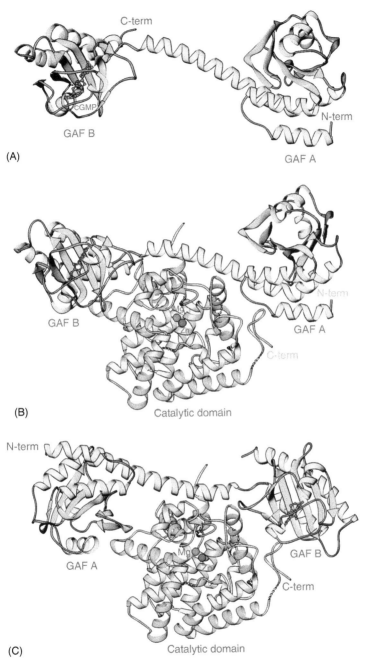

FIGURE 30.5 (See color insert following page 274.) Ribbon diagram of PDE2A domains. (A) The isolated regulatory GAF domain. cGMP binds only to GAF B despite GAF A having similar folding. (B) and (C) Modeling on potential assembly of the GAF and catalytic domains. Two models have an orientation difference of about 180° (switching left to right and then 180° rotating about the horizontal axis).

catalytic pocket. However, GAF A or GAF B, depending on the models, interacts with the H-loop at the active site of the catalytic domain, implying an allosteric mechanism of regulation.

30.2.5 PDE3B

PDE3 hydrolyzes both cAMP and cGMP with K_m values of 0.1–0.5 μM for both cAMP and cGMP.[3] It is also known as a cGMP-inhibited PDE because cGMP may physiologically compete with cAMP at the catalytic site and impede cAMP hydrolysis. The structure of the catalytic domains of PDE3B is highly similar to PDE4 (Figure 30.3 and Figure 30.4D).[20] Superposition between PDE3B and PDE4D2 yields an RMS deviation of 1.2 Å for the backbone atoms of 290 comparable residues (674–688, 692–933, 734–755, 802–864, 866–910, 919–1007, 1059–1072 in PDE3B). The main differences are observed in the N-terminus and the insertions of two large fragments. The N-terminal helix H1 of PDE3B shows positional shifts of as much as 10 Å from those of PDE4D2. An insertion of 44 AAs between helices H6 and H7 is unique to PDE3 (Figure 30.3) and was predicted by sequence alignment. However, an insertion of 47 residues between helices H15 and H16, which also exists in PDE1B, is revealed only by the structural comparison (Figure 30.3). Since both insertions are far away from the active site of PDE3, their roles in the catalysis are not clear.

30.2.6 PDE9A

Among cGMP-specific PDEs, PDE9 has the highest affinity for cGMP with K_m values of about 70 nM for cGMP and 230 μM for cAMP.[34,44] However, the conformation of the catalytic domain of PDE9A (Figure 30.6) is most closely related to that of cAMP-specific PDE4, instead of cGMP-specific PDE5.[31] The residues after 208 of PDE9A2 can be superimposed very well over those of PDE4D2, as shown by an RMS deviation of 1.5 Å

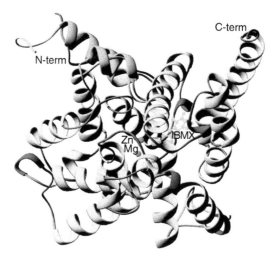

FIGURE 30.6 (See color insert following page 274.) Ribbon presentation for the catalytic domain of PDE9A2. The green ribbons are the PDE4D2 loops from superposition.

for the superposition of Cα atoms of 288 comparable residues between PDE9A2 and PDE4D2. The structural variation is observed mainly in the N-terminal region, in which residues 181–207 of PDE9A2 or 79–114 of PDE4D2 have a totally different three-dimensional arrangement and are not superimposable (Figure 30.3 and Figure 30.6). In addition, PDE9A2 shows a new 3_{10}-helix for residues 440–443, an extension of 10 residues in C-terminus of helix H16, and an α-helix (residues 273–276) corresponding to a 3_{10}-helix in PDE4D2 (Figure 30.3).

30.2.7   PDE5A

PDE5 is another family of cGMP-specific enzymes with K_m values of 3 μM for cGMP[45] and >40 μM for cAMP.[3] The catalytic domain of PDE5A1 has an overall folding topology similar to that of PDE4D2 (Figure 30.7).[18,26,29] However, only 12 helices of PDE5 are comparable with those of other PDE families. The superposition between unliganded PDE5A1 and PDE4D2 yields an RMS deviation of 1.5 Å for the Cα atoms of 257 comparable residues. Two regions show dramatic differences in secondary and tertiary structures. The N-terminal regions (residues 535–566 in PDE5A1 and 79–113 in PDE4D2) have a totally different three-dimensional arrangement although both contain two helices (Figure 30.7).

Another region showing conformational differences is the H-loop at the active site (residues 660–683 in PDE5A1 or 207–228 in PDE4D2).[26] The H-loops of all other PDE families contain two short α-helices and have a uniform and comparable conformation. In contrast, the H-loop in the catalytic domain of unliganded PDE5A1 and its complexes with inhibitors 3-isobutyl-1-methylxanthine (IBMX), sildenafil, and icarisid II show four different conformations. In the structure of the unliganded PDE5A1, the H-loop contains a few turns and majority of its residues exist in coil conformation (Figure 30.7). The binding

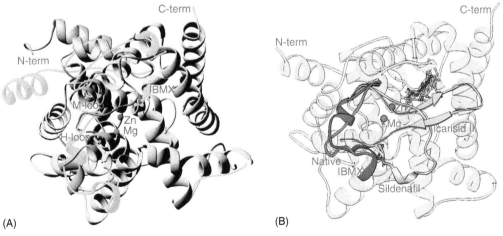

(A)   (B)

FIGURE 30.7 (See color insert following page 274.) Ribbons diagram of the catalytic domain of PDE5A1. (A) Structural comparison between PDE5A1–IBMX and PDE4D2–IBMX. Common structures are drawn in cyan. Loops in different conformations are shown in green for PDE5A1 and gold for PDE4D2. (B) Inhibitor-induced conformational changes of the PDE5A1 catalytic domains. H-loops as well as the inhibitors are represented by colors of purple (unliganded), blue (IBMX), green (sildenafil), and gold (icarisid II).

of IBMX converts the H-loop into two short α-helices (residues 663–667 and 671–675) and makes the Cα atoms in H-loop moved as much as 7 Å from those in the unliganded PDE5A1 structure. In the PDE5A1–sildenafil structure, the H-loop shows dramatic changes in both secondary and tertiary structures. Upon binding of sildenafil, the H-loop is turned into a β-turn and a 3_{10}-helix (residues 667–670 and 672–675) and the whole loop migrates as much as 24 Å to cover the active site (Figure 30.7). The most dramatic change of the H-loop occurs in the structure of PDE5A1–icarisid II complex, in which the H-loop is converted to two β strands (residues 662–666 and 675–679) and migrates as much as 35 Å to close up completely the binding pocket. None of the four conformations of the H-loop in PDE5 is superimposable with those in other PDE families. The closest is the H-loop in the PDE5A1–IBMX structure, which contains two short α-helices as other PDE families do. However, the Cα atoms of helix H8 in PDE5A1–IBMX migrate as much as 7 Å from the corresponding residues in PDE4D2.

30.2.8 SUMMARY OF THE CATALYTIC DOMAIN STRUCTURES

1. A core catalytic domain with about 300 AAs, starting helix H3 (residues 115 to 410 in PDE4D2, Figure 30.3), can be defined on the basis of the structural comparison among seven PDE families. This core domain of PDE families, except the H-loop in PDE5, has similar folding and is superimposable among PDE families. The large fragment insertions of (1) 44 residues between H6 and H7 in PDE3B and (2) 38 and 47 residues between H15 and H16 in PDE1B and PDE3B (Figure 30.3) are located at considerable distance from the active site and are unlikely to directly interact with the substrate. Thus, these fragments may impact catalytic activity via allosteric mechanism or play a role in the interaction with regulatory domains of PDE or other proteins.

2. The active sites of the PDE families 1, 2, 3, 4, 7, and 9, and possibly other structurally unknown PDEs have uniform and comparable conformations. Thus, substrate differentiation between cAMP and cGMP in these PDE families is likely determined by variation of amino acid types and subtle conformational differences of the active site residues. PDE5 appears to be a unique family of PDE and shows four different conformations for the H-loop at the active site. Thus, the hydrolysis of the phosphodiester bond of nucleotides would be achieved via an allosteric mechanism in PDE5.

30.3 HYDROLYSIS MECHANISM

The crystal structures of the PDE4D/4B–AMP and PDE5A–GMP complexes have been determined and show how the products interact with the active sites of PDEs.[18,22,25] In the structures of the PDE4D2–AMP and PDE4B–AMP complexes,[18,22,25] the AMP phosphate group coordinates with both metal ions and forms hydrogen bonds with His160 and Asp318 (Figure 30.8). The adenine of AMP adopts an *anti* conformation and orients to the hydrophobic pocket made up of residues Tyr159, Leu319, Asn321, Thr333, Ile336, Gln369, and Phe372. It forms three hydrogen bonds with the side chains of Gln369 and Asn321 and stacks against Phe372. The ribose of AMP has a configuration of C3′ *endo* puckering and makes van der Waal's contacts with residues His160, Met273, Asp318, Leu319, Ile336, Phe340, and Phe372 of PDE4D2. AMP binding does not dramatically change the conformations of the active site, as revealed by the comparison with the unliganded PDE4D2. However, Asn321 changes its side chain conformation to form two hydrogen bonds with the adenosine ring of AMP.

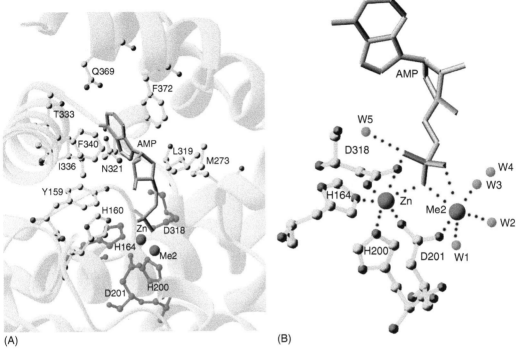

FIGURE 30.8 (See color insert following page 274.) (A) AMP binding at the active site of PDE4D2. The metal binding residues are shown in purple. (B) Binding of the metal ions. Dotted lines label the interactions involving the metals and AMP. Me2 is the second metal. (From Ref. [25]. With permission.)

The structural studies have led to a proposal for the mechanism of hydrolysis of the cyclic phosphodiester bond by PDE.[25] The divalent metals in both unliganded and AMP-bound structures have six coordinations that are arranged in an octahedral configuration. However, the phosphate oxygen atoms of AMP replace two water molecules and a water–OH bridging two divalent metal ions (Figure 30.8B). On basis of this observation, bridging water–OH was proposed to serve as the nucleophile to attack the phosphorus (Figure 30.9).[25] The

FIGURE 30.9 A proposed mechanism for hydrolysis of the phosphodiester bond of cyclic nucleotides by PDE. Amino acid numbers are for PDE4D2. (From Ref. [25]. With permission.)

metal-binding residue Asp318 that also forms a hydrogen bond with the bridging water–OH may activate the attacking nucleophile. The interactions of AMP phosphate with His160 and two metal ions, which are predicted to exist in the PDE–cAMP complex, may polarize the phosphodiester bond and thus assist the hydrolysis. In addition, His160 may serve as a proton donor for the completion of the phosphodiester bond hydrolysis. This proposal is consistent with the early model that the bridging hydroxide ion serves as the nucleophile on the basis of the quantum mechanical calculations.[46] The absolute conservation of His160 and the metal-binding residues in PDE families and the binding similarity between AMP and GMP suggest that this mechanism may be universal for all class I PDEs.

30.4   DISSECTION OF INHIBITOR SELECTIVITY

Extensive structural studies have revealed detailed interactions of inhibitors at the active site of PDEs and led to identification of subpockets or elements that are important for the recognition of nonselective and family-selective inhibitors.[18–31] The structures of PDE3, 4, 5, 7, and 9 in complex with the nonselective inhibitor IBMX showed two common elements for recognition of all inhibitors: a hydrogen bond with an invariant glutamine and stacking against a highly conserved phenylalanine (Gln369 and Phe372 in PDE4D2, Figure 30.2).[20,26,30] These two elements form a core of a common subpocket for binding of all inhibitors. Because of its hydrophobic nature, this subpocket is also called a hydrophobic clamp.[27] The structures of PDE4 and PDE5 in complex with various family-selective inhibitors confirm the existence of the common pocket and also show additional subpockets that contribute to differentiation of family-selective inhibitors.[18,20,24,27,28] An important subpocket or so-called "Q-pocket"[27] is around the area for binding of the cyclopentyloxy group of rolipram. This pocket contributes residues for direct interactions with inhibitors and also scaffolding residues for the stabilization of conformations of the interacting residues, and may play key roles in the determination of inhibitor selectivity.[24,27]

The kinetic studies on PDE7 and PDE4 and their mutants provide further insight into inhibitor selectivity. PDE7 was originally known as a rolipram-insensitive PDE family. Mutations of PDE7 residues to those of PDE4 and vice versa showed that sensitivity of the mutants to PDE4 inhibitors can be mutually switched.[30] The single mutations of S373Y, S377T, and I412S in PDE7 or Y329S in PDE4 produced a several fold gain or loss of sensitivity to rolipram and other PDE4 inhibitors (Figure 30.10). Double mutations of S373Y/S377T, S373Y/I412S, and S377T/I412S further amplified the sensitivity of PDE7 to PDE4 inhibitors, whereas the triple mutation of S373Y/S377T/I412S in PDE7 engineered a PDE4-like enzyme (Figure 30.10). These mutagenesis experiments suggest that multiple elements must work together to determine the inhibitor selectivity.[30]

The argument of multiple determinants for inhibitor selectivity is supported by the structures. First, the role of an invariant glutamine (Gln369 in PDE4D2, Figure 30.11A) was recognized by its hydrogen bond with the inhibitors in the various PDE–inhibitor structures.[20,23,24,27] The side chain amide group of the glutamine showed opposite orientations in the structures of PDE4 and PDE5[18,25] and was proposed to play a key role in substrate specificity.[18] Thus, the conformation and positioning of the glutamine is likely to be critical for inhibitor selectivity. Second, scaffolding residues that stabilize the conformation and position of the invariant glutamine are also critical. For example, Gln369 in PDE4D2 is stabilized by a hydrogen bond with Tyr329, and the mutation of Y329S in PDE4D2 or of

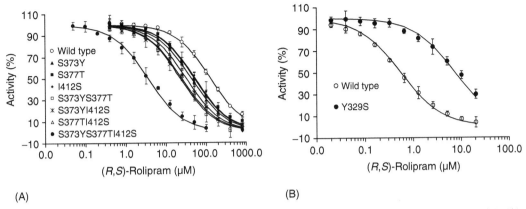

FIGURE 30.10 Inhibition on cAMP activity of the catalytic domains of wild type PDE7A1 (A) and PDE4D2 (B) as well as their mutants by PDE4 inhibitor (R,S)-rolipram. (From Ref. [30]. With permission.)

S373Y in PDE7 significantly change the inhibitor selectivity (Figure 30.10).[30] Third, residues that impact shape and size of the binding pocket will play roles in determining inhibitor selectivity. An example is Ile412 in PDE7A1, which corresponds to Ser368 in PDE4D2 and thus makes a smaller pocket in PDE7A (Figure 30.11). The I412S mutation contributes about twofold of sensitivity of PDE7A to PDE4 inhibitors.[30] In short, the structural and mutagenesis studies suggested that at least three factors play roles in inhibitor selectivity: (1) the conformation and positioning of the invariant glutamine, (2) residues scaffolding the glutamine, and (3) residues affecting shape and size of the binding pockets.[30] Other residues such as Asn321 in PDE4D2, which shows high variation across PDE families (Table 30.2), may also play a key role in the inhibitor selectivity, although this has yet to be addressed experimentally.

30.5 PERSPECTIVE

Although structural studies have been extensively performed, only preliminary knowledge on substrate specificity and inhibitor selectivity has been obtained. The multiple conformations of the H-loop at the active site of PDE5 suggest an allosteric mechanism for the hydrolysis of nucleotides. In contrast, other known structures of PDE families show rigid and comparable active sites, suggesting a more competitive mechanism of catalysis. The structures of the catalytic domains have shown that the conformations of the active sites of PDE families cannot be simply categorized according to their cAMP- or cGMP-specificity. Thus, mechanisms for the recognition of cAMP and cGMP are expected to be complicated and each family of PDEs may have its characteristic pattern. To elucidate substrate specificity, crystal structures of representative PDE families (cAMP-, cGMP-, and dual specificity) in complex with both cAMP and cGMP or their analogs are essential. In addition, structures of full-length PDEs or fragments containing both regulatory and catalytic domains are necessary for the illustration of regulation on the catalytic activity although expression of large quantity of functional full-length PDE proteins are challenging at present.

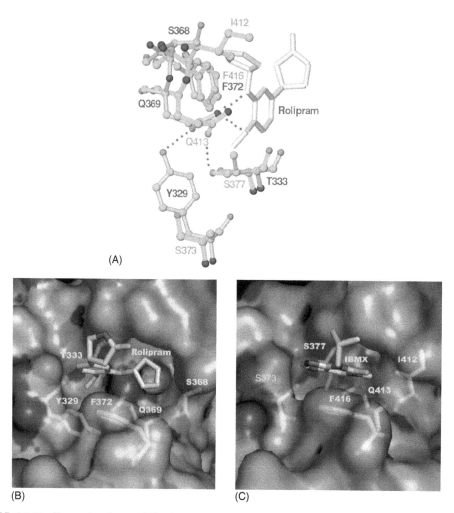

FIGURE 30.11 (See color insert following page 274.) Inhibitor binding in PDE4 and PDE7. (A) Superposition of rolipram-binding residues of PDE4D2 (cyan sticks and blue labels) over those of PDE7A1 (green sticks and red labels). Dotted lines represent hydrogen bonds. The cyclopentanyl ring of rolipram will clash with the side chain of Ile412 if Gln413 forms hydrogen bonds with the oxygen atoms of phenylmethoxy of rolipram. (B) Surface presentation of the subpocket for the cyclopentanyl group of rolipram in PDE4D2. (C) Surface presentation of the same binding pocket in PDE7A1. It is clear that the size and shape of the subpocket for the cyclopentanyl group are different. (From Ref. [30]. With permission.)

Inhibitor selectivity is an essential issue for the improvement of drug efficacy and minimization of side effects. However, it remains a mystery even for the PDE4 family, which has been most extensively studied. A multidisciplinary approach appears to be necessary for a complete illustration of the inhibitor selectivity. First, structures of all 11 PDE families and their complexes with nonselective and selective inhibitors are needed to show conformational differences of the inhibitor-binding pockets and to identify potential elements in the determination of inhibitor selectivity. Second, mutagenesis experiments may show contribution of

each element to the inhibitor selectivity and lead to identification of a characteristic combination of recognition elements. Finally, structure-based development of novel selective inhibitors and an analysis of quantitative structure–activity relationship will confirm and extend our understanding of inhibitor selectivity.

ACKNOWLEDGMENT

We thank the support of NIH (GM59791 to H.K).

REFERENCES

1. Torphy, T.J. 1998. Phosphodiesterase isozymes: Molecular targets for novel antiasthma agents. *Am J Respir Crit Care Med* 157:351.
2. Soderling, S.H., and J.A. Beavo. 2000. Regulation of cAMP and cGMP signaling: New phosphodiesterases and new functions. *Curr Opin Cell Biol* 12:174.
3. Mehats, C., et al. 2002. Cyclic nucleotide phosphodiesterases and their role in endocrine cell signaling. *Trends Endocrinol Metab* 13:29.
4. Houslay, M.D., and D.R. Adams. 2003. PDE4 cAMP phosphodiesterases: Modular enzymes that orchestrate signalling cross-talk, desensitization and compartmentalization. *Biochem J* 370:1.
5. Castro, A., et al. 2005. Cyclic nucleotide phosphodiesterases and their role in immunomodulatory responses: Advances in the development of specific phosphodiesterase inhibitors. *Med Res Rev* 25:229.
6. Goraya, T.A., and D.M. Cooper. 2005. Ca^{2+}–calmodulin-dependent phosphodiesterase (PDE1): Current perspectives. *Cell Signal* 17:789.
7. Barnette, M.S., and D.C. Underwood. 2000. New phosphodiesterase inhibitors as therapeutics for the treatment of chronic lung disease. *Curr Opin Pulmonary Med* 6:164.
8. Movsesian, M.A. 2000. Therapeutic potential of cyclic nucleotide phosphodiesterase inhibitors in heart failure. *Expert Opin Investig Drugs* 9:963.
9. Truss, M.C., et al. 2001. Phosphodiesterase 1 inhibition in the treatment of lower ruinary tract dysfunction: From bench to bedside. *World J Urol* 19:344.
10. Liu, Y., et al. 2001. Cilostazol (pletal): A dual inhibitor of cyclic nucleotide phosphodiesterase type 3 and adenosine uptake. *Cardiovasc Drug Rev* 19:369.
11. Huang, Z., et al. 2001. The next generation of PDE4 inhibitors. *Curr Opin Chem Biol* 5:432.
12. Corbin, J.D., and S.H. Francis. 2002. Pharmacology of phosphodiesterase-5 inhibitors. *Int J Clin Pract* 56:453.
13. Rotella, D.P. 2002. Phosphodiesterase 5 inhibitors: Current status and potential applications. *Nat Rev Drug Discov* 1:674.
14. Giembycz, M.A. 2002. Development status of second generation PDE4 inhibitors for asthma and COPD: The story so far. *Monaldi Arch Chest Dis* 57:48.
15. Spina, D. 2003. Theophylline and PDE4 inhibitors in asthma. *Curr Opin Pulm Med* 9:57.
16. Lipworth, B.J. 2005. Phosphodiesterase-4 inhibitors for asthma and chronic obstructive pulmonary disease. *Lancet* 365:167.
17. Manallack, D.T., R.A. Hughes, and P.E. Thompson. 2005. The next generation of phosphodiesterase inhibitors: Structural clues to ligand and substrate selectivity of phosphodiesterases. *J Med Chem* 48:3449.
18. Zhang, K.Y., et al. 2004. A glutamine switch mechanism for nucleotide selectivity by phosphodiesterases. *Mol Cell* 15:279.
19. Iffland, A., et al. 2005. Structural determinants for inhibitor specificity and selectivity in PDE2A using the wheat germ in vitro translation system. *Biochemistry* 44:8312.

20. Scapin, G., et al. 2004. Crystal structure of human phosphodiesterase 3B: Atomic basis for substrate and inhibitor specificity. *Biochemistry* 43:6091.
21. Xu, R.X., et al. 2000. Atomic structure of PDE4: Insight into phosphodiesterase mechanism and specificity. *Science* 288:1822.
22. Xu, R.X., et al. 2004. Crystal structures of the catalytic domain of phosphodiesterase 4B complexed with AMP, 8-Br-AMP, and rolipram. *J Mol Biol* 337:355.
23. Lee, M.E., et al. 2002. Crystal structure of phophodiesterase 4D and inhibitor complex. *FEBS Lett* 530:53.
24. Huai, Q., et al. 2003. Three dimensional structures of PDE4D in complex with roliprams and implication on inhibitor selectivity. *Structure* 11:865.
25. Huai, Q., J. Colicelli, and H. Ke. 2003. The crystal structure of AMP-bound PDE4 suggests a mechanism for phosphodiesterase catalysis. *Biochemistry* 42:13220.
26. Huai, Q., et al. 2004. Crystal structures of phosphodiesterases 4 and 5 in complex with inhibitor IBMX suggest a conformation determinant of inhibitor selectivity. *J Biol Chem* 279:13095.
27. Card, G.L., et al. 2004. Structural basis for the activity of drugs that inhibit phosphodiesterases. *Structure* 12:2233.
28. Card, G.L., et al. 2005. A family of phosphodiesterase inhibitors discovered by cocrystallography and scaffold-based drug design. *Nat Biotechnol* 23:201.
29. Sung, B.J., et al. 2003. Structure of the catalytic domain of human phosphodiesterase 5 with bound drug molecules. *Nature* 425:98.
30. Wang, H., et al. 2005. Multiple elements jointly determine inhibitor selectivity of cyclic nucleotide phosphodiesterases 4 and 7. *J Biol Chem* 280:30949.
31. Huai, Q., et al. 2004. Crystal structure of phosphodiesterase 9 in complex with inhibitor IBMX. *Proc Natl Acad Sci USA* 101:9624.
32. Martinez, S.E., et al. 2002. The two GAF domains in phosphodiesterase 2A have distinct roles in dimerization and in cGMP binding. *Proc Natl Acad Sci USA* 99:13260.
33. Smith, K.J., et al. 1996. Determination of the structure of the N-terminal splice region of the cyclic AMP-specific phosphodiesterase RD1 (RNPDE4A1) by ^{1}H NMR and identification of the membrane association domain using chimeric constructs. *J Biol Chem* 271:16703.
34. Fisher, D.A., et al. 1998. Isolation and characterization of PDE9A, a novel human cGMP-specific phosphodiesterase. *J Biol Chem* 273:15559.
35. Wang, P., et al. 2003. Identification and characterization of a new human type 9 cGMP-specific phosphodiesterase splice variant (PDE9A5) differential tissue distribution and subcellular localization of PDE9A variants. *Gene* 314:15.
36. Zhu, J., E. Mix, and B. Winblad. 2001. The antidepressant and antiinflammatory effects of rolipram in the central nervous system. *CNS Drug Rev* 7:387.
37. Torphy, T.J., et al. 1992. Coexpression of human cAMP-specific phosphodiesterase activity and high affinity rolipram binding in yeast. *J Biol Chem* 267:1798.
38. Barnette, M.S., et al. 1996. Association of the anti-inflammatory activity of phosphodiesterase 4 (PDE4) inhibitors with either inhibition of PDE4 catalytic activity or competition for [3H]rolipram binding. *Biochem Pharmacol* 51:949.
39. Laliberte, F., et al. 2000. Conformational difference between PDE4 apoenzyme and holoenzyme. *Biochemistry* 39:6449.
40. Michaeli, T., et al. 1993. Isolation and characterization of a previously undetected human cAMP phosphodiesterase by complementation of cAMP phosphodiesterase-deficient *Saccharomyces cerevisiae*. *J Biol Chem* 268:12925.
41. Hetman, J.M., et al. 2000. Cloning and characterization of PDE7B, a cAMP-specific phosphodiesterase. *Proc Natl Acad Sci USA* 97:472.
42. Laxman, S., A. Rascon, and J.A. Beavo. 2005. Trypanosome cyclic nucleotide phosphodiesterase 2B binds cAMP through its GAF-A domain. *J Biol Chem* 280:3771.

43. Martins, T.J., M.C. Mumby, and J.A. Beavo. 1982. Purification and characterization of a cyclic GMP-stimulated cyclic nucleotide phosphodiesterase from bovine tissues. *J Biol Chem* 257:1973.

44. Soderling, S.H., S.J. Bayuga, and J.A. Beavo. 1998. Identification and characterization of a novel family of cyclic nucleotide phosphodiesterases. *J Biol Chem* 273:15553.

45. Fink, T.L., et al. 1999. Expression of an active, monomeric catalytic domain of the cGMP-binding cGMP-specific phosphodiesterase (PDE5). *J Biol Chem* 274:24613.

46. Zhan, C., and F. Zheng. 2001. First computational evidence for a catalytic bridging hydroxide ion in a phosphodiesterae active site. *J Am Chem Soc* 123:2835.

31 Bench to Bedside: Multiple Actions of the PDE3 Inhibitor Cilostazol

Junichi Kambayashi, Yasmin Shakur, and Yongge Liu

CONTENTS

31.1 Introduction 627
31.2 Pharmacology and Mechanistic Studies 628
    31.2.1 Medicinal Chemistry 628
    31.2.2 Dual Inhibition of PDE3 and Adenosine Uptake 628
    31.2.3 Antithrombotic Effect of Cilostazol 630
    31.2.4 Effect on Skin Bleeding Time 632
    31.2.5 Effect on Vasculature 633
    31.2.6 Effect on Lipid Metabolism 634
    31.2.7 Neuronal Protective Effect 634
    31.2.8 Summary 635
31.3 Clinical Applications 636
    31.3.1 Current Marketing Authorization Status 636
    31.3.2 Recent Patient Exposure 636
    31.3.3 Safety and Adverse Events 636
    31.3.4 Approved Indications 637
        31.3.4.1 Ischemic Symptoms Due to Peripheral Arterial
                 Occlusive Disease 637
        31.3.4.2 Intermittent Claudication 637
        31.3.4.3 Prevention of Stroke Recurrence 638
    31.3.5 Potential New Indications 639
        31.3.5.1 Prevention of Thrombosis after Coronary Stenting 639
        31.3.5.2 Prevention of Restenosis after Percutaneous
                 Transluminal Coronary Angioplasty with or without Stent 640
        31.3.5.3 Attenuation of Restenosis in Diabetic Patients 641
        31.3.5.4 New Indications Related to Vasodilatory Effect of Cilostazol 642
31.4 Summary 642
Acknowledgment 642
References 642

31.1 INTRODUCTION

It is well established that cAMP plays an important role in the regulation of many processes within the cardiovascular system, including platelet activation, vascular smooth muscle, and cardiomyocyte cell function. Intracellular levels of cAMP are controlled by the opposing

627

actions of the adenylate cyclases, which are responsible for cAMP synthesis and by phospho-diesterases (PDEs), which are responsible for its degradation. As the sole mechanism respon-sible for the breakdown of cAMP, PDEs have long been regarded as a target for pharmacotherapy. Of the 11 PDE gene families, PDE3 is one of the major forms expressed in the cells of the cardiovascular system and has been shown to play a significant role in modulating platelet and vascular smooth muscle cell functions.[1–5] In the early 1980s, in a quest to identify a new class of antiplatelet drug, several selective PDE3 inhibitors demon-strating potent antiplatelet and vasodilatory activities were identified by Otsuka. Cilostazol was selected from a number of analogs for further development because of its potent anti-platelet and vasodilatory activities and its minimal cardiac effects.

Cilostazol (Pletaal) was approved for the treatment of various ischemic symptoms caused by peripheral arterial occlusive disease in Japan in 1988[6] and shortly thereafter in several other Asian countries. Peripheral arterial occlusive disease is a progressive disease caused by narrowing or occlusion of peripheral arteries, particularly those in the leg, causing various ischemic symptoms such as leg pain on walking (intermittent claudication) and gangrene, leading to limb loss.[7] Peripheral arterial occlusive disease is considered to be a manifestation of systemic atherosclerosis comparable to ischemic stroke and myocardial infarction and affects approxi-mately 8–12 million people in the United States alone. In 1999 and 2001, cilostazol (Pletal) was approved in the United States, UK, and Ireland, respectively, for treatment of the symptoms of intermittent claudication due to peripheral arterial occlusive disease, with a contraindication in patients with congestive heart failure (CHF) of any severity.[6] In addition to the treatment of peripheral arterial occlusive disease, including intermittent claudication, the clinical application of cilostazol has been explored in several other therapeutic areas: recently cilostazol has been approved in Japan and several other Asian countries for the reduction of stroke recurrence and has also been evaluated for use in the prevention of restenosis in postpercutaneous transluminal coronary angioplasty, with promising results (see details in Section 31.3.5.2).

PDE3 inhibition was initially considered to be cilostazol's sole mechanism of action. However, recent studies have revealed inhibition of adenosine uptake to be another critical pharmacological action of cilostazol.[8,9] This dual inhibitory property of cilostazol may explain its favorable and unique pharmacological profile: potent antiplatelet and vasodilatory effects, with minimal cardiac effect.[10,11] Since its launch more than 15 years ago, cilostazol has been of benefit to many patients worldwide and its clinical applications have continued to expand. We believe that the success of cilostazol in its clinical application is due to its dual inhibitory effect on PDE3 and adenosine uptake.[8,9] In the following sections, we will sum-marize the results of preclinical studies involving cilostazol and highlight the efficacy, safety, and clinical utility of this drug.

31.2 PHARMACOLOGY AND MECHANISTIC STUDIES

31.2.1 MEDICINAL CHEMISTRY

Cilostazol (6-[4-(1-cyclohexyl-1H-tetrazol-5-yl)butoxy]-3,4-dihydro-2(1H)-quinolinone) (see Figure 31.1) is a 2-oxo-quinoline derivative, which was synthesized in 1983 by Otsuka Pharmaceutical Company, Tokushima, Japan.[12] It is an odorless white to off-white crystal-line powder, practically insoluble in water, but soluble in dimethylsulfoxide.

31.2.2 DUAL INHIBITION OF PDE3 AND ADENOSINE UPTAKE

Cilostazol is a selective and potent inhibitor of the PDE3 enzyme family that hydrolyzes cAMP and cGMP and is comprised of two subfamilies, PDE3A and PDE3B.[13] The inhibitory

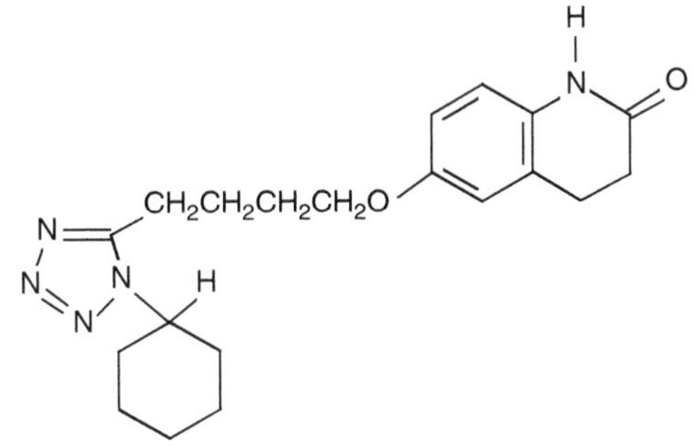

FIGURE 31.1 Chemical structure of cilostazol.

potency of cilostazol for both PDE3 subtypes is similar, with IC_{50} values for the recombinant enzymes being around 0.5 μM.[13,14] The PDE3 subtypes show both distinct and overlapping tissue and cellular distributions.[15] For example, PDE3A is expressed in platelets and cardiomyocytes, and PDE3B in adipocytes, hepatocytes, pancreatic β-cells, macrophages, and T cells. Both subtypes have been detected in vascular smooth muscle cells. Other than its potent effects on PDE3, cilostazol has been reported to have a weaker inhibitory effect on PDE5 (IC_{50} of 4.4 μM),[14] although other studies using similar substrate concentrations (Shakur et al., unpublished observations) have found no effect of this drug on recombinant PDE5 at concentrations as high as 10 μM. The recombinant PDE5 for these studies was expressed in Sf9 insect cells and mammalian COS-7 cells, respectively. Whether differences in the recombinant protein produced using these expression systems account for the observed differences in inhibition by cilostazol is not known. Additionally, unlike the PDE3 inhibitor, milrinone, which has been associated with decreased mortality in patients with CHF, cilostazol is not an inhibitor of PDE4. The IC_{50} values for PDE4 are >100 and 16 μM for cilostazol and milrinone, respectively.[13]

An important characteristic of cilostazol that distinguishes it from other PDE3 inhibitors, such as milrinone, is its ability to inhibit adenosine uptake (IC_{50} 5–10 μM) in a wide range of cell types, including endothelial cells, vascular smooth muscle cells, erythrocytes, platelets, and cardiac ventricular myocytes.[10,16] In contrast, milrinone has been found to have no significant inhibitory effect on adenosine uptake at concentrations as high as 100 μM.[16] Assessment of the relative potency of cilostazol for inhibition of PDE3 compared to adenosine uptake is not possible based on a comparison of the IC_{50} values, since values for the former were calculated using a cell-free assay and those for the latter in intact cells. Nevertheless, studies have demonstrated that the inhibition of adenosine uptake is a significant feature in the overall pharmacology of this drug and will be discussed later.

The chronic use of certain PDE3 inhibitors, such as milrinone, has been associated with proarrhythmic activities, possibly caused by excessive increases in intracellular cAMP in the heart in patients with severe CHF.[17–19] At the time of its approval, cilostazol, being a PDE3 inhibitor, was considered to belong to the same class of compounds as milrinone and consequently, even though its effect on mortality in CHF patients is unknown, its use was contraindicated in patients with heart failure. However, new evidence suggests cilostazol's ability to inhibit adenosine uptake is an important feature that may counteract the undesirable cardiac effects observed with other PDE3 inhibitors, such as milrinone.[9,11]

As with other adenosine uptake inhibitors, clinically relevant concentrations of cilostazol (~3 μM) have been shown to increase rabbit cardiac interstitial and circulatory adenosine levels.[11] This increase in extracellular adenosine has the favorable consequence of enhancing antiplatelet[10] and vasodilatory[11] effects, while diminishing the positive inotropic response

caused by PDE3 inhibition in the heart.[11] The differential effect of adenosine on PDE3 inhibition in different cell types can be explained by differences in the distribution of adenosine receptor subtypes and their coupling to G_i or G_s proteins. In cardiac ventricular myocytes and pacemaker cells, the G_i-coupled adenosine A_1 receptor is the predominant form[20] and acts to inhibit cAMP production. A_1 receptors may also directly inhibit cAMP-dependent protein phosphorylation.[21] Thus, elevation of cardiac adenosine levels caused by inhibition of adenosine uptake would be expected to diminish the effect of PDE3 inhibition on the heart. This was demonstrated by studies in isolated rabbit hearts, which showed that cilostazol induced a smaller increase in contractility compared to milrinone and that A_1 antagonism increased the cardiotonic effect of cilostazol.[22,11] On the other hand, in platelets and smooth muscle cells, the G_s-coupled A_2 receptor predominates. In this case, activation of the A_2 receptor would be expected to enhance the increases in cAMP caused by inhibition of PDE3. The interaction between PDE3 inhibition and adenosine on cardiac ventricular myocytes, vascular smooth muscle cells, and platelets is summarized in Figure 31.2. The evidence is consistent with the interpretation that the dual inhibition of PDE3 and adenosine uptake contributes to the overall efficacy and safety of cilostazol.

31.2.3 ANTITHROMBOTIC EFFECT OF CILOSTAZOL

Various *in vitro* studies have established that cilostazol inhibits, in a dose-dependent fashion, aggregation of platelets from several species including human, dog, rabbit, guinea pig, rat, and mouse.[23–27] The inhibition was observed against aggregation induced by a wide range of platelet activators such as ADP, collagen, epinephrine, arachidonic acid, and thrombin receptor–activating peptide. In addition to the effect of cilostazol to inhibit platelet aggregation, it has also been shown to disperse human and rabbit platelet aggregates that were formed in response to ADP, collagen, arachidonic acid, or epinephrine.[23] Cilostazol differs from aspirin in that it can inhibit both primary and secondary aggregations as well as shear stress–induced platelet aggregation.[28,29]

The inhibitory effects of cilostazol on platelet aggregation have also been demonstrated in various *ex vivo* and *in vivo* studies. Orally administered cilostazol significantly reduced *ex vivo* platelet aggregation induced by ADP, collagen, and arachidonic acid.[30,31] In various animal studies, cilostazol has been shown to inhibit thrombotic occlusion: one study utilized a vascular graft implanted in canine femoral artery[32]; in another study, cilostazol inhibited reocclusion after canine coronary arterial thrombolysis with recombinant tissue-type plasminogen activator.[33] In a porcine carotid artery injury model, with electrical stimulation, cilostazol significantly prolonged the time to total occlusion and decreased thrombus weight.[34] In other studies, cilostazol reduced thrombosis formation in a cyclic flow reduction model in rabbits[35] and in mice, it prevented death due to pulmonary thromboembolism produced by intravenous injection of ADP or collagen.[23] In a randomized, double-blind crossover study in patients comparing the inhibitory effects of cilostazol, aspirin, and ticlopidine on platelet aggregation *ex vivo*, it was demonstrated that cilostazol was more effective than aspirin in inhibiting ADP-induced platelet aggregation and better than ticlopidine in inhibiting arachidonic acid–induced platelet aggregation.[36] In patients who had undergone coronary artery bypass surgery, cilostazol was found to significantly suppress platelet aggregation induced by ADP and arachidonic acid.[37]

Evidence that PDE3 inhibition plays an important role in the antithrombotic effect of cilostazol comes from the observations that cAMP levels in washed human and rabbit platelets were increased by cilostazol in a dose-dependent manner.[26] Additionally, in the presence of forskolin or PGE_1, the effect of cilostazol on intraplatelet cAMP was enhanced as was its inhibitory effect on platelet aggregation.[22,25,26,38] However, PDE3 inhibition is not the

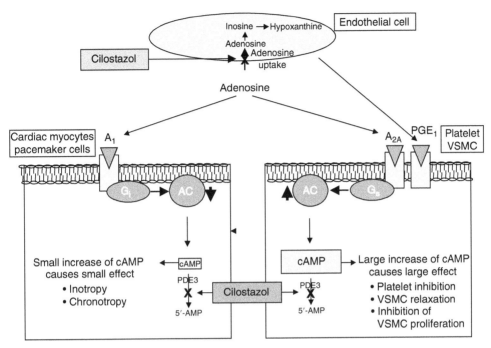

FIGURE 31.2 Cilostazol's pharmacology involves the interaction of PDE3 inhibition and adenosine. Adenosine is produced through the metabolism of ATP to ADP and AMP, and further de-phosphorylation by 5'-nucleotidase. Once in the extracellular space, adenosine is rapidly taken up into the endothelial cells by the adenosine uptaker (or transporter), a nucleoside-specific facilitated diffusion carrier. Within the cell, adenosine is efficiently degraded by adenosine deaminase to inosine and subsequently to hypoxanthine. Inhibition of adenosine uptake by cilostazol increases the interstitial adenosine level. Adenosine level is high in cardiac and working skeletal muscle due to the high turnover rate of ATP, and it is further increased under conditions of restricted flow, such as ischemia. In cardiac myocytes and pacemaker cells, adenosine acts on A_1 receptors (the predominant adenosine receptor subtype in these cells) to inhibit adenylate cyclase (AC) via the G_i coupling, thus reducing the cAMP level. Subsequent inhibition of cAMP degradation by cilostazol through the PDE3 inhibition results in a small increase in cAMP and therefore in correspondingly small cardiac functional changes, i.e., small effect on positive inotropy and chronotropy.[16] On the other hand, in platelets and vascular smooth muscle cells (VSMC), adenosine stimulates A_{2A} receptors (the predominant adenosine receptor subtype in these cells) to activate AC via the G_s protein, resulting in an elevation of cAMP. This in turn has synergistic effects on cAMP levels when PDE3 is also inhibited.[10] Thus, the combination of effects on adenosine uptake and PDE3 is more effective at inhibiting platelet activation than either inhibition of adenosine uptake or PDE3 alone.[10] In VSMC, the increase in cAMP induces cell relaxation and inhibition of cell proliferation. PGE_1 released from EC has also been shown to synergize with cilostazol in the inhibition of platelet activation.[39]

only mechanism involved in cilostazol's inhibitory effect on platelet aggregation, especially under *in vivo* conditions. Recent studies provide evidence that changes in extracellular adenosine levels brought about by inhibition of adenosine uptake also contribute to this inhibitory effect. A study by Sun et al.[10] showed that cilostazol inhibited the uptake of adenosine generated during platelet activation, causing a measurable increase in extracellular adenosine levels. The inhibitory effect of cilostazol was significantly reversed by ZM241385, an A_{2A} adenosine receptor antagonist, and by adenosine deaminase.[10] Regarding the antiplatelet

effect, cilostazol has been found to be 50-fold more potent in *in vivo* studies than in *in vitro* studies, suggesting that it may have other actions, such as increasing PGE_1 release from endothelium.[39]

The potency of cilostazol's inhibition of adenosine uptake is moderate when compared with known adenosine uptake inhibitors, such as dipyridamole.[40] Thus, a combination of cilostazol with dipyridamole may further enhance the antiplatelet effect. Indeed, several recent studies from our laboratories have shown that dipyridamole synergistically increased the antiplatelet effect of cilostazol both *in vitro*,[41] *ex vivo*,[41] and *in vivo*.[35,42] Whether a novel combination therapy can be developed requires further investigation.

31.2.4 EFFECT ON SKIN BLEEDING TIME

One of the major side effects associated with antiplatelet treatments, such as aspirin and clopidogrel, is increased bleeding risk. The molecular events regulating thrombosis are considered in large part to be similar to those underlying hemostasis, and so it is interesting that in animal and clinical studies cilostazol effectively inhibits thrombosis without increasing bleeding risk. Bleeding time in rabbits was increased by aspirin, but not by cilostazol[35] and, in humans, several studies have reported no increase in bleeding risk in patients treated with cilostazol.[43–46] For example, Wilhite et al.[44] (see Figure 31.3) showed that aspirin and clopidogrel significantly prolonged bleeding time individually and to a greater extent in combination, whereas cilostazol did not. Moreover, when cilostazol was added to aspirin, clopidogrel, or a combination of the two, there was no additional increase in bleeding time. Therefore, cilostazol can be used safely in combination with other antiplatelet agents such as aspirin or clopidogrel. It is challenging to fully explain the dissociation between the effect on platelets and bleeding time and at this point we can only speculate on the reason for this. In both thrombosis and hemostatic plug formation, platelet activation per se may be very similar, but the triggering agonist, i.e., collagen, ADP, thrombin, etc., may be different. More importantly, there may be significant differences in the local environment at the thrombotic site and the bleeding site. At the thrombotic site, it is likely that PGI_2, PGE_1,

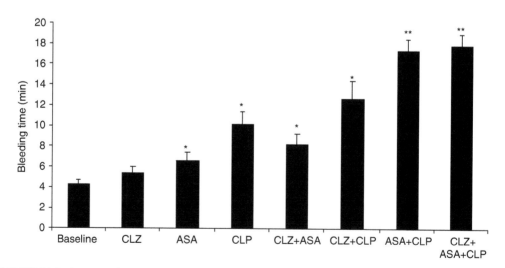

FIGURE 31.3 A comparison of skin bleeding time after treatment with cilostazol, aspirin, and clopidogrel. (Data were obtained from the study by Wilhite et al.[44]) Patients were treated with aspirin (ASA, 325 (Ref. [44]) mg daily), clopidogrel (CLP, 75 mg daily), cilostazol (CLZ, 100 mg twice daily), or a combination for 14 days. $*p < 0.05$ vs. baseline. $**p < 0.05$ vs. all single agents and vs. cilostazol+ASA ASA and cilostazol+CLP.

and adenosine are released from the damaged or ischemic endothelial cells. Since all these mediators are stimulants of adenylate cyclase, it is possible that cilostazol may synergize with these mediators in elevating intraplatelet cAMP. This idea is supported by the finding that the inhibitory effect of cilostazol on platelet aggregation was potentiated by the presence of endothelial cells and that this potentiation was reversed by aspirin.[39] In contrast, mediators such as PGI_2, PGE_1, and adenosine may not be released at the site of hemostatic plug formation as the endothelial damage is mechanical rather than pathological (inflammation or ischemia). Differences in the local environment (presence of PGI_2, PGE_1, and adenosine) at the sites of thrombosis may also explain why the antiplatelet effect of cilostazol is much more potent *in vivo* than *ex vivo,* or *in vitro.* It has been demonstrated in patients with thrombotic disease that cilostazol does not inhibit the production of PGI_2,[47] whereas aspirin is known to inhibit PGI_2 synthesis. Further studies are required to provide a better understanding of the mechanism by which cilostazol inhibits thrombosis without affecting hemostasis, and may help toward the design of novel antithrombotic agents in the future.

31.2.5   EFFECT ON VASCULATURE

The vasodilatory activity of cilostazol may contribute its therapeutic activity in intermittent claudication and other diseases. Although cilostazol has broad vasodilatory effects, the magnitude is not uniform. For example, vasodilation of the vertebral and femoral arteries is greater than in renal arteries[23] and cilostazol dilated large cerebral arteries without affecting regional cerebral blood flow in humans.[48] Also, in intermittent claudication patients treated with cilostazol, resting ankle-brachial index was increased,[32] suggesting an improved circulation to the leg. While these studies looked at large vessels, we studied blood flow in the tissue of rabbit gastrocnemius muscle. Cilostazol had no effect on resting blood flow in the muscle tissue, but it augmented blood flow in electrically stimulated contracting muscle,[49] suggesting that the vasodilatory effect of cilostazol was selective to contracting muscle. The selective effect of cilostazol on exercise muscle may be mediated by an increase in adenosine production in contracting muscle.[49]

PDE3 inhibitors reduce vascular smooth muscle cell proliferation,[1] thus it is not surprising that cilostazol has been shown to have antiproliferative effects. Cilostazol inhibits rat and human vascular smooth muscle cell proliferative responses to a variety of growth factors, including platelet-derived growth factor, insulin, insulin-like factor 1 and serum, indicating that this effect is nonspecific for growth factors.[50–52] Cilostazol has been shown to down-regulate the cell cycle transcription factor E2F,[53] mitogen-activated protein kinase, and cell cycle–related cyclins and their associated kinases.[54] Other factors that may contribute to the antiproliferative effect of cilostazol include inhibition of growth factor expression in smooth muscle cells,[55] reduction in platelet-derived growth factor release as a consequence of antiplatelet effects, and suppression of P-selectin-mediated platelet activation and platelet–leukocyte interaction.[56]

In contrast to its inhibitory effects on vascular smooth muscle cells and their growth factors, cilostazol has been found to stimulate endothelial cell growth and increase endothelial cell-specific growth factors.[57] Circulating vascular endothelial growth factor was significantly increased in cilostazol-treated, but not in pentoxifylline- or placebo-treated nondiabetic intermittent claudication patients.[58] The induction of vascular endothelial cell monocyte chemoattractant protein-1 (thought to be an initial event in the development of atherosclerotic lesions) was inhibited by cilostazol.[59] Neutrophil adhesion to endothelial cells was inhibited by cilostazol, suggesting that it may reduce neutrophil infiltration and superoxide production, both of which may trigger atherosclerotic lesions.[60]

In vivo, cilostazol has been shown to suppress neointimal formation in the balloon-injured carotid artery of the rat[61] and after stenting in dogs.[62] In the rat, some of these effects have

been attributed to the release of vascular hepatocyte growth factor, which preferentially enhanced endothelial cell but not smooth muscle cell growth.[63] Cilostazol has been shown to suppress atherosclerotic lesion formation in mice through reduction of superoxide and TNF-α formation, thereby reducing NF-κB activation/transcription, VCAM-1/MCP-1 expression, and monocyte recruitment.[64] Several promising prospective, randomized clinical studies on the prevention of coronary restenosis following percutaneous coronary intervention and stent implantation will be discussed later in the clinical section.

31.2.6 EFFECT ON LIPID METABOLISM

Various studies suggest that cilostazol may protect against atherosclerosis by decreasing levels of atherogenic proteins and increasing those of antiatherogenic proteins. In studies on intermittent claudication patients, plasma levels of triglycerides were decreased and high-density lipoprotein cholesterol and apolipoprotein A_1 were increased following 12 weeks of cilostazol treatment,[65] and remnant lipoprotein concentrations, which were found to be significantly elevated in these patients, were reduced by cilostazol.[66] In patients with type 2 diabetes, treatment with cilostazol was associated with decreased plasma triglycerides, apolipoproteins (B, CII, CIII, E) and remnant-like lipoprotein particle cholesterol and with an increase in high-density lipoprotein cholesterol and apoliporotein AII.[67–72] A decrease in plasma triglycerides was also observed, following treatment for 6 months with cilostazol, in type 2 diabetic patients with peripheral vascular disease. Additionally, an increase in the levels of the omega-3 fatty acid, docohexaenoic acid, was also observed in these patients.[73]

Although the mechanism of this action of cilostazol has not been fully elucidated, there is evidence to suggest that cilostazol may exert its effects on lipid metabolism by increasing lipoprotein lipase activity: cilostazol increased cardiac and plasma lipoprotein lipase in diabetic rats.[74] Activation of the capillary-bound fraction of lipoprotein lipase from rat heart has been observed following incubation with the catalytic subunit of protein kinase A.[75] It is therefore possible that the effect of cilostazol on cardiac lipoprotein lipase may be mediated via this cAMP-dependent pathway. Lipoprotein lipase production and release from adipocytes has also been shown to be regulated by cAMP.[76,77] The cAMP level in adipocytes is regulated by several receptors including α- and β-adrenoreceptors and the adenosine A_1 and A_2 receptors.[78–80] In the case of the adenosine receptors, the inhibitory effect of the A_1 receptor is thought to predominate over the activating effect of the A_2 receptor.[78] These studies suggest the potential for cilostazol, via its dual inhibitory actions on PDE3 and on adenosine uptake, to affect lipoprotein lipase production and release from the adipocyte. Whether any effect of cilostazol on adipocyte lipoprotein lipase production or release would be of a stimulatory or inhibitory nature would ultimately depend on the relative potency for inhibition of adenosine uptake vs. PDE3 inhibition in this cell type, and requires further studies. Other studies suggest that cilostazol may inhibit the production of cytokines (TNF-α and IL-6) that have inhibitory effects on lipoprotein lipase activity.[81,82] Cilostazol has been shown to decrease lipopolysaccharide (LPS)-induced release of TNF-α from human umbilical endothelial cells and to reduce levels of IL-6 in intermittent claudication patients.[58,64,83] Finally, studies in rats suggest that cilostazol may lower plasma triglycerides by inhibiting the rate of triglyceride secretion.[84]

31.2.7 NEURONAL PROTECTIVE EFFECT

Cilostazol has been shown to protect neurons against ischemia-induced cell death. Using magnetic resonance imaging (MRI), Lee et al. showed that cilostazol, given postischemia,

reduced cerebral infarct size following a 2 h ischemia of the middle cerebral artery and 24 h reperfusion.[85] In a similar model, infarct size was significantly reduced by cilostazol, but not by aspirin and clopidogrel.[64] Moreover, only cilostazol decreased DNA fragmentation observed in the cortical tissue. Investigation of the mechanism of this neuronal protection demonstrated that cilostazol ameliorated the neuronal damage by suppression of apoptotic cell death via maxi-K channel opening[86] and that the suppression of apoptosis was mediated by increased casein kinase 2 phosphorylation and decreased PTEN (phosphatase and tensin homolog deleted from chromosome 10) phosphorylation.[86] Similar findings have also been observed in a neuroblastoma cell line SK-N-SH.[43] In addition, Choi et al. have demonstrated that cilostazol decreased Bax protein and cytochrome *c* release and increased the level of Bcl-2 protein in the ischemic and reperfused cortical region in rats, lending further support to the finding that cilostazol has a protective effect against apoptosis.[87] In this study, the cAMP level was significantly elevated in the cortex following treatment with cilostazol, suggesting that PDE3 inhibition plays a role in these effects. The neuronal protective effects observed for cilostazol, together with its well-known antiplatelet and vasodilatory effects, support the clinical findings that cilostazol prevents the recurrence of stroke.[88]

31.2.8 SUMMARY

With ongoing preclinical and clinical research, it has become evident that cilostazol possesses a broad spectrum of pharmacological actions (summarized in Figure 31.4). While PDE3 inhibition undoubtedly plays an important role, the additional inhibition of adenosine uptake

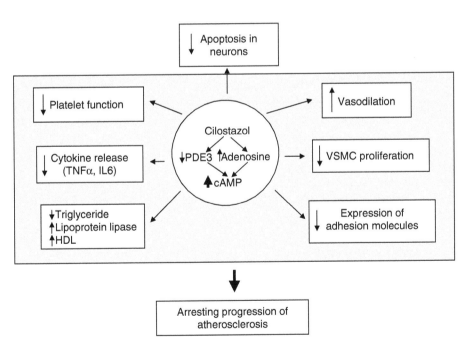

FIGURE 31.4 A summary of the biological actions of cilostazol. The effects of cilostazol on platelets, the vasculature and inflammatory cells as shown in the shaded area may lead to the reduction and arrest of the progression of atherosclerosis, the cause of most cardiovascular diseases, such as heart attack, stroke, and peripheral arterial occlusive disease. An effect on inhibiting apoptosis in neurons is also shown. IL6, interleukin 6; HDL, high-density lipoprotein; and VSMC, vascular smooth muscle cell.

has proven to be an integral part of cilostazol's overall effect. The combination of cilostazol's inhibitory effects on PDE3 and adenosine uptake contributes to its unique profile of beneficial and safe treatment/prevention of various diseases, especially those related to atherosclerosis. Atherosclerosis is a multifactorial disease usually involving damage to the inner arterial wall (endothelial cells), lipid deposition, proliferation of vascular smooth muscle cells, and accumulation of platelets, which eventually results in blockage of the artery leading to heart attack, stroke, and peripheral artery occlusive disease. Most of the pharmacological actions listed in Figure 31.4 are essential elements in the treatment of the symptoms arising from atherosclerosis and in slowing disease progression. Finally, another important feature of cilostazol is the relatively lower risk of bleeding complications compared to other antiplatelet agents, which is an important benefit in its clinical applications.

31.3 CLINICAL APPLICATIONS

31.3.1 CURRENT MARKETING AUTHORIZATION STATUS

Cilostazol has been marketed as Pletaal in Japan since 1988 for the treatment of various ischemic symptoms due to peripheral arterial occlusive disease. Marketing approval for ischemic symptoms has also been obtained in South Korea (1990), Thailand (1992), Philippines (1994), People's Republic of China (1996), Argentina (1997), Indonesia (1997), Taiwan (2000), Hong Kong (2001), Pakistan (2003), Egypt (2003), and Vietnam (2005). Additionally, marketing approval for intermittent claudication has been obtained in the United States (1999), UK, and Ireland (2000) and is pending in other European countries. Cilostazol has also been approved for prevention of recurrence of cerebral infarction in Japan (2003), Korea (2003), Thailand (2003), Argentina (2004), Indonesia (2004), Philippines (2004), Hong Kong (2004), and Pakistan (2004).

31.3.2 RECENT PATIENT EXPOSURE

The total number of patients exposed to cilostazol in sponsored studies conducted in Asia and the United States during January 2004 to January 2005 was approximately 3347 (Periodic Safety Update Report for Pletal, January 16, 2004 through January 15, 2005).

The number of patients exposed to the marketed formulation worldwide has been calculated using the number of milligrams sold from January 2004 to 2005. The total number of milligrams sold worldwide was 27,878,036,800 mg. The defined daily dose of cilostazol is 200 mg, so patient exposure is calculated to be approximately 139,390,184 patient days and 381,891 patient years.

31.3.3 SAFETY AND ADVERSE EVENTS

Adverse events associated with cilostazol were assessed in eight placebo-controlled studies involving 2274 intermittent claudication patients in the United States and the UK.[89] Minor adverse effects occurred more frequently with cilostazol than placebo. In order of frequency, the most common adverse effects reported included headache, diarrhea, and palpitations. These symptoms rarely required discontinuation of the drug, as they were mild to moderate in severity and often transient. The side effects of cilostazol in the Japanese stroke trial in 1052 patients mirrored those seen in intermittent claudication clinical trials.[88] The adverse drug reactions of cilostazol in worldwide postmarketing experience have been reported to the Food and Drug Administration (FDA) since 1999. There have been no serious findings leading to changes in the US labeling of cilostazol.

31.3.4 Approved Indications

31.3.4.1 Ischemic Symptoms Due to Peripheral Arterial Occlusive Disease

In the early 1980s, pilot clinical studies in Japan revealed that cilostazol had significant antiplatelet[30] and vasodilatory activities.[32,90] Subsequently, cilostazol was evaluated in small-scale pilot studies for the treatment of ischemic leg ulcers due to peripheral arterial occlusive disease and favorable results were obtained with an optimal dose regimen of 100 mg twice a day. Furthermore, in a multicenter study involving 177 peripheral arterial occlusive disease patients with ischemic leg ulcers, it was demonstrated that cilostazol was superior to placebo in reducing the size of the ulcer.[91] Cilostazol was subsequently approved for this indication, which is one of the most severe forms of peripheral arterial occlusive disease, and also for use in treatment of other more minor ischemic symptoms including rest pain, intermittent claudication, and cold sensation associated with peripheral arterial occlusive disease. There are two distinct disease entities in peripheral arterial occlusive disease that have similar ischemic symptoms; one is due to atherosclerosis obliterans, and the other is thromboangitis obliterans (also called Burger's disease). The incidence of the latter is relatively higher in Japan and neighboring countries, in comparison with Western countries. Treatment of the ischemic symptoms caused by thromboangitis obliterans has been very difficult, but cilostazol has been found to be effective in patients with thromboangitis obliterans in the above mentioned clinical studies.

31.3.4.2 Intermittent Claudication

Since the approval of cilostazol in Japan, attempts have been made to obtain regulatory approval for its use in the treatment of critical limb ischemia in the United States and European Union. However, there has been controversy over the use of size reduction of the ischemic ulcer as an endpoint of critical ischemia, because of spontaneous healing, involvement of infection, and the placebo effect. Several factors, including the fact that most patients with critical leg ischemia have been treated using a multidisciplinary approach, including surgical revascularization and the recommendation to use leg amputation as an endpoint for the treatment of critical leg ischemia, made it difficult to conduct large controlled studies for cilostazol on this indication. Therefore, intermittent claudication was selected as an endpoint for the US and EU studies. Intermittent claudication, the most frequent symptom of peripheral arterial occlusive disease, is the result of poor oxygenation of the muscles of the lower extremities and is experienced typically as an aching pain, cramping or numbness in the calf, buttock, hip, thigh or arch of the foot. The symptoms are induced by walking or exercise and are relieved by rest. Intermittent claudication is the characteristic symptom of stage II peripheral arterial occlusive disease and the severity may be measured objectively, using standardized treadmill testing. Though the incidence of intermittent claudication is very high, there has been no established drug treatment except pentoxifylline and its effectiveness has been questioned by some investigators.[7] Under these circumstances, randomized, double-blind, placebo-controlled 12–24-week trials were conducted in patients with moderate to severe intermittent claudication. Trial results demonstrated that cilostazol significantly increased walking distances on the treadmill and improved quality of life in comparison with placebo.[65,92–94] Additionally, a large comparative 24-week trial showed that cilostazol, 100 mg twice daily, was significantly more effective than pentoxifylline, 400 mg three times daily.[95] In this study, the effect of pentoxifylline was not significantly different from that of placebo (see Figure 31.5). Based on these controlled clinical trials, cilostazol has been approved for the treatment of intermittent claudication in the United States, UK, and Ireland. Prior to approval, the FDA advisory committee had raised some concerns regarding

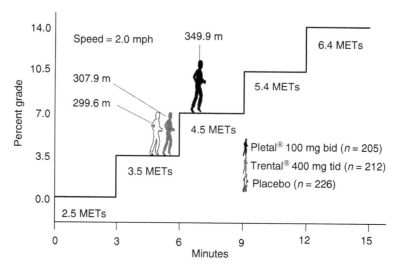

FIGURE 31.5 Comparison of the effect of cilostazol (Pletal), pentoxifylline (Trental), or placebo on maximal walking distance in patients with intermittent claudication. For a detailed description of the study, please see Ref. [95]. In this study, patients walked on a calibrated treadmill at a constant speed of 2 mph. Walking was started on a level treadmill, with the incline being increased 3.5% every 3 min. Patients who took cilostazol 100 mg bid for 24 weeks had a maximal walking distance of 349.9 m, which was significantly longer than that for patients taking placebo or pentoxifylline 400 mg tid. Moreover, cilostazol-treated patients were able to walk through more stages of the treadmill test than those treated with placebo or pentoxifylline. Thus, the maximal walking distance for the cilostazol group was achieved at an intensity equivalent to 4.5 METs. This intensity is 29% greater than the intensity achieved by either pentoxifylline or placebo. MET (metabolic equivalent term; 1 MET = basal aerobic oxygen consumption = 3.5 ml O_2/kg/min) is an accepted treadmill workload variable that is used to compare exercise capacity.

the long-term cardiac safety of cilostazol. The concern was based upon the clinical outcome of the PROMISE trial that had revealed that chronic use of the PDE3 inhibitor milrinone increased mortality in patients with CHF.[17] At that time the additional mechanism of action, i.e., inhibition of adenosine uptake, had not been elucidated, and consequently cilostazol was classified as belonging to the same group as milrinone. Cilostazol has not been shown to increase cardiovascular mortality in clinical trials in the United States and safe long-term use has been demonstrated in several Asian countries.[89] In the clinical trials conducted in the United States, exercise-limiting CHF was an exclusionary criterion, so that relatively few patients with CHF (and none with severe CHF) participated in the trials. The drug's long-term effect on mortality in this group is, therefore, unknown. Also, the safety data base from Japan and other Asian countries did not provide specific data on the safety of cilostazol in patients with CHF. Thus, cilostazol was approved, with a contraindication against CHF of any severity.

31.3.4.3   Prevention of Stroke Recurrence

In Japan and other Asian countries, cilostazol has been tested in several disease conditions where administration of an antiplatelet agent is considered to be beneficial. Among them, one large study on the prevention of recurrence of stroke was initiated in 1992 in Japan.[88] This study, called Cilostazol Stroke Prevention Study (CSPS), enrolled a total of 1052 eligible patients who had suffered cerebral infarction at 1 to 6 months prior to the enrollment.

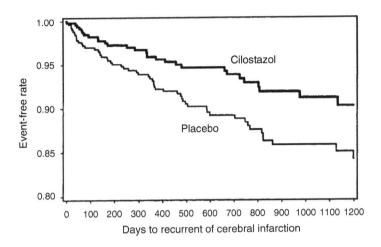

Number of patients at risk:

Cilostazol	526	421	386	364	327	284	248	219	174	151	129	103	78
Placebo	526	456	429	403	364	297	264	232	204	177	155	116	96

FIGURE 31.6 Effect of cilostazol on the recurrence of cerebral infarction. Oral administration of cilostazol (100 mg twice daily) (526 patients) or placebo (526 patients) was randomly assigned to the patients. The proportions of patients who did not experience recurrence of cerebral infarction (event-free) over a period of 1200 days are shown as Kaplan–Meier plots. The numbers of patients at risk at different time points during the trial are shown below the figure. A beneficial effect of cilostazol was clearly shown. (Reproduced from Ref. [88]. With permission.)

Patients were randomly assigned to cilostazol (100 mg twice daily), or placebo, with a primary endpoint of prevention of recurrent stroke. The study demonstrated that cilostazol significantly reduced the relative risk (41.7%; 95% confidence limits, 9.2% to 62.5%) of the recurrence in comparison with placebo treatment ($p = 0.0150$), as shown in Figure 31.6.[88] Based on this data, cilostazol was approved for the secondary prevention of stroke in Japan in 2003. The extent of risk reduction for stroke is remarkable in comparison to that estimated in other studies for aspirin (18% to 25%), or clopidogrel plus aspirin (8.9%),[96] but is similar to the 33% to 36% benefit ascribed the combination of aspirin plus dipyridamole in the European Stroke Prevention Study (ESPS) and ESPS-2 studies.[97,98] The robust benefit of cilostazol in this particular indication may be due to a combination of its antiplatelet effect and its vasodilatory actions. It is interesting to note that dipyridamole also inhibits adenosine uptake and produces vasodilation of cerebral arteries in humans.[99,100] Whether the combination of cilostazol and aspirin would have a greater effect on prevention of stroke reoccurrence compared to cilostazol alone is an interesting question that remains to be addressed.

31.3.5 POTENTIAL NEW INDICATIONS

31.3.5.1 Prevention of Thrombosis after Coronary Stenting

Coronary stenting has become the established treatment for coronary artery disease and is widely utilized in interventional cardiology. A coronary stent is a small, coiled wire-mesh tube, which is inserted into a narrowed or completely occluded coronary artery during angioplasty and expanded using a small balloon. The use of adjunctive antiplatelet therapy

has been found to prevent subacute thrombosis, which usually occurs within a month after stenting. A combination of aspirin and ticlopidine/clopidogrel has been administered for this purpose. The antiplatelet effect of cilostazol alone or combination with aspirin has been investigated for the prevention of thrombosis after coronary stenting. In two small, prospective, nonrandomized studies,[101,102] cilostazol 200 mg daily appeared to be as safe and effective as aspirin following stenting. In another study, Park et al. compared the effect of cilostazol 200 mg with ticlopidine 500 mg in 490 patients undergoing elective stent implantation.[103] At the 30 day follow-up, rates of subacute thrombosis and myocardial infarction were similar in the two groups, indicating that in the prevention of subacute thrombosis, cilostazol is at least as effective as ticlopidine. Since then, there have been several studies to compare the effect on subacute thrombosis between cilostazol and ticlopidine. Hashiguchi et al. performed a meta-analysis of four pooled studies comparing the effectiveness and safety of aspirin plus cilostazol and aspirin plus ticlopidine.[104] The results indicated that there were no differences between the drugs with regard to their effectiveness and safety for a 1-month period as an adjunctive therapy after coronary stenting. A similar meta-analysis was reported, comparing cardiac events including subacute thrombosis, among various adjunctive pharmacotherapies. The study demonstrated that neither the combination of clopidogrel and aspirin nor cilostazol and aspirin can be statistically distinguished from ticlopidine plus aspirin for the prevention of 30 day poststent cardiac adverse events.[105] However, in contrast to these meta-analyses, a randomized trial conducted by Sekiguchi et al. demonstrated that ticlopidine group showed significantly less subacute thrombosis after stenting compared with the cilostazol group.[106] Also, clopidogrel was compared with cilostazol in a relatively large study (nearly 700 patients) with regard to efficacy and safety after coronary stenting.[107] The results of this study also indicated that cilostazol is as safe and effective as clopidogrel in preventing thrombotic complications after stenting. The advantage of cilostazol over other treatments currently in use for this indication may be its reduced effect on skin bleeding time, giving it a better risk–benefit profile.

31.3.5.2   Prevention of Restenosis after Percutaneous Transluminal Coronary Angioplasty with or without Stent

Coronary angioplasty is a common and often successful treatment for ischemic heart disease. However, up to 50% of angioplasty patients soon develop significant restenosis, a narrowing of the artery caused by migration and growth of smooth muscle. Stents introduced into the coronary artery, to keep it open after angioplasty, considerably reduce the incidence of restenosis, but 10 to 50% of patients receiving a stent still develop restenosis.[108] Cilostazol has been evaluated for the prevention of coronary artery restenosis following percutaneous coronary intervention since the late 1990s. Between 1997 and 1999, the results of six controlled studies in Japan were published, which compared the effect of aspirin and cilostazol alone on restenosis in patients with various types of coronary intervention, such as percutaneous transluminal coronary angioplasty with or without stent.[109–114] Even though the number of patients in each study was relatively small (<100), all the studies showed statistically significant lower restenosis rates compared to aspirin alone. An additional four controlled studies have compared aspirin plus cilostazol and aspirin plus ticlopidine in patients with percutaneous transluminal coronary angioplasty with stents.[103,115–117] These studies were also carried out in Japan and Korea, and three out of the four clearly demonstrated lower restenosis rate in patients with cilostazol. In 2001, a large multicenter, controlled study was initiated in the United States to compare the effect of cilostazol and placebo in patients with a coronary stent.[118] A total of 705 patients were randomly assigned to either cilostazol (354 patients) or placebo (351 patients) following successful stent implantation. Restenosis

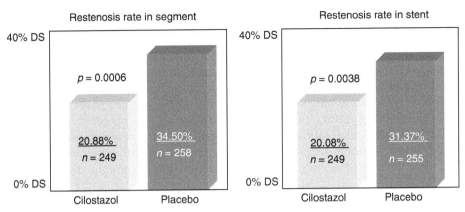

FIGURE 31.7 Cilostazol reduces coronary stent restenosis in a randomized, double-blind, placebo-controlled trial (CREST, Cilostazol for Restenosis Trial).[119] Patients who had successful coronary stent implantation received in addition to aspirin, either cilostazol (100 mg twice daily) or placebo. Restenosis was determined by quantitative coronary angiography at 6 months. The analysis segment included the stented segment, and 5 mm proximal and distal to the stent. DS, diameter stenosis. Cilostazol treatment caused a 39.5% (95% confidence interval: 34.9% to 44.2%) and a 36% (95% confidence interval: 31.6% to 40.6%) reduction of restenosis rate in segment and in stent, respectively. For details of the study, please see Ref. [119].

was determined by quantitative coronary angiography at 6 months and the results have recently been published.[119] Restenosis, defined as 50% narrowing, occurred in 22.0% of patients in the cilostazol group and in 34.1% of the patients in the placebo group ($p < 0.01$), a 35.4% risk reduction. Figure 31.7 shows more details on restenosis at both the segment and the stent. There was no difference in bleeding, rehospitalization, target vessel revascularization, myocardial infarction, or death between the groups. These outcomes clearly demonstrated the excellent risk–benefit ratio of cilostazol in those cardiac patients.

31.3.5.3 Attenuation of Restenosis in Diabetic Patients

Recently, two studies have clearly demonstrated the favorable long-term effect of cilostazol in patients with type 2 diabetes.[50,120] Both studies used the intima-media thickness of the carotid artery as a surrogate for progression of atherosclerosis in diabetic subjects. In the first study, a total of 89 subjects were randomly assigned to the cilostazol (43) or control (46) groups. After the observation period of 3.2 ± 0.5 years, intima-media thickness increased by 0.18 ± 0.19 mm in the control group, whereas intima-media thickness did not change in the cilostazol group ($p < 0.01$). Moreover, in the control group, 2 out of 46 subjects showed symptomatic brain infarctions and 10 out of 34 subjects who were without infarct-like regions, assessed by standard brain MRI examination, showed silent brain infarctions after the observation period. On the other hand, no subjects in the cilostazol group showed silent brain infarction, or stroke during the study period. Similarly in the second study, carotid intima-media thickness showed a significantly greater increase in the control group than in the cilostazol group (0.12 ± 0.14 mm vs. 0.04 ± 0.02 mm, $p < 0.05$) during the observation period of 2.6 ± 0.17 years. Similar benefits were observed in the CREST trial.[119] Although the number of patients was relatively small in these studies, they clearly indicated that cilostazol attenuated progression of atherosclerosis in diabetic patients. We believe that these favorable results are due to a combination of antiplatelet and vascular smooth muscle antiproliferative effects.

31.3.5.4   New Indications Related to Vasodilatory Effect of Cilostazol

The vasodilatory effect of cilostazol has been explored in various pathological conditions. Rajagopalan et al.[121] demonstrated in a placebo-controlled study with 40 patients with Raynaud's syndrome that cilostazol improved various parameters of the syndrome, including increased brachial artery diameter. This effect remains to be confirmed in larger studies.

The effect of cilostazol in patients with vasospastic angina pectoris has been investigated in a small-scale controlled study.[122] A total of 30 vasospastic angina pectoris patients and 10 normal subjects were randomly divided into three groups: no antiplatelet group, aspirin group, and cilostazol group. Various parameters characteristic to vasospastic angina pectoris were significantly improved only in the cilostazol group, suggesting that cilostazol may improve coronary vascular endothelial function and coronary hemodynamics in patients with vasospastic angina pectoris.

The effect of cilostazol on cerebral hemodynamics in normal subjects has also been examined in a double-blind, randomized, crossover pilot study, which demonstrated that cilostazol dilated the superficial temporal artery, without any change in global or regional cerebral flow.[48] This suggests that cilostazol may be an interesting candidate for future clinical trials of delayed cerebral vasospasm subsequent to subarachnoidal hemorrhage.

31.4   SUMMARY

After more than 20 years of intensive preclinical and clinical research, it is clear that cilostazol's mechanism of action involves more than just the inhibition of PDE3. While the discovery of cilostazol and its successful clinical applications have contributed to a better understanding of the role of PDE3 in biological systems, the unique pharmacology of cilostazol as a dual inhibitor of PDE3 and adenosine uptake distinguishes it from other PDE3 inhibitors that have been tested and found to be largely unsuccessful in the clinic. To date, cilostazol has proven to be a useful therapeutic agent in a wide range of diseases involving vascular abnormalities and thrombosis. In addition to its usefulness as a therapeutic agent for the diseases discussed above, continuing basic and clinical research on cilostazol may reveal its potential for the treatment of additional disease conditions for which there is currently an unmet medical need.

ACKNOWLEDGMENT

We thank Mr. Simon Lockyer for proof-reading the manuscript.

REFERENCES

1. Polson, J.B., and S.J. Strada. 1996. Cyclic nucleotide phosphodiesterases and vascular smooth muscle. *Annu Rev Pharmacol Toxicol* 36:403–427.
2. Hidaka, H., H. Hayashi, H. Kohri, et al. 1979. Selective inhibitor of platelet cyclic adenosine monophosphate phosphodiesterase, cilostamide, inhibits platelet aggregation. *J Pharmacol Exp Ther* 211:26–30.
3. Lindgren, S.H., T.L. Andersson, E. Vinge, and K.E. Andersson. 1990. Effects of isozyme-selective phosphodiesterase inhibitors on rat aorta and human platelets: Smooth muscle tone, platelet aggregation and cAMP levels. *Acta Physiol Scand* 140:209–219.
4. Vigdahl, R.L., J. Mongin, Jr., and N.R. Marquis. 1971. Platelet aggregation. IV. Platelet phosphodiesterase and its inhibition by vasodilators. *Biochem Biophys Res Commun* 42:1088–1094.

5. Yamazaki, H., T. Motomiya, N. Mashimo, T. Asano, and H. Hidaka. 1978. Platelet aggregation and cyclic nucleotide phosphodiesterase activity in arteriosclerotic patients. *Thromb Haemost* 39:158–166.

6. Ikeda, Y., T. Sudo, and Y. Kimura. 2002. Cilostazol. In *Platelets*, ed A.D. Michelson, 817–823. Academic Press, San Diego, CA.

7. Hiatt, W.R. 2001. Medical treatment of peripheral arterial disease and claudication. *N Engl J Med* 344:1608–1621.

8. Liu, Y., Y. Shakur, M. Yoshitake, and J.J. Kambayashi. 2001. Cilostazol (Pletal): A dual inhibitor of cyclic nucleotide phosphodiesterase type 3 and adenosine uptake. *Cardiovasc Drug Rev* 19:369–386.

9. Kambayashi, J., Y. Liu, B. Sun, Y. Shakur, M. Yoshitake, and F. Czerwiec. 2003. Cilostazol as a unique antithrombotic agent. *Curr Pharm Des* 9:2289–2308.

10. Sun, B., S.N. Le, S. Lin, et al. 2002. New mechanism of action for cilostazol: Interplay between adenosine and cilostazol in inhibiting platelet activation. *J Cardiovasc Pharmacol* 40:577–585.

11. Wang, S., J. Cone, M. Fong, M. Yoshitake, J. Kambayashi, and Y. Liu. 2001. Interplay between inhibition of adenosine uptake and phosphodiesterase type 3 on cardiac function by cilostazol, an agent to treat intermittent claudication. *J Cardiovasc Pharmacol* 38:775–783.

12. Nishi, T., F. Tabusa, T. Tanaka, T. Shimizu, and K. Nakagawa. 1985. Studies on 2-oxoquinoline derivatives as blood platelet aggregation inhibitors. IV. Synthesis and biological activity of the metabolites of 6-[4-(1-cyclohexyl-1*H*-5-tetrazolyl)butoxy]-2-oxo-1,2,3,4-tetrahydroquinoline (OPC-13013). *Chem Pharm Bull (Tokyo)* 33:1140–1147.

13. Shakur, Y., M. Fong, J. Hensley, et al. 2002. Comparison of the effects of cilostazol and milrinone on cAMP–PDE activity, intracellular cAMP and calcium in the heart. *Cardiovasc Drugs Ther* 16:417–427.

14. Sudo, T., K. Tachibana, and K. Toga, et al. 2000. Potent effects of novel anti-platelet aggregatory cilostamide analogues on recombinant cyclic nucleotide phosphodiesterase isozyme activity. *Biochem Pharmacol* 59:347–356.

15. Shakur, Y., L.S. Holst, T.R. Landstrom, M. Movsesian, E. Degerman, and V. Manganiello. 2001. Regulation and function of the cyclic nucleotide phosphodiesterase (PDE3) gene family. *Prog Nucleic Acid Res Mol Biol* 66:241–277.

16. Liu, Y., M. Fong, J. Cone, S. Wang, M. Yoshitake, and J. Kambayashi. 2000. Inhibition of adenosine uptake and augmentation of ischemia-induced increase of interstitial adenosine by cilostazol, an agent to treat intermittent claudication. *J Cardiovasc Pharmacol* 36:351–360.

17. Packer, M., J.R. Carver, R.J. Rodeheffer, et al. 1991. Effect of oral milrinone on mortality in severe chronic heart failure. The PROMISE Study Research Group. *N Engl J Med* 325:1468–1475.

18. Packer, M. 1998. Treatment of chronic heart failure. *Lancet* 340:92–95.

19. Thadani, U., and D.M. Roden. 1998. FDA Panel report: January 1998. *Circulation* 97:2295–2296.

20. Dobson, J.G.J., and R.A. Fenton. 1998. Cardiac physiology of adenosine. In *Cardiovascular biology of purines*, eds. G. Burnstock, J.G.J. Dobson, B.T. Liang, and J. Linden, 21–39. Boston, MA: Kluwer Academic Publishers.

21. Narayan, P., R.M. Mentzer, Jr., and R.D. Lasley. 2000. Phosphatase inhibitor cantharidin blocks adenosine A(1) receptor anti-adrenergic effect in rat cardiac myocytes. *Am J Physiol Heart Circ Physiol* 278:H1–H7.

22. Cone, J., S. Wang, N. Tandon, et al. 1999. Comparison of the effects of cilostazol and milrinone on intracellular cAMP levels and cellular function in platelets and cardiac cells. *J Cardiovasc Pharmacol* 34:497–504.

23. Kimura, Y., T. Tani, T. Kanbe, and K. Watanabe. 1985. Effect of cilostazol on platelet aggregation and experimental thrombosis. *Arzneimittelforschung* 35:1144–1149.

24. Uehara, S., and A. Hirayama. 1989. Effects of cilostazol on platelet function. *Arzneimittelforschung* 39:1531–1534.

25. Tani, T., K. Sakurai, Y. Kimura, T. Ishikawa, and H. Hidaka. 1992. Pharmacological manipulation of tissue cyclic AMP by inhibitors. Effects of phosphodiesterase inhibitors on the functions of platelets and vascular endothelial cells. *Adv Second Messenger Phosphoprotein Res* 25:215–227.

26. Kariyazono, H., K. Nakamura, T. Shinkawa, T. Yamaguchi, R. Sakata, and K. Yamada. 2001. Inhibition of platelet aggregation and the release of P-selectin from platelets by cilostazol. *Thromb Res* 101:445–453.
27. Matsumoto, Y., K. Marukawa, H. Okumura, T. Adachi, T. Tani, and Y. Kimura. 1999. Comparative study of antiplatelet drugs in vitro: Distinct effects of cAMP-elevating drugs and GPIIb/IIIa antagonists on thrombin-induced platelet responses. *Thromb Res* 95:19–29.
28. Minami, N., Y. Suzuki, M. Yamamoto, et al. 1997. Inhibition of shear stress–induced platelet aggregation by cilostazol, a specific inhibitor of cGMP-inhibited phosphodiesterase, *in vitro* and ex vivo. *Life Sci* 61:L-9.
29. Tanigawa, T., M. Nishikawa, T. Kitai, et al. 2000. Increased platelet aggregability in response to shear stress in acute myocardial infarction and its inhibition by combined therapy with aspirin and cilostazol after coronary intervention. *Am J Cardiol* 85:1054–1059.
30. Yasunaga, K., and K. Mase. 1985. Clinical effects of oral cilostazol on suppression of platelet function in patients with cerebrovascular disease. *Arzneimittelforschung* 35:1186–1188.
31. Woo, S.K., W.K. Kang, and K.I. Kwon. 2002. Pharmacokinetic and pharmacodynamic modeling of the antiplatelet and cardiovascular effects of cilostazol in healthy humans. *Clin Pharmacol Ther* 71:246–252.
32. Yasuda, K., M. Sakuma, and T. Tanabe. 1985. Hemodynamic effect of cilostazol on increasing peripheral blood flow in arteriosclerosis obliterans. *Arzneimittelforschung* 35:1198–1200.
33. Saitoh, S., T. Saito, A. Otake, et al. 1993. Cilostazol, a novel cyclic AMP phosphodiesterase inhibitor, prevents reocclusion after coronary arterial thrombolysis with recombinant tissue-type plasminogen activator. *Arterioscler Thromb* 13:563–570.
34. Kohda, N., T. Tani, S. Nakayama, et al. 1999. Effect of cilostazol, a phosphodiesterase III inhibitor, on experimental thrombosis in the porcine carotid artery. *Thromb Res* 96:261–268.
35. Li, H., J. Cone, M. Fong, J. Kambayashi, M. Yoshitake, and Y. Liu. 2005. Antiplatelet and antithrombotic activity of cilostazol is potentiated by dipyridamole in rabbits and dissociated from bleeding time prolongation. *Cardiovasc Drugs Ther* 19:41–48.
36. Ikeda, Y., M. Kikuchi, H. Murakami, et al. 1987. Comparison of the inhibitory effects of cilostazol, acetylsalicylic acid and ticlopidine on platelet functions ex vivo. Randomized, double-blind cross-over study. *Arzneimittelforschung* 37:563–566.
37. Tanemoto, K., Y. Kanaoka, and M. Kuinose. 2000. Assessment of antithrombotic agents using the platelet aggregation test. *Cur Ther Res* 61(11):798–806.
38. Ikeda, Y. 1999. Antiplatelet therapy using cilostazol, a specific PDE3 inhibitor. *Thromb Haemost* 82:435–438.
39. Igawa, T., T. Tani, T. Chijiwa, et al. 1990. Potentiation of anti-platelet aggregating activity of cilostazol with vascular endothelial cells. *Thromb Res* 57:617–623.
40. Eisert, W.G. 2002. Dipyridamole. In *Platelets*, ed. A.D. Michelson, 803–815. New York: Academic Press.
41. Liu, Y., J. Cone, S.N. Le, et al. 2004. Cilostazol and dipyridamole synergistically inhibit human platelet aggregation. *J Cardiovasc Pharmacol* 44:266–273.
42. Nomura, S., N. Inami, T. Iwasaka, and Y. Liu. 2004. Platelet activation markers, microparticles and soluble adhesion molecules are elevated in patients with arteriosclerosis obliterans: Therapeutic effects by cilostazol and potentiation by dipyridamole. *Platelets* 15:167–172.
43. Kim, J.S., K.S. Lee, Y.I. Kim, Y. Tamai, R. Nakahata, and H. Takami. 2004. A randomized crossover comparative study of aspirin, cilostazol and clopidogrel in normal controls: Analysis with quantitative bleeding time and platelet aggregation test. *J Clin Neurosci* 11:600–602.
44. Wilhite, D.B., A.J. Comerota, F.A. Schmieder, R.C. Throm, J.P. Gaughan, and R.A. Koneti. 2003. Managing PAD with multiple platelet inhibitors: The effect of combination therapy on bleeding time. *J Vasc Surg* 38:710–713.
45. Tamai, Y., H. Takami, R. Nakahata, F. Ono, and A. Munakata. 1999. Comparison of the effects of acetylsalicylic acid, ticlopidine and cilostazol on primary hemostasis using a quantitative bleeding time test apparatus. *Haemostasis* 29:269–276.
46. Mallikaarjun, S., W.P. Forbes, and S.L. Bramer. 1999. Interaction potential and tolerability of the coadministration of cilostazol and aspirin. *Clin Pharmacokinet* 37 (Suppl 2):87–93.

47. Nagakawa, Y., Y. Omaki, and H. Orimo. 1986. Effects of cilostazol (OPC-13013) on arachidonic acid metabolism. *Jpn Pharmacol Ther* 14:6319–6324.

48. Birk, S., C. Kruuse, K.A. Petersen, O. Jonassen, P. Tfelt-Hansen, and J. Olesen. 2004. The phosphodiesterase 3 inhibitor cilostazol dilates large cerebral arteries in humans without affecting regional cerebral blood flow. *J Cereb Blood Flow Metab* 24:1352–1358.

49. Wang, S., J. Cone, M. Fong, M. Yoshitake, J. Kambayashi, and Y. Liu. 2005. Cilostazol increases tissue blood flow and interstitial adenosine levels in stimulated rabbit gastrocnemius muscle [abstract]. *J Mol Cell Cardio* 38:868.

50. Takahashi, S., K. Oida, R. Fujiwara, et al. 1992. Effect of cilostazol, a cyclic AMP phosphodiesterase inhibitor, on the proliferation of rat aortic smooth muscle cells in culture. *J Cardiovasc Pharmacol* 20:900–906.

51. Shimizu, E., Y. Kobayashi, Y. Oki, T. Kawasaki, T. Yoshimi, and H. Nakamura. 1999. OPC-13013, a cyclic nucleotide phosphodiesterase type III, inhibitor, inhibits cell proliferation and transdifferentiation of cultured rat hepatic stellate cells. *Life Sci* 64:2081–2088.

52. Hayashi, S., R. Morishita, H. Matsushita, et al. 2000. Cyclic AMP inhibited proliferation of human aortic vascular smooth muscle cells, accompanied by induction of p53 and p21. *Hypertension* 35:237–243.

53. Kim, M.J., K.G. Park, K.M. Lee, et al. 2005. Cilostazol inhibits vascular smooth muscle cell growth by downregulation of the transcription factor E2F. *Hypertension* 45:552–556.

54. Rybalkin, S.D., and K.E. Bornfeldt. 1999. Cyclic nucleotide phosphodiesterases and human arterial smooth muscle cell proliferation. *Thromb Haemost* 82:424–434.

55. Kayanoki, Y., W. Che, S. Kawata, Y. Matsuzawa, S. Higashiyama, and N. Taniguchi. 1997. The effect of cilostazol, a cyclic nucleotide phosphodiesterase III inhibitor, on heparin-binding EGF-like growth factor expression in macrophages and vascular smooth muscle cells. *Biochem Biophys Res Commun* 238:478–481.

56. Inoue, T., T. Uchida, M. Sakuma, et al. 2004. Cilostazol inhibits leukocyte integrin Mac-1, leading to a potential reduction in restenosis after coronary stent implantation. *J Am Coll Cardiol* 44:1408–1414.

57. Morishita, R., J. Higaki, S.I. Hayashi, et al. 1997. Role of hepatocyte growth factor in endothelial regulation: Prevention of high D-glucose-induced endothelial cell death by prostaglandins and phosphodiesterase type 3 inhibitor. *Diabetologia* 40:1053–1061.

58. Lee, T.M., S.F. Su, C.H. Tsai, Y.T. Lee, and S.S. Wang. 2001. Differential effects of cilostazol and pentoxifylline on vascular endothelial growth factor in patients with intermittent claudication. *Clin Sci (Lond)* 101:305–311.

59. Nishio, Y., A. Kashiwagi, N. Takahara, H. Hidaka, and R. Kikkawa. 1997. Cilostazol, a cAMP phosphodiesterase inhibitor, attenuates the production of monocyte chemoattractant protein-1 in response to tumor necrosis factor-alpha in vascular endothelial cells. *Horm Metab Res* 29:491–495.

60. Omi, H., N. Okayama, M. Shimizu, et al. 2004. Cilostazol inhibits high glucose-mediated endothelial-neutrophil adhesion by decreasing adhesion molecule expression via NO production. *Microvasc Res* 68:119–125.

61. Ishizaka, N., J. Taguchi, Y. Kimura, et al. 1999. Effects of a single local administration of cilostazol on neointimal formation in balloon-injured rat carotid artery. *Atherosclerosis* 142:41–46.

62. Kubota, Y., K. Kichikawa, H. Uchida, et al. 1995. Pharmacologic treatment of intimal hyperplasia after metallic stent placement in the peripheral arteries. An experimental study. *Invest Radiol* 30:532–537.

63. Aoki, M., R. Morishita, S. Hayashi, et al. 2001. Inhibition of neointimal formation after balloon injury by cilostazol, accompanied by improvement of endothelial dysfunction and induction of hepatocyte growth factor in rat diabetes model. *Diabetologia* 44:1034–1042.

64. Lee, J.H., G.T. Oh, S.Y. Park, et al. 2005. Cilostazol reduces atherosclerosis by inhibition of superoxide and tumor necrosis factor-alpha formation in low-density lipoprotein receptor-null mice fed high cholesterol. *J Pharmacol Exp Ther* 313:502–509.

65. Elam, M.B., J. Heckman, J.R. Crouse, et al. 1998. Effect of the novel antiplatelet agent cilostazol on plasma lipoproteins in patients with intermittent claudication. *Arterioscler Thromb Vasc Biol* 18:1942–1947.

66. Wang, T., M.B. Elam, W.P. Forbes, J. Zhong, and K. Nakajima. 2003. Reduction of remnant lipoprotein cholesterol concentrations by cilostazol in patients with intermittent claudication. *Atherosclerosis* 171:337–342.

67. Tamai, T., A. Shimada, H. Maeda, et al. 1992. Reduction of low density lipoprotein by cilostazol in the non-insulin dependent diabetic patient. *Jap Pharmacol Ther* 20:241–248.

68. Watanabe, N., Y. Ishikawa, Y. Kitagawa, et al. 1996. Effects of cilostazol on the lipid metabolism in patients with hypertriglyceridemia. *Jap Pharmacol Ther* 24:127–132.

69. Ishikawa, M., Y. Yamada, C. Hirose, D. Tujino, K. Hoshi, and N. Saitou. 1997. The effects of cilostazol on serum lipid metabolism and ASO in NIDDM with hypertriglyceridemia. *Ther Res* 18:198–204.

70. Mishima, Y., A. Kuyama, M. Ando, M. Kibata, and T. Ishioka. 2000. Effects of cilostazol on serum lipoprotein concentrations and particle size of low-density lipoproteins in patients with dyslipidemic NIDDM. *J Jap Atherosclerosis Soc* 27:17–22.

71. Takayoshi, T., O. Shinichi, A. Ryuzo, et al. 2001. Effect of cilostazol on lipid, uric acid and glucose metabolism in patients with impaired glucose tolerance or type 2 diabetes mellitus: A double-blind, placebo-controlled study. *Clinical Drug Investig* 21:325–335.

72. Ikewaki, K., K. Mochizuki, M. Iwasaki, R. Nishide, S. Mochizuki, and N. Tada. 2002. Cilostazol, a potent phosphodiesterase type III inhibitor, selectively increases antiatherogenic high-density lipoprotein subclass LpA-I and improves postprandial lipemia in patients with type 2 diabetes mellitus. *Metabolism* 51:1348–1354.

73. Nakamura, N., T. Hamazaki, H. Johkaji, et al. 2003. Effects of cilostazol on serum lipid concentrations and plasma fatty acid composition in type 2 diabetic patients with peripheral vascular disease. *Clin Exp Med* 2:180–184.

74. Tani, T., K. Uehara, T. Sudo, K. Marukawa, Y. Yasuda, and Y. Kimura. 2000. Cilostazol, a selective type III phosphodiesterase inhibitor, decreases triglyceride and increases HDL cholesterol levels by increasing lipoprotein lipase activity in rats. *Atherosclerosis* 152:299–305.

75. Oscai, L.B., R.A. Caruso, and W.K. Palmer. 1986. Protein kinase activation of heparin-releasable lipoprotein lipase in rat heart. *Biochem Biophys Res Commun* 135:196–200.

76. Motoyashiki, T., M. Fukamachi, T. Morita, H. Shiomi, and H. Ueki. 1998. Involvement of adenosine in vanadate-stimulated release of lipoprotein lipase activity. *Biol Pharm Bull* 21:889–892.

77. Ranganathan, G., D. Phan, I.D. Pokrovskaya, J.E. McEwen, C. Li, and P.A. Kern. 2002. The translational regulation of lipoprotein lipase by epinephrine involves an RNA binding complex including the catalytic subunit of protein kinase A. *J Biol Chem* 277:43281–43287.

78. Londos, C., D.M. Cooper, W. Schlegel, and M. Rodbell. 1978. Adenosine analogs inhibit adipocyte adenylate cyclase by a GTP-dependent process: Basis for actions of adenosine and methylxanthines on cyclic AMP production and lipolysis. *Proc Natl Acad Sci USA* 75:5362–5366.

79. Larrouy, D., A. Remaury, D. Daviaud, and M. Lafontan. 1994. Coupling of inhibitory receptors with G$_i$-proteins in hamster adipocytes: Comparison between adenosine A1 receptor and alpha 2-adrenoceptor. *Eur J Pharmacol* 267:225–232.

80. Hadri, K.E., A. Courtalon, X. Gauthereau, A.M. Chambaut-Guerin, J. Pairault, and B. Feve. 1997. Differential regulation by tumor necrosis factor-alpha of beta1-, beta2-, and beta3-adrenoreceptor gene expression in 3T3-F442A adipocytes. *J Biol Chem* 272:24514–24521.

81. Bullo, M., P. Garcia-Lorda, J. Peinado-Onsurbe, et al. 2002. TNFalpha expression of subcutaneous adipose tissue in obese and morbid obese females: Relationship to adipocyte LPL activity and leptin synthesis. *Int J Obes Relat Metab Disord* 26:652–658.

82. Greenberg, A.S., R.P. Nordan, J. McIntosh, J.C. Calvo, R.O. Scow, and D. Jablons. 1992. Interleukin 6 reduces lipoprotein lipase activity in adipose tissue of mice *in vivo* and in 3T3-L1 adipocytes: A possible role for interleukin 6 in cancer cachexia. *Cancer Res* 52:4113–4116.

83. Kim, K.Y., H.K. Shin, J.M. Choi, and K.W. Hong. 2002. Inhibition of lipopolysaccharide-induced apoptosis by cilostazol in human umbilical vein endothelial cells. *J Pharmacol Exp Ther* 300: 709–715.

84. Maeda, E., G. Yoshino, K. Nagata, et al. 1993. Effect of cilostazol on triglyceride metabolism in rats. *Curr Ther Res* 54:420–424.

85. Lee, J.H., Y.K. Lee, M. Ishikawa, et al. 2003. Cilostazol reduces brain lesion induced by focal cerebral ischemia in rats—an MRI study. *Brain Res* 994:91–98.
86. Lee, J.H., K.Y. Kim, Y.K. Lee, et al. 2004. Cilostazol prevents focal cerebral ischemic injury by enhancing casein kinase 2 phosphorylation and suppression of phosphatase and tensin homolog deleted from chromosome 10 phosphorylation in rats. *J Pharmacol Exp Ther* 308:896–903.
87. Choi, J.M., H.K. Shin, K.Y. Kim, J.H. Lee, and K.W. Hong. 2002. Neuroprotective effect of cilostazol against focal cerebral ischemia via antiapoptotic action in rats. *J Pharmacol Exp Ther* 300:787–793.
88. Gotoh, F., H. Tohgi, S. Hirai, et al. 2000. Cilostazol stroke prevention study: A placebo-controlled double-blind trial for secondary prevention of cerebral infarction. *J Stroke and Carebrovasc Dis* 9:147–157.
89. Otsuka America Pharmaceutical Inc. NDA 20-863: Cilostazol. 1997.
90. Ohashi, S., M. Iwatani, Y. Hyakuna, and Y. Morioka. 1985. Thermographic evaluation of the hemodynamic effect of the antithrombotic drug cilostazol in peripheral arterial occlusion. *Arznei-mittelforschung* 35:1203–1208.
91. Mishima, Y., S. Tanabe, T. Sakaguchi, T. Katsunma, A. Kusaba, and H. Uchida. 1986. Evaluation of OPC-13013 in chronic arterial occlusion. *J Clin Exp Med* 139:133–157 (in Japanese).
92. Dawson, D.L., B.S. Cutler, M.H. Meissner, and D.E.J. Strandness. 1998. Cilostazol has beneficial effects in treatment of intermittent claudication: Results from a multicenter, randomized, prospective, double-blind trial. *Circulation* 98:678–686.
93. Money, S.R., J.A. Herd, J.L. Isaacsohn, et al. 1998. Effect of cilostazol on walking distances in patients with intermittent claudication caused by peripheral vascular disease. *J Vasc Surg* 27:267–274.
94. Beebe, H.G., D.L. Dawson, B.S. Cutler, et al. 1999. A new pharmacological treatment for intermittent claudication: Results of a randomized, multicenter trial. *Arch Intern Med* 159:2041–2050.
95. Dawson, D.L., B.S. Cutler, W.R. Hiatt, et al. 2000. A comparison of cilostazol and pentoxifylline for treating intermittent claudication. *Am J Med* 109:523–530.
96. CAPRIE Steering Committee. 1996. A randomised, blinded, trial of clopidogrel versus aspirin in patients at risk of ischaemic events (CAPRIE). CAPRIE Steering Committee. *Lancet* 348:1329–1339.
97. European Stroke Prevention Study. ESPS Group. 1990. *Stroke* 21:1122–1130.
98. Diener, H.C., L. Cunha, C. Forbes, J. Sivenius, P. Smets, and A. Lowenthal. 1996. European Stroke Prevention Study. 2. Dipyridamole and acetylsalicylic acid in the secondary prevention of stroke. *J Neurol Sci* 143:1–13.
99. Hwang, T.L., A. Saenz, J.J. Farrell, and W.L. Brannon. 1996. Brain SPECT with dipyridamole stress to evaluate cerebral blood flow reserve in carotid artery disease. *J Nucl Med* 37:1595–1599.
100. Ito, H., T. Kinoshita, Y. Tamura, I. Yokoyama, and H. Iida. 1999. Effect of intravenous dipyridamole on cerebral blood flow in humans. A PET study. *Stroke* 30:1616–1620.
101. Ochiai, M., T. Isshiki, S. Takeshita, et al. 1997. Use of cilostazol, a novel antiplatelet agent, in a post-Palmaz-Schatz stenting regimen. *Am J Cardiol* 79:1471–1474.
102. Yoshitomi, Y., S. Kojima, T. Sugi, M. Yano, Y. Matsumoto, and M. Kuramochi. 1998. Antiplatelet treatment with cilostazol after stent implantation. *Heart* 80:393–396.
103. Park, S.W., C.W. Lee, H.S. Kim, et al. 1999. Comparison of cilostazol versus ticlopidine therapy after stent implantation. *Am J Cardiol* 84:511–514.
104. Hashiguchi, M., K. Ohno, R. Nakazawa, S. Kishino, M. Mochizuki, and T. Shiga. 2004. Comparison of cilostazol and ticlopidine for one-month effectiveness and safety after elective coronary stenting. *Cardiovasc Drugs Ther* 18:211–217.
105. Schleinitz, M.D., I. Olkin, and P.A. Heidenreich. 2004. Cilostazol, clopidogrel or ticlopidine to prevent sub-acute stent thrombosis: A meta-analysis of randomized trials. *Am Heart J* 148:990–997.
106. Sekiguchi, M., H. Hoshizaki, H. Adachi, S. Ohshima, K. Taniguchi, and M. Kurabayashi. 2004. Effects of antiplatelet agents on subacute thrombosis and restenosis after successful coronary stenting: A randomized comparison of ticlopidine and cilostazol. *Circ J* 68:610–614.

107. Lee, S.W., S.W. Park, M.K. Hong, et al. 2005. Comparison of cilostazol and clopidogrel after successful coronary stenting. *Am J Cardiol* 95:859–862.
108. Fattori, R., and T. Piva. 2003. Drug-eluting stents in vascular intervention. *Lancet* 361:247–249.
109. Take, S., M. Matsutani, H. Ueda, et al. 1997. Effect of cilostazol in preventing restenosis after percutaneous transluminal coronary angioplasty. *Am J Cardiol* 79:1097–1099.
110. Kunishima, T., H. Musha, F. Eto, et al. 1997. A randomized trial of aspirin versus cilostazol therapy after successful coronary stent implantation. *Clin Ther* 19:1058–1066.
111. Sekiya, M., J. Funada, K. Watanabe, M. Miyagawa, and H. Akutsu. 1998. Effects of probucol and cilostazol alone and in combination on frequency of poststenting restenosis. *Am J Cardiol* 82: 144–147.
112. Yamasaki, M., K. Hara, Y. Ikari, et al. 1998. Effects of cilostazol on late lumen loss after Palmaz–Schatz stent implantation. *Cathet Cardiovasc Diagn* 44:387–391.
113. Tsuchikane, E., A. Fukuhara, T. Kobayashi, et al. 1999. Impact of cilostazol on restenosis after percutaneous coronary balloon angioplasty. *Circulation* 100:21–26.
114. Tsuchikane, E., T. Kobayashi, and N. Awata. 2000. The potential of cilostazol in interventional cardiology. *Curr Interv Cardiol Rep* 2:143–148.
115. Ochiai, M., K. Eto, S. Takeshita, et al. 1999. Impact of cilostazol on clinical and angiographic outcome after primary stenting for acute myocardial infarction. *Am J Cardiol* 84:1074–1076.
116. Tanabe, Y., E. Ito, I. Nakagawa, and K. Suzuki. 2001. Effect of cilostazol on restenosis after coronary angioplasty and stenting in comparison to conventional coronary artery stenting with ticlopidine. *Int J Cardiol* 78:285–291.
117. Makutani, S., K. Kichikawa, H. Uchida, et al. 1999. Effect of antithrombotic agents on the patency of PTFE-covered stents in the inferior vena cava: An experimental study. *Cardiovasc Interv Radiol* 22:232–238.
118. Douglas, J.S., W.S. Weintraub, and D. Holmes. 2003. Rationale and design of the randomized, multicenter, cilostazol for RESTenosis (CREST) trial. *Clin Cardiol* 26:451–454.
119. Douglas, J.S., Jr., D.R. Holmes, Jr., D.J. Kereiakes, et al. 2005. Coronary stent restenosis in patients treated with cilostazol. *Circulation* 112:2826–2832.
120. Mitsuhashi, N., Y. Tanaka, S. Kubo, et al. 2004. Effect of cilostazol, a phosphodiesterase inhibitor, on carotid IMT in Japanese type 2 diabetic patients. *Endocr J* 51:545–550.
121. Rajagopalan, S., D. Pfenninger, E. Somers, et al. 2003. Effects of cilostazol in patients with Raynaud's syndrome. *Am J Cardiol* 92:1310–1315.
122. Watanabe, K., S. Ikeda, J. Komatsu, et al. 2003. Effect of cilostazol on vasomotor reactivity in patients with vasospastic angina pectoris. *Am J Cardiol* 92:21–25.

32 Reinventing the Wheel: Nonselective Phosphodiesterase Inhibitors for Chronic Inflammatory Diseases

Mark A. Giembycz

CONTENTS

32.1 Introduction 649
32.2 Phosphodiesterase 4 Inhibitors in Asthma and COPD 650
32.3 Serious Adverse Events of Phosphodiesterase 4 Inhibitors 651
32.4 Dual-Selective Phosphodiesterase Inhibitors..651
    32.4.1 Phosphodiesterase 1/4 Inhibitors...651
    32.4.2 Phosphodiesterase 3/4 Inhibitors...653
    32.4.3 Phosphodiesterase 4/5 Inhibitors...654
    32.4.4 Phosphodiesterase 4/7 Inhibitors...655
    32.4.5 Nonselective Phosphodiesterase Inhibitors 658
32.5 Concluding Remarks..659
Acknowledgments 659
References 659

32.1 INTRODUCTION

An enzyme that hydrolyzes the 3'-ribose phosphate bond of cAMP to the catalytically inactive 5'-adenosine monophosphate was identified more than 40 years ago.[1] Since that time 11 molecularly, biochemically, and immunologically distinct enzyme families—collectively known as cyclic nucleotide phosphodiesterases (PDEs) that selectively degrade cyclic purine nucleotides have unequivocally been identified.[2–5] Phosphodiesterases that act on cyclic pyrimidine monophosphates have also been discovered but they have received relatively little attention and, instead, most investigators have focused on those enzymes that hydrolyze cAMP and cGMP for which functionally important second messenger roles have been established.

Over the last 15 years there has been considerable interest in the cAMP-specific, or PDE4, family of enzymes as intracellular targets that could be exploited to therapeutic advantage for a multitude of diseases associated with chronic inflammation.[6–13] Although drug targeting of

PDE4 is based on a conceptually robust hypothesis,[14] dose-limiting side effects, of which nausea and vomiting are the most common and troublesome, have hampered their clinical development. A fundamental challenge that still is to be met by the pharmaceutical industry is to synthesize compounds with an improved therapeutic ratio given that the adverse effects of PDE4 inhibitors represent an extension of their pharmacology.[15] Several strategies (not described here) are being considered to dissociate the beneficial from detrimental effects of PDE4 inhibitors[6,14,16] with some degree of success.[7,17–19] However, compounds with an optimal pharmacophore have not yet been reported. An alternative approach of improving the therapeutic ratio and safety of PDE4 inhibitors may lie in the synthesis of compounds with broader PDE selectivity. Of the 11 PDE families that have been unequivocally identified, dual-selective compounds that inhibit PDE4 as well as PDE1, PDE3, PDE5, or PDE7 could offer potential opportunities to enhance clinical efficacy. Moreover, despite the persuasive rationale of developing selective PDE4 inhibitors, second generation, nonselective compounds might, paradoxically, offer the best approach to alleviate inflammation.

In this chapter, the possibility that dual-selective and nonselective PDE inhibitors could provide an improved approach to alleviate chronic inflammation is discussed. Although many diseases fall within this category including asthma, chronic obstructive pulmonary disease (COPD), atopic dermatitis, psoriasis, lupus, rheumatoid arthritis, and multiple sclerosis, the discussion herein will focus on the role of PDE inhibitors in the treatment of chronic airways inflammation.

32.2 PHOSPHODIESTERASE 4 INHIBITORS IN ASTHMA AND COPD

Despite encouraging data from some clinical trials in asthma and COPD,[7,17,19–28] the current generation of PDE4 inhibitors including cilomilast (*Ariflo*) and roflumilast (*Daxas*) is hampered by a low therapeutic ratio. This limitation became clear early on in the development of these compounds, with nausea, diarrhoea, abdominal pain, vomiting, and dyspepsia being the most common adverse events documented (Figure 32.1). Unfortunately, these unwanted actions represent an extension of the pharmacology of PDE4 inhibitors and are typical of first generation compounds such as rolipram and nonselective PDE inhibitors in general.[29]

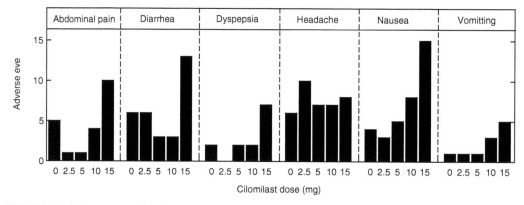

FIGURE 32.1 Percentage of adverse events produced by the PDE4 inhibitor cilomilast reported in all controlled asthma and COPD clinical trials. The number of subjects who received cilomilast or placebo were: 0 ($n = 1802$), 2.5 mg ($n = 72$), 5 mg ($n = 366$), 10 mg ($n = 437$), and 15 mg ($n = 2475$). (Data taken from Ref. [32].)

32.3 SERIOUS ADVERSE EVENTS OF PHOSPHODIESTERASE 4 INHIBITORS

Documentation of serious toxicities resulting from the administration of PDE4 inhibitors is relatively sparse when compared to inhibitors of other cAMP PDE families. However, of current concern in the deployment of PDE4 inhibitors is their potential to produce arteritis. Indeed, the development of arteritis has been identified by the US Food and Drug Administration (FDA) as a significant safety issue requiring rigorous monitoring in clinical trials of PDE4 inhibitors[30] and this concern is echoed in studies where the toxicology of PDE4 inhibitors is evaluated.[31] Arteritis[32] is characterized by inflammation, haemorrhage, and necrosis of blood vessels, and may be irreversible in animals.[33,34] Mechanistically, arteritis is thought to result from hemodynamic changes produced by excessive and prolonged vasodilatation of specific vascular beds, although the means by which PDE4 inhibitors cause certain vessels to become targets of inflammation is unknown.[35] In nonhuman primates, studies with PDE4 inhibitors generally have not identified pathologies, including arteritis, similar to those reported in other species used for toxicology, and this has lead to a view that arteriopathies may be nonprimate-specific.[36,37] Indeed, rats and dogs have an increased susceptibility to drug-induced vascular lesions because of the common occurrence of arteriopathies in these species.[38–42] Consistent with this hypothesis, cilomilast is reported not to produce medial necrosis of mesenteric arteries in primates unlike comparable studies performed in rodents where medial necrosis of mesenteric arteries is often precipitated.[43] However, a recent comprehensive toxicological study found that a PDE4 inhibitor, SCH 351591, produced, in Cynomolgus monkeys, acute to chronic inflammation of small- to medium-sized arteries in many tissues and organs.[31] These findings of arteriopathy in primates, previously thought to be resistant to toxicity, have serious implications for human risk. Indeed, it is noteworthy that Merck in 2003 abandoned the development of its lead PDE4 inhibitor due to an incidence of colitis, and were concerned that it was secondary to arteritis.[44] As COPD is a chronic disease requiring long-term therapy, a wide margin of safety will be needed because toxicity cannot be adequately monitored, especially as presentation of mesenteric ischemia is vague in humans and diagnostic tools are poor. However, comfort can be derived from the knowledge that no clinically relevant pathological effects have been produced in patients treated for many years with theophylline, which produces medial necrosis of mesenteric vessels in rats.[45,46] Moreover, mice deficient in *PDE4A*, *PDE4B*, and *PDE4C* are not reported to develop arteritis (reviewed in Ref. [13]). Thus, PDE inhibitor-induced vascular lesion may be species-specific.

It should be noted that a detailed toxicological study published in 2005 found that subacute administration of a PDE4 inhibitor, BYK169171, did not cause in rats a primary vasculitis of the splanchnic vasculature (i.e., the blood vessels of the mesentery, spleen, and liver).[35] Instead, a nonpurulent mesenteritis was induced that was followed, in some but not all animals, by a segmental necrotizing panarteritis.[35] This toxicological profile is thus distinct from that evoked by selective inhibitors of PDE3 (see below).

32.4 DUAL-SELECTIVE PHOSPHODIESTERASE INHIBITORS

One means of improving the therapeutic ratio and safety of PDE4 inhibitors may lie in the synthesis of compounds with broader PDE selectivity. Of the 11 PDE families that have been unequivocally identified, dual-selective compounds that inhibit PDE4 and either PDE1, PDE3, PDE5, or PDE7 currently offer potential opportunities to enhance clinical efficacy.

32.4.1 Phosphodiesterase 1/4 Inhibitors

PDE1 is a generic term that describes a family of enzymes whose activity is dependent upon the presence of Ca^{2+} and calmodulin. In humans, PDE1s are encoded by three genes

(*PDE1A, PDE1B, PDE1C*) with further complexity arising from differential mRNA splicing. PDE1A and PDE1B preferentially hydrolyze cGMP whereas PDE1C degrades both cAMP and cGMP with high affinity. Like all other PDE isoenzymes, their expression is regulated both transcriptionally and posttranslationally. In airways smooth muscle from humans and other species, PDE1 activity accounts for a significant proportion (30% to 80%) of the cyclic nucleotide hydrolytic activity,[47–49] yet its function is unknown.

In 1997, PDE1 was implicated in human vascular smooth muscle proliferation[50,51] and it is tempting to speculate that one or more PDE1 isoenzymes could subserve the same function in airway myocytes. Rybalkin et al. found that PDE1C was markedly induced in proliferating but not quiescent smooth muscle cells derived from human aorta and, like human airway myocytes, this enzyme was the predominant PDE in these cells.[50] Subsequently, it was established that induction of PDE1C correlated with cell cycle progression and that inhibition of this enzyme with either 8-methoxymethyl-3-isobutyl-1-methylxanthine (a modestly selective PDE1 inhibitor) or antisense oligonucleotides directed against PDE1C significantly reduced mitogenesis.[51] As neither zaprinast nor sildenafil (selective PDE5 inhibitors) were antimitogenic in the same system, it was concluded that proliferation, due to PDE1C induction, involved degradation of cAMP rather than cGMP.[51] Collectively, these data indicate that induction of PDE1C lowers cAMP in vascular myocytes and so relieves an endogenous break, allowing mitogenesis to proceed unhindered. Given that airway remodeling is characteristic of asthma and COPD,[52,53] dual-selective inhibitors of PDE1C and PDE4 may target proliferating airways smooth muscle cells, and so retard the remodeling process and, at the same time, arrest inflammation (via PDE4 inhibition). In addition, extensive remodeling of the pulmonary vasculature is also prevalent in COPD (see below and Ref. [54]). Accordingly, if PDE1 also plays a central role in regulating the proliferation of pulmonary vascular myocytes then inhibition of this enzyme could equally lead to a reduction in pulmonary vascular resistance (see section on PDE5 below). Unfortunately, there are few reports of selective PDE1 inhibitors that can be used to rigorously test this possibility, although Schering-Plough has identified certain tetracyclic guanines that may be useful pharmacological tools[55] (Figure 32.2).

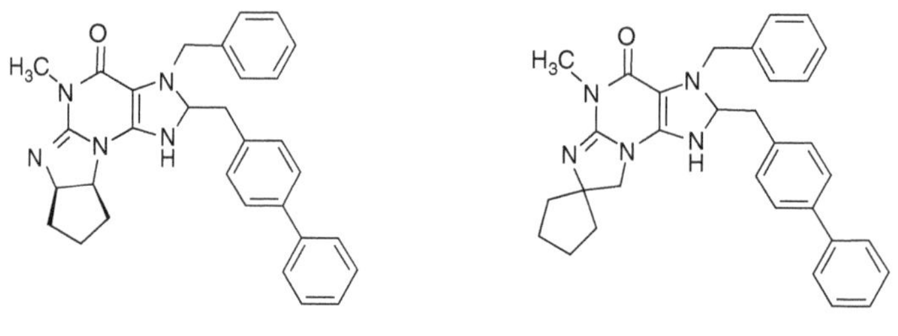

	IC$_{50}$ (nM)					
Compound	PDE1	PDE2	PDE3	PDE4	PDE5	PDE5/PDE1 Ratio
1	0.07	2100	3500	700	305	4357
2	0.6	1800	7000	200	200	333

FIGURE 32.2 Structure of two inhibitors of PDE1 based on the tetracyclic guanine template and their isoenzyme selectivity. **1**, (6a*R*,9a*S*)-2-(biphenylylmethyl)-5,6a,7,8,9,9a-hexahydro-5-methyl-3-(phenylmethyl)cyclopent[4,5]imidazo[2,1-b]purin-4(3H)-one; **2**, 5′-Methyl-2′-(biphenylylmethyl)-3′-(phenylmethyl)spiro [cyclopentane-1,7′(8′*H*)-[3*H*]imidazo[2,1-b]purin]-4(5′*H*)-one.

32.4.2 PHOSPHODIESTERASE 3/4 INHIBITORS

Phosphodiesterase 3 is a cAMP-specific enzyme that is ubiquitously distributed across many cells and tissues.[56,57] Two distinct, but related, genes that encode PDE3 isoenzymes have been identified and are designated PDE3A and PDE3B.[56–59] Interest in PDE3 as a target for the treatment of asthma and COPD stemmed primarily from the finding that selective inhibitors promote bronchodilatation in humans.[60–63] Therefore, conceptually, inhibitors that block both PDE3 and PDE4 should exhibit both anti-inflammatory and bronchodilator activity and so have superior efficacy over compounds that block only PDE4. In addition, inhibition of PDE3 may also have desirable effects on the function of certain pro-inflammatory and immune cells especially during concurrent PDE4 inhibition. For example, T-lymphocytes, macrophages, monocytes, epithelial and endothelial cells, which are sources of PDE4, also express PDE3.[6] *In vitro* studies have shown that while PDE3 inhibitors generally have little or no effect on T-cell proliferation or on IL-2 generation, they significantly enhance the effect of PDE4 inhibitors.[64,65] Similar results have been reported for TNFα release from human alveolar macrophages.[66]

Several dual-selective inhibitors of PDE3 and PDE4 have been developed and evaluated in humans including zardaverine, ben(z)afentrine, tolafentrine, and pumafentrine although there is a dearth of peer-review data to gauge the clinical progress of these molecules.[67,68] The most advanced compound in clinical trials for both asthma and COPD was Altana's pumafentrine. However, development of that compound was discontinued in late 2002 as initial phase II trials data did not meet the expectation with regard to duration of action.[69] It is implied in Altana's 2003 annual report that research is now focused on the active metabolite, hydroxypumafentrine, although reference is made only to compounds in preclinical development and so the active metabolite too may also have been abandoned.

Although the scientific rationale is clear for developing hybrid PDE3/PDE4 inhibitors (i.e., such compounds should exhibit both anti-inflammatory and bronchodilator activity), there may be major safety concerns with this approach. Selective PDE3 inhibitors were developed in the 1980s as safer alternatives to cardiac glycosides for the treatment of dilated cardiomyopathy. Acutely, beneficial effects on myocardial contractility and on vascular smooth muscle tone were reported but, for reasons that are still not clear, chronic treatment resulted, paradoxically, in a significant increase in mortality.[70] These findings are worrisome as many individuals with COPD also have right heart failure secondary to pulmonary hypertension[71] and so PDE3/PDE4 inhibitors would, presumably, be contraindicated in this clinical setting. Moreover, the long-term effect of a dual-selective PDE inhibitor of this combination in individuals with (mild) COPD in whom cardiac function is normal is unknown.

Another potential cause for concern is arteritis. There is an extensive literature on PDE3-inhibitor-induced arteriopathy in laboratory animals with the splanchnic vessels and coronary arteries of rats and dogs, respectively, being the most susceptible to lesions.[72] Whether arteriopathies evoked by PDE3/PDE4 inhibitors are also to be expected in humans with *chronic* use, as would be required in asthma and COPD, is unknown but it is worth re-stating that both theophylline and the selective PDE3 inhibitor, cilostamide, are used clinically and no evidence to date has emerged that these compounds produce vascular lesions. Nevertheless, whether clinically meaningful bronchodilatation can ultimately be achieved in humans with an acceptable degree of vasodilatation is largely unexplored but coincident headache is not uncommon with current PDE3 or PDE3/PDE4 inhibitors at bronchodilator doses.[63,73–75]

The fact that pumafentrine progressed to phase II clinical trials may suggest that systemic exposure or cardiovascular activity can be minimized. Conceivably, airway-selective compounds could be realized through formulation, route of administration, or by exploiting the fact that PDE3 and PDE4 inhibitors may interact synergistically to reduce human airways smooth

muscle tone[47] and elicit anti-inflammatory activity.[76] Indeed, it seems entirely logical, because of synergy in target cells, to predict that proportionally lower doses of dual-selective inhibitors will be required to elicit the same degree of therapeutic benefit as a PDE4 inhibitor alone. The same argument can also be made with regard to the evocation of adverse events.

Selective targeting of either PDE3A or PDE3B could also offer a more novel approach to reduce cardiovascular toxicity. Indeed, when PDE3 heterogeneity was first appreciated, the terms cardiac-type and adipocyte-type PDE3 were used to describe what are now called PDE3A and PDE3B, respectively. This crude taxonomy is based on the finding that PDE3A is expressed primarily in cardiac and vascular myocytes whereas adipocytes, amongst other cells, are rich in PDE3B. While this differential expression pattern might lead one to hope that the cardiovascular and bronchodilator actions of PDE3 inhibitors are mediated by PDE3A and PDE3B, respectively, this convenient distinction is complicated by the finding that human cardiac myocytes and airways smooth muscle cells express mRNA for both isoforms (author's unpublished observations). Moreover, in many cell types where there is coincident gene expression, PDE3A usually predominates.[77] Whether PDE3A and PDE3B mediate nonoverlapping functions in the same tissue is unexplored although logic dictates that this is highly likely. Thus, it is possible that the cardiovascular actions of PDE3 inhibitors can be dissociated from their bronchodilator and anti-inflammatory activities.[65] Fortuitously, subtype-selective compounds are beginning to emerge that could be employed to assess this possibility[78] (Figure 32.3).

32.4.3 PHOSPHODIESTERASE 4/5 INHIBITORS

Chronic generalized alveolar hypoxia occurs in human diseases associated with decreased ventilation such as COPD. Indeed, many people with COPD have concurrent pulmonary hypertension that is believed to be due to hypoxic pulmonary vasoconstriction (HPV). Although the increase in pulmonary artery pressure is usually mild to moderate, some individuals suffer from right heart failure secondary to severe pulmonary hypertension. In addition, there can be extensive remodeling of the pulmonary vasculature in COPD with all

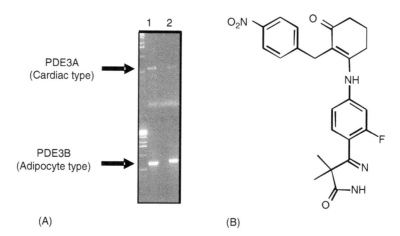

(A)   (B)

FIGURE 32.3 (A) Identification in human airway myocytes (lane 1) and ventricular myocardium (lane 2) of mRNA transcripts for PDE3A and PDE3B using gene-specific generic primers (PCR performed at 36 cycles of amplification). (B) A vinylologous amide pyrazolone, which is 33 times selective for PDE3B vs. PDE3A (IC$_{50}$ values: 150 vs. 4.5 nM, respectively).

vessel layers (intima, media, adventitia) affected, which is minimally reversible with supplemental oxygen.[79]

There is good evidence that sildenafil, an inhibitor of the cGMP-specific, or PDE5, isoenzyme family, decreases pulmonary vascular resistance in humans with pulmonary hypertension[80–90] and in volunteers with HPV.[91] This is seen in short- and long-term dosing trials (see Ref. [87]). Other PDE5 inhibitors (tadalafil, vardenafil) have also been shown to reduce pulmonary blood pressure in humans with pulmonary hypertension,[92] indicating that this effect is likely to be a characteristic of this class of drugs. This finding is important because many of the newer generation PDE5 inhibitors (e.g., tadalafil) have, in humans, a significantly longer half-life than sildenafil[93] and would be expected to exert a more prolonged depressor action in the pulmonary vasculature. Ideally, a drug designed to reduce pulmonary vascular resistance should have little or no effect on systemic blood pressure or cardiac output, and sildenafil and tadalafil seemingly exhibit this desirable property.[82,83,92,94–96] Curiously, vardenafil does not exhibit selectivity for the pulmonary vasculature being almost equiactive on systemic and pulmonary blood pressure.[92] This effect of vardenafil is unlikely to be due to the inhibition of a PDE that is insensitive to sildenafil and tadalafil, and may reflect an interaction with an unknown target that can regulate systemic vascular tone.

Selective targeting of the pulmonary vasculature by PDE5 inhibitors may reflect uniform and abundant expression of PDE5 throughout the muscularized vascular tree,[97] high nitrergic drive to pulmonary vessels, either through tonic release of endothelium-derived NO (although endothelium-dependent relaxation is apparently impaired in pulmonary arteries from subjects with COPD[98] and eNOS levels are lower in heavy smokers[99]) or through a relatively denser nitrergic innervation. Regardless, the NO axis is activated and appears to play an important role in regulating the tone of hypertensive pulmonary vessels,[54,95,97] which may explain the efficacy of PDE5 inhibitors in HPV but not systemic hypertension. NO is also believed to be involved in mitogenesis of pulmonary vascular smooth muscle and it is interesting that Wharton et al.[100] reported an antiproliferative action of a PDE5 inhibitor, sildenafil, on myocytes derived from the human pulmonary artery.

In addition to relaxing pulmonary vascular smooth muscle, Charan[101] has published case reports describing beneficial effects of sildenafil on lung function. In two patients with erectile dysfunction who had concurrent COPD, oral sildenafil produced a rapid (within 1 h of administration) and long-lasting improvement in FEV_1 and FVC (24% and 12%, respectively), indicative of direct bronchodilatation.[101] There is also evidence in guinea-pig models of asthma and COPD that sildenafil can suppress pulmonary inflammation and airways hyperreactivity following allergen and LPS challenge.[102] Thus, dual-selective inhibitors of PDE4 and PDE5 could act at multiple levels in COPD, evoking beneficial effects on inflammation, airways and vascular smooth muscle tone, and possibly remodeling.

32.4.4 PHOSPHODIESTERASE 4/7 INHIBITORS

In 1993, Micheali et al.[103] reported the development of a highly sensitive functional screen for the isolation of cDNAs that encode cAMP PDEs by complementation of defects in a strain of the yeast *Saccharomyces cerevisiae* that lacks both endogenous cAMP PDE genes. Three groups of cDNAs were isolated from a human glioblastoma cDNA library using this technology. Two of those genes were closely related to the *Drosophila* dunce cAMP PDE (i.e., PDE4-like), whereas the third encoded an enzyme that readily degraded cAMP but with characteristics distinct from all other known PDEs. At that time, the new PDE was given the name high-affinity cAMP-specific phosphodiesterase 1 (HCP-1). Primary sequence analysis established that HCP-1 shared significant homology with a stretch of 300 amino acids located toward the carboxy terminus that constitute the catalytic domain of other mammalian cAMP

PDEs.[103] Within this highly conserved region, HCP-1 exhibited the highest and lowest degrees of homology to PDE4 (35% identity; 51% similarity) and PDE2 (24% identity; 37% similarity) family members, respectively. However, given that the homology in the catalytic domain between PDE4 variants varies between 85% and 95% and that cAMP hydrolysis by HCP-1 was insensitive to the selective PDE4 inhibitors, Ro-20,1724 and rolipram, it was concluded that the novel enzyme represented the first member of a previously unknown cAMP PDE family and was designated PDE7.[103]

Two genes (*PDE7A*, *PDE7B*) have been identified in the mouse, rat, and human that encode PDE7 isoenzymes.[103] The products of these two genes are not distributed evenly and are believed to subserve nonoverlapping, tissue-specific functions.[104–107] Thus, PDE7A is abundantly expressed in the lung and immune system[108] whereas PDE7B is enriched in pancreas, brain, heart, thyroid, and skeletal muscle.[104,105,109] Screening of murine and human skeletal muscle DNA libraries has identified three splice variants of PDE7A (PDE7A1, PDE7A2, PDE7A3) derived from the same gene. All human pro-inflammatory and immune cells that have been studied express mRNA for PDE7A1 and PDE7A2 at an approximate ratio of 4:1.[108] However, irrespective of the method used for detection, PDE7A2 has never been found at the protein level despite unequivocal identification of PCR products corresponding to this transcript.[108] Very little is known about PDE7A3 or its cellular distribution, although studies by Glavas et al.[110] identified this splice variant in human CD4[+] T-lymphocytes after costimulation with anti-CD3/CD28 antibodies. Given that PDE7A3 and PDE7A1 are probably regulated by the same promoter,[111] it is possible that the expression pattern of both transcripts is similar.

Despite the discovery of PDE7 more than a decade ago, there were, until 2003, surprisingly few reports of selective inhibitors. Now, several compounds have been described (Figure 32.4) including BRL 50481[112] and BMS-586353[113] that are suitable for *in vitro* and *in vivo* pharmacological testing (reviewed in Refs. [114,115]). The advent of selective inhibitors has allowed an assessment of the functional role of PDE7A in several cells and tissues. Excitement in this area initially was fueled by a report in 1999 that IL-2 production by, and the subsequent

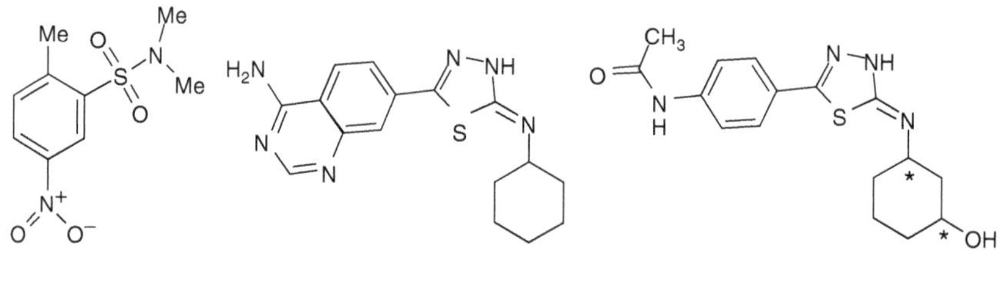

	IC$_{50}$ (μM)		
Compound	PDE4	PDE7	PDE4/PDE7
3	62	0.36	238
4	92.2	0.0039	23667
5	40.8	0.085	480

FIGURE 32.4 Structure of some PDE7 inhibitors and their isoenzyme selectivity relative to PDE4. 3, BRL 50481 (*N*,*N*,2-trimethyl-5-nitrobenzene sulfonamide); 4, 7-[(5*Z*)-5-(cyclohexylimino)-4,5-dihydro-1,3,4-thiadiazol-2-yl]-2-methylquinazolin-4-amine; 5, *N*-(4-{(5*Z*)-5-[(3-hydroxycyclohexyl)imino]-4,5-dihydro-1,3,4-thiadiazol-2-yl} phenyl) acetamide. *Chiral center.

proliferation of, anti-CD3/anti-CD28-stimulated human T-lymphocytes was associated with induction of PDE7A1, and that delivery to these cells of antisense oligonucleotides directed against all PDE7A mRNA transcripts prevented these responses.[110,116] However, subsequent investigations with small molecule PDE7 inhibitors have not corroborated these findings. Thus, BRL 50481 has no effect on IL-15-induced proliferation of human CD8$^+$ T-lymphocytes.[112] Similarly, proliferation and Th1 (IL-2, IFNγ, TNFα) and Th2 (IL-4, IL-5, IL-13) cytokine production evoked by ligation of CD3/CD28 is preserved in T-lymphocytes taken from *PDE7A* knockout mice and from wild-type animals treated with BMS-586353.[113] It is unclear why these results do not concur with data reported by Li et al.,[116] but it seems unlikely to be species related or to a redundant mechanism in mice that compensates for the deficiency in *PDE7A*. Two additional possibilities may account for the discrepancy. In the study reported by Smith et al.,[112] CD8$^+$ T-lymphocytes were isolated from other leukocytes by negative immunoselection using a mixture of antibodies against CD11b, CD16, CD19, CD36, CD56, and CD4. Although the same methodology was used by Li et al.,[116] antibodies against CD25 and HLA-DR were also employed, which will remove all activated and proliferating T-cells. Thus, it is possible that naive T-cells are regulated differently by PDE7A when compared to their activated, proliferating counterparts. Alternatively, the use of naked antisense oligonucleotides, as used by Li et al.,[116] may not have targeted specifically the mRNA of interest or, alternatively, evoked toxic effects that were sequence nonspecific.[117]

Although inhibition of PDE7A with BRL 50481 does not attenuate the proliferation of human T-cells per se, it significantly augments the antimitogenic and cAMP-elevating activities of rolipram.[112] Similarly, the suppression by PDE4 inhibitors of TNFα release from LPS-stimulated human monocytes and lung macrophages is significantly enhanced by BRL 50481.[112] Collectively, these data are reminiscent of the behavior of PDE3 inhibitors in human T-cells[65] and demonstrate that PDE7A can suppress pro-inflammatory and immune cell function if PDE4 is inhibited concomitantly.

Of additional interest is the finding that culture of human monocytes resulted in a time-dependent upregulation of PDE7A1 and conferred functional sensitivity to BRL 50481 (i.e., LPS-induced TNFα release was significantly inhibited).[112] Moreover, in monocytes in which PDE7A1 was up-regulated, the inhibition of TNFα release evoked by the PDE4 inhibitor, rolipram, and other cAMP-elevating agents was enhanced in a purely additive manner. These data imply that PDE7 inhibitors alone may regulate the responsiveness of monocytes and possibly other pro-inflammatory and immune cells under circumstances when PDE7A is highly expressed. In this respect, many cytokines relevant to the pathogenesis of chronic inflammatory diseases signal, in part, through a PKC-dependent mechanism[118] and it is known that the human PDE7A1 promoter is activated by phorbol esters.[111] Moreover, PDE7A1 expression in human monocytes treated with phorbol myristate acetate is significantly increased relative to time-matched control cells (S.J. Smith and M.A. Giembycz, unpublished observations).

Although inhibition of PDE7A has little or no demonstrable anti-inflammatory activity under normal conditions, mice deficient in the *PDE7A* gene respond to immunization with a significantly enhanced antibody response when compared to wild-type animals.[113] Thus, PDE7A may play a central role in cAMP/PKA signaling processes such as B-lymphocyte function.[113]

On balance, the ubiquitous distribution of PDE7A across human immune and pro-inflammatory cells has provided the impetus to design selective small molecule inhibitors and to evaluate their potential as novel anti-inflammatory drugs. However, of the limited studies thus far reported, inhibitors of PDE7 are remarkably inactive *in vitro* and *in vivo* on functional responses (e.g., T-cell proliferation, cytokine output) that may be considered pro-inflammatory. At best,

they interact synergistically (normally additively) with PDE4 inhibitors and, in this respect, their behavior is reminiscent of PDE3 inhibitors in the same experimental settings. Nevertheless, PDE7A may play prominent functional roles under circumstances where it is up-regulated or in responses that have not been empirically investigated such as antibody production.[113] In addition, based on results obtained with BRL 50481, the possibility that greater efficacy could be achieved with hybrid inhibitors of PDE7 and PDE4 is a concept worthy of further investigation and is the focus of several patent filings.[119,120]

32.4.5 Nonselective Phosphodiesterase Inhibitors

The nonselective PDE inhibitor, theophylline, was originally identified by Kossel in Berlin in 1888 and was synthesized 12 years later by Boehringer. However, more than two decades elapsed before the bronchodilator activity of theophylline was realized[121] and a further 56 years before a possible mechanism of action was suggested.[1] Since the early 1950s, theophylline and some related alkylxanthines have been used widely in the treatment of asthma and, more latterly, COPD although recent international guidelines recommend that these drugs be relegated to third-line therapy for both indications.[122] The primary use of theophylline is as a bronchodilator and significant improvements in lung function are evoked at plasma concentrations above 10 μg/ml (55 μM).[122] The biochemical basis of theophylline-induced bronchodilatation is due primarily to inhibition of PDE3 in airways smooth muscle, although the compound is not very potent.[123] Indeed, *in vitro* studies have found that the EC_{50} of theophylline for the relaxation of isolated human tracheal smooth muscle is ~70 μM,[123–125] which equates to a plasma concentration of 32 μg/ml assuming that 60% of the drug is bound to plasma proteins.[126] Theophylline has also been reported to suppress a number of inflammatory indices in asthma and COPD[127–134] at low doses (5 μg to 10 μg/ml) that produce only a modest, but probably significant, inhibition of PDE (see Ref. [135]). Moreover, the concurrent administration of theophylline to patients with mild to moderate asthma who are not adequately controlled by inhaled corticosteroids provides equivalent or even superior control than doubling the dose of inhaled corticosteroids.[136–138] Despite demonstrable therapeutic benefit in asthma, it is the low therapeutic ratio of theophylline that is a major cause for concern as adverse events tend to occur when the plasma concentration exceeds 20 μg/ml (110 μM). The most common adverse events are attributable to PDE4 inhibition and include nausea, vomiting, gastrointestinal irritation, and headache. Other, more serious, side effects such as central nervous system (CNS) stimulation, diuresis, and cardiac arrhythmias occur at higher concentrations and are due primarily to antagonism of adenosine A_1-receptors.[122] Higher plasma concentrations still may precipitate convulsions and even death.[122] Thus, it is tempting to speculate that second generation, nonselective PDE inhibitors that are devoid of activity at adenosine receptors may have a superior therapeutic activity over theophylline and compounds that selectively target PDE4. Improved activity may be realized for several reasons. First, it is well established that inhibitors of PDE4 can act synergistically in many pro-inflammatory and immune cells with compounds that block both PDE3[64–66] and PDE7.[112] Thus, lower doses of a nonselective compound may be more efficacious than a PDE4 monotherapy, whereas adverse events attributed to inhibition of PDE4 in nontarget tissues are reduced. The opportunities for synergy between other PDE families would also be favored. Second, regardless of synergy, concurrent inhibition of multiple PDEs would be expected to exert clinically relevant effects, not induced by selective PDE4 inhibition, on other processes that contribute to the pathogenesis of asthma or COPD including airways remodeling (PDE1-regulated), endothelial cell permeability (PDE2-regulated), mast cell stabilization (PDE3-mediated) airways smooth muscle tone (PDE3-regulated), and remodeling of the pulmonary vasculature (PDE5-regulated).

It is noteworthy that nonselective compounds might also be associated with an increased risk of adverse events over PDE4 inhibitors, either through mechanisms of synergy in nontarget tissues or by virtue of inhibiting multiple PDEs. Clinical experience with theophylline suggests that anti-inflammatory activity and associated clinical benefit can be realized at an acceptable therapeutic ratio (see above). It has been suggested that the effects of low-dose theophylline (5 to 10 μg/ml) are unrelated to PDE inhibition.[139,140] However, even at 5 μg/ml (27 μM) and taking into account plasma protein binding, theophylline will still inhibit, albeit modestly, many PDEs by up to ~25% depending on the isoenzyme.[135] If functional synergy is also seen due to inhibition of two or more distinct PDE family members, then a compelling argument can be made that PDE inhibition accounts for the mechanism of action of theophylline.

32.5 CONCLUDING REMARKS

The decision by the pharmaceutical industry to develop second generation PDE4 inhibitors for the treatment of asthma and COPD (and other chronic inflammatory disorders) is based on a conceptually robust hypothesis that is now supported by a wealth of preclinical data. Therefore, it is highly likely that if approved and also shown to be potentially disease-modifying in clinical trials, PDE4 inhibitors will offer physicians a novel class of drugs to treat patients in whom lung function is compromised by emphysema or bronchitis. Nevertheless, further major refinements in the design or use of PDE4 inhibitors are still necessary if major clinical benefit is to be realized. Of the possibilities currently available, dual-selective inhibitors that target PDE4 and either PDE1, PDE3, PDE5, or PDE7 may represent a novel approach worthy of empirical investigation. Moreover, the synthesis of nonselective compounds, devoid of activity at adenosine receptors, may paradoxically prove to be an even better therapeutic strategy at enhancing efficacy with a more favorable therapeutic ratio. Indeed, it would be ironic if, after 20 years of research into selective PDE inhibitors for airways inflammatory diseases, it was deemed necessary, on the balance of available evidence, to reinvent the wheel with the development of nonselective compounds.

ACKNOWLEDGMENTS

M.A.G. is an Alberta Heritage Foundation Senior Scholar and is funded by the Canadian Institutes of Health Research and the Alberta Lung Association.

REFERENCES

1. Butcher, R.W., and E.W. Sutherland. 1962. Adenosine 3′,5′-phosphate in biological materials. I. Purification and properties of cyclic 3′,5′-nucleotide phosphodiesterase and use of this enzyme to characterize adenosine 3′,5′-phosphate in human urine. *J Biol Chem* 237:1244.
2. Soderling, S.H., S.J. Bayuga, and J.A. Beavo. 1998. Identification and characterization of a novel family of cyclic nucleotide phosphodiesterases. *J Biol Chem* 273:15553.
3. Beavo, J.A. 1988. Multiple isozymes of cyclic nucleotide phosphodiesterase. *Adv Second Messenger Phosphoprotein Res* 22:1.
4. Beavo, J.A. 1995. Cyclic nucleotide phosphodiesterases: Functional implications of multiple isoforms. *Physiol Rev* 75:725.
5. Conti, M., and S.L. Jin. 1999. The molecular biology of cyclic nucleotide phosphodiesterases. *Prog Nucleic Acid Res Mol Biol* 63:1.
6. Torphy, T.J. 1998. Phosphodiesterase isozymes: Molecular targets for novel antiasthma agents. *Am J Respir Crit Care Med* 157:351.

7. Giembycz, M.A. 2001. Cilomilast: A second generation phosphodiesterase 4 inhibitor for asthma and chronic obstructive pulmonary disease. *Expert Opin Investig Drugs* 10:1361.
8. Huang, Z., et al. 2001. The next generation of PDE4 inhibitors. *Curr Opin Chem Biol* 5:432.
9. Doherty, A.M. 1999. Phosphodiesterase 4 inhibitors as novel anti-inflammatory agents. *Curr Opin Chem Biol* 3:466.
10. Souness, J.E., D. Aldous, and C. Sargent. 2000. Immunosuppressive and anti-inflammatory effects of cyclic AMP phosphodiesterase (PDE) type 4 inhibitors. *Immunopharmacology* 47:127.
11. Dyke, H.J., and J.G. Montana. 2002. Update on the therapeutic potential of PDE4 inhibitors. *Expert Opin Investig Drugs* 11:1.
12. Burnouf, C., and M.P. Pruniaux. 2002. Recent advances in PDE4 inhibitors as immunoregulators and anti-inflammatory drugs. *Curr Pharm Des* 8:1255.
13. Houslay, M.D., P. Schafer, and K.Y. Zhang. 2005. Phosphodiesterase-4 as a therapeutic target. *Drug Discov Today* 10:1503.
14. Giembycz, M.A. 2000. Phosphodiesterase 4 inhibitors and the treatment of asthma: Where are we now and where do we go from here? *Drugs* 59:193.
15. Duplantier, A.J., et al. 1996. Biarylcarboxylic acids and amides: Inhibition of phosphodiesterase type IV versus [^{3}H]rolipram binding activity and their relationship to emetic behavior in the ferret. *J Med Chem* 39:120.
16. Souness, J.E., and S. Rao. 1997. Proposal for pharmacologically distinct conformers of PDE4 cyclic AMP phosphodiesterases. *Cell Signal* 9:227.
17. Compton, C.H., et al. 2001. Cilomilast, a selective phosphodiesterase-4 inhibitor for treatment of patients with chronic obstructive pulmonary disease: A randomised, dose-ranging study. *Lancet* 358:265.
18. Torphy, T.J., et al. 1999. Ariflo (SB 207499), a second generation phosphodiesterase 4 inhibitor for the treatment of asthma and COPD: From concept to clinic. *Pulm Pharmacol Ther* 12:131.
19. Giembycz, M.A. 2002. Development status of second generation PDE4 inhibitors for asthma and COPD: The story so far. *Monaldi Arch Chest Dis* 57:48.
20. 2004. Roflumilast: APTA 2217, B9302-107, BY 217, BYK 20869. *Drugs RD* 5:176.
21. Rabe, K.F., et al. 2005. Roflumilast—an oral anti-inflammatory treatment for chronic obstructive pulmonary disease: A randomised controlled trial. *Lancet* 366:563.
22. Giembycz, M.A. 2006. An update and appraisal of the cilomilast phase III clinical development programme for chronic obstructive pulmonary disease. *Br J Clin Pharmacol* 62:138.
23. Lagente, V., et al. 2005. Selective PDE4 inhibitors as potent anti-inflammatory drugs for the treatment of airway diseases. *Mem Inst Oswaldo Cruz* 100 Suppl 1:131.
24. Brown, W.M. 2005. Cilomilast GlaxoSmithKline. *Curr Opin Investig Drugs* 6:545.
25. Soto, F.J., and N.A. Hanania. 2005. Selective phosphodiesterase-4 inhibitors in chronic obstructive lung disease. *Curr Opin Pulm Med* 11:129.
26. Lipworth, B.J. 2005. Phosphodiesterase-4 inhibitors for asthma and chronic obstructive pulmonary disease. *Lancet* 365:167.
27. Jeffery, P. 2005. Phosphodiesterase 4-selective inhibition: Novel therapy for the inflammation of COPD. *Pulm Pharmacol Ther* 18:9.
28. Gamble, E., et al. 2003. Antiinflammatory effects of the phosphodiesterase-4 inhibitor cilomilast (Ariflo) in chronic obstructive pulmonary disease. *Am J Respir Crit Care Med* 168:976.
29. Howell, R.E., W.T. Muehsam, and W.J. Kinnier. 1990. Mechanism for the emetic side effect of xanthine bronchodilators. *Life Sci* 46:563.
30. Chowdhury, B.A. 2003. Memorandum. Overview of the FDA background materials for NDA 21–573, Ariflo (cilomilast) Tablets, 15 mg, for the maintenance of lung function (FEV1) in patients with chronic obstructive lung disease (COPD). Available at: http://www.fda.gov/ohrms/dockets/ac/03/briefing/3976B1_02_B-FDA-Tab%201.pdf
31. Losco, P.E., et al. 2004. The toxicity of SCH 351591, a novel phosphodiesterase-4 inhibitor, in Cynomolgus monkeys. *Toxicol Pathol* 32:295.
32. GlaxoSmithKline. 2003. SB 207499 (Ariflo, Cilomilast)—New Drugs Application (21–573). Pulmonary and Allergy Drug Products Advisory Committee. Transcript of Hearing held on 5

September 2003 at the Holiday Inn Gaithersburg, Gaithersburg, Maryland, USA. Available at: http://www.fda.gov/ohrms/dockets/ac/03/transcripts/3976T1.doc

33. Savage, C.O. 2002. The evolving pathogenesis of systemic vasculitis. *Clin Med* 2:458.
34. Davies, D.J. 2005. Small vessel vasculitis. *Cardiovasc Pathol* 14:335.
35. Mecklenburg, L., et al. 2006. Mesenteritis precedes vasculitis in the rat mesentery after subacute administration of a phosphodiesterase type 4 inhibitor. *Toxicol Lett* 163:54.
36. Rao, R.R., et al. 1985. Pre-clinical toxicity studies on the new nitroimidazole 1-methylsulphonyl-3-(1-methyl-5-nitroimidazole-2-yl)-2-imidazolidinone. *Arzneimittelforschung* 35:1692.
37. Jones, T.R., et al. 1998. Effects of a selective phosphodiesterase IV inhibitor (CDP-840) in a leukotriene-dependent non-human primate model of allergic asthma. *Can J Physiol Pharmacol* 76:210.
38. Ruben, Z., et al. 1989. Spontaneous disseminated panarteritis in laboratory beagle dogs in a toxicity study: A possible genetic predilection. *Toxicol Pathol* 17:145.
39. Bishop, S.P. 1989. Animal models of vasculitis. *Toxicol Pathol* 17:109.
40. Slim, R.M., et al. 2002. Effect of dexamethasone on the metabonomics profile associated with phosphodiesterase inhibitor-induced vascular lesions in rats. *Toxicol Appl Pharmacol* 183:108.
41. Robertson, D.G., et al. 2001. Metabonomic assessment of vasculitis in rats. *Cardiovasc Toxicol* 1:7.
42. Larson, J.L., et al. 1996. The toxicity of repeated exposures to rolipram, a type IV phosphodiesterase inhibitor, in rats. *Pharmacol Toxicol* 78:44.
43. GlaxoSmithKline. 2003. SB 207499 (Ariflo, Cilomilast)—New Drugs Application (21–573). Pulmonary and Allergy Drug Products Advisory Committee. Nonclinical Findings. Available at: http://www. fda.gov/ohrms/dockets/ac/03/briefing/3976B1_01_C-Glaxo-Nonclinical%20-Findings.pdf
44. Data Monitor. 2003. Press Release: Merck discontinues development of PDE4 inhibitor compound. Available at: http://www.datamonitor.com/~90c3cffcaf0546f7ad089c97929631c8~/companies/company/?pid = E915DFB3-6FBF-4C67-AE8F-29EB1160FC4C&nid = 816AB3F2-982C-44B3-887C-3451D89A6362&type = NewsWire&article = 1
45. Nyska, A., et al. 1998. Theophylline-induced mesenteric periarteritis in F344/N rats. *Arch Toxicol* 72:731.
46. Collins, J.J., et al. 1988. Subchronic toxicity of orally administered (gavage and dosed-feed) theophylline in Fischer 344 rats and B6C3F1 mice. *Fundam Appl Toxicol* 11:472.
47. Torphy, T.J., et al. 1993. Identification, characterization and functional role of phosphodiesterase isozymes in human airway smooth muscle. *J Pharmacol Exp Ther* 265:1213.
48. Giembycz, M.A., and P.J. Barnes. 1991. Selective inhibition of a high affinity type IV cyclic AMP phosphodiesterase in bovine trachealis by AH 21–132. Relevance to the spasmolytic and anti-spasmogenic actions of AH 21–132 in the intact tissue. *Biochem Pharmacol* 42:663.
49. Torphy, T.J., et al. 1991. Role of cyclic nucleotide phosphodiesterase isozymes in intact canine trachealis. *Mol Pharmacol* 39:376.
50. Rybalkin, S.D., et al. 1997. Calmodulin-stimulated cyclic nucleotide phosphodiesterase (PDE1C) is induced in human arterial smooth muscle cells of the synthetic, proliferative phenotype. *J Clin Invest* 100:2611.
51. Rybalkin, S.D., et al. 2002. Cyclic nucleotide phosphodiesterase 1C promotes human arterial smooth muscle cell proliferation. *Circ Res* 90:151.
52. Jeffery, P.K. 2001. Remodeling in asthma and chronic obstructive lung disease. *Am J Respir Crit Care Med* 164:S28.
53. Aoshiba, K., and A. Nagai. 2004. Differences in airway remodeling between asthma and chronic obstructive pulmonary disease. *Clin Rev Allergy Immunol* 27:35.
54. Naeije, R., and J.A. Barbera. 2001. Pulmonary hypertension associated with COPD. *Crit Care* 5:286.
55. Ahn, H.S., et al. 1997. Potent tetracyclic guanine inhibitors of PDE1 and PDE5 cyclic guanosine monophosphate phosphodiesterases with oral antihypertensive activity. *J Med Chem* 40:2196.
56. Kelly, J.J., and M.A. Giembycz. 1994. Current status of cyclic nucleotide phosphodiesterase isoenzymes. In *Methylxanthines and phosphodiesterase inhibitors in the treatment of airway disease.* Eds. J. Costello, and P.J. Piper, 27–80. London: Parthenon Publishing.

57. Shakur, Y., et al. 2001. Regulation and function of the cyclic nucleotide phosphodiesterase (PDE3) gene family. *Prog Nucleic Acid Res Mol Biol* 66:241.

58. Meacci, E., et al. 1992. Molecular cloning and expression of human myocardial cGMP-inhibited cAMP phosphodiesterase. *Proc Natl Acad Sci USA* 89:3721.

59. Taira, M., et al. 1993. Molecular cloning of the rat adipocyte hormone-sensitive cyclic GMP-inhibited cyclic nucleotide phosphodiesterase. *J Biol Chem* 268:18573.

60. Myou, S., et al. 1999. Bronchodilator effect of inhaled olprinone, a phosphodiesterase 3 inhibitor, in asthmatic patients. *Am J Respir Crit Care Med* 160:817.

61. Bardin, P.G., et al. 1998. Effect of selective phosphodiesterase 3 inhibition on the early and late asthmatic responses to inhaled allergen. *Br J Clin Pharmacol* 45:387.

62. Leeman, M., et al. 1987. Reduction in pulmonary hypertension and in airway resistances by enoximone (MDL 17,043) in decompensated COPD. *Chest* 91:662.

63. Fujimura, M., et al. 1995. Bronchodilator and bronchoprotective effects of cilostazol in humans in vivo. *Am J Respir Crit Care Med* 151:222.

64. Robicsek, S.A., et al. 1991. Multiple high-affinity cAMP-phosphodiesterases in human T-lymphocytes. *Biochem Pharmacol* 42:869.

65. Giembycz, M.A., et al. 1996. Identification of cyclic AMP phosphodiesterases 3, 4 and 7 in human CD4$^+$ and CD8$^+$ T-lymphocytes: Role in regulating proliferation and the biosynthesis of interleukin-2. *Br J Pharmacol* 118:1945.

66. Schudt, C., H. Tenor, and A. Hatzelmann. 1995. PDE isoenzymes as targets for anti-asthma drugs. *Eur Respir J* 8:1179.

67. Ukena, D., et al. 1995. Effects of the mixed phosphodiesterase III/IV inhibitor, zardaverine, on airway function in patients with chronic airflow obstruction. *Respir Med* 89:441.

68. Foster, R.W., et al. 1992. Trials of the bronchodilator activity of the isoenzyme-selective phosphodiesterase inhibitor AH 21–132 in healthy volunteers during a methacholine challenge test. *Br J Clin Pharmacol* 34:527.

69. Altana Press Release. 2002. ALTANA confirms its forecast for the whole year. Available at: http://www.altana.com/root/index.php?page_id=202&cms_press_id=176

70. Movsesian, M.A. 2003. PDE3 inhibition in dilated cardiomyopathy: Reasons to reconsider. *J Card Fail* 9:475.

71. Naeije, R. 2003. Pulmonary hypertension and right heart failure in COPD. *Monaldi Arch Chest Dis* 59:250.

72. Joseph, E.C. 2000. Arterial lesions induced by phosphodiesterase III (PDE III) inhibitors and DA$_1$ agonists. *Toxicol Lett* 112–113:537.

73. Yamashita, K., et al. 1990. Increased external carotid artery blood flow in headache patients induced by cilostazol. Preliminary communication. *Arzneimittelforschung* 40:587.

74. Wilsmhurst, P.T., and M.M. Webb-Peploe. 1983. Side effects of amrinone therapy. *Br Heart J* 49:447.

75. Brunnee, T., et al. 1992. Bronchodilatory effect of inhaled zardaverine, a phosphodiesterase III and IV inhibitor, in patients with asthma. *Eur Respir J* 5:982.

76. Hatzelmann, A., et al. 1996. Enzymatic and functional aspects of dual-selective PDE3/4 inhibitors. In *Phosphodiesterase inhibitors*, First ed., Eds. C. Schudt, G. Dent, and K.F. Rabe, 147–160. London: Academic Press.

77. Maurice, D.H., et al. 2003. Cyclic nucleotide phosphodiesterase activity, expression, and targeting in cells of the cardiovascular system. *Mol Pharmacol* 64:533.

78. Edmondson, S.D., et al. 2003. Benzyl vinylogous amide substituted aryldihydropyridazinones and aryldimethylpyrazolones as potent and selective PDE3B inhibitors. *Bioorg Med Chem Lett* 13:3983.

79. Timms, R.M., F.U. Khaja, and G.W. Williams. 1985. Hemodynamic response to oxygen therapy in chronic obstructive pulmonary disease. *Ann Intern Med* 102:29.

80. Galie, N., et al. 2005. Sildenafil citrate therapy for pulmonary arterial hypertension. *N Engl J Med* 353:2148.

81. Prasad, S., J. Wilkinson, and M.A. Gatzoulis. 2000. Sildenafil in primary pulmonary hypertension. *N Engl J Med* 343:1342.

82. Wilkens, H., et al. 2001. Effect of inhaled iloprost plus oral sildenafil in patients with primary pulmonary hypertension. *Circulation* 104:1218.
83. Michelakis, E., et al. 2002. Oral sildenafil is an effective and specific pulmonary vasodilator in patients with pulmonary arterial hypertension: Comparison with inhaled nitric oxide. *Circulation* 105:2398.
84. Michelakis, E.D., et al. 2003. Long-term treatment with oral sildenafil is safe and improves functional capacity and hemodynamics in patients with pulmonary arterial hypertension. *Circulation* 108:2066.
85. Chockalingam, A., et al. 2005. Efficacy and optimal dose of sildenafil in primary pulmonary hypertension. *Int J Cardiol* 99:91.
86. Lepore, J.J., et al. 2005. Hemodynamic effects of sildenafil in patients with congestive heart failure and pulmonary hypertension: Combined administration with inhaled nitric oxide. *Chest* 127:1647.
87. Steiner, M.K., et al. 2005. Pulmonary hypertension: Inhaled nitric oxide, sildenafil and natriuretic peptides. *Curr Opin Pharmacol* 5:245.
88. Sastry, B.K., et al. 2002. A study of clinical efficacy of sildenafil in patients with primary pulmonary hypertension. *Indian Heart J* 54:410.
89. Sastry, B.K., et al. 2004. Clinical efficacy of sildenafil in primary pulmonary hypertension: A randomized, placebo-controlled, double-blind, crossover study. *J Am Coll Cardiol* 43:1149.
90. Bharani, A., et al. 2003. The efficacy and tolerability of sildenafil in patients with moderate-to-severe pulmonary hypertension. *Indian Heart J* 55:55.
91. Zhao, L., et al. 2001. Sildenafil inhibits hypoxia-induced pulmonary hypertension. *Circulation* 104:424.
92. Ghofrani, H.A., et al. 2004. Differences in hemodynamic and oxygenation responses to three different phosphodiesterase-5 inhibitors in patients with pulmonary arterial hypertension: A randomized prospective study. *J Am Coll Cardiol* 44:1488.
93. Gupta, M., A. Kovar, and B. Meibohm. 2005. The clinical pharmacokinetics of phosphodiesterase-5 inhibitors for erectile dysfunction. *J Clin Pharmacol* 45:987.
94. Ghofrani, H.A., et al. 2002. Combination therapy with oral sildenafil and inhaled iloprost for severe pulmonary hypertension. *Ann Intern Med* 136:515.
95. Michelakis, E.D. 2003. The role of the NO axis and its therapeutic implications in pulmonary arterial hypertension. *Heart Fail Rev* 8:5.
96. Jackson, G., et al. 1999. Effects of sildenafil citrate on human hemodynamics. *Am J Cardiol* 83:13C.
97. Sebkhi, A., et al. 2003. Phosphodiesterase type 5 as a target for the treatment of hypoxia-induced pulmonary hypertension. *Circulation* 107:3230.
98. Dinh-Xuan, A.T., et al. 1991. Impairment of endothelium-dependent pulmonary-artery relaxation in chronic obstructive lung disease. *N Engl J Med* 324:1539.
99. Barbera, J.A., et al. 2001. Reduced expression of endothelial nitric oxide synthase in pulmonary arteries of smokers. *Am J Respir Crit Care Med* 164:709.
100. Wharton, J., et al. 2005. Antiproliferative effects of phosphodiesterase type 5 inhibition in human pulmonary artery cells. *Am J Respir Crit Care Med* 172:105.
101. Charan, N.B. 2001. Does sildenafil also improve breathing? *Chest* 120:305.
102. Toward, T.J., N. Smith, and K.J. Broadley. 2004. Effect of phosphodiesterase-5 inhibitor, sildenafil (Viagra), in animal models of airways disease. *Am J Respir Crit Care Med* 169:227.
103. Michaeli, T., et al. 1993. Isolation and characterization of a previously undetected human cAMP phosphodiesterase by complementation of cAMP phosphodiesterase-deficient *Saccharomyces cerevisiae*. *J Biol Chem* 268:12925.
104. Hetman, J.M., et al. 2000. Cloning and characterization of PDE7B, a cAMP-specific phosphodiesterase. *Proc Natl Acad Sci USA* 97:472.
105. Sasaki, T., et al. 2000. Identification of human PDE7B, a cAMP-specific phosphodiesterase. *Biochem Biophys Res Commun* 271:575.
106. Hoffmann, R., S. Abdel'Al, and P. Engels. 1998. Differential distribution of rat PDE-7 mRNA in embryonic and adult rat brain. *Cell Biochem Biophys* 28:103.

107. Sasaki, T., J. Kotera, and K. Omori. 2002. Novel alternative splice variants of rat phosphodiesterase 7B showing unique tissue-specific expression and phosphorylation. *Biochem J* 361:211.
108. Smith, S.J., et al. 2003. Ubiquitous expression of phosphodiesterase 7A in human proinflammatory and immune cells. *Am J Physiol Lung Cell Mol Physiol* 284:L279–L289.
109. Gardner, C., et al. 2000. Cloning and characterization of the human and mouse PDE7B, a novel cAMP-specific cyclic nucleotide phosphodiesterase. *Biochem Biophys Res Commun* 272:186.
110. Glavas, N.A., et al. 2001. T cell activation up-regulates cyclic nucleotide phosphodiesterases 8A1 and 7A3. *Proc Natl Acad Sci USA* 98:6319.
111. Torras-Llort, M., and F. Azorin. 2003. Functional characterization of the human phosphodiesterase 7A1 promoter. *Biochem J* 373:835.
112. Smith, S.J., et al. 2004. Discovery of BRL 50481 [3-(*N,N*-dimethylsulfonamido)-4-methylnitrobenzene), a selective inhibitor of phosphodiesterase 7: *in vitro* studies in human monocytes, lung macrophage and CD8$^+$ T-lymphocytes. *Mol Pharmacol* 66:1679.
113. Yang, G., et al. 2003. Phosphodiesterase 7A-deficient mice have functional T cells. *J Immunol* 171:6414.
114. Giembycz, M.A., and S.J. Smith. 2006. Phosphodiesterase 7 as a therapeutic target. *Drugs Future* 31:207.
115. Giembycz, M.A., and S.J. Smith. 2006. Phosphodiesterase 7A: A new therapeutic target for alleviating chronic inflammation? *Curr Pharm Des* 12:3207.
116. Li, L., C. Yee, and J.A. Beavo. 1999. CD3- and CD28-dependent induction of PDE7 required for T cell activation. *Science* 283:848.
117. Stein, C.A. 2001. The experimental use of antisense oligonucleotides: A guide for the perplexed. *J Clin Invest* 108:641.
118. Kontny, E., et al. 2000. Rottlerin, a PKC isozyme-selective inhibitor, affects signaling events and cytokine production in human monocytes. *J Leukoc Biol* 67:249.
119. Hatzelmann, A., et al. 2004. Phthalazinone derivatives useful as PDE4/7 inhibitors. WO 02/085906. Altana Pharma AG.
120. Pitts, W.J., et al. 2002. Dual inhibitors of PDE7 and PDE4. WO 02/088079. Bristol-Myers Squibb.
121. Schultze-Werninghaus, G., and J. Meier-Sydow. 1982. The clinical and pharmacological history of theophylline: First report on the bronchospasmolytic action in man by S.R. Hirsch in Frankfurt (Main) 1922. *Clin Allergy* 12:211.
122. Barnes, P.J. 2003. Theophylline: New perspectives for an old drug. *Am J Respir Crit Care Med* 167:813.
123. Cortijo, J., et al. 1993. Investigation into the role of phosphodiesterase IV in bronchorelaxation, including studies with human bronchus. *Br J Pharmacol* 108:562.
124. Goldie, R.G., et al. 1986. In vitro responsiveness of human asthmatic bronchus to carbachol, histamine, β-adrenoceptor agonists and theophylline. *Br J Clin Pharmacol* 22:669.
125. Finney, M.J., J.A. Karlsson, and C.G. Persson. 1985. Effects of bronchoconstrictors and bronchodilators on a novel human small airway preparation. *Br J Pharmacol* 85:29.
126. Guillot, C., et al. 1984. Spontaneous and provoked resistance to isoproterenol in isolated human bronchi. *J Allergy Clin Immunol* 74:713.
127. Pauwels, R., et al. 1985. The effect of theophylline and enprofylline on allergen-induced bronchoconstriction. *J Allergy Clin Immunol* 76:583.
128. Ward, A.J., et al. 1993. Theophylline—an immunomodulatory role in asthma? *Am Rev Respir Dis* 147:518.
129. Mapp, C., et al. 1987. Protective effect of antiasthma drugs on late asthmatic reactions and increased airway responsiveness induced by toluene diisocyanate in sensitized subjects. *Am Rev Respir Dis* 136:1403.
130. Sullivan, P., et al. 1994. Anti-inflammatory effects of low-dose oral theophylline in atopic asthma. *Lancet* 343:1006.
131. Jaffar, Z.H., et al. 1996. Low-dose theophylline modulates T-lymphocyte activation in allergen-challenged asthmatics. *Eur Respir J* 9:456.

132. Kraft, M., et al. 1996. Theophylline: Potential antiinflammatory effects in nocturnal asthma. *J Allergy Clin Immunol* 97:1242.
133. Lim, S., et al. 2001. Low-dose theophylline reduces eosinophilic inflammation but not exhaled nitric oxide in mild asthma. *Am J Respir Crit Care Med* 164:273.
134. Culpitt, S.V., et al. 2002. Effect of theophylline on induced sputum inflammatory indices and neutrophil chemotaxis in chronic obstructive pulmonary disease. *Am J Respir Crit Care Med* 165:1371.
135. Dent, G., and K.F. Rabe. 1996. Effects of theophylline and non-selective xanthine derivatives on PDE isoenzymes and cellular function. In *Phosphodiesterase inhibitors*. Eds. C. Schudt, G. Dent, and K.F. Rabe, 41–64. London: Academic Press.
136. Evans, D.J., et al. 1997. A comparison of low-dose inhaled budesonide plus theophylline and high-dose inhaled budesonide for moderate asthma. *N Engl J Med* 337:1412.
137. Ukena, D., et al. 1997. Comparison of addition of theophylline to inhaled steroid with doubling of the dose of inhaled steroid in asthma. *Eur Respir J* 10:2754.
138. Lim, S., et al. 2000. Comparison of high dose inhaled steroids, low dose inhaled steroids plus low dose theophylline, and low dose inhaled steroids alone in chronic asthma in general practice. *Thorax* 55:837.
139. Cosio, B.G., et al. 2004. Theophylline restores histone deacetylase activity and steroid responses in COPD macrophages. *J Exp Med* 200:689.
140. Ito, K., et al. 2002. A molecular mechanism of action of theophylline: Induction of histone deacetylase activity to decrease inflammatory gene expression. *Proc Natl Acad Sci USA* 99:8921.

33 Medicinal Chemistry of PDE4 Inhibitors

Jeffrey M. McKenna and George W. Muller

CONTENTS

33.1 Introduction 667
33.2 Rolipram and Its Clinical Limitations 668
33.3 Clinical PDE4 Inhibitors..670
    33.3.1 Current Clinical Inhibitors 670
    33.3.2 Discontinued Clinical Inhibitors..677
33.4 General Structural Types for Inhibition of PDE4 682
    33.4.1 Rolipram as a Starting Point...683
    33.4.2 Theophylline as Inspiration 686
    33.4.3 Alternative Sources of Inspiration 688
33.5 PDE4 Subtype–Selective Inhibition 690
References 691

33.1 INTRODUCTION

The nucleotides cyclic adenosine $3',5'$-monophosphate (cAMP) and cyclic guanosine $3',5'$-monophosphate (cGMP) are key second messengers involved in a variety of cellular responses initiated either by the actions of a hormone or a neurotransmitter.[1,2] Two enzymatic processes control the intracellular levels of these two nucleotides: the first of these is through regulation of their syntheses (achieved through the actions of adenylate or guanylate cyclase); and the second is by way of their hydrolysis catalyzed by phosphodiesterases (PDEs). The latter enzyme superfamily has been targeted by the pharmaceutical industry for the development of inhibitors for various therapeutic uses,[3,4] the most well known being PDE5 inhibition for erectile dysfunction.[5] There are at present 11 known families of PDEs that control the levels of cAMP or cGMP by catalyzing their hydrolysis to AMP or GMP, respectively. The 11 family members are classified according to differences in sequence, substrate specificity, cofactor requirements, and response to known inhibitors.[6,7] The PDE4 family of enzymes is one of the key PDEs responsible for controlling intracellular cAMP levels (the other cAMP-specific PDEs are PDE7 and PDE8).[8] Members of the PDE4 family are predominantly responsible for hydrolyzing cAMP within immune cells[9] and cells in the central nervous system (CNS).[10] PDE4 enzymes have been reported to selectively hydrolyze cAMP ($K_m = 0.2–4.0\ \mu M$) and have a low affinity for cGMP ($K_m > 1000\ \mu M$) in *in vitro* systems.[11] In a similar fashion to all mammalian PDE superfamily proteins, PDE4 has a conserved catalytic domain near its C-terminus; components of the regulatory domain are located in both the N-terminus and extreme C-terminus.[4,8] The catalytic domain of all PDE4s includes two consensus metal binding sites for zinc and (most likely) magnesium, and the inclusion of these metal ions within the active site are crucial to confer catalytic activity.[12] The PDE4

superfamily includes four different gene products or isozymes termed PDE4A through PDE4D and each of these has additional splice variants.[13-16] The resulting proteins are differentially expressed among various tissues and cells[17,18] and so it would seem possible for inhibitors to exhibit quite different pharmacology, as a consequence of the specificity profile taken together with its tissue distribution. This chapter will discuss the medicinal chemistry of selective small-molecule PDE4 inhibitors with a focus on those inhibitors that have reached the clinic. The reader is directed toward the other chapters in this book for further discussions of the four subtypes and current opportunities that are exploited in this area.

As described in earlier chapters, inhibiting the hydrolysis of cAMP and thus increasing the intracellular level of cAMP has wide-ranging pharmacological consequences that are of interest to the pharmaceutical industry. The most heavily investigated area is the clinical application of PDE4 inhibitors in respiratory diseases and within this area there continues to be a focus on asthma and chronic obstructive pulmonary disease (COPD).[19] However, PDE4 inhibitors have also been shown to have utility in a variety of inflammatory conditions[20] and autoimmune diseases,[21] and they have also demonstrated potential clinical utility within the CNS.[22] Although the first clinical trials undertaken to assess a selective PDE4 inhibitor in humans was in a CNS indication (*vide infra*), most of the effort over the past decade has been directed toward use of this class of inhibitor to diseases of the airways with a lesser effort focusing on learning and memory, though this focus is changing. It is also interesting to note that inhibitors of PDE4 were first directed toward asthma even though they were shown to be less-effective bronchodilators when compared with β_2-adrenoreceptor agonists (which also induce cAMP accumulation). However, it is the possibility that PDE4 inhibitors affect the inflammation underlying chronic airway diseases that has provided the drive and focus for many investigators to develop this class of compound as a respiratory therapy.[23-26] More recently, evidence has been gained which could direct this class of drug toward use as immunomodulatory agents,[27] in allergic disorders[28] or in autoimmune diseases.[29] The impetus for these latter indications was provided by an early report demonstrating that PDE4 inhibition could inhibit the production of TNF-α in activated human monocytes and peripheral blood mononuclear cells (PBMC).[30] The recent clinical validation of TNF-α as a target in autoimmune diseases following the approval of Enbrel, Humira, and Remicade has further focused efforts on the development of orally active TNF-α inhibitor small-molecule equivalents of these protein-based therapies. Although there still remain questions concerning the long-term use of these injectable protein-based drugs, the primary concern is associated with an increased incidence of re-exacerbation of latent tuberculosis or other infectious diseases that drive the search for alternatives.[31,32] With the concerns surrounding injectable protein-based treatments, it can be appreciated that a small molecule oral TNF-α inhibitor might be the next step in this therapeutic arena.

33.2 ROLIPRAM AND ITS CLINICAL LIMITATIONS

Any discussion of the medicinal chemistry of PDE4 inhibitors has to begin with rolipram (1), the key prototypical PDE4 inhibitor, because of the large number of programs that have used its key pharmacophore in their design and discovery strategies.[33] Rolipram was one of the initial selective PDE4 inhibitors reported (along with the less potent Ro 20-1724).[34] Rolipram is a selective potent inhibitor of the PDE4 enzyme isolated from the U937 cell line (IC$_{50}$ ~ 400 nM) and a potent TNF-α inhibitor (IC$_{50}$ ~ 150 nM) in lipopolysaccharide (LPS)-stimulated PBMC (Figure 33.1).

Rolipram was evaluated clinically for efficacy within the CNS, and specifically as a possible therapy for depression.[35] Early, open-label, studies suggested good efficacy.[36] However, later studies indicated the compound to be less efficacious and to have nausea associated with it.[37] Rolipram was also assessed in animal studies for the treatment of Parkinson's disease, and

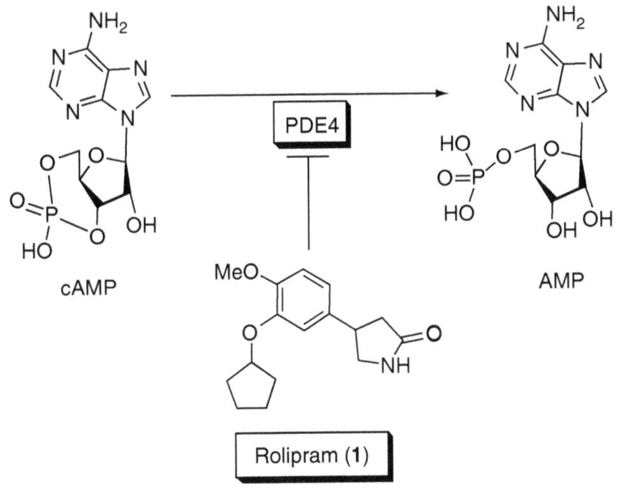

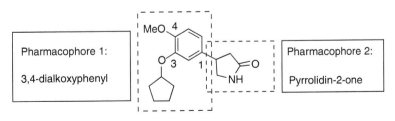

FIGURE 33.1 cAMP is converted to AMP by PDE4 and that activity can be inhibited by rolipram (**1**).

initial studies were promising; however, clinically no improvements were seen and, again, nausea was associated with its dosing.[38,39] Consequently, rolipram's clinical development was discontinued because of nausea, vomiting, and general intestinal side effects associated with its dosing.[36,40,41] The dog and ferret have been used subsequently to measure the emetic effects of rolipram and more recent PDE4 inhibitors in research environments.[42]

Rolipram (**1**) from the perspective of a medicinal chemist is a simple molecule. It can be viewed as containing two pharmacophores, a 3-cyclopentoxy-4-methoxyphenyl group and a pyrrolidin-2-one, a functional group also referred to as a γ-lactam (Figure 33.2). Even though rolipram (**1**) is a highly emetic agent, it has served as a key starting point for the design of many new PDE4 inhibitors. Specifically, many researchers still utilize the 3,4-dialkoxyphenyl pharmacophore of rolipram, a moiety that is responsible for both its PDE4 specificity and potency.[43] The lack of affinity of rolipram (**1**) for either the dual-activity PDEs (1, 2, 3, 10, and 11) or the cAMP-specific PDEs 7 and 8 has been ascribed to its interaction (via the 3,4-dialkoxyphenyl motif) with the side chain of the conserved glutamine residue.[44-46] This key interaction is mediated through a hydrogen-bond to the side-chain amide nitrogen atom of glutamine residue. In PDE4, the glutamine carboxamide is in a unique environment as this side-chain amide is locked and unable to rotate; also the close interaction of rolipram with N321 (asparagine) and Y159 (tyrosine) may be another factor aiding this selectivity. The reader is directed toward the Chapter 6 for further discussions on the crystal structure of the PDE4 enzyme and inhibitor. Rolipram continues to be used in research as one of the gold standards in PDE4 inhibitor. This widespread use of rolipram (**1**) as the benchmark for so many comparisons can be explained by three contributing factors: it is readily prepared in the laboratory; it is PDE4 selective; and it has strong potency both *in vitro* and *in vivo*.

FIGURE 33.2 (±)-Rolipram (**1**).

A recent hypothesis within the PDE4 area is that inhibition of PDE4D is responsible for the widespread emetic effects seen with PDE4 inhibitors.[47] Rolipram is highly emetic, and exhibits a minor selectivity in favor of PDE4D in its subtype inhibition profile, with the inhibition of PDE4C being weakest among the four subtypes.[48] Another explanation for rolipram's emetic activity is a two binding site hypothesis.[49] Rolipram (**1**) has been shown to bind to PDE4 enzyme in both a low-affinity mode, i.e., the low-affinity rolipram-binding site (LARBS), and a high-affinity mode, i.e., high-affinity rolipram-binding site (HARBS). The low-affinity site activity has been reported to correlate with the inhibition of cAMP hydrolysis, and the high-affinity site activity correlates more closely with the emetic side effects.[50,51] As a consequence of this hypothesis, a goal of many medicinal chemistry efforts has been to design PDE4 inhibitors with improved LARBS and HARBS ratios. However, recent data would suggest that this hypothesis may be too simple. Although the driving forces behind inhibitor–conformer interactions are still to be uncovered, it is clear that PDE enzymes have only a single binding site for rolipram, namely the catalytic unit, where the rolipram serves as a competitive inhibitor. Nonetheless, in light of these observations, *in vitro* model systems have been set up to identify these conformational changes in the enzyme.[52,53] Other data are suggestive of the alternate conformations that are due to either differing phosphorylation,[54] the formation of complexes of the PDE4 protein with other proteins, such as PDE4A4 with SH3-domain containing proteins[55] and XAP2, or as a consequence of exposure to thiol reagents.[56] Indeed, given the multiplicity of PDE4 isoforms it may well be that there are a large number of ways in which modification of specific isoforms or interaction with proteins could trigger a conformational change in the catalytic unit. Nevertheless, the avoidance of emetic responses and its apparent correlation with LARBS allows the medicinal chemist a pragmatic pathway forward using this correlation in an effort to optimize their series based on this differing binding affinity seen under specific conditions.

33.3   CLINICAL PDE4 INHIBITORS

Following rolipram's clinical withdrawal, numerous clinical studies have been conducted with selective PDE4 inhibitors[57] but, as yet, none of these selective PDE4 inhibitors have reached the marketplace. Success in this arena has been limited by clinical side effects associated with the PDE4 inhibitors, of which emesis is one that has been dose-limiting. The many PDE4 inhibitor research programs have resulted in the discovery of a large number of selective PDE4 inhibitors.[58–65] In addition to the emetic effects, more recently an inflammatory syndrome has been identified as another dose-limiting toxicity observed in rodents.[66] This newer side effect, together with the standard safety issues associated with drug development, will have to be monitored closely in all PDE4 medicinal chemistry programs.

The drugs described within Section 33.3.1 represent the most clinically advanced PDE4 inhibitors. In reference to these compounds, and looking forward, it is the success of these drugs that could determine the future interest in PDE4 as a molecular target from within the pharmaceutical sector. However, the therapeutic potential of PDE4 inhibition is so large that even without clinical success of a current agent it is likely that there will still be a continued interest in this arena. This interest could focus on isozyme-selective inhibitors or allosteric inhibitors with differing selectivity profiles that could be pursued.

33.3.1   Current Clinical Inhibitors

Roflumilast (**2**; Altana/Tanabe; BY-217; Daxas) is in clinical development for both asthma and COPD.[19,67,68] Its registration dossier was presented for review in Europe in 2004[69] but its NDA filing has been delayed because of a slower-than-anticipated enrollment in extensive

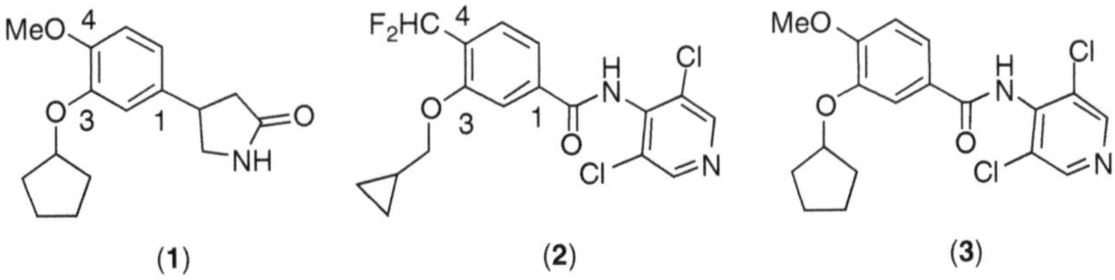

FIGURE 33.3 Rolipram (**1**), roflumilast (**2**), and piclamilast (**3**).

Phase 3 trials. In July of 2005, Pfizer returned all rights for the development of the drug to Altana, reversing the agreement of April 2002 in which the two companies had defined a codevelopment strategy for roflumilast.[70] The compound was originally disclosed in 1995, and at that time it represented a new class of PDE4 inhibitor, a so called second-generation PDE4 inhibitor.[71] Roflumilast (**2**) does contain the 3,4-dialkoxyl pharmacophore found within rolipram (**1**), and it is also closely related to the Rhone-Poulenc-Rorer PDE4 inhibitor, piclamilast (**3**) (Figure 33.3). Roflumilast varies only in the alkyl groups of the 3,4-dialkoxyl pharmacophore when compared to piclamilast (**3**). The Rhone-Poulenc-Rorer compound, piclamilast (**3**), is discussed later in the chapter and contains the 3-cyclopentoxy-4-methoxyphenyl group of rolipram, whereas in roflumilast (**2**) the 3-cyclopentoxy group is replaced by a 3-cyclopropylmethoxy group and the 4-methoxy group is replaced by a 4-difluoromethoxy group. It is worth noting that roflumilast has a far superior bioavailability and is less plasma protein-bound when compared with piclamilast.[72]

Roflumilast is one of the more potent PDE4 inhibitors known, having a PDE4 IC_{50} of 0.8 nM, which is about 100-fold more potent than cilomilast (**4**, *vide infra*), and it should be noted that it is nonselective with respect to the PDE4 subtypes (Figure 33.4).[72] Roflumilast has good oral bioavailability (absolute bioavailability, *F*, of 79%) in humans[73] and has a plasma half-life of 15.7 h, which supports its clinical oral dosing paradigm of 0.5 mg once daily. The pyridine unit of roflumilast is metabolized, creating an active metabolite, its *N*-oxide (**5**) (Figure 33.5), a compound that is reported to show a similar potency to its parent molecule.[72]

Together, these two compounds (roflumilast and its *N*-oxide) display linear pharmacokinetics and have been shown to lack drug–drug interactions with erythromycin, budesonide, or salbutamol. Further adding to the drug–drug interaction data, neither food nor cigarette smoke has been shown to exert any significant effect on the pharmacokinetics of roflumilast.[74] Preclinically, roflumilast and its *N*-oxide were efficacious in allergen-induced early airway reaction (EAR) in guinea pigs, and both were eightfold more potent than the closely related piclamilast, and 32-fold more potent than rolipram.[75] Moreover, in an antigen-induced late inflammatory response in brown Norway rats, roflumilast was again the most efficacious of the compounds studied. Roflumilast and its close analog piclamilast were

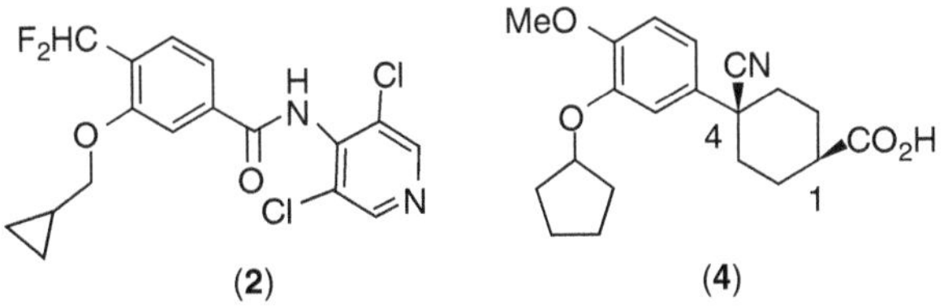

FIGURE 33.4 Roflumilast (**2**) and cilomilast (**4**).

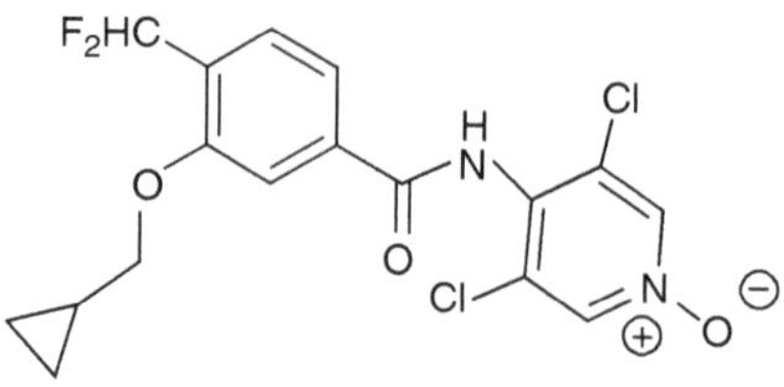

FIGURE 33.5 Roflumilast *N*-oxide (**5**).

significantly more potent when compared to rolipram and cilomilast in the inhibition of LPS-induced release of TNF-α in PBMC.[72]

Cilomilast (**4**; GlaxoSmithKline; SB-207499; Ariflo) is currently undergoing late Phase 3 studies in COPD,[19,76] having received an approvable action letter from the US Food and Drug Administration (FDA) in 2003.[77,78] Cilomilast (**4**) is also structurally related to rolipram (**1**) (Figure 33.6), containing the exact 3-cyclopentoxy-4-methoxyphenyl moiety found in rolipram. Cilomilast differs from rolipram in that it now contains three pharmacophoric groups, namely its carboxylic acid, a nitrile, and a 3,4-dialkoxyphenyl moiety rather than the two groups mentioned for rolipram. Intriguingly, the evolution of this compound relied upon rolipram, a fact that was clearly outlined in the first report of the compound in 1998.[79] This systematic study of structure–activity relationships sought to optimize a series of PDE4 inhibitors for their inhibition of monocyte-derived PDE4 catalytic activity versus their ability to compete for the high-affinity [3H]-rolipram-binding site (HARBS).

Modifications about rolipram (**1**) identified the cyclopentanone analog (**6**) to contain an appropriately positioned hydrogen-bond acceptor; this modification effectively replaced rolipram's γ-lactam with a cyclopentanone. Thereafter, the closely related cyclopentanecarboxylic acid derivative (**7**) was also identified as an analog of interest containing a second possible replacement for the lactam. Both of these viable replacement ring systems possess

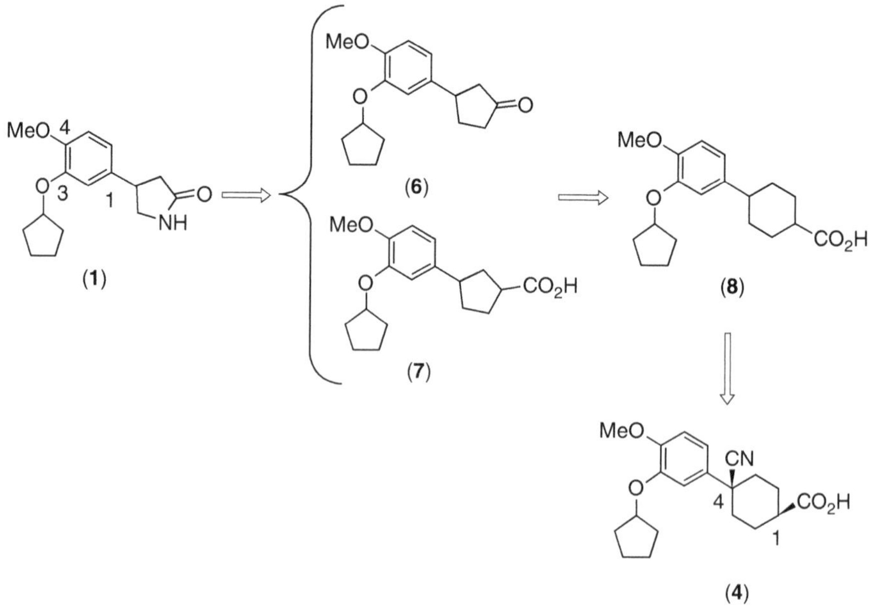

FIGURE 33.6 The evolution of cilomilast (Ariflo; **4**) from rolipram (**1**).

conformational flexibility and so in an effort to rigidify the two pharmacophores, a 1,4-substituted cyclohexane system was investigated, as exemplified by (8). This 1,4-substituted cyclohexane spacer had the ability to define a rigid pharmacophore, avoid chirality, but provide some flexibility in the positioning of the necessary functional groups. Having previously noted that the inclusion of a carboxylic acid was possible, SAR studies centered upon this achiral 1,4-carboxylic acid, and with the final identification of the benzylic nitrile in a *cisoid* relationship to the acid, cilomilast (4) was born. The drug is a selective PDE4 inhibitor as it is essentially inactive against PDEs 1, 2, 3, 5, and 7.[80] Interestingly, cilomilast is approximately 10-fold selective for the PDE4 isoform PDE4D having an IC_{50} of 12 nM (IC_{50} of PDE4A, 115 nM; PDE4B, 86 nM; PDE4C, 308 nM, respectively).[6,80] It is reported that cilomilast (4) causes much less nausea or gastric acid–secretion when compared with rolipram (1),[81] and it has also been suggested that limiting the brain exposure to PDE4 inhibitors may decrease their emetic effects. Cilomilast is extensively metabolized, the major pathways for its biotransformation are through *ortho*-dealkylation (specifically *ortho*-decyclopentylation); 3-hydroxylation of the cyclopentyl ring; and formation of an acyl glucuronide.[82] The drug is extensively cleared in the urine with approximately 90% of the drug excreted in this fashion, about 1% of the original dose does remain unchanged following oral dosing. The drug is rapidly absorbed, has excellent bioavailability ($F = 96\%$) in humans, and a plasma elimination half-life is 7 h.[83,84] There is little possibility for drug–drug interactions with cilomilast, as the only cytochrome P450 implicated in its hydroxylation is CYP2C8, a cytochrome P450 with few inhibitors or substrates.[85] As previously noted with roflumilast, cigarette smoke does not affect the pharmacokinetics of cilomilast, and the compound exhibits linear pharmacokinetics, which are unaffected by age or food.[84] Cilomilast has been shown to be effective in numerous preclinical animal models, reducing ovalbumin-induced contraction of guinea pig isolated tracheal strips,[86] blocking early and late phase reactions in an allergen-induced bronchospasm model, and also decreasing production of TNF-α from LPS-stimulated human monocytes adoptively transferred to BALB/c mouse peritoneum.[87] Cilomilast has also been assessed in a model of allergic dermatitis (BALB/c mice sensitized to toluen-2,4-diisocyanate); this study demonstrated that pretreatment with cilomilast significantly reduced swelling, granulocyte influx, and IL-1β secretion.[88] Also in animal studies, cilomilast was reported to cause a periarteritis in both rats (30 mg/kg/day) and mice (200 mg/kg/day), a side effect that has also been noted for rolipram.[66,89] As a result of these data, the FDA recommended that during clinical trials, gastrointestinal toxicity be monitored separately from other adverse effects.[90] It is not clear at this time whether cilomilast will eventually receive FDA approval.

Tetomilast (9; Otsuka; OPC-6535) is in clinical development for use in the treatment of ulcerative colitis (UC)[91] and COPD.[92] A Phase 3 study was initiated in mid-2003 for UC, and in 2004, a Phase 2 trial was initiated in COPD. The drug mediates its effects as a result of its dual inhibition of PDE4 and superoxide production. Extravascular recruitment of neutrophils plays an important role in the tissue damage associated with inflammatory conditions such as UC and COPD. Neutrophil activation is controlled by a number of intracellular pathways, but particularly through cAMP-dependent protein kinase (PKA) that stimulates the synthesis of PDE4 and so increases conversion of cAMP to AMP. Tetomilast (9) was first reported in 1995 as a lead candidate following a structure–activity relationship study (Figure 33.7).[93] Tetomilast is an indirect analog of rolipram even though it contains the 3,4-dialkoxyphenyl pharmacophore found within rolipram (1). In this case, it is exemplified as a 3,4-diethoxyphenyl group instead of the 3-cyclopentoxy-4-methoxyphenyl group found in rolipram. The original lead (10) for this drug was identified in a screening campaign as a 2,5-biarylthiazole. The replacement of the thiazole with alternative heteroaromatic rings did not improve activity and thus the thiazole was the optimal central ring. A methodical optimization

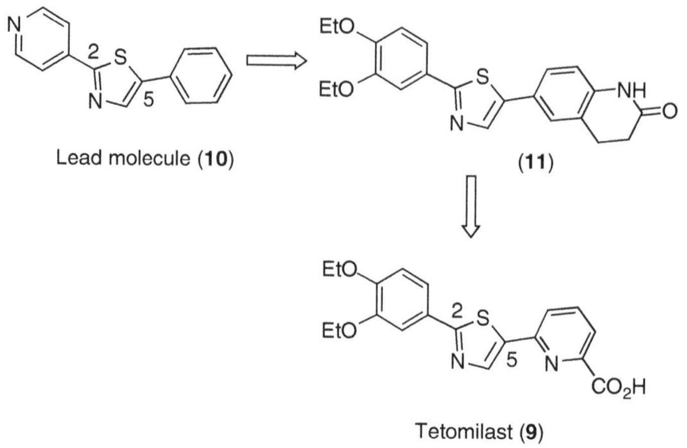

Lead molecule (10)   (11)

Tetomilast (9)

FIGURE 33.7 Tetomilast (**9**) from its initial lead compound (**10**).

identified the best group at C2, and insertion of various spacers between the thiazole and the C2-dialkoxyphenyl ring reduced the potency. Optimization of the C4 aromatic ring had initially identified a 6-(3,4-dihydro)-quinolin-2-one as the preferred group, as exemplified by (**11**). However, aqueous solubility was a problem with these derivatives, and the introduction of a carboxylic acid was the answer to this challenge. The preclinical development of tetomilast demonstrated the ability to inhibit LPS-induced TNF-α production in a dose-dependent fashion in human PBMC and human whole blood (HWB).[94] In other studies, anesthetized, ventilated pigs were infused with live *Pseudomonas aeruginosa* to induce sepsis, and tetomilast (**9**) significantly reduced acute lung injury as indicated by an 18-fold increase in bronchoalveolar lavage protein and neutrophil content; this resulted in significant improvement in arterial oxygenation.[95] Hartley strain guinea pigs were used to evaluate tetomilast in a COPD model in which the drug was given orally (10 mg/kg) once daily over a 4-week period. Significant improvements in specific airway resistance and peak expiration flow were observed when compared with vehicle control. The dextran sulfate sodium (DSS)-induced chronic colitis rat model was used to assess the effects of the compound in inflammatory bowel disease. Lower doses of the compound were administered in this model (1 mg/kg orally once daily) where reductions in diarrhea and rectal bleeding were observed when compared with vehicle control.[96]

CC-10004 (Celgene Corporation) has advanced into Phase 2 trials in asthma and psoriasis. The structure of this compound has yet to be revealed, though it has been reported to be of the same structural type as their other reported PDE4 inhibitors and it is a potent PDE4 inhibitor (PDE4 of IC_{50}, ~74 nM).[97] CC-10004 inhibits TNF-α production in LPS-activated human PBMC with an IC_{50} of 77 nM and in LPS-activated HWB with an IC_{50} of 110 nM. CC-10004 was reported to induce relaxation of guinea pig tracheal rings with an EC_{50} of 310 nM. It was also reported to be orally efficacious in the ovalbumin-induced allergic asthma model and inhibit LPS-induced lung neutrophilia in the rat. Intratracheal dosing in the LPS-induced rat lung neutrophilia model lowered the EC_{50} dose by 30- to 100-fold. It was reported to have a 30-fold improvement in therapeutic index, relative to cilomilast in a ferret model of lung neutrophilia versus the emetic effect observed. In a Phase 1 trial, CC-10004 was reported to be well tolerated (at doses of 20 mg once daily). Pharmacokinetic analysis demonstrated that a C_{max} that was sixfold greater than the PDE4 IC_{50} at a 10 mg dose and that a C_{max} was 14-fold greater than the PDE4 IC_{50} at the 20 mg dose. In 1998, Celgene reported on a novel series of selective PDE4 inhibitors.[98] During the structure–activity relationship studies, it was

noted that the 3-ethoxy substitution was generally more potent than the corresponding 3-cyclopentoxy substitution (found in rolipram). Inhibition of TNF-α in activated human PBMC with these analogs correlated with their PDE4 inhibitory activity as expected for a series of PDE4 inhibitors.

AWD-12-281 (**12**; Elbion/GlaxoSmithKline) is a potent, selective PDE4 inhibitor designed to act locally upon inflammatory cells. AWD-12-281 (**12**) contains the 3,5-dichloro-4-pyridyl moiety found in both roflumilast (**2**) and piclamilast (**3**) but not the 3,4-dialkoxyphenyl pharmacophore of rolipram (**1**). It is interesting to note that the substituted indole of AWD-12-281 appears isosteric with the 3,4-dialkoxyphenyl moiety found in roflumilast (**2**) and rolipram (**1**). AWD-12-281 is one of the few clinically advanced PDE4 inhibitors that do not contain the 3,4-dialkoxyphenyl group or a closely related mimic. The compound is currently in Phase 2 clinical trials for the treatment of COPD and in Phase 1 clinical trials for the treatment of atopic dermatitis.[99] The compound can be administered via powder inhalation, as a nasal spray, or as a topical cream to both minimize systemic exposure and limit the side effects observed with oral PDE4 compounds currently in clinical trials.[100] AWD-12-281 has shown exceptional safety and tolerability which is potentially because of its low dissolution rate, its rapid metabolism, and high plasma protein binding. The compound undergoes rapid glucuronidation on the phenol within its indole nucleus and has a high clearance from plasma.[101] The compound has a low–relative affinity for HARBS, and it has been shown to suppress the production of various cytokines in LPS-stimulated PBMC, including IL-2, IL-5, and TNF-α. In side-by-side studies aimed at the assessment of its emetic potential, it has been shown that AWD-12-281 (**12**) (Figure 33.8) demonstrated a lower potential for emetic episodes than cilomilast (**4**) (in ferrets) or roflumilast (**2**) (in pigs).[102]

HT-0712 (**13**; Inflazyme/Helicon; IPL-455903) is being investigated for use as a nootropic agent (a learning or memory enhancer).[103] Helicon disclosed in late in 2004 that a multidose phase 1 study was to be undertaken[104] and thereafter Inflazyme announced in mid-2005 that the drug was safe and well tolerated, and importantly without emetic effects, in a single, ascending dose study in 50 healthy volunteers.[105] From a structural perspective, HT-0712 (**13**) (Figure 33.9) has commonalities with rolipram (**1**), as it contains the 3-cyclopentoxy-4-methoxyphenyl pharmacophore. However, instead of the γ-lactam present in rolipram its second pharmacophore is a ring-expanded δ-lactam that is further benzylated at the α-carbon. Preclinically, the compound was efficacious (at 0.1 mg/kg i.p.) in a murine model of long-term memory deficit (the CBP $+/-$ mutant model).[106]

EHT-202 (ExonHit Therapeutics) is in Phase 1 clinical trials for evaluation in the treatment of Alzheimer's disease.[107] The compound (whose structure has not been disclosed) is a dual inhibitor of PDE4 and GABA-A. In November 2004, ExonHit reported that the compound

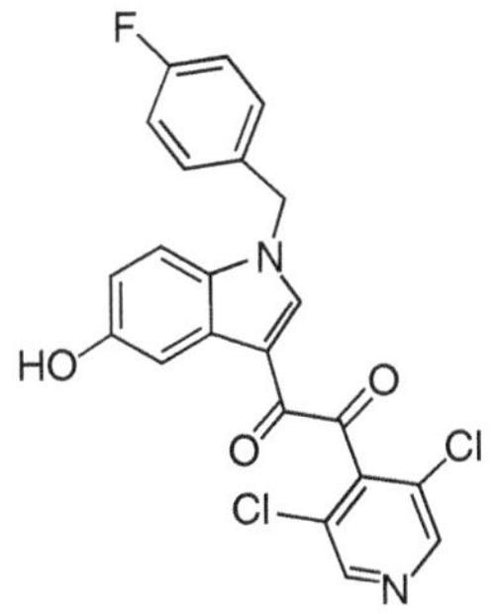

FIGURE 33.8 AWD-12-281 (**12**).

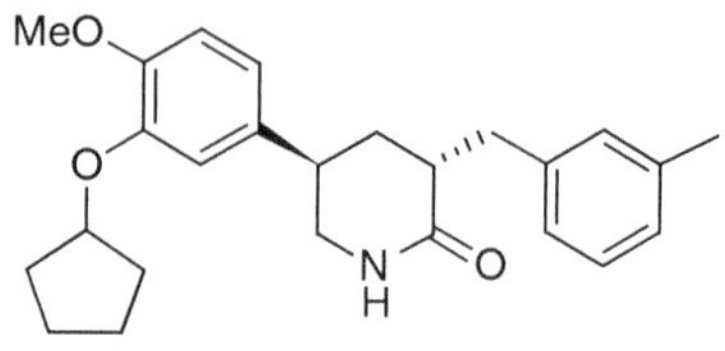

FIGURE 33.9 HT-0712 (**13**).

was orally active; was efficacious *in vivo* in protecting cells from ischemia; improved cognitive performance in aged rats; and at a dose of 10 mg/kg was mildly anxiolytic.[108]

GRC-3886 (Glenmark/Forest/Teijin) in March of 2005 successfully completed phase 1 clinical single- and multiple-dosing studies in the UK for asthma and COPD.[109] Glenmark also indicated that WHO had granted the compound the international nonproprietary name of Oglemilast. Preclinically, GRC-3886 has been demonstrated to inhibit all 4 PDE4 subtypes (IC_{50} of PDE4A, 1.3; of PDE4B, 2.5; of PDE4C, 6.2; PDE4D, 1.7 nM), its structure has not been disclosed, but it is reported to be a novel PDE4 inhibitor with a >7000-fold selectivity for PDE4 over the other PDE (PDE 1 to 11) isoforms. GRC-3886 has been shown to inhibit LPS-induced serum TNF-α production with an ED_{50} of 0.52 mg/kg dosed orally by gavage in Sprague–Dawley rats. It also reversed Freund's complete adjuvant (FCA)–induced acute hyperalgesia in Sprague–Dawley rats with an ED_{50} of 0.85 mg/kg (p.o.) and produced significant amelioration of collagen-induced arthritis (CIA) in DBA/1J mice at a dose 10 mg/kg/day (p.o.) comparable to that produced by valdecoxib at 1 mg/kg/day (p.o.).[110]

ONO-6126 (**14**; Ono Pharmaceutical Co. Ltd.) is in Phase 2 clinical trials for bronchial asthma and COPD.[111] ONO-6126 (Figure 33.10) is a close structural analog of cilomilast thus containing the 3-cyclopentoxy-4-methoxyphenyl pharmacophore of rolipram and cilomilast and its second ring is reminiscent of the *cis*-4-cyano-cyclohexane carboxylic acid of cilomilast. Preclinical data associated with the compound were presented at the 98th American Thoracic Society meeting in 2002, which demonstrated its *in vivo* antiinflammatory effects with reduced emetic liabilities.[112] The compound was also reported in an abstract to have demonstrated suppressive effects on TNF-α release following repeated dosing in healthy Japanese male subjects.[113]

Mesopram (**15**; Schering AG; SH-636) (Figure 33.11) as the name implies is a close structural analog of rolipram containing the 3,4-dialkoxyphenyl pharmacophore. The molecule has a 3-propoxy group in place of the 3-cyclopentoxy group present in rolipram. Schering last reported on this compound in 2003,[114] indicating that mesopram alleviated experimental colitis in mice, a TNF-α-mediated disease where the biologic TNF-α inhibitors have had success. Mesopram was also shown to ameliorate the symptoms during experimental autoimmune encephalomyelitis (EAE), data that could support its use in a clinical setting as a treatment for multiple sclerosis.[115]

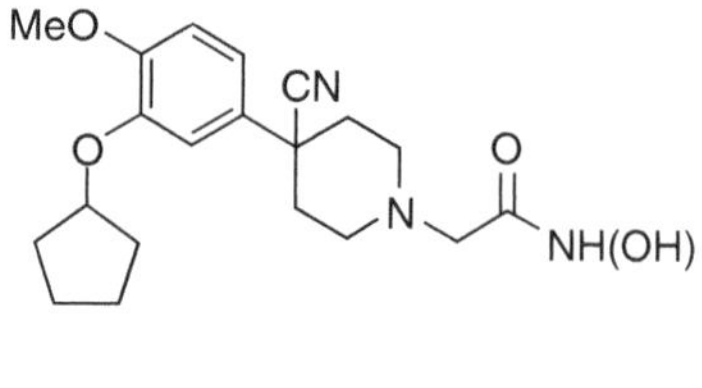

FIGURE 33.10 ONO-6126 (**14**).

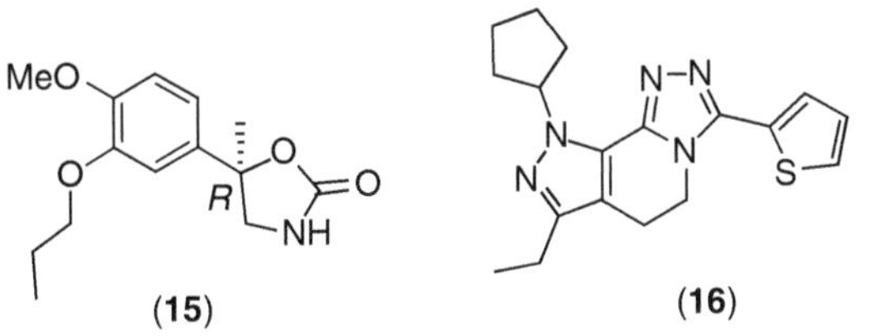

FIGURE 33.11 Mesopram (**15**) and tofimilast (**16**).

Tofimilast (**16**; Pfizer; CP-325366) (see Figure 33.11) is not a rolipram analog and can be considered more of a cAMP- or xanthine-type analog; little data are available at this stage with regard to its development. The compound was originally patented in 1996,[116] and a large- scale process reported for its synthesis in 2001.[117]

From this overview of current PDE4 inhibitors currently in development for clinical use and whose structures have been disclosed, it can be seen that the pharmacophore displayed by rolipram has been incorporated into a large number of these compounds. Thus, rolipram has had a lasting effect on the medicinal chemistry field of PDE4 inhibitors that continues to this day.

33.3.2 Discontinued Clinical Inhibitors

Piclamilast (**3**; Rhone-Poulenc-Rorer; RP-73401) was originally reported in 1994.[118] Initial SAR quickly identified 2,6-disubstitution to be preferred in the nondialkoxyphenyl ring. Further refinement and optimization through variation of this arene ring yielding hetero-cyclic-based amides identified the 2,6-dichloro-4-pyridine unit to be most potent (Figure 33.12). This compound, and others within the benzamide series, showed the same preferences for the 3,4-dialkoxy bearing ring as had previously been elucidated,[119] so that the most preferred dialkoxyphenyl moiety was the 4-methoxy-3-cyclopentoxyphenyl motif as is found in rolipram (**1**). The development of this compound (RP-73401) proceeded until 1996 when it was halted in Phase 2. Piclamilast is a very potent PDE4 inhibitor with a PDE4 IC_{50} of ~1 nM isolated from a variety of cell types, and is 19,000 times selective for PDE4 compared to other PDE isoenzymes.[120] As a result of its low bioavailability because of an extensive first-pass effect in the liver[121] and amide hydrolysis,[122] the compound was preferentially administered by inhalation.[123] It is worth noting that piclamilast is metabolized exclusively by CYP2B6, and hydroxylation only occurs at the 3-position of the cyclopentyl ring, which is also a site of biotransformation noted with cilomilast (**4**; *vide infra*). In a 2 year inhalation rat carcinogenicity study, piclamilast (**3**) was found to cause tumors of the nasal olfactory region.[124] The compound is a potent suppressor of eosinophil function,[120] caused

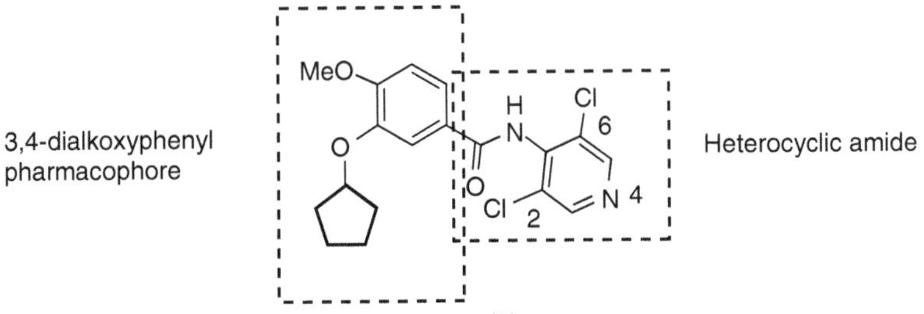

3,4-dialkoxyphenyl pharmacophore

Heterocyclic amide

FIGURE 33.12 Piclamilast (**3**).

significant bronchodilation,[123,125] and an inhibitor of anti-CD3-induced IL-4 release; this latter effect was seen to correlate with the inhibition of a low-affinity PDE4 isoform.[126] Most recently, this compound has been studied for its ability to modulate various immune responses. It should be noted how structurally similar roflumilast (2) is to this compound, with the only differences being in the ether alkyl substituents. As can be seen later in the chapter in Section 33.4.3, a number of groups used this compound, (3), as the basis for developing their own PDE4 programs.

IC-485 (ICOS Corporation), whose structure has yet to be disclosed, entered clinical development as a potential treatment for COPD, but in March 2005 results from a Phase 2 study were released, indicating that the drug had failed to meet its primary endpoint of improving lung function (in 258 patients comparing three doses of IC-485 to placebo).[127]

Filaminast (17; Wyeth-Ayerst; WAY-PDA-641) again falls into the rolipram analog realm and varies from rolipram in the replacement of the pyrrolidin-2-one with an oxime-carbamate (Figure 33.13). Its clinical development was also halted in Phase 2 and previously it had been shown to be a selective PDE4 inhibitor that was 10 times more potent than rolipram (1) having an IC_{50} of 50 nM. The compound was tested in dogs for typical PDE inhibitor side effects and although no effects on heart rate or cardiac contractility were observed (up to 3 mg/kg), emesis, though less severe when compared with rolipram, was seen at similar doses to that of rolipram.[128]

Atizoram (18; Pfizer; CP-80633) differs from the more typical PDE4 inhibitor in the 3,4-dialkoxyphenyl pharmacophore through the inclusion of a lipophilic [2.2.1]-bicyclohep-tane in place of the more often seen cyclopentane as the 3-alkyl ether linkage, but atizoram should still be considered to be rolipram analog (see Figure 33.13). The compound was demonstrated to be emetic,[129] which ultimately limited its clinical development for asthma. The PDE4 inhibitor was moderately potent and nonselective across the PDE4 subtypes,[130] was a bronchodilator,[131] and was shown preclinically to be an effective inhibitor of TNF-α in mice.[132] The compound is of specific interest because when it was applied bilaterally over eight days to 20 atopic dermatitis patients, the compound was shown to be efficacious. This was the first such clinical efficacy demonstrated for a PDE4 inhibitor in the treatment of atopic dermatitis.[133]

Lirimilast (19; Bayer; BAY-19-8004) was halted in development stage in mid-2001[134] with Bayer stating that the compound had not met the desired target product profile. The compound was a selective PDE4 inhibitor (PDE4 IC_{50} of ~67 nM) that was not selective for any particular PDE4 subtype (Figure 33.14). As expected, the compound exhibited a broad profile of antiinflammatory activity in animal models of COPD, asthma, and was an effective bronchodilator.[135] The first disclosure from Bayer of the benzofurans as PDE4 inhibitors was in late 1994 in the patent literature and at that time this series represented quite a structural deviation from rolipram (1).[136] In Phase 1 clinical studies, the compound displayed linear pharmacokinetics with a half-life of 25 h along with the dose-related–side

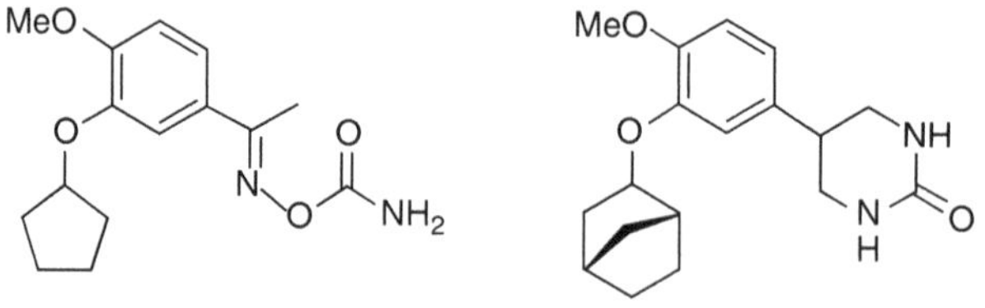

(17)   (18)

FIGURE 33.13 Filaminast (17) and atizoram (18).

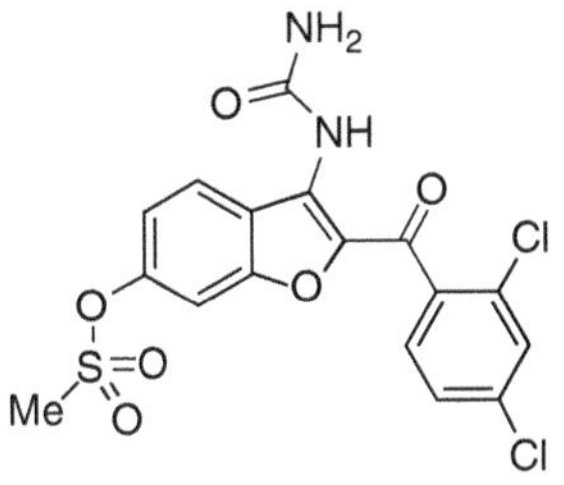

FIGURE 33.14 BAY-19-8004 (**19**).

effect of nausea, which, as for cilomilast was transient and early in the dosing regimen. A report of the compound in a 1-week study for the treatment of COPD patient has been published, in which forced expiratory volume in 1 sec (FEV$_1$) and sputum cell number were not affected.[137]

D-4418 (**20**; Chiroscience now Celltech/Schering-Plough) was stopped in Phase 2 clinical trials for the treatment of asthma and other respiratory disorders. As can be seen from the structure, D-4418 is a benzamide in which the Chiroscience group has used 8-methoxy-quinoline (in a strategy related to the benzofuran cyclization described in the later Section 33.4.1) as an isostere for the 3,4-dialkoxyphenyl pharmacophore of piclamilast (**3**). In an *in vitro* setting, D-4418 (**20**) was modestly potent (PDE4 IC$_{50}$ of ~170 nM). In the guinea pig, the compound was found to have good bioavailability ($F = 62\%$) and when dosed at 30 mg/kg, it showed significant inhibition of eosinophil influx and hyperreactivity. When dosed at 60 mg/kg, it was not emetic in the ferret.[138] Chiroscience further optimized D-4418 and identified a more potent (PDE4 IC$_{50}$ of 51 nM), bioavailable and efficacious compound (**21**; SCH 365351).[139] Further investigations, primarily from a pharmacokinetic perspective, identified SCH 351591 (**22**), the *N*-oxide of (**21**), to be of greater interest (Figure 33.15).[140] This compound is equipotent to (**21**), is not PDE4 subtype selective, and displays better pharmacokinetics when compared with (**21**). SCH 351591 (**22**) was recently shown to be efficacious in antigen-induced–guinea pig or primate asthma models.[141] Although the compound was found to be better tolerated in the ferret than cilomilast (**4**), it has been documented that SCH 351591 (**22**) produces, in Cynomolgus monkeys, acute to chronic inflammation of small- to medium-sized arteries in many tissues and organs.[142]

CDP-840 (**23**; Celltech/Merck Frosst) evolved from a SAR program using rolipram (**1**) as its starting point.[143] The compound was well tolerated and found to reduce antigen-induced bronchoconstriction and pulmonary eosinophilic inflammation in animal models and attenuate late-phase asthmatic response to allergen challenge in patients; however, it was discontinued from clinical development on the basis of the data accumulated from three separate Phase 2a studies in a patient population with mild-to-moderate asthma.[144] CDP-840 (**23**) is a

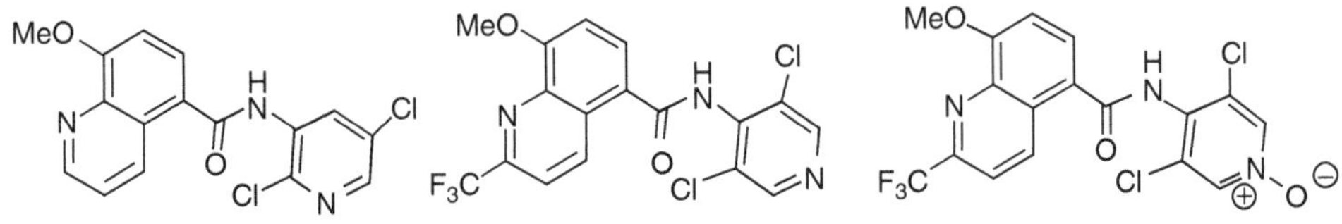

D-4418 (**20**)   SCH 365351 (**21**)   SCH 351591 (**22**)

FIGURE 33.15 D-4418 (**20**) and SCH 365351 (**21**), and SCH 351591 (**22**).

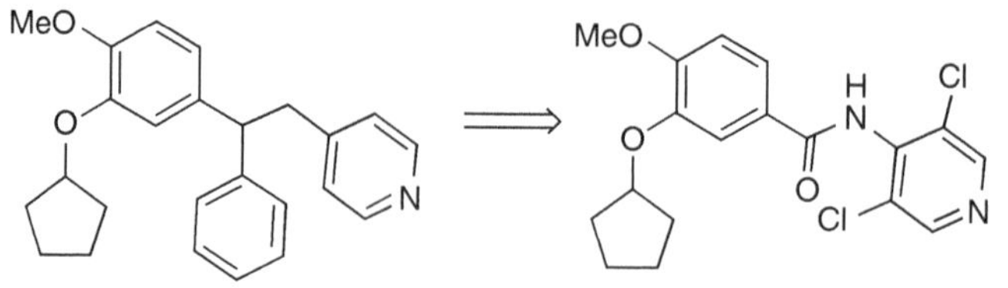

FIGURE 33.16 CDP-840 (**23**) and piclamilast (**3**).

rolipram analog containing the identical dialkoxyphenyl group, although it is possible to identify this compound to an equal extent with piclamilast (**4**), differing in replacement of the amide unit with a simple aliphatic two-atom spacer to a 4′-pyridyl group (Figure 33.16). However, even with this done, CDP-840 has a second phenyl group at the benzylic position of the alkoxyphenyl ring. This second phenyl ring of CDP-840 has no substituents, and it is hydroxylated *in vivo*, in the rat, in its *para*-position. It should be noted that its *para*-chloro derivative is more metabolically stable. It is interesting to note that this particular phase 1 oxidation is not seen in other species, including humans, where the major circulating metabolite was seen to be the pyridinium glucuronide.[145] CDP-840 was potent (PDE4 IC$_{50}$ of 4.3 nM), but lost substantial activity in the LPS-induced TNF-α HWB assay (IC$_{50}$ of ~16 μM). Thus, metabolic stability and whole blood assay activity were two aspects of CDP-840 (**23**) that required improvement, and were the focus of follow-up derivatives.

L-826,141 (**26**; Celltech/Merck Frosst) was targeted as a therapy for asthma and COPD. The evolution of this compound from CDP-840 (**23**; Figure 33.17) first relied upon the reduction of metabolism of the dialkoxyphenyl unit.[146] The first synthesized analogs included a series of *para*-chlorophenyl compounds, and then using this substructure, modifications of both ether groups were made resulting in the quick identification of the *bis*-difluoromethyl ether combination (**24**). The focus, once this had been accomplished, was the reoptimization

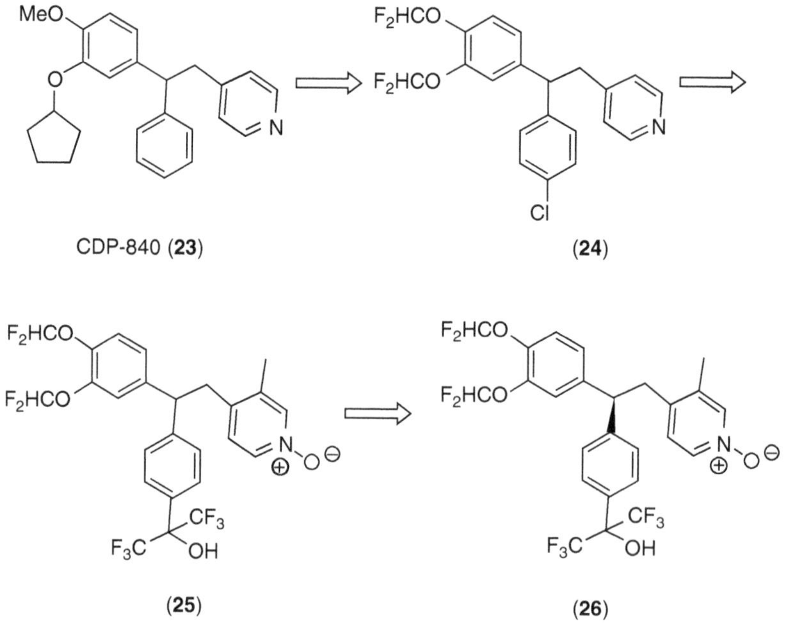

CDP-840 (**23**)   (**24**)

(**25**)   (**26**)

FIGURE 33.17 The evolution of L-826,141 (**26**) from CDP-840 (**23**).

of the aromatic ring neighboring the dialkoxyphenyl group, where a few hundred analogs were prepared. Many of these analogs showed good enzyme potency, but it was the results from the LPS-induced HWB TNF-α inhibition assay where the activity of CDP-840 had suffered that finalized the substituent in the phenyl ring. The simple *para*-hydroxymethyl derivative was most interesting, and SAR studies based on this compound identified tertiary carbinols as the superior group through a combination of enzyme inhibition potency, the HWB TNF-α inhibition assay and pharmacokinetics. Finally, the *bis*-trifluoromethyl carbinol was selected, and oxidation of the pyridine to its *N*-oxide yielded L-791,943 (**25**). The unfortunate aspect of this compound was an excessively long half-life and so the introduction of a metabolic labile substituent (soft spot) was the strategy employed to rectify this drawback.[147] A multitude of positions for this soft spot were tested and eventually a single methyl group introduced at the *meta*-position within the pyridyl ring was chosen; this racemate was resolved to give L-826,141 (**26**). This compound is a potent PDE4 inhibitor (IC_{50} of 1.3 nM), inhibits PDE4B and PDE4D more potently (~threefold) than PDE4A or PDE4C, and is potent in the HWB assay (IC_{50} of ~300 nM).[148] The compound had a shorter half-life in rats, monkeys, and dogs but still had good bioavailability ($F = 60\%$) in the rat. The faster metabolism of this compound was driven through the soft spot, generating the corresponding benzylic alcohol and explained the improved pharmacokinetics. Finally, the compound was not emetic in ferrets at 30 mg/kg.

CI-1018 (**27**; Parke-Davis now Pfizer) was under the development for the treatment of asthma and COPD (Figure 33.18). CI-1018 (**27**) is clearly not related to rolipram (**1**); it is a benzodiazepine that was identified following screening of a library that contained a number of cholecystokinin-receptor antagonists. The initial lead was modified (a methyl added, and 2-indolyl replaced with 4-pyridyl) to give CI-1018 (PDE4 IC_{50} of ~1.14 μM).[149] The SAR demonstrated the need for a five-membered fused ring and the necessity for the (R) absolute stereochemistry. This compound was further optimized for greater PDE selectivity, potency in the *in vitro* inhibition of TNF-α release from HWB assay, and its emetic profile. Accordingly, CI-1044 (**28**) (see Figure 33.18) was identified (PDE4 IC_{50} of ~270 nM), and the compound was well tolerated in the ferret at 40 mg/kg.[150] More recently, CI-1044 was reported to inhibit plasma TNF-α release in LPS-stimulated whole blood of healthy volunteers and COPD patients.[151]

YM-976 (**29**; Yamanouchi) was first disclosed within the patent literature in 1996[152] and thereafter entered the clinic. The lead for this series was identified during random screening and optimization generated a compound that has similarity to nitraquazone (*vide infra*, **56** Section 33.4.2), having a chlorine atom as the nitro-replacement (Figure 33.19). The compound is potent but a nonsubtype selective PDE4 inhibitor (PDE4 IC_{50} of 2.2 nM) and is 1000-fold selective for PDE4 over PDE1, 2, 3, and 5. The drug displays a high affinity for the high-affinity rolipram-binding site (IC_{50} of 2.6 nM) but with a much improved ratio relative to rolipram, and interestingly it has also been reported that the compound does not

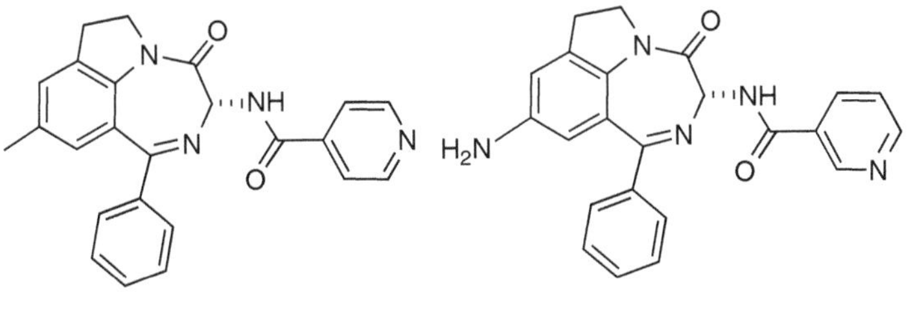

FIGURE 33.18 CI-1018 (**27**) and CI-1044 (**28**).

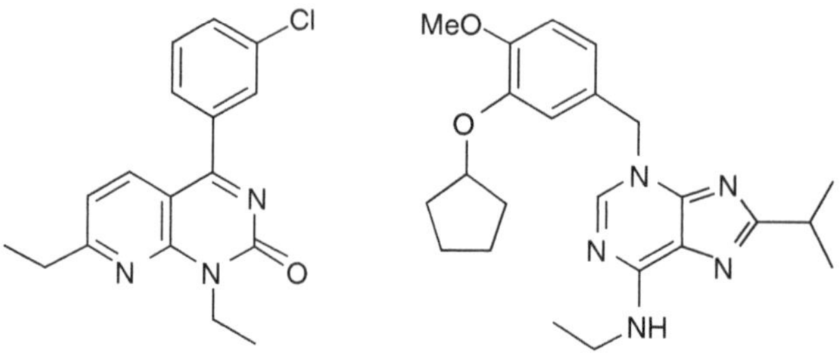

FIGURE 33.19 YM-976 (**29**; Yamanouchi) and V-11294A (**30**; Napp).

enter the rat brain.[48,153] In preclinical models of antigen-induced eosinophil infiltration in mice, rats, or ferrets, the compound was shown to be efficacious with ED_{50} of 3.6, 1.7, and 1.2 mg/kg, respectively. The ferret also provided data on the emetic potential of the drug; it was efficacious at 1.2 mg/kg but no emetic episodes were reported at 10 mg/kg.[154]

V-11294A (**30**; Napp) is a xanthine that has been assessed in healthy volunteers for its pharmacokinetic and pharmacodynamic effects.[155] The compound was bioavailable, having a plasma half-life of greater than 7 h. It produces *in vivo* a single active metabolite (by way of N-de-ethylation), which is equipotent to the parent (see Figure 33.19). V-11294A was also free of emetic episodes at doses up to 300 mg. This dose was also used to demonstrate *ex vivo* that the compound could inhibit TNF-α production in whole blood following LPS-stimulation that persists for at least 24 h. Preclinically, the compound was demonstrated to be a selective PDE4 inhibitor (PDE4 IC_{50} of 405 nM), capable of inhibition of TNF-α synthesis in human monocytes. In ferrets, the compound was not emetogenic at doses of 30 mg/kg (p.o.) despite reaching plasma concentrations 10-fold the IC_{50} for PDE4.[156] The compound retains the same 3,4-dialkoxyphenyl unit found in rolipram, is a second-generation compound, and yet obviously also has theophylline as inspiration for its design (*vide infra*, **31**, Section 33.4.2 and Figure 33.20). The paucity of information about this compound would seem to suggest that its development has been discontinued.

MEM-1414 (Memory/Roche) was selected as a clinical candidate for the treatment of Alzheimer's disease, however, in April of 2005 a company update stated that the development of MEM-1414 was to be stopped.[157] Later in 2005, Memory reacquired all rights to the compound by an amendment to the agreement. The companies continue to work on PDE4 inhibitor backup compounds.[158]

33.4 GENERAL STRUCTURAL TYPES FOR INHIBITION OF PDE4

From the above discussion, it is evident that the archetypal PDE4 inhibitor, rolipram, provided many investigators with a structural starting point for PDE4-inhibitor design. This is particularly true for the development of cilomilast (**4**) and roflumilast (**2**), which, as

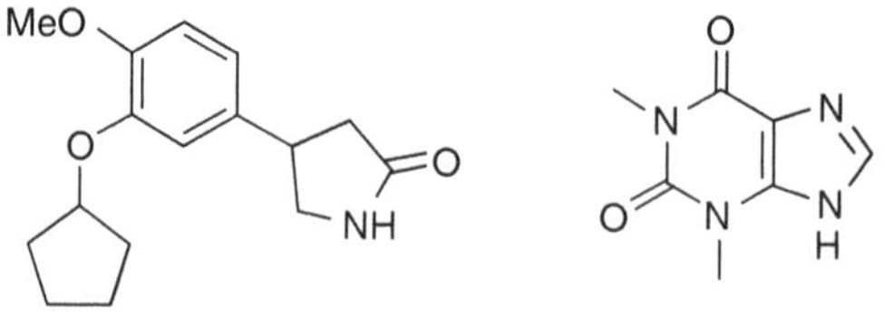

FIGURE 33.20 (±)-Rolipram (**1**) and theophylline (**31**).

we have discussed, are the most advanced in clinical development. Having said this, the drive to identify new scaffolds for PDE4 inhibition with the aim to reduce the specific PDE4 side effects has not slowed over time.[159] Another molecule that has been used as a template for the design of the PDE4 inhibitor is the nonselective PDE inhibitor, theophylline (31), whose pharmacology, in part, at least, appears to be mediated by PDE4 inhibition. At clinically attainable concentrations, theophylline also acts as a nonselective adenosine-receptor antagonist; so it is ironic that it should serve as a template for improved analogs as it is a nonselective inhibitor of PDEs and has such a crossover in its pharmacology.[160]

Theophylline is a purine and may be thought of as an adenosine analog in this context. These two molecules and the design strategies based around them will be discussed; this together with the key information gained from other inhibitors will provide an overview of the current structural modifications, which have taken forward the inhibitor design and medicinal chemistry associated with PDE4.

33.4.1 ROLIPRAM AS A STARTING POINT

Medicinal chemists had the good fortune during the investigation of the structure–activity relationship of rolipram to be able to approach it in modular fashion. This simple fact has allowed the rapid identification of key structural motifs in new inhibitors and the development of key logical structure–activity relationships to be built around them.

As we have noted already, modification of the 3,4-dialkoxyphenyl moiety has been largely restricted to minor changes of the alkyl chain of the aryl ether with a particular focus on the 3-alkoxy moiety. However, more radical changes are possible as first noted during investigations of the dialkoxyphenyl motif by the Glaxo group.[161] They demonstrated that the cyclic methylenedioxyphenyl-, the ethylenedioxyphenyl- and cyclohexylidene acetal analogs related to rolipram, all lost activity when compared with rolipram.

These results guided them to a successful replacement of the dialkoxyphenyl aromatic by using a benzofuran and in doing so opened the door to the general concept of cyclizing onto one of the phenolic oxygen atoms to achieve new classes of PDE4 inhibitor (Figure 33.21). This research identified a benzofuran series through cyclization effectively onto the 4-methoxy group (thus retaining the 3-cyclopentyloxy functionality). The direct analog (32) of rolipram (1) was essentially equivalent in potency in its inhibition of PDE4 (IC_{50} of ~260 nM). It was also noted that reduction of the resulting benzofuran to its dihydrofuran derivative yielded a less potent molecule. In an interesting extension to this, the Rhone-Poulenc-Rorer group combined the benzamide of piclamilast (3) with the cyclization strategy to yield two series of inhibitors, one based on benzofuran[162] and the other on benzimidazole[163] as the aromatic pharmacophore. The benzofuran used by the Rhone-Poulenc-Rorer group differed from

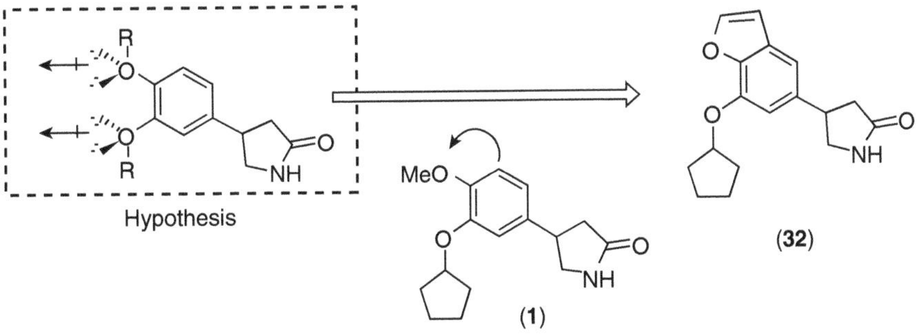

FIGURE 33.21 Optimum orientation of the lone pairs driving the cyclization hypothesis.

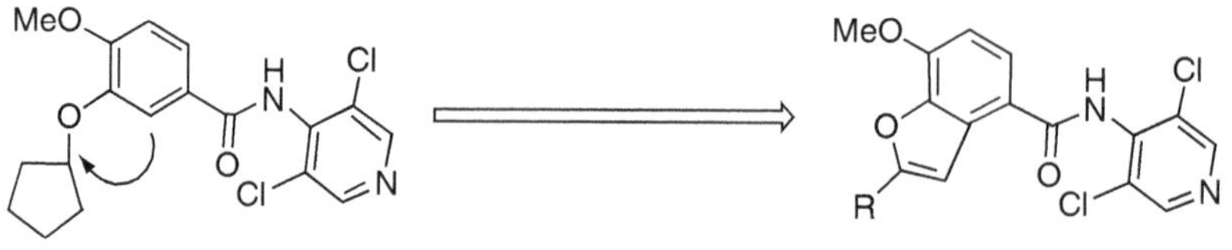

FIGURE 33.22 The alternate cyclization strategy to a benzofuran.

that previously reported, as it now involved the cyclization onto the 3-position of the aromatic ring leaving the 4-methoxy group intact (Figure 33.22).

The resulting 7-methoxybenzofuran (**33**) was found to be a potent replacement for the 3,4-dialkoxyphenyl pharmacophore in the benzamide series. This new pharmacophore was also transposed onto a series of rolipram pyrrolidinone analogs. This dual combination resulted in hybrid compounds that contained changes for both components of rolipram, and it was most interesting that this benzofuran replacement was not universally applicable and equally there was no indication, a priori, if the interchange would be beneficial. Darwin Discovery have also developed several series of benzofuran derived inhibitors, as exemplified by (**34**) and (**35**), that they have disclosed within the patent literature.[164,165] They have disclosed some data related to these benzofurans, showing compound (**36**) to be the most preferred analog, having a PDE4 IC$_{50}$ of 1.6 nM and showing good activity in a guinea pig skin eosinophilia model at 10 mg/kg across a range of mediators.[166] They have also indicated that the benzofuran can be further modified to a pyridofuran, as in (**37**).[167] The compounds have a wide variety of functional groups all in the 2-position of the benzofuran. Kyowa Kakko have also been active in the benzofuran arena, with patent applications claiming benzofuran-2-carboxamides, such as (**38**) as PDE4 inhibitors.[168] A modification of note within the benzofuran derivatives is the spiro-fused tetrahydrofurans reported by Byk Gulden (Altana), as indicated by structure (**39**), again based on modification at the 2-position of the benzofuran.[169] These compounds are reported as potent inhibitors (PDE4 IC$_{50}$ < 10 nM), and it should be noted how different this is when compared with the cyclohexylidene acetal reported by Glaxo (K_i ~ 63.9 μM). The last reported interest within the benzofuran area is that from Memory Pharmaceuticals. Again using the 7-methoxybenzofuran as a base, they report that 4-amino derivatives, such as (**40**), have utility in the treatment of cognitive and memory impairment (Figure 33.23).[170]

The second bioisosteric replacement for the 3,4-dialkoxyphenyl moiety identified by the Rhone-Poulenc-Rorer group was the 7-methoxybenzimidazole.[163] The regiochemistry between the pendant methoxy and benzamide units of these PDE4 inhibitors was identical to the previously noted benzofurans. Additional SAR studies revealed that the C2-position of the benzimidazole could be modified, leading them to potent, efficacious, and orally bioavailable compounds. In short, the benzimidazole conserved the necessary recognition elements for PDE4 inhibition.

The identification of a nonoxygenated isosteric replacement of the 3,4-dialkoxyphenyl ring has proven to be more difficult to achieve. The first such disclosure identified a series of indoles as potent PDE4 inhibitors in which the indole itself was serving as the isosteric group.[171] Initially, poor *in vivo* results were obtained with the compounds and these results were ascribed to poor absorption, but later use of the pyridine-*N*-oxide (within the dichloropyridine benzamide unit) overcame this apparent poor absorption problem to generate efficacious orally active compounds (**41**, Figure 33.24).[172] More recently, Pfizer has disclosed a series of compounds in which the dialkoxyphenyl motif has been replaced by an

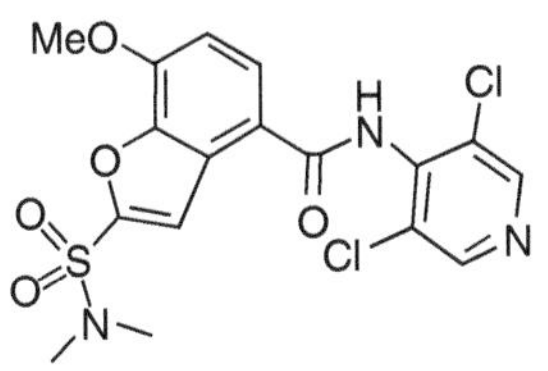

FIGURE 33.23 Benzofurans (**34–40**) from Darwin Discovery, Kyowa Kakko, and Byk Gulden and Memory Pharmaceuticals.

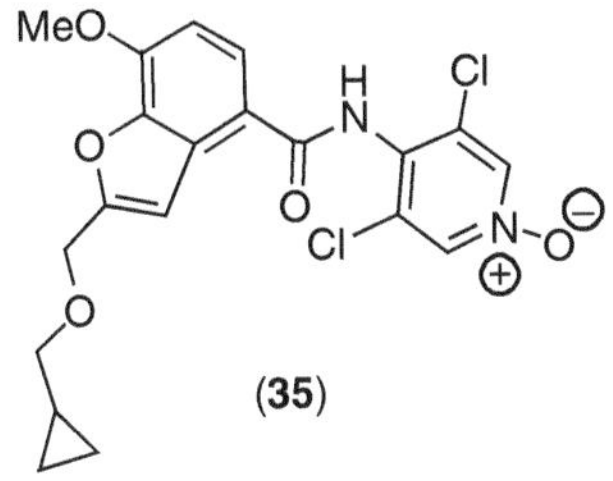

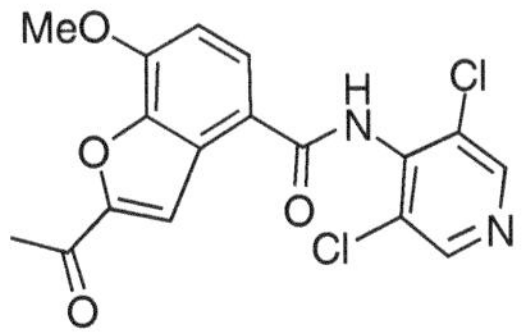

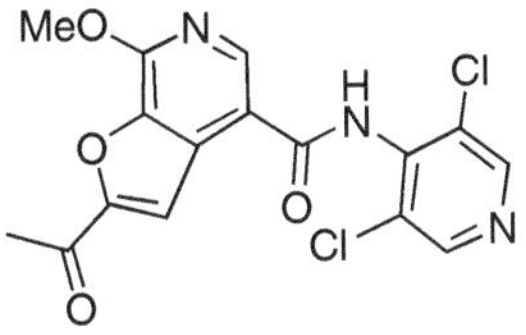

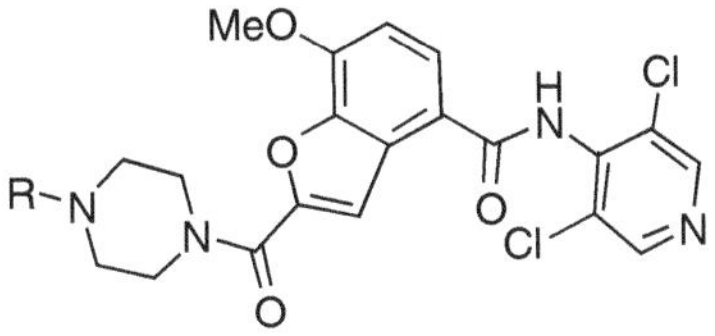

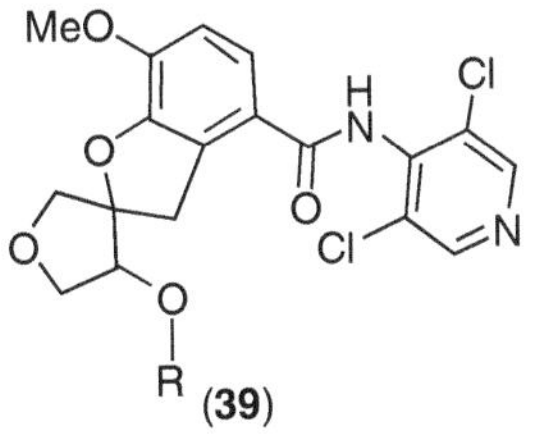

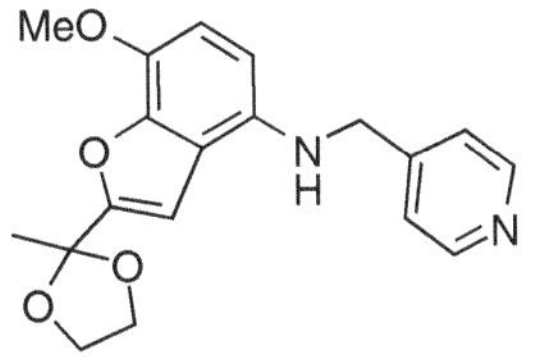

indazole (**42**). These compounds were claimed to be efficacious for the treatment of asthma and COPD and to be inhibitors of TNF-α release, however, no biological data were disclosed.[173]

The replacement of the γ-lactam of rolipram (**1**) has also led to the discovery of some interesting analogs and structure–activity relationships. The lactam can be ring-expanded (*vide infra*, (**12**); HT-0712), and equally can be transformed into a cyclic urea (an imadazolidinone);[174,175] both changes retain potency for the inhibition of PDE4, thereby demonstrating the importance of the 3,4-dialkoxyphenyl pharmacophore. This later series of compounds ultimately yielded atizoram (*vide infra*, (**18**); Pfizer; CP-80633) and further studies by the same group showed that the lactam ring of rolipram could also be substituted by an oxindole.[176] These oxindoles were characterized as compounds that had a low affinity for HARBS.

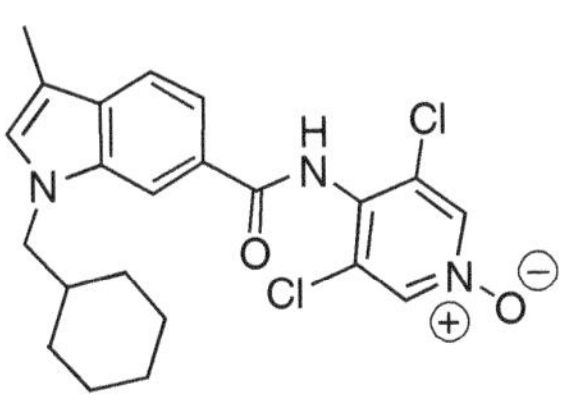

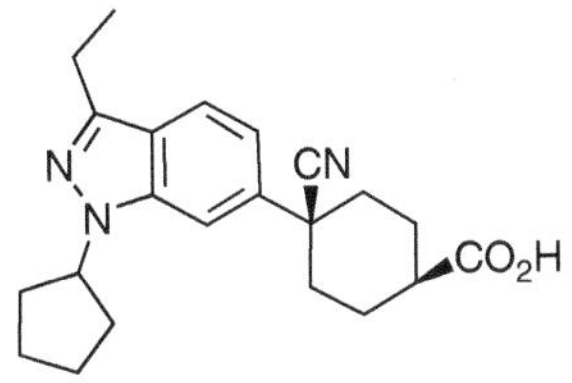

FIGURE 33.24 Nonoxygenated isosteric replacement of the 3,4-dialkoxyphyenyl pharmacophore; indole (**41**) and indazole (**42**).

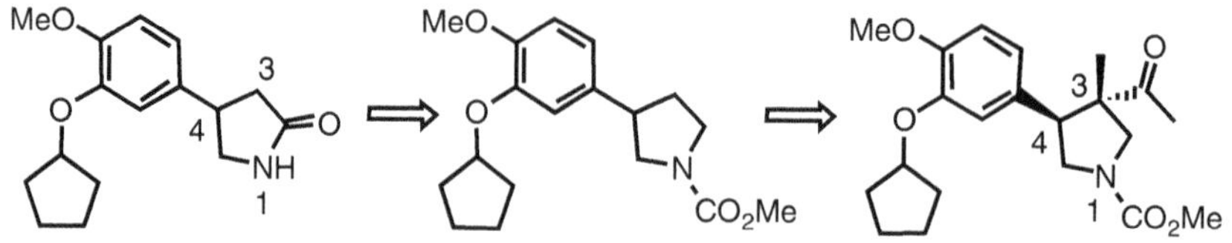

FIGURE 33.25 Realization of 3,4-disubstituted pyrrolidine (**44**) from rolipram (**1**).

Removal of the lactam carbonyl from rolipram (**1**) yielded the corresponding 4-aryl-pyrrolidine.[177] Initial indications were that a 10-fold improvement on rolipram potency could be achieved if the pyrrolidine was capped as either a *t*-butyl or phenyl carbamate (methyl carbamate shown; (**43**) Figure 33.25). A more comprehensive structure–activity relationship study followed.[178] A series of 3,4-disubstituted pyrrolidines were described, and the importance of the *trans* relative stereochemistry between C3 and C4 was clearly demonstrated (between the aryl and hydrogen-bond acceptor). Further refinement of this 3,4-disubstituted pyrrolidine series was realized when the possibility of a conformational switch within the framework of the inhibitor was elucidated.[179] The inclusion in the C3-position of the pyrrolidine ring of a hydrogen-bond acceptor, a methyl ketone, which mimicked the lactam carbonyl of rolipram, and a methyl group (creating a quaternary stereogenic centre) created potent analogs (PDE4 IC_{50} of ~1 nM). These two modifications when combined, as in (**44**), represented an overall 100-fold increase in potency when compared to the analogous 4-aryl pyrrolidine compound (**43**). The rationale presented for the improvement was that the methyl group increased the conformational freedom of the molecule and thereby allowed access to the conformer in which the hydrogen-bond acceptor at C3 could best mimic the position of the carbonyl of rolipram. It is interesting to note that ICOS has reported a further modification, specifically reduction of the pyrrolidine methyl ketone; this yielded an interesting pair of secondary alcohols (IC-86518 and IC-86521), which have equivalent PDE4 potency but significantly different emetic potential ($ED_{50} > 50$ and 0.5 mg/kg, respectively).[180]

33.4.2 Theophylline as Inspiration

The nonspecific PDE inhibitor theophylline (**31**) is quite distinct from rolipram (**1**) in structure (Figure 33.20). It is equally different in potency of inhibition for PDE4 with IC_{50} values in the millimolar (mM) range rather than the inhibitory potency for rolipram, which is in the nanomolar (nM) range. It is also an adenosine-receptor antagonist with potency in the micromolar (μM) range.[160] The obvious challenges presented to the medicinal chemists when initiating a drug discovery program using theophylline as a starting point are to be able to derive selectivity and potency as their optimization programs progress.

Although much work has been carried out using theophylline as a starting point, much of it was conducted before the separation of the PDE isozymes. However, an example of the improvements made in this area is the work carried out by Miyamoto et al.[181] with the report of various 1,3-dialkyl xanthines. The improvements in these analogs are seen at the level of PDE4 selectivity and potency, but the compound still exhibited adenosinergic activity. Further modification, specifically the inclusion of a 7-oxopropyl group, reduced the adenosinergic activity while maintaining PDE4 inhibitory activity, as exemplified in denbufylline (**45**),[182] which was limited in the development by its poor pharmacokinetics (Figure 33.26).[183] Napp further optimized the purine-based inhibitors through addition of the 3,4-dialkoxyphenyl motif found within rolipram to yield V-11294A and has since reported[184] modification of

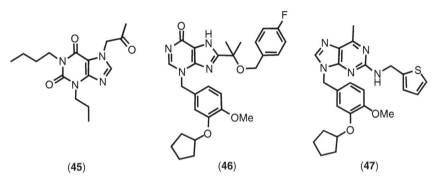

FIGURE 33.26 Denbufylline (**45**), Napp's modified purine series (**46**), and an example from Mitsubishi, Tokyo (**47**).

the purine core that yields much enhanced potency for inhibition of PDE4. These compounds are selective with respect to PDE3 and PDE5 with inhibition IC_{50}s with (**46**) that are >10 μM compared with its PDE4 IC_{50} of 0.34 nM.

A novel class of purine-based derivatives, exemplified by (**47**), have been reported by Mitsubishi, Tokyo.[185] This extensive patent covered 1200 compounds, all of which were benzylated in the purine nucleus, and these benzyl groups contained the 3,4-dialkoxyphenyl pharmacophore of rolipram. The compounds were claimed for the treatment of asthma.

Cipamfylline (**48**; BRL-61063) is also a xanthine analog, discovered by SmithKlineBeecham,[186] and licensed to Leo Pharmaceuticals. This compound selectively inhibited PDE4 in human lung (IC_{50} of 1.27 μM) and TNF-α secretion *in vitro*.[187] It is currently in a Phase 2 clinical trial for atopic dermatitis. Arofylline (**49**; LAS-31025, Almirall Prodesfarma) is interesting in this context as it has an aromatic group appended at N3, whereas most xanthines have alkyl groups in this position. It was initially reported to be in clinical trials for the treatment of asthma but was discontinued in Phase 3 because of undesirable side effects. Almirall took their lead for a series of 2,5-dihydropyrazolo[4,3-c]quinolin-3-one from nitraquazone and arofylline.[188] The lipophilic group at N2 was crucial for the inhibition of PDE4 activity, and the cyclopentyl derivative (**50**; shown, Figure 33.27) from within the series exhibited PDE4 inhibition potency comparable with rolipram. It was distinguished by having an improvement within the HARBS and PDE4 ratio to >100. Its minimum emetic dose in dogs was >3 mg/kg following i.v. dosing.

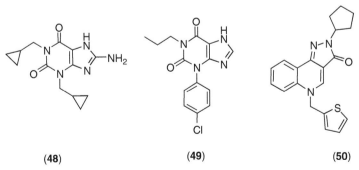

FIGURE 33.27 Cipamfylline (**48**), arofylline (**49**), and a 2,5-dihydropyrazolo[4,3-c]quinolin-3-one (**50**).

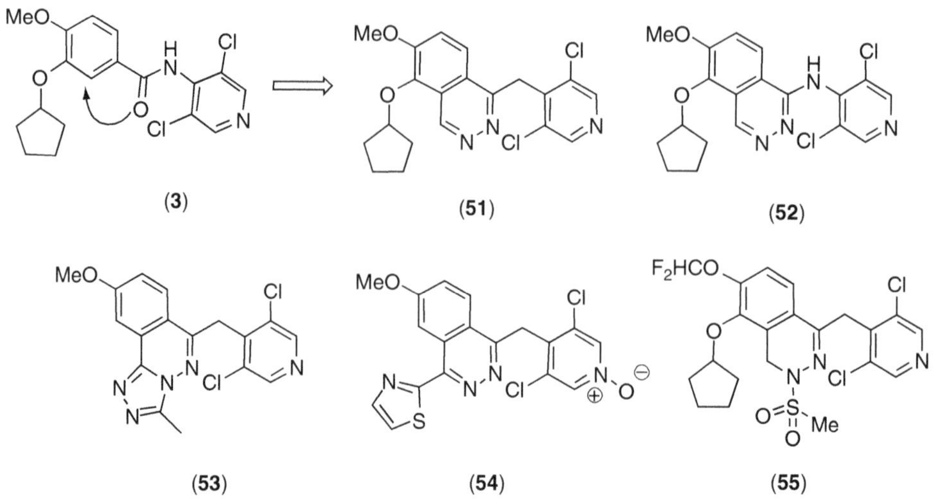

FIGURE 33.28 Various phthalazines (**51–55**) evolved from piclamilast (**3**).

33.4.3 Alternative Sources of Inspiration

Having detailed some of the modifications made with rolipram (**1**) and theophylline (**31**) as the point of attack in the generation of new PDE4 inhibitors, piclamilast (**3**; RP73401) has also inspired medicinal chemists to create alternative scaffolds. In an effort to achieve this, cyclization between the 2-position of the 3,4-dialkoxyphenyl aromatic ring and the carbonyl of the dichloro-pyridine benzamide was undertaken. The phthalazines emerged as viable inhibitors, in which the π-bond of the carbonyl was mimicked by the imino-substructure of the phthalazine ring. The dichloropyridine unit was introduced with either a C- or an N- as a spacer linking (**51** and **52**) the heteroaromatic ring to the phthalazine core.[189] The compounds shown, (**51**) and (**52**), were both potent PDE4 inhibitors having PDE4 IC_{50} values of 53 and 59 nM, respectively. Compound (**51**) was also studied for emetic behavior in the dog, where it displayed an $ED_{50} > 10$ μmol/kg (i.v.). This work has spawned numerous series, including synthesis of more elaborate tricyclic phthalazines (**53**),[190] mono-methoxyphthalazines (**54**),[191] and the related 3,4-dihydroisoquinoline series (**55**) (Figure 33.28).[192]

Another alternative, and equally effective in retrospect, has been the approach to identify a new template for a PDE4 inhibitor by mimicking the pharmacophore presented by nitra-quazone or TVX 2706 (**56**).[193] Researchers from Pfizer[194] and Syntex[195] both used this molecule to identify novel PDE4 inhibitors (Figure 33.29). At Pfizer, the identification of CP-77059 (**57**) followed the development of a hypothesis that the nitro-group attached at the

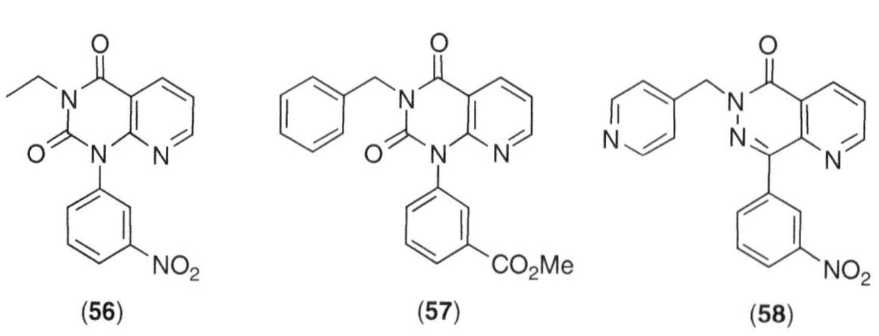

FIGURE 33.29 Nitraquazone (**56**), CP-77059 (**57**), and Syntex pyridazinone (**58**).

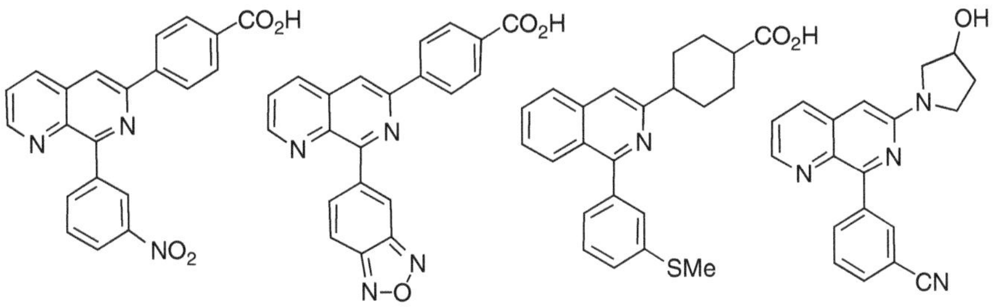

(59)   (60)   (61)   (62)

FIGURE 33.30 1,7-naphthyridines and related series from Novartis.

meta-position mimicked the amide of rolipram within the binding site and so represented a key element for recognition and inhibition. The replacement of this nitro-group was possible, but was only really effective when a carbomethoxy group was employed. Syntex produced potent compounds, bringing together the isosteric benzene–pyridine replacement and the pyrimidone–pyridazinone interchange as in compound (**58**), however, the increased potency was accompanied by the concomitant increase in side effects (emesis), which ultimately limited the development of these compounds.

The series of naphthyridines generated, developed, and reported by Novartis also follow the use of nitraquazone (**56**) as the structural lead molecule.[196] Indeed, most of the SAR for this series was carried out with the *meta*-nitro-phenyl ring at C8, a feature in common with nitraquazone. These 1,7-naphthyridines are PDE4D selective, displaying potencies for PDE4D with IC$_{50}$s as low as 1 nM, and when substituted in the 6-position with 4'-benzoic acid, they were orally active (**59**). All the compounds tested were inactive against PDE3, and it should be noted that they were not selective in regard to their inhibition of the PDE4 catalytic activity versus their ability to inhibit HARBS. The compound furthest advanced within this series appears to be NVP-ABE-171 (**60**), which is PDE4D-selective (PDE4 IC$_{50}$s: A, 602; B, 34; C, 1230; D, 1.5 nM) and has been shown to have better *in vivo* activity in a rodent model of lung inflammation than cilomilast.[197] There have been two further reports related to this class of compound by Novartis within the patent literature, claiming isoquinolines (**61**) and naphthyridines (**62**) (Figure 33.30).[198,199]

Although they may appear to be unrelated, a series of triazolopyridopyridazines (**63**) from Almirall Prodesfarma (Figure 33.31) also contained a *meta*-substituted aromatic ring as seen with the naphthyridines from Novartis (**59**) and nitraquazone (**56**).[200] It is evident that either nitraquazone or the phthalazine disclosures (see (**53**), Figure 33.28) could have provided the impetus for the series.

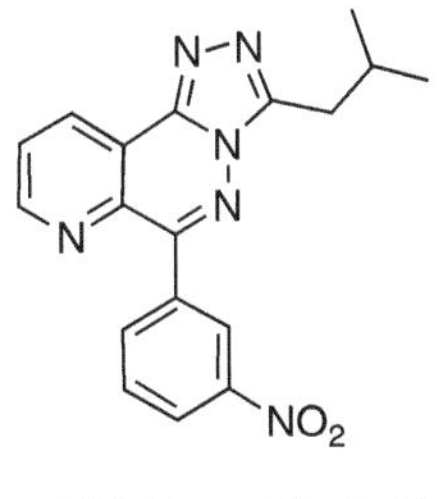

FIGURE 33.31 Triazolopyridopyridazines (**63**) from Almirall Prodesfarma.

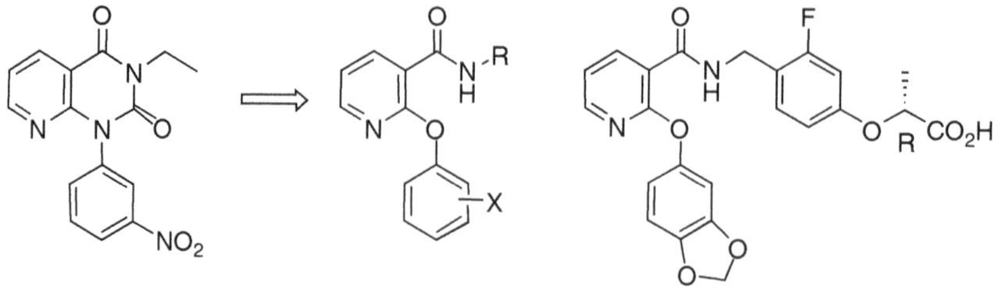

FIGURE 33.32 Ring opened nicotinamide ethers from Pfizer.

Nonetheless, the compounds blocked TNF-α secretion and, as a whole, the series were significantly less emetic in dogs than in rolipram. The introduction within the nucleus of the nitrogen atom increased the *in vitro* potency 30-fold, with the *meta*-nitro increasing the potency a further order of magnitude (PDE4 IC_{50} of ~40 nM).

Pfizer, in a second optimization based upon nitraquazone (56), approached the problem quite differently and unearthed a series of nicotinamide ethers (64) (Figure 33.32). The strategy followed a ring-opening rationale, and the PDE4 inhibitors that were generated were easily synthesized, which facilitated a rapid optimization process.[201]

The structure–activity relationship established that a wide variety of groups were tolerated within the amide unit but that the 2-ether was far more sensitive, with *meta*-substituted phenyl ethers being preferred. There have been numerous patent filings around this class of inhibitor,[202–204] including pyrimidine-based inhibitors;[205] however, a recent disclosure highlighted CP-671305 (65), a nicotinamide derivative, as the most advanced compound from this series.[206] It is interesting to note that this compound is also PDE4D-selective, having approximately 100-fold preference for the inhibition of this isozyme (PDE4 IC_{50}s: A, 310; B, 287; C, 3858; D, 3 nM) and is orally active in rat, dog, and monkey (*F*: 43%–80%; 45%; 26%, respectively).

33.5   PDE4 SUBTYPE–SELECTIVE INHIBITION

The pursuit of subtype-selective inhibitors by pharmaceutical research groups has been spurred on by the belief that subtype-selective inhibitors could overcome some, if not all, of the clinical limitations observed within the class as a whole. It has been suggested that the nausea and emesis seen with many inhibitors are associated with PDE4D inhibition[47] and hence suppressing this activity could be beneficial. As the PDE4 family is comprised of four gene products (PDE4A through PDE4D), a variety of options exist for the generation of selective inhibitors. To further complicate this picture, the four PDE4 isoforms can be further subdivided into three categories: a long form, a short form, and a super-short form.[13–16] The four subtypes are differentially expressed among tissue and cells.[17,18] It is clear that PDE4C is predominantly localized within the testis, skeletal muscle, human fetal lung, and within cerebellum (temporal cortex and granule layer) of the human brain. In contrast to PDE4C, the remaining three isoforms, namely PDE4A, B, and D are widely distributed in immune and inflammatory cells. Data from gene knockout studies have shown that PDE4B$^{-/-}$ mice[207] exhibit a profound attenuation in the ability of LPS to generate TNF-α which was not mirrored in the PDE4D$^{-/-}$ mice.[208] The PDE4D$^{-/-}$ mice were unique in a lack a muscarinic acetylcholine receptor function. When one adds to this the data generated at SmithKlineBeecham, namely a correlation between suppression of human inflammatory cell function and PDE inhibition, indicating that PDE4A or PDE4B, and not PDE4D, is the important target,

a hypothesis for a selective PDE4A or PDE4B inhibitor in lung diseases such as asthma and COPD emerges. Whether the specific targeting of a PDE4 subtype will have the beneficial effects postulated remains to be determined, and provides the pharmaceutical industry a keen focal point. The intervention of crystallography[44–46,209–212] and the realization of subtype-selective inhibitors bound within them are foreseen, and the results are awaited with anticipation. Importantly, the *in vivo* pharmacological data from these selective compounds will facilitate a correlation of the inhibition with either a beneficial therapeutic benefit or the adverse side effect liability associated with the subtype inhibition. The data have not been collected, but it will be of interest to determine if lack of brain exposure is enough to improve the therapeutic index.

REFERENCES

1. Conti, M., and S.-L.C. Jin. 2000. The molecular biology of cyclic nucleotide phosphodiesterases. *Prog Nucleic Acid Res Mol Biol* 63:1.
2. Mehats, C., et al. 2002. Cyclic nucleotide phosphodiesterases and their role in endocrine cell signaling. *Trends Endocrinol Metab* 13:29.
3. Manganiello, V.C., and E. Degerman. 1999. Cyclic nucleotide phosphodiesterases (PDEs): Diverse regulators of cyclic nucleotide signals and inviting molecular targets for novel therapeutic agents. *Thromb Haemostasis* 82:407.
4. Houslay, M.D., P. Schafer, and K.M.J. Zhang. 2005. Keynote review: Phosphodiesterase-4 as a therapeutic target. *Drug Discov Today* 10:1503.
5. Carson, C.C., and T.F. Lue. 2005. Phosphodiesterase type 5 inhibitors for erectile dysfunction. *BJU Int* 96:257.
6. Torphy, T.J. 1998. Phosphodiesterase isozymes: Molecular targets for novel antiasthma agents. *Am J Respir Crit Care Med* 157:351.
7. (a) Soderling, S.H., and J.A. Beavo. 2000. Regulation of cAMP and cGMP signalling: New phosphodiesterases and new functions. *Curr Opin Cell Biol* 12:174; (b) Beavo, J.A., and L.L. Brunton. 2002. Cyclic nucleotide research—still expanding after half a century. *Nat Rev Mol Cell Biol* 3:710.
8. Conti, M., et al. 2003. Cyclic AMP-specific PDE4 phosphodiesterase as critical components of cyclic AMP signaling. *J Biol Chem* 278:5493.
9. Burnouf, C., and M.-P. Pruniaux. 2002. Recent advances in PDE4 inhibitors as immunoregulators and antiinflammatory drugs. *Curr Pharm Des* 8:1255.
10. Renau, T.E. 2004. The potential of phosphodiesterase 4 inhibitors for the treatment of depression: Opportunities and challenges. *Curr Opin Investig Drugs* 5:34.
11. Beavo, J.A., et al. 1994. Multiple cyclic nucleotide phosphodiesterases. *Mol Pharmacol* 46:399.
12. Francis, S.H., I.V. Turko, and J.D. Corbin. 2001. Cyclic nucleotide phosphodiesterases: Relating structure to function. *Prog Nucleic Acid Res Mol Biol* 65:1.
13. Beavo, J.A., M. Conti, and R.J. Heaslip. 1994. Multiple cyclic nucleotide phosphodiesterases. *J Mol Pharmacol* 46:399.
14. Houslay, M.D., and D.R. Adams. 2003. PDE4 cAMP phosphodiesterases: Modular enzymes that orchestrate signaling crosstalk, desensitization, and compartmentalization. *Biochem J* 370:1.
15. Bolger, G., et al. 1993. A family of human phosphodiesterase homologous to the dunce learning and memory gene product of *Drosophila melanogaster* are potential targets for antidepressant drugs. *Mol Cell Biol* 13:6558.
16. Houslay, M.D., M. Sullivan, and G.B. Bolger. 1998. The multienzyme PDE4 cyclic adenosine monophosphate-specific phosphodiesterase family: Intracellular targeting, regulation, and selective inhibition by compounds exerting antiinflammatory and antidepressant actions. *Adv Pharmacol* 44:225.
17. Bolger, G.B. 1994. Molecular biology of the cyclic AMP-specific cyclic nucleotide phosphodiesterases: A diverse family of regulatory enzymes. *Cell Signal* 6:851.
18. Engels, P., K. Fichtel, and H. Lubbert. 1994. Expression and regulation of human and rat phosphodiesterase type IV isogenes. *FEBS Lett* 350:291.

19. Lipworth, B.J. 2005. Phosphodiesterase-4 inhibitors for asthma and chronic obstructive pulmonary disease. *Lancet* 365:167.
20. Banner, K.H., and M.A. Trevethick. 2004. PDE4 inhibition: A novel approach for the treatment of inflammatory bowel disease. *Trends Pharmacol Sci* 25:430.
21. Souness, J.E., and M. Foster. 1998. Potential of phosphodiesterase type IV inhibitors in the treatment of rheumatoid arthritis. *IDrugs* 1:541.
22. O'Donnell, J.M., and H.-T. Zhang. 2004. Antidepressant effects of inhibitors of cAMP phosphodiesterase (PDE4). *Trends Pharmacol Sci* 25:158.
23. Spina, D. 2003. Phosphodiesterase-4 inhibitors in the treatment of inflammatory lung disease. *Drugs* 63:2575.
24. Barnes, P.J. 2000. New drugs in asthma. *Nat Rev Drug Disc* 3:831.
25. Spina, D. 2000. The potential of PDE4 inhibitors in asthma or COPD. *Curr Opin Investig Drugs* 1:204.
26. Jeffrey, P. 2005. Phosphodiesterase 4-selective inhibition: Novel therapy for the inflammation of COPD. *Pulm Pharmacol Ther* 18:9.
27. Souness, J.E., D. Aldous, and C. Sargent. 2000. Immunosuppressive and antiinflammatory effects of cyclic AMP phosphodiesterase (PDE) type 4 inhibitors. *Immunopharmacology* 47:127.
28. Crocker, I.C., and R.G. Townley. 1999. Therapeutic potential of phosphodiesterase 4 inhibitors in allergic diseases. *Drugs Today (Barc)* 35:519.
29. Bielekova, B., et al. 2000. Therapeutic potential of phosphodiesterase-4 and -3 inhibitors in Th1-mediated autoimmune diseases. *J Immunol* 164:1117.
30. Semmler, J., H. Wachtel, and S. Enders. 1993. The specific type IV phosphodiesterase inhibitor rolipram suppresses tumor necrosis factor-α production by human mononuclear cells. *Int J Immunopharmacol* 15:409.
31. Baugh, J.A., and R. Bucala. 2001. Mechanisms for modulating TNFα in immune and inflammatory disease. *Curr Opin Drug Disc Dev* 4:635.
32. Palladino, M.A., et al. 2003. Anti-TNFα therapies. The next generation. *Nature Rev Drug Disc* 2:736.
33. Schwabe, U., et al. 1976. 4-(3-cyclopentyloxy-4-methoxyphenyl)-2-pyrrolidinone (ZK62711): A potent inhibitor of adenosine cyclic 3',5'-monophosphate phosphodiesterase in homogenates and tissue slices from rat brain. *Mol Pharmacol* 12:900.
34. Gruenman, V., and M. Hoffer. 1975. 4-benzyl-2-imidazolidinones from N-[(1-cyano-2-phenyl)ethyl] carbamates. US Patent 3,923,833.
35. Bobon, D., et al. 1998. Is phosphodiesterase inhibition a new mechanism of antidepressant action? A double blind double-dummy study between rolipram and desipramine in hospitalized major and/or endogenous depressives. *Eur Arch Psychiatry Neurol Sci* 238:2.
36. Horowski, R., Y. Sastre, and M. Hernandez. 1985. Clinical effects of the neurotropic selective cAMP phosphodiesterase inhibitor rolipram in depressed patients: Global evaluation of the preliminary reports. *Curr Ther Res* 38:23.
37. Scott, A.I., et al. 1991. In-patient major depression: Is rolipram as effective as amitriptyline?. *Eur J Clin Pharmacol* 40:127.
38. Casacchia, M., et al. 1983. Therapeutic use of a selective cAMP phosphodiesterase inhibitor (rolipram) in Parkinson's disease. *Pharmacol Res Commun* 15:329.
39. Parkes, J.D., et al. 1984. Rolipram in Parkinson's disease. *Adv Neurol* 40:563.
40. Barnette, M.S., et al. 1995. The ability of phosphodiesterase IV inhibitors to suppress superoxide production in guinea pig eosinophils is correlated with inhibition of phosphodiesterase IV catalytic activity. *J Pharmacol Exp Ther* 273:674.
41. Zeller, E., et al. 1984. Results of a phase II study of the antidepressant effect of rolipram. *Pharmacopshychiatry* 17:188.
42. Duplantier, A.J., et al. 1996. Biarylcarboxylic acids and amides: Inhibition of phosphodiesterase type IV versus [³H]Rolipram binding activity and their relationship to emetic behavior in the ferret. *J Med Chem* 39:120.
43. Schneider, H.H., et al. 1986. Stereospecific binding of the antidepressant rolipram to brain protein structures. *Eur J Pharmacol* 127:105.

44. Zhang, K.Y., et al. 2004. A glutamine switch for nucleotide selectivity for phosphodiesterases. *Mol Cell* 15:279.

45. Mallanack, D.T., R.A. Hughes, and P.E. Thompson. 2005. The next generation of phosphodiesterase inhibitors: Structural clues to ligand and substrate selectivity of phosphodiesterases. *J Med Chem* 48:3449.

46. Card, G.L., et al. 2004. Structural basis for the activity of drugs that inhibit phosphodiesterases. *Structure* 12:2233.

47. Robichaud, A., et al. 2002. Deletion of phosphodiesterase 4D in mice shortens α_2-adrenoreceptor-mediated anesthesia, a behavioral correlate of emesis. *J Clin Invest* 110:1045.

48. Aoki, M., et al. 2001. Studies on mechanisms of low emetogenicity of YM976, a novel phosphodiesterase type 4 inhibitor. *J Pharmacol Exp Ther* 298:1142.

49. Souness, J.E., and S. Rao. 1997. Proposal for pharmacologically distinct conformers of PDE4 cyclic AMP phosphodiesterases. *Cell Signal* 9:227.

50. Barnette, M.S., et al. 1995. Inhibitors of phosphodiesterase IV (PDE IV) increase acid secretion in rabbit isolated gastric glands: Correlation between function and interaction with a high-affinity rolipram binding site. *J Pharmacol Exp Ther* 273:1396.

51. Schmiechen, R., H.H. Schneider, and H. Wachtel. 1990. Close correlation between behavioral response and binding in vivo for inhibitors of the rolipram-sensitive phosphodiesterase. *Psychopharmacology* 102:17.

52. Laliberte, F., et al. 2000. Conformational difference between PDE4 apoenzyme and holoenzyme. *Biochemistry* 39:6449.

53. Liu, S., et al. 2001. Dissecting the cofactor-dependent and independent bindings of PDE4 inhibitors. *Biochemistry* 40:10179.

54. (a) Alvarez, R., et al. 1995. Activation and selective inhibition of a cyclic AMP-specific phosphodiesterase, PDE-4D3. *Mol Pharmacol* 48:616; (b) Hoffmann, R., et al. 1998. cAMP-specific phosphodiesterase HSPDE4D3 mutants which mimic activation and changes in rolipram inhibition triggered by protein kinase A phosphorylation of ser-54: Generation of a molecular model. *Biochem J* 333:139.

55. McPhee, I., et al. 1999. Association with the SRC family tyrosyl kinase LYN triggers a conformational change in the catalytic region of human cAMP-specific phosphodiesterase HSPDE4A4B. Consequences for rolipram inhibition. *J Biol Chem* 274:11796.

56. (a) Kelly, J.J., P.J. Barnes, and M.A. Giembycz. 1996. Phosphodiesterase 4 in macrophages: Relationship between cAMP accumulation, suppression of cAMP hydrolysis and inhibition of [3H]R-(−)-rolipram binding by selective inhibitors. *Biochem J* 318:425; (b) Bolger, G.B., et al. 2003. Attenuation of the activity of the cAMP-specific phosphodiesterase PDE4A5 by interaction with the immunophilin XAP2. *J Biol Chem* 278:33351.

57. Giembycz, M.A. 2005. Phosphodiesterase-4. *Proc Am Thorac Soc* 2:326.

58. Lombardo, L.J. 1995. Phosphodiesterase-IV inhibitors: Novel therapeutics for the treatment of inflammatory diseases. *Curr Pharm Des* 1:255.

59. Cavalla, D., and R. Frith. 1995. Phosphodiesterase IV inhibitors: Structural diversity and therapeutic potential in asthma. *Curr Med Chem* 2:561.

60. Norman, P. 1999. PDE4 inhibitors 1999. *Expert Opin Ther Patents* 9:1101.

61. Martin, T.J. 2001. PDE4 inhibitors—a review of the recent patent literature. *IDrugs* 4:312.

62. Montana, J.G., and H.J. Dyke. 2001. Phosphodiesterase 4 inhibitors. *Ann Reports Med Chem* 36:41.

63. Norman, P. 2002. PDE4 inhibitors 2001. Patent and literature activity 2000—September 2001. *Expert Opin Ther Patents* 12:93.

64. Odingo, J. 2005. Inhibitors of PDE4: A review of recent patent literature. *Expert Opin Ther Patents* 15:773.

65. Huang, Z., et al. 2001. The next generation of PDE4 inhibitors. *Curr Opin Chem Biol* 5:432.

66. Larson, J.L., et al. 1996. The toxicity of repeated exposures to rolipram, a type IV phosphodiesterase inhibitor, in rats. *Pharmacol Toxicol* 78:44.

67. van Schalkwyk, E., et al. 2005. Roflumilast—an oral once-daily phosphodiesterase 4 inhibitor, attenuates allergen-induced asthmatic reactions. *J Allergy Clin Immunol* 116:292.

68. Rabe, K.F., et al. 2005. Roflumilast—an oral antiinflammatory treatment for chronic obstructive pulmonary disease: A randomised controlled trial. *Lancet* 366:563.

69. Altana, A.G., press release, February 13, 2004.

70. Altana, A.G., press release, July 1, 2005.

71. Amschler, H., et al. 1995. Fluoroalkoxy-substituted benzamides and their use as cyclic nucleotide phosphodiesterase inhibitors. PCT WO9501338.

72. Hatzelmann, A., and C. Schudt. 2001. Antiinflammatory and immunomodulatory potential of the novel PDE4 inhibitor roflumilast in vitro. *J Pharmacol Exp Ther* 297:267.

73. Timmer, W., et al. 2002. The new phosphodiesterase 4 inhibitor roflumilast is efficacious in exercise-induced asthma and leads to suppression of LPS-stimulated TNF-α ex vivo. *J Clin Pharmacol* 42:297.

74. Reid, P. 2002. Roflumilast. *Curr Opin Investig Drugs* 3:1165.

75. Brundschuh, D.S., et al. 2001. In vivo efficacy in airway disease models of roflumilast, a novel orally active PDE4 inhibitor. *J Pharmacol Exp Ther* 297:280.

76. Compton, C.H., et al. 2001. Cilomilast, a selective phosphodiesterase-4 inhibitor for treatment of patients with chronic obstructive pulmonary disease: A randomized, dose-ranging study. *Lancet* 358:265.

77. GlaxoSmithKline, press release, October 27, 2003.

78. See the following website for the slides presented at the PADAC in connection with cilomilast: http://fda.gov/ohrms/dockets/ac/03/slides/3976S1_02_FDA-Ariflo_files/slide0002.htm.

79. Christensen, S.B., et al. 1998. 1,4-Cyclohexanecarboxylates: Potent and selective inhibitors of phosphodiesterase 4 for the treatment of asthma. *J Med Chem* 41:821.

80. Torphy, T.J., et al. 1997. Molecular basis for an improved therapeutic index of SB 207499 (Ariflo), a second-generation phosphodiesterase 4 inhibitor. Abstract. *Eur Respir J* 10:S313.

81. Torphy, T.J., et al. 1999. Ariflo (SB 207499), a second generation phosphodiesterase 4 inhibitor for the treatment of asthma and COPD: From concept to clinic. *Pulmon Pharmacol Ther* 12:131.

82. Giembycz, M.A. 2001. Cilomilast: A second generation phosphodiesterase 4 inhibitor for asthma and chronic obstructive pulmonary disease. *Exp Opin Investig Drugs* 10:1361.

83. Zussman, B.D., et al. 2001. Bioavailability of the oral, selective phosphodiesterase 4 inhibitor cilomilast. *Pharmacotherapy* 21:653.

84. Zussman, B.D., et al. 2001. An overview of the pharmacokinetics of cilomilast (ariflo), a new, orally active phosphodiesterase 4 inhibitor, in healthy young and elderly volunteers. *J Clin Pharmacol* 41:950.

85. Murdoch, R.D., et al. 2004. Lack of pharmacokinetic interactions between cilomilast and theophylline or smoking in healthy volunteers. *J Clin Pharmacol* 44:1046.

86. Underwood, D.C., et al. 1998. Antiasthmatic activity of the second-generation phosphodiesterase 4 (PDE4) inhibitor SB 207499 (ariflo) in guinea pig. *J Pharmacol Exp Ther* 287:988.

87. Griswold, D.E., et al. 1998. SB 207499 (ariflo), a second generation phosphodiesterase 4 inhibitor, reduces tumor necrosis factor α and interleukin-4 production in vivo. *J Pharmacol Exp Ther* 287:705.

88. Baumer, W., et al. 2002. Effects of the phosphodiesterase 4 inhibitors SB 207499 and AWD 12-281 on the inflammatory reaction in a model of allergic dermatitis. *Eur J Pharmacol* 446:195.

89. Slim, R.M., et al. 2003. Apoptosis and nitrative stress associated with phosphodiesterase inhibitor-induced mesenteric vasculitis in rats. *Toxic Pathol* 31:638.

90. See the following FDA website: http://www.fdaadvisorycommittee.com/FDC/AdvisoryCommittee/Committees/PulmonaryAllergy + Drugs/090503_Ariflo/090503_ArifloP.htm

91. Otsuka Maryland Research Institute Inc., press release, July 23, 2003.

92. See company website: http://www.otsuka.com/OMRI/OMRIRespiratory.htm

93. Chihiro, M., et al. 1995. Novel thiazole derivatives as inhibitors of superoxide production by human neutrophils: Synthesis and structure–activity relationships. *J Med Chem* 38:353.

94. Nagamato, H., et al. 2004. OPC-6535 inhibits human and porcine monocyte tumor necrosis factor-α production in vitro and in vivo. *Gastroenterol* 126 (Suppl 2): Abs. W1090.

95. Bloomfield, G.L., et al. 1997. OPC-6535, a superoxide anion production inhibitor, attenuates acute lung injury. *J Surg Res* 72:70.

96. O'Mahony, S. 2005. Tetomilast. *IDrugs* 8:502.
97. Schafer, P. 2005. Therapeutic index of CC-10004. *Asthma and COPD*, SMI, London, UK, April, 2005.
98. Muller, G.W., et al. 1998. Thalidomide analogs and PDE4 inhibition. *Bioorg Med Chem Lett* 8:2669.
99. Draheim, R., U. Egerland, and C. Rundfeldt. 2004. Antiinflammatory potential of the selective phosphodiesterase 4 inhibitor *N*-3,5-dichloro-pyrid-4-yl)-[1-(4-fluorobenzyl)-5-hydoxy-indole-3-yl]-glyoxylic acid amide (AWD 12-281), in human cell preparations. *J Pharmacol Exp Ther* 308:555.
100. Marx, D., et al. 1999. The pharmacological activity of AWD 12-281, a potent phosphodiesterase 4 (PDE4) inhibitor for the treatment of allergic rhinitis and asthma. *Pneumologie* 53:443.
101. See a presentation at http://www.elbion.de/pdf/Runfeldt-SRI-Conf2004.pdf.
102. Kuss, H., et al. 2003. In vivo efficacy in airway disease models of *N*-(3,5-dichloropyridin-4-yl)-[1-(4-fluorobenzyl-5-hydroxy-indole-3-yl]-glyoxidic acid amide (AWD 12-281), a selective phosphodiesterase 4 inhibitor for inhaled administration. *J Pharmacol Exp Ther* 307:373.
103. Rose, G.M., et al. 2005. Phosphodiesterase inhibitors for cognitive enhancement. *Curr Pharm Des* 11:3329.
104. Helicon Therapeutics Inc., press release, December 10, 2004.
105. See http://www.inflazyme.com/corporate_profile.shtml.
106. Bourtchouladze, R., et al. 2003. A mouse model of Rubinstein–Taybi syndrome: Defective-long-term memory is ameliorated by inhibitors of phosphodiesterase 4. *Proc Natl Acad Sci* 100:10518.
107. ExonHit Therapeutics, press release, December 4, 2003.
108. ExonHit Therapeutics, press release, November 22, 2004.
109. Glenmark Pharmaceuticals Ltd, press release, August 24, 2005.
110. Gullapalli, S., et al. 2005. GRC3886—a selective PDE4 inhibitor with potential effect in rheumatoid arthritis. *7th World Congress on Inflammation, Melbourne.*
111. Ono Pharmaceuticals, status of development pipeline, February 5, 2004.
112. Takeda, H. 2002. Pharmacological profile of ONO-6126, a novel phosphodiesterase 4 (PDE4) inhibitor. *98th Int Conf Am Thorac Soc Atlanta.*
113. Furuie, H., et al. 2003. Suppressive effect of novel phosphodiesterase 4 (PDE4) inhibitor ONO-6126 on TNF-α release was increased after repeated oral administration in healthy Japanese subjects. Abstract. *Eur Respir J* 22 (Suppl 45): Abst. 2257 see also https://www.ersnetsecure.org/public/prg_congres.abstract?ww_i_presentation = 11244
114. Loher, F., et al. 2003. The specific type-4 phosphodiesterase inhibitor mesopram alleviates experimental colitis in mice. *J Pharmacol Exp Ther* 305:549.
115. Dinter, H., et al. 2000. The type IV phosphodiesterase inhibitor mesopram inhibits experimental autoimmune encephalomyelitis in rodents. *J Neuroimmunol* 108:136.
116. Duplantier, A.J., and K. Cooper. 1996. Preparation of tricyclic 5,6-dihydro-9H-pyrazolo[3,4-c]1,2,4-triazolo[4,3-a]pyridines as inhibitors of phosphodiesterase (PDE) type IV and the production of tumor necrosis factor (TNF), PCT WO9639408.
117. Urban, F.J., et al. 2001. Process research and large-scale synthesis of a novel 5,6-dihydro-(9H)-pyrazolo[3,4-c]1,2,4-triazolo[4,3-a]pyridine PDE-IV inhibitor. *Org Proc Res Dev* 5:575.
118. Ashton, M.J., et al. 1994. Selective type IV phosphodiesterase inhibitors as antiasthmatic agents. The synthesis and biological activities of 3-(cyclopentyloxy)-4-methoxybenzamides and analogues. *J Med Chem* 37:1696.
119. Marivet, M.C., et al. 1989. Inhibition of cyclic adenosine-3′,5′-monophosphate phosphodiesterase from vascular smooth muscle by rolipram analogues. *J Med Chem* 32:1450.
120. Souness, J.E., et al. 1995. Suppression of eosinophil function by RP 73401, a potent and selective inhibitor of cyclic AMP-specific phosphodiesterase: Comparison with rolipram. *Br J Pharmacol* 115:39.
121. Stevens, J.C., et al. 1997. Human liver CYP2B6-catalysed hydroxylation of RP 73401. *J Pharmacol Exp Ther* 282:1389.
122. Cassidy, K.C., et al. 2000. Quantitation of *N*-(3,5-dichloropyri-4-yl)-3-cyclopentyloxy-4-methoxy-benzamide and 4-amino-3,5-dichloropyridine in rat and mouse plasma by LC/MS/MS. *J Pharm Biomed Anal* 22:869.

123. Raeburn, D., et al. 1994. Antiinflammatory and bronchodilator properties of RP 73401, a novel and selective phosphodiesterase type IV inhibitor. *Br J Pharmacol* 113:1423.

124. Pino, M.V., et al. 1999. Toxicologic and carcinogenic effects of the type IV phosphodiesterase inhibitor RP 73401 on the nasal olfactory tissue in rats. *Toxicologic Pathology* 27:383.

125. Naline, E., et al. 1996. Effects of RP 73401, a novel, potent, and selective phosphodiesterase type 4 inhibitor, on contractility of human, isolated bronchial muscle. *Br J Pharmacol* 118:1939.

126. Souness, J.E., et al. 1999. Suppression of anti-CD3-induced interleukin-4 and interleukin-5 release from splenocytes of mesocestoides corti-infected BALB/c mice by phosphodiesterase 4 inhibitors. *Biochem Pharmacol* 58:991.

127. ICOS Corporation, press release, March 23, 2005.

128. Heaslip, R.J., et al. 1994. Phosphodiesterase inhibition, respiratory muscle relaxation and brochodilation by WAY-PDA-641. *J Pharmacol Exp Ther* 268:888.

129. Watson, J.W., et al. 1995. Anti-anaphylactic and side effect activity of the PDEIV inhibitor CP-80,633 in the ferret. Abstract. *J Allergy Clin Immunol* 95:850.

130. Cohan, V.L., et al. 1996. *In vitro* pharmacology of the novel phosphodiesterase type 4 inhibitor, CP-80633. *J Pharmacol Exp Ther* 278:1356.

131. Turner, C.R., et al. 1996. In vivo pharmacology of CP-80,633, a selective inhibitor of phosphodiesterase 4. *J Pharmacol Exp Ther* 278:1349.

132. Cheng, J.B., et al. 1997. The phosphodiesterase type 4 (PDE4) inhibitor CP-80,633 elevates plasma cyclic AMP levels and decreases tumor necrosis factor-α (TNF-α) production in mice: Effect of adrenalectomy. *J Pharmacol Exp Ther* 280:621.

133. Hanifm, J.M., et al. 2001. Type 4 phosphodiesterase inhibitors have clinical and in vitro anti-inflammatory effects in atopic dermatitis. *J Invest Dermatol* 107:51.

134. Bayer, A.G., press release, June 15, 2001.

135. Sturton, G., and M. Fitzgerald. 2002. Phosphodiesterase 4 inhibitors for the treatment of COPD. *Chest* 121:192S.

136. Fischer, R., et al. 1994. Benzofuranyl- and thiophenyl-alkanecarboxylic acid derivatives, EP 0623607.

137. Grootendorst, D.C., et al. 2003. Efficacy of the novel phosphodiesterase-4 inhibitor BAY 19-8004 on lung function and airway inflammation in asthma and chronic obstructive pulmonary disease (COPD). *Pulm Pharmacol Ther* 16:341.

138. Buckley, G.M., et al. 2002. 8-Methoxyquinoline-5-carboxamides as PDE4 inhibitors: A potential treatment for asthma. *Bioorg Med Chem Lett* 12:1613.

139. Billah, M., et al. 2002. 8-Methoxyquinolines as PDE4 inhibitors. *Bioorg Med Chem Lett* 12:1617.

140. Billah, M., et al. 2002. Synthesis and profile of SCH351591, a novel PDE4 inhibitor. *Bioorg Med Chem Lett* 12:1621.

141. Billah, M.M., et al. 2002. Pharmacology of *N*-(3,5-dichloro-1-oxido-4-pyridinyl)-8-methoxy-2-(trifluoromethyl)-5-quinoline carboxamide (SCH 351591), a novel, orally active phosphodiesterase 4 inhibitor. *J Pharmacol Exp Ther* 302:127.

142. Losco, P.E., et al. 2004. The toxicity of SCH 351591, a novel phosphodiesterase-4 inhibitor, in Cynomolgus monkeys. *Toxicol Pathol* 32:295.

143. Alexander, R.P., et al. 2002. CDP840. A prototype of a novel class of orally active antiinflammatory phosphodiesterase 4 inhibitors. *Bioorg Med Chem Lett* 12:1451.

144. Celltech, press release, February 1, 1996.

145. Li, C., et al. 2001. Investigation of the in vitro metabolism profile of a phosphodiesterase-IV inhibitor, CDP-840: Leading to structural optimization. *Drug Metab Dispos* 29:232.

146. Guay, D., et al. 2002. Discovery of L-791,943: A potent, selective, nonemetic and orally active phosphodiesterase-4 inhibitor. *Bioorg Med Chem Lett* 12:1457.

147. Frenette, R., et al. 2002. Substituted 4-(2,2-diphenylethyl)pyridine-*N*-oxides as phosphodiesterase-4 inhibitors: SAR study directed toward the improvement of pharmacokinetic parameters. *Bioorg Med Chem Lett* 12:3009.

148. Claveau, D., et al. 2004. Preferential inhibition of T helper1, but not T helper2, cytokines *in vitro* by L-826,141 [4-{2-(3,4-*bis*-difluoromethoxyphenyl)-2{4-(1,1,1,3,3,3-hexafluoro-2-hydroxypropan-2-yl)-phenyl]-ethyl}-3-methylpyridine-1-oxide], a potent and selective phosphodiesterase 4 inhibitor. *J Pharmacol Exp Ther* 310:752.

149. Pascal, Y., et al. 2000. Synthesis and structure–activity relationships of 4-oxo-1-phenyl-3,4,6,7-tetrahydro-[1,4]diazepino[6,7,1-hi]indoles: Novel PDE4 inhibitors. *Bioorg Med Chem Lett* 10:35.

150. Burnouf, C., et al. 2000. Synthesis, structure–activity relationships, and pharmacological profile of 9-amino-4-oxo-1-phenyl-3,4,6,7-tetrahydro-[1,4]diazepino[6,7,1-hi]indoles: Discovery of potent, selective phosphodiesterase type 4 inhibitors. *J Med Chem* 43:4850.

151. Ouagued, M., et al. 2005. The novel phosphodiesterase 4 inhibitor, CI-1044, inhibits LPS-induced TNF-α production in whole blood from COPD patients. *Pulm Pharmacol Ther* 18:49.

152. Takayama, K., et al. 1996. Novel naphthyridine derivative and medicinal composition thereof, PCT WO9606843.

153. Aoki, M., et al. 2000. A novel phosphodiesterase type 4 inhibitor, YM976 (4-(3-chlorophenyl)-1,7-diethylpyrido[2,3-d]pyrimidin-2(1H)-one), with little emetogenic activity. *J Pharmacol Exp Ther* 295:255.

154. Aoki, M., et al. 2000. Effect of a novel antiinflammatory compound, YM976, on antigen-induced eosinophil infiltration in the lungs in rats, mice, and ferrets. *J Pharmacol Exp Ther* 295:1149.

155. Gale, D.D., et al. 2002. Pharmacokinetic and pharmacodynamic profile following oral administration of the phosphodiesterase (PDE) 4 inhibitor V11294A in healthy volunteers. *Br J Clin Pharmacol* 54:478.

156. Gale, D.D., et al. 2003. Pharmacology of a cyclic nucleotide phosphodiesterase type 4 inhibitor, V11294. *Pulm Pharmacol Ther* 16:97.

157. Memory Pharmaceuticals, press release, April 15, 2005.

158. Memory Pharmaceuticals, press release, August 18, 2005.

159. Dal Piaz, V., and M.P. Giovannoni. 2000. Phosphodiesterase 4 inhibitors, structurally unrelated to rolipram, as promising agents for the treatment of asthma and other pathologies. *Eur J Med Chem* 35:463.

160. Mann, J.S., and S.T. Holgate. 1985. Specific antagonism of adenosine-induced bronchoconstriction in asthma by oral theophylline. *Br J Clin Pharmacol* 19:685.

161. Stafford, J.A., et al. 1994. Structure activity relationships involving the catachol subunit of Rolipram. *Bioorg Med Chem Lett* 1855:4.

162. McGarry, D.G., et al. 1999. Benzofuran based PDE4 inhibitors. *Bioorg Med Chem* 7:1131.

163. Regan, J., et al. 1998. 2-substituted-4-methoxybenzimidazole-based PDE4 inhibitors. *Bioorg Med Chem Lett* 8:2737.

164. Dyke, H., et al. 1999. Benzofuran-4-carboxamides and their therapeutic use, PCT WO9940085.

165. Dyke, H., et al. 1997. Benzofuran carboxamides and their therapeutic use, PCT WO9744337.

166. Buckley, G., et al. 2000. 7-methoxybenzofuran-4-carboxamides as PDE 4 inhibitors: A potential treatment for asthma. *Bioorg Med Chem Lett* 10:2137.

167. Dyke, H., et al. 1997. Furopyridine derivatives and their therapeutic use, PCT WO9964423.

168. Ohshima, E., et al. 1999. Benzofuran derivatives, PCT WO09916768.

169. Martin, T., et al. 1999. Benzamides with tetrahydrofuranyloxy substitutents as phosphodiesterase 4 inhibitors, PCT WO09964414.

170. Schumacher, R.A., A.T. Hopper, and A. Tehim. 2004. 6-Amino-1H-indazole and 4-aminobenzofuran compounds as phosphodiesterase 4 inhibitors, PCT WO2004009557.

171. Hulme, C., et al. 1998. The synthesis and biological evaluation of a novel series of indole PDE4 inhibitors I. *Bioorg Med Chem Lett* 8:1867.

172. Hulme, C., et al. 1998. Orally active indole *N*-oxide PDE4 inhibitors. *Bioorg Med Chem Lett* 8:3053.

173. Marfat, A. 1999. Therapeutically active compounds based on indazole bioisostere replacement of catechol in PDE4 inhibitors, PCT WO9923076.

174. Koe, B.K., et al. 1990. Effects of novel catechol ether imidazolidinones on calcium-independent phosphodiesterase activity, [3H]Rolipram binding, and reserpine-induced hypothermia in mice. *Drug Dev Res* 21:135.

175. Saccomano, N.A., et al. 1991. Calcium-independent phosphodiesterase inhibitors as antidepressants: [3-(bicycloalkoxy)-4-methoxyphenyl]-2-imidazolidinones. *J Med Chem* 34:291.

176. Masamune, H., et al. 1995. Discovery of micromolar PDE IV inhibitors that exhibit much reduced affinity for the [³H]Rolipram binding site: 3-norbornyloxy-4-methoxyphenylmethylene oxindoles. *Bioorg Med Chem Lett* 5:1965.

177. Feldman, P.L., et al. 1995. Phosphodiesterase type IV inhibition. Structure–activity relationships of 1,3-disubstituted pyrrolidines. *J Med Chem* 38:1505.

178. Stafford, J.A., et al. 1995. Phosphodiesterase type IV (PDE IV) inhibition. Synthesis and evaluation of a series of 1,3,4-trisubstituted pyrrolidines. *Bioorg Med Chem Lett* 5:1977.

179. Stafford, J.A., et al. 1995. Introduction of a conformational switching element on a pyrrolidine ring. Synthesis and evaluation of (R*,R*)-($\pm$)-methyl 3-acetyl-4-[3-cyclopentyloxy)-4-methoxyphenyl]-3-methyl-1-pyrrolidinecarboxylate, a potent and selective inhibitor of cAMP-specific phosphodiesterase. *J Med Chem* 38:4972.

180. Odingo, J. 2003. American Chemical Society, New York City, September 7–11.

181. Miyamoto, K., et al. 1994. Selective tracheal relaxation and phosphodiesterase-IV inhibition by xanthine derivatives. *Eur Pharmacol (Mol Pharmacol Soc)* 267:317.

182. Goring, J.E. 1980. Xanthine derivatives, a process for their preparation and their use in pharmaceutical compositions, EP0018136.

183. Nicholson, C.D., S.A. Jackman, and R. Wilke. 1989. The ability of denbufylline to inhibit cyclic nucleotide phosphodiesterase and its affinity for adenosine receptors and the adenosine re-uptake site. *Br J Pharmacol* 97:889.

184. Chasin, M., P. Hofer, and D. Cavalla. 2001. Novel hypoxanthine and thiohypoxanthine compounds, PCT WO0111967.

185. Tanaka, T., et al. 1999. Purine derivatives and medicine containing the same as the active ingredient, PCT WO9924432.

186. Buckle, D.R., et al. 1994. Inhibition of cyclic nucleotide phosphodiesterase by derivatives of 1,3-*bis*(cyclopropylmethyl)xanthine. *J Med Chem* 37:476.

187. Kaplan, J.M., et al. 1995. Effect of TNF-α production inhibitors BRL 61063 and pentoxifylline on the response of rats to poly I:C. *Toxicology* 95:187.

188. Crespo, M.I., et al. 2000. Synthesis and biological evaluation of 2,5-dihydropyrazolo [4,3-c]quinolin-3-ones, a novel series of PDE 4 inhibitors with low emetic potential and antiasthmatic properties. *Bioorg Med Chem Lett* 10:2661.

189. Napoletano, M., et al. 2000. The synthesis and biological evaluation of a novel series of phthalazine PDE4 inhibitors I. *Bioorg Med Chem Lett* 10:2235.

190. Napoletano, M., et al. 2000. Benzazine derivatives as phosphodiesterase 4 inhibitors, PCT WO0021947.

191. Napoletano, M., et al. 2001. Phthalazine PDE4 inhibitors. Part 2: The synthesis and biological evaluation of 6-methoxy-1,4-disubstituted derivatives. *Bioorg Med Chem Lett* 11:33.

192. Napoletano, M., et al. 2000. Phthalazine derivatives as phosphodiesterase 4 inhibitors, PCT WO0005219.

193. Glaser, T., and J. Traber. 1984. TVX 2704—a new phosphodiesterase inhibitor with antiinflammatory action biochemical characterization. *Agents Actions* 15:341.

194. Lowe, J.A., et al. 1991. Structure–activity relationship of quinazolinedione inhibitors of calcium-independent phosphodiesterase. *J Med Chem* 34:624.

195. Wilhelm, R.S., et al. 1993. Benzo and pyrido pyridazinone and pyridazinthione compounds with PDE IV inhibiting activity, PCT WO9307146.

196. Hersperger, R., et al. 2000. Palladium-catalyzed cross-coupling reactions for the synthesis of 6,8-disubstituted 1,7-naphthyridines: A novel class of potent and selective phosphodiesterase type 4D inhibitors. *J Med Chem* 43:675–682.

197. Trifilieff, A., et al. 2002. Pharmacological profile of a novel phosphodiesterase 4 inhibitor, 4-(8-benzo[1,2,5]oxadiazol-5-yl-[1,7]naphthyridine-6-yl)-benzoic acid (NVP-ABE171), a 1,7-naphthyridine derivative, with antiinflammatory activities. *J Pharm Exp Ther* 301:241.

198. Denholm, A., et al. 2003. Naphthyridine derivatives, their preparation and their use as phosphodiesterase isoenzyme 4 (PDE4) inhibitors, PCT WO03039544.

199. Denholm, A., et al. 2004. [1,7]Naphthyridines as PDE4 inhibitors, PCT WO04055013.

200. Gracia Ferrer, J., et al. 1999. 1,2,4-Triazolo[4.3-b]pyridazine derivatives and pharmaceutical compositions containing them, PCT WO9906404.
201. Vinick, F.J., et al. 1991. Nicotinamide ethers: Novel inhibitors of calcium-independent phosphodiesterase and [³H]Rolipram binding. *J Med Chem* 34:86.
202. Chambers, R.J., T.V. Magee, and A. Marfat. 2002. Ether derivatives useful as inhibitors of PDE4 isozymes, PCT WO02060896.
203. Chambers, R.J., T.V. Magee, and A. Marfat. 2002. Preparation of nicotinamide biaryl derivatives as inhibitors of PDE4 isozymes, PCT WO2002060875.
204. Marfat, A., and M.W. McKechney. 2002. Thiazolyl, oxazolyl, pyrrolyl, and imidazolyl-acid amide derivatives useful as inhibitors of PDE4 isozymes, PCT WO02060898.
205. Chambers, R.J., T.V. Magee, and A. Marfat. 2001. Pyrimidine carboxamides useful as inhibitors of PDE4 isozymes, PCT WO200157025.
206. Kalgutkar, A.S., et al. 2004. Disposition of CP-671,305, a selective phosphodiesterase 4 inhibitor in preclinical species. *Xenobiotica* 34:755.
207. Jin, S.-L.C., and M. Conti. 2002. Induction of cyclic nucleotide phosphodiesterase PDE4B is essential for LPS-activated TNF-α responses. *Proc Natl Acad Sci* 99:7628.
208. Mehats, C., et al. 2003. PDE4D plays a critical role in the control of airway smooth muscle contraction. *FASEB J* 17:1831.
209. Lee, M.E., et al. 2002. Crystal structure of phosphodiesterase 4D and inhibitor complex. *FEBS Lett* 530:53.
210. Huai, Q., et al. 2003. Three-dimensional structures of PDE4D in complex with rolipram and implications on inhibitor selectivity. *Structure* 11:865.
211. Xu, R.X., et al. 2004. Crystal structures of the catalytic domain of phosphodiesterase 4B complexed with AMP, 8-Br-AMP, and rolipram. *J Mol Biol* 337:355.
212. Card, G.L., et al., 2005. A family of phosphodiesterase inhibitors discovered by cocrystallography and scaffold-based drug design. *Nat Biotechnol* 23:201.

Index

A

Acetobacter xylinum, 212
Acute promyelocytic leukemia (APL) model, 576
Adaptor protein 2 (AP2), 116
Adenosine 3′,5′-cyclic monophosphate (cAMP), 3, 10, 62, 79, 255
  anchored phosphodiesterases and localized pools of, 378–379
Adenylate cyclase (AC8) expression, 44
Adipose triglyceride lipase (ATGL), 87
Adrenocorticotropic hormone (ACTH), 66
Affi-10 gel, 83
Aggrenox, 11
AIPL1, 183
AKAP450, 119–120
A-kinase activating proteins (AKAPs), 10, 113, 117–119, 383–384
  anchoring of, 379–383
  as phosphodiesterase anchoring proteins
    AKAP450, 384–385
    mAKAP, 385
  signal complexes organized by, 385–386
A1016–K1052 residues, 589
Alzheimer disease, 214, 549
Amantadine, 41
Amino acid sequence, analysis of, 37–38, 107
Amyloid precursor protein (APP), 547–549
Anabaena adenylyl cyclase, 137, 245, 584
Angiotensin-converting enzyme (ACE) activity, 511
Anti-CD3/anti-CD28-stimulated human T-lymphocytes, 657
Anti–glial fibrillary acidic protein (GFAP) antibody, 231
Aplysia kurodai, 100
Arcuate nucleus (ARC), 348
Arg728, 614
Ariflo, 11
βARKct expression, 367
Arofylline, 687
Asn321, 612, 618
Asp201, 608
Asp318, 608
Asp430, 614

Asp764, 142
Asp439 carboxylate side chain forms, 64
Aspirin, 630, 632–633, 640
Atherosclerosis, 636
Atizoram, 678
ATP-binding cassette transporter (ABCR), 181
Atrial natriuretic factor (ANF) treatment, 45
Atrial natriuretic peptide (ANP), 66, 154
  signaling, 575
AtT20 neuroendocrine cells, 39
AWD-12-281, 275

B

BALB/c mouse peritoneum, 673
BAY 60-7550, 69
BAY 60-7750, 59–60
BAY 73-6691, 232
B-cell lymphocytic leukemia (B-CLL), 561
Ben(z)afentrine, 653
Benzimidazol, 683
Benzofuran, 683–684
Bile acid glycochenodeoxycholate (GCDC), 574
Bipolar affective disorder, 224–225
BMPR2, 502
BMS-586353, 600, 656
Bos taurus, 25
BRL 50481, 656, 658
BRL-50481, 600
BTPDE1A1, 25
BtPDE2A1, 60
BtPDE1B1, 36
Budesonide, 671
Burger's disease, 637
BYK169171, 651

C

Caenorhabditis briggsae (CAE71587), 215
Caenorhabditis elegans, 4, 100, 206
Caenorhabditis elegans (NP_490787), 215
Ca^{2+}-induced–Ca^{2+} release (CICR), 396
Calcineurin, 36, 40–41
Calcitonin gene-related peptide, 524

Calcium and calmodulin (CaM), binding
        complex of, 35
Calcium-channel blockers, 41
Calmodulin, 39
m-calpain, 37
Calpain, 38
CaM kinase II, 36
CaM kinase kinase, 36
cAMP regulatory elements (CREs), 102
cAMP response element binding protein
        (CREB) transcription factors, 508
cAMP response element modulator (CREM)
        gene, 577
Cardiac excitation-contraction coupling, 396
Cardiac fibroblasts, 366
    PDE activities in, 367–368
Cardiac hypertrophy, 11, 155
    role of cyclic AMP, 367
Cardiac natriuretic peptides signaling, 398
Cardiac nitric oxide signaling, 397–398
Cardiovascular systems, PDE1 isoforms of,
        44–45
C4 aromatic ring, 674
Catechol, 595
Catecholamines, 397
Catecholamine-stimulated lipolysis, 89
Caveolae signaling platforms, 85
CC-10004, 674
CDP-840, 679–680
CEM-ICR27 line cells, 567
Central nervous system (CNS) diseases, 11
CG5411 gene, 304–305
CG8279 gene, 304
CG10231 gene, 305–306
CG14940 gene, 304
CG32648 gene, 305
cGMP, *see* Guanosine 3′,5′-monophosphate
cGMP-binding protein PDE (cG-BPP), 132
cGMP-gated ion channel (CNGB1), 181
cGMP signaling, 398
Chlorambucil, 563
Chloramphenicol acetyl transferase (CAT), 110
Cholecystokinin (CCK), 348
Cholesterol biosynthesis, 85
C767-H781 residues, 589
Chromosome 2q32, 42
Chronic obstructive pulmonary disease
        (COPD), 11, 155, 668
Chronic renal disease, 155
Chymotrypsin, 60
α-chymotrypsin, 37
CI-1018, 681
Cialis, 11
Cilomilast, 207, 595, *see* Ariflo

Cilostamide, 58, 68, 70, 80
Cilostazol, 11, 80–81
    clinical applications
        approved indications, 637–639
        marketing status, 636
        patient exposure, 636
        potential indications, 639–642
        safety and adverse events, 636–637
    inhibitory effect on platelet aggregation,
        631–632
    pharmacology and mechanistic studies
        antithrombotic effects, 630–632
        dual inhibition of PDE3 and adenosine
            uptake, 628–630
        on lipid metabolism, 634
        medicinal chemistry, 628
        on neuronal protective effect, 634–635
        on skin bleeding, 632–633
        on vasculature, 633–634
Cilostazol Stroke Prevention Study (CSPS), 638
Ciona intestinalis, 215
Cipamfylline, 687
CL 316,243, 89–90
Clathrin, 116
Clopidogrel, 632
Conditioned avoidance responding (CAR),
        in rodents, 250
Coronary stenting, 639
COS7 cells, 111
CP-77059, 688
CP-671305, 690
CREST trial, 641
C-terminal catalytic (C) domain, 135
C-terminal portion, of PDE10A protein, 239
C-terminal prenylation, 175
C-type natriuretic peptide (CNP), 371
Cyclic AMP
    and proapoptotic signaling in heart cells, 367
    role in cardiac hypertrophy, 367
Cyclic GMP binding, 137–138, 150
Cyclic nucleotide phosphodiesterases (PDEs), 3
    and amino acid sequence relationships, 5
    anchoring of, 10
    catalytic domain structures, 584–589
    class I, 4, 6–7
    class II, 4, 6
    class III, 4
    compartmentation of signaling, 398–399
    dissection of inhibitor selectivity, 620–621
    hydrolysis mechanism, 618–620
    hydrophobic clamp, 593–594
    inhibitors, 11, 14
    mammalian cyclic nucleotide isoforms,
        20–24

methods to study compartmentalization in
    intact myocytes, 399–401
nomenclature systems of, 9, 36
nucleotide selectivity, 589–593
pathophysiological role of cAMP
    compartmentation, 407–408
PDE2 enzymology
  catalytic site inhibitors, 59–60
  kinetics, 56–59
PDE1 family, of enzymes
  inhibitors, 40–41
  kinetic properties, 38–39
  primary structure and activation
      by Ca^{2+}/CaM, 36–38
  regulation by phosphorylation, 39–40
PDE3 gene family
  crystal structure of PDE3B, 83
  insulin-induced formation of
      macromolecular complex containing
      PDE3B, 85–86
  kinetic properties and inhibitor sensitivities,
      80–81
  role and regulation of PDE3B, 86–88
  role in diabetes, 88–91
  structural organization, 81–83
  subcellular localization, 85
  tissue-and cell specific expression
      and function, 83–85
PDE2 gene organization and splice
    variants
  gene organization, 60
  splice variants, 60
PDE1 genetics
  PDE1A variants, 42
  PDE1B variants, 42–43
  PDE1C variants, 43
PDE2 physiological roles
  atrial natriuretic peptide and adrenal
      steroidogenesis, 66–67
  in fibroblasts, 69
  in immunology, 71–72
  in inhibiting calcium currents in hearts, 69–70
  in localization of tissues, 70–71
  in neuronal NO-cGMP pathways, 68–69
  platelet aggregation, 67–68
  role in chloride transport, 69
  role in endothelial cells, 67
  role in olfactory epithelium, 71
PDE2 structure
  allosteric cGMP-binding pocket, 62–64
  catalytic domain, 65
  regulatory segment, 61–62
PDE1 tissue specific expressions
  of cardiovascular system, 44–45

immune system, 47–48
nervous system, 45–47
testis and sperm, 48–49
potential for structure based rational drug
    design, 601–602
regulation of cardiac functions
  by cAMP pathways, 397
  cardiac excitation–contraction coupling, 396
  by cGMP pathways, 397–398
retinal, 10
role in cyclic nucleotide compartmentation
  and hormone specificity, 401–403
  specific roles of isoforms in, 403–407
structural basis of inhibitor potency
  IBMX binding, 595
  PDE3 inhibitors, 597–598
  PDE4 inhibitors, 595
  PDE5 inhibitors, 595–597
structural basis of inhibitor selectivity
  PDE4 and PDE5, 598–600
  PDE4 and PDE7, 600–601
  PDE5 and PDE6, 600
  PDE4B and PDE4D, 601
structure of family of
  of catalytic domain, 618
  PDE4, 608–612
  PDE2A, 614–616
  PDE5A, 617–618
  PDE7A, 612
  PDE9A, 616–617
  PDE1B, 613–614
  PDE3B, 616
in vitro specificity, 11–12
in vivo selectivity, 12–15
Cyclic nucleotide phosphodiesterases (PDEs),
    of protozoa, *see* Protozoal
    phosphodiesterases (PDEs)
Cyclic nucleotide phosphodiesterases (PDEs), role
    in apoptosis
  antiapoptotic effects of cAMP, cGMP, and
      phosphodiesterase inhibitors
    cAMP signaling protects hepatocytes from
        Fas and Bile acid-induced apoptosis,
        574–575
    cGMP signaling and apoptosis in endothelial
        cells, colon tumor cell lines, and
        hepatocytes, 575
    and induction of differentiation in myeloid
        cells, 575–577
    of PDE4 and PDE5, 573–574
  proapoptotic effects of phosphodiesterase
      inhibitors
    in B-cell chronic lymphocytic leukemia,
        564–565

cAMP-induced lymphoid cell death,
561–563
induced EPAC activation, 571
interrelationship between glucocorticoid and
cAMP/PDE4 inhibitor-mediated
apoptosis, 567–569
in models of acute lymphoblastic leukemia
and diffuse large B-cell lymphoma,
565–567
other mechanisms, 569–571
specificity of, 571–572
and regulation of physiologic apoptotic
signaling, 577
Cyclic nucleotide phosphodiesterases (PDEs), role
in heart failure and hypertension
role of PDE3
in cardiac and vascular smooth muscle,
486–487
role of isoforms in regulation of cyclic-
nucleotide-mediated signaling, 487–489
use of inhibitors in treatment of heart failure,
489–490
role of PDE5
in cardiac and vascular smooth muscles,
492–493
isoforms and regulation of cyclic nucleotide-
mediated signaling, 493–494
use of inhibitors, 494–495
Cyclic nucleotide phosphodiesterases (PDEs), role
in vascular cyclic nucleotide signaling
arterial structure-function
cyclic nucleotides and their effectors,
443–444
development and vascular remodeling,
441–443
PDE expression in cardiovascular cells,
444–447
regulated expression and function
of arterial cells
in PDE1 variants, 447–449
in PDE3 variants, 449–451
in PDE4 variants, 453–455
Cyclic nucleotide phosphodiesterases (PDEs), role
in vascular diseases
cGMP-induced proliferation and effect on
PDE expression in the human
permanent endothelial cell line EAhy
926, 425–427
distribution in various endothelial cells
aortic endothelial cells, 419–420
human endothelial cells, 420–423
pulmonary endothelial cells, 420
effectiveness of PDE inhibitors in Chicken
Embryo CAM, 435

involvement in migration and proliferation
steps in VEGF-Stimulated HUVEC
cAMP accumulation in VEGF-treated
HUVEC, 430–431
cell cycle protein phosphorylation and
expression in VEGF-stimulated
HUVEC, 432–434
effect of PDE expression in HUVEC,
427–430
VEGF-induced HUVEC cycle progression,
432
VEGF-induced HUVEC migration,
434–435
VEGF-induced HUVEC proliferation,
431–432
mediation of cAMP–cGMP crosstalk in BAEC,
423–424
phenotypic expression in BAEC, 424–425
role in growth of vascular smooth muscles
PDE1, 473–475
PDE3, 475–476
PDE5, 475
role in regulation of vascular smooth muscles
PDE1, 471
PDE3, 472–473
PDE5, 471
role of cAMP and cGMP in vascular smooth
muscle, 466–467
role of phosphodiesterase inhibitors
atherosclerosis and restenois, 478–479
nitrate tolerance, 476–477
pulmonary hypertension, 477–478
with structurally and functionally different
phosphodiesterases
in the cross talk between Ca^{2+} and cyclic
nucleotides, 467–468
in cross talk between cAMP and cGMP
signaling, 468–470
as a negative feedback regulator of cAMP
signaling, 470
as a negative feedback regulator of cGMP
signaling, 470–471
Cyclic nucleotide phosphodiesterases (PDEs),
study using fruitfly models, *see*
Drosophila melanogaster
6-[4-(1-cyclohexyl-1*H*-tetrazol-5-yl)butoxy]-3,4-
dihydro-2(1*H*)-quinolinone, *see*
Cilostazol
3-cyclopentoxy-4-methoxyphenyl group, 669
3-cyclopentoxy-4-methoxyphenyl moiety, 672
Cynomolgus monkeys, 651
CYP2B6, 677
Cyp40 protein, 107
Cys386, 61

D

D-4418, 679
DA-8159, 368
Danio rerio, 206, 215
Deprenyl, 41
Dextran sulfate sodium (DSS)-induced chronic colitis rat model, 674
3,4-dialkoxyphenyl pharmacophore, 685
3,4-dialkoxyphenyl ring, 684
Dictyostelium discoideum, 41, 222
Differentiation-inducing factor-1 (DIF-1), 41
3,5-dimethyl-1-phenyl-1H-pyrazole-4-carboxylic acid ethyl ester (PhPCEE), 601
Dipyridamole, 58, 178, 245
Dipyridamole inhibition, 71
DISC1 gene, 112
Disrupted in schizophrenia (DISC1) protein, 112–113
DMPDE "32948", 4
dnc gene, 304
Dorsomedial hypothalamus (DMH), 348
Drosophila, 4
Drosophila melanogaster, 100, 111, 206, 215
Drosophila melanogaster phosphodiesterases
  biochemistry and pharmacology of
    Dunce, 311–312
    PDE1, 312–314
    PDE6, 312–314
    PDE11, 312–314
  expressions in adult model, 306–307
  gene information of phosphodiesterases
    CG5411, 304
    CG8279, 304
    CG10231, 305–306
    CG14940, 304
    CG32648, 305
    Dunce, 304
  inhibitor sensitivities of, 314–315
  as model organism, 301–302
  phenotypic evidence for functional conservation of, 315–317
  protein sequence alignments of, 307
  structural features of
    PDE1, 307
    PDE6, 307–310
    PDE8, 310
    PDE9, 310
    PDE11, 310–311

E

EGTA, 40
EHT-202, 675–676
Enoximone, 80, 228
Erectile dysfunction (ED)
  drugs, 40
  treatment of, 133
ERK5, 119
ERK5-PDE4D3-Epac1-mAKAP complex, 119–120
Erythro-9-(2-hydroxy- 3-nonyl)-adenine (EHNA), 56, 59, 68–70, 228, 573
Erythromycin, 671
Escherichia coli, 83
  direct oxygen sensor (EcDOS), 212
  Fh1A, 137
  FhlA, 245
  FhlA protein, 584
Ethylene diaminetetraacetate (EDTA), 143
European Stroke Prevention Study (ESPS), 639
Excitation–contraction coupling (ECC), 397
Expressed sequence tag (EST) database, 19, 205, 238
Extracellular matrix (ECM) proteins, 365
Extracellular signal-regulated kinase (ERK), 540

F

Familial hemiplegic migraine (FHM), 521
FEZ1 protein, 112
Filaminast, 595, 678
FKBP52 protein, 107
Fluorescence resonance energy transfer (FRET), 67
Forskolin, 69, 561, 564, 573–574
Free fatty acids (FFA), 88

G

GAF-AB x-ray crystal model, 58
GAF domains, 4, 6, 55–56, 58, 61, 64, 173–174, 211, 247
GAF functions, 138–141
Gel filtration chromatography, 85, 181
GenBank, 9, 19, 25, 199, 222–223, 239
Geranylgeranyl group, 175
Ghrelin binding, 352
Glioblastoma multiforme, 41
Gln369, 612, 618, 620–621
Gln775, mutagenesis of, 145
Gln817, 145
Glomerulosa cells, 66
Glu278, 614
Glucagon-like peptide-1 (GLP-1), 348
Glucon-like peptide 1 (GLP-1), 80

Glucose-dependent insulinotropic polypeptide (GIP), 354
Glutamine residue, 244
GLUT-4 translocation, 80, 86
Glyceryl trinitrate (GTN), 520
Golgi apparatus, 198
Golgi retention, 110
G-protein-coupled receptor, *see* Rhodopsin
G-protein receptor associated kinase2 (GRK2), 113
G-protein signaling-9 (RGS9), 169
G-protein β-subunit (Gβ5), 169
Granulocyte–macrophage colonystimulating factor (GM-CSF), 48
Granulosa cells, 102
GRC-3886, 676
Green fluorescent protein (GFP), 110
GTP-binding protein-coupled receptor (GPCR)-coupled signaling, 577
Guanine nucleotide exchange factor (Epac2), 67, 79
Guanosine 3′,5′-monophosphate, 3, 10, 62, 221, 255
Guanylate cyclase activating proteins (GCAPs), 169
Guanylyl cyclase (GC), 45

H

H89, 119
H-89, 575–576
Hartley strain guinea pigs, 674
HCc catalytic sequence, 4
HD(X)2H(X)4N motif, 83
Heat shock protein (HSP)-90, 86
HEK293 cells, 213, 228
Hek293 cells, 117
High-affinity cAMP-specific phosphodiesterase 1 (HCP-1), 655
High-affinity rolipram-binding site (HARBS), 105, 670
Hill coefficient, 58
His164, 608
His200, 608
His617, 142
His653, 142
Histamine, 520
H-loop influences, on PDE5, 146–147
$HNX_2HNX_NE/D/QX_{10}HDX_2HX_{25}E$, 6
Homo sapiens, 9
Hop1 protein, 107
Hsp90, 107
HsPDE2A3, 60
HsPDE4A1, 9

HT-0712, 275
HT29 cells, 575
Human and murine PDE7A1, 196–199
Human PDE7A2, 199
Human PDE7A3, 199
human *PDE9A* gene, 227
Human PDE8A Unigene Hs.9333, 210
Human umbilical vein endothelial cells (HUVECs), 67
Huntington's disease, 248
HUT78 lymphoma cells, 199
Hypoxic pulmonary vasoconstriction (HPV), 503, 654

I

IC-485, 678
IC_{50} values, 12, 40, 58–59, 62, 70, 207, 227, 564, 629, 686
Ile336, 612, 618
Ile422, 64
Ile458, 62
Imidazole, 591
Immune system, PDE1 isoforms of, 47–48
Inducible cAMP early repressor (ICER), 577
Inducible cyclic AMP early repressor (ICER), 367
Inhibitory γ-subunit, 175–176
Injections, of 8-bromo-cAMP or cAMP-elevating agents, 350
Insulin, 80, 86, 88, 633
Insulin receptor tyrosine kinase (IRTK), 88
Interleukin-1β (IL-1β), 69
Intermittent claudication, 13
Intravenous glucose tolerance tests (IVGTT), 354
Isobutylmethylxanthine, 40
3-isobutyl-1-methylxanthine (IBMX), 14, 58, 69, 143, 146–147, 177, 195, 200, 207, 228, 617
Isoproterenol, 69, 86, 89

J

JAK–STAT pathway, 350
Janus kinase (JAK), 86
JNK3 kinase, 116
Jurkat cells, 102, 568

K

KT5720, 119, 574
KT5823, 573

L

L-826,141, 680–681
Lateral hypothalamus (LH), 348
L-butathione-[S,R]-sulfoximine, 573
Leber's congenital amaurosis, 183
Lectin, 71
Leptin, 80, 86
Leu232, 614
Leu319, 618
Leu389, 614
Leukemic cell line (L1210 cells), 560
Levitra, 11
Lipopolysaccharide (LPS), 102
Lipopolysaccharide (LPS)-stimulated PBMC, 668
Lipoprotein lipases, 634
Lirimilast, 678
Lixazinone, 80
Low-affinity rolipram-binding sites (LARBS), 105, 670
L-type Ca^{2+} channels (LTCCs), 396
LY-83583, 69
LY294002, 574

M

Macrophage colony-stimulating factor (M-CSF), 48
Male erectile dysfunction, 608
Maturation-promoting factor (MPF), 84
MC3T3-E1, 214
Melanin-concentration hormone (MCH), 348
α-melanocyte-stimulating hormone (a-MSH), 348
MEM1018, 552
MEM1091, 552
MEM-1414, 682
MERCK1, 83
Mesopram, 589, 595, 676
Met337, 612
Met357, 612
Metal binding sites for zinc and magnesium, 667
Methionine, 135, 224
8-methoxymethyl-IBMX, 40, 48
8-methoxymethyl-3-isobutyl-1-methylxanthine, 652
Methylxanthines, 563
Met381 of PDE8A1, 209
M138/G139 residues, 174
Michaelis–Menten function, 56, 58
Michaelis–Menten kinetics, 206
Migraine syndrome, 519–520
Milrinone, 573, 629
Mitochondria, 168
Mitogen-activated protein (MAP) kinase, 84

MK-801, 68–69, 547
MMGL, 110–111
MMPDE1A1, 25
Mono-MAC6, 102
Motapizone, 68
Multalin, 5
Mus musculus, 25
Myeloid zinc-finger-1 (MZF-1), 239
Myomegalin, 110–111
Myosin light chain kinase, 36, 45

N

N6-benzoyl-cAMP, 574
Nervous system, PDE1 isoforms of, 45–47
Neuronal NO-cGMP pathways, 68–69
NF-kB p65–p50 complex, 213
N496–F508 residues, 589
N^G-monomethyl arginine (L-NMMA), 69
NIH 3006, 83
Nitraquazone, 687–689
Nitric oxide (NO), 519, 522–524
Nitric oxide–sensitive guanylyl cyclase, 69
Nitroglycerin (NTG), 44
NKFDE motif, 64
N[KR]X(5–24)FX(3)DE motif, 64
N-methyl-D-aspartate (NMDA), 68–69, 248
    receptor–mediated cyclic AMP signaling, 542
Nomenclature system, of PDEs, 9, 36
Nonprotein-encoding regions, 19
Nonselective PDE inhibitors, 40
N-terminal portion, of PDE10A protein, 239
N-terminal regulatory domain, 36
N-tosyl-L-lysinechloromethyl ketone (TLCK), 511
NUDEL protein, 112

O

Olfactory cyclic nucleotide-gated channel (CNGA2), 232
OMIM 180071, 182
OMIM 180072, 182
OMIM 180073, 182
OMIM 604392, 183
ONO-6126, 676
Oocytes, 84

P

Papaverine, 58, 207, 245
Paraventricular nuclei (PVN), 348
Parkinson's disease, 668
PAS domain, 211–212

PD98059, 103, 574
PDE2
  enzymology
    catalytic site inhibitors, 59–60
    kinetics, 56–59
  gene organization and splice variants
    gene organization, 60
  physiological roles
    atrial natriuretic peptide and adrenal
      steroidogenesis, 66–67
    in fibroblasts, 69
    in immunology, 71–72
    in inhibiting calcium currents in hearts,
      69–70
    in localization of tissues, 70–71
    in neuronal NO-cGMP pathways, 68–69
    platelet aggregation, 67–68
    role in chloride transport, 69
    role in endothelial cells, 67
    role in olfactory epithelium, 71
  structure
    allosteric cGMP-binding pocket, 62–64
    catalytic domain, 65
    regulatory segment, 61–62
PDE4
  alternative sources for inhibition, 688–690
  antidepressant effects of inhibitors
    clinical studies of, 544
    preclinical models, 542–544
    roles of subfamilies of, 544–545
  inhibitors of
    clinical inhibitors, 670–677
    discontinued clinical inhibitors, 677–682
  and memory
    aging and Alzheimer's disease, 550–551
    inhibitors in memory models, 545–549
    long-term potentiation, 549–550
    role of subfamilies of, 551–552
  in signaling pathways associated with
    depression and memory
    N-Methyl D-Aspartate receptor–mediated
      signaling, 542
    noradrenergic and serotonergic signaling,
      541–542
  structure and pharmacology
    enzyme structure, 540
    inhibitor interactions, 540–541
  subtype-selective inhibition, 690–691
PDE5, molecular characteristics of
  characteristics of gene encoding, 134–135
  different conformers of, 148–149
  as drug target in treatments, 155
  effects of inhibition, 371
  inhibitor specificity, 148

  ligand interactions, 144–146
  physical characteristics
    crystal structures of, 147
    domain organization, 135–136
    metal-binding sites, 142–143
    regulatory domain, 136–141
    structural features, 141–142
    substrate or inhibitor binding, 143–147
  regulatory mechanism on cGMP signaling
    and action of inhibitors
    in erectile dysfunction and pulmonary
      hypertension, 154–155
    impact of concentration and affinity of
      cGMP receptors on cGMP signaling,
      152–153
    negative feedback pathway, 153–154
    short-term regulation of cGMP signaling,
      150–152
  role in migraine
    induction of headache through inhibition,
      526–531
    migraine pathophysiology and pain
      pathway, 520–522
    migraine syndrome, 519–520
    in pain pathway of migraine, 525–526
    signaling molecules associated with
      migraine, 522–525
  structural basis of inhibitor potency of, 595–597
  tissue distribution, 133–134
PDE6
  allosteric role of cGMP-binding GAF domains
    of PDE6, 179–180
  catalytic properties of, 177
  interacting proteins
    glutamic acid–rich protein-2 (GARP2), 181
    prenyl binding protein/PDEd (PrBP/δ),
      181–182
  pharmacology
    effects of PDE inhibitors on photoreceptor
      and retinal physiology, 177–178
    potency and selectivity of PDE inhibitors for
      PDE6 *in vitro,* 178–179
  regulation of, by G-Protein, Transducin,
    180–181
  retinal diseases due to, 182–183
  structure of catalytic and inhibitory
    subunits of
    catalytic subunits, 173–175
    inhibitory γ-subunit, 175–176
  tissue and subcellular distribution of, 171
PDE8A
  biochemical characteristics, 206–207
  biological functions
    PDE8A in testis, 214

PDE8B in brain, 214
  in various cell types, 214–215
  gene organization and alternative
    transcripts, 208–210
  localization of, 210–211
  nonmammalian, 215
  pharmacological characteristics, 207–208
  structure
    motifs, 213–214
    PAS domain, 211–213
    REC domain, 211
PDE8A1, 206
PDE9A
  cloning, 222–223
  crystal structure, 228–229
  enzymatic profiles and sensitivity to
    inhibitors, 227
  genomic organization and splice variants,
    223–227
  inhibitors of, 232–233
  in situ hybridization analysis of, in
    rat brain, 230–231
  subcellular localization of variants of,
    231–232
  tissue-expression patterns in humans and
    mice, 229–230
PDE10A
  activity and function of
    enzymatic and kinetic properties, 244–245
    physiological functions and therapeutic
      implications, 249–251
    physiological regulation of cyclic
      nucleotides, 248–249
    regulation of, 245–248
  cloning of, 238
  genomic organization of, 241
  localization of, 241–244
  primary structure and splice variants, 238–239
  transcriptional start site, 239–241
PDE11A
  detection and localization of proteins of,
    263–264
  enzymatic properties, 257–259
  genomic organization, 264–269
  other protein features, 261
  physiological role of, and effects of
    inhibition, 269–271
  primary structure, 256–257
  sensitivity to known phosphodiesterase
    inhibitors, 259–261
  single nucleotide polymorphisms, 269
  tissue distribution of mRNA, 261–263
PDE1A gene, 42
PDE10A gene, 247

PDE7A transcripts
  PDE7A1, 196–199
  PDE7A2, 199
  PDE7A3, 199
PDE3B, role of
  in energy balance regulation
    dysregulation of cAMP in the
      hypothalamus, 352
    hypothalamic effects of cAMP on food
      intake, 349–350
    regulation of feeding at hypothalamus, 348
    regulation of food intake by Leptin,
      350–352
    working model for hypothalamic control of
      food intake involving regulation of
      cAMP, 352
  in regulation of insulin secretion, 352–357
PDE7B1 expression, 200
PDE1B genes, 42–43
PDE7B transcripts, 200–201
PDE1C, 10–11
PDE1C genes, 43
PDE4D5-D556A, 114
PDE4 enzymes, cellular functions of
  associations between PDE4 isoforms and AKAPs
    ERK5 associates with PDE4D3, 119
    functional consequences of the
      ERK5–PDE4D3-Epac1-mAKAP
      interaction, 119
    interaction of PDE isoforms with
      AKAPs in the Centro some, 119–120
    interactions between PDE4D3,
      mAKAP, and Epac1, 119
    interactions between PDE4D3 and
      mAKAP, 118–119
  cyclic AMP regulation of, 102
  dimerization, multimerization, and
      intramolecular domain interactions of
    dimerization of, with pharmacological
      effects, 105
    and high-affinity rolipram-binding site,
      105–106
    interactions between UCR1 and UCR2,
      104–105
  interaction of PDE4A5 with the
      immunophilin XAP2, 106–107
    functional and pharmacological
      implications of, 107–108
  interaction of PDE4D isoforms with
      myomegalin, 110–112
  interaction of PDE4 isoforms with protein
      tyrosine kinases
    functional and pharmacological
      implications of, 109–110

interaction with SRC, FYN, and
LYN, 109
interaction of PDE4 isoforms with the
disrupted in schizophrenia (DISC1)
protein, 112–113
phosphorylation by ERK kinases, 103–104
phosphorylation by PKA, 102–103
role of PDE4 isoforms in regulating signaling
through the β-adrenergic receptor
with β-arrestins, 113–114
functional implications of the PDE4D5–β-
arrestin-β2AR interaction, 117–118
role of RACK1 in the PDE4D5–β-arrestin
interaction, 116–117
role of the PDE4D5 amino-terminal
region, 116
role of the PDE4D5–β-arrestin interaction on
phosphorylation of β_2 AR, 114–116
role of the PDE4A1 amino-terminal region in
membrane association, 110
transcriptional regulation of *PDE4* genes as a
feedback mechanism in signaling,
101–102
PDE4 from knockout mice, physiological
functions
and central nervous system functions
antidepressant effects of rolipram, 334–335
hypnotic action of adrenoceptors, 335–336
differential role in airway inflammation of
asthma, 328–329
involvement of PDE4D in modulation of
excitation-contraction coupling in
heart, 336–338
PDE4 from knockout mice, physiological
functions
role of PDE4B
in control of neutrophil recruitment to
lung, 333–334
regulation of PDE4D for gonadotropin-
dependent pattern gene expression in
ovarian follicle, 326–327
role of PDE4B, 329–333
role of PDE4D in control of airway smooth
muscle tone, 327–328
role of PDE4D in neonatal survival and
growth, 324–326
PDE3 gene family
crystal structure of PDE3B, 83
insulin-induced formation of macromolecular
complex containing PDE3B, 85–86
kinetic properties and inhibitor sensitivities,
80–81
role and regulation of PDE3B, 86–88
role in diabetes, 88–91

structural organization, 81–83
subcellular localization, 85
tissue-and cell specific expression and
function, 83–85
PDE3/PDE4 inhibitors, 653
Pentoxifylline, 637
Peptide KRpTIRR, 569
Peripheral blood mononuclear cells (PBMC), 668
Phe340, 612
Phe372, 612, 618
Phe438, 64
Phe831, 614
Phenylalanine, 593
9-(6-phenyl-2-oxohex-3-yl)-2-(3,4-
dimethoxybenzyl)-purin-6-one
(PDP), 59
Pheochromocytoma PC12, 66
Phorbol myristate acetate (PMA), 88
Phorbol-12-myristate-13-acetate (PMA), 71
Phosphatidylcholine, 103
Phosphatidylinositol-3-kinase (PI3K) p85, 86
Phosphodiesterase inhibitors
adverse events of, 651
in asthma and COPD, 650–651
dual selective
1/4 inhibitors, 651–653
3/4 inhibitors, 653–654
4/5 inhibitors, 654–655
4/7 inhibitors, 655–658
nonselective phosphodiesterase
inhibitors, 658–659
Phospholamban (PLB), 396
Phosphoprotein phosphatase inhibitor
calyculin A, 40
Phosphoprotein phosphatase (PP)-2A, 86
Phosphorylated peptides, 40
Phosphorylation, 36, 39, 86–87
S13, 118
at S54, 103–104
Photoreceptive cells, in pineal gland, 171
Phototransduction, 167
Phytohemagglutinin, 71
Piclamilast, 595, 671, 677, 688
PI3K-PDE3B-cAMP signaling
pathway, 351
Pituitary adenylate cyclase-activating
polypeptide (PAKAP), 577
Platelet aggregation, 67–68, 79
PLB985 AML cells, 577
PLX513, 566
Poly(ADP-ribose)polymerase (PARP)
cleavage, 569
PP5A protein, 107
Prolactin releases, 72

Proline-23, 176
Pro-opiomelanocortin (POMC)
    neurons, 348
Prostacyclin (PGI2), 67
Protein Data Bank, 584, 608–609
Protein kinase B (PKB), 86
Proteolysis, 36
Protozoal phosphodiesterases (PDEs)
  DdPDE1
    cell biology of, 284–285
    DdPDI of, 285
  DdPDE2, 286–287
  DdPDE3, 287–288
  DdPDE4, 288
  DdPDE5, 288–289
  DdPDE6, 289
  DdPDE7, 289
  fungal class I
    Candida spp. PDE2, 282–283
    Saccharomyces cerevisiae ScPDE2,
        281–282
  fungal class II
    biological role of, 281
    Candida spp. PDE1, 280
    other fungi, 280–281
    Saccharomyces cerevisiae ScPDE1,
        278–280
    Schizosaccharomyces pombe SpPDE1, 280
  kinetoplastida
    Leishmania major, 293
    LmjPDEA of *Leishmania major*, 291
    Plasmodium (PfPDE1), 295
    TbrPDEA of *Trypanosoma brucei*, 290–291
    TcrPDEC of *Trypanosoma cruzi*, 294
    Trypanosoma brucei, 291–293
    Trypanosoma cruzi, 293–294
Protozoans, 4
Proximal renal tubules, 133
Pseudomonas aeruginosa, 674
Pulmonary arterial hypertension, 11
  and bone morphogenetic protein receptor
        mutation and modifying factors
        (PDE5), 501–502
  hypoxia-induced pulmonary vasconstriction
        and phosphodiesterase inhibition,
        503–504
  and PDE5 adaptor proteins
    interaction of PDEγ with PDE6, 512–513
    PDEγ and mitogenic signaling, 513–515
  phosphodiesterase activity in chronic
        hypoxic models of
    angiotensin II and PDE5 expression,
        511–512
    chronic hypoxia, 504–508

cross talk regulation between PDE3 and
        PDE5, 511
    cyclic nucleotides, 504–508
    role of Nuclear Factor-kB and PDE5
        Expression, 510–511
    role of PDE3, 508–509
    role of PDE5, 509–510
    vascular reactivity, 504–508
  use of PDE5 inhibitor sildenafil, 502–503
Pulmonary microvascular endothelial cells
    (PMVECs), 575
Pumafentrine, 653
Purkinje cells, 230–231
Purkinje neurons, 133
Pyrazole, 595

Q

Q-pocket, 620

R

RACK1, 116–117
5′ rapid amplification of cDNA ends
    (RACE), 239
Rattus norvegicus, 25
Raynaud's syndrome, 155
REC domain, 211
Receptors for activated C-kinase 1
    (RACK1), 541
Renal ducts, 133
Restenosis, 640–641
RGS9 anchoring protein (R9AP), 169
Rhodopsin, 168
Rhone-Poulenc-Rorer compound,
    see Piclamilast
Riboprobes, 70–71
RIP-PDE3B mice, 354, 356–357
mRNA expression, 7, 19
RNase protection, 43
RNPDE1A1, 25
RnPDE2A2, 60
RO 20-1724, 70
Rod photoreceptors function, 166
Roflumilast, 595, 670–671
Rolipram, 104–106, 109, 118, 196, 540, 612,
    683–686
  clinical limitations of, 668–670
  3,4-dialkoxyphenyl pharmacophore of, 669
  lack of affinity of, 669
  pyrrolidone groups of (R)- and (S)-of, 612
RP-73401, *see* Piclamilast
RPMI 8392 cells, 565

RPMI-8392 cells, 563
R755–S798 residues, 589
RU486, 568

S

S579A, 103
Saccharomyces cerevisiae, 655
Salbutamol, 671
Sarcoendoplasmic reticulum Ca^{2+}-ATPase
        (SERCA), 396
S49 cells, 561, 568
SCH51866, 40, 227–228
SCH351591, 651
Schizophrenia, 251
Schizosaccharomyces pombe SpPDE1, 280
S579D, 103
S54D mutation, 103
Second messengers, *see* Adenosine 3′,5′ -cyclic
        monophosphate (cAMP); Guanosine
        3′,5′-monophosphate
Ser92, 136
Ser102, 136, 139–140, 148, 151, 153
Ser120, 39
Ser138, 39
Ser273, 83
Ser368, 612
Ser424, 64
Ser428, 88
Serine 296/302, 86
Serotonin (5-HT), 524
Serotonin inhibitor, 68
Severe combined immunodeficient
        (SCID) mice, 577
Sf9 insect cells, 83, 629
Sildenafil, 40–41, 132, 143–144, 146–147,
        150, 152–155, 227, 368–371, 502, 589,
        596, 652, 655, *see* Viagra
    methylpiperazine of, 597
S578 in HsPDE2A, 58
Skene periurethral glands, 133
SKF38393, 550
SKF94836, 505
SM00471, 174
SMART database, of protein domains, 60
Smooth muscle cells (SMCs), 44
Sodium nitroprusside, 69
SRC protein tyrosine kinase, 109
Stroke, 11
Superose 6 chromatography, 86
Sus scrofa, 25
S454 –V471 residues, 587
SW480 colon carcinoma cell line, 575
S373Y, 621

T

Tadalafil, 132, 143, 146, 152–154, 597, 655,
        see Cialis
Takifugu rubripes, 215
T cell proliferation, 198
Tetomilast, 673
Tetraodon nigrovirdis, 215
Tetratricopeptide repeat (TPR) domain, 106
Theophylline, 40, 81, 651, 686–688
Thick ascending limb (THAL), 69
Thr-96, 222
Thr-235, 222
Thr-260, 222
Thr333, 612, 618
Thr488, 64
Thr492, 62
Ticlopidine, 630
TNF-α, 67, 574, 668, 682, 685
Tofimilast, 677
Tolafentrine, 653
Toll-like receptors (TLRs), 102
TPAS1, 110
Transaortic constriction (TAC), 368
Trequinsin, 207
Triazolopyridopyridazines, 689
Trifluoperazine, 40
Triton X-100, 71
Trypanosoma brucei, 614
Trypanosoma cruzi, 6
TVX 2706, 688
Type 2 diabetes (T2D), 81
Tyr159, 612, 618
Tyr329, 612, 620
Tyr448, 614
Tyr481, 62
Tyr719, 614

U

U0126, 547
U-937 cell line, 576
U937 cells, 71, 103
Udenafil, 132
UK122764, 143
Upstream conserved sequences (UCRs) 1 and 2,
        100, 104–105, 112
US Food and Drug Administration (FDA),
        651, 672

V

V-11294A, 682, 686
Val484, 63

Val660–His684, 142
Vardenafil, 40–41, 132, 143–144,
        146, 152–154, 227–228, 589,
        596, 655, *see* Levitra
    ethylpiperazine in, 597
Vascular smooth muscle cells (VSMC),
        80, 102
Ventromedial hypothalamus
        (VMH), 348
VETKKVTSSGVLLL sequence, 103
Viagra, 11
Vinpocetine, 40, 563, 573

W

WEHI-7.1 cells, 567
Western blotting, 48

X

Xanthane, 595
Xanthine, 208
XAP2 protein, 106–107

Y

YM-976, 681–682
Y329S, 620

Z

Zaprinast, 132, 177–178, 227–228,
        526, 575, 652
Zardaverine, 653
Zebra fish, 4
Zn^{2+} binding, 6
Zydena, 11, 132